MOLECULAR BIOTECHNOLOGY

MOLECULAR BIOTECHNOLOGY
Principles and Practices

CHANNARAYAPPA

Principal, Professor and Head
Department of Genetics
Vydehi Institute of Biotech Sciences
Bangalore, India

CRC is an imprint of the Taylor & Francis Group,
an informa business

Universities Press (India) Private Limited

Registered Office
3-5-819 Hyderguda, Hyderabad 500 029 (A.P.) India
email: info@universitiespress.com

Distributed in India, Pakistan, Nepal, Myanmar (Burma), Bhutan, Bangladesh and Sri Lanka by
Orient Longman Private Limited
Registered Office
3-6-752 Himayatnagar, Hyderabad 500 029 (A.P.) India

Other Offices
Bangalore / Bhopal / Bhubaneshwar / Chennai
Ernakulam / Guwahati / Hyderabad / Jaipur / Kolkata
Lucknow / Mumbai / New Delhi / Patna

Distributed in the rest of the world by
CRC Press LLC, Taylor and Francis Group,
6000 Broken Sound Parkway, NW, Suite 300, Boca Raton, FL 33487, USA

© Universities Press (India) Private Limited 2007

Cover and book design
© Universities Press (India) Private Limited 2007

All rights reserved. No part of the material may be reproduced or utilized
in any form, or by any means, electronic or mechanical, including photocopying,
recording, or by any information storage and retrieval system,
without written permission from the copyright owner.

First published in India by
Universities Press (India) Private Limited 2007

ISBN 13: 978 1 4200 5157 5
ISBN 10: 1 4200 5157 1

Set in Optima 10/13 by
OSDATA, Hyderabad

Printed in India by
Orion Printers Private Limited, Hyderabad 500 004

Published by
Universities Press (India) Private Limited
3-5-819 Hyderguda, Hyderabad 500 029 (A.P.) India.

To

The authors who provided me information and inspiration

Contents

	Preface	ix
PART I: Introduction to Biotechnology		**1–35**
1.	Biotechnology: Scope and Importance	2
2.	Biosafety and Good Laboratory Practices	17
PART II: Advanced Techniques in Molecular Biology		**36–124**
3.	Techniques of Cell Fractionation and Centrifugation	37
4.	Chemical Synthesis of Nucleic Acids	58
5.	DNA Chip Technology and its Potential Applications	74
6.	Bioinformatics in Biotechnology	94
PART III: Working with Nucleic Acids		**125–361**
7.	Isolation of Nucleic Acids	126
8.	Measuring Nucleic Acid Concentration and Purity	166
9.	Electrophoretic Techniques	191
10.	DNA Sequencing	232
11.	Genetic Maps and Marker Analysis	247
12.	Polymerase Chain Reaction (PCR)	288
13.	*In Situ* Hybridization	317
PART IV: Recombinant DNA and Genetic Engineering		**362–547**
14.	Fundamentals of Recombinant DNA Technology	363
15.	Enzymes in Molecular Cloning	382
16.	Gene Constructs and Cloning Vectors	401
17.	DNA Libraries	441
18.	Molecular Biology of Gene Transfer Systems	473
19.	Selection and Screening of Recombinant Molecules	511
PART V: Applications of Biotechnology		**548–644**
20.	Genetic Engineering of Microorganisms	549
21.	Genetic Engineering of Animals	581
22.	Genetic Engineering in Plants	612

PART VI: Working with Proteins — 645–785

23. Protein Purification Techniques — 646
24. Protein Detection and Estimation — 663
25. Protein Fractionation Techniques — 679
26. Immunochemical Techniques — 734

PART VII: Bacterial and Mammalian Cell Culture — 786–868

27. Biology of Bacteria — 787
28. Cultivation of Mammalian Cells *In vitro* — 836

PART VIII: *In Vitro* Plant Cell Culture and Crop Improvement — 869–1037

29. Plant Cell Culture Laboratory and Requirements — 870
30. Plant Culture Media, Preparation, and Culture Initiation — 887
31. Micropropagation — 905
32. Cultures of Organized Tissues — 925
33. Culture of Unorganized Tissues — 944
34. Cryopreservation and Distribution of Clonal Material — 980
35. Measurement of Plant Cell Growth and Cytological Analysis — 992
36. Protoplast Fusion and Somaclonal Variation — 1003
37. Application of Plant Cell, Tissue and Organ Culture — 1026

PART IX: Environmental Biotechnology — 1038–1154

38. Biotechnology in Pollution Control — 1039
39. Biodiversity and Genetic Conservation — 1073
40. Bioenergy Fuel from Biomass — 1107
41. Regulatory Aspects of Using Genetically-Modified Organisms — 1118
42. Intellectual Property Rights and Socio-Legal Aspects of Biotechnology — 1136

Appendices — 1155

Index — *1183*

Preface

Biotechnology is an important tool that can be applied to various economic sectors such as the production of food crops, livestock management, human health care, chemical industries, and environmental management. Many universities, understanding the importance of biotechnology research and the need for qualified manpower to exploit such technologies, have started undergraduate and master's degree programs. With the burgeoning number of such courses, experts have realized the urgent need for developing a suitable curriculum, possibly in the form of a model textbook aimed at providing undergraduate and postgraduate students with a strong base in this emerging and highly promising interdisciplinary area.

Molecular Biotechnology: Principles and Practices, is designed to balance between two important aspects of the science. The first aspect is the principles of molecular biology, which constitutes the theoretical knowledge pertaining to molecular biotechnology. The second aspect is practices in molecular biology, or the experimental approach to the study of biological processes. This book can serve as a textbook for both undergraduate and postgraduate students of molecular biology and biotechnology. It can also be used as a laboratory reference book in most research laboratories. The salient feature of this book is that it covers a wide range of molecular techniques in biotechnology and provides a source of information to readers at all levels. Concise and straightforward explanations of both theory and techniques associated with molecular/recombinant technologies are very few in literature. Most research articles in the field discuss either theory or techniques individually, but rarely explain both together and adequately. This makes life difficult for the student, teacher or researcher, who is new to the subject. Although several books on biotechnology have been published in the last decade, most are either very shallow or cover few areas in depth. Rarely do they cover the broad spectrum of topics which would provide enough information for understanding the subject or provide simple protocols for execution of molecular biology experiments. Realizing this deficiency, I have made an attempt to explain the basic concepts in biotechnology and the detailed steps of some important experiments in relatively simple terms. In my opinion, this book can help the reader to easily understand the subject and also execute the experiments very efficiently.

The book is divided into nine sections, containing 42 chapters and an appendix. In recent years, both the amount of molecular biology knowledge and its rate of growth have exploded. It is im possible to keep pace with this development and include references to all the work exhaustively. Therefore, in this book, only a few representative methods, which can provide standard protocols and general information on those techniques, are presented under each section. Alternative protocols provided for each technique make it more suitable for different laboratory conditions and systems used. Sources where additional information can be obtained are sufficiently quoted at the end of each chapter. Each chapter is adequately illustrated with computer graphics in an attempt to convey the practical as well as theoretical aspects of the techniques. It should, therefore be well suited for both the lecture room and the laboratory bench. The illustrations are adequately labeled and explains step-by-step, the procedure to be followed at each level of the process.

I dedicate this work to the authors of the books, which I have referred to as a source of information. They have given me the inspiration to write this book. Writing a book requires a lot of effort, patience and dedication without really gaining much benefit or appreciation. However, in my opinion, the benefits obtained by the readers is priceless. My personal experience of reading books is that it not only enriched my knowledge, but that it also helped me choose the right path in life. I hope this book will prove to be useful to the readers. I invite their valuable comments and suggestions for improvements in future editions.

I would like to thank many people for the encouragement and inspiration they have given me while I was writing this book. I especially thank the staff and management of Universities Press (India) Pvt. Ltd. for their editorial assistance and the fine production of this book. I want to thank Prof. M. Udaya Kumar, HOD, Crop Physiology, UAS; Mr. A. M. Veerabhadraiah, Associate Professor, Soil Chemistry, UAS; Dr. K. V. Devaraj, former Vice-Chancellor, UAS, Bangalore; Dr. Balakrishna Gowda, Associate Professor and Dr. R. Uma Shaanker, Associate Professor, UAS, Bangalore; Dr. K. V. Janardhan, HOD, Biotechnology, Vydehi Institute of Biotech Sciences, Bangalore; Dr. J. N. Vinay, Postdoctoral fellow, Arizona State University, USA, for their help and inspiration while writing this book.

Channarayappa
crayappa@rediffmail.com

PART – I
Introduction to Biotechnology

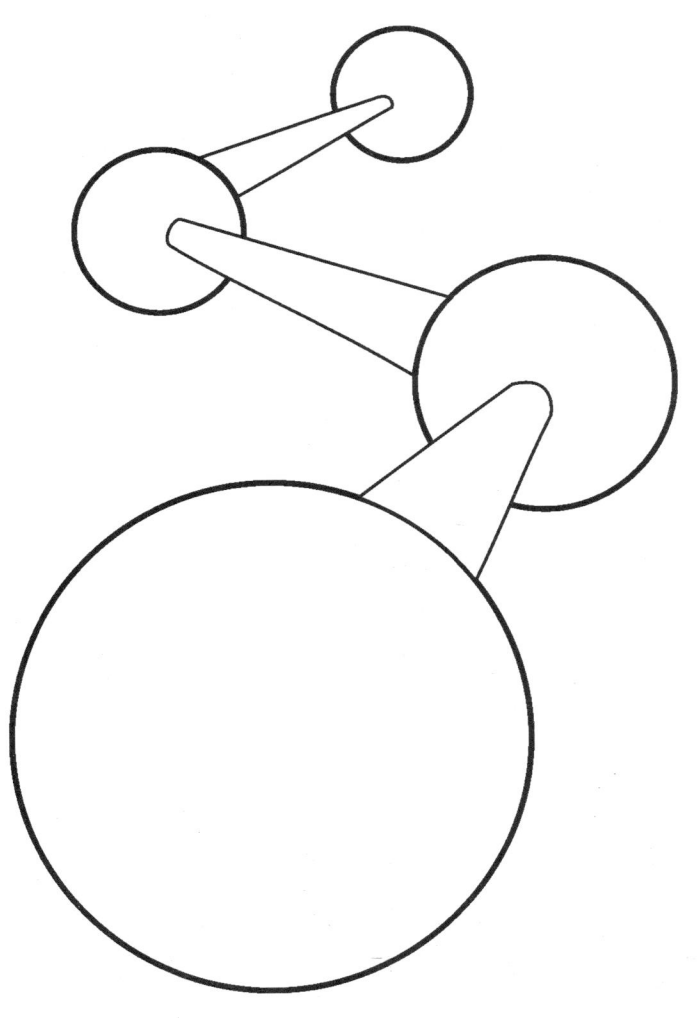

Biotechnology: Scope and Importance

I INTRODUCTION

The recent advances in science and technology have driven the biological sciences into a new era. All of this can probably be traced back to when Anton van Leeuwenhoek, a Dutch dry-goods dealer, ground the first microscope lens. Through his newly invented glass, he discovered a previously unseen cellular world. The second half of the 21st century was a truly exciting time for molecular biologists. Many inventions, including new methods for analyzing proteins, DNA and RNA, fueled an explosion of information and enabled scientists to understand cells and multicellular organisms at the molecular level. Now we have molecular blueprints (genomic sequences) for many organisms. The main objective of modern science, however, is to know the nature of genetic material and to find the answers to questions like: Which genes determine specific characters? How do they get switched on and off spatially and temporally? How do we correct genetic defects? How can we best manipulate genomes?

A. WHAT IS BIOTECHNOLOGY?

The field of biotechnology, which emerged as a new discipline, was a result of the fusion of biology and technology. Biology is the science of all living organisms or their components, whereas technology deals with the physical–chemical properties and techniques (Chapters 3–13) applied to the production of biological products/services. The emergence of biotechnology has been possible mainly due to the revolutionary discoveries made in these two areas. Biotechnology has been defined in many ways by many organizations. Biotechnology may be broadly defined as *"the controlled use of selected/manipulated biological systems or processes for the production of abundant/novel products or services"*. Therefore, the area covered under biotechnology is very vast and the techniques involved are widely divergent. Biotechnology can be applied in areas as diverse as agriculture, animal husbandry, medicine, environment, industry, and biological conservation.

Biotechnology is multidisciplinary by its very nature and encompasses several disciplines of basic sciences (e.g., genetics, biochemistry, molecular biology, chemistry, microbiology, immunology, cell and tissue culture, and physiology), engineering (processing technology, biochemical engineering, electronics and physical sciences) and also other disciplines like sociology, economics, politics, law and ethics (Figure 1.1).

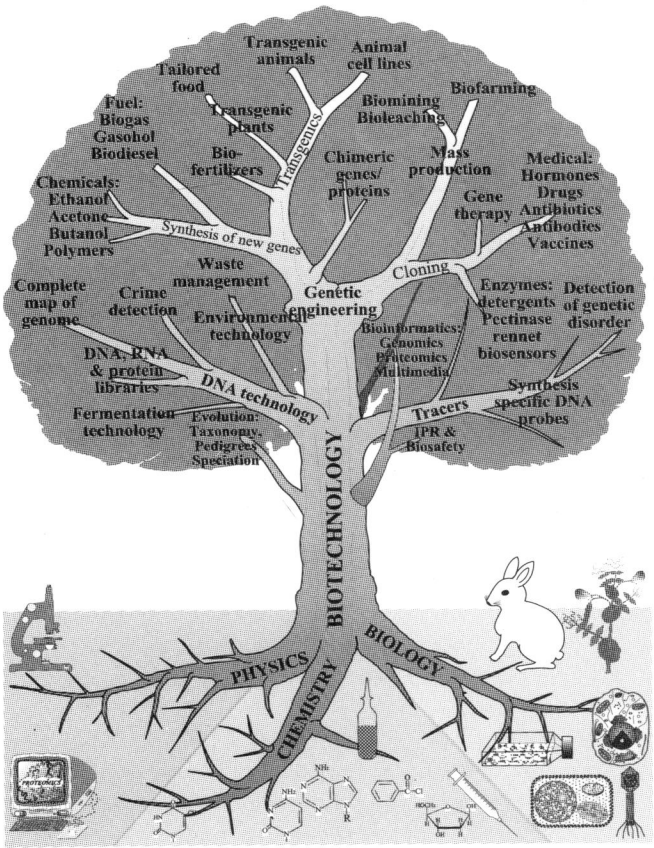

Figure 1.1 *Biotechnology tree: Evolution of the field of biotechnology and some important areas of biotechnology applications. The three main roots represent the importance of basic science knowledge*

B. WHEN DID BIOTECHNOLOGY BEGIN?

Although the term biotechnology is a recent development, its origin can be traced back to prehistoric times. Humans have been altering the genetic composition of plants for millennia – retaining seeds from the best crops and planting them in the following years, breeding and cross- breeding varieties to make them taste sweeter, grow bigger, last longer, etc. In this way, early agriculturists transformed the wild tomato (*Lycopersicon*), from a fruit the size of a peanut to today's giant, juicy and fleshy tomato. From a weedy plant called *teosinte* with an "ear" barely an inch long has emerged our foot-long ears of sweet, nutrient-rich, yellow corn. Man has also selected hundreds and thousands of new crop varieties by selection and hybridization. In ancient scripts, it has been documented that humans employed microorganisms as early as 5000 BC for making wine, vinegar, yogurt, leavened bread, etc. The discovery that fermentation converted fruit juice into wine, milk into cheese and yogurt, and solutions of malt and hops into beer seems to have set in motion the study of biotechnology. The early animal breeders soon realized that different physical traits could be either magnified or lost by mating the appropriate pairs of animals, thereby engaging in the traditional manipulations of biotechnology. However, the use of microorganisms for the production of chemicals on a commercial scale begun during the First World War, and has recently been more fully exploited due to the advancement of modern biotechnology.

C. MODERN BIOTECHNOLOGY

Modern biotechnology is innovative and quite different from the conventional practices. Traditional breeders made crosses only between related organisms whose genetic composition was compatible (genetically closer). Doing it this way involved the transfer of tens of thousands of genes (many genes were not required) after years of long selection procedure. By contrast, today's genetic engineers can transfer just a few desirable genes at a time, between species that are distantly related or not related at all. In other words, scientists can extract a desirable gene from virtually any living organism and insert it into virtually any other organism. They can put human gene into a plant or microorganism or in any other combination. For example, they can put a rat gene into lettuce to make a plant that produces vitamin C or insert a microbial toxin gene into cotton plants to make it insect-resistant. All this genetic manipulation became possible by the discovery of techniques of gene splicing and recombinant DNA technology (see Chapters 7–19). The engineered organisms which scientists produce by transforming genes between species are called "transgenic" organisms. Transgenic animals and several dozen transgenic food crops are currently in the market. Most of these crops are engineered to help farmers deal with age-old agriculture problems: good seeds, insects, diseases, nutrient composition, stress tolerance, etc (see Chapters 20–22).

The beginning of modern biotechnology can be traced back to 1865, when Gregor Mendel published the results from his experiments conducted on the garden pea on the inheritance of seven different physical traits. This and many other studies eventually led to the concept of the gene as the basic unit of heredity. Over the next century, many other researchers with sophisticated techniques and instruments contributed to the growth of modern biotechnology (Table 1.1).

Table 1.1 *Chronology of some major developments in biotechnology*[*]

Before 6000 BC	Yeasts used to make wine and beer
About 4000 BC	Yeasts were used for making leavened bread
1866	Mendel published his research findings, experiments conducted on the garden pea, which led to the concept of the gene as the basic unit of heredity.
1869	Friedrich Miescher isolated nuclein, later shown to be DNA, from the nuclei of white blood cells.
1885	*E. coli* bacterial cells are identified and grown under controlled conditions.
1902	Archibald Garrod's report that the human disease "Alkaptonuria" behaves as a Mendelian recessive trait led to the suggestion that enzymes are encoded by genes.

Continues...

Continued...

Table 1.1 *Chronology of some major developments in biotechnology**

1910	Thomas Hunt Morgan showed the first evidence of the presence of genes in chromosomes. He used microorganisms to treat sewage.
1912–14	Large-scale production of acetone, butanol and glycerol using bacteria.
1917	Karl Ereky coined the term "biotechnology".
1940	George Beadle and Edward Tatum hypothesized the concept of "one-gene-one-enzyme".
1943	Penicillin was produced on an industrial scale.
1944	Avery, Macleod and McCarty demonstrated that DNA, not protein, carries hereditary information. Penicillin was produced on a large scale for the first time.
1952	Alfred Hershey and Martha Chase demonstrated that the genes of a bacteriophage are made of DNA and are capable of directing the synthesis of new bacteriophage proteins.
1953	Watson and Crick determined the structure of the DNA double helix. They drew from the work of other scientists to propose that the molecule is an alpha double helix structure in which the two strands are both complementary and antiparallel to one another.
1955–65	The role of tRNA, mRNA, and rRNA as well as of DNA and RNA polymerases in gene function was elucidated.
1957	The Central Dogma, which states that hereditary information flows from DNA to RNA to protein, was put forth by Francis Crick and George Gamov.
1961	Marshall Nirenberg and Har Gobind Khorana correctly translated the genetic code.
1962	Uranium was mined with the aid of microbes (Canada).
1967	DNA ligase was isolated and identified.
1970	Stewart Lin and Werner Arber identified the first restriction endonucleases. In the same year researchers discovered the enzyme reverse transcriptase, which catalyzes the reaction in which DNA is transcribed from an RNA template.
1972	Khorana and co-workers synthesized an entire tRNA gene. Paul Berg created the first recombinant DNA molecule.
1973	Stanley Cohen, Herbert Boyer and colleagues constructed a functioning plasmid, containing genes, which confer resistance to both tetracycline and streptomycin; i.e., the establishment of recombinant DNA technology.
1976	Techniques were developed to determine the sequence of DNA.
1977	Somatostatin became the first human hormone to be synthesized by a bacterial cell as a result of transformation with human DNA.
1981	The use of monoclonal antibodies for diagnosis was approved in the USA.
1983	Approval was granted for the use of insulin produced by genetically-engineered microbes (GEMs).
1984	Animal interferons, produced by GEMs, were approved of for the protection of cattle against diseases.
1988	The polymerase chain reaction (PCR) method was published.
1990	Approval was granted in the USA for a trial of human somatic cell gene therapy.
To date	The field of biotechnological research enormously expanded and is still expanding exponentially.

* *Collected from different books and journals.*

D. GENE TECHNOLOGY IS A BASIC TOOL FOR BIOTECHNOLOGY

Scientists continue to find new ways to insert desirable genes for specific traits into the DNA of different biological systems (plants, animals and microorganisms). A field of promise and a subject of debate, genetic engineering is changing the food we eat and the world we live in. Many scientists envision a cornucopia (similar to the "*kalparuksha*" or "*Kamadenu*"): the mass production of rare plants; highly variegated and long shelf-life flowers; tomatoes and broccoli produced with pharmaceutical compounds and industrial chemicals; vaccine-producing bananas; vitamin-enriched rice, sweet potatoes, and cassava to help the malnourished poor and vegetable oils so loaded with therapeutic ingredients that doctors "prescribe" them for patients at risk of cancer, heart disease, diabetes, etc; cheaper and safer fuel; clean environment, etc. The possibilities are endless.

Overall, gene manipulation has provided novel solutions to experimental problems in biology; these solutions, in turn, have led to novel products. Most biotechnology companies and research institutes make use of gene technology or genetic engineering, which mainly involves recombinant DNA and gene cloning. This technology allows the splicing of a DNA molecule at desired places to isolate a specific DNA segment and then inserting it into another DNA molecule at a desired position (Chapter 15). The resultant product is called "recombinant DNA" and the technique often called "genetic engineering". Using molecular techniques, we can isolate and clone a single copy of a gene or a DNA segment into an indefinite number of copies with similar properties. This became possible due to the identification and modification of different kinds of vectors (a self-replicating DNA molecule is called "vector", e.g., plasmid, phage or virus) with suitable properties (Chapter 16), and which can be mobilized to a suitable host, where they reproduce along with the host. The vector carrying the inserted DNA will also replicate faithfully through the vector DNA. This technique is called "gene cloning" (Chapter 14). A gene is a part of a chromosome and is responsible for a specific character or trait of an organism/cell. Genes produce their phenotypic effects by specifying the amino acid sequences of specific proteins. There are also various biological tools that are used to carry out manipulation of genetic material and cells.

The gene cloning technique has had a tremendous impact on all areas of molecular biology and, consequently, on biotechnology. Recombinant DNA technology broadly involves: (1) the isolation of a specific DNA (Chapter 7), (2) the selection of vectors (Chapter 16), (3) the preparation of a chimeric DNA (Chapter 14), (5) cloning of the chimeric DNA, and (6) the screening of recombinants (Chapter 19).

II WHAT MAKES BIOTECHNOLOGY A POWERFUL TECHNOLOGY?

Biotechnology is a unique and powerful technology, since it helps humans in many ways, for example, (i) mass production, (ii) the generation of novelty, and (iii) better service.

A. OVERPRODUCTION OF CELLULAR COMPONENTS

For commercial use or in academic studies, the determination of the structure, function or utility of a protein demands that adequate amounts of purified material are available. This is not always an easy task, particularly when the protein is normally present in very low levels in the cell mass. Genetic engineering provides a means of generating sufficient material. For example, 5 mg of somatostatin was first isolated from half a million sheep brains and a small amount of epidermal growth factor from 40,000 gallons of human urine. After the advent of gene cloning, the same amount of material was obtained from a few liters of bacterial culture. This principle has been applied to a wide range of cellular proteins and was the basis for many of the biotechnology start-up companies in the world. Over-production need not be restricted to proteins. It is possible to raise the levels of most intracellular components, provided that they are not toxic to the producing organism. This can be done by cloning all the genes for a particular biosynthetic pathway and over-expressing them. Alternatively, it is possible to shut down particular metabolic pathways and thus re-direct particular intermediates towards the desired end product.

B. GENERATION OF NOVELTY

By and large, gene manipulation has provided novel solutions to experimental problems in biology. These solutions have led to creation of novel products. Gene manipulation has been used to permanently modify the germ cells of animals ("transgenesis"), e.g., the production of 'supermice' which are extra-large as a result of the over-production of the human growth hormone. Transgenic animals that over-express foreign proteins and secrete them in milk have been developed. Novel therapies for various human diseases have been also created. Transgenic plants carrying genes resistant to various stresses have also been produced. However, the development of these products has also raised some novel problems. Some of these benefits have been discussed in the following sections.

C. BETTER SERVICES

Biotechnology can also be of better service to human beings, other organisms and to the ecosystem by providing solutions to many natural and man-made problems. For example, biotechnology can provide solutions to environmental pollution, it can help to monitor and conserve biodiversity, and reduce the rapid loss of natural resources by providing alternative solutions. It can also aid in the overall improvement of biological safety and in maintaining the ecological balance (see the following sections for more details).

III APPLICATIONS OF BIOTECHNOLOGY

Biotechnology has rapidly emerged as an area of activity that will have a potential impact on virtually all domains of human welfare – food processing, human health, agriculture animal improvement, and environmental protection (Table 1.2). As a result, it plays a very important role in employment, production, trade, economics and economy, human health, conservation of biodiversity and even in the socio-economical and political status of a nation. This is clearly reflected in the emergence of numerous biotechnology companies throughout the world.

Table 1.2 *A list of areas in which biotechnology is making significant contributions*

Pharmaceutical industries	Agriculture
Human health care products	Horticulture and floriculture
Animal improvement	Environment
Dairy and animal husbandry	Renewable energy and biofuels
Secondary metabolites	Population control
Food processing and beverage	Microbial-mining or leaching
Crime detection and parentage disputes	Fisheries and aquaculture
Embryo-cloning and gene therapy	Forestry
Intellectual property rights and trade	Biodiversity conservation

The importance of biotechnology to human welfare is becoming increasingly more important. The products of DNA research, ranging from proteins to engineering organisms, have a wide range of applications. The following are some of the contributions of biotechnology to overall development of human welfare.

A. MEDICAL BIOTECHNOLOGY

1. Diagnosis of biological disorders

The production of biological reagents for the diagnosis of biological disorders and infectious diseases is a major industry. On a global scale, the revenue from sales is estimated at many billions of dollars. There are three types of diagnostic reagents: biochemical reagents for assaying specific enzymes, antibodies for detecting specific proteins, and a recent development, nucleic probes (Chapters 12–13).

Molecular analysis of genetic disorders: There are several hundred genetic diseases in man, which are the result of mutations. For many of these genetic diseases there is no definitive treatment, although in some cases human gene therapy may become possible. DNA research helps in understanding many diseases or genetic disorders at the molecular level, such as sickle cell anemia, thalassemias, familial hypercholesterolemia, etc.

Laboratory diagnostic applications: By using recombinant DNA techniques, many diseases can be diagnosed, e.g., AIDS, hepatitis B, etc. In the case of most diseases, diagnosis is the detection of a specific microorganism. For example, a patient with a disorder of the gastrointestinal tract could be infected with more than 15 types of microorganisms. By using the right probe, a particular infectious microbe can be accurately detected from stool samples without any cultivation and cytological work.

Prenatal diagnosis of diseases: The DNA collected from the amniotic fluid can be used to predict the risk of developing genetic diseases (e.g., sickle cell anemia and many other genetic defects) and this can be done by using DNA probes.

2. Hybridoma technology

This produces large quantities of monoclonal antibodies, which are used for the diagnosis of various diseases, e.g., venereal diseases, hepatitis B, viral diseases, cancer, etc (Chapter 26).

3. Genetically-engineered microbes

Vaccines: Conventional viral vaccines consist of inactivated, virulent strains or live, attenuated strains, but they are not without their problems (e.g., there is a danger of vaccine-related disease). Many problems can be overcome by recombinant vaccines, which are cleaner and safer (e.g., human hepatitis B virus, *E. coli* vaccines for pigs, rabbits, etc.)

GEMs can be used to produce DNA probes that can be used for diagnosing diseases such as *kala-azar*, sleeping sickness, malaria, etc. They can also be a source of valuable drugs like human insulin, human interferon, human and bovine growth hormones, etc (Chapter 20).

4. Advanced therapeutic techniques

Production of proteins in abundance: Using recombinant DNA techniques, several proteins have been produced in abundance for therapeutic purposes. These include insulin, growth hormone, erythropoietin, interferons, blood-clotting factors, vaccines and superoxide dismutase.

Gene therapy can be used to cure various genetic diseases like Huntington's chorea and cystic fibrosis.

Artificial insemination and sexing: Genetic procedures can be used to treat infertility as well as produce babies of a specified sex (by artificial insemination with X or Y carrying sperms, prepared by sperm separation techniques).

5. Application to forensic medicine

Finger-printing and forensics: In paternity disputes, the parents can be identified by using DNA or auto-antibody finger-printing. Using the same technique, criminals can be identified very accurately by blood or semen stains, hair roots, etc., collected from a crime scene and compared with the suspect's DNA.

B. INDUSTRIAL BIOTECHNOLOGY

An industrial biotechnology process that uses microorganisms for producing a commercial product typically has three operational stages.
(1) *Upstream processing:* preparation of a raw material so that it can be used as a food source for the target microorganism.
(2) *Fermentation and transformation*: Growth of the desirable microorganism in a fermenter (usually a bioreactor with >100 liters), which produces the desired product by fermentation/biotransformation process.
(3) *Downstream processing*: This involves the purification of the desired product either from the cell medium or the cell-mass. Some of the important areas of commercialization of biotechnology are listed below.

1. Production of organic compounds

Many useful compounds, such as ethanol, lactic acid, glycerine, citric acid, gluconic acid, acetone, etc., can be produced by using various microorganisms, mainly bacteria.

2. Secondary metabolites

Many antibiotics can be produced (e.g., penicillin, streptomycin, erythromycin, mitomycin, cycloheximide, etc.), as they are generally produced by fungi, bacteria and actinomycetes as secondary metabolites.

3. Biotransformation

This is the transformation of toxic compounds to less toxic or non-toxic compounds. By means of biotransformation, less useful and cheaper compounds can be converted into more useful and valuable ones – e.g., steroid hormones from sterols, sorbose from sorbitol – using microorganisms or immobilized enzymes in a fermenter.

Enzymes like proteases, lipases, α–amylases can be produced by fungi, bacteria, etc., and these enzymes are used in the detergent, textile, leather, dairy and pharmaceutical industries.

4. Biomass production

Single-cell proteins (SCP) are generally a total microbial biomass freed from toxins and contaminants, if any, and are obtained from bacteria, yeasts, fungi or algae for use as human food and as animal feed (in the form of supplements).

5. Biofuels

Biofuels, mainly ethanol and biogas, are produced from cheap, less useful and abundant substrates, e.g., sugarcane bagasse, wood, cow-dung and other animal excreta, leaf litter and agricultural wastes. Biofuels are generally produced through fermentation by microorganisms (Chapter 40).

6. Bioleaching

Biomining or bioleaching is the microbial (mainly bacterial) process of mineral extraction through leaching from low-grade ores, e.g., copper, gold, uranium, etc (Chapter 20).

7. Immobilization of enzymes

The immobilization of enzymes results in their efficient functioning and repeated industrial applications. Enzymes may be immobilized in various types of matrices, and used for conversion of different substrates into products which can be easily purified.

8. Engineering of proteins/enzymes

This means changing the primary structure of the existing proteins/enzymes by genetic engineering. The protein/enzyme can be modified into a more efficient prototype, e.g., with substrate specificity, temperature sensitivity, etc. This has been successfully carried out for T4 lysozyme, trypsin, subtilisin, lactate dehydrogenase, etc.

By engineering proteins, immunotoxins (by joining a natural toxin with a specific antibody) can be produced which can destroy specific cell types. Therefore, it can be effectively used in the treatment of particular diseases, e.g., cancer.

C. ANIMAL BIOTECHNOLOGY

1. *In vitro* fertilization and embryo transfer

In vitro fertilization and embryo transfer have been successfully applied both in humans and in animals. In the case of human beings with the help of biotechnology, some couples suffering from infertility can bear children (Chapter 21).

The rapid multiplication of superior genotypes in animals can be achieved by hormone-induced superovulation, *in vitro* fertilization and/or embryo splitting followed by embryo transfer into a foster mother.

2. Transgenic animals

Transgenic animals can be created for increased milk, rapid growth rate, resistance to diseases and the production of valuable proteins in the milk/serum/urine. Many transgenic animals carrying novel characteristics have been developed in mice, pigs, chicken, rabbits, cattle, sheep and fish.

Transgenic fish: Atlantic salmon grow more slowly during the winter, but engineered salmon carrying modified growth-hormone genes from other fish, reach market size in about half the normal time.

Pharmaceutical-producing animals: Scientists are using biotechnology to insert genes into cows and sheep, so that the animals produce beneficial pharmaceuticals in their milk.

D. PLANT BIOTECHNOLOGY

1. Diagnostics

The accurate identification of viruses is critical in the prediction of plant diseases in annual crops, for the prevention of infection in planting stock, in monitoring disease-control methods and in diagnosing diseases in plants held in quarantine. Biotechnology is a useful tool in identifying viruses.

2. Plant cell culture

Clonal propagation: A very high proportion of plant material can be multiplied by clonal propagation through meristem culture. For example, many fruit and forest trees (e.g. teak) are very slow-growing and their growth can be increased manifold by means of biotechnological protocols (Chapters 29–37).

Embryo culture: Embryo culture is used to rescue unviable hybrids and haploid plants from inter-specific hybrids. This is one of the most important techniques in plant breeding programs.

Virus-free plants: Meristem culture is generally combined with thermotherapy/cryotherapy. This procedure is usually followed to recover virus- and pathogen-free stocks of clonal crops. It is very useful in clonal crops and germplasm exchange.

Anther culture: The rapid isolation of homozygous lines can be achieved by chromosome doubling of haploids produced through anther culture. This technique has been successfully used in the development of variety, e.g., in rice and wheat.

Somaclonal variation: This refers to the isolation of stable somaclonal variants with improved yield and other desirable traits such as resistance to diseases, cold, herbicides, metal toxicity, salt and other abiotic stresses. Many examples of the successful isolation of genotypes to create novel and heritable traits have been reported for many crops.

3. Transgenic plants

Transgenic plants have been established in many crop varieties by using transfer techniques for various characteristics, which include insect resistance, protection against viruses, herbicide resistance, storage protein improvement, secondary metabolites, etc (Chapter 22).

Herbicide resistance: Farmers routinely spray herbicides to kill weeds (they also kill normal crops!). However, biotech crops can carry special "tolerance" genes that help them withstand the spraying of chemicals that kill nearly every other kind of plant.

Insect resistance: Some biotech varieties make their own insecticides. For example, Bt transgenics, which is a gene borrowed from a common soil bacterium, *Bacillus thuringiensis*, (Bt for short). Bt genes code for toxins that are considered harmless to humans but lethal to certain insects. When these insects feed on Bt plants, the toxin attacks their digestive tracts, and they die within a few days.

Disease resistance: Many crops have been genetically engineered to resist diseases. For example, squash and papaya have been genetically engineered to resist viral diseases. Potatoes have been transformed with the genes of bees and moths to protect the crops from the potato blight fungus, and grapevines have been altered with silkworm genes to make the vines resistant to Pierce's disease, which is spread by insects.

Generation of novel food content of plants: It is widely recognized that gene manipulation techniques have revitalized the biotechnology industry. Altered starch content, oil and amino acid composition in storage proteins of seeds are a few of the many transgenic food crops developed.

Pigmentation in transgenic plants: Plants are widely used for ornamental purposes; therefore, considerable attempts have been made to develop varieties exhibiting new colors or pigmentation patterns. The pigmentation in flowers is mainly due to three classes of compounds: the flavonoids, the carotenoids and the betalains. Several flavonoids genes have been cloned and many-hued color patterns have been created.

4. Molecular mapping

Molecular markers, e.g., RFLPs, RAPD, AFLP, QTLs, etc., have been extensively used both for linkage mapping and for the mapping of quantitative trait loci. These techniques are very powerful tools for the indirect selection of quantitative traits and for several other important applications (Chapter 11).

E. BIODIVERSITY CONSERVATION

The term "biodiversity" epitomizes a new name for species richness occurring as an interacting system in a given habitat. Biodiversity can be measured at the gene level, the species level or at the ecosystems level. The conservation of biodiversity is vital for our future survival and quality of life. Due to human interference (increased use of agricultural land, environmental pollution due to industrialization and mining) hundreds and thousands of species have been lost and thousands more are on the verge of extinction. Thus, biodiversity conservation is becoming one of the priority areas of human activities. India is rich in both fauna and floral biodiversity. In India, over 1, 15,000 species of plants and animals have already been identified and described. India is also an important center of diversity and origin of over 167 important cultivated plant species and domestic animals. There is an urgent need for biodiversity conservation throughout the world (Chapter 39).

Gene Bank and Plant Conservation: To feed the ever-growing human population, agriculture has been extended to large areas and, as a result, most of the cropping area is occupied by a few high-yielding crops (monoculture). Consequently, our natural resources are rapidly disappearing and the genetic diversity of many species is either decreasing or lost. Therefore, to save the valuable and threatened species, germplasm conservation is essential. The gene bank (germplasm) has undertaken the challenge of conserving the gene pool of many species. Conservation efforts are taking place at different levels: international, national and local. During the last two decades, many regional and international genetic resource centers (GRCs) have been set up in different countries. For example, the International Rice Research Institute (IRRI), Manila, has collected about 25,000 varieties of rice germplasm. Similarly, at the Maize and Wheat Improvement Center (Mexico) (CIMMYT), more than 12,000 varieties have been conserved.

F. EVOLUTION

Recombinant DNA technology is of great service in bridging several missing links in the evolution. This has done by amplifying the DNA (by PCR) from the archeological samples of extinct animals.

G. ENVIRONMENTAL BIOTECHNOLOGY

1. Sewage treatment

Microorganisms which are isolated from the natural environment or genetically engineered are used for efficient sewage treatment and deodorization of human excreta (Chapter 38).

2. Biodegradation of oil spills

Oil spills and automobile exhaust are major sources of environmental pollution in recent times, particularly in industrial areas. A strain of *Pseudomonas putida* has been used to treat environmental pollution. Many other efficient strains are in the developmental stages.

3. Biodegradation of xenobiotics and industrial effluents

Many industrial wastes and commercial products contain toxic materials. Most of them are resistant to natural degradation and are gradually accumulating in the environment. This biomagnification of toxins is a serious threat to human health and environment. Even though microbes in nature can biodegrade them, many man-made compounds are too complex for the natural microorganisms to break down. Now, many microorganisms with the capability to detoxify genobiotics have been developed by means of genetic engineering.

4. Biological control: An alternative to chemical control of plant diseases and pests

The biocontrol of diseases and insect pests, by using viruses, bacteria, amoebae, fungi, etc, is environmentally friendly and circumvents the use of pesticides, which cause environmental pollution and pose health hazards.

H. INTELLECTUAL PROPERTY RIGHTS AND FARMER'S RIGHTS

In recent years, many private companies and individuals have developed many novel products and techniques. Since they spend substantial amount of money and manpower, they have made a case to establish a legislation to protect their inventions from piracy (Chapter 42).

Intellectual property rights (IPR): These include patents, trade secrets, trademarks, and copyrights, which can be protected through a variety of laws in different countries.

Farmer's rights: In recent years there have also been discussions about the protection of farmers' traditional knowledge (farmers' rights) so that they are not exploited by the industries or any other organizations.

I. COMMERCIAL POTENTIAL OF BIOTECHNOLOGY

The ultimate aim of biotechnological research is the development of commercial products. Biotechnology appears to have an unlimited commercial potential to develop a wide range of valuable and novel products/services, which, generally, cannot be achieved by conventional methods. It deals with virtually all aspects of human existence. Its practical applications are ever-growing and it has now become a multibillion-dollar industry. In 1985, there were over 400 biotechnology companies in the United States, and today there are over 900 biotechnology companies in USA alone and about 1200 worldwide. In mid-1991, there were over 130 biotechnologically-developed pharmaceuticals; and the present rate of development indicates that, globally, biotechnological products can exceed more than $100 billion/year by the year 2010. In India alone it is estimated that this figure will rise to around $5 billion by the year 2010.

This new discipline clearly demands very high expertise and skill, and continued support in terms of heavy funding and dedicated efforts. This gives developed countries a dramatic edge over developing countries, both with regard to skilled workers, funding for research and the exploitation of market potential.

IV SAFETY OF BIOTECH PRODUCTS

Biotechnology will undoubtedly provide unprecedented benefits to humankind (Chapters 20–22). Although it is very exciting and important to emphasize the positive aspects of new advances, there are also many concerns and consequences (possible hazardous effects on society and environment) that must be addressed.

A. BIOTECH FOOD

Risks exists everywhere in our normal food supply; for example, about a hundred people die each year from peanut allergies. However, with genetically-engineered foods, we can substantially minimize these risks by doing rigorous testing. Generally, transgenic products go through more testing than any of the other foods we eat; there would be rigorous screening for potential toxins and allergens and assessment of the nutrient composition and levels. For example, a biotechnology company launched a project to insert a gene from the Brazil nut into a soybean, to make a protein rich in one essential amino acid. Because the Brazil nut is known to contain an allergen, the company also tested the product for adverse human reaction, with the idea that the transgenic soybean might accidentally enter the human food supply. When tests showed that humans would react to the modified soybeans, the project was abandoned. However, if similar types of allergenic proteins were present in a new food developed by conventional methods, they would not have been detected.

So far, none of the current biotechnology products have been implicated in allergic reactions or any other healthcare problem in people. In fact, biotechnology foods have some health benefits. Corn damaged by insects often contain high levels of fumonisins, toxins made by fungi that are carried on the backs of insects and that grow in the wounds of the damaged corn. Laboratory tests have linked fumonisins with the incidence of cancer in animals and humans. Studies show that most Bt corn have lower levels of fumonisins than conventional corn damaged by insects.

B. EFFECT ON THE ENVIRONMENT

The main concern about genetically-engineered crops is the issue of "gene flow", i.e., the movement of genes via pollen and seeds from one population of plants to another (Chapter 41). The probability of the effect on non-target organisms by genetically-engineered plants was realized as a result of a study suggesting that Bt corn pollen harmed monarch butterfly caterpillars. The pollen dusted on milkweed leaves causes a certain percentage of mortality to monarch caterpillars and this led some environmentalists to believe that Bt corn is dangerous to wildlife. However, the pollen densities from Bt corn rarely reach damaging levels on milkweed, and the chances of a caterpillar finding Bt pollen doses are negligible. Actually, butterflies are safer in a Bt cornfield than they are in conventional cornfield, because conventionally used chemical pesticides kill not just caterpillars but most insects in the field. Many scientists argue that plants offer an environmentally friendly alternative to pesticides, which tend to pollute surface- and groundwater and harm wildlife. The use of Bt varieties has dramatically reduced the amount of pesticides applied, for example, to cotton crops.

Perhaps a major concern to be considered has to do with insect evolution. Crops that continuously make Bt may hasten the evolution of insects impervious to the Bt. To delay the evolution of resistant insects, special measures for farmers who grow Bt crops have been devised. Farmers must plant a moat or 'refuge' of conventional crops near their engineered crops. The idea is to prevent two resistant bugs from mating. The few insects that emerge from Bt fields resistant to the insecticide would mate with their non-resistant neighbors living on conventional crops nearby; the result could be offspring that are susceptible to Bt.

Given the risks, many ecologists suggest that the industry should build up the extent and rigor of its testing and that governments should strengthen their regulatory regimens to more fully address environmental effects.

V SUGGESTED READING

Anonymous (1984) *Commercial Biotechnology: An International Analysis,* Office of Technology Assessment, US Congress, US Government Printing Office, Washington DC.

Anonymous (1987) *New Developments in Biotechnology—Background Paper: Public Perceptions of Biotechnology,* Office of Technology Assessment, US Congress, US Government Printing Office, Washington, DC.

Bulock JD and Kristiansen B (1987) *Basic Biotechnology,* Academic Press, London.

Dubey RC (1993) *A Textbook of Biotechnology,* S Chand & Company Ltd, New Delhi.

Gibbs DF and Greenhalgh ME (1983) *Biotechnology, Chemical Feed Stocks and Energy Utilization,* Francis Pinter (Publ.), London.

2

Biosafety and Good Laboratory Practices

I INTRODUCTION

Safety is of primary importance in the laboratory and many molecular techniques can often be hazardous. Therefore, good, explicit laboratory guidelines are necessary to carry out experiments. The main purpose of this chapter is to give some idea of the risks involved while doing laboratory work, the legal recommendations and the essential requirements to be employed, to prevent mishaps and finally the way to conduct oneself in the laboratory environment. The collection of laboratory safety principles that follows is certainly not comprehensive. It does not list all the hazardous and safety considerations; therefore, this chapter cannot be considered as a safety manual. The individual must thoroughly understand all the safety considerations and perform laboratory operations with maximum care. This chapter describes only some general and specific safety principles that are commonly followed when working in the molecular biology laboratory.

A close attention to safety aspects is an integral part of all laboratory procedures and national legislation impose legal requirements on those person(s) planning or carrying out such procedures. However, it remains the responsibility of the researcher to ensure that the procedure(s) followed are carried out in as safe a manner as possible and that all necessary safety instructions and regulations are implemented. In view of the rapid improvements in laboratory techniques and the introduction of new instruments and chemicals, *the researcher is urged to review and evaluate the information provided by the manufacturer, for each reagent, piece of equipment or other agents*, for instructions on usage, added warnings, and precautions. In particular, workers in recombinant DNA laboratories must be aware of safe methods of handling radioisotopes, protection from chemical hazards, and containment of biohazardous materials. When working with any hazardous substance, proper protection procedures should be always followed. In this chapter, the different biohazardous agents used, their nature of action, disposal methods and safety regulations are briefly explained.

II HANDLING AND CARE OF GENETICALLY-MODIFIED LIFE FORMS

Humans have been evaluating and modifying microorganisms, plants and animals through conventional approaches – such as selection of elite strains, breeding and mutagenesis – for centuries without paying much attention to biosafety. However, the recent surge in molecular techniques that enable us to genetically modify and manipulate genomes by inserting genes from distantly related organisms, are of concern to many of us. Although, recombinant DNA technology has many advantages, at the same time it also warrants the possibility of future biohazards. Therefore, foreign genes expressed in various organisms (genetically-modified organisms, GMOs) raise various concerns in the scientific community as well as in public.

A. MICROORGANISMS

There are four main aspects that all researchers working with microorganisms should be familiar with: (i) the precautions that should be taken to prevent contact with all or any cultured microorganisms handled in the laboratory; (ii) an understanding of the basic process of infection and the potential of different organisms to cause disease which will help the worker to anticipate and prevent unnecessary hazards; (iii) the potential danger(s) of genetically-engineered microorganisms and their dissemination in the environment; and (iv) a thorough knowledge about the handling, identification, disposal of all pathogens used in the laboratory which is essential for prevention and rectifying the problems associated with laboratory microorganisms.

There are many reports of presumed laboratory-acquired infections. It is apparent that the first step in selecting safe procedures is determining what organisms are pathogenic, and how infectious they are to human beings. Unfortunately, the problem is not so simple and it may be simpler to *regard all microorganisms as presenting some degree of hazard in one way or another*. The safest way to approach work with live microorganisms is to make the following assumptions: (1) every microorganism used in the laboratory is *potentially hazardous*; (2) every culture fluid contains potentially pathogenic organisms; and (3) every culture fluid contains toxic substances.

Microorganisms can transmit infections by different ways and means. The actual processes whereby microbial cultures are aerosolized and the laboratory techniques predispose to this are as follows: (1) the bursting of a liquid film, (2) the mixture of gas and liquid, (3) vibrations, (4) falling drops, (5) the "string of beads", (6) centrifugal force, (7) sizzling, (8) electrostatic effects and (9) gross splashes.

1. Safe handling and management of accidents

The aim of safe handling of laboratory cultures is to prevent whatever possible contact can be foreseen between viable microorganisms and the worker. Contact with organisms may be either by direct handling or exposure to cultures or contaminated material or air.

Any accident, such as spilled cultures, cuts and abrasions, should be reported and immediately treated. If a cut or abrasion is received in the laboratory, make sure that suitable first-aid treatment is obtained. A spilled culture should be flooded with a suitable disinfectant solution, which should be left for 15–30 min. before cleaning up. The spilled area must be properly identified by posting a notice to prevent other workers from coming into contact with the spillage. Broken glass should be collected with forceps and placed in an autoclavable sharps container.

2. Physical Protection

Undoubtedly, the most important ways of providing protection are: (1) the use of safety cabinets (Figure 2.1); (2) protective clothing; (3) good laboratory design; and (4) medical aspects. There are four biosafety levels of physical containment, and the level for a particular procedure is dependent upon the perceived safety risk as outlined in the guidelines. BL1 (Biosafety level 1) is the least stringent containment level and BL4 is the most stringent. All the levels require adherence to good microbiological practices, decontamination of work surfaces, and decontamination of wastes prior to disposal. Higher levels of containment require more stringent conditions, such as biological safety cabinets, the use of surgical masks, negative pressure rooms, etc. In addition to physical containment, there are various levels of biological containment as defined by the National Institute for Health, USA (NIH) guidelines. These depend upon the viability and transmission potential. The current NIH, Department of Biotechnology (DBT, India),

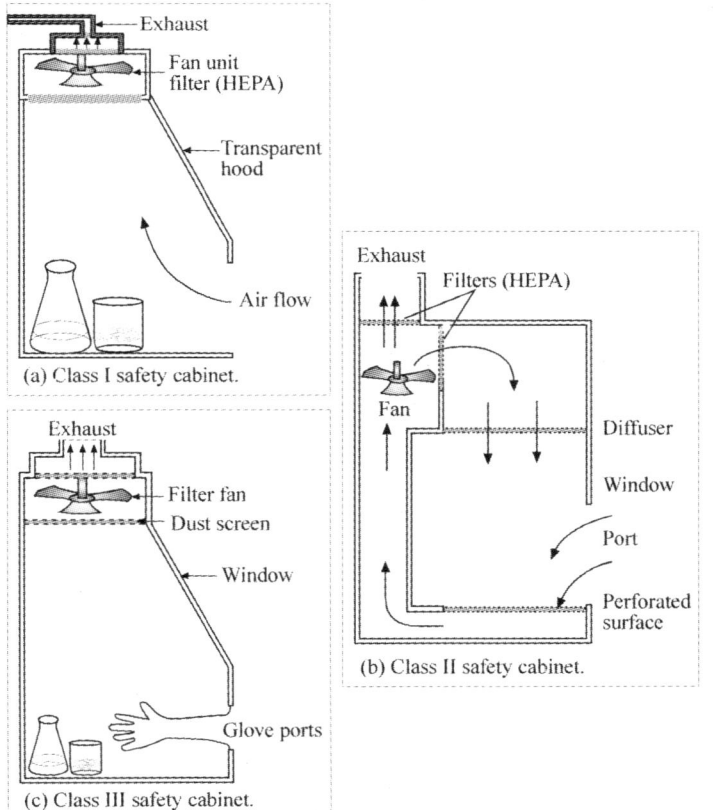

Figure 2.1 *Safety cabinets for prevention of airborne exposure: (a) a class I cabinet, (b) a class II safety cabinet, and (c) a class III safety cabinet*

Genetic Engineering Approval Committee (GEAC, India), and the Ministry of Environment and Forests (MoEF, India) guidelines, adhere to BL1 physical containment procedures, which are summarized below:
1. Follow standard microbiological practices.
2. Keep doors and windows closed during the laboratory session to prevent contamination from air currents. Access to the laboratory should be limited when experiments are in progress.
3. At the beginning and termination of each laboratory session, wipe work-tops with a disinfectant solution.
4. Decontaminate all wastes prior to disposal. If they are to be moved prior to decontamination, they must be placed in leak-proof containers.
5. Mouth pipetting is prohibited. Pipettes must be plugged with cotton; never blow infectious material out of pipettes. Mechanical pipetting devices are preferred.
6. Eating, drinking, smoking, and cosmetic applications are prohibited. Food may be stored in cabinets or refrigerators designated for this purpose only.
7. A sink for hand-washing must be available within the laboratory.
8. Wash hands with detergent and dry them with paper towels upon entering and prior to leaving the laboratory.
9. Minimize the creation of aerosols. Procedures which can produce aerosols include: grinding, blending, sonicating, re-suspending, inserting a hot loop into a culture, centrifugation, forceful ejection of fluid from a pipette or syringe, opening a tube containing a lyophilized agent, releasing the vacuum on a freeze dryer, opening a tube which may cause implosion, etc.
10. Use an alcohol-moistened cotton plug around the stopper and needle when removing a syringe and needle from a rubber-stoppered vacuum bottle. Expel the excess fluid and bubbles from a syringe vertically into a cotton plug moistened with disinfectant. It is recommended to avoid the use of hypodermic needles and syringes specifically for handling recombinant microorganisms.
11. Do not place contaminated instruments, such as inoculating loops, needles and pipettes, on the work-tops. Loops and needles should be sterilized by incineration, and pipettes should be disposed of in the designated receptacles.
12. Upon entering the laboratory, place coats, books, and other things in specified locations – not on the work-tops. However, while working, wear a lab coat or apron to protect clothing from contamination or accidental discoloration by staining solutions.
13. Keep hands away from mouth, nose, eyes and face. This may prevent self-inoculation.
14. An insect and rodent control program must be in effect. Windows must be fitted with fly screens.
15. Carry cultures in a test-tube rack when moving around the laboratory.
16. Closed shoes should be worn at all times in the laboratory setting.
17. The laboratory space must be easily accessible for cleaning. Work-tops must be impervious to water and resistant to acids, alkalis, organic solvents, and moderate heat. The laboratory furniture must be sturdy.
18. Before centrifuging, inspect the tubes for cracks and the inside of the trunnion cup (rotor bucket) for any spillage. Use centrifuge trunnion cups with screw-caps or their equivalent. Never over-fill the tubes with the sample.
19. All accidents, including minor cuts, abrasions, and spills of culture or reagents, must be reported to the safety authority and treated immediately. Spilled cultures or broken culture tubes should be covered immediately with paper towels and then saturated with a disinfectant solution. After 15 min. of reaction time, remove the towels and dispose of them in the recommended way.

20. Masks, safety goggles, and laboratory coats should be worn if aerosols might be formed or splattering of fluids is likely to occur.
21. On completion of the session, place all cultures and materials in the designated disposal bins.

3. Guidelines for recombinant organisms

When genetic material was successfully transferred from one organism into another in which it did not naturally occur, it was immediately apparent that these techniques could offer substantial advantages in terms of specificity and speed over the traditional methods of, for example, microbial or plant breeding. However, many prominent scientists and concerned citizens questioned the possibility of these techniques generating new hazards. Some concerns were that the creation of new genetically-modified microorganisms (GM) would upset the ecological equilibrium and the transfer of DNA from microorganisms to animals or human cells would cause cancer or new diseases and pose health hazards. The publication of these hypothetical dangers initiated the DNA debate, and the formulation of guidelines for recombinant DNA research or application was established in many countries. These activities introduced the concept of *biological containment* and a regulatory system developed based on these scientific concepts.

4. Infectious microorganisms

Infectious microorganisms are classified into four risk groups in the increasing order of risk, based on the following parameters: (i) the pathogenicity of the agent; (ii) the mode(s) of transmission and host range of the agent; (iii) the availability of effective preventive treatments or curative medicines; and (iv) the capability to cause diseases in human/animal strains. These parameters may be influenced by various factors, such as: immunity, the density and movement of the host population, the presence of vectors for transmission and the standards of environmental hygiene (Ghosh, 1995).

B. TRANSGENIC ANIMALS

The development of transgenic animals with commercial values – e.g., increased growth rate, improved production efficiency, disease-resistant breeds, resistance to various ecological stresses – is gaining importance throughout the world. The commercial application of transgenic animals will, however, require satisfactory answers to environmental and human safety issues.

Before the entry of transgenes into the human food chain, a satisfactory assessment must be conducted on the effects of the transgenic constructs in human body when consumed. Therefore, there is a need to formulate guidelines for experimentation and the controlled use of transgenic animals.

C. TRANSGENIC PLANTS

Transgenic plants have been developed and tested the world over to generate varieties for various characters. The safety questions associated with such plants are diverse and it is difficult to ascertain their consequences immediately. Therefore, there is a pressing need to frame rules and guidelines to assess their effect on the environment and society. Biosafety rules deal with transgenic plants and experimental protocols have been formulated for research work in India; however, protocols to address safety issues are still to be evolved, since not much data is available. The reader can refer to Chapter 41 for more details on the biosafety of transgenic plants.

III CHEMICAL HAZARDS

Most of the chemicals (both organic and inorganic) used in molecular laboratories are toxic/ genotoxic, or flammable (some explosive), or both. Every possible precaution should be taken to avoid swallowing, inhaling, or spilling of chemical compounds. Many chemicals are readily absorbed through the skin and can be just as toxic when absorbed as when swallowed. Other toxic substances may act synergistically, that is, a mixture or combination of two or more chemicals may be more toxic than would be expected from the toxicity of each component. When an unfamiliar chemical need to be used, it is always wisest to read the instructions that come with the product and also to look it up in the *Merck Index* before using it. It is difficult to formulate specifications for molecular biology laboratories. This is because laboratories vary depending upon the type of work, the kind of organisms, the chemicals used, and the facilities available. However, certain aspects are rather universal in nature and researchers are obligated to take utmost care to minimize the risks from laboratory contamination and wastes.

If any material is spilled on the skin, wash it off immediately with large volumes of water. Do not taste any chemicals, solutions, or other material unless specifically instructed to do so. Every possible precaution should be taken to keep the air free from contaminants. Among other measures, the following will help achieve this.
1. All the operations (transfers, weighing, diluting, preparing solutions, etc.) related to chemicals should be performed in a standard fume hood (Figure 2.2).
2. Replace all tops and stoppers on containers immediately after use.

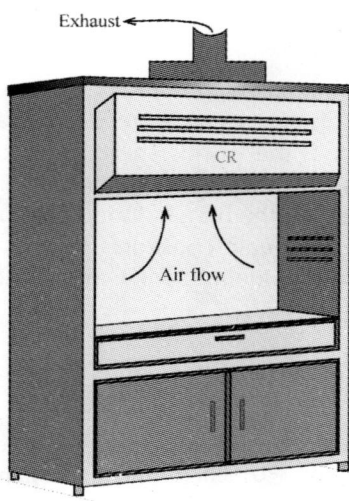

Figure 2.2 *Chemical fume hood*

3. Unless specifically required to do so, do not remove chemicals or reagents from the areas designated for their storage and measurement.
4. Never use more than the quantities indicated. This will help to ensure successful experiments as well as minimize evaporation of hazardous substances.
5. When disposing of any liquids in the sink pour them directly into the drain and flush them down with large quantities of water. If chemicals are not permitted to be poured into the drain, follow the proper disposal methods recommended.
6. When distilling or heating a mixture under efflux, ensure that the water is flowing through the condenser so that even volatile liquids are condensed.
7. There should not be any open flames in the laboratory when large quantities of flammable liquids are used.
8. When strong corrosive substances are used (e.g., strong acids and bases), extra precaution is required. The following precautions should be observed: (a) if any corrosive materials comes into contact with the skin, wash it off immediately with large quantities of water; (b) use only the necessary quantities; (c) wear eye protection when working with corrosive substances; (d) when mixing a strong acid with water, always add the acid very slowly to the water while stirring the mixture.

IV RADIOISOTOPES IN BIOLOGICAL SCIENCES

Isotope or tracer techniques have contributed greatly to the development of the biological sciences. Radioactive tracers for hydrogen, carbon, sulfur, phosphorus and many other elements became readily available and a very large number of compounds labeled with tracers are now produced. Tracer techniques were appreciably facilitated because of the ease with which radioactive materials could be detected. The use of radioactive tracers, however, involves a health hazard due to the experimenter being exposed to the radiation emitted by the radioisotopes –either as a source external to the body or as an internal source if the material is ingested. In designing safe protocols for the use of radioactivity, the importance of common sense, based on an understanding of the general principles of isotopic decay and the importance of continuous monitoring (e.g. Geiger counter can be used), cannot be overemphasized. In addition, it is also critical to take into account the rules, regulations, and limitations imposed by each specific institution.

The radioactivity of a given substance is measured in terms of its ionizing activity. A Curie (Ci) by definition is the amount of radioactive material that will produce 3.4×10^{10} disintegration (ions) per second (Appendix 8, Table 3).

A. SOME APPLICATIONS OF RADIOACTIVE TRACERS IN BIOLOGY

The main uses of radioactive tracers in biology are for: (1) metabolic studies and enzyme assay involving the use of labeled substrates; (2) the determination of trace levels of nucleic acids, amino acids, sugars and steroids in fluids by radioisotope dilution analysis; (3) the determination of tracer elements by radio-activation analyses.

All these generally require the ultimate determination of small amounts (mCi–µCi) of radioactivity. After the irradiation, it is generally necessary to chemically separate the low levels of activity associated with the trace elements from the high levels of activity associated with the matrix. Most of the activity presented is unwanted and usually ends up as radioactive waste.

B. RADIATION HAZARDS AND PROTECTION

Radiation hazards can arise from radioisotope sources that are external to the body as well as from radioisotopes that have been ingested and are distributed. External hazards, either closed or open sources, may be present in handling radioactive materials. Internal hazards arise in handling all open sources and may be considerable even with µCi level sources, depending upon the isotope involved. Even if the isotope is one of low radiochemical toxicity, such as tritium, care must be exercised over long periods to ensure that chronic ingestion of small amounts of activity does not lead to a significant build-up of ingested material. The following series of precautions that must be observed in a tracer laboratory are very similar to the normal precautions for work in any molecular laboratory.

C. THE GENERAL PRECAUTIONS TO BE FOLLOWED IN A TRACER LABORATORY

1. Know the rules: Be sure that each individual is authorized to use each particular isotope and uses it in an authorized work area.
2. Mouth operations are not allowed in the laboratory; other prohibited activities are pipetting by mouth, smoking, drinking, chewing, licking of labels, applying cosmetics and mouth glass-blowing, etc.
3. While working with radioisotopes, all operations must be performed under radioisotope protection shields, such as a Plexiglas shield (Figure 2.3).
4. Protective clothing must be worn, the very minimum being a laboratory coat and either rubber or disposable plastic gloves. Laboratory coats should be worn in the laboratory at all times and the gloves when handling open radioactive sources.
5. Work must be conducted in a confined area so that if there is a spill of labeled material, the area of contamination is known and contained. Such confined areas should be in fume cupboards when using radioactive gases and powders, and on the bench in large paper lined trays for liquids. The areas designated for radioactive work must be covered with absorbent paper backed with an impervious material, for easy and rapid clean up in case of spillage.
6. All radioactive liquids must be double contained. For this purpose, vessels containing radioactive solutions must be placed in containers packed with absorbing materials such as vermiculite or cotton-wool.
7. Protective gloves must be worn when operating light switches, counting and monitoring equipment or opening drawers.
8. On removing or putting on gloves, particularly those that have been used several times, it is important not to touch the outside of the gloves with the hands. Disposable gloves are preferable.
9. Hands must be washed and monitored before leaving the laboratory. It is important to realize that *radioactive contamination cannot be neutralized*. It can only be placed in a safe place and allowed to decay.
10. Accidents and spills should be reported to the appropriate agency.
11. All the material used for radioactive work should be properly labeled. It is only common courtesy to alert co-workers of the existence of anything radioactive. Yellow hazard tape printed with the international symbol should be used.

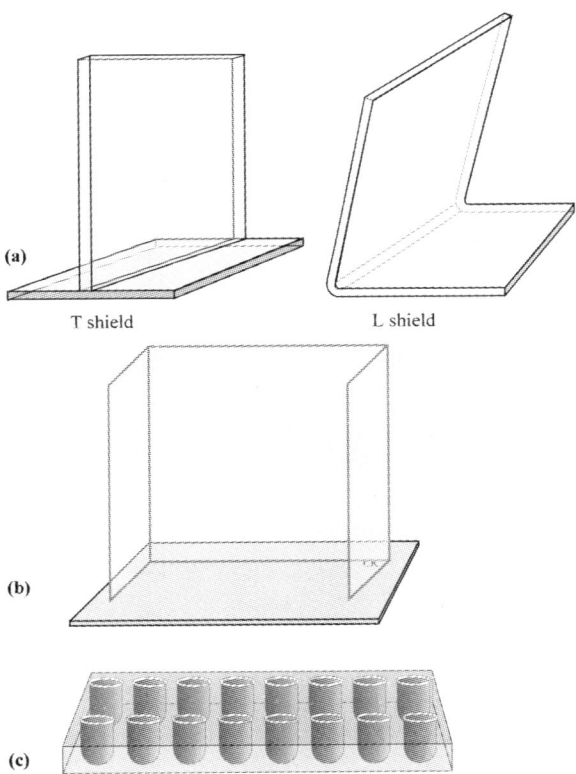

Figure 2.3 *Plexiglas shields for working with radioisotopes (a) Two portable shields (T and L designs) and (b) a bench top shield. (c) Tube rack for samples in microcentrifuge tubes. Plexiglas should be atleast 0.25 inch thick.*

12. Radioactivity should be monitored early and often. Portable radiation detection monitors are essential for every laboratory that uses radioactivity.

D. DECONTAMINATION OF APPARATUS AND PERSONAL DECONTAMINATION

Most radioactive contamination is easily removed from the apparatus by using a detergent, especially the surface-active types such as Decon–90. Some materials, such as [^{32}P], can be strongly adsorbed onto the glassware but can usually be removed with concentrated acids. Concentrated HCl will remove phosphate, but more stubborn contamination can be removed by soaking the apparatus in a chromic acid solution. Chelating agents, such as a 5% (w/v) solution of EDTA, are also useful. Highly-contaminated apparatus, which such procedures have failed to decontaminate successfully, may either be stored to permit radioactive decay or discarded as radioactive waste. Any contamination of the skin should be washed thoroughly with soap/detergent.

V BIOLOGICAL HAZARD SAFETY

Antibodies, sera and cells (particularly, but not exclusively, those of human and primate origin) pose a significant biological hazard. All such materials, whatever be their origin, may harbor human pathogens and should be handled as potentially infectious material in accordance with local and national guidelines. Any recombinant DNA work associated with protocols is likely to require permission from the relevant regulatory body and users must consult the local safety officer before embarking upon this type of work.

VI LABORATORY EQUIPMENTS AND MAINTENANCE

In this chapter, we will discuss some of the widely used, but often poorly-described, techniques or bits of information which are helpful or necessary for successfully using biotechnological procedures. Much of this information is well known in laboratories that practice molecular biology or molecular genetics, but is transmitted almost entirely by word-of- mouth. Therefore, there is a need to record this information, particularly for beginners who are not familiar with the procedures to maintain a good laboratory for molecular biology work and need some basic information to avoid contaminating the experiments. A few basic techniques for cleaning and sterilizing the labware and liquids are also briefly explained.

A. GLASS AND PLASTIC CONTAINERS

In a molecular biology laboratory, various kinds of glassware and plastics are used (Figures 2.4 and 2.5). The variation in physical and chemical properties of glass is due to Na_2O and other impurities that are added to SiO_2 in order to lower glass viscosity and thereby facilitate the manufacture of glass products. Generally, a lower concentration of impurities leads to a harder glass that is more resistant to chemicals and scratching. This hard glass also has a lower thermal coefficient of expansion, so it is less likely to break when shocked by a temperature change.

The three plastics most commonly used in place of glass are *polyethylene*, *polypropylene*, and *polystyrene*. The inexpensive polyethylene and polystyrene have low melting temperatures and melt when autoclaved. Polypropylene does not melt when autoclaved, although it does become more brittle and may craze after repeated autoclaving. Polyethylene, which is pliable, is useful in the manufacture of squeeze bottles that dispense water, detergents, ethanol, and other mild solvents. Solutions and biological materials, such as nucleic acids, proteins, and viruses, can be stored in either hard glass or polypropylene containers. Often, very small quantities of almost any charged substance will bind to glass and plastic. Although siliconizing the container will help, storing sub-microgram quantities of nucleic acids, proteins, and the like is risky.

All glassware should be sterilized by autoclaving. Borosilicate glass (e.g., Pyrex) or factory-washed soda-glass apparatus needs no special treatment before being used, other than normal washing up. New, unwashed soda glass should be soaked in HCl overnight to partially neutralize the alkali contained in the glass. A great deal of sterilized plasticware is available. Autoclaving is appropriate for some, but not all, plasticware, depending on the type of plastic. All the procedures commonly used in molecular cloning can be carried out in glassware or

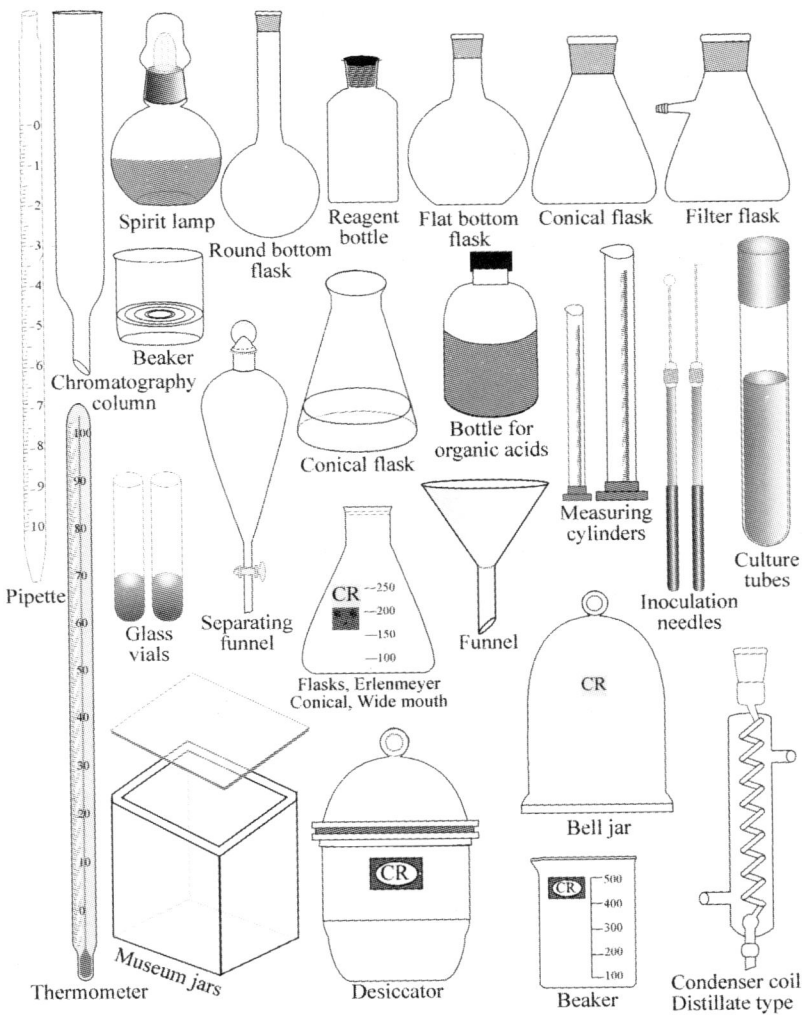

Figure 2.4 *Commonly used laboratory glassware*

plasticware prepared in this way; there is no significant loss of material by adsorption onto the surfaces of the containers. However, for certain procedures, it is advisable to use glassware or plasticware that has been coated with a thin film of silicone. A simple procedure for siliconizing small items such as pipettes, tubes, and beakers is available in the *Maniatis Manual* (Appendix E).

B. WASHING OF LABWARE

All experiments in molecular biology stipulate that all items of glassware and other equipments are scrupulously clean before use. This is particularly important for reactions involving proteins

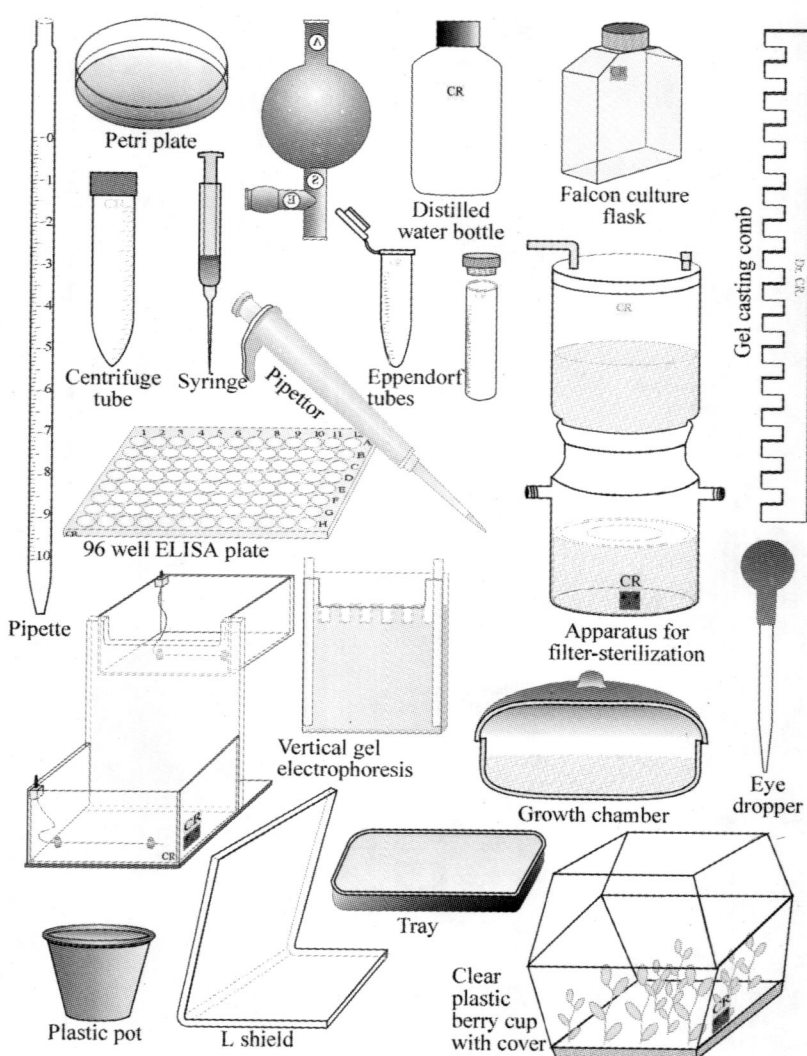

Figure 2.5 *Commonly used plasticware in a molecular biology laboratory*

and nucleic acids, which are usually carried out in small volumes; very small quantities of nucleic acids can be absorbed onto the surface of the glassware used. This problem can be significantly reduced if small polypropylene test-tubes are used. If these are not easily available, the glassware being used should be coated with silicon. There are two essential requirements for molecular biology or tissue culture work. The laboratory ware that is used to prepare reagents and that comes into contact with cell suspensions must be clean and sterile.

Usually, washing the glassware with hot water and soap (or washing compounds, such as sodium metasilicate, trisodium phosphate or synthetic detergents) followed by thorough rinsing in tap water, and then in distilled water, is sufficient. Occasionally, more rigorous means are called for; e.g., washing in ethyl alcohol containing 5% HCl or "cleaning solution" (technical potassium dichromate in commercial concentrated sulfuric acid). In addition, the glassware selected should be free from cracks and stains.

Method-1: Washing of glassware

Materials

(i) Good quality detergent; (ii) steady water supply through well-regulated taps; (iii) distilled or double-distilled water; (iv) various sizes of bottle-brushes; (v) pipette washing jars with racks.

Procedure

1. Immediately after use, rinse the glassware with tap water to remove the medium, cells, etc. Fill the containers (such as beakers, flasks, reagent-bottles) with tap water and completely submerge the test-tubes and pipettes in tap water.
2. Using a brush, scrub the glassware with a detergent and warm tap water. Thoroughly rinse all traces of detergent from the glassware with warm tap water. Then rinse the glassware again, a minimum of 3 times, with distilled water and double-distilled water.
3. Wash the pipettes by dipping them repeatedly into a washing jar filled with diluted detergent. Then, transfer the pipettes to another washing jar filled with flowing warm tap water and rinse the pipettes until all traces of detergent have been removed. Finally, rinse at least 3 times each with distilled water and double-distilled water.

Note

1. Alternatively, automatic washing machines may be used.

Method-2: Washing pipettes

Remarkably, no uniform or best solution seems to exist to the problem of washing and sterilizing pipettes. If pipettes are being used for the transfer of dangerous pathogens, it is best to use disposable pipettes, which are placed in jars of suitable disinfectant and sterilized before the final disposal.

Procedure

1. As soon as the pipettes are used, soak them in soapy water.
2. Pre-rinse the pipettes. Fill a pipette washer basket one-half to two-thirds full with pipettes. Drain out the soapy water for 5 min. in an automatic rinsing container. Drain the basket of pipettes very thoroughly, shaking them rather vigorously.
3. Soak the pipettes in a cleaning solution. The cleaning solution, dichromate in concentrated sulfuric acid, is still the most satisfactory cleanser. Very carefully, lower the basket of rinsed pipettes into the cleaning solution and cover with a plastic wrap on top as the bubbles start to burst. Soak them for a minimum of 3 hours.
4. Rinse the pipettes; then, lift the basket above the acid and let it drain completely. Quickly put the basket into an automatic rinsing container. The rinse will be more effective if the basket is shaken a little from time to time during the rinsing. After rinsing, put the basket of wet pipettes on a sink drain ledge for at least 15 min. Wash the pipettes with distilled water at least once.

5. Dry and sterilize the pipettes. Load the empty pipette cans not more than two-thirds full with pipettes. Close the cans and bake at 150°C overnight to dry and sterilize.

C. STERILIZATION OF LABWARE

Most molecular biology experiments require scrupulously clean and sterile conditions. Laboratory ware can be sterilized by different methods (Table 2.1), depending upon the kind of material to be sterilized. Sterilization implies the destruction or removal of all life forms. Although chemicals may sometimes be used to sterilize objects, the principal methods for achieving sterilization employ physical agents such as heat, radiation, and filtration.

The various species of microorganisms differ in their susceptibility to physical and chemical agents. In the spore-forming species, the growing vegetative cells are much more susceptible than the spore forms; bacterial spores are extremely resistant. The many processes and substances used as antimicrobial agents manifest their activity in one of several ways. A thorough knowledge of the mode of action of a particular agent may make it possible to predict the conditions under which it will function most effectively as well as the kinds of microorganisms it will be most effective against.

Table 2.1 *Methods of sterilization*

Method	Time and temperature	Use	Mode of action	Disadvantages
Dry heat	1 hr at 170°C in a hot-air sterilizer	Glassware	Oxidizes bacterial components	Cannot be used for certain media
Intermittent or fractional sterilization ("tyndalization")	Free-flowing steam at 100°C for 30 min. on each third consecutive day	Media, tissues, liquids, solids	Heat, shock the material in first 30 min. activates spore germination, kills in subsequent heating cycles	Tedious, time-consuming
Moist heat	15 to 30 min. at 121°C under 15 psi	Glassware media, liquids, solids, metals	Coagulates proteins of microorganisms	Some nutrients are unstable
Filtration	Need millipore filter vacuum, etc.	Media, liquids	Microorganisms cannot pass through the filter pores	Glassware cannot be sterilized
Toxic gases (cold sterilization)	Ethylene oxide under slight pressure	Large surface areas, equipment	Kills all type of life forms	Flammable, sometimes hard to remove traces
Radiation	Ultraviolet (UV) or ionizing (γ or x-rays)	Media, liquids, solids, pharmaceuticals	Cause mutation, interferes with metabolism	Does not penetrate glassware, access to radioactive core required

Method-1: Sterilization of laboratory glassware with heat

The killing effect of heat on microorganisms has long been acknowledged. Heat is fast, reliable, and relatively inexpensive, and it does not introduce chemicals into a substance, as disinfectants sometimes do. Above the maximum physiological temperatures, any biochemical change in the cell's molecules result in its death. Heat also repels water, and since all organisms depend on water, this loss may be lethal. The killing rate of heat may be expressed as a function of time and temperature (Table 2.2).

Table 2.2 *Different temperature environments used in heat sterilization*

Heat (°C)	Temperature environments and control of organisms
160	Spores killed in 2 hr in hot-air oven or 1 hr in hot oil.
121	Spores killed in 15–30 min. in autoclave.
100	Spores killed in 2 hr or more in boiling water; 30 min./day for 3 days in fractional sterilization.
82	Pathogenic bacteria killed in 3 sec. in pasteurization.
72	Pathogenic bacteria killed in 15–17 sec. in pasteurization (71.6°C).
63	Pathogenic bacteria killed in 30 min. in pasteurization (62.9°C).
37	Normal human body temperature (37°C).
5	Refrigerator.
-10	Home freezer (–10°C).

Materials

(i) Autoclave, (ii) sterile indicators, (iii) cotton, (iv) cheesecloth, (v) aluminum foil, (vi) tape, (vii) autoclave bags, (viii) pipette canisters with stainless steel caps.

Procedure

1. Prepare the items to be autoclaved as follows:
   a. Tighten the caps on reagent bottles, then loosen a turn.
   b. Plug the flasks, graduated cylinders, etc., with cotton-filled cheesecloth plugs and cover the plugs securely with aluminum foil.
   c. Plug the pipettes with cotton and burn off the excess cotton with a Bunsen burner. Place the pipette, either wrapped in paper or placed in metal cylinders, in the canisters. Then place the caps on the canisters.
2. Tape a sterile indicator to the glassware and canisters, and autoclave at 121°C for 20 min. Usually, a fast exhaust and a drying cycle are preferable.
3. After exposure to this temperature, the slowly-cooled, wrapped dishes may be removed and stored in the desk until needed.

Method-2: Sterilization of liquid reagents

Reagents such as water, normal saline, and phosphate-buffered saline can be sterilized by autoclaving. Many other reagents, however, must be sterilized by membrane filtration since they contain heat-sensitive components.

Materials

(i) Autoclavable reagent bottles with caps, (ii) autoclave, (iii) cool normal saline, (iv) sterile membrane filtration units, 0.22 µm or 0.45 µm (millipore), (v) disposable syringes and needles, (vi) boiling distilled water.

Sterilization by autoclaving

Procedure: (e.g., water, normal saline, PBS, etc.)

1. Fill the reagent bottles until they are two-thirds full of liquid. Twist the cap tightly, and then loosen the cap one turn. Tape the sterile indicator on the bottles.
2. Place the bottles in a steel tray with sufficiently high sides to catch any spilled liquid, if bottles should break in the autoclave.
3. Autoclave at 121°C for 20 min. by slow exhaust.
4. After autoclaving, let the bottles cool and then tighten the caps.

Sterilization by membrane filtration

Some fluids or liquid culture media are very labile and cannot be subjected to heating without altering their chemical nature. Fluids like toxins, sugars, and amino acids are made sterile by passing them through various types of filters designed to retain bacteria. The filter is a mechanical device for removing microorganisms and bigger particles from a solution. Several types of filters (e.g., inorganic filters, organic filters and membrane filters, etc.) are available for use in the laboratory. For air filtration, *high-efficiency particulate air* (HEPA) filters are generally used. Filtration can be achieved either by the application of negative pressure (i.e., suction), or positive pressure.

Procedure

1. Carefully remove the sterilized filter units from the autoclaving bag and put the needle (with its cap still in place) onto the unit.
2. Fill the syringe with boiling water, attach the filter unit, remove the needle cap, and push a large volume of water through the filter. Then, recap the needle, remove the filter unit, fill the syringe with cool normal saline, and pass the saline through the filter unit, (this procedure removes any toxic wetting agents that are present in the filters).
3. Fill the syringe with the reagent (media, buffers, antiserum, etc.) and pass it through the filter unit into a sterile tube.
4. Finally, disassemble/dismantle the filter unit to check whether the filter is intact. Occasionally, small cracks appear in the filters. If this occurs, re-filter the reagent through another filter unit. Do not pull back on the syringe barrel while the filter unit is attached or the filter will tear.

Notes

1. The growth media obtained from cultured cells usually need to be pre-filtered to remove any debris that can clog the 0.22 µm and 0.45 µm filters. This can be done by passing the media through filters of decreasing pore size or, alternatively, centrifuging the media at 12,000 g for 20 min. prior to filtering.

2. When membrane filters are used to sterilize solutions containing low concentrations of macromolecules, considerable losses may occur due to adherence of the macromolecules to the filters. Treating the filters by flushing with small amounts of serum or protein (1% BSA) prior to the passage of the biologically active solution reduces the likelihood of such losses.

Method-3: Pasteurization

Pasteurization is not the same as sterilization. Its purpose is to reduce the bacterial population of a liquid, such as milk, and to destroy organisms that may cause spoilage and human diseases; spores are not affected by pasteurization. One method for milk pasteurization, called the *holding method*, involves heating the milk to 62.9°C for 30 min. Although thermophilic bacteria thrive at this temperature, they cannot grow at body temperature. Two other methods of pasteurization are the *flash pasteurization method* carried out at 71°C for 15 sec., and the *ultra-pasteurization method* done at 82°C for 3 sec.

Method-4: Hot oil

This method is generally followed by dentists and physicians. They use hot oil at 160°C for 1 hr, for the sterilization of instruments. Hot oil does not rust metals, and minimal corrosion takes place. However, once hot oil sterilization is done, the instruments must be cleaned and dried before storage.

Method-5: Sterilization by irradiation

Ultraviolet Light: One type of radiant energy, *ultraviolet light* (UV), is useful for controlling microorganisms. UV light has a wavelength between 100 and 400 nm, with the energy at about 265 nm most destructive to bacteria. UV light causes cross-linking between thymine molecules in nucleic acids and replication is impaired.

Other type of radiation useful for destroying microorganisms includes x-rays and gamma rays. Both have shorter wavelengths than the UV light, and can create ions. For this reason, these radiations are called *ionizing radiation*. The ions quickly combine with and destroy proteins and nucleic acids causing death.

Another form of energy, the *microwave*, has a wavelength longer than that of UV light. In a microwave oven, the microwaves are absorbed by the water molecules. The molecules are set into high-speed motion, and the heat of friction is transferred to foods, which become hot rapidly.

A particular form of light energy can also be used for sterilization. When concentrated by sophisticated devices, light energy forms a *laser beam*. The term "laser" stands for light amplification by stimulated emission of radiation. The microorganisms can be destroyed in a fraction of a second, but the laser must reach all parts of the material to effect sterilization.

Method-6: Ultrasonic vibrations

Ultrasonic vibrations are high-frequency sound waves beyond the range of the human ear. When propagated in fluids, ultrasonic vibrations cause the formation of microscopic bubbles, or cavities, and the water appears to boil. Some observers call this "cold boiling". The cavities rapidly collapse and send out shock waves, and the microorganisms in the fluid are quickly disintegrated by the external pressures. The formation and implosion of the cavities is known as *cavitation*. Ultrasonic vibrations are valuable in research for breaking open tissue cells and obtaining their parts for study. A device called the *cavitron* which uses the principle of ultrasonic vibrations, is used by dentists to clean teeth.

VII CONTROL OF MICROBES BY CHEMICAL AGENTS

The chemical control of microorganisms is used in diverse areas. The physical agents for controlling microorganisms are generally intended to achieve *sterilization*. This implies the destruction of all forms of life. Chemical agents, by contrast, rarely achieve sterilization. Instead, they are expected to destroy only the pathogenic organisms in an object. The selective process of destroying pathogens is called *disinfection*. If the term is applied to a lifeless object, such as a tabletop, the chemical agent is known as a *disinfectant*. However, if it is a living object, such as a tissue, the chemical is an *antiseptic*. Both antiseptics and disinfectants usually have *bacteriocidal* (kills bacteria) properties, but occasionally they have *bacteriostatic* (temporarily prevents further multiplication) properties as well.

Selection of antiseptics and disinfectants: To be useful as an antiseptic or disinfectant, a chemical agent must have certain properties. The first prerequisite is that it must be able to kill microorganisms. It should also be non-toxic to animals or humans, soluble in water and have a substantial shelf-life. It should also be useful in a very diluted form, act quickly, should not separate on standing, should penetrate well; and it should not corrode instruments. Of course, in addition, it should be relatively inexpensive.

Important chemical agents: The chemical agents currently in use for controlling microorganisms range from very simple substances, such as halogen ions, to very complex compounds, such as detergents. *Halogens* are a group of highly-reactive elements whose atoms have seven electrons in the outer shell. Two halogens, chlorine and iodine, are commonly used for disinfection. *Phenol and phenolic compounds* (phenolics) have played a key role in disinfection practices; phenol remains the standard against other antiseptics and disinfectants. Some metals are called *heavy metals* (mercury, copper sulfate, and silver nitrate) because of their large atomic weights and complex electron configurations. Heavy metals react very well with proteins, particularly with the protein's sulfhydryl groups (–SH), and they are believed to bind protein molecules together by forming bridges between the groups. However, heavy metals are not sporicidal. *Alcohols* are effective skin antiseptics and are valuable disinfectants for medical instruments. The preferred alcohol for practical use is ethyl alcohol. Ethyl alcohol is active against vegetative bacterial cells but it has no effect on spores. It denatures proteins and dissolves lipids, an action that may lead to the disintegration of the cell membranes. *Ethyl alcohol* is also a strong dehydrating agent. Usually, a 5–80% alcohol solution is recommended, since water prevents rapid evaporation and assists penetration into the tissues. *Isopropyl alcohol*, or rubbing alcohol, has high bactericidal activity in concentrations as high as 99%. *Methyl alcohol* is toxic to the tissues and is rarely used.

Other chemical agents: Apart from disinfectants and antiseptics, there are some other chemicals that can be used for sterilization. Some of these agents are formaldehyde (*formalin*), ethylene oxide, gluteraldehyde, hydrogen peroxide, soaps and detergents (anionic–cationic detergents, quaternary ammonium compounds), acids (benzoic, salicylic and undecylenic acids), and organic acids (e.g., lactic and acetic acids, propionic acid).

Micropipetting

In almost all laboratories, micropipetting techniques are used to transfer very small quantities of liquid samples. The mastery of these techniques is important for good results in all of the experiments that follow. Before using micropipettes, one should be familiar with metric units of measurement and their conversions (see Appendix 8, Table 2). Some of the important precautions one must take while using micropipettes are listed below.

1. Never rotate the volume adjustor beyond the upper or lower range of the micropipettor.
2. Always ensure that the end of the micropipettor be aligned with the tip, otherwise the precision position of measurement will be altered.
3. Never touch the tip to other objects or put down a filled micropipettor on a bench.
4. Never immerse the barrel of the micropipettor in fluid.
5. Keep the micropipettor tip away from any flames.
6. Do not detach or reassemble the micropipettor yourself, let the authorized service personnel do the servicing when necessary.
7. Always use standard size micropipette tips and, depending on the size of the sample volume to be transferred, use the appropriately-sized micropipette tip and micropipettor.
8. To expel the sample into a reaction tube, after releasing the sample, touch the tip on the inside wall of the reagent tube.

VIII SUGGESTED READING

Alcamo IE (1997) *Fundamentals of Microbiology*, The Benjamin/Cummings Publishing Company (an imprint of Addison Wesley Longman, Inc.).

Anonymous (1993) *Biosafety in Microbiogical and Biomedical Laboratories*, 3rd Edn, US DHHS, Public Health Service, Centers for Disease Control and Prevention, Atlanta, Georgia and NIH, Bethesda. Maryland.

Anonymous, *NIH Guidelines for Recombinant DNA Technology Research*, National Institute of Health, USA.

Cartledge TG (1992) Introduction to microbial growth and cultivation, In: *In Vitro Cultivation of Microorganisms* (Ed. Cartledge, TG), Butterworth-Heinemann Ltd, Oxford.

Darlow, HM (1972) Safety in the microbiological laboratory: An introduction, In: *Safety in Microbiology* (Ed. Shapton DA and Board RG), Academic Press.

Ghosh PK (1995) *Biotech Industry Guide*, Biotech Consortium India Limited, New Delhi.

Harrigan WF (1998) *Laboratory Methods in Food Microbiology*, Academic Press, CA. http://www.apnet.com.

Hellman A, Oxman MA and Pollack R (Eds.) (1973) *Biohazards in Biological Research*, Cold Spring Harbor Laboratory, NY.

Tolin SA and Vidaver AK (1989) Guidelines and regulations for research with genetically-modified organisms: A view from academe, *Annu. Rev. Phytopath.*, **27**:551–581.

Wilson K and Walker J (2000) *Practical Biochemistry: Principles and Techniques* (5th Edn.), Cambridge University Press, UK.

Winnacker EL (2003) *From Genes to Clones*, Panima Publishing Corporation. (Appendix F, pp 549–567).

PART – II
Advanced Techniques in Molecular Biology

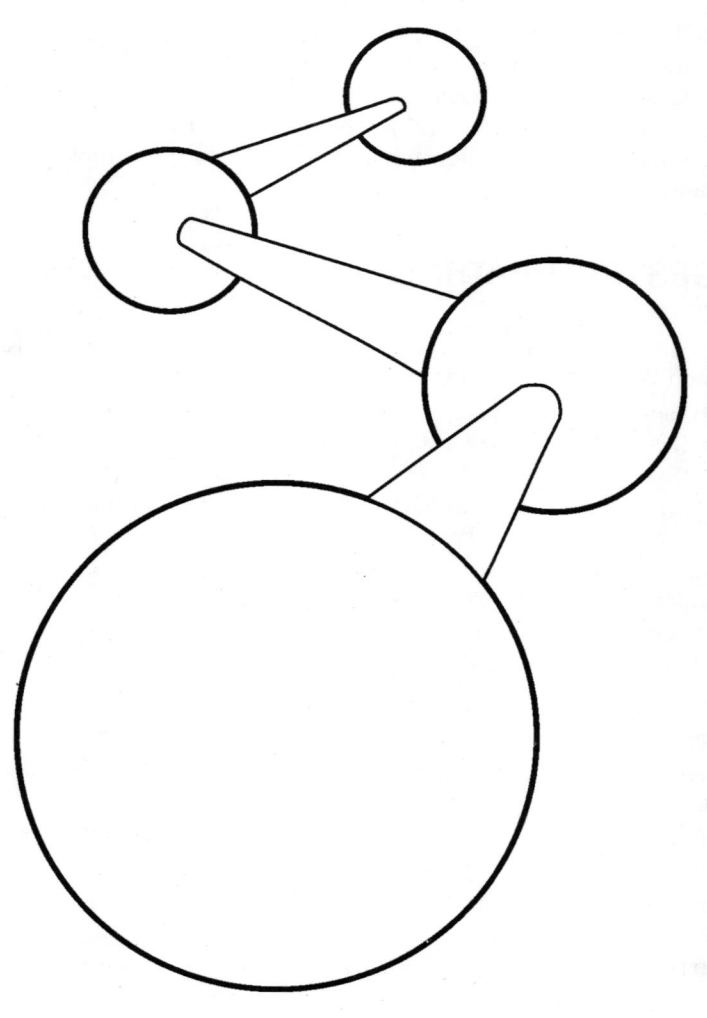

3

Techniques of Cell Fractionation and Centrifugation

I INTRODUCTION

Cell fractionation essentially proceeds in two consecutive stages: (1) homogenization, which disrupts the tissue and releases the cellular components into the resultant homogenate, and (2) centrifugation, which separates the individual components within the homogenate according to density, size and shape. The ultimate objective of the cell fractionation procedure and the nature of the starting material are important parameters in defining homogenization as a semi-empirical method of cell disruption. In order to understand the complex nature of biological membranes and organelles within a cell, the plasma membrane (and cell walls in plants) must be disrupted. The challenge of *in vitro* work is to choose methods of cell disruption and incubation of subcellular fractions, cells, tissues or organs, that are appropriate to the particular investigation and which minimizes artifacts. The ideal homogenate comprises all the desired cellular components in abundance, in an unaltered morphological and metabolic state. In practice, the fractionation of selected material often produces components with at least one altered characteristic.

II HOMOGENIZATION

A. HOMOGENIZATION MEDIUM

Cell disruption is usually performed in either a slightly hypo-osmotic or an iso-osmotic medium to preserve morphological integrity. Sucrose is commonly used as an osmoticum to prevent subcellular organelles or vesicles from adverse swelling or shrinkage. Mannitol and sorbitol have also been used in cases where sucrose has been found to interfere with the biochemical analysis of the subcellular component. Although the homogenization medium is usually aqueous in nature, non-aqueous media, such as ether/chloroform, have been used to isolate subcellular organelles.

Chelating agents (EDTA or EGTA) may be added to the homogenization medium to remove divalent cations, such as magnesium and calcium, which are required by the membrane proteases. The disruption of plant tissue cells is very difficult, because they are encapsulated within a rigid cell wall that may be strengthened by the lignin. Plant cells also often release phenols that can have several deleterious effects on plant enzymes. Therefore, it is important to remove the phenolic compounds as quickly as possible and this is best achieved by using adsorbents (e.g., borate, germinate, sulfites and mercaptobenzothiazole) that bind to the phenols or quinones. Cell or tissue fractionation is invariably performed at 4°C, in order to reduce the activity of membrane proteases. Other compounds used in the homogenization media include disulfide-reducing agents such as 2ME, DTT, reduced glutathione or cysteine.

Plant tissues also contain active lipolytic enzymes, which can attack the lipid components of membranes directly and disrupt organelles. Each individual homogenization medium is different and may contain specific inhibitors for enzymes such as phospholipases or phosphatases; glycerol is often added to the isolation medium in order to inhibit phosphatidic acid phosphatase. Ethanolamine and choline chlorides are commonly used to inhibit phospholipase D. The buffering capacity of the homogenization medium for plant-cell fractionation must be high in order to offset the release of organic acids by the disrupted vacuole, which constitutes a substantial volume of the cell.

B. PROCEDURE OF CELL AND TISSUE DISRUPTION

Different cell disruption methods are used to release proteins from cell or tissues. Generally, cell disruption is accompanied by liquid/solid separations and, in some cases, the solubilization of the released protein. Usually, both physical and mechanical processes are employed, although some chemical effects are utilized. Table 3.1 summarizes the prevalent methodologies of cell disruption. The release of materials from mammalian cell mass is generally easier than from bacterial, yeast and plant cells.

In order to minimize membrane protease activity and the denaturation of proteins by excessive heat, all stages of cell fractionation must be performed at 4°C. All the media and apparatus should be pre-cooled and maintained at this low temperature throughout the procedure. Enzyme activities may be lost if there is foaming of the homogenate in the high-speed blender. The following physical and non-physical homogenizing procedures impose different degrees of physical stress upon the plant tissue.

Table 3.1 *Cell disruption methods*

Mechanical
    Liquid shear:
        Ultrasonic, agitation, liquid extrusion
    Solid shear:
        Grinding, pressure

Non-mechanical
    Desiccation:
        Freeze-drying, air-drying, solvent-drying

Continues...

Continued...

Table 3.1 *Cell disruption methods*

Lysis:
    Physical
        Osmotic shock, pressure release (French press), freeze/thaw
    Chemical
        Detergents, glycine
    Enzymatic
        Antibiotics, lysozyme, phage, others

1. Physical methods of cell disruption

(a) Solid shear methods of cell disruption

The disruption of the cells or tissue results from the shearing forces generated between the cells and a solid abrasive. These techniques are severe and tend to damage large subcellular organelles. The tissue is placed in a ceramic mortar and a heavy, round-ended pestle is used to grind the material in the presence of a small amount of abrasive in the form of coarse sand, or fine silica sand and alumina for more delicate tissue. Plant tissue is sometimes frozen quickly in liquid nitrogen and then ground using a mortar and pestle.

(b) Liquid shear methods of cell disruption

These methods of cell and tissue disruption rely on the shearing forces generated between the tissue and liquid medium. The material is placed in a pre-cooled, capped mixing container with the homogenization medium and subjected to efficient blending for a short and defined period of time. The cutting blades in a blender are oriented at different angles to each other to enhance the mixing of the homogenate and rotate at a high speed but for only a short period of time, because considerable shearing forces can be generated in that short time. In tissue homogenizers (Figure 3.1), the tissue is ground by the relatively mild shearing forces generated by an upward and downward rotation of a pestle within a glass cylinder. The Potter–Elvehjem homogenizer comprises of a power-driven Teflon, Pyrex glass or Lucite pestle and a glass cylinder. The Dounce homogenizer operates in an identical manner to that of the Potter–Elvehjem homogenizer.

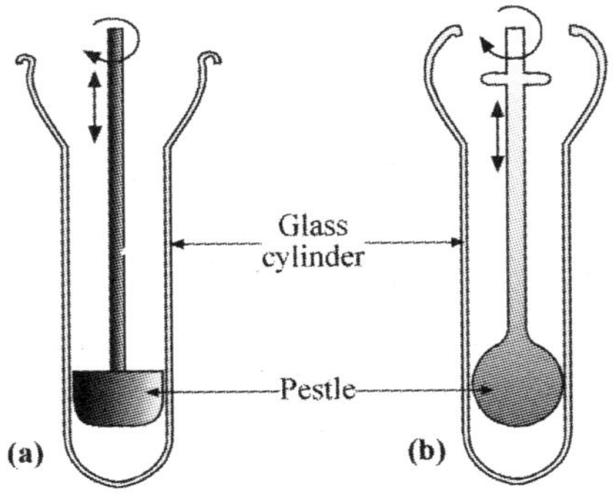

Figure 3.1 *(a) The motor-driven Potter–Elvehjem homogenizer; (b) The Dounce homogenizer*

2. Non-physical methods of cell disruption

Organic solvents, such as chloroform/methanol mixtures, are commonly used to dissolve membrane lipids and release the integral proteins and subcellular components. However, organic solvent or *mineral acid* extraction does not preserve morphological or metabolic integrity. *Chaotropic anions,* such as potassium thiocyanate, potassium bromide and lithium diiodosalicylate, are believed to enhance the solubility of hydrophobic groups in an aqueous environment by increasing the disorder of the water molecules around the group. This facilitates the transfer of the apolar groups to the aqueous phase, thereby weakening the hydrophobic interactions, which enhance the stability of most membranes, multimeric proteins and the native conformation of biological macromolecules. *Detergents* solubilize the integral membrane proteins by interacting with the phospholipid bilayer. They form mixed micelles of the various components of the membrane and the detergent.

Enzymatic digestion of the cell wall of the plant cell results in protoplasts, which may then be disrupted by mild-shearing forces generated by the Potter–Elvehjem or the Dounce homogenizer. The isolation of fragile organelles requires tailor-made conditions and the formulation of the homogenization medium tends to arise by intuitive empirical design. The general protocol is fairly well established but the precise components and their concentration must be established for individual plant species.

III ACIDS AND BASES

A. ACIDITY: TOTAL vs pH

Two interpretations are possible for the term "acidity (reaction) of a solution": (a) the *total acidity* or entire amount of acid present, regardless of the kind or strength; and (b) the *strength of acid* or the quantity of hydrogen ions present, regardless of the total amount of that acid present. Total acidity is frequently referred to as "titrable" acidity and is measured by means of titration with a standard alkali and expressed in terms of the alkali. For example, to neutralize 10 ml of 0.1 N HCl or CH_3COOH (acetic acid), 10 ml of 0.1 N NaOH is required.

When considering the strength of an acid, one is not concerned with the amount of acid but rather with the number of hydrogen ions [H^+] present in an aqueous solution. This number depends upon the extent of dissociation of the specific acid, and the measure of the [H^+], or the concentration of hydrogen ions, is termed pH. In pure water and dilute solutions, the product of the hydrogen ion and hydroxide ion concentration (known as the ion product of water, K_w) is constant: $[H^+][OH^-] = K_w = 10^{-14}$ (at 25°C).

B. IONIZATION OF ACIDS

The simple definition of an *acid* as a substance that ionizes to furnish a proton (proton donors), and of a *base* as a substance that ionizes to provide a hydroxyl ion (proton acceptors), is not particularly useful in many biological situations. According to the Bronstead definition, acids dissociate to a proton and to the conjugate base of the acid, whereas bases accept a proton to become an acid.

$$HA \rightleftharpoons H^+ + A^-$$

The conjugate base of the acid HA is A^-. The base formed A^- need not carry only a single negative charge, and in some cases it is a neutral molecule. The weak acid HA can exist as a positively or negatively charged molecule as well as a neutral molecule; it can be both an acid and a base. For example, the bicarbonate ion, HCO_3^-, is a weak acid, but it is also the conjugate base of carbonic acid, H_2CO_3.

Strong acids, e.g., HCl, are almost completely dissociated in aqueous solution. Therefore, the molar concentration of the proton formed is equal to the molar concentration of acid that was present.

$$HCl \longrightarrow H^+ + Cl^-$$

The acidity of aqueous solutions of strong acids is seldom expressed in terms of the hydrogen ion concentration, but rather as the pH of the solution. As originally defined by Sorenson, the pH of a solution is equal to the negative logarithm of the hydrogen ion activity.

$$pH = \log_{10} \frac{1}{aH^+} = -\log aH^+$$

In dilute solutions, which are those encountered in most biochemical situations, the activity will approach the concentration. Therefore, the expression becomes

$$pH = \log \frac{1}{[H^+]} = -\log[H^+]$$

where the use of brackets indicates molar concentrations. It must be stressed that the pH values are logarithmic functions of the hydrogen ion concentration and that a solution of pH 3 has a hydrogen ion concentration (10^{-3} M), ten times that of a solution with pH 4 (10^{-4} M).

C. MEASUREMENT OF pH IN A BUFFER SOLUTION

Water has many chemical and physical properties, not least of which is a very slight tendency to ionize as hydroxonium ions (H_3O^+) and hydroxyl ions (OH^-). When an H_2O molecule binds a proton to form H_3O^+ it is acting as a base, whereas when it forms OH^- it is acting as an acid. The equilibrium constant for H_2O ionization is

$$K_{eq} = \frac{[H^+][OH^-]}{[H_2O]} = 1.8 \times 10^{-16} M$$

where concentrations are expressed in molarity.

H_2O has been shown to be 55 M (1l of water weigh 1000 g and has a molecular mass of 18; therefore, the molarity of water equals 1000/18 = 55.6 M). Thus,

$$[1.8 \times 10^{-16}][55] = [H^+][OH^-]$$

$$1 \times 10^{-14} M^2 = [H^+][OH^-]$$

where $1 \times 10^{-14} M^2$, known as the ion product of water (K_w) is constant.

Electrolytes that react with water to increase the proportion of OH^- are bases, whereas acids increase the proportion of H^+. Pure water is electrically neutral, i.e., $[H^+]=[OH^-]=10^{-7}M$.

The pH of a solution can be determined in a number of ways. A convenient term for expressing concentrations of H^+ is as its negative logarithm (pH), i.e., pH = $-\log [H^+]$. An estimation of the pH can be obtained by the use of indicators that change color at different pH, but more accurate measurements are made with the use of a pH meter and a glass electrode. The pH meter measures the difference between the potential of the glass electrode (E_g) and the potential of a reference electrode (E_{ref}). The relationship between this difference at 25°C is

$$pH = \frac{E_g - E_{ref}}{0.0591}$$

The value measured is that of the hydrogen ion activity, not of the hydrogen ion concentration. A discussion of the theory and use of the glass electrode is beyond the scope of this section. It is very difficult to measure the potential difference, $E_g - E_{ref}$, accurately enough to standardize solutions. What is done in practice is to calibrate the pH meter with solutions of known pH.

D. THE HENDERSON–HASSELBATCH EQUATION

A rearranged form of the mass law, called the Henderson–Hasselbatch equation, can calculate the pH of buffers. In the buffer solution dissociation, two equations are important:

HA $\rightleftharpoons$ $H^+ + A^-$ (weakly dissociated)

BA $\rightleftharpoons$ $B^+ + A^-$ (highly dissociated).

The K_a was previously defined as

$$\frac{[H^+][A^-]}{[HA]} = K_a$$

where the brackets indicate molar concentration (moles per litre).

Therefore $\quad [H^+] = K_a \dfrac{[HA]}{[A^-]}$

If the logarithm of both sides of this equation is taken and the equation multiplied by -1

$$\log[H^+] = \log K_a + \log\frac{[HA]}{[A^-]}$$

$$-\log[H^+] = -\log K_a - \log\frac{[HA]}{[A^-]}$$

Therefore, by previous definitions

$$pH = pK_a + \log\frac{[Salt]}{[Acid]} \quad \text{or} \quad pH = pK_a + \log\frac{[\text{proton acceptor}]}{[\text{proton donor}]}$$

E. METHODS OF MEASURING pH

1. Colorimetric

This method is based on the fact that certain dyes change color with a change in acidity (or alkalinity). The color change takes place because the dye dissociates into a colored ion and a non-colored, or differentially colored, ion. The degree of dissociation is a function of the $[H^+]$; every shade corresponds to a definite pH (Figure 3.2). Specific dyes have definite ranges over which they are sensitive. The first dyes used in colorimetric measurements were of plant or animal origin (litmus from a lichen and cochineal from an insect). However, many modern indicators are synthetic products.

2. Potentiometric

When two metal electrodes are placed in a solution of an acid, a potential (electromotive force, e.m.f.) is set up because of the reaction between the metal and the surrounding electrolyte. By exactly opposing this force with one of a known value, these differences in potential are measurable in terms of millivolts, and the values may be converted into pH values upon proper calibration. The e.m.f. is a function of the number of the hydrogen ions present.

3. pH Electrodes

Nowadays, the pH electrode/pH meter is one of the most basic items of equipment found in biology laboratories. The pH electrode depends upon the ion exchange in the hydrated layers formed on the electrode surface. There are different types of electrodes, some of which are briefly described below.

 (a) *Hydrogen*: A platinum electrode, coated with platinum black and saturated with gaseous hydrogen, is frequently used. Owing to the cumbersome equipment required, its application is limited in biological systems.

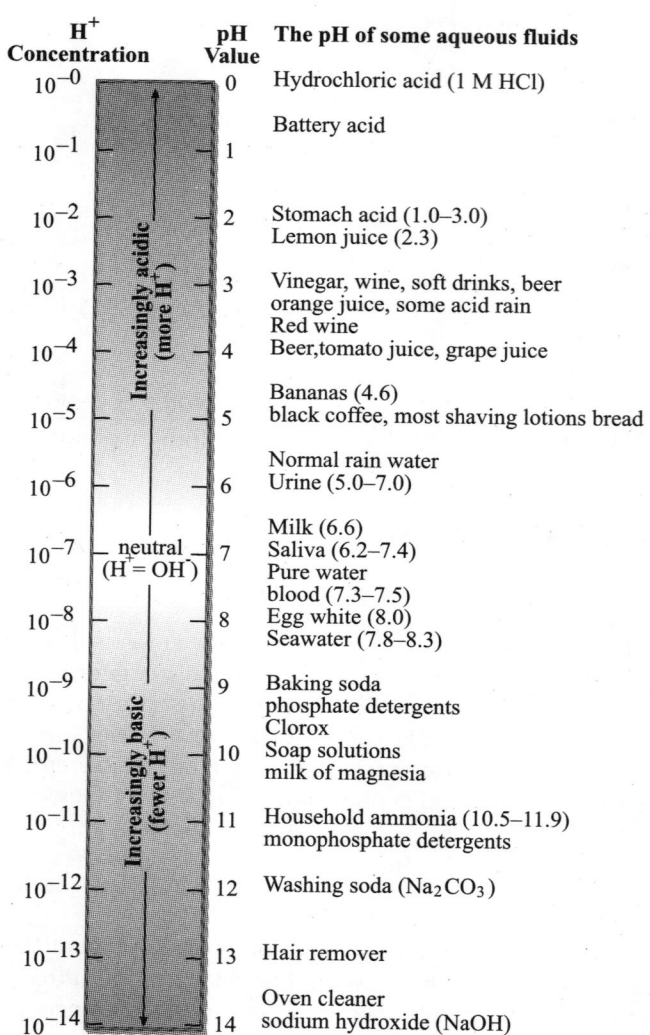

Figure 3.2 *The pH scale in which a litre of fluid is assigned a number according to the number of hydrogen ions in it. The scale ranges from 0 (most acidic) to 14 (most basic). A change of only 1 on the scale means a tenfold change in the H^+ concentration*

(b) *Quinhydrone*: Quinhydrone, in the presence of a gold or platinum electrode, is often used. Its application is limited because it is readily affected by certain organic materials. In addition, values above a pH of 8.0 cannot be determined with accuracy.

(c) *Glass*: Perhaps the most convenient and accurate way of determining pH is by using a glass electrode. This apparatus has become extremely popular because of the development of a rugged glass electrode. It consists of a thin glass membrane that is differentially permeable to hydrogen ions, and precise, accurate measurements can be made with relative ease. The glass electrode can be used in any type of biological system.

The pH electrode depends upon the ion exchange in the hydrated layers formed on the surface of the glass electrode. Glass is composed of a silicate network amongst which metal ions are coordinated to oxygen atoms, and it is these metal ions that exchange with H^+. The glass electrode acts like a battery, whose voltage depends on the H^+ activity of the solution in which it is immersed. The size of the potential (E) due to H^+ is given by the equation

$$E = 2.303 \frac{RT}{F} \log \frac{[H^+]_i}{[H^+]_o},$$

where $[H^+]_i$ and $[H^+]_o$ are the molar concentrations of H^+ inside and outside the glass electrode. In practice, $[H^+]_i$ is fixed and is generally 10^{-1} because the electrode contains 0.1 M HCl. Since pH = $-\log[H^+]$, it follows that the developed potential is directly proportional to the pH of the solution outside the electrode.

(d) *pH meters:* pH meters are available in a variety of shapes and sizes, which are suitable for many different applications. The typical pH meters use a combination glass electrode. This consists of an inner pH electrode, which is in contact with the sample through pH-sensitive glass, and an outer KCl-containing reference electrode in contact with the sample through a porous glass plug. The pH-sensitive glass and the plug can both become clogged. Clogging causes sluggish response of the electrode as well as decreased sensitivity. This can be dealt with by immersing the electrode overnight in a solution of 1 mg/ml pepsin in 0.1 N HCl. If the clogging is not very severe, the electrode can be soaked for the same amount of time in a solution of 0.1 N HCl. In addition a brief acetone rinse often removes hydrophobic substances.

There are two basic controls available on a pH meter. One regulates the gain of its amplifier, and the other is an offset adjustment. These may be disguised under a variety of names like "temperature", which is a gain control, and "buffer adjust," which is an offset control. There may even be several knobs regulating the same function. Because pH electrodes lose sensitivity with age, the gain control will often have to be increased to allow the pH meter to be accurate at all pH levels. To adjust the pH meter so that it accurately measures at all pH levels, the following facts should be noted: (1) the pH meter measures and indicates a voltage produced by the electrode; (2) the gain control changes the amplification of the voltages; and (3) the offset adds a fixed constant to all voltages and pH levels. If the temperature setting required differs greatly from the ambient temperature, the electrode may be clogged. In such cases, the electrode should be cleaned. When it is no longer possible to adjust the electrode because it is too insensitive, it should be replaced.

The calibration process necessitates the use of two solutions with widely differing pH. Usually calibration is first carried out with a pH 7 buffer, followed by a pH 4 buffer (if the sample is expected to be acidic) or a pH 9 buffer (if the sample is expected to be basic). Once the pH electrode is calibrated it can simply be immersed in the solution to be measured and the pH can be estimated quickly and accurately.

Comments

1. Some electrodes give false readings with Tris-buffers.
2. Some electrodes have a rubber plug at the top. This should be open during use in order to allow pressure equilibration, and closed when not in use to minimize the evaporation of KCl from the reference electrode.

3. Most pH electrodes need to be filled with saturated KCl. The levels should be checked periodically and more added if necessary.
4. In storage, the probe should be immersed in saturated KCl. It should then be rinsed with water. Do not wipe off remaining drops because a static charge may thus be given to the probe. Remove the drop by touching it to the side of the rise vessel.
5. If excessive KCl crystals have formed in the bottom of the probe, dissolve them by warming the probe under running water and replacing the saturated KCl solution with a new saturated KCl solution. Several heat–chill cycles should suffice to solubilize KCl crystals in a probe.

F. pH AND BUFFER SOLUTIONS

Almost every biological process is pH-dependent; a small change in pH produces a large change in the rate of the process. This is true not only for the many reactions in which the H^+ ion is a direct participant, but also for those in which there is no apparent role for H^+ ions. The enzymes catalyze cellular reactions with characteristic pK_a values. The buffers are the weak acids and their conjugate bases. Buffers are aqueous systems that tend to resist changes in the pH when small amounts of acid (H^+) or base (OH^-) are added. A buffer system consists of a weak acid and its conjugate base. Buffering results from two reversible reaction equilibriums occurring in a solution of nearly equal concentrations of a proton donor and its conjugate proton acceptor. Whenever H^+ or OH^- is added to a buffer, the result is a small change in the ratio of the relative concentrations of the weak acid and its anion and, thus, a small change in the pH. The decrease in concentration of one component of the system is exactly balanced by an increase in the other. Thus, the sum of the buffer components does not change, only their ratio does.

Since biochemical reactions occur in an aqueous environment, and for the most part in an environment which is maintained close to neutrality, a thorough understanding of the properties of acids and bases and of a buffered system must be established. As these subjects are adequately dealt with in general chemistry and quantitative analysis courses, only a brief review will be presented in this chapter.

G. BUFFERS USED IN CHEMICAL REACTIONS

For a weak acid to be used as buffer in an *in vitro* biochemical reaction, it should, of course, have a pK_a close to that of the desired pH to maximize its buffering capacity. It should be nontoxic to the biochemical reaction being studied, colorless, and free of any UV absorption in the region where proteins and nucleic acids absorb. A good buffer is also one in which changes in temperature, concentrations or ionic strength have a minimal effect on dissociation. The second dissociation of phosphoric acid is in a good pH range, but phosphate solutions are easily precipitated by divalent cations, and phosphate is often a product or reactant in biochemical reactions. Tris [tris (hydroxymethyl) aminomethane] is often used above pH 7, but its dissociation is rather temperature-dependent, and other complex amines such as HEPES and PIPES are now used.

Solving problems using the Henderson–Hasselbath equation

1. Calculate the pK_a of lactic acid, given that when the concentration of lactic acid is 0.010 M and the concentration of lactate is 0.087 M, the pH is 4.80.

$$pH = pK_a + \log\frac{[\text{Lactate}]}{[\text{Lactic acid}]}$$

$$pK_a = pH - \log\frac{[\text{Lactate}]}{[\text{Lactic acid}]}$$

$$= 4.80 - \log\frac{0.087}{0.010} = 4.80 - \log 8.7 = 4.80 - 0.94 = \mathbf{3.86}$$

2. Calculate the pH of a mixture of 0.1 M acetic acid and 0.2 M sodium acetate, where the pK_a of acetic acid is 4.76.

$$pH = pK_a + \log\frac{[\text{Acetate}]}{[\text{Acetic acid}]}$$

$$= 4.76 + \log\frac{0.2}{0.1} = 4.76 + 0.301 = \mathbf{5.06}$$

3. Calculate the ratio of the concentrations of acetate and acetic acid required in a buffer system of pH 5.30, where the pK_a of acetic acid is 4.76.

$$pH = pK_a + \log\frac{[\text{Acetate}]}{[\text{Acetic acid}]}$$

$$\log\frac{[\text{Acetate}]}{[\text{Acetic acid}]} = pH - pK_a$$

$$= 5.30 - 4.76 = 0.54$$

$$\frac{[\text{Acetate}]}{[\text{Acetic acid}]} = \text{antilog}\, 0.54 = \mathbf{3.47}$$

When making up solutions, the following guidelines should be observed.
1. Use the highest grade of reagents whenever possible.
2. Prepare all the solutions with the highest quality of distilled water available.
3. Autoclave the solutions when possible. If the solution cannot be autoclaved and you wish to store it, sterilize it by filtration through a 0.22 μm filter.

4. Check the pH meter carefully, using freshly-prepared solutions of standard pH, before adjusting the pH of the buffers.
5. Always label the containers with the name of the solution, the percentage or concentration, the name of the experimenter and the date. For highly basic solutions, such as 1 M NaOH, be sure to use plastic containers – glass will be corroded by the bases.
6. Store solutions cold whenever it is possible.

IV CENTRIFUGATION

A. BASIC CONCEPTS OF SEDIMENTATION THEORY

Centrifugation is a simple operation used in almost all biological separations. The principle behind centrifugation is that two particles in suspension, which have different masses or densities, will settle to the bottom of a tube at different rates. It fractionates by applying a centrifugal field to the mixture to be separated instead of gravity (Figure 3.3). Centrifugation is used for two basic purposes: as a preparative technique to separate one type of material from others, and as an analytical technique to measure physical properties (e.g., molecular weight, density, shape, and equilibrium binding constants) of macromolecules.

A centrifuge is a device for separating particles from a solution. In biology, the particles are usually cells, subcellular organelles, or large molecules, which are generally referred to as "particles" to simplify the terminology. The physical parameters, which determine the extent of fractionation, apply equally to such diverse particles as macromolecules and cells. If a solution of large particles is left to stand, then the particles will tend to sediment under the influence of gravity. For a given particle, the rate or velocity at which it sediments is proportional to the force applied, such that the particles sediment more rapidly when the force applied is greater than the gravitational force of the earth. Therefore, the basis of centrifugation separation techniques is to exert a larger force than that exerted by the earth's gravitational field. Although the nature of the particles (molecular mass, shape and density) may place restraints on the centrifu-

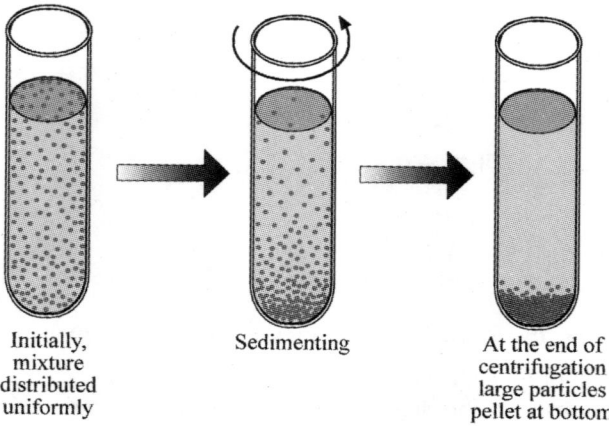

Figure 3.3 *Preparative centrifugation*

gation conditions that can be used, if the sedimentation of a spherical particle reaches a constant value, the net force on the particle is equal to the force resisting its motion through the liquid. This resisting force is called "fractional" or "drag force". From Stokes' law it can be calculated that the sedimentation rate, v, of a particle is given by

$$v = \frac{d^2(\rho_p - \rho_m)}{18\mu} \times g$$

From this equation, it can be seen that: (1) the sedimentation rate of a given particle is proportional to the square of the diameter (d) of the particle. (2) The sedimentation rate is proportional to the difference between the density of the particle and the density of the liquid medium, ($\rho_p - \rho_m$). (3) The sedimentation rate is zero, when the density of the particle is equal to the density of the liquid medium. (4) The sedimentation rate decreases as the viscosity (μ) of the liquid medium increases. (5) The sedimentation rate increases as the force field (g) increases.

The force field relative to the earth's gravitational field (RCF), exerted during centrifugation, is defined by the equation

$$\text{RCF} = \frac{\omega^2 r}{980},$$

where r is the distance between the particle and the center of rotation in cm; the rotor speed ω in rad/sec can be calculated from the equation

$$\omega = \frac{\text{rev}}{\text{min}} \times \frac{2\pi}{60} = \frac{\text{rev}}{\text{min}} \times 0.10472$$

The sedimentation velocity per unit of centrifugal force is called the sedimentation coefficient, s:

$$s = \frac{1}{\omega^2 r} \times \frac{dr}{dt},$$

where dr/dt is the rate of movement of the particle in cm/sec. The sedimentation coefficients are usually expressed in Svedbergs (S), equivalent to 10^{-13} sec. Thus, a particle whose sedimentation coefficient is measured at 10^{-12} sec, i.e., 10×10^{-13} sec, is said to have a sedimentation coefficient of 10 S.

B. CENTRIFUGATION METHODS

There are two main types of centrifugation techniques: preparative centrifugation and analytical centrifugation techniques. Preparative centrifugation techniques are concerned with the actual separation, isolation and purification of, for example, whole cells, subcellular organelles, plasma membranes, polysomes, ribosomes, chromatin, nucleic acids, viruses, etc. In contrast, analytical centrifugation techniques are devoted mainly to the study of pure, or virtually pure, macromolecules or particles. They are primarily concerned with the study of the sedimentation characteristics of biological macromolecules and molecular structures and utilize specially designed rotors and detector systems. Such studies yield information from which the purity, relative molecular mass and shape of the material may be deduced. In this chapter, three main types of centrifugal fractionation, namely, (1) differential pelleting (differential centrifugation), (2) rate-zonal density-gradient sedimentation, and (3) isopycnic density-gradient sedimentation, are briefly discussed.

1. Differential centrifugation

Of these techniques, differential pelleting is the most commonly used method for fractionating material according to size and density. This technique is only suitable for materials with markedly different sedimentation rates. In this method, the material to be fractionated (e.g., cell suspension, cell homogenate, etc.) is initially distributed uniformly throughout the sample solution, which is the sole occupant of the centrifuge tube. A particular centrifugal field is chosen over a period of time in order to pellet. In the early spins, only the largest particles' sediments (or pellet) go to the bottom of the centrifuge tube. Such particles may be whole cells, fragments of cell wall, nuclei, or grains of starch. Later, chloroplasts and mitochondria, tonoplast, endoplasmic reticulum, etc., are pelleted. At the highest speeds, and these might be tens of thousands of revolutions per minute, microsomes and ribosomes, even dissolved protein molecules, can be pelleted.

After, each centrifugation, the pellet is enriched with the larger particles of the mixture (Figure 3.4). However, the pellet obtained always consists of a mixture of the different types of particles, and it is only the most slowly sedimenting component of the mixture that remains in the supernatant liquid which can be purified by a single centrifugation. The amount of contamination in the pellet can be reduced by washing it, or by subjecting it to repeated steps of resuspension in the suspension medium and centrifugation. However, this inevitably reduces the yields obtained. Another difficulty with this technique is that there are significant differences in the size of the same organelle within a single cell type.

2. Rate-zonal density gradient centrifugation

The efficiency of the fractionation of particles according to size and shape can be improved markedly by using rate-zonal density-gradient centrifugation through a density gradient. In rate-zonal centrifugation, the particles have to reach sedimentation equilibrium. Unlike differential centrifugation, where the sample is distributed throughout the medium, in rate-zonal centrifugation, the sample is initially present only on top of the gradient as a narrow band. Centrifugation subsequently proceeds for a fixed period of time so that the particles separate into a series of bands in accordance with parameters, such as the centrifugal field, size and shape of the particle and difference in density between the particle and the suspending medium. The technique can yield data about the molecular weight, density, shape and purity of a given group of molecules. There are several materials that can serve as density-gradient supports for the preparation of gradients (1–1.4 g/ml). These include *cesium chloride*, *sucrose* (sugar), *Ficoll* and *percoll*. Percoll is a colloid (finely-divided particles suspended in an uniform medium) of silica particles, 15–30 nanometers in diameter, which have been coated with polyvinyl pyrrolidone (PVP), thus increasing their stability as a colloid and eliminating their toxicity to cells. Sucrose can be used in this model experiment, but cannot be used to separate cells and some cell organelles.

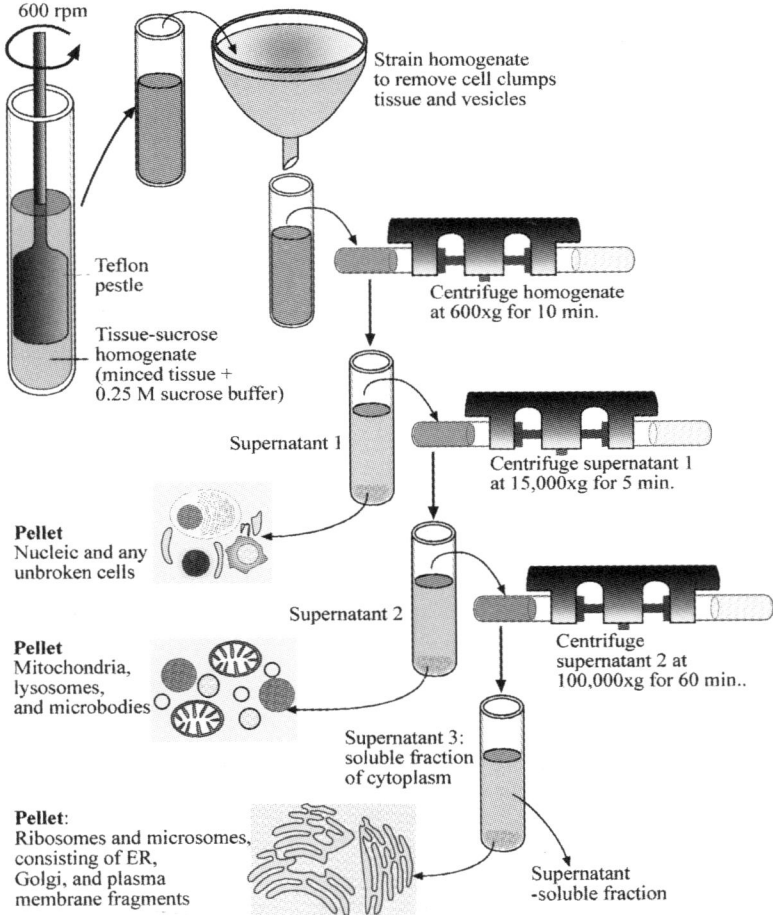

Figure 3.4 *General procedure for cell fractionation by differential centrifugation. Repeated centrifugation at progressively higher speeds will fractionate homogenates of cells into their components*

Density gradients can be constructed in centrifuge tubes as discontinuous or continuous gradients. A discontinuous gradient is constructed by loading a centrifuge tube with layers of varying densities of the separating medium. The sample to be separated is added to the top layer of the medium and centrifuged at high speed. This leads to the components of the sample sedimenting and forming a band at the interface of the layer, which matches their density (Figure 3.5). A continuous gradient is formed when a medium with a uniform density is subjected to a centrifugal force.

3. Isopycnic sedimentation

The third method for separating particles is *isopycnic sedimentation,* in a gradient whose maximum density exceeds that of the particles. This is an equilibrium technique in which particles

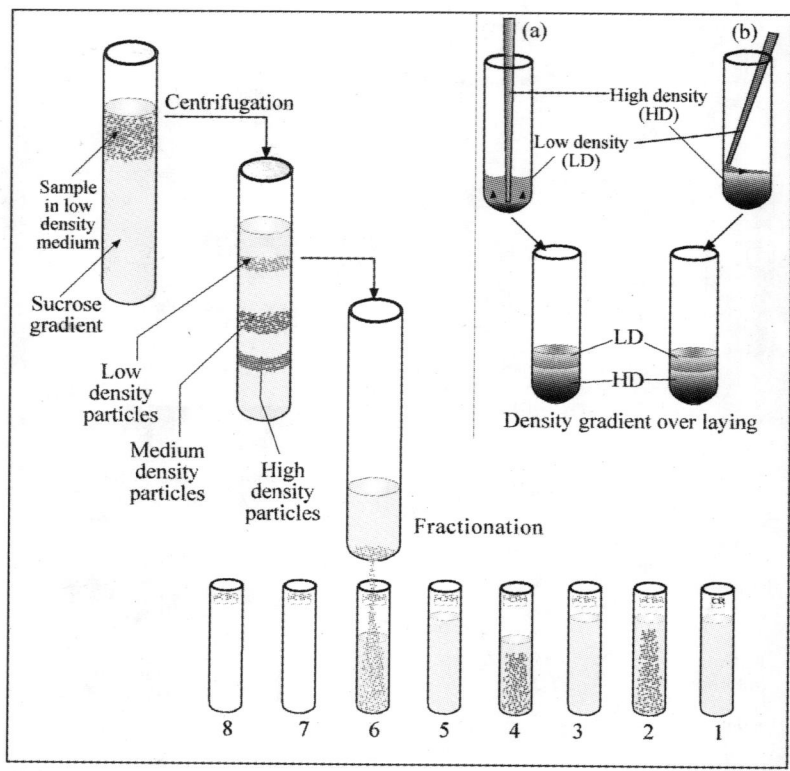

Figure 3.5 *Schematic diagram of rate-zonal and isopycnic centrifugation. Subcellular component fractionations. Insert: different techniques of overlaying the gradients in a tube: (a) light end first; (b) dense end first*

are separated on the basis of their buoyant densities, independently of the time of centrifugation and of the size and shape of the particles. It is important to realize that the effective buoyant density of any particle is a function of the actual density of the particle and its degree of hydration. Therefore, the choice of media is extremely important, because the high viscosity of the medium can lead to the poor resolution of bands. For example, the density of non-hydrated DNA is close to 2.0 g/cm^3, but its observed buoyant density can vary from 1.7 g/cm^3 to 1.1 g/cm^3, depending on the water activity of the gradient medium. Isopycnic centrifugation in the salts of alkaline metals (e.g., CsCl and NaI) has been widely used for the separation, fractionation and purification of macromolecules and nucleoproteins. Isopycnic sucrose gradients have been used widely for the fractionation of cell organelles like mitochondria and nuclei, whereas the more osmotically-inert polysaccharides (e.g., Ficoll) have been used for the separation of cells.

The basic concept of this method involves the generation of a density along the length of a tube, such that its density near the bottom is greater than that of the densest particles in the mixture to be fractionated, while at the meniscus the density is less than that of the highest particles. After an appropriate period of centrifugation, the sample particles will have moved along the gradient until they reach a region where the density of the surrounding medium and the buoyant densities of the particles are equal. The sedimentation of these particles will then stop and particles of different densities will finally be localized in different regions of the gradient; i.e., they will be fractionated according to their buoyant densities. If DNA is present in the centrifuged CsCl solution, it will move to a position of equilibrium in the gradient equivalent to its buoyant density (Figure 3.6). For this reason, this technique is also called isopycnic (same density) centrifugation.

Caesium chloride centrifugation is an excellent means of removing RNA and proteins in the purification of DNA. The density of DNA is typically slightly greater than 1.7 g/cm^3, while the density of RNA is more than 1.8 g/cm^3; proteins have densities less than 1.3 g/cm^3. In CsCl solutions of appropriate density, the DNA bands near the center of the tube, the RNA pellets to the bottom, and the proteins float near the top. Single-stranded DNA is denser than double-helical DNA. The irregular structure of randomly-coiled ssDNA allows the atoms to pack to-

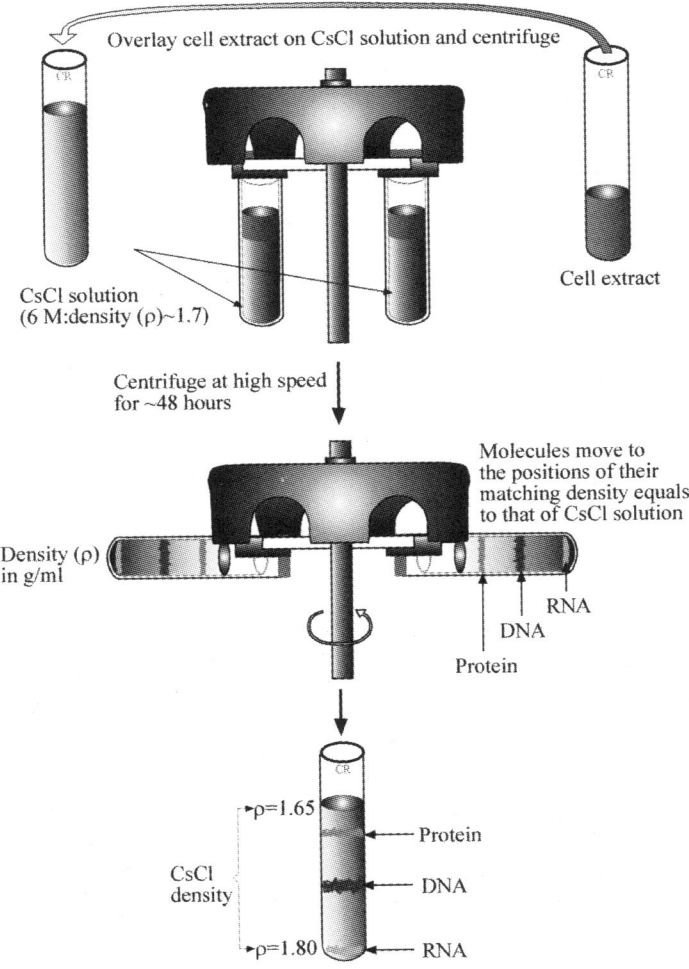

Figure 3.6 *Isopycnic density gradient ultracentrifugation used to isolate DNA*

gether through van der Waals interactions. These interactions compacts the molecule into a smaller volume than that occupied by the hydrogen-bonded double helix.

The techniques mentioned thus far can be used for fractionating particles on either a preparative or an analytical scale. However, in order to obtain really accurate quantitative data, it is necessary to use the purpose-built analytical ultracentrifuge.

4. Ultracentrifugation

Ultracentrifugation was developed by a Swedish biochemist Svedberg (1923). The principle is based on the generation of centrifugal force to as high as 600,000 g (earth's gravity $g = 9.81$ ms^2) that allows the sedimentation of particles or macromolecules. The rate at which the sedimentation occurs in ultracentrifugation primarily depends on the size and shape of the particles or macromolecules. It is expressed in terms of *sedimentation coefficient*(s) and is given by the formula

$$S = \frac{v}{\omega^2 r},$$

where v = Migration (sedimentation of the molecule)
ω = Rotation of the centrifuge rotor in radians/sec
r = Distance in cm from the center of rotor.

Very detailed manuals for the operation of the ultracentrifuge and rotors are generally provided by the manufacturer along with the centrifuges. However, some important details are hard to find among the great mass of information. In addition, there are a few tricks to loading tubes that are not mentioned in standard literature.

(a) Filling and balancing tubes

1. Balance the tubes within 0.5 g. For aqueous solutions the eye is a good estimator, but with denser solutions it is necessary to use a balance.
2. Fill the non-capped swinging bucket tubes to within 4 mm of the top. The thin-wall capped tubes for angle rotors must be completely filled.
3. Tubes with aluminum caps present several problems, some of which will be overcome with the new heat-sealable tubes. For tubes with aluminum caps, it is best to attach the cap to the tube using a vise. The balance tubes must be filled with solutions of the same density as the sample tubes and then balanced by weighing. Use a 25 ml syringe with an 18-gauge needle to complete filling a tube with mineral oil; then, screw the Allen screw into the top and invert the tube. There should not be more than 0.1 ml of air bubbles.

(b) Loading rotors

1. A 12-place rotor can be balanced with 2, 3, 4, 5, 6, 7, 8, 9, 10, or 12 equally-filled tubes. For swinging bucket rotors, the use of all buckets, whether or not they have tubes inside, are recommended. Always put the buckets on the hanger that has the same number and never use buckets from a different rotor.
2. The samples must be isolated from the vacuum of the centrifuge chamber or they will evaporate and unbalance the rotor. Check all the O-rings and gaskets, including the O-ring in the angle rotor handle; they should be slightly greasy and not gritty. When tightening screws to form seals, do not tighten to the point that metal-to-metal contact is made.

(c) Operation of ultracentrifuge

Before using the centrifuge, the experimenter should know all the main rules of operation, the control measures, the peculiarities and the "reefs" in the performance of an experiment. They should be checked personally before every run.

(d) Making a centrifuge run

1. The appropriate rotor should be seated freely on the drive shaft.
2. *Loading the sample* (Figure 3.7): The samples should be placed in an appropriate centrifuge tube that fits well into the rotor used. The samples should be well balanced by weighing them along with a centrifuge tube, and tubes with equal weights should be loaded exactly opposite to each other in a rotor.

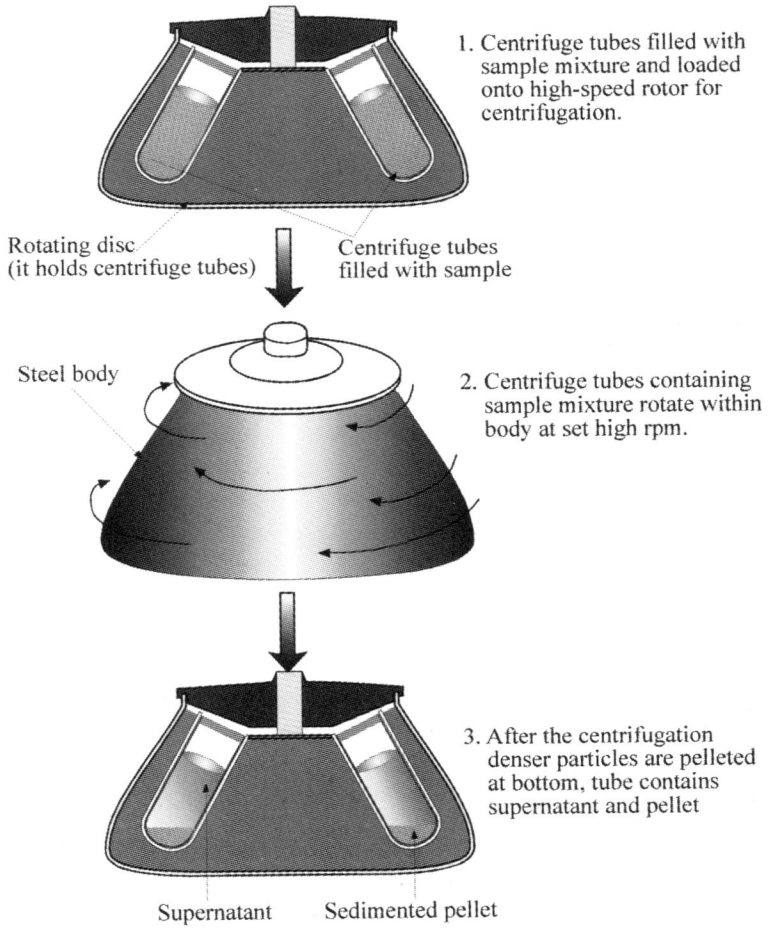

Figure 3.7 *Different steps in the process of centrifugation with fixed angle rotors*

3. *Vacuum system*: In order to avoid friction, air is pumped out from the rotor chamber. The rotor chamber is tightly sealed by a rubber O-ring (R), which is clamped by a heavy, free-lying chamber lid. After closing the chamber door, the vacuum pump can be turned on.
4. *Temperature setting*: It is best to expose the rotor with the tubes to the pre-selected temperature for several hours. Therefore, while selecting the operating conditions, it is advisable to fix upon ambient temperature or upon the temperature of a cold-room or a refrigerator so that the rotor can be placed there beforehand.
5. *Setting the rotor speed*: The speed selector dial must be set to a pre-selected value. The speed indicator's accuracy lies within ±500 rpm. The exact speed value can be determined by reference to the revolution counter
6. *Setting of the run time*: The optimum time required for the centrifugation should be set before the start of the centrifugation.
7. *Accelerating and braking the rotor*: The acceleration rate dial determines both the acceleration rate and the deceleration rate; however, the latter mode operates only when the brake button is depressed.
8. *Starting a run*: If the light of the indicator lamp "speed normal" is on, it indicates that the centrifuge has been correctly prepared for the start of the run. This can be effected by pressing the push button "START/RUN".
9. *Termination of a run*: When the rotor has come to a complete halt, the light "SPEED NORMAL" goes off. Later, the cooling system ("REFRIG" button) and vacuum can be turned off. The rotor chamber is then filled with atmospheric air and, after the characteristic hiss has ceased, the chamber door can be opened and the rotor removed.

C. TYPES OF CENTRIFUGES

Centrifuges may be classified into four major groups: the small, bench-top centrifuges; large-capacity refrigerated centrifuges; high-speed refrigerated centrifuges; and ultracentrifuges of two types, preparative and analytical.

D. CARE OF CENTRIFUGES AND ROTORS

For the complete novice who has no experience of using a particular centrifuge or a particular rotor, it is very important to read the manufacturers' instructions and, if possible, consult someone who can give advice on practical details. For example, rotors from other models should not be used and the balancing of tubes, running under an imbalanced rotor, causes the centrifuge to deteriorate faster. It is also important to contain the sample during centrifugation, particularly as the samples are often biohazardous or radioactive. Also, the rotors of all ultracentrifuges run in a high vacuum to minimize heating due to air resistance and so it is especially important to isolate the sample, to prevent evaporation, by capping the tubes securely. The function of the cap is threefold. First, it ensures that the sample is contained within the tube; second, it prevents the sample being exposed to the high vacuum surrounding the rotor; and third, it supports the top of the tube and prevents deformation during centrifugation. Another problem is that of spillage of sample material into the rotor buckets. The rotors must always be rinsed thoroughly with distilled water and left to drain after use.

Obviously, it is important that the users are aware of the capabilities and limitations of these instruments, since such factors determine the procedures to be used for particular separations and enable one to predict the likely result of the experiment. In the case of the rotor's speed, capacity and geometry are the main variable parameters, whereas tubes may vary in their ability to withstand centrifugal forces and their resistance to solvents. Centrifuge rotors can be broadly divided into five types: (a) fixed-angle, (b) swing-out, (c) vertical tube rotors, (d) zonal, and (e) elutriator rotors. Detailed descriptions of the routine maintenance necessary for individual rotors of all types are given in the manufacturers' instruction manuals.

Sample containers: Centrifuge tubes and bottles are manufactured in a range of different sizes, in varying thicknesses and rigidity, and from a wider variety of materials including glass, cellulose esters, polyallomer, polycarbonate, polyethylene, polypropylene, kynar, nylon and stainless steel. The correct choice of sample container is important in order to achieve the desired degree of separation of particles from a sample mixture. The types of containers used will depend on the nature and volume of the sample to be centrifuged, the type of rotor used, the available centrifuge, the centrifugal forces to be withstood, its chemical resistance (transparent or opaque), and whether they will be sliced or punctured for post-centrifugation analyses. Glass centrifuge tubes are usually suitable only for centrifugation at low speeds, because they disintegrate in higher centrifugal fields. Centrifuge tubes and bottles should always be filled to the correct level, and the maximum permitted rotor speeds, depending on the particular container, should be observed. Ideally, the tubes used in fixed-angle and vertical tube ultracentrifuge rotors should be completely filled, in order to support the tube against the very high centrifugal forces generated. The capping and sealing of tubes used in vertical tube rotors and most fixed-angle rotors are also important, because the tubes have to withstand the large upward hydrostatic force generated during centrifugation by the liquid in the tube.

V SUGGESTED READING

Birnie GD and Rickwood D (1978) *Centrifugal Separations in Molecular and Cell Biology*, Butterworth, London.

Griffiths A (2000) Centrifugation techniques, In: *Practical Biochemistry, Principles and Techniques*, Eds. Wilson K and Walker J, Cambridge University Press.

Rickwood D and Birnie GD (1978) Introduction: Principles and practices of centrifugation, In: *Centrifugal Separations in Molecular and Cell Biology*, Eds. Birnie GD and Rickwood D, Butterworth and Co (Publishers) Ltd.

Van Holde KE, Curtis Jonson W and Shing HO, P (1998) *Principles of Physical Biochemistry*, Prentice Hall, Upper Saddle River, NJ.

Wilson K and Walker J (2000) *Practical Biochemistry Principles and Techniques*, Cambridge University Press, UK.

4

Chemical Synthesis of Nucleic Acids

I INTRODUCTION

One of the experimental techniques that played a key part in the rapid development of genetic engineering is the chemical synthesis of oligonucleotides of defined structure. The chemistry of deoxyoligonucleotide synthesis has been the subject of study for many years. The knowledge and experience accumulated during these years by using the Khorana's phosphodiester approach and the Letzinger's phosphotriester method have led to the development of the first reliable and relatively easy manual process of deoxyoligonucleotide (oligonucleotides) synthesis. Caruthers and co-workers, using a modified version of phosphotriester chemistry, worked out a method for synthesizing oligonucleotides on polymer support. Later, a rapid and routine automated synthesis based on solid-phase phosphoramide chemistry was developed (Horvath et al. 1987). Now, it has become a routine tool in the molecular biology laboratory. Oligonucleotide lengths are usually stated as numbers; e.g., '10mer', '18mer', '25mer', etc.

Solid-phase synthesis methods have made the construction of short, defined-sequence DNA fragments accessible to all workers and it is often an essential part of the process of isolating a single gene from libraries containing tens of thousands to millions of gene fragments. Owing to the unique chemical structure of DNA, a piece of DNA (oligonucleotides) with a particular sequence will recognize and bind only to another piece with a complementary sequence, ignoring all other pieces of DNA with other sequences in a complex mixture. Therefore, small pieces of chemically-synthesized DNA can be used as unique molecular probes to isolate gene-sized pieces of DNA that contain the appropriate complementary sequence. The correct sequence for the DNA probes is often deduced from the amino acid sequence of the corresponding protein by translating (using the genetic code dictionary) a protein sequence into a DNA sequence. Compared to other ways in which genes can be obtained, the synthetic method offers new potentials. Synthetic oligonucleotides have significantly contributed to the methodologies of molecular cloning and DNA characterization. These readily-available, single-stranded DNA oligonucleotides may be used to assemble whole genes or gene fragments, to amplify specific DNA fragments, to introduce specific mutations into isolated DNAs, as DNA hybridization probes, as adjuncts to gene sequencing, or as linkers to facilitate gene cloning.

II PHOSPHORAMIDITE CHEMISTRY

A. PRINCIPLE

The oligonucleotides of any specified sequence can be synthesized by a series of chemical reactions (Figure 4.1). Phosphoramidite chemistry is currently the accepted method of oligonucleotide synthesis. The general strategy involves the sequential addition of a nucleotide to the 5′ end of oligonucleotide units, as nucleoside phosphoramidite derivatives to a nucleoside covalently attached to an insoluble resin. The key to the precise ordering of nucleotide additions is the use of chemically-modified nucleotides (phosphoramidites) containing an activated phosphoester group at the 3′ carbon, but one that is also chemically blocked from reaction at the 5′ end (Figure 4.2). The blockage of the 5′ nucleotide carbon assures that the phosphoramidites cannot react with each other but only with the reactive 5′ end of the growing oligonucleotide chain. The addition of nucleotides begins with a single glass-bead-bound nucle-

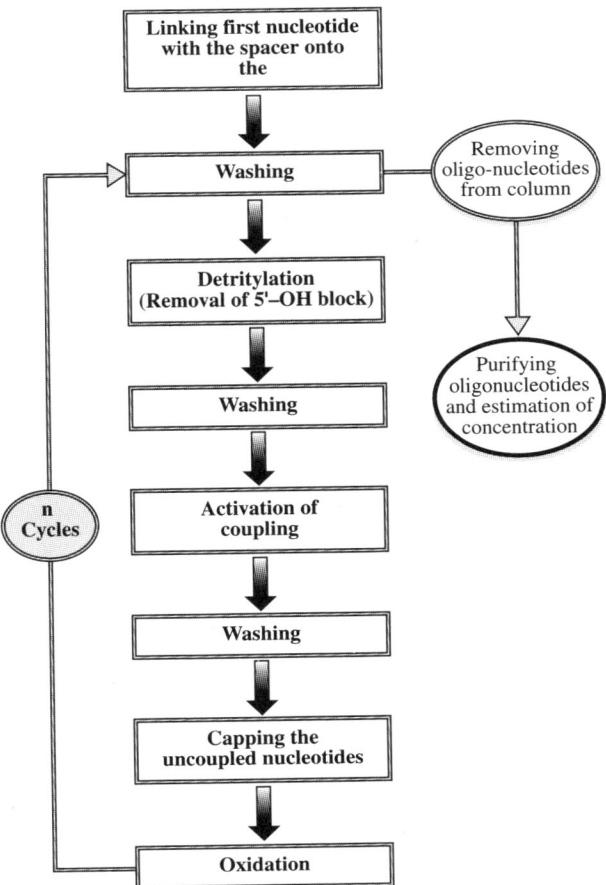

Figure 4.1 *Flow chart for solid-phase oligonucleotide synthesis by using phosphoramidites*

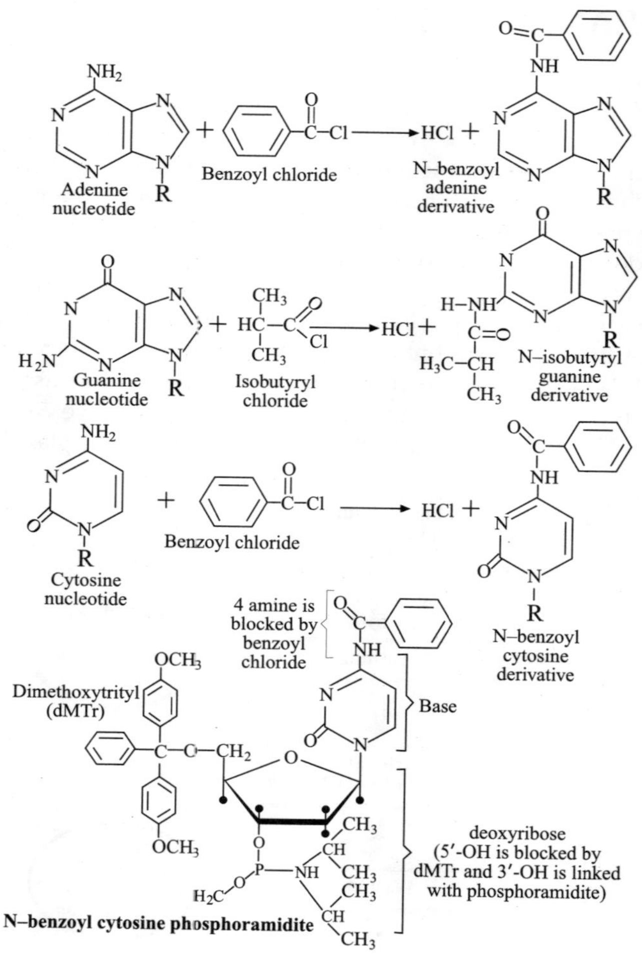

Figure 4.2 *Blocking of reactive amine groups on the phosphoramidite nucleosides*

otide, either A, C, G or T, depending on the desired 3' terminal nucleotide. The excess reagents, starting materials and byproducts are removed after each step by filtration. After the desired oligonucleotide has been formed, it is freed of all blocking groups, hydrolyzed from the resin, and purified by gel electrophoresis.

B. GENERAL APPROACH TO SOLID-PHASE OLIGONUCLEOTIDE SYNTHESIS

Currently, the phosphoramidite method of chemical DNA synthesis is the procedure of choice. Chemical synthesis takes place in the 3'→5' direction (reverse of the biological polymerization direction). The steps involved in cycle are shown in Figure 4.3. The multi-step synthesis proceeds as described below.

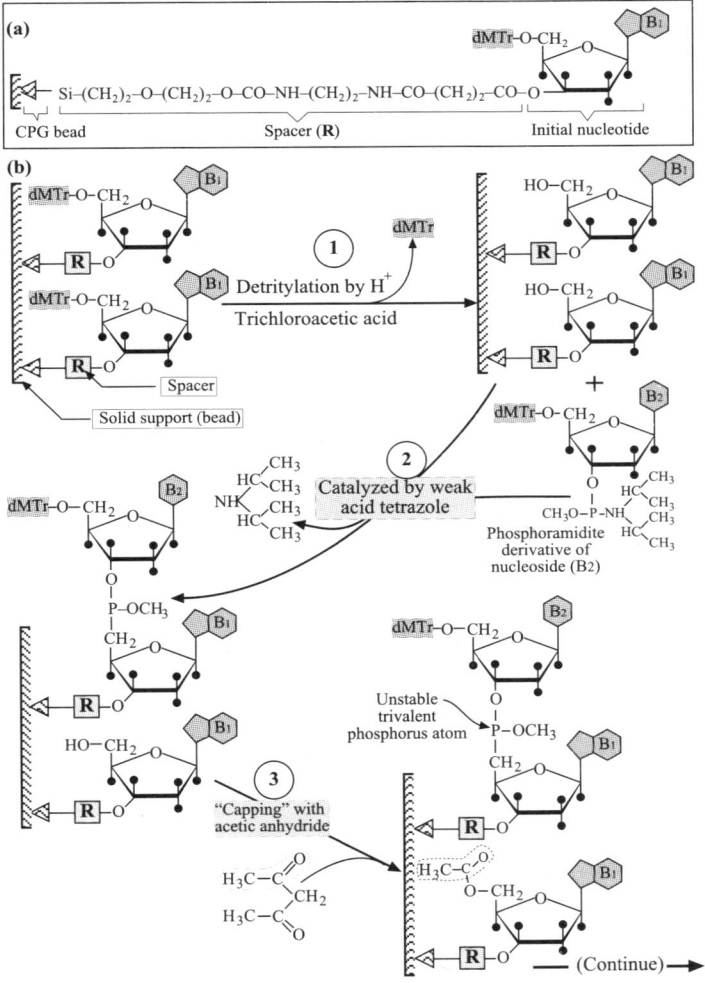

Figure 4.3 (a) Starting complex for the chemical synthesis of a DNA strand, (b) DNA synthesis initiation reaction: synthesis starts by using the phosphoramidite nucleosides

1. Columns containing a first immobilized nucleotide

All solid-phase method cycles start with one nucleoside immobilized via its 3′–OH group and the coupling of the next nucleoside to the 5′–OH of the first. The 3′–OH group of sugar is covalently attached to an insoluble and inert resin or matrix, typically either controlled pore glass (CPG) or silica bead, with a long alkyl amine spacer-chain (Figure 4.3 insert). The growing chain remains bound during the reaction series. The chain is thus synthesized from 3′→5′ end. The eventual product has free OH groups at both its 3′ and 5′ ends. Thus, for each synthesis, one of four prepared columns, corresponding to one of the four glass-bound nucleosides that will represent the terminal 3′ nucleotide in the completed oligonucleotide, must be selected.

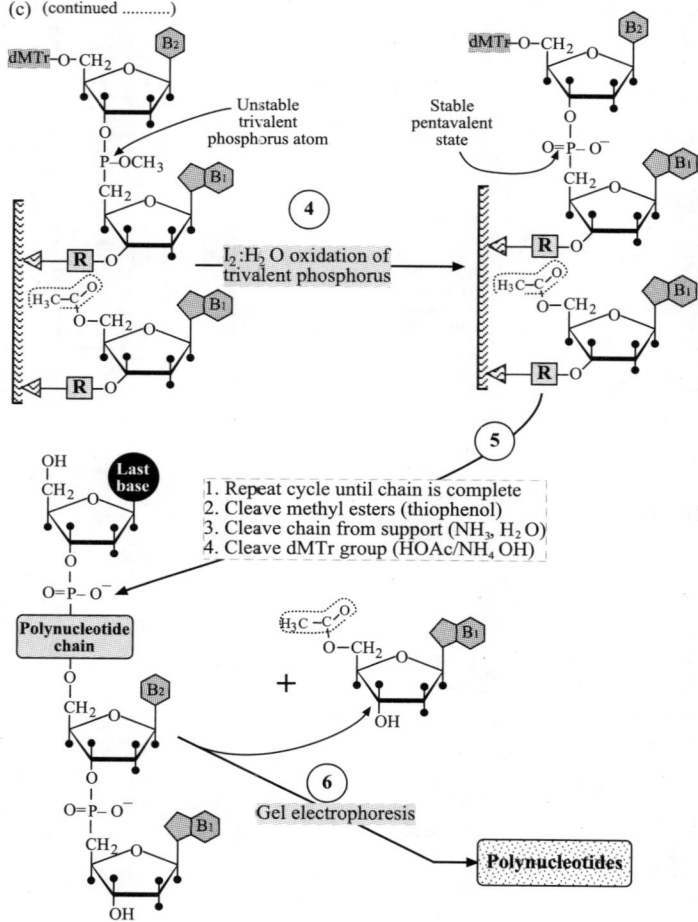

Figure 4.3c *(c) Synthesis of DNA using the phosphoramidite chemistry (reaction extension and termination)*

2. Protection of reactive 5'–OH and amine groups

All the incoming nucleosides must be blocked at 5'–OH. The 5'–OH protecting groups are common to all synthetic strategies. All the reactions for the synthesis cycle start with the nucleoside blocked at the 5' carbon by a protecting dimethoxytrityl (DMTr) group and amino groups protected with N–benzoyl or N–isobutyryl derivatives. The acid-labile 'dimethoxytrityl' group (4,4'–dimethoxytriphenylmethyl; abbreviated as DMT or DMTr) is commonly used. The 'pixyl' group (9–phenylxanthen–9–yl) also serves as an alternative for 5'–OH protection.

The experimental conditions, which result in the formation of a phosphodiester link between nucleosides, will lead to undesirable side reactions unless we block them. The exocyclic amino groups on three of the four bases (adenine, guanine, and cytosine) require protection, as do the free OH groups on the phosphates (Figure 4.2).

3. Deprotection or Detritylation

The 5'–OH protecting group must be removed prior to each nucleoside addition. The DMTr is rapidly removed by protic acids, such as trichloroacetic acid or dichloroacetic acid. The intense orange color of the dimethoxytrityl cation released allows the estimation of the amount of nucleoside deprotected. However, caution should be issued that the use of protic acids for this deprotection does lead to problems with 'depurination'.

4. Coupling reaction

The efficiency of the coupling step is critical to the success of a synthesis; that is, the coupling of the activated 3' phosphoramidite to the deprotected 5' ends of the oligonucleotide. In the coupling step, the second base (B–2) is added in the form of a nucleoside phosphoramidite derivative whose 5'–OH bears a dMTr blocking group so it cannot polymerize with itself. The presence of a weak acid, such as tetrazole, activates the phosphoramidite, and it rapidly reacts with the free 5'–OH of B–1, forming a dinucleotide linked by a phosphite group. Chemical synthesis thus takes place in the 3'→5' direction.

5. Capping

This involves the re-blocking of any oligonucleotide 5' ends that were not reacted with the phosphoramidites, as in step 2. The unreacted free 5'–OHs of B–1 (usually only 2–6% of the total) are blocked from further participation in the polymerization process by acetylation with acetic anhydride in step 3, referred to as capping.

6. Oxidation of the phosphorus to the pentavalent state

The phosphite linkage between B–1 and B–2 is highly reactive and it is oxidized by iodine water or aqueous iodine (I_2) to form the desired more stable phosphate group. This step (step 4) completes the cycle.

7. Cycle repetition

This consists of the repetition of steps 1 to 4, until the desired full-length oligonucleotide is achieved.

8. Deblocking

This is an alkaline treatment to achieve cleavage of the cyanoethyl groups attached to the phosphodiester bonds, linking the individual nucleotides and hydrolysis of the terminal 3' ester bond, to release the oligonucleotide. When the chain is complete, it is cleaved from the support with NH_4OH, which also removes the N–benzoyl and N–isobutyryl protecting groups from the amino functions on the A, G, and C residues (step 5).

9. Purification

The newly-synthesized oligonucleotide product is present in a mixture of prematurely-terminated chains and cleaved chains. The freed oligonucleotides can be purified (step 6) by gel electrophoresis, or by HPLC, RP–HPLC, etc (refer to Chapter 9).

C. AUTOMATED DNA SYNTHESIS

Commercial oligonucleotide synthesizers automate the procedure by introducing the reagents onto the column according to a microprocessor-controlled, pre-programmed sequence. Generally, DNA synthesizers consist of a set of valves and pumps that are programmed to introduce, in the correct order, specified nucleotides and the reagents required for the coupling of each consecutive nucleotide to the growing chain. Its mechanical design permits rapid, simple, and reliable operation at a relatively low cost. This automatic instrument allows the synthesis of deoxyoligonucleotides with an excellent yield ($\geq 97\%$/cycle) and at a high speed (<10 min/cycle). The accuracy of the nucleotide additions must be extremely high for oligonucleotides that are <30 to 40 nucleotides in length: for example, consider the case where nucleotides are added with an accuracy of 99% (0.99^n), where n is the chain length in number of nucleotides. So, for an oligonucleotide of length 64, only about 0.99^{64} or 52% of the final product would represent the correct sequence. Although the accuracy of this order is attainable in many commercial machines, a practical chain-length limit of between 100 and 200 nucleotides is currently the maximum achievable on most systems.

Since the 5' end of the oligonucleotide will have a free –OH after deblocking, the oligonucleotide can be easily labeled using T4 polynucleotide kinase, which catalyzes the transfer of ^{32}P–labeled phosphate from the gamma position of ATP to the 5' end of the chain. By this technique, however, the specific activity of the probe (radioactivity per microgram of DNA) decreases with chain length, since only one ^{32}P–labeled phosphate group can be transferred per chain.

1. Synthesizer design

The general design of the instrument (Horvath et al. 1987) is shown in the schematic diagram (Figure 4.4). The primary reactor is a flow-through column, and the reagent and solvent deliveries are controlled by a series of zero-dead volume, pneumatically-actuated diaphragm valves. To move reagents and solvents from the reservoirs, positive argon pressure is used rather than a mechanical pump. The valves are automatically operated by a control unit. A separate reactor, the activation vessel, is designed for the activation of each phosphoramidite prior to coupling. If mixed probes are being synthesized, the activation vessel is also used for the accurate premixing and the simultaneous activation of two, three, or four different phosphoramidites.

(a) *Primary reaction vessel:* The vessel that houses the silica support on which the oligonucleotide is attached is similar to the one used in the gas-phase sequenator. The cartridge is formed from three Pyrex cylinders housed in a metal jacket. The silica resides in the central cavity of the middle cylinder and is held in place by porous Teflon membranes that also provide the seals between the glass cylinders.

(b) *Activation vessel:* The deoxynucleoside 3'–phosphoramidites are activated by a reaction with tetrazole in an activation vessel, before they are delivered into the primary reaction vessel. The pre-reaction vessel is a conical Pyrex flask with an internal volume of about 3 ml and is fitted with Teflon tubing connectors. The nucleoside solution is delivered into the flask, the tetrazole solution is added, and the solutions are thoroughly mixed by bubbling argon through it before it is transferred from the flask to the primary reaction vessel. In each synthetic cycle, the pre-reactor is automatically cleaned by solvent washes.

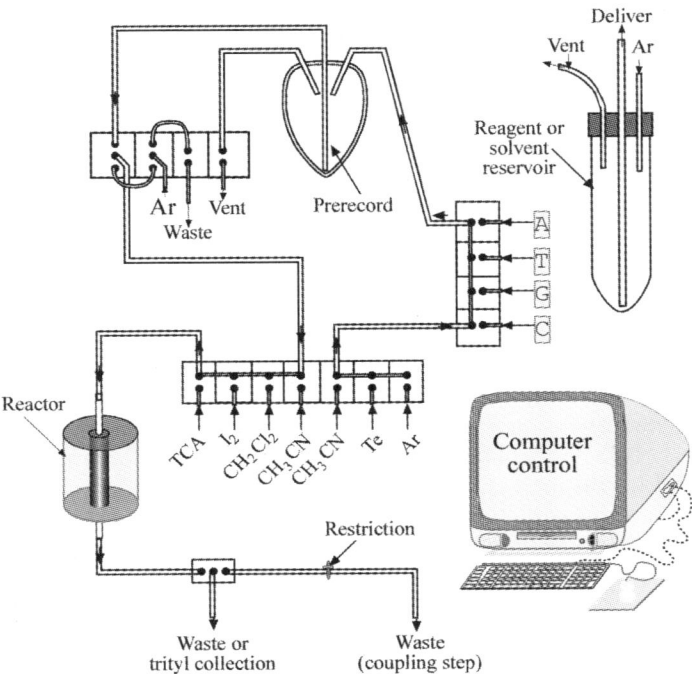

Figure 4.4 *A schematic diagram of an automatic DNA synthesizer. (Redrawn from Horvath et al. 1978)*

(c) *Delivery valves:* The valves that control the liquid and gas movements through the instrument are zero-dead volume, pneumatically-actuated diaphragm valves. They have been modified in two ways: the FEP–Teflon diaphragms used in the earlier valves have been replaced by 0.02-inch-thick Kalrez diaphragms and these diaphragms are held closed by the piston of a solenoid unit mounted as an integral part of the diaphragm. This allows the reduced pressure inside the piston chamber to lift the diaphragm away from the surface of the valve block.

(d) *Reagent/solvent storage:* All the reagents and solvents are stored in Pyrex bottles, (reservoirs) protected from light. All tubing and bottle cap surfaces exposed to chemical vapors or liquid are made of either Teflon or Kalrez.

(e) *Argon supply*: Two 200-cubic-foot cylinders supply argon to the reagent/solvent reservoirs and the reaction vessel. This supply is passed through a molecular sieve filter to eliminate water vapor and is distributed through a manifold where it is reduced by four low-pressure, low-flow gas regulators to the 3–8 psig level required.

(f) *Waste collection:* The effluents from either reaction vessel can be directed into a Pyrex collection bottle through the appropriate delivery valve. During the delivery, the activated nucleoside from the primary vessel is first directed through an adjustable needle valve that precisely regulates the flow rate. This valve is fitted with a Kalrez O-ring. The effluent from the primary vessel during and immediately after the detritylation step is directed into a fraction collector, in order to facilitate the quantitative monitoring of each coupling reaction.

(g) *Control unit:* The valve operations are controlled by a microcomputer interfaced to the valves through a Hewlett–Packard Model 3497 system controller or its equivalent. The system software allows automatic operation with a manual override of any valve function. The synthesizer programs can be customized to effect modifications of the valve and timing sequencing of any or all steps in them. Normally, however, all the user has to do is type in the base sequence of deoxyoligonucleotide to be synthesized. This sequence can include from one to four bases to be added at each cycle of the synthesis, allowing for the synthesis of heterogeneous or mixed probes.

Materials

Protected nucleosides: 5'–O–dimethoxytrityl (dMTr) deoxythymidine, 5'–O–dMTr–N–benzoyldeoxycytidine, 5'–O–dMTr–benzoyldeoxy–adenosine, and 5'–O–dMTr–N–isobutyryldeoxyguanosine. Chloro–N,N–dimethylaminomethoxyphosphine and the nucleoside 3'–N, N–diisopropylaminophosphoramidite of the protected deoxynucleosides of each of the four bases can be prepared as in the published protocols (Beaucage and Caruthers, 1981). Solvents and reagents can be purified or dried as described below. Dichloromethane should be dried over 3 Å molecular sieves and passed through an activated alumina column prior to use. Acetonitrile should be distilled over calcium hydride and sorted over 3 Å molecular sieves. Tetrahydrofuran should be distilled over sodium spheres (1 mol/liter) in the presence of benzophenone (0.125 mol/liter). 2,6–lutidine, double-distilled over p–toluenesulfonyl chloride and calcium hydride and stored over 4 Å molecular sieves. Iodine, resublimed; trichloroacetic acid; 1 H–tertrazole, resublimed.

2. Synthesizer operation

In order to optimize the conditions for routine synthesis, the solvents and reagents for each individual step of the synthesis cycle are prepared as listed in Table 4.1.

Table 4.1 *General protocol for DNA synthesis*

Reagent or solvent	Function
CH_2Cl_2	Wash
3% TCA/CH_2Cl_2	5'–detritylation
CH_2Cl_2	Wash
CH_2CN	Wash
Activated nucleoside (Nu + Te)	Coupling
CH_2CN	Wash
I_2/H_2O: 2,6–lutidine:THF	Oxidation
CH_2CN	Wash

(a) Synthesis start-up

The synthesis of the deoxyoligonucleotide with a defined sequence begins by loading the proper, silica-bound, 3'-end deoxynucleoside into the primary reactor. The derivation of the polymer support (Fractosil–500, Merck) was made according to the published protocol of Matteucci and Caruthers (8). The synthesis then automatically proceeds in the 3'→5' direction, by adding one deoxynucleotide per cycle. The synthesis cycle includes three reactions (Figure 4.3)–5'–detritylation, coupling (activation and condensation), and oxidation.

(b) Detritylation

The 5′–hydroxyl of the silica-bound deoxynucleoside (or deoxyoligonucleotide) is protected with the acid-labile dimethoxytrityl group. For automated synthesis, the 3% trichloroacetic acid (TCA)/ dichloromethane solution proves to be an excellent reagent. The detritylation reaction is completed in less than a minute, and, because of the subsequent fast and efficient dichloromethane wash to remove excess reagent, the depurination is negligible. The dimethoxy cations give a bright orange color in solution with lamdamax 498 nm, and they are automatically collected and used for the quantitative monitoring of the coupling yield of each previous cycle.

(c) Coupling

5′– and base-protected deoxynucleoside 3′–N, N–diisopropylaminophosphoramidites are used for the sequential addition of the four bases to the silica-bound deoxynucleoside (or deoxyoligonucleotide). While the column is being washed with anhydrous acetonitrile, the selected deoxynucleoside 3′–phosphoramidite: acetonitrile solution is delivered to the activation vessel and mixed with a 1H–tetrazole: acetonitrile solution. The activation is extremely rapid. In mixed base positions, the proper deoxynucleoside 3′–phosphoramidite solutions are delivered into the activation vessel one after the other in a well-defined molar ratio and then mixed with the 1H–tetrazole solution. By applying positive argon pressure in the activation vessel and by opening the proper valves, the activated deoxynucleoside 3′–phosphoramide solution passes through the column and the condensation is complete in 3 min. and 30 sec. The flow rate and, thus, the reaction time can be finely adjusted with a needle valve that is built into the outlet line of the column. Using only a 10-fold molar excess of the deoxynucleoside 3′–phosphoramidites, the coupling yields are ≥97%. The carefully-prepared, protected deoxynucleoside 3′–N,N–diisopropyl–aminophosphoramidites are stable in an anhydrous acetonitrile solution for more than a week. After coupling, both the activation vessel and the column are washed with acetonitrile.

(d) Oxidation

The 3′–5′ phosphite triester bond, formed in the previous coupling step, is oxidized to a phosphate triester by using I_2 in a water: 2,6–lutidine: tetrahydrofuran solution. This oxidation is complete in 30 sec. Then, the I_2 solution can be removed from the column using a 55-sec acetonitrile wash. The detailed protocol of the DNA synthesizer is given in Table 4.2. One synthesis cycle involves 24 steps, and, as indicated, some of these steps have double functions in order to shorten the cycle time. The total cycle, i.e., the addition of one base, takes 9 min. 24 sec.

(e) Deprotection and cleavage from the support

Once the synthesis cycle described above is repeated the desired number of times, the silica-bound, protected deoxyoligonucleotides removed from the primary reactor (column), are cleaved from the silica support and freed of the protecting groups. The deprotection procedure starts with treating the silica-bound reaction product with a thiophenol: triethylammonium dioxane solution at room temperature for 45 min. The thiophenoxide ions remove the methyl groups from the internucleotide phosphotriester bonds. The silica is then extensively washed with methanol in order to completely remove the thiophenol reagent and dried under reduced pressure. This step is followed by a treatment with concentrated ammonium hydroxide at room temperature for 3 hr to cleave the base-protected diester oligonucleotides from the support. After separating the silica and the supernatant by centrifugation, the supernatant containing the product is treated with additional concentrated ammonium hydroxide at 50°C for up to 12 hr to remove the base-protecting groups, namely, N–benzoyl groups from deoxycytosines and deoxyadenosines and N–isobutyryl groups from deoxyguanosines. The ammonium hydroxide solution is removed by vacuum drying, and finally the last 5'–end dimethoxytrityl group is cleaved by using 80% acetic acid solution for 15 min. The reaction mixture containing the crude, unprotected deoxyoligonucleotide is then lyophilized to dryness.

Table 4.2 *Steps involved in each cycle of the automated DNA synthesis**

Steps	Function	Duration (sec.)
1.	Deliver CH_2Cl_2 to reactor	20
2.	Flush argon through reactor	6
3.	Deliver 3% TCA/CH_2Cl_2 to reactor (5' detritylation)	55
4.	Flush argon through reactor	6
5.	Deliver CH_2Cl_2 to reactor; Pressurize nucleotide (Nu) and tetrazole (Te) reservoirs	20
6.	Flush argon through reactor	6
7.	Deliver CH_3CN to reactor; Deliver Nu to pre-reactor	13
8.	Deliver CH_3CN to reactor; Flush argon through pre-reactor to push Nu into pre-reactor (clean the line)	3
9.	Deliver CH_3CN to reactor; Deliver Te to pre-reactor	19
10.	Deliver CH_3CN to reactor; Flush argon through reactor to push Te into pre-reactor (clean the line)	3
11.	Deliver CH_3CN to reactor; Bubble argon into pre-reactor to mix Nu and Te solution (activation)	8
12.	Flush argon through reactor	5
13.	Push activated nucleotide from pre-reactor to the inlet of reactor	2
14.	Deliver activated nucleotide to reactor (condensation)	210
15.	Flush argon through reactor	3
16.	Deliver CH_3CN to pre-reactor (wash)	5
17.	Push CH_3CN from pre-reactor to reactor	35
18.	Deliver CH_3CN to reactor; Deliver CH_3CN to pre-reactor (wash)	4
19.	Deliver CH_3CN to reactor; Push CH_3CN from pre-reactor to waste	37
20.	Flush argon through reactor	5
21.	Deliver I_2/H_2O: 2,6–lutidine:THF	30
22.	Flush argon through reactor	9
23.	Deliver CH_3CN to reactor	55
24.	Flush argon through reactor	5

Total cycle time: 9 min. 24 sec.

(* Source: Horvath et al., 1978)

(f) Purification

For purification of the crude deoxyoligonucleotides, either preparative polyacrylamide gel electrophoresis or preparative reverse-phase HPLC have been used. In practice, mixed probes, single probes, and gene fragments with high guanosine content and/or with long chains (≥30) are usually purified by preparative gel electrophoresis. Other oligonucleotide chains, such as primers for deoxy DNA sequencing, are usually purified by HPLC.

Method-2: Getting the primers ready for use in PCR

Oligonucleotide synthesizers are lovely machines that synthesize oligonucleotides of whatever sequence required. The oligos are synthesized 3' to 5' from the 3' nucleotide that is bound to silica beads in a column. When chemical synthesis is completed, the oligos have to be freed from the silica beads. The protective groups that run the length of the oligo also need to be removed. This can be done as described below.

Procedure

1. Take some concentrated ammonia hydroxide and keep it in the refrigerator. At the time of use, take 4–5 ml out of the bottle and place it in a smaller bottle.
2. Carefully open the column with a screwdriver. Gently tap the beads into a 1.5 ml Eppendorf tube.
3. Pipette 1 ml of NH_4OH into the tube. Then, finger flick to suspend the beads and let it stand or slowly rotate at room temperature for 1 hr. Pipette off the aqueous layer and put into a sterile, glass tube with a screw-on cap that screws down tight.
4. Repeat step 3 two more times, although the yields do not improve much.
5. Incubate the solution for 12–15 hr at 55°C.
6. Let the sample cool to room temperature, then open tube. Dry the sample using a stream of nitrogen, argon or air.
7. Suspend the "curd" on the bottom of the vial in 500 µl of sterile, distilled water.
8. Make a 1/300 dilution and determine the absorbance at 260 and 280 nm using the spectrometer.
9. The purity of the preparation is assessed on the basis of the 260/280 ratio; values near 1.7 are considered good.
10. To determine the concentrations, use the following formula. DNA at a concentration of 33 µg/ml has an absorbance of 1 OD at 260 nm

    $$\text{Concentration } (\mu M) = (A_{260} \times D \times 33{,}000) \div (N \times 330),$$

    where A_{260} is the Spec reading at 260 nm, D is the dilution of primer stock, and N is the number of base pairs of the primer. This gives a concentration in µM. The 33,000 is the conversion for the Spec reading to µg/l and the 330 in the denominator is the atomic mass unit for a single nucleotide.
11. Make 200 µl of a 100 µM primer solution and store the remainder at –20° or –80°C. From the 100 µM solution, make 500 µl of a 10 µM working primer solution.

D. PRACTICAL LIMITATION

Oligonucleotides have a few inherent problems; some oligonucleotides are more prone to synthesis failure than others. Apart from the difficulties imposed by the coupling step yield, reagent impurities, and side reactions as the chain length increases, depurination may impose a serious limitation. Adenine-rich sequences will be subject to more losses during synthesis. A further problem arises when oligonucleotide probes that are constructed from amino acids have more than one codon: serine and arginine have six each, proline and threonine have four each, etc. This means that the 'true' probe is one of many possible oligonucleotides. The yield of the oligonucleotides also decreases as the length of the oligonucleotide increases.

III APPLICATION OF CHEMICALLY SYNTHESIZED OLIGONUCLEOTIDES

The ease of customizing reaction cycles in automated, solid-phase DNA synthesizers allow for the efficient and site-specific introduction of chemical modifications in oligonucleotides. For the synthesis of oligonucleotides, automated synthesizers make use of 5'-protected dNTPs anchored to a controlled pore-glass support at the 3'-end via a chemically cleavable linker. This approach circumvents the need to purify synthetic intermediates at each step of the reaction cycle, thereby saving enormous amounts of time.

Synthetic oligodeoxyribonucleotides (henceforth referred to as oligonucleotides) have been incorporated into a number of molecular biology protocols in recent years. Reductions in synthesis time and cost have made oligonucleotides available to almost any laboratory. Oligonucleotides have been used in the construction of synthetic gene sequences, for the addition of new restriction enzymes for either primary nucleotide sequence determination or complementary DNA (cDNA) synthesis. Oligonucleotides have also been extensively developed as probes in various hybridization techniques. Some of the applications of chemically synthesized oligonucleotides are listed in Table 4.3.

Table 4.3 *Application of chemically-synthesized oligonucleotides*

Purpose		Description
DNA amplification	:	Primers for DNA amplifications by PCR: Specific primers, random primers, nested primers, degenerate primers, universal primers (e.g., Universal primer R, GTATCACGAGGCCCTT) etc.
		cDNA synthesis: Poly-T oligonucleotides can be used in the synthesis of complimentary DNA directly from RNA, using the reverse transcriptase enzyme.
DNA sequencing	:	Sequencing: both by enzymatic and PCR-based sequencing; oligonucleotide primers are used.

Continued...

Purpose	Description
Antisense DNA/RNA	: Antisense RNA or DNA technology has been extensively used to inhibit specific gene expression using short oligonucleotide sequences. Antisense RNA was also used to control RNA viruses.
Insertion of point mutations	: Synthetic adapters are used for introduction of point mutations of one or a few base pairs addition or deletions.
Insertions of restriction sites	: Restriction sites can be introduced into primers and amplified (linker DNA). The linker DNA primers are very useful in cloning experiments.
Oligonucleotide probes	: Short or long oligonucleotide probes are used in Southern and Northern blotting to detect corresponding complimentary strands.
In situ hybridization	: Radioisotope- or non-radioisotope-labeled nucleic acid probes are extensively used.
Construction of genes	: The partial of a complete functional gene can be constructed by using complimentary and overlapping oligonucleotides.
DNA chip technology	: Oligonucleotide microarrays are prepared for gene expression profiling.

A. CHEMICALLY SYNTHESIZED GENES

For the synthesis of a double-stranded gene or a gene fragment, each complementary stand has to be made separately. The production of short genes (60–80 bp) is technically straightforward and can be accomplished by synthesizing the complementary strands and then annealing them. However, for the production of total genes or longer sequences (>300 bp) special strategies must be invoked, because the coupling efficiency of each cycle during chemical DNA synthesis is never 100%. One method for building a synthetic gene requires the initial production of a set of overlapping, complementary oligonucleotides, each of which is between 20 and 60 nucleotides long. The sequences of the strands are planned so that after annealing, the two end segments of the gene are aligned to create blunt ends. Each internal section of the gene has complementary 3' and 5' terminal extensions that are precisely designed to the base pair with an adjacent section (Figure 4.5a). After the gene sequence is assembled, the gaps are sealed by a T4 DNA ligase.

An alternative way to prepare a full-size gene is to synthesize a specified set of overlapping oligonucleotides (40 to 100 nt). After the 3' and 5' extensions (6 to 10 nt) are annealed, large gaps still remain, but the base-paired regions are both long enough and stable enough to hold the structure together. The duplex is completed and the gaps are filled by enzymatic DNA synthesis with E. coli DNA polymerase I. This enzyme fills the gaps by extending the 3' ends (DNA synthesis). After the enzymatic synthesis is completed, the nicks are sealed with T4 DNA ligase (Figure 4.5b). Table 4.4 lists some of the genes that have been chemically synthesized.

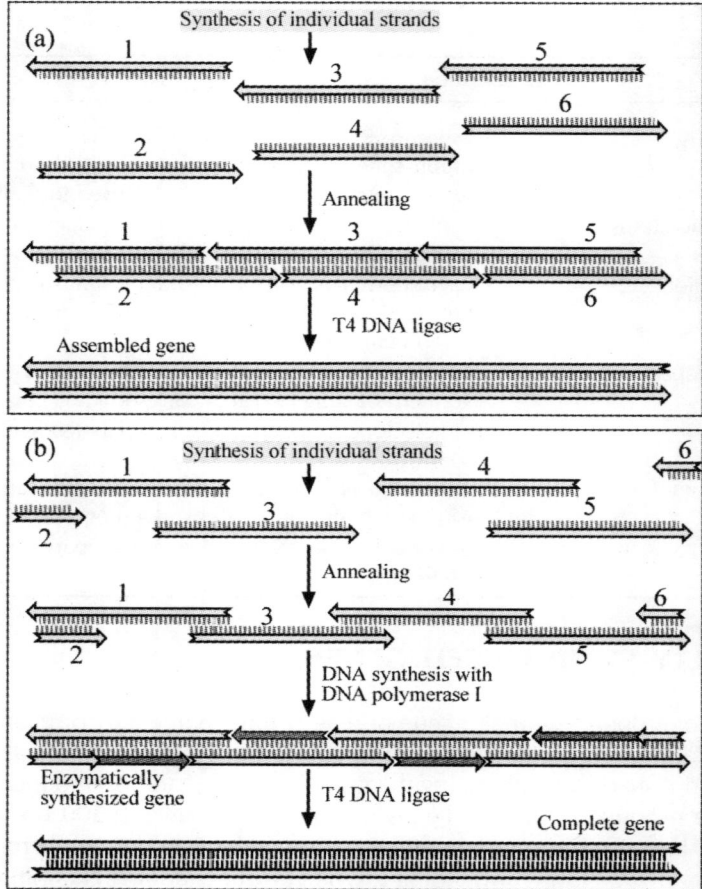

Figure 4.5 (a) Assembly of a synthetic gene from short oligonucleotides (20 to 60-mers) are synthesized chemically. Their sequences are designed so that they form a double-stranded tandem array after annealing. T4 DNA ligase is used to seal the nicks and produce an intact version of the gene, (b) Assembly and in vitro enzymatic DNA synthesis of a gene

Table 4.4 Some chemically synthesized genes

Gene	Size (bp)	Gene	Size (bp)
Tissue plasminogen activator	1610	RNase T1	324
Rhodopsin	1057	Bovine intestinal Ca-binding protein	298
c-Ha-ras	576	Connective tissue activating peptide III	280
α–Interferon	542	Hirudin	226
γ–Interferon	453	tRNA	126
Lysozyme	385	Secretin	81
RNase A	375	Proenkephalin	77
Cytochrome b_5	330		

IV SUGGESTED READING

Caruthers MH (1985) Gene synthesis machines: DNA chemistry and its uses, *Science*, **230**:281–285.

Chow F, Kempe T and Palm G (1981) Synthesis of oligodeoxyribonucleotides on silica gel support, *Nucleic Acids Res.*, **9**:2807–2817.

Dorper T and Winnacker EL (1983) Improvements in the phosphoramidite procedure for the synthesis of oligonucleotides, *Nucleic Acids Res.*,**11**:2575–2584.

Ferretti L, Karnik SS, Khorana HG, Nassal M, and Oprian DD (1986) Total synthesis of a gene for bovine rhodopsin, *PNAS*, USA, **83**:599–603.

Garret RH and Grisham CM (1999) *Biochemistry*, Saunders College Publishing.

Glick BR and Pasternak JJ (1996) *Molecular Biotechnology*, Panima Publishing Corporation, New Delhi, Bangalore.

Horvath SJ, Firca JR, Hunkapiller T, Hunkapiller MW and Hood L (1987) An automated DNA synthesizer employing deoxynucleotide 3'–phosphoramidites, *Methods in Enzymology*, **154**: 314–326.

Khorana HC, Agarwal KL, Besmer P, Buchi H, Caruthers MH, Cashion PJ, Fridkin M, Jay E, Kleppe K, Kleppe R, Kumar A, Loewen PC, Miller RC, Minamoto K, Panet A, Raj-Bhandary UL, Ramamoorthy B, Sekiya T, Takeya T and Sande JH (1976) Total synthesis of the structural gene for the precursor of a tyrosine suppressor transfer RNA from *Escherichia coli*, 1. General Introduction, *J. Biol. Chem.*, **251**:565–570.

Walker JM and Gingold EB (1993) *Molecular Biology and Biotechnology*, Panima Educational Book Agency.

Winnacker EL (2003) *From Genes to Clones*, Panima Publishing Corporation.

5

DNA Chip Technology and its Potential Applications

I INTRODUCTION

Despite many years of biochemical advancement, there are still thousands of genes in eukaryotic cells and quite a few in bacteria that we know nothing about. A number of molecular biology techniques have been developed and many of these techniques can detect even a single nucleotide change in a genome. Although they are very sensitive and well standardized, their applications have many limitations. Some of the limitations include cost, being labor-intensive and time-consuming, and only a few samples can be processed at a time. The molecular biology techniques used thus far can be used to monitor the expression of only a single gene at a time. In addition to this, analyzing changes in entire genome or a larger part of a genome is very important for diagnosing probable changes among trillions of nucleotides in a genome. Unfortunately, at present no technique can do analysis at the genome scale. On the other hand, the rapid advancement of genome-scale sequencing has driven the development of methods to exploit this information by characterizing biological processes in new ways. The knowledge of the coding sequences of virtually every gene in an organism, for instance, encourages the development of technology to study the expression of all of them at once. To this end, a variety of techniques have evolved to rapidly and efficiently monitor the transcript abundance for all of an organism's genes.

Important refinements of the technology underlying DNA libraries, PCR, and hybridization have come together in the development and use of the wealth of sequence information. The analytical techniques that exploit the phenomenon of sequence-specific hybridization between nucleic acids have been the workhorses of modern molecular biology. One way is to exploit the parallelism inherent in using arrays of bound DNA as analytical tools. One of the most powerful tools that has recently bridged the gap between sequence information and functional genomics is *DNA microarray* (or DNA microchip, biochip, genechip, etc.) *technology*. Microarray assays arose out of biochemical experiments on solid surfaces. These DNA chips can gather genetic information at twenty-five times the rate of traditional methods involving PCR and gel electrophoresis, at a fraction of their cost. The method is based on hybridization and principally corresponds to the classical Southern or Northern blot techniques, with the difference that in array hybridization the "probe" is immobilized and the free nucleic acid is the "target". The chip-based approaches utilize high-density molecular arrays to examine biochemical samples (Figure 5.1). Mixtures of DNA or RNA isolated from biological sources are labeled enzymatically by incorporating nucleotides bearing reporter tags and hybridized to microarrays. Hybridization reactions yield heteroduplexes between the individual components of the fluorescent sample (*probe*) and a complementary sequence (*target*) on the chip surface. Owing to the fact that each target element or 'feature' is chemically homogenous and occupies a known location, the identity and quantity of each component in the fluorescent mixture can be ascertained by measuring the fluorescence intensity at each position on the microarray.

Solid surface, micro-scale assays permit the use of sophisticated fluorescence detection technology, including confocal laser scanning and charge-coupled device (CCD) imaging, which allow high-speed, quantitative data gathering. Microarrays are prepared by various synthesis or deposition strategies. Synthesis strategies typically produce microarrays consisting of groups of oligonucleotides ranging in size from 10–25 bases, while chips prepared by micro-deposition technologies are usually composed of 0.5–2.0 kb cDNAs amplified by the PCR.

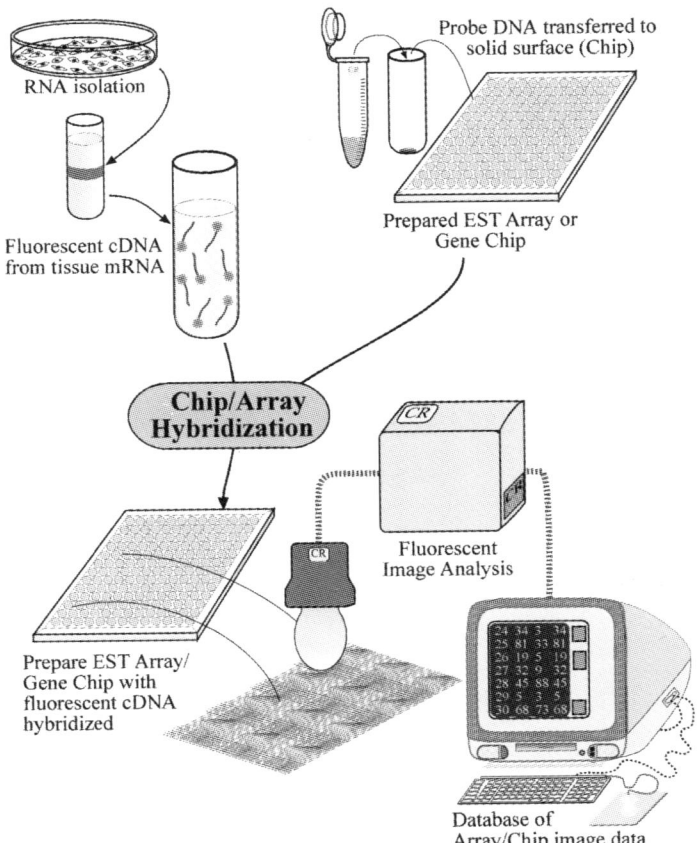

Figure 5.1 *The principle steps in producing and utilizing microarray. Microarray are used to monitor the expression of thousands of genes simultaneously*

Arrays enable the study of literally thousands of genes in a single experiment. The potential importance of arrays is enormous and has been highlighted by the number of publications. In this chapter, a brief outline of the salient features of this technology and some of its potential applications are discussed.

II MICROCHIPS

A. WHAT ARE DNA MICROCHIPS?

DNA chips are simply glass surfaces bearing arrays of DNA fragments at discrete addresses, at which the fragments are available for hybridization. In the context of molecular biology, the word "*array*" is normally used to refer to a series of DNA or protein elements firmly attached in a regular pattern, to some kind of solid supportive medium. A DNA array is often used interchangeably with a gene array or microarray. The DNA arrays on the chip are hybridized to a complex sample of fluorescently labeled DNA or RNA in solution (Figure 5.2). These DNA chips are commonly used either to monitor expression of the arrayed genes in mRNA populations

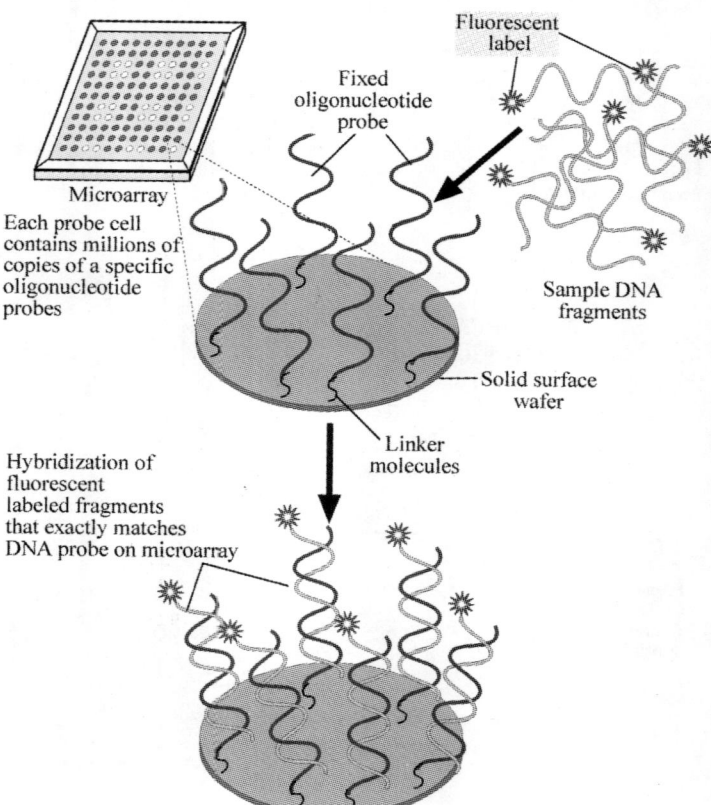

Figure 5.2 *Hybridization of fluorescent labeled DNA or RNA samples to fixed unlabeled probes*

DNA Chip Technology and its Potential Applications   77

from living cells or to detect DNA sequence polymorphisms or mutations in genomic DNA. The DNA chip format has an advantage over larger flexible membranes in that it reduces the diffusion necessary for complex hybridization probes to find complementary-arrayed DNAs.

B. DNA CHIP FORMATS

Any known DNA sequence, from any source, can be used in a microarray. DNA chip technologies are distinguished by: (1) the sizes of the arrayed DNA fragments, (2) the methods of arraying, (3) the chemistries and linkers for attaching the DNA to the chip, and (4) the hybridization and detection methods. The DNA elements that make up the DNA arrays can be oligonucleotides, partial gene sequences, or full-length cDNA. Thus far, two general types of microarray formats have been developed: *DNA fragment-based microarrays* (cDNA-array formats) (Figure 5.3 (a)) and *oligonucleotide-based microarrays* (*in situ*-synthesized oligonucleotide arrays). These two formats are illustrated in Figure 5.3 (b) and (c).

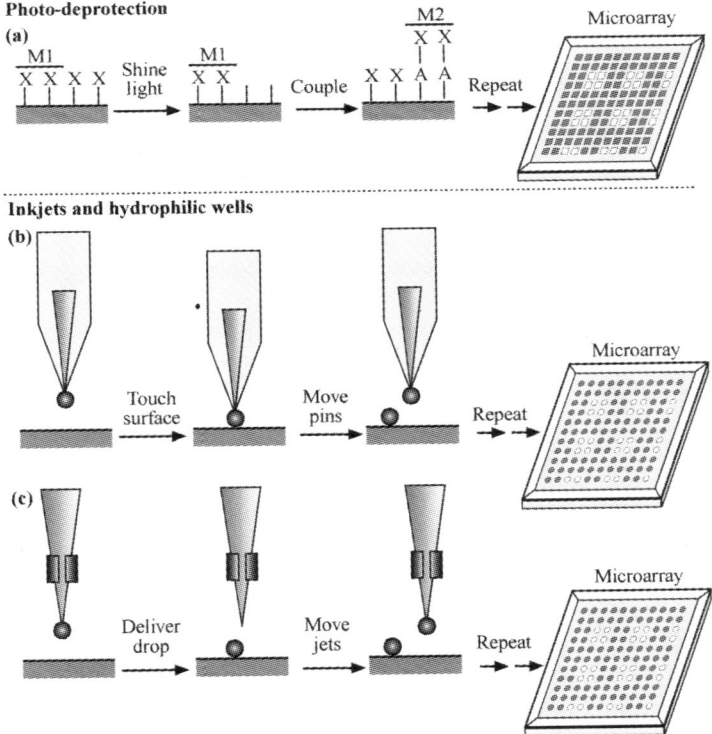

Figure 5.3   *Three approaches to microarray manufacturing are depicted (a) Photolithography: a glass wafer modified with photolabile protecting groups (X) is selectively activated by shining light through a photomask (M1). The wafer is then flooded with a photoprotected DNA base (A–X), resulting in spatially defined coupling on the chip surface. A second photomask (M2) is used to deprotect defined regions of the water. Repeated deprotection and coupling cycles enable the preparation of high-density oligonucleotide microarray, (b) Mechanical microspotting; a small volume of probe sample is spotted by a robotic control system and (c) Ink-jetting: a small amount of probe sample is printed through piezoelectric fitting in a ink jet printer*

The term *"probe"* designates the recognition nucleic acid on the DNA microarray and *"target"* stands for the nucleic acid of interest, whose identity/abundance is being determined. The simplest form of DNA chip technology has been to spot denatured cDNA molecules onto a glass microscope slide using a computer-driven robotic arm to position the probe material precisely.

C. DNA CHIP MANUFACTURING TECHNOLOGY

Currently, two major DNA chip technologies dominate the market. The first is based on an *in situ* synthesis of oligonucleotides on the chip by means of *photolithography* or *layered ink-jet printing* of "C", "G", "A", or "T" molecules. The second method is based on *robot spotting* of pure cDNA fragments on the chip surface. In addition, several other chip technologies, which are still in the developmental and testing phases, are expected to be commercially released in the next few years. Each technology has its unique strengths and weakness that may make it more suitable for some specific chip applications.

1. Oligonucleotide Microarrays

Synthetic DNA arrays can be made either by synthesizing the individual oligos in separate tubes and then attaching them, in a regular array to a flat substrate, or the oligos can be synthesized *in situ* in regular arrays. The former method allows for a post-synthetic purification of the DNA before attachment but becomes impractical as the number of different oligos increases into tens or hundreds of thousands. The most efficient strategy for oligonucleotide microarray manufacturing involves DNA synthesis on solid surfaces using combinatorial chemistry methods (refer to Chapter 4). Glass is currently used as the preferred synthesis substrate, because of the inert chemical properties of this material, the ability to chemically derivatize the surface, and its low level of intrinsic fluorescence. As it is impossible to purify full-length reaction products from the shorter precursors, '*in situ*' synthesis strategies have limitations as far as the oligonucleotide length is concerned. On the other hand, considering that DNA is synthesized at a large number of sites in parallel and that sequence data can be taken directly from databases without the physical handling of colonies, the synthesis methods have a significant advantage over deposition strategies when very high density microarrays are required.

(a) Photolithographic patterning and photo-deprotection

Of the three approaches currently used to manufacture oligonucleotide microarrays, the light-directed deprotection method is the most effective strategy for generating very high-density chips. This technique forms the basis of all modern semiconductor ("computer chip") manufacture. It works by shining a light through a photolithography mask, which is much like a photographic negative, onto a light-sensitive surface. In this method, modified nucleotide phosphoramidites bearing a photo-labile protecting group are used, allowing light to be used as the activating agent in the synthesis reaction (Figure 5.3). A single round of synthesis involves light-directed deprotection, followed by nucleotide coupling. Photolithography masks are used to control the regions of the chip designated for illumination. If the surface of the chip is flooded with one of the four bases, it results in the selective coupling of that base to each deprotected region on the synthesis surface. Considering that a separate deprotection and coupling cycle is required for each of the four possible bases, the maximum number of cycles needed to produce all of the possible 20-mer primers would be 80 (i.e., 4×n, where n is the length of the oligonucleotide). This method currently allows for very high-density (250,000 features/cm^2) microarrays to be manufactured; however, current coupling efficiencies impose a limit of ~25 bases to these chips. Beyond a 100 cycles, the accumulation of incomplete synthesis products becomes a problem. Recent advances in related photolithographic technologies, such as those involving 'acid resists', may increase the efficiency of *in situ* synthesis so that oligonucleotides containing >25 bases are feasible. Generally speaking, photolithography is a relatively poor prototyping technology and a rather good production technology. One principal disadvantage of this method is that a significant amount of design work and cost is associated with the mask design. However, once a set of photolithographic masks have been made, a large number of chips can be produced at a reasonable cost.

In DNA photolithography, UV light is honed through holes in the masks in order to direct a parallel, step-wise, synthesis of oligonucleotides. At each step in the synthesis, oligonucleotides that, for example, require a cytosine in the next position, are deprotected by light directed to the appropriate positions by a mask. The chip is then flooded with activated cytosine nucleotides, which couple to the deprotected positions. Uncoupled cytosines are then washed away, another mask is applied, and the deprotection and coupling steps are carried out with adenosine, for example. The repetition of this cycle ~70 times, with seventy different masks, allows the synthesis of the complete array of thousands of 25-mer oligonucleotides in tandem.

(b) Inkjets and hydrophilic wells

Other *in situ* manufacturing technologies can be used to make oligonucleotide microarrays. For example, ink-jet printing technology is being used to direct the delivery of phosphoramidites to specified addresses. In this method, a glass surface is coated with a light-sensitive hydrophobic material. The synthesis areas are then prepared by photolithographic etching and the phosphoramidites are delivered to the resulting hydrophobic 'wells' to allow chemical coupling. The wafer is then washed with an oxidizer and a deblocking agent to complete a round of coupling. Although only modest densities (10,000 features/cm^2) are attainable using this technology, the simplicity of ink-jetting may make oligonucleotide microarrays accessible to a large number of researchers in the near future.

The technique uses the print heads of commercial piezoelectric ink-jet printers to deliver reagents to individual spots on the array. A piezoelectric ink-jet head (Figure 5.4) consists of a small reservoir with an inlet port and a nozzle at the other end. One wall of the reservoir consists of a thin diaphragm with an attached piezoelectric crystal. When a voltage is applied

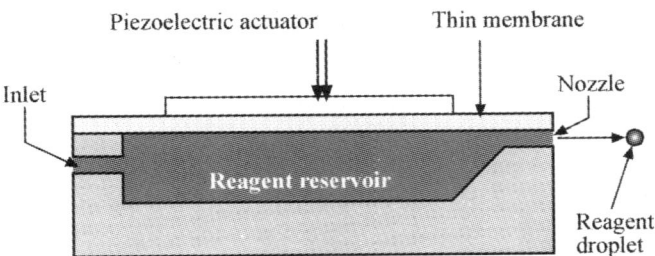

Figure 5.4 *A piezoelectric ink-jet. When voltage is applied to the piezoelectric actuator it contracts laterally, which pushes the membrane into the reagent reservoir and results in the ejecting of a droplet from the nozzle. The reservoir refills via capillary action through the inlet. (Redrawn from Blanchard 1998)*

to the crystal, it contracts laterally, thus deflecting the diaphragm and ejecting a small drop of fluid from the nozzle. The reservoir then refills by means of capillary action through the inlet. One, and only one, drop is ejected for each voltage pulse applied to the crystal, thus allowing complete control over when a drop is ejected. These devices can deliver thousands of drops per second (150,000 per min.) and can be controlled by a computer. Ink-jets can be used in two ways to make arrays. The nucleoside monomers and activating agent (tetrazole) can be simultaneously delivered by separate ink-jets to couple different nucleosides to different spots. The common oxidation, deprotection, and washing steps are performed by dipping the array in the appropriate reagents. Otherwise, the deprotection reagent can be delivered to the selected spots on the array, which is then treated with one of the four nucleosides. The coupling, oxidation and washing steps are again performed uniformly across the whole array surface. Since a computer controls the pattern of reagents as the array is being made, it is as easy to make 10 arrays with different sequences as it is to make 10 identical arrays. This flexibility is perhaps the main advantage of the ink-jet approach. Achieving high density with the ink-jet approach requires one more trick. Two drops of a liquid applied too closely together on a surface will tend to spread into each other and mix (e.g., 40 picolitre drops need 600 microns spacing). One way to engineer patterns in the surface chemistry of the array is to produce spots of a relatively hydrophilic character, surrounded by hydrophobic barriers. There are several ways to engineer this – the use of a laser printer to lay down a "solvent-repellant grid" of pigment on the array surface or the use of fluorinated alkyl silanes, which covalently couple to glass and present an extremely hydrophobic surface.

The oligonucleotide approaches to gene-expression monitoring are advantageous, because they allow the user to avoid gene regions that are repetitive or homologous to other known genes. Such oligonucleotides are designed on the basis of sequence information alone, which removes the need to maintain cDNA clones. Typically, 20 pairs of oligonucleotides are arrayed to represent each gene; this improves the quantification, specificity and reliability of the gene-expression data. Each oligonucleotide that matches the gene sequence is paired with a second mismatched oligonucleotide that differs only in the central nucleotide. During data analysis, one can sort out spurious hybridization by excluding oligonucleotides that fail to hybridize more strongly than corresponding mismatched oligonucleotides.

Hybridization samples for Affymetrix gene-expression chips are typically prepared from 3–5 µg of mRNA, or 5–50 µg of total RNA, in a two-step process.

2. cDNA Microarrays

This simple and elegant array format developed at Stanford University is extensively used at Incyte Inc., and has been widely adapted by academic investigators. These microarrays are formed by robotically depositing large and specific fragments of DNA (e.g., cDNA) at indexed locations (DNA spots are 50–150 µM diameter) onto a coated support matrix. Nylon membranes are used by most off-the-shelf array providers like Clontech Laboratories Inc., Genome systems Inc., and Research Genetics Inc. Microarrays such as those produced by Affymetrix Inc., Incyte pharmaceuticals Inc., and many do-it-yourself (DIY) arraying groups use glass wafers or slides. The fragments can originate from a variety of sources including anonymous cDNA clones, EST clones, anonymous genomic clones, or DNA amplified from open reading frames (ORFs) found in sequenced genomes. With currently available technology, it is feasible to array up to 10,000 spots per 3.24 cm^2. Incyte currently prepares DNA chips that bear 10,000 spots per 3.24 cm^2.

Once produced, the microarrays are hybridized with fluorescently-labeled cDNA probes obtained from control and experimental tissues (diseased or drug-treated). After removing the unbound probe, only the fluorescent probe that has hybridized to fragments on the microarrays excited by light remains. The fluorescent signal emitted from each spot is a reflection of the abundance of the corresponding sequence in the original probe. As the detection is fluorescence-based, the signal output is very sensitive, and individual mRNA species can be detected as a threshold of 1 part in 100,000 to 1 part in 500,000. In addition, the signal output has a broad dynamic range, so both weak and strong signals can be monitored on the same microarray. Thus, quantitative information about mRNA levels for a large number of genes can be collected concurrently. A dual-color probe can be hybridized to a single Incyte DNA chip and scanned separately for each color, to assess differential gene expression in an internally-controlled manner.

Microarray technology is ideally suited for making pair-wise comparisons of samples. Two fluorescent tags, with different excitation and emission optima, can be used to label two distinct probes (e.g., two mRNA populations from physiologically or genetically distinct samples). The two probes are mixed and allowed to hybridize to the same microarray, thereby eliminating any differences associated with the hybridization process. For each DNA fragment in the microarray, the ratio of fluorescence emission at the two wavelengths reflects the ratio of the abundance of that sequence in the two probes. A ratio of two or greater is generally accepted as significant.

Microarrays containing large DNA segments, such as cDNA, are generated by physically depositing small amounts of each DNA of interest onto known locations on glass surfaces. So far, two advanced liquid delivery technologies have been extensively used. The first consists of the mechanical microspotting approaches that utilize pins, tweezers, capillaries and other pronged devices to deposit small quantities of DNA onto known addresses using motion-control systems. The rate and accuracy by which the pre-synthesized material can be delivered to the surface depends on a large number of parameters, including the precision of the robotic control system and the micro-machining technology used to manufacture the printing devices. The most convenient printing substrates are standard 1×3 inch microscope slides bearing chemically-modified surfaces that allow DNA attachment. A variety of surface-coupling chemistries, derived in part from oligonucleotide microarray experiments, have recently been employed including polylysine coatings and reactive aldehydes.

Modern gridding robots can easily print up to 100 chips in a single session. Recent advances in microspotting technology, such as the availability of high-precision printer heads containing 32 pins, allow for the preparation of 100 microarrays containing >10,000 features in ~12 hr. Further improvements in microspotting technologies will eventually permit the automated fabrication of cDNA microarrays containing 100,000 features. High-density DNA printing, together with suitable fluorescence detection equipment, will enable efficient and inexpensive genome analysis of any eukaryotic organism.

The second approach to cDNA microarray fabrication utilizes ink-jet nozzles to deliver concentrated DNA solutions to glass surfaces. This technology relies on the so-called piezoelectric effect, whereby certain materials such as ceramic expand when exposed to an electric potential. Piezoelectric fittings, tightly attached to glass capillaries, allow the selective contraction of the capillaries in an electrically-controlled manner. Piezojets, attached to suitable motion control systems, are capable of delivering 100 picoliter droplets at a rate of ~10,000 droplets per second. Several companies, including Incyte Pharmaceuticals and Protogene, have proprietary ink-jetting technologies. The recent preparation of high-density cDNA microarrays for human gene expression analysis demonstrates the viability of ink-jet technology.

One major advantage of the deposition technologies, relative to the synthesis approaches, is the capacity to prepare microarrays of virtually any molecule of interest including genomic DNAs, antibodies, lipids, carbohydrates, small molecules, and so on. At the current printing densities, it would be possible to prepare microarrays of bacterial artificial chromosomes (BACs) sufficient to cover the entire 3×10^9 bp of the human genome. The microarrays of human BACs might find use in a variety of genomics experiments including gene mapping, DNA fingerprinting and promoter identification. Although microarrays of long DNA molecules are unsuitable for examining sequences at single nucleotide resolution, microarrays of genomic DNA can be used for massive, parallel analysis of genomes at a single-gene resolution. One of these applications is outlined in the next section.

II PHYSICAL CHEMISTRY OF MICROARRAY HYBRIDIZATION

Several parameters influence the rate of heteroduplex formation during the hybridization reaction: (1) the concentration of the target (immobilized DNA), (2) the concentration of the probe (labeled DNA, RNA), (3) sequence composition of the heteroduplexes (AT and GC content), and (4) salts and temperature.

A. CONCENTRATION OF TARGET AND PROBE DNA

When the target concentration is ~10-fold greater than its cognate species in the probe mixture, pseudo-first-order reaction kinetics ensue, such that the hybridization rate is determined largely by the probe concentration. Thus, a two-fold increase in probe concentration produces a two-fold increase in the signal. Pseudo-first-order conditions greatly improve quantitation by minimizing the effects of minor differences in the target concentration caused by imprecision in microarray manufacturing methods. Hybridization reactions, which involve target concentrations that are equal to or less than the probe concentration, display second-order hybridization kinetics. Under these conditions, small differences in the concentration of immobilized DNA (target) can have a large effect on the rate of hybridization and the absolute signal. Efficient synthesis and micro-deposition strategies, used in conjunction with robust coupling chemistries, permit the preparation of microarrays that have a high target concentration and, therefore, display pseudo-first-order kinetics over a wide range of sample concentrations. Most of the published assays with oligonucleotide and cDNA microarrays utilize chips that contain a surplus of immobilized target DNA relative to the probe.

B. SALTS AND TEMPERATURE

The presence of monovalent cations, such as sodium, increases the rate of heteroduplex formation by shielding the negatively charged phosphate backbones that would otherwise hinder base-pairing interactions between the target and probe molecules. Typically, sodium ion concentrations of ~1 molar are used for both oligonucleotide and cDNA microarray experiments. Temperature exerts a positive effect on hybridization if the temperature is sufficiently below the melting temperature (m) of the heteroduplexes. Hybridization temperatures of

25–42°C and 55–70°C are typically used for oligonucleotide and cDNA microarray experiments, respectively. The optimal salt concentration and hybridization temperature must be empirically determined for a given set of experimental conditions.

C. SEQUENCE COMPOSITION

Sequence composition is the parameter over which the experimenter has least control, and is a much greater concern with oligonucleotides than cDNAs. Pairs of GC form three hydrogen bonds, whereas AT pairs form only two; consequently, an increased duplex stability is observed in GC-rich regions. When these stable regions are ~50 bases long, they are assumed to serve as 'nucleation' sites from which the heteroduplex formation spreads. Considering that thousands of different heteroduplexes are being formed in a single microarray hybridization, it is critical to select a set of 'compromise' reaction conditions that gives the optimal signal to noise discrimination for as many of the heteroduplex probes as possible. Introducing the appropriate positive and negative control sequences on the array, as well as spiking in a number of positive controls empirically, achieves this.

The physical chemistry of hybridization to oligonucleotide microarrays is clearly different from that of the cDNA microarrays. As previously mentioned, hybridization involving oligonucleotides are more sensitive to the GC content of individual heteroduplexes than experiments involving longer DNA sequences. To minimize this problem, agents that equalize the binding energies of GC and AT base pairs, such as tetramethylammonium chloride (TMAC), can be added to the hybridization reaction. As the base composition issues are much less important with 0.5–2.0 kb sequences, the use of TMAC is an unnecessary precaution for cDNA microarray experiments. Single-base mismatches can have a pronounced impact on the hybridization reassociation of short sequences (i.e., oligonucleotides) and relatively little effect on cDNA hybridization. This fact can be exploited in resequencing and diagnostic applications of oligonucleotide microarrays. In general, the issues of hybridization specificity and cross-hybridization are important and complex issues and must be carefully considered and controlled in each of the applications involving microarrays of oligonucleotides and cDNAs.

Method-1: cDNA microarrays

Preparation of oligonucleotides

Synthesis of oligonucleotides

The oligonucleotides can be synthesized with a 394 DNA/RNA synthesizer (refer to the earlier chapter).
1. Synthesize the oligonucleotides, fluorescently labeled at the 5' terminus, by using FAM (6–carboxyfluorescein) and HEX (hexachlorinated analogue of FAM) amidites. The synthesis of oligonucleotides for immobilization starts with 3–methyluridine, located at the 3' end.
2. Label the oligonucleotides and DNA for hybridization also at the 3' end as follows. Incubate 100 pmol of an oligonucleotide at 37°C for 1 hr in 20 μl of the reaction mixture containing 100 mM cacodilate buffer, pH 6.8, 1 mM $CaCl_2$, 0.1 mM DTT, 200 pmol of tetramethylrhodamine (TMR)-conjugated dUTP, and 10 units of terminal deoxynucleotidyl transferase.

Another procedure for labeling oligonucleotide containing the 5' amino group is by using an excess of the N–hydroxysuccinimide ester of 5–carboxy TMR (molecular probes) in dimethyl sulfoxide, with 50 mM sodium borate buffer (pH 9.0,) at 60°C for 30 min.
3. Purify the oligonucleotides on NAP–5 columns that have been pre-packed with Sephadex G–25.

Preparation of sample DNA

Double-stranded (ds) DNA preparation

1. The dsDNA sequences from the gene of interest can be synthesized by PCR amplification. Initially, dsDNA should be amplified using genomic DNA, primers for genomic DNA and standard PCR conditions. In a 1 ml Eppendorf tube, to set up a 100 μl reaction mix, add the following:
    A PCR buffer containing 40 mM KCl, 20 mM MOps (pH 8.0), 2 mM $MgCl_2$
    1 ng of genomic DNA
    100 mM each of dNTPs
    0.3 mM each of amplification primers
    2.5 units of AmpliTaq DNA polymerase
    Reaction conditions: 30 cycles of 40 s at 93°C, 30 s at 67°C, and 30 s at 72°C.

Synthesis of shorter dsDNA fragments by nested primers

1. Transfer 5 μl of PCR products from the previous amplification to another reaction mixture, for amplification with nested primers. Then, amplify the DNA with nested primers for 25 cycles (30 s at 90°C, 30 s at 50°C, and 20 s at 72°C).
2. After amplification, purify the PCR fragments with a *QIAquick* PCR Purification Kit (Quiagen Chatsworth, CA) as suggested in the manufacturer's protocol.

Single-stranded (ss) DNA Preparation

1. The ssDNA can be prepared by single primer re-amplification (Ausubel *et al.*, 1994). As per the 100 μl PCR reaction, transfer 100 ng of PCR-amplified dsDNA into a 1 ml Eppendorf tube. Then, amplify this with fluorescently-labeled primers (e.g., 5'–HEX.labeled primers) for 25 cycles (30 s at 90°C, 30 s at 50°C, and 20 s at 72°C).
2. Purify the single-stranded amplification products by electrophoresis in 8% polyacrylamide gel.
3. Cut out the fragment and elute it in 1 M $LiClO_4$ and then precipitate it in 10 volumes of 2% $LiClO_4$ in acetone.

DNA fragmentation

1. Fragment the DNA by the same chemical reactions used in the Maxam–Gilbert sequencing method (see Chapter 10). Dissolve 3 μg of PCR-amplified dsDNA in 20 μl of 80% formic acid and incubate it at room temperature for 10 min.
2. Stop the reaction by adding 20 volumes of cold 2% $LiClO_4$ in acetone and cool for 10 min at –20°C.
3. Precipitate the DNA by spinning it down in a micro-centrifuge for 3 min.
4. Wash the pellet twice with acetone and air dry it.
5. Dissolve it in 100 μl of 1 M piperdine and incubate it at 95°C for 30 min.
6. Extract by chloroform and precipitate it in 2% $LiClO_4$ in acetone.

7. Dephosphorylate the resulting products twice, using 1 unit of shrimp alkaline phosphatase in 10 µl of the buffer, provided by the manufacturer, for 1 hr at 37°C.
8. Inactivate the phosphatase by heating it at 95°C for 15 min.
9. Label the fragments with 1 nmol of TMR-conjugated dUTP, using terminal deoxynucleotidyl transferase, as described above.

D. ARRAY PRINTING

Arrays are typically printed on one of two types of support matrices: nylon membranes or glass wafers or slides. Although standard microscope slides may be used, they must be prepared to facilitate the bonding of the DNA to the glass. Several different coatings have been successfully used, including silane and lysine. The coating of the slides can easily be carried out in the laboratory. Once the support matrix has been prepared, the DNA elements can be applied by several methods. Affymetrix, Inc. has developed a unique photolithographic technology for attaching oligonucleotides to glass wafers. More often than not, DNA is applied by either non-contact or contact printing. Non-contact printers use thermal, solenoid, or piezoelectric technology to spray aliquots of solution onto the support matrix and may be used to produce slide or membrane based arrays; Cartesian Technologies Inc. has developed the nQUAD technology for use in its PixSgs printers. The system couples a syringe pump with the microsolenoid valve, a combination that provides the rapid quantitative dispensing of nanoliter volumes (down to 4.2 nL) over a variable volume range. A different approach to non-contact printing uses a solid pen-and-ring combination (Figure 5.5).

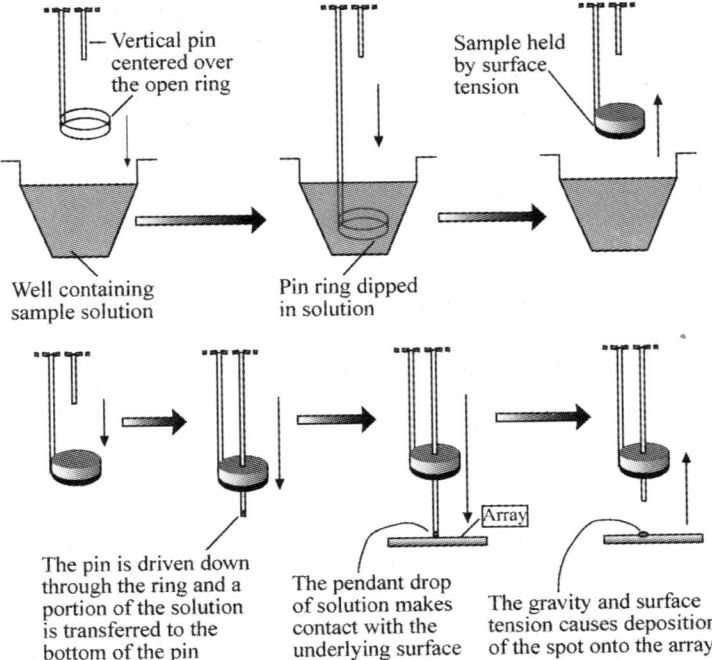

Figure 5.5 *Array printing by pin and ring system. (Redrawn from Rockett and Dix 1999)*

1. cDNA microchip manufacturing on glass surface

Polyacrylamide gel provides a maximal capacity of 50 mM for 3D-immobilization. This is over a 100 times higher than the 2D-immobilization capacity of a glass surface. Another advantage of the gel support is that more than 70% of the oligonucleotides can participate in the hybridization process and they can be well spaced from each other. This prevents interference between different molecules of the oligonucleotides and DNA during hybridization. Thus, it facilitates the efficient discrimination of perfect duplexes from mismatched ones.

1. Prepare a glass-immobilized gel matrix of elements by polymerization of 20 μm thin polyacrylamide (8% acrylamide/0.28% bisacrylamide) gel on a glass surface treated by Bind-silane (LKB).
2. Remove strips of the gel in X–Y directions with a scribing machine, forming an array of the gel elements of size 40 × 40 μm or 100 × 100 μm and spaced 80 or 200 μm apart, respectively. Alternatively, photolithography and the laser evaporation technique can be used to remove the gel from the space between the microchip elements.
3. Activate the polyacrylamide gel by substituting some amide groups with hydrazide groups by hydrazine-hydrate treatment.
4. Hydrophobize the glass space between the gel elements by treatment with Repel-Silane.
5. Synthesize the oligodeoxynucleotides for immobilization with 3′-terminal 3-methyluridine, activated by oxidizing with $NaIO_4$, to produce dialdehyde groups for coupling with the hydrazide groups of the gel.
6. Transfer the solution of activated oligonucleotides on to the micromatrix element with a specially devised one-pin robot (Figure 5.6). The robot performs 240 transfers per hr of activated oligonucleotide solutions onto the activated micromatrix gel elements. The reproducibility of the transfer is ±8%. The pin temperature should be kept at dew point to prevent drop evaporation or water condensation on the pin. Microchips prepared in this way usually sustain 15 rounds and, in some cases, 50 rounds of hybridization.
7. After the transfer, decrease the matrix temperature and condense the water on the gel.
8. Cover the fully-swollen gel matrix with mineral oil and keep it at 20°C for 48 hr for quantitative oligonucleotide immobilization; then, wash out the oil using ethanol and distilled water.
9. Dry the microchips. These can be kept at 4°C for a year before use.

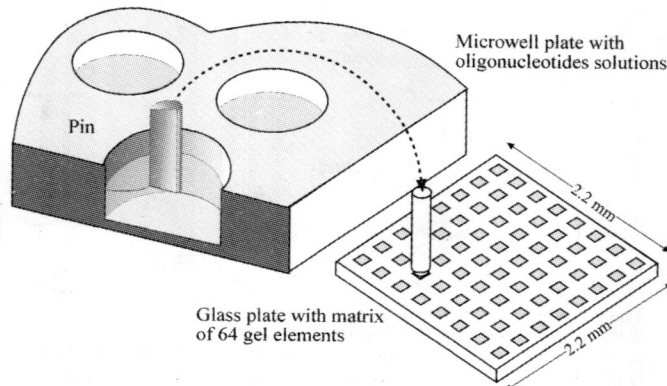

Figure 5.6 *The scheme of one-pin robot for manufacturing oligonucleotide microchips (Redrawn from Yershow et al. 1996)*

E. HYBRIDIZATION

Oligonucleotide hybridization probes, under appropriate experimental conditions, enable the discrimination of nucleic acid sequences that differ by as little as a single nucleotide. This sequence specificity has led to the development of many hybridization techniques to detect target sequences, starting from one nucleotide to many differences. The effective temperature stability of the duplexes formed with gel-immobilized oligonucleotides depends on their concentration. The difference in melting temperature due to differing GC contents can become a problem when high-stringency experiments are performed. This can be somewhat mitigated through the use of tetramethylammonium chloride, which tends to stabilize AT bp and bring their melting point closer to that of GC bp. Steric hindrance between the glass surface and the DNA can reduce hybridization yields if the linker molecule between the glass and the DNA is too short.

The detection of hybridization is generally done by labeling the sample either radioactively, with a chemiluminescent reporter, or with a fluorescent dye, then detecting the presence of the label on the array after hybridization and washing. The labeling can be done by making copies of the sample with labeled PCR primers, or incorporating labeled nucleotides, or by direct chemical means.

1. All the procedures should be performed on a Peltier thermostable (working stage from −5°C to +60°C). The hybridization of a microchip with fluorescently-labeled DNA (1–5 pmol) or oligonucleotides (5 pmol) should be carried out at 5°C for 30 min. in 2–5 μl of the hybridization buffer (1 M NaCl, 1 mM EDTA, *1% Tween* 20, 5 mM sodium phosphate, pH 7.0) for oligonucleotides, and in 0.1X buffer for DNA.
2. Then wash the microchip with 100 μl of the hybridization buffer at 5°C for 10 s and cover it with 5 μl of the same buffer.
3. For ethidium bromide (EtdBr) staining, hybridize 1 mM solution of unlabeled DNA with the microchip and stain with 1 mM EtdBr in 0.1X buffer for 5 min. at 5°C.
4. Contiguous stacking hybridization can be carried out in two steps. After the initial hybridization of the microchip with PCR-amplified ssDNA, as described above, hybridize the microchip with 1 μl of a mixture of pentamers (5 pmol of each) in 1X buffer for 5 min. at 4°C. Wash off the hybridized 5-mers from the microchip at 20°C for 2 min. in 1X buffer and repeat the hybridization under the same conditions with the other mixtures of 5-mers.

F. IMAGE ANALYSIS SYSTEM

To monitor fluorescence signals on either the FAM or TMR and HEX a two-wavelength fluorescence microscope can be used (Yershov et al., 1996). It includes a 350-W high-pressure mercury lamp, Phloem opaque with interference excitation and barrier filters, special optics, and a CCD camera (TE/CCD512SF; Princeton Instruments, Trenton, NJ). The X3 objective with a 0.4 numerical aperture allows the illumination of an object field up to 7 mm in diameter. It projects a 2.7 × 2.7 mm area of the microchip on the 8.1 × 8.1 mm matrix of a Peltier-cooled CCD camera. The exposure times vary from 0.4 to 30 s, with a readout time of about 1.3 s; variations in the sensitivity within the object area should not exceed 5%. The system allows work with a X1.7 objective and the same aperture for analyzing a 5 × 5 mm microchip area. Two sets of filters, for FAM or HEX and TMR, can be quickly changed for the analysis of these fluorophores. The image of the microchip on the CCD camera can be displayed and analyzed on a PC using specially developed software. The system provides real-time and simultaneous monitoring of hybridization kinetics at different temperatures for all microchip elements. Parallel two-color monitoring of the hybridization can be carried out with a DNA sample labeled with two different fluorophores (Figure 5.7). For printing, a linear transformation, which brings the highest pixel values to the same level for all images, should be used. For digital estimation, the image of the microchip element should be fully covered by a 'square' (total signal, S) twice the size of the element. Then, a frame (total signal, F) with an area equal to that of the square should be constructed around the square. The signal of the element (E) can be calculated as $E = S - F$.

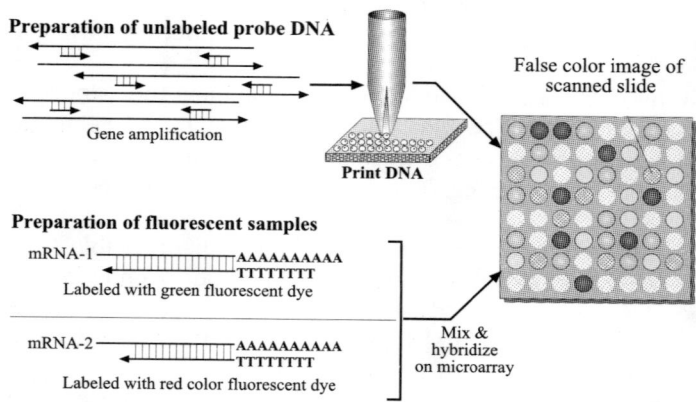

Figure 5.7 *The principle steps in producing and utilizing DNA microarrays (Redrawn from Kehoe et al. 1999)*

G. ADVANTAGES AND DISADVANTAGES OF THE TWO MICROARRAY METHODS

1. Advantages

1. Oligonucleotide-based microarrays, particularly those produced by photolithography methods, have very high densities (250,000 oligonucleotides per cm^2) and are inherently more consistent from array to array in comparison to DNA fragment-based microarrays.
2. The inherent potential of misplacing clones when handling the thousands of clones required to construct DNA fragment-based microarrays can also be avoided with oligonucleotide-based microarrays, because the oligonucleotides are synchronized *in situ* on the basis of gene sequence obtained directly from the sequence databases.
3. Because the sequence that hybridizes in oligonucleotide-based microarrays is very short (i.e., 20–25 nucleotides), the hybridization reactions are very sensitive to single-nucleotide mismatches, unlike DNA fragment-based microarrays.
4. Since oligonucleotide-based microarrays are very sensitive to single-nucleotide mismatches, these microarrays are particularly well-suited to genotyping or resequencing applications.
5. Oligonucleotide-based microarrays are, on the whole, more sensitive and exhibit a broader dynamic range and permit the simultaneous analysis of two complex probes.

6. Only oligonucleotide-based microarrays can provide a genome-scale analysis of gene-expression patterns.
7. One significant implication of DNA microarray technology is that many identical arrays can be produced to serve as a common experimental platform among researchers.

2. Disadvantages

1. The major disadvantage is that their construction depends heavily on the availability of accurate sequence data.
2. Conversely, sequence information is not required for the construction of DNA fragment micro-arrays, and this type of microarray is more appropriate for the majority of plant species.

III   APPLICATION OF DNA CHIP TECHNOLOGY

The main advantage of DNA chips is the ability to assay many combinations of genes simultaneously. Whether assaying genomic DNA or mRNA, the unique features of DNA chips are its high throughput and the fact that it addresses broader biological questions. Outlined in Table 5.1 are several areas where DNA chip technology has been exploited. A few methods where this technology has been applied are also discussed below.

Table 5.1 *Applications of microchip technology*

1. Diagnosis of genetic disorders
    a. Predetermining and preventing diseases before spell disorder
    b. Nucleotide polymorphisms (single-nucleotide polymorphism, SNP)
    c. Rapid diagnosis of infections
2. Application in basic work
    a. Cell-cycle variations
    b. Screen populations for polymorphic "expression fingerprints"/allelic and evolutionary sequence comparison
    c. Mutation detection
    d. Variation in gene expression in different tissues due to developmental and physiological processes.
    e. Detection of reduced transcript accumulation
    f. Gene expression patterns in cell organelles
    g. Assigning functions
    h. New gene discovery
    i. Genome mapping
3. Pattern of gene expression
    a. Developmental stage
    b. In pathogenic tissues
    c. Drug treatment
    d. In stressful environmental conditions

Continues...

Continued...

Table 5.1 *Applications of microchip technology*

---

4. Designing new and better drugs
   a. Targeting of abnormally active or inactive or non-specific to time and tissue genes
5. Detecting the genes of disease-causing parasites
6. Determining the sequence of the genetic materials of any organism
7. Mutation detection with oligonucleotide microarrays
8. Genotyping by hybridization to oligonucleotide microarrays
9. Shared data management and analysis
   a. Transcription accumulation studies
   b. DNA microarrays and proteomics
   c. Molecular bar coding and reverse genetics

A. SOME AVAILABLE METHODS OF MICROARRAY APPLICATIONS

Method-1: Gene Expression Patterns

Using DNA chip technology, it is possible to determine the differences in levels of gene expression between normal and disease samples. To detect changes in patterns of gene expression that may occur during disease states, an array assay can be used to monitor gene expression in different cells and tissues (Figure 5.8).

a. Use a DNA chip spotted with cDNA from diseased tissue. This array format developed by companies (Table 5.2) is a combination of DNA base pairs from a cell, or tissue sample, of interest. The DNA arrays are printed onto a glass slide. This array contains both the target cDNA and the control DNA in different quadrants of the slide.
b. To probe this chip, use fluorescent cDNA probes on normal and diseased tissue; for example, benign hyperplasia prostate tissue (red and green probes can be used).
c. Scan the hybridized array slides to identify the genes expressed in diseased cells from normal cells. If two-colored probes are used, the results will be: green samples contain genes expressed in diseased cells, red samples contain genes expressed in normal cells and yellow indicates genes expressed in both cell types.
d. The results can interpret the increase in gene expression of disease-related genes and decrease in the expression of other genes. This demonstrates the difference in gene expression between the normal and diseased states.

Method-2: Human diagnostics

The DNA chip array assay has tremendous potential in human diagnostics for monitoring changes in gene expression within specific cell types, tissue types, physiological states, and the genetic composition of the patient, etc. A patient sample can be examined for changes in gene expression that may provoke a cancerous growth or diseased state. The efficacy of therapeutic treatment, such as protease inhibitors for AIDS patients, can be monitored with specifically-designed DNA chips.

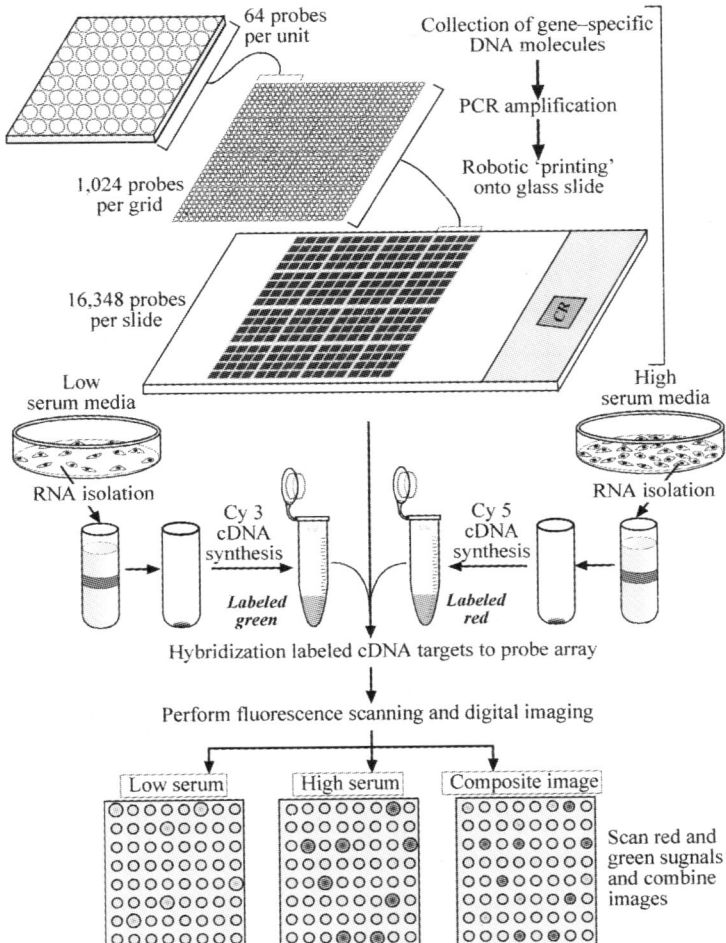

Figure 5.8 *Global changes in gene expression can be monitored using cDNA microchips containing known ESTs*

Method-3: To determine tissue-specific gene expression

a. Probe a DNA chip spotted with genes from a large variety of human tissues with many types of human cyclin mRNA species, each tagged with a different fluorescent color.
b. Results may show a variety of human tissues expressing different cyclin types at different levels; for example, cyclin A is expressed at high levels in the brain, thyroid, retina, prostate, and pancreas.

Method-4: DNA chip technology in studying functional genomics

After examining differences in gene expression, the next step will be to look at the function of these differentially expressed genes. Presently, researchers use fully-sequenced organisms, such

Table 5.2 *DNA Chip Manufacturers*

Company	Data	Array format
Affymetrix, Santa Clara, CA	Expression-profiling disease management and polymorphism analysis	On-chip photolithographic synthesis of oligos onto silicon
Axys Pharmaceuticals, San Diego, CA	Novel gene identification, candidate gene validation, diagnostics and mapping	Off-set printing on array
Brax, Cambridge, UK	Diagnostics, expression profiling, and novel gene identification	Short, synthetic oligos synthesized off-chip
Clontech, Palto Alto, CA	Differential gene expression for cellular pathways and targeted research	cDNA immobilized onto nylon membranes
Genetic MicroSystems, Woburn, MA	Highly parallel genomic research and drug discovery	Surface tension spotting onto slides using pin-and-ring technology to create uniform samples
Genome Systems, St. Louis, MO	Differential gene expression	cDNA spotted onto nylon membranes
Hyseq, Sunnyvale, CA	Universal sequencing chip, expression profiling, novel gene identification, polymorphism analysis	DNA sample printed onto HyGnostics and gene-discovery membranes
Luminex Austin, TX	DNA and protein diagnostics, haplotype determination	Oligos coupled to fluorescent-coded microspheres
Nanogen, San Diego, CA	Diagnostics and short tandem repeat identification	Oligos captured onto electroactive spots on silicon
Protogene Laboratories, Palo Alto, CA	Expression profiling, and polymorphism analysis	On-chip synthesis of oligos printed onto glass
The German Cancer Institute, Heidelberg, Germany	Expression profiling and diagnostics	Protein-nucleic acid macrochip with on-chip synthesis of probes

as yeast, to conduct gene disruption studies. A gene of interest is disrupted or knocked-out in the yeast genome. The fitness of this mutant is monitored, using a variety of selective growth conditions, to determine the significance of the disrupted gene.

Method-5: Gene Mutation Patterns

DNA array analysis has been widely used to detect genetic mutations. Mutations within the BRCA1 exon 11 genes have been detected from breast and ovarian cancers. At the NIH, the Collins groups developed DNA arrays of 96,000 oligonucleotides to screen for 24 heterozygous mutations within exon 11 of the hereditary breast and ovarian cancer gene. The results showed a promising start on a 5,592 bp gene known to have mutations within 22 exons. A diagnostic DNA chip able to detect all mutations within the BRCA–1 gene, requiring 400,000 probes, is currently under investigation.

Method-6: Sequencing by Hybridization

To detect a gene expressed at levels as low as one copy per cell, Hyseq, in Sunnyvale, California, has developed the "sequencing by hybridization" technology. This sequencing by hybridization technology can identify target genes without prior sequence knowledge. A series of short, overlapping probes are hybridized and bioinformatically reconstructed to recreate the DNA sequence. An extension of their technology yields a universal sequencing chip that is an excellent tool for both diagnostics as well as gene discovery.

Method-7: Mutant Yeast Molecular Bar Coding

In this method, a library of mutant yeast strains are created by deleting a unique gene or DNA sequence from the yeast genome; in each yeast strain this deleted area is tagged. The yeast strain is then grown under selective growth conditions. Then, the deletion tags from surviving strains are PCR amplified and these products are hybridized to a DNA chip. The hybridization pattern demonstrated yeast strains indicating which deleted genes are critical for survival of the strain. The viability of the yeast with a particular deletion gives researchers clues to the function of the disrupted gene.

IV SUGGESTED READING

Ausubel FM, Brent R, Kingston RE, Moore DD, Seidman JG, Smith JA, and Struhl K (Eds) (1999) *Short Protocols in Molecular Biology*, 4th Edn, John Wiley and Sons, Inc.

Blanchard A (1998) Synthetic DNA arrays, *Genetic Engineering*, **20**:111–121.

Brown PO and Bostein D (1999) Exploring the new world of the genome with DNA microarray, *Nature Genetics*, **21**:33–37, http://genetics.nature.com.

Debouck C and Goodfellow PN (1999) DNA microarrays in drug discovery and development, *Nature Genetics*, **21**:48–50, http://genetics.nature.com.

Duggan DJ, Vittner M, Chen Y, Meltzer P and Trent JM (1999) Expression profiling using cDNA microarrays, *Nature Genetics*, **21**:10–14, http://genetics.nature.com.

Kehoe DM, Villand P and Somerville S (1999) DNA microarrays for studies of higher plants and other photosynthesis organisms, *Trends Pl. Sci.*, **4**(1):33–41.

Schena M, Shalon D, Davis RW and Brown PO (1995) Quantitative monitoring of gene expression patterns with a complementary DNA microarray, *Science*, **270**:467–470.

Southern E, Mir K and Shchepinov M (1999) Molecular interactions on microarrays, *Nature Genetics*, **21**:5–9.

6

Bioinformatics in Biotechnology

I INTRODUCTION

Molecular biology studies genes and proteins, which are macromolecules consisting of several thousand atoms and molecules. The genomic sequence data of various organisms, including humans, have now reached such a level of sophistication that it is quite common for a large stretch of DNA to be sequenced and for that sequence to be manipulated or stored in a computer database. This accomplishment has led to a rapid build-up of the sub-disciplines of genomics, bioinformatics, and proteomics. Since the human genome-sequencing project begun in 1990, we have been flooded with biological data at an unprecedented rate. For example, as of August 2000, the GenBank repository of nucleic acid sequences contained 8,214,000 entries and the protein sequence data contained 88,166. On an average, the amount of information stored in these databases is doubling every 15 months. Since it is not possible to draw such a complex picture as a genome sequence with paper and pen, biology has to marry with informatics (information technology) to give birth to a new discipline called *"bioinformatics"*; this has given rise to a whole new area in molecular biology. *Bioinformatics is the science of managing and analyzing biological information*. This can also be called *computational biology*, as computer applications are being used. Thus, bioinformatics is an integration of mathematical, statistical and computer methods to analyze biological, biochemical and biophysical data. This science develops computer databases and algorithms for the purpose of speeding up and enhancing biological research.

In recent years, biotechnology has made outstanding progress, allowing scientists to modify genetic structure in order to develop, for example, new drugs or pest-resistant plants. This process has been greatly assisted by the science of bioinformatics, which is a collective name used for computer approaches in fields such as molecular biology, biotechnology, medicine and agriculture. A number of large-sequence facilities are now fully automated and sequences can be downloaded automatically to those databases from robotic workstation servers. Surprisingly, this accumulation of genetic information has been well matched by developments in computer hardware and software. Database sequence-similarity searching is now carried out thousands of times each day by researchers worldwide.

II APPLICATION OF BIOINFORMATICS AT MOLECULAR ORGANIZATION

The information needed for the organization and functioning of life begins from an atom to a cell, to an organism and beyond the organism (Figure 6.1). Therefore, it is very important to understand the structure, order and physical–chemical properties of molecules at various levels of organization. This knowledge will certainly utilize the computational power of computers in biological systems at different levels of molecular organization and functions. The potential application of bioinformatics in biological systems is briefly discussed below.

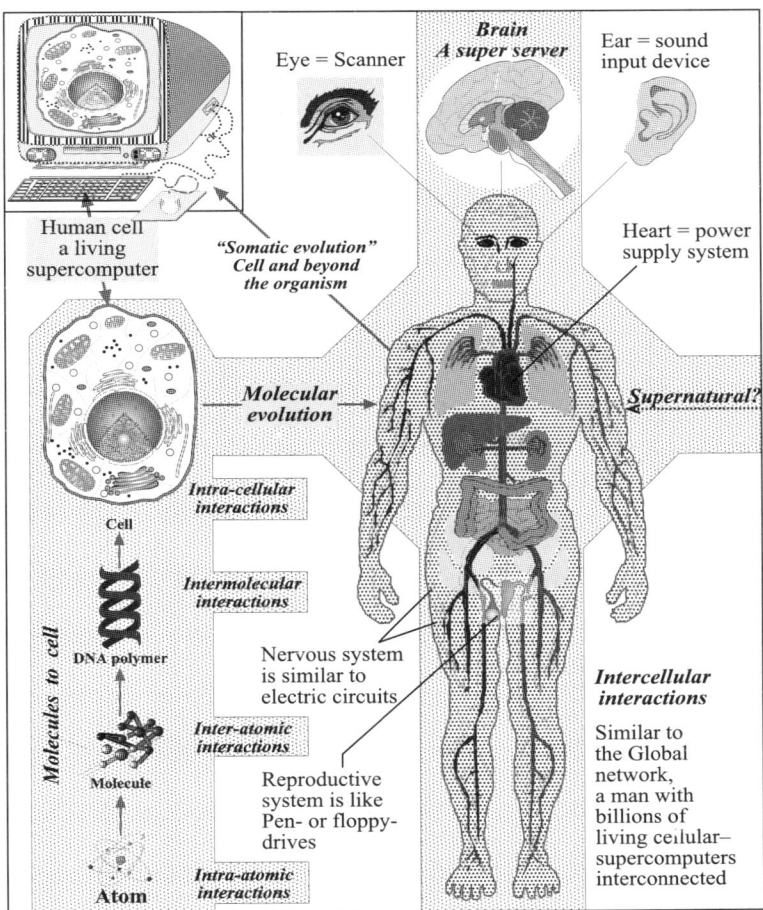

Figure 6.1 *A hypothetical concept which draws parallelism between the recent revolution in the field of bioinformatics and molecular evolution that creates cellular organisms. In parallel to global electronic networking, a highly sophisticated and well structured communication system can be seen in multicellular organisms such as human body*

A. INTRA-ATOMIC INTERACTIONS

The structural, physical and chemical properties of the atom itself is the fundamental force behind all the biological activities on the earth and, probably, in the universe. All the atoms, except hydrogen, contain three subatomic particles called protons (positively charged), neutrons (neutral) and electrons (negatively charged). These three subatomic particles vary in their number and organization between the atoms. Thus, there are different atoms on the earth and the elemental properties (*intra-atomic* interactions) are dictated by the composition of these subatomic particles, which, in turn, also provide a basis for the generation of *inter-atomic* (between the atoms) interactions.

B. INTER-ATOMIC INTERACTIONS

Inter-atomic interactions are the basis for the formation of molecules and their interactive properties with other molecules. Thus, inter-atomic interactions determine the structural and functional ability of a molecule. The structural organization and functions of all biomolecules including water, amino acids, proteins, nucleic acids, sugars, lipids, etc., are determined by intermolecular interactions. Therefore, to understand, for example, how the protein structure might catalyze a reaction, we need to understand enough about chemical-reaction mechanisms to develop a hypothesis about how the reaction might work, given the shape of the protein and the location of various amino acids with specific intermolecular interactions. Protein association is often mediated by the electrostatic (intra-atomic interactions) properties of the protein structure; interactive molecules can be drawn together over considerable distances by strong electrostatic potentials. Within protein structures, hydrogen bonds and other inter-atomic interactions confer structural stability (Figure 6.2). It is a must for anyone to know about molecular specificity in practical applications of biochemistry – designing small molecules or peptide drugs, understanding the molecular basis of disease and immunity, etc. Providing detailed information on those molecules is beyond the scope of this book; therefore, I suggest that readers refer any standard textbook on biochemistry and organic chemistry. Information on inter-atomic forces, such as: covalent bonds; hydrogen bonds; hydrophobic and hydrophilic interactions; Van der Waals forces; charge–charge, charge–dipole, and dipole–dipole interactions; repulsive forces; relative strength of inter-atomic forces; etc, will be very useful in understanding their complexity and for the application of bioinformatics.

C. CELL STRUCTURE, FUNCTION AND PERPETUATION

The information available at the cellular level will be of a higher order of molecular interactions, but still basic to the multicellular organism. The cell is a structural and functional unit of all living organisms. It is a highly-organized structure and contains many subcellular structures made up of diverse micro- and macro-molecules. The chemical and physical properties generated after the assembly of various molecules into a very complex body (subcellular structure), gives the cell a *"life"*. The assembly and interactions of diverse molecules in the cell provides the capabilities to function, repair, respond to the internal and external stimuli, and perpetuate faithfully for number of generations by itself (cell division). The understanding of the subcellu-

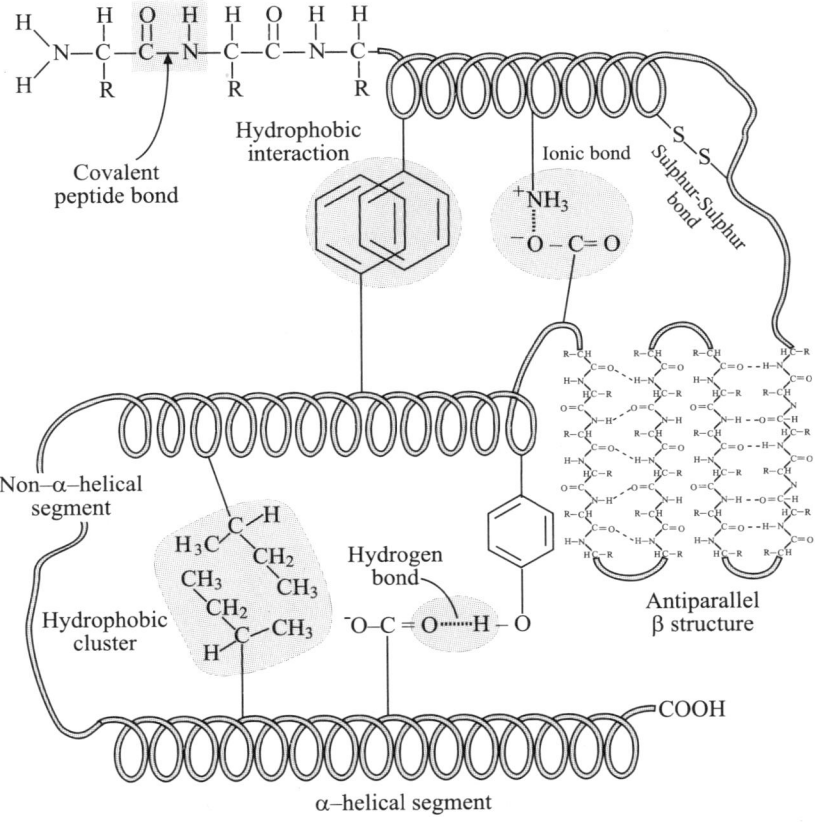

Figure 6.2 *A hypothetical globular protein having several types of covalent and non-covalent side-chain interactions*

lar functions and the cellular interactions with the environment is one of the fundamental aspects of molecular biology.

On a logical basis, when comparisons are made between a cell and a computer, many paralle-lisms can be inferred. The cell appears to be a highly-advanced, living supercomputer (Figure 6.1). Apparently, computer hardware, software, and its accessories function and appear to resemble the architecture and functions of the cell and its components. The levels of information storage, processing and retrieval in a cell from DNA to RNA then translation into a protein, appear similar to the functioning of processor language, assembly language and application language programs in a computer system. The nucleus can be compared to the hard disc where information is permanently stored, and the RNA-ribosomal complex on the endoplasmic reticulum is like a RAM where RNA turnover takes place, and so on. The cell organelles, for example mitochondria, function like a power supply system at the cell level. The cell uses a *parallel computing* system for information processing and problem solving, instead of solving them one at a time (as in silicon-based computers), employing multiple processors to solve each sub-problem simultaneously. Thus the cells can process information much faster (up to 10^{20} operations sec) and more efficiently than the man-made supercomputer (which has a process maximum of 10^{14} operations/sec). Another very unique function of a cell is that it can duplicate by itself and can respond better to environmental changes; these features are missing in silicon-based computers.

D. INTER-CELLULAR INTERACTIONS IN MULTICELLULAR ORGANISMS

Multicellular organisms are made up of many cell types. For example, the human body consists of more than 100 billion individual cells. The multicellular organization in human beings can be compared to a global computer network system or something of an even higher order; it is like connecting more than 100 billion cells under single server. The cells in the human body are interlinked by the vascular system and the neural network. The brain is a powerful bio-processor and a central regulator of all the coordinated functions in a body; thus, the brain can be called a *"superserver"* (which makes advanced servers) created by nature during the millions of years of the evolutionary process. The neural network in the human body is a highly advanced and very complex one, which can transude and transmit signals more quickly and efficiently than the modern electronic communication systems. The sensory organs, like the eyes, ears, mouth, skin, etc, can be equated to the input/output devices in a computer, such as the camera/scanner, sound-cord/speaker, keyboard, printer, etc.

E. EXTRA-SOMATIC EVOLUTION

Among the higher organisms, human beings are considered to be the most highly-evolved organisms *"extra-somatically"* on the earth. Humans have developed the ability to understand and regulate other organisms and the environment in which they live. Physical abilities such as strength, size, shape, color, etc, are all due to the result of *"somatic evolution"* (related to genetics). Somatic evolution is, perhaps, the single most important mechanism that has created all the organisms on the earth. However, when humans are compared with other organisms, the physical abilities of humans are not extraordinary. Even the metabolic activities are based on the same physical and chemical principles. But, what makes humans unique is the development of the human brain and its analytical abilities (e.g., memory, information-processing abilities, etc). The extraordinary processing capacity of the human brain might have been responsible for extra-somatic evolution. Thus, human beings have been able to harvest the benefits of nature better than any other organism on earth. They have also been able to develop a sense of conservation for contingency and for future use. The development of agricultural practices, the establishment of verbal and literary languages, social and cultural activities, educational and scientific investigations, harvesting physical and chemical energy to perform work, the development of human welfare, the realization of ecological and biological balance on the earth, etc, are some of the important components of extra-somatic evolution. To maintain continued dominance over other organisms and the environment, humans have always tried to invent newer and better technologies to manipulate other organisms and the environment for their continued benefit. Among the new inventions, the development of computers and electronic communication system is one of the recent accomplishments.

III BIOINFORMATICS ACCESS TOOLS

Initially, in molecular genetics, computers were mainly used to store and disseminate DNA sequence information through institutional or government-sponsored computer links connecting academic research centers. More recently, however, the Internet has become a new type of electronic tool for molecular genetic research (Table 6.1). The rapid growth in the number of Internet users and the application of information technology to molecular genetics have been due to the introduction of a powerful graphical interface called the *World Wide Web* (*www*). It permits the display of files as multimedia objects that are linked together by hypertext addresses. Software programs, called web browsers, allow individual Internet users to interface with the web through personal computers that are electronically connected to the Internet. Individual web-sites are web-pages that can be located by unique Internet addresses. A web-page contains text, graphics, and marked hypertext links that are encoded in the *HTML* (Hyper Text Markup Language) language. Web-page files must be stored on dedicated Internet-connected computers that function as web servers. A web "home" page is the assigned starting point for a set of related web-sites (Table 6.2).

Table 6.1 *Some important bioinformatics access tools*

Access tools	Service they provide
Telnet	A protocol that let users log onto a remote computer directly through the Internet. Telnet allows someone using a computer full access to log into another computer over the Internet, assuming s/he has logging in privileges on that computer as well.
FTP	FTP stands for file transfer protocol, and it is the name of a program used for file transfer between computers with full access to each other, assuming that the user has logging in privileges on both the local and remote computers. Anonymous FTP is a common practice, whereby anyone on the Internet may transfer files from a remote system with the user ID "anonymous" and an arbitrary password.
Gopher	Gopher is a user-interface program that makes FTP and other types of connections for computer users when they select an item in a menu. It is an easy way to get material off the Internet without having to know where the material exists.
Archie	It helps people to locate items in thousands of FTP archives around the world. It is the first information-retrieval system developed for the Internet.
Veronica	Veronica is a very rodent-oriented net-wide index to computerized archives. Veronica searches through hundreds of Gopher holes looking for any thing that matches the keyword supplied by the user, and assembles a list of gopher servers that contains items of interest. Veronica checks only the titles of gopher items, not their contents.
Wide Area Information Servers	The idea behind WAIS is to make anonymous FTP archives more accessible by indexing their contents for easy searching and browsing. WAIS servers are used as back-end engines for gopher servers.
World Wide Web	The web is yet another tool for gathering useful information from the Internet. This system merges the power of worldwide networks, hypertext, and multimedia. The web looks like a document that users can open and read, but selecting certain words via the mouse or keyboard causes other documents to be retrieved and opened for inspection.

Table 6.2 *Nucleic acid and protein database resources available on the World Wide Web*

Database or resource	Information available	URL (uniform resource locator)
General DNA sequence database		
EMBL	European genetic database	http://www.ebi.ae.uk
GenBank	A US genetic database, GenBank contains sequence data from the translated coding regions from DNA sequences	http://ncbi.nlm.nih.gov
DDBJ	Japanese genetic database	http://ddbj.nig.ac.jp
Protein sequence database		
Swiss-Prot	A European protein-sequence database, it is an annotated protein-sequence database. The Swiss-Prot protein knowledge-base consists of sequence entries. Sequence entries are composed of different line types, each with their own format.	http://expasy.hcuge.ch/sprot/sprot-top.html
TREMBL	European protein sequence database	http://www.ebi.ac.uk/pub/database/trembl
PIR	A US protein information resource, source, it is a division of the NBRF, which is affiliated to Georgetown University Medical Center, established in 1984. PIR, in collaboration with MIPS and JIPID, produces the PIR-International.	http://www.nbrf.georgetown.edu/pir
MOWSE	Peptide mass mapping and sequencing	http://srs.hgmp.mrc.ac.uk/cgi-bin/mowse
ProFound	Peptide mass mapping and sequencing	http://prowl.rockefeller.edu/cgi-bin/ProFound
PeptIdent	Peptide mass mapping and sequencing	http://www.expasy.ch/tools/peptident
MASCOT		http://www.matrixscience.com/
FindMod	Post-translational modification	http://www.expasy.ch/tools/findmod/
SEAQUEST	Uninterpreted MS/MS searching	http://fields.scripps.edu/sequest/
FASTA Search Programs	Protein and nucleotide database searching	http://fasta.bioch.virginia.edu/

Continues...

Continued...

Table 6.2 *Nucleic acid and protein database resources available on the World Wide Web*

Database or resource	Information available	URL (uniform resource locator)
Cleaved radioactivity of phosphopeptides	Protein-phosphorylation site mapping	http://fasta.bioch.virginia.edu/crp
Protein structure databases		
PDB	Brookhaven protein database is the single worldwide archive of structural data of biological macromolecules. The PDB makes data available in two formats: the legacy PDB flat format and the newer mmCIF data format.	http://www.pdb.bnl.gov http://www.rcsb.org/pdb/
NRL-3D	Protein structure database	http://www.gdb.org/Dan/proteins/nrl3d.html
Genome project databases		
Human Mapping Database, Johns Hopkins, USA		http:/gdbwww.gdb.org
DbdbEST (cDNA and partial sequences)		http:/www.ncbi.nih.gov
Genethon Genetic maps based on repeat markers		http://www.genethon.fr
Whitehead Institute (YAC and physical maps)		http://www-genome.wi.mit.edu

The Internet network has grown, both in terms of the number of networked computer servers and the amount of information that can be accessed through test-searching programs, such as Gopher. The Internet today is a vast collection of linked high-speed computers that serve as regional centers for user access through smaller *local area networks*, called LANs. Through the use of routing protocols, established at the regional network servers, a single user can send or receive files by connecting to Internet with a dial-up telephone link to an *Internet service provider* (ISP), or by an Ethernet-type connection with a LAN (Figure 6.3). On the Internet, aside from the use of e-mail and online journal publications, the primary research-related activities are: (1) DNA sequence analysis, using sequence-alignment programs; (2) searching integrated databases to locate biological resources and reagents; and (3) interactive communication with specialized biological sciences user groups. To further emphasize the utility of these molecular genetic tools on the web, laboratory practicum 9 describes how searching DNA sequences databases can be used as a dry-lab method to identify functionally-related genes.

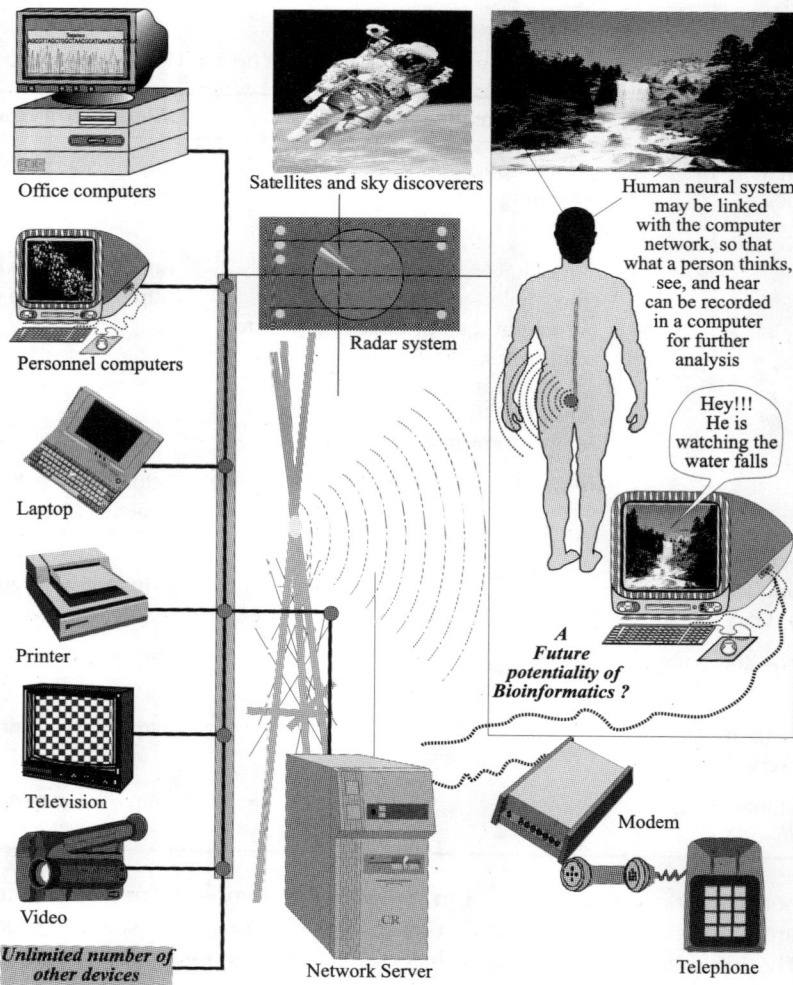

Figure 6.3 *Global computer network connection and potentiality of the bioinformatics in future*

IV THE ROLE OF BIOINFORMATICS IN BIOTECHNOLOGY

As mentioned earlier, bioinformatics actually represents an extension of *computational biology*, which has existed as a field of investigation for decades. The recent development of appropriate algorithms (sequence of commands for solving complex, highly-specific problems) and modern computing power have, however, accelerated the rate at which computational biology has progressed. Further, this also deals with special themes in molecular biology, like evolutionary genomics, functional genomics and structural genomics. Bioinformatics produces databanks, such as collections of protein sequences for biology and biotechnology, as well as computer software for analysis. Traditionally, the primary task of bioinformatics was the collection and maintenance of sequence information in public interconnected databases and to compare the DNA sequence data erupting from the genome-sequencing projects. Today, one of the major challenges in bioinformatics is to handle, maintain, and integrate the growing flood of sequence data of DNA and protein sequences and also of DNA chip-based multiparallel expression analysis or genome fingerprinting. The bioinformatics effort covers four areas of activities: (1) organization of the experimental data into a systematic database by arranging independent data into a systematic database which links independent data to genes, pathways and clones (2) support to individual projects in data collection and evaluation, (3) linking the results to the comprehensive database of plants, and (4) development of novel methods for the interpretation of data. Bioinformatics also applies its methodology for analysis to allow for the prediction of information based on known data and the deduction of information from large-scale sampling (data mining).

The users, such as biotechnologists and academic biologists, may choose to access these data over the Internet, using publicly available databanks. Bioinformatics currently deals with several main types of biological data which are discussed below.

Sequences and structure of genes and proteins: Nucleotide sequences are the simplest way to represent a macromolecule. The structure of genes that code for the sequence of amino acids in proteins is produced in this form by genome sequencing groups. Protein sequences are usually obtained from computer-based translation of genomic data. In prokaryotes, gene counting is a relatively straightforward task, since it has simple genomes and relatively fewer genes (about 3,600 protein coding genes). However, understanding the eukaryotic genome is a much more complex endeavor. As mentioned previously, in addition to protein-coding sequences, many eukaryotic genomes contain repetitive DNA (non-coding) and do not correlate between gene number and genome size. The alternative use of promoters, exons, and transcription-termination sequences during the processing of primary mRNA transcripts further complicates the process. Consequently, a single stretch of DNA sequence may participate in coding for the amino acid sequence of more than one protein. In addition, DNA rearrangements further aggravate attempts at precisely defining what constitutes a gene, and exacerbates the difficulties associated with developing accurate gene counts. Thus, the use of computers in computational biology is highly recommended.

3-D molecular structures: These are obtained by physical measurement (NMR, x-ray, etc) combined with computer modeling.

Genome structures and functions: The genome of an organism is composed of its entire genetic material. Information on the structure and function of a genome is a basic description that has increased exponentially, especially since the onset of genome projects, like the human genome sequencing programs. The data are currently organized into a small number of large public databanks, available through the Internet.

A. APPLICATION OF BIOINFORMATICS IN THE LABORATORY

The use of computers for biology starts in the laboratory; for instance, to plan how a DNA molecule will be cut and tailored with the several hundreds of enzyme reagents available. In order to carry out the relatively simple task of cutting out a precise fragment of a DNA piece, it is necessary to find one or two enzymes that cut somewhere near the ends of the desired piece, but will not cut the fragment itself. One such enzyme may cut a piece of DNA into a few or into several hundred fragments, depending on the sequence of the DNA piece. A computer can enumerate all the possible fragments that can be obtained, and suggest enzyme combinations, and protocols for the experiment. A more sophisticated task is the characterization of a gene sequence that is obtained from an experiment. To this end, the biologist performs a database search on several of the publicly-accessible and frequently-updated databanks.

B. DNA SEQUENCE ANALYSIS PROGRAMS AND DATABASES

In the early 1980s, molecular biologists used the Internet as a mechanism to search nucleotide databases. Using a *file transfer protocol* (FTP), and on-site DNA sequence algorithms such as the University of Wisconsin's *Genetics Computer Group* (GCG) programs, researchers were able to perform sophisticated DNA sequence analyses using UNIX computers located at their own institution. However, early versions of GCG and other programs had two major limitations — difficulty in mastering the necessary on-line command routines that were required to utilize the full potential of these sequence analysis programs; and the fact that the analysis had to performed using data terminals that were directly linked to a large institutional main-frame computer.

However, in just a few years, the graphical interface of the web transformed DNA sequence analyses into a simple point-and-click task that could be performed by any laboratory, office, or home computer connected to the Internet. There are numerous ways to access public-domain DNA sequence databases, but one of the most convenient methods is to use an Internet connection to search *GenBank* with the *Basic Local Alignment Search Tool* (BLAST). GenBank is maintained by the National Center for Biotechnology Information (NCBI), a division of the National Library of Medicine located on the NIH campus in Bethesda, Maryland. GenBank is a member of the International Nucleotide Sequence Database Collaboration, an organization that includes the European Molecular Biology Laboratory (EMBL) at the European Bioinformatics Institute (EBI) at Cambridge, UK, and the DNA DataBank of Japan (DDBJ). The GenBank, EMBL, and DDBJ databases freely exchange DNA sequence files on a daily basis to maintain a comprehensive set of all known sequences. The number of entries in the collective GenBank databases have been doubling every two years.

BLAST is a sequence homology algorithm that uses statistical calculations to identify significant sequence matches between a known sequence provided by the user, called a query, and DNA sequences in the GenBank database. The search parameters for a database query are chosen by the user and ultimately determine the type of matches that will be scored as significant. The BLAST homology search programs were developed by David Lipman and his colleagues at NCBI and can be accessed and executed through the web. Detailed explanations about the default BLAST search parameters, as well as statistical considerations used for determining significant homology values, can be found through the NCBI web-site. The simplest GenBank sequence comparison uses the BLAST program *blastn* to compare a nucleotide query sequence to all the nucleotide sequences in GenBank. Table 6.3 describes five of the BLAST search programs that can be used to compare DNA or protein sequences with any of the available NCBI databases (Table 6.4). The results from BLAST searches are typically used to design follow-up laboratory experiments that are capable of assigning biological relevance to the most significant nucleotide or amino acid homologies.

C. BLAST PROGRAMS AND DOCUMENTATION

NCBI currently supports two versions of BLAST, which are free of charge: BLAST 2.0 and Position-Specific Iterated BLAST (PSI-BLAST). BLAST 2.0 is the standard version of BLAST, which allows a user to search a sequence database with a nucleotide or protein sequence of interest. BLAST 2.0 places gaps into the query and target sequences, so that separate areas of similarity between the two sequences can be returned as one hit. PSI-BLAST is an iterative BLAST search, which is optimized for finding distantly related sequences. Other BLAST programs (Table 6.3) are also available from the NCBI web-page.

With "QBLAST" all the searches submitted to the servers are issued with a request ID (RID). When the button is clicked, a separate status screen will appear which will tell the user the time the search was submitted, and once the search is completed the results will automatically appear in the new screen.

The easiest way to access the BLAST suite of programs is through the NCBI World Wide Web site, at http://www.ncbi.nlm.gov/BLAST/. All versions of BLAST are accessible from this site, and can be used to query all sequence databases available at the NCBI. All documentation, which includes an overview of BLAST, BLAST frequently-asked questions (FAQs), a "What's New" page, the BLAST manual, and a list of references, is also available here.

Table 6.3 *BLAST search programs*

Program	Query	Database sequence	Comments
BLASTP	Amino acid sequence	Protein sequence	Can be run in standard mode or in a more sensitive iterative mode (PSI-BLAST), which uses the previous search results to build a profile for subsequent rounds of similarity searching.
BLASTN	Nucleotide (both strands)	Nucleotide sequence	Parameters optimized for speed, not sensitivity; not intended for finding distantly related coding sequences. Automatically checks complementary strand of query.
BLASTX	Nucleotide (six-frame translation)	Protein sequence	Very useful for preliminary data containing potential frame-shift errors (ESTs, HTGs, and other "single-pass" sequences).
TBLASTN	Protein sequence	Nucleotide sequence (six-frame translation)	Essential for searching protein queries against the EST database. Often useful for finding undocumented open-reading frames or frame-shift errors in the database sequences.
TBLASTX	Nucleotide (six-frame translation)	Nucleotide (six-frame translation)	Should be used only if BLASTN and BLASTX produce no results. Restricted for search against EST, STS, HTGS, GSS, and *Alu* databases.

Cn3D

The stand-alone software for viewing structures in three-dimension (3-D), Cn3D, is a helper application for web browsers that allows the viewing of 3-D structures from NCBI's Entrez retrieval service. Cn3D is a visualization tool for biomolecular structures, sequences and sequence alignments. It helps to correlate structure and sequence information. Cn3D displays structure–structure alignments along with their structure-based sequence alignments, so as to emphasize which regions of a group of related portions are most conserved in structure and sequence.

D. INTEGRATED BIOLOGICAL RESOURCE DATABASES

One of the most exciting molecular genetic applications of the web has been the ability to exploit hypertext file-linking functions, using strategies that create Internet entry points for integrated resource databases. The basic idea of these linked websites has been to enable researchers to place DNA sequence information into a proper biological context. By linking gene names, GenBank sequence files, MedLine abstracts, and investigator-generated research databases into common biological themes using extensively annotated web pages, it is now possible for scientists, clinicians, and interested web visitors to find valuable experimental data through the Internet.

Table 6.4 *Representative NCBI databases that can be queried with BLAST programs using the web interface*

Type	Name	Description
Nucleotide	nr	All non-redundant GenBank +EMBL + DDBJ +PDB sequences (no ESTs)
	month	All new or revised nr sequences released in the last 30 days
	dbest	All non-redundant GenBank + EMBL + DDBJ EST nucleotide sequences
	dbsts	All non-redundant GenBank + EMBL + DDBJ STS sequences
	htgs	High-throughput genomic sequences (direct nucleotide sequences)
	epd	Eukaryotic promoter database
Protein	Swissprot	Recent major release of the SWISS-PROT protein sequence database
	pdb	Sequences from protein structures in Brookhaven Protein Data Bank
	kabat	Kabat's database of sequences of immunological interest
	yeast	*S. cervesiae* protein sequences

For example, the Genome Net project in Japan is an integrated research database resource, which is operated jointly by the Kyoto University and the University of Tokyo. *GenomeNet* is a basic science resource consisting of database retrieval tools and hypertext-linked databases that have been organized so that a researcher can better exploit database information. GenomeNet represents one of several new types of molecular genetic research tools that utilize complex search algorithms to "mine" existing nucleic acid and protein databases for potential novel relationships of biological relevance.

E. BIONET NEWS GROUPS FOR MOLECULAR GENETIC RESEARCH

So far, we have only described the ways that molecular genetic researchers use the Internet to obtain information. The Internet is also an important means of communication between individuals (E-mail) and between groups of users with a common interest. Internet newsgroups function as electronic bulletin boards, and have been around since the inception of the Internet network. Newsgroups were originally created by computer scientists as an efficient mechanism to exchange technical information. The participants in Internet newsgroups can read current announcements, post questions, answer questions, or search the archives for specific topics. The newsgroup postings are organized as threads of related sub-topics, which permit users to carry on an electronic conversation in the context of the entire forum. The benefit of participating in an Internet newsgroup is that it provides a medium for global communication on related topics. However, the usefulness of a newsgroup to its members depends completely on the quality and diversity of the postings, and so some newsgroups use monitoring protocols to maintain proper "netiquette" behavior.

Newsgroups specifically dedicated to biological research are administered by an organization called BIOSCI, presently located at Stanford University. BIOSCI was initiated through grants by the Department of Energy and the National Science Foundation, but since 1996, it has depended on web advertising revenue and company sponsorship to cover overhead expenses. The BIOSCI home page provides a convenient entry point for newsgroup users to begin participating in the fora of their choice. One of the most popular BIOSCI newsgroups, called "bionet.molbio.-methods–reagnts," attracts more than 1000 molecular biological research postings a month. Table 6.5 lists some of the other topics that are represented by the BIOSCI newsgroups on the web.

Table 6.5 *Representative BIOSCI newsgroups that can be accessed through the BIOSCI WW home page*

Newsgroup	Topic
Amyloid	Forum for Alzheimer's disease disorders including prion diseases
Arabidopsis	Newsgroup for the *Arabidopsis* Genome Project
Bioforum	Discussions about biological topics lacking dedicated news group
Bionews	General announcements of widespread interest to biologists
Bio-Matrix	Applications of computers to biological databases
Biotechniques	Discussions of articles in the journal *Biotechniques*
Cell-Biology	Discussions about cell biology and cancer research at the cellular level
EMBL-Databank	Messages to and from the EMBL database staff
GenBank-Bb	Messages to and from the GenBank database staff
Genstructure	Genome and chromatin structure and function
Protein-Analysis	Discussions about research on proteins and SWISS-PROT databank

F. PATTERN RECOGNITION IN DNA AND PROTEIN SEQUENCES

The large amount of information derived from genome sequencing and protein sequencing needs computer software technology for storage, retrieval, analysis of data and determination of structural and functional properties. Useful information for a novel DNA or protein sequence can be derived from databanks. For example, the lucine-rich lucine zippers and the zinc-finger family of proteins is expected to bind DNA to DNA and to probably regulate transcription, while a member of the protein kinase family is expected to phosphorylate certain amino acids in certain proteins, including itself in some cases. A newly-found DNA sequence may well be similar to one or more already known sequences and such similarities are likely to provide strong clues as to the likely identity and function of the sequence in question. Searching for sequence similarities requires the use of computers, sequence databases and appropriate algorithms.

1. Sequence Analysis

Sequence data is the most abundant type of biological data that is available electronically. Sequence analysis techniques can be applied to DNA and RNA (nucleotide) sequences or to protein (amino acid) sequences. Pair-wise sequence comparison is the most essential technique in computational biology. It allows us to do everything from sequence-based database searching, to building evolutionary trees and identifying characteristic features of protein families, to creating homology models. Besides, it is also the key to larger projects including comparing genomes, exploring the sequence determinants of protein structure, connecting expression data to genomic information, and much more. The types of analysis that can be done with sequence data are: (i) Knowledge-based single sequence analysis for sequence characteristics, (ii) Pair-wise sequence comparison and sequence-based searching. (iii) Multiple-sequence alignment. (iv) Sequence motif discovery in multiple alignments. (v) Phylogenetic inference.

To understand how DNA and protein sequences are informative, we should know something about the chemistry of DNA and proteins. The readers are recommended to read some standard textbooks and some of the later chapters in this book for information on the chemistry and techniques used for the sequence analysis of nucleic acids and proteins. Many molecular biology research projects involve the cloning and sequencing of a gene or other DNA sequences of interest. The task of collecting, assembling, and correcting the sequence data are generally performed with the help of computer programs; software packages are able to assemble, correct the sequence, and analyze the sequence in greater details. The handling of this task by the program varies greatly and there is no single ideal package.

(a) Sequence reading

The task of sequence reading depends on the sequencing protocols used. In many laboratories, the sequence is generated on an autoradiograph, from which the sequence is read. Automated gel readers are now available, which scan an autoradiograph and read the sequence directly into a computer file. These devices correct the problems in the gels, enhance the quality of the autoradiograph, and produce a sequence reading with the difficult regions highlighted. These data can be transferred to the computer using a communications link.

Once a large chunk of DNA has been mapped and sequenced, the task of understanding its function begins. Many programs can be used to explore the sequence of genes and other biologically important features. A feature is a sequence pattern with some functional significance, such as ORFs, start-and-stop codons, splice sites, exon/intron number and boundaries, signal sequences, sequence that are bound by proteins (regulatory sequences), restriction sites and coding and non-coding sequences. *Genefinders* are programs that identify all the open reading frames in unannotated DNA. They use a variety of approaches to locate the genes, but the most successful ones combine both content-based and pattern-recognition approaches. The GRAIL family of programs, developed at Oak Ridge National Laboratories, uses a neural network to combine evidence from seven different statistical measures of DNA content; the subsequent versions measure additional features to better exploit these different types of data. Some other commonly-used programs in gene finding are GENSCAN, PROCRUSTES and Gene-Wise.

(b) Sequence alignment

The comparison of protein and DNA sequences is one of the foundations of bioinformatics. Sequence alignment attempts to align two or more sequences, such that regions of structural or functional similarity between the molecules are highlighted by asking whether our unknown sequence is in any way similar to known sequences and, ideally, to sequences of known function or structure. We can identify the unknown sequence and predict its structure and function. There are various types of software available on the web to perform a wide range of different types of sequence alignment. Sequence alignment is a non-trivial computational task, and the choice of the program and program parameters have a significant impact on the sensitivity of the outcome. In describing sequence comparisons, several different terms are commonly used. Sequence identity, sequence similarity, and sequence homology are the most important of these terms. What really matters when evaluating a sequence alignment is whether a given alignment is random or meaningful. If the alignment is meaningful, then we have to assess how meaningful it is, which can be done by constructing a scoring matrix.

(i) Global and local sequence alignment: The alignment of two sequences could reflect a common evolutionary origin or could try to represent common structural roles, which might not always be congruent with evolutionary history. Alignments are generally restricted to describing the most common mutations: insertions, deletions and single-residue substitutions. Insertions or deletions are represented by null characters while substitutions are represented by the alignment of two different letters. The global sequence alignment necessitates the complete alignment of the input sequences, whereas the local alignment aligns only their most similar segments. The method adopted depends upon whether the sequences are presumed to be related over their entire lengths or to share only isolated regions of homology.

(ii) Pair-wise sequence alignment: There are two well-known algorithms for calculating the optimal alignment between two sequences. The Needleman–Wunsch algorithm calculates a global alignment between two sequences – the best alignment including all of the shortest sequences. The Smith–Waterman algorithm returns the best local alignments between two sequences. ALIGN is a simple utility for comparing global alignments and is part of the FASTA software distribution. To run ALIGN and any of the other FASTA programs, the sequence data has to be in FASTA format. This format is very flexible, and it is one of the most commonly used formats for sequence analysis programs. So far, the most popular tool for searching sequence database is a program called BLAST. BLAST is the algorithm at the core of most online sequence-search servers. It performs pair-wise comparisons of sequences, seeking regions of local similarity, rather than optimal global alignments between whole sequences. Another method for local sequence alignment is the FASTA algorithm.

(iii) Multiple sequence alignment: One of the goals of biology has been the creation of taxonomy for living things, a method of organizing species in terms of their relationships to one another. Multiple sequence alignment techniques are most commonly applied to protein sequences; ideally, they are statements of both evolutionary and structural similarity among the proteins encoded by each sequence in the alignment. We know that proteins with closely related functions are similar in both sequence and structure from organism to organism, and that sequence tends to change more rapidly than structure in the course of evolution. In multiple alignments, generated from sequence data alone, regions that are similar in sequence are usually found to be superimposable in structure as well. ClustalW is a very good program that also happens to be free, well supported, capable of dealing with large sequence alignments and available for MacIntosh, Windows and various Unix systems. There is also the graphical user interface–ClustalX.

(iv) Phylogenetic analysis: Phylogenetic inference is the process of developing hypothesis about the evolutionary relatedness of organisms, based on their observable characteristics. Traditionally, phylogenetic analyses have been based on the *gross anatomy of species*. Sequence-based phylogenies are quantitative and can provide valuable, scientifically valid evidence to support theories of evolutionary history, when they are based on sufficient amounts of data. One of the earliest means to understand algorithms for tree drawing is the *pair-wise distance method*. This method produces a rooted tree. The algorithm is initialized by defining a matrix of distances between each of sequences in the input set. The sequences are then clustered according to distance; in effect, building the tree from the branches down to the roots. Phylogenetic trees can also be built based on *neighbor joining*, which is another distance matrix method. It eliminates any possible error that can occur when the UPGMA method is used. UPGMA produces trees in which the branches that are closest together by absolute distance are placed as neighbors in the tree. This assumption places a restriction on the topology of the tree, which can lead to incorrect tree construction under some conditions.

Phylogenetic trees based on *maximum parsimony*: A more widely-used algorithm for tree drawing is called parsimony. Parsimony is related to Occam's Razor, a principle formulated by the medieval philosopher William of Ockham that states that the simplest explanation is probably the correct one. Parsimony searches among the sets of possible trees to find the one requiring the least number of nucleic acid or amino acid substitutions to explain the observed differences between sequences. The *maximum likelihood* methods also evaluate every possible tree topology, given a starting set of sequences. These methods are probabilistic; that is, they search for the optimal choice by assigning probabilities to every possible evolutionary change at informative sites, and by maximizing the total probability of the tree. The maximum likelihood methods use information about amino acid or nucleotide substitution rates, analogous to the substitution matrices that are used in multiple sequence alignment.

There is a variety of phylogenetic analysis software available for many operating systems. One of the most extensively used is maintained by Dr Joe Felsenstein, the author of the PHYLIP package, and is accessible from the PHYLIP web-page (http://evolution.genetics.washington.edu/phylip.html).

(v) Coding region identification: The coding region is defined as the region that is translated by a ribosome into a polypeptide. It is delimited by sequences that function in the initiation and termination of transcription and translation. In prokaryotes, the coding region is a single open reading frame (ORF), or continuous stretch of sequence that contains no termination codons. Eukaryotic genes, however, are commonly organized as exons and introns and, hence, may comprise of several disjointed ORFs. The initiation codon, usually an AUG, signals the start of translation, while a termination codon in the same reading frame marks the end of the translated region. Any sequence between an initiation and termination codon is thus an ORF, and potentially encodes protein. As a result of the use of the triplet code for translation, any particular DNA sequence has six potential reading frames that can encode proteins: three on one strand of the DNA proceeding in the 5'→3' direction, and another three, also proceeding in the 5'→3' direction on the complementary strand. Thus, there are six possible readings in which to look for the ORFs that encode the protein of interest.

V VISUALIZING PROTEIN STRUCTURE AND COMPUTING STRUCTURAL PROPERTIES

The study of the 3-D structures of proteins gives more information than primary sequence analysis. The visualization of structure and measurement of structural properties are important tools for molecular and structural biologists. Developing the 3-D structure of a protein and analyzing its shape in detail can reveal the location of catalytic sites and interaction sites. It can also help to identify targets for the site-directed mutagenesis studies that are so often used to arrive at a detailed characterization of a protein's functional chemistry. There are many specialized analysis programs in protein structure literature. Software have been used for analyzing and modeling protein structure, for such purposes as visualization and plotting, geometric and surface property analysis, classification, analysis of intermolecular interactions and solvent interactions, and computation of some physico-chemical properties. For all-purpose molecular structure modeling, the easiest-to-use tools are still commercial packages such as MSI's Quanta and Insight, Tripos' SYBYL, and others.

To work with protein sequence and structure, a working knowledge of protein chemistry is essential. Proteins often perform their functions using standard organic reaction mechanisms, mediated by amino acids and small organic molecules (cofactors) that bind to the protein, or by metal ions. To understand how the protein structure might catalyze a reaction, it is essential to know enough about organic reaction mechanisms to develop a hypothesis about how the reaction might work, given the shape of the protein and the location of various amino acids. Protein association is often mediated by the electrostatic properties of the protein structure; inter-atomic interactions and molecular shapes are the basis of the specificity of intermolecular interactions. Therefore, the visualization of the protein structure and the computation of structural functions are important in understanding molecular specificity in practical applications of biochemistry, in designing small-molecule or peptide drugs, understanding the molecular basis of disease and immunity, or delving into the specific molecules concerned in sending molecular signals between the cells and through the body.

A. WEB-BASED PROTEIN STRUCTURE TOOLS

The most important source of information about protein structure is the protein databank (PDB). In addition to being an entry point to structural data itself, the PDB website (http://www.rcsb.org/pdb) contains links to many tools database for applying them to individual protein structures as the database search continues. Information from the database is made available through the Protein Structure Explorer interface. Each protein molecular structure can be viewed using 3-D display tools, such as RasMol and the Java QuickPDB viewer. PDB files and file headers can be viewed in HTML format and downloaded in a variety of formats. The links to the protein structure classification databases CATH, FSSP and SCOP, are provided, along with the tools CE and VAST, which search for structures based on structural alignment. The average geometric properties, including dihedral angles, bond angles, and bond lengths, can be displayed in a tabular format with extremes and the deviations noted. The sequences can be viewed and labeled according to the secondary structure and the sequence information downloaded in the FASTA format. We can also go directly to the page for a particular protein of interest by entering that protein's four-letter PDB code in the Explore box on the PDB's main page. From the individual protein page generated by the Structure Explorer, the PDB provides a menu of links through which to connect to other tools.

B. PROTEIN ANALYSIS

For analyzing a protein sequence, one is required to search the protein database for proteins that have similar sequences (Text box 6.1). This is the best way to learn about the structure and function of a new sequence. A similar sequence may imply homology (common ancestry) and, therefore, may imply a common function. If the sequence is very similar (>50% matching residues) to a protein whose three-dimensional structure is known, then one can try to build a model for the new sequence based on the coordinates of the known 3-D structure. Once two sequences have been aligned, we can look for similarity among 3-D structures. This can be obtained by comparing the patterns of hydrophobicities along the two sequences. If the two proteins have folded in similar way, it means that the corresponding residues of the two proteins have the same attraction or repulsion to water.

Text box 6.1: BASIC BLAST SEARCH PROCEDURE

The use of the **BLAST** (Basic Local Alignment Search Tool) is to identify an unknown sequence. It has manty advantages: (1) through BLAST we can access the BLAST server via the NCBI (National Center for Biotechnology Information) at NIH (National Institute of Health); (2) the correct BLAST search tool can be chosen to compare a known sequence to sequences in the databases at NCBI; (3) probable sequence can be identified from the sequence data provided.

The NCBI maintains many DNA and protein sequence databases, which are submitted by scientists. NCBI makes these data available to the public over the Internet. Each sequence in a database has an accession number, information relevant to the sequence, and the sequence itself. The computer program "BLAST" has been developed by Altschul et al. (1991) and Schuler *et al.* (1994). The blastn program is used to align DNA sequence of interest and the blastp program searches the databases for amino acid sequences (the sequence of interest is called the ***query*** sequence). The blastx program compares a nucleotide query sequence, translated in all six reading frames against a protein sequence database.

Procedure

Assume you have sequenced a fragment of DNA and used a computer program to translate that DNA sequence into a sequence of amino acids. The blast search query DNA sequence or the blast search query protein sequence can be used.

Continues...

Continued...

Search with blastn

1. Go to the NCBI website at <http://www.ncbi.nlm.nih.gov/> and select "BLAST" from the menu line at the top of the screen.
2. Select "Standard nucleotide–nucleotide BLAST [blastn]" under the "Nucleotide BLAST" option and a new screen will appear.
3. Type or paste your query DNA sequence into the big "Search" box. Use the default settings (i.e., you want "nr" in the "Choose database" box, and further down the page, under the "Options for advanced blasting" section, the "low complexity" box should be checked for "Choosing filters." Click on the "BLAST!" button to start the search.
4. A new screen will appear that states "your request has been successfully submitted and put into the Blast Queue." Below the line that reads "The request ID is XXX-XXX" click on the "Format!" button. The "BLAST Search Results" screen will appear next. When the search is completed, the results will be displayed.
5. Scroll down to "Distribution of # Blast Hits on the Query Sequence." This shows a graphical view of all the sequences that are similar to the query sequence. They are color-coded bars, with red indicating sequences that are very similar and black indicating sequences that are less similar. By placing the mouse over a colored bar, the name of the sequence can be bought to display.
6. Scroll down past the box with the colored bars to a list of the "Sequences producing significant alignments." On the left side of the list are links, clicking on which will take you to the sequence file in the database (gb = GenBank, emb = European Molecular Biology, etc.) and the accession number of the sequence. In the sequence file, the name of the gene, the source of the DNA, the title of relevant journal articles, the translated protein sequence, the DNA sequence, and much more information can be seen.
7. On the right-hand side of the list, there is a "Score" and an "E value" for each sequence. The score indicates the degree of similarity between that sequence and the query sequence; the higher the score, the better the match. The lower the E values, the higher the probability that the match between the two sequences is not due to random chance. The E value for good matches is usually expressed as a negative exponent (e.g., e^{-134}). An E value of 0 indicates identity or near-identity of the database sequence with the query sequence. Short sequences of identical matches will give higher E values than longer sequences of identical matches because there is a greater probability of shorter sequences matching by chance than longer sequences.
8. Below the list of similar sequences, the alignment of each of the database sequences to the query sequence itself can be found. Below the name of each sequence, the similarity scores and E values are presented, as well as the percentage of identity between the two DNA sequences.

Search with blastp

Follow the same steps of the previously explained procedure using the amino acid sequence as the query sequence to search the databases for similar amino acid sequences.

C. SECONDARY STRUCTURE PREDICTION

The amino acid sequence of a protein (primary structure) contains interesting information in and of itself. One protein sequence can be compared and contrasted with the sequence of other proteins to establish its relationship, if any, to known protein families, and to provide information about the evolution of biochemical function. Several methods have been developed to predict the positions of α-helices, β-strands and turns from the amino acid sequences based on physical, chemical and structural properties. They are: (i) observing physical properties of amino acids, (ii) finding probabilities of amino acid distributions as determined by x-ray crystallography, (iii) on-sequence patterns typical of the various types of secondary structure segments, (iv) neutral nets, and (v) artificial intelligence approaches. The accuracy of the prediction can be enhanced by using an alignment of several homologous sequences and a combination of different methods. In some cases, when the protein is known to contain exclusively α-helices or β-strands, the prediction is more accurate for these secondary elements.

D. HYDROPHOBICITY PATTERNS

An important property of amino acid chains is their hydrophobicity. In the hydrophobicity scale, positive numbers are assigned to amino acids with apolar (hydrophobic) side chains and negative numbers to those with polar (hydrophilic) side chains. In membrane proteins, the side chains in contact with lipids are hydrophobic. The α-helix is an important component of integral membrane proteins. The portion of the protein that is embedded in the membrane is α-helical because the α-helix provides maximum number of hydrogen bonds which, in turn, reduces the hydrophilic nature of peptide linkages.

E. DETECTION OF MOTIFS

It is possible to detect motifs in a new protein based on those previously identified in other proteins. The presence of such motifs can indicate a similar function or a similar 3-D structure. PROSITE, a pattern database, contains more than 338 patterns. A pattern is a way to describe which string of the amino acid code is sufficient to identity the motif; these patterns recognize functional motifs. The programs available match the pattern against a protein sequence and decide whether it contains the motif represented by the pattern.

F. POST-TRANSLATIONAL MODIFICATION SITES

There are many possibilities for post-translational modifications to proteins since there are plenty of reactive hydroxyl, sulphahydral, carboxy, immidazole, guanidinium and tyrosol groups present in amino acid side chains. However, there are some frequently observed modifications whose sequence specificity is known: glycosylation; disulphide bonds, proteolytic cleavage, phosphorylation, etc.

G. SUPER-SECONDARY STRUCTURES

The sequence segments have been identified in a variety of proteins that are believed to fulfill the same function and probably have the same 3-D structure. These motifs may occur in proteins that have no apparent similarity in the rest of their sequences: calcium-binding proteins (EF hand), helix-turn-helix; zinc-finger, etc.

H. PROTEOMICS

The advent of large-scale DNA sequencing has transformed biological research in a short span of time. With the discovery of most human and other organisms' genes, it is now apparent that a 'factory approach' to address biological problems is desirable if we are to achieve a comprehensive understanding of complex biological processes. In this section, the crucial contribution of proteomics to our understanding of biology and medicine, through the global analysis of gene products, is discussed.

The term "*proteomics* or *proteome*" refers to the total protein complement expressed by the genome, and generally means the high-throughput systematic separation, identification and characterization of proteins (the term "proteomics" was first coined in 1995). Proteomics attempts to analyze all the proteins in a cell from the point of view of their individual functions and understand how the interaction of specific proteins with other cellular components (e.g., nucleic acids, membranes, organelles, etc.) affects the function of those proteins. The concept of proteomics is fundamentally different from genomics – the genome of a cell is essentially static and well defined in all the cells of an organism, the proteome, however, continually changes in response to external and internal conditions. For example, *E. coli* expresses different proteins and, thus, has a different proteome depending on the type of media it grows. Similarly, during multicellular mammalian development, specialized cells express different proteins, but with characteristic proteomes, representing diverse tissues. Based on the expression of proteins in specialized cell of a tissue, various proteomic projects have been developed (Table 6.6).

Table 6.6 *Major proteomic research projects*

Project type	Nature of investigation
Characterization of sample protein	Alignment of amino acid sequence of newly discovered protein with known proteins. Comparisons of a nucleotide sequence of a gene with known genes. Translation of a nucleotide sequence into an amino acids sequence.
Estimation of protein expression levels	Estimate relative amounts of specific proteins in various cells/tissues types of an organism.
Protein interactions	Reveal different macromolecular complexes in which specific proteins interact.
Predict post-translational modifications	From amino acid sequence data, identify putative candidate residues for phosphorylation and potential sites for linked oligonucleotides, etc.

Continues...

Continued...

Identify the processed proteins	Compare length of protein with known, larger molecular weight precursors.
Establish phylogenetic relationships	Make comparative analysis for protein sequences among diverse organisms.
Understand the 3-D structure of proteins (i.e. protein folding)	Configure three-dimensional structural information of proteins with whole genome sequences to deduce the function of proteins and their regulation. Identify novel protein folding motifs in proteins from unusual microorganisms (e.g. thermophilic, acidophilic bacteria).
Protein trafficking	Find out the cellular trafficking of proteins, particularly regions such as ER, Vesicular and membranes.
Disease treatment	Elucidate the specific protein interactions that establish onset of disease states.

As pointed out above, a major tool for proteomics is in silico-biology, making comprehensive use of the data derived from the genome-sequencing efforts. However, the information which can be deduced from the genome data about their encoded proteins is rather limited. For example, transcript expression profiles usually do not reflect the actual protein expression profiles. Combining PCR and proteome analysis showed that the correlation between changes in mRNA levels and altered protein abundance was only 0.48. There is a strong demand for genome-wide elaboration of protein expression profiles. The limitation of proteomics is the lack of genome-wide experimental approaches for the analysis of protein expression, post-translational modifications, 3-D structure, enzymatic activity, etc. A protein database, in which all available information is integrated, is being constructed for many prokaryotes and some model eukaryotes, including *Arabidopsis*.

A broad definition of proteomics involves many different areas of study. These include protein–protein interaction studies, protein modifications, protein function, and protein localization studies, to name a few. The aim of proteomics is not just to identify all the proteins in a cell, but also to create a complete 3-D map of the cell indicating where the proteins are located. The fulfillment of these goals certainly require contributions from many disciplines, such as molecular biology, biochemistry and bioinformatics.

The growth of proteomics is a direct result of the advances made in large-scale nucleotide sequencing of expressed sequence tags and genomic DNA. Protein identification (by MS or Edman sequencing) relies on the presence of some form of database for the given organism. The majority of DNA and protein sequence information has been accumulated over the last 5–10 years. The genome sequencing of 45 microorganisms has been completed and that of 170 more is underway. At present, the sequencing of 5 eukaryotic genomes have been completed: *Arabidopsis thaliana, Saccharomyces cerevisiae, Schizosaccharomyces pombe, Caenorhabditis elegans* and *Drosophila melanogaster*. In addition, the rice, mouse and human genomes are near completion.

1. Potentiality of Proteomics

With the accumulation of vast amounts of DNA sequences in databases, researchers now realize that merely having the complete sequences of genomes is not sufficient to elucidate the biological function. A cell is normally dependent upon a multitude of metabolic and regulatory pathways for its survival. There is no strict linear relationship between genes and the protein complement or the 'proteome' of a cell. Proteomics is complementary to genomics because it focuses on the gene products, which are the active agents in the cells. Thus, proteomics directly contributes to drug development, as almost all drugs are directed against proteins. The existence of an open reading frame (ORF) in genomic data does not necessarily imply the existence of a functional gene. Despite the advances in bioinformatics, it is still difficult to predict genes accurately from genomic data. Although the sequencing of related organisms will ease the problem of gene prediction through comparative genomics, the success rate for the correct prediction of the primary structure is still low. This is particularly true in the case of small genes or genes with little or no homology to other known genes. Therefore, the verification of a gene product by proteomics methods is an important first step in 'annotating the genome'. Any modifications of the proteins that are not apparent from the DNA sequence, such as isoforms and post-transcriptional and post-translational levels, make proteomics distinct and a valuable tool.

2. Types of Proteomics (Figure 6.4)

(1) Protein-expression proteomics: The quantitative study of protein expression between samples that differ by some variable is known as expression proteomics. In this approach, the protein expression of the entire proteome, or of sub-proteomes, between samples can be compared. Information from this approach could identify novel proteins in signal transduction or identify disease-specific proteins.

(2) Structural proteomics: The goal of the study of proteomics is to map the structure of protein complexes or the proteins present in a specific cellular organelle – this is known as *"cell mapping"* or *structural proteomics*. Structural proteomics attempts to identify all the proteins within a protein complex or organelle, and locate and characterize all protein–protein interactions. The isolation of specific sub-cellular organelles or protein complexes by purification can greatly simplify the proteomic analysis. This information will help piece together the overall architecture of cells and explain how the expression of certain proteins gives a cell its unique characteristics.

(3) Functional proteomics: "Functional proteomics" is a broad term for many specific, directed proteomics approaches. In some cases, specific sub-proteomes are isolated by affinity chromatography for further analysis. This could include the isolation of protein complexes or the use of protein ligands to isolate specific types of proteins. It allows us to study a selective group of proteins and characterize them, and also provides important information about protein signaling, disease mechanisms or protein–drug interactions.

3. Basic technology for proteomics

The growth of proteomics has been based on the advances made in protein technologies. Many new technologies have emerged and the old ones have been improved in particular areas; for example, protein separation to protein identification. However, there is a vital need for the

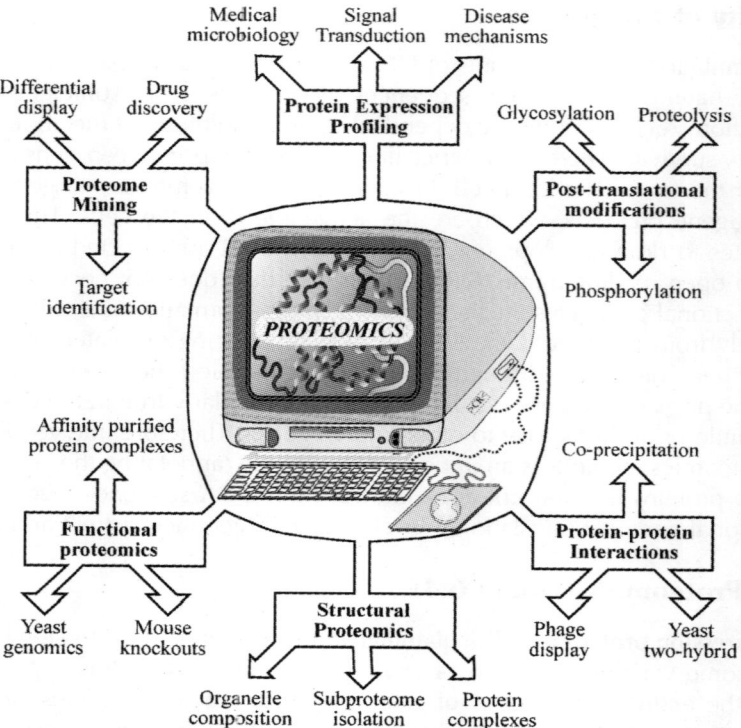

Figure 6.4 *Types of proteomics and their applications to biological sciences (Redrawn from Graves and Haystead, 2002)*

development of more sophisticated and new technologies for proteomics to reach its full potential. A typical proteomics experiment can be broken down into the following categories: (i) the separation and isolation of proteins from a cell line, tissue, or organism, (ii) the acquisition of protein structural information for the purposes of protein identification and characterization, and (iii) database utilization. Currently, the prevailing operational definition for proteomics is the combined use of mass spectrometry (MS) with two-dimensional polyacrylamide gel electrophoresis (2D-PAGE) analysis to study gene expression at the protein level. At present, 2D-PAGE is still the most comprehensive, quantitative, high-resolution method for displaying proteins, while MS is the most sensitive method available to identify proteins (see Chapters 23–26).

4. Applications of proteomics technology

Proteomics has unique and significant advantages as an important complement to the genomics approach, and the applications for this technology are readily apparent. However, many types of information cannot be obtained from the study of genes alone. For example, proteins, not genes, are responsible for the phenotypes of cells. It is impossible to elucidate the mechanisms of disease, aging, and the effects of the environment solely by studying the genome. Only through the study of proteins can protein modifications be characterized and the targets of drugs identified. For convenience, the applications of proteomics can be discussed under the following subheadings.

(a) Annotation of the genome: One of the applications of proteomics will be to identify the total number of genes in a given genome. This *"functional annotation"* of a genome is necessary because it is still difficult to accurately predict genes from genomic data. The exon–intron structure of most genes cannot be accurately predicted by bioinformatics. To achieve this goal, genomic information will have to be integrated with data obtained from protein studies in order to encompass the existence of a particular gene.

(b) Protein expression mapping: The analysis of mRNA expression by various methods has become increasingly popular. These methods include the serial analysis of gene expression (SAGE) and DNA microarray technology. However, the analysis of mRNA is not a direct reflection of the protein content in the cell. Consequently, many studies have now shown a poor correlation between mRNA and protein-expression levels. The formation of mRNA is only the first step in series of events resulting in the synthesis of a protein (Figure 6.5). During mRNA processing, polyadenylation, and mRNA editing, many different protein isoforms can be generated from a

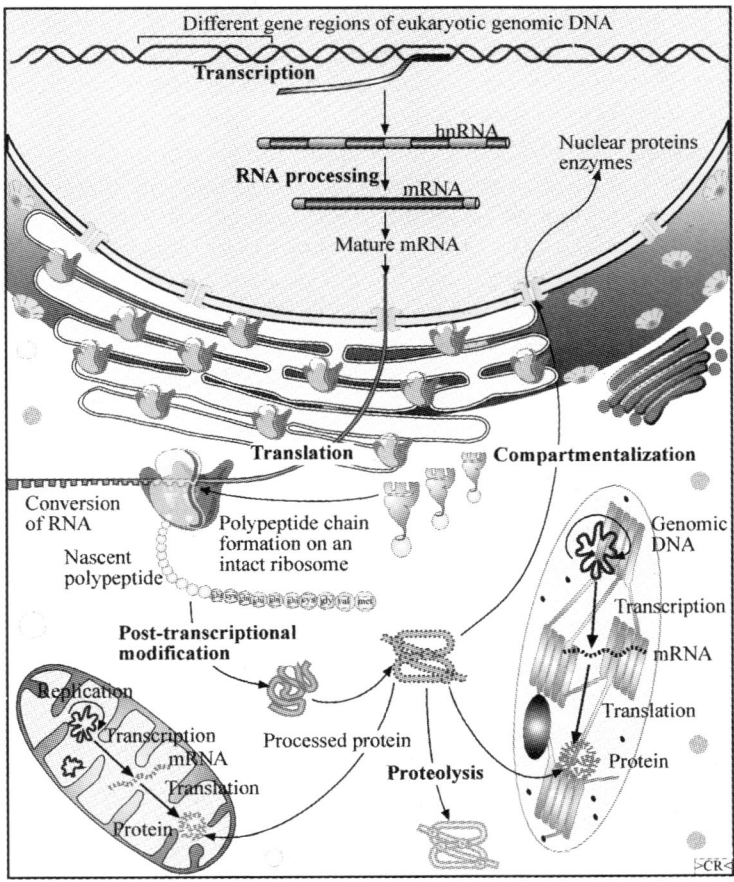

Figure 6.5 *Schematic representation of genetic information copying from DNA to RNA, processing of RNA, translation of RNA into protein, processing of nascent polypeptide to functional protein, compartmentalization and degradation in the cell*

single gene. Proteins, having been formed, are subject to post-translational modification. It has been estimated that up to 200 different types of post-translational protein modifications exist. The average number of protein forms per gene was predicted to be one or two in bacteria, three in yeast, and three or more in humans. Therefore, it is clear that the tenet of "one gene, one protein" is an oversimplification.

Protein-expression mapping is a quantitative study of global changes in protein expression in tissues, cells or body fluids, using 2-D gels and image analysis. This method has the advantage of direct determination of protein abundance and the detection of post-translational modification, such as glycosylation or phosphorylation, which result in a shift in mobility; mass spectrometry may be used for the subsequent characterization of the protein of interest. Due to the fact that thousands of proteins are imaged in one experiment, a picture of the protein profile of the sample at a given point in time is obtained, facilitating comparative proteome analysis.

(c) Protein function: The sequence data suggests that no function has been assigned for about one-third of the sequences in organisms for which the genomes have been sequenced. The complete identification of all proteins in a genome will aid the field of structural genomics, in which the ultimate goal is to obtain 3-D structures for all proteins in a proteome. This is essential because the functions of many proteins can only be inferred by examination of their 3-D structure, when it exists in its native form.

(d) Protein modifications: One of the important applications of proteomics will be the characterization of post-translational protein modifications. Proteins get modified post-translationally in response to a variety of intracellular and extracellular signals. For example, protein phosphorylation is an important signaling mechanism and the disregulation of protein kinases or phosphatases can result in oncogenesis. By using a proteomics approach, any changes in the modifications of the many proteins expressed by a cell can be analyzed simultaneously.

(e) Protein localization and compartmentalization: Protein localization is very important for normal cell function. Proteomics aims to identify the sub-cellular location of each protein. This information can be used to create a 3-D protein map of the cell, providing novel information about protein regulation.

(f) Protein-complex identification: A great deal can be inferred about a protein's function by studying the proteins with which it associates, its cellular location and any changes in these parameters introduced by stimuli. The processes of cell growth, programmed cell death, and the decision to proceed through the cell cycle, are all regulated by signal transduction through protein complexes. The direct measurement of protein–protein interactions by the purification of protein complexes and MS-identification has recently proved to be very valuable in discovering novel signaling proteins and in elucidating the components of the spliceosome complex. The systematic use of gene tagging with peptide sequences that enable affinity purification, sub-cellular fractionation, affinity purification and MS-identification could deliver a map of many protein complexes and their cellular locations. This would effectively be a physical map of the cell that would increase in depth over time and could incorporate different cell types and conditions; it could even be oriented towards proteins of interest.

5. The future of proteomics

There is a growing realization of the importance of proteomics in the life sciences industry. This is evidenced by the increasing investment in this area by pharmaceutical and genomics companies and the proliferation of numerous companies dedicated to proteomics technology. This increased attention and investment should fuel the continued rapid advance of proteomics technology. The improvement of sample solubilization, sub-cellular fractionation, protein detection and the commercialization of narrower range pH gradients, should improve the sensitivity, resolution and representation of proteins using 2-D gels. MS technology has also been continually improving in sensitivity and ease of use. It is anticipated that the next generation of this technology will see another order of magnitude increase in sensitivity.

VI INTERNET-BASED MULTIMEDIA IN BIOTECHNOLOGY

Internet-based multimedia has the potential to make effective communication of scientific information more insightful and long enduring; e.g., the structure–function relationships in biological macromolecules. Multimedia plays a vital role, especially since the software and hardware needed for creation and viewing of multimedia content for the Internet are widely available and affordable. Multimedia formats can deliver more information to a wider audience at lower cost. Thus, an increasing number of life scientists can and do use multimedia materials both in the laboratory and at home.

A. WHAT IS MULTIMEDIA?

The term multimedia implies a combination of visual and audio components and can be used to communicate in more than one way. The visual component can be text, bitmapped images, digital video (e.g. Quick Time™ or Real video™ or Dynamic HTML), or some other visual components (e.g., those produced by a graphics plug-in, Java™ applet or Virtual Reality Modeling Language, VRML, 3-D immersive view). The audio components typically consist of either soundbites downloaded as playable files or a narration continuously streamed to the user.

In multimedia, if the end user who views a multimedia application is allowed to control what and when the elements are delivered; it is called *"interactive multimedia"*. If the end user is provided with a structure of linked elements through which he can navigate, it is called *"Hypermedia"*. Multimedia projects may or may not be interactive; the users can sit back and watch, just as if they are watching a movie or television. Multimedia projects can be *"non-linear"* also, in which case the users are given navigational control to scan through the content at will and can result in interactive multimedia. This is called **"scripting"** or **"story-boarding"**; i.e., the message one wants to convey, describing the parameters of the project, the artwork, and the programming.

Multimedia elements are put together into a project using software tools called *"authoring tools"*; these software tools are designed to manage individual multimedia elements and provide user interaction. In addition to providing a method for the user to interact with the project, most authoring tools also offer facilities for creating and editing text and images, and have extensions to derive video disc players, videotape players, and other relevant hardware peripherals. Sounds and movies are usually created with editing tools dedicated to these media, and then the elements are imported into the authoring system for playback. The sum of what gets played back and how it is presented to the viewer is called the *"human interface"*. This interface is just as much the rules for what happens to the user input as it is the actual graphics on the screen.

What makes multimedia so special? The ever-increasing amount of scientific data and lecture material that is appearing online clearly reveals the importance of the Internet for the delivery of information (Table 6.7). Also, placing scientific information onto a website that allows anyone with access to a web browser to download this material and browse through it is compelling. Although, the assimilation of a text read from a computer screen is significantly less efficient than the assimilation of text read from a hard copy, we can benefit from watching and listening to someone presenting a seminar, rather than just reading the text of the presentation. Thus, much of the thrust towards the incorporation of multimedia components into text-based web content has been targeted at improving content assimilation and creating an efficient distance-learning paradigm. In scientific communities, the aim is to produce multimedia presentations that retain user interest by means of audio-visual stimuli or interactivity. The most manifest form of this technique is on-line digital video. This may be an animated depiction of the cell cycle or an animated molecular graphics presentation, but is more commonly limited to 'talking heads'; e.g., a single photograph of someone giving a lecture.

Table 6.7 *Examples of multimedia-based internet delivery in the life sciences*

Source	URL	Description
University of Pennsylvania Oncolink	http://goldwein1.xrt.upenn.edu/tour1-144.ram	A cancer information site that employs multimedia components
Multimedia Medical Reference Library	http://www.med-library.com/medlibrary/	A variety of multimedia-based biomedical resources
Multimedia index	http://www.a.utokyo.ac.jp/dennocyte/data/data-e/intro.html	Graduate School of Agricultural and Life Sciences, Faculty of Agriculture at the University of Tokyo
Virtual Frog Dissection Kit	http://www.itg.lbl.gov/ITG.hm.pg.docs/dissect/dissect.html	Part of the 'whole frog' project from Lawrence Berkeley National Laboratory
University of California at Berkeley Online classroom	http://www.cmil.unex.berkeley.edu/online/tat2/html/octop.html	Online learning facility that includes bioscience material
Guidelines for Multimedia Courseware Developers in Higher Education	http://ibis.nott.ac.uk/guidelines	Guidelines for producing multimedia teaching content from the University of Nottingham Bioinformatics Research Group
Java simulation of chemically-induced skin cancer in mice	http://www.public.usit.net/wiarda/scientific/cancersim.html	From Dorothea Wiarda's Group
Home page of the Multimedia Educational Repository for Learning and Online Teaching	http://meriot.csuchico.edu/	Resources and links for those interested in the creation of multimedia teaching materials

Continues...

Continued...

Source	URL	Description
Japanese National Institute of Multimedia Education	http://www.nime.ac.jp/index-e.html	Various resources for the creation of online and offline multimedia learning capability
Hemoglobin Lab Java applet	http://www.cdl.edu/HemoglobinLab/hblab.html	California State University for Distributed Learning
NewMedia Centers home page	http://www.csulb.edu/gc/nmc/	A non-profit organization helping institutions of higher education enhance teaching and learning through the use of new media

B. GENERAL APPLICATIONS OF MULTIMEDIA

Multimedia has many applications, five of which are briefly explained below.

1. Multimedia Tele-teaching system: Distance learning is an application of multimedia communication, wherein a professor is able to teach students located in distant sites on the computer network. Multimedia involves the use of audio, text, graphics, data, and video information. Multimedia material is transmitted to all the participants while the audio stream is generated in real time by the professor.

2. Virtual classroom and multimedia: The virtual classroom is a teaching and learning environment located within a computer-mediated communication (CMC) system. Classroom activities such as online exams, quiz, grad-book, which are rendered "Virtual", are provided by the electronic information exchange system. Participation is generally asynchronous; i.e., the virtual classroom participants may dial in at any time, around the clock, and from any location in the world accessible by a telephone, public X.25 data network, or the Internet.

3. Multimedia in medical applications: The medical environment is one of the earliest adapters of multimedia systems. This is due to overwhelming time constraints in many medical areas in conjunction with the readiness developed within BERKOM, which has provided distributed access to medical databases via high-speed networks. A major goal that was achieved within the projects was the integration of low-cost end systems in a usually high-end environment. This makes it possible to install the end systems in every surgery, providing access to x-rays stored in a central or a decentralized database. End systems can be upgraded to support audio and video communication, in order to provide a CSCW environment for easy cooperation between doctors. High-definition x-rays as well as real-time audio/video communication drive the need for high-speed networks. Current prototypes use the BERKOM test-bed. The high bandwidth costs also favor solutions based on ISDN technology.

4. Multilingual video on demand: This involves providing close captioning and multi–language support for a video lecture presentation; e.g., a presentation in English supported by any other language.

5. Documentary: In a protein documentary for example, the scientist shares his/her excitement about the structure through an audio track that is ideally synchronized with appropriate views on the molecule. In this way, the viewer does not have to shift focus but starts, stops and replays segments until s/he understands the interaction being demonstrated.

VII SUGGESTED READING

Aebersold R, Rist B and Gygi SP (2000) Quantitative proteomics analysis: Methods and applications, *Ann. N. Y. Acad. Sci.*, **919**:33–47.

Andersen JS and Mann M (2000) Functional genomics by mass spectrometry, *FEBS Lett.,* **480**:25–31.

Appel RD and Hochstrasser DF (1999) Computer analysis of 2-D images, *Methods Mol. Biol.,* **112**:363–381.

Ausubel FM, Brent R, Kingston RE, Moore DD, Seidman JG, Smith JA, and Struhl K (Eds) (1999) *Short Protocols in Molecular Biology*, 4th Edn, John Wiley and Sons, Inc.

Baxevanis AD and Qualette BFF (1999) *Bioinformatics: A Practical Guide to the Analysis of Genes and Proteins, Molecular Medicine Today,* **5**:332.

Blackstock WP and Weir MP (1999) Proteomics: Quantitative and physical mapping of cellular proteins, *Trends Biotechnol.,* **17**:121–127.

Burley *et al.* (1999) Structural genomics: Beyond the human genome project, *Nat. Genet.,* **23**:151–157.

Gibas C and Jambeck P (2001) *Developing Bioinformatics Computer Skill,* Shroff Publishers and Distributors Pvt. Ltd., Mumbai, Calcutta.

Graves PR, Haystead TAJ (2002) Molecular biologist's guide to proteomics, *Microbiology and Molecular Biology Reviews,* **66**:39–63.

Higgins D and Taylor W (2000) *Bioinformatics: Sequence, Structure, and Databanks*, Oxford University Press.

Miguel AA and Sander C (1997) Bioinformatics: From genome data to biological knowledge, *Current Opinion in Biotechnology,* **8**:675–683.

Quinn GB, Taylor A, Wang H-P and Bourne PE (1999) Development of internet-based multimedia applications. *TIBS* **24**:321–324.

Wang JH and Hewick RM (1999) Proteomics in drug discovery. *DDT* **4**:129–1333.

PART – III
Working with Nucleic Acids

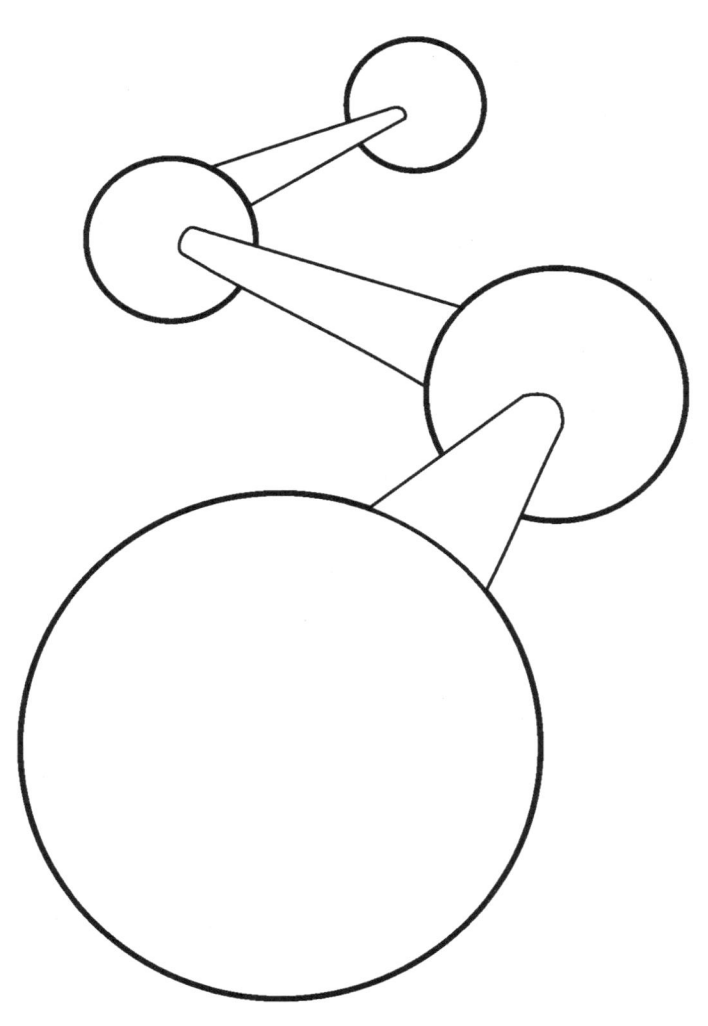

7

Isolation of Nucleic Acids

I INTRODUCTION

The isolation of nucleic acids is a fundamental process required for most molecular biology work, such as genomic characterization, gene mapping, sequencing, DNA amplification, construction of DNA libraries, etc. The amount of DNA in an organism, the organization of DNA in a cell and the sub-cellular components vary enormously from organism to organism. Thus, the process of isolating DNA from different organisms is very complex and there is no single protocol available that suits all cell types. It is very important to have a thorough knowledge about the physical and chemical properties of DNA and its organization in a cell. Even though a number of DNA isolation protocols have been developed, in this chapter I have listed only few standard methods. Therefore, I request the readers to refer to specific protocols reported elsewhere in literature for their individual requirements.

II NUCLEIC ACIDS

Nucleic acids (NA) are biological macromolecules that have numerous roles; they participate as essential intermediates in virtually all aspects of cellular metabolism. They have very important biological roles, as elements of heredity, agents of genetic information transfer, and being a structural part of functional units, etc. Nucleic acids are linear polymers of nucleotides. The two basic types of NA are deoxyribonucleic acid (DNA) and ribonucleic acids (RNA) and all hereditary information resides in these nucleic acids. In most organisms, genes are segments of DNA molecules, but in a few phages and in many animal- and plant-viruses, RNA is the genetic material. A great deal of what we know about the nature of genes and gene expression has been obtained from knowledge of the structure of DNA or RNA.

A. COMPONENTS OF NUCLEIC ACIDS

DNA and RNA are polynucleotides – that is, polymers of nucleotides. Each nucleotide has three components – a nitrogen base, a sugar and a phosphate (Figure 7.1). The complete digestion of NA liberates the nitrogenous bases of purines and pyrimidines, a five-carbon sugar, and phosphoric acid in equal amounts.

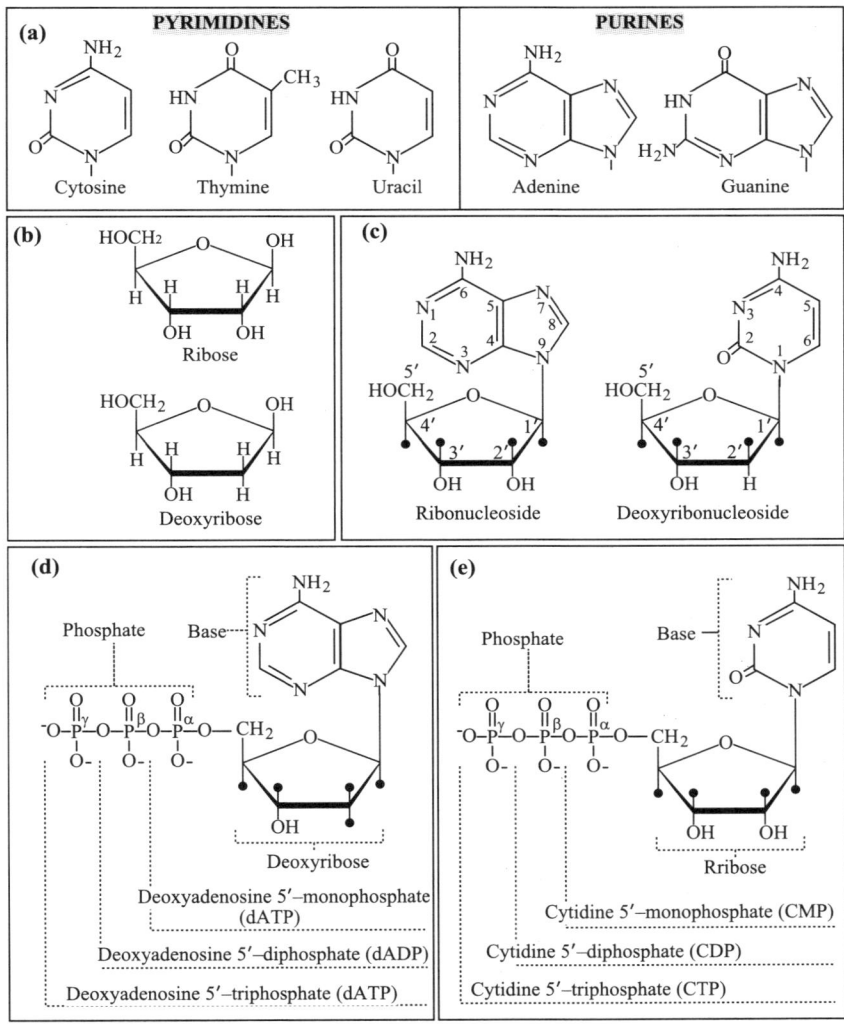

Figure 7.1 The chemical structures of nucleic acid subunits: (a) Pyrimidine and purine bases; (b) Ribose and Deoxyribose sugars; (c) Ribonucleoside (e.g. Adenosine), deoxyribonucleoside (e.g. deoxycitdine); (d) Deoxyribonucleotides (e.g. deoxyadenosine triphosphate); and (e) Ribonucleotides (e.g. Cytosine triphospahte).

1. **A cyclic five-carbon sugar:** The pentose sugar is ribose in ribonucleic acid (RNA), and deoxyribose in deoxyribonucleic acid (DNA). The structures of ribose and 2'–deoxyribose differ only in the absence of a 2'–OH group in deoxyribose. The bulky 2'–OH group in RNA can form hydrogen bonds and makes RNA more susceptible to chemical and enzymatic degradation than DNA.

2. **A purine or pyrimidine base** is attached to the 1'–carbon atom of the sugar by an N-glycosidic bond. The bases are the purines: adenine (A) and guanine (G), and the pyrimidines:

cytosine (C), thymine (T), and uracil (U). DNA and RNA both contain A, G, and C; however, T is usually found only in DNA, and U is usually found only in RNA.

3. **A phosphate** is attached to the 5'–carbon of the sugar by the DNA phosphoester linkage. This phosphate is responsible for the strong negative charge of both nucleotides and nucleic acids. A base linked to a sugar is called a nucleoside; thus, a nucleotide is a nucleoside phosphate.

The nucleotides in NA are covalently linked by a second phosphoester bond that joins the 5'–phosphate of one nucleotide and the 2'–OH group of the adjacent nucleotide. The bond between the phosphate and the 3'– and 5'–carbon atoms is called a *phosphodiester bond* (Figure 7.2). The resulting polynucleotide is an alternating sugar-phosphate backbone with one 3'–OH terminus and one 5'–phosphate terminus. A typical RNA molecule is a single-stranded polyribonucleotide. In contrast, DNA contains two polydeoxynucleotide strands wrapped round one another to form a double-stranded helix.

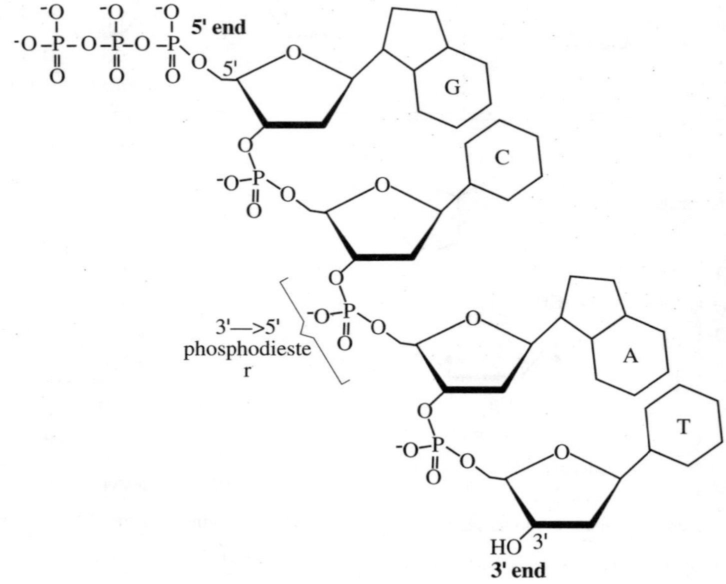

Figure 7.2 *Phosphodiester bonds and the covalent structure of four nucleotide long single-stranded DNA*

B. THE DOUBLE HELIX

X-ray diffraction studies have indicated that the common form of DNA, called B–DNA, is an extended chain with a highly-ordered structure: DNA is helical, with the nucleotide bases stacked and their planes separated by 3.4 nm. Chargaff had carried out a chemical analysis of the molar content of the bases in DNA molecules isolated from many organisms – in each case [A] = [T] and [G] = [C]. Watson and Crick combined the physical and chemical data and deduced the model showing that the two strands are coiled about one another to form a double-stranded helix (Figure 7.3). The sugar–phosphate backbones follow a helical path at the outer edge of the molecule, and the bases are in a helical array in the central core. The bases of one strand

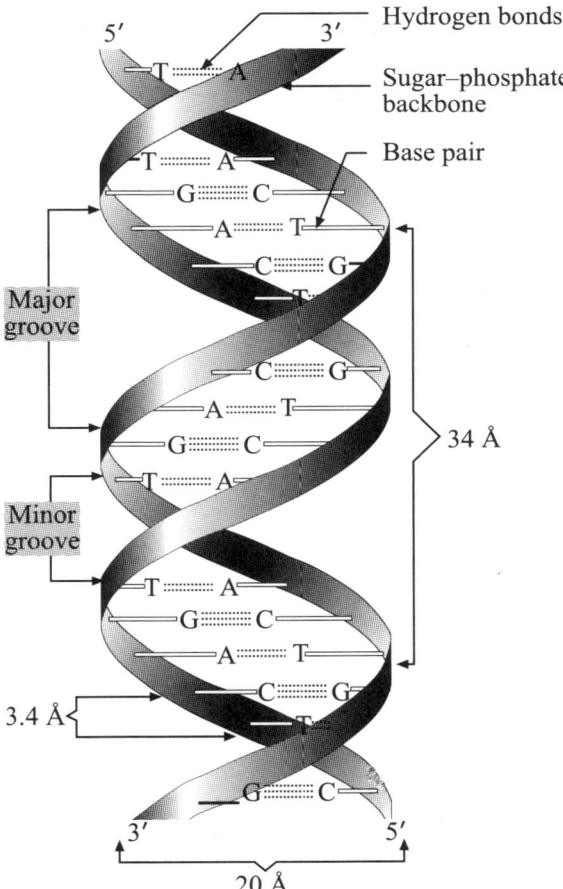

Figure 7.3 *A schematic view of the three dimensional structure of DNA double helix*

are hydrogen-bonded to those of the other strand (complementary to each other) to form the purine–pyrimidine base pairs AT and GC (Figure 7.4). Because each pair contains one two-ringed purine (A or G) and one single-ringed pyrimidine (T or C), the length of each pair is about the same and the helix can fit into a smooth cylinder. Otherwise, the helix would have either too thick (if both are purines A and G) or too thin (if both are pyrimidine, C and T) strands.

The two bases in each base pair lie in same plane, and the plane of each pair is perpendicular to the helix axis. The base pairs are rotated with respect to each other to produce 10 pairs per helical turn. The diameter of the double helix is 20 nm and the DNA is a right-handed helix. This means that each strand appears to follow a clockwise path, moving away from a viewer looking down the helical axis. The DNA helix has two external helical grooves, a deep wide one (the *major groove*) and a shallow narrow one (the *minor groove*). Most of the specific DNA-binding proteins bind in the major groove. The two polynucleotide strands of the DNA double helix are anti-parallel. Thus, in a linear double-strand helix, there is one 3'-OH and one 5'-P terminus at each end of the helix. By convention, the sequence of bases of a single

130 MOLECULAR BIOTECHNOLOGY: PRINCIPLES AND PRACTICES

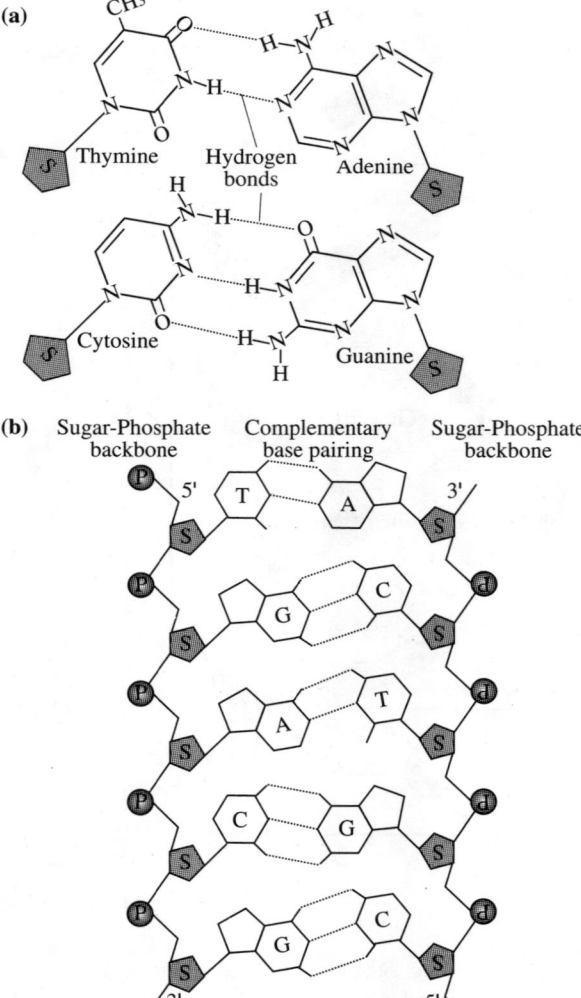

Figure 7.4 *(a) Hydrogen bonding between the adenine-thymine (A:T) and guanine-cytosine (G:C) base pairs (b) The base pairing between two DNA chains, S=deoxyribose sugar and P=phosphate group*

chain is usually written with the 5'–P terminus at the left; for example, GCT denotes the trinucleotide P–5'–GCT–3'–OH, which may also be written as pGpCpT.

C. CIRCULAR AND SUPERHELICAL DNA

The intact chromosomes of most prokaryotes and plasmids are circular. A circular molecule may be a covalently-closed circle, which has one or more interruptions (nicks) in one or both strands. Generally, covalently-closed circles are twisted and such a circle is said to be a *supercoiled* (that means another form of coiling *superimposed* on that found in the linear DNA). The DNA helix can be *right-handed* or *left-handed*. The right-handed or positive coil means that when one is looking down the helical axis, the coil follows a clockwise path and moves away from the viewer; if the path is counter-clockwise, the coil is left-handed or negative.

The two ends of a linear DNA helix can be brought together and joined in such a way that each strand is continuous. In doing so, one of the ends is rotated 360 degrees with respect to the other to produce some unwinding of the double helix and then the ends are joined; the resulting covalent circle, if the hydrogen bonds reform, twists in the opposite sense (opposite to unwinding direction) to form a twisted circle, to relieve strain. Such a molecule looks like a figure 8 (i.e., has one crossover point or *node*). If instead, it is twisted 720 degrees before joining, the resulting superhelical molecule has two nodes. The reason for the twisting is, in the case of a 720-degree unwinding of the helix, 20 bp must be broken (because the helix has been unwound two turns). To maintain a right-handed (positive) helical structure with 10 bp per turn, the DNA deforms in such a way that the underwinding is rewound and compensated for by a negative (left-handed) twisting of the circle. Similarly, the initial rotation might instead be in the direction of overwinding, in which case the joined circle twists in the opposite direction, forming a positive superhelix. Generally, naturally-occurring superhelical DNA molecules are initially underwound and, hence, form negative superhelices. Furthermore, the degree of twisting is about the same for all molecules; one negative twist is produced per 200 bp, or 0.5 twists per turn of the helix. In bacteria, the underwinding of superhelical DNA is a result of introduced turns into pre-existing circles by an enzyme called *DNA gyrase*. Three other arrangements that could counteract the strain of underwinding are possible: (1) The number of base pairs per turn of the helix could change; (2) All of the underwinding could be taken up by having one or more large, single-stranded bubbles; (3) The underwinding could be taken up in part by superhelicity and in part by bubbles. If a circular molecule were made that was not underwound, transient breakage and remaking of hydrogen bonds (breathing) would introduce compensating transient negative twists.

D. THE STRUCTURE OF RNA

A typical cell contains about 10 times as much RNA as DNA. Generally, RNA exists as a single-stranded polynucleotide, except in the case of a few viruses where it is double-stranded. The three major types of RNA that exist in a cell are ribosomal RNA, transfer RNA, and messenger RNA. All of these molecules superficially resemble single-stranded DNA, in that single-stranded regions are interspersed with intermolecular, double-stranded regions (Figure 7.5). This chapter briefly describes the different protocols used for the isolation of DNA and RNA.

III SOURCES OF DNA FOUND IN DIFFERENT ORGANISMS

The organization and kind of genetic material varies, depending upon the type of organism (Figure 7.6). The prokaryotic cells contain bacterial chromosomes and some cells with plasmid or phage DNAs. In eukaryotes, more than one type of genomic material can be isolated (nuclear,

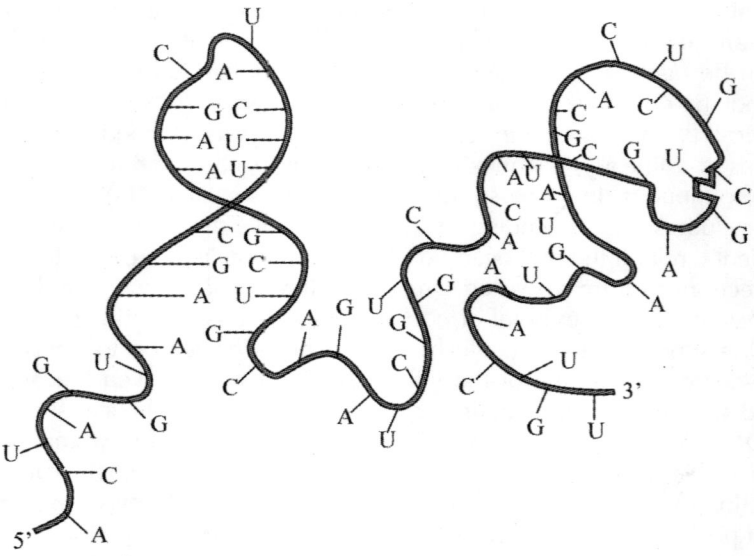

Figure 7.5 *Base pairing within a single strand of RNA showing a possible folded structure*

mitochondrial, and chloroplast). On the other hand, viral genomes have very diverse genomic DNA structures and categories.

A. NUCLEAR DNA

The DNA in the eukaryotic nucleus exists as a nucleoprotein complex called chromatin. The proteins of chromatin fall into two classes: histones and non-histone chromosomal proteins. Histones are abundant structural proteins, whereas non-histones are represented by only a few copies of each of the many diverse proteins involved in genetic regulation. Five distinct histones aggregate to form an octomeric core structure (nucleosome), around which the DNA helix is wound. A higher order of chromatin structure is created when the nucleosomes, in their characteristic beads-on-a-string motif, are wound in the fashion of solenoid with six nucleosomes per turn. This solenoid filament then forms long DNA loops of variable lengths; such loops are arranged radially about the circumference of a single turn to form a miniband unit of the chromosome.

B. ORGANELLAR DNA

In addition to nuclear DNA, the eukaryotes also contain organellar DNA. Mammalian cells have mitochondria, whereas plant cells contain both mitochondria and plastids. Both of these organelles are self-replicating and have their own genomic DNA. The *mitochondria* of both animals and plants contain DNA (mtDNA) of varying sizes and proportion in the cellular DNA. It varies in size from 15–2400 kb, depending upon the organism. However, the number of mitochondria per cell is large, albeit relative to the nuclear DNA; usually, the mtDNA still

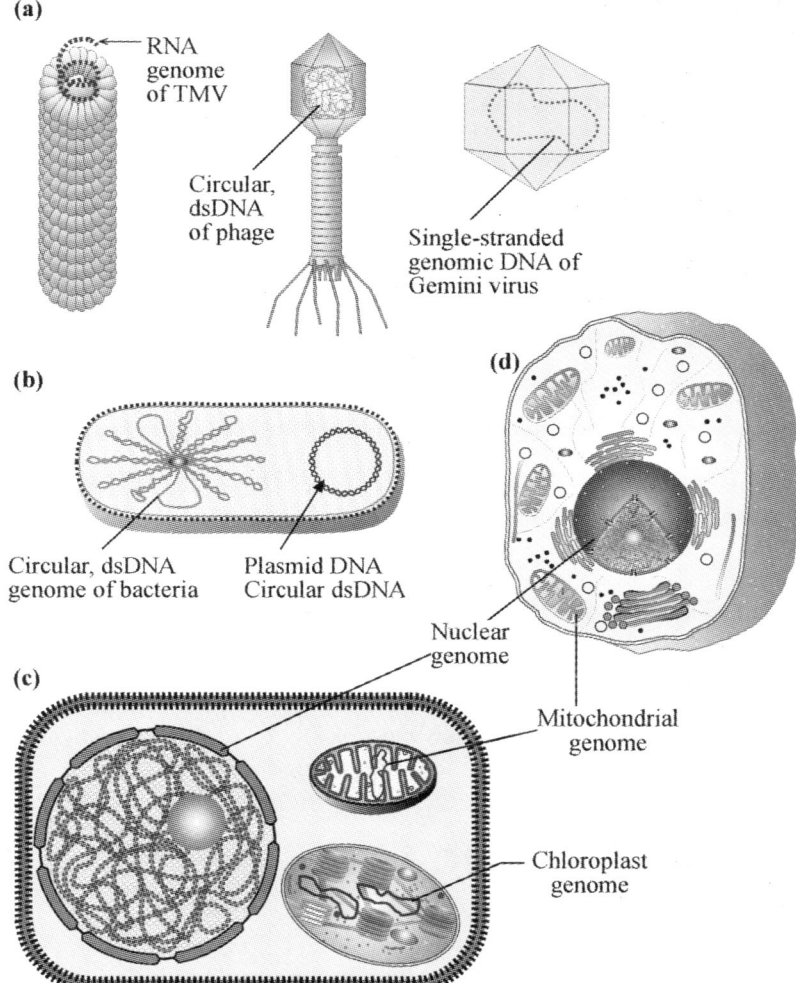

Figure 7.6 *Schematic view of the structure and location of genomic DNA of (a) Viral genomes, (b) the E. coli chromosome with a plasmid DNA, (c) Plant nuclear chromosome with mitochondrial and chloroplast genomes, (d) Animal nuclear chromosome with mitochondrial genomes*

represents less than 1% of the cellular DNA (except in the case of yeast where it has as high as 18%). The mtDNA is usually circular, but in some organisms it also exists in linear form. The chloroplast genomes (cpDNA) are relatively large, usually ranging from 120 kb to 210 kb. This is comparable to the size of a large bacteriophage like T4 (119 kb). Each cell may have several *chloroplasts* and each chloroplast may have several circular DNA molecules, so that the cpDNA may make up to 14% of the cellular DNA in lower eukaryotes. In higher eukaryotes, however, the cpDNA comprises only a small proportion of the total DNA in the cell.

C. BACTERIAL GENOMIC AND PLASMID DNA

In prokaryotes, the genomic DNA is generally a single, tightly-coiled circular molecule and very few proteins are associated with the DNA double helix. The genomic DNA in prokaryotes, which are loosely clumped in a cytoplasm and do not have any membrane, are called nucleoids. Many bacteria, in addition to their genomic DNA, also harbor self-replicating, double-stranded extrachromosomal DNA molecules called plasmids.

D. VIRAL DNA

Many viruses, however, have diverse forms of genetic material. Their genomes can be DNA or RNA, either single-stranded or double-stranded, and circular or linear. The viral genetic material is simply covered with a protein coat and, in a few cases, a lipid membrane (enveloped viruses). They can be present in a nucleus or in the cytoplasm of the cell or can also be present outside the cell.

The application of molecular biology techniques to the analysis of complex genomes depends on the ability to prepare pure, high-molecular-weight DNA. The DNA content of some cells probably exceeds 10^9 g/mol, making them the largest linear covalent structures in nature. Genomic DNA are seldom isolated intact, because these molecules are highly sensitive to shear forces and prone to DNases. Care should be taken to keep both of these factors to a minimum during the isolation of DNA. The conventional purification techniques, those described in this chapter, can approach up to 10^7 g/mol.

This section begins with protocols for the isolation of DNA from various sources of cells and organisms, including plant viruses. For each type of DNA, a brief theoretical discussion and step-wise explanation of different methods for DNA isolation are provided.

IV ISOLATION OF NUCLEIC ACIDS

The nucleic acid (NA) to be manipulated must be extracted from its source, either from an intact organism or from cells. The NA thus isolated must be purified before it can be used. The degree of purity needed is determined by the goals of the experiment. One of the most common methods for extracting and purifying nucleic acids uses *phenol* to extract NA in large- or small-scale procedures, from complex extracts or from simpler *in vitro* manipulations.

Various versions of phenol extraction methods have been developed. The primary function of phenol is to remove the proteins from the aqueous solution containing NA. The extraction sample usually contains many kinds of nucleases that could damage the DNA, while others could simply interfere with later manipulations. EDTA (ethylenediaminetetraacetic acid) is often added; it is a chelating agent that binds Mg^{2+}, which is required for nucleases to act on the DNA. Only highly-purified phenol must be used, because the oxidized form of phenol can damage the DNA. Mixing the phenol with the sample under particular conditions favors the dissociation of the proteins from the nucleic acids. The sample is then centrifuged. Centrifugation yields two phases: a lower organic phenol phase carrying the protein, and the less dense aqueous phase, containing the NA. Often, some phenol extraction protocols include *chloroform,* which denatures proteins, removes lipids, and improves the efficiency of the extractions. To reduce the foaming caused by chloroform, *isoamyl alcohol* is usually added. The extraction of DNA from cells or organisms should be carried out as quickly as possible in ice with refrigerated buffers to minimize the activity of any nuclease present in the cells that can degrade the DNA.

Precipitating nucleic acids: During NA extraction, cloning, or many other projects, it is often necessary to concentrate the DNA samples or change the solvent in which a nucleic acid is dissolved. Often, the DNA isolated by phenol extraction contains trace amounts of phenol, which could disrupt the activity of enzymes in subsequent manipulations if it is not purified further. This purification can be achieved by ethanol precipitation, isopropanol precipitation, or by many other methods. The most versatile is probably *ethanol precipitation*, because it can concentrate both DNA and RNA and purify the DNA after the phenol extractions. Basically, combining the DNA sample with salt and ethanol, at −20°C or lower, precipitates the DNA. The precipitated salt of the nucleic acid is then sedimented by centrifugation, the ethanol supernatant is removed, and the NA pellet is resuspended in a buffer.

A. ISOLATION OF PLASMID DNA

Many methods have been developed to purify plasmid DNA from bacteria. These methods invariably involve three steps: (i) The selective propagation of bacterial cells in culture, (ii) the harvesting and lysis of the bacteria, and (iii) the purification of plasmid DNA by deproteinization and precipitation. Three of the most popular procedures for the small-scale isolation of plasmid DNA from bacteria are the alkaline lysis method, the boiling method, and precipitation with CTAB. Figure 7.7 explains the general steps of plasmid DNA preparations.

1. Growth of bacterial culture

Plasmids are always purified from cultures that have been inoculated with a single bacterial colony selected from an agar plate. Many of the currently used plasmid vectors replicate to such a high copy number that they can be purified in a large yield from cultures grown to the log phase in a standard LB medium. However, in those cases where plasmids grow poorly because of their large size or due to the foreign sequences they carry, it may be worthwhile to explore alternative ways to grow and treat the bacterial culture (see Chapter 27).

2. Harvesting and lysis of the bacteria

Bacteria are removed from centrifugation and lysed by any one of a large number of methods, including treatment with non-ionic or ionic detergents, organic solvents, alkalis, or heat. These methods are dictated by three factors: the size of the plasmid, the strain of *E. coli*, and the technique that is to be used to purify the plasmid DNA.

3. Purification of plasmid DNA

All the methods in common use exploit the relatively small size and covalently closed circular nature of plasmid DNAs. For many years, equilibrium centrifugation in CsCl–ethidium bromide gradients has been the method of choice to prepare large amounts of plasmid DNA. However, this process is expensive and time-consuming, and many alternative methods have been

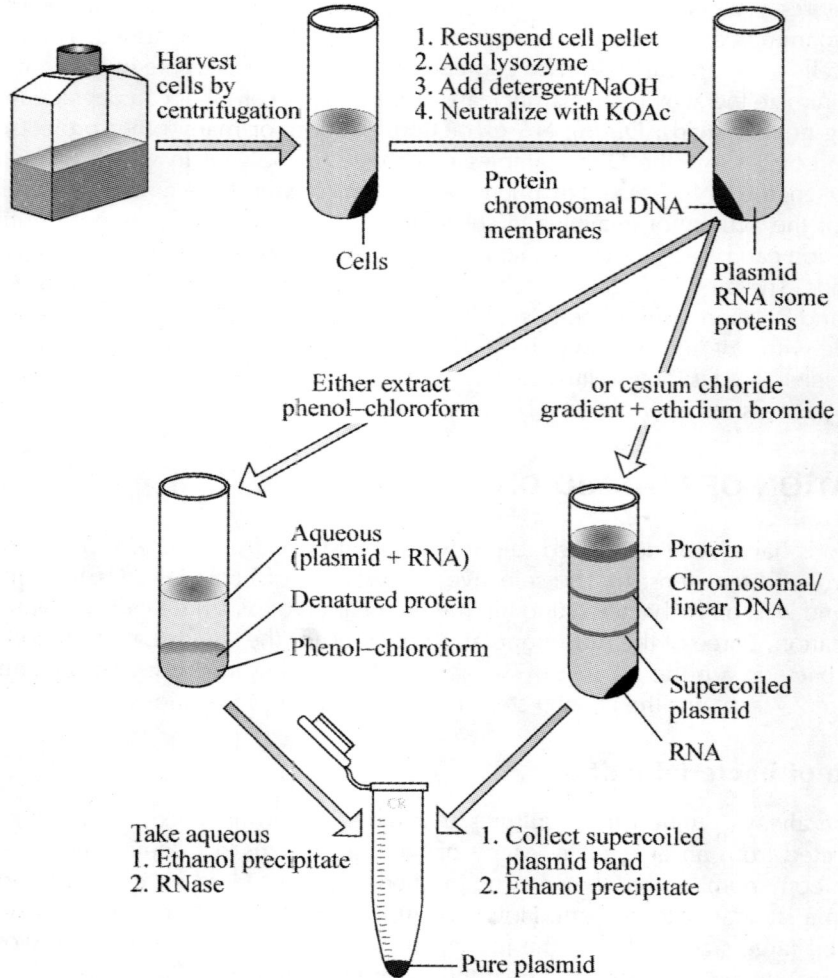

Figure 7.7 *General scheme for plasmid preparation*

developed, the majority of which involve the use of ion-exchange or gel-filtration chromatography or differential precipitation to separate plasmid and host DNAs. The best of them, differential precipitation with polyethylene glycol, has recently been improved to the point where it yields plasmid DNA of an extremely high purity. The following procedures are presented: (1) a rapid isolation of small amounts (miniprep) of DNA, which is not very pure. (2) a scaled-up protocol for large amounts of DNA of high purity.

Materials

Bacterial strains and plasmids (*E. coli* strains DH5α and MC 1061) are commonly used as hosts for the propagation of the plasmids. Plasmids pSP65, or pGEM–1 (Promega Corp, Madison, WI), or recombinant clones derived from other plasmids can be used.

TELT Solution (100 ml): 50 mM Tris–HCl, pH 8.0; 62.5 mM Na_2EDTA; 2.5 M LiCl; 4% (vol/vol) Triton X–100

TE Buffer (100 ml): 10 mM Tris, 1 mM EDTA, pH 8.0

LB Broth (see Appendix I)

Host strain with plasmid.

Ampicillin (50 mg/ml)–Store solution at –20°C

GTE Buffer (100 ml): 50 mM Glucose, 25 mM Tris.Cl pH 8.0, 10 mM EDTA, autoclave

Lysis solution: 0.2 N NaOH, 1% (w/v) SDS

5 M Potassium acetate pH 4.8 (29.5 ml Glacial acetic acid, Adjust pH to 4.8 with conc. KOH; Adjust volume to 100 ml with ddH_2O)

TBE Buffer pH 8.3 (10 X): 0.89 M Tris base, 0.89 M Boric acid, 0.02 M $Na_2EDTA.2H_2O$, ddH_2O to 1 litre

Tracking Dye/Loading Buffer: 40% (w/v) sucrose or glycerol, 0.25% bromophenol blue, TBE (1x)

1:1 (v/v) Phenol:Chloroform; Ethanol 70% and 95%; 3 M Sodium acetate; RNase A (10 mg/ml stock); Lysozyme; Isopropanol; ss–phenol; Chloroform; Agarose; Methylene Blue Plus™; Ethidium bromide (10 mg/ml)

Microfuge tubes, Gloves, Goggles, Microfuge, Pipettors, Centrifuge tubes (50 ml), Bench-top microcentrifuge (termed as "Microfuge"), Bench-top clinical centrifuge (e.g., IEC Model HNS–II, International Equipments Co., MA), Refrigerated high-speed centrifuge (e.g., Sorvall RC–5B, Du Pont Co), Variable-volume pipettors 1–50, 50–200, 200–100 µl (Micropipette, VWR Corp., PA) Polaroid Film 667, UV white light, Transilluminator or hand-held UV lamp, Electrophoresis cell, Power supply, Microwave, Camera.

Plasmid isolation—a miniprep

Harvesting and lysis of bacteria

1. Transfer a single bacterial colony into 2 ml of LB medium containing the appropriate antibiotic (e.g., ampicillin, 60µg/ml) in a loosely-capped 15-ml tube. Incubate the culture overnight at 37°C with vigorous shaking.
2. Pour 1.5 ml of the culture into a microfuge tube. Centrifuge at 12,000g for 30 seconds at 4°C in a microfuge. Store the remainder of the culture at 4°C.
3. Remove the medium by aspiration, leaving the bacterial pellet as dry as possible and proceed with one of the small-scale DNA preparations outlined below.

Method-1: Alkaline lysis "Miniprep" method

The bacterial cells that grow in the LB medium with an antibiotic carry the plasmids that have the antibiotic resistance gene, because *E. coli*, which is sensitive to this antibiotic, would not be able to grow by itself. In the preparation of plasmid DNA by an "alkaline lysis" method, SDS is added to denature proteins, rupturing the bacterial cell walls, and NaOH is added to denature both chromosomal and plasmid DNAs. However, in the pH range of 12.0–12.6, small, covalently-closed circular DNA (such as plasmid DNA) does not denature, although high molecular weight chromosomal DNA does forming an insoluble network on renaturation, and enabling one to selectively extract the soluble plasmid DNA. Potassium acetate neutralizes the solution, thereby creating an environment conducive to plasmid DNA reannealing. Bacterial proteins and DNA that have now precipitated out of the solution are removed by centrifugation. The plasmid DNA that remains in the solution is then purified and concentrated by the addition of ethanol, and the resulting precipitate is resuspended in an appropriate volume of buffer.

Procedure

1. Resuspend the bacterial pellet in 100 µl of ice-cold Solution I by vigorous vortexing. Solution I (50 mM glucose; 25 mM Tris.Cl pH 8.0; 10 mM EDTA pH 8.0) prepared in batches of 100 ml, autoclave and store at –4°C.
2. Add 200 µl of freshly-prepared Solution II (0.2 N NaOH; 1% SDS). Close the tube tightly, and mix the contents by inverting the tube rapidly five times. Ensure that the entire surface of the tube comes into contact with the Solution II. Do not vortex. Store the tube on ice.
3. Add 150 µl of ice-cold Solution III (5 M potassium acetate 60 ml; glacial acetic acid 11.5 ml; H_2O 28.5 ml). Close the tube and vortex it gently in an inverted position for 10 seconds to disperse Solution III through the viscous bacterial lysate. Store the tube on ice for 3–5 min..
4. Centrifuge at 12,000g for 5 min. at 4°C in a microfuge. Then, transfer the supernatant to a fresh tube.
5. *Optional*: Add an equal volume of phenol:chloroform and mix by vortexing. After centrifuging at 12,000g for 2 min. at 4°C in a microfuge, transfer the supernatant to a fresh tube.
6. Precipitate the double-stranded DNA with two volumes of ethanol at room temperature, and mix by vortexing. Allow the mixture to stand for 2 min. at room temperature.
7. Centrifuge at 12,000g for 5 min. at 4°C in a microfuge.
8. Remove the supernatant by gentle aspiration. Place the tube in an inverted position on a paper towel to allow all the fluid to drain away. Remove any drops of fluid adhering to the walls of the tube.
9. Rinse the pellet of double-stranded DNA with 1 ml of 70% ethanol at 4°C. Remove the supernatant as described in step 8, and allow the pellet to dry in the air for 10 min.
10. Redissolve the nucleic acids in 50 µl of TE (pH 8.0) containing DNase-free pancreatic RNase (20 µg/ml), and vortex briefly. Store the DNA at –20°C.

Method-2: Boiling "Miniprep" method

The boiling method is based on the observation that when bacteria – the cell walls of which have been removed by lysozyme treatment – are lysed with a non-ionic detergent, boiled for a brief period, and cooled, the chromosomal DNA forms a debris. The clot is pelleted by centrifugation, leaving the soluble plasmid DNA in the supernatant. The boiling might also inactivate the bacterial DNases and facilitate the isolation of plasmid DNA in an undegraded form.

Procedure

1. Inoculate 3 ml of sterile LB medium containing 100 µg/ml of the appropriate antibiotic. Incubate it at 37°C overnight in a shaking water bath. Three millilitres of such a plasmid-carrying culture, grown overnight, typically yields 4 to 6 µg of plasmid DNA.

2. Pour the grown culture into two plastic microcentrifuge tubes (do not use glass, as the DNA will adhere to the walls of the tube). Spin it for 2 min. in a microcentrifuge.
3. Decant the supernatant completely, and resuspend the cell pellets in 0.5 ml of sucrose solution containing 8% sucrose, 5% Triton X–100, 50 mM EDTA, and 50 mM Tris, pH 8.0.
4. Add 35 µl of a freshly-prepared solution of lysozyme (10 mg/ml in 10 mM Tris, pH 8.0), and mix by shaking it gently.
5. Place the tube in a boiling water bath for 60 seconds and cool on ice.
6. Centrifuge for 15 min. in a microcentrifuge.
7. Remove the pellets by gently teasing them out of the tube with a toothpick, leaving the clear supernatant behind.
8. Add 25 µl of 2.5 M NaCl and 0.6 ml of cold isopropanol to the supernatant. Mix thoroughly and incubate it for 15 min. in a dry-ice/ethanol bath.
9. Centrifuge it again in the microcentrifuge for 15 min.
10. Decant and discard the supernatant. Resuspend the plasmid pellets in 50 µl of sterile TE buffer (10 mM Tris, 1 mM EDTA, pH 8.0). Add 6 µl of 2.5 M NaCl and 1.25 µl of cold ethanol. Incubate it in a dry-ice/ethanol bath for 10 min., and centrifuge for 10 min. Then, decant and discard the supernatants.
11. Cover the tubes with Parafilm, and punch several holes through the covering. Place the tubes in a small beaker in a desiccator containing silica gel. Create a vacuum by attaching the desiccator to an aspirator, and allow the pellet to dry in this fashion for 10 to 15 min.

Method-3: Phenol–chloroform lysis extraction method

Procedure

1. Inoculate 3 ml of sterile LB medium containing 100 µg/ml of the appropriate antibiotic. Incubate it at 37°C, overnight, in a shaking water bath. Three millilitres of such a plasmid-carrying culture, grown overnight, typically yields 4 to 6 µg of plasmid DNA.
2. Resuspend the bacterial pellet thoroughly by vortexing it in 200 µl of TE buffer. Then, place it in a boiling water bath for 1 minute.
3. Add 200 µl of TNE and vortex it for 15 seconds. Add 400 µl of a 1:1 phenol–chloroform mixture and vortex vigorously for 15 second. Spin at full speed in a microfuge for 15 seconds. Then, spin at full speed in a microfuge for 15 min. at room temperature.
4. Transfer the upper aqueous layer to another 1.5 ml microfuge tube. Add an equal volume of chloroform vortex for 10 seconds, and spin at room temperature for 30 seconds to separate the two layers. Transfer the upper aqueous layer to another 1.5 ml microfuge tube. Precipitate the plasmid DNA by mixing it with an equal volume of ice cold 2–propanol. Spin it for 15 min. at 4°C in a microfuge to pellet down the precipitate.
5. Dissolve the pellet in 50 µl of TE and, if the intention of the screening is to verify cloning, subject 5 µl of the DNA preparation to electrophoresis on a 0.6% (w/v) agarose gel to carry out a preliminary identification of recombinant clones containing inserts. To the remaining DNA solution of each positive clone identified in this manner, add 4 µl of 3M sodium acetate and add water to bring the volume up to 400 µl. Add 800 µl of ethanol and mix well. Store it at –70°C for 30 min. Then, pellet down the precipitate by centrifugation in a microfuge at 4°C for 15 min.

6. Dissolve the pellet in 20 μl of TE. The yield of plasmid DNA is about 2 μg/1.5 ml of bacterial culture. If the DNA preparation is intended for sequence analysis, remove RNA by ribonuclease treatment and purify the plasmid DNA by phenol–chloroform extraction followed by ethanol precipitation as described in method-2.

Method-4: Large-scale preparation of plasmid DNA

At present the best method for obtaining up to 1 mg of highly pure plasmid DNA among the procedures just described above is the ethidium bromide–CsCl method. However, all these procedures cannot be easily scaled up to yield substantially larger quantities of plasmid. An alternative method, described herein, is readily adaptable to any quantity of DNA because it uses precipitation steps to purify the plasmid DNA. This method, however, has its own drawbacks (e.g., it often yields plasmid DNA contaminated with chromosomal DNA and tRNA). However, these problems can be solved if additional purification steps are planned.

Enhancing the plasmid yield

A reproducibly high plasmid yield (2–5 mg/500 ml culture) can be achieved by growing the bacteria in a rich medium as described below:

1. Grow a 30 ml culture of the bacterial strain carrying a desirable plasmid to the lag phase.
2. Inoculate 500 ml of LB or Terrific Broth medium (37°C) containing appropriate antibiotic in a 2-litre flask with 25 ml of the late-log-phase culture. Incubate the culture for 2.5 hours at 37°C on a rotary shaker (300 rpm).
3. *Optional*: Add 2.5 ml of a solution of chloramphenicol (34 mg/ml in ethanol) to give a final concentration of 170 μg/ml.
4. Incubate the culture for a further 12–16 hours at 37°C on a rotary shaker (300 rpm).

Harvesting

1. Centrifuge 500 ml of the culture at 4000 rpm for 15 min. at 4°C. Discard the supernatant. Then, place the open centrifuge bottle in an inverted position to allow all the supernatant to drain away.
2. Resuspend the bacterial pellet in 100 ml of ice-cold STE (0.1 M NaCl, 10 mM Tris.Cl, pH 8.0, 1 mM EDTA pH 8.0).
3. Collect the bacterial cells by centrifugation, as described in step 1.

Lysis by sodium dodecyl sulfate

1. Resuspend the washed bacterial pellet from a 500-ml culture in 10 ml of an ice-cold solution of 10% sucrose in 50 mM Tris.Cl (pH 8.0). Transfer the suspension to a 30-ml plastic screw-cap tube.
2. Add 2 ml of a freshly-prepared solution of lysozyme (10 mg/ml in 10 mM Tris.Cl pH 8.0).
3. Add 8 ml of 0.25 M EDTA (pH 8.0) and mix the suspension by inverting the tube several times. Place the tube on ice for 10 min.
4. Add 4 ml of 10% SDS. Immediately stir the contents of the tube with a glass rod, so as to disperse the SDS evenly throughout the bacterial suspension. Be as gentle as possible to minimize any shearing of the liberated DNA.
5. As soon as the stirring is completed, add 6 ml of 5 M NaCl (final to 1 M). Again, use a glass rod to stir the contents of the tube gently but thoroughly. Place the tube on ice for at least 1 hour.

6. Remove any high-molecular-weight DNA and bacterial debris by centrifugation at 30,000 rpm for 30 min. at 4°C. Carefully transfer the supernatant to a 50 ml disposable plastic centrifuge tube and discard the pellet.
7. Extract the supernatant once with phenol:chloroform and once with chloroform.
8. Transfer the aqueous phase to a 250-ml centrifuge bottle and add 2 volumes of ethanol at room temperature. Mix well and let it stand at room temperature for 1–2 hours.
9. Recover the nucleic acids by centrifugation at 5000g for 20 min. at 4°C, preferably in a temperature-controlled horizontal rotor.
10. Discard the supernatant and wash the pellet with 70% ethanol at room temperature and centrifuge as in step 9. Discard as much of the ethanol as possible, and then invert the centrifuge bottle on a pad of paper towels to allow the last of the ethanol to drain away. Dry the centrifuge bottle briefly in a vacuum desiccator, but do not allow the pellet to dry out completely.
11. Dissolve the DNA in 3 ml of TE (pH 8.0) and store at –20°C or purify the plasmid DNA by precipitation with polyethylene glycol (PEG).

Purification of plasmid DNA by precipitation with PEG

1. Transfer the nucleic acid solution to a 15 ml Corex tube, and add 3 ml of an ice-cold solution of 5 M LiCl. Mix well and then centrifuge the solution at 10,000 rpm for 10 min. at 4°C.
2. Transfer the supernatant to a fresh 30-ml Corex tube. Add an equal volume of isopropanol and mix well. Recover the precipitated nucleic acids by centrifugation at 10,000 rpm for 10 min. at room temperature.
3. Decant the supernatant carefully, and invert the open tube to allow the last drops of supernatant to drain away. Rinse the pellet and the walls of the tube with 70% ethanol at room temperature. Drain off the ethanol, and use a Pasteur pipette attached to a vacuum line to remove any beads of liquid that adhere to the walls of the tube. Place the inverted, open tube on a pad of paper towels for a few min. at room temperature to allow the last traces of ethanol to evaporate.
4. Dissolve the pellet in 500 µl of TE (pH 8.0) containing DNases-free pancreatic RNase (20µg/ml). Transfer the solution to a microfuge tube and store at room temperature for 30 min.
5. Add 500 µl of 1.6 M NaCl containing 13% (w/v) polyethylene glycol (PEG 800) and mix well. Recover the plasmid DNA by centrifugation at 12,000g for 5 min. at 4°C in microfuge.
6. Remove the supernatant by aspiration. Dissolve the pellet of plasmid DNA in 400 µl of TE (pH 8.0). Extract the solution once with phenol, once with phenol:chloroform, and once with chloroform.
7. Transfer the aqueous phase to a fresh microfuge tube. Add 100 µl of 10 M ammonium acetate, and mix well. Add 2 volumes of ethanol, and store the tube for 10 min. at room temperature. Recover the precipitation of plasmid DNA by centrifugation at 12,000g for 5 min. at 4°C in a microfuge.
8. Remove the supernatant by aspiration. Add 200 µl of 70% ethanol at 4°C, vortex briefly, and then centrifuge at 12,000g for 2 min. at 4°C in a microfuge.
9. Remove the supernatant by aspiration, and store the open tube on the bench until all visible traces of ethanol have evaporated.

10. Dissolve the pellet in 500 µl of TE (pH 8.0). Measure the OD_{260} of a 1:100 dilution (in TE pH 8.0) of the solution, and calculate the concentration of the plasmid DNA (1 OD_{260} = 50 µg of plasmid DNA/ml). Store the DNA in aliquots at –20°C.

B. ISOLATION OF BACTERIAL CHROMOSOMAL DNA

The isolation of high-molecular-weight chromosomal DNA is often the first step in molecular cloning. In prokaryotes, the chromosomal DNA is associated with very few proteins and is clumped very loosely in the cytoplasm, which is called the *nucleoid*. The chromosomal DNA of *E. coli* is a large circular molecule containing approximately 3×10^6 base pairs. The DNA is attached to the plasma membrane at several points. Large DNA is very sensitive to mechanical shear, which causes random breaks in the phosphate bonds of the molecule. Therefore, the isolation of chromosomal DNA with a minimum number of breaks is a difficult task. However, if the extraction is performed carefully, large fragments of chromosomal DNA can be obtained with an average fragment length of 10^5–20^5 bp.

The resuspended cells are first mixed with EDTA. The DNA being extracted is protected from DNase degradation since the complexed form of Mg^{2+} cannot be utilized by the enzyme. The addition of the ionic detergent, Sarkosyl, dissolves the cell membrane and denatures many proteins. RNase is also present to degrade high-molecular-weight RNA. The proteolytic enzyme (protease) is added to the cell lysate in order to digest proteins that are free in solution or bound to the DNA. RNases and proteases are exceptionally stable enzymes and they remain active in the presence of denaturing detergents, such as Sarkosyl, and at high temperatures. Eventually, the RNase will be degraded by the protease. In the presence of salts, DNA and RNA precipitate from solutions containing high percentages of isopropanol or ethanol, while smaller molecules, such as sugars and amino acids, remain in solution.

Method-1: Preparation of bacterial genomic DNA – Basic protocol

Materials

Tris buffer: 10 mM Tris-Cl (pH 8.0), 0.1 mM EDTA

RNase in Tris buffer: Pancreatic RNase A, 10 mg/ml, in TE buffer, preheated to 80°C for 10 min. to inactivate DNases

5 M NaCl solution; EDTA Buffer; Sarkosyl solution; Protease in Tris buffer; 91–100% isopropanol (at –20°C).

Hand gloves; Test tubes 15 ml; Pipettes 5 and 10 ml; Pipette pumps; Beakers or flasks 50 and 250 ml; Horizontal gel electrophoresis apparatus with DC power supply; Automatic micropipettes with tips; Water bath (45°C); Recombinant DNA visualization system (white light with methylene Blue Plus™ staining).

Procedure

Extracting DNA from cells

1. Label the test tube containing 5 ml of resuspended cells.
2. Add 2 ml of EDTA buffer to the cell suspension.

3. Add 0.5 ml of RNase solution. Tightly cap it and mix by inverting the tube several times.
4. Let the cells incubate for 5 min. at room temperature.
5. Add 0.5 ml of the Sarkosyl solution.
6. Add 1 ml of Protease solution. Tightly cap it and gently mix by slowly inverting the tube 3 times.
7. Incubate the tube for 20 min. in a 45°C water bath.
8. Add 0.6 ml of 5 M NaCl solution and mix by inverting the tube..
9. Slowly pour the viscous DNA solution into a clean 50 ml beaker.

Spooling DNA

1. Carefully overlay the viscous DNA solution with the specified volume of ice-cold isopropanol. Let the isopropanol slowly stream down the inside wall of the tube; isopropanol is less dense than water so it will be the upper layer.
2. Submerge the end of a glass rod just below the interface of the isopropanol and the aqueous DNA solution.
3. Swirl the rod several times in a circular motion to spool out the DNA.
4. Remove the rod to see if the precipitate is being collected. The precipitate will appear semi-transparent and gelatinous in texture.
5. Continue swirling the rod to collect the DNA from the solution. Allow the end of the rod to occasionally touch the bottom of the tube.
6. Note the appearance of the spooled DNA. As the DNA adheres to the rod, its initial gelatinous texture (as it appeared in step 5) will become more compact and fibrous in appearance.

Redissolve spooled DNA

1. Remove the rod from the tube, ensuring that the excess isopropanol drips off.
2. Add 2 ml of Tris Buffer to a clean test tube.
3. Submerge the coated end of the rod into the buffer.
4. Swirl the rod several times to dislodge some of the DNA.
5. Cover the test tube, with the rod still inside, with a plastic wrap, Parafilm or foil to prevent evaporation.
6. Allow the DNA to rehydrate at room temperature. High-molecular-weight DNA can take several days to completely rehydrate and dissolve. Check the tube to see if the DNA precipitate has completely dissolved.

Method-2: Preparation of E. coli DNA

The isolation of DNA from *E. coli* poses relatively few problems. With sufficient care, it is possible to isolate very pure DNA with an average fragment size above 10 kb. The procedure (Schleif and Wensink, 1981) is given below. It yields adequate DNA for cloning by recombinant DNA technology or for other procedures.

Materials

For additional materials refer method-1 (Basic protocol). Lysozyme solution: dissolve at concentration of 2 mg/ml in 0.15 M NaCl, 0.1 M EDTA, pH 8.0.

Procedure

1. Grow the cells overnight to the stationary phase in 5 ml of YT medium.
2. Centrifuge 3×1.5 ml of cells in the Eppendorf centrifuge.
3. Resuspend the combined cell pellets from the 4.5 ml of cell culture in 15 µl of freshly-prepared lysozyme solution.
4. Incubate it at 37°C for 10–20 min. until the cells just begin to lyse. Lysis is indicated by an increase in viscosity.
5. Freeze in dry-ice ethanol. Then, add 125 µl of 1% SDS, 0.1M NaCl, 0.1M Tris–HCl, pH 9.0 to the frozen solution. After adding, stir the mixture as the cells thaw out.
6. Add 150 µl of phenol that has been saturated with the buffer and then mix it well.
7. Remove the upper aqueous phase that contains the DNA, and add 300 µl of 95% ethanol. Mix and freeze it at –20° or –70°C for 30 min., and centrifuge in the Eppendorf centrifuge for 5 min.
8. Resuspend the DNA pellet in 100 µl of 0.1 × SSC buffer (see Appendix 3). The low concentration of salt promotes dissolving of the DNA. After the DNA has dissolved, add 5 µl of 20 × SSC buffer.
9. Add 4 µl RNase T1 at 0.8 mg/ml and 3 µl RNase A at 2 mg/ml and incubate at 37°C for 30 min.
10. Add 100 µl of phenol saturated with SSC, vortex, and centrifuge it for 3 min. Remove the aqueous layer and precipitate the DNA in ethanol.

Note

1. Unless otherwise specified, all the operations should be carried out at 0°C with solutions that have been chilled to 0°C.

C. ISOLATION OF EUKARYOTIC NUCLEAR DNA

The isolation of nuclear DNA from eukaryotic cells is at the heart of molecular biology. The protocol (Figure 7.8) varies depending upon the cell types and the organisms. The isolation of nuclear DNA from plant tissues requires a different protocol from that required for mammalian nuclear DNA. This is because plant cells are surrounded by rigid cell walls and plant tissues often contain a variety of secondary metabolites that can damage the DNA. Thus, DNA isolation from plants can present some particular difficulties. The methods presented here makes it possible to extract purified high-molecular-weight (>50 kb) plant DNA, without the use of expensive equipment and time-consuming procedures; furthermore, tissues as small as individual ovules and embryos, or even small pieces of tissue from various parts of the same plant, can be used. The DNA isolation methods discussed here represent the total cellular DNA (nuclear, chloroplast, and mitochondrial). The extraction procedures for plant DNA in general must accomplish the following: (1) The cell walls must be broken in order to release the cellular constituents. (2) The membrane must be disrupted, so that the DNA is released into the extraction buffer. (3) The DNA must be protected from the endogenous nucleases. (4) Shearing of the DNA must be minimized. DNA in solution can be broken by exposure to turbulence due to drawing through small orifices. (5) Enzyme inhibiting polysaccharides are often present in the purified DNA particularly in higher plants, and this has to be removed.

Isolation of Nucleic Acids

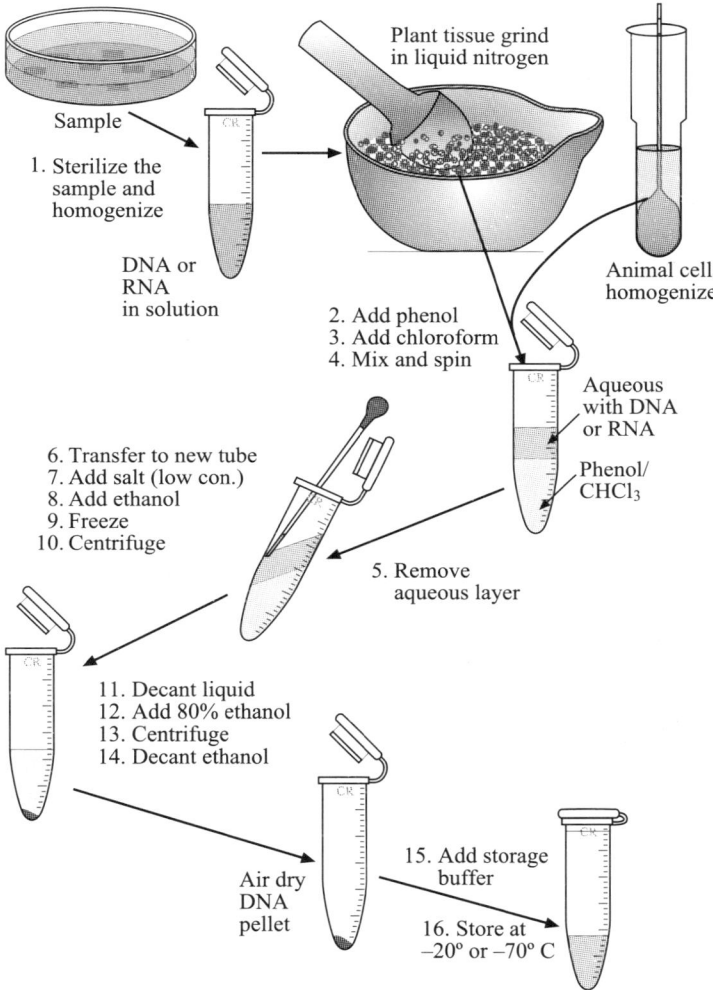

Figure 7.8 *Nucleic acid extraction and precipitation*

Method-1: Preparation of plant-nuclear DNA using CsCl centrifugation–Basic protocol

Materials

Extraction buffer: 100 mM Tris.Cl pH 8.0, 100 mM EDTA, 250 mM NaCl, 100 μg/ml proteinase K

TE buffer: 10 mM Tris–HCl pH 7.6, 1 mM EDTA

Cold, sterile water; Liquid nitrogen; 10% (w/v) N–lauroylsarcosine (Sarkosyl); Isopropanol; Cesium chloride; 10 mg/ml ethidium bromide; CsCl–saturated isopropanol; Ethanol;

3 M sodium acetate; pH 5.2; Beckman JA–14; JA–20 or JA–21; and VTi80 rotors or their equivalents.

Procedure

1. Harvest 10–50 g of fresh plant tissues. In order to reduce the starch content, use younger plants and place them in a dark place 1–2 days prior to harvest.
2. Rinse with cold, sterile water, blot dry, and freeze with liquid nitrogen. Grind to a fine powder with mortar and pestle.
3. Transfer the frozen powder to a 250 ml centrifuge bottle, immediately add 5–10 ml extraction buffer/g plant tissue, and stir gently to mix. Add 10% (w/v) Sarkosyl to a final concentration of 1%, and incubate for 1–2 hr at 55°C. In all the subsequent steps, the solutions should be vortexed or mixed vigorously.
4. Centrifuge it for 10 min. at 5000g in a JA–14 rotor, at 4°C. Save the supernatant and repeat this step if necessary to remove any debris.
5. Add 0.6 vol of isopropanol and mix. If the precipitate is not visible, place it for 30 min. at −20°C. Centrifuge it for 15 min. at 7500g in a JA–14 rotor, at 4°C. Discard the supernatant.
6. Resuspend the pellet in 9 ml of TE buffer, add 9.7 g solid CsCl, mix, and incubate it for 30 min. on ice. Centrifuge it for 10 min. at 7500g in a JA–20 rotor, at 4°C, and save the supernatant. The Sarkosyl phase, which may form on the top of the solution, can be removed by filtering the supernatant through two layers of cheesecloth.
7. Add 0.5 ml of 10 mg/ml ethidium bromide and incubate it for 30 min. on ice. Centrifuge it for 10 min. at 7500g, at 4°C. Transfer the supernatant into two 5 ml quick-seal ultracentrifuge tubes and seal them.
8. Centrifuge for 4 hr at 525,000g in a VTi80 rotor, at 20°C, or overnight at 300,000g, at 20°C. Collect the DNA band using a 15-G needle and syringe.
9. Remove the ethidium bromide by repeatedly extracting the DNA with CsCl-saturated isopropanol.
10. Add 2 vol of water and 6 vol of 100% ethanol, mix, and incubate it for 1 hr at −20°C. Centrifuge it for 10 min. at 7500g in a JA–20 rotor, at 4°C.
11. Resuspend the pellet in TE buffer, reprecipitate by adding 1/10 vol of 3 M sodium acetate and 2 vol of 100% ethanol. Repeat the centrifugation. Air dry and resuspend in the TE buffer. The yield should be 10–40 μg DNA (50-kb length)/g fresh plant tissue.

Method-2: Preparation of plant-nuclear DNA using CTAB

The non-ionic detergent cetyltrimethylammonium bromide (CTAB) can be used to liberate and complex with total cellular nucleic acids (nuclear, chloroplast, and mitochondrial DNA) from a wide array of plant genera and tissue types. The DNA preparations, thus obtained, are suitable for analysis using restriction enzymes and Southern blotting techniques without purification on CsCl gradients.

Materials

EB (extraction buffer): 1% CTAB (w/v), 50 mM Tris.Cl (pH 8.0), 50 mM EDTA (pH 8.0), 0.7 M NaCl, 1 mM 1,10–O-phenanthroline, 1% β-mercaptoethanol.

Precipitation buffer (PB): 1% CTAB (w/v), 50 mM Tris.Cl (pH 8.0), 50 mM EDTA (pH 8.0). Autoclave and store at room temperature.

RNase stock solution: 1 mg/ml RNase A, 100 U/ml RNase T1. The solution should be heated near to boiling point for at least 10 min. to destroy any DNases and stored frozen.

CTAB (cetyltriethylammoniumbromide) stock: 10% (w/v). As it takes some time to dissolve, heat gently under stirring to accelerate the process. Autoclave and store at room temperature.

Liquid nitrogen; Chloroform; Isopropyl alcohol; 80% Ethanol; 15 mM ammonium acetate pH 7.5.

Cleaned and autoclaved mortar, pestle, and metallic spatula; 2.2–ml Eppendorf tubes; 10-100-, and 1000-μl pipettes with corresponding tips; Centrifuge tubes 50 ml; Water bath 65°C; Centrifuge.

Procedure

Preparation of sample

1. The plant material (callus, leaves, seedlings, etc.) used in this protocol should be air dried or freeze dried. The tissue can be air dried in the laboratory by spreading the cut shoots on a sheet of absorbent paper and turning this daily, it usually dries in 5–7 days. This dried tissue can be saved for use at a later date by sealing it in a tightly closed jar and freezing.
2. Alternatively, harvest the fresh plant material and seal it in plastic bags. Freeze the samples by putting them in a –70°C environment, or immersing them in liquid nitrogen.

DNA Extraction

1. Grind 1 g of dried (air-dried or freeze-dried) plant tissue to a fine powder in a mortar with a pestle.
2. Mix the powder with 25 ml of EB buffer in a 50 ml, capped centrifuge tube.
3. Place the tube in a 65°C water bath; incubate it there for 1 hr. Several times during the hour, mix the tube's contents by inversion.
4. Remove the tube from the water bath and allow it to cool for several minutes on the bench before proceeding to the next step.
5. Add 20 ml chloroform to the tube, cap and mix by inversion until the contents are thoroughly mixed. When mixed, the extraction buffer and the chloroform will form a thick emulsion.
6. Centrifuge the tube at 3,500g for 10 min. to break the emulsion and separate the tube contents into two phases.
7. Upon removal from the centrifuge, the contents of the tube form three distinct layers. At the bottom is a green layer of chloroform. Often, bits of plant material are pelleted under this layer. In the middle is an interface (whitish or yellowish in color) consisting largely of denatured protein and bits of leaf tissue. At the top of the tube is the straw-colored aqueous layer. This top layer contains the majority of the DNA in the preparation. Using a pipette, remove this top layer to a small flask or beaker, without disturbing the interface.
8. What remains in the centrifuge tube is the interface and the organic (chloroform) waste.
9. To the aqueous phase in the flask, add 2/3 of isopropyl alcohol (i.e., 34% of tube's contents). And mix by swirling the contents. The DNA will precipitate to form a cottony mass.

10. Using a glass rod or a Pasteur pipette, transfer the DNA to a clean flask. Add 10 ml of 80% ethanol and 15 mM ammonium acetate, and swirl to wash.
11. After about 20 min., transfer the precipitated DNA to a microcentrifuge tube and centrifuge briefly to drive the DNA to the bottom of the tube. Using a Pasteur or capillary pipette, remove the residual ethanol. Allow the DNA to dry in the uncapped tube for about 10 min. on the bench top.
12. Add 0.75 ml of TE buffer to the DNA to dissolve the precipitate and store it at 4°C.
13. Treat with RNase.

Method-3: Preparation of plant-genomic DNA for PCR analysis

The following method (based on Dewards et al., 1991) describes the rapid extraction of small amounts of plant genomic DNA, which are suitable for PCR analysis. This method can be used for a variety of plant species and it does not require the liquid nitrogen grinding or phenol/chloroform extraction steps.

Materials

Extraction buffer: 200 mM Tris.Cl pH 7.5, 250 mM NaCl, 25 mM EDTA, 0.5% SDS.

TE buffer: 10 mM Tris–HCl pH 7.6, 1 mM EDTA.

Leaf tissue; Microcentrifuge, Eppendorf tubes, Disposable grinders (Bel-art products: Scienceware, Pequannock, NJ), Isopropanol, Ethanol.

Procedure

1. In a sterile Eppendorf tube collect a pinch out of the disc of leaf material and place it directly into the tube. This ensures uniform sample size and also reduces any possible contamination arising from handling the tissue.
2. Macerate the tissue in the original Eppendorf tube by using disposable grinders at room temperature, without buffer, for 15 seconds.
3. Add 400 µl of the extraction buffer and vortex the sample for 5 seconds. This mixture can be left at room temperature until all the samples have been extracted (>1 hr).
4. Centrifuge the extract at 13,000 rpm for 1 min. and transfer 300 µl of the supernatant to a fresh Eppendorf tube.
5. Add 300 µl of isopropanol to the supernatant and mix. Leave the supernatant at room temperature for 2 min.
6. Centrifuge the supernatant at 13,000 rpm for 5 min. Vacuum-dry the pellet, dissolve it in 100 µl of 1X TE, and store it at 4°C. This will remain stable for more than a year.

Method-4: Universal and rapid salt extraction of high-quality genomic DNA

This method (Aljanabi and Martinez, 1997) is very simple, fast and universally applicable to extract high-quality megabase genomic DNA from different organisms. It does not require expensive and environmentally hazardous reagents and equipments and the amount of tissue required is only ~50–100 mg. The quality of DNA extracted is suitable for carrying out PCR-based reactions (RAPD), restriction digestion, Southern blot, and cloning techniques.

Materials

Salt homogenizing buffer: 0.4 M NaCl, 10 mM Tris.Cl pH 8.0, 2 mM EDTA pH 8.0.

Procedure

1. Homogenize 50–100 mg of tissue (from plant, animal, insect, etc) in 400 µl of sterile salt homogenizing buffer, using a Polytron Tissue Homogenizer, for 10–15 s.
2. Add 40 µl of 20% SDS (2% final concentration) and 8 µl of 20 mg/ml proteinase K (400 µg/ml final concentration) and mix well.
3. Incubate the samples at 55–65°C for at least 1 hr or overnight.
4. Add to each sample 300 µl of 6 M NaCl (NaCl saturated in water).
5. Vortex the samples for 30 s at maximum speed and centrifuge the tubes for 30 min. at 10,000g.
6. Transfer the supernatant to fresh tubes, add an equal volume of isopropanol to each sample and mix well.
7. Incubate the samples at –20°C for 1 hr. Then centrifuge them for 20 min., at 4°C, at 10,000g.
8. Wash the pellet with 70% ethanol, vacuum dry and resuspend it in 300–500 µl sterile distilled water.
9. The purity of DNA, determined from the A_{260}/A_{280} ratio, will be average to 1.78. Generally, there is no RNA contamination or DNA degradation. The yield of DNA ranges from 500–800 ng/mg fresh weight. The amount of tissue can be scaled up to more than 10-fold without any loss in the quantity or quality of the DNA.

D. PLANT VIRAL NUCLEIC ACID PRECIPITATION

The genome of viruses is defined by its state in the mature virion, and varies between virus families. Unlike the genomes of true organisms, the virus genome can consist of DNA or RNA, which may be double- or single-stranded, circular or linear, depending upon the type of virus. In some viruses, the genome consists of a single molecule of nucleic acid (*monopartite*), which may be linear or circular, but in others it is segmented or diploid (*bipartite*). Single-stranded viral genomes are described as positive sense, negative sense or ambi-sense. In recent years, the development of new methods for the genetic modification of plants has promoted intensive efforts to produce plants with greater resistance to viral diseases and other beneficial qualities. The most significant contribution of a plant virus to the development of systems for the constitutive and transient expression of foreign genes in plants has been that of the 35S promoter sequence of CaMV. The first of these was the demonstration of resistance to plant virus attack by expression of the TMV CP gene from tobacco plants. Since then, resistance has been induced by transformation with the CP genes of many plant viruses in different host plants. More recently, other viral genes including the replicase and movement protein genes have been used to bring about resistance in plants.

The study of viral genomes involves virus adaptation, purification, DNA isolation, restriction enzyme analysis, genome amplification, cloning and sequencing. Very pure virus preparations are essential for chemical, physical, and certain biochemical studies and many biological investigations depend upon the availability of at least partially purified preparations. It must be stressed here that no two viruses are exactly alike and, consequently, there are about as many purification procedures as there are viruses. In this chapter, only a few representative nucleic acid isolation procedures, which represent the plant viruses with genome-type DNA or RNA, are briefly described.

Protocol–1: Isolation of Caulimovirus and DNA extraction (Covey et al., 1998))

Each member of the caulimovirus group has a circular dsDNA genome of approximately 8 kbp, which is encapsidated in a spherical, naked nucleocapsid of approximately 50 nm in diameter. The outer layer contains 420 protein subunits arranged with $T=7$ icosahedral symmetry. The DNA is associated with internally disposed domains of the coat protein. Other proteins may also be present in the virion. Caulimoviruses characteristically produce subcellular inclusion bodies in infected tissues that contain most of the virions found in cells, embedded in an apparently random manner. The replication of caulimoviruses involves the alternation of genomes as DNA and RNA forms, progeny virion DNA being generated by reverse transcription of a terminally redundant, genome-length RNA utilizing a virus-encoded polymerase.

The main problem to overcome in purifying caulimoviruses from infected tissue is the disruption of the two types of inclusion bodies containing the virions. The virion isolation method, described here, was designed to liberate virions from inclusion bodies by incubation of plant extracts in urea. The solubilization of cells and the prevention of virus aggregation is enhanced by the inclusion of the non-ionic detergent Triton X-100. Polyphenoloxidase activity is minimized by the reducing agent sodium sulfite. The liberated virions are purified by differential centrifugation. The virion DNA is released from the purified virus by digestion with proteinase K.

Materials

Celite abrasive; Tip-flattened sterilized glass rod; Sodium phosphate buffer pH 7.2; Solid sodium sulfite (0.75% in extraction buffer); Solid urea (1 M); 10% Triton X-100; DNase buffer (50 mM Tris–Cl pH 7.5, 5 mM $MgCl_2$); Phenol:chloroform:isoamyl alcohol (25:24:1); 10 mg/ml Proteinase K in TE with 1% SDS; 2 mg/ml DNase I, in DNase buffer; 2 mg/ml Pancreatic ribonuclease A (RNase A) in TE; 10% SDS; TE solution; 0.5 M $MgCl_2$; 30% Polyethylene glycol 6000 (PEG); Refrigerated ultracentrifuge; Bottom-drive blender cooled to 4°C; Muslin cloth; Rubber policeman.

Procedure

1. Harvest 100g of turnip leaves, systematically infected with CaMV. Select younger leaves, up to ca. 10 cm in length. Wash and lightly dry leaves.
2. Place the leaves in the blender with 200 ml 0.5 M PBS pH 7.2, and 0.75 g sodium sulfite per 100 ml homogenate. Blend to fine fragments in the cold.
3. Pour the homogenate into a beaker adding 6 g urea and 25 ml of 10% Triton X-100, both per 100 ml homogenate. Stir it with a magnetic stirrer at 4°C, overnight.
4. Centrifuge the extracted homogenate at 4000g for 10 min. at 4°C.
5. Gently pour the supernatant through four layers of muslin and then distribute the green liquid into tubes for pelleting virus by ultracentrifugation at 70,000g for 2 hr at 4°C in an ultracentrifuge.
6. Pour off the supernatant and resuspend the pellets by dispersing them in sterile distilled water per tube for 1–2 hr, using a rubber policeman occasionally.

7. Pool the resuspended pellets and centrifuge them twice in a microcentrifuge to remove particulate matter.
8. Pellet the virions from the supernatant by ultracentrifugation at 136,000g at 4°C for 1 hr.
9. Resuspend the pellets, each in 1 ml of DNase buffer. The virion yield can be assessed at this point by UV spectrophotometry. A suspension of CaMV virions of 1 mg/ml has an OD_{260} of 7. The virions can be stored at 4°C or −20°C.
10. To isolate the virion DNA, the purified virions are first treated with DNase I (10 µg/ml, for 10 min. at 37°C; reaction stopped by addition of EDTA to 1 mM) to remove any fragments of plant DNA. The virions are disrupted by adding proteinase K to a final concentration of 0.5 mg/ml with 1% SDS in TE and incubating for 15 min. at 37°C. The DNA is purified from the lysed mixture by phenol:chloroform extraction (minimum twice). Then, the DNA is concentrated by ethanol-precipitation, collected, and quantified by absorbance at 260 nm.

Protocol–2: Tobamovirus isolation and RNA extraction (Chapman, 1998)

The tobamoviruses produce rigid, rod-shaped particles, with dimensions of approx 300 × 18 nm, and form one of the most extensively studied groups of plant viruses. The members of this group infect a wide range of angiosperms, and individual members frequently have wide experimental host ranges. The best characterized of the tobamoviruses is the type member tobacco mosaic virus (TMV). The TMV genomic material is a single- and positive-stranded RNA, which is about 6395 nucleotides long. TMV was the first virus to be purified, in 1935 (Stanley 1935), and since then many methods have been developed for its purification. The method described below is simple and does not require ultracentrifugation. The purification is dependent upon the process of virion precipitation in the presence of the hydrophilic polymer polyethylene glycol (PEG).

Materials

0.5 M Phosphate buffer: Prepare a 0.5 M solution of disodium hydrogen orthophosphate and adjust the pH to 7.2 with 0.5 M potassium dihydrogen orthophosphate.

Virion extraction buffer: Add 1% (v/v) 2-mercaptoethanol to the 0.5 M phosphate buffer described above.

10 mM phosphate buffer: Prepared by a 50-fold dilution of the 0.5 M phosphate buffer described above.

5X RNA extraction buffer: 0.5 M sodium chloride, 5 mM EDTA, 5% (w/v) SDS, 0.1 M Tris.Cl (pH 8.0).

5 M Sodium chloride; 20% (w/v) PEG (mol wt 8000); 3 M sodium acetate (pH 5.0), treated with 0.1% DEPC; Distilled water treated with 0.1% DEPC; Butanol; Chloroform; Absolute ethanol; Phenol:chloroform; Acid-washed sand; Miracloth; Waring-type blender; Cheesecloth; Polypropylene and polycarbonate centrifuge tubes; Microcentrifuge tubes; Sterile 1.5-ml microcentrifuge tubes; Liquid nitrogen; mortar, and pestle.

Procedures

Virus purification

1. Collect 20 g of systemically-infected leaf tissue displaying infection symptoms. Using a pestle and mortar, with a little acid-washed sand to aid homogenization, grind the leaf tissue in 60 ml of virion extraction buffer. Start grinding with a small amount of buffer and progressively add more. Continue grinding until the tissue is well macerated.
2. Filter the homogenate through two layers of Miracloth into the polypropylene centrifuge tubes. Squeeze as much of the liquid as is possible out of the material retained by the Miracloth without contaminating the filtrate with particles.
3. Add butanol (0.8 ml/10 ml of filtrate) drop-wise to the filtrate, while swirling the contents of the tube. Cap the tubes and incubate them at room temperature for 15 min. Every few minutes, mix the contents of the tube by inversion. Chlorophyll and other coagulated material should collect in the upper organic phase.
4. Centrifuge the tube contents at 10,000g for 30 min. at 12°C. Recover the lightly-pigmented aqueous phase and avoid taking any of the pelleted material or the upper, organic layer. If any of this undesired material is carried over, the aqueous phase should be centrifuged again in fresh tubes for 15 min. Then, filter the clarified extract through two layers of Miracloth into fresh centrifuge tubes.
5. Add 20% of the PEG solution to give a final concentration of 4%. Mix the contents of the tube by inversion and incubate it on ice for 15 min. Periodically mix the contents of the tube. The solution should turn cloudy as the virus precipitates.
6. Pellet the virus by centrifugation at 10,000g for 15 min. at 4°C. This should yield a whitish pellet that may be contaminated with some traces of pigmented material. Decant the supernatant and centrifuge briefly to collect residual liquid at the bottom of the tube. Pipette off the residual liquid.
7. Dissolve the pellet in 8 ml of 10 mM phosphate buffer. Centrifuge it at 10,000g for 15 min. at 4°C. Then, transfer the supernatant to a fresh tube.
8. Add 1.7 ml of 5 M NaCl and 2.42 ml of 20% PEG to the supernatant. Mix the contents in the tubes and incubate it on ice for 15 min. Pellet the virus by centrifugation at 10,000g for 15 min. at 4°C; this should yield a white viral pellet. Decant the supernatant; centrifuge briefly and pipette off the residual liquid.
9. Dissolve the pellet in 2 ml of 10 mM phosphate buffer. Divide the solution between two 1.5-ml microcentrifuge tubes. Centrifuge the tubes in a microcentrifuge at 13,000g for 30 s and pipette the supernatant into fresh microcentrifuge tubes.
10. This procedure should yield at least 20 mg of virus. To determine the yields, prepare dilutions of small allocates of the preparation and measure the absorbance at 260 and 280 nm. An A_{260}/A_{280} ratio of about 1:19 is expected for TMV; however, this ratio varies between different tobamoviruses.

Viral RNA Extraction

1. Dilute an aliquot of the virion preparation to 10 mg/ml with a 10 mM phosphate buffer. Pipette 0.8 ml of this dilution into a 2 ml microcentrifuge tube and add 0.2 ml of 5X RNA extraction buffer. Add 1 ml of phenol:chloroform, vortex briefly until an emulsion is formed, and then centrifuge it in a microcentrifuge tube at 13,000g for 5 min. at room temperature.

2. Collect the upper aqueous phase from the third phenol:chloroform extraction. Add an equal volume of chloroform, and vortex it briefly to form an emulsion. Separate the phases by microcentrifugation.
3. Collect the upper aqueous phase, which should contain about 0.7 ml, and divide it equally between two 2-ml microcentrifuge tubes. To each tube, add 0.1 vol of 3 M sodium acetate and 2.5 vol of ethanol. Mix and incubate the sample at –20°C for 15 min. to precipitate the RNA.
4. Pellet the RNA by microcentrifugation at 13,000g for 15 min. at 4°C. This should yield a visible white pellet. Decant the supernatant, centrifuge briefly to collect the residual liquid at the bottom of the tubes, and pipette this off without disturbing the pellet. Dry the pellet under a vacuum for few min. Add 0.2 ml of DEPC-treated water and dissolve the pellet by gently vortexing.
5. This procedure should yield about 200 µg of RNA. Then, prepare dilutions of the dissolved RNA and measure the absorbency at 260 and 280 nm; the preparation should have an A_{260}/A_{280} ratio of 2.0. The RNA yield can be estimated by assuming that a 40 µg/ml solution of RNA has an absorbance of 1.

V RNA EXTRACTION AND ANALYSIS

Isolating intact, clean RNA is important in cloning experiments and is essential to analyzing gene expression (Figure 7.9). A typical mammalian cell contains about 10–15 µg of RNA, 80–85% of which is rRNA. Most of the remaining 15–20% consists of a variety of low-molecular-weight species (tRNA, small nuclear RNA, etc). These RNAs are of a defined size and sequence and can be isolated in a virtually pure form by gel electrophoresis, density gradient centrifugation, or anion-exchange or high-performance liquid chromatography. In contrast, mRNA, which makes up between 1% and 5% of the total cellular RNA, is heterogeneous in both size and sequence. However, most of the eukaryotic mRNAs carry at their 3' termini a tract of polyadenylic acid residues that is generally long enough to allow the mRNAs to be purified by an affinity chromatograph on oligo (dT)-cellulose.

The critical factor in all RNA experiments is the isolation of full-length RNA (undegraded RNA). Most cell types contain relatively high levels of RNase activity, which is normally located in the vacuoles. These ribonucleases are very stable and are highly active in the absence of co-factors. During the RNA extraction procedure, the RNA should be protected against these endogenous and external sources of RNase. The cells are packed with RNases and, thus, most methods to isolate RNA rely on a chaotropic agent to rapidly disrupt and denature the proteins. The RNA can be isolated from the other cellular macromolecules in this mix by virtue of its differential precipitation in the presence of acetate salts or its greater density. The first step in all RNA isolation protocols involves lysing the cell in a chemical environment, which results in the denaturation of ribonuclease. The RNA is then fractionated from the other cellular macromolecules. The cell type from which the RNA is to be isolated, and the eventual use of that RNA, will determine which of the procedures will be appropriate.

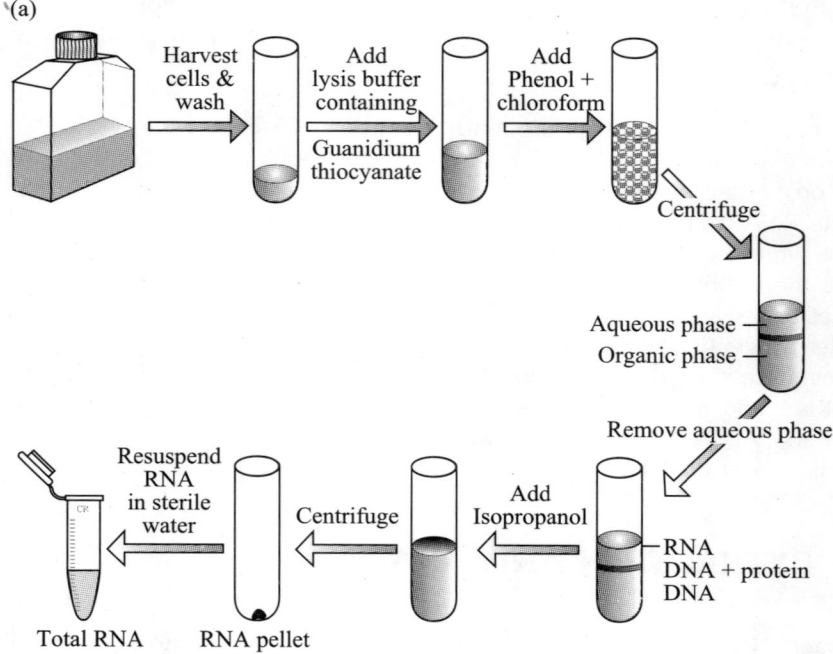

Figure 7.9 *General protocol for isolation of total RNA (refer text for more details)*

Reducing ribonuclease contamination

To avoid RNase contamination from skin and dust, always wear latex or vinyl gloves when carrying out the following procedures and frequently change these gloves. Non-disposable plastic ware should be treated by rinsing them with 0.1 M NaOH/1 mM EDTA solution, followed by ribonuclease-free water. Glassware should be baked at temperatures higher than 240°C for 4 hr or overnight. Alternatively, plastic and glassware can be treated with diethyl pyro-carbonate (DEPC), which is an efficient inactivator of ribonucleases. The glassware, solutions and water should be treated by 0.2% (v/v) DEPC solution and followed by autoclaving, to remove any residual traces of DEPC. Those solutions, which cannot be autoclaved, have to be prepared either by using DEPC-treated water or filter sterilization to remove the RNases.

Methods available

Most of the methods employed for the isolation of RNA from mammalian (*Method 1*) cells use the protein denaturant *guanidine isothiocyante* (GITC) in an initial cell or tissue extraction, followed by ultracentrifugation through a cesium chloride gradient, to separate the RNA from other macromolecules (Chargwin *et al.*, 1979). Although, this method yields pure RNA, it is limited by the difficulty in the simultaneous extraction of RNA from multiple samples and CsCl contamination of RNA. For these reasons, this method has largely been superseded by the single-step extraction method of Chomczynski and Sacchi (1987). This method uses acid GITC as the initial denaturant, followed by an acid phenol–chloroform extraction step to separate the RNA from other cellular contaminants. Even though the RNA extraction methods used for yeast (*Method 3*) and *E. coli* (*Method 4*) are similar to those used for mammalian cells, yeast cells requires an extra step to disrupt the tough cell wall before extraction. The protocols for the isolation of poly(A)$^+$ RNA (*Method 5*) depend on the fact that most eukaryotic mRNAs end in a stretch of adenylate residues, the poly(A) tail. The mRNA can be removed by hybridizing the poly(A) tails of mRNA to immobilized oligo(dT) on solid support (*Method 6*).

Method-1: Preparation of cytoplasmic RNA from tissue culture cells

Materials

Lysis buffer, ice-cold: 0.1 M dithiothreitol, 4 M guanidine isothiocyanate, 0.5% (v/v) N-lauryl sarcosine, 20 mM sodium acetate, pH 4.0

Phosphate buffer saline (PBS), ice-cold; 20% SDS; 20 mg/ml proteinase K; 25:24:1 phenol/chloroform/isoamyl alcohol; 24:1 chloroform/isoamyl alcohol; 3 M sodium acetate, pH 5.2 (DEPC–treated); 100% ethanol; 75% ethanol/25% 0.1 M sodium acetate, pH 5.2; DEPC-treated water.

Beckman JS-4.2 rotor or equivalent; Ice bath; UV spectrophotometer and cuvettes; Vortex mixer; Water bath.

Procedure

For monolayer cultures

1. Wash the cultures three times with 1 ml of ice-cold PBS for each 10 cm dish. Scrape the cells into a small volume of ice-cold PBS and transfer them into a centrifuge tube on ice. Centrifuge it for 5 min. at 300g, discard the supernatant, and keep it cold. Alternatively, trypsinize to remove the cells from the flask, and wash the cells twice with cold PBS.

For suspension cultures

1. Centrifuge the suspension cultures for 5 min. at 300g, discard the supernatant, and resuspend it in one-half the original volume of ice-cold PBS. Centrifuge and discard the supernatant.
2. Resuspend the cells in 375 µl of ice-cold lysis buffer and incubate it for 5 min. on ice; then, microcentrifuge it for 2 min. at 4°C. Transfer the supernatant into a clean tube containing 4 µl of 20% SDS and vortex it immediately.
3. Add 2.5 µl of 20 mg/ml proteinase K, and incubate it for 15 min. at 37°C.
4. Extract with 400 µl phenol/chloroform/isoamyl alcohol by vortexing for 1 min. Microcentrifuge it, transfer the upper phase to a clean tube, and repeat the extraction. Extract with 400 µl chloroform/isoamyl alcohol, and transfer the upper phase to a clean tube.
5. Add 40 µl of 3 M sodium acetate, pH 5.2, and 1 ml of 100% ethanol. Mix these and precipitate the mixture for 15–30 min. on ice, or overnight at –20°C.
6. Microcentrifuge it for 15 min. at 4°C and rinse the pellet with 1 ml of 75% ethanol/5% 0.1 M sodium acetate, pH 5.2. Dry and resuspend it in 100 µl of DEPC-treated water. Dilute 10 µl in 1 ml alkaline water and determine A_{260} and A_{280}. Store the remainder at -70°C. The yields are usually 30–100 µg RNA for a confluent 10 cm dish or 10^7 lymphoid cells.

Method-2: Isolation of total and polysomal RNA from plant tissues

The following procedure has been applied to a wide variety of plant tissues, including adult leaves and roots, seedlings grown under dark or light conditions, and callus and suspension cells, invariably yielding large amounts of undegraded total RNA suitable for *in vitro* translation and cDNA synthesis.

Materials

RBA extraction buffer: 100 mM LiCl, 1% SDS, 100 mM Tris–NaOH (pH 9.0), 10 mM EDTA. This can be stored at room temperature for a maximum of 6 months.

Distilled phenol with 0.1% hydroxyquinoline: This can be stored at –20°C for a maximum of 6 months.

Chloroform (pro-analysis)

8 M LiCl: The solution should be filtered through a sterile Whatman 1 MM after autoclaving and keeping it overnight at room temperature. The Whatman paper can be autoclaved wrapped in aluminum foil (this can be stored at –20°C for a maximum of 6 months).

2 M LiCl: This can be prepared from the filtrated 8 M LiCl (it can be stored at +4°C for a maximum of 1 month).

80% ethanol; Double-distilled sterile water.

Procedure

1. Harvest the material in plastic tubes filled with liquid N_2 and placed in a container filled with liquid N_2. Determine the fresh weight of the material
2. Prepare a 1:1 mixture of RNA extraction buffer and phenol with hydroxyquinoline, and heat the mixture to 90°C in a water bath in a fume hood.
3. Grind the material in liquid N_2 in a pre-cooled mortar and pestle until a fine homogeneous powder is obtained.
4. Transfer the frozen powder to an Erlenmeyer flask of suitable size kept in an ice–salt mixture, pre-cooled at –80°C, with the aid of a metal or plastic spatula cooled in liquid N_2.
5. Add 2 ml of the well-mixed phenol/extraction buffer per gram of fresh weight of plant material, and swirl the flask vigorously. Intermittently heat the water bath (90°C) until a milky suspension is obtained, which is devoid of clumps of frozen material. The final temperature of this mixture should be 25–30°C. Place the Erlenmeyer flask on a gyrator shaker at 300 rpm for 5 min. at room temperature.
6. Add 1 ml of chloroform per gram fresh weight of plant material and continue shaking for 15–30 min. at room temperature.
7. Transfer the milky suspension to the centrifuge tubes, and centrifuge it at 20,000g for 30 min. at 25°C.
8. Remove the aqueous upper phase with a pipette and transfer it into an Erlenmeyer flask. Add 1 ml of chloroform per gram fresh weight of plant material and place the mixture on a shaker for 15 min. at 300 rpm.
9. Transfer the mixture to chloroform-resistant glass or plastic tubes and centrifuge it for 15 min. at 12,000g at 25°C.

10. Remove the aqueous phase as described in step 8. Then, transfer it to a measuring cylinder and subsequently transfer it into centrifuge tubes. Add 1/3 volume of 8 M LiCl, mix well and precipitate the RNA for 16–48 hr at 4°C.
11. Centrifuge it at 12,000g for 30 min. at 4°C. Wash the resulting RNA pellet once with 2 M LiCl at 4°C and twice with 80% ethanol, and vacuum-dry. Dissolve the pellet in double-distilled water and store it at –20°C. Generally, up to 20 mg of RNA per ml can be dissolved.

Notes

1. An extraction with phenol only results in a significant loss of poly(A)⁻containing RNA. This problem is eliminated by the combined chloroform/phenol extraction.
2. Sometimes, the RNA preparations may be contaminated with polysaccharides. The most convenient method to remove this contamination is by dissolving the pellet in a small volume of water, heating it at 60°C for 5 min., and centrifuging at 40,000g for 30 min. The resulting gel-like pellet contains most of the impurities and there is almost no loss of RNA.

Method-3: Small-scale extraction of yeast RNA

Materials

Absolute and 70% ethanol; AE buffer: 50 mM sodium acetate; 10 mM EDTA, pH 5.3; Phenol:AE: phenol re-equilibrated with AE buffer; Phenol:chloroform; 10% SDS; 3 M sodium acetate, pH 5.3; TE buffer: 1 mM EDTA, 10 mM Tris.Cl pH 7.5; Bench-top centrifuge; Collecting tubes; 15 ml Centrifuge tubes; Liquid nitrogen; Microcentrifuge tubes; 65°C water bath; Vortex mixer; Spectrophotometer.

Procedure

1. Warm up the AE buffer, 10% SDS and phenol: AE to 65°C.
2. From each extraction, set up the following 1.5 ml microcentrifuge tubes: tube A containing 40 μl 10% SDS; tube B containing 200 μl phenol:chloroform; tube C containing 40 μl 3 M sodium acetate, pH 5.3.
3. Harvest 10 ml of yeast culture (OD600 = ~0.5) by centrifugation at 5000 g for 30 sec at 4°C.
4. Discard the supernatant and drain the tube on a paper tissue.
5. Resuspend the cell pellet in 400 μl AE buffer, and transfer the contents to tube A.
6. Add 440 μl phenol:AE, tighten the screw-cap, and vortex for 30 sec.
7. Place the tube in a 65°C water bath for 3 min., vortex for 1 min., replace in the 65°C water bath for 3 min., and re-tighten the screw-cap.
8. Freeze the sample in liquid nitrogen for 30 sec, and then place it in the 65°C water bath for 3 hr.
9. Centrifuge the sample in a microcentrifuge for 10 min.
10. Transfer the aqueous phase (about 420 μl) to tube B, vortex for 10 sec, and then leave the tube at room temperature for 5 min. Centrifuge the sample for 2 min. in a microcentrifuge.
11. Transfer the aqueous phase (about 400 μl) to tube C.
12. Add 1 ml of pre-cooled ethanol, vortex for 10 sec, and leave the tube at –20°C for 2–3 hr or overnight.
13. Centrifuge the sample for 15 min. in a microcentrifuge.

14. Discard the supernatant and gently rinse the precipitate with 1 ml of 70% ethanol.
15. Discard the 70% ethanol, centrifuge the sample at 10,000g for 5 sec, remove the remaining supernatant carefully with a 20–200 μl pipette, and air-dry the precipitate.
16. Add 30–50 μl of TE, place it in a 65°C hot water bath for 5 min., vortex sample for 10 sec, centrifuge it briefly, and store the sample at −70°C.

Method-4a: Preparation of bacterial (*E. coli*) RNA – Basic protocol

This procedure delivers high-quality RNA from *E. coli* or cyanobacteria, which is suitable for Northern blotting, S1 mapping, and primer extension.

Materials

100 ml of *E. coli* culture; Stop buffer; STET lysing solution; Buffered phenol; Chloroform; 3 M sodium acetate, pH 6; 0.2 M and 10 mM vanadyl-ribonucleoside complex (VRC); 1:1 buffered phenol/chloroform; DEPC-treated water; Cesium chloride, solid; CsCl cushion: 5.7 M CsCl in 100 mM EDTA, pH 7; 10% and 70% ethanol, ice-cold.

Beckman JA-14 and JA-17 rotors or equivalents; 15-ml polypropylene tube; Beckman TL-100 ultracentrifuge with a TLA-100.3 rotor and 13×51 mm polycarbonate centrifuge tubes.

Procedure

1. Grow a 100 ml culture of *E. coli* to a log phase (stop growth by adding 1/20 vol stop buffer, if RNA is not used for primer extension or S1 nuclease analysis) and place it on ice.
2. Centrifuge the cells for 5 min. at 5500g in JA-14 rotor, at 4°C. Resuspend the pellet in 2 ml of STET lysing solution, add 100 μl of 0.2 M VRC, and transfer it into a 15-ml polypropylene tube.
3. Add 1 ml buffered phenol, vortex it for 1 min.; then, add 1 ml chloroform and vortex it for 1 min. Centrifuge it for 10 min. at 10,000g in a JA-17 rotor, at 4°C, and collect the top aqueous phase (avoid the interphase).
4. Add 1/10 vol of 3 M sodium acetate and 2 vol of ice-cold 100% ethanol. Centrifuge it for 10 min. at 10,000g, at 4°C, and resuspend the pellet in 2 ml of 10 mM VRC.
5. Extract twice with 1:1 phenol/chloroform and reprecipitate as in step 4.
6. Resuspend the pellet in 2 ml of DEPC-treated water, add 1 g of solid CsCl and dissolve it completely. Layer 2.25 ml of this solution onto a 0.75 ml CsCl cushion in a 13×51 mm TLA-100.3 polycarbonate tube. Centrifuge it for 1 hr at 280,000g, at 20°C.
7. Carefully remove the DNA at the interface and then remove the upper CsCl layer with a sterile Pasteur pipette. Pour off the remaining supernatant, mark the position of the RNA pellet, and wipe the walls of the centrifuge tube with a tissue (Figure 7.10). Re-suspend the pellet in 0.36 ml of DEPC-treated water and transfer it to a 1.5-ml microcentrifuge tube.
8. Immediately add 1/10 vol of 3 M sodium acetate and 2.5 vol of ice-cold 100% ethanol. Precipitate it for 20 min. at 70°C, and microcentrifuge for 5 min. at high speed, at 4°C. Add 1 ml of ice-cold 70% ethanol to the RNA pellet and microcentrifuge it for 5 min. at high speed, at 4°C.
9. Air-dry the pellet and dissolve it in 200 μl of DEPC-treated water. Quantify by measuring the A_{260} and A_{280}. Adjust it to 4 μg/μl finally and keep it at −70°C for long-term storage or store as an ethanol precipitate.

Isolation of Nucleic Acids 159

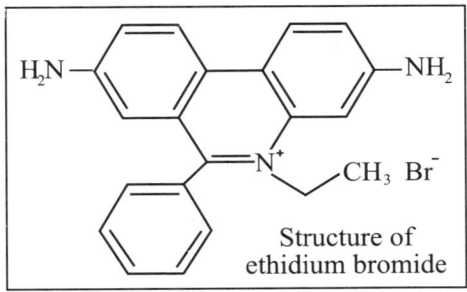

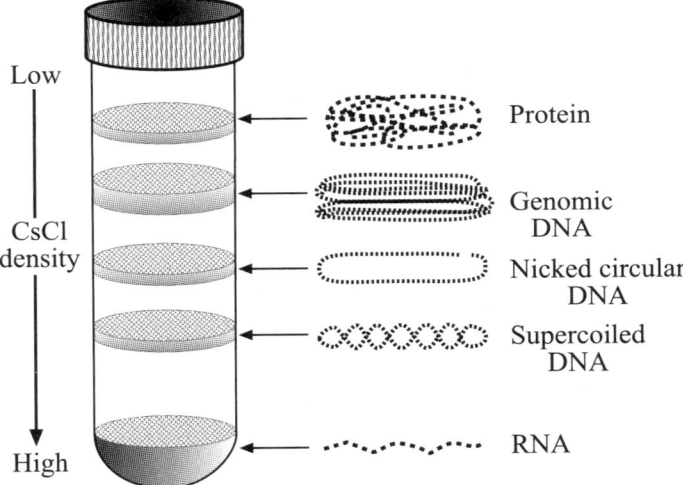

Figure 7.10 *Cesium chloride density gradients can be used to separate genomic DNA and plasmid DNA based on differential buoyant densities resulting from the amount of ethidium bromide absorbed. The chemical structure of ethidium bromide is shown*

Method-4b: Rapid isolation of RNA from *E. coli*

Procedure

1. Centrifuge the cells from a 10 ml gram-negative bacteria culture for 10 min. at 12,000g, at 4°C. Resuspend it in 10 ml of protoplasting buffer, add 80 μl of 50 mg/ml lysozyme, and incubate it for 15 min. on ice.
2. Centrifuge the protoplasts for 5 min. at 5900g, at 4°C. Resuspend them in 0.5 ml gram-negative lysing buffer, add 15 μl of DEPC, mix, and transfer the mixture to a microcentrifuge tube. Incubate it for 5 min. at 37°C, and chill on ice.
3. Add 250 μl of saturated NaCl, mix, and incubate it for 10 min. on ice.
4. Microcentrifuge for 10 min. at high speed, at room temperature or 4°C. Remove the supernatant to two clean microcentrifuge tubes. To each tube, add 1 ml of ice-cold 100% ethanol and precipitate it for 30 min. on dry ice or overnight at −20°C.

5. Microcentrifuge it for 15 min. at high speed, at 4°C, rinse the pellet in 500 µl of ice-cold 70% ethanol, and air-dry. Then, dissolve it in 100 µl of DEPC-treated water, dilute 10 µl into 1 ml alkaline water, and determine the A_{260} and A_{280}. Store the RNA at –70°C.

Method-5: Extraction of polysomes and polysomal RNA

The following procedure has been extracted from the published protocol of Vries *et al.* (1988) and gives better yield and quality of polysomal RNA.

Materials

Polysome buffer (PB): 1% Nonidet P40 (Shell), 50 mM $MgCl_2$, 25 mM EGTA, 50 mM Tris–NaOH (pH 9.0), 250 mM NaCl. This can be stored at 4°C for a maximum of 1 month.

Gradient buffer (GB): 10 mM $MgCl_2$, 5 mM EGTA, 10 mM Tris–NaOH (pH 8.5), 50 mM NaCl. This can be stored at 4°C for a maximum of 1 month

GB + 0.2 M EDTA

75% w/v sucrose in GB: Dissolve 590g of sucrose (pro-analysis) in 410 ml of GB to obtain a solution with a density of 1.2806 and 1.4396 at 20°C. The resulting volume is approximately 750 ml. After dissolution of all the sucrose check the refractive index and add 0.1% v/v of DEP, stir slowly overnight and heat at 60°C for 1 hr (this can be stored at 4°C for a maximum of 2 months).

RNA-extraction buffer; 8 M LiCl; phenol; chloroform; ethanol 96% and 70% (see solutions for total RNA).
The solutions should be made RNases-free, as indicated for the extraction of total RNA.

Procedure

1. Harvest the material in liquid N_2 and determine the fresh weight.
2. Grind the plant material in liquid N_2, using a mortar and pestle, until it becomes a fine powder.
3. Transfer the frozen powder to a second mortar (pre-cooled at 0 to 4°C) and add 4 ml/g of plant material in ice-cold polysome buffer (PB). Gently suspend the powder with a pestle until the slurry does not contain any frozen material.
4. Centrifuge it for 10 min. at 0°C and 27,000g. Filter the supernatant solution through a 3G-1 glass filter into a measuring cylinder in a cold room.
5. Transfer the filtrate into Type 42.1 or Type 65 (Beckmann) polycarbonate tubes and fill them to approximately 80%. Underlayer the filtrate with 2 ml (Type 65) or 5 ml (Type 42.1) of 60% w/v sucrose in gradient buffer (GB) with a 2-mm, inner-diameter Hamilton needle fitted to a 10 ml syringe and fill up tubes with the filtrate. Centrifuge this for 3 hr (Type 42.1) at 0°C and 40,000 rpm or 105 min. (Type 65) at 49,000 rpm. After centrifugation, decant the tubes carefully and place the inverted tubes on sterile tissues to drain off the remaining liquid from the opalescent polysome pellets. The pellets can now either be resuspended directly or be quickly frozen in liquid N_2 and stored at –80°C.
6. Prepare linear 10–40% w/v sucrose gradients in GB, in polymer SW28 or SW40 tubes (Beckmann), and keep it at 4°C.

7. Carefully resuspend the polysome pellets by pipetting or low-speed vortexing in GB to a final concentration of approximately 2.5 mg ml-1 of polysomal RNA (60 A_{260} units), which is about 5 mg ml-1 of polysomes.
8. Carefully load 16–20 mg or up to 5 mg of polysomes on SW28 or SW40 gradients, respectively. Centrifugation for 30 min. at 0°C and 40,000 rpm (SW40Ti) without a break, or 70 min. at 25,000 rpm (SW28), is sufficient to yield a clear separation between the residual monosomes and most of the contaminating hnRNA and polysomes >100 S. Usually, it is sufficient to completely monitor only 1 out of 6 gradients by A_{280} or A_{260} extinction in the flow cell and to fractionate the remainder after identifying the monosome peak and polysomes >100 S.

   The polysome-containing fractions are then collected and precipitated overnight at –20°C by the addition of 1/10 volume of 3 M sodium acetate pH 7.0 and 2 volumes of 96% ethanol. The precipitate is collected by centrifugation at 20,000g for 30 min. at 4°C, washed once with 70% ethanol and dried under a vacuum. The precipitated polysomes can now be extracted to yield polysomal RNA or can be EDTA-released to remove any residual traces of hnRNA.
9. The extraction of RNA from ethanol-precipitated polysomes or monosomes is performed by the phenol/chloroform/LiCl procedures (steps 5–11) as described for the extraction of total RNA. Before the RNA extraction, the polysomes or monosomes are first dissolved to approximately 1 mg/ml. The extent of cell disruption determines the final polysome yield.

Method-6: Purification of Poly(A)$^+$ RNA

Only a small fraction of the RNA extracted from eukaryotic cells (usually <5%) is mRNA. The mRNAs of eukaryotes generally differ from other RNA molecules (rRNA, tRNA and small RNAs) by possessing a characteristic polyadenylic acid [poly(A)] tail at 3' termini. This property permits the efficient separation of mRNAs by affinity chromatography on using immobilized complimentary strands of either oligothymidilic [oligo-(dT)] or oligouridilic (oligo-U) acids. Differential adhesion to, and elution from, these affinity ligands is effected by changing salt concentrations. A cellulose matrix is used to give a sufficiently slow flow rate to complete the poly(A) tail binding after two column runs. This is an essential step when preparing mRNA, which should be used as a template for the construction of cDNA libraries. Poly(A)$^+$ RNA usually yields better results than total RNA when analyzed by northern hybridization or nuclease-S1 assays. Figure 7.11 illustrates the general protocol for poly(A)$^+$ isolation.

Materials

Preparation of oligo(dT)–cellulose as described by Gilham (1964)

1. Suspend 0.5–1.0 g of oligo(dT)–cellulose in 0.1 N NaOH.
2. Wash the column with 10 ml of 5 M NaOH and rinse with water. Pour a column of oligo(dT)-cellulose (0.5–1.0 ml packed volume) in a sterile Dispocolumn (Bio-Rad) or a Pasteur pipette, plugged with sterile glass-wool that has been traced with diethyl pyrocarbonate [DEPC] and autoclaved. Wash the column with 3 column volumes of sterile water.
3. Equilibrate the oligo(dT) column with 10 to 20 ml of 1X column-loading buffer until the pH of the effluent is less than 8.0 (~7.5).

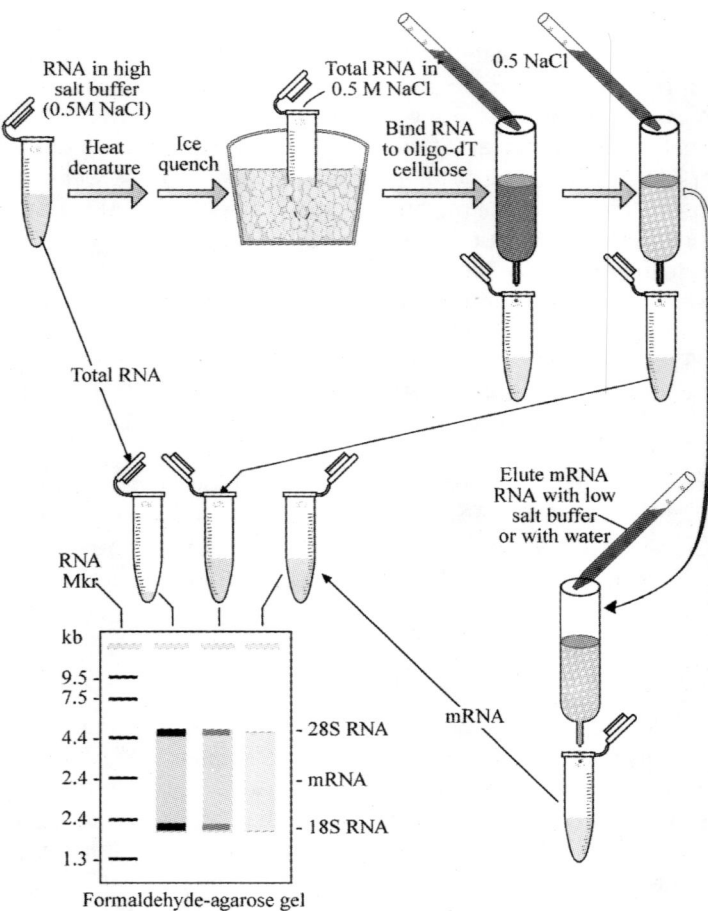

Figure 7.11 *Purification of total RNA from tissue culture cells using guanidinium thiocynate and oligo-dT cellulose*

Commercially obtained: Oligo-dT(12–18) Cellulose (Pharmacia, Type 7).

1X Loading buffer: 0.05 M sodium citrate, 0.5 M NaCl, 1.0 mM EDTA, 0.1% SDS.
2X Loading buffer: 0.1 M sodium citrate, 1.0 M NaCl, 2.0 mM EDTA, 0.2% SDS.
0.1X Salt loading buffer: 0.05 M sodium citrate, 0.1 M NaCl, 1.0 mM EDTA, 0.1% SDS.
Elution buffer: 0.05 M Tris.HCl (pH 7.5), 1.0 mM EDTA.
Washing buffer: 0.1 M NaOH, 5.0 mM EDTA.
3 M Sodium acetate (pH 5.2)
96% ethanol, kept at –20°C.

Procedure

1. Equilibrate 0.2 mg of oligo-dT cellulose in 3 ml of loading buffer and pour to form a 0.4 ml column.

2. Wash the column with 1.5 ml each of: (a) sterile water; (b) NaOH–EDTA; (c) sterile water.
3. Prepare the RNA by overnight ethanol precipitation. Thoroughly drain the ethanol from the RNA pellet and then resuspend it in a buffer. Measure the absorbance at 260 and 280 nm. The A_{260}/A_{280} should be about 2.
4. Adjust the RNA solution to 5–10 mg/ml with elution buffer and heat for 5 min. at 65–70°C, quick-cool, and mix it with an equal volume of 2X buffer and cool to room temperature.
5. Apply 1 ml volume to the oligo-dT column.
6. Collect the flow-through from the column. Re-heat, cool and re-apply it to the column.
7. The second flow-through is the first poly(A)⁻ fraction; it will typically contain 95% of the poly(A)⁻ RNA. Collect the remaining poly(A)⁻ RNA with 10 column volumes of loading buffer followed by 5 volumes of 0.1 M salt buffer. Monitor the OD of the elute at 260 nm; the final 0.1 M salt washes should show no absorbance at 260 nm.
8. Elute poly(A)⁺ RNA with 0.5 ml fractions of up to 10 column volumes of elution buffer, and assess the yield by spectrophotometry. About 5–10% of total RNA can be expected in this fraction.
9. Pool poly(A)⁻ fractions and the ethanol precipitate by adding sodium acetate to 300 mM and 2.5 volumes of 96% ethanol. Keep it at –20°C for >2 hr.
10. Separately collect the first poly(A)⁺ RNA fraction and a pool of the remaining fractions. Add sodium acetate to 0.3 M and 2.5 volumes of 96% ethanol, and place at –20°C to precipitate. Incubate it overnight at –20°C or 30 min. on dry ice/ethanol.
11. Centrifuge it for 30 min. at 304,000g, at 4°C, to pellet the (very dilute) RNA. Discard the ethanol and air-dry the pellet. Resuspend it in 150 µl of RNase-free TE buffer and pool the samples. Check the quality of the RNA by heating 5 µl for 5 min. at 70°C and fractionating on a 1% agarose gel. Approximately 1% of the input RNA should be recovered as poly(A)⁺ RNA. It should appear as a smear from ~20 kb down, with the greatest intensity in the 5- to 10-kb range.

Notes

1. Washing with NaOH/EDTA cleans and regenerates the column and this should be carried out before each reuse. If sterility is maintained, the column can be kept for several weeks at 4°C.
2. If the recovery of poly(A)⁺ RNA exceeds 20% of the total, a second cycle of extraction should be carried out.
3. The A_{260} will increase as the poly(A)⁻ RNA washes through the column; thereafter, it will decrease. The washing stage is complete when the A_{260} returns to a background level. An A_{260}/A_{280} ratio of pure RNA will be between 1.8 and 2.0.

Method-7: Purification of mRNA using oligo(dT) immobilized on magnetic beads

Materials

2X Binding buffer: 1 M LiCl, 2 mM EDTA, 20 mM Tris.Cl pH 7.5.

Deionized, RNase-free water; DEPC; Dynabeads® oligo(dT)25; 1X Elution buffer: 2 mM EDTA; RNA sample; RNase inhibitor.

1X Washing buffer: 0.15 M LiCl, 1 mM EDTA, 10 mM Tris.Cl pH 7.5.

Magnetic particle concentrator (Dynal); Microcentrifuge tubes; Sterile RNase-free microtubes, pipettes and pipette tips; 65°C water bath or heating block.

Procedure

1. Adjust the volume of 75 µg total RNA to 100 µl with distilled DEPC-treated water or with 1X elution buffer.
2. Heat to 65°C for 2 min. to disrupt any secondary structure, providing optimal conditions for hybridization.
3. Transfer 0.2 ml (1.0 mg) of Dynabeads® oligo(dT)25 from the stock tube into a microcentrifuge tube in the Dynal MPC magnet. After 30 s, remove the supernatant and wash once with 100 µl of 2X binding buffer. Remove the supernatant again with the aid of the Dynal MPC magnet.
4. Transfer the tube to another rack and add 100 ml of 2X binding buffer.
5. Add the total RNA to the bead solution and mix it gently. Let the mixture stand to hybridize for 3–5 min. at room temperature.
6. Place the tube in the Dynal MPC® magnet for 30 s and remove the supernatant.
7. Remove the tube and wash the beads 2X with 200 µl of washing buffer. Be sure to remove all of the supernatant after the final washing step.
8. Add the desired amount of 1X elution buffer. Heat to 65°C for 2 min., and place the tube immediately in the DynalMPC® magnet. Transfer the eluted mRNA to a new RNAse-free tube. The Dynabeads® oligo(dT)25 remain bound to the magnet. If the eluted mRNA is not to be used immediately, it should be stored at −70°C. RNasin/DTT or other RNase-inhibitors may be added to the solution.

Notes

1. Regeneration of Dynabeads® oligo(dT)25 for reuse: Resuspend the used Dynabeads® oligo(dT)25 in 200 µl 0.1 M NaOH, and heat to 65°C for 2 min. Place the tube in the Dynal MPC® magnet for at least 30 s and remove the supernatant. Repeat steps 1–3 twice. Resuspend the Dynabeads® oligo(dT)25 in storage buffer, place the tube in the Dynal MPC® magnet for at least 30 s and remove the supernatant. Repeat this washing step until the pH of the supernatant is less than pH 8.0 (~3X). Resuspend the Dynabeads® oligo(dT)25 in the desired volume of storage buffer. Finally, store the Dynabeads® oligo(dT)25 at 4°C.

VI SUGGESTED READING

Aljanabi SM and Martinez I (1997) Universal and rapid salt-extraction of high quality genomic DNA for PCR-based techniques, *Nucl. Acids Res.*, **25**: 4692–4693.

Anant S and KN Subramanian (1992) Isolation of low molecular weight DNA from bacteria and animal cells, *Methods Enzymol.*, **216**:20–29.

Birnboim HC and Doly J (1979) A rapid alkaline extraction procedure for screening recombinant plasmid DNA, *Nucl. Acids Res.*, **7**:1513.

Boom R, Sol CJA, Salimans MMM, Jansen CL, Wertheim van Dillen PME and van der Noordaa J (1990) Rapid and simple method for purification of nucleic acids, *J. Clinical Microbiol.*, **28**:495–503.

Challen C (1996) Purification of mRNA using oligo(dT) immobilized on magnetic beads (p21) In: *Gene Transcription: RNA Analysis*, (Ed) K Docherty, John Wiley and Sons.

Chapman SN (1998) Tobamovirus isolation and RNA extraction, In: *Methods in Molecular Biology*, vol. **81**: Plant virology protocols: from virus isolation to transgenic resistance, (Ed) GD Foster and SC Tylor, Humana Press Inc., Totowa, NJ.

Chirgwin JM, Przybyla AE, MacDonald RJ and Rutter WJ (1979) *Biochemistry*, **18**:5294–5299.

Chomczynski P and Sacchi N (1987) *Analyt. Biochem.*, **162**:156–159.

Covey SN, Nadia RJ, Al-Kaff NS and Turner DS (1998) Caulimovirus isolation and DNA extraction, In: *Methods in Molecular Biology*, vol. **81**: Plant virology protocols: From virus isolation to transgenic resistance, (Ed) GD Foster and SC Tylor, Humana Press Inc., Totowa, NJ.

Dewards K, Johnstone C and Thompson C (1991) A simple and rapid method for the preparation of plant genomic DNA for PCR analysis, *Nucleic Acids Research*, **19**:1349.

Liechtenstein C, Draper J (1985) Genetic engineering of plants, In: Glover DM (Ed) *DNA Cloning*, vol. II, IRL Press, Oxford, pp67–119.

Murray MG, Thompson WF (1980) Rapid isolation of high-molecular-weight plant DNA, *Nucl. Acids Res.*, **8**:4321–4325.

Sambrook J, Fristch EF and Maniatis (1989) *Molecular Cloning: A Laboratory Manual* (2nd Edn), Cold Spring Harbor Laboratory, New York.

8

Measuring Nucleic Acid Concentration and Purity

I INTRODUCTION

The reliable quantitation of nanogram and microgram amounts of DNA and RNA in solution is an essential procedure for researchers in molecular biology. There are two types of methods commonly used to measure the amount of nucleic acids in a preparation. If the sample is pure, spectrophotometric measurement of the amount of UV irradiation absorbed by the bases is simple and accurate (optical methods). If the amount of DNA or RNA is very small, or if the sample contains significant quantities of impurities, the amount of nucleic acid can be estimated from the intensity of fluorescence emitted by ethidium bromide or the Hoechst stain. The quantity and purity of nucleic acids can be estimated also by gel electrophoresis and dot- and slot-blot methods.

II QUANTITATION OF NUCLEIC ACIDS

A. SPECTROPHOTOMETRIC METHODS

Both DNA and RNA absorb ultraviolet (UV) light so efficiently that absorbance spectroscopy can be used as an accurate, rapid, and non-destructive method to determine concentrations as low as 2–5 µg/ml. The absorption of the sample is measured at several different wavelengths to assess the purity and concentration of nucleic acids. The nitrogenous bases in oligonucleotides have an absorption maximum at approximately 260 nm. Using a 1-cm light path, the extinction coefficient for DNA at this wavelength is 20. Based on this extinction coefficient, the absorbance at 260 nm in a 1 cm quartz cuvette of a 50 µg/ml solution of double-stranded DNA is equal to 1 (see equation 2.1). This value varies slightly with the (G+C)% of the nucleic acid, but such variation rarely needs to be considered in molecular biology.

Absorbance is also useful as a measure of the purity of nucleic acid preparations. The absorbance spectrum of double- to single-stranded structures and the increase in absorbance (*hyperchromacity*) during the transition from double- to single-stranded forms are both fairly accurate measures of purity. Proteins absorb maximally at approximately 280 nm, mainly due to tryptophan residues. Therefore, the ratio of 260/280 is a measure of the purity of a DNA preparation and it should fall between 1.65 and 1.85 unless the DNA has a very bizarre (G+C)%. A lower value suggests protein contamination. If phenol, which has one λ max at 270 nm, is contaminating the DNA preparation, then the A_{260} will be abnormally high, leading to an overestimation of the DNA concentration.

Materials

Equipment

DNA sample to be quantitated; Calf thymus DNA standard solutions; Matched 1-ml quartz cuvettes (1-cm path length); 1X TE buffer; Microcentrifuge tubes; Single- or dual-beam UV spectrophotometer.

Procedure

1. Prepare a dilution of chromosomal DNA by adding 20 μl of DNA solution to 0.98 ml of TE buffer in a microcentrifuge tube. Mix it well by inverting the tube.
2. Allow 20 min. for the UV lamp in the spectrophotometer to warm up. Then, set the wavelength of the spectrophotometer to 260 nm. Add the TE buffer to the blank. Measure the absorbance of the DNA dilution and write it down.
3. Repeat this procedure at 280 nm.
4. Calculate the concentration of DNA dilution, assuming that DNA at a concentration of 50μg/ml has an OD of 1 at 260 nm. Calculate the DNA concentration in the original DNA solution.

    Label the tube containing the DNA with the date, contents, concentration, and the initials of a person who has carried out the isolation, and then cover the label with transparent tape for preservation (this is not necessary if a permanent marker pen is used). Finally, store the DNA in a freezer box at −20°C.

    $$\text{DNA concentration} = (OD_{260}) \times (\text{dilution factor}) \times (50\ \mu g\ \text{DNA/ml}\ OD_{260}\ \text{unit}). \tag{2.1}$$

5. *Total yield:*
    # Mg DNA recovered = (DNA concentration in μg/ml) × (total volume in ml). (2.2)
6. The expected yield of DNA: Use the following values for *E. coli*: the MW of the *E. coli* chromosome is 3.1×10^9; the Avogadro's number is 6.02×10^{23} molecules /mole; the culture volumes are 1.2 ml; in the late log phase, there are approximately 2 chromosomes cell.

    E. coli expected yield:

    #chromosomes = Vol (ml) × (cell/ml) × (#chromosomes/cell) (2.3)

    $$\text{Expected yield} = \frac{3.1 \times 10^9\ \text{g/mol}}{6.02 \times 10^{23}\ \text{chromosomes/mol}} \times \text{chromosomes} \tag{2.4}$$

For other organisms:
7. The % recovery of DNA for both *E. coli* and other organisms.

$$\% \text{ recovery} = \frac{\text{actual yield}}{\text{expected yield}} \times 100 \tag{2.5}$$

8. The number of haploid genomes of, for example, *Drosophila* DNA (ignoring the contribution of mtDNA is as calculated below). The number of Daltons per haploid of *Drosophila* genome is 1.1×10^{11}

$$\text{Number of haploid genomes} = \frac{\text{g of DNA recovered}}{\text{g/haploid genome}} \tag{2.6}$$

$$\text{g/haploid genome} = \frac{1.1 \times 10^{11} \text{g/mol}}{6.02 \times 10^{23} \text{molecules/mol}} \tag{2.7}$$

To determine the concentration (C) of the DNA present, use the A_{260} reading in conjunction with one of the following equations:

$$\text{Single-stranded DNA} = C \text{ (pmol/}\mu\text{l)} = \frac{A_{260}}{10 \times S} \tag{2.8}$$

$$C \text{ (}\mu\text{g/ml)} = \frac{A_{260}}{0.027}$$

$$\text{Double-stranded DNA} = C \text{ (pmol/}\mu\text{l)} = \frac{A_{260}}{13.2 \times S} \tag{2.9}$$

$$C \text{ (}\mu\text{g/ml)} = \frac{A_{260}}{0.020}$$

$$\text{Single-stranded RNA} = C \text{ (}\mu\text{g/ml)} = \frac{A_{260}}{0.025} \tag{2.10}$$

$$\text{Oligonucleotide} = C \text{ (pmol/}\mu\text{l)} = A_{260} \times \frac{100}{1.5N_A + 0.71N_C + 1.20N_G + 0.84N_T}$$

where S represents the size of the DNA in kilobases and N is the number or residues of base A, G, C, or T.

The calculations are based on the Lambert–Beer law,
$A = ECl$, where A is the absorbance at a particular wavelength, C is the concentration of DNA, l is the pathlength of the spectrophotometer cuvette (typically 1 cm), and E is the extinction coefficient.

9. Use the A_{260}/A_{280} ratio and the readings at A_{230} and A_{325} to estimate the purity of the nucleic acid sample

For quantitating the amount of DNA or RNA, the readings should be taken at wavelengths of 260 nm and 280 nm. The reading at 260 nm allows the calculation of the concentration of nucleic acid in the sample. An OD of 1 corresponds to approximately 50 µg/ml for double-stranded DNA, 40 µg/ml for single-stranded DNA and RNA, and ~20 µg/ml for single-stranded oligonucleotides. The ratio between the readings at 260 nm and 280 nm (OD_{260}/OD_{280}) provides an estimate of the purity of the nucleic acid. Pure preparations of DNA and RNA have OD_{260}/OD_{280} values of 1.8 and 2.0, respectively.

An absorbance at 230 nm reflects the contamination of the sample by phenol or urea, whereas an absorbance at 325 nm suggests contamination by particulate and dirty cuvettes. Light scatter at 325 nm can be magnified 5-fold at 260 nm.

Note

1. If there is contamination with protein or phenol, the OD_{260}/OD_{280} will be significantly less than the values given above and an accurate quantitation of the amount of nucleic acid will not be possible.

B. FLUORESCENT DYE-BASED METHOD

If the quantity of DNA is very low (<250 µg/ml) or heavily contaminated with other substances that absorb UV irradiation, a spectrophotometric assay may not be possible. Ethidium bromide (EtBr) binds to DNA by interaction, so that the total fluorescence is proportional to the DNA mass. A quick way to estimate the amount of DNA in such samples is to utilize the UV-induced fluorescence emitted by the EtBr molecules intercalated into the DNA. As the amount of fluorescence is proportional to the total mass of the DNA, the quantity of DNA in the sample can be estimated by comparing the fluorescent yield of the sample with that of a series of standards; as little as 1–5 µg of DNA can be detected by this method.

The fluorescence of ethidium bromide is increased about 50-fold when it is intercalated into the DNA. Since, the fluorescence excited by UV light and the emission is in the visible spectrum, very simple methods can be used to detect DNA. In contrast to the UV absorbance, however, this test destroys the DNA. A spectrofluorometer could be used to quantitate the increased fluorescence of an ethidium bromide solution caused by the presence of DNA. However, it will be considerably easier and sufficiently precise for most practical purposes to make a set of standards and to visually compare them to the unknown sample. The comparisons can be made directly by observing the fluorescence with a shielded eye or indirectly with photographs of the sample and standards.

Method-1: Ethidium bromide fluorescent quantitation of dsDNA

Materials

Stock buffer: 0.01 M Tris–HCl, pH 7.4; 20 mM NaCl; 1 mM EDTA; 1 µg/ml ethidium bromide (this stock buffer remains viable for months if stored in the refrigerator).

Procedure

1. Stretch commercial plastic wrap over a transilluminator or over a ring, such as the rim of a Petri dish with the bottom removed.
2. In an orderly array, spot the samples with a micropipettor of DNA (1–5 µl) and 2 µl of a series of DNA standards (0.5–20 µg/ml, prepared by serial dilution) onto the plastic.

3. Add an equal volume of TE buffer (pH 8) containing 2 µg/ml of EtBr to each spot. Mix it with a micropipettor.
4. Illuminate the spots with a short-wave UV light.
5. Observe the DNA in the 2 sets after the transilluminator is turned on. Use suitable eye protection when looking at the UV light. Plastic safety glasses are commonly used, but it is better to have a large 6 mm thick Plexiglas shield that will protect both the eyes and the skin.
6. Record the fluorescence of the sets by the photographic procedures detailed in the section on gel electrophoresis. A visual interpretation of the fluorescence, from the unknown sample into the series containing the standard, is reasonably accurate.

Notes

1. Two common substances, such as SDS and DEAE-cellulose contaminants, can effectively quench the fluorescence. Whenever there is a doubt, add a known quantity of DNA to a drop containing the sample in order to test that the increase in fluorescence is as expected.
2. The EtBr drop test can be used for both single- and double-stranded DNA.
3. The drop of the stock solution need not be put directly on a transilluminator. Often, it is more convenient to put the drops in the top or bottom half of a plastic Petri plate and then to invert the plate with a deft twist of the wrist in such a way that the drops are not disturbed. The inverted plate can then be put on an upward-shining UV source.

C. GEL ELECTROPHORESIS METHOD

Agarose gel electrophoresis can be used to resolve the different molecular configurations of a DNA molecule as well as to separate DNA fragments of different lengths. Electrophoresis through mini-gels provides a rapid and convenient way to measure the quantity of DNA and to analyze its physical state at the same time. The latter is the method of choice if there is a possibility that the samples may contain significant quantities of RNA.

Method-1: Mini-gel method

Materials

Loading buffer; Horizontal 0.8% agarose gel; DNA samples and Marker DNA; Gel electrophoresis systems with power supply; Ethidium bromide, Gel viewing UV light box, photography equipment.

Procedure

1. Mix 2 µl of the DNA sample with 0.4 µl of gel-loading buffer IV (bromophenol blue only) and load the solution into a slot in a 0.8% agarose mini-gel containing EtBr (0.5 µg/ml).
2. Mix 2 µl of each of a series of standard DNA solutions (0, 2.5, 5, 10, 20, 30, 40, and 50 µg/ml) with 0.4 µl of gel-loading buffer IV.
3. Follow the procedure used for pouring and preparing gels described in the earlier methods.
4. Load the samples into the wells of the gel. The standard DNA solutions should contain a single species of DNA approximately the same size as the expected size of the unknown DNA.
5. Carry out the electrophoresis until the bromophenol blue has migrated approximately 1–2 cm.

6. After running it either overnight at 20 V or for ~ 2 hr at 200V, stain the gel with 1 µg/ml EtBr for 20 min., and then destain the gel in water or by immersing it for 5 min. in an electrophoresis buffer containing 0.01 M $MgCl_2$.
7. Photograph the gel using short-wavelength UV irradiation. Compare the intensity of fluorescence of the unknown DNA with that of the DNA standards and estimate the quantity of DNA in the sample.
8. When running plasmid DNA, prepare a semi-log plot of the plasmid size (plot kb on the log scale) against the distance migrated (on the linear scale). Use the plot to calculate the size of plasmid in the sample.

Method-2: Agarose plate method

Any contaminants present in the DNA sample can either contribute to or quench the fluorescence. To avoid these problems, the DNA samples and standards can be spotted onto the surface of a 1% agarose slab gel containing EtBr (0.5 µg/ml). Allow the gel to stand at room temperature for a few hours, so that small contaminating molecules have the chance to diffuse away.

Procedure

1. Prepare the petri dishes containing 10 ml of 1% (w/v) agarose (with 0.5 µg/ml of EtBr) in a TEA buffer.
2. On the agarose, spot 5 µl of DNA from a stock solution of a known concentration (20 µg/ml), as well as serial dilutions in the TE buffer (pH 8) of the stock solution (spot 5 µl) of stock and 1:2, 1:4, 1:8, 1:16, and 1:32 dilutions.
3. Spot 1, 2, and 5 µl of the DNA sample on the agarose plate.
4. Allow about 30 min. for the spots to soak into the agarose. This step facilitates the dispersion from the DNA of any contaminating fluorescent or quenching material that may be in the preparation. Leaving the plate overnight will allow more of the interfering material to diffuse into the agarose.
5. Place the petri dish on the transilluminator and photograph it. By comparing the fluorescence intensities, one can determine the approximate concentration of the DNA sample.

III CLEANING DNA

DNA preparations are frequently contaminated with substances that inhibit enzymes, degrade DNA, or cause artifactual rates of DNA migration in gels or a variety of other problems. In some cases, simple methods can be used to overcome these problems. DNA degradation due to nuclease contamination can be overcome by phenol extraction, followed by ethanol precipitation. An alternative response is to add EDTA to 1 mM in order to chelate the divalent cations that are almost always required by nucleases and contaminating organisms. Another common problem, particularly with DNA preparations from eukaryotic cells, is the presence of high molecular weight carbohydrates. These can inhibit enzymatic reactions and can usually be eliminated by pelleting the carbohydrate at 15,000g for 15 min. at 4°C.

The DNA can be cleaned by column chromatography with DEAE-cellulose or hydroxyapatite (Figure 8.1). In these chromatographic methods, the DNA is selectively bound to the

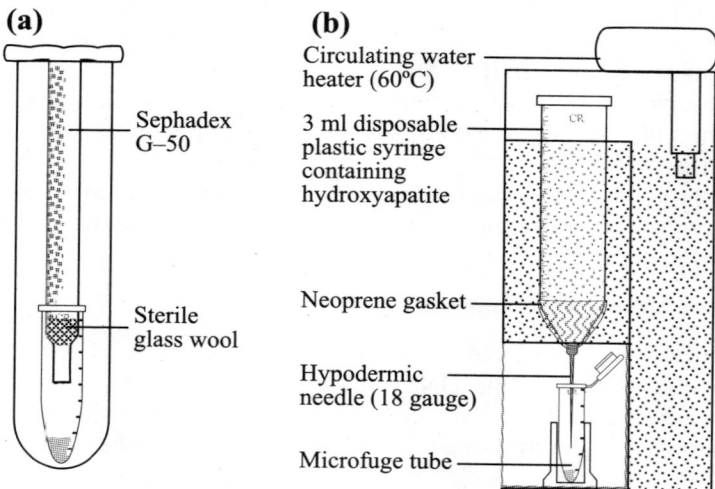

Figure 8.1 *Gel-filtration chromatography: (a) Sephadex chromatography and (b) Apparatus for hydroxyapatite chromatography*

column material; contaminants are washed away and then the DNA is eluted. DEAE chromatography is easier and yields cleaner DNA, but has the drawback that longer DNA binds more tightly and is selectively lost. The virtue of hydroxyapatite chromatography is that it works with much larger DNA. Its drawbacks are that it yields slightly less clean DNA and that this DNA is in a highly concentrated phosphate solution. The phosphate cannot be removed by ethanol precipitation of the DNA because the phosphate also precipitates. Dialysis is usually the best method, but phosphate dialysis is very slow. Occasionally, the volume of the solution is small enough for the phosphate to be removed by Sephadex gel filtration.

Method-1: Cleaning by DEAE

The method described below gives about a 40% recovery of DNA 6 kb long, and an almost 100% recovery of DNA that is 1 kb or less in length.

Procedure

1. Use washed pre-swollen, microgranular DEAE-cellulose (DE 52, Whatman) without pre-washing. It binds approximately 25 µg of DNA/ml of column volume. Place a substantial wad of siliconized glass wool into the shoulder of a Pasteur pipette.
2. Soak the DEAE in 10 mM Tris-HCl, pH 7.5, 0.3 M NaCl for at least 5 min. Then pipette about 1 ml of the slurry into the Pasture pipette mounted on a ring stand.
3. Wash the column with at least 4 volumes of the same solution. Allow the solution to flow under the force of gravity.
4. Load the DNA onto the column in a solution that has the same ionic strength.
5. Wash the column 5 times with 2 ml of the same solvent. Elute the DNA with 2.5 ml of 1 M NaCl, 10 mM Tris-HCl, pH 7.4.
6. Dilute this DNA solution with water to 0.2 M NaCl, add 2 vol of ethanol, and precipitate the DNA.

Notes

1. If the DEAE-cellulose is eluted with the DNA, it can be removed by a short centrifugation in an Eppendorf microcentrifuge.
2. For most applications, it is useful to determine the lowest salt concentration, which will elute the DNA sample from the column. Often, the column can be washed with 0.4 M salt and the DNA eluted with 0.6 M salt.

Method-2: Cleaning by hydroxyapatite

Procedure

1. Prepare the phosphate buffer stock (PB), 1.2 M, by mixing 2 vol of 1.2 M Na_2HPO_4 with 1 vol of 1.2 M NaH_2PO_4 and then adjust it to pH 7.0 by adding one or the other of the 1.2 M salt solutions.
2. Hydroxyapatite (HAP, DNA grade, Bio-Rad) is adequate for this procedure after the fine particles (fines) are removed. The fines will clog the column and reduce the flow rate. To remove the fines, suspend the dry HAP in 0.12 M PB by swirling and then let it settle for approximately 2 min. Remove the supernatant by aspirating with a Pasteur pipette. Repeat this twice and add 0.12 M PB to achieve a total volume that is approximately twice the volume of the settled HAP.
3. Swirl this HAP preparation and add 0.5 ml of the resulting slurry to a water-jacketed (60°C) column with a diameter of approximately 1 cm and a sintered glass filter at its bottom. Let the HAP sit for 1 min. in the column; then stir it and remove the liquid by applying pressure to the top of the column. This can be done with a syringe attached to a needle that sticks through a rubber stopper which is, in turn, pressed into the top of the column.
4. Add the DNA in at least 1 ml of 0.012 M PB, pH 7.0, and let it drain through the HAP with no pressure applied. If the volume of the DNA is large, stir the mixture. When the draining is complete, apply pressure with a syringe to remove all liquid. On all subsequent washes and elutions, remove all the liquid in a similar fashion.
5. Elute single-stranded DNA with 2 vol of 2 ml 0.12 M PB, 60°C, or elute both single- and double-stranded DNA with 2 vol of 2 ml 0.4 M PB, at 60°C.
6. Combine the 2 vol of 2 ml and remove the phosphate by dialysis against a volume of 1 litre. Dialyze twice against 0.2 M NaCl, TE, pH 8 for 2 h and then twice for at least 1 day against the same solution. Then, ethanol-precipitate the DNA. The 0.2 M NaCl accelerates the dialysis rate of the highly-charged phosphate, presumably because it shields the charge interactions between the phosphate and dialysis membrane.

Note

1. The binding capacity of DNA-grade HAP is approximately 0.6 mg of DNA/ml of bed volume. However, this capacity tends to vary with different commercial lots.

Method-3: Precipitation and size fractionation of DNA with PEG

Polyethylene glycol (PEG, Carbowax 5000, Union Carbide) precipitates DNA in a nearly quantitative manner, provided that the initial DNA concentration is greater than 50 µg/ml. This method is inexpensive, has virtually unlimited capacity, does not damage the DNA, and is easy to perform. It also has the very useful additional feature that it can be manipulated to selectively precipitate different sizes of DNA molecules, utilizing different concentrations of PEG, and separated with only about 5% cross-contamination. The same method can also be used to separate plasmid DNA from chromosomal DNA.

The procedure described below is used for separating 2 EcoRI restriction fragments that are approximately 440 and 5000 base pairs' length. Whenever possible, DNA manipulations are performed at 0°C in order to minimize unwanted chemical and enzymatic degradation. However, the cohesive ends remaining after EcoRI endonuclease digestion allow the fragments to aggregate at 0°C. Therefore, size fraction by PEG precipitation is performed at 37°C.

Procedure

1. Heat the sample to 65°C for 10 min. to inactivate the EcoRI enzyme.
2. Dilute the sample 10-fold with TE buffer in order to reduce the concentration of Mg^{2+} to below 0.5 mM. However, the resulting solution must contain at least 50 μg DNA/ml.
3. Add 5 M NaCl to give a final concentration of 0.55 M.
4. Add and dissolve 6 g PEG per 100 ml and incubate it at 37°C for 20 hr.
5. Centrifuge it at 10,000 rpm at room temperature in a Sorvall GSA rotor for 10 min. The supernatant contains the 440 base pair fragment.
6. To precipitate the 440 bp fragment from the supernatant, add and dissolve PEG to make a final concentration of 18% and incubate it at 0°C for at least 2 hr.
7. Centrifuge it at 10,000 rpm at 0–4°C in a Sorvall GSA rotor.
8. Resuspend the pellet in a TE buffer.
9. Precipitate the DNA in ethanol and resuspend in TE or any another desired buffer.

IV STORING DNA

With proper precautions, clean DNA can be stored for long periods of time with little degradation. Storing it at 4°C in 10 mM Tris, pH 8.0, 1 mM EDTA (TE) for 6 months produces less than 1 single-strand break (nick) per 3×10^5 base pairs of super-coiled DNA. A pH of 8 is used because the DNA deamination rate is lower than it would be if the DNA were stored at pH 7. The presence of EDTA eliminates free divalent cations, which are required for the activity of most nucleases. The EDTA also inhibits the growth of microorganisms, which may synthesize large quantities of nuclease and/or shift the sequence of the DNA to the sequence of the contaminating organism. DNA can even be stored with a lower nicking rate if it is kept frozen in the above solution. However, it appears that every cycle of freezing and thawing introduces some nicks. This can be a serious problem since most –20°C freezers are equipped with automatic defrost timers.

V RNA ANALYSIS

The measurement of the RNA concentration and purity is an important step after the extraction of RNA from cells/tissue. The efficiency of protocol and experimentation mainly depends upon the RNA yield and the purity of the RNA. Isolated RNA may be contaminated by other macromolecules, such as DNA or polysaccharides. This contamination can be removed by enzyme digestion or centrifugation through a dense CsCl solution. However, the major problem is that of degradation of the RNA, and so it would be advantageous to be able to make a rapid assessment of the extracted RNA. The most convenient method of doing this is by gel electrophoresis. Separating the RNA by electrophoresis in a 1–1.4% agarose gel, followed by ethidium bromide staining, offers a quick and simple assessment of the quality of the RNA. However, the RNA molecules do tend to aggregate and the secondary structure of RNA can hide the presence of nicks in the RNA. Thus, for a more definitive analysis it is better to use gel electrophoresis under denaturing conditions. A number of denaturing systems can be used including urea, formamide, methyl mercuric hydroxide.

Several methods are available to check RNA preparations for degradation; four of these procedures will be briefly reviewed and discussed.

A. QUANTITATION OF RNA BY UV SPECTROPHOTOMETERY

Method-1: UV Spectrophotometry

Ultraviolet spectrophotometry can be used to assess both the purity and recovery of nucleic acids. A UV spectrophotometer is required; double-beam and scanning spectrometers speed up the analysis. Nucleic acid bases absorb at around 260 nm. The OD_{260} of a 50 µg/ml solution of double-stranded nucleic acid is approximately 1 A unit (1-cm light path). For single-stranded nucleic acid (such as RNA), 1 A unit corresponds to 40 µg/ml.

The ratio of the OD_{260} to the OD_{280} gives a measure of the purity of the nucleic acid sample. Protein and phenol (another common contaminant) absorb maximally at around 289 nm. For DNA, the OD_{260}/OD_{280} ratio should approach 2, while, for RNA, a value of 1.8 to 1.9 indicates purity.

B. FRACTIONATION OF RNA BY SIZE

A number of methods are used to fractionate RNA molecules of particular size. Two commonly used methods are electrophoresis through agarose gels and sedimentation through sucrose gradients. Both methods have been used to estimate the sizes of the mRNAs that code for particular proteins and enrich the populations of mRNA to be used for cDNA cloning of species of interest. Electrophoresis through agarose gels gives better separation of different size molecules of RNA, but recovery of RNA from sucrose gradients is much more efficient.

Method-1: Electrophoresis of RNA through agarose gels

Before fractionation, the RNA is treated with methylmercuric hydroxide, a reagent that reacts primarily with the imino–bonds of uridine and guanosine in RNA. Because these bonds may be involved in Watson–Crick base pairing, methylmercuric hydroxide is an effective denaturing agent that disrupts all secondary structures in the RNA. Methylmercuric hydroxide is extremely toxic and also volatile. Therefore, all operations of solutions containing concentrations of methylmercuric hydroxide in excess of 10^{-2} M should be carried out in a chemical hood.

Materials

Methylmercury gel running buffer 1X: 50 mM boric acid; 5 mM $Na_2B_4O_7 \cdot 10H_2O$; 10 mM Na_2SO_4, if necessary, adjust the pH to 8.1. Caution: buffers that contain nitrogen bases, EDTA, or chloride ions should not be used because these compounds form complexes with methylmercuric hydroxide. 2X gel loading buffer: 25 µl of 1 M methylmercuric hydroxide, 500 µl of 4X methylmercury gel-running buffer, 200 µl of 100% glycerol, 275 µl water, 0.2% (w/v) bromophenol blue.

Procedure

1. Prepare the gel (1% agarose for RNA molecules 1 kb or larger in size; 1.4% agarose for smaller species of RNA). Dissolve the agarose in 1X methylmercury gel running buffer.
2. Mix equal volumes of 2X gel-loading solution and the solution of RNA (up to 10 µg may be loaded per standard 0.6 cm slot).
3. Load the samples and run the gel at 5–6 V/cm for 12–16 hr. Methylmercuric hydroxides adds to the gel but not to the methylmercury gel-running buffer.
4. After electrophoresis, the RNA may be stained by incubating the gel for 30–45 min. in 0.1 M ammonium acetate containing 0.5 µg/ml ethidium bromide. The ammonium ions convert methylmercury to a charged non-volatile complex and enhance binding of ethidium bromide to the RNA.

Recovery of RNA from agarose gels

1. Prepare and run the gel as described above, except use low-melting-temperature agarose.
2. After electrophoresis, soak the gel in 0.1 M dithiothreitol for 30–40 min.
3. Cut the gel into slices approximately 3 mm in width. The stained tracks containing fragments of DNA of known size, 18S and 28S human rRNAs, or 9S rabbit β-globin mRNA may be used as molecular-weight standards. The sizes of these RNAs are 6333, 2366, and 710 nucleotides, respectively. Alternatively, mixtures of RNAs of known sizes can be purchased from any supplier.
4. Transfer each gel slice to a polypropylene tube, and add at least 4 vol of 0.5 M ammonium acetate (pre-heated to 65°C), so that the gel slice dissolves completely. Otherwise, the agarose will be carried over into the aqueous phase during the subsequent extractions with phenol and chloroform.
5. Heat the samples at 65°C until the agarose is completely dissolved, and vortex the samples well.
6. Extract the samples with phenol equilibrated with Tris.Cl (pH 7.6) at room temperature. Separate the phases by centrifugation at 2000g for 10 min. at 4°C. During extraction with organic solvents, agarose becomes a powder and forms a layer at the interface during centrifugation.
7. Re-extract the aqueous phase at least twice more with chloroform. Repeated extractions with chloroform may be required to remove the agarose completely.
8. Transfer the aqueous phase to a fresh tube, and add 0.1 vol of 3 M sodium acetate (pH 5.2). Mix well and add 3 vol of ethanol, and store the solution for at least 1 hr at –70°C.
9. Recover the RNA by centrifugation at 12,000g for 10 min. at 4°C. Carefully discard the supernatant, and wash the pellet with 70% ethanol. Re-centrifuge briefly, and allow the RNA pellet to dry in the air.

10. Redissolve the RNA in a small volume (5–10 µl) of water, add 3 vol of ethanol, mix well, and store the preparation at −70°C until it is needed. To recover the RNA, add 3M sodium acetate (pH 5.2) to a final concentration of 0.3 M, mix well, and centrifuge at 12,000g for 5 min. at 4°C, in a microfuge.

Method-2: RNA size fractionation by sucrose gradient centrifugation

A convenient method for the fractionation of up to 1 mg of RNA is to separate the RNA according to size by centrifugation through a sucrose gradient. The different fractions then yield the different size classes of RNA. The sucrose fraction procedure, described below, appears to separate RNA according to its molecular weight. The DMSO treatment disaggregates the RNA. Sterilize all solutions either by adding diethylpyrocarbonate to 0.1% and indicating at 35–45°C for several hours, at which time no odor of DEPC should remain, or by autoclaving a 10-fold concentrated buffer solution (10X buffer: 1 M NaCl; 10 mM EDTA; 5 mM Tris-HCl pH 7.5; 5% SDS) and a stock of distilled water. The distilled water and concentrated buffer can then be mixed and combined with RNase-free sucrose to give the desired solutions. Sterilizing buffers containing sucrose will, of course, lead to caramelized solutions.

Making the sucrose gradients

Procedure

1. Place 6.3 ml of 15% and 5.9 ml of 30% sucrose solutions into the left and right chambers, respectively, of a DEPC-treated plastic gradient maker. The gradient maker is of the same design as that used to make gradient acrylamide gels for protein separations. Be sure to eliminate bubbles, which may form in the channel between the two wells. These bubbles can be prevented by prefilling the channel with 15% sucrose in the buffer.
2. Place a sterile Teflon magnetic stirring flea in the right, that is, the 30% chamber, and use a magnetic stirrer to force it to stir the solution in that chamber.
3. Adjust the plastic tubing outlet from the right chamber so that it touches the lip of a DEPC-treated SW40 tube about 3–5 mm from its top. Open the outlet valve and when the 30% solution begins to flow, open the channel between the two chambers. The flow rate should allow the centrifuge tube to be filled in 15–20 min. If the gradient forms too quickly, excessive mixing of the layers will result, which can lead to a loss of resolution in the gradients.
4. It is useful to let the gradients sit at room temperature for about 1 hr before loading a sample for centrifugation. This will remove any homogeneities in the gradient.

Preparing the RNA

Procedure

1. First, pellet the RNA to be fractionated. Pellet it from an ethanol solution in a siliconized tube, drain briefly, use a sterile Q-tip to remove excess ethanol, and then resuspend the pellet before it dries. The resuspension should be in 99% DMSO, 10 mM Tris-HCl pH 7.5. Up to 100 µg of RNA can be solubilized in 0.2 ml of this DMSO solution. If a larger amount of RNA is solubilized in this volume, aggregation usually becomes a serious problem. If the RNA does not dissolve at this concentration, it is best to dilute the solution with more buffer.

2. Heat the RNA in the DMSO solution for 5 min. at 37°C. Then dilute it 4-fold with 5 mM Tris-HCl pH 7.5; 1 mM EDTA; 0.5% SDS; and heat at 65–70°C for 1 min. Finally, quick-cool the solution in ice before layering it onto a gradient.

Sedimenting and collecting the RNA

Procedure

1. A general guide to centrifugation conditions is given by the sedimentation of 18S and 28S rRNA at 25°C in these gradients. At 38,000 rpm for 16 hr in an SW40 rotor, the 28S RNA will pellet and 18S RNA will be within the last one-fifth of the gradient. At 30,000 rpm for 16 hr, the 28S RNA will be about three-quarters of the way down the gradient and 18S will be in the middle. Centrifuging at 28,000 rpm for 14 hr or 40,000 rpm for 6 hr, will put the 18S RNA about one-third to one-half of the way down the gradient.
2. Collect the 0.5 ml fractions and follow the RNA profile by its UV absorbance. The RNA can be recovered by adding Na^+-acetate to 0.25 M and precipitating it with ethanol as usual. Carrier RNA may have to be added to achieve the efficient precipitation of RNA that is at very low concentrations. The RNA, thus fractionated, can be translated *in vitro* protein-synthesizing systems and is an efficient template for cDNA synthesis with reverse transcriptase.

C. *IN VITRO* TRANSCRIPTION AND TRANSLATION

Method-1: In vitro transcription

Many vectors are commercially available, including an SP6, T7 or T3 promoter upstream and downstream of a multiple-cloning site. These vectors permit the transcription of both sense RNA copies (cRNAs) of inserted cDNA sequences, and antisense RNA copies that can be used to quantitate RNA by *in situ* hybridization and RNase protection assays. The vector SP64T contains 5′ and 3′ untranslated sequences from the *Xenopus* globin gene flanking the cloning site (Figure 8.2). The presence of these sequences has been shown to increase the stability of the resultant cDNA in microinjected *Xenopus* oocytes and to increase the transcription efficiency in most systems.

Materials

Absolute ethanol; 7 M Ammonium acetate; Chloroform; m7G(5′)ppp(5′)G cap structure analog (10 mM); Distilled water; Dithiothreitol (DTT) (500 mM); Linearized DNA at 1–2 μg/ml; Phenol:chloroform (1:1); RNase inhibitor, 38 U/μl; RNase- and DNase-free bovine serum albumin (BSA) 2.5 mg/ml; Scintillation fluid; 10 mM NTPs (ATP, CTP, GTP, UTP) adjusted to pH 7.0; SP6 RNA polymerase, T7 RNA polymerase or T3 RNA polymerase at 40–60 U/μl; Trichloroacetic acid (TCA) (10% w/v); tRNA; [α–^{32}P]–UTP (30 TBq/mmol, 800 Ci/mmol).

10X Transcription buffer: 0.4 M Tris.Cl pH 8.0, 60 mM $MgCl_2$, 20 mM spermidine HCl, 50 mM NaCl.

GFC filters; Microcentrifuge; Micropipettor; Perspex radiation shield; RNase-free 1.5 ml microcentrifuge tubes and pipette tips; Scintillation counter; Vacuum filtration apparatus; Water bath at 37°C.

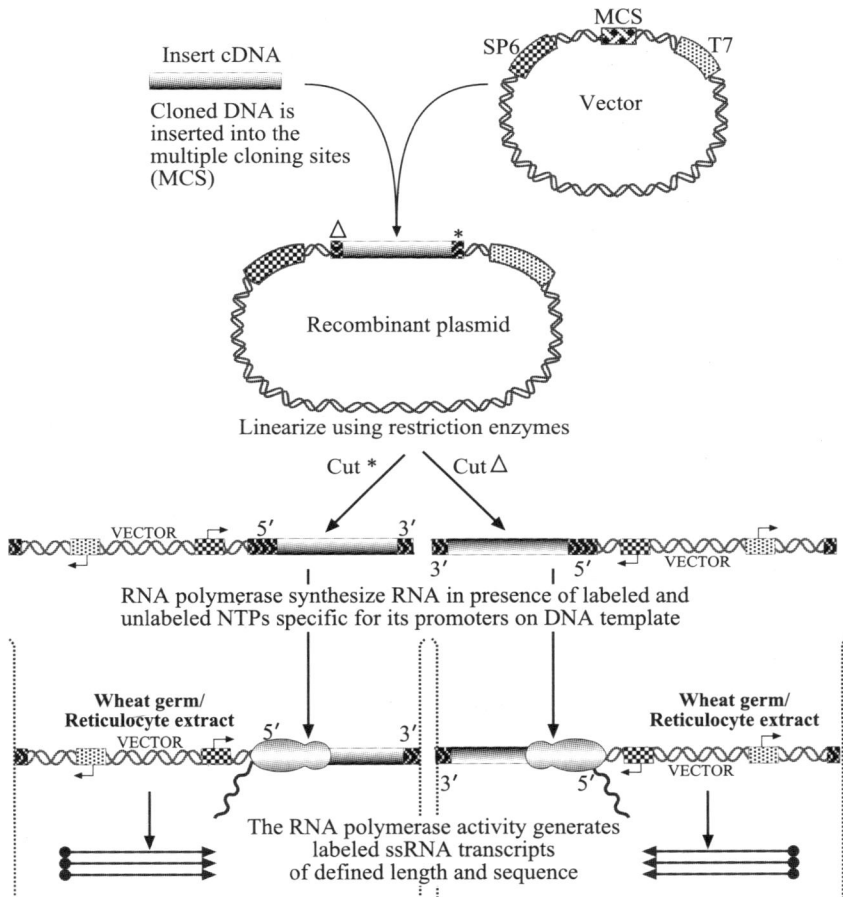

Figure 8.2 *The diagrammatic illustration of in vitro transcription reaction*

Procedure

1. To a microcentrifuge tube at room temperature, add the following, in the given order: 12.5 µl of distilled water; 5 µl DTT; 5 µl BSA; 5 µl each NTPs; 1 µl GTP; 1 µl [α-^{32}P]–UTP (optional); 2.5 µl m7G(5′) ppp (5′)G-cap structure analog; 5 µl of 10X transcription buffer; 1 µl RNase inhibitor; 1 µl linearized DNA; 1 µl RNA polymerase (SP6, T7 or T3).
2. Mix them gently and incubate the mixture at 37°C for 60 min.
3. Add 150 µl of distilled water and 200 µl of phenol:chloroform. Mix them and microcentrifuge for 5 min. Then, remove the aqueous phase and place it in a fresh microcentrifuge tube.
4. Back-extract the phenol phase with 100 µl of distilled water. Mix and centrifuge it as described in step 3. Remove the aqueous layer and combine it with the aqueous phase from step 3.
5. Add 300 µl of chloroform to the aqueous phase, mix and centrifuge for 5 min. in a microcentrifuge tube, measure the volume and precipitate the RNA by adding 0.1 vol of 7 M ammonium acetate and 2.5 vol of absolute ethanol. Leave it overnight at −70°C.

6. Centrifuge in a microcentrifuge for 20 min. at 4°C to pellet the RNA. Remove the ethanol and wash the pellet in 70% ethanol. Centrifuge as before, remove the ethanol and dry the pellet.
7. Redissolve the RNA in 10 µl distilled water.

Notes

1. The high concentration of spermidine can cause precipitation of DNA. Hence, the DNA should be almost the last reagent to be added.
2. The production of transcripts can be verified by removing 2 µl of the reaction and running it on a 1–1.4% agarose gel for 10–15 min. at 100 V.

Method-2: In vitro translation of mRNA

Valuable information regarding the biosynthesis of a protein can be gathered using simple *in vitro* translation systems. The *in vitro* translation of isolated mRNA into its corresponding polypeptides can be accomplished in a cell-free system derived from rabbit reticulocytes or wheat-derived extracts. By comparing incorporation levels to identify the RNA standards of good quality, the integrity of mRNA can be checked. The *in vitro* system is very sensitive to mRNA degradation, resulting in an immediate drop in incorporation level. However, although this method is very fast (3–4 hr) it is also expensive, necessitating radioactively-labeled amino acids. In addition, if impurities of the RNA (polysaccharides, salt) are present, this will result in reduced incorporation and may lead to inadvertently discarded RNA preparations. The cell-free system derived from rabbit reticulocytes is prepared by the lysis of blood from rabbits stimulated to produce high reticulocyte levels by drug-induced anemia. The reticulocyte mRNA is removed from the lysate by treatment with micrococcal nuclease. Since this enzyme is dependent on Ca^{2+}, the subsequent addition of EGTA to chelate this cation will prevent degradation of any exogenous mRNA added to the system for translation. The translated products are usually labeled with ^{35}S-methionine, although other amino acids may be substituted if desired. This standard *in vitro* translation system will synthesize polypeptide products without any post-translational modifications, such as signal peptide cleavage or glycosylation. Wheat germ lysate is an alternative, endogenous RNA-free, commercially available (BCL, Amersham) system, in which *in vitro* translations can also be performed.

Materials

Rabbit reticulocyte lysate, either prepared from anemic rabbits (Merrick, 1983) or purchased in commercial kits (e.g., Amersham N90). Store lysates strictly at –70°C or under liquid nitrogen until use; Purified mRNA or total RNA: usually 0.5–1 µg or 5 µg, respectively; ^{35}S-methionine; KM: 2 M KCl, 10 mM $MgCl_2$; CP: 0.2 M creatinine phosphate; AA: 3 mM L-leucine; 3 mM L-valine; 2 mM all other amino acids except methionine.

Procedure

1. Spin the RNA down from the ethanol and resuspend it at 1 µg/ml in water. Include two control tubes: one without added RNA and another with 1 µg of TMV genomic RNA, or other RNA giving product of known size.
2. Heat the RNA to 65°C for 5 min. (to denature it) and add to the lysate mix.
3. Mix 1 µl each of KM, CP and AA with 16 µl of lysate.

4. Add 10–100 µCi of ^{35}S-methionine.
5. Add the RNA and incubate it at 30°C for 60–90 min.
6. Place the tubes on ice while 1 µl aliquots are taken for scintillation counting and/or samples are taken for SDS-PAGE analysis. Store the samples at –70°C for longer periods.

D. ELECTROPHORESIS METHODS

Method-1: Electrophoresis in polyacrylamide gels

RNA samples of 2–10 µg can be subjected to electrophoresis through either cylindrical or slab 2.5% polyacrylamide gels in 40 mM Tris-HCl, 40 mM NaH_2PO_4, 2 mM EDTA, pH 7.5, 0.2% SDS, 10% glycerol and subsequently scanned at 260 nm with an IDCO gel scanner. Although laborious, this is the most reliable method for checking RNA integrity, since it will show very limited degradation resulting in a shift in relative abundance between the 25 and 16 S ribosomal RNA peaks.

Method-2: Non-denaturing agarose gel electrophoresis of RNA

The RNA samples (approximately 1 µg) can be subjected to electrophoresis through 1.0–1.5% agarose gels (in 40 mM Tris, 20 mM hydrogen acetate, 2 mM EDTA, pH 8.1) and conveniently stained with EtBr (1 µg/ml). This method is very fast and allows a gross indication of the presence of both ribosomal RNA peaks.

Materials

Sample buffer: 50% (w/v) glycerol, 1 mM EDTA pH 8.0, 0.25% (w/v) Bromophenol blue, 0.25% (w/v) xylene cyanol FF.

10X TBE: 0.9 M Tris base, 0.9 M boric acid, 25 mM EDTA.

Agarose; 0.1 M NaOH; Molecular weight markers; 5 mg/ml Ethidium bromide solution; Double-distilled water; Microwave; Camera; Pyrex vessel; Gel electrophoresis apparatus; UV light box; Tissue paper.

Procedure

1. Prior to the electrophoresis, soak the electrophoresis tank, gel tray and comb in 0.1 M NaOH for at least 6 hr.
2. Remove the gel tray and comb from the tank and rinse them with autoclaved double-distilled water. Dry the edges of the tray with tissue paper and seal with tape.
3. Add 1.5 g of research-grade agarose to 100 ml 1X TBE electrophoresis buffer. Melt the agarose in a Pyrex vessel by boiling for 2 min. Ensure completion by gently swirling the solution and checking for the presence of any undissolved agarose.
4. Cool the gel solution by running cold water over the vessel containing the gel until it is at approximately 50°C. Add 10 µl 5 mg/ml EtBr to the gel mixture and mix by gentle swirling.
5. Pour the gel solution into the gel tray until the solution touches the top of the teeth of the comb. Leave the gel to set for 30 min. at room temperature.
6. Rinse the electrophoresis tank with autoclaved double-distilled water and fill the tank with 1X TBE to a level that will cover the gel.

7. Remove the comb from the gel and place the gel into the tank.
8. The total RNA of between 0.1–1.0 µg/µl can be loaded on to the gel with 1/10 vol of the sample buffer. Electrophoresis the RNA samples together with any required molecular weight markers at 20–25°C for 1 hr at 80 V.
9. Visualize the RNA bands by placing the gel on a UV light box and record the image either using an image-capture device or by photography.

E. CHARACTERIZATION OF RNA OR TRANSCRIPT MAPPING

The following six methods, which are commonly used in the characterization of RNA, are described below. 1) Northern hybridization, 2) Dot and slot hybridization, 3) Mapping RNA using nuclease S1 or ribonuclease, 4) Primer extension, 5) Solution hybridization, 6) Filter hybridization, and 7) *In situ* RNA hybridization.

Method-1: Northern hybridization (RNA Blotting)

By using Northern hybridization method, the size and amount of specific mRNA molecules in preparations of total or poly(A)$^+$ RNA are determined. The RNA molecules are separated according to size by electrophoresis through a denaturing agarose gel, and transferred to activated cellulose, nitrocellulose, glass or nylon membranes. The RNA of interest is then located by hybridization with radio-labeled DNA or RNA followed by autoradiography. The RNA separation on agarose gels is performed in the presence of formaldehyde to keep the nucleic acid in a denatured, single-stranded form. The RNA molecules can be transferred to nylon or nitrocellulose membranes in a manner analogous to Southern blotting. No denaturation step is necessary before the transfer, as the RNA is already single-stranded. Note that the formaldehyde gels are brittle and require careful handling.

The attachment of the denatured RNA to nitrocellulose is presumed to be non-covalent but is essentially irreversible. A wide variety of probes may be used to detect the RNA transferred to nitrocellulose filters or nylon membranes. This includes double-stranded DNA labeled by nick translation, single-stranded DNA prepared by primer extension of an oligonucleotide annealed to a recombinant M13 bacteriophage, radio-labeled synthetic oligonucleotides, and RNA synthesized *in vitro* with prokaryotic DNA-dependent RNA polymerase. The protocol presented for Northern hybridization also works well with most types of positively-charged nylon membranes, but may not be optimal for any particular brand.

Materials

RNA samples: 10 µg total or 1 µg poly(A)$^+$ RNA per track.

RNA loading buffer: 50% sucrose with 0.25% bromophenol blue.

MEA buffer: Make up 10X MEA as follows: 200 mM Morpholinopropane sulphuric acid (MOPS)-acetate (pH 7.0), 10 mM EDTA, and 50 mM sodium acetate.

Formaldehyde (Formalin): Formalin is a 37% solution of formaldehyde in water, approximately 12.3 M.

Gel running buffer: 1X MEA with 1.85% formaldehyde (or 50% formalin). It should be a total volume of 1 litre for a medium-sized flatbed.

Gel: 1.2% agarose in 1X MEA with 3.7% formaldehyde (or 10% formalin).

Markers: Denatured (i.e., single-stranded) markers, such as the BRL RNA ladder (5μg) and *E.coli* rRNA (2 μg) should be run.

Blotting equipment and materials as for Southern blotting; 5 mg/ml ethidium bromide (EtBr) in water; 20X SSC; Ultrapure agarose

Procedure

1. Prepare a 1.2% agarose in a 1X MEA buffer supplemented with 3.7% formaldehyde.
2. Make a gel-running buffer of 1X MEA with 1.85% formaldehyde.
3. Prepare the RNA sample in 25% formaldehyde and 1X MEA. Heat it to 65°C for 5–15 min. and then chill it on ice. Centrifuge the samples for 5 seconds to deposit all of the fluid in the bottom of the microfuge tubes. Add 2 μl of sterile, DEPC-treated formaldehyde gel-loading buffer. Add 1 μl EtBr (optional). Up to 30μg of RNA may be analyzed in each lane of the gel.
4. Before loading the sample, pre-run the gel for 5 min. at 5 V/cm.
5. Add 10% by volume (approx. 2 μl) of the RNA loading buffer to the sample and load onto the agarose gel.
6. Electrophorese it at 50 V overnight (approx. 2.5 V/cm). If it is done any faster, it is likely to cause smearing of the RNA. Recirculate the buffer (positive to negative) using a peristaltic pump, and place the entire apparatus in a fume hood to remove any poisonous fumes.
7. At the end of the run, if the EtBr is not added in the loading buffer, stain the gel with EtBr (0.5 μg/ml in 0.1 M ammonium acetate) for 30–40 min.
8. Photograph the gel under UV light (the rRNA species should be visible) and then soak the gel in 10X SSC for 1 hr before blotting.
9. Cut the nitrocellulose or nylon membrane to cover the gel, pencil the details on the bottom and remove the bottom right-hand corner for orientation. Soak the nitrocellulose first in sterile water, and then in 20X SSC. The nylon may be placed directly into the 20X SSC. Soak 4 pieces of 3MM paper in 20X SSC, and briefly wash the gel. Assemble the blot as indicated in Figure 8.3.
10. Leave it overnight at room temperature. Then, remove the nitrocellulose, float it on the surface of 2X SSC (do not immerse) and air dry it for 15 min. Bake it at 80°C for 2 hr. Verify the transfer of the rRNA under UV illumination.
11. Store it in a sealed bag until ready to probe.

Hybridization and autoradiography

The conditions for pre-hybridization, hybridization, and washing of RNA immobilized on filters are essentially the same as those used for DNA.

1. Pre-hybridize the filter for 1–2 hr in

either (at 42°C)	or (at 68°C)
50% formamide	
5X SSPE	6X SSC
2X Denhardt's reagent	2X Denhardt's reagent
0.1% SDS	0.1% SDS

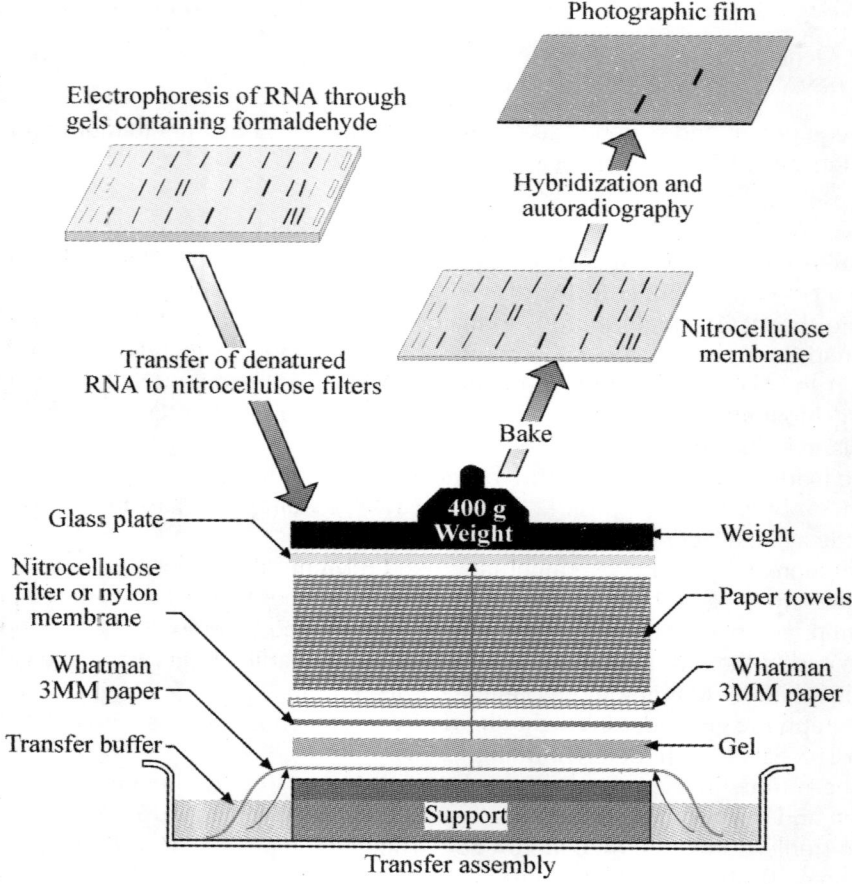

Figure 8.3 Northern blot (similar to Southern blot) analysis. RNA isolated from cells (from a specific cell type or grown in a specific condition) can be probed with labeled cDNA or RNA molecules to find out the presence/absence of a particular RNA molecule of interest

2. Add the denatured radio-labeled probe directly to the pre-hybridization fluid. To detect low-abundance mRNAs, use at least 0.1 μg of probe whose specific activity exceeds 2×10^8 cpm/μg. Continue the incubation for 16–24 hr at the appropriate temperature.
3. Wash the filter for 20 min. at room temperature in 1X SSC, 0.1% SDS, followed by three washes of 20 min. each at 68°C in 0.2X SSC, 0.1% SDS.
4. Establish an autoradiography by exposing the filter for 24–48 hr to X-ray film (Kodak XAR-2 or its equivalent) at −70°C with an intensifying screen.

Method-2: Dot- and slot-blot hybridization of RNA

Dot hybridizations were originally performed by spotting a small sample of the RNA preparation onto dry nitrocellulose, which was then dried, hybridized with a specific ^{32}P-labeled DNA or RNA probe, and exposed to X-ray film. Although accurate quantitation was not feasible because of the large and variable size of the spots, it was nevertheless possible in many cases to obtain a good idea of the intensity of specific gene expression in specific tissues or cultured cells. Recently, this technique has been improved in two ways:

1. Filtration manifolds have been designed to accept a large number of samples and to deposit the nucleic acids onto the nitrocellulose in a fixed pattern, which allows the results to be quantitated by scanning densitometry. The filtration manifold consists of a Lucite block containing a number of slots into which the samples are applied. The manifold fits onto a suction platform containing a sheet of nitrocellulose, onto which the samples are deposited. These manifolds are commercially available.
2. A method involving slot blots has been developed to measure the concentration of a specific RNA in unfractionated cytoplasm prepared from freshly-harvested or frozen cultured cells or animal tissues. This improvement eliminates the tedious task of purifying the RNA from a large numbers of samples.

Procedure

Slot hybridization of RNA

1. Wet a piece of nitrocellulose (0.45 μm pore size) briefly in water and soak it in 20X SSC for 1 hr at room temperature. Meanwhile, clean the manifold carefully with 0.1 N NaOH and then rinse it well with sterile water.
2. Place two sheets of heavy, absorbent paper, previously wetted with 20X SSC, on the top of the vacuum unit of the sample wells cut into the upper section of the manifold. Smooth away any air bubbles trapped between the upper section of the manifold and the nitrocellulose. Clamp the two parts of the manifold together, and connect the vacuum unit to a vacuum line.
3. Fill all of the slots with 10X SSC, and apply gentle suction until all of the fluid has passed through the nitrocellulose filter. Turn off the vacuum and refill the slots with 10X SSC.
4. Mix the RNA (dissolved in 10 μl of H_2O) with:
   100% formamide 20 μl
   Formaldehyde (37%) 7 μl
   20X SSC 2 μl
   Incubate the mixture for 15 min. at 68°C, and then cool the samples on ice. Up to 20 μg of RNA may be applied to each slot of the manifold.
5. Add 2 volumes of 20X SSC to each of the samples.
6. Apply gentle suction to the manifold until the 10X SSC in the slots has passed through the nitrocellulose filter. Turn off the suction.
7. Load the sample into the slots, and then apply gentle suction. After all the samples have passed through the filter, rinse each of the slots twice with 1 ml of 10X SSC.
8. After the second rinse has passed through the filter, continue the suction for 5 min. to completely dry the nitrocellulose filter.
9. Remove the nitrocellulose filter from the manifold and allow it to dry completely at room temperature. Bake the filter for 2 hr at 80°C in a vacuum oven.
10. Hybridize the filter to a radio-labeled probe as described earlier. By using single-stranded probes radio-labeled to a specific activity of $>5 \times 10^8$ cpm/μg, it is possible to detect mRNAs that are present at approximately 5 copies per cell, if 20 μg of total cellular RNA is applied to a single slot.

Method-3: Ribonuclease protection assay

This procedure, first described by Zinn et al. (1983) has largely superseded the nuclease S1 protection and, for many applications, is the method of choice. In this method, a cDNA corresponding to the mRNA of interest is cloned in the antisense orientation into one of a number of vectors that contain a polylinker adjacent to a promoter sequence for a bacteriophage RNA polymerase. The plasmid DNA is then purified and the radio-labeled antisense transcript is generated *in vitro,* using purified bacteriophage RNA polymerase. This probe is then hybridized with the RNA sample to generate RNA/RNA duplexes, which are visualized after treating the hybridization reaction with ribonucleases (RNases) that are specific for single-stranded RNA. The main advantage of this approach is that it is extremely sensitive. Since the probe is synthesized by a polymerization reaction, it can be radio-labeled to a very high specific activity. The nature of the probe synthesis reaction also ensures that the probes are always made to a defined specific activity. The RNase protection assay is also more sensitive than the Northern blotting and it also offers a considerable safety advantage over Northern blotting since it is not necessary to carry out electrophoresis in large amounts of formaldehyde.

The key to the success of the RNase protection is the ability of the RNases to degrade single-stranded RNA specifically. Unlike S1 and *Mung* bean nucleases, the RNases are sequence-specific, and it is necessary to use a mixture of enzymes (e.g., RNases A and T1) to ensure efficient degradation. Furthermore, RNases show less double-stranded nuclease activity than the S1 and *Mung* bean nucleases and, consequently, lower RNA degradation and lower background. However, a potential disadvantage of the RNase protection procedure is the need to generate specific subclones. This assay uses RNA made from a phage promoter as a probe. It is an alternative to S1 analysis and produces high-specific-activity probes.

Materials

5X transcription buffer; 200 mM DTT; 3NTP mix (ATP, UTP, and GTP at 4 mM each); [α-^{32}P]CTP (10 mCi/ml, 400 to 800 Ci/mmol); Placental ribonuclease inhibitor (e.g., RNAsin from Promega Biotec); 1 mg/ml template DNA; Bacteriophage RNA polymerase; 2.5 mg/ml DNase I (RNase-free); 10 mg/ml tRNA; DEPC-treated water; 25:24:1 phenol/chloroform/isoamyl alcohol; 2.5 M ammonium acetate; 100% ethanol; 75% ethanol/25% 0.1 M sodium acetate, pH 5.2 (DEPC-treated); 20% (w/v) SDS; 20 mg/ml proteinase K (store at −20°C); 40 µg/ml ribonuclease A; 2 µg/ml ribonuclease T1; RNA loading buffer.

10X Hybridization buffer: 400 mM Pipes, 10 mM EDTA, 4. M NaCl, pH 6.4.

Procedure

Preparation of radio-labeled RNA probes

1. At room temperature, mix in an autoclaved microcentrifuge tube (20 µl total): 4 µl 5X transcription buffer; 1 µl 200 mM DTT; 2 µl 3NTP mix; 10 µl [α-^{32}P]CTP (10 mCi/ml, 400 to 800 Ci/mmol); 1 µl placental ribonuclease inhibitor (20 to 40 U); 1 µl 1 mg/ml template DNA (50 µg/ml final); 1 µl bacteriophage RNA polymerase (5 to 10 U).
2. Incubate the reaction mix 30 to 60 min. at 40°C for SP6 RNA polymerase, or 37°C for T7 and T3 RNA polymerases.
3. Add 5 µg or 10 U RNase-free DNase I and incubate it for 15 min. at 37°C.

4. Add 2 µl of 10 mg/ml tRNA and water to make up 50 µl. Extract with phenol/chloroform/isoamyl alcohol.
5. Add 200 µl of 2.5 M ammonium acetate and 750 µl of 100% ethanol to the aqueous phase. Mix and place it for 15 min. on ice. Then, microcentrifuge it for 15 min. at 4°C.
6. Dissolve the pellet in 50 µl water and reprecipitate it twice as described in step 5. Rinse the final pellet with 75% ethanol/25% 0.1 M sodium acetate, pH 5.2. Dry the pellet and dissolve in 100 µl hybridization buffer. Count 1 µl of labeled probes for radioactivity. Use the probe the day it is prepared or within a few days, if stored at 4°C.

RNase protection assay

1. Ethanol-precipitate RNA and dissolve it in 30 µl of hybridization buffer containing 5×10^5 cpm of the probe RNA. Denature it for 5 min. at 85°C. Transfer this to the desired hybridization temperature (30°C to 60°C) and incubate it for >8 hr or overnight.
2. Add to each reaction 350 µl ribonuclease digestion buffer containing 40 µg/ml ribonuclease A and 2 µg/ml ribonuclease T1. Incubate it for 30 to 60 min. at 30°C.
3. Add 10 µl of 20% SDS and 2.5 µl of 20 mg/ml proteinase K. Incubate it for 15 min. at 37°C.
4. Extract with 400 µl phenol/chloroform/isoamyl alcohol. Transfer the aqueous phase to a clean tube containing 1 µl of 10 mg/ml yeast tRNA and add 1 ml ethanol. Precipitate, dry the pellet, and redissolve it in 3 to 5 µl of RNA loading buffer.
5. Denature it for 3 min. at 85°C, and analyze it on a denaturing polyacrylamide/urea (sequencing) gel. Finally, autoradiograph it overnight with an intensifying screen.

Method-4: Primer extension method for measurement of RNA levels

This is a highly specialized application, which like the nuclease S1 map and RNase protection assays, allows the accurate determination of transcriptional start-points or splice junctions. Primer extension is used to map and quantitate the 5' termini of RNA and to detect the precursors and processing intermediates of mRNA. The basis of the procedure is very simple. The test RNA is hybridized with an excess of a single-stranded DNA primer radio-labeled at its 5' terminus. Reverse transcriptase is then used to extend this primer to produce the cDNA complementary to the RNA template. The length of the resulting end-labeled cDNA, as measured by electrophoresis through a polyacrylamide gel under denaturing conditions, reflects the distance between the end-labeled nucleotide of the primer and the 5' terminus of the RNA. The yield of cDNA is approximately proportional to the concentration of the target sequences in the mRNA preparation. This technique is the method of choice for measuring RNA steady-state levels for a number of reasons. It allows direct quantitation of levels without the difficulties of RNA blotting and it provides a precise definition of the transcription start site for the introduced gene in the transformed tissue.

If desired, the end-labeled cDNA can be extracted from the gel and sequenced by the Maxam-Gilbert method. It is also possible to sequence the cDNA directly by including dideoxyribonucleoside triphosphates in the primer-extension reaction. Either the ssDNA or denatured dsDNA can be used as primers. The fragments of double-stranded DNA used as primers are usually 75–150 nucleotides in length and are radio-labeled at both ends by phosphorylation or by the incorporation of radio-labeled dNTPs at a recessed 3' terminus. Single-stranded DNA primers, which are preferred for experiments of this type, are synthetic oligonucleotides about 30–40 nucleosides in length. It is important to use primers whose target sequences are located within 100 nucleotides of the 5' terminus of the mRNA. The main limitation to this method is that it is extremely insensitive, and prone to considerable background problems. Furthermore, the 5' ends of mRNAs are often highly structured; with the result that the reverse transcriptase does not copy these sequences.

Materials

TE buffer: 10 mM Tris.Cl; 1 mM EDTA pH 7.5.

2.5X Kinase buffer: 125 mM Tris.Cl pH 9; 25 mM $MgCl_2$; 12.5 mM DTT; 2.5 mM spermidine; 0.25 mM EDTA (use within two weeks).

5X Annealing buffer: 1.25 M KCl; 10 mM Tris pH 7.9; 1 mM EDTA.

PE mix: 10 mM $MgCl_2$; 5 mM DTT; 20 mM Tris, pH 8.3; 0.33 mM of each dNTPs; 100 µg/ml actinomycin-D.

Formaldehyde sequencing dyes: 98% deionized formamide; 1mM EDTA; 0.3% xylene cyanol; 0.3% bromophenol blue.

Vanadyl is prepared as described in Maniatis *et al.*, 1982.

Procedure

^{32}P 5'-end-labeling of oligonucleotide synthetic primers

1. One should plan to use 0.1 pmol primers for each reaction. Here is an example of a kinasing reaction for about 50 reactions. In an Eppendorf tube, dry 5 µl (10 pmol) $^{32}P-\gamma-ATP$ (5000 Ci/mmol) and then add:

   2 µl DNA primer (5 pmol)
   2 µl 2.5 X kinase buffer
   1 µl polynucleotide kinase
   Incubate the reaction at 37°C for 15 min.

2. Ethanol precipitation:
   (1) Add 75 µl TE + 54 µl 5 M NH_4OAc + 20 µg yeast RNA carrier + 350 µl ethanol, chill to –70°C for 30 min. and centrifuge for 15 min. at 4°C.
   (2) Dissolve pellet in 90 µl TE + 10 µl 3M NaOAc pH 6 + 250 µl ethanol, chill to –70°C for 30 min., 15 min. cold spin, 95% ethanol wash, dry.

3. Dissolve the final pellet in 50 µl TE (about 0.1 pmol/µl) and store at 4°C.

Primer extension reactions

4. For annealing, mix together on ice:

   2–10 µg total RNA (in 3–5 µl)
   ^{32}P 5'-end-labeled oligo primer (0.1 pmol)
   2 µl 5X annealing buffer
   1 µl 30 mM vanadyl
   Bring it to 10 µl with TE. Incubate it at the optimal annealing temperature for 1–8 hr (spin down condensed water in the tubes at least once every hour).

Primer extension reaction

1. For each annealing reaction tube, prepare 23 µl of PE mix + approximately 0.5 µl (10 units) of AMV reverse transcriptase. Mix them and add 23.5 µl to each tube, mix gently (not vortex), then incubate them at 37–42°C for 45 min. to 1 hr.
2. Ethanol precipitation: To each tube add 300 µl of ethanol and vortex, incubate them for 20 min. at −70°C, spin for 15 min. at 4°C, then wash with 70% ethanol, and let them dry.
3. Load and run the gel: Dissolve each pellet (by vortex) in 3 µl TE buffer, and then add 6 µl of formamide sequencing dyes. Heat it at 85°C for 5 min. and load onto an 8% sequencing gel, and run until the xylene cyanol dye is 16–20 cm from the top. After drying, wrap the gel with saran wrap to avoid contamination of the screen and cassette.
4. Expose the gel to X-ray film to determine the size of the extended product with a screen at ~70°C.

Note

1. For each primer/RNA, the optimal annealing temperature should be empirically tested. As a thumb rule, calculate the Tm-10°C of a primer (Tm = 4°C for each G or C and 2°C for each A or T), then try the primer extension reactions at several annealing temperatures around Tm-10°C. Also, primers with lower GC content require a longer annealing time.

Method-5: In situ RNA hybridization in plant tissues

The study of gene products using biochemical and molecular techniques often requires tissue samples containing a considerable amount of the target molecule. This presents a difficulty because many developmentally interesting genes are expressed either in a minority of cells in complex tissues, or for only brief periods of time during the differentiation of an organism or tissue. *In situ* hybridization is the most direct way of examining the modulation of gene expression during development at the individual cell level. This technique has been used primarily in animal tissues to determine the location of a particular mRNA or to map the chromosomal positions of cloned DNA. This approach has also been used as a diagnostic tool for the detection of cells infected with viruses. More recently, *in situ* hybridization has been applied to the localization of specific RNAs in plant tissues. Several methods have been used to detect cellular mRNA. Blotting analysis, for example, measures the overall expression levels of mRNA in the tissue, although sacrificing spatial resolution and is inefficient in the examination of heterologous tissues where individual cells express the gene of interest. In contrast, *in situ* hybridization allows the determination of the precise spatial distribution of the gene. Several *in situ* hybridization protocols have been optimized for the localization of particular mRNAs in plants (Meyerowitz, 1987).

VI SUGGESTED READING

Docherty K (1996) Primer extension analysis, In: *Gene Transcription: Analysis: Essential Techniques*, (Ed) Docherty K, John Wiley and Sons, Chichester.

Dunsmuir P, Bond D, Lee K, Gidoni D, and Townsend J (1989) Stability of introduced genes and stability in expression, In: *Plant Molecular Biology Manual*, (Ed) Gelvin SB, Schilperoort RA, and Verma DPS, Kluwer Academic Publishers.

Jones P (1996) Non-denaturing agarose gel electrophoresis of RNA, p9, In: *Gene Transcription: RNA Analysis,* (Ed) Docherty K, John Wiley and Sons.

Lehrach H, Diamond D, Wozney JM and Boedtker H (1977) RNA molecular weight determination by gel electrophoresis under denaturing conditions, a critical reexamination, *Biochemistry,* **16**:4743–4751.

Merric WC (1983) Translation of exogenous mRNAs in reticulocyte lysates, *Methods Enzymol.,* **101**:606–615.

Pelham HRB and Jackson RJ (1976) An efficient mRNA-dependent translation system from reticulocyte lysates, *Eur. J. Biochem.,* **67**:247–256.

Schleif RF and Wensink PC (1981) *Practical Methods in Molecular Biology*, Springer Verlag Inc, New York.

Shennan KIJ (1996) *In vitro* transcription, p 31, In: *Gene Transcription: RNA Analysis*, (Ed) Docherty K, John Wiley and Sons.

Thomas PS (1980) Hybridization of denatured RNA and small DNA fragments transferred to nitrocellulose, *Proc. Natl. Acad. Sci. USA,* **77**:5201–5205.

Zyskind JW and Bernstein SI (1989) *Recombinant DNA Laboratory Manual*, Academic Press, Inc.

9

Electrophoretic Techniques

I INTRODUCTION

Most biological macromolecules carry an electrical charge and can thus move in an electric field. For example, phosphate groups in sugar–phosphate backbone of nucleic acids are negatively charged at neutral pH. The side chains of four of the amino acids are highly ionized and are, therefore, charged at the near neutral pH values which exist in most biological systems *in vivo*. Glutamic acid and aspartic acid are negatively charged, while lysine and arginine are positively charged. Histidine is also positively charged but the charge is weak. Therefore, the proteins are zwitterions (a molecule with both positive and negative charges) and the net charge of a protein at a particular pH is determined by its content of acidic and basic amino acids, and by their degree of ionization at that pH. Thus, at nearly all pH values, proteins carry a net charge, which is either negative or positive depending on the pH value. These electrical properties of proteins are exploited in the various techniques of electrophoresis described in this chapter.

The versatility and resolving capacity of these techniques has resulted in their becoming the most popular tools in the cell and molecular biologist's arsenal and also an indispensable aid to the analytical biochemist.

II THEORY OF ELECTROPHORESIS

A. PRINCIPLES

The term "electrophoresis" is derived from Greek and means "carried by electricity". Thus, *electrophoresis is defined as the migration of a charged particle in an electric field*. Many important biological molecules, like amino acids, peptides, proteins, nucleotides and nucleic acids, posses ionizable groups and, therefore, at any given pH, exist in solution as electrically-charged species – either as cations (+) or anions (–). Under the influence of an electric field, these charged particles will migrate either to the cathode or to the anode, depending on the nature of their net charge. Under conditions of constant velocity, the driving force on a particle is the product of the charge on the particle and the applied field strength. This force is counteracted by the frictional resistance of the separation medium, which is proportional to its shear velocity. While

Stocks' law is obeyed in a free solution, the condition is complicated if a gel medium is used and, consequently, frictional resistance also depends on additional factors like gel density and particle size.

Gel electrophoresis, being a simple method to separate molecules according to their size, has made gels useful for a large number of purposes. Two DNA fragments that fall within the length range of 1–50,000 bp can be resolved on gels, provided that their length differs by at least 1%. This remarkably high resolution is one of the reasons gel electrophoresis is so useful. However, no single gel can separate molecules with such high resolution over such a broad range of lengths. However, the reproducibility of gel resolution depends on variables such as the gel concentration, buffer, ionic strength, pH, and the voltage drop per unit distance along the gel. Some less obvious variables, such as the temperature of the gel and the length of time the gel has been allowed to polymerize, are also influencing factors. A wide variety of buffers can be used for electrophoresis.

B. ELECTROPHORETIC MOBILITY: EFFECTS OF pH AND IONIC STRENGTH

An important concept in electrophoresis is *electrophoretic mobility*, which is *defined as the velocity of the particle per unit field strength*. The choice of applied voltage and separation path length, which determine the field strength, together with the time of the run, are important parameters in optimizing an electrophoretic separation. For example, the net charge on a protein is dependent upon the pH of the medium; the pH exerts a profound influence on protein mobility during electrophoresis. If the operative pH is the same as the pI of the protein, it will not migrate during electrophoresis. At pH values below its isoelectric point (pI), the protein will move towards the cathode, while at pH values above its pI, it will migrate towards the anode. The *ionic strength* of the separation medium also exerts a major influence on electrophoretic mobility. Separation media of low ionic strength permit higher rates of migration than do those of higher ionic strength (higher ionic strength results in sharper zones of separation). During every electrophoretic separation, electric energy is transformed into heat, and this is termed as *Joule heating*. This can result in deleterious effects, such as enzyme denaturation, increased diffusion of protein molecules, and even damage to the instrument itself. The choice of buffer strength is crucial here, as the higher its ionic strength, the greater its conductivity, and the greater the amount of heat generated. The electrical resistance of the gel increases as the moving boundary migrates through the gel, due to a decrease in conductance. In contrast to a constant current, at a constant applied voltage, the current and Joule heating will decrease with time, but at the expense of increased separation and reduced resolution.

C. ZONE ELECTROPHORESIS

In case of proteins, as proteins are charged at any pH other than their pI, they will migrate in an electric field at a rate that is dependent on their charge density. *Charge density is defined as the ratio of charge to mass* – the higher the ratio, the faster the molecules migrate. Thus, if an electric field is applied to a solution of proteins, the different molecules will migrate at different rates, depending on their charge density. Therefore, if the sample is initially present as a narrow zone, proteins of different mobilities will travel as direct zones and thus separate during electrophoresis. This approach is known as *zone electrophoresis* or *gel electrophoresis*.

D. EQUIPMENT

The equipment required for electrophoresis consists basically of two items: a power pack and an electrophoresis unit. Electrophoresis units are available for running either vertical or horizontal gel systems. The power pack supplies a direct current in an appropriate buffer, which is essential to maintain a constant state of ionization of the molecules being separated. Any variation in the pH would alter the overall charge and, hence, the mobility of the molecules being separated.

E. SUPPORT MEDIA

Although it is possible to carry out zonal electrophoresis in free solution, the adverse effects of diffusion and convection current caused in a free solution could be minimized by stabilizing the support medium. This was achieved by carrying out electrophoresis on a porous mechanical support, which was wetted in the electrophoresis buffer and in which electrophoresis of buffer ions and samples could occur. The support medium cuts down convection currents and diffusion, so that the separated components remain as sharp zones. Of all the media tried, two have proved outstanding for electrophoresis – polyacrylamide and agarose gels. Use of agarose or polyacrylamide gel is the standard method used to separate, identify and purify DNA fragments. The technique is simple, rapid to perform, and capable of resolving fragments of DNA that cannot be separated properly by other methods. Furthermore, the location of DNA within the gel can be determined directly by staining with low concentrations of the fluorescent intercalating dye ethidium bromide bands containing as little as 1–10 μg of DNA, which can be detected by direct examination of the gel in UV light. If necessary, these bands of DNA can be recovered from the gel and used for a variety of cloning purposes. Polyacrylamide gels are most effective for separating small fragments of DNA (5–500 bp). It has a high resolution power; fragments of DNA that differ in size by as little as 1 bp can be separated from one another. Although agarose gels have a lower resolving power than polyacrylamide gels, they have a greater range of separation. DNAs from 200–50 kb in length can be separated on agarose gels of various concentrations. Each of these types of gel electrophoresis will be discussed briefly in the following sections.

1. Agarose Gels

Agarose, which is extracted from seaweed (purified polysaccharide from rhodophyte "red" algae of the genera *Gelidum* and *Gracilaria*) and is composed of modified galactose residues, is a linear polymer. There are many special grades and chemically modified forms of agarose, with different melting and gel properties, available commercially. A combination of strength and macroporacity makes agarose gels indispensable for electrophoresis of very large macromolecules, particularly nucleic acids. The gelling properties are attributed to both inter- and intramolecular hydrogen bonding within and between the long agarose chains. The polymeric chain of agarose is built of alternating 1,3-linked β-D-galactopyranose and 1,4-linked 3,6-anhydro-α-L- galactopyranose units (Figure 9.1). As mentioned earlier, gel formation proceeds by the hydrogen bonding of bundles of filaments into a three-dimensional network. The pore size in the gel is controlled by the initial concentration of agarose; large pore sizes are formed

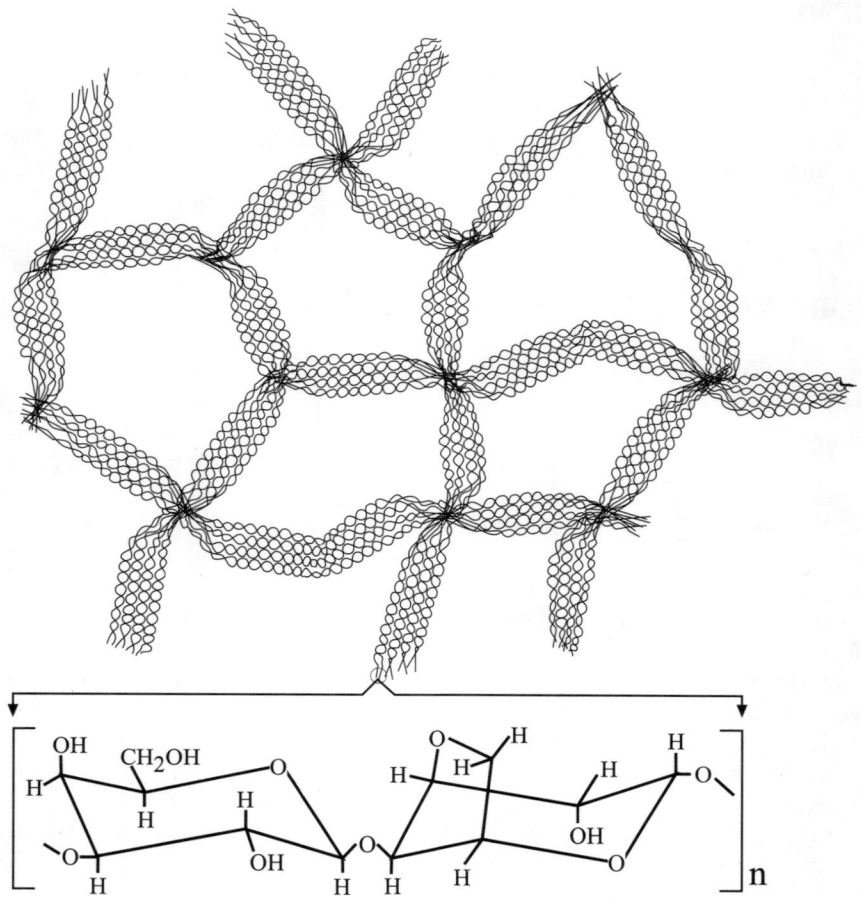

Figure 9.1 *Schematic representation of the gel structure and partial structure of gel forming polymer of agarose (alternating 1,3-linked β-D-galactose and 1,4-linked, 3,6-anhydro-α-L-galactose)*

from low concentrations and smaller pore sizes are formed from higher concentrations. Several agarose preparations form strong gels at as low a concentration as 0.3%. Agarose gels are used for the electrophoresis of both proteins and nucleic acids. Agarose gels form transparent solutions (melt) at 84–96°C, and low melt agar at 70°C. Agarose is sold in different purity grades, based on the sulphate concentration – the lower the sulphate content, the higher the purity.

Agarose gels are cast by melting the agarose in the presence of the desired buffer until a clear, transparent solution forms. The melted solution is then poured into a mold and allowed to harden. Upon hardening, the agarose forms a matrix, the density of which is determined by the concentration of the agarose. When an electric field is applied across the gel, DNA, which is negatively charged at neutral pH, migrates toward the anode. The rate of migration is determined by a number of parameters.

(a) Factors affecting the rate of DNA migration in agarose gels

(i) Molecular size of the DNA

Molecules of linear double-stranded DNA, which tend to become oriented in an electric field in an end-on position, migrate through the gel matrices at rates that are inversely proportional to the log10 of the number of base pairs. The larger molecules migrate more slowly because of greater frictional drag and because they find their way through the pores of the gel less efficiently than smaller molecules.

(ii) Agarose concentration

A linear DNA fragment of a given size migrates at different rates through gels containing different concentrations of agarose. There is a linear relationship between the logarithm of the electrophoretic mobility of DNA (μ) and the gel concentration (r), which is described by the equation:

$$\log \mu = \log \mu_o - K_r^t$$

where μ_o is the free electrophoretic mobility of DNA and K_r is the retardation coefficient, a constant that is related to the properties of the gel and the size and shape of the migrating molecules. Thus, by using gels of different concentrations (Table 9.1), it is possible to resolve a wide size range of DNA molecules.

Table 9.1 *Range of separation in gels containing different concentrations of agarose*

Agarose concentration (% [w/v])	Range of DNA molecules (kb)
0.3	5–60
0.6	1–20
0.7	0.8–10
0.9	0.5–7
1.2	0.4–6
1.5	0.2–3
2.0	0.1–2

(iii) Conformation of the nucleic acids

The super-helical circular (form I), nicked circular (form II), and linear (form III) DNAs of the same molecular weight migrate through agarose gels at different rates. The relative mobilities of the three forms depend primarily on the agarose concentration in the gel, but they are also influenced by the strength of the applied current, the ionic strength of the buffer, and the density of the super-helical twists in the form I DNA. Under some conditions, form I DNA migrates faster than form III DNA; under other conditions, the order is reversed. A long dsDNA is rather flexible; it is propelled through the pores of a gel in a snake-like fashion. On the other hand, a single-stranded chain of the same length folds into a random coil such that its migration in the gel is hindered. In this case, the denatured DNA moves at a slower rate on electrophoresis than the native one.

(iv) Applied voltage

At low voltages, the rate of migration of the linear DNA fragments is proportional to the voltage applied. However, as the electric field strength is raised, the mobility of high-molecular-weight fragments of DNA increases differentially. Thus, the effective range of separation in agarose gels decreases as the voltage is increased. To obtain the maximum resolution of DNA fragments that are greater than 2 kb in size, the voltage should not be more than 5 V/cm (length is between the electrodes not within the gel itself).

(v) Direction of the electric field

The DNA molecules that are larger than 50–100 kb in length migrate through agarose gels at the same rate, if the direction of the electric field remains constant. However, if the direction of the electric field is altered periodically, the DNA molecules are forced to change course. This is because the larger molecules of DNA take longer to realign themselves to the new direction of the field.

(vi) Base composition and temperature

The electrophoretic behavior of DNA in agarose gels is not significantly affected either by the base composition of the DNA or the temperature at which the gel is run. Thus, in agarose gels, the relative electrophoretic mobilities of different DNA fragments do not change between 4–30°C.

(vii) Presence of intercalating dyes

Ethidium bromide, a fluorescent dye that is used to detect DNA in agarose and polyacrylamide gels, reduces the electrophoretic mobility of linear DNA by about 15%. The dye intercalates between stacked base pairs, extending the length of linear and nicked circular DNA molecules.

(viii) Composition of the electrophoresis buffer

The electrophoretic mobility of DNA is affected by the composition and ionic strength of the electrophoresis buffer. In the absence of ions, electrical conductance is minimal and the DNA migrates very slowly, if at all. In buffers of high ionic strength electrical conductance is very efficient and significant amounts of heat are generated. Different buffers are available for electrophoresis of native double-stranded DNA (Table 9.2). These contain EDTA and Tris-acetate (TAE), Tris-borate (TBE), or Tris-phosphate (TPE) at a concentration of approximately 50 mM (pH 7.5–7.8).

Table 9.2 *Commonly used electrophoresis buffers*

Buffer	Working solution	Concentrated stock solution
Tris-acetate (TAE)	1X: 0.04 M Tris-acetate 0.001 m EDTA	50X: 242 g Tris base 57.1 ml glacial acetic acid 100 ml 0.5 M EDTA (pH 8.0)

Continues...

... *Continued*

Buffer	Working solution	Concentrated stock solution
Tris-phosphate (TPE)	1X: 0.09 M Tris-phosphate 0.002 M EDTA	10X 108 g Tris base 15.5 ml 85% phosphoric acid (1.679 g/ml) glacial acetic acid 40 ml 0.5 M EDTA (pH 8.0)
Tris-borate (TBE)	0.5X: 0.045 M Tris-borate 0.001 M EDTA	5X: 54 g Tris base 27.5 g boric acid 20 ml 0.5 M EDTA (pH 8.0)
Alkaline	1X: 50 mN NaOH 1 mM EDTA	1X 5 ml 10 N NaOH 2 ml 0.5 M EDTA (pH 8.0)

(ix) Apparatus used for agarose gel electrophoresis

There are a wide variety of designs for the hardware of agarose gel electrophoresis and almost all of them work equally well. However, there are two fundamental differences in design. One is the electrical connection between the buffer in the gel and in the reservoirs at the positive and negative poles. Most systems permit direct contact between the gel and the reservoir and some other use a paper wick. The second fundamental design difference is between horizontal (Figure 9.2) and vertical (Figure 9.3) gels.

The most commonly used configuration is the horizontal slab gel, which is superior to the vertical gels in several respects. Horizontal slab gels are usually poured onto a glass plate or plastic tray that can be installed on a platform in the electrophoresis tank. Electrophoresis is carried out with the gel submerged just beneath the surface of the buffer. The resistance of the gel to the passage of electric current is almost the same as that of the buffer, and so a considerable fraction of the applied current passes along the length of the gel. The horizontal gel apparatus has several other advantages: samples can be easily loaded; the gel can be stained in the apparatus after electrophoresis; vacuum grease does not need to be used; the DNA migration can be easily followed during electrophoresis; gel formation is very simple; and the apparatus is fairly inexpensive. The gel box is made of Plexiglas and has one or two removable sides, so that after electrophoresis the gel can be gently removed from the box.

(b) Preparation of an agarose gel

1. Seal the edges of a clean, dry, glass plate (or open ends of a gel-casting tray) with paper adhesive tape or autoclave tape so as to form a mold. Set the mold on a horizontal leveled surface.
2. Prepare sufficient electrophoresis buffer (Table 9.2) to fill the electrophoresis tank and to prepare the gel. Add the correct amount of powdered agarose and a measured quantity of electrophoresis buffer into a clean glass container. The volume of buffer should not occupy more than 50% of the container, and the same batch of buffer should be used for both gel preparation and gel running.

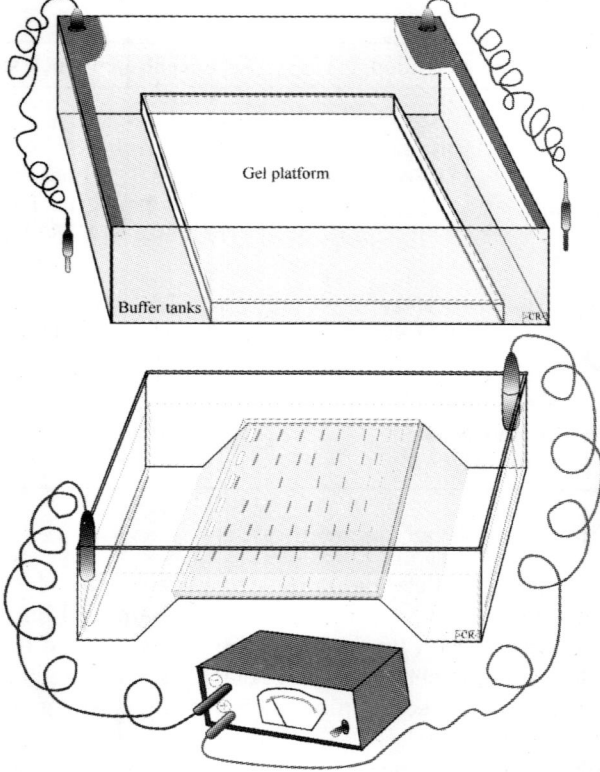

Figure 9.2 *Diagram of horizontal gel electrophonesis apparatus, two models*

3. Melt the agarose in the buffer. Loosen the cap of the container and heat the slurry in a boiling-water bath, or in a microwave, or by autoclave until the agarose dissolves. The gel slurry should be boiled just enough to completely dissolve the agarose.
4. Allow the melted agarose solution to cool to 55–60°C and, if desired, add ethidium bromide (0.5 μg/ml) and mix well. Care should be taken when using high concentrations of agarose (2% and above); cool the solution quickly to 70°C and pour the gel immediately. The higher the agarose concentration, the quicker the gel hardens.
5. Use a Pasteur pipette to seal the edges of the mold with a small quantity of agarose solution. Allow the seal to set.
6. Position the comb 0.5–1.0 mm above the plate so that complete wells are formed when the agarose is added.
7. Pour the remainder of the warm agarose solution into the mold. The gel should be between 3–5 mm thick. Make sure there are no air bubbles around the teeth of the comb or in the gel.
8. After the gel is completely set (30–45 min.), carefully remove the comb and the autoclave tape. Very low concentrations of agarose gels (below 0.55%) may require longer hardening periods and/or hardening at low temperatures. Preclude any damage to the wells by lifting one end of the comb until the first tooth is free of the gel, and then holding

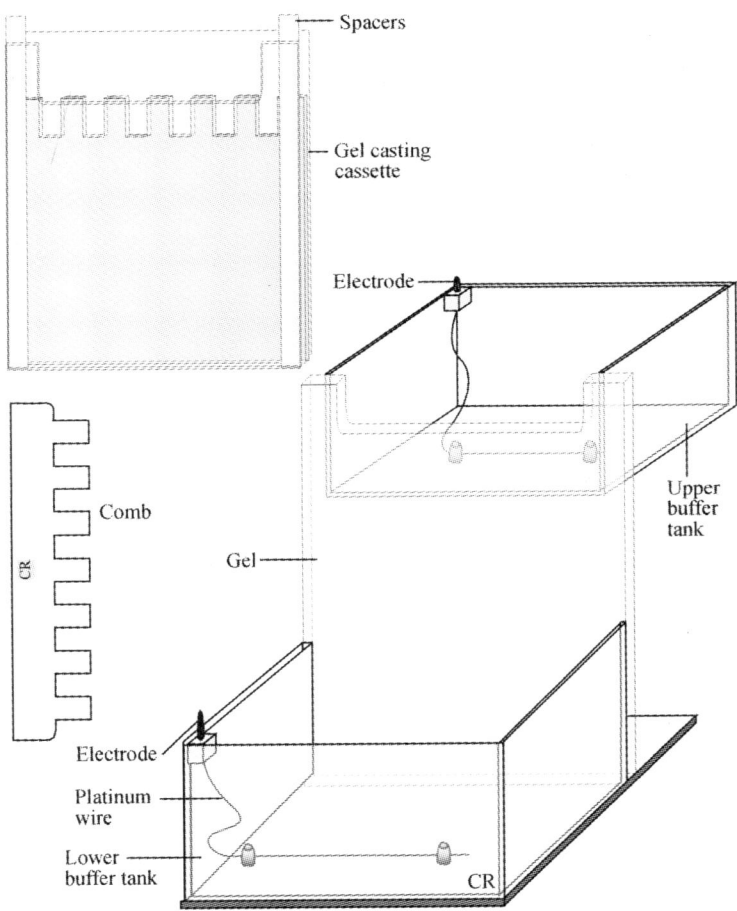

Figure 9.3 *Diagram of vertical slab gel electrophoresis apparatus*

that end of the comb fixed while lifting the other end of the comb. This minimizes the vacuum in the space between the gel wells and the comb. Then, mount the gel in the electrophoresis tank.
9. Add just enough electrophoresis buffer to cover the gel to a depth of about 1 mm and make sure that all the wells are filled. The gel may be used after a 10-min. soaking. The gels may be stored for several days at room temperature and can be kept clean and wet by covering the box with a plastic sheet.
10. Fill each reservoir with about 300 ml of buffer; more or less buffer can be used. If the buffer touches the solder connection between the platinum wire and the banana plug connector, the electrode reactions will soon unsolder the connection. On the other hand, too little buffer in the tank will provide poor buffering capacity and the electrode reactions will change the pH drastically. In turn, this pH change can have dramatic effects on the hardness of the agarose.

11. Mix the DNA samples with the gel-loading buffer (Table 9.3). The salt concentration in the sample should be less than 0.1 M to minimize any smearing of the DNA bands. The glycerol in the gel-loading buffer allows the sample to settle at the bottom of the well. Slowly load the mixture into the slots of the submerged gel using a disposable micropipettor or a Pasteur pipette (Figure 9.4). Ensure that no air bubbles are left in the wells, as bubbles in the wells will distort the electric field in the gel and may result in distorted band patterns.
12. Close the lid of the gel tank and attach the electrical leads, so that the DNA will migrate toward the anode (red lead). Apply a voltage of 1–5 V/cm (measured as the distance between the electrodes). If the leads have been attached correctly, bubbles will not be generated at the anode and cathode (due to electrolysis) and, within a minute, the bromophenol blue should migrate from the wells into the body of the gel. Run the gel until the bromophenol blue (co-migrate with ~0.3 kb dsDNA) and xylene-cyanol FF (co-migrate with ~ 4 kb dsDNA) have migrated the appropriate distance through the gel. Ethidium bromide migrates toward the cathode (opposite to that of the DNA), therefore restaining of gel may sometimes be necessary.

Table 9.3 *Gel-loading buffers (tracking dyes)*

Buffer types	6X Buffer	Storage temperature
I	0.25% bromophenol blue 0.25% xylene cyanol FF 40% (w/v) sucrose in water	4°C
II	0.25% bromophenol blue 0.25% xylene-cyanol FF 15% Ficoll (Type 400) in water	Room temperature
III	0.25% bromophenol blue 0.25% xylene-cyanol FF 30% glycerol in water	4°C
IV	0.25% bromophenol blue 40% (w/v) sucrose in water	4°C
V	Alkaline loading buffer 300 mN NaOH 6 mM EDTA 18% Ficoll (Type 400) in water 0.15% bromocresol green 0.25% xylene-cyanol FF	4°C

13. For a 100 ml, 1% agarose gel, electrophoresis at 40 mA for about 10 hr is adequate for most purposes. The DNA migration rate can be manipulated by changing the ionic strength of the buffer or by changing the current; lowering the ionic strength or increasing the current will increase the migration rate. The rate of migration can be estimated by the migration of the bromophenol blue dye, which in such a gel co-migrates with double-stranded DNA of about 0.3 kb. When connecting the electrodes, remember that the DNA will migrate to the positive pole. If too high a current is used the gel will heat up, usually in a non-uniform way, causing the DNA migration rates to vary in different parts of the gel.

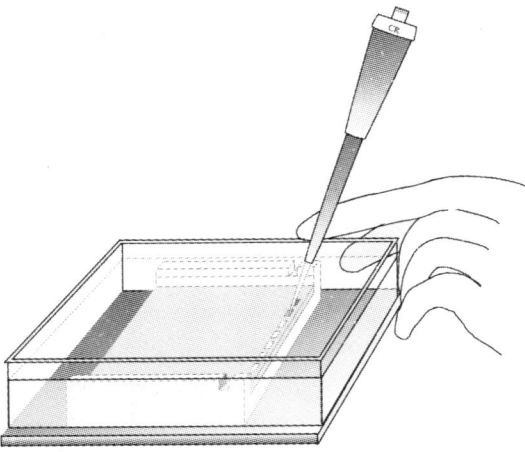

Figure 9.4 *Loading samples on to the well in horizontal gel electrophoresis*

14. Turn off the electric current, and remove the leads and lid from the gel tank. If ethidium bromide staining is not enough then re-stain the gel, examine the gel by UV light and photograph it.

(c) Staining DNA in agarose gels

The most convenient method to visualize DNA in agarose gels is staining with the fluorescent dye, ethidium bromide. Ethidium bromide contains a planar group that intercalates between the stacked bases of the DNA. UV radiation at 254 nm is absorbed by the DNA and transmitted to the dye; radiation at 302 nm and 366 nm is absorbed by the dye itself. In both cases, the energy is re-emitted at 590 nm in the red-orange region of the visible spectrum. Because the fluorescent yield of the ethidium bromide:DNA complexes is much higher than that of the unbound dye, small amounts of DNA can be detected in the gels. Ethidium bromide can be used to detect both single- and double-stranded nucleic acids (DNA and RNA). However, the affinity of the dye for single-stranded nucleic acid is relatively low and fluorescent yield is comparatively poor.

Ethidium bromide is usually incorporated into the gel with electrophoresis buffer at a concentration of 0.5 µg/ml. Although the electrophoretic mobility of linear double-stranded DNA is reduced by approximately 15% in the presence of the dye, the ability to examine the gel directly under UV illumination during or at the end of the run is often a great advantage. However, some people feel that sharper bands are obtained when the gel is run in the absence of the dye. In this case, after electrophoresis is completed, the gels can be stained by soaking them for 30–45 min. in a solution of ethidium bromide solution (0.5 µg/ml) at room temperature. The detection of smaller amount of DNA (<10 µg) can be made easier if the background fluorescence is reduced by de-staining. Unbound ethidium bromide can be reduced by soaking the gels in water or 1 mM $MgSO_4$ for 20 min. at room temperature. Ethidium bromide solution is usually prepared as a stock solution of 10 mg/ml in water, which is stored at room temperature in the dark.

(d) Photography

Photographs of gels may be made using transmitted or incident UV light. Most commercially available UV light sources emit UV light at 302 nm. The fluorescent of ethidium bromide:DNA complexes is considerably greater at this wavelength than at 366 nm and slightly less than at short wavelength (254 nm) light. The most sensitive film is Polaroid Type 57 or 667 (ASA 3000). With an efficient UV light source (>2500μW/cm^2), a Wratten 22A filter, and a good lens (f=135 mm), an exposure of a few seconds is sufficient to obtain images of bands containing as little as 10 μg of DNA. With long exposure time and a strong UV light source, the fluorescence emitted from as little as 1 μg of DNA can be recorded on film.

2. Polyacrylamide Gel Electrophoresis

Electrophoresis in acrylamide gels is generally called as PAGE, which is an abbreviation for polyacrylamide gel electrophoresis. Acrylamide is a white crystalline monomer whose structure is $CH_2=CHCONH_2$. Dry acrylamide if stored in a dark vessel, especially if in the cold, can be used within a year. Acrylamide is toxic; therefore, solutions should be prepared under a hood, using rubber gloves. N,N'-Methylene-bis-acrylamide (referred to here as 'Bis' for short) $CH_2(NHCOCH=CH_2)_2$ is used to cross-link the linear polyacrylamide chains (Figure 9.5). Ammonium persulfate is used as an initiator of the polymerization process; it decomposes gradually in aqueous media and, thus, only freshly-prepared solutions should be employed. Photopolymerization is an alternative method that can be used to polymerize acrylamide gels. The ammonium persulphate and TEMED are replaced by riboflavin and, when the gel is poured, it is placed in front of a bright light for 2 to 3 hr. The photodecomposition of riboflavin generates a free radical that initiates polymerization.

(a) Purity of acrylamide

The chemical purity of acrylamide and Bis is very important. General laboratory-grade reagents are likely to be impure, containing breakdown products such as acrylic acid, polyacrylic acid, ammonia and β',β'',β'''-nitrilo-tris-proprionamide. Prior to using these compounds for electrophoresis they should be recrystallized, using chloroform for acrylamide and acetone in the case of Bis. It is also important to realize that the stocks of acrylamide, whatever their initial purity, will breakdown on prolonged storage. The changed acrylic acid, in particular, is deleterious to most electrophoretic techniques. The breakdown particles can be easily removed by mixing the acrylamide solution with 1 g of a suitable ion-exchange resin (e.g., amberlite MB-1 monobed resin) for at least 1 hr. The resin can be removed by filtration and the resulting solution used for gel polymerization.

(b) Polymerization catalysts

Gel polymerization is usually initiated with ammonium persulfate and the reaction is accelerated by the addition of the catalyst, N,N,N',N''–tetramethylethylenediamine (TEMED); $(CH_3)_2NCH_2CH_2N(CH_3)_2$ is a colorless liquid, which is 6.7 M when undiluted. Strictly speaking, TEMED is not an initiator of polymerization but is a powerful catalyst of the process. This reaction is more efficient than the alternative method of UV activated polymerization with riboflavin. The optimum temperature for gel polymerization is in the range of 25–30°C. Polymerization using pre-cooled acrylamide solutions (0–4°C) is recommended only to prolong the polymerization time.

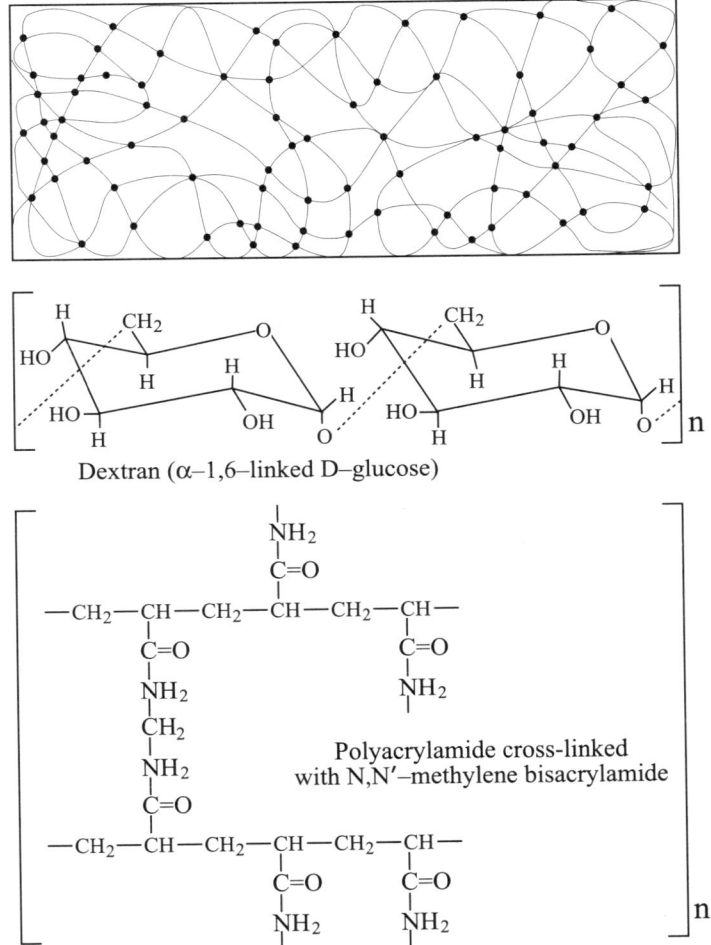

Figure 9.5 *Schematic representation of the gel structures of point cross-linked polymers such as polyacrylamide and dextran (Sephadex)*

Some of the factors that influence the polymerization process are that water-dissolved oxygen may be regarded as a bi-radical, capable of interrupting the chain reaction of free-radical polymerization of acrylamide. Also, oxygen present in the air when in contact with a solution of monomers can prevent polymerization of acrylamide and, in order to avoid this, the surface of the solution is overlaid with a 3–5 mm layer of degassed water or iso-butanol.

(c) Pore size

Acrylamide gels are defined in terms of the total percentage of acrylamide present, and the pore size in the gel can be varied by changing the concentrations of both the acrylamide and the bis-acrylamide. Very dilute gels (<2.5%) are mechanically unstable, thereby limiting the maximum effective pore size to about 80 mm.

(d) Gel configuration and concentration

Cylindrical rod gels: The original procedures for performing gel electrophoresis in polyacrylamide gels were developed using cylindrical gels polymerized in glass tubes. The tubes which are used normally have an internal diameter of 3–5 mm and gels varying in length from about 5 to 15 cm can be prepared.

Slab gels: Flat slab gels are the method of choice for electrophoresis in one dimension, as many samples can be electrophoresed simultaneously on the same gel under identical conditions, thereby increasing the reproducibility and comparability of the separation patterns. There are two types of slab gels, based on the type of apparatus that holds them: vertical slab gels and horizontal slab gels.

Gel concentration: *Homogenous gels*: theoretically, there is a gel concentration which is optimal for the resolution of a given pair of proteins; however, there is no single gel concentration which will give the maximum separation of the components of a more complex protein mixture.

Gradient gels: The effective separation by polyacrylamide gels can be extended using gels containing a linear or non-linear concentration gradient. The average pore size of these gels decreases with increasing gel concentration, so that there is an effective band-sharpening effect during electrophoresis. Thus, gradient polyacrylamide gels have the dual advantages of separating proteins with a wider range of sizes, i.e., both separation and resolution is improved. Most separations carried out using linear polyacrylamide gradients as their gel are relatively straightforward. However, for a particularly difficult separation problem, it may be necessary to resort to the use of convex, concave or more complicated gradient shapes rather than a linear one.

(e) Gel preparation

Polyacrylamide gels are more tedious to prepare and run than agarose gels. However, acrylamide gels have some advantages over agarose gels: (i) their resolving power is so great that they can separate molecules of DNA with as little as 0.2% differences (i.e., 1 bp in 500 bp, Table 9.4); (ii) they can accommodate larger quantities of DNA (up to 10 μg of DNA in a single slot of 1 cm×1 mm) than agarose gels; and (iii) the DNA recovered from polyacrylamide gels is extremely pure. Two types of polyacrylamide gels are commonly used: non-denaturing polyacrylamide gels and denaturing polyacrylamide gels.

When a series of runs is planned, it is advisable to prepare, in advance, a concentrated (30–40%) aqueous solution of acrylamide mixed in definite proportions with Bis. This solution may be stored for weeks in the refrigerator. The stock buffer solution and TEMED may also be stored for a prolonged time in the cold, while the ammonium persulfate solution should be freshly prepared before a run.

Non-denaturing polyacrylamide gels (*Native* or *buffer gels*) are used for the separation and purification of a particular protein (often an enzyme) and fragments of double-stranded DNA. These gels are poured and run in 1X TBE at low voltage (1–8 V/cm) to prevent the denaturation of small fragments of DNA by the heat generated by the passage of the electric current. Most species of double-stranded DNA migrate through non-denaturing acrylamide gels at a rate that is inversely proportional to the log10 of their size. However, their electromobility is also affected by their base composition and sequence. The kinks formed at specific sequences in double-stranded DNA can affect mobility up to 10% in exactly the same size molecules.

Denaturing polyacrylamide gels are used for the separation and purification of single-stranded fragments of DNA. These gels are polymerized in the presence of a denaturing agent (e.g., urea or, less frequently, formamide) that suppresses base pairing in nucleic acids (alkalis cannot be used as denaturing agents because they deaminate acrylamide, and methylmercuric hydroxide cannot be used since it inhibits polymerization). The denatured DNA migrates through these gels at a rate that is almost completely independent of its base composition and sequence. Some of the uses of denaturing polyacrylamide gels are: the isolation of radio-labeled DNA probes, the analysis of the products of nuclease-S1 digestions, and the analysis of the products of DNA sequencing reactions.

Table 9.4 *Effective range of separation of DNAs in polyacrylamide gels*

Acrylamide (%[w/v])[a]	Separation range (bp)	Xylene cyanol FF[b]	Bromophenol blue[b]
3.5	1000–2000	460	100
5.0	80–500	260	65
8.0	60–400	160	45
12.0	40–200	70	20
15.0	25–150	60	15
20.0	6–100	45	12

[a] bis-acrylamide is 1/30th the concentration of acrylamide
[b] The numbers are the approximate sizes of double-stranded DNA (bp), which co-migrates with dye

(f) Preparation of non-denaturing polyacrylamide gels

Most vertical electrophoresis tanks obtained from commercial sources are constructed to hold glass plates of 20 × 40 cm in size, but the size can vary depending upon the availability of suitable tanks. The spacers vary in thickness from 0.5–2.0 mm, depending on the quantity of DNA to be loaded. The thicker the gel, the hotter it will become during electrophoresis; overheating results in 'smiling' bands of DNA and other problems. Therefore, thinner gels are preferred, since they produce sharper and flatter bands of DNA.

Materials

1. Solutions

    30% Acrylamide: Dissolve 29 g Acrylamide and 1 g N,N'-methylenebisacrylamide in 100 ml water. Heat the solution to 37°C to dissolve the chemicals.

    Prepare 1X TBE: 89 mM Tris-borate, 2 mM EDTA (pH 8.0), TBE is usually made and stored as a 5X stock solution. The pH of the buffer should be approximately 8.3.

    Prepare 10% Ammonium persulfate: Dissolve 1 g Ammonium persulfate in 10 ml of water. The solution may be stored at 4°C for several weeks.

2. *Prepare the glass plates and spacers for pouring the gel:* If necessary, clean them with KOH/methanol, which is prepared by adding ~5 g of KOH pellets to 100 ml of methanol. Then, wash the glass plates and the spacers in a warm detergent solution and rinse them well, first in tap water and then in deionized water. Hold the plates by the edges so that the oils from the fingers are not deposited on the working surfaces of the plates. Rinse the plates

with ethanol and set them aside to dry. The glass plates must be free of grease spots to prevent air bubbles from forming in the gel. Treat one surface of each plate with a silicon solution. This prevents the gel from adhering too tightly to both plates and reduces the possibility that the gel will tear when it is removed from the mold after the electrophoresis is completed.

3. There are many types of electrophoresis apparatus available commercially. Typically, the two plates are of slightly different size and one of them is notched. Lay the larger plate flat on the bench and arrange the spacers at each side parallel to the two edges. A couple of minute dabs of petroleum jelly (or silicone vacuum grease) help to keep the spacer bars in position. Lay the inner (notched) plate in position, resting on the spacer bars. Bind the entire length of the two sides and bottom of the plates with a gel-sealing tape to make a watertight seal. Alternatively, clip the sides of the gel cassette with bulldog clips.

4. After determining the size of the glass plates and the thickness of the spacers, calculate the volume of acrylamide solution required.

5. Wearing gloves, perform the following manipulations over a tray so that any spilled acrylamide solution will not spread over the bench.
    a. Add 35 µl of TEMED (N,N,N',N'-tetramethylethylenediamine) to each 100 ml of acrylamide solution. Mix the solution by swirling.
    b. Draw the solution into the barrel of a 50 ml syringe. Invert the syringe and expel any air that has entered the barrel. Introduce the nozzle of the syringe into the space between the two glass plates. Expel the acrylamide solution from the syringe, filling the space almost to the top. Keep the remaining acrylamide solution at 4°C to reduce the rate of polymerization. If the plates were clean, there should be no trapped air bubbles, and if they were sealed well, no leaks. If any air bubbles have formed, empty the gel mold and re-pour the gel, after thoroughly re-cleaning the glass plates.
    c. Lay the glass plates against a test-tube rack at an angle of approximately 10°. This decreases the chance of leakage and minimizes distortion of the gel. If a bottom spacer is used, be sure to remove it before fixing it to the electrophoresis apparatus.

6. Immediately insert the appropriate comb, being careful not to trap any air bubbles under the teeth. The tops of the teeth should be slightly higher than the top of the glass. Clamp the comb in place with a bulldog paper clip. If necessary, use the remaining acrylamide solution to fill the gel mold completely. Check that no acrylamide solution is leaking from the gel mold.

7. Allow the acrylamide to polymerize for 60 min. at room temperature, adding additional acrylamide solution if the gel retracts significantly. When polymerization is complete, a schleren pattern will be visible just beneath the teeth of the comb.

8. Carefully remove the comb, and immediately rinse out the wells with water. Using a razor blade or a scalpel, remove the electrical tape from the bottom of the gel.

9. Attach the gel to the electrophoresis tank, using large bulldog paper clips on the sides and three-prong clamps on the shoulders. The notched plate should face inward towards the buffer reservoir.

10. Fill the reservoirs of the electrophoresis tank with a 1X TBE. Use a bent Pasteur pipette or syringe needle to remove any air bubbles trapped beneath the bottom of the gel.

11. Use a Pasteur pipette to flush out the wells with 1X TBE.

12. Mix the DNA samples with the appropriate amount of 6X gel-loading buffer. Load the mixture into the wells using a drawn-out glass micropipette or Hamilton syringe. Long,

disposable micropipette tips can also be used for this purpose. Draw up the sample into the loading device, and then insert the tip of the device into the well. This should be done quickly, in a single movement, since the DNA sample tends to dribble out of the loading device after the tip is immersed in the electrophoresis buffer.

13. Connect the electrodes to a power pack (positive electrode connected to the bottom reservoir). Non-denaturing polyacrylamide gels are usually run at voltage gradients between 1 V/cm and 8 V/cm.
14. Run the gel until the tracking dyes have migrated the desired distance. Then, turn off the electric power, disconnect the leads, and discard the electrophoresis buffer from the reservoirs. Detach the glass plates, and use a scalpel or razor blade to remove the electrical tape. Lay the glass plates on the bench. Using a thin spatula, lift a corner of the upper glass plate and check that the gel remains attached to the lower plate. Pull the upper plate smoothly away and remove the spacers.
15. Use any of the methods described on the following pages to detect the positions of bands of DNA in the polyacrylamide gel.

Notes

1. It is essential to wash out the wells thoroughly as soon as the comb is removed. Otherwise, small amounts of acrylamide solution trapped by the comb will polymerize in the well, producing irregularly-shaped surfaces that give rise to distorted bands of DNA.
2. It is important to use the same batch of electrophoresis buffer in both of the reservoirs and in the gel. Small differences in ionic strength or pH produce buffer fronts that can greatly distort the DNA migration.
3. If any remnants of unpolymerized acrylamide remain in the wells, diffuse, wavy bands of DNA will be observed.
4. Raise the loading device as the sample is loaded into the well. The tip of the device should always be above the level of the sample in the well. Do not attempt to expel the entire sample from the loading device, since this almost always produces air bubbles that blow the sample out of the well or obstruct the electric flow.
5. If electrophoresis is carried out at a higher voltage, differential heating in the center of the gel may cause bowing of the DNA bands or even melting of the strands of small DNA fragments.

F. DETECTION OF DNA IN POLYACRYLAMIDE GELS

The most convenient stain for visualizing DNA in gels is ethidium bromide. Both agarose and acrylamide gels can be stained by immersing them in about 400 ml of 0.5–1 µg/ml of ethidium bromide. Stocks of ethidium bromide can be kept for months at 4°C, at concentrations between 100–1 mg/ml.

1. Staining with ethidium bromide

1. Gently submerge the gel and its attached glass plate in the staining solution (0.5 µg/ml ethidium bromide in a 1X TBE). Use just enough staining solution to cover the gel completely. After staining it for 30–45 min. at room temperature, remove the gel, using the

glass plate as a support. Carefully blot any excess liquid from the surface of the gel with a pad of paper towels; do not use absorbent paper. Cover the gel with a piece of Saran Wrap, and try to avoid creating air bubbles or folds in the Saran Wrap.
2. To photograph the gel, place a piece of Saran Wrap on the surface of an UV transilluminator. Invert the gel and place it on the transilluminator. Remove the glass plate, leaving the gel attached to the Saran Wrap and then photograph the gel.

Note

1. Owing to the fact that polyacrylamide quenches the fluorescence of ethidium bromide, it is not possible to detect bands that contain less than about 10 ng of DNA using this method.

2. Autoradiography

(a) Unfixed, wet gels

1. Wrap the gel, together with its supporting glass plate, in Saran Wrap (try to avoid creating air bubbles or folds in the Saran Wrap). To align the gel and the film, attach adhesive dot labels marked with radioactive ink onto the surface of the Saran Warp. Cover the labels with scotch tape. This prevents any possible contamination of the film holder or intensifying screen with radioactive ink.
2. Invert the gel and expose it to X-ray film (Kodak XAR-2 or its equivalent) as follows: In a darkroom, tape the sealed gel to a piece of X-ray film cut to the same size as the glass plate. The plate serves as a weight to ensure good contact between the Saran Wrap and X-ray film. Do not use a metal film cassette, which may break the glass and crush the gel, instead, wrap the gel and film in light-tight aluminum foil. Expose the film for the appropriate period of time at room temperature or at $-70°C$ with an intensifying screen.

Note

1. If any bands of radioactive DNA are to be recovered from the gel, the gel should not be fixed or dried.

(b) Fixed, Dried gels

1. Immerse the gel, together with its attached glass plate, in 7% acetic acid for 5 min. Remove the gel from the fixative by carefully lifting the glass plate from the fluid.
2. Rinse the gel briefly in deionized water. Remove any excess fluid from the surface of the gel with a pad of paper towels. Do not use absorbent paper, to which the gel will stick.
3. Wrap the gel and glass plate in Saran Wrap, and establish an autoradiograph as described above. Alternatively, dry the gel onto a piece of Whatman 3MM paper using a commercial gel dryer. Drying the gel is necessary only when the gel contains DNA labeled with weak β-emitting isotopes, such as ^{35}S; when small amounts of ^{32}P-labeled DNA are used, long exposures (longer than 24 hr) are necessary to obtain an adequate autoradiographic image.

Note

1. Try to minimize the movement of fluid across the surface of the gel during fixation. The aim is to keep the gel attached to its supporting glass plate. In most cases, the gel can then

be carefully unfolded and restored to its original shape. To avoid problems, some workers use a piece of plastic mesh to hold the gel in place during fixation.

III RECOVERY AND PURIFICATION OF DNA FRACTIONATED ON GELS

Although many methods have been developed over the years to recover DNA from gels, none has proved entirely satisfactory. Some of the problems are: the presence of inhibitors of enzymatic reactions; inefficient recovery of large fragments of DNA; inefficient recovery of small amounts of DNA; and inability to recover a number of different fragments simultaneously. Of the many techniques available to recover DNA from gels, the following three are the most reliable. Although these three techniques collectively meet all of the desired criteria, no single technique meets them all.

Method-1: Electroelution into dialysis bags

This method allows the recovery of a wide size range of DNAs, but it is a very inconvenient technique. Therefore, it should only be used when it is necessary to isolate large fragments of DNA (>5 kb in length) that are inefficiently recovered by other techniques. The DNA band can be detected in the gel either by ethidium bromide staining and illumination with long-wavelength UV light or by autoradiography. Electroelution into dialysis bags is by far the most inconvenient of these techniques; however, it is also most effective for the recovery of large fragments of DNA (>5 kb).

Procedure

1. Digest an amount of DNA that will yield about 1 µg of the fragment(s) of interest. Separate the fragments by electrophoresis through an agarose gel of the appropriate concentration that contains 0.5 µg/ml ethidium bromide. Locate the band of interest with a hand-held UV lamp.
2. Using a sharp scalpel or razor blade, cut out a slice of agarose containing the band of interest, and place it on a trough-like spatula that has been wetted with 1X TAE electrophoresis buffer.
3. After cutting out the band, photograph the gel as described earlier so that there is a record of which band was eluted.
4. Seal one end of a piece of dialysis tubing with a dialysis clip. Fill the dialysis bag to overflowing with 1X TAE. Then, holding the neck of the bag, use the spatula to transfer the slice of agarose into the buffer-filled bag.
5. Allow the gel slice to sink to the bottom of the bag. Remove most of the buffer, leaving just enough to keep the gel slice in constant contact with the buffer. Then clip the bag just above the gel slice; avoid trapping air bubbles.
6. Immerse the bag in a shallow layer of 1X TAE in an electrophoresis tank. When the current is switched on, the DNA is electroeluted out of the gel and onto the inner wall of the bag. The processes can be conveniently monitored with a hand-held, long-wavelength UV light if sufficient DNA is available.

7. Reverse the polarity of the current for 1 min. to release the DNA from the wall of the bag. Recover the bag from the electrophoresis chamber, and gently massage the side of the bag where the DNA has accumulated to remove the DNA from the wall. The process can be followed with a hand-held UV lamp.
8. Open the bag, and carefully transfer the buffer surrounding the gel slice to a fresh disposable plastic tube. Using a Pasteur pipette, wash out the bag with a small quantity of 1X TAE, and add the wash to the tube.
9. Stain the gel slice by immersing it in 1X TAE containing ethidium bromide (0.5 µg/ml) for 30–45 min. at room temperature. Examine the stained slice by UV illumination to confirm that all the DNA has been eluted.
10. Purify the DNA either by passing it through DEAE-Sephracel or by extraction with organic solvents.

Method-2: Low-melting-temperature agarose gels

The recovery of DNA from gels cast with low-melting-temperature agarose is less reproducible than the other methods, but this method has the advantage that certain enzyme reactions (e.g., digestion with restriction enzymes and ligation) can be carried out directly in the melted gel. A number of types of agarose are available, in which hydroxyethyl groups have been introduced into the polysaccharide chain. This substitution causes the agarose to gel at approximately 30°C and to melt at around 60°C – well below the melting temperature of most double-stranded DNAs. These properties have been exploited to develop a simple technique for the recovery of DNA from gels.

Procedure

1. Pour a gel containing the appropriate concentration of low-melting-temperature agarose, as described in the earlier section. When the gel has cooled to room temperature, transfer it and its supporting glass plate to a horizontal surface in a cold room, to ensure that the gel sets completely.
2. Mix the samples of DNA with a gel-loading buffer, load it onto the gel, and carry out electrophoresis at 4°C to ensure that the gel does not melt during the run.
3. Locate the band of interest using a hand-held UV lamp. This minimizes any radiation damage to the DNA.
4. Using a sharp scalpel or razor blade, cut out a slice of agarose containing the band of interest and transfer it into a clean, disposable plastic tube.
5. After cutting out the band, photograph the gel as described earlier so that there is a record of which band was eluted.
6. Add about 5 vol of 20 mM Tris-HCl (pH 8.0), 1 mM EDTA (pH 8.0) to the slice of agarose. Close the top of the tube and incubate for 5 min. at 65°C to melt the gel.
7. Cool the solution to room temperature, and then add an equal volume of phenol, equilibrated to pH 8.0 with 0.1 M Tris-HCl. Vortex the mixture for 20 s, and then recover the aqueous phase by centrifugation at 4,000g for 10 min. at 20°C. The white substance found at the interface is powdered agarose. Re-extract the aqueous phase once with phenol:chloroform and once with chloroform.
8. Transfer the aqueous phase to a polystyrene centrifuge tube, and add 0.2 vol of 10 M ammonium acetate and 2 volumes of ethanol at 4°C. Store the mixture for 10 min. at room temperature, and then recover the DNA by centrifugation (e.g., at 5,000 rpm for 20 min. at 4°C in a Sorvall RT 6000 centrifuge). Usually, at this stage, the DNA is pure enough to

serve as a substrate for restriction endonucleases, ligases, etc. However, if necessary, the DNA can be further purified by passing it through DEAE-Sepharcel.

Note

1. Because low-melting-temperature agarose remains fluid at 37°C, enzymatic manipulations can be carried out by adding portions of the melted gel slice directly to the reaction mixture.

Method-3: Extracting DNA fragments from polyacrylamide gels

A number of methods have been used for extracting DNA from PAGE gels. None of them is entirely satisfactory, but the best in use at the present time are given below. If enzymes do not operate properly on gel-extracted DNA, they can usually be used successfully if bovine serum albumin is included in the reaction mix at a concentration of 250 µg/ml. The best method to isolate DNA from polyacrylamide gels is the "crush and soak" technique, originally described by Maxam and Gilbert (1977). This elution method works very well for smaller DNA fragments. The recovery rates begin to fall as the fragments become larger than several thousand base pairs. The DNA obtained is of very high purity and is free of contaminants that inhibit enzymes or are toxic to transfected or microinjected cells.

Method

Elution buffer: 0.5 M ammonium acetate; 10 mM magnesium acetate; 1 mM EDTA (pH 8.0); 0.1% SDS.

"Crush and soak" method

1. Run the polyacrylamide gel as described previously. Locate the DNA band of interest by autoradiography or by examination of ethidium bromide-stained gels.
2. Using a sharp scalpel or razor blade, cut out the segment of the gel containing the band of interest. Do not attempt to remove the gel from the Saran Wrap before cutting; instead, cut through both the gel and the Saran Wrap together and then peel the small piece of gel containing the DNA from the Saran Wrap.
3. Transfer the gel slice to a microfuge tube. Use a disposable pipette tip to crush the slice against the wall of the tube or crush it with a siliconized glass rod. Over-crushing causes problems at later stages, so don't overdo it.
4. Calculate the approximate volume of the slice and add 1–2 vol of elution buffer to the microfuge tube (<0.5 ml is preferred).
5. Close the tube and incubate it at 37°C on a rotating wheel or rotary platform. Small fragments of DNA (<500 bp) are eluted in 3–4 hr; larger fragments take 12–16 hr (incubate it overnight, if possible).
6. Separate the liquid from the fragments of acrylamide by centrifuging the slurry at 12,000 g for 1 min. at 4°C in a microfuge. Transfer the supernatant to a fresh microfuge tube, being careful to avoid transferring any fragments of polyacrylamide.
7. Add an additional 0.5 vol of elution buffer to the pellet of polyacrylamide, vortex briefly, and re-centrifuge. Combine the two supernatants.
8. Remove any remaining fragments of polyacrylamide by passing the supernatant through a disposable plastic column or a syringe barrel containing a Whatman GF/C filter or packed siliconized glass wool.

9. Add 2 vol of ethanol at 4°C and store the solution on ice for 30 min. Recover the DNA by centrifugation at 12,000g for 10 min. at 4°C in a microfuge.
10. Redissolve the DNA in 200 µl of TE (pH 7.6), add 25 µl of 3 M sodium acetate (pH 5.2), and reprecipitate the DNA with 2 vol of ethanol as described in step 9.
11. Do several cycles of ethanol precipitation to completely remove the contaminants.
12. Carefully rinse the pellet once with 70% ethanol, and redissolve the DNA in TE (pH 7.6) to a final vol of 10 µl.
13. Check the amount and quality of the DNA fragment by polyacrylamide gel electrophoresis.

Note

1. The recovery efficiency can be substantially increased by including 5 µg of carrier tRNA in steps 4 and 9.

Method-4: The 'freeze–squeeze' method for recovering DNA from agarose gel slice

The *'freeze–squeeze'* procedure to recover DNA from gel slices is illustrated in Figure 9.6a. This method involves freezing the agarose band and centrifuging out the liquid, including the DNA. The DNA is recovered in a small enough volume, so that it can be used directly and does not need to be precipitated. This method is faster, easier, and gives greater recovery than other methods.

Method-5: Recovery of DNA from agarose or polyacrylamide gels by isotachophoresis

The isotachophoresis method can be used for the elution of DNA from both agarose and polyacrylamide gel electrophoresis (Figure 9.6b)

Method-6: Use of Agarase enzyme

The enzyme agarase is used to digest the gel matrix, and the DNA is purified by phenol extraction.

IV GEL ELECTROPHORESIS-BASED MOLECULAR TECHNIQUES

A. SOUTHERN BLOT HYBRIDIZATION

Identifying Specific DNA Sequences by Southern Blotting: Any given DNA fragment is unique solely by virtue of its specific nucleotide sequence. The only practical way to find a specific DNA sequence among a vast population of heterogenous DNA fragments is to exploit its sequence specificity to identify it. In 1975, Ed Southern invented a technique capable of doing just that. The basic principle of the Southern blot, as it is done currently, is illustrated in Figure 9.7. In this method, the whole (genomic) DNA is digested with one or more restriction enzymes and the resulting fragments are separated according to fragment length on an agarose gel. The DNA is then denatured *in situ* and the fragments are blotted onto the solid support (usually a nitrocellulose filter or nylon membrane) either by pressure, capillary transfer, or by

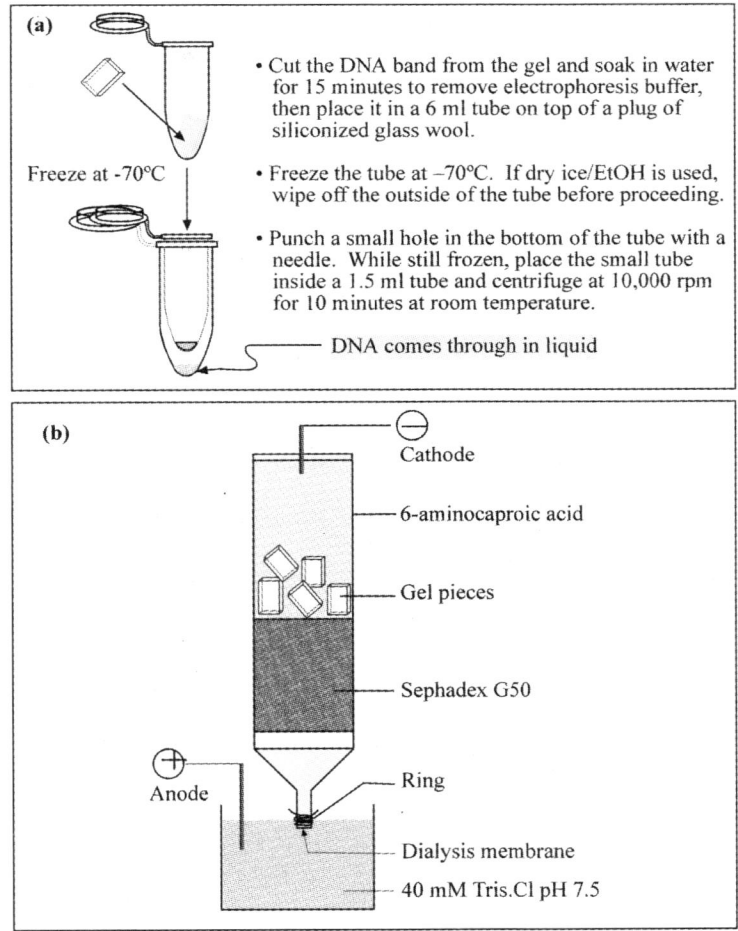

Figure 9.6 *DNA recovery from gels (a) The 'freeze–squeeze' method for recovering DNA from a slice of an agarose gel, (b) Recovery of DNA from agarose or polyacrylamide gel slices by isotachophoresis*

transferring the fragments onto the filter electrophoretically. The relative positions of the DNA fragments are preserved during their transfer to the filter. The DNA attached to the filter is hybridized to radio-labeled DNA or RNA, and autoradiography is used to locate the positions of bands complementary to the probe.

The amount of genomic DNA needed to generate a detectable hybridization signal depends on a number of factors, including the proportion of the genome that is complementary to the probe, the size of the probe and its specific activity, and the amount of genomic DNA transferred to the filter. Under optimal conditions, the method is sufficiently sensitive to detect less than 0.1 µg of DNA complementary to a probe that has been radio-labeled with ^{32}P. Because the strength of the signal is proportional to the specific activity of the probe and inversely proportional to its length, Southern hybridization reaches its optimum sensitivity when very short

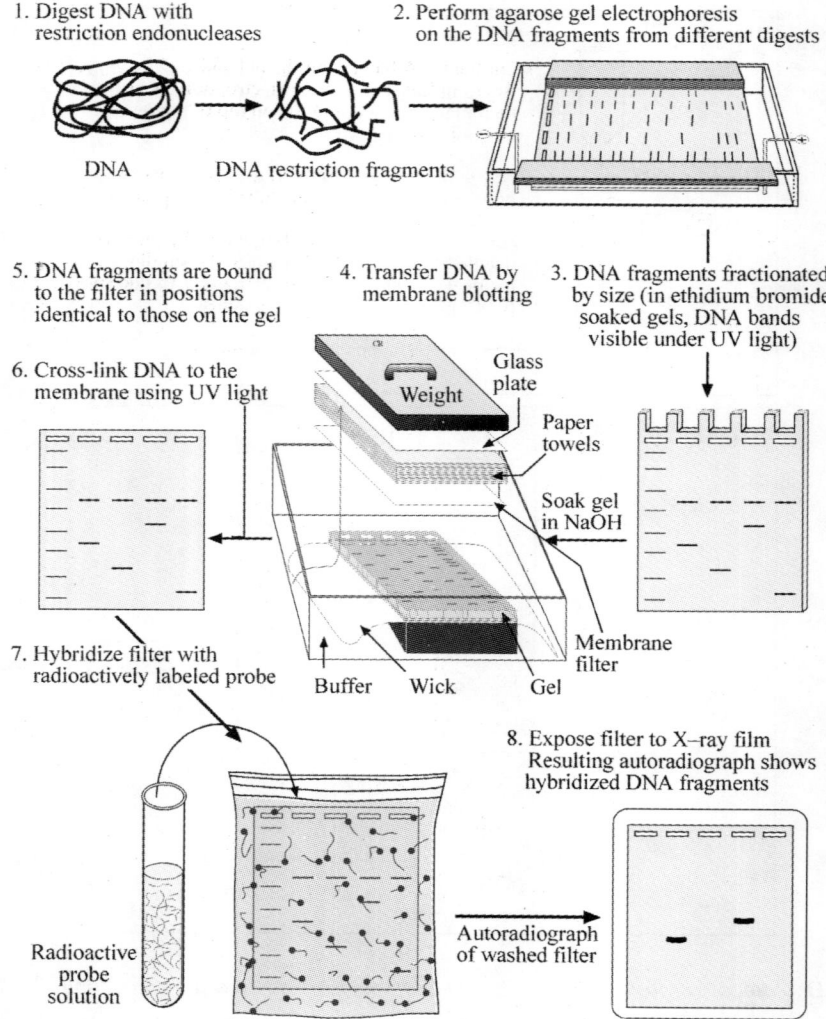

Figure 9.7 *Membrane blotting and hybridization using the Southern blot technique*

probes are used. To obtain a detectable signal with oligonucleotide probes, it is therefore necessary to use radio-labeled oligonucleotides to the highest specific activity possible, to increase the amount of target DNA on the filter, and to expose the autoradiograph for several days.

1. Separation of restriction fragments by agarose gel electrophoresis

Materials

20X SSC (Standard Saline Citrate): 174 g/l NaCl; 88.2 g/l sodium citrate; pH 7.0.

Nitrocellulose or nylon: Wear gloves to handle membranes.

Clingfilm, Saran Wrap or similar material; Whatman 3MM filter paper; Paper towels; Glass plates; Denaturing solution; 0.5 M NaOH, 1.5 M NaCl.

Neutralization solution: 0.5 M Tris–HCl, pH 6.8; 1.5 M NaCl.

Depuration solution: 0.25 M HCl.

Procedure

1. Digest an appropriate amount of DNA with one or more restriction enzymes. For Southern analysis of mammalian genomic DNA, approximately 10 µg of DNA should be loaded into each slot of the gel when probes of standard length (>500 bp) and high specific activity (>10^9 cpm/µg) are used. To detect single-copy sequences, 30–50 µg of DNA are needed when oligonucleotides are used as probes. Proportionately less DNA may be used when the sample contains higher concentrations of the sequences of interest (Note 1).
2. At the end of the digestion, add the appropriate amount of gel-loading buffer and separate the fragments of DNA by electrophoresis through an agarose gel (for genomic DNA, a 0.7% gel cast in 0.5X TBE may be used).
3. After the electrophoresis is completed, photograph the gel as described earlier. Place a transparent ruler alongside the gel, so that the distance that any band of DNA has migrated can be read directly from the photographic image.

Note

1. The concentrations of DNA in preparations of high-molecular-weight genomic DNA are often so low that it is necessary to carry out restriction digestions in relatively large volumes. The chief problem encountered during the digestion of high-molecular-weight DNA is the unevenness of digestion caused by variations in the local concentrations of DNA.

2. Transfer of DNA from agarose gels to solid supports

There are three methods to transfer DNA fragments from agarose gels to solid supports (nitrocellulose filters or nylon membranes):

(a) Electrophoretic transfer

This method is not practical when nitrocellulose is used as the solid support, because of the high ionic strengths of the buffers that are required to bind nucleic acids to these filters. These buffers conduct electric current very efficiently, and it is necessary to use large volumes to ensure that the suffering power of the system does not become depleted by electrolysis. Until recently, electrophoretic transfer was carried out with supports such as diazobenzyloxymethyl (DBM)- or o-minophenyl-thioether (APT)-cellulose, which bind nucleic acids efficiently at low ionic strength. Recently, however, charged nylon membranes are commonly used; nucleic acids as small as 50 bp will bind on these membranes in buffers of very low ionic strength.

Single DNA and RNA can be transferred directly; however, the fragments of double-stranded DNA must first be denatured *in situ*. The gel is then neutralized and soaked in an electrophoresis buffer (1X TBE), before being mounted between porous pads aligned between parallel electrodes in a large tank of buffer. The time required for the complete transfer depends on the size of the DNA fragments, the porosity of the gel, and the strength of the applied field.

As the electrophoretic transfer requires comparatively large electric currents, it is often difficult to maintain the electrophoresis buffer at a temperature compatible with the efficient transfer of DNA. Owing to these and other problems, electrophoretic transfer is recommended to be used only when transfer by capillary action or under vacuum is inefficient (e.g., PAGE is used).

(b) Vacuum transfer

DNA and RNA can be transferred rapidly and quantitatively from gels under a vacuum. Several vacuum transfer devices are now available commercially, in which the gel is placed in contact with a nitrocellulose filter or nylon membrane supported on a porous screen over a vacuum chamber. The buffer, drawn from an upper reservoir, elutes nucleic acids from the gel and deposits them on the filter or membrane. Vacuum transfer is more efficient than the capillary method and is extremely rapid; DNA that have been partially depurinated and denatured with alkali are quantitatively transferred within 30 min. from gels of normal thickness (4–5 mm) and normal agarose concentration (<1%). If carried out carefully, vacuum transfer can result in a two- to three-fold enhancement of the hybridization signal obtained from Southern transfers. All of the commercially available apparatuses work well, as long as care is taken to ensure that the vacuum is applied evenly over the entire surface of the gel. It is also important not to apply too much vacuum during transfer. When the vacuum exceeds 60 cm of water, the gels become compressed and the efficiency of the transfer is reduced.

(c) Capillary transfer

In the capillary transfer method, DNA fragments are carried from the gel in a flow of liquid and deposited on the surface of the solid support. The liquid is drawn through the gel by capillary action that is established and maintained by a stack of dry, absorbent paper towels (Figure 9.8). The rate of the DNA transfer depends on the size of the DNA fragments and the concentration of agarose in the gel. Small fragments of DNA (<1 kb) are transferred almost quantitatively from a 0.7% agarose gel within 1 hr; larger fragments are transferred more slowly and less efficiently (e.g., 15 kb length requires >18 hr). The efficiency of transfer of large DNA fragments is determined by the fraction of molecules that escape from the gel before it becomes dehydrated. This dehydration problem can be alleviated by a partial hydrolysis of the DNA prior to the capillary transfer. The DNA in the gel is exposed to a weak acid (which causes partial depurination),

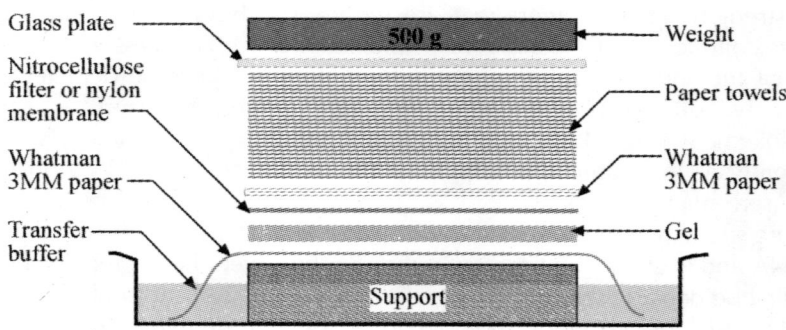

Figure 9.8 *Capillary transfer of nucleic acids from agarose gels to solid supports*

followed by a strong base (which hydrolyzes the phosphodiester backbone at the depurination site). The resulting fragments of DNA (~1 kb in length) can be transferred rapidly from the gel with high efficiency. However, it is important not to let the depurination reaction proceed too far; otherwise, the DNA is cleaved into small fragments that are too short to bind efficiently to the solid support. Depurination can also cause the fuzzy appearance of the bands on gels, because of the increased diffusion of DNA during the transfer. When the target DNA fragments are present in high concentrations, the capillary method can be used to transfer the DNA simultaneously and rapidly from a single gel to two nitrocellulose filters or nylon membranes (Smith and Summers, 1980).

Method-1: Capillary transfer of DNA to nitrocellulose filters:

Procedure

1. After electrophoresis, transfer the gel to a glass-baking dish and trim away any unused areas of the gel with a razor blade. Cut off the bottom left-hand corner of the gel; this serves to orient the gel during the succeeding operations.
2. Denature the DNA by soaking the gel for 45 min. in several volumes of 1.5 M NaCl, 0.5 N NaOH with constant, gentle agitation. The gel should be completely submerged under the buffer.
3. Rinse the gel briefly in deionized water, and then neutralize it by soaking it for 30 min. in several volumes of a solution of 1 M Tris (7.4), 1.5 M NaCl at room temperature with constant, gentle agitation. Change the neutralization buffer and continue soaking the gel for a further 15 min.
4. While the gel is in the neutralization solution, wrap a piece of Whatman 3MM paper around a piece of Plexiglas, or a stack of glass plates, to form a support that is longer and wider than the gel. Place the wrapped support inside a large baking dish. Fill the dish with the transfer buffer (10X SSC or 10X SSPE) until the level of the liquid reaches almost to the top of the support. After the 3MM paper on the top of the support becomes thoroughly wetted, smooth out all the air bubbles with a glass rod. The binding of the DNA to the nitrocellulose depends on the ionic strength of the transfer buffer. The smaller the DNA fragments, the higher the ionic strength required. Alternatively, nylon membranes, which bind small DNA fragments more efficiently than nitrocellulose filters, may be used. The efficiency of retention of the small DNA fragments also depends on the pore size of the filter. For the efficient retention of DNA fragments less than 300 nucleotides, a pore size of 0.2 microns is used.
5. Using a fresh scalpel or a paper cutter, cut a piece of nitrocellulose filter about 1 mm larger than the gel in both dimensions. Use gloves and blunt-ended forceps to handle the filter. A nitrocellulose filter that has been touched by greasy hands will not moisten.
6. Float the nitrocellulose filter on the surface of a dish of deionized water until it is wetted completely from beneath, and then immerse the filter in a transfer buffer for at least 5 min. Using a clean scalpel blade, cut off a corner from the nitrocellulose filter to match the corner cut from the gel.
7. Remove the gel from the neutralization solution and invert it, so that its underside is now facing upwards. Place the inverted gel on the support so that it is centered on the wet 3MM paper, and ensure that there are no air bubbles between the 3MM paper and the gel.
8. Surround, but do not cover, the gel with Saran Wrap or Parafilm. This serves as a barrier which prevents liquid flowing directly from the reservoir onto paper towels placed on top of the gel. If these towels are not precisely stacked, they tend to droop over the edge of the

gel and may touch the support. This type of short-circuiting is a major cause of inefficient transfer of DNA from the gel to the filter.

9. Place the wet nitrocellulose filter on top of the gel, so that the cut corners are aligned. One edge of the filter should extend just over the edge of the line of slots at the top of the gel. Do not move the filter once it has been applied to the surface of the gel. Make sure that there are no air bubbles trapped between the filter and the gel.
10. Wet two pieces of 3MM paper (cut to exactly the same size as the gel) in 2X SSC and place them on top of the wet nitrocellulose filter. Smooth out any air bubbles with a glass rod.
11. Cut a stack of paper towels (5–8 cm high) just smaller than the 3MM papers and place them on the 3MM papers. Put a glass plate on top of the stack and weight it down with a 500 g weight. The objective is to set up a flow of liquid from the reservoir through the gel and the nitrocellulose filter, so that the fragments of denatured DNA are eluted from the gel and deposited on the nitrocellulose filter.
12. Allow the transfer of DNA to proceed for 8–24 hr. As the paper towels become wet, they should be replaced.
13. Remove the paper towels and the 3MM papers above the gel. Turn over the gel and the nitrocellulose filter and lay them, gel-side-up, on a dry sheet of 3MM paper. Mark the positions of the gel-side-up, on a dry sheet of 3MM paper. Mark the positions of the gel slots on the filter with a very soft lead pencil or a ballpoint pen.
14. Peel the gel from the filter and discard it. Soak the filter in 6X SSC for 5 min. at room temperature. This removes any pieces of agarose sticking to the filter. To assess the efficiency of the DNA transfer, the gel may be stained for 45 min. in a solution of ethidium bromide (0.5 µg/ml in water) and examined under UV light.
15. Remove the filter from the 6X SSC and allow any excess fluid to drain away. Place the filter flat on a paper towel to dry, for at least 30 min. at room temperature.
16. Sandwich the filter between two sheets of dry 3MM paper. Fix the DNA to the filter by baking it for 30 min. to 2 hr at 80°C, in a vacuum oven. Over-baking can cause the filter to become brittle.
17. Hybridize the DNA immobilized on the filter to a ^{32}P-labeled probe, as described in Chapter 13.

Method-2: Transfer of DNA from agarose gels to nylon membranes

As there are many inherent problems associated with nitrocellulose papers as a solid-phase, alternative supporting matrices have been explored. This problem has been solved by the introduction of various types of nylon membranes that bind nucleic acids irreversibly and are far more durable than nitrocellulose filters. Immobilized nucleic acids can, therefore, be hybridized sequentially to several different probes without damaging the membrane. Furthermore, because nucleic acids can be immobilized on nylon in buffers of low ionic strength, the transfer of nucleic acids from gels to nylon membranes can be carried out electrophoretically. This can be useful when capillary or vacuum transfer of DNA is inefficient; e.g., when fragments of DNA are transferred from polyacrylamide gels. The sole disadvantage of nylon membranes is a tendency to give increased levels of background hybridization, especially with RNA probes. This problem can be overcome by using increased amounts of blocking agents.

Two basic types of nylon membranes are available commercially: unmodified nylon and charge-modified nylon. Many different types of charged nylon membranes are available that vary in the type of charge, the method used to apply it, and the density of the nylon mesh. Both double- and single-stranded nucleic acids may be applied to nylon membranes in a wide range of solvents. However, each manufacturer provides specific instructions for the transfer of nucleic acids to their particular type of charged nylon membrane. The general protocol given below works well with most types of positively-charged nylon membranes, but may not be optimal for any particular brand.

The nylon membranes must be treated to immobilize the DNA after it has been transferred. The DNA becomes fixed to the nylon membranes if it is thoroughly dried or if it is exposed to low-dose UV irradiation (254 nm) after the transfer of the nucleic acid. The latter method, although a nuisance to set up, is preferred since it is much quicker and the nucleic acid becomes covalently attached to the membrane. In addition, there are reports that DNA transferred to nylon membranes in buffers of high ionic strength and immobilized with UV irradiation yields greatly enhanced hybridization (Khandjian, 1987).

Capillary transfer of DNA to nylon membranes under natural conditions

1. Fractionate the DNA by gel electrophoresis, and process the gel as described earlier.
2. While the gel is in the neutralization solution, prepare the nylon membrane as follows:
   a. Using a fresh scalpel or a paper cutter cut a piece of membrane about 1 mm larger than the gel in both dimensions. Use gloves and blunt-ended forceps to handle the membrane.
   b. Float the membrane on the surface of a dish of deionized water until it is wetted completely from beneath, and then immerse the membrane in a transfer buffer (10X SSC or 10X SSPE) for at least 5 min. Using a clean scalpel blade, cut off a corner from the membrane to match the corner cut off from the gel.
3. Transfer the denatured DNA from the gel to the membrane by capillary action as described earlier.
4. To fix the DNA to the membrane, either place the dried membrane between two pieces of 3MM paper and bake the membrane for 30 min. to 2 hr at 80°C in a vacuum or conventional oven, or expose the side of the membrane carrying the DNA to a source of UV irradiation (254 nm).

Capillary transfer of DNA to nylon membranes under alkaline conditions

The ability of positively charged nylon membranes to bind denatured DNA under alkaline conditions has advantages in some circumstances; e.g., when using probes that are complementary to inverted repeat sequences. However, the level of background hybridization is almost always significantly higher when the membrane has been exposed to high concentrations of alkali for extended periods of time. This problem can often be overcome by increasing the concentration of blocking agents in the hybridization stage.

Procedure

1. After electrophoresis, transfer the gel to a glass-baking dish and trim away any unused areas of the gel with a razor blade. Cut off the bottom left-hand corner of the gel; this serves to orient the gel during the succeeding operations.
2. Denature the DNA by soaking the gel for 15 min. in a denaturation/transfer solution (0.4 N NaOH, 1 M NaCl) with constant, gentle agitation. Change the solution and continue to soak the gel for a further 20 min. with gentle agitation.

3. While the gel is soaking in the denaturation/transfer solution, prepare the nylon membrane as follows:
   a. Using a fresh scalpel or a paper cutter cut a piece of membrane about 1 mm larger than the gel in both dimensions. Use gloves and blunt-ended forceps to handle the membrane.
   b. Float the membrane on the surface of a dish of deionized water until it is completely wetted from beneath, and then immerse the membrane in a denaturation/transfer solution for at least 5 min. Using a clean scalpel blade, cut off a corner from the membrane to match the corner cut off from the gel.
4. Transfer the denatured DNA from the gel to the membrane by capillary action.
5. Peel the gel from the membrane and discard it. Soak the membrane in 0.5 M Tris-Cl (pH 7.2), 1 M NaCl for 15 min. at room temperature. This neutralizes the membrane and removes any pieces of agarose adhering to it.
6. Remove the membrane from the neutralizing solution and allow any excess fluid to drain away. Place the membrane flat on a paper towel to dry for at least 30 min. at room temperature.
7. To fix the DNA to the membrane, either place the dried membrane between two pieces of 3MM paper, and bake the membrane for 30 min. to 2 hr at 80°C in a vacuum or conventional oven, or expose the side of the membrane carrying the DNA to a source of UV irradiation (254 nm).

3. Hybridization of radio-labeled probes to immobilized nucleic acids

There are many methods to hybridize radioactive probes in solution to nucleic acids immobilized on solid supports, such as nitrocellulose filters or nylon membranes. These methods differ in the following respects: (1) the solvent and temperature used (e.g., 68°C in aqueous solution or 42°C in 50% formamide), (2) the volume of the solvent and length of hybridization (large volumes for periods as long as 3 days or minimal volumes for periods as short as 4 hr), (3) the degree and method of agitation (continuous or stationary shaking), (4) the use of agents, such as Denhardt's reagent or BLOTTO, to block the non-specific attachment of the probe to the surface of the solid matrix, (5) the concentration of the labeled probe and its specific activity, (6) the use of compounds, such as dextran sulfate or polyethylene glycol, which increase the rate of reassociation of nucleic acids, (7) the stringency of the washing following the hybridization.

(a) Hybridization of radio-labeled probes to nucleic acids immobilized on solid support

Although the method given below deals with RNA or DNA immobilized on nitrocellulose filters, only slight modifications are required to adapt the procedure to nylon membranes. These modifications are noted at the appropriate places in the text.

Pre-hybridization solutions

 For the detection of low-abundance sequences
  6X SSC (or 6X SSPE)
  5X Denhardt's reagent
  0.5% SDS
  100 µg/ml denatured, fragmented salmon sperm DNA
or
  6X SSC (or 6X SSPE)
  5X Denhardt's reagent

0.5% SDS
100 µg/ml denatured, fragmented salmon sperm DNA
50% formamide

For the detection of moderate- or high-abundance sequences:
6X SSC (or 6X SSPE)
0.05X BLOTTO

or

6X SSC (or 6X SSPE)
0.05X BLOTTO
50% formamide

Hybridization solution for nylon membranes

6X SSC (or 6X SSPE)
0.5% SDS
100 µg/ml denatured, fragmented salmon sperm DNA
50% formamide (if hybridization is to be carried out at 42°C)

Procedure

1. Prepare the appropriate pre-hybridization solution for the task at hand. Approximately 0.2 ml of pre-hybridization solution will be required for each square centimeter of nitrocellulose filter or nylon membrane. The pre-hybridization solution should be filtered through a 0.45 µM disposable cellulose-acetate filter.
2. Float the nitrocellulose filter or nylon membrane containing the target DNA on the surface of a tray of 6X SSC (or 6X SSPE) until it becomes thoroughly wetted from beneath. Submerge the filter for 2 min.
3. Slip the wet filter into a heat-sealable bag. Add 0.2 ml of pre-hybridization solution for each square centimeter of nitrocellulose filter or nylon membrane. Squeeze as much as possible from the bag, and then seal the open end of the bag with the heat sealer. Incubate the bag for 1–2 hr, submerged in a water bath, set at the appropriate temperature (68°C for aqueous solvents; 42°C for solvents containing 50% formamide). Heating the pre-hybridization solution to the correct temperature before adding it to the bag will prevent the formation of small bubbles on the surface of the filter as the temperature of the solution increases.
4. If the radio-labeled probe is double-stranded, denature it by heating it for 5 min. at 100°C. Single-stranded probes need not be denatured. Chill the probe rapidly in ice water.
5. Working quickly, remove the bag containing the filter from the water bath. Open the bag by cutting off one corner with scissors. Add the denatured probe to the pre-hybridization solution, and then squeeze as much air as possible from the bag. Reseal the bag with the heat sealer, so that as few bubbles as possible are trapped in the bag. To avoid any radioactive contamination of the water bath, the resealed bag should be sealed inside a second, non-contaminated bag. When using nylon membranes, the pre-hybridization solution should be completely removed from the bag and immediately replaced with the hybridization solution. The probe is then added and the bag is resealed.
6. Incubate the bag submerged in a water bath, set at the appropriate temperature, for the required period of hybridization.

7. Wearing gloves, remove the bag from the water bath and immediately cut off one corner. Pour out the hybridization solution into a disposable container, and then cut the bag along the length of three sides. Remove the filter and immediately submerge it in a tray containing several hundred millilitres of 2X SSC and 0.5% SDS at room temperature. Do not allow the filter to dry out at any stage during the washing procedure.
8. After 5 min., transfer the filter to a fresh tray containing several hundred millilitres of 2X SSC and 0.1% SDS and incubate it for 15 min. at room temperature with occasional gentle agitation. If short oligonucleotides are used as probes, washing should be carried out only for brief periods (1–2 min.) at appropriate temperatures.
9. Transfer the filter to a flat-bottomed plastic box containing several hundred ml of fresh 0.1X SSC and 0.5% SDS. Incubate the filter for 30 min.–1 hr at 37°C with gentle agitation.
10. Replace the solution with fresh 0.1X SSC and 0.5% SDS, and transfer the box to a water bath set at 68°C for an equal period of time. Monitor the amount of radioactivity on the filter using a hand-held mini-monitor. The parts of the filter that do not contain DNA should not emit a detectable signal. There should be no detectable signal on the mini-monitor from filters containing mammalian DNA that have been hybridized to single-copy probes.
11. Briefly wash the filter with 0.1X SSC at room temperature. Remove most of the liquid from the filter by placing it on a pad of paper towels.
12. Place the damp filter on a sheet of Saran Wrap. Apply adhesive dot labels marked with radioactive ink to several asymmetric locations on the Saran Wrap. These markers serve to align the autoradiograph with the filter. Cover the labels with scotch tape.
13. Cover the filter with a second sheet of Saran Wrap, and expose the filter to an X-ray film to obtain an autoradiographic image. The exposure time should be determined empirically. However, single-copy sequences in mammalian genomic DNA can usually be detected after 16–24 hr of exposure at –70°C with an intensifying screen.

(b) Hybridization of radio-labeled oligonucleotides to genomic DNA

Oligonucleotide probes as short as 17 nucleotides in length may be used to detect single-copy sequences in restriction digests of eukaryotic genomic DNA that have been transferred to solid supports. Duplexes with a single base-pair mismatch are significantly less stable and dissociate at a lower temperature than their perfectly-matched counterparts. It has, therefore, been possible to use oligonucleotides of defined sequences to probe fetal DNA for the presence of specific point mutations that cause conditions such as sickle-cell anemia, certain thalassemias, etc. The methods used when hybridizing with oligonucleotide probes are similar to those described earlier in this chapter.

4. Removal of radio-labeled probes from solid phase by washing

The probes become irreversibly bound if the nitrocellulose filters and nylon membranes are allowed to dry. Therefore, every effort should be made to ensure that the solid supports remain wet at all stages during the hybridization, washing, and exposure to X-ray film.

Removing probes from nitrocellulose filters

1. Heat several hundred ml of 0.05X SSC, 0.01 M EDTA (pH 8.0) to boiling pont. Remove the fluid from the heat and add SDS to a final concentration of 0.1%. Immerse the filter in the hot elution buffer for 15 min.

2. Repeat step 1 with a fresh batch of boiling elution buffer.
3. Rinse the filter briefly in 0.01X SSC at room temperature. Drain most of the liquid from the filter by placing it on a pad of paper towels.
4. Sandwich the damp filter between two sheets of Saran Wrap, and apply it to an X-ray film to check that all the probe has been removed.
5. The filter may now be dried, wrapped loosely in aluminum foil, and stored under vacuum at room temperature until needed.

Removing probes from nylon membranes

1. Either immerse the membrane in several hundred ml of 1 mM Tris-Cl (pH 8.0), 1 mM EDTA (pH 8.0), 0.1X Denhardt's reagent or immerse the membrane in 50% formamide, 2X SSPE for 1 hr at 65°C.
2. Rinse the membrane briefly with 0.1X SSPE at room temperature. Remove most of the liquid from the membrane by placing it on a pad of paper towels.
3. Sandwich the damp membrane between two sheets of Saran Wrap, and apply it to an X-ray film to check that all the probe has been removed.
4. The membrane may now be dried, wrapped loosely in aluminum foil, and stored under vacuum at room temperature until needed.

B. Analysis of RNA by Northern Blotting

If DNA fragments can be transferred from agarose gels to nitrocellulose or nylon membranes for hybridization studies, one might expect that RNA molecules separated by gel electrophoresis could be similarly transferred and analyzed. Indeed, such RNA transfers are used routinely in molecular genetic laboratories. These RNA blots are called Northern blots, in recognition of the fact that the procedure is the mirror image of the Southern blotting technique. The northern blot procedure is essentially identical to that used for Southern blot transfers; however, RNA molecules are more sensitive to degradation by RNases. Thus, extra precautions must be taken to prevent any contamination of the solutions with RNase. RNase molecules are extremely stable enzymes; they are not inactivated by boiling or other treatments that would totally destroy most enzymes. Most RNA molecules contain a considerable secondary structure. Therefore, they must be kept denatured during electrophoresis if one wants to separate them on the basis of size. Denaturation is accomplished by adding formaldehyde or some other chemical denaturant to the loading buffer and the electrophoresis-running buffer. After the transfer to an appropriate membrane, the RNA blot is hybridized to either RNA or DNA probes, just as with Southern blots.

The Northern blot hybridization procedure is extremely useful in studies of gene expression. It can be used to determine whether a particular gene is transcribed in all tissues of an organism or only in certain tissues. It can also be used to study the temporal pattern of the expression of individual genes during their growth and development. The ease of Northern blotting has led to its widespread use. The main advantages of this procedure are that it is possible to determine the total size of a transcript, and the filters can be screened under a variety of stringencies to detect transcripts from gene families or related organisms. Furthermore, since the target material is immobilized, it is possible to screen the filter repeatedly with different probes. Finally, the ability to conduct the hybridization procedure on an immobilized surface allows the use of non-radioactive probe methods. However, there are a number of technical limitations that prevent its universal application: (a) it requires high quality RNA samples, (b) since the procedure detects the entire RNA molecule, it is necessary to use agarose gels; agarose gels are poor at resolving high-molecular-weight species and these are also difficult to size accurately, (c) filter hybridization is not as efficient as solution hybridization, with the result that Northern blots are generally reckoned to have about 10% of the sensitivity of an equivalent solution method.

C. STANDARD WESTERN BLOT PROTOCOL

The high specificity and affinity of ligands, such as antibodies and lectins, make these reagents very sensitive tools for the identification and characterization of proteins separated by gel electrophoresis. Polyacrylamide gel electrophoresis has been used as an important tool for the separation and characterization of proteins. Since many functional proteins are composed of two or more sub-units, individual polypeptides are separated by electrophoresis in the presence of the detergent SDS, which denatures the proteins. Protein samples can be reacted with an appropriate specific anti-serum prior to electrophoresis; the resulting immunoprecipitate can then be recovered and the reactive proteins analyzed by electrophoresis. Alternatively, the sample, after removal of the immunoprecipitate, can be analyzed to determine which proteins have been depleted or removed. This procedure is known as *immunodeletion*. It is also possible to use antibodies to precipitate protein components following electrophoresis; this procedure is known as *immunofixation*. However, the restrictive nature of most polyacrylamide gels makes this technique unsatisfactory. This is due to the slow rate of diffusion of antibodies or other large-probe molecules. The problems associated with immunofixation were overcome by the development of blotting techniques, in which the separated proteins are transferred to membranes, and individual proteins can be detected by using specific antibodies. This transfer of proteins from acrylamide gels to membranes is called *Western blotting* and is performed by using an electric current – so-called *"electro blotting"*. After the transfer, a specific protein of interest is identified by placing the membrane with the immobilized proteins in a solution containing an antibody to the protein. Non-bound antibodies are then washed off the membrane. The presence of the primary antibody is detected by placing the membrane in a solution containing a "secondary" antibody, which is conjugated to either a radioactive isotope or an enzyme that produces a visible product when the proper substrate is added (Figure 9.9).

1. Electro blotting apparatus

Protein transfer can be achieved by contact diffusion, capillary or vacuum methods. A wide variety of electro blotting apparatus has been described and many systems are available commercially. There are two basic designs; the first configuration uses a *vertical buffer tank*. In this type of blotter, the gel and the blotting membrane are held in intimate contact by a sandwich of filter papers and sponge pads held under pressure in a frame fixed between the electrodes. The second configuration, known as *semi-dry blotting*, employs flat-plate electrodes arranged in a horizontal apparatus, providing a homogenous electrical field, with a short inter-electrode distance and using small amounts of buffer. The gel matrix is sandwiched between two stacks of filter papers, wetted with transfer buffer, which are in direct contact with the plate electrodes.

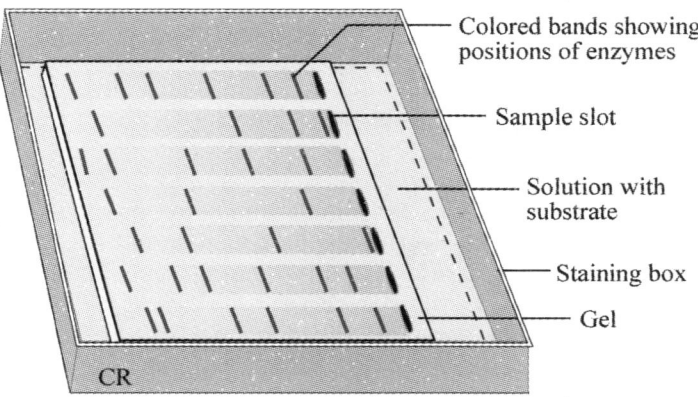

Figure 9.9 *Enzymatic staining of proteins in gel after gel electrophoresis*

2. Blotting matrices

Several types of filters with different properties are available for Western blotting of proteins. Nitrocellulose, the most popular matrix as it does not require derivatization prior to use, is compatible with most general protein stains, is relatively inexpensive and has a high protein-binding capacity (249 µg/cm^2, pore range from 0–0.45 µm). The proteins are bound to the nitrocellulose membranes and are adsorbed by a combination of hydrophobic and electrostatic interactions. But proteins bind covalently to diazo papers (e.g., diazobenyloxymethyl cellulose and diazophenylthioether cellulose), which need to be activated prior to use. More recently, hydrophobic polyvinylidene difluoride (PVDF) membranes have become available; these membranes have a high mechanical strength and are compatible with most Western blotting protocols.

3. Transfer buffers

The choice of the transfer buffer is very important, since it influences both the efficiency of elution from the polyacrylamide matrix and the retention by the blotting matrix. The composition and pH of the buffer determines the elution and the protein retention, influenced by additives such as detergents and methanol. The most commonly-used buffer system for electroblotting of proteins from SDS-PAGE gels is that originally described by Towbin *et al.*: 25 mM Tris, 92 mM glycine, pH 8.3. Methanol (10–20%, v/v) is often added to this buffer as it removes SDS from protein-SDS complexes and increases the affinity of binding of the proteins to the nitrocellulose. However, methanol acts as a fixative and reduces the efficiency of protein elution, so extended transfer times must be used. Several alternative buffer systems, not containing glycine, can be used when the blotted proteins are to be characterized chemically.

4. Blocking

After the proteins have been transferred to the membrane and before probing with a specific ligand, all the unoccupied binding sites on the support must be blocked. Bovine serum albumin (3–5% in PBS for 1 hr) is the most commonly-used blocking agent, although other proteins including other animal sera, ovalbumin, hemoglobin, casein, and gelatin have been used. A solution of non-fat dried milk (3% (w/v) in PBS) has become popular as a blocking agent and usually results in very low background staining. Note that if detergents (e.g., Tween 20, polyethylene sorbitan monolaurate) are used as blocking agents, then nylon membranes should be used.

Obviously, the Western blot procedure is a very powerful tool for identifying and characterizing specific gene-products. Many variations exist among the protocols for running Western blots. These variations represent all aspects of the procedure from the choice of gel buffers and membranes to transfer conditions, blocking and processing steps. The following procedure is written as a general protocol for use with standard Tris-Glycine SDS-PAGE gels and nitrocellulose membranes. Optimization for particular situations may require modifying the conditions described.

Protocols–1: Running gel and transfer of proteins to solid surface

Materials

Equipment: Electrophoretic transfer apparatus; Plastic bag sealer; Rocking apparatus.

Reagents: Amido black stain, 0.2% in 7% acetic acid. NGS (normal goat serum), heat inactivated at 56°C for 60 min., filtered through a 0.22 µm filter.

Transfer buffer: 12 g Tris base, 57.65 g glycine, 4 litres distilled water, after Tris and glycine are dissolved, add 1 litre methanol. pH to 8.3 with HCl.

Destaining buffer (optional): 30% methanol and 10% acetic acid in distilled water.

Blot buffer: 5% (w/v) Carnation non-fat dry milk in PBS (BLOTTO buffer)

Phosphate-buffered saline (PBS).

PBS-Tween buffer: 0.05% Tween-20 in PBS.

Primary antibody: antisera made against protein(s) of interest; this may be monoclonal or heterologous antisera.

Secondary antibody: Biotinylated goat anti-primary IgG, Avidin-horseradish peroxidase (HRP).

Substrate: 30% hydrogen peroxide diluted to 1% (1:30) in H_2O

Color indicator stock solution: 4-chloro-1-napthol (3 mg/ml H_2O).

Procedure

In advance

1. Prepare the polyacrylamide-SDS gel for the separation of proteins (see previous section). Obtain an antibody against the protein sequence of interest.
2. While the gel is running, soak all the fiber pads, filter papers, and transfer membrane in the transfer buffer. Ensure that no bubbles are trapped in the filters or fiber pads. (Note: PVDF membranes require pre-wetting in 100% methanol before soaking them in the transfer buffer).

Western blot

Electrophoretic transfer in polyacrylamide gel to NC filter

1. After the electrophoresis, cut off the bottom right-hand corner of the gel. This will ensure that the gel is oriented correctly in the transfer apparatus. Equilibrate/rinse both the gel and the NC filter in transfer buffer for 5–15 min.
2. Assemble the transfer cassette according to the manufacturer's instructions. Place the NC filter against the gel on a flat surface. Smooth the filter over the gel by rolling it with a 5 ml glass pipette to remove all the air bubbles. Ensure that the gel is oriented, so that after transfer the lanes will appear on the membrane in the desired order.
3. Wrap a piece of 3MM filter paper (pre-wetted with transfer buffer) around the gel/filter to make a sandwich. Keep it wet and avoid bubbles.
4. The paper/gel/filter/paper sandwich is placed in an electrophoretic transfer apparatus, as suggested by the manufacturer, with the gel facing the cathode.
5. Place it in the buffer tank and fill with transfer buffer to cover the gel.
6. Connect the tank to the power supply and run the electrophoretic transfer, as suggested by the manufacturer. Some common conditions are constant voltage, setting starting voltage at a point which gives a starting current of 200 mA (the current will drift down during the transfer). Transfer from a 1 mm-thick mini-gel, in the range of 8–12% acrylamide, is usually complete in about 90 min. The optimal transfer time should be determined experimentally.
7. After the transfer is complete, remove the filter and gel from the unit. Then, discard the gel.
8. Briefly float the filter in amido black stain, until the molecular weight standards become visible. Mark the standard positions.
9. Wash the NC briefly in 100 ml water. Use a destaining buffer if necessary.
10. Incubate the filter in a blot buffer for 1 hr at 37°C.
11. Wash in PBS-Tween buffer for 1 hr at room temperature.
12. Seal the filter inside a plastic bag with a heat seal apparatus. The seal should be as close as possible to filter to keep the inside volume small.
13. Cut off a corner of the bag to use as a buffer entry port. Then, close the hole with a dialysis clamp.

5. Immuno-staining of proteins

Antibody reaction

1. Mix 100 µl NGS, the primary antibody (dilution will vary with antibody, good starting points are 1–2 µg/ml for purified antibody or 1:500 dilution for raw serum) in 10 ml of blot buffer. Add it to the bag with the filter. Incubate it for 2 hr at room temperature (or overnight at 4°C) with rocking.
2. Wash the filter in a shallow dish with four changes of 75 ml of PBS-Tween buffer for a total of 300 ml over 30 min.; a minimum 5 min. in each wash solution. Discard each wash.
3. Add the biotinylated goat anti-rabbit IgG 40 µl in 10 ml blot buffer with 100 µl NGS (0.05–1 µg/ml for chromogenic detection, 0.01–0.2 µg/ml for chemiluminescence) to the bag. Incubate it for 1 hr at room temperature with rocking.
4. Wash 3 times for a minimum of 5 min. each time in a wash solution. Discard each wash.

5. Briefly rinse the membrane in substrate buffer (or PBS or Tris without Tween 20) to remove residual detergent – 10 s is sufficient.
6. Add avidin-HRP (40 µl in 10 ml of blot buffer with 100 µl of NGS). Do not pour the substrate solution directly onto the membrane. Incubate it for 30 min. at room temperature with rocking.
7. Wash 3 times for a minimum of 5 min. each time in a wash solution. Discard each wash.
8. Prepare the color indicator/substrate. Mix 3 ml color indicator stock solution with 9 ml PBS. Add 150 µl of 1% hydrogen peroxide, and use immediately.
9. Incubate it at room temperature with rocking, until a purple color develops and the background begins to be detectable (1–10 min.).
10. Transfer it to the filter and wash in another container containing water to halt color development. Air-dry and store it in a plastic bag in the dark.

6. Data analysis

1. Identify and number the lanes on the membrane or its photocopy to aid analysis.
2. The molecular size standards will not be visible, unless pre-stained or colored protein size markers are used. The blot is a mirror image of the gel depending on the side of the gel placed against the nitrocellulose membrane. Determine the lanes which contain the protein bands that have reacted with the antibodies. Generally, most lanes will contain one predominant band, although some lanes may have additional faint bands.
3. Determine the molecular weight of a protein and variability as explained earlier.

D. ANALYSIS OF NUCLEIC ACIDS BY DOT BLOT HYBRIDIZATION

This method provides a rapid determination of the presence of specific RNA or DNA in samples. A direct transfer of RNA or DNA to "slots" or "dots" on the NC is made (Figure 9.10), and the spotted filter is then hybridized with a labeled probe. This method can also be used on partially-degraded RNA samples with good semi-quantitative results.

Requirements

Equipment: Dot blot apparatus (e.g., Bio-Rad or Schleicher and Schuell); Vacuum source.

Reagents: 20X SSC buffer; TE buffer

Dye: 0.25% bromophenol blue in 25% Ficoll

Procedure

In advance

Prepare the DNA or RNA samples (1–2 µg), as described earlier. The sample should be in water or TE buffer. Prepare a ^{32}P-labeled probe by nick translation or end-labeling (see Chapter 13).

Preparation of the blot with the apparatus

1. Mix 10 ml of 20X SSC with 10 ml of water.

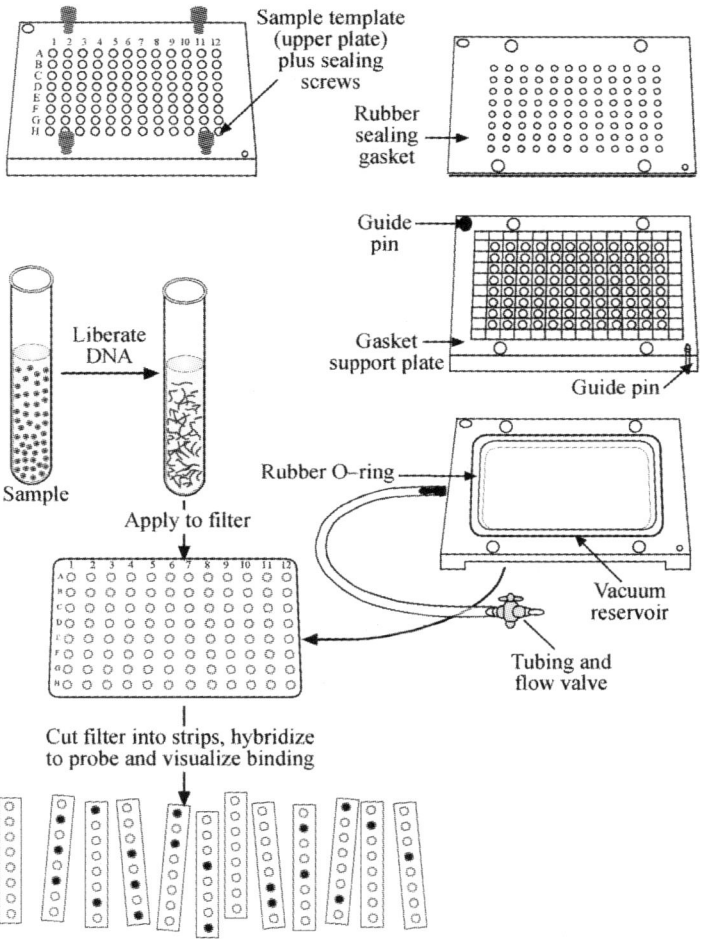

Figure 9.10 *Diagram of a typical "dot blot" methodology and the apparatus vacuum manifold*

2. Wet one piece each of NC and Whatman 2MM paper (each cut to the size of the apparatus) in the mixture from step 1. The edges of the filters may have to be cut to fit around the papers in the apparatus.
3. Place the Whatman 3MM filter in the apparatus first, and then place the NC on top of the paper. Be sure to clean apparatus thoroughly before and after use.
4. Put a lid on the apparatus and latch it in place. Hook it up to the vacuum source.
5. Rinse the wells with 100 µl of 20X SSC.
6. Place 2 µl of dye in several outer wells as position markers.
7. *For RNA dot blot*: Place 1–2 µg of RNA in 50 µl of 20X SSC. Add to the wells.
   For DNA dot blot: Place 1–2 µg of DNA in 10 µl of TE buffer. Heat the sample to 95°C for 5 min. Add 40 µl of 20X SSC to the samples, and add this to the wells. For semi-quantitative analysis, serial 1:1 dilutions of samples added across rows can be used to yield a concentration curve.

8. Rinse each well with 100 µl of 20X SSC under a vacuum.
9. Turn off the vacuum. Take off the top of the apparatus and remove the filter.
10. Bake the NC filter between two fresh Whatman 3MM filters in a vacuum oven at 80°C for 2 hr.
11. Hybridize the filter with the radio-labeled probe, as described in Chapter 13.

Without apparatus

1. Wet the NC filter with 20 ml of a 1:1 mixture of 20X SSC and water.
2. Blot the filter with dry Whatman 3MM paper filters.
3. In a defined pattern of rows and columns, spot small, uniform-volume DNA or RNA samples in 20X SSC buffer, as prepared in step 7, on the filter. Mark the additional spots with dye for orientation.
4. Perform steps 10 and 11, given above.

E. PULSED-FIELD GEL ELECTROPHORESIS

The agarose gel methods for DNA, described above, can fractionate DNA of 60 kb or less. The introduction of pulsed-field gel electrophoresis (PFGE) and the further development of variations on the basic technique now mean that DNA fragments up to 2×10^3–10^6 kb can be separated. This technique, therefore, allows the separation of whole chromosomes by electrophoresis. The method basically involves electrophoresis in agarose, where two electric fields are applied alternately at different angles for a defined time period (e.g., 60 s). Activation of the first electric field causes the coiled molecules to be stretched in the horizontal plane and they start to move through the gel (Figure 9.11). The interruption of this field and the application of the second field force the molecule to move in the new direction. Since there is a length-dependent relaxation behavior when a long-chain molecule undergoes conformational change in an electric field, the smaller a molecule, the quicker it realigns itself with the new field and is able to continue moving through the gel. Larger molecules take longer to realign. In this way, with continual reversing of the field, the smaller molecules draw ahead of the larger molecules and separate according to size. There are many improved versions of PFGE available, but detailed description of these techniques is beyond the scope of this chapter.

V SUGGESTED READING

Alwine JC, Kemp DJ and Stark GR (1977) Method for detection of specific RNAs in agarose gels by transfer to diazobenzyloxymethyl paper and hybridization with DNA probes, *Proc. Natl. Acad. Sci. USA*, **74:**5350–5354.

Bollag DM, Rozycki MD, and Edelstein SJ (1996) *Protein Methods*, (2nd Edn) NY: Wiley-Liss, 195–227.

Docherty K (1996) (Eds) *Gene Transcription Analysis: Essential Techniques*, John Wiley and Sons, Chichester.

Dunn DJ (1993) *Gel Electrophoresis of Proteins*, Bios Scientific Publishers.

Harlow E and Lane D (1988) *Antibodies: A Laboratory Manual*, Cold Spring Harbor, NY: Cold Spring Harbor Laboratory Press, 471–510.

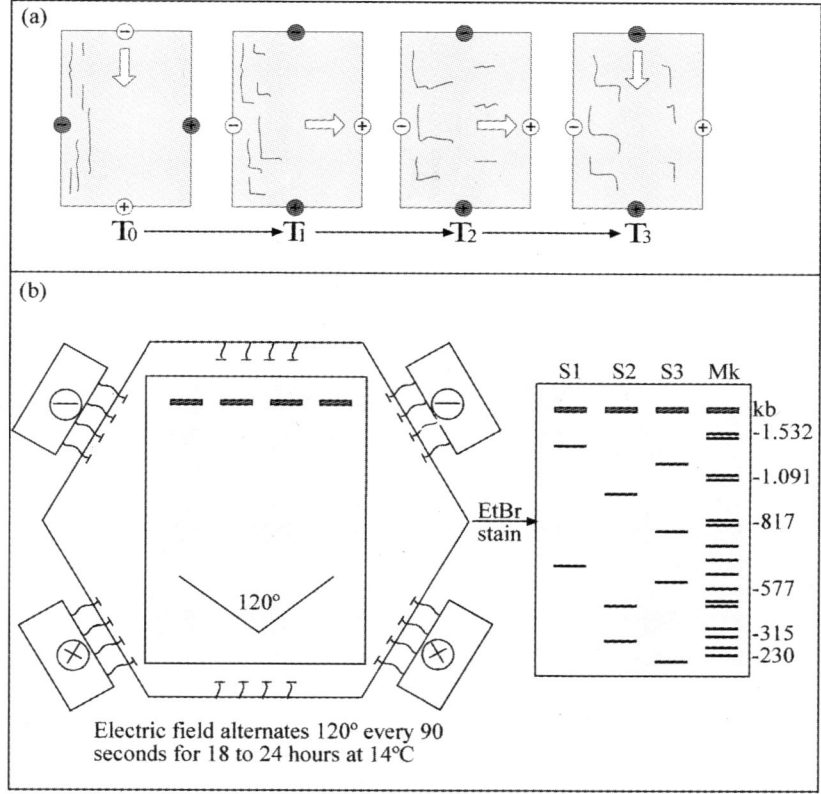

Figure 9.11 *Pulsed field gel electrophoresis (PFGE) is based on the principle that large DNA molecules migrate at different rates through an alternating-angle electric field. (a) Larger DNA molecules that are migrating parallel to the electric field in one direction take more time (T) to reorient in an alternating electric field than do similar molecules. (b) CHEF pulsed-field gel systems use a hexagonal gel box that alters the angle of the fields relative to the agarose gel*

Meinkoth J and Wahl G (1984) Hybridization of nucleic acids immobilized on solid supports, *Anal. Biochem.*, **138**:267–284.

Olszewska E and Jones K (1988) Vacuum blotting enhances nucleic acid transfer, *Trends Genet.*, **4**:92–95.

Sambrook J, Fritsch EF, Maniatis T (1989) *Molecular Cloning: A Laboratory Manual*, (2nd Edn) Cold Spring Harbor Lab, Cold Spring Harbor, 9.31–9.46.

Southern EM (1975) Detection of specific sequences among DNA fragments by gel electrophoresis, *J. Mol. Biol.*, **98**:503–517.

Tautz D and Rez M (1983) *Analytical Biochemistry*, **132**:14.

Wahl GM, Stern M and Stark GR (1979) Efficient transfer of large DNA fragments from agarose gels to diazobenzyloxymethyl-paper and rapid hybridization by using dextran sulfate, *Proc. Natl. Acad. Sci. USA*, **76**:3688–3687.

10

DNA Sequencing

I INTRODUCTION

Although the restriction maps and RFLPs provide valuable information regarding the base sequence of a DNA molecule, the information on a DNA sequence is not complete until the base sequence of the DNA molecule is complete. Determination of the order or sequence of bases along a length of DNA is one of the central techniques in molecular biology. The precise usage of codons, information regarding mutations and polymorphisms, and the identification of gene-regulatory control sequences can only be elucidated by analyzing DNA sequences. Sequencing is often a necessary component of a cloning project, while in other cases it is the desired end point and the sequences are used in taxonomic, ecological, or evolutionary studies. Technological advances have facilitated the sequencing of entire genomes, which will further revolutionize both basic and applied knowledge of gene structure and function. The development of computerized databases of DNA and protein sequences allows the construction of the structure and function of the proteins and their secondary structures.

The two basic and rapid sequencing techniques in current use are: the enzymatic method of Sanger sequencing or *chain termination* or the *dideoxy method,* and the chemical degradation technique (*base-specific chemical cleavage method*) of Maxam and Gilbert. In addition to these basic methods, several automated methods are also extensively used. Although the two basic techniques are very different in principle, they generate a separate population of radio-labeled oligonucleotides that begin from a fixed point and terminate randomly at a fixed residue or combination of residues. Since each base in the DNA has an equal chance of being the variable terminus, each population consists of a mixture of oligonucleotides whose lengths are determined by the location of a particular base along the length of the original DNA. These populations with heterogenous fragments of DNA are then resolved by electrophoresis into as little difference as one base length. When the DNA mixtures are loaded into the adjacent lanes of a sequencing gel, the order of nucleotides along the DNA can be read on an autoradiographic image of the gel, since the DNA molecules are labeled with radioactive ^{32}P.

II MAXAM AND GILBERT CHEMICAL METHOD OF SEQUENCING

The scientists, Allan Maxam and Walter Gilbert, developed a method to determine the sequence of bases in a DNA molecule. The Maxam and Gilbert method, also called the chemical-cleavage method, depends on the ability of certain chemicals to selectively degrade particular bases within a molecule. This is less frequently used today than the other methods, partly because the chemicals used are toxic and partly because the methods are more labor intensive. The primary advantage of this method over the Sanger method is that the DNA sequences are obtained from the original DNA molecule and not from a synthesized copy. Thus, one can analyze DNA modifications such as methylation and study DNA secondary structure and the interaction of proteins with DNA using this method.

The technique has five basic steps: (1) One end of each strand of the DNA molecule to be sequenced is radioactively labeled and the strands are separated from one another. (2) A particular base, or pair of bases, is chemically modified; e.g., some of the guanine bases or some of the cytosine bases are modified. (3) The modified bases are removed from the DNA backbone. (4) The backbone is cleaved wherever a base is missing. (5) The resulting single-stranded DNA fragments are size-separated by polyacrylamide gel electrophoresis. This method is often used for sequencing small fragments of DNA, such as oligonucleotides (Figure 10.1).

Once many copies of the DNA to be sequenced have been radio-labeled at one end, they are denatured and divided into four samples or aliquots. Various chemical treatment of each of the four aliquots causes modification as depicted in Table 10.1.

Table 10.1 *Chemical modifications and cleavage sites in the Maxam–Gilbert method*

Base	Base-modifying reagent	Specific modification*
G	Dimethyl sulfate	Methylation of N7 with dimethyl sulfate at pH 8.0 makes the C_8–C_9 bond specifically susceptible to cleavage by base
A+G	Piperdine formate	Piperdine formate at pH 2.0 weakens the glycosidic bond of adenine and guanine by protonating nitrogen atoms in the purine rings, resulting in depurination
C+T	Hydrazine	Hydrazine opens a pyrimidine ring which recycles in a five-membered form that is susceptible to removal.
C	Hydrazine + NaCl	In the presence of 1.5 m NaCl only cytosine reacts appreciably with hydrazine
A>C	NaOH at 90°C	1.2 N NaOH at 90°C results in a strong cleavage at A and a weaker cleavage at C

* Hot (90°C) piperdine (1M in H_2O) is used to cleave the sugar-phosphate chain of DNA at the sites of chemical modifications.

The chemical reaction in each aliquot is controlled so that it remains incomplete. This means that only a few target bases are modified in any particular DNA molecule. All of the base modified aliquots are then treated with a second chemical, which causes the DNA backbone to break

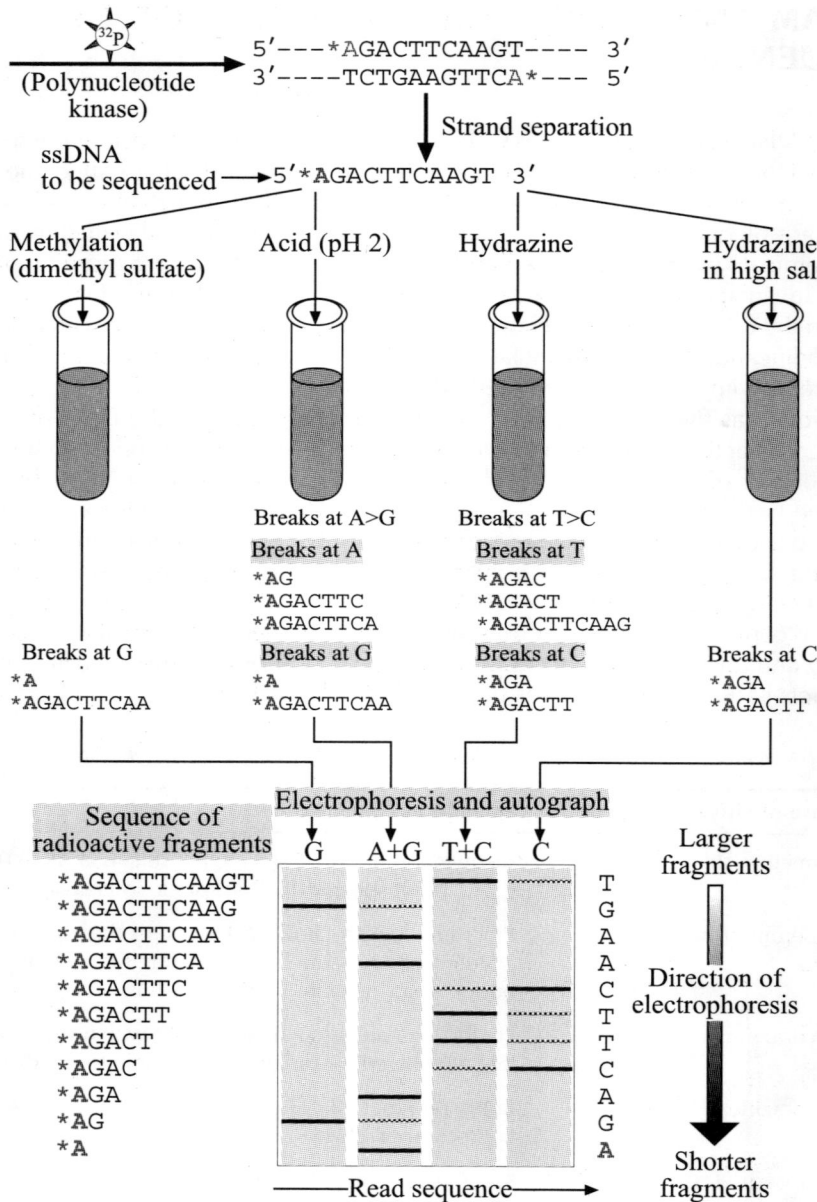

Figure 10.1 *DNA sequencing by the Maxam and Gilbert techniques illustrated*

wherever there is a modified base. This breakage results in the production of a series of DNA fragments of varying lengths. All fragments, regardless of the presence or absence of the radio-label, separate on the basis of size by electrophoresis on a polyacrylamide gel. Band patterns form on the gel when it is exposed to autoradiography.

III DNA SEQUENCING BY DIDEOXYNUCLEOTIDE CHAIN TERMINATORS

The dideoxy sequencing method capitalizes on the ability of the DNA polymerase to use 2′, 3′-dideoxynucleoside triphosphate as substrates. 2′, 3′ ddNTPs differ from conventional dNTPs in that they lack a hydroxyl residue at the 3′ position of deoxyribose. The ddNTPs can be incorporated by the DNA polymerase enzyme into a growing DNA chain through their 5′ triphosphate groups. However the absence of a 3′-hydroxyl residue prevents the formation of a phosphodiester bond with the succeeding dNTP. Thus, further extension of the growing DNA chain is impossible. So, when a small amount of one ddNTP is included with the four conventional dNTPs in a reaction mixture for DNA synthesis, there is competition between extension of the chain and infrequent, but specific, termination. Since the incorporation of ddNTPs rather than dNTP is a random event, the reaction will produce new molecules varying widely in length, but all terminating at the same type of base. The oligonucleotide chain lengths are determined by the distance between the terminuses of the primer used to initiate DNA synthesis and the sites of premature termination. By using the four different ddNTPs (ddGTP, ddCTP, ddATP, and ddTTP) in four separate enzymatic reactions, populations of oligonucleotides are generated that terminate at positions occupied by every A, C, G, or T in the template strand. The four samples are then denatured by heating and loaded next to each other on a polyacrylamide gel for electrophoresis (Figure 10.2).

A. REAGENTS USED IN THE SANGER METHOD OF DNA SEQUENCING

1. Primers

In enzymatic sequencing reactions, the priming of DNA synthesis is achieved by the use of a synthetic oligonucleotide complementary to a specific sequence on the template strand. The problem of obtaining primers can be solved by using a "universal" primer that anneals to the vector sequences flanking the target DNA. The universal primer is typically 15–29 nucleotides in length and anneals to the sequences immediately adjacent to the *Hind*III site in the polycloning region of the bacteriophage M13mp18 and the *Eco*RI site in M13mp19. In addition, several companies sell primers that have been designed to allow the sequencing of target DNAs cloned into a variety of restriction sites in different plasmids.

2. Templates

The two types of DNA which can be used as templates in the Sanger method of sequencing are: pure ssDNA and dsDNA that has been denatured by heat or alkali. The best results are obtained from ssDNA templates, which are usually isolated from recombinant bacteriophage M13 particles. Dideoxy sequencing can also be easily carried out using double-stranded DNA, if it is first denatured with alkali or heat. The dideoxy sequencing of a double-stranded template is particularly useful when DNA sequencing is the only rapid method available for verifying a particular plasmid construction. However, for large-scale sequencing projects, use of a single-stranded DNA vector system is recommended. Two factors are critical here: the quality of the template DNA and the type of DNA polymerase that is used.

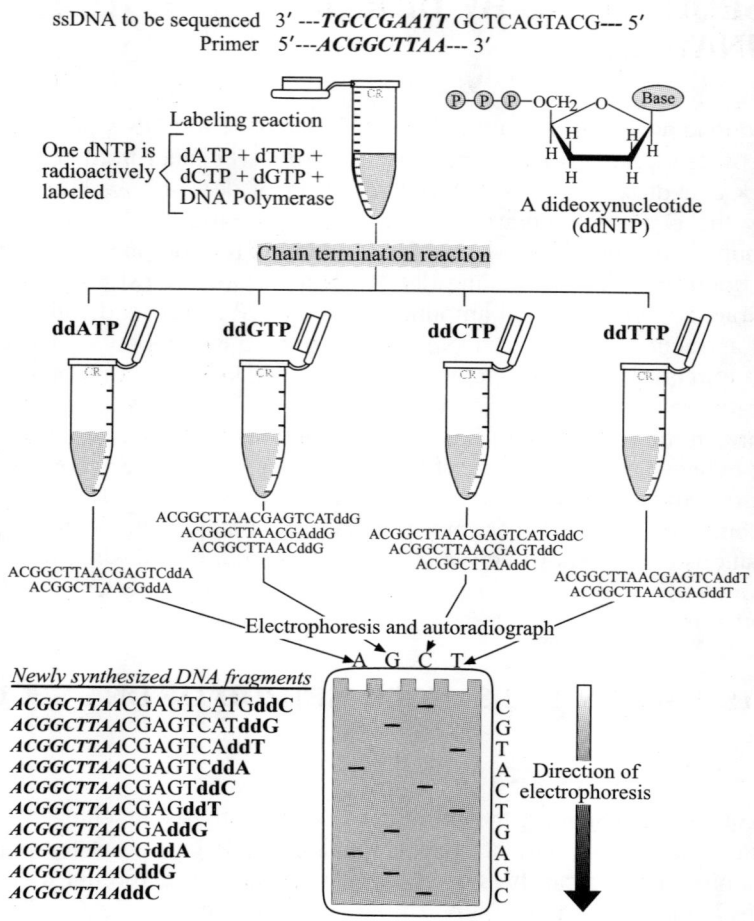

Figure 10.2 *DNA sequencing by using didcoxynucleotide triphosphates as chain terminators*

3. DNA polymerases

Several different enzymes are commonly used for dideoxy-mediated sequencing, including the Klenow fragment of *E. coli* DNA polymerase I, reverse transcriptase, bacteriophage T7 DNA polymerases that have been modified to eliminate 3'→5' exonuclease activity, and the thermostable DNA polymerase isolated from *Thermus aquaticus* (*Taq* DNA polymerase). The properties of these DNA polymerases differ greatly in ways that can considerably affect the quantity and quality of the DNA sequence obtained from chain-termination reactions.

4. Radio-labeled dNTPs

Since the beginning of DNA sequencing, all DNA sequencing has been carried out with [α-^{32}P]dNTPs. However, the strong β particles emitted by ^{32}P create two problems. First, because of scattering, the bands on the autoradiograph will be larger than the bands on the gel; this creates the problem of reading the sequences correctly. Secondly, the decay of ^{32}P cause the radiolysis of the DNA in the sample. However, ^{32}P offers the advantage of short exposure times. The introduction of [^{35}S]dATP greatly alleviates both of these problems because [^{35}S]dATP emits weak β particles. This allows the unambiguous determination of several hundred nucleotides of a DNA sequence from a single reaction set.

5. Analogs of dNTPs

DNA sequences with dyad symmetry (a high G+C content) can form intra-strand secondary structures that are not fully denatured during electrophoresis. This can cause an anomalous pattern of migration, in which adjacent bands get compressed to the point where they are difficult to read. The compressed regions of gels can usually be resolved by using nucleotide analogs, such as dITP (deoxy inosine triphosphate) or 7-deaza-dGTP. These analogs pair weakly with conventional bases and are good substrates for DNA polymerases. Any compression that is not resolved either by dITP or 7-deaza-dGTP can almost always be cleared up by determining the sequence of both strands of the DNA.

B. SETTING UP DIDEOXY-MEDIATED SEQUENCING REACTIONS

1. Preparation of single-stranded DNA

Small-scale preparations of single-stranded bacteriophage M13 or phagemid DNAs should be isolated as described for methods used to prepare plasmid DNA (see Chapter 8).

2. Preparation of primers

Many companies supply universal primers of several different lengths (15–26 bases) as lyophilized powders, which should be dissolved at a concentration of 2 µg/ml in 10 mM Tris-Cl (pH 7.4), 5 mM NaCl, 0.1 mM EDTA (pH 8.0). If the primers are supplied in solution form, they should be stored at –20°C. The primers synthesized in the laboratory generally need not be purified by gel electrophoresis before use. However, it is often necessary to experiment with laboratory-synthesized primers to determine the optimum ratio of primer to template in the sequencing reaction. Laboratory-synthesized primers should be dissolved at a concentration of 2 µg/ml, as described above, and stored at –20°C.

3. Setting of chain-extension/chain-termination reaction

The competition between chain elongation and termination can be determined by adjusting the ratio of dNTP to ddNTP in each of the four sequencing reactions. Each of the polymerases used for sequencing has a different affinity for both dNTPs and ddNTPs. Thus, the optimization of ratios between dNTPs and ddNTPs are necessary for any particular task. Each new batch of reagents should be checked by setting up a series of test reactions with a standard template and primer and a range of concentrations of dNTPs and ddNTPs.

(a) Stock solutions of dNTPs and ddNTPs

The following (Table 10.2) stock solutions are used to prepare the working solutions of dNTPs and ddNTPs for sequencing reactions using the Klenow fragment of *E. coli* DNA polymerase I or Sequenase.

Table 10.2 Stock solutions of dNTPs and ddNTPs in TE buffer

Stock solutions of dNTPs		Stock solutions of ddNTPs	
0.5 mM dATP	302 µg/ml	0.5 mM ddATP	5.7 µg/ml
0.5 mM dCTP	284 µg/ml	0.5 mM ddCTP	5.4 µg/ml
0.5 mM dGTP	304 µg/ml	0.5 mM ddGTP	5.7 µg/ml
0.5 mM dTTP	291 µg/ml	0.5 mM ddTTP	5.5 µg/ml

(b) Denaturing polyacrylamide gels

Both the nucleotide number and the accuracy of the nucleotide sequence are determined by the quality of the polyacrylamide gels used to display the radio-labeled DNA fragments generated by chemical and enzymatic sequencing reactions. Different electrophoresing times and optimal conditions, between 300 and 500 nucleotides of reliable sequence, can be obtained from a denaturing polyacrylamide gel. The concentration of acrylamide used to prepare the gel depends on the size of the DNA fragments that are to be analyzed. To read sequence within 50 nucleotides of the 5' terminus of the primer, the gels should be cast with high (12–20%) concentrations of acrylamide. The sequences lying between 25 and 400 nucleotides from the terminus of the primer can be determined from gels containing 6% acrylamide. More distant sequences can be determined from gels cast with 4 or 5% polyacrylamide.

(c) Preparation of buffer-gradient polyacrylamide gels

The method given below describes how to pour and run sequencing gels that contain increasing concentrations of buffer toward the base of the gel. The sequencing gels should be made at least 2 hours before use and can be made up to 24 hours before they are needed.

1. Prepare the following stock solutions:

    40% Acrylamide solution

Acrylamide (DNA-sequencing grade)	380 g
N, N'–methylenebisacrylamide	20 g
Distilled H$_2$O to 600 ml	

    Heat the solution to 37°C to dissolve the chemicals, and adjust the volume to 1 litre with distilled water. Filter the solution through a nitrocellulose filter, and store in dark bottles at room temperature.

    5X TBE

Tris base	54 g
Boric acid	27.5 g
0.5 M EDTA (pH 8.0)	20 ml
Deionized H$_2$O to 1 litre	

10% Ammonium persulfate

    Ammonium persulfate   1 g

    H_2O to 10 ml

The solution may be stored at 4°C for several weeks.

6% Acrylamide/urea top solution

    40% acrylamide solution   75 ml

    5X TBE   50 ml

    Urea (ultrapure)   230 g

Adjust the volume to 500 ml with deionized water, and filter the solution through a nitrocellulose filter (e.g., Nalge, 0.45 µm size). The 6% acrylamide/urea top solution may be stored for several weeks at 4°C.

6% Acrylamide/urea bottom solution

    40% acrylamide solution   30 ml

    5X TBE   100 ml

    Urea (ultrapure)   92 g

    Sucrose   20 g

    Bromophenol blue   10 mg

Filter the solution by a nitrocellulose filter (e.g., Nalge, 0.45 µm size). The 6% acrylamide/urea bottom solution may be stored for several weeks at 4°C.

2. Prepare the number of glass plates required to accommodate the sequencing reactions. If it is necessary to remove the old silicon from the plates, swab them with KOH/methanol, which is prepared by adding ~5 g of KOH pellets to 100 ml of methanol.

    Then wash the glass plates and spacers in warm detergent solution, and rinse them thoroughly in tap water, followed by deionized water. Hold the plates by their edges, so that material from the gloves or oils from your hands do not become deposited on the working surfaces of the plates. Rinse the plates with ethanol and set them aside to dry. The plates must be cleaned meticulously to ensure that air bubbles do not form when the gel is poured.

    Treat one surface of each plate with silicone solution. This prevents the gel from sticking tightly to both plates and reduces the possibility that the gel will tear when it is removed from the mold after the electrophoresis is completed.

3. There are many types of electrophoresis apparatus available commercially, and the arrangement of the glass plates and spacers differs slightly from manufacturer to manufacturer. In all cases, the aim is to form a watertight seal between the plates and spacers, so that the unpolymerized gel solution does not leak out. Typically, the two plates are of slightly different size and one of them is notched. The spacers are made of thin (usually 0.4 mm) flexible plastic.

    Lay the larger plate flat on the bench and arrange the two spacers in place along the sides. A couple of minute dabs of petroleum jelly help to keep the spacers in position during the next steps. Ensure that there is no dust on the plates, and then lay the smaller plate in position, resting on the spacers. Clamp together one side of the plates with several large (5-cm length) bulldog binder clips. Bind the entire length of the other side and the bottom of the plates with gel-sealing tape to make a watertight seal. Take particular care with the bottom corners of the plates, since these are the places where leaks most often occur. Transfer the bulldog clips to the sealed side of the plates, and bind the remaining side of the plates with gel-sealing tape.

4. Place the comb into the open end of the gel mold and test to see that it fits snugly. Remove the comb and lay the empty gel mold on the bench.
5. Cover the working area with plastic-backed protective paper, as it is almost impossible to pour sequencing gels without dripping acrylamide solutions onto the bench. Prepare a sequencing gel as described below. The preparation of the gel must be done without interruption.

For a 20-cm × 40-cm sequencing gel

1. Place 10 ml of 6% acrylamide/urea bottom solution in a small Erlenmeyer flask.
2. Place 35 ml of 6% acrylamide/urea top solution in a 100 µl beaker.
3. Add 40 µl of 10% ammonium persulfate to the 6% acrylamide/urea bottom solution. Mix the solution by rapidly swirling the flask.
4. Add 120 µl of 10% ammonium persulfate to the 6% acrylamide/urea top solution. Mix the solution by rapid swirling.
5. Add 50 µl of TEMED to the top solution and mix. Draw 7 ml of the solution into a 25-ml pipette fitted with a pipettor bulb. Draw the remainder of the solution into a 50-ml hypodermic syringe.
6. Add 15 µl of TEMED to the bottom solution and mix. Draw 7 ml of the bottom solution into the same 25-ml pipette used for the top solution. Allow 2–3 air bubbles to pass upward through the acrylamide solution in the pipette. This establishes a crude buffer gradient.
7. Slowly pour the solution from the pipette down one side of the gel mold while holding the mold at an angle of approximately 45° to the horizontal. To avoid producing air bubbles, pour the solution in a continuous stream. When the pipette is nearly empty, lower the mold slowly to the horizontal position. Quickly resume pouring the gel using the top solution in the 50-ml syringe. Take care that no air bubbles form and that the bottom solution is not pushed up the opposite side of the gel mold by the incoming top solution. This can be avoided by tilting the gel at an angle and pouring the top solution down the lower side.
8. Carefully examine the gel for air bubbles. If any are present in the buffer gradient toward the bottom of the gel, it may be necessary to pour another gel. Bubbles in the upper portion of the gel can sometimes be moved with a thin spacer to a position where they will not interfere with the migration of the DNA samples. However, this is possible only when the full width of the gel is not to be used for loading samples. The presence of bubbles is a sure sign that the gel plates were not thoroughly cleaned before the mold was assembled.
9. Lay the mold down at an angle, so that the top of the mold rests on a support about 5-cm high. This reduces the hydrostatic pressure at the base of the mold and prevents leaks and bowing of the gel plates.
10. Immediately insert the flat side of a shark's-tooth comb about ~ 0.5 cm into the gel solution. Insert both ends of the comb into the fluid to an equal depth so that the flat surface is level when the gel is standing in a vertical position. Clamp the comb into position using bulldog binder clips. Use the remaining 6% acrylamide/urea top solution in the hypodermic syringe to form a bead of acrylamide across the top of the gel. Allow the gel to polymerize for 1–2 hr.
11. Wash out the 25-ml pipette and the syringe so that they do not become clogged with polymerized acrylamide.

12. After 45 min. of polymerization, examine the gel for the presence of a schlieren line just underneath the flat surface of the comb. This is a sign that polymerization has occurred satisfactorily. When the polymerization is complete, the gel can be stored for up to 24 hr at room temperature. To prevent dehydration during storage, leave the comb in the gel and surround the top of the gel with paper towels dampened with 1X TBE, and cover the paper towels with Saran Wrap.

4. Loading and running gradient sequencing gels

Prepare an oven or a water bath set at 80°C before starting the gel, because it is required to denature the sequencing reactions just before they are loaded onto the gel.

1. When the polymerization of the gel is complete, use damp paper towels to wipe away any dried polyacrylamide/urea from the outside of the gel mold. Carefully remove the comb from the top of the gel, and strip the electrical tape from the bottom of the gel mold.
2. Attach the gel mold to the electrophoresis apparatus with bulldog binder clips or plastic-coated laboratory clamps. Depending on the design of the electrophoresis apparatus, it may be necessary to clamp plastic-covered metal plates to the gel mold to ensure even diffusion of the heat produced during electrophoresis. Fill the top reservoir with 0.5X TBE and the bottom reservoir with 1X TBE.
3. Remove the cover from the microtiter plate containing the completed sequencing reactions. Incubate the open plate in the oven for 10 min. at 80°C.
4. While the plate is incubating, fill a 10-ml hypodermic syringe fitted with a 22-gauge needle with 1X TBE. Squirt the TBE across the submerged, flat-loading surface of the gel to remove any adhering fragments of urea and polyacrylamide.
5. Reinsert the comb with its teeth just sticking into the loading surface of the gel. Remove the comb, and once again wash out the slots of the gel with 1X TBE.
6. Remove the microtiter plate from the 80°C oven or water bath, and immediately transfer it onto packed ice. Keep the plate at 0°C until the samples have been loaded onto the gel. This retards renaturation of the template and the radio-labeled strands.
7. Load 1–2 µl of each sequencing reaction onto adjacent slots of the gel. Keep a record of the order of the templates and load the samples in every reaction set in the same order. TCGA is the best order, because the two tracks (G and C) that suffer the most from abnormal patterns of migration are then adjacent to one another and can be easily compared. In addition, if the gel has been loaded in the order TCGA, the sequence of the complementary strand (3'→5') can be read by flipping the autoradiograph over and reading the gel from the bottom.
8. When all the samples are loaded, connect the electrodes to the power pack and the electrophoresis apparatus (anode to the bottom reservoir). The gel should be run at constant power (35–40 W for a 20- × 40-cm gel; ~1700 V) for the time required to achieve the optimal resolution of the sequence of interest. The time required can be estimated by monitoring the migration of the marker dyes in the formamide/EDTA/XC/BPB gel-loading buffer (formamide 10 ml, xylene cyanol FF 10 mg, bromophenol blue 10 mg, 0.5M EDTA (pH 8.0) 200 µl).

Depending on the distance between the sequence of interest and the oligonucleotide primer, a second loading of the sequencing samples can be applied to the gel approximately 15 min. after the bromophenol blue from the first loading buffer has migrated from the gradient gel

(~3 hours). The sequence obtained from the first loading will be more distal to the primer, whereas that obtained from the second loading should be denatured at 80°C for 2–3 min. just before loading.

C. AUTORADIOGRAPHY OF SEQUENCING GELS

1. At the end of the electrophoresis run, turn off the power and disconnect the sequencing apparatus from the power pack. Dispose of the electrophoresis buffer, and then remove the gel mold from the apparatus.
2. Lay the gel mold flat on plastic-backed protective bench paper, with the smaller plate uppermost. Remove any remaining pieces of electrical tape. Using the end of a metal spatula, slowly pry apart the plates of the mold. With luck, the gel will remain attached to the lower plate. If it does not, place the plate back on the gel, invert the plates, and try again.
3. When the glass plates have been separated, cut off the bottom corner of the side of the gel that was loaded first. This serves to orient the gel during the subsequent manipulations. Transfer the gel to a shallow bath containing 10% methanol and 10% acetic acid in water. This fixes the gel and removes the urea and sucrose, which would otherwise prevent the gel from drying completely and cause it to stick to the autoradiographic film. There is no need to agitate the fluid while the gel is being fixed.

    The time required for fixation varies according to the thickness of the gel. The same batch of fixation fluid can be used to fix several gels.
4. Pick up the supporting plate by its edges, and lift the gel out of the fixation fluid. Take care that the gel does not slide off the plate. When the plate is clear of the fixation fluid, hold the gel in place with a gloved hand and tilt the plate to allow excess fixation fluid to drain away.
5. Lay the plate (gel uppermost) on a piece of protective paper. Remove any folds or distortions by gently kneading the gel with gloved fingers. Wipe away the excess fixation fluid from the glass plate with paper towels, avoiding touching the surface of the gel.
6. Place a piece of Whatman 3MM paper on top of the gel. The paper should be slightly larger (2–3 cm) than the gel in both length and width, and should be centered over the gel. Apply a gentle pressure so that the gel becomes firmly attached to the rough surface of the paper. Hold the paper in place with one hand and pick up the supporting glass plate with the other. Quickly flip the plate over, and lay it down on a dry piece of protective paper.
7. Slide the piece of protective paper to the edge of the bench. Take hold of the leading edge of the 3MM paper and pull it downward while moving the glass plate slowly toward the edge of the bench. The gel will stick to the 3MM paper as it is peeled from the glass plate.
8. Lay the 3MM paper on two other pieces of 3MM paper of the same size. Cut a piece of Saran Wrap slightly longer and wider than the gel and lay it on top of the gel. Try to avoid any creases and bubbles; this is best done with the help of another person. Hold the corners of the Saran Wrap and pull it outward so that it is tightly stretched. Lower the stretched Saran Wrap onto the surface of the gel. Once the Saran Wrap has touched the gel, do not attempt to remove it since this will cause the gel to tear.
9. Using a paper cutter, trim all three pieces of 3MM paper and the Saran Wrap so that they are nearly the same size as the gel.

10. Dry the gel for 30–40 min. under a vacuum on a commercial gel dryer set at 80°C.
11. Remove the gel from the dryer and peel off the Saran Wrap. The dried gel should feel smooth to the touch but not sticky. To orient the autoradiograph, attach a small adhesive label marked with radioactive ink to the 3MM paper in the space created by cutting off the bottom corner of the gel.
12. Establish an autoradiograph by exposing the gel to an X-ray film (Kodak XAR-2, XAR-5, or their equivalent) for 16–24 hr at room temperature. If possible, use spring-loaded metal cassettes to ensure a direct contact between the entire surface of the dried gel and the film emulsion.
13. Develop the autoradiograph, and read the sequence of the DNA as described in next section.

D. READING THE SEQUENCE

Reading the sequences from gels is not as easy as it appears and is an acquired skill. The following list of tips may help to simplify the process and minimize problems.

1. As soon as the autoradiograph is developed, label it with the date and the names of the templates. Mark each set of sequencing reactions clearly.
2. Ensure that you can distinguish between the left and the right sides of the gel. The image of the radioactive ink should appear at the bottom of the sequencing reaction that was loaded on the first track of the gel.
3. When searching for a correspondence between the new sequence and one that is already known, look for obvious "signatures", such as homopolymeric runs or alternating purine and pyrimidines (e.g., GTGTGT). Once found, these signatures can be used to quickly locate the sequence of interest.
4. An unknown sequence should be read and recorded at least twice, preferably by different people. The two readings should then be compared and discrepancies should be resolved, if necessary, by further sequencing.
5. If the gel has been loaded in the order TCGA, the sequence of the complementary strand ($3' \rightarrow 5'$) can be read by flipping the autoradiograph over and reading the gel from the bottom.
6. The following guidelines are useful when reading gels:
   — Single C bands are generally weaker than single bands of the three other nucleotides.
   — The first A in a homopolymeric run of As is generally stronger than the rest.
   — The first C in a homopolymeric run of Cs is usually much weaker than the second.
   — G bands are weak when they are preceded by a T band.

IV SEQUENCING DNA BY PCR

DNA fragments can be amplified by the polymerase chain reaction (PCR) directly from genomic or cloned DNA. In this method, a specific segment of DNA can be amplified by one million-fold or more using *Taq* DNA polymerase (see chapter on PCR) on other thermophilic DNA polymerases. The PCR eliminates the need to prepare large amounts of DNA for the sequencing and sub-cloning steps. The conventional PCR requires primers of known sequences that flank the region to be amplified. Several techniques have been developed for sequencing dsDNA produced by the PCR. The asymmetric PCR allows the preferential amplification of only one of the complementary DNA strands, so that ssDNA template is produced for dideoxy sequencing.

V AUTOMATED DNA SEQUENCING

Most eukaryotic chromosomes are much larger than those of bacteria or yeast, so complete DNA sequencing is not feasible without the use of instruments that partly automate the process. Large-scale DNA sequencing is well under way in a number of model organisms. Recent advances in dye-terminator chemistry have led to the development of automated sequencing methods that involve the use of dideoxynucleotides labeled with different fluorochromes. The principle behind automated large-scale DNA sequencing is shown in Figure 10.3.

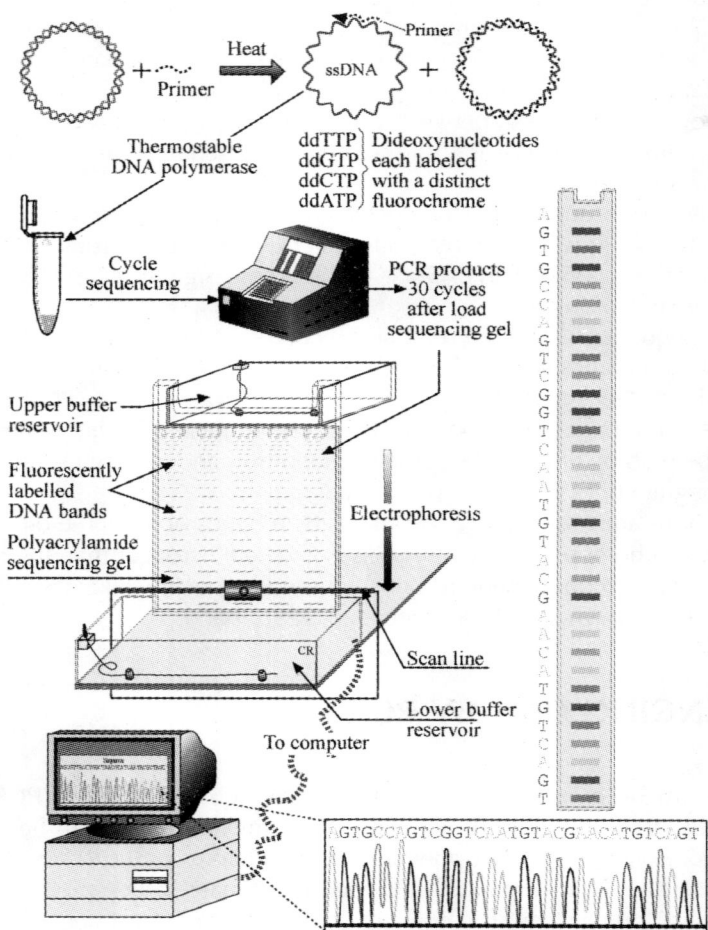

Figure 10.3 *Schematic diagram of an automated fluorescence-based DNA sequencer uses dideoxy chain termination reactions to generate labeled reaction products that are size-fractioned by gel electrophoresis*

This method is essentially the same as that of Sanger's dideoxy chain termination method, except the method of gel reading. In automated DNA sequencing, the nucleotides that terminate synthesis are labeled with different fluorescent dyes (G, black; A, green; T, red; C, blue). Because the colors distinguish the products of DNA synthesis that terminate with each nucleotide, the products of all the synthesis reactions can be put together in the same tube and separated by electrophoresis in a single lane. The fluorescent dye that each band contains is excited by laser light, and the color of the fluorescence is automatically read by a photocell and recorded in a computer. In addition to real-time detection, the lengths of sequence that may be analyzed are in excess of 500 bp. Further improvements are likely to be made in electrophoresis by using capillary electrophoresis. The consequence of automated sequencing and the incorporation of PCR-cycle sequencing has substantially decreased the time needed to undertake sequencing projects.

VI   ANALYZING DNA SEQUENCE DATA

Genome-sequencing projects that generate substantial amounts of data will require appropriate computer assistance to analyze them (Doolittle 1990, Gribskov and Devereux 1991). Software packages are available for all common laboratory computer systems and, depending on the size of the computer, can analyze the sequences in greater or lesser detail. Sequence data can be feed into the computer in several ways. Gels can be read visually by the scientist, digitized and entered, or scanned by automated laser scanners, which can enter the data directly. Some software has been developed that can interpret ambiguities in the autoradiograph. The automated methods also increase the speed and help to minimize any clerical errors that may occur. The computer can compare readings from several sequencing runs, search for and identify overlaps and contiguities between runs, compare results from sequencing the complementary strands of the DNA, and identify any possible errors. Once the sequence data have been entered into the computer and possible errors are corrected, the next step is to analyze the data.

The DNA-sequence databanks are expanding rapidly and are important resources for the research community. There are three major DNA-sequence databases – the DNA Data Bank of Japan (DDBI), the European Molecular Biology Laboratory Nucleotide Sequence Data Library (EMBL), and the Gene Bank Genetic Sequence Data Bank (GenBank). The entries of DNA sequences are growing every day in an exponential fashion. Great care is taken to ensure that the data entered are accurate, but errors are apparently common in published accounts in both the data banks and scientific journals.

VII   SUGGESTED READING

Ambrose BJB and RC Pless (1987) DNA sequencing: Chemical methods, *Methods Enzymol.*, **152**:522–538.

Ausubel FM, Brent R, Kingston RE, Moore DD, Seidman JG, Smith JA, and Struhl K (Eds) (1999) *Short Protocols in Molecular Biology* (4th Edn) John Wiley and Sons, Inc.

Barness WM (1987) Sequencing DNA with dideoxyribonucleotides as chain terminators: Hints and strategies for big projects, *Methods Enzymol*, **152**:538–556.

Davis CG, Dibner MD, and Battery JF (1986) *Methods in Molecular Biology*, Elsevier Science Publishing Co.

Doolittle RF (Ed) (1990) Molecular evolution: Computer analysis of protein and nucleic acid sequences, *Methods Enzymol.*, Vol. **183**, Academic, New York.

Garret RH and Grisham CM (1999) *Biochemistry*, Saunders College Publishing.

Glick BR and Pasternak JJ (1996) *Molecular Biotechnology*, Panima Publishing Corporation, Bangalore, New Delhi.

Gribskov M and J Devereux (Eds) (1991) *Sequence Analysis Primer,* Stockton Press, New York.

Hoy MA (1994) *Insect Molecular Genetics,* Academic Press, Inc.

Hunkapiller T, Kaier RJ, Koop BF and Hood L (1991) Large-scale and automated DNA sequence determination, *Science*, **254**:59–67.

Innis MA, Gelfand DH, Sninsky JJ and White TJ (Eds) (1990) *PCR Protocols*, Academic Press, Inc.

Peters P (1994) *Biotechnology: A Guide to Genetic Engineering,* Wm. C. Brown Publishers.

Sambrook J, Fritsch EF and Maniatias T (Eds) (1989) *Molecular Cloning: A Laboratory Manual* (2nd Edn) Cold Spring Harbor Laboratory Press, Cold Spring Harbor, NY.

Steinberg M, Guyden J, Calhoun D, Staiano-Coico L and Coice R (1993) Recombinant DNA technology, *Concepts and Biomedical Applications*, PTR Prentice Hall, Inc.

Strickberger MW (1999) *Genetics,* Prentice-Hall of India Private Limited.

Wilson K and Walker J (Eds) (2000) *Practical Biochemistry: Principles and Techniques* (5th Edn) Cambridge University Press.

11

Genetic Maps and Marker Analysis

I INTRODUCTION

Mendel's laws of inheritance (1900) and the chromosome theory of inheritance (Sutton, 1903) laid the foundation of classical genetics. These and other discoveries led to the concept that characters (traits) are controlled by factors (now called 'genes') that are passed on from parents to the children generation after generation with high fidelity. Further, it was revealed that chromosomes are the main vehicles of heredity and genes are linearly arranged on the chromosomes. The first genetic map was developed by Sturtevant (1913), on a chromosome of *Drosophila*. Since then, a number of genetic maps have been developed in several plant species including man and bacteria. The main purpose of gene mapping is for assigning a gene to a locus on a chromosome. By mapping the relative positions of many genes and other markers, it is possible to generate a chromosome map or a map of the entire genome. Gene mapping has been used to help understand and manipulate the inheritance of biological traits, particularly by using genetic linkage to associate the inheritance of one trait with another and by correlating differences in phenotype with differences in chromosome structure. Three kinds of markers – *morphological* (traits), *biochemical or protein-based markers* (e.g., storage proteins and isozymes), and *molecular* (DNA) – have been used in the construction of genetic maps.

Morphological markers: Conventional linkage maps are constructed based only on those genes which produce distinct phenotypic (morphological) effects. This technique is also called *naked-eye polymorphisms (NEP)*, in which inheritance can be followed with the naked eye. This method is very quick, simple and does not need any laboratory equipment. Although this method is very effective in gene mapping, it has many limitations: (1) The number of such genes is very few when compared to the total number of genes, (2) the conventional method of mapping is tedious and time-consuming, (3) mapping of genes governing quantitative traits (quantitative trait loci, QTL) is difficult, and (4) morphological markers are highly influenced by the environment, developmental stage, and some pleiotropic effects.

Protein-based markers: These markers are detected as electrophoretic variants of proteins. For example, the detection of polymorphisms based on *isozyme analysis*. Isozymes are variant forms of an enzyme, usually detectable through electrophoresis due to differences in their net electrical charges. These were the first to be used, leading to the development of many of the principles applicable to molecular markers. However, the limited number of good isozyme loci in most species has shifted the focus to DNA marker analysis.

Molecular markers, commonly referred to as DNA markers, are numerous and represent a milestone in genetics by providing the capacity for the complete coverage of nuclear, mitochondria and chloroplast genomes. Due to the many limitations in morphological and isozyme analysis, attempts have recently been focused on using molecular markers (e.g., nucleic acids and proteins) for linkage mapping. The mapping of DNA-sequence polymorphisms in genomic DNAs is one of the most significant developments in molecular biology. Essentially, there are three types of gene maps: *genetic* (*linkage*) *maps*, *cytogenetic maps* and *physical* (*molecular*) *maps*. Generally, three types of molecular markers are used for marker analysis: (i) isozymes, (ii) restriction fragment-length polymorphisms (RFLPs), and (iii) PCR-based molecular markers. There are many other variants, such as Random amplified polymorphic DNAs (RAPDs), Amplified fragment-length polymorphism (AFLP), Amplicon length polymorphism (ALP), Arbitrarily-primed PCR (AP-PCR), Allele-specific PCR (AS-PCR), DNA amplification fingerprinting (DAF), Specific amplicon polymorphism (SAP), Sequence-characterized amplified region (SCAR), Single-strand conformation polymorphism (SSCP), Microsatellite simple sequence length polymorphism (SSLP), Simple sequence repeats (SSR), Sequenced tagged sites (STS), End sequence tagged (EST), and Variable length nucleotide tandem repeats (VNTR). However, in this chapter, only few representative basic methods have been discussed.

II GENETIC OR LINKAGE MAPPING

A. CONSTRUCTION OF GENETIC LINKAGE MAPS BASED ON DNA BREAKS

There are several mapping techniques, which exploit this principle by artificially introducing chromosome breaks (Figure 11.1), including *radiation hybrid mapping* (where chromosome breaks are introduced by irradiating somatic cell hybrids) and *HAPPY mapping* (where genomic DNA is sheared by vortexing or sonication). In each case, physical mapping technology is then applied to detect linkages: the presence or absence of two markers on the same DNA fragment is assessed by hybridization, PCR, or detection of the gene product. After repeated experimentation, the degree of linkage is calculated as the frequency with which markers are separated. Radiation hybrid maps are calibrated in *centiRays* (cRx, where x is the dosage of X-rays in Rads), 1 cR being equivalent to a 1% frequency of separation. The common method of linkage mapping, however, exploits the natural chromosome breakpoints, which arise due to the crossing-over (homologous recombination) during meiosis.

B. CONSTRUCTION OF LINKAGE MAPS BASED ON GENETIC RECOMBINATION

On the basis of studies in garden peas, Gregor Mendel (1850) formulated some of the fundamental principles concerning the mechanisms of heredity. First, he proposed that the units of inheritance were discrete, occurs as pairs, and can exist in alternative forms. Second, only one

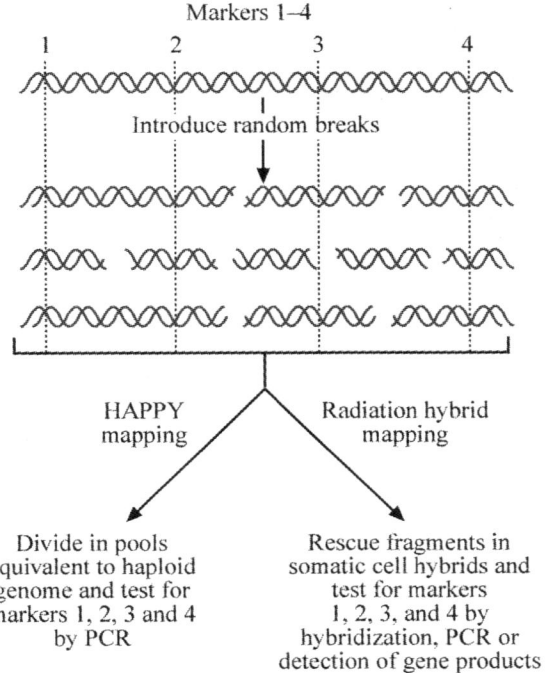

Figure 11.1 *Genetic linkage mapping using artificially introduced break points*

member of each pair of genes is distributed to a sex cell. Third, the pairs of genes assort independently of one another, with the result that all possible genetic combinations will result if a large number of population from a single genetically-appropriate mating are sampled (Figure 11.2). The loci carried on the same chromosome are said to be linked to each other. There are as many *linkage groups (l)* as there are autosomes in the haploid set plus sex chromosomes. For example, *Drosophila* has five linkage groups (2n=8; l=3 autosomes+X+Y), whereas humans have 24 linkage groups (2n=46; l=22 autosomes+X+Y). Prokaryotes and viruses, which usually have a single chromosome, are discussed elsewhere.

Theoretically, if genetic linkage were complete (loci locked together on a chromosome), then all genes along any chromosome would be passed on to the sex cells as an intact block, with no new genetic combinations on the chromosomes formed during the meiotic process (Figure 11.3a). However, in most instances, linkage is not complete. Exchanges occur between gene sites during meiosis, thereby creating new combinations of genes (Figure 11.3b). Because recombination tends to occur with increasing frequency as the distance between two specific gene loci increases, the percentage of recombination can be used to represent a measure of distance (map distance) between the two genes. Thus, by analyzing recombination frequencies among the progeny of parents that are heterozygous, for a number of linked genes, a genetic map that places the genes in a linear array can be constructed.

The distance between the loci represents the frequency of the occurrence of recombination and is not equivalent to a precise physical distance. However, approximations for the value of percentage recombination and the number of base pairs of DNA have been deduced by

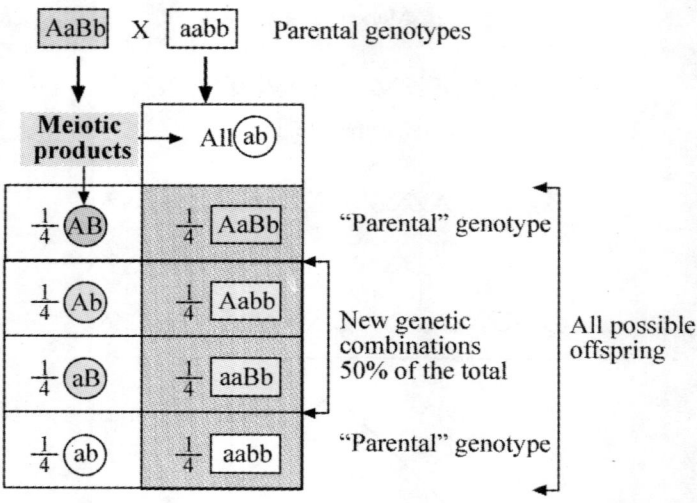

Figure 11.2 *Independent assortment. When an organism with a doubly heterozygous genotype (AaBb) is crossed with an organism with a doubly homozygous recessive genotype (aabb), then 50% of the offspring will have the parental genotypes and remaining 50% will have new genotype combinations*

correlating physical chromosome maps with genetic maps. In gene-mapping terminology, 1 map unit is equal to 1 *centimorgan* (cM) and 1 cM is equal to 1% recombination which, in humans, is roughly equivalent to 1×10^6 bp of DNA. Some important points to be noted about genetic linkage and gene mapping are as follows. First, to keep count of the occurrence of new genetic combinations, the genetic arrangement of one of the parents must be at least doubly heterozygous; i.e., AB/ab (coupling or *cis* configuration) or Ab/aB (repulsion or *trans* configuration). Second, the recombination between two genes occurs regardless of the arrangement of the genes. Third, 0% recombination denotes a complete linkage; and a value of 50% recombination between two loci means that the genes are either on different chromosomes or so far apart on the same chromosome that they cannot be mapped. To resolve the problem of two genes being far apart but on the same chromosome, the genes that lie between the original two genes must also be mapped to determine whether or not all of them make up a single linkage group.

Genetic maps have both theoretical and practical importance. For example, in the tomato plant, the close linkage of a gene for a character (phenolic compound) to another gene (disease-resistance) that causes a genetic disease can be used as a diagnostic tool. In genotypes where this linkage has been established, an individual found to be carrying the allele of a phenolic compound can also carry the disease-resistance gene. This information can help breeders to select disease-resistant seedlings at an early stage, based on the presence of phenol. To generate detailed genetic maps for eukaryotic organisms, many genes, each with at least two alleles, must be identified. Then, genetically-defined crosses must be carried out and the frequency of recombinants among large number of offspring must be tallied. The results indicate the extent of linkage between genes. Thus, high-quality genetic maps can be produced using multipoint (two or more linked gene pairs) crosses.

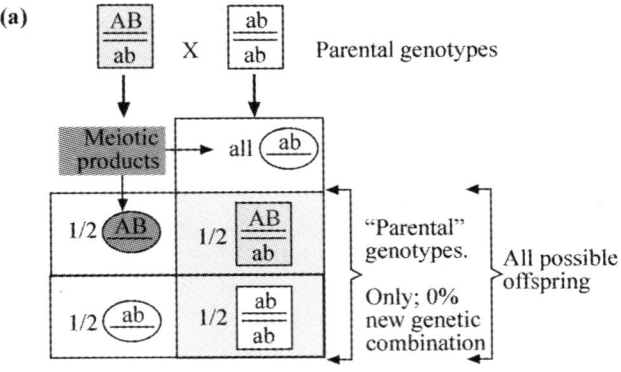

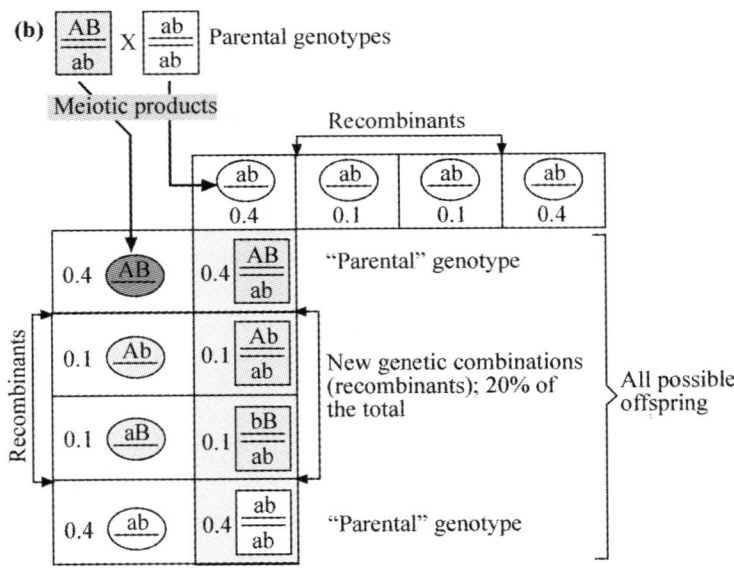

Figure 11.3 *(a) complete linkage, (b) incomplete linkage. During meiosis in this example, 20% (i.e. 0.1+0.1=0.2) of the products result from a recombinations(s) event between A and B loci. Recombination occurs regardless of the genotype of the parents (Redrawn from Glick and Pasternak, 1996)*

III CYTOGENETIC MAPPING

Cytogenetic maps can be constructed by correlating the phenotype with observable chromosome rearrangements and deficiencies. This type of map has a low resolution and is applicable only to few species, e.g., mammals and *Drosophila*, which display either natural or artificially inducible, reproducible, chromosome-banding patterns. Cytogenetic maps of simple genomes are used to locate regions that differ in their physical properties. In addition to differential banding, *in situ* hybridization is extensively used for cytogenetic mapping (for more details on methods of *in situ* hybridization refer to Chapter 13).

IV PHYSICAL OR MOLECULAR MAPPING

Physical maps reveal the sites for restriction enzymes, sites and sizes of deletions, AT-rich regions, etc. A physical or molecular map is created by ordering cloned fragments of genomic DNA and is calibrated in real units (base pairs). The physical map has the highest resolution and is the ultimate aim of molecular mapping. The advent of molecular markers has allowed the three types of maps to become progressively integrated. Polymorphic molecular markers, such as restriction fragment-length polymorphisms (RFLP), and sequence-tagged sites (STSs) containing minisatellite DNA, can be used both as genetic markers in linkage analysis as well as nucleic acid probes to identify physical clones. Also, *in situ* hybridization allows the physical clones to be precisely assigned to the chromosome bands on cytogenetic maps. Molecular markers are based on two basic techniques: (i) Southern blot hybridization and (ii) Polymerase chain reaction (PCR). This chapter describes some of the current approaches and techniques used for RFLP and PCR-based genetic mapping.

A. HETERODUPLEX MAPPING

This is a superior technique for the comparative analysis of nucleotide changes in a gene of different species. A *heteroduplex* is a double-stranded nucleic acid molecule, in which the two strands are not entirely complementary. For example, in a DNA heteroduplex one strand may contain one allele of a gene, while the other strand may have another corresponding allele. Such heteroduplex molecules are prepared *in vitro* by mixing two denatured DNA preparations and later allowing them to renature at renaturation conditions. Many of the double-stranded molecules thus produced will be heteroduplexes, depending on the amount of nucleotide complementarity. In such heteroduplexes the segment of wild type strand, which is missing in the mutant strand, will remain unpaired and can be seen as a loop under an electron microscope.

By using this technique, changes in mutant DNA segments corresponding to wild type DNA segments can be identified and mapped. Thus, this technique is known as *heteroduplex mapping*. It permits us to determine the size and location of a deleted region in a gene. By using this technique, the presumed deletion mutants of r^{II} locus were mapped by Benzer and shown to contain deletions in the r^{II} region. This technique was later extended to DNA:RNA heteroduplexes also, and it permitted the discovery and determination of the introns and exons of eukaryotic genes.

B. RESTRICTION DIGESTION AND RESTRICTION FRAGMENT MAPPING

The technology of recombinant DNA relies heavily on restriction enzymes, which are discussed in detail in Chapter 15. The restriction digestion analysis of DNA has been generally referred to as *fingerprinting, profiling, typing, genotyping,* or *identity testing*. DNA fingerprinting is useful for forensic identification, determination of family relationships, genome-linkage mapping, antenatal diagnosis, localization of disease loci, determination of genetic variation, molecular archaeology, and epidemiology. Since DNA is the medium of heredity, it directly reflects the relatedness or phylogeny of the sample material. Because DNA chemistry (replication, packaging, methylation, gene expression, etc) varies between cell types and individual cells, inherited base pair mutations at every cell division that are themselves stably inherited, even single cells from an individual, can ultimately be characterized. Usually, variations in the DNA base sequence are exploited to produce a DNA fingerprint, which highlights the uniqueness of a particular sample.

The RFLP technique depends on the natural variation in DNA base sequences. Homologous restriction fragments of DNA (obtained after *restriction digestion*), which differ in size or length, can be used as genetic markers to follow chromosome segments through genetic crosses. Since there are enormous amounts of DNA in cells of higher organisms (e.g., plants, mammals, etc), no two organisms are likely to have an identical DNA base sequence. Thus, there is tremendous amount of DNA variation present in natural populations of higher organisms. The natural variation in DNA sequences can be detected in several ways. Among the many methods available, RFLP is the most commonly used technique. This new technology promises to revolutionize some areas of plant genetics and plant breeding in addition to the above-said applications. RFLP analysis involves: (i) DNA isolation, (ii) DNA cleavage by restriction enzymes, (iii) Electrophoresis of resulting fragments, (iv) Southern transfer of separated fragments to membrane support, (v) Radioactive labeling of suitable probes, (vi) Hybridization of probes to membrane-supported fragments, and (vii) Print detection as a banding pattern on X-ray film.

RFLP analysis can make use of single-copy probes or repetitive probes. Both DNA and random genomic libraries have been used as sources of probes for RFLP mapping. Variant banding patterns, or fingerprints, on X-ray film are the result of polymorphic fragments detected by a particular probe. The identification of DNA bands by simple ethidium bromide staining is possible only when the number of fragments produced is low enough to resolve distinctly on the gel (e.g., chloroplast DNA produces about 40 fragments when digested with *Eco*RI). When the genomic DNAs of plants and mammals are used, they generate thousands of fragments making fragment separation difficult. This necessitates the use of labeled (radioactive or non-radioactive) probes to detect the sequence of interest on the gel. The probe DNA may hybridize to multiple tandem-repetitive or hypervariable minisatellites and produce complex fingerprint patterns. Alternatively, the probes may be locus-specific for individual hypervariable loci, which produce simpler patterns by detecting alleles from single or even multiple loci (Figure 11.4).

The polymorphism in the size of restriction fragments is a consequence of the variation in the distribution of restriction sites. This can occur through the loss or gain of sites by base substitution, insertions or deletions, inversions, translocations, or transpositions. The position of the sites can change due to insertions, deletions or genomic rearrangements. The number of fragments produced and the sizes of each fragment will reflect the distribution of restriction sites in the DNA. The fragments produced will thus be specific for each target DNA/restriction enzyme combination and can be used as a "fingerprint" specific for a given target DNA (or DNA of an organism). The variation in restriction sites does not necessarily reflect a functional difference in the sequence examined. Accordingly, the restriction fragments are well suited to be genetic markers. If the individuals were crossed, the F_1 would contain both segments and would be recognized as being heterozygous. This illustrates the co-dominant nature of these nuclear markers. These fragments would then segregate in a backcross or F_2 generation in a simple Mendelian fashion. By chi-square or χ^2-analysis, the linkage relationships can be

254 MOLECULAR BIOTECHNOLOGY: PRINCIPLES AND PRACTICES

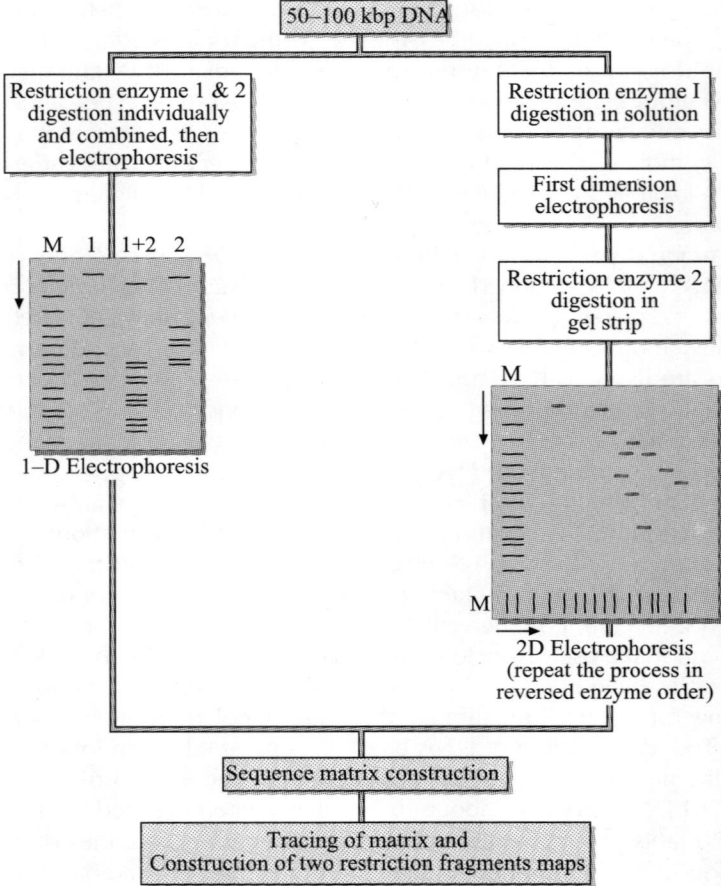

Figure 11.4 *The general scheme for the construction of two DNA restriction maps based on 1-D electrophores, two-enzyme digestions, two 2D electrophoresis, and sequence matrix. M, DNA fragments as size markers; arrows denote the direction of electrophoresis*

determined for a theoretically unlimited number of fragments. Extensive work has been done in developing linkage maps and using RFLPs to map various genetic diseases in man. In plants, the linkage analysis of DNA markers has been done for maize, tomato, peas and lettuce.

1. Storing and Handling of Restriction Enzymes

Before commencing of restriction digestion or RFLP analysis, it is very important to know the nature and procedure to handle the restriction enzyme(s), because restriction enzymes are at the heart of RFLP analysis. Restriction enzymes, like many other enzymes, are most stable at cold temperatures and lose activity if warmed for any length of time. The guidelines listed below are recommended for handling restriction enzymes.

1. Always store the enzymes in a non-frost-free freezer that maintains a constant temperature between –10°C and –20°C.

2. When a large shipment of an enzyme is received, split it into several smaller aliquots of 50–100 µl in 1.5 ml tubes. Use a permanent marker on a tape to clearly identify the aliquots with the enzyme type, concentration in units/µl, and date received. Use up one aliquot before starting another.
3. Remove the restriction enzymes from the freezer directly onto crushed or cracked ice in an insulated ice bucket or cooler. Keep the enzymes on ice at all times during the preparation, and return to the freezer immediately after use.
4. Preferably, set up the restriction digests on ice. Then, there is little loss of enzyme activity during the time taken to set up the reaction.
5. *Storing DNA and restriction buffer*: Purified DNA is generally stored in the refrigerator (approximately 4°C). DNA can be kept at –10°C to –20°C for long-term storage of several months or more. However, repeated freezing and thawing damages the DNA. The restriction buffer is best kept frozen and is not affected by freeze–thawing.

2. Buffers

Several types of buffers are used in this course, as groups of restriction enzymes operate under different conditions of salt and pH. For optimal activity, several different buffers would be needed for the enzymes used in this course. To simplify procedures, we use a "compromise" restriction buffer – a universal buffer that is a compromise between the conditions preferred by various enzymes.

For the digests of genomic DNA to be analyzed by Southern blotting and hybridization, the amount of DNA to use per track on the Southern depends on the genome size of an organism (e.g., for higher eukaryotes, 20–25 µg should be used). Different restriction enzymes cleave optimally in different buffers, but all require the presence of divalent ions, usually Mg^{2+}, in a buffered salt solution. The addition of BSA stabilizes enzyme activity. Manufacturers often supply buffers optimized for their enzymes, but the buffers given in Table 11.1 will do for most enzymes. Restriction enzymes are supplied in buffered 50% glycerol and should be kept at –20°C and handled on ice for as short a time as possible. Units are defined as the amount of enzyme required for cleaving a given DNA substrate in a given time. The digests should be set up so as to give 2–5 fold over the digestion of test DNA.

Table 11.1 *Restriction enzyme buffers*

Buffer strength	NaCl	Tris-HCl (pH 7.5)	$MgCl_2$	DTT
Low salt	0	10 mM	10 mM	1 mM
Medium salt	50 mM	10 mM	10 mM	1 mM
High salt	100 mM	50 mM	10 mM	1 mM

These should be made up as sterile 10X stocks and kept frozen in aliquots.

One-For-All: This is used for most restriction enzymes (10X): 100 mM Tris-acetate pH 7.5, 100 mM magnesium acetate, and 500 mM potassium acetate.

3. Restriction enzyme digestion

The product of the action of a restriction enzyme on a DNA sample is called *restriction digestion*. The volumes required for restriction enzyme digestion vary, depending on the concentration of the DNA and on the capacity of the wells in the gel. A typical restriction digest might contain 5 μg of DNA in 20 μl of restriction mix.

Procedure

1. On ice, take DNA and add sterile distilled water to 17 μl.
2. Add 2 μl of the appropriate 10X buffer.
3. Add 1–5 units of the restriction enzyme. Do not add more than 10% by volume to the reaction, as the glycerol in the enzyme stock (50%) will inhibit cleavage if its concentration exceeds 5% in the reaction. Spin the tubes briefly in a microfuge or in a preparatory centrifuge.
4. Transfer it to a constant-temperature water bath (usually 37°C, Note 1 and Note 2) for 1–3 hr (at least 20 min.). After several hours, the enzymes lose their activity, and the reactions may be stored in freezer (–20°C) until one is ready to continue.
5. Inactivate the enzyme by heating to 65°C for 5 min. (Note 2), or by adding 20 mM EDTA.
6. The DNA is now ready for electrophoresis. For long-term storage, the samples should be refrigerated or frozen.
7. Analyze the restriction digestion by the electrophoresis of an aliquot on an agarose minigel. Include an uncut control.
8. The cut DNA may require purifying away from the BSA and restriction enzyme, and resuspending in a different buffer for subsequent procedures. This is achieved by extracting once with phenol/chloroform, once with chloroform, and precipitating it with salt and ethanol before resuspension.

Notes

1. The given buffers will do for all the commonly used enzymes, except for a small number of enzymes which work best in 20 mM KCl. Restriction enzyme digestion buffer: 10 mM Tris–HCl, pH 8.0, 10 mM $MgCl_2$ and 1 mM DTT and is incubated at 30°C.
2. Note that the enzymes isolated from thermophyllic bacteria (e.g., *Taq* 1) are inactive at 37°C, and have temperature optima at 60–80°C. These reactions should therefore be carried out at 65°C. The enzymes cannot be denatured by heating.

4. DNA can be mapped using restriction endonucleases

Double-stranded DNA by its own unique sequence characterizes each and every molecule of DNA. It is known that restriction enzymes cleave the DNA only in the vicinity of particular sequences of base pairs, known as restriction sequences. This means that the cleavage sites on a DNA molecule are constant: every time a DNA molecule X is treated with enzyme Y – under the same conditions of time, concentration, and temperature – the same fragments will be produced. Therefore, the DNA molecules can be characterized by the sizes of the fragments, which are produced as a result of cleavage by any particular enzyme. And we also know that the resultant fragments can be separated and their sizes determined by agarose-gel electrophoresis. The band patterns generated by this procedure can be used to prepare *restriction maps* for different DNA molecules. Restriction maps are schematic representation of DNA molecules and show the locations of recognition sequences for various restriction enzymes.

DNA is polymorphic and polymorphisms can be used as genetic markers. Any two members of the same species have the same basic arrangement of genes on their chromosomes. Furthermore, the base sequence of any particular gene in different members of the same species is basically the same. However, it is most likely that the actual base sequence of these two genes differ slightly. This slight difference provides an example of a *genetic polymorphism*, a variant in the base pair sequence of either the coding or the non-coding portion of a DNA molecule that occurs with a frequency of 1% or more. A polymorphism in the coding sequence of a gene may lead to differences in the gene's polypeptide product. Non-coding polymorphisms, which occur at approximately 1 nucleotide in every length of 250 nucleotides, do not appear to have any phenotypic effects; however, they can be used as genetic markers. The presence of a genetic polymorphism can be detected by using a variety of restriction endonucleases. The variable fragment lengths resulting from genetic polymorphisms are called *RFLPs*.

The general steps involved in RFLPs are as follows: (i) isolate the total genomic DNAs from several strains or related species, (ii) digest the DNA samples with a selected restriction enzyme(s), (iii) resolve the digested fragments through electrophoresis, (iv) transfer the resulting gel patterns to a suitable solid support and expose them to a suitably-labeled, appropriate DNA probe under conditions favoring DNA:DNA hybridization (Southern blotting), (v) wash the membranes at high stringency to remove the free and non-specifically bound probes, and (vi) detect the bands to which the probe are hybridized by filming them through radioautography or other means (gel-documentation system).

The band pattern of the RFLPs generated will mainly be determined by the following factors: (1) the species/strain specific differences in the DNA sequences, (2) the kind of restriction enzyme used, and (3) the probe employed to detect the specific DNA sequence. The genetic polymorphism is detected by the RFLPs due to the following changes in the DNA of the organism: (i) variations in the base sequence of recognition sites of the restriction enzymes, and (ii) variability of fragment length due to the relatively large addition/deletions in the genomic DNA (results in differential movement of the band on the gel). The DNA probes may be obtained from: (i) genomic libraries, (ii) cDNA libraries, or (iii) chromosome-specific libraries (e.g., addition/substitution lines, flow-sorted chromosomes, chromosome-specific repeated sequences or micro-dissected chromosomes). A copy of single-gene sequences are the best probes.

The RFLPs have several unique advantages: (1) the number of RFLP loci in a genome is very large, so that even very small segments of the chromosomes can be mapped (even a cistron could be mapped), (2) mapping does not necessarily depend on gene function, thus phenotypically/biochemical non-detectable genes can be identified, (3) quantitative trait loci can also be mapped, which is virtually impossible through conventional technique, (4) it is much faster than conventional linkage mapping, and (5) fewer individuals (25–50 individuals/F_2 generation) need to be studied.

5. Applications of RFLPs

The RFLPs have many applications, some of which are: (i) gene(s) that are known to be linked with an RFLP locus can be isolated and characterized, (ii) finger-printing of strains/varieties for their genetic polymorphisms, (iii) linkage mapping of quantitative trait loci, (iv) identification of the most important loci affecting a quantitative trait, (v) highly-efficient indirect selection for tightly-linked quantitative trait loci and even for those oligo-genes a direct selection which may otherwise be either difficult or costly, (vi) determination of chromosome segments alterations, which is likely to yield the best results, (vii) establishing the relationships among various strains/species, and (viii) understanding the identity and function of thus far 'mysterious' polygenes. RFLP maps have been constructed for many organisms (crop species, mammals, microorganisms, etc.).

6. RFLP protocols

Method-1: Standard protocol for the construction of RFLP of plasmid genomic DNA

Materials

Horizontal agarose gel electrophoresis apparatus, well-forming combs, gel mold, etc.

40X TAE buffer; Ethidium bromide; gel-loading solution

Procedure

1. Prepare a restriction enzyme (RE) digest of the DNA samples to be analyzed. Check the progress of the digestion with a mini-gel (optional).
2. If the RE digest is satisfactory, run the samples with agarose gel. The agarose content used will vary with the size of desired DNA fragments. For example, 0.8% agarose gel can resolve 1–25 kb. For larger fragments (20–100 kb), use 0.5% agarose. For small fragments (0.2–2 kb), prepare 1.5% agarose.
3. In a separate Eppendorf add one-fifth of 1 vol of loading buffer and the RE digest. Heat them to 65°C for 2 min. and load up to 35 µl per well on the gel.
4. Run the gel for 4 hr to overnight at 40 V in 1X TAE buffer with ethidium bromide in the gel. The resolved restriction fragments can be viewed on a UV-transilluminator box and photographed.
5. The sizes of the cleavage products for the standard are known and can be used to estimate the size of unknown DNA fragments, by establishing a size calibration curve on semilog graph paper. The size (in kilobases) is plotted on the linear abscissa to make this curve. A comparison of the distance migrated in an experiment with co-migrated fragment size standards allows a reasonably accurate size determination (± 5%) of the fragment of interest.

Method-2: Construction of restriction fragment maps of 50–100 kilobase DNA

The construction of restriction fragment maps obtained from plasmid, cosmid, and YACs is a key step in many molecular biology experiments – particularly those involving sequencing, such as in the genome consisting of 10^8–10^9 base pairs (bp), which is usually broken down to the level of cosmids or YACs with the size range of 30–300 kb. The following procedure consists of a two-enzyme digestion, two-dimensional (2D) electrophoresis, as well as the construction of a sequence matrix based on the data of lambda phage DNA (linear or circular DNA) as an example to illustrate this approach (Figure 11.5). Using a two-enzyme digestion, 2D electrophoresis procedure, all the restriction fragments in 50- to 100-kb DNA can be individually resolved and displaced on a 2D plane. This 2D gel pattern, with appropriate markers, provides a fixed set of x, y coordinates for each fragment obtained from the single and double digestion as well as the relationship between all the doubly-digested restriction fragment (second digestion) and their corresponding parental restriction fragments (first digestion).

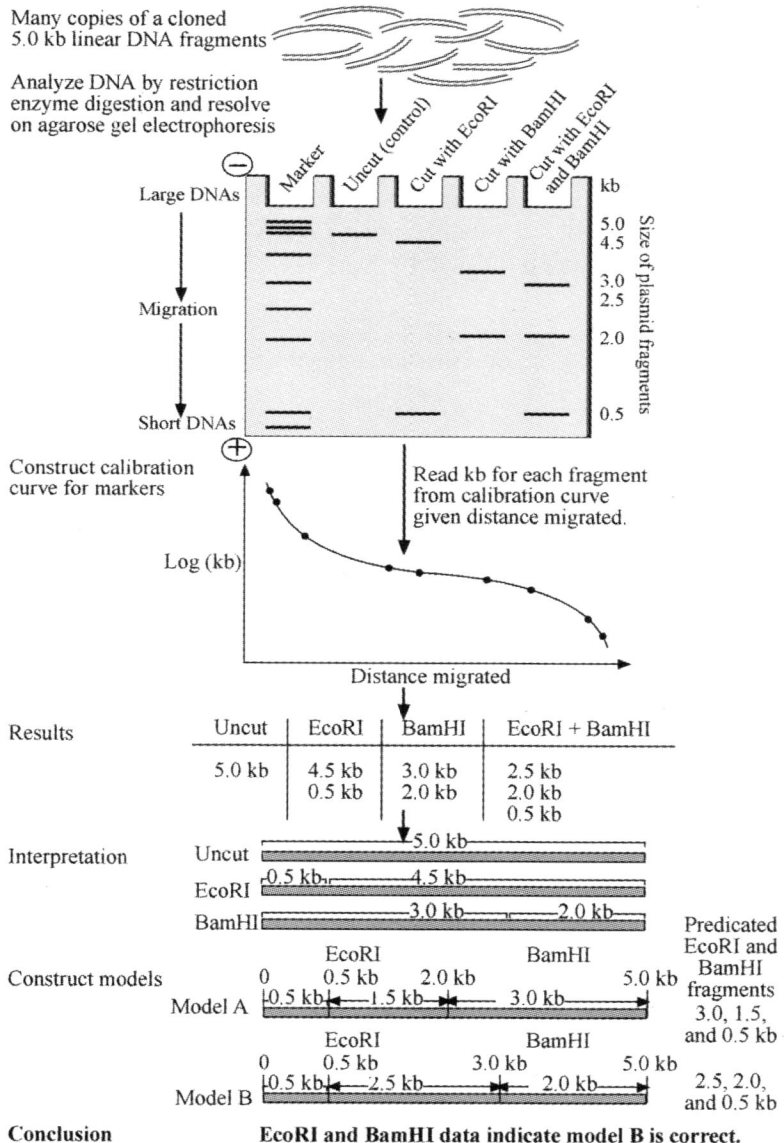

Figure 11.5 *Construction of a restriction map for EcoRI and BamHI in a DNA fragment (Redrawn from Genetics by Russell PJ, Benjamin/Cummings, 1998)*

Through the two-enzyme–2D analysis, the relationship between the singly-digested fragments and the doubly-digested fragments of both forward and backward order can be experimentally and unambiguously determined. The connection of one doubly-digested fragment to one singly-digested fragment from the first enzyme, as well as to the other singly-digested fragment from the second enzyme, provides the unambiguous overlapping relationship that is the cornerstone of the mapping process. Based on the above reasoning, a sequence matrix is constructed from the one-dimensional (1D) electrophoresis pattern and two 2D electrophoresis patterns. Each doubly-digested fragment in the sequence matrix defines the overlap of two adjacent fragments produced after a two-enzyme digestion. In the construction of the restriction map, each doubly-digested fragment contains the information for defining the overlap of the parental, singly-digested fragment obtained from the first enzyme as well as the parental, singly-digested fragment obtained from the second enzyme. Therefore, a continuous overlapping can be produced when all the doubly-digested fragments are placed in a proper order. The following protocol is based on the paper published by Yi M et al. (1993).

Methods

Restriction enzyme digestion: Choice of restriction enzyme pairs chosen for this demonstration is

EcoRI/BamHI, EcoRI/HindIII.

EcoRI reaction buffer: 100 mM Tris (pH 7.75), 50 mM NaCl, 10 mM $MgCl_2$.

BamHI reaction buffer: 100 mM Tris (pH 7.75), 100 mM NaCl, 10 mM $MgCl_2$.

HindIII reaction buffer: 50 mM Tris (pH 8.0), 50 mM NaCl, 10 mM $MgCl_2$.

TE buffer: 10 mM Tris, 1 mM EDTA (pH 8.0).

Procedure

1. Digest λ phage DNA with EcoRI and BamHI or HindIII (each with 2.5 units/µg DNA) separately in their respective reaction buffer at 37°C, overnight.
2. Digest 1.5 µg of λ phage DNA with EcoRI together with BamHI (each with 2.5 units/µg DNA).
3. Extract the DNA solutions with the same volume of phenol once and the same volume of chloroform twice.
4. Add 1/9 vol of 3 M sodium acetate, pH 7.0.
5. Precipitate the DNA fragments overnight, by adding 2.5 vol of 95% (v/v) ethanol at –20°C.
6. Centrifuge the DNA precipitates at 14,000 rpm at 4°C for 10 min.
7. Rinse the DNA pellets with 70% ethanol twice, and 95% ethanol once.
8. Dry the pellets in a vacuum and redissolve the DNA pellets in the TE buffer.

One-Dimensional Electrophoresis

Reagent: TPE: 0.08 M Tris-phosphate, 0.08 M EDTA (pH 7.7).

Procedure

1. Load 1.0 µg of EcoRI fragments of λ phage DNA, 1.0 µg of BamHI fragments of λ phage DNA, and 1.5 µg of EcoRI/BamHI doubly-digested fragments of λ phage DNA separately onto different lanes in a horizontal 25 × 20 × 0.46 cm 1% agarose gel slab.
2. Load the mixture of 0.5 µg of λ/HindIII fragments and 1.6 µg of 1-kb DNA ladder fragments on the gel slab at the first lane as size markers.
3. Perform electrophoresis at 40 V for 20 hr in TPE buffer at room temperature.
4. Stain the gel with ethidium bromide, and take a photograph.
5. Determine the molecular sizes of all the fragments by comparison with the size markers.

Two-dimensional electrophoresis: First dimension

Procedure

1. Load 1.5 µg of *Eco*RI fragments of λ phage DNA onto a horizontal 25 × 20 × 0.46 cm gel slab containing 1% ultrapure agarose. The sample well is 5 mm wide and 1 mm thick.
2. Load 0.5 µg of λ/*Hind*III fragments and 1.6 µg of 1 kb DNA ladder fragments as size markers.
3. Perform electrophoresis in a TPE buffer at 40 V for 20 hr at room temperature.

Second enzyme *in situ* digestion in gel

1. Cut the DNA-containing portion of the gel into a 16 × 0.8 cm strip.
2. Equilibrate the gel strip with *Bam*HI reaction buffer by shaking it at 100 rpm at room temperature for 1 hr.
3. Put the gel strip into a 21 × 1.25 cm tube (total volume: 26 ml) containing *Bam*HI reaction buffer, 325 µg/ml of molecular biology-grade bovine serum albumin (BSA) and a 4 mM concentration of molecular biology-grade spermidine.
4. Rotate the tube gently at 37°C for 1 hr and add 2000 units of *Bam*HI to the tube.
5. Rotate the tube gently at 37°C for 20 hr.
6. Equilibrate the gel strip with TPE buffer by shaking it at room temperature for 1 hr.

Two-dimensional electrophoresis: Second dimension

1. Fuse the gel strip with a 1% agarose gel slab by adding melted agarose to fill the gap.
2. Load the size markers in a well located at the end of the fused gel strip.
3. Perform the second-dimensional electrophoresis, with a direction perpendicular to the first one, at 35 V for 20 hr, in TPE buffer at room temperature.
4. Stain the gel slab with ethidium bromide and take a photograph.
5. Determine the molecular sizes of all the fragments by comparison with the size markers.
6. In cases in which *Bam*HI is used as first restriction enzyme and *Eco*RI is used as the second enzyme, the procedure is the same as described above except that the enzyme order is reversed, together with their corresponding reaction buffers (Figure 11.6).
7. The same procedure can be used for the *Eco*RI/*Hind*III pair with the corresponding buffers.

Construction of the matrix

1. Set the vertical column on the very left side of the matrix (Figure 11.7). The upper section contains *Eco*RI singly-digested fragments, and the lower section contains *Bam*HI singly-digested fragments listed in Table 11.2. The order of arrangement is always from the smallest to the largest fragments, from top to bottom.
2. Set the dividing horizontal row between the upper and lower sections with the *Eco*RI and *Bam*HI doubly-digested fragments listed in Table 11.2. The order of arrangement is always from the smallest to the largest fragments, from left to right.
3. Define the locations of E (for *Eoc*RI fragments overlapping with *Bam*HI fragments in the doubly-digested fragments) B (for *Bam*HI fragments overlapping with *Eco*RI fragments in doubly-digested fragments) and S (for fragments that are not digested by the second enzyme or located at the end). The S fragments arising from the end do not exist in the digestion or cellular DNA.

262 MOLECULAR BIOTECHNOLOGY: PRINCIPLES AND PRACTICES

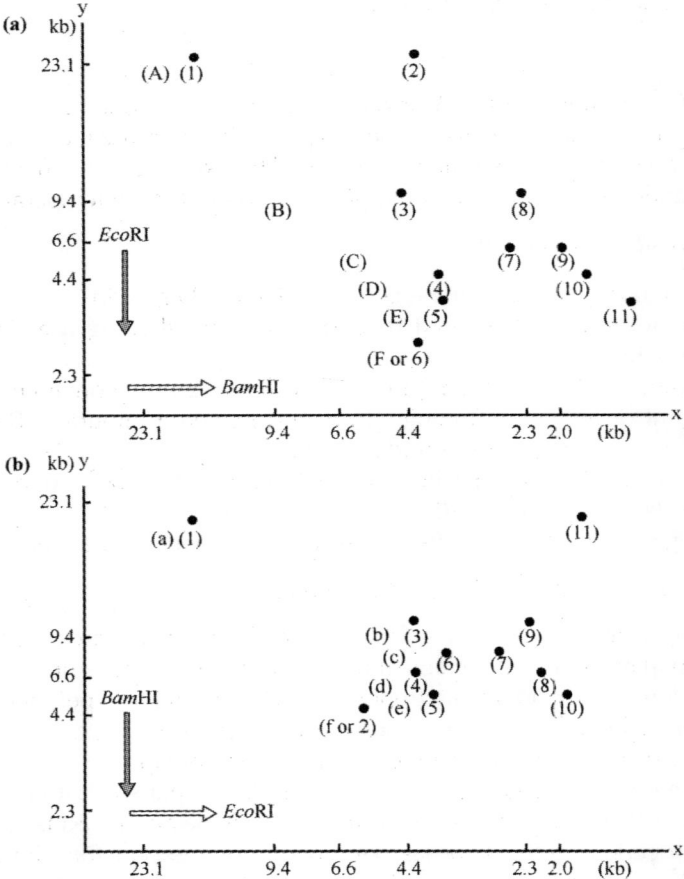

Figure 11.6 *Diagrammatic interpretation of two-dimensional (2D) gel electrophoresis pattern of λ phage DNA (a) 1.5μg of DNA digested by EcoRI→BamHI (fragments A–F) and (b) 1.5μg of DNA digested by BamHI→EcoRI (fragments a–f). The single spots (F and f), do not have a cutting site for the second enzyme BamHI and EcoRI, respectively. The vertical arrow denotes the direction of the first dimension, while the horizontal arrow denotes the direction of the second-dimension electrophoresis (Redrawn from: Yi M et al, 1993).*

Construction of the restriction fragment maps

1. As shown in Figure 11.7, connect E, B, and S locations throughout the matrix by starting the trace with the smallest fragment at the top of the left column (e.g., 3.65 kb). After reaching S1, go down vertically (passing the 3.70-b fragment in the medial row) to meet B (B1) in the lower section, move horizontally to another B (B2) and then vertically upward (passing the 3.37 kb fragment in the medial row) to E (E1, then E2), and so on. The tracing is carried out alternatively between a horizontal movement followed by a vertical one and vice versa. In the meantime, when the fragment is reached, the order of this fragment in its section is determined according to the numerical order in the tracing. For example, the B reached by

Genetic Maps and Marker Analysis 263

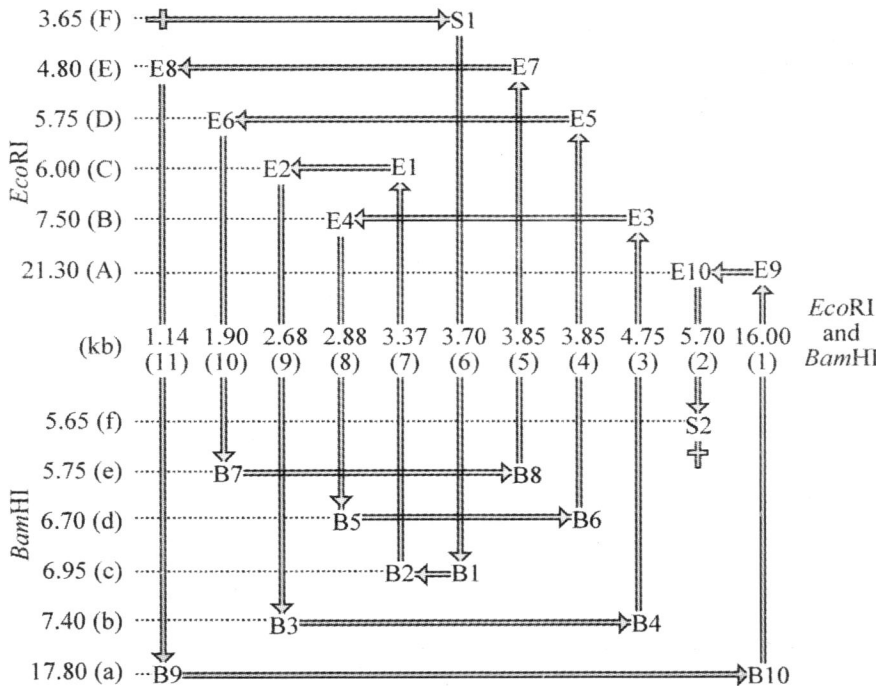

Figure 11.7 *Sequence matrices for construction of the linear restriction fragment map of lambda phage DNA. The vertical column at the left shows the size of all the EcoRI-digested fragments in the upper half and all the BamHI-digested fragments in the lower half. The horizontal row shows all the sizes of the doubly digested fragments as listed in table 11.2 "E" and "B" denote EcoRI- and BamHI-digested fragments respectively. "S" denotes a fragment that is not digested by the second enzyme. The numerical subscripts denote the linear order for the restriction fragments in the map*

tracing in the lower section of the matrix is ordered as B1, and the second B reached is ordered as B2. When tracing goes through the first turn in the upper section, it determines the order of E1 and E2. The next tracing goes to the lower section again, giving the order of B3 and B4, and so on.

2. Construct two restriction fragment maps simultaneously from the sequence matrix. In Figure 11.7, the top section of the map is for the ordering of the λ DNA *Eco*RI restriction fragments. The bottom section is for the ordering of *Bam*HI fragments. The construction of these two maps is based on the tracing in Figure 11.6. Following the tracing procedure described above, starting from the 3.65 kb fragment (F) at the top of the left column in the upper section, the tracing reaches S1 first. The 3.65 kb fragment (F), which is the singly-digested parental fragment of S1, is determined as the first fragment in the *Eco*RI map. The tracing goes down vertically in the matrix to reach B1 and B2 in the lower section. The singly-digested 6.95 kb fragment (c) shown in the left column, which is the parental fragment of B1 and B2, is now determined as the first fragment in the *Bam*HI map. Then the tracing goes up vertically to reach E1 and E2, which originate from the 6.0 kb fragment

(C), and is now the second fragment in the *Eco*RI map. Continuing the tracing, the parental fragment of B3 and B4, the 7.40 kb (b) fragment, will be denoted as the second fragment in the *Bam*HI map. The next tracing will add 7.5 kb (B) to the *Eco*RI map as the third fragment. The tracing in the sequence matrix represents the order and the overlap of the fragments.

Table 11.2 *Molecular sizes of singly-, doubly- and single followed by second-enzyme digested restriction fragments from λ phage DNA.*

1–D§ *Eco*RI	2–D¶	1–Dᵃ *Bam*HI	2–D•	1–D≠ *Eco*RI and *Bam*HI
21.30 (A)	(1, 2)	17.80 (a)	(1, 11)	16.00 (1)
7.50 (B)	(3, 8)	7.40 (b)	(3, 9)	5.70 (2)
6.00 (C)	(7, 9)	6.95 (c)	(6, 7)	4.75 (3)
5.75 (D)	(4, 10)	6.70 (d)	(4, 8)	3.85 (4)
4.80 (E)	(5, 11)	5.75 (e)	(5, 10)	3.85 (5)
3.65 (F)*	(6)	5.65 (f)*	(2)	3.70 (6)
				3.37 (7)
				2.88 (8)
				2.68 (9)
				1.90 (10)
				1.14 (11)

§=Fragment sizes obtained from 1-D electrophoresis after singly digested by *Eco*RI.
¶=Probable fragment sizes obtained from 2-D gel, after 1-D bands digested by second enzyme and electrophoresed (*Eco*RI→*Bam*HI).
ᵃ=Fragment sizes obtained from 1-D electrophoresis after singly digested by *Bam*HI.
•=Probable fragment sizes obtained from 2-D gel, after 1-D bands digested by second enzyme and electrophoresed (*Bam*HI→*Eco*RI).
≠=Fragment sizes obtained from 1-D electrophoresis after combined digestion of λ phage DNA (*Eco*RI + *Bam*HI).
*=Fragments (F and f digestions does not contain a cleavage site for second enzymes).

As shown in the constructed λ DNA restriction maps (Figure 11.8), each fragment from enzyme 1 is bound by two restriction sites of enzyme 1 (except the end fragments S) and is ordered by the two overlapping fragments generated by the second enzyme. For example, the order of the 6.0 kb E fragment (for the *Eco*RI mapping) is defined by E1 and E2 overlapping B2 and B3 generated by *Bam*HI. The numerical determinants in Figures 11.6 and 11.7 (such as E1-10, B1-10, and S1-2) are the overlapping orders of the fragments in the map. These orders are not defined by the polarity (5' end and 3' end) of the DNA strands in the duplex. Therefore, these orders can be reverted depending on the choice of polarity.

Notes

1. All the restriction fragments obtained from the first/single digestion must be completely retained and resolved in the 1-D electrophoresis and all the doubly-digested fragments must be retained and resolved within the 2D gel.
2. The correct choice of a restriction enzyme pair is crucial. To get the appropriate enzyme pair, the DNA sample should be singly digested by 8–10 enzymes. From the 1D electro-

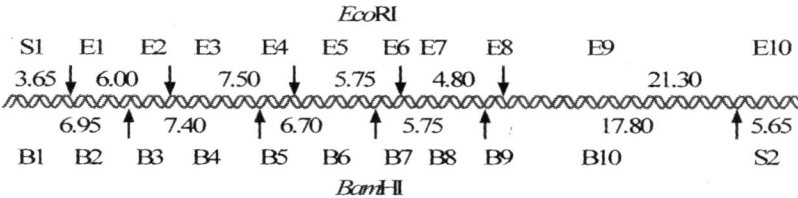

Figure 11.8 *The linear map of λ phage DNA constructed from EcoRI and BamHI restriction fragments*

phoresis pattern, the enzymes that produce 6–10 fragments in a single digestion should be chosen. If too few fragments are produced, the obtained map of the restriction sites will not be useful. If too many fragments are produced, an overcrowding in the 2D gel pattern may be observed, leading to overlapped and unresolved DNA spots.

3. Some of the restriction enzymes either exhibit poor activity in the gel or are rather expensive. Currently, the preferred list of acceptable enzymes includes *Eco*RI, *Eco*RV, *Hind*III, *Bst*EII, *Sal*I, *Sty*I, and *Xho*II.

7. Construction of RFLP genetic maps

The application of RFLP markers in the construction of genetic mapping is a little different from the previous applications. For example, a genetic cross in a diploid organism in which two homozygous parents are crossed to produce an F_1 and the F_1 is selfed to produce an F_2 progeny. It is understood that the F_1 hybrid from two such parents will have one set of chromosomes from each of the parents and that the two members of a homologous chromosome pair will be different from one another to the degree in which the parents differ in the DNA base sequence. When the F_1 plant undergoes meiosis to produce gametes, its chromosomes will undergo recombination by crossing over. This recombination process will form gamete chromosomes, which are mosaics with segments from each of the two parental chromosomes, and no two chromosomes will have an identical array of segments. In F_2 plants produced by selfing the F_1, homologous pairs will contain different mosaics of parental chromosome segments. This recombination process is the basis of conventional genetic mapping, which depends on two observations: (i) chromosome segments, which are not on the same chromosome, will undergo random recombination; (ii) chromosome segments on the same (homologous) chromosome will undergo recombination according to a function, which depends on their physical distance apart. Chromosome segments that are close together (closely linked) will undergo recombination less commonly than will segments that are further apart. Thus, the genetic distance or map distance is defined as a function of the recombination which occurs during gamete formation.

As discussed earlier in this chapter, previously the only way to follow chromosome segments and observe their recombination during gamete formation was to observe the phenotype caused by the action of genes which happened to be on the segments. By observing a phenotype, one could infer the behavior of the respective chromosome segments during meiosis and homologous recombination. Following the chromosome segments in this indirect manner we can construct fairly detailed genetic maps of several plant species. However, the method is cumbersome and time consuming. This problem is resolved by RFLP markers, a way to directly follow chromosome segments during recombination and greatly simplify the construction of genetic maps. Instead of looking at the phenotype caused by the presence of a gene on a chromosome segment, one looks directly at an RFLP marker on the segment itself. Thus, one is looking directly at the genotype of the plant, rather than indirectly, through the phenotype produced by gene action. The inheritance pattern of RFLP markers is similar to the way conventional gene markers function and can be analyzed by conventional Mendelian methods. The maps of RFLP markers can be constructed in the same fashion as can maps of conventional markers.

8. Advantages of RFLP markers over conventional genetic markers

When conventional genes and RFLP markers are compared, it is apparent that RFLP markers have many advantages for the construction of genetic maps. Some of the most obvious of these are:

(a) *Natural variation*: To construct a genetic map by following phenotypic variation, the genes used to follow the chromosome segments must have different alleles of the gene in question. Often, one cannot depend on the natural variation of conventional markers, and variants will have to be produced by mutagenesis. RFLP mapping, on the other hand, utilizes naturally-existing variation in DNA sequences and the induction of mutant-alleles is not necessary.

(b) *Developmental stage of the plant organ:* To infer the segregation of chromosomal segments by conventional genetic marker analysis, one has to depend on gene expression. This can leads to many complications. For example, genes may be expressed only in certain tissues, or at certain developmental stage or under certain environmental conditions. In some cases, scoring for a conventional gene may be expensive or time consuming. Since RFLP markers directly detect the RFLP variation in the chromosomal DNA, irrespective of the developmental stage and type of plant part, a small amount of DNA can be isolated from a sample and used for RFLP analysis.

(c) *Phenotypic effect*: RFLP markers are reflections of the natural variation present in the DNA sequence of organisms. In the vast majority of cases, they will have no phenotypic effect at all. In natural populations, an unlimited number of RFLP markers are present in any point; in contrast, phenotypic markers present only in few cases. These facts have important consequences for the construction of genetic maps.

(d) *Environmental effect*: Conventional genes depend on gene expression for scoring. The DNA sequence comprising the gene must be transcribed into mRNA, the mRNA must be translated to protein, and the protein must assume the proper configuration and localization to exert its effect. The function of the protein is often due to its enzymatic activity or its ability to bind to some cell component. All these stages can be, and often are, affected by the environment. Since the analysis of RFLP does not depend on gene expression, it is not affected by the environment.

(e) *Interaction with other genes*: Since conventional gene markers depend on gene expression to be scored, other genes often modulate their phenotypic expression. One of the simplest cases is dominance; many mutations of common genes are recessive. The recessive types are scored only if plants are homozygous, for the mutant alleles and heterozygotic plants cannot be distinguished from homozygotic wild-type plants. The RFLP markers are co-dominant, and the three possible genotypes for one marker can always be distinguished. In addition, the conventional markers will often display different phenotypes in different genetic backgrounds due to gene interactions. The RFLP markers are independent of each other and maintain their identity in any genetic background.

9. Construction of genetic maps using RFLP markers in plants

Before the construction of an RFLP map, a series of steps must be well-defined. For some important crop species (e.g., rice, maize, tomato, wheat, etc) RFLP maps have already been developed and anyone can use these available maps without having to construct new ones.

(a) Select the parental lines

The parent lines used for RFLP mapping must be genetically divergent enough to exhibit sufficient RFLP, but should not be too distant as to cause sterility of the progeny. The presence of some agronomic traits that can segregate in a cross are highly desirable. Since one does not know *a priori* how much RFLP variation will be present in a crop species, it is necessary to do a survey using a random selection of cloned probes. One should utilize any available information to select the plants to be surveyed. The survey should include cultivars, introductions and also existing systematic or evolutionary studies. After selecting a range of plants to be surveyed, the DNA is isolated from individual plants of each of the lines, restriction digested, and screened for polymorphisms by the standard method. Each accession can then be scored for the RFLP alleles present, and accessions showing a usable amount of polymorphism can be selected for crossing.

(b) Produce mapping population

The selected parent plants are crossed to produce an F_1 plant(s). A mapping population can be generated by selfing the F_1 to produce F_2, which is then scored for segregation of the RFLPs, or by backcrossing the F_1 to one of the parents and observing segregation in the first backcross generation. The inclusion of the F_2 population will provide more information than the backcross population of a comparable size. A mapping population of about 50 F_2 or backcross plants is sufficient for a fairly detailed map.

(c) Score RFLPs in the mapping population

Once the mapping population is obtained, the DNA is isolated from each individual plant in the population. One should remember that the chromosomes of each plant in the mapping population contain a unique array of parental chromosome segments/genes. To establish an RFLP map, it is necessary to determine which parent's chromosome segments are present in each plant in the mapping population. The unique pattern of chromosomal segments would be destroyed by recombination in the F_2 or backcross plants that are allowed to reproduce sexually. Therefore, every effort should be made to keep the individual plants in the mapping population alive so that repeated DNA extractions can be made, and a large amount of DNA can be accumulated. Mapping proceeds by sequentially scoring RFLPs in the individual plants of the mapping population by the following series of procedures.

(i) Screening for polymorphism: The probes are selected sequentially from the library and tested against the parents in an effort to determine which restriction enzymes will detect a polymorphism between the parents. The number of enzymes depends on the amount of variation present in the parent plants (e.g., in maize one or two enzymes were used to detect polymorphisms with most probes, but in rice 11 enzymes were used and about 75% of the probes were polymorphic).

(ii) Scoring: After a polymorphic probe/enzyme combination is detected, it can then be scored in the mapping population. To achieve this, a series of agarose gels must be prepared and filters prepared from them for DNA hybridization (called F_2 survey or mapping filters). There must be one filter set for each restriction enzyme to be used in the mapping project, and each filter set must contain digested DNA from each individual plant in the mapping population (e.g., one filter with *Eco*RI and another with *Pst*I digested). If a polymorphism between the parent plants has been detected, with a given probe/enzyme combination, then that probe will be scored on the corresponding mapping filter set. For example, "A" probe is found to be polymorphic between the parents using *Pst*I, then it could be scored on the *Pst*I mapping filter set. In an F_2 mapping population scored for a single RFLP, only three types of plants can be present. There are two homozygotes (each parental types) and the heterozygotes.

(iii) Linkage analysis: The data accumulated from scoring the mapping population sequentially with probes from the library is used to construct the linkage map. Linkage analysis is based on the degree to which probes tend to co-segregate. The first probe scored in a mapping study may not provide information on linkage, but beginning with the second probe one can determine whether any linkage is indicated. If the second probe is linked to the first one, they will tend to co-segregate (i.e., F_2 plants, which are homozygous or heterozygous for the first probe, will tend to also be homozygous or heterozygous, respectively, for the second probe). If no linkage is indicated, the distribution of homozygotes and heterozygotes for the first two probes will tend to be random. Statistical tests, such as a chi-square analysis, can be used to test for randomness of segregation and, thus, linkage. The first few probes, since they are randomly selected, are unlikely to be linked and hence unlikely to show co-segregation. However, as one sequentially adds probes, the linkages will be detected. Initially, there may be many more linkage groups than there are chromosomes, but the two numbers will tend to converge as more markers are added. As each probe is screened, it is tested for possible linkages to all the other markers, which have been mapped before. It is evident that a great amount of data is rapidly accumulated in an RFLP mapping project, and a computer is required for efficient data storage and analysis. The program, LINKAGE–1 (Sulter *et al.*, 1983) is used in genetic mapping and Mapmaker software (Lander *et al.*, 1987) for most mapping projects.

10. RFLP maps and conventional maps

The main objective of the RFLP mapping project is the production of a "saturated" map, which is a map with RFLP markers spaced every 10–20 *centimorgans* (cm) over all the chromosomes. On such a map, any conventional gene will be within a few cm of an RFLP marker. Since conventional plant genetic maps are about 1500 cm, at least 150 well-spaced markers are necessary for a saturated map.

   (a) *Rationalization of RFLP maps with conventional maps*: After the construction of RFLP maps, the immediate step is to associate the linkage groups of RFLP markers with the conventional genetic map (available). This can be done in several ways, depending on the particular situation.

   (b) *Placing conventional genes on the RFLP map*: Once the RFLP map is constructed, conventional genes can be placed on the map. In order to do this, it is necessary to have a mapping population in which both the conventional gene and the RFLP markers are segregated. By determining which RFLP markers show co-segregation with the conventional marker, one can place the conventional marker on the map.

11. Uses of RFLPs in plant breeding

The construction of an RFLP map can be very advantageous in systematic and evolutionary studies. However, in plant breeding, an RFLP map is not useful by itself unless it is used in conjunction with the analysis of conventional markers. One way in which RFLP maps can be used to supplement regular plant breeding protocols is by utilizing indirect selection. This procedure is more useful when one wants to select a conventional gene (the direct selection of a gene would be expensive, difficult or time consuming) by indirect selection. With indirect selection, the gene of interest is selected along with more closely linked RFLP markers. If the RFLP markers are indeed closely linked, they will remain associated with the gene of interest during segregation. This allows one to select for the RFLP marker with the confidence that the conventional gene will also be present, since only relatively rare recombination events would separate the two. An indirect selection for an RFLP marker, rather than a direct selection for the gene, can have advantages in several plant-breeding scenarios. For example, if one is trying to introgress a recessive gene into a cultivar by using a backcross-breeding program.

Backcross protocols involve alternate backcross and selection phases. If a gene is dominant it can be directly selected, but recessive genes will not be expressed in any of the backcross plants, and it would be necessary to carry out progeny testing (probably by selfing the backcross plants to test for the presence of the recessive gene). However, a recessive allele could be indirectly selected by selecting for a linked RFLP marker. No progeny testing is necessary and the process would be greatly speeded up. In some cases, the indirect selection for dominant genes could also be profitable in a backcross program, in which plants can be scored for a linked RFLP marker while in the juvenile stage, since only a little DNA needs to be isolated. The plants not bearing the desired allele could be weeded out early, saving space and expense. RFLP analysis is also useful when two or more genes have the same phenotypic effect.

Quantitative traits: Most traits of agronomic significance in plants are controlled by a number of genes scattered over the chromosome(s). Each of the individual genes of such a polygenic system may contribute either a small positive or negative effect to the trait of interest. Since the quantitative traits do not show a clear dominance, the phenotype has a large environmental component. All these characteristics taken together make quantitative traits very difficult to analyze. Thus, one is forced to use biometrical methods and extensive testing over different years and in different environments in an effort to advance toward the desired state. Since one can use RFLP markers to simultaneously follow the segregation on all chromosome segments during a cross, the basic idea is to look for a correlation between the quantitative trait of interest and the specific chromosome segments marked by the RFLPs. If correlations do exist, then the chromosome segment must be involved in the quantitative trait (one or more genes determining the trait must be on that chromosome segment). The difficult part in the procedure is establishing the correlations between the trait and the specific chromosome segments. Once the specific chromosome segments are implicated in the trait, specific chromosome segments with a positive effect on a quantitative trait can be selected from a population of plants and incorporated into single plants with a high level of efficiency. This is possible because of the ability to score for several RFLP markers simultaneously in a single plant, since RFLP analysis is free from environmental influence or gene interactions.

12. Restriction landmark genomic scanning (RLGS)

This technique was introduced first time by Hatada et al. (1991) for the genomic DNA analysis of higher organisms. It is based on the principle that restriction enzyme sites can be used as landmarks. It employs direct labeling of genomic DNA at the restriction site and 2D electrophoresis to resolve and identify these landmarks. The technique has proven its utility in the genome analysis of closely-related cultivars, for obtaining polymorphic markers that can be cloned by the spot target method, and has also been used as a new fingerprinting technique for rice cultivars.

C. DNA TYPING

The method of fingerprinting has been extensively used in the forensic laboratory. The principle is that no two individuals have exactly the same fingerprints, so fingerprints left at the crime scene are important evidence in a criminal investigation. For DNA typing – the use of DNA analysis to identify an individual – scientists may use highly polymorphic markers scattered throughout the genome. Each marker consists of a restriction fragment within which are short, identical segments of DNA tandemly arranged, head to tail. The differences between individuals result from a great variation in the number of tandem repeats, which are called variable number of tandem repeats, or VNTRs (also called minisatellite sequences). The use of VNTRs as markers is illustrated in Figure 11.9. For example, individual A is heterozygous for two alleles at a VNTR locus flanked by restriction sites. One allele has three copies of the repeat sequence and the other allele has ten copies. Individual B is homozygous for a 6-repeat allele. When the genomic DNA is digested with the enzyme, restriction fragments of different sizes are produced because of the VNTRs. The results of the digest can be visualized by the use of a probe for the particular repeat sequence at the marker locus. In this way, two DNA bands would be detected for individual A and one DNA band would be detected for individual B.

D. PCR-BASED DNA MAPPING

DNA markers linked to important economic traits are rapidly becoming an integral part of plant-breeding programs, to help in the identification and selection of desirable genotypes in segregating populations. The application of DNA markers in genetic mapping, genetic diagnostics, molecular taxonomy, and evolutionary studies has been well established. The most commonly used DNA markers are restriction fragment-length polymorphism (RFLPs). However, the detection of RFLPs by DNA blot hybridization is laborious and incompatible with applications requiring high throughput. Conventional RFLP requires relatively large amounts of DNA, while Southern blotting and hybridization need radioactive isotopes and autoradiography. These factors make conventional RFLP analysis slow and expensive. The development of new methods based on PCR amplification of genomic DNA and detection of the differences in the PCR products have become popular. Genetic tests based on the PCR are simple to perform and, using PCR, particular fragments of DNA can be amplified to large quantities. The amplified portions of the DNA are defined by short (16–20 base), synthetic DNA molecules (primers) that match the ends of the DNA to be amplified. However, target DNA sequence information is required to design specific primers. The PCR-based markers can be utilized in much the same way as the RFLPs. In some cases, RFLP markers are actually converted to PCR-based markers.

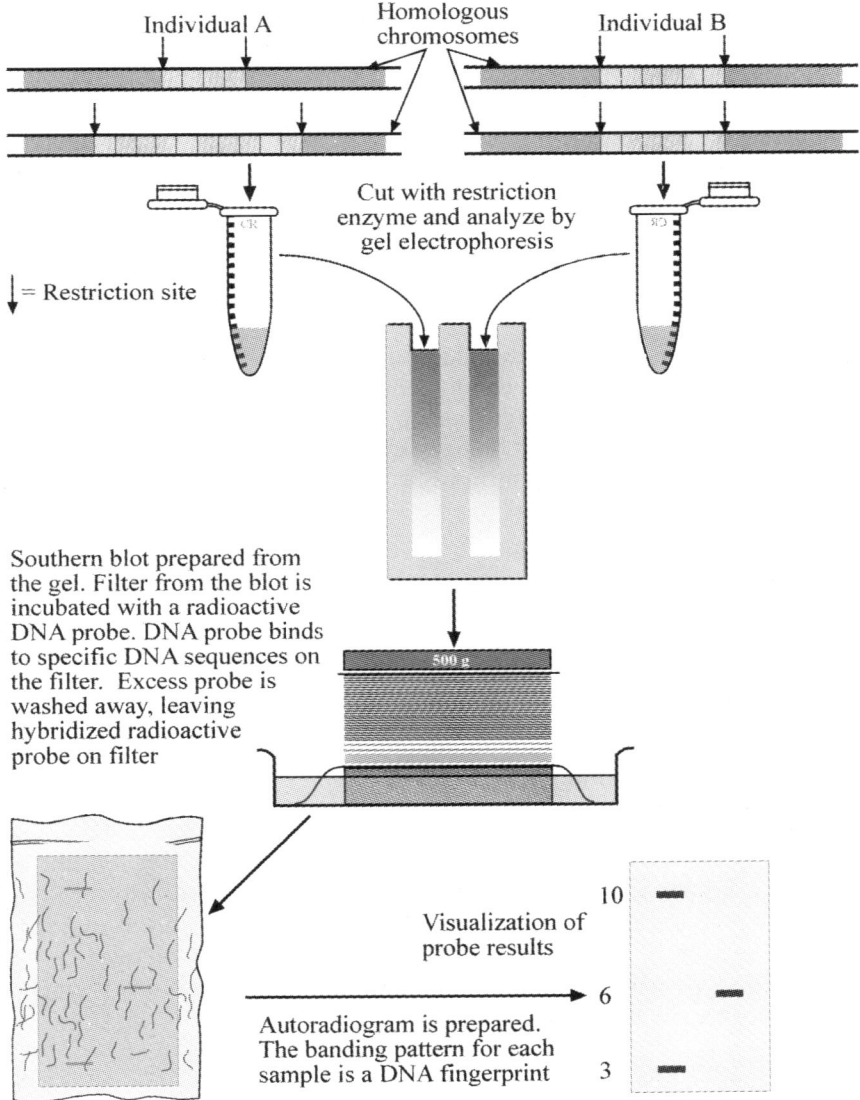

Figure 11.9 *Diagrammatic illustration of the concept involved in using variable number of tandem repeats (VNTRs) as DNA markers*

(1) RFLP-markers converted into PCR-based markers

(a) Sequence tagged sites (STS)

RFLP probes that are specifically linked to a desired trait can be converted into PCR-based STS markers, based on the nucleotide sequence of the probe giving polymorphic band patterns, to obtain a specific amplicon. STS is a short stretch of genomic sequence that can be detected by PCR. Using this technique, the tedious hybridization procedures involved in RFLP analysis can be eliminated. Each STS is mapped to a specific site as a landmark in the genome. In mammalian genomes a large number of STS have been produced by analyzing RFLP markers, microsatellite, expressed sequences, and fragments of yeast artificial chromosome (YAC) inserts. STS is important in converting a genetic map to a physical map. Based on the sequence information of STS, specific genes can also be isolated. This approach is extremely useful for studying the relationship between various species. When these markers are linked to some specific traits – for example, the stem-rust gene in barley or the powdery mildew resistance gene – they can be easily integrated into plant-breeding programs for marker-assisted selection of the trait of interest.

The STS marker system has all the advantages of the PCR techniques. A limitation of STS is that considerable effort and time are needed to generate DNA sequence data for designing suitable primers. Besides, the fragments amplified by the primers may not show length variation between individuals. However, this problem is circumvented by the restriction digestion of the PCR product, which detects polymorphism in the target site of a restriction enzyme internal to the two primer binding sites (this method is called cleaved amplified polymorphic sequence, *CAPS*).

(b) Allele-specific associated primers (ASAPs)

To get an allele-specific marker, a specific allele is sequenced and specific primers are designed for amplifying the DNA template to generate a single fragment at stringent annealing temperatures. These markers tag the specific alleles in the genome and are more or less similar to SCARs.

(c) Expressed sequence tag markers (EST)

This technique was introduced by Adams et al. (20). The ESTs are obtained by partial sequencing of random cDNA clones. Once generated, they are useful in cloning specific genes of interest and synteny mapping of functional genes in various related organisms. ESTs are used in full genome sequencing and mapping programs of a number of organisms and for identifying active genes, thus helping in identification of diagnostic markers. Moreover, an EST that appears to be unique helps to isolate new genes. In many crops, thousands of functional cDNA clones are being converted into EST markers.

(d) Single-strand conformation polymorphism (SSCP)

This is a powerful and rapid technique for gene analysis, particularly for the detection of point mutations and the typing of DNA polymorphisms. SSCP can identify the heterozyogosity of DNA fragments of the same molecular weight and can even detect the changes of a few nucleotide bases, as the mobility of the ssDNA changes with change in its GC content due to its conformational change. To overcome problems of reannealing and complex banding patterns, an improved technique called asymmetric-PCR SSCP was developed, wherein the denaturation step was eliminated and a large amount of the sample could be loaded for gel electrophoresis, making it a potential tool for high-throughput DNA polymorphism. It was found useful in the detection of heritable human diseases.

(2) Multi-locus probes

(a) Repetitive DNA

The discovery of repetitive DNA in genomes (30–90%) of many organisms is one of the major steps in molecular genetics. Repetitive DNAs are highly polymorphic in nature. Repetitive DNA regions contain genetic loci comprising of several hundred alleles, differing from each other with respect to length, sequence, or both, interspersed ubiquitously in tandem arrays. The repetitive DNA regions play an important role in absorbing mutations (buffering) in the genome. Of the mutations that occur in the genome, only inherited mutations play a vital role in evolution or polymorphism. Thus, repetitive DNA and mutational forces functional in nature together form the basis of a number of marker systems that are useful for various applications in plant genome analysis. The markers belonging to this class are both hybridization-based and PCR-based.

(b) Microsatellites and minisatellites

The terms microsatellites and minisatellites were introduced by Litt and Lutty (1989) and Jeffrey (1985), respectively. Both are multi-locus probes creating complex banding patterns, and are usually non-species specific occurring ubiquitously. They essentially belong to the repetitive DNA family. *Microsatellites* are a class of repeat sequences (ranging from 1–6 bp long) that are comprised of tandem repeats of short, core sequences dispersed throughout the genome. When these core microsatellite sequences are hybridized to restriction enzyme-digested genomic DNA, they simultaneously detect several hyper-variable loci. The specific oligonucleotide for simple sequence repeats or *"minisatellites"* (about 11–60 bp long) are another class of repeats that can also detect high levels of polymorphism at multiple loci. The microsatellite assay is very useful as it is sufficient to merely separate the amplification products by electrophoresis. However, the main limitation of this assay is the cost of establishing polymorphic primer sites and the investment in synthesizing the oligonucleotide. Microsatellite markers can detect higher levels of allelic variation than other markers (RFLP or RAPD).

(3) Microsatellite and minisatellite sequences converted into PCR-based markers

(a) Sequence-tagged microsatellite site markers (STMS)

This technique includes DNA polymorphism, using specific primers designated from the sequence data of a specific locus. The primers complementary to the flanking regions of the simple-sequence repeat loci yield highly polymorphic amplification products. Polymorphism appears because of variations in the number of tandem repeats (VNTR loci) in a given repeat motif. Tri- and tetranucleotide microsatellites are more popular for STMS analysis because they present a clear banding pattern after PCR and gel electrophoresis. The di- and tetranucleotide repeats are present mostly in the non-coding regions of the genome, while 57% of the trinucleotide repeats are shown to reside in or around the genes. A very good relationship between the number of alleles detected, and the total number of simple repeats within the targeted microsatellite DNA has been observed.

(b) Direct amplification of minisatellite DNA markers (DAMD-PCR)

This technique was introduced by Heath et al. (1993) and has been explored as a means of generating DNA probes which are useful for detecting polymorphism. The DAMD-PCR clones can yield individual-specific DNA fingerprinting pattern and, thus, have potential as markers for species differentiation and cultivar identification.

(c) Inter simple sequence repeat markers (ISSR)

This technique was first reported by Zietkiewicz et al. (1994). Primers based on microsatellites are used to amplify inter-SSR DNA sequences. Here, various microsatellites anchored at the 3'-end are used for amplifying genomic DNA, which increase their specificity. These are mostly dominant markers, though occasionally a few of them exhibit co-dominance. An unlimited number of primers can be synthesized for various combinations of di-, tri-, tetra- and penta-nucleotides, etc, with an anchor made up of a few bases and which can be exploited for a broad range of applications in plant species.

(4) Other repetitive DNA-type markers

(a) Transposable elements

A large number of transposable elements have been identified in many organisms, including humans and plants. However, only a few have been exploited as molecular markers. Evolutionarily, they have contributed to the genetic differences between species and individuals by playing a role in retrotransposition events promoting unequal crossing-over. Retrotransposition-mediated fingerprinting has been shown to be an efficient fingerprinting method for the detection of genetic differences between different species.

(b) Alu-repeats

The method of fingerprinting genotypes using semi-specific primers was developed to complement repetitive DNA elements, called "Alu-repeats", in human genome analysis. Alu-repeats are a class of randomly-repeated interspersed DNA, preferentially used for Alu-PCR as they reveal considerable levels of polymorphism. These representatives of short–long interspersed nuclear elements are known as SINES and LINES respectively. Alu elements are approximately 300 bp in size and suggested to be originated from special RNA species that have been reintegrated. These repeats have been studied mostly in humans, while their function in other organisms remains largely unexplored.

(c) Repeat complementary primers

Primers complementary to other repetitive sequence elements were also successfully used for the generation of polymorphisms. Some examples are introns/exons splice junctions, tRNA genes, 5sRNA genes and Zn-finger protein genes. One of the advantages of these new strategies is that they are more amenable to automation than the conventional hybridization-based procedures.

(5) Arbitrary Sequence Markers

(a) RAPD Analysis

Williams *et al.* (1990) developed random amplified polymorphic DNA (RAPD) technology, which has since been widely used for the genetic analysis of biological systems. RAPD analysis has been most useful in those cases where molecular markers, such as isozymes and RFLPs, are not available. However, these genetic markers behave as dominant traits since a single band is either amplified or not amplified at one or more loci. Different alleles at a locus are rarely amplified; the quantification of the amplified DNA band is currently not possible due to an inherent variation in the efficiency of the amplification reaction between DNA samples. The careful optimization of the reaction conditions, the utilization of positive and negative controls, and testing the reproducibility of the amplification reaction are necessary before drawing any conclusions from the results obtained.

This method uses 10-base primers (10-mer) of arbitrary sequences to amplify discrete DNA fragments, using genomic DNA as a template. The RAPD markers are generated by the amplification of random DNA segments with single primers of arbitrary nucleotide sequence. RAPD markers can be used for genetic mapping applications as well as for genetic diagnostics. The assay is non-radioactive, requires only nanogram quantities of DNA, and is applicable to a broad range of species. A detailed experimental protocol for RAPD assays and applications, emphasizing their use for genetic analysis in plants, is briefly discussed.

To perform a RAPD assay, a single oligonucleotide of an arbitrary DNA sequence is mixed with genomic DNA in the presence of a *Taq* polymerase, dNTPs, a suitable buffer. It is then subjected to temperature cycling conditions typical of the PCR. The products of the reaction depend on the sequence and length of the oligonucleotide, as well as the reaction conditions. At an appropriate annealing temperature during the thermal cycle, the single primer binds to sites on opposite strands of the genomic DNA that are within an amplifiable distance of each other, and a discrete DNA segment is produced. In practice, the DNA amplification reaction is repeated on a set of DNA samples with several different primers, under conditions that result in several amplified bands from each primer. The polymorphic bands are noted – for example, between the parents of a cross – and the polymorphisms can be mapped in a segregating population. Often, a single primer can be used to identify several polymorphisms, each of which maps to a different locus (Figure 11.10).

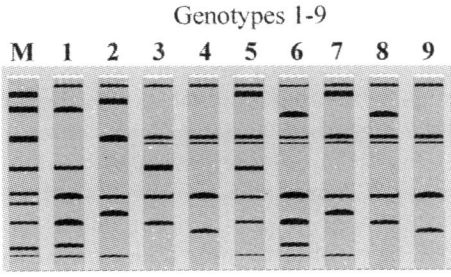

Figure 11.10 *Schematic illustration of RAPDs generated through the use of random primers in PCR*

(b) Applications of RAPD assay

(i) Genome mapping

The RAPD assay has been widely used for the genetic analysis of biological systems. When a RAPD marker is detected as a DNA segment amplified from one parent in a genetic cross but not from the other parent (Mendelian fashion), the marker can be followed in the segregating progeny and can be assigned to a locus in a genetic map. The advantages of the RAPD technique over the RFLP analysis are its ability to detect extensive polymorphism, its simplicity, speed, and the need for only small amounts of genomic DNA. The RAPD technology has greater potential for revealing minor changes in the genomes of related genotypes compared to isozymes and RFLPs. When using RAPD markers for genetic maps, certain types of populations, such as back-crosses and recombinant inbred lines, are preferred to F_2 populations. This is because the RAPD techniques tend to provide only dominant genetic markers. A given RAPD band does not indicate whether its respective locus is homozygous or heterozygous. If it is necessary to identify heterozygous regions, two closely-linked RAPD markers, each amplified from a different parent, may be used as a pair. It is also possible to quantitate the intensity of RAPD bands using standard densitometric techniques. A RAPD band derived from a heterozygous region will have half the intensity of a band derived from a homozygous region.

(ii) DNA fingerprinting, varietal identification and germplasm diversity

The ease in detecting sequence polymorphisms using RAPD markers makes them attractive for the study of individual identity and taxonomic relationships. RAPD analysis has been used for varietal identification, phylogenetic relationships, fingerprinting, germplasm evaluation and the assessment of genetic diversity in germplasm collections. Artificial variation represents a potential problem in surveys of genetic variation in natural populations, and must be distinguished from true polymorphism for the application of RAPD to be both accurate and reliable. Some properties of the RAPD fragments cannot be easily determined, such as the homology of equal molecular weight bands shared between individuals (homologous or homoplastic characters), or the origin of the fragments (nuclear or cytoplasmic). The presence of spurious bands and the continuum of band intensities create additional complications in using RAPD technology for germplasm evaluation and diversity studies. These properties of the RAPD markers can reduce their utility in determining genetic relationships within and between populations.

(iii) Gene tagging

RAPD analysis has been used to tag several important traits, such as bacterial speck resistance in tomato, nematode resistance in tomato, downy mildew resistance in lettuce, and rust resistance in the common bean. Despite the general utility of RAPDs for gene tagging, there are several potential shortcomings, which reduce their effectiveness for marker-aided selection or map-based cloning. Sequence-tagged sites (STSs), and sequence-characterized amplified regions (SCARs), have been developed in an effort to make PCR-based markers more reliable. To transform a linked RAPD into a SCAR marker, the RAPD fragment is cloned, the two ends of the fragment are sequenced, and specific pairs of primers (20–30 mers) complementary to the DNA sequence of the RAPD fragment are synthesized. The advantages of the SCARs over the RAPD markers are that they detect only a single locus, their amplification is less sensitive to reaction conditions, and they have the potential to be converted into co-dominant markers.

(iv) The use of mapped RAPD markers in different genetic crosses

A RAPD marker mapped in one cross may be non-polymorphic in another cross involving different parents. To use the same marker in the second cross, it may be necessary to find a polymorphism at the locus using another procedure. For example, if the band can be amplified from both parents, the DNA could be cleaved with restriction enzymes, either before or after amplification, to identify a restriction site polymorphism lying between the priming sites. The sequence-based differences in secondary structure could be detected on a gradient denaturing gel or by the single-stranded conformation polymorphism (SSCP) method.

(v) Population genetics

DNA markers can be used to measure similarity among individuals in natural or artificial populations within a species. To compare the relative efficiencies of RAPD vs RFLP markers, consider that a single RAPD primer will usually amplify several independent genetic loci, but can be used to identify the presence of only one allele. Only the presence or absence of a RAPD band can be detected. In contrast, the RFLP probes can usually be used to assay only one or a few loci of complementary nucleotide sequences, but can also be used to detect multiple alleles at each locus due to the variations in the sizes of the RFLP fragments.

(vi) Molecular taxonomy

Any amplification done with RAPD primers is extremely sensitive to single base changes in the primer-target site. This feature suggests that RAPDs should be highly useful for phylogenetic analysis among closely related individuals, but less useful for the analysis of genetically diverse individuals.

Method-1: A general protocol for RAPD analysis

Materials

Taq enzyme (AmpliTaq from Perkin-Elmer Cetus; Source DNA: plant genomic DNA; Oligo-nucleotide primers (randomly-generated oligonucleotides of 10 bases comprising 50–70% G+C; Amplification buffer: 10 mM Tris.Cl (pH 8.3), 50 mM KCl, 2 mM $MgCl_2$, 0.001% (w/v) gelatin; dNTPs: 100 µM each of dATP, dGTP, dCTP, and dTTP.

Procedure

1. In an Eppendorf tube add the following: 2.5 µl of amplification buffer (10x), 100 µM each of four dNTPs, 0.2 µM primer, 25 µg of genomic DNA, and 0.5 unit of AmpliTaq polymerase, and make up the volume to 25 µl with distilled water. Overlay a drop of mineral oil.
2. Perform the amplification in a thermal cycler programmed for 45 cycles (often 35 cycles are sufficient) of 1 min. at 94°, 1 min. at 36°, and 2 min. at 72°, using the fastest available transitions between each temperature.
3. Analyze the amplification products by electrophoresis in 1.4% agarose gels and detect any staining the gels with ethidium bromide.

Re-amplification of RAPD bands for use as hybridization probes

1. Excise the bands of interest visualized with a low-intensity UV light, by punching them with a glass capillary.

2. Place the agarose plugs containing the DNA in 100 µl of 10 mM Tris.Cl (pH 7.5), 0.1 mM EDTA.
3. Heat the sample at 94°C to dissolve the agarose.
4. Re-amplify 1 µl aliquots of several dilutions under standard conditions with the same primer that was originally employed to generate the band.
5. Check the samples for purity by running them on agarose gel to identify the pure samples containing only the desired band.

Documentation of results

The RAPD gels are easily photographed under the standard conditions used for ethidium bromide-stained gels. Alternatively, the gels can be imaged with an image analyzer and stored electronically. Multi-point maps and LOD scores can be calculated using the Mapmaker program.

(c) RAPD analysis using plant samples

The following RAPD protocol is written based on the procedure developed by Landry BS (1993) for plant DNA.

Method-1: DNA microextraction protocol for plant leaf samples

The DNA microextraction protocol, described below, works for many species of plants and can be successfully used for marker-assisted selection in plant-breeding programs. Crude DNA extracts from both the parental lines and the segregating populations are generally suitable for the RAPD analysis.

Solutions

Digestion buffer (100 ml): 4 g Mannitol (4%), 4 g Sorbitol (4%), 0.11 g $CaCl_2$ (10 mM), 0.21 g 2-[N-Morpholino] ethane-sulfonic acid (potassium salt), MES, (10 mM), pH 6.2. Filter-sterilize the buffer. The enzymes should be added only before digestion, depending on the volume needed in the following concentrations: Macerase 1%, Cellulysin 5%. Dissolve, filter-sterilize, and aliquot 200µl into 1.5 ml microfuge tubes.

Washing buffer (100 ml): 0.61 g Trizma base (50mM), 3.72 g $Na_2EDTA.2H_2O$ (100mM), 5.50 g Sorbitol (5.5%), pH to 8.0 with HCl.

Lysis buffer (100 ml): 0.61 g Trizma base (50 mM), 3.72 g $Na_2EDTA.2H_2O$ (100mM), pH to 8.0 with HCl.

Sodium Sarkosyl solution (10 ml): 0.5 g Na Sarkosyl (5%).

Procedure

1. Using a common paper punch, collect a small leaf disk from a young leaf or a fully-developed cotyledon and place it in a 1.5 ml microcentrifuge tube containing 200 µl of freshly-prepared enzyme digestion buffer.
2. Vacuum-infiltrate the solution into the leaf tissues for 30 min. (at 25 mm Hg). Ensure that the leaf disks are in the solution.
3. Release the vacuum from the vacuum chamber, close the tube immediately, and incubate it at room temperature in the dark, overnight (18 h).

4. Place the tube on ice, and gently tap the side of the tube to facilitate the release of protoplasts. Centrifuge it at 300 g, 4°C for 5 min. to pellet the protoplasts.
5. Remove the supernatant and discard the solution. Add 200 µl of washing buffer and resuspend the protoplasts by tapping the tube gently. Centrifuge it at 300 g, 4°C for 5 min.
6. Resuspend the protoplasts in 160 µl of lysis buffer at room temperature. Add 40 µl of sodium sarkosyl solution to lyse the protoplasts and the nuclei by inverting the tube several times.
7. Pellet the debris by centrifugation at 16,000 g, at room temperature, for 15 min.
8. Recover the supernatant; add 90 µl of ammonium acetate solution and 200 µl of isopropanol. Invert the tube several times and incubate it at room temperature for 15 min.
9. Collect the DNA pellet by centrifugation at 16,000 g, at room temperature, for 15 min. Discard the supernatant and add 500 µl of 70% ethanol. Invert the tube and centrifuge it at 16,000 g, at room temperature, for 5 min. Remove the supernatant and then dry the pellet in a vacuum.
10. Resuspend the DNA pellet in 50 µl of TE buffer plus RNase (10 µg/ml) and incubate it at 55°C for 10 min. to help resuspension of the DNA and RNA digestion. The presence of the RNA helps to precipitate a low amount of DNA; therefore, the RNA digestion can be carried out at the end of the procedure.
11. Centrifuge at it 16,000 g for 2 min. to pellet the debris and transfer the supernatant to a new tube. Store it at –20°C. This protocol yields 0.25–1.00 µg of high-molecular-weight total DNA and a large quantity of RNA as well.

Method-2: PCR reaction with RAPD primers

Solutions

PCR reaction buffer (1X) for 20 reactions: add 50µl of 10X Salt buffer, 50µl of 10X Nucleotide buffer, 5µl of 20µM stock RAPD Primer (0.2 µM), 2µl of *Taq* Polymerase (0.5 U/reaction), 352 µl of distilled water. Prepare prior to use and keep on ice.

10X salt buffer: 100 mM Tris-HCl pH 8.2, 500 mM KCl, 20 mM $MgCl_2$, 0.02% Autoclaved Gelatin. Prepare 100 ml of the 10X buffer in advance, aliquot 1 ml to each tube, and store them at 4°C.

10X nucleotide buffer (1 ml): Mix 20 µl each of 100 mM stock of each dNTPs (dATP, dTTP, dGTP, dCTP) and adjust the volume to 1 ml by adding 920 µl of distilled water.

Procedure

1. Prepare the PCR reaction buffer and distribute 23 µl into each 0.5 ml thin-walled tube (Cetus/Perkins–Elmer), on ice. The PCR reaction buffer contains all the reagents for the PCR reaction, as explained earlier (*Taq* polymerase, RAPD primer (10-mer), nucleotide (4 dNTPs), and PCR buffer) except the plant DNA template.
2. Add 2 µl (1/10 of total vol) of plant DNA. Mix it well and then spin down the content for a few seconds at medium speed in a microcentrifuge. Layer the solution with 25 µl of mineral oil, then place the tubes into the temperature cycler.
3. Thermocycler setting for 45 cycles of PCR
   One cycle as follows:
       30 s at 94°C (ramping 1.0 sec)

10 min. at 42°C (ramping 2.0 sec)
Forty-five cycles as follows:
   1 s at 50°C (ramping 5.0 sec)
   45 s at 72°C (ramping 1.0 sec)
   5 s at 94°C (ramping 1.0 sec)
   30 s at 42°C (ramping 2.0 sec)
One cycle as follows:
   5 min. at 72°C (ramping 1.0 sec)

The ramping value represents the time (in seconds) required to increase or decrease the temperature by 1°C.

(d) Electrophoretic analysis of the amplified product and selection of useful primers

Generally, half the RAPD reaction is loaded onto the gel. The amplification products were analyzed by electrophoresis in 1.4% agarose gels and detected by staining them with ethidium bromide. Good quality results can be obtained if the RAPD amplification products are analyzed by mixed gel electrophoresis: 3.0% Nusieve–agarose (3:1), for 3 h at 5 Volts/cm in the TAE buffer system and stained with ethidium bromide before UV photography. The useful marker-amplified loci are generally bright bands. They are present in one parental line and completely absent in the other parental line; faint bands do not segregate properly.

(e) Segregating analysis and genetic mapping

The populations most suitable for RAPD mapping are haploid or doubled haploid lines derived from the microspores of an F_1 plant or recombinant inbred lines derived from repeated selfing of individual plants of an F_2 population. The dominance-recessive relationship of alleles does not interfere with the segregation analysis in these populations. Backcross populations can be used, but the useful RAPD markers are those that will permit the amplification of a DNA fragment only from the recurrent parent. There is a loss of genetic information when F_2 populations are used to map RAPD markers, since heterozygotes cannot be separated from homozygotes. The polymorphic DNA bands are scored for their presence or absence.

(f) Some important variables of the RAPD assay

It is not clear why some combinations of target DNA sequences and primers are more sensitive to variation than others, and why some primers produce highly reproducible bands while others do not. RAPD analysis is more prone to artificial variation than the standard PCR procedure because of its generally low stringency conditions for amplification. To obtain reproducible DNA amplification, it is important to establish precise protocols and maintain consistent reaction conditions, such as:

1. It is very important to always begin with the standard conditions.
2. The most important variable is the concentration of genomic DNA. Because different DNA extraction procedures produce DNA of widely different purity, it may be necessary to optimize the amount of DNA used in the RAPD assay. Too much DNA may result in smears and too little DNA gives irreproducible patterns. At least 850 haploid genome equivalents per 25 µl reaction volume are required for reproducibility (<285 genome equivalents some bands will be lost). In general, at a genomic DNA concentration of 1 µg/ml, reproducible amplification is achieved with the DNA from a wide variety of organisms.

3. For reproducibility, it is important to note that both the magnesium ion concentration and the annealing temperature affect the relative intensity of the amplified bands. As the magnesium concentration increases, some DNA segments are amplified more efficiently while others are less amplified. The annealing temperature in the thermal cycler profile is typically set at 36°C in the thermal cycler; 40°C is too high to obtain a good amplification with many 10-base primers.
4. Primer concentrations between about 0.1 and 2.0 µM are optimal. A deoxynucleotide triphosphate concentration of 100 µM for each of the four bases is adequate for generating RAPDs. At lower concentrations, the intensity of the bands in the gel becomes progressively weaker. The recommended concentration of *Taq* polymerase enzyme is 20 units/ml; but it is worthwhile to optimize for the DNA of different species.
5. The primer sequence should contain at least 9 bases and 4 G+C bases.
6. Keeping the reaction mixture free of contaminating DNA is essential. The results can even vary between thermal cyclers manufactured by different companies.
7. Contamination: The problem of contaminated amplification reactions is a serious concern for both investigative and diagnostic applications of PCR. Rigorous precautions must be taken to minimize and monitor potential contamination, including good laboratory practices, storing, and using reagents in small aliquots, the use of positive displacement pipettes, physical separation of the reaction preparation from the analysis and the use of blank reactions with no template DNA as negative controls.

(6) Some variations in the RAPD Technique

(a) DNA amplification fingerprinting (DAF)

In this technique, single arbitrary primers as short as 5 bases have been employed to amplify DNA using PCR. In a spectrum of products obtained, the simple patterns are useful as genetic markers for mapping, while the more complex patterns are useful for DNA fingerprinting. The band patterns are reproducible and can be analyzed using PAGE and silver staining. DAF requires the careful optimization of parameters; however, it is extremely amenable to automation and the fluorescent tagging of primers for the early and easy determination of amplified products.

(b) Arbitrary primed polymerase chain reaction (AP-PCR)

This is a special case of RAPD, wherein discrete amplification patterns are generated by employing single primers of 10–50 bases in length in the PCR of genomic DNA (Welsh and McClelland, 1991). In the first two cycles, annealing is done under non-stringent conditions, and the final products are structurally similar to the RAPD products. This variant of RAPD is not very popular as it involves autoradiography. Recently, it has been simplified by separating the fragments on agarose gels and using ethidium bromide staining for visualization.

(7) Sequence characterized amplified regions for amplification of specific band (SCAR)

In this technique, the RAPD marker termini are sequenced and longer primers are designed (22–24 base long) for the specific amplification of a particular locus (Michelmore *et al*. 1991).

These are similar to STS markers, both in construction and application. The presence or absence of the band indicates a variation in the sequence. These have better reproducibility than the RAPDs. The SCARs are usually dominant markers, but some of them can be converted into co-dominant markers by digesting them with tetra-cutting restriction enzymes, and polymorphism can be studied either by denaturing gel electrophoresis or SSCP. Some of the advantages of SCAR are: map-based cloning, as they can be used to screen pooled genomic libraries by PCR, physical mapping, locus specificity, etc. The SCARs also allow comparative mapping or homology studies among related species, thus making it an extremely adaptable technique.

(8) Cleaved amplified polymorphic sequences (CAPs)

These polymorphic patterns are generated by the restriction enzyme digestion of PCR products. Such digests are compared for their differential migration during electrophoresis (Jarvis et al., 1994). The PCR primer for this process can be synthesized based on the sequence information available in the databank of genomic or cDNA sequence or cloned RAPD bands. These markers are co-dominant in nature.

(9) Randomly amplified microsatellite polymorphisms (RAMPO)

In this PCR-based strategy, genomic DNA is first amplified using arbitrary (RAPD) primers. The amplified products are then electrophoretically separated and the dried gel is hybridized with microsatellite oligonucleotide probes. Several advantages of oligonucleotide fingerprinting, RAPD, and microsatellite-primed PCR are thus combined, these being the speed of the assay, the high sensitivity, the high level of variability detected and the non-requirement of prior DNA sequence information. This technique has been successfully employed in the genetic fingerprinting of many agricultural crops.

(10) Amplification fragment length polymorphism (AFLP)

AFLP are DNA fragments with different nucleotide sequences, of which a large number of copies have been synthesized by PCR. AFLP can be used for DNAs of any origin or complexity and the fingerprints are produced without any prior knowledge of the sequence, using a limited set of genetic primers. Only the DNA sequence interval (several thousand nucleotides) between the sites where one or more base pairs or polynucleotide (15–35 bp) or oligonucleotide (2–10 bp) primers anneal to a DNA template is so amplified (Figure 11.11). AFLPs are highly heritable, polymorphic, apparently selectively neutral, and are nearly ubiquitous in plant tissue. AFLP is used as a new technique for DNA fingerprinting. This technique is based on selective PCR amplification of the restriction fragments from the total digest of genomic DNA. This method involves three steps: (i) restriction of the DNA and ligation of the oligonucleotide adapters, (ii) selective amplification of sets of restriction fragments, and (iii) gel analysis of the amplified fragments. AFLPs are resolved through agarose or polyacrylamide gels; the amplified DNA is generally stained with ethidium bromide or another non-specific stain for DNA. Typically, 60–100 restriction fragments are amplified and detected on denaturing polyacrylamide gels. The reproducibility of AFLP is ensured by using restriction-site-specific adapters and adapter-specific primers, with a variable number of selective nucleotides, under stringent amplification conditions.

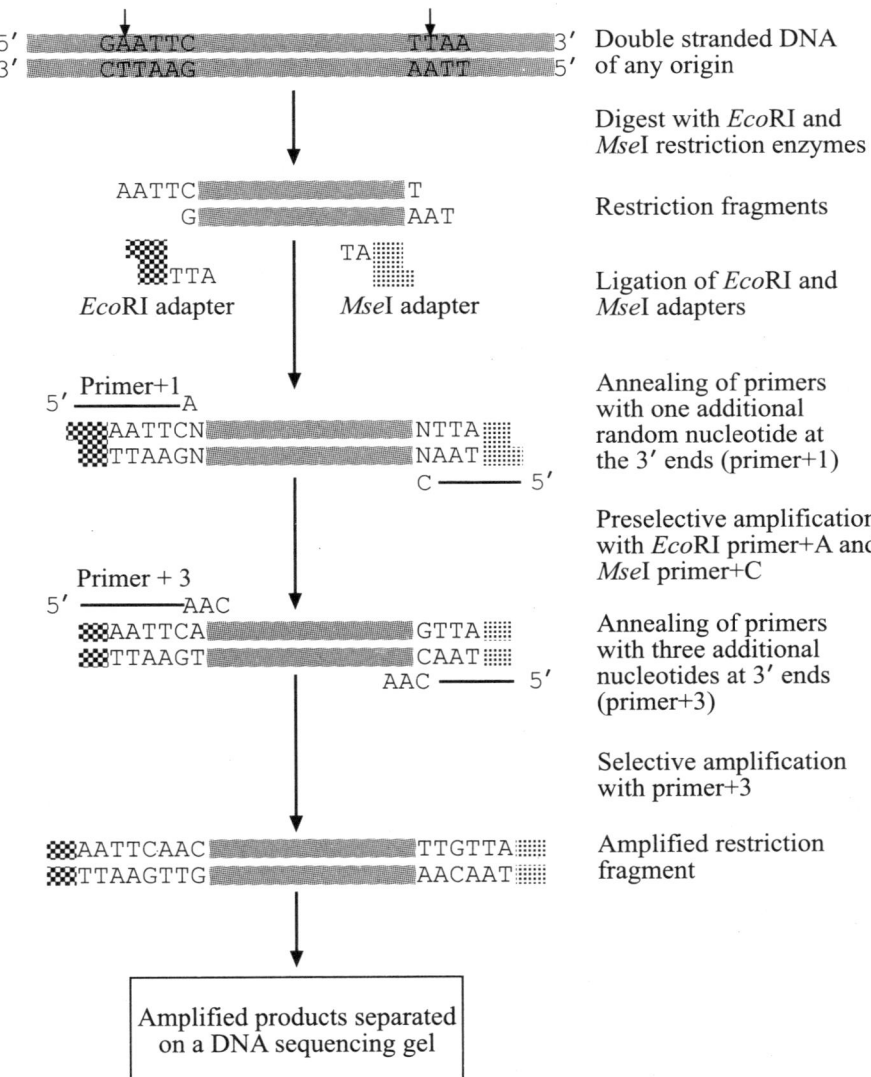

Figure 11.11 *Principle steps involved in the AFLP mapping. Arrows indicates restriction endonuclease cutting sites*

AFLP analysis depicts unique fingerprints, regardless of the origin and complexity of the genome. Since most AFLP fragments correspond to unique positions on the genome, they can thus be exploited as landmarks in genetic and physical mapping. AFLPs are extremely useful as tools for DNA fingerprinting and also for cloning and mapping of variety-specific genomic DNA sequences. Similar to RAPDs, the bands of interest obtained by AFLP can be converted into SCARs. Thus, AFLP is a versatile technique, lending itself to a variety of applications.

V PHYSICAL MAPPING OF GENOMES

In the physical mapping of genomes, the distance between genes are shown in base pairs (bp) rather than map units. The preparation of physical maps requires the cloning of many pieces of DNA and the determination of the size (in bp) of the DNA segments, the genes they contain, and their relative locations along a chromosome. Thus, the physical mapping of genomes primarily involves preparing chromosome-specific libraries either by using pulsed-field gel electrophoresis (see Chapter 9) or by automated chromosome sorting. Physical mapping at the end provides a complete sequencing of whole genomes.

VI APPLICATIONS OF MOLECULAR MARKERS IN CROP IMPROVEMENT

Molecular markers have provided useful information in understanding the genomic architecture of many species, their evolutionary relationships and in manipulating genomes to increase the efficiency of various conventional crop improvement programs. Fortunately, molecular markers offer great potential to overcome some of the limitations (availability of number of markers, environmental and pleiotropic effects) inherited by morphological and isozyme markers. Molecular maps have provided a wealth of information on the genetic architecture of crop plant genomes, recombination, and the cytogenetic-genetic distance along chromosomes. The maps offer new opportunities to apply marker-assisted selection, map-based cloning of genes, determining synteny relationship, and introgression of alien germplasm. Some of the applications are discussed below.

1. *Development of integrated saturated genetic maps*: There are a large number of molecular markers and they have provided an excellent opportunity to develop saturated genetic maps and to integrate genetic, cytological and molecular maps. Comprehensive molecular maps have now become available for many crops. For example, in rice, more than 2300 DNA markers have been mapped; similarly, 600 markers have been mapped to seven chromosomes of barley, and 1000 markers on the tomato genetic map.
2. *Gene tagging and marker assisted selection*: The process of locating genes of interest via their linkage to molecular markers is referred to as gene tagging. A molecular marker very closely linked to a gene can act as a "tag", which can be used in the indirect selection of the gene (often desirable to tag agronomically valuable traits) in breeding programs. A large number of genes for disease and insect resistance and for several other characters have been tagged with molecular markers. More effective ways of obtaining linked markers utilize near-isogenic lines (NILs). These NILs have been produced for many crop plants by repeated cycles of backcrossing and selection. The development of saturated molecular maps, the tight linkage of target genes with the molecular markers, and the conversion of these markers to PCR-based markers have made it feasible to use marker-aided selection (MAS) in plant-breeding programs.
3. *Quantitative trait loci mapping*: A number of economically important traits, such as yield, quality and tolerance to various abiotic stresses (drought, salinity, submergence, etc), are quantitative in nature. The genetic differences affecting such traits are controlled by a

relatively large number of loci, each of which can make a small positive or negative contribution to the final phenotypic value of the traits. Such loci are termed 'quantitative trait loci' (QTLs). The genes controlling such traits, called polygenes or minor genes, also follow the laws of Mendelian inheritance, but are largely influenced by their environment. It is now possible to map major QTLs on chromosomes. The number of QTLs varies depending upon the trait, the nature of parents, and the phenotypic variance. The minor effects of QTLs can also be detected by the marker method depending upon: (i) the map distance from the nearest marker to the QTL: the closer a QTL is to a marker, the less the effects of QTL can be detected; (ii) the size of the segregating population: the larger the population size, the more likely it is that the effects of lesser QTL will reach statistical significance; (iii) heritability: the lower the heritability, the less likely it is that a QTL will be detected.

4. *Comparative mapping and synteny relationship*: Molecular markers have revealed that many distantly related species have co-linear maps for a portion of their genomes. These markers have provided a new and unexpected wealth of information on the *synteny* relationships among plant genomes. This means the use of a common set of DNA probes to detect and map homologous sequences across a sexually high degree of conservator in terms of copy number and the homology of low-copy probes, the linkage and the locus order. Comparative mapping has shown that a locus for liguleless has been conserved among rice, wheat and maize. Similarly, parallel linkage has been observed between genes for resistance to leaf rust and prolamines in wheat, oats and maize. Based on comparative mapping, the species with a smaller genome size (such as rice *vs* maize and wheat) could be used to accelerate the positional cloning of orthologous genes. Once the gene in the source species has been cloned and sequenced, this information may be used to quickly isolate the orthologous gene in the target species.

5. *Chromosome walking and map-based cloning of genes*: The genetic maps saturated with molecular markers provide opportunities for the isolation of plant genes, based on the map position of a phenotype. The possibilities of cloning plant genes based upon map position has been greatly enhanced by the continuing development of yeast artificial chromosome (YAC) and bacterial artificial chromosome (BAC) libraries and accelerated DNA sequencing. Map-based cloning consists of four major steps: (i) the development of high-resolution molecular linkage maps in the region of interest, (ii) the determination of the physical distance that needs to be covered in a chromosome walk, (iii) chromosome walking using various libraries containing large segments of the genome, and (iv) verification that the target gene has been isolated by using transformation.

6. *DNA fingerprinting of crop germplasm*: In the past, the genetic composition of the germplasm of crops and their wild relatives have been based on reproductive biology, morphological traits, pedigree records, breeding behavior, *in situ* and *ex situ* conservation of agricultural traits, chromosome structure, and behavior and protein markers. The DNA data revealed by the RFLPs are being routinely used for taxonomic, genetic and phylogenetic information. The variation in DNA sequences has made molecular markers a valuable descriptor. The repetitive and arbitrary DNA markers are used in the genotyping of cultivars. Microsatellite lines – $(CT)_{10}$, $(GAA)_5$, $(AAGG)_4$, $(AAT)_6$, $(GATA)_4$, $(CAC)_5$ – and minisatellites have been employed in DNA fingerprinting for detecting genetic variation, identifying cultivars, and genotyping. Fingerprinting is important in plant variety protection, registration, certification and patents. DNA markers could improve the management of plant genetic resources, particularly to assess genetic diversity in the collections. DNA markers are much

more powerful tools to determine genetic diversity than protein and morphological markers. In future, DNA-based descriptors will be valuable tests for providing information on genetic identity. Recently, using these markers, the sex of dioeceous plants has been detected. The microsatellite probe $(GATA)_4$ is found to reveal sex-specific differences in Southern analysis, and can be used as a diagnostic marker in this system where male and female plants are undistinguishable until they flower.

7. *DNA fingerprinting of pathogen populations*: In the past, pathogen populations have been characterized primarily by virulence analysis on varieties carrying different genes. It is difficult to precisely characterize diversity and phylogenetic relationships based on virulence analysis. Moreover, pathotyping is laborious, time consuming, and the disease reaction is influenced by the environment. The differential hosts may possess more than one resistance gene or unknown resistance factor. Molecular markers offer new opportunities to describe genetic variation and understand the evolution of pathogen populations. Lineage analysis has been used to determine the relationship among various pathotypes of blast pathogen.

8. *Rapid and precise identification of pest populations using molecular probes*: The availability of nucleic acid probes has made it possible to screen large samples of germplasm for the diagnosis of disease-causing organisms attacking crop plants. The nucleic acid spot hybridization method is routinely used to detect the presence of viruses in the tomato and potato germplasm. This technique has been used to detect the tomato leaf-curl virus in tomato leaves and also in the viruliferous whiteflies which vector the virus.

9. *Predicting heterosis based on molecular diversity*: The prediction of hybrid performance is important for hybrid crop-breeding programs. The availability of molecular maps offer an opportunity to predict hybrid performance and thus economize the time, labor and efforts needed in field screening. In crop plants, one of the classical examples of the several useful traits that are maternally inherited is cytoplasmic male sterility (CMS). It is now well established that the genes for CMS are located on the mitochondrial genome. Conventional methods require extensive backcrossing to develop isogenic lines and, later, identifying CMS sources based on test crosses with known restorer lines. The identification of CMS lines, based on molecular probes, could enhance the efficiency of breeding hybrid crop varieties. Kiang et al. (1993) identified and cloned a 4.5 kb DNA fragment from mtDNA of the CMS line of *Lolium perenne* which did not hybridize to sequences in the mtDNA of fertile lines. This technique could be used as a marker to distinguish between fertile and sterile lines.

10. *Understanding alien introgression and genome differentiation*: Wide hybridization, involving the transfer of useful genes from wild to cultivated species, has been one of the important plant breeding methods to widen the gene pool of crop plants. The limited recombination between the chromosome of cultivated and wild species has been the major barrier deterring the transfer of alien genes. Monitoring alien introgression by using morphological and isozyme markers is not efficient, particularly when the two genomes are highly divergent. Fortunately, molecular markers in combination with fluorescence *in situ* hybridization techniques have enabled the precise detection of the introgression of alien chromosome segments and the characterization of sexual hybrids, somatic hybrids, cybrids, and transgenic plants.

VII FURTHER READING

Ausubel FM, Brent R, Kingston RE, Moore DD, Seidman JG, Smith JA, and Struhl K (1997) *Short Protocols in Molecular Biology* (4th Edn), Wiley and Sons, Inc.

Burke J. F (Ed) (1996) *PCR: Essential Techniques*, John Wiley and Sons, Chichester.

Fickett JW (1996) Finding genes by computer—The state-of-the-art, *Trends Genet.*, **12**:316–320.

Landry BS (1993) DNA mapping in plants (pp269–285) In: *Methods in Plant Molecular Biology and Biotechnology*, CRC Press Inc.

Hatada I, Hayashizaki Y, Hirotsune S, Komatsubara H and Mukai T (1991) *PNAS, USA,* **88**:9523–9527.

Heath DD, Iwama GK and Devlin RH (1993) *Nucleic Acids Res.*, **21**:5782–5785.

Joshi SP, Ranjekar PK and Gupta VS (1999) Molecular markers in plant genome analysis, *Current Science*, **77**:230–240.

Litt M and Luty JA (1989) *Am. J. Hum. Genet.*, **44**:397–401.

Michelmore RW, Paran I and Kesseli RV (1991) *PNAS, USA,* **88**:9828–9832.

Primrose SB (1995) *Principles of Genome Analysis*, Blackwell Science, Oxford.

Sambrook J, Fritsch EF and Maniatias T (1989) *Molecular Cloning: A Laboratory Manual* (2nd Edn) Cold Spring Harbor Laboratory Press, Cold Spring Harbor, NY.

Singer M and Berg P (1991) *Genes and Genomes*, Blackwell Science, Oxford.

Singh BD (1983) *Plant Breeding: Principles and Methods*, Kalyani Publishers.

Stuber CW (1995) Mapping and manipulating quantitative traits in maize, *Trends Gent.*, **11**:477–481.

Welsh J and McClelland M (1991) *Nucleic Acids Res.*, **19**:861–866.

Williams JGK, Kubelik AR, Livak KJ, Rafalski JA and Tingey SV (1990) DNA polymorphisms amplified by arbitrary primers are useful as genetic markers, *Nucl. Acids Res.*, **18**:6531.

Williams JGK, MK Hanafey, JA Rafalski, and SV Tingey (1993) Genetic analysis using random amplified polymorphic DNA markers, *Methods in Enzymology*, **218**:704–141.

Yi M, Ichikawa N and P.O.P Ts'O (1993) Construction of restriction fragment maps of 50- to 100-kilobase DNA, *Methods in Enzymology*, **218**:651–671.

Zabeau M (1993) *European Patent Application*, No. 0534858AI.

Zietkiewiez E, Rafalski A and Labuda D (1994) *Genomics,* **20**:176–183.

12

Polymerase Chain Reaction (PCR)

I INTRODUCTION

One of the most important developments in applied molecular genetics was the invention of an amazingly simple DNA amplification strategy called the polymerase chain reaction (PCR) developed in April 1983 by Kary Mullis. He was awarded the Nobel Prize in Chemistry in 1993 for this invention. PCR has become a core technique for molecular biologists in an astonishingly short time. It is a technique that mimics nature's way of replicating DNA. The PCR reaction is a rapid procedure for the *in vitro* enzymatic amplification of a specific segment of DNA and is capable of increasing the amount of a target DNA sequence in a sample by synthesizing many copies of the DNA segment. PCR is carried out in discrete cycles and each cycle of amplification can double the amount of target DNA. The target DNA is exponentially amplified, such that after n-cycles there is 2^n times as much target as was present initially. PCR has a tremendous variety of applications, and has been adopted as an essential research tool because it can take a minute sample of genetic material and duplicate enough of it for a study.

II POTENTIALITY OF PCR

Specificity of the PCR reaction: *Amplification of unique sequences*: Since all DNA polymerases require primers to initiate strand synthesis, the target for PCR amplification can be specified by designing primers to anneal to particular unique DNA sequences. The most efficient primers for specific amplification reactions are 17–30 nucleotides in length, as this represents a sequence unlikely to be repeated by chance in the unique sequence DNA of higher eukaryotes. Some loss of specificity may occur if the primers anneal at lower temperatures, as mismatches are tolerated and then become stabilized on the template by primer extension. Specificity can also be increased by the use of nested primers, where products from a particular amplification are subsequently amplified with a second set of primers which flank the same target site but are internal to the original primers. Any spurious products amplified in the first reaction by mispairing are unlikely to also possess the correct sites for the internal primers, so only the genuine products will be amplified in the second reaction (*nested PCR*). Unique sequence amplifications have many applications, one of them being as a diagnostic technique to detect polymorphisms and unknown mutations.

Amplification of related sequences: Not all PCR applications aim to generate specific products. It is also possible for the PCR to identify families of related sequences or to amplify the DNA from one species based on sequence information from another. In such cases, specific primers may be designated around a highly-conserved domain, allowing products with differing internal sequences to be identified. Another approach is to use degenerate oligonucleotide primers (DOP-PCR), a mixture of primers with alternative nucleotides at certain positions. This strategy, known as homology screening, involves the design of primers around a conserved domain but incorporating all the known sequence variations in the family. Another use of degenerate primers is to amplify a specific target sequence corresponding to a known protein, when only the polypeptide sequence of the protein is known. In this case, the use of DOP-PCR reflects the degeneracy of the genetic code.

Amplification of unrelated sequences: A major application of PCR is the identification of microorganisms in, for example, infected tissue of contaminated food at restaurants. PCR can be used as a simple diagnostic test, allowing the unambiguous identification of pathogenic or non-pathogenic microorganisms based on their nucleotide sequence (Figure 12.1). An added advantage is that many species can be assayed at one time by using multiple sets of primers in the same reaction (multiplex PCR), each of which generates a specific diagnostic-sized product.

Amplification of arbitrary sequences: In certain cases, the aim of the PCR is not to generate one or more specific product or related products, but to produce a collection of purely arbitrary sequences which are used as the basis of further analysis. The two major applications of this strategy are cDNA cloning and gene mapping. The amplification is achieved using arbitrary primers, i.e., short primers (9–15 nt), which anneal to many sites and amplify a random collection of unrelated products. The arbitrary amplification of cDNA produces a library of partial cDNA fragments, or ESTs, which can be used in the same way as sequence-tagged sites for genome mapping. ESTs can be used to characterize differentially expressed genes and also to produce markers for genetic and physical mapping. RAPDs are randomly amplified polymorphic DNA markers used extensively to map plant genomes (refer to Chapter 11 for more details).

PCR mutagenesis: Primer design can be used not only to amplify sequences but also to alter them. By this method, point mutations can be introduced *in vitro* and in the end-modification of PCR products to facilitate further manipulation. In PCR mutagenesis, the primers are designed to mismatch with their target, and the conditions are chosen so that annealing is still permitted. Amplification thus introduces the mutation into the amplified product. For generating PCR products with unique sequences at the ends, novel sequences are added to the 5' primer ends, which do not pair with the template DNA, but allow extra sequences to be added on to each end of the PCR product. This has an advantage of making the addition of restriction sites, universal primer binding sites for sequencing, and bacteriophage promoters to facilitate *in vitro* transcription possible.

Downstream applications: In addition to the above applications, PCR products are required in large quantities for numerous downstream applications. PCR products are inserted into cloning vectors in different ways, such as Blunt-end cloning, T-vectors, linker primers, and dinucleotide/trinucleotide sticky-end cloning (DISEC/TRISEC).

The applications of PCR in research

PCR can be applied to a diverse array of both basic and applied problems (Table 12.1). The protocols for these methods are available in several books (Inns et al. 1990, Erlich et al. 1989,

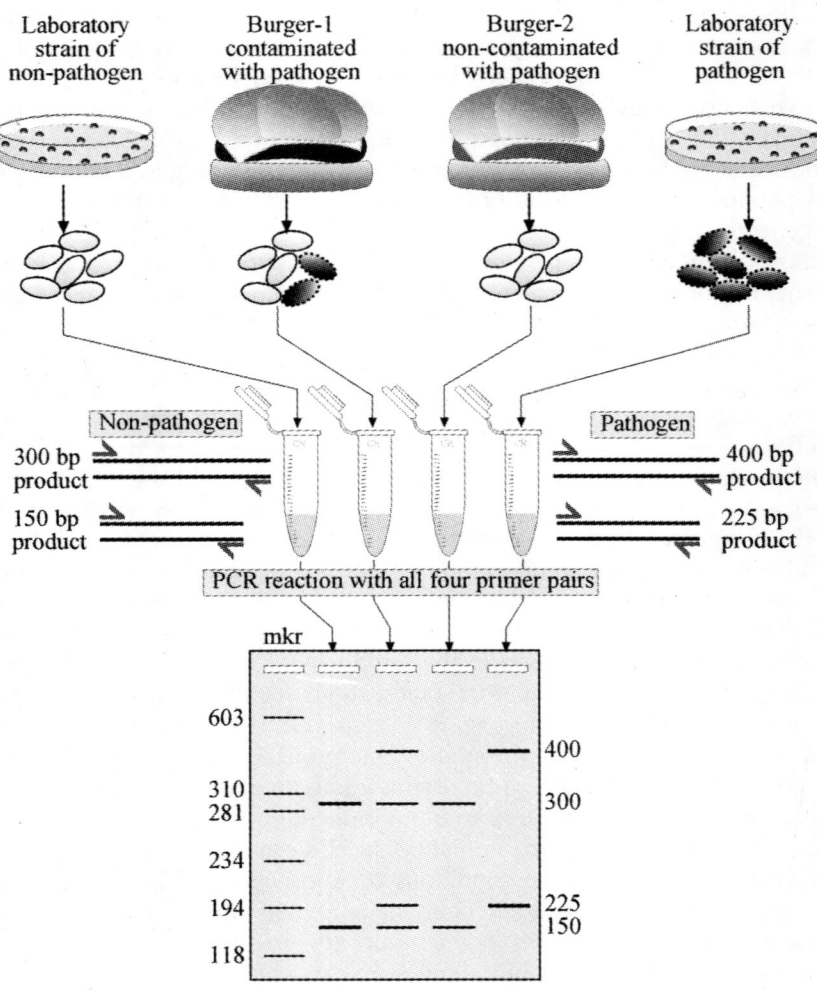

Figure 12.1 *Multiplex PCR can be used simultaneously to screen for pathogenic strains of E.coli using strain-specific primer pairs*

Ausubel *et al.* 1991) as well as many individual papers in journals. In addition, many different PCR applications are described in the journals of *BioTechniques* and *PCR Methods and Applications*.

Table 12.1 *Problems and potential of PCR applications*

Problems and applications	PCR technique(s)
Amplify ancient DNA, fossil samples	Standard PCR
Amplify mRNA	RNA PCR
Identification of sequences *in situ*	*In situ*-PCR
Chromosome walking	Inverse PCR; SSP-PCR
Cloning a gene	Blunt-end cloning; sticky-ended cloning, anchored PCR, PCR with degenerate primers, SSP-PCR
Constructing a genetic map	AP-PCR, RAPD, Inverse PCR
Constructing a phylogeny	Standard PCR with primers having polylinkers for cloning and sequencing, asymmetric PCR and sequencing, PCR-RFLP
Detecting gene expression and quantitation	RNA-PCR
Detecting mutations, point mutations, deletions, additions, inversions, translocations, etc.	Standard PCR, RAPD, AP-PCR, PCR-RFLP
Detecting pathogens *in situ*	Standard PCR
Detecting integrated DNA in transgenic organisms	Standard PCR, ligation-mediated PCR, inverse PCR, anchored PCR
Engineering DNA	
Introduction of restriction sites into DNA fragments	Attach sequences to the 5' end of primers and conduct the standard PCR
Introduction of deletion or addition mutations	Standard PCR
Label DNA with radioisotopes or non-radioisotopes for sequencing, probes or isolating DNA strands onto a column	Standard PCR
Assemble overlapping DNA segments to make synthetic DNA	Standard PCR
Introduce substitutions, deletions, or insertions in product DNA	Alter primer sequence when synthesizing, then standard PCR
Evolutionary analyses	Standard PCR, RAPD, AP-PCR, DNA sequencing, PCR-RFLP
Identifying strains, races, or biotypes	Standard PCR, RAPD, AP-PCR, PCR-RFLP?
Identifying upstream and downstream sequences	Inverse PCR, SSP-PCR
Monitoring environmental dispersal of individuals	Standard PCR, RAPD, AP-PCR
Sequencing a gene	Asymmetric PCR to produce ssDNA, Dideoxynucleotide chain-termination sequencing method with *Taq* polymerase
Forensic investigations	PCR, PCR-RFLP

Data in the table is collected from various sources.

III THE BASIC PRINCIPLE

The principle of the PCR technique is that of an exponential increase in the amount of template (target) DNA. A target DNA sequence to be amplified is chosen first. It is not necessary to know the entire nucleotide sequence of the target DNA, but two small stretches of DNA of known sequence that flank this target segment must be found and used to design two synthetic oligonucleotide primers. These two oligonucleotides are used as primers for a series of reactions that are catalyzed by a DNA polymerase. These oligonucleotides typically have different sequences and are complementary to the sequences that lie on opposite strands of the template DNA and flank the segment of DNA that has to be amplified. The template DNA is first denatured by heating (94°C) in the presence of a large molar excess of each of the two oligonucleotides and the four dNTPs. The reaction mixtures is then cooled to a temperature (55°C) that allows the oligonucleotide primers to anneal to their target sequences, after which the annealed primers are extended (72°C) with DNA polymerase. The cycle of denaturation, annealing, and DNA synthesis is then repeated many times (Figure 12.2). Because the products of one round of amplification serve as templates for the next, each successive cycle essentially doubles the amount of the desired DNA product. It is this fact that accounts for the exponential increase in the PCR product with the cycle number.

After the first few cycles, the major product is fragments of double-stranded DNA, whose termini is defined by the 5' termini of the oligonucleotide primers and whose length is defined by the distance between the primers. If the amount of the target exactly doubles with each cycle, as few as 20 cycles will generate about a million times more target sequence than was present initially. In addition, longer DNA molecules are generated during the reaction. The products of a successful first round of amplification are heterogeneously-sized DNA molecules, whose lengths may exceed the distance between the binding sites of the two primers. In the second round, these molecules generate DNAs of defined length that will accumulate in an exponential fashion in later rounds of amplification and will form the dominant products of the reaction (Figure 12.3).

IV STANDARD PCR PROTOCOL AND OPTIMIZATION OF AMPLIFICATION

Although the PCR reaction is relatively simple, the biochemical and kinetic parameters that affect PCR specificity and sensitivity are actually quite complex and it can sometimes be troublesome to define a reaction condition that generates reproducible results. The following PCR protocols are basic protocols which work with a variety of DNAs.

A. COMPONENTS OF PCR

The first step is to get the initial PCR(s) to work. The rule about which parameters affect PCR is simple: *everything*! However, some of the most important parameters that effect PCR amplification are described below.

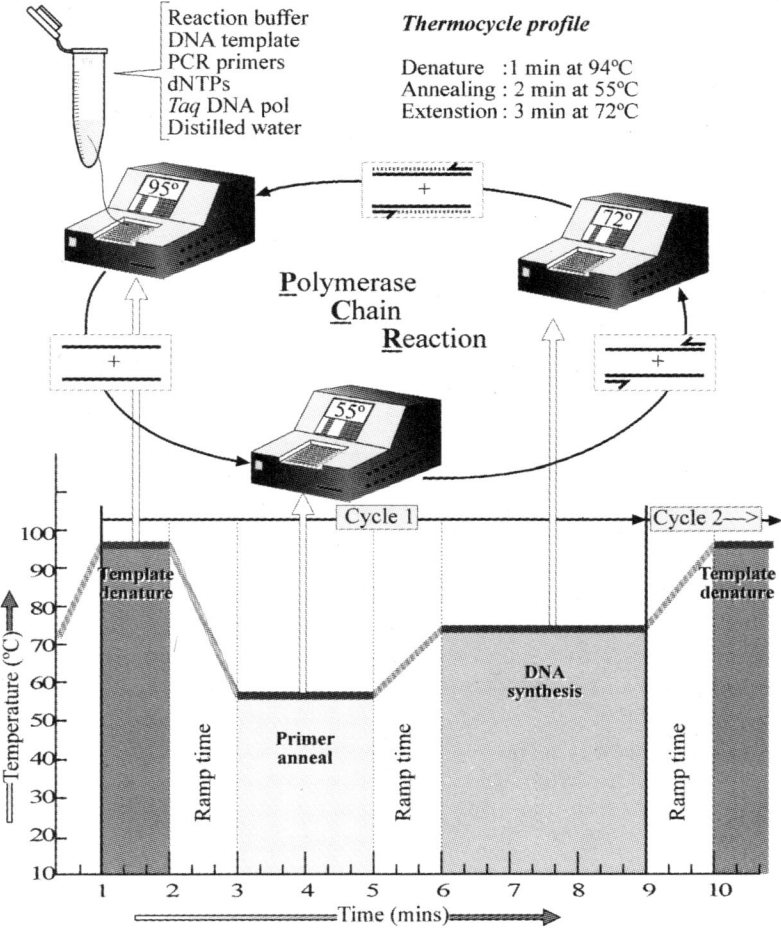

Figure 12.2 *The temperature profile of a PCR cycle is controlled by the thermal cycler program. Each PCR cycle requires three temperature steps to complete one round of DNA synthesis*

1. Oligonucleotide primers

The specificity and amplification of the PCR depends critically upon the **_primers_**. Primer design is an art. While designing the primers, the quality of primers depends on the length, the nucleotide composition, and the primer–template match. The approach to the selection of efficient and specific primers remains somewhat empirical; there are no discrete rules that will guarantee the choice of an effective primer pair. However, the majority of primers can be made to work, and the following information will help in their design.

The oligonucleotides used for priming the PCR should be at least 13 nucleotides long and can be as long as 80 nucleotides. In most cases, 18–24 nucleotides length is preferred. The longer the primer, the higher the annealing temperature can be, and the greater the specificity. However, long primers need purification, otherwise primer dimerization and non-specific am-

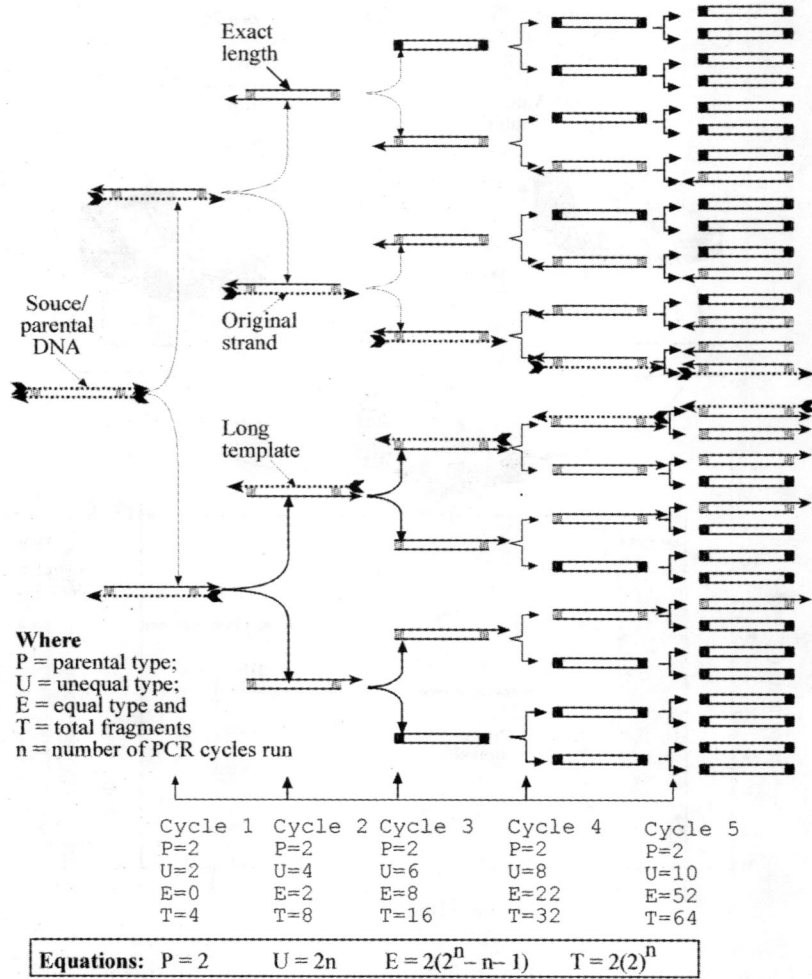

Figure 12.3 *Polymerase chain reaction: different types of amplification products generated at the end of "n" number cycles*

plification may occur. The ideal primer has roughly equal numbers of each nucleotide, without internal repeats or self-homology, and a G+C content of about 45–55%. The optimal distance between the primers is very application specific, but for most diagnostic PCR assays, it is best when the opposing primers are spaced 150–500 bp apart. Usually, oligonucleotides are used at concentrations between 0.1–1.0 µM in PCR. This is usually sufficient for at least 30 amplification cycles.

2. Buffers used for PCR

The standard reaction buffer for PCR contains 50 mM KCl, 10 mM Tris.HCl (pH 8.3) and 1.5 mM MgCl. When incubated at 72°C, the pH of the reaction drops by more than a full unit, producing a buffer whose pH is approximately 7.2. The presence of divalent cations is critical. Mg^{2+} ions are superior to manganese, and calcium ions are ineffective. Because the optimal concentration of Mg^{2+} is quite low (1.5 mM), it is important that the preparation of template DNA does not contain high concentrations of chelating agents, such as EDTA, or of negatively-charged ionic groups, such as phosphates. Therefore, the DNAs to be used as templates should be dissolved in 10 mM Tris.HCl (pH 7.6), 0.1 mM EDTA (pH 8.0).

The standard buffer works well for a wide range of templates and oligonucleotide primers, but it may not be optimal for any particular combination. Thus, the conditions given should be regarded as a point of departure to explore modifications and potential improvements. In particular, the concentration of Mg^{2+} should be optimized whenever a new combination of target and primers is first used or when the concentration of dNTPs or primers is altered. The dNTPs are the major source of phosphate groups in the reaction, and any change in their concentration affects the concentration of the available Mg^{2+}.

3. Choice of polymerase enzymes

The enzymatic basis for PCR amplification is a DNA polymerization reaction that extends the annealed primers in the standard 5′–3′ direction. The original PCR protocol used the Klenow fragment of *E. coli* DNA polymerase I to perform the primer extension reaction; however, this meant that fresh enzymes had to be added after each round of denaturation because this enzyme is easily heat-inactivated. A related problem with using an *E. coli* DNA polymerase is that the optimal activity level of the enzyme is 37°C, which greatly limits the specificity of the reaction owing to degenerate primer annealing at this low temperature. Both of these problems can be solved by switching to a DNA polymerase isolated from a thermophilic bacterium. The first commercially available thermostable DNA polymerase for PCR came from the thermophilic eubacterium, *Thermus aquaticus*. Two forms of *Taq* DNA polymerases are now available: the native enzyme purified from *Thermus aquaticus* and a genetically-engineered form of the enzyme synthesized in *E. coli* (AmpliTaq™). Both forms of the polymerase carry a 5′→3′ polymerization activity, but they lack a 3′→5′ exonuclease activity (Table 12.2).

Table 12.2 *Thermostable DNA polymerases commonly used in PCR*

DNA polymerase	Source	Exonuclease activity
Pfu	*Pyrococcus furiosus*	3′→5′ (proofreading)
Pfu (exo-)	*Pyrococcus furiosus*	No
Psp	*Pyrococcus furiosus* Sp.GB-D	3′→5′ (proofreading)
Psp (exo-)	*Pyrococcus furiosus* Sp.GB-D	No
Pwo	*Pyrococcus* Sp.	3′→5′ (proofreading)
Taq (native and/or recombinant)	*Thermus aquaticus*	5′→3′
Taq, N-terminal deletion	*Thermus aquaticus*	No
Tbr	*Thermus brocianus*	5′→3′
Tfl	*Thermus flavus*	–
Tli	*Thermus litoralis*	3′→5′ (proofreading)
Tli (exo-)	*Thermus litoralis*	No
Tma	*Thermus maritima*	3′→5′ (proofreading)
Tth	*Thermus thermophilus*	5′→3′

4. Deoxyribonucleotide triphosphates

dNTPs (including dITP, 7-deaza-dGTP and ddNTPs) are purchased or prepared at saturating concentrations (200 µM for each dNTP) in 0.1 mM EDTA. A stock solution of dNTPs (50 mM) should be adjusted to pH 7.0 with 1 N NaOH, to ensure that the pH of the final reaction does not fall below 7.1. The actual concentrations should be determined spectroscopically, aliquoted into 0.5 ml lots in sterile tubes and can be stored at –20°C up to 1 year. The four dNTPs should be used at equivalent concentrations to minimize misincorporation errors.

5. Source DNA or target sequences

One of the most appealing features of PCR is that the quantity and quality of the DNA sample to be subjected to amplification need not be high. A single cell, or crude lysates prepared by simply boiling cells in water, or specimens with an average molecular length of only a few hundred base pairs are usually adequate for successful amplification. The DNA containing the target sequences can be added to the PCR mixture in single- or double-stranded form. Although the size of the DNA is generally not a critical factor, the amplification is improved if very high-molecular-weight DNAs (e.g., genomic DNAs) are digested with a rarely cutting restriction enzyme (e.g., *Sal*I of *Not*I). The concentration of target sequences in the template DNA obviously varies according to the circumstances and is often not under the control of the experimenter. However, it is worthwhile setting up a series of control reactions that contain decreasing amounts of known target sequences (1.0, 0.1, 0.001 µg, etc) to check that the amplification reaction is functioning at the required sensitivity.

6. Amplification reaction and specificity

There are two important fundamental aspects of the PCR amplification: one is the enormous amplification achieved, and the other is the specificity of the PCR. The PCR amplification procedure tolerates many variations from the ideal reaction conditions. In particular, a degree of mispriming will be tolerated, as long as the 3'-endmost two to three bases do not contain a mismatch. The target sequence is defined by the flanking primers and the specificity derives from the specific hybridization of the primers under the annealing conditions set for the thermal cycle. The fact is that the length of the target sequence is limited to less than a few kilobases. This is especially likely when degenerate primers are employed. However, an unintended point of hybridization will often be irrelevant because it is not likely that a second primer molecule will hybridize near enough for amplification to occur.

7. Cycling parameters

PCR is performed by incubating the samples at three different temperatures corresponding to the three steps (denaturation, annealing, and extension) in a cycle of amplification (Figure 12.2). This cycling can be accomplished either manually with preset water baths or automatically with thermal cyclers available from several manufacturers. In a typical reaction, the dsDNA is denatured by briefly heating the sample to 90–95°C. The primers are allowed to anneal to their complementary sequences by briefly cooling them to 40–60°C, followed by heating to 70°C to extend the annealed primers with the *Taq* polymerase. It is often helpful to precede the first cycle with an initial denaturation step for 3 min. at 93°C. The specificity can be improved by adding the *Taq* polymerase at an elevated temperature, rather than prior to the first denaturation step. The time of incubation at 70–75°C varies according to the length of target being amplified. The optimum number of cycles will depend mainly upon the initial concentration of the target DNA when the other parameters are optimized. Too many cycles can increase the amount and complexity of non-specific background products. Of course, too few cycles give a low product yield.

8. Setting of a standard amplification reaction

Procedure

1. Set up the general PCR reaction in a 0.5 ml microfuge tube as described below, as in Table 12.3. Mix well and centrifuge it.
   10X PCR buffer: 600 mM Tris-HCl (pH 9.5, room temperature), 150 mM $(NH_4)_2SO_4$ and 20 mM $MgCl_2$. Overlay with mineral oil (100 µl).
   Note: The above given standard *conditions work for a wide range of templates and oligonucleotide primers, but they may* not be optimal for any particular combination.
2. Heat the reaction mixture for 5 min. at 94°C, to denature the DNA completely.
3. While the reaction mixture is still at 94°C, add 0.5 µl of *Taq* DNA polymerase (5 units/µl). *Taq* DNA polymerase is supplied in a storage buffer containing 50% glycerol. This solution is very viscous and is difficult to pipette with accuracy. The best method is to centrifuge the tube containing the enzyme at 12,000g for 10 s at 4°C in a microfuge, and then to withdraw the required amount of enzyme.

Table 12.3 *PCR reagents needed for a 100 µl of reaction mixture*

Reagent	Volume	Final concentration
10x PCR buffers	10 µl	1X
10 mM dNTPs mix	2 µl	Each dNTP 0.2–1.25 mM
Primer 1	Variable	0.1–1.0 µM
Primer 2	Variable	0.1–1.0 µM
Template DNA	Variable	0.1–1.0 µg/100 µl
Taq DNA polymerase	0.5 µl	2.5 units/100 µl
DMSO[a]	Variable	5%
Glycerol[a]	Variable	10%
PMPE[a]	Variable	1%
Sterile water	Variable	–
Final volume	100 µl	–

[a] Substitute with other enhancer agents as available

4. Overlay the reaction mixture with 100 µl of light mineral oil. This prevents any evaporation of the sample during repeated cycles of heating and cooling. In the new PCR machines this problem is overcome by maintaining uniform temperature on the microfuge tops.
5. Carry out the amplification as described below. The typical conditions for denaturation, annealing, and polymerization are given in Table 12.4.

Transfer the samples to –20°C for storage. If the PCR is run overnight, then set the thermocycler to 4°C at the end of the last cycle, so that the next day the early morning samples can be transferred to –20°C for storage.

Table 12.4 *Cycling procedure*

File #	Cycle	Temperature (°C)	Time	No. of Cycles
First cycle:				1
	Denaturation	94	5 min	
	Anneal	50	2	
	Elongate	72	3	
Subsequent cycles:				30–35
	Denature	94	1 s	
	Anneal	50–65	2 min	
	Elongate	72	3 min	
Last cycle:				1
	Denature	94	1	
	Anneal	50	2	
	Final elongation	72	10 min	

6. Carefully withdraw a sample of the amplified DNA from the reaction mixture from the bottom of the tube and analyze it by gel electrophoresis, Southern hybridization, or DNA sequencing. If necessary, the oil can be removed from the sample by extraction with 150 µl of chloroform. The aqueous phase, which contains the amplified DNA, forms a micelle near the meniscus. This micelle can be transferred to a fresh tube with an automatic micropipettor. However, in new machines, the adding of oil is avoided by uniformly heating the tubes at the top also, so there will not be any condensation of water under the caps.

Precautions

1. The timing of individual steps should begin only after the reaction mixtures have reached the required temperatures. Between 30–60 s are required for the reaction mixture to reach the desired temperature after the temperature shift. It is important to adjust the length of the incubation steps to compensate for these reactions.
2. The annealing temperature (50°C) chosen here is a compromise. The amplification is more efficient if the annealing is carried out at lower temperatures (37°C), but the amount of mispairing is very high. At higher temperatures (55°C), the specificity of the amplification reaction is increased, but its overall yield will be reduced.
3. The longer the distance between the primers, the longer the time required for the complete synthesis of the entire target sequence. The time given above is optimized for a target sequence of about 500 bases long.
4. The number of amplification cycles depends on the concentration of target DNA in the reaction mixture. At least 25 cycles are required to amplify single-copy target sequences in mammalian genomic DNA, just enough to detect on agarose or polyacrylamide gels. Also, the amount of *Taq* polymerase usually become limiting after 25–30 amplification cycles.

B. OPTIMIZATION OF PCR REACTION

In order to get the most out of the PCR, optimization is always necessary. PCR reactions are not 100% efficient, even when using cloned DNA and primers of defined sequence. Usually, the reaction conditions must be varied to improve the efficiency. If the reaction is not optimal, the PCR often generates a smear of products on a gel rather than a defined band. The usual parameters to be varied include the annealing temperature and Mg^{2+} concentration; too low annealing temperature favor mispairing. Contamination is another aspect to be controlled.

1. Enhancing PCR sensitivity

The method for optimization uses a logical system of changes in pH, magnesium and potassium chloride. The change in pH can alter the activity of the DNA polymerase and the ability of the DNA strands to denature; the magnesium concentration affects the specificity of the DNA polymerase reaction; while the KCl affects enzyme activity and strand denaturation. The optimal magnesium concentration varies with each new sequence, but is usually between 1–4 mM.

2. Hot start PCR

The 'hot start' protocol featured here uses wax-embedded components dried in trehalose, which stabilizes biological molecules when dried. In this case, the PCR primers are mixed with the trehalose solution, spotted onto polypropylene thread and then coated in wax. These beads can be stored and added to a 15 µl PCR reaction mix. As the temperature increases the wax will melt and the primers will be released. These will anneal at 55°C and then be extended at 72°C, thus ensuring the specificity of the procedure.

3. Nested PCR

If mispriming occurs in a PCR due to sequence similarities between the target and related DNA, an aliquot of the products can be re-amplified with a second set of primers which have sequences 3' to the original set, i.e., nested (Figure 12.4a). Nested PCR enhances the PCR specificity, but re-amplification with a new set of primers internal to the fragment amplified by the first pair generates a shorter PCR product. If on the first round of PCR some non-specific products have been produced, giving a smear or a number of bands, using nested PCR should ensure that only the desired product is amplified from this mixture as it should be the only sequence present containing both sets of primer-binding sites. A further modification of this method has been called 'drop in drop out' PCR, in which the two sets of primers are added simultaneously. The primers are designed such that the outer set amplifies at a much higher melting temperature than the inner set and the inner set has a low annealing temperature. Cycling begins at high temperatures, allowing only the outer primers to initiate synthesis. After several rounds the annealing temperature is dropped, allowing the inner set to prime as well. After few rounds of PCR, when the shorter product accumulates, the denaturation is also dropped so that products of the outer primers can no longer melt to form templates, which causes the accumulation of products from the inner primers only.

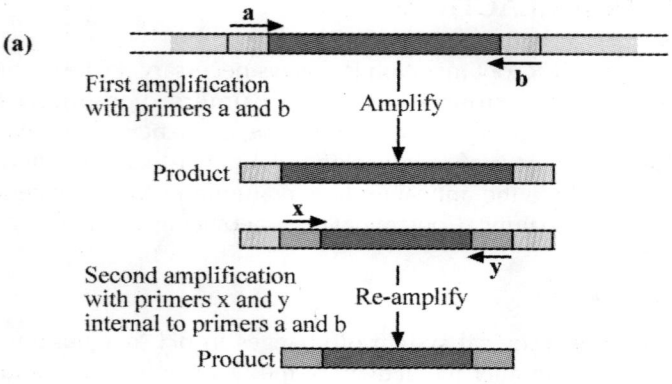

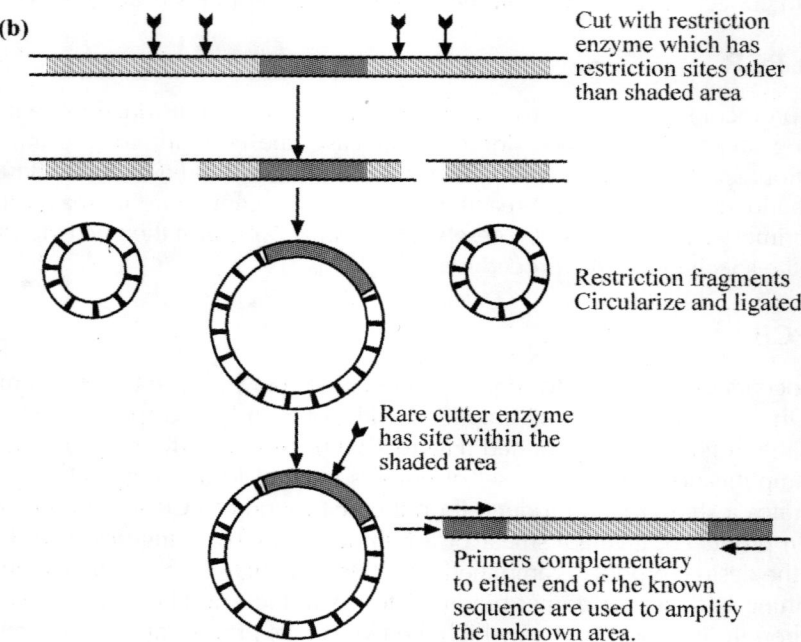

Figure 12.4 *Modification techniques for PCR reaction (a) Nested PCR and (b) Inverse PCR*

4. Touchdown PCR

The principle behind touchdown PCR is to initiate synthesis at very high annealing temperatures, which permit only perfectly matched primer–template hybrids to form. Later, the annealing temperature is dropped in a stepwise fashion with each cycle. As the amplification begins to accumulate in alternative cycles, high-temperature annealing becomes much less critical for specificity. The products are unlikely to have any sites for mispairing. The benefit of decreasing the annealing temperature in touchdown PCR is to increase specificity without compromising the yield.

5. Anchored and Inverse PCR

The main limitation of PCR is to know the primer sequences surrounding the target DNA. If the area to be amplified is not known, a way around the need for a flanking primer at the other end is to use an adaptation known as anchored PCR. For example, a restriction fragment may only have a known sequence to which a specific primer can be made at one end of the target area. Using terminal deoxynucleotidyl transferase, the other end of the fragment can be given a poly G tail (the anchor). A poly C primer can now act as the primer to the specific primer.

An alternative technique of amplifying into unknown regions when a known sequence is in one area, but amplification of the DNA on either side of this area is desired is given below. Initially, the target DNA is cut with a restriction enzyme, which does not cut within the known sequence, but at unknown points on either side. The resulting DNA fragments can then circularize as the single-strand overhangs, left by the restriction enzyme and anneal to one another. The circle is joined by the addition of a ligase and then re-opened using another rarer cutting restriction enzyme, which cleaves only within the known sequence (Figure 12.4b). The known area is now at both ends of the template and amplification is normal in relation to the original sequence, because the area has been inverted by the circularization process.

6. Elimination of contamination

Owing to the fact that PCR is a very sensitive technique and can copy a DNA sequence from very small amounts of DNA (theoretically from a single molecule), it is crucial to adopt meticulous laboratory techniques to prevent any contamination of the laboratory, supplies, and equipment with the target DNA. The DNA contamination in PCR reactions can come from unlikely sources; indeed, recombinant *Taq* polymerase is frequently contaminated with DNA. It is important to remove this if the PCR is to be performed on microbiological DNA samples. Multiple controls are required for the PCR; the choice of controls will facilitate the detection of contamination and false positives.

C. PROBLEMS AND SOME SIMPLE SOLUTIONS

A normal PCR amplification product on gel gives a sharp, bright, and tight band of the correct size from each sample reaction and no bands at all in a negative control. Unfortunately, this is not always the case – there are many other problems, both inherent and extraneous.

1. No PCR product

If there is no PCR product, it is advisable to repeat the experiment. Check the starting template DNA concentrations and dilute them if necessary. Try to lower the annealing temperature in the PCR cycle and run 5–10 cycles at very low annealing and extension temperatures (42°C and 68°C for instance), then run another 30–40 cycles at higher temperatures. Check the primer concentrations; if there are paint bands, then replace the tubes and run it again for another 10–20 cycles.

2. Bright bands in the well of the gel

These are almost certainly caused by over-amplification of the PCR product. To redress this problem, first dilute the template DNA and then raise the annealing temperature (>55°C).

3. Smearing of PCR products or multiple bands

Use a low concentration of template DNA and increase the annealing temperature by 2–5°C. Standardize the optimum magnesium concentrations by amplifying at varying concentrations. Then, decrease the number of cycles. The formation of multiple bands can be checked by using only one of the primers in the same reaction, in order to determine if primers exist along with the artifacts.

4. Bands in the blank

Check all the solutions carefully and change the solutions if there is any uncertainty. Wash the pipettes thoroughly and check for any contamination.

V ADVANCES AND EXTENSIONS TO BASIC PCR STRATEGY

The applications of PCR are very extensive and diverse, and are still growing. The PCR technique has already been found very useful in the diagnosis of genetic disorders, the detection of nucleic acid sequences of pathogenic organisms in clinical samples, the genetic identification of forensic samples, analysis of allelic sequence variations, analysis of RNA transcript structure, direct sequencing of amplified DNA, detection of gene expression, engineering of gene expression, analysis of mutations in activated oncogenes, evolutionary analysis, etc. In addition, as discussed below, PCR amplification is being used to carry out a variety of tasks in molecular cloning and the analysis of DNA, including: (1) generating specific sequences of cloned double-stranded DNA for use as probes, (2) producing probes specifically for uncloned genes by the selective amplification of particular segments of cDNA. (3) making libraries of cDNA from small amounts of mRNA, (4) creating large amounts of DNA for sequencing, (5) analyzing mutations, (6) chromosome crawling/walking, and (7) identifying unknown genes, called differential display.

A. SOME MODIFICATIONS OF PCR

The standard PCR protocol has been extensively modified, based on the type of target sequence used and the nature of product required, for example, if we are interested in amplifying DNA from an organism for which there is little genetic information available, or if we want single-stranded rather than double-stranded DNA, or if we want to find a DNA sequence upstream or downstream of a specific gene. Some solutions to these problems have been achieved by modifying the types and numbers of primers used in the PCR. Other newer applications of PCR are also being developed rapidly (Table 12.5).

Table 12.5 *Modifications of the PCR using different types of primers*

PCR type and primer sizes	Potential application
Standard: Paired, 15–30 nucleotides (nt) Each	DNA amplified on target DNA, for which sequence information is available
Anchored: One known primer, the second is made	Synthesis of cDNA with the known primer is carried out using mRNA. For example, a poly(G) tail is made by synthesizing a primer with a poly(C) sequence, which allows the amplification of a second DNA strand that is complementary to the cDNA.
Arbitrary: Single, 18–30 nt, arbitrary sequence	PCR carried for regions of DNA internal to regions to where arbitrary primers (M13 sequencing primer, M13 reverse sequencing primer or T3 sequencing primer) anneal on opposite strands. One or more DNA fragments will be produced and these can be used to generate genome maps or discriminate among individuals, populations, or species.
Asymmetric: Paired, 10–30 nt primers in a 1:50 to 1:100 ratios	Amplify single-stranded DNA for sequencing.
DAF-PCR: Single, 5 nt, arbitrary sequence	DNA amplification fingerprinting (DAF-PCR) by the amplification of genomic DNA using a single, short primer of arbitrary sequence.
Degenerate: Multiple types, 15–30 nt	Amplify DNA from genes related to those for which the sequence is known, or for which only part of the sequence is known, or for members of a gene family. The degeneracy of the DNA code for amino acids determines how many primer types are included in the reaction.
Inverse: Paired, 15–30 nt, inverse orientation	Amplify regions of DNA of unknown sequence flanking a known sequence; used for identifying upstream and downstream sequences, chromosome walking, library screening, determining insertion sites of transposable elements, etc. The primers are oriented so that the DNA synthesis occurs away from the core DNA with a known sequence.
PCR-RFLP: Paired, 18–30 nt	Genomic DNA is amplified by standard PCR, and then cut with a restriction enzyme. The banding patterns are visualized on a gel after staining with ethidium bromide. Standard primers are used; if none are available, sequences can be obtained from DNA cloned in a genomic library.

Continues...

. . . Continued

PCR type and primer sizes	Potential application
Quantitation of mRNA	Several methods are known: (1) two different cDNAs are amplified and the absolute level of one is calculated if the other is known, (2) the sample is spiked with a known amount of control DNA and target, and the control DNA are amplified and compared to estimate the amount of target DNA.
RAPD-PCR: Single, 10 nt, random sequence	Random amplified polymorphic DNA-PCR. Amplify regions of DNA that are flanked by the random primer sequences. Multiple DNA fragments may be produced and used as markers for genome mapping or identifying individuals, populations, or species.
RNA amplification: 18–22 nt	The mRNA is reverse-transcribed and the cDNA is amplified by PCR using primers 18–22 nucleotides long.
SSP-PCR	Single-specific-primer PCR amplifies dsDNA when sequence information is available for one end only. A second generic vector primer is used at the unknown end. The primers are used in a PCR after ligating the unknown end to a generic vector. This method can also be used to obtain sequence data and restriction site data in unknown regions outside the known sequence.

Protocols for some modified PCR methods

At present, PCR offers the fastest and most accurate tool for verifying transgenics, as well as determining the changes in a particular gene sequence resulting from mutations or tissue culture. PCR has been used in combination with RAPD technology to confirm the presence of transforming DNA in single cells and protoplasts, and to follow them as recipient cells undergo mitotic division. A modification of PCR technology, known as "inverse PCR", has been used to determine the T-DNA copy number in transgenic plants generated by *Agrobacterium*-mediated transformation. Another modification, known as reverse transcriptase (RT)-PCR, has been used to detect antisense transcripts in transgenic plants. In this chapter, the application of standard PCR assay in plants (Method–1) is briefly discussed.

Method-1: PCR amplification of plant DNA

Materials

Absolute ethanol; 10 mM ATP; 10 ml Chloroform-isoamyl alcohol (24:1); 10 ml CTAB-2% w/v CTAB, 1.4 M NaCl, 0.2% v/v mercaptoethanol, 20 mM EDTA, 100 mM Tris (pH 8). 10 mM dNTP solutions; 10X DNA polymerase buffer; DNase-free RNase; 50 mM DTT; 70% Ethanol; 10X Kinase buffer; 10X lambdaexo buffer (670 mM glycine-KOH pH 9.4,

25 mM MgCl$_2$, 500 µg/ml BSA); Phenol-chloroform-isoamyl alcohol (26:24:1); Sterile distilled water; T4 polynucleotide kinase; *Taq* DNA polymerase; TE buffer (50 mM Tris-HCl pH 8, 10 mM EDTA).

Equipment: Corex tube; Centricon-3 microconcentrator; Centrifuge; Electrophoresis equipment; Eppendorf tube; Ethidium bromide-stained agarose gel; GeneAmp PCR System 9600 thermal cycler; Incubator; Liquid nitrogen; Microcentrifuge tubes (0.5 ml); Micropipettor plus tips; Oligo DNA synthesizer; Pestle and mortar; UV light source; Vacuum drier.

Procedure

DNA sample preparation

1. Grind about 0.5 g of leaf material in a pestle and mortar in liquid nitrogen to a fine powder, and then transfer it to a preheated (60°C) solution of 10 ml CTAB in a Corex tube.
2. Incubate the buffer–tissue mixture at 60°C for 30 min., with occasional swirling.
3. Extract the buffer–tissue mixture twice with 10 ml chloroform-isoamyl alcohol (24:1) to remove the chlorophyll and other hydrophobic components, and to denature and separate the proteins from the DNA.
4. Resolve the phases by centrifuging the sample at 4000 rpm for 10 min. at room temperature.
5. Remove the top aqueous phase to a new tube and precipitate the nucleic acids by adding 20 ml cold (–20°C) absolute ethanol. Leave the sample to stand on ice for 20 min. and then pellet the nucleic acids by centrifugation at 2400 rpm for 5 min.
6. Wash the pellet with 70% ethanol and air-dry it for 10 min.
7. Resuspend the pellet in 0.5 ml 1X TE buffer and transfer it to an Eppendorf tube.
8. Add DNase-free RNase (1 U/ml) and incubate the sample at 37°C in a water bath for 1 hr to digest the RNA.
9. Repeat steps 3–7 above and store the DNA at –20°C until needed. Before storage, run 2 µl of the sample in an ethidium bromide-stained agarose gel to observe the DNA.

Amplification reaction

1. Automated amplification of a targeted DNA segment can be achieved with phosphorylated primer 1 and unphosphorylated primer 2. The procedure can be carried out in 50 µl vol composed of:
   5 µl dNTP mix (10 mM of each dATP, dCTP, dGTP and dTTP in equal ratio)
   5 µl of 10X DNA polymerase buffer
   200 µg of genomic DNA
   250 µg of phosphorylated primer
   250 µg of unphosphorylated primer
   2 U of *Taq* DNA polymerase
   made up to a total volume of 50 µl with sterile distilled water.
2. Prepare the sample in 0.5 ml microcentrifuge tubes without the *Taq* polymerase, then place the sample in the heat block.
3. Heat the sample to 94°C for 5 min. (to denature DNA), then add *Taq* polymerase.
4. Perform the cycling reaction: 94°C for 30 s (denaturation step), 50°C for 30 s (annealing) and 72°C for 1 min. (extension).
5. After 30 cycles, hold at the extension step for 10 min. at 10°C until it is retrieved for analysis.

6. To check whether the reaction was successful, electrophorese 5 µl of the PCR products on an 0.8% agarose gel containing ethidium bromide and visualize under UV light to observe the bands.
7. Analyze the PCR products by DNA sequencing.

Notes

1. CTAB disrupts the membrane so that the DNA is released into the extraction buffer, while EDTA protects the DNA from endogenous nuclease by binding magnesium ions, generally considered a necessary cofactor for most nucleases.
2. λ exonuclease digestion to generate ssDNA template: Combine: 45 µl DNA, 6 µl 10X λ exonuclease buffer, 1 µl λ nuclease (5 U/ml) and make up to 60 µl with ddH$_2$O. Incubate it at 37°C for 1 hr. To remove the buffer and enzyme, add 130 µl TE and transfer to a Millipore 100000 NMWL filter; spin at 4500 rpm in a variable speed microcentrifuge for 5 min.; add 200 µl distilled water and spin for a further 4 min.; Pipette the sample from the top of the filter to a new Eppendorf tube and make it up to 60 µl. Do not let the filter dry out.

B. REVERSE TRANSCRIPTASE-MEDIATED PCR

The amplification of double-stranded DNA by PCR is not limited to genomic DNA targets; it can also be applied to double-stranded cDNA that has been synthesized from RNA using a reverse transcription reaction. Sensitive methods for the detection and analysis of RNA molecules are an important aspect of most cell/molecular biology studies. The adaptations of PCR technology have provided a breakthrough in this area. Reverse transcription followed by PCR (RT-PCR), leading to the amplification of specific RNA sequences in cDNA form, is a sensitive means for detecting RNA molecules, a means for obtaining material for sequence determination, and a step in cloning a cDNA copy of the RNA. Its advantages as a preparative technique over conventional library-based cDNA cloning methods include speed, the requirement for only small amounts of target material, and tolerance of large amounts of contaminating rRNA and tRNA, allowing whole cellular RNA to be used as the source. Arbitrary RT-PCR, using either random hexamers or longer arbitrary primers and oligo-dT primers, which hybridize to the polyadenylate tails of mRNA molecules, can be used to generate representative pools of *expressed sequence tags* (ESTs) for further analysis. A second application is the identification of differentially-expressed genes. In this technique, known as differential display PCR or mRNA fingerprinting, the ESTs are amplified from different sources and compared side-by-side using electrophoresis. Differentially-expressed genes are identified as extra or missing bands on the gel, and this can be a sensitive approach for detecting regulated gene products. Various strategies can be adopted for first-strand cDNA synthesis: the reverse transcriptase reaction can be primed by the downstream PCR primer annealed to the RNA, or by random hexamers, or by an oligo(dT) primer at the poly(A) tail of mRNA. The second strand of the cDNA is synthesized by the *Taq* DNA polymerase during the first cycle of the PCR. Because of the speed with which RT-PCR can be carried out, it is an attractive approach for obtaining a specific cDNA sequence for cloning.

Method-1: Amplification of DNA by reverse transcription of mRNA

The first strand of cDNA generated by reverse transcription of mRNA can be used as a template for PCR, as described below.

Target sequences

The target sequences in this method are mRNA molecules isolated from different organs or cell types. Use autoclaved tubes and solutions, wherever possible, and wear gloves to minimize nuclease contamination from the fingers.

Reverse transcriptase amplification reaction

1. Set up a reaction to synthesize the first strand of cDNA. In a sterile 0.5 ml microfuge tube, mix the constituents given below in the following order.

Amplification buffer (10X)	2 µl
dNTPs (each at a concentration of 10 mM, pH 7.0)	2 µl
Oligo(dT) 12–18 (100 µg/ml) (see note 1)	1 µl
Placental RNase inhibitor	20 units
Template mRNA (see note 2)	1–2 µg
50 mM $MgCl_2$	1 µl
H_2O to a final volume of	20 µl
Murine reverse transcriptase	100–200 units

   Incubate the reaction for 30 min. at 37°C.
   10X Amplification buffer: 500 mM KCl; 100 mM Tris .Cl (pH 8.3 at room temperature); 15 mM $MgCl_2$; 0.1% gelatin.

2. Heat the reaction at 95°C for 5 to 10 min. in a water bath and then quickly chill on ice. The heat treatment denatures the RNA–cDNA hybrid and inactivates the reverse transcriptase.

3. Add the following into the sterile 0.5 ml tube containing heat treated reverse transcriptase reaction.

Upstream oligonucleotide primer (i.e., primer complementary to the original mRNA)	10–50 pmoles
Downstream oligonucleotide primer (i.e., primer complementary to the first strand of cDNA)	10–50 pmoles
1X amplification buffer to a final volume of 79 µl	
Taq DNA polymerase (5 units/µl)	1–2 units

   The total amount of "upstream" primer in the amplification reaction should be 10–50 pmoles. The presence of excess oligonucleotide primers in the reaction often leads to the amplification of undesirable non-target sequences. Thus, the optimum concentration of primers should be standardized by preliminary experiments.

4. Overlay the reaction mixture with 100 µl of light mineral oil. This prevents evaporation of the sample during the repeated cycles of heating and cooling.
5. Carry out the desired number of cycles of amplification as described earlier (usually around 20–50 cycles). The number of PCR cycles required depends on the abundance of the target.
6. A thermal cycle profile that works well for amplification of a target of more than 500 bp is: denaturing for 30 s at 95°C; cooling over 1 min. to 55°C; annealing primers for 30 s at 55°C; heating over 30 s to 72°C; extending the primers for 30 s at 72°C and heating over 1 min. to 95°C, and so on.

Notes

1. 10–50 picomoles (pmoles) of a synthetic oligonucleotide, complementary to a desired sequence of the mRNA, may be used in place of oligo(dT) as a primer. The amount of $MgCl_2$ in the amplification buffer needs to be supplemented with additional $MgCl_2$ in the reverse transcriptase reaction mixture.
2. It may be helpful to heat treat the RNA sample at 90°C for 5 min. and quickly chill on ice before adding it to the reaction; presumably this breaks up aggregates and some secondary structures that may inhibit the cDNA priming step.

C. PCR AS A QUANTITATIVE TECHNIQUE

A quantitation of the PCR analysis would provide us with more information and ever since the discovery of PCR it has always been considered important that the technique should be quantitative. This is particularly true in the case of diagnostics where it is necessary to determine precise amounts of DNA in, for example, infected fluids and also in the study of gene expression. Although PCR is extensively used to amplify and detect minute quantities of nucleic acids, it has not been routinely used to quantitate the concentration of the sequences of interest in the original sample of nucleic acids. This is because the efficiency of the amplification process is greatly affected by small differences in any one of a number of variables. These include: (i) the quality and concentrations of the reagents used (polymerase enzyme, dNTPs, primers and template DNA); (ii) the conditions used for denaturation, annealing, and primer extension; (iii) the rate at which the temperature is changed from one step to another within each amplification cycle; and (iv) the priming efficiency of the primers, which is affected by several variations, including the propensity of the oligonucleotides to form self-priming hairpin loops. PCR is so sensitive to small differences in conditions that sets of almost identical amplification reactions can vary up to tenfold in the final yield. Therefore, it is impossible to draw any firm conclusions about the initial concentration of the target sequence from the yield of amplified product in a standard PCR. Recently, however, methods have been devised to measure the overall efficiency of individual PCRs. If we know the amount of amplified products generated, it is then a simple matter to calculate the original concentration of the sequences of interest.

The method involves the co-amplification of two templates: the desired target sequence and a control template that is added in known amounts to a series of amplification reactions. Although the control template uses the same primers as the target sequence, it must be distinguishable from the target sequence following amplification. For example, the control template might be an RNA generated by the *in vitro* transcription of a mutated version of the target sequence that either contains a novel restriction site or lacks a naturally occurring site. The amplified products of the control and target sequences can then be differentiated by restriction enzyme digestion. Alternatively, the control template might differ from the target sequence in size, for example, by deletion of a small segment of the sequences that lie between the two primers.

When the aim of the experiment is to measure the concentration of mRNA, the best control template is a transcript that can be copied by reverse transcriptase and then co-amplified with the target sequences. When measuring the concentration of DNA, the best control template is a fragment of DNA. In either case, the control template should differ from the target sequence as little as possible in order to minimize systematic differences in the efficiency of amplification. Because the control template and target sequences are present in the same amplification reaction and use the same primers, the effect of the variables mentioned earlier is nullified. Therefore, the relative amounts of the two amplified products reflect the relative concentrations of the control and target sequences in the original reaction mixture.

Method–1a: Quantitation of DNA concentration

Procedure

1. Using the guidelines given above, design and prepare a control template suitable for the task at hand. Measure the concentration of the control template as carefully as possible, preferably by spectrophotometry. Alternatively, estimate the amount of control template by ethidium bromide-mediated fluorescence.
2. Make a series of tenfold dilution (in water) containing from 10 µg/ml to 1 fg/ml of the control template. After use, these dilutions should be stored at –70°C for use in step 6.
3. Set up a series of amplification reactions, as described before (DNA templates), or phage (RNA templates). In addition to the usual ingredients, this series of reactions should contain:

    Tenfold dilutions of control template (step 2)   10 µl
    [α–^{32}P]dCTP (sp.act. 3000 Ci/mmole)   10 µCi

4. Carry out a series of PCR for the appropriate number of cycles as described in Table 12.4. Analyze the products of the reactions as described in step 5, a or b, below.
5. a) When using a control template that differs from the target sequence in size, analyze the sizes of the amplified products in a 20 µl aliquot of each of the reactions by gel electrophoresis and autoradiography. Recover the amplified bands of the control template and target sequences from the gel, and measure the amount of radioactivity in them in a liquid scintillation counter. Calculate the relative amounts of the two radiolabeled DNAs in each of the PCRs.
   b) When using a control template that contains a novel restriction site or lacks a naturally occurring site, heat the samples to 94°C for 5 min. following the final round of amplification. Allow the samples to cool to room temperature and then digest a 20 µl aliquot of each of the reactions with the appropriate restriction enzyme. Analyze the sizes of the amplified DNA fragments by gel electrophoresis and autoradiography, and quantitate the amount of radioactivity in each of them as described in the step above (Note 2).
6. Examine the results to determine the concentration of control template that yields approximately the same amount of amplified product as the target sequence. Set up a second series of amplification reactions containing a narrower range of concentrations of the control template. In addition to the usual ingredients, this series of reactions should contain incremental dilutions of the control template:

    (1:10, 2:10, 3:10, 4:10, etc.)   10 µl
    [α–^{32}P]dCTP (sp.act. 3000 Ci/mmole)   10 µCi (Note 1)

7. Carry out a series of PCRs for the appropriate number of cycles as described in Table 12.4.
8. Analyze the radio-labeled products of the amplification reactions as described in step 5. For each amplification reaction, measure the ratio of the yield of the amplified control template to the yield of amplified target sequence. Plot this ratio against the amount of control

template added to each amplification reaction. From the resulting straight line, determine the equivalence point (i.e., the amount of control template that gives exactly the same quantity of amplified product as the target sequence in the reaction). Calculate the concentration of the target sequence in the original sample.

Notes

1. Do not reduce the concentration of unlabeled dCTP in the reaction mixture to increase the specific activity of the precursor pool.
2. The aim of the denaturation and reannealing step is to allow efficient formation of heteroduplexes consisting of one strand of the control template and one strand of the target sequence. When equal concentrations of the two templates are present in the original reaction mixture, the reannealed products will consist of 50% heteroduplexes and 25% of each type of homoduplex. Only one of the two types of homoduplexes will be susceptible to digestion with the appropriate restriction enzyme. Which of the two types of homoduplexes will be susceptible to cleavage depends on whether the control template carries an extra restriction site or lacks a site that is present in the target sequence. Heteroduplex formation is far less efficient when the target and control templates differ significantly in size.

Method–1b: Quantitating viral DNA molecules relative to host cellular sequences

Materials

Cells or tissue sample; Proteinase digestion buffer; 20 mg/ml proteinase K; Phenol buffered with 50 mM Tris.Cl/10 mM EDTA, pH 7.4; 24:1 chloroform/isoamyl alcohol; 10 mM ammonium acetate; 100% ethanol (ice cold) and 70% ethanol; TE buffer, pH 7.5; Reaction mix cocktail; Mineral oil; 0.8 U/μl *Taq* DNA polymerase; Oligonucleotide primers for hybridization; Screw-cap microcentrifuge tubes, autoclaved; Microcapillary pipettes; automated thermal cycler.

Procedure

1. Place the cells or tissue sample in a screw-cap microcentrifuge tube. Add ~100 μl proteinase digestion buffer per ~2×10^6 cells and 20 mg/ml proteinase K to 100 μg/ml. Incubate the sample overnight at 50°C.
2. Mix the sample gently. Add 100 μl buffered phenol, mix gently, add 100 μl of 24:1 chloroform/isoamyl alcohol, and again mix gently. Microcentrifuge it for 5 min. and transfer the aqueous phase to a new microcentrifuge tube.
3. Back-extract organic phase with proteinase digestion buffer. Microcentrifuge it for 5 min. and add the aqueous phase to the aqueous phase in step 2.
4. Extract the aqueous phase twice with equal volumes of 24:1 chloroform/isoamyl alcohol, centrifuging each time to separate the phases.
5. Top the aqueous phase, add 10 M ammonium acetate to 2.5 M final and 2.5 vol cold 100% ethanol. Mix and place it on dry ice for 30 min. Microcentrifuge it for 15 min. at 4°C, wash the pellet with 70% ethanol, microcentrifuge it again, and dry the pellet.
6. Resuspend the pellet in TE buffer, pH 7.5 (100 μl for sample prepared from 2×10^6 cells at 5–100 ng/μl). Store the DNA at 4°C.
7. Estimate the DNA concentration by running the aliquot on agarose gel alongside the standard DNA or by ethidium bromide dot quantitation.

8. Prepare one tube containing 110 and 90 ng DNA from uninfected cells or tissue, and several tubes containing 100 ng DNA. Use these tubes to make a set of 10-fold serial dilutions of the sequence of interest as follows: add a known amount (e.g., 20,000 molecules) of DNA containing the sequences of interest to the stock DNA and mix. Then add 1/10 of this material to the stock DNA and mix. Repeat this several times more until the tube contains ≤10 molecules of the sequence of interest. Add 1/10 of the material from that tube to another tube containing 90 ng of DNA. The final volume of each tube should be ≤71 µl.
9. For each amplification reaction, prepare a screw-cap microcentrifuge tube containing 24 µl of the reaction mix cocktail and enough sterile distilled water for 100 µl final (after the addition of DNA and *Taq* DNA polymerase). Mix and overlay each reaction with ~100 µl mineral oil.
10. Open only those tubes that will contain equivalent samples. Add 100 ng of the sample DNA to each appropriate tube, close the tubes, and briefly microcentrifuge the tubes to mix.
11. Heat-denature the samples for 1 min. at 94°C in a water bath in an automated cycler.
12. Open the tubes (containing equivalent samples) and add 5 µl of 0.8 U/µl *Taq* DNA polymerase. Close the tubes and repeat steps 10–12 with a nested set of equivalent samples.
13. Briefly microcentrifuge the tubes. Cycle the tubes once for 2 min. at 55°C (reannealing) and once for 3 min. at 72°C (extension).
14. Cycle the tubes 29 times for 1 min. at 94°C, 2 min. at 55°C, and 3 min. at 72°C. Extend the cycling for an additional 7 min. at 72°C. Store the completed reactions at 4°C.
15. Electrophorese the aliquots (1/10 reaction) on the appropriate agarose or polyacrylamide gel. Include the lanes with DNA molecular weight markers that will be visible both upon ethidium bromide staining and by autoradiography. Stain and photograph the gel. The DNA fragment corresponding to the PCR product from the host DNA should be readily visible and measures the efficiency of each amplification reaction.
16. Transfer it to a nitrocellulose/nylon membrane and UV cross-link the DNA to the filter. Prehybridize and hybridize with the end-labeled oligonucleotide specific to the sequence of interest. Analyze it by using autoradiography.
17. For more quantitative analysis, strip filter previous probe by boiling in water for 15 min. (if necessary) and hybridize with probe specific for host single-copy sequence. Quantitate the signals by densitometric scanning and compute the standard curve from the dilution series, normalizing it to the host sequence signals. Determine the number of molecules in the experimental samples by interpolation from the standard curve. The standard curve will give a linear relationship between the log of the autoradiographic signal and the log of the amount of DNA.

D. MUTAGENESIS

A complete match between the primer and the target DNA is not necessary, provided allowance is made for this in determining the conditions for primer annealing. Therefore, the technique can be used to engineer specific mutations in the amplified sequence by constructing a primer that corresponds to the mutated sequence, rather than to the original.

E. SEQUENCING

By reducing the amount of one of the two primers, it is possible to arrange for a preferential amplification on one of the strands, resulting in the preparation of effective ssDNA, which can then be used directly for sequencing. The other amplification primer can then also serve as a primer for the sequencing reactions. The preferential amplification of one strand in this way is known as *asymmetric PCR*. Single-stranded material can also be obtained by tagging one of the primers and then using an affinity purification method to separate the tagged strand.

F. LIGASE CHAIN REACTION

The *ligase chain reaction* (*LCR*) or *ligation amplification reaction* (***LAR***) is an *in vitro* method for amplifying nucleic acids, which involves a cyclical accumulation of the ligation products of two pairs of complementary oligonucleotides (Figure 12.5). The LCR uses oligos that completely cover the target sequence (Barany 1991). One set of oligonucleotides is designed to be completely complementary to the left half of a sequence being sought, and a second set matches the right half. Both sets are added to the test sample, along with ligase cloned from thermophilic bacteria. If the target sequence is present, the oligos blanket their respective halves of that stretch with their ends barely abutting at the center. The ligase interprets the break

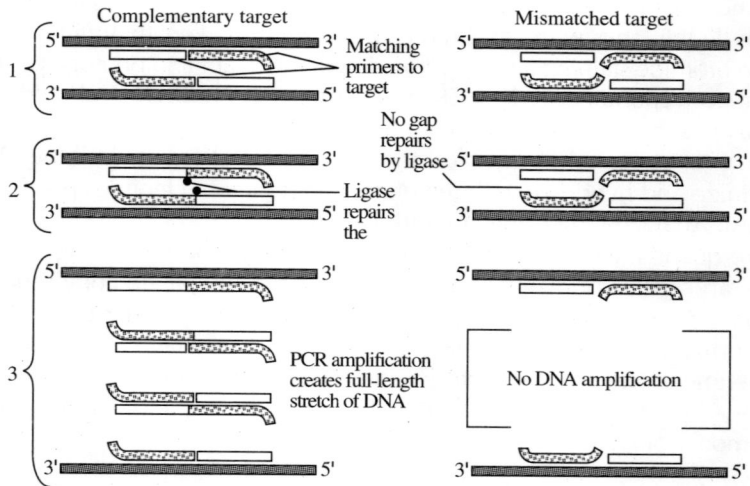

Figure 12.5 *The ligation chain reaction (LCR) uses primes that anneal to a specific complementary sequence on the target DNA to be amplified. In this method two primers that are completely cover the target sequence; one to the left half and the other primer matches the right of a sequence. Both primers are added to the target DNA along with ligase. If the target sequence is present, the primers blanket the DNA fragment with the ends of the primers abutting in the center. The ligase interprets the tiny break between these ends as a nick in need of repair and ligates them together, which creates a full-length stretch of DNA complementary to the target. Amplification subsequently is achieved in a manner similar to the PCR (Redrawn from Hoy MA, 1994)*

between the ends as a nick in need of repair and produces a covalent bond. This creates a full-length segment of DNA complementary to the target sequence. The solution is then heated to separate the new full-length strand from the original strand, and both then serve as targets for additional oligos. After sufficient cycles of amplifications, the full-length ligated oligos are identified in one of several ways, using radioisotopes, fluorescence, or immunological methods. The LCR is particularly useful because the ligase will not join the adjoining oligos if the oligos and the target are not perfectly matched. Thus, this procedure is highly specific. However, PCR is faster and requires fewer copies of the target DNA than the LCR. Combining both the PCR and LCR takes advantage of the best attributes of both reactions.

G. METHODS USED FOR AMPLIFICATION OF UNKNOWN SEQUENCES

The PCR reaction is limited to the amplification of DNA lying between two defined primers, and therefore knowledge of a sequence prior to amplification is essential. However, it is often necessary to characterize the DNA flanking a region for which the sequence is known, for example, to clone a full-length cDNA starting with an EST, or to examine the regulatory elements which lie upstream of an amplified gene segment. *Inverse PCR* (or *inside-out PCR*) allows the amplification of flanking sequences in genomic DNA. If the DNA is digested with a restriction endonuclease, the target sequence for a given PCR will be embedded in a larger fragment of DNA containing both 5' and 3' flanking regions. This fragment can be circularized using DNA ligase and the flanking regions amplified using the same primer pairs which generated the original product, but instead facing outwards so that they amplify the remainder of the circle. Intra-molecular secondary structures such as stem-loops (*panhandle PCR*) and internal bubbles (*vectorette PCR*) for strand-specific priming has been used for the addition of linkers to the ends of linear DNA products. If the gene-specific primer is biotynlated, the products can be captured with streptavidin (capture PCR). A similar strategy is used to produce full-length cDNAs from expressed sequence tags. This technique, termed as *rapid amplification of cDNA ends* (RACE) utilizes a gene-specific primer and an oligo-dT primer which hybridizes to the polyadenylate tail of the cDNA to specifically amplify the 3'-end (*3' RACE*). The 5'-end can be amplified in a similar way by adding an artificial tail to the second cDNA strand, using terminal transferase. Another method which differs from the previous one is *asymmetric PCR* (*single-stranded PCR*), where one of the primers is added in great excess, so that after a limited number of rounds of normal PCR amplification, one of the primers becomes depleted and the reaction switches to a linear accumulation of single strands.

H. REALTIME PCR

PCR technique has been further improved as "**Real time PCR**" to measure the number of copies of a gene or an RNA molecule in tissues. This improvement makes PCR much more accurate than the conventional PCR technique. There are many instances where it is important to know the number of copies of a particular nucleic acid sequence in cells or tissues. For example, in case of medical diagnostics, this may be very useful to assess the level of a viral infection or to determine the number of copies of an amplified oncogene in tumor cells. In cell biology, it is important to monitor the changes in level of a specific mRNA during embryogenesis. Although various conventional methods (blotting techniques, PCR, etc.) are extremely sensitive and can detect very low levels of a nucleic acid sequences, these techniques are not consistent and vary greatly even among the replicate samples. For accurate quantitative estimation of specific nucleic acid sequences real time PCR has been developed. This technique involves, monitoring the amount of DNA product at every cycle of the PCR reaction. It is very rapid and can be completed in 2–3 hours.

Methodologies

Many methods are available to monitor accurately the production of new copies of the target DNA molecule during the PCR reaction. Most of these techniques require a modified thermal cycler, which is adapted by incorporating a spectrophotometer to measure fluorescence in the PCR reaction. Fluorescence is the key to following the rate of amplification of the target DNA molecule.

One of the methods is to simply measure the fluorescence produced by dyes binding to dsDNA molecules. In this protocol the dye fluoresce only when bound to double-stranded DNA and can be used when the target is a cDNA molecule that has been copied from mRNA. This method is simple and cheap, but is more susceptible to errors. It can be used to analyze native DNA molecules (e.g. the level of a DNA virus in a tissue).

Other methods require the construction of specialized DNA probes that are complementary to a part of the target DNA sequence. The probes carry fluorescent reporter molecules that will only fluoresce under certain conditions. One of the probe-based method utilizes two short probes that are synthesized complementary to adjacent regions of the PCR target (Figure 12.6). One probe has a fluorochrome attached to its 3' end and the other a different fluorochrome attached to its 5' end. The fluorochromes are chosen so that the 3' fluorochrome emits a wavelength that excites the 5' fluorochrome. Only fluorescence from the 5' fluorochrome is monitored (Figure 12.6). During the annealing stage of the PCR cycle the labeled probes anneal to their target sequences. In the presence of an appropriate wavelength of light the 3' fluorochrome is excited and emits light that excites the 5' fluorochrome. The light emitted by the 5' fluorochrome is then detected. This only happens when the two fluorochromes are held very close together in the DNA duplex. During the extension phase the two probes are displaced from the complex and fluorescence is no longer observed. Hence, the intensity of the fluorescence during the annealing phase is proportional to the number of copies of the target DNA that are present at the end of the previous cycle of the PCR reaction.

Two other methods use related technologies. The molecular beacons method employs fluorochromes attached to a single probe. They also fluoresce only when the probe is bound to the target. However, the chemistry employed is very different. In the probe hydrolysis or Tacman method fluorescence is only seen during the extension phase of the PCR cycle. Here a single probe carries both a fluorochrome and a molecule that quenches fluorescence. The probe is degraded by the 5' exonuclease activity of the *Taq* polymerase during the extension phase, thus separating the fluorochrome from the quencher. This allows fluorescence to be emitted.

In each reaction cycle, the quantity of product doubles until the reaction slows down as the concentration of precursors decrease. Initially, the level of product is too low to be detected. Therefore to determine the number of target molecules at the start of the reaction a threshold fluorescence level, **Tc**, is set that is just above the limit of detection of fluorescence in the system. The higher the number of target molecules present initially the fewer cycles will be required to reach Tc. Tc is recorded for a number of parallel reactions. To obtain the number of target molecules present at the start of the reaction, the Tc value is interpolated into a

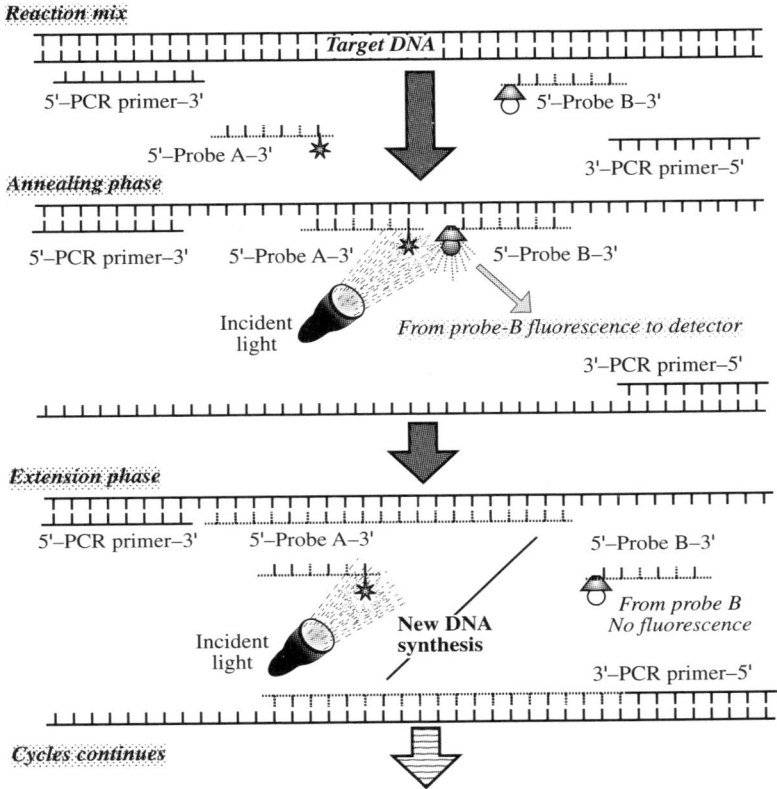

Figure 12.6 *Real time PCR for quantification of number of gene copies (or RNA molecules) in tissues. Probe A is tagged with a fluorochrome that is different from the fluorochrome tagged to probe B. After the denaturation of the target DNA, it becomes single stranded. At the annealing phase, the two single-stranded PCR primers and probes hybridize with their complementary sequences in the target DNA. In the presence of exciting wavelengths of light, fluorochrome A excites and emits short range fluorescence. Since the two probes are held together in the hybrid the fluorescence emitted from probe A can reach the fluorochrome of probe B and induce it to emit its own fluorescence. It is this wavelength that the detector is programmed to detect. During the extension phase, both probes detach from the hybrid and the fluorochromes are no longer close enough together to interact, thus fluorochrome B ceases to emit its fluorescence. All the components recycle during consecutive cycles*

calibration curve obtained by measuring Tc for a series of reaction mixtures containing different numbers of the target molecule.

VI FURTHER READING

Ausubel FM, Brent R, Kingston RE, Moore DM, Seldman JG, Smith JA, and Struhl K (1997) *Short Protocols in Molecular Biology* (4th Edn) Wiley and Sons, Inc.

Ausubel FM, Brent R, Kingston RE, Moore DM, Seldman JG, Smith JA, and Struhl K (1991) The polymerase chain reaction, *Current Protocols in Molecular Biology,*. Greene Publ. Assoc., New York.

Barany F (1991) The ligase chain reaction in a PCR world, *PCR Methods Appl.*, **1**:5–6.

Burke JF (Ed) (1996) *PCR Essential Techniques*, John Wiley and Sons, Chichester.

Carbonari M, Sbarigia D, Cibati M and Fiorilli M (1993) Optimization of PCR performance, *Trends Genet.*, **9**:42–43.

Dallman MJ and Porter ACG (1991) Semi-quantitative PCR for the analysis of gene expression, pp. 215–224, In: *PCR: A Practical Approach*, MJ McPherson, Quirke P and Taylor GR, (Eds) IRL, Press. Oxford.

Ellingboe J and Gyllensten Eds. 91992) *The PCR Techniques: DNA Sequencing,*. Eaton Publ., Natick. MA.

Engelke DR, Krikos RA, Bruck ME and Ginsburg D (1990) Purification of *Thermus aquaticus* DNA polymerase expressed in *Escherichia coli, Anal. Biochem.*, **191**:396–500.

Erlich HA, Ed. (1989) *PCR Technology, Principles and Applications of DNA Amplification*, Stockton Press, New York.

Foley KP, Leonard MW and Engel JD (1993) Quantitation of RNA using the polymerase chain reaction, *Trends Genet.*, **9**:380–386.

Innis MA, Gelfand DH, Sninsky JJ and White TJ (Eds) (1990) *PCR Protocols: A guide to Methods and Applications,* Academic Press, San Diego, CA.

Kaufman DL and Evans GA 91990) Restriction cleavage at the termini of PCR products, *BioTechniques*, **8**:304–306.

Landegren U (1993) Ligation-based DNA diagnostics, *BioEssays*, **15**:761–765.

McPherson MP, Quirke P and Tayor GR (Eds) (1991) *PCR: A Practical Approach*, IRL Press, Oxford.

Mullis KB (1990) The unusual origins of the polymerase chain reaction, *Scientific American*, **262**:56–65.

Olsen DB, Wunderlich G, Uy A and Eckstein F (1993) Direct sequencing of polymerase chain reaction products, *Methods Enzymol.*, **21**:79–92.

Ragot M and Hoisington DA (1993) Molecular markers for plant breeding: comparisons of RFLP and RAPD genotyping costs, *Theor. Appl. Genet.*, **86**:975–984.

Rychlik W, Spencer WJ and Rhoads RE (1990) Optimization of the annealing temperature for DNA amplification *in vitro, Nucleic Acids Res.*, **18**:6409–6412.

Sambrook J, Fritsch EF and Maniatias T (1989) *Molecular Cloning: a Laboratory Manual* (2nd Edn) Cold Spring Harbor Laboratory Press, Cold Spring Harbor, NY.

Stingey SV and del Tfo JP (1993) Genetic analysis with random amplified polymorphic DNA markers, *Plant Physiol.*, **101**:349–352.

Welsh J and McClelland M (1990) Fingerprinting genomes using PCR with arbitrary primers, *Nucleic Acids Res.*, **18**:7213–7218.

Williams JGK, MK Hanafey, JA Rafalski, and SV Tingey (1993) Genetic analysis using random amplified polymorphic DNA markers,. *Methods in Enzymol.*, **218**:704–141.

Zilberberg N and Gurevitz M (1993) Rapid isolation of full-length cDNA clones by "inverse PCR": Purification of a scorpion cDNA family encoding α-neurotoxins, *Anal. Biochem.*, **209**:203–205.

Internet Resources

http://www.promega.com/amplification/amptech.html
http://www.genome.wi.mit.edu
http://www.alkami.com/primers
http://bioinformatics.weizonann.ac.il/mb/bioguide/pcr/contents.html

13

In Situ Hybridization

I INTRODUCTION

In situ hybridization (ISH) is a powerful technique that enables the visualization of nucleic acid probes on target tissues, cells, organelles, cytoplasm, nuclei and chromosomes, so that the location of the nucleic acid (DNA and RNA) can be determined *in vivo*. The method is based on the formation of double-stranded hybrid molecules, which form between a DNA or RNA target sequence and the complementary, single-stranded labeled probe. There are many different types of ISH. The ability to detect nucleic acids *in situ* enables the: (i) construction of physical maps of chromosomes; (ii) analysis of chromosomes structure and aberrations; (iii) investigation of function and evolution of chromosomes and genomes; (iv) determination of the spatial and temporal expression of genes; (v) identification and characterization of viruses, viral sequences and bacteria in tissues; (vi) sex determination; (vii) localization of transformation sequences and oncogenes; and (viii) analysis of neurotransmitter messages. It has also made an important contribution in determining the spatial and temporal expression of genes. ISH also plays an important role in medical diagnosis and plant-breeding programs.

Understanding ISH requires knowledge of molecular biology, genetics, immunochemistry and histochemistry. Although the first ISH procedures made use of radioisotopes, the ISH now followed largely uses nucleic acid probes that are modified with stable, non-radioactive labels. Since its introduction in the late 1970s, the impact of non-radioactive ISH in biochemical research and diagnosis has been tremendous. This has been the result of continuous improvements in the different steps that are essential for successful ISH; i.e., specimen preparation, probe selection and labeling, cytochemical detection, and microscopy.

There are many common features in the variety of methods, which have been published in the last 30 years. A typical *in situ* hybridization (ISH) protocol includes the following steps (Figure 13.1).

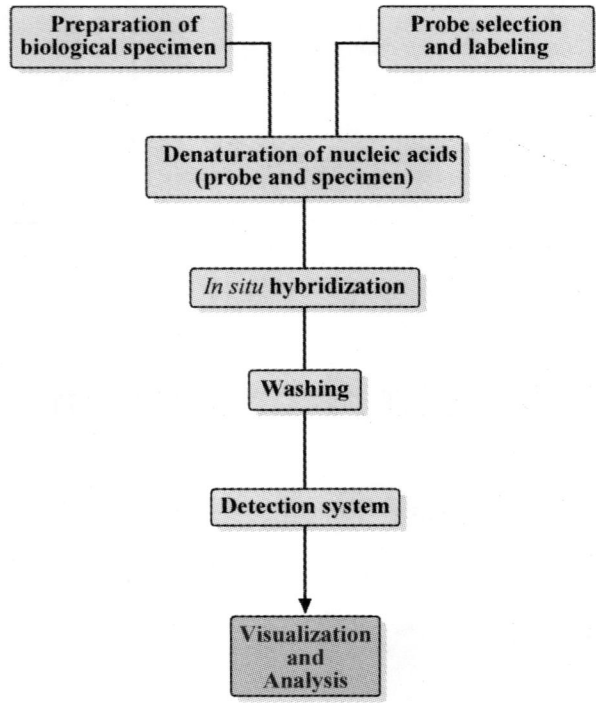

Figure 13.1 *Flow chart of an outline of the in situ hybridization procedure*

II A STANDARD *IN SITU* PROTOCOL

A. PREPARATION OF MATERIAL

For successful ISH experiments the material must be adequately preserved and the target sequences and tissue morphology maintained. In addition, the cell/tissue has to be permeable to probes and detection reagents.

1. Tissue fixation

Fresh material should be used wherever possible. Once harvested, the material should be fixed quickly to minimize endogenous nuclease activity and other degradation processes. Rapid fixation is particularly important when RNA is to be detected, because RNA is very sensitive to the degrading activity of RNase. Two categories of fixatives are predominantly used: cross-linking fixatives (e.g., gluteraldehyde, formaldehyde) or protein-precipitating fixatives (e.g., ethanol or methanol mixed with acetic acid in the ratio 3:1). The fixative chosen depends on the material and probe being used, the method of imaging the probe hybridization sites, and the level of sensitivity required.

Optimal fixation procedures should preserve morphology while ensuring the accessibility of the target RNA or DNA sequences. Some useful fixatives for RNA–RNA ISH are 4% paraformaldehyde and PLPD (phosphate buffer, lysine, periodate, dichromate). Fixation with EAA (ethanol acetic acid 3:1 v/v) decreases the intensity of the signal and formaldehyde– containing fixatives such as formalin or AFE (2% acetic acid, 4% formaldehyde, 85% absolute ethanol) are not very suitable for ISH. Many protocols include digestion with pronase or proteinase K to optimize the accessibility of the target sequence, but this is more important for paraffin-processed tissue sections than for frozen sections or cell smears. Immediately after this digestion, the cells may be incubated in paraformaldehyde to cross-link the exposed target sequence to the remaining cell components, with a resulting two-fold increase in sensitivity. For DNA–DNA ISH, paraformaldehyde and formalin can be employed successively. In the digoxigenin detection method, it is essential to fix with 4% paraformaldehyde for at least 16 hr, to allow for a more aggressive treatment with proteinase K, which reduces the background.

2. Preparation of slides

The ISH material should be mounted on glass slides, as chromosome and nuclei spreads do not adhere well to unclean slides and consequently can be lost during ISH. The preparation of chromic acid washed slides, which are suitable for DNA–DNA ISH, is described in Method 1. The sectioned material for light microscopy should be mounted onto slides coated with either poly-L-lysine or activated 3-aminopropyltriethoxy-saline (APES). For RNA–RNA ISH, special precautions must be taken to remove the RNase. The slides and cover slips are pretreated to avoid non-specific adhesion of probe and ensure the adhesion of cells. The pretreatment protocol employed can contribute to background to a variable degree, especially when bovine serum albumin (BSA) is used.

Method-1: Preparation of chromic acid cleaned slides

Procedure

1. Place the slides in chromium trioxide solution in 80% (w/v) sulfuric acid for at least 3 hr at room temperature.
2. Wash the slides in running water for 1 min.
3. Rinse the slides thoroughly in distilled water.
4. Air dry and place the slides in 100% ethanol. Remove and dry slides just prior to use.

Method-2: Pretreatment of slides and cover slips

Procedure

Poly-L-lysine-coated slides

1. Place the slides in concentrated nitric acid for 30 min.
2. Rinse in deionized water for up to 2 hr; then air dry.
3. Soak the slides in acetone for 15 min.; then bake at 180°C for 2 hr.
4. After cooling, add a small drop (8 µl) of poly-L-lysine to each slide and, using a baked cover slip, draw out into a film to cover the slide.
5. Dry the slides overnight on a 40°C hotplate.

3-Aminopropyltriethoxy-silane (APES)-coated slides

1. Place the slides in concentrated nitric acid for 30 min.
2. Rinse in deionized water for up to 2 hr; then air dry.
3. Soak the slides in acetone for 10 min. and then bake at 180°C for 2 hr.
4. Dip the slides into 2% (v/v) APES in acetone.
5. Rinse thoroughly in acetone and air dry.
6. Activate APES by placing slides in 2.5% (v/v) gluteraldehyde in 1X PBS for 1 hr.
7. Wash the slides in RNase-free deionized water and air dry.

Cover slips

1. Immerse cover slips in 0.1 M HCl for 20 min., rinse in absolute ethanol and air dry.
2. Siliconize with dimethyldichlorosilane and bake at 180°C for 2 hr.

3. Cell spreading

The highest-quality chromosome preparations are needed for DNA–DNA ISH. In particular, the remains of cytoplasm and other cellular material will severely reduce the ISH signal and generate high levels of background. Ideally, there should be little or no contact between cells, but the density should not be so low that the cells are difficult to find. Good chromosome preparations should appear with high contrast when examined dry on a microscope slide, using phase-contrast microscopy. Plant tissue containing dividing cells can be used – usually young root-tip meristems are chosen, but other tissues are also usable. The metaphase index in any meristem can be increased using ice water or spindle-inhibiting drugs prior to fixation. The cell wall of plants presents problems that do not occur with animal cells by limiting probe penetration and causing high levels of background. Plant material is usually digested with enzymes to remove the cell wall after fixation. Two methods for preparing mitotic chromosome preparations from plant meristems (*Method-1 and Method-2*) and a basic method for preparing animal cell smears (*Method-3*) are described below.

Method-1: Mitotic chromosome preparation from plant meristems

Materials

Fixative (freshly prepared): 3 parts 100% ethanol or methanol to 1 part glacial acetic acid.

10X enzyme buffer (pH 4.8): 40 mM citric acid; 60 mM sodium citrate; dilute 1:10 in water for use.

2X enzyme solution: 2% (w/v) cellulase (1.8% (w/v) dry powder from *Aspergillus niger*, Calbiochem, 0.2% (w/v) 'Onozuka' (RS), 20% (v/v) pectinase; make up in 1X enzyme buffer. Store in aliquots at –20°C.

Procedure

Accumulation of metaphases and fixation

1. To accumulate metaphases, excised root tips, buds or other meristems are treated with one of the following:
   (i) Aerated distilled water at 0°C for 24 hr (ideal for cereals).

(ii) 0.01–0.05% (w/v) colchicine for 3–6 hr at room temperature or 16–24 at 4°C.
(iii) 2 mM 8-hydroxyquinoline for 1–2 hr at room temperature followed by 1–2 hr at 4°C (suitable for dicotyledonous)
2. The material is then instantly fixed in freshly-prepared fixative for at least 10 hr at room temperature. Some workers transfer the material into 100% ethanol at this stage.
3. Fixed material can be stored for up to 3 months at −20°C prior to making chromosome preparations, by either the squashing or dropping method.

Chromosome preparations

(a) Squashing

1. Wash the material for 3–5 min. in 1X enzyme buffer to remove the fixative.
2. Transfer the material into 1X or 2X enzyme solution and digest wall material at 37°C until the material is soft, but ensure that the morphology remains (usually 1–2 hr). Adjust the time and the concentration of the enzyme to suit the material; usually, the longer the material has been fixed, the longer the digestion that is needed.
3. Wash the material in 1X enzyme buffer for at least 15 min.
4. Transfer the material into 45% aqueous acetic acid for 1–3 min.
5. Make chromosome preparations in 45% acetic acid on chromic acid-washed slides. Using only the meristematic tissue, remove as much of the other tissue as possible. Apply a clean cover slip to the material in a minimal amount of liquid. Gently disperse the material between the glass slide and cover slip by tapping the cover slip gently with a needle and squash the cells using thumbnail pressure.
6. Place the newly-spread slide onto dry ice for 5–10 min. or immerse it in liquid nitrogen, then flick off the cover slip with a razor blade.
7. Allow the slide to air dry. After the preparations, the slides should be screened and only top-quality spread preparations selected. The spread preparations can be stored desiccated for up to 1 month in the fridge or freezer (−20°C).

(b) Dropping

1. Wash material for 3–5 min. in 1X enzyme buffer to remove the fixative. Remove as much of the non-meristematic tissue as possible.
2. Transfer the material into 2X enzyme solution in a 1.5 ml microcentrifuge tube and digest wall material at 37°C, until the material is soft and breaks up easily. The material is gently dispersed with a pipette. Adjust the time and concentration of the enzyme to suit the material.
3. Centrifuge it for 3 min. at 800 g, discard the supernatant and add 1 ml of fresh 1X enzyme buffer. Re-suspend the pellet with a pipette and leave for 1 min.
4. Repeat step 3 twice.
5. Centrifuge it at 800 g for 3 min., discard the supernatant, add 1 ml of fresh fixative, and re-suspend the pellet with a pipette.
6. Repeat step 5 twice, then centrifuge it for 3 min. at 800 g, discard the supernatant and re-suspend the pellet in 50–100 μl of fresh fixative.
7. Drop 10–20 μl of cell suspension onto a chromic acid-washed glass slide from a height of 5 cm, and blow on it gently.

8. Allow the slide to air dry. After the preparations, the slides should be screened and only top-quality spread preparations selected. The spread preparations can be stored desiccated for up to 1 month in the fridge or freezer (−20°C).

Method-2: Preparation of cell smears for ISH

1. Wash the cells in PBS, count, and smear onto glass slides.
2. For RNA, fix in 4% paraformaldehyde (pH 7.0) in PBS at 4°C for 60 min., rinse in PBS for 5 min., dehydrate in a series of graded alcohol, and store dry at −20°C until required.
3. For DNA, fix it in 4% paraformaldehyde in PBS for 16–24 hr, rinse it twice in 0.1 M Tris pH 7.4 for 5 min., immerse it twice in 0.25% (v/v) Triton X-100, 0.25% (v/v) Nonidet P40 in 0.1 M Tris pH 7.4 for 5 min., and wash it twice in 0.1 M Tris pH 7.4 for 5 min.
4. Incubate it for 10 min. at 37°C in 100 µg/ml proteinase K in 50 mM Tris, 5 mM EDTA, and wash it twice in 0.1 M Tris with 2 mg/ml glycine for 5 min.
5. Immerse it for 15 s in 20% (v/v) acetic acid at 40°C and wash it twice for 5 min. in 0.1 M Tris.

Method-3: Tissue sectioning

To localize cellular DNA and RNA or to detect bacteria, viruses and viral sequences, the material is normally sectioned. The material for sectioning can be fixed and embedded in paraffin wax or resin or can be rapidly frozen and cryosectioned. The embedding medium must preserve the target sequences and maintain good morphological structure. The choice of the embedding medium depends on the material used and the sensitivity of detection required. The sections are transferred to glass slides for light microscopy or coated gold grids for electron microscopy.

Paraffin sections: Paraffin-embedded material preserves the tissue structure excellently and enables the collection of serial sections. Paraffin sections are routinely used in pathology and are suitable for ISH experiments.

Resin sections: Material for examination at high resolution using either light microscopy (LM) or EM is embedded in acrylic resins following fixation with cross-linking fixatives such as glutaraldehyde. Following polymerization, ultra-thin sections of the material are cut. Acrylic, rather than epoxy resins are used because they are hydrophilic and give better access to the probe and detection reagents.

Cryosections: This is a quick way to obtain sections; the material can be frozen, sectioned and hybridized with a nucleic acid probe on the same day. The material can be fixed either before or after freezing and sectioning. For mRNA detection, the fixation step usually takes place before freezing to inactivate the RNase and reduce diffusion of the target sequences.

Method-3a: Fixing and paraffin-embedding material

Materials

Fixative (freshly prepared): 4% (w/v) paraformaldehyde in 1X PBS, pH 7.2.

Graded ethanol series in saline: prepare 30%, 50%, 75%, 85% (v/v) ethanol in 0.85% (w/v) NaCl, 95% (v/v) ethanol in water, and 100% ethanol. De-gas this using a vacuum pump and chill it prior to use.

Procedure

1. Fix the plant material at 4°C overnight.
   (a) The tissue is dissected and placed immediately in the fixative. If the material floats, it must be vacuum-infiltrated by submerging it under the surface of the fixative and then applying a vacuum. The material should sink once the vacuum is released. The fixative should then be renewed since volatile components are lost during the vacuum treatment.
   (b) Cell suspensions can be spun down (200 g for 5 min.) and re-suspended in molten 1% (w/v), low-melting point agarose in 1X PBS at 55°C. Once solidified, the agar block should be refixed for 1 hr at room temperature and treated like a tissue block.
2. Fix the material in 0.85% (w/v) NaCl for 30 min. on ice.
3. Dehydrate the plant material in the graded ethanol series (30–100% ethanol) for 90 min. each on ice and 100% ethanol overnight at 4°C.
4. Replace the 100% ethanol with fresh 100% ethanol and leave it for 1 hr at room temperature. Then place it in 1 part of 100% ethanol to 1 part of Histo-Clear for 1 hr at room temperature, followed by 3 × 1 hr in 100% Histo-Clear at room temperature.
5. Embed the material in wax chippings; leave it at 40°C overnight. Transfer it to molten wax at 60°C and change the molten wax twice a day for 3 days. This can be reduced empirically, depending on the tissue; inadequately-embedded material will not section properly.
6. Place the material into flexible plastic moulds containing molten paraffin wax. Float the mould on water to solidify the wax. The nucleic acids within the wax blocks are stable and the blocks can be stored at 4°C for months or years.
7. Once the material is ready for sectioning, the paraffin wax block is trimmed to give a trapezoid face with the longer of the parallel sides mounted to strike the microtome blade first. The sections are typically cut to 10 μm thickness and ribbons of serial sections are floated on water on poly-L-lysine-coated slides. The slides are placed on a hotplate at 40°C for a few minutes to allow the sections to expand. The water is then drained off and the slides left on the hotplate overnight to dry.
8. The sections should be dewaxed by rinsing twice in Histo-Clear for 10 min. each.
9. Rehydrate the sections in the graded ethanol series (100% down to 30% ethanol) for 1 min. in each, followed by incubation in 1X PBS for 5 min. prior to ISH.

Method-3b: Embedding and sectioning material in acrylic resin for ISH

Materials

Fixative buffer (pH 6.9): A = 0.1 M Na_2HPO_4; B = 0.1 M KH_2PO_4. Mix 3 parts A to 2 parts B.

Fixative: 2% (v/v) glutaraldehyde (EM grade); 0.2% (v/v) saturated aqueous picric acid, made up in fixative buffer.

Procedure

1. Fix the material for 2 hr at room temperature in fixative and wash for 2–5 min. in fixative buffer.
2. Dehydrate in 10%, 20%, 30%, 50%, 70%, 90%, 100%, 100% ethanol; 10 min. in each.
3. Embed in LR White resin (medium grade; London Resin Company by replacing ethanol with LR White in the following ratios: ethanol-LR White 3:1 (30 min.), 1:1 (30 min.) 1:3 (30 min.) then 100% LR White for 2 days, changing the solution five times.

4. Polymerize at 65°C for 15 hr.
5. The block containing the material is trimmed to give a trapezoid face with the long axis mounted to strike the knife-edge first. The sections are floated onto a solution of 1% (v/v) benzyl alcohol in water to re-expand the sections after the compression caused by sectioning. If a diamond knife is used, care must be taken because benzyl alcohol can dissolve the knife's cement. The sections are typically cut to 0.1–0.25 µm thickness.

Sections for glass slides

Glass slides are treated with poly-L-lysine or Vectabond (Vector Laboratories) according to the manufacturer's instructions. Vectabond is very satisfactory, giving good section survival and low levels of background. The sections are transferred to the slides with a metal loop (or 2 mm-diameter hole EM grid).

Sections for EM grids

The sections are picked up on fine-gold mesh grids (e.g., 400 mesh) prepared as described in the earlier section.

Method-3c: Cryosectioning plant material for ISH and examination by LM

Materials

Fixative buffer (pH 6.9): A = 0.1 M Na_2HPO_4; B = 0.1 M KH_2PO_4. Mix 3 parts A to 2 parts B.

Fixative: 2% (v/v) glutaraldehyde (EM grade); 0.2% (v/v) saturated aqueous picric acid, made up in fixative buffer.

Cryoprotectant and mountant: e.g., O.C.T. compound (Tissue-Tek, Agar Scientific).

Procedure

1. Fix the material for 2 hr at room temperature in fixative and wash for 2–5 min. in a fixative buffer.
2. Blot-dry and immerse the material as far as possible.
3. Freeze at –20°C.
4. Mount the frozen tissue-block onto a cryostat microtome. Cut thick sections (10–20 µm) at –14°C. Ensure that the steel blade is sharp and correctly aligned.
5. Pick up the sections on a coated slide. Allow the sections to thaw and air-dry before using it for ISH.

4. Whole-mount preparations

For the study of whole organisms, organs or undisrupted cells, material up to 1–2 mm in diameter can be fixed, using precipitating or cross-linking fixatives, and the ISH protocol conducted on the whole material or on fresh vibratome sections. This approach is becoming increasingly popular, particularly for the detection of mRNA. Whole-mount material may have probe and detection reagent penetration problems, so the plant cells should be pretreated with cell wall digesting enzymes (e.g., cellulase and pectinase). If antibodies are used to detect the probe hybridization sites (e.g., anti-digoxigenin), the background is reduced by pre-absorbing the antibodies on the control material (not used for the experiment) to remove any non-specific antibody activity. The principal advantage of this technique is that the topography of the tissue is retained. In combination with the confocal microscope or deblurred optical sections, this technique can be used to produce 3-D information.

1. Pretreatment of material

Before ISH, the material is pretreated to reduce non-specific probe hybridization to non-target nucleic acids and to reduce non-specific interactions with proteins or other components that may bind the probe. These steps also assist probe and detection reagent penetration and maintain the stability of the target sequences.

(a) *RNase treatment*: For the detection of DNA sequences the material is typically treated with RNase A (digests single-stranded RNA) to remove cytoplasmic and nuclear RNA, thus preventing hybridization with the probe. This is particularly important if the sequence being detected is transcribed.

(b) *Acetylation*: Acetylation neutralizes positively charged molecules, such as basic proteins, and prevents non-specific binding of the probe to the slide when poly-L-lysine-coated slides are used. Acetylation also removes endogenous biotin, which can otherwise cause background signals if a biotinylated probe is used. This pretreatment is optional, although it is often used for RNA–RNA ISH and rarely used for DNA–DNA ISH.

(c) *Permeabilization*: The use of enzymes to digest proteins (e.g., pronase E, proteinase K or pepsin/HCl) helps with probe and detection reagent accessibility, a process called permeabilization. The enzymes probably act by unmasking nucleic acids from the associated proteins; this is particularly necessary when the proteins have been cross-linked with the fixative glutaraldehyde or formaldehyde. This step can be omitted for detecting repeated sequences in cell spreads. When used on cell spread material, the concentrations should be lower than that used on sectioned material. For example, the range in proteinase K concentration can be from 0.01 µg ml^{-1} (cell spreads and whole mounts) to 0.5 µg ml^{-1} (cryosections) and from 1–5 µg ml^{-1} (resin sections). If there is a problem with the reagent penetration, the concentrations can be raised, and if the cell or tissue morphology is poor the concentration should be lowered.

(d) *Prehybridization fixation and material drying*: The prehybridization fixation is designed to prevent/reduce diffusion and loss of cellular RNA or DNA. It also stabilizes the chromosomes prior to the rigorous denaturation procedures, which can lead to DNA loss. After fixation, the material is dehydrated in an ethanol series and air-dried. Although this step is not essential, applying the probe to desiccated material ensures that the probe is not diluted by any residual prehybridization solutions. The dehydration step is omitted if the material has been embedded in acrylic resin, because of adverse reactions of the resin with alcohol.

B. DENATURATION AND STRINGENCY

1. Stringency

ISH reactions exploit the kinetics of nucleic acid duplex formation. Nucleic acid duplexes form by hydrogen bonding between two complementary nucleic acid strands. Each nucleic acid strand is a linear, unbranched polymer consisting of a sugar-phosphate backbone with a nucleotide linked to each sugar residue. The stability of the duplex under defined conditions can be determined by calculating its melting temperature (T_m). For duplexes longer than 250 bp in solution, the T_m can be calculated using equation(1).

$$T_m = 0.41(\%GC) + 16.6 \log M - \frac{500}{n} - 0.61(\% \text{ formamide}) + 81.5 \quad (1)$$

where T_m = 'melting' temperature (°C), (%GC) = percentage of guanine and cytosine in the probe sequence (if unknown, this is usually taken to be 45% for cereals and 40% for humans), M = the concentration of monovalent cations (Na$^+$) in the hybridization solution (mol^{-1}), n = probe length in base pairs (e.g. 250–500 bp), % formamide = concentration of formamide expressed as (v/v) percentage. T_m for RNA–DNA hybrids is 10–15°C higher. T_m for RNA–RNA hybrids is 20–25°C higher.

The stability of the hybrid nucleic acid depends on a number of factors: 1) the proportion of guanine and cytosine (%GC), 2) the length of the hybrid nucleic acid, 3) the environment of the nucleic acid hybrids, 4) the type of nucleic acid hybrid, and 5) the presence of mismatched hybrids.

The term *hybridization stringency* refers to the degree of mismatch tolerated by a specific set of hybridization conditions and is usually given in terms of the minimal percent base match required for duplex formation. This terminology suggests that the stability of the duplex bears a simple relationship to the average fraction of base paired residues. In fact, the stability of the duplex depends greatly on the arrangement of paired bases. For example, the duplexes (A) and (B) both exhibit 70% homology (14 out of 20 matched bases) to their complementary strands, but (B) will be more stable than (A) under the same reaction conditions because the matched bases in (B) are in a contiguous block, while the matched bases in (A) are interspersed.

Duplex (A) $\begin{bmatrix} 5\text{'-CATAGGCTTGATTCAGAAGC-3'} \\ 3\text{'-GTCTCAGATCTGAGCCTCCG-5'} \end{bmatrix}$

Duplex (B) $\begin{bmatrix} 5\text{'-CCTAGGCTTGATTCAGACGC-3'} \\ 3\text{'-ATGTCCGAACTAAGTCCTGG-5'} \end{bmatrix}$

Nevertheless, as a rule, there is a reduction in the duplex melting temperature (T_m) equivalent to between 0.5° and 1.4°C for every 1% mismatch in base sequence. The stringency with which an ISH experiment is carried out determines the approximate percentage of nucleotides that are correctly matched in the probe and target duplex, and is calculated using the following equation:

$$\text{Stringency (\%)} = 100 - Mf(t_m - t_a) \quad (2)$$

where Mf = mismatch factor (1 for probes longer than 150 bp, 5 for probes shorter than 20 bp), t_m = calculated melting temperature (°C; T_m, Equation (1)) and t_a = temperature (°C) at which the ISH mixture or washing conditions were used.

Stringency is controlled by the temperature, ionic conditions and concentration of helix-destabilizing molecules in the hybridization mix and post-hybridization washing solutions. Under conditions of high stringency only probes with high similarity to the target sequence will be stable enough to remain hybridized. As stringency is lowered the number of mismatched nucleotides that remain hybridized increases.

(a) *Salt*: The presence of salts strongly influences the rate of hybridization, an effect governed primarily by the cationic concentration and which is more sensitive to divalent than monovalent cations. The mechanism of the action is thought to involve neutralization of the negatively-charged phosphate backbone, with a resulting decrease in the polarity of the molecule(s). Any changes in salt concentration alter the stability of the duplex, which is reflected by changes in the melting temperature.

(b) *Temperature:* The rate of hydrogen-bond formation between complimentary strands accelerates with temperature, but is later offset by the reverse reaction (breaking of hydrogen bonds) as temperatures approach the melting temperature (T_m) of the duplex.

(c) *Denaturant:* Chemical denaturants alter the stringency of hybridization by interfering with hydrogen bonding. Denaturants that have been used for this purpose are urea, ethylene glycol, sodium perchlorate, and tetraethyl or methylammonium chloride.

However, the most commonly-used denaturant is formamide, which acts to destabilize duplexes, so that only segments that are properly base paired are stable in solutions containing high concentrations of the denaturant. Stringency is more easily controlled by the inclusion of formamide than by salt concentration or temperature. Formamide, an organic solvent, reduces the thermal stability of double-stranded polynucleotides. It reduces the T_m of DNA–DNA duplexes in a linear fashion by 0.72°C for each 1% increment in formamide concentration. The relationship between the T_m of RNA-containing duplexes and the formamide concentration is non-linear, and optimum conditions are best found empirically. Formamide can be used to increase the stringency of the hybridization and washing steps without having to use very high temperatures, which could prove damaging to the tissues, or affect the adhesion of tissue sections to the slides.

2. Denaturation

For DNA–DNA ISH, both the probe and target sequences should be denatured to make them single stranded prior to ISH. Ribopores are single-stranded but because they sometimes form intramolecular duplexes within similar regions of their sequence, they too are usually denatured. The conditions used to denature RNA or DNA probes are usually about 30°C above the estimated T_m of the duplex. The degree of denaturation is less critical for the probe than it is for the target DNA, where there is a narrow window between adequate denaturation and DNA loss. A major area of variation between laboratories is the denaturation procedure for target DNA. The widely-used procedures involve acids – alkali or formamide – in ionic buffers at various temperatures, concentrations and durations. For non-radioactive probes in which the label is attached to the probe via an ester linkage, alkali treatment must not be used since it will remove the label by alkali hydrolysis. In some laboratories, the slides are denatured separately from the probe, in others the target and probe are denatured separately from the slide, while in still others the target and probe are denatured together.

C. NUCLEIC ACID *IN SITU* HYBRIDIZATION

The basic principles of duplex formation are derived from the reassociation of the denatured probe and the heterologous nucleic acid sequence mix. In practice, the annealing of two nucleic acid strands in the laboratory is much less precise than the natural base pairing of nucleic acids that occurs in the intact cell. The renaturation (hybridization) of DNA or RNA duplexes depends on the salt concentration, temperature, and formamide concentration. It is also influenced by the pH and the characteristics of the probe used. The hybridization buffer contains substances which reduce the background; some of these factors are briefly discussed below. The maximum rate of DNA renaturation is at a temperature between 16–32°C below the melting temperature (T_m). For DNA–DNA ISH, a temperature of 25°C below T_m is recommended. Due to the higher thermal stability of RNA–DNA and RNA–RNA hybrids, higher temperatures can be employed for RNA ISH. In the pH range 5–9, the rate of renaturation is relatively independent of the pH. For hybridization involving RNA, it is preferable to chose mildly acidic conditions, because of the susceptibility of RNA to alkaline hydrolysis. Most mixes include agents like citrate or EDTA to complex divalent cations, which strongly stabilize the duplex DNA. With decreasing salt concentrations the electrostatic repulsion between two DNA strands of duplex DNA increases and the duplex is destabilized.

The precise amount of target nucleic acid available for ISH is difficult to assess because of the unknown effects of the nucleic acid conformation and its interactions with associated molecules, particularly proteins. However, in most ISH experiments, the probe concentration is in excess of the target concentration and the reaction follows first-order kinetics. The rate of hybridization depends on the probe length, complexity of sequence and concentration. In general, long probes may result in a slower rate because of limited diffusion into the material. The hybridization rate is increased by using dextran sulfate. It is probable that when one strand is immobilized *in situ*, the rate of hybridization is reduced by a factor of 7–10 over the time taken for DNA duplexes to form in solution.

D. POST-HYBRIDIZATION WASHING

The washing steps are intended to remove probes hybridized to sequences with incomplete homology, but to conserve the binding of probes in perfectly matched hybrids. Post-hybridization washes are usually carried out in a slightly more stringent solution than the hybridization mixture to denature and remove weakly-bound probes. The degree to which mismatches decrease the stability of duplexes is inversely related to the probe length. In short probes, such as synthetic oligonucleotides, the effect of single mismatches on the hybrid stability is pronounced. The stringency can be varied by alterations in the salt concentration, temperature, or formamide concentration. An additional step is often included for RNA–RNA ISH. RNA probes tend to be 'sticky', producing high level of background signal. This is most effectively removed by a post-hybridization RNase A digestion. RNase A only removes single-stranded, and hence unhybridized, RNA, leaving the nucleic acid duplexes intact. The higher stringency of the washing procedure leads to improved signal–noise ratio, with a low background. If the stringency of the wash is too high, even perfectly-matched hybrids become destabilized, resulting in a decrease of both the signal and background.

E. DETECTION OF THE ISH SITES OR SIGNALS

After the probe hybridization and stringent washing, the probe hybridization sites are detected. The methods of detection and visualization depend on the type of label incorporated into the probe.

1. Detection of radioactively labeled probes

Radioactively-labeled probe hybridization sites are usually detected by autoradiography. The slides are first coated with a radiation-sensitive emulsion, which is allowed to dry to give maximum contact between emulsion and specimen. The emulsion is usually 3–4 µm thick, representing a good compromise between sensitivity (increase with film thickness) and resolution (decrease with film thickness). The signal is formed by the interaction of ß-particles, emitted from the radioisotope, with atoms in the emulsion. The energy of the interaction reduces silver halides in the emulsion to metallic silver, thereby generating a latent image. The latent image is developed and fixed using standard photographic procedures. The silver grains are visible by light- and dark-field microscopy or electron microscopy.

For the microscopic assessment of the radioactive label signal, the slides are dipped in liquid autoradiographic emulsion. After development, they are counter-stained with haematoxylin/eosin. For the rapid detection of specific RNAs in a small number of cells in smears, some optimizing methods are applied to replicate sections or smears. In one of the most sensitive non-radioactive detection methods, the DNA probe is tagged with digoxigenin, a glycoside, which is recognized by a specific antibody (raised against it as a hapten). The antibody is conjugated to alkaline phosphatase, which can be detected by an enzyme-linked color reaction.

2. Detection of non-radioactively labeled probes

Most probe labels are immunogenic and detectable by antibodies raised against the label. Biotin has two detection systems, either an anti-biotin antibody or the biotin-(strept)avidin system.

(a) Immunochemistry: The signal-generating system, which enables the sites of probe hybridization to be visualized, may be conjugated to the primary antibody in a one-step detection method. Alternatively, a two-step detection method may be used, in which a second antibody carries the signal-generating system. The latter method is generally more sensitive than the former because secondary antibodies, each carrying the signal-generating system, can potentially bind to each primary antibody molecule. For mercury-labeled probes an additional detection step is required since mercury itself is not immunogenic. After hybridization the probe is detected by first making the mercury react with a ligand containing an immunogenic group such as trinitrophenol (Tnp). The Tnp is detected using anti-Tnp as the primary antibody. A second antibody containing the signal-generating system allows detection of the probe hybridization sites. One advantage of this system is that the presence of mercury in the probe does not interfere with probe hybridization and a high degree of mercury incorporation is possible. Mercury-labeled probes are potentially more sensitive than other non-radioactively labeled probes.

(b) Biotin-(strept)avidin system: Avidin, a glycoprotein extracted from egg white, has a high affinity for biotin. The association constant is about 10^6 times greater than that for antibody-antigen association constants. The first step in the detection of biotin is the addition of avidin conjugated to a signal-generating system. The signal may then be amplified using biotinylated anti-avidin (an antibody raised against avidin, conjugated to biotin), followed by a further layer of avidin conjugated to the signal-generating system. Since avidin can bind up to 4 biotin molecules, the potential for amplification is high. An alternative to biotin is streptavidin, derived from the bacterium *Streptococcus avidin*. Streptavidin is uncharged and so non-specific electrostatic binding may be reduced. In addition, the molecule lacks carbohydrate groups, so non-specific binding to lectins is prevented.

F. SIGNAL-GENERATING SYSTEMS

Next to specimen and probe preparation methods, the success of the ISH technique has been the result of the availability of a large variety of radioisotopic and cytochemical detection methods. Visualization of the probe hybridization site depends on a signal-generating system, which is conjugated to the antibody or (strept)avidin. Signal-generating systems in addition to non-radioactive systems fall into three main groups: (i) fluorochromes, (ii) enzymes and (iii) metals. The availability of different systems has enabled the simultaneous detection of several differently-labeled nucleic acid sequences. Using light microscopy, different probes may be distinguished via fluorochromes with different spectral characteristics; different enzymes that produce different colored end products or a combination of both. The three main classes of signal-generating systems are described below.

1. Fluorochromes

Fluorochromes are visualized by excitation with light of the appropriate wavelength (excitation wavelength) and imaging of the emitted fluorescence (emission wavelength) using appropriate light filters. A wide variety of fluorochromes are available that can be attached to (strept)avidin or to antibodies. The fluorescent properties of some of the more commonly-used fluorochromes are outlined in Table 13.1. In some cases, fluorescent detection is preferred over enzymatic systems because of better spatial resolution, the ability to quantify fluorescent signals by photon counting, and the greater potential for the simultaneous detection of more than one probe. New fluorochromes, based on cyanins that emit in the infrared region, can now extend the spectral range still further. Fluorochromes that can be excited at the same wavelength of light but emit light at different wavelengths, allowing for easy detection of different labels at the same time, are also being developed.

Table 13.1 *Fluorochromes used for ISH probe detection and DNA counter-staining*

Fluorochromes	Max excitation wavelength (nm)	Max emission wavelength (nm)	Color of fluorescence
(a) *Signal-generating systems*			
Coumarin AMCA[a]	350	450	Blue
Fluorescein[a] FITC	495	515	Green
R-phycoerythrin	450–570	575	Red
Rhodamine[a]	550	575	Red
Rhodamine600 TRITC	575	600	Red
Texas red	595	615	Red
Ultralight680	red	680	Far red
Cy 5	648	665	Far red
Cyt	640	705	Far red
(b) *DNA counterstains*			
Chromomycin A3	430	570	Yellow
DAPI	355	450	Blue
Hoechst 33258	356	465	Blue
Propidium iodide	340, 530	615	Red

[a] Available conjugated directly to nucleotides

2. Enzyme-mediated reporter systems

These systems work by catalyzing the precipitation of a visible product at the hybridization site. Many enzymes commonly used in immunocytochemistry are available conjugated to either (strept)avidin or an antibody; these are listed in Table 13.2. The most common are horseradish peroxidase and alkaline phosphatase. The choice of the enzyme depends on the material examined. Some tissues contain endogenous enzyme activity which, if not blocked, can lead to unacceptable background labeling and confusing results. For tissues in which endogenous peroxidase activity is a problem, the activity may be blocked with periodate and borohydrate, sodium nitroferricyanide or phenyl hydrazine. Endogenous alkaline phosphatase activity may be blocked by incubating the sections in levamisole. The major advantages of using enzyme-mediated detection systems are the stability of the signal and the simplicity and low cost of the light microscope needed to visualize the signal.

Table 13.2 *Enzymes used in signal-generating systems*

Enzyme	Enzyme substrate	Embedding	Bright-field	Reflection contrast	Fluorescence
A-Pase	N-ASMX-P +Fast Red TR	Aqueous	Red	Yellow	+ (Red)
A-Pase	N-ASMX-P +Fast Blue BN	Aqueous	Blue	Yellow	+ (Red)
A-Pase	N-ASGR-P +Fast Blue BN	Aqueous	Green/grey	Red	–
A-Pase	BCIP+NBT	Aqueous	Blue/purple	Orange/yellow	–
A-Pase	N-ASMX-P + New Fuchsin	Aqueous	Red	Yellow	+ Red
A-Pase	NBT/BCIP/INT	Aqueous	Brown	White	–
HPO	H_2O_2 + AEC	Aqueous	Red	Yellow	–
HPO	H_2O_2 +Chloronaphthol	Aqueous	Purple	White/yellow	–
HPO	H_2O_2 + DAB	Aqueous	Brown	White	–
HPO	H_2O_2 + TMB	Organic	Green	Pink/red	–
GlO	PMS + NBT	Aqueous	Blue/purple	Yellow	–

* (*Abbreviations used*: A-Pase = alkaline phosphatase; HPO = horseradish peroxidase; GlO = glucose oxidase; AEC = aminoethyl-carbazole; BCIP = bromo-chloroindolyl phosphate; DAB = diaminobenzidine; N-ASGR-P = naphthol–ASGR-phosphate; N-ASMX-P = naphthol-ASMX-phosphate; NBT = Nitro blue tetrazolium; PMS = phenozine methosulphate; NBT/BCIP/INT substrate = product name Dako; TMB = tetramethylbenzidine. (Source: Speel *et al.* 1995)

3. Metal receptor systems for light and electron microscopy

The most-widely used metal for ISH is colloidal gold, which is available conjugated to (strept)avidin and antibodies. Colloidal gold can be visualized at both the light and electron microscope levels. Detection sensitivity using gold probes at the light microscope level can be increased either chemically by silver enhancement or by visualization using reflection-contrast microscopy. At the electron microscope level, colloidal gold offers a number of advantages over enzyme-mediated electron-dense precipitates (e.g., HRPO/DAB). Colloidal gold is available in a number of discrete sizes (e.g., 1, 5, 10, 15, and 20 nm) and, by using a combination of sizes to detect different probes, more than one sequence can be detected simultaneously. In addition, the discrete spherical structure of each particle of colloidal gold enables some degree of signal quantification and a higher resolution of signal detection. Other metal reporter systems (including ferritin and haemocyanin) are available but are less widely used, possibly because their sizes are less precisely controlled than colloidal gold.

4. Multiple-sequence detection

Multiple-labeling strategies are very important in determining the relationship of sequences to each other, to identify chromosomes simultaneously with a new probe sequence, and to localize many probes simultaneously. Multiple-labeling experiments employ more than one probe, each of which is differently labeled. The simplest multiple-labeling experiments involve incorporating different fluorochrome-conjugated nucleotides directly into each probe. This experiment is simple because after the probe hybridization and washing, the different sites of probe hybridization can be identified immediately by their fluorescence color. Some multiple labeling combinations – e.g., a biotinylated probe and a digoxigenin-labeled probe – require two different detection systems. The optimum signal-generating systems for multiple labeling are fluorochromes. When choosing the fluorochromes, ensure that the fluorescent spectra for excitation and emission do not overlap because too much spectral overlap will blur the distinction between the probe signals.

Many options are available now for elegant multiple labeling, by detecting the sites of probe hybridization with several different fluorochromes. By carefully choosing fluorochromes with complementary excitation and emission spectra, and using more than one label in some probes, up to 20 different YAC clones have been simultaneously localized on the same human chromosomes to form 'chromosomal barcodes'. Detecting many sequences requires combinatorial labeling of probes and also an epifluorescence microscope equipped with a digital camera and computer software. Multiple-labeling strategies can also be conducted on material examined by electron microscopy. Alternatively, different probes can be hybridized to the same material separately, and the sites of hybridization of each probe can be compared with the first probe in photographs or by aligning and displaying the images together using digital imaging.

G. IMAGING SYSTEMS AND ANALYSIS OF SIGNAL

Visualization methods aim to localize the ISH signal with the maximum sensitivity and spatial resolution possible. Many of the limitations of ISH can be tackled not only by improvements in experimental protocols but also by careful choice of visualization methods. For example, low signal strength can be compensated for by sensitive visualization techniques (e.g., low light-sensitive cameras) as well as by experimental amplification of the signal. Low contrast can be improved by changes in the microscopy technique and the careful use of filters. Other problems can be tackled by data analysis; e.g., high backgrounds or low signal resolution can be overcome by statistical data analysis.

1. Bright-field microscopy

(a) Before ISH: Light microscopy (LM) is important for prescreening the slides before the hybridization steps to ensure that the quality of the material is high. In particular, the chromosome preparations, small nuclei, and tissue sections need to be checked before the ISH procedure to ensure that the quality of the preparation is suitably high; a poor slide will give poor results!

(b) Radioactive label detection: Transmitted LM is the preferred method for visualizing ISH sites, following autoradiography. The material can be visualized by staining or by phase contrast. Contrast filters are normally used to increase the contrast between silver grains in the emulsion, while filters complementary in color to the stain are used to image the material. Dark-ground microscopy, in which silver grains appear as bright dots on a black background, can be useful for examining large clouds of silver grains.

(c) Detection of ISH signal generated by enzyme-mediated reporter systems: To visualize the colored products precipitated by enzyme-mediated reporter systems, transmitted LM is usually used. As with radioactive label detection, the contrast of the signal can be enhanced by using filters and material counter-stains.

2. Reflection-contrast microscopy

In comparison with bright-field and fluorescence microscopy, the advantage of reflection-contrast microscopy (RCM) is the high sensitivity and spatial information that can be obtained with the use of specifically localized, non-fading enzyme precipitates that exhibit reflectance properties. This type of microscopy uses white light to illuminate the slide from above (epi-illumination) and a special reflection contrast objective lens to focus polarized light onto the specimen. Where signal, material and background have different reflectance properties, a high-contrast image is generated. No counter-staining is needed, since unstained cells and tissue sections can already be visualized on the basis of their reflectance properties. This method gives better contrast than transmitted LM and is extremely sensitive, enabling the detection of low-level ISH signals. The sensitive RCM technique has its own intrinsic types of artifacts, e.g., dirt particles and small amounts of reaction product that are precipitated as a result of the non-specific binding of cytochemically-introduced reagents.

3. Epifluorescence microscopy

The principle behind fluorescence microscopy is that a photon of a particular wavelength (excitation wavelength) excites an electron in the fluorochromes, making it jump into an outer electron shell. The excited electron is unstable and on returning to its ground (stable) state, loses energy, which is emitted as light (fluorescence). According to Stoke's Law the wavelength of emitted light is always longer than the excitation wavelength. The light source for epifluorescence microscopy is usually an ultrahigh-pressure mercury vapor lamp, 50 or 100 W) which emits ultraviolet, visible and infrared light. The microscope has excitation filters to select the correct wavelengths of light for the particular fluorochrome (Table 13.3). The selected wavelengths are focused on the specimen using the objective lens. The excitation filter may be a band-pass filter that only transmits light within a narrow, defined range of wavelengths or a short-pass filter, which only transmits light below a certain wavelength. The emitted light is imaged through a long-pass barrier filter, which only transmits light longer than a certain wavelength. In order for the microscope to operate successfully, an additional filter is needed between the excitation filter and the long-pass filter; this is called the chromatic-beam splitter. It is positioned between the excitation filter and the specimen at 45° and reflects short, excitation wavelength light onto the specimen. The fluorescent light, emitted from the specimen at a longer wavelength, is almost fully transmitted through the chromatic-beam splitter to the long-pass filter. The chromatic beam splitter is important as it allows the excitation filter and the long-pass filter to lie in different light paths above the specimen.

Table 13.3 *Filters used to visualize fluorochromes by epifluorescence microscopy*

Fluorochromes	Excitation color	Excitation filter	Chromatic-beam splitter	Barrier filter	Fluorescence color
DAPI	Ultraviolet	G365 or BP340–380	CBS420	LP420	Blue
FITC	Blue/violet	BP450–490	CBS510	LP520	Green
Texas red	Green	BP536–556 or BP515–560	CBS580	LP590	Red
Propidium iodide	Green	BP536–556 or BP515–560	CBS580	LP590	Red

The fluorochromes are readily bleached by epifluorescence microscopy, although the speed of labeling can be reduced using anti-fade reagents. Epifluorescence microscopy has progressed enormously with the introduction of high-performance interference filters in which the filter effect is based on light interference rather than absorption. The transmission and reflectance properties of these filters are determined by the thickness, number, composition and sequence of deposition of the layers. These interference filters can be designed to transmit or block certain wavelengths of light very specifically and effectively. Another important advance is the development of dual- and triple- band filter sets, which enable the simultaneous visualization of two or three fluorochromes, respectively.

4. Confocal microscopy

Conventional epifluorescence microscopy can suffer from a distorted image generated from out-of-focus information interfering with the image. This is particularly true when the signal is imaged within intact cells or cell layers. The confocal scanning optical microscope or confocal microscope removes much of the out-of-focus information, enabling non-invasive optical sectioning. A complete series of optical sections through biological material enables a study of the three-dimensional distribution of the ISH signal in biological material. The confocal microscope scans the specimen with a spot of laser light focused through a conventional epifluorescence microscope. The excitation wavelength of the light excites fluorochromes used in the signal-generating systems. The emitted light passes through the confocal aperture that is confocal with the 'in-focus' information. Most of the light that passes through the aperture is in focus, while the light from above and below the focal plane is defocused at the aperture and prevented from reaching the photomultiplier. The photomultiplier, receiving mainly focused light, displays the image point by point on to an appropriate high-resolution monitor. The technique offers certain advantages: (1) The brief period of fluorochromes excitation required for producing a confocal image (~2 s) reduces the photobleaching. (2) The digital images recorded by the confocal microscope can be stored and recombined, enabling the alignment of different images.

5. Recording light microscope images

(a) Photographic camera systems: Recording fluorescent images photographically can be difficult because extended exposure times can bleach the fluorochromes. Typical exposures on films may vary from 10 s–2 min. depending on the film used. Some of the real advantages in using conventional camera systems are: (1) The choice of film can improve the contrast between the signal and counter-stain, (2) long exposure times can accumulate the signal, and (3) the total amount of information that can be stored on photographic film is extremely large. Recent developments in color print film make it the preferred choice to record color images with good contrast, resolution and sensitivity.

(b) Digital imaging: Recording images electronically using video or low-light cameras has the advantage of image processing. Optical sections can be taken through the material and out of-focus information removed using complex algorithms. Contrast and intensity changes are easily and quickly performed and artificial colors can be assigned to sectors to enhance the contrast. Different images can be overlaid and compared directly, and relative sizes of the ISH signal calculated and compared, enabling some degree of signal quantification. One of the major uses of digital images has been to pseudocolor chromosomes after a multicolor fluorescence ISH approach based on combinatorial probe labeling. Digital imaging can also be used to remove known artifacts or to improve the presentation or display of data. All electronically accumulated images are recorded at low resolution compared to photographic films. Many digital cameras are now commercially available with a range of resolutions along with good image-processing software.

6. Transmission electron microscopy (TEM)

Imaging ISH by electron microscopy is relatively straightforward and requires no further experience beyond conventional procedures. Block staining of embedded material with osmium or uranyl acetate is probably not possible before ISH because these stains can inhibit the ISH reaction and, consequently, the contrast of the material at relatively low kV. However, a low kV can be damaging to material and support films and should be used cautiously. Uranyl acetate and/or lead citrate staining after ISH can improve the contrast of the image, but care should be taken not to over-stain or the *in situ* signal may be difficult to distinguish.

7. Quantitation of signal

Ideally, knowledge of the precise location and quantity of ISH signal is required. As the detection sensitivity improves both requirements are increasingly achieved. ISH still represents the best way to determine the number of chromosomal locations for a sequence and the location of any nucleic acid within a particular cell. However, the amount of signal is more difficult to determine and quantification techniques are still in their infancy. There is still no method to determine the copy number of a nucleic acid sequence from the strength of the ISH signal.

The quantification of radioactive *in situ* signal can be assessed by statistically counting the number of silver grains per unit area of the tissue. This method has been used for the chromosomal assignment of genes. However, there are many factors that affect the response of the emulsion to the radiation emitted by the radioactive label and so care must be taken in interpreting the results. For the quantitation of non-radioactively generated signals various methods have been used. In particular, digital imaging techniques, such as the cooled CCD camera, enable the assessment of quantitative data on signal intensities.

H. THE CHOICE OF PROBE

Nucleic acid sequences are usually detected using labeled nucleic acids. In principle, a probe can be RNA or double- or single-stranded DNA. Double-stranded probes have a number of disadvantages in comparison with single-stranded probes. For use, the probes must be denatured; and renaturation may occur in solution, thus reducing the concentration of the probe available for hybridization. The tendency of double-stranded probes to form long concatenates in solution limits their tissue penetration. RNA probes are preferable to conventional nick-translated DNA probes because they are easy to prepare and can increase the sensitivity of the detection method. Furthermore, DNA–RNA or RNA–RNA hybrids are thermally more stable than DNA–DNA hybrids, which allows for more stringent conditions. Single-stranded RNA probes are particularly effective for ISH. Using SP6 or T7 cloning vectors, transcription (with the appropriate bacterial polymerase) results in the synthesis of labeled RNA. The versatility of the vector allows the transcription of RNA from almost any desired DNA sequence, which can be labeled to high specific activity. An additional RNase digestion step can be included to reduce the background by removing the excess probe, which may be bound non-specifically.

Hybrid formation in solution is proportional to the single-stranded probe length. For ISH, the probes have to diffuse into the dense matrix of cells and chromosomes. It is usually desirable to reduce the probe length and this can be achieved by measured exposure to alkali conditions (RNA probes). The best probe length for ISH is about 100–300 bases. The use of short synthetic DNA probes permits the choice of conditions for highly selective binding of probe, but the overall stability of the duplexes formed is poor. However, the longer probes (>1 kb) may have tissue penetration problems. The reannealing rate is also proportional to the probe concentration. But high probe concentrations increase non-specific binding and it is advisable to test the optimal concentration for each probe. Alternatively, oligonucleotide sequences can be synthesized *de novo* or an organism's total genome used.

1. Cloned nucleic acids: To amplify a specific DNA sequence by cloning, the DNA is inserted into a vector and both vector and insert are amplified inside the appropriate host cells. The amplified DNA is then extracted. Some commonly-used vectors include bacterial plasmids, bacteriophages and cosmids. Yeast artificial chromosomes (YACs), which accept up to 1 Mbp of foreign DNA, are increasingly being used for the long-range physical mapping of chromosomes including by ISH. The methods for cloning DNA sequences, maintaining bacterial stocks, and isolating DNA and RNA can be found in Chapters 9, 10, 17–25.

2. Synthetic oligonucleotides: Synthetic oligonucleotides are short nucleotide sequences, usually between 10–50 bp long, prepared using a DNA synthesizer (Chapter 4). Their major advantage is that they can be tailor-made to hybridize to specific sequences. Oligonucleotide probes can be used to detect genes, repeat sequences and RNA. Unlabeled oligonucleotides have been used as primers for directly labeling DNA on chromosomes, a process called primed *in situ* labeling.

3. Genomic DNA probes: The total genomic DNA can be labeled and used as a probe (genomic probe) to discriminate the genomic origins of chromosomes in hybrid plants or to identity individual chromosomes in cell fusion hybrids. Differentiation between the two genomes is hugely improved when the genomic probe is hybridized in the presence of an excess concentration of unlabeled total genomic DNA from the other genome (blocking DNA). The blocking DNA hybridizes to common sequences in the probe, on the chromosomes, and to molecules in the cytoplasm, thereby preventing hybridization of the probe to these sequences. Only sequences specific to the target are available for *in situ* probe hybridization.

4. PCR-generated probes: PCR has enabled DNA sequences to be specifically and directly amplified. The method can use cloned DNA sequences, employing suitable primers that flank the insert. An important use of PCR amplification methods will be to amplify specific sequences, or class of sequences, from the total genomic DNA or DNA isolated from flow-sorted chromosomal material.

I. PROBE LABELING METHODS

1. The choice of label

Nucleic acid labels fall into two broad categories, radioactive and non-radioactive. The most sensitive detection methods employ radioactive labels. Autoradiographic detection of labeled probes depends on the isotope used and the specific activity, which has to be high enough to permit detection after hybridization within a reasonable exposure time and has to have a good signal–noise ratio. Several isotopes are available for radioactive labeling. The use of [^{32}P] allows rapid detection of the signal, yet cellular localization is sub-optimal because of the long path-length of the [^{32}P] ß–rays. Autoradiographs of high quality and improved cellular localization employ ^{35}S-labeled probes; [^{35}S] emits ß-rays of much shorter path-length. One of the major problems using [^{35}S] is the non-specific binding of the label to cells or sections. It has been suggested that pre-hybridization in the presence of non-labeled thio α UTP reduces this non-specific binding. It is also possible to label nucleic acids with tritium or by iodination with [^{125}I]iodine. Tritium-labeled probes give accurate intracellular localization, but low-abundance target molecules may require exposures of up to 100 days. This may result in high background and is unacceptably long for most purposes.

The disadvantages of the radiographic method – safety hazards, instability of the label, and long exposure times – can be avoided by using non-radioactive labels such as biotin, acetylaminofluorene, fluorescent labels, mercurated and sulphonated probes, or digoxigenin. One of the most sensitive methods employs digoxigenin as label. Random priming with digoxigenin has the advantage of high labeling efficiency. ISH procedures employing this method achieve high sensitivity, good intracellular localization, and low background. Enzymatic labeling methods usually result in the higher incorporation of modified nucleotides and can, therefore, generate the most sensitive probes. Non-radioactive detection also has the advantage of simultaneous detection of several DNA sequences through double ISH.

(a) Radioactive labels: The radioactive labeling of nucleic acids is usually achieved by enzymatically incorporating nucleotides containing ^{32}P, ^{35}S, ^{125}I or ^{3}H. The labeled nucleotide is essentially identical to its unlabeled counterpart. The labeled probe is not susceptible to steoric hindrance during hybridization and this may be one reason why radioactive probes are more sensitive than non-radioactive probes. The particular isotope that is chosen will depend upon the application since there is an inverse relationship between sensitivity and resolution (Table 13.4). For example, the high emission energy (E_{max}) of ^{32}P makes it suitable for detecting

sequences in less than 7 days, but gives a wide scattering of silver grains in photographic emulsions during autoradiography and has low signal resolution. In contrast, the weak β-emission of ^{3}H results in a high signal resolution but it can take at least 2 weeks to expose the photographic emulsion. Ultimately, the choice of isotope depends on the particular requirements of the experiment in question.

Table 13.4 *Properties of radioisotopes commonly used to label nucleic acids of ISH*

Isotope	Particle(s) emitted	Emax (MeV)	Half-life	Resolution[a] (μm)	Approximate exposure time
^{3}H	β	0.018	12.4 years	0.5–1	2–12 weeks
^{125}I	β	0.004	60 days	1–10	12–20 days
	γ	0.035			
^{35}S	β	0.167	87.4 days	10–15	12–20 days
^{32}P	β	1.71	14.3 days	20–30	2–5 days
^{33}P	β	0.25	28 days	15–20	10–20 days

[a] Scatter of silver grains in emulsion around the point source

(b) Non-radioactive labels: In recent years, radioisotopes are being replaced by non-radioisotopic labels. In general, non-radio-labeled probes may be safer and specialized laboratories are not necessary. Non-radio-labeled probes can also be stored for long periods, the hybridization sites can be detected quickly and several sequences can be detected simultaneously by using different probe labels. Two types of non-radioactive ISH procedures can be distinguished: the direct and the indirect methods.

In the direct method, the label that has been incorporated into the nucleic acid can be visualized directly once the ISH has been completed. For example, chemically coupled fluorochrome tetramethyl rhodamine isothiocyanate (TRITC) onto the 3'-terminus of RNA and fluorochromes-labeled nucleotides, or fluorescein isothiocyanate (FITC), rhodamine (TRITC), 7-amino-4-methyl-coumarin-3-acetic acid (AMCA)-linked deoxynucleotide triphosphates (dNTPs), which can be enzymatically incorporated into the nucleic acid, are becoming extremely important because of the simplicity of detection. However, they may not be sensitive enough to detect low- and single-copy sequences. The availability of an anti-fluorescein antibody conjugated to fluorescein enables the signal to be amplified and may increase the sensitivity of detection.

With indirect methods the label to be incorporated into the probe cannot be directly visualized; instead, a second molecule, the reporter molecule, is attached to the probe label after hybridization and enables visualization of the probe hybridization sites. A range of non-radioactive labels can be incorporated into nucleic acids (Table 13.5).

Table 13.5 *Some examples of non-radioactive labels that can be incorporated into nucleic acids*

(Photo)biotin	biotin–11–dUTP, biotin–16–dUTP, biotin–14–dATP, and biotin–11–dCTP.
(Photo)digoxigenin	digoxigenin–11–dUTP or digoxigenin–11–UTP, derivatives of photodigoxigenin.

Continues...

... *Continued*

2-Acetylaminofluorene (AAF)
Sulphone groups
Mercury
Bromodeoxyuridine (BrdU)
Fluorochromes
Dinitrophenol (Dnp)

2. Enzymatic labeling methods

Enzymatic labeling procedures can be divided into those that result in uniformly labeled probes (*In vitro* transcription, nick translation, oligolabeling) and those that label the end of a nucleic acid strand (end-labeling). Uniformly-labeled probes are most commonly used as they incorporate more than end-labeling procedures. However, end-labeled probes may be more sensitive because of a reduction in steoric hindrance between the probe and the target DNA.

(a) In vitro transcription of riboprobes

The sequence of interest is cloned into a vector (e.g., Bluescript from Stratagene) containing the bacteriophage RNA polymerase promoter sequences (e.g., T3, T7 or SP6 RNA polymerase). In the presence of the appropriate RNA polymerase and labeled (radioactive or non-radioactive) and unlabeled nucleotides, single-stranded RNA probes (riboprobes) are synthesized. The high specificity of a bacteriophage polymerase for its promoter enables these vectors to be used to generate single-stranded RNA probes complementary to the coding (sense) or non-coding (anti-sense) strands. For the detection of single-stranded mRNA, the coding strand provides a useful negative control for the specificity of the positive, i.e., anti-sense probe. Prior to *in vitro* transcription the vector is linearized, preferably using a restriction enzyme that generates blunt or protruding 5'-termini.

Method-1: Synthesizing highly radioactive RNA complementary to DNA

It is frequently useful to synthesize highly radioactive RNA complementary to a DNA sample. This is usually done so that the labeled RNA preparation can be hybridized to the DNA and then the unhybridized RNA can be efficiently and selectively removed. The RNase will digest the unhybridized RNA but will digest very little, if any, of the RNA that is in a duplex. The radioactive RNA copy of DNA can be easily synthesized using *E. coli* RNA polymerase to non-specifically transcribe the DNA into RNA. The method described below will yield [^{3}H]-labeled complementary RNA with a specific activity of approximately 10^7 cpm/µg. [^{32}P]-labeled precursors will yield specific activities about 10-fold higher.

Procedure

1. Vacuum-dry the following in a siliconized 13 × 100 mm glass tube or a 1.5 ml polypropylene microfuge tube: 0.6 nmol of [^{3}H]–rUTP and [^{3}H]–rCTP (about 40 and 25 Ci/mmol, respectively) and 15 nmol of rATP and rGTP.
2. Ethanol-precipitate 0.5 µg of DNA and re-suspend it in 50 µl of 1X nick translation buffer and 5 µl of 1.5 M KCl.

3. Re-suspend the nucleotide triphosphates in 50 µl of the DNA solution and incubate at 37°C for 5 min. Add 2 units of *E. coli* RNA polymerase. Assay the reaction at 0 min. and at 40 min. intervals thereafter. After the 160 min. sample, let the reaction continue and use the TCA precipitation method to assay the extent of RNA synthesis.
4. The incorporation of the label into the RNA should be linear for several hours. When the incorporation has stopped, add 50 µg of carrier RNA and cool to 25°C. Add 1 µg of DNase I (RNase-free) and incubate it at 25°C for 30 min.
5. Add 50 µl of TE pH 8, phenol-extract twice, and re-extract the 2 phenol phases with 100 µl of TE, pH 8. Remove the aqueous phases and re-extract the phenol phase a second time. Combine all 3 aqueous phases from the extractions.
6. Wash a 7.5 ml bed volume Sephadex G-100 column with 20 ml of TE, pH 8, 0.5% (vol/vol) diethylpyrocarbonate. Load the aqueous phase extract on the column and collect the excluded volume as described for nick translation.

Method-2: Synthesizing radio-labeled DNA complementary to RNA

In some experiments it is useful to synthesize radioactive DNA complementary to RNA. This can be done by utilizing the poly-A tail, which is present on most eukaryotic RNAs. The RNA is first annealed to oligo-dT, which then serves as a primer for elongation by the avian myeloblasts virus (AMV) reverse transcriptase. By providing the reverse transcriptase with highly-labeled triphosphates the DNA can be radio-labeled to a very high specific activity (Figure 13.2). One useful feature of this reaction is that the common contaminant of mRNA preparations, rRNA, is a very poor substrate because it lacks a poly-A tail.

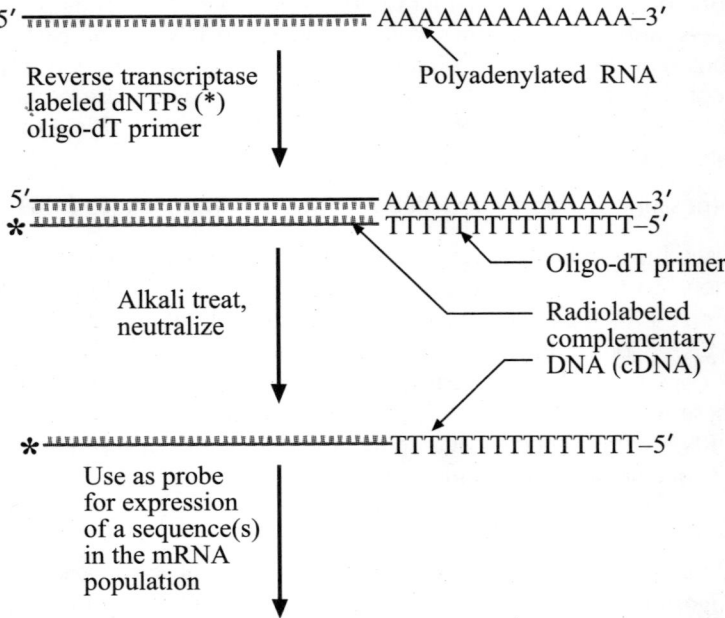

Figure 13.2 *Complementary DNA (cDNA) probes are created from any RNA using reverse transcriptase to polymerize a cDNA from the RNA template using labeled deoxyribonucleotide triphosphates (* indicates the radiolabeled cDNA strand)*

Requirements

Salts (5X): 250 mM Tris-HCl pH 8.2; 50 mM $MgCl_2$; 500 mM KCl; 2 mM DTT. Store them frozen at –20°C in 20 µl aliquots. Do not freeze.

Primer (10X): 100 µg/ml oligo-dT12–18. Store it frozen.

Triphosphates (5X): 5 mM each of dGTP, dATP, and dTTP.

Assay buffer: 400 µg/ml bovine albumin; 10 mM EDTA; 2% SDS.

Stop buffer: 10 mM EDTA; 0.1% SDS; 0.2 M Na^+–acetate pH 5.0.

Procedure

1. Vacuum-dry 0.5 nmol of alpha-$[^{32}P]$-dCTP in a 1.5 ml polypropylene microfuge tube or in a siliconized 13 × 100 mm glass tube.
2. Add the following while the tube is on ice: 11.5 µl sterile distilled water; 5 µl 5X salts; 0.5 µl 200 mM sodium pyrophosphate; 2.5 µl 10X primer; 5 µl 5X triphosphates; 0.5 µl of 1 mg/ml poly-A^+ RNA.
3. Incubate at 42°C for 5 min. and then add a saturating amount of AMV reverse transcriptase and incubate for 1 hr. The enzyme is commercially available, and the saturating amount must be determined empirically for each batch of enzyme. The quantity of the primer oligo-dT is sufficient to saturate all the poly-A template in the reaction.
4. Take assay samples of the reaction at 0, 20, 40, and 60 min. by transferring 0.5 µl samples to 200 µl of assay buffer on ice. For 1 hr put the reaction tube on ice and follow the standard TCA precipitation protocol, to assay the incorporation. The reaction usually is completed in 1 hr. The addition of more enzymes rarely causes additional incorporation, whereas the addition of more RNA occasionally leads to more incorporation. Approximately 20% of the label will be incorporated when the reaction has stopped.
5. Add 200 µl of the stop buffer, 30 µg of carrier tRNA, 200 µl of 5 M NH_4^+–acetate, and 3 vol of 95% ethanol, –20°C. Store it overnight at –20°C.
6. Centrifuge it for 10 min. at 10,000 rpm and carefully discard the highly radioactive supernatant. Re-suspend the pellet in 300 µl TE, pH 8, and add 10 µl of 10 M NaOH. Incubate it at 100°C for 2 min. to hydrolyze the RNA and then transfer it to ice. Neutralize it by adding 9 µl of 10 M HCl and then small volumes of 2 M HCl until the pH is 7. Vortex between additions of acid and measure the pH by spotting 1 µl onto pH indicator sticks.
7. Add 30 µl of carrier tRNA and 2 vol of 95% ethanol at –20°C, mix and store overnight at –20°C.
8. Pellet the DNA as in step 5; wash the pellet by swirling it with 500 µl of 70% ethanol, –20°C, and centrifuge. Discard the supernatant; dry the walls of the tube with a cotton swab (Q-tip), and vacuum dry the pellet. Re-suspend the pellet in 200 µl TE, pH 8. Assay the fraction of label that is incorporated into the DNA by TCA precipitation.

(b) End-labeling of nucleic acids

Method-1: Radio-labeling the 5' ends of RNA or DNA

Both RNA and DNA can be efficiently labeled by replacing their terminal 5' phosphate with $^{32}PO_4$. The 5' ends of single-stranded structures and the protruding, flush or indented 5' ends of the double-stranded structures can be labeled. This method of labeling is useful for general purposes, but is particularly valuable when the label must be at the end of the molecules. Such end-labeling is required in some sequencing procedures, for e.g., in the Smith and Birnstiel (1976) method of restriction site mapping, and when nucleic acids, like 5 S RNA and tRNA, are difficult to label by other *in vitro* methods.

Procedure

1. The specific activities obtained by this method are, of course, directly related to the number of 5' ends and therefore to the size of the nucleic acid. For this reason higher molecular weight RNA is usually hydrolyzed under conditions that yield 5' hydroxyl ends on fragments that have an average length of approximately 50 to 100 nucleotides. This is done by incubating it at 90°C for 30 min. in 5 mM glycine pH 9.5, 10 µM EDTA, and 100 µM spermidine in a 1.5 ml polypropylene microfuge tube.
2. If it is flush-ended or a 5'-indented dephosphorylated DNA is to be labeled, denature it by heating at 100°C for 10 min. in the same buffer. If the DNA to be labeled has less than a few thousand nucleotides of sequence information and the amount of DNA to be labeled is more than 1 µg, then the DNA can renature and thus prevent efficient kinasing. To reduce this renaturation, quench the denatured DNA on ice and immediately do the kinase reaction. Duplex-kinased DNA is frequently the desired end product. The denatured DNA usually reassociates to form duplexes during the kinase and subsequent reactions of most procedures.
3. Begin the kinase reaction by adding 45 µl of the RNA or DNA (1–50 pmol of ends) to γ-[^{32}P]-ATP that is in molar excess to the number of 5'-ends and never less than 1 µM. The nucleic acid solution should not contain ammonium ion because even small amounts of this ion will inactivate the enzyme. If necessary, precipitate the nucleic acid with ethanol and wash the pellet with 95% ethanol. Add 5 µl of 100 mM $MgCl_2$, 50 mM DTT, and 500 mM Tris-HCl pH 9.5. Add 2 units of T4 polynucleotide kinase and incubate it at 37°C for 30 min. Assay the incorporation by TCA precipitation.
4. Remove the unincorporated label by chromatography on DEA-cellulose, by Sephadex gel filtration, or by three ethanol precipitations. Generally, the unincorporated label can be removed by several ethanol precipitations. The first precipitation of the nucleic acid is done in ammonium acetate solution instead of NaCl solution, because the ammonium ion inactivates the enzyme. Add 200 µl 2.5 M ammonium acetate, mix and add 1 µl 1 mg/ml tRNA and 750 µl 95% ethanol. Mix the tube by inverting several times, then put the tube in a dry ice-ethanol bath for 15 min. Centrifuge the tube for 15 min. in a microfuge. After this, remove the supernatant with a Pasteur pipette and re-suspend the DNA in 250 µl of 0.25 M NaCl. Add 500 µl of 95% ethanol, chill it in dry ice-ethanol for 15 min., and then centrifuge it for 15 min. in the microfuge. Wash the DNA pellet to remove any remaining salt. This wash should be done by adding 1 ml of 95% ethanol, inverting the tube several times, and recentrifuging it.
5. Any DNA that has phosphate on its 5' ends must also be dephosphorylated prior to the kinase reaction. For single-stranded DNA or for 5' protruding-end double-stranded DNA the dephosphorylation reaction can be done at 37°C.
6. Because of its secondary structure, tRNA is not reproducibly labeled by the above method and therefore should first be treated with bacterial alkaline phosphatase (BAP). This treatment dephosphorylates the 5' ends of RNA prior to its phosphorylation by kinase. The BAP is dissolved at 0.5 units/ml in 0.1 M Tris pH 8.3, and is dialyzed against this buffer to

completely remove ammonium ions, which inhibit kinase. The dialyzed BAP can be stored at 4°C for at least 5 months without loss of activity. Add 10^{-4} units of enzyme per picomole of 5'ends to the nucleic acid in 0.1 M Tris pH 8.3, 60°C and incubate it for 30 min. This BAP reaction removes about 95% of the terminal 5' phosphates. After the reaction, extract it twice with water-saturated phenol and once with chloroform and then ethanol-precipitate.

Method-2: End-labeling of 3'-OH termini of single- or double-stranded DNA

The enzyme terminal deoxynucleotidyl transferase (TdT) is an unusual DNA polymerase that catalyzes the addition of nucleotides on to the 3'-OH termini of double- and single-stranded DNA molecules in a template-independent reaction. The enzyme strongly prefers to use DNA with protruding 3'-termini as acceptors. Blunt or recessed 3'-termini can also be used, although less efficiently, provided that buffers of low ionic strength containing Co^{2+}, Mg^{2+} or Mn^{2+} are used. The extent of labeling depends on the number of 3'-OH groups initially present, since these serve as the initiation sites for the nucleotide addition (Figure 13.3). Large DNA fragments can be cleaved (e.g., with restriction enzymes or DNase) to increase the number of 3'-OH termini that are available for labeling. The enzyme will accept modified nucleotides (radioactive

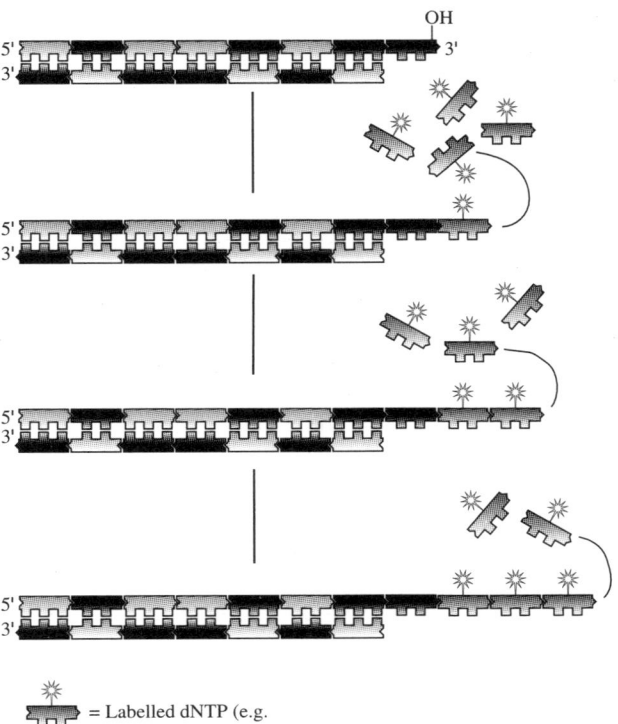

Figure 13.3 *The end-labeling reaction. The reaction uses the enzyme terminal deoxynucleotidyl transferase (TdT). The enzyme use 3'-termini as acceptors. The reaction is template independent and the reaction mixture contains only the modified dNTP so that the tail produced becomes completely labeled*

or non-radioactive nucleotides), and if the reaction mixture contains only the modified dNTP, a 'tail' can be produced that is completely labeled. The enzyme is particularly useful for labeling oligonucleotide probes (<100 bp) for which nick translation and oligolabeling methods are not suitable. The speed of the end-labeling reaction is rapid, and the kits that are available have optimized conditions to enable the length of the 'tail' to be easily controlled.

Materials

5X DNA tailing buffer: 1 M potassium cacodylate pH 7.2, 0.125 M Tris.Cl pH 6.6, 1.25 mg ml^{-1} BSA.

Cobalt chloride: 25 mM.

dATP: 10 mM dATP in 100 mM Tris.Cl pH 7.5.

Labeled nucleotides: Either 1 mM digoxigenin-11-dUTP; or 1 mM biotin-11-dUTP; or 1 mM fluorescein-11-dUTP

Terminal deoxynucleotidyl transferase (TdT): 10–15 units µl^{-1}

Procedure

1. In a 1.5 ml microfuge tube place the following: 4 µl of 5X DNA tailing buffer, 4 µl of cobalt chloride, 1 µl of labeled nucleotide, 1 µl of dATP, µl of DNA equivalent to 250–400 ng, y µl of water. Make up the total volume to 19 µl.
2. Add 1 µl of TdT, mix gently and centrifuge briefly.
3. Incubate for 15 min. at 37°C or 2–3 hr at room temperature.
4. Ethanol-precipitate and re-suspend the DNA in 10–20 µl 1X TE.

(c) DNA labeling by nick translation

The most widely-employed technique for preparing labeled probes is by nick translation. The nick translation reaction employs two enzymes, DNase I and *E. coli* DNA polymerase, which incorporate labeled nucleotides along both strands of the DNA duplex (Figure 13.4). The polymerase does this by binding to a nick in the DNA, excising the nucleotide to the 5′ side of the nick, and then, while moving down the DNA strand, replicating that nucleotide using a triphosphate from the reaction mixture. By a series of these reaction steps the polymerase moves the nick down the DNA strand, replicating nucleotides from the DNA with labeled bases from the triphosphates included in the reaction. Usually, only one labeled nucleotide is used. The most critical parameter in the reaction is the activity of DNase I, which 'nicks' the dsDNA. The technique is rather delicate in that too little nicking of the DNA leads to inefficient incorporation of the label, whereas too much nicking leads to a DNA product that is too short for most experiments. Now there are commercially-available enzyme mixtures of *E. coli* DNA polymerase I and DNase I which are optimized to produce >50% incorporation of label into DNA in 60–90 min., giving probe lengths of about 200–400 bp.

Method-1: Radio-labeling of DNA by nick translation

The conditions described below will yield [^{32}P]-labeled DNA with a specific activity of approximately 10^8 cpm/µg or [^{3}H]-labeled DNA with approximately 10^7 cpm/µg. Iodine-labeled triphosphates are now commercially available (New England Nuclear) and are reported to give specific activities of 10^9 cpm/µg with the procedure described below. The labeled product has

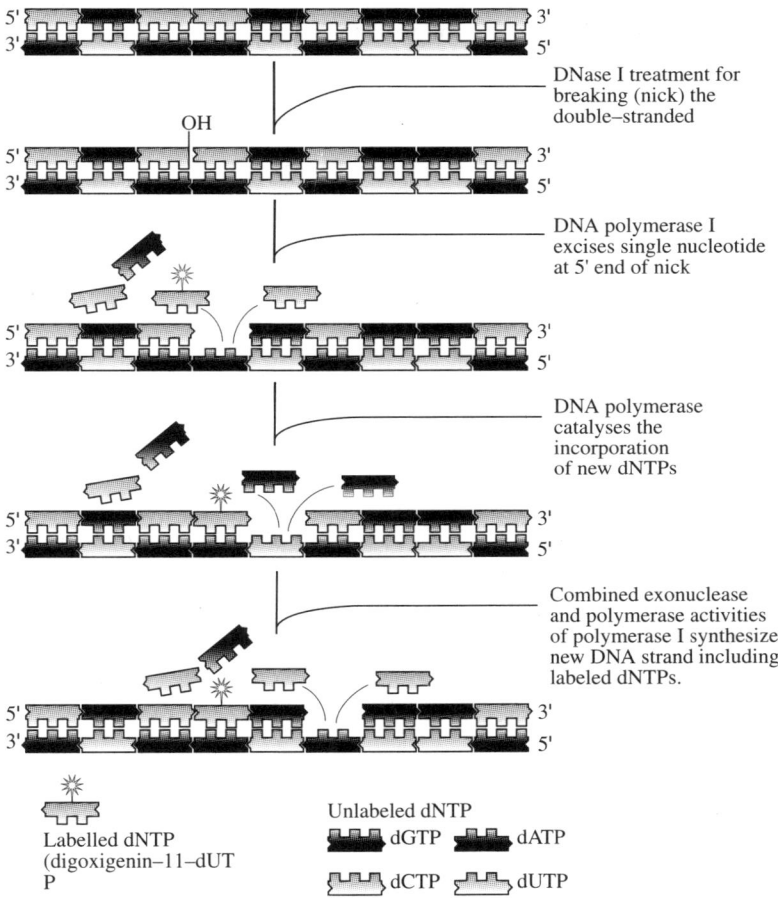

Figure 13.4 *Probe labeling by nick translation reaction*

an average single-strand size of approximately 600 nucleotides and is satisfactory for filter and solution hybridization reactions. Higher molecular weight products with somewhat lower specific activities can be obtained by lowering the DNase concentration 10-fold as long as the DNA polymerase I is not contaminated by significant endonucleolytic activity. The procedure given is for [^{32}P]-labeling of DNA. For [^{3}H]-labeling, use labeled dCTP and dTTP for reasons of cost and specific activity. Change the mixture of unlabeled triphosphates accordingly.

Materials

Nick translation buffer (10X): Mix 2.5 ml of 1 M Tris pH 7.8 and 0.45 ml of 1 M MgCl$_2$ and autoclave. Then add 40 µl of 98% 2-mercaptoethanol, 2.5 mg of albumin, and finally sterile water to yield a volume of 5 ml. This stock solution can be stored, tightly capped, at 4°C for many months without loss of effectiveness; freezing will give a precipitate. The final concentrations of the components in this buffer are as follows: 500 mM Tris.Cl pH 7.8; 90 mM $MgCl_2$; 100 mM 2-mercaptoethanol; 500 µg/ml bovine albumin.

Stop buffer: 10 mM EDTA; 0.5% SDS; 10 mM Tris–HCl pH 7.4; 0.2 M NaCl.

Procedure

1. Vacuum dry 0.2 nmol or 30 µCi of $\alpha[^{32}P]$-dCTP (>300 Ci/mmol) in a siliconized 13 × 100 mm glass tube or a 1.5 ml polypropylene microfuge tube.
2. Add the following solutions and make up the volume to 50 µl with distilled water:
   – 5 µl of 10X nick translation buffer;
   – 4 µl of 100 µM dATP, dGTP and dTTP. This yields 0.4 nmol of each unlabeled triphosphate in the final reaction mixture.
   – 0.2 µg DNA. This gives approximately 0.1 nmol of each nucleotide in a 50% G+C DNA. Up to 15 µl DNA in TE pH 8 can be added without inhibiting the reaction.
3. Dilute DNase stock into a 1X nick translation buffer at 4°C. The DNase stock, 1.0 mg/ml, is stored in 10 µl aliquots at –20°C in microfuge tubes. The appropriate dilution is crucial to the size and specific activity of the product. It must be empirically determined for each batch of frozen aliquots and for each lot of DNA polymerase. The latter is important because endonucleolytic activity frequently contaminates commercial DNA polymerase I preparations. If the polymerase has usual Worthington product, a 1:46,000 dilution is fine. Start the dilution by adding 1 ml of 1X nick translation buffer, at 4°C, to the 10 µl frozen aliquot of DNase.
4. Incubate the reaction mixture at 15°C for 10 min. and then add 1 unit of DNA polymerase I and µl of diluted DNase I. At 90 min. store the reaction on ice and wait for the assay results.
5. Assay the reaction at 0, 30, 60, and 90 min. by trichloroacetic acid (TCA) precipitation of approximately 0.1 µl of the reaction mixture. Add the sample to 200 µl of 2% SDS, 10 mM of EDTA, and 400 µg/ml of BSA. Mix and then spot 20 µl of the mixture onto a GF/C glass filter (Whatman) as a measure of the total counts in the sample. Then add 2 ml of 5% TCA and follow the TCA precipitation procedure. The reaction should plateau by 60–90 min. At the plateau, between 20–90% of the radio-label should be incorporated. Occasionally the addition of more polymerase will increase the incorporation to a new plateau, but usually it is not worth the effort.
6. Stop the reaction by adding 150 µl of stop buffer and then 15 µg of carrier DNA. Pass the reaction mix through a Sephadex G-100 column in TE to remove the unincorporated triphosphates, the salts, and other contaminants. The 8 ml Bio-Rad Econo columns are convenient for this. Collect 0.5 ml (20 drop) fractions with a Gilson Micro-Fraction collector. If the fractions are collected in a convenient 72-hole polypropylene Gilson tray, a hand monitor using the homemade lead shield can assay the profile of radioactivity eluted. Make the shield by punching a 1-cm hole in a thin sheet lead.

Method-2: Labeling of DNA with non-radioactive nucleotides by nick translation

Materials

10X nick translation buffer: 0.5 M Tris.Cl pH 7.8, 0.05 M $MgCl_2$, 0.5 mg ml^{-1} BSA.

Unlabeled nucleotide mixture: dCTP, dGTP, and dATP individual nucleotides. Make 0.5 mM solution of each nucleotide in 100 mM Tris.Cl pH 7.5, and prepare a 1:1:1 mixture.

Labeled nucleotide:

For digoxigenin: Mix digoxigenin-11-dUTP (1 mM stock solution) and dTTP (1 mM stock) to a final concentration of 0.35 mM digoxigenin-11-dUTP and 0.65 mM dTTP.

For biotin: Use 0.4 mM biotin-11-dUTP (powder in 100 mM Tris.Cl pH 7.5).

DNA polymerase I/DNase I: 0.4 units μl^{-1} (Gibco BRL).

Procedure

1. In a 1.5 ml microfuge tube place the following: 5 µl of 10X nick translation buffer; 5 µl of unlabeled nucleotide mixture and either 1 µl of digoxigenin-11-dUTP-dTTP mixture or 2.5 µl of biotin-11-dUTP or 2.5 µl of fluorochrome-labeled nucleotide mixture, 1 µl of 100 mM DTT, X µl of DNA equivalent to 1 µg, y µl of water. Make up a total volume equal to 45 µl.
2. Add 5 µl of DNA polymerase I/DNase I solution, mix gently and centrifuge briefly.
3. Incubate at 15°C for 90 min.
4. Add 5 µl of 0.3 M EDTA pH 8.0, to stop the reaction.
5. Add 5 µl of 3M sodium acetate (or 5 µl of 4 M LiCi) and 150 µl of ice-cold 100% ethanol.
6. Precipitate the DNA in the freezer (–20°C) overnight or on dry ice for 1–2 hr.
7. Spin the tubes at –10°C for 30 min., 12,000 g.
8. Discard the supernatant and then wash the pellet by adding 0.5 ml of ice-cold 70% ethanol and then spinning for 5 min. as above (step 7).
9. Discard the supernatant and leave the pellet until it is dry.
10. Re-suspend the DNA in 1X TE: 10 µl for genomic probes and 10–30 µl for cloned probes.

(d) Random-primed probe labeling

Oligolabeling of DNA uses the Klenow fragment of *E. coli* DNA polymerase and a mixture of all possible oligonucleotides (usually hexanucleotides) that serve as primers for the DNA polymerase. Since practically all sequence combinations are represented in the hexanucleotide primer mix, the primers bind to the template, on average, every at 80–100 bases. The random primers are annealed to denatured DNA to form short double-stranded regions which can then be filled in using DNA polymerase I and dNTPs. The new strand can be synthesized by Klenow fragmentations of DNA polymerase in a reaction mixture containing labeled and unlabeled nucleotides (Figure 13.5). The main advantages of this method are that it requires less than 200 ng of DNA, both dsDNA or ssDNA can be used as a template for the reaction, and oligolabeling can be used on small fragments of DNA (100–500 bp). However, the main disadvantages are that dsDNA must be denatured, and circular DNA is not efficiently labeled and should be linearized. Typically hexanucleotide primers are used, but longer oligonucleotide primers (e.g., 9–14 bp) are now being employed. Longer primers allow the primer concentration to be greatly reduced because they have increased binding efficiency to template DNA.

Materials

10X hexanucleotide reaction mixture (in 10X buffer): 0.5 M Tris.Cl pH 7.2, 0.1 M $MgCl_2$, 1 mM DTT, 2 mg ml^{-1} BSA, 62.5 A_{260} units ml^{-1} random hexanucleotides.

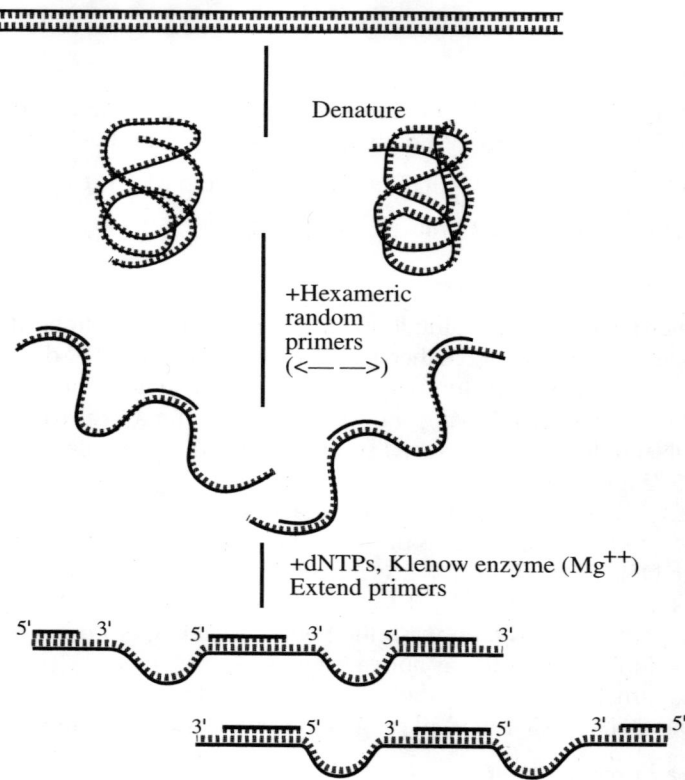

Figure 13.5 *Labeling of probes by the random primer technique*

Unlabeled nucleotide mixture: dCTP, dGTP and dATP individual nucleotides. Make 0.5 mM solution of each nucleotide in 100 mM Tris.Cl pH 7.5, and prepare a 1:1:1 mixture.

Labeled nucleotide:

For digoxigenin: Mix digoxigenin-11-dUTP (1 mM stock solution) and dTTP (1 mM stock) to a final concentration of 0.35 mM digoxigenin-11-dUTP and 0.65 mM dTTP.

For biotin: Use 0.4 mM biotin-11-dUTP (powder in 100 mM Tris.Cl pH 7.5).

For fluorochromes-labeled nucleotides: Use either fluorescein-11-dUTP or rhodamine-4-dUTP (1 mM stock) and dTTP (1 mM stock). Mix 1:1 fluorochrome dUTP:dTTP for use.

Klenow enzyme: 6 units μl^{-1}.

Procedure

1. Denature linearized DNA (50–200 ng) in boiling water for 5 min., and then chill on ice for 5 min.

2. In a 1.5 ml microfuge tube place the following: 3 μl of unlabeled nucleotide mixture, 1.5 μl of labeled nucleotide, 2 μl of 10X hexanucleotide reaction mixture, X μl of DNA equivalent to 1 μg, y μl of water. Total volume should be equal to 19 μl.
3. Add 1 μl of Klenow enzyme, mix it gently and centrifuge briefly.
4. Incubate it at 37°C for 6–8 hr or overnight.
5. Add 2 μl of 0.3 M EDTA pH 8.0, to stop the reaction.
6. Add 2 μl of 3 M sodium acetate (or 2 μl of 4 M LiCi) and 60 μl of 100% ethanol.
7. Precipitate the DNA in the freezer (–20°C) overnight or on dry ice for 1–2 hr.
8. Spin the tubes at –10°C for 30 min., 12,000 g.
9. Discard the supernatant and then wash the pellet by adding 0.5 ml of ice-cold 70% ethanol and then spinning it for 5 min. as above (step 8).
10. Discard the supernatant and leave the pellet until it is dry.
11. Re-suspend the DNA in 1X TE: 10 μl for genomic probes and 10–30 μl for cloned probes.

(e) Polymerase chain reaction (PCR)

A modification of oligolabeling is the PCR. The reaction employs a heat-stable DNA polymerase *Taq* polymerase. The enzyme exhibits highly processive $5' \rightarrow 3'$ polymerase activity. (A detailed explanation is available in Chapter 12.) *Taq* polymerase has been shown to accept modified nucleotides (e.g., radioactive nucleotides, digoxigenin- or biotin-labeled nucleotides) as substrates. Therefore, PCR can be used not only to amplify DNA but also produce large quantities of labeled-probe DNA that is suitable for ISH. PCR may become the standard method to label cloned DNA sequences of less than 4 kbp.

Materials

10X PCR buffer: 100 mM Tris.Cl pH 8.3, 500 mM KCl, 30 mM $MgCl_2$, 0.1% gelatin.

Unlabeled nucleotides: 2.5 mM dATP, dCTP, dTTP, dGTP solution of each nucleotide in 100 mM Tris.Cl pH 7.5.

Labeled nucleotides: Either 1 mM digoxigenin-1-dUTP, or 0.4 mM biotin-11-dUTP, or 1 mM fluorescein-11-dUTP, or 1 mM rhodamine-4-dUTP.

M13 primers: (a) M13 reverse sequencing primer (17 bases) from Pharmacia, μM solution supplied. (b) DNA: Miniprep DNA should be diluted 1:100 in water; use 3 μl per reaction.

Procedure

1. In a 1.5 ml microfuge tube mix: 5 μl of 10X PCR buffer, 2 μl each of dATP, dCTP, dGTP, 3.25 μl of dTTP, 1.75 μl of labeled nucleotide, 9 μl of M13 single-strand primer, 2 μl of M13 reverse primer, 3 μl of DNA, 23.5 μl of water. Total volume should be 49.5 μl.
2. Put some grease around the bottom of the tubes and place them in a temperature cycling machine. Run program 1.

Step	Temp. (°C)	Time (min)
1. Denaturation	91	5
2. Annealing	47	5

3. Remove each tube individually and add 0.5 µl of *Taq* DNA polymerase, mix and return it to the machine.
4. When the enzyme has been added to all the tubes, overlay each tube with 50 µl of mineral oil and start program 2.

Step	Temp. (°C)	Time (min)
1. Synthesis	72	1
2. Denaturation	91	1
2. Annealing	47	1

5. Remove the PCR product from under the mineral oil and ethanol-precipitate.
6. Re-suspend the pellet in 20–30 µl of 1X TE.
7. Check the PCR product by agarose gel electrophoresis. The DNA incorporating labeled nucleotides will migrate slower than the unlabeled DNA.

(f) Chemical method of oligonucleotide synthesis

Oligonucleotides of any specified sequences can be synthesized by a series of reactions in which nucleotides are added sequentially to the 5′-end of an oligonucleotide. Techniques are available for the synthesis of short fragments of DNA of a specified base sequence. Synthetic DNA is widely used in molecular genetics, especially in genetic engineering, but also in basic research. The procedures for synthesis of DNA can be completely automated so that an oligonucleotide of 30–35 bases can be easily made in a few hours and oligonucleotide fragments can be joined enzymatically using DNA ligase.

DNA is synthesized in a solid phase procedure in which the first nucleotide in the chain is fastened to an insoluble porous support (such as silica gel with particles of about 50 µM in size). The overall procedure, the chemical details of which need not concern us here, is shown in Chapter 4. Several chemical steps are needed for the addition of each nucleotide. After each step is completed, the reaction mixtures are flushed out of the solid support and the series of reactions repeated for the addition of the next nucleotide. Once the desired length is achieved, the oligonucleotide is removed from the solid-phase support and purified to eliminate the by-products and contaminants.

Synthetic DNA molecules are widely used for various purposes in genetic engineering; for instance, as a probe to detect, via nucleic acid hybridization, specific DNA sequences. We will describe later how synthetic DNA is used in a procedure called site-directed mutagenesis to create mutations at specific locations on the genome. Finally, synthetic DNA is employed extensively as a source of DNA primers for PCR (refer to Chapter 4 for more details).

Checking incorporation of labels: After labeling, it is advisable to check the incorporation of the labeled nucleotides. For radionucleotides this can be measured using a scintillation counter. Biotin or digoxigenin label incorporation can be measured using a dot blot assay, while the incorporation of fluorochrome-labeled nucleotides can be measured by placing a small amount of probe on a glass slide and examining for fluorescence in an epifluorescence microscope with suitable filters.

3. Chemical labeling

1.2–Acetylaminofluorene: dsDNA, ssDNA and RNA can be chemically labeled with 2-acetylaminofluorene (AAF), which is highly immunogenic. AAF is introduced into the nucleic acid by reacting it with N-acetoxy-2-acetylaminofluorene (N-Aco-AAF). The main site of AAF introduction is the C-8 position of the guanine residues. The degree of modification can be controlled by varying the ratio of N-Aco-AAF to the nucleic acid, although a degree of modification of 5–10% is considered sufficient. Lower degrees of modification (e.g., about 2%) result in a weaker signal, whereas with higher degrees of modification steoric factors may interfere with antibody binding during the detection steps. The reaction is simple and rapid (20–30 min.), producing highly stable probes; however, its use has perhaps been limited by the toxicity of AAF, which is carcinogenic.

(a) Sulphonation: ssDNA probes can be chemically labeled by the insertion of a sulphone group at the C-6 of cytosine residues. The reaction is catalyzed by sodium bisulphite. The resulting sulphone group is stabilized by the substitution of the amine group on C-4 of cytosine with methoxyamine to produce a sulphonate derivative of cytosine that is highly immunogenic. Approximately 10–15% of the cytosine residues become sulphonated during this reaction. The reaction is simple, and can be performed on unpurified DNA, although DNA less than 100 bp is not efficiently labeled. The reaction is usually conducted overnight but can be performed in 4–6 hr by increasing the temperature of the reaction to 42°C and increasing the amount of methoxyamine. A kit is now available from FMC Bioproducts (Chemiprobe kit) for labeling in this way.

(b) Mercuration: Mercury can be incorporated at the C-5 position of the pyrimidine bases (RNA: cytosine, uracil, DNA: cytosine) in nucleic acids. By varying the incubation time (usually 8–16 hr), the degree of mercury modification can be manipulated. Within 8 hr, 30–40% of the uracil and cytosine residues are modified in the RNA probes and 40–50% of the cytosine bases in DNA. The toxicity of the procedure has limited its use.

(c) Photolabeling: A number of light-sensitive compounds are now available that can be used to label DNA and RNA; these include photobiotin and photodigoxigenin. These labeling methods have, however, made little impact on ISH although there is no theoretical reason why they cannot be used.

4. Primed *in situ* labeling

As an alternative to the labeling of nucleic acid followed by probe hybridization *in situ*, a method has been developed called primed *in situ* labeling (PRINS). PRINS involves first annealing a specific DNA sequence to the material; the hybridized DNA then serves as a primer for the incorporation *in situ* of labeled nucleotides by either DNA polymerase or reverse transcriptase. Cloned DNA, synthetic oligonucleotides and PCR products have all been shown to serve as primers for the *in situ* labeling reaction. The direct labeling of the chromosome after primer hybridization has been used to label chromosomes and to localize mRNA.

J. SENSITIVITY

For RNA probes, the reported maximum sensitivity is 20 mRNA copies per cell. Lynn *et al.* (4) report the detection of sequences that comprise about 0.05% of the total RNA with asymmetric probes. The sensitivity of probes labeled with digoxigenin has been reported as 1–2 genomes/cell for the human papilloma virus (HPV) genomes.

K. CONTROLS

Controls should be incorporated with all ISH experiments. At all times, Northern and Southern blots can be carried out to determine the specificity of probe hybridization to the target sequences. To establish the specificity of the hybridization several types of controls can be applied, like prior tissue digestion with RNase and DNase to remove the target sequence; competition with unlabeled probes; comparison of the signals obtained with adequate and inadequate probes; comparison of ISH with immunohistochemical analysis of the encoded protein; and a comparison of the observed result with the known or supposed localization of the mRNA within suitable test tissues. The best biological, 'negative' controls are cells, which are very similar to the cells studied, yet known to lack the target sequence. A test system comprising positive and negative cell lines is very useful for the optimization of ISH.

L. TROUBLESHOOTING

1. Poor signal

One of the major problems often encountered during ISH is getting no or poor signals. When this happens it is difficult to determine the source of the problem. When there is some signal, however faint or 'dirty', improving the signal is a matter of adjusting the parameters to reduce background, non-specific hybridization or cross-hybridization, or to increase the signal strength. It is therefore essential that the initial experimental parameters be chosen to obtain an *in situ* signal alongside suitable controls.

(a) *Probe:* To test the probe, a small quantity is bound to nitrocellulose or a charged nylon membrane and tested for the incorporation label. Radioactively-labeled probes can be tested by scintillation counting. For non-radioactively-labeled probes the result gives a semi-quantitative measure of the label incorporation into the probe. Ideally, when testing non-radioactive probes, the same detection reagents as those used for the ISH experiment should be used to verify the effectiveness of the detection reagents.

(b) *Stringency*: The ISH experiment should initially be conducted at low stringency (ideally 70–80%) probe and target similarity to encourage the binding of the probe to the target sequence *in situ*. The signal strength levels of background and cross-hybridization to non-target sequences can be assessed at this stringency. Increasing the stringency of hybridization and post-hybridization washing has the effect of removing non-specific hybridization and weakly-bound probes, but can also reduce the signal strength. When using non-radioactively-labeled probes the signal can be amplified to compensate for this.

(c) *DNA denaturation conditions*: When there is no ISH signal, it is probably because of a problem in DNA denaturation. Chromosomal DNA has a narrow window at which optimal denaturation occurs. The denaturation conditions depend on the type of sequence and the extent to which it protected by DNA-associated proteins. If the denaturation conditions are not adequate, then the target DNA will not separate into single strands and the probe DNA will be unable to hybridize to the sequence *in situ*. Conversely, if the material is over-denatured, excessive DNA loss from the material will occur. Over-denaturation is usually clearly seen after ISH, as the chromosomes appear ragged and ghost-like. If the counter-staining of chromosomes is weak and there is little or no ISH signal, then the results are suggestive of over-denaturation of the chromosomal DNA.

2. Too much background/non-specific hybridization

Increasing the stringency is one of the many ways to reduce background hybridization. Others include material pretreatment, increased washing steps, or the addition of unlabeled nucleic acids to the probe mixture. All the methods can be used separately or in combination.

(a) *Increased pretreatments*: RNase digestion, to remove RNA, is an important pretreatment when detecting DNA sequences because RNA contribute significantly to background signal. Pepsin or proteinase K pretreatments can be increased to remove cytoplasmic proteins. The acetylation of amino groups can reduce non-specific electrostatic binding of the probe. Acetylation can be used to reduce the levels of endogenous biotin, which is important if the probes are labeled with biotin.

(b) *Increased washing steps*: The background can be reduced by increasing the number and duration of the washing steps or by incorporating harsher detergents. However, care must be taken not to damage or lose the material with overzealous washing conditions.

(c) *Removal of 'sticky' non-specific binding sites*: If the background signal is not removed by increased stringency, pretreatments and/or washing, then the problem may relate to the non-specific binding of the probe or detection reagents to endogenous components in the system. This can be tested by conducting an ISH experiment without any labeled probe and examining the background labeling. The effect of probe-binding molecules, which contribute significantly to background labeling, can be reduced by adding an excess of unlabeled DNA or tRNA to the probe hybridization mix.

(d) *Detection systems for non-isotopic labels with the use of antibodies or (strept)avidin*: Non-specific binding of these molecules to the material can contribute to background labeling and can be reduced using a high concentration of a protein solution, which may preferentially bind to these 'sticky' sites. In addition, reducing the concentration of detection reagents may reduce some background.

3. Patchiness

Almost any slide with ISH shows patchiness of signal. The patchiness probably arises during tissue preparation, especially during tissue squashing or spreading. In particular, the cytoplasmic debris around chromosomes inhibits ISH. This could be because the DNA is heavily protected by the cytoplasm-preventing probe and detection reagent access, DNA denaturation, or a combination of the above. Therefore, cell-spread preparations should be as clean as possible with little interfering cytoplasm. Patchiness of signal can also arise if the reagents have not been well mixed or if air bubbles form under the coverslips during the incubation steps.

III DETAILED PROTOCOLS FOR ISH

The following protocols provide a radioactive method for ISH of RNA target sequences, employing asymmetric RNA probes (Figure 13.6), and a non-radioactive method for ISH of DNA target sequences, employing random-primed digoxigenin DNA probes (Figure 13.7).

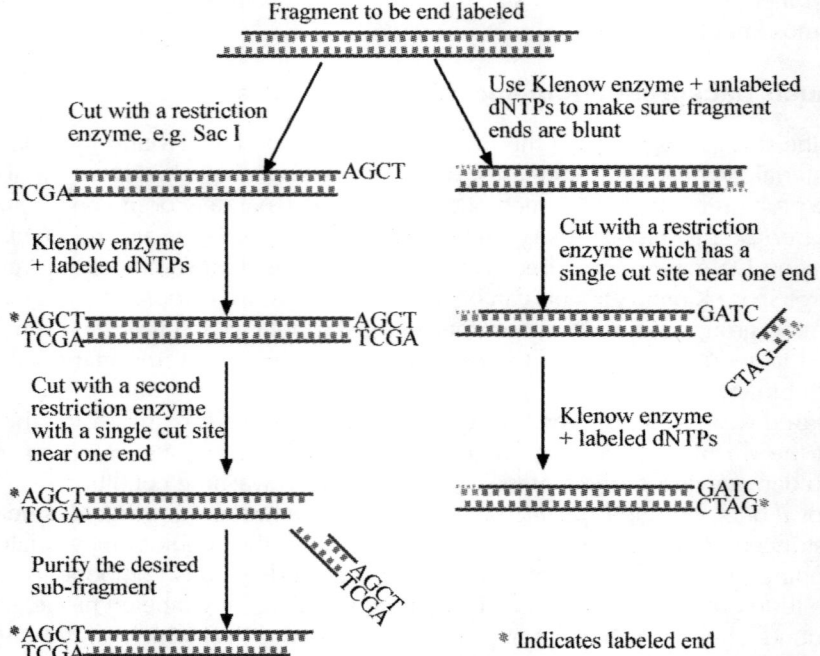

Figure 13.6 *Asymmetric end labeling of DNA fragments*

Method-1: Basic protocol for ISH with radioactive-labeled asymmetric RNA probes

Procedure

1. Linearize the plasmid with the appropriate restriction enzyme.
2. Generate a labeled RNA probe following the supplier's protocol (Transprobe, Pharmacia). At room temperature, use 1 µg of the DNA template and 100 µCi (3.7 MBq) [α-^{32}P]UTP specific activity 800 Ci/mmol (29.6 GBq µmol) (Amersham) or 100 µCi (3.7 MBq) [α-^{35}S]SUTP at 400 Ci/mmol (14.8 GBq/µmol) (Du Pont/NEN). Redissolve the ethanol-precipitated probes in a hybridization buffer containing 50–70% deionized formamide, 2X SSC, 0.5X Denhardt's solution, 1 mM EDTA pH 7.0 and 200 µg/ml sonicated salmon sperm DNA. Incubate the hybridization mixture at 85°C for 5 min. and measure a small sample in a liquid scintillation counter.
3. Apply to each slide 8 µl of the hybridization mixture with 200,000–400,000 cpm. (total) per slide, cover with a siliconized coverslip (19 mm diameter), and seal by immersion in mineral oil. Allow the hybridization to proceed for 7–18 hr at a temperature suitable for the probe sequence (40–60°C).
4. Remove the mineral oil by several rinses in chloroform. Allow the coverslips to float off in 2X SSC at room temperature. Incubate the slides in 50–70% deionized formamide and 0.1X SSC for 60 min. (one bath change) at hybridization temperature. Rinse it in 2X SSC at room temperature for 5 min. Remove the single-stranded RNA by digestion with 50 µg/ml RNase

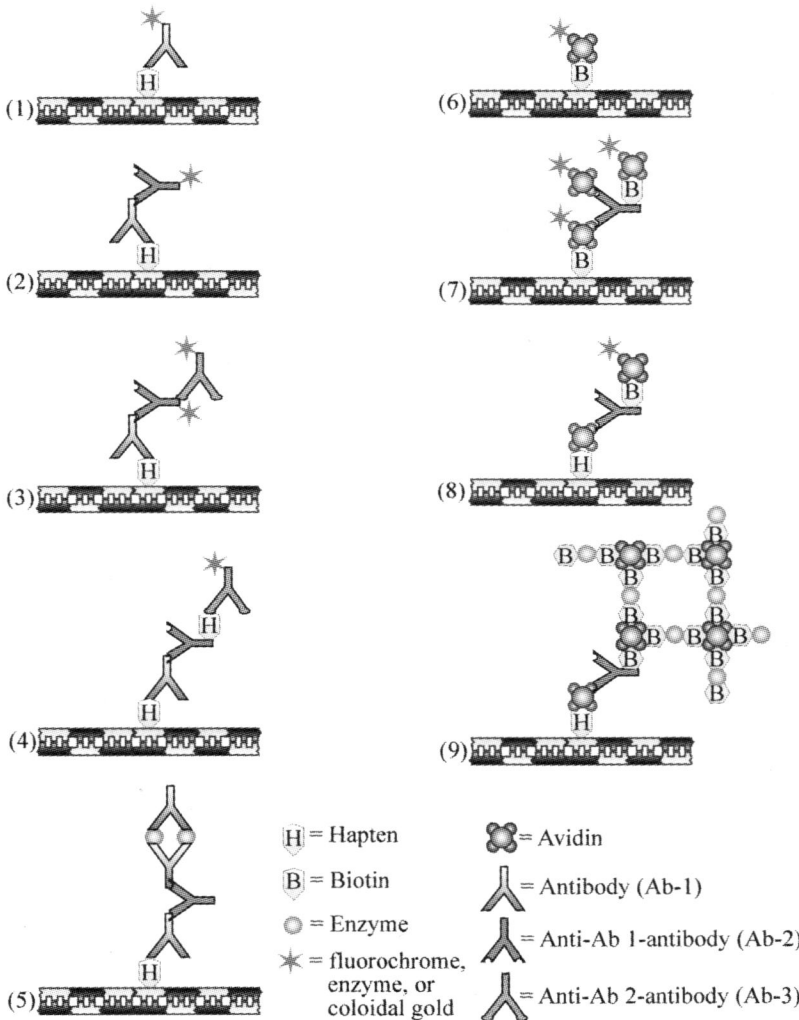

Figure 13.7 *Cytochemical detection and amplification systems that are frequently used for ISH. Hapten=biotin, digoxigenin, dinitrophenyl, fluorescein isothiocynate (FITC)*

A in 2X SSC at 37°C for 60 min. Incubate the slides in 50–70% deionized formamide and 0.1 SSC at hybridization temperature for 15 min., and finally rinse in 0.1X SSC for 5 min. at room temperature.

5. Dehydrate the slides and dip them in Ilford K5 autoradiographic film emulsion, diluted 1:1 with 1% glycerol. Store them in a dry and light-tight box containing silica gel as a desiccant for 5 days at −70°C. Alternatively, expose them to preflashed Fuji Rx X-ray film for 24–48 hr and subsequently dip them in liquid emulsion. Develop and counter-stain them with haematoxylin eosin in the usual way.

Notes

1. All solutions should be treated with 0.1% DEPC to destroy ribonucleases. They must then be autoclaved to inactivate DEPC and prevent the carboxymethylation of RNA.
2. [α–^{32}P]NTP that is more than 10 days old can substantially inhibit transcription.

Method-2: Basic protocol for ISH with digoxigenin-labeled DNA probes

Procedure

1. Linearize the plasmid with the appropriate restriction enzyme.
2. Label 0.2 µg DNA with digoxigenin-11-UTP by random priming, following the recommended protocol provided by the supplier (Boehringer). Remove the unincorporated nucleotides using Geneclean.
3. Add 140 ng/ml of the probe to a hybridization buffer containing 2X SSC, 5% (w/v) dextran sulfate, 50% deionized formamide and 0.2% low-fat skimmed milk. Apply 50 µl of the hybridization mix per slide, cover it with Gelbond (ICN), and seal the edges with nail varnish. Denature the target DNA and probe at 90°C for 10 min. and allow the hybridization to proceed for 16 hr in a humidified chamber at a temperature appropriate for the sequence used as a probe – in a protocol suitable for HPV detection – at 42°C.
4. Remove the Gelbond and wash the slides for 10 min. in 2X SSC at room temperature. Then wash the slides for 20 min. in 2X SSC at 60°C, for 10 min. in 0.2X SSC at room temperature, and finally for 20 min. in 0.2X SSC at 42°C.
5. In a minor modification to the supplier's protocol, process the slides (at room temperature) as follows: Wash for 5 min. in buffer 1 (0.1 M Tris pH 7.5, 0.15 M NaCl). Immerse them for 30 min. in buffer 1 supplemented with 20% sheep serum. Incubate them in a humidified box with buffer 1 and a 1:5000 dilution of anti-digoxigenin antibody for 20 min. Wash twice in buffer 1 for 20 min. Place them in buffer 2 (0.1 M Tris pH 9.5, 0.1 M NaCl, 5 mM MgCl$_2$) for 60 min. Apply buffer 2 and 0.33 mg/ml nitro-blue tetrazolium salt and 0.17 mg/ml 5-bromo-4-chloro-3-indolyl phosphate. Stop the reaction after 16 hr by immersion in 20 mM Tris pH 7.5, 5 mM EDTA and mount in Aquamount (BDH).

Method-3: Basic protocol for ISH to cellular RNA using fixed sections or cells

The ISH to cellular RNA is used to determine the cellular localization of specific messages within complex cell populations and tissues.

Materials

Solutions

Glass slides containing paraffin sections or cells. Dewaxing/ rehydration (dehydration) series: 3 staining dishes of xylenes, 2 staining dishes of 100% ethanol, and 1 staining dish each of 95%, 70%, and 50% ethanol. 0.2 M HCl; 2X SSC, 70°C; 1X and 3X PBS; Pronase, predigested and lyophilized (optional). 2 mg/ml of glycine in 1X PBS (optional). 4% (w/v) PFA fixative, freshly prepared at room temperature. 10 mM DTT in 1X PBS, freshly prepared at 45°C; Blocking solution, prepared immediately before use at 45°C; Triethanolamine buffer (0.1 M TEA), freshly prepared; Acetic anhydride; [^{35}S]UTP-labeled riboprobes; S-riboprobe competitor; 50 mM DTT; Hybridization mix A; Moist chamber solution A;

Wash solutions A, B, and C; RNase digestion solution; 50%, 70%, and 95% ethanol/0.3 M ammonium acetate solutions; 100% ethanol.

Equipment

Two sets of slide racks; 10 glass staining dishes; 45°, 55°, and 50°C water baths; Slide box with desiccant; 100°C heating block or water bath; 45°C incubator; Moist chambers; 4 glass staining dishes.

Procedure

1. Prepare the dewaxing/ rehydration series and 0.2 N HCl, while the slides containing the specimen sections are warmed to room temperature. Start preheating 2X SSC to 70°C.
2. Dewax the slides in dishes by three changes in xylenes, 2 min. each (this step is not necessary for slides containing cell smears).
3. Rehydrate them in dishes through the following regimen: 100% ethanol–twice, 2 min. each; 95% ethanol–2 min.; 70% ethanol–2 min.; 50% ethanol–2 min.
4. Denature the specimens for 20 min. at room temperature in 0.2 N HCl.
5. Heat-denature for 15 min. at 70°C in 2X SSC. Rinse them for 2 min. in 1X PBS.
6. Postfix the specimens in freshly-prepared 4% PFA fixative for 5 min. at room temperature. Block fixation should be for 5 min. in 3X PBS. Rinse them twice, 30 s each time, in 1X PBS.
7. Equilibrate the specimens in 10 mM DTT prepared in 1X PBS 10 min. at 45°C in a water bath.
8. Block with freshly-prepared blocking solution 30 min. at 45°C in a water bath covered with aluminum foil.
9. Rinse them twice, 2 min. each time, in 1X PBS at room temperature.
10. Equilibrate the specimens for 2 min. in freshly-prepared TEA buffer. Transfer the slide rack to fresh TEA buffer and add acetic anhydride to 0.25% final. Mix them quickly and incubate for 5 min.
11. Block the specimens for 5 min. in 2X SSC.
12. Dehydrate the specimens through 50%, 70%, 95%, and 100% ethanol (twice), 2 min. each, at room temperature.
13. Air-dry the specimens (or dry in a desiccator), ensuring that the specimens are absolutely dry before proceeding. Store the specimens in slide box with a desiccant overnight at –70°C.
14. Centrifuge the ethanol-precipitated antisense and sense ^{35}S-labeled riboprobes, as well as the S-riboprobe competitor. Dry the pellets. Dissolve each pellet in 5 μl sterile 50 mM DTT. Add 2.5 μl (half a reaction) of the S-riboprobe competitor to both the antisense and sense riboprobe.
15. Heat the dissolved probes to 100°C for 3 min. in a water bath.
16. Immediately add enough hybridization mix A to obtain a 0.3 μg/ml of the final probe concentration. Mix well and measure 1 μl (expected counts 1×10^5 cpm/μl). Place the tubes in a water bath at 45°C (hybridization temperature).
17. Set up the hybridizations by carefully spreading an appropriate amount (e.g., 20 μl/20 mm^3) of the probe on the specimens.
18. Place the specimens in a moist chamber containing the humidified chamber solution A and incubate them at 45°C for an appropriate hybridization time. Perform a series of hybridizations from 30 min–4 hr (equilibrate and seal the humidified chamber carefully).

19. During the last hour of hybridization, prepare and preheat the wash solutions A, B, and C.
20. Start washing the slides by dipping them one at a time in 100 ml wash solution A at 55°C. Immediately place them in a slide rack in a staining dish filled with wash solution A.
21. Incubate them twice, 15 min. each time, in wash solution A at 55°C (this removes much of the radioactivity). Incubate them twice, 15 min. each time, in wash solution B at 55°C. Incubate them twice, 2 min. each time, in wash solution C at room temperature. Do not allow the slides to dry during any of the hybridization and washing steps.
22. Add 500 µl of RNase digestion solution per slide, covering all the specimens, and place the slides in a moist chamber. Incubate them for 15 min. at room temperature.
23. Wash the slides twice, 30 min. each time, in wash solution C at 50°C with gentle shaking. Wash the slides twice, 30 min. each time, in wash solution A at 50°C with gentle shaking. Wash the slides twice, 5 min. each time, in 2X SSC at room temperature.
24. Dehydrate through the following regimen (2 min. each): 50% ethanol/0.3 M ammonium acetate; 70% ethanol/0.3 M ammonium acetate; 95% ethanol/0.3 M ammonium acetate; 100% ethanol.
25. Air-dry slides. Expose the slides at least overnight against a film and then perform emulsion autoradiography.

Method-4: In situ PCR and hybridization to detect low-abundance nucleic acids

This method is a novel approach for detecting low-abundance nucleic acid targets in nuclear and cytoplasmic regions by the amplification of specific target sequences using an *in situ* polymerase chain reaction (ISPCR). This is followed by *in situ* reverse transcription (if the target sequence is RNA) and ISH of the amplified sequences.

Materials

Slides containing fixed specimens; 0.3% H_2O_2 in PBS (prepare freshly); PBS; 1 mg/ml proteinase K; Rinse buffer (RNase free DNase solution without DNase); DEPC-treated water; 10X AMV/MoMuLV reaction buffer: 100 mM Tris.Cl pH 8.3, 500 mM KCl, 15 mM $MgCl_2$, Prepare fresh; 10 mM 4dNTP mix: 10 mM each dNTP in TE buffer, pH 7.5 (store at −20°C); 40 U/µl RNasin; 20 µM downstream primer; 20 U/µl avian myeloblastosis virus (AMV) or Moloney murine leukemia virus (MoMuLV) reverse transcriptase or 20 U/µl SuperScript II; 10 M DTT; 2 H_2O_2 25 µM forward and reverse primers; 1 M Tris.Cl pH 8.3; 1 M KCl; 100 mM $MgCl_2$; 0.01% (w/v) gelatin; 5 U/µl *Taq* DNA polymerase; 100% ethanol; 2X SSC; 55°C, 92°, 95°, and 105°C heating blocks accommodating glass slides; Moist chamber; 42°C incubator; 20 × 60-mm glass coverslips; Clear nail polish or varnish; Thermal cycler accommodating glass slides.

Procedure

1. Incubate the slides with fixed specimens for 5–120 s on a 105°C heating block.
2. Incubate the slides overnight at 37°C or room temperature in 0.3% H_2O_2 in PBS to inactivate endogenous peroxidase activity, and then wash once with PBS.
3. Dilute 1 ml of 1 mg/ml proteinase K in 150 ml PBS (6 µg/ml final). Immerse the slides in this solution and incubate them at room temperature. After 5 min., examine the cells under a microscope at 400X; if the majority of cells of interest exhibit small round "peppery dots", proceed immediately to step 4. Otherwise, continue incubation up to 60 min., examining

the slide at 400X at 5 min intervals and proceeding immediately to step 4 at the point where cell surface bubbles appear.
4. Heat the slides for 2 min in a 90°C heating block to inactivate the proteinase K, then rinse for 10 s in PBS and 10 s in water. Allow the slides to air dry. To reverse transcribe RNA targets, proceed with step 5 or step 7; to amplify DNA targets, proceed to step 10.
5. Add 10 µl of RNase-free DNase solution to each well of the slides. Incubate them overnight at 37°C in a moist chamber.
6. Rinse the slides once with rinse solution, then twice with DEPC-treated water and allow them to air dry.
7a. If you are using AMV or MoMuLV reverse transcriptase: Make up the following reverse transcription cocktail (20 µl total volume):
2 µl 10X AMV/MoMuLV RT buffer (1X final)
2 µl 10 mM 4dNTP mix (1 mM final)
0.5 µl 40 U/µl RNase (1 U/µl final)
1.0 µl 20 µM downstream primer (1 µM final)
0.5 µl 20 U/µl AMV or MoMuLV reverse transcriptase (0.5 U/µl final)
8 µl of DEPC-treated water.
7b. If you are using SuperScript II reverse transcriptase: Make up the following reverse transcriptase cocktail (20 µl total volume):
4 µl 5X reaction buffer (supplied with enzyme, 1X final)
2 µl 10 mM 4dNTP mix (1 mM final)
0.5 µl 40 U/µl RNase (1 U/µl final)
1.0 µl 20 µM downstream primer (1 µM final)
0.5 µl 20 U/µl SuperScript II (0.5 U/µl final)
1.2 µl 0.1 M DTT (6 mM final)
4.8 µl of DEPC-treated water.
8. Add 10 µl of the reverse transcription cocktail to each well of the slides and carefully cover each well with a 20 × 60-mm glass coverslip. Incubate them 1 hr at 42°C or 37°C in a moist chamber.
9. Incubate the slides for 2 min on a 92°C heating block. Remove the coverslips and wash the slides twice with water.
10. Make up the following amplification cocktail (100 µl total volume):
5 µl 25 µM forward primer (1.25 µM final)
5 µl 25 µM reverse primer (1.25 µM final)
2.5 µl 10 mM 4dNTP mix (200 µM each dNTP final)
1.0 µl 1.0 M Tris.Cl, pH 8.3 (10 mM final)
5.0 µl 1.0 M KCl (50 mM final)
2.5 µl 100 mM $MgCl_2$ (2.5 mM final)
10 µl 0.01% gelatin (0.001% final)
2 µl 5 U/µl *Taq* DNA polymerase (0.1 U/µl final)
66 µl water.
11. Layer 8µl (for 3-wel slide) of the ISPCR amplification cocktail onto each well using a 20 µl micropipettor, so that the whole surface of the well is covered with the solution.
12. Place a 20 × 60-mm glass coverslip over each slide and carefully seal the edge of the coverslip to the slide with clear nail polish or varnish.
13. Incubate the slides for 90 s on a 92°C heating block then transfer them to a thermal cycler.

14. Carry out PCR using the following amplification cycles or optimized conditions.
    30 cycles: 30 s   94°C (denaturation)
    1 min   ~45°C (annealing)
    1 min   72°C (extension)
    Final step: indefinitely   4°C (hold).
15. Remove the slides from the thermal cycler and soak them in 100% ethanol for 5 min. to dissolve the nail polish. Pry off the coverslips using a razor blade or other fine blade. Scratch off any remaining nail polish so that fresh coverslips can be placed evenly in the hybridization/detection steps.
16. Incubate the slides for 1 min. on a 92°C heating block, and then soak them for 5 min. in 2X SSC at room temperature. The slides can be stored for 2–3 weeks at 4°C.

Hybridization and detection of ISPCR-amplified target material

1. Prepare the following hybridization mix:
    2 µl 200 pM probe (pM final)
    50 µl deionized formamide (50% final)
    10 µl 20X SSC (2X final)
    10 µl 100X Denhardt's solution (10X final)
    10 µl 10 mg/ml sonicated salmon sperm DNA (1 mg/ml final)
    1 µl 10% SDS (1% final)
    7 µl water.
2. Add 10 µl of the hybridization mix to each well of the slides containing ISPCR-amplified nucleic acids. Cover the wells with coverslips and heat the slides for 5 min. on a 95°C heating block.
3. Incubate the slides for 2–4 hr at 48°C in a moist chamber.
4. Wash the slides twice in 100 mM Tris.Cl pH 7.5/150 mM NaCl, 15 min. each time at room temperature, then wash once more in alkaline phosphatase substrate buffer for 5 min. at room temperature.
5. Preheat 50 ml of alkaline phosphatase substrate buffer to 37°C in a Couplin jar. Add 200 µl of 75 mg/ml NBT and 166 µl of 50 mg/ml BCIP. Mix well, then incubate the slides in this solution at 37°C until the desired level of signal is achieved (10 min–2 hr), then stop the reaction by rinsing the slides in several changes of deionized water.
6. Stain the slides for 5 min. at room temperature in either 0.2% Gills hematoxylin (if peroxidase-based color development was used) or in 1% nuclear fast red stain (if alkaline phosphatase-based color development was used). Rinse the slides in several changes of tap water.
7. Mount the slides by applying one drop mounting medium per well and covering with a coverslip. View them immediately, taking care not to disrupt the coverslip, or allow the mounting medium to dry overnight at room temperature.

IV APPLICATIONS OF ISH

In situ hybridization can be used to detect DNA sequences, which are part of normal or abnormal genes, to study gene expression by mRNA detection, or to identify the chromosomal location of DNA sequences. The method is particularly useful if target sequences are distributed in a non-random way in tissues, for the visualization of heterogeneity and the study of cell differentiation. The method described for RNA–RNA hybridization is generally applicable to the rapid screening of a small number of cultured cells for the expression of oncogene mRNA. Under the right circumstances, the problem of extracting rare RNAs in sufficient amounts for detection and the intracellular or intranuclear localization of RNA is possible. Similar methods have been used to visualize homeotic gene expression in developing larvae. DNA–DNA ISH is also suitable for the detection of viral genomes in sections of routinely-processed archival paraffin blocks of human tissues.

V SUGGESTED READING

Lang SM, Wyllie AH and Conkie D (1992) *In situ* hybridization, In Freshney, RI (Ed) *Animal Cell Culture: A Practical Approach* (2nd Edition), IRL.

Leitch AR, Schwarzacher T, Jackson D and Leitch IJ (1994) *In situ Hybridization: A Practical Guide*, Bios Scientific Publishers.

Lynn DA, Angerer LM, Bruskin AM, Klein WH and Angerer RC (1983) Localization of a family of mRNAs in a single cell type and its precursors in Sea urchin embryos, *Proc. Natl. Head. Sci.*, **80**: 2656.

Manuelidis L, Langer-Safer PR and Ward DC (1982) High-resolution mapping of satellite DNA using biotin-labeled DNA probes, *J. Cell Biol.*, **95**:619–625.

Meyerowitz EM (1987) *I- situ* hybridization to RNA in plant tissue, *Plant Mol. Biol. Rep.*, **5**:242–250.

Pinkel D, Straume T and Gray JW (1986) Cytogenetic analysis using quantitative, high-sensitivity, fluorescence hybridization, *Proc. Natl. Acad. Sci. USA*, **83**:2934–2938.

Raikhel NV, Bednarek SY and Lerner DR (1989) *In situ* RNA hybridization in plant tissues, In, Gelvin SB, Schilperoort RA, and Verma DPS (Eds) *Plant Molecular Biology Manual*, Kluwer Academic Publishers.

Speel EJM, Ramaekers FCS and Hopman AHN (1995) Cytochemical detection systems for *in situ* hybridization, and the combination with immunocytochemistry: Who is still afraid of red, green and blue? *Histochemical Journal,* **27**: 833–858.

PART – IV
Recombinant DNA and Genetic Engineering

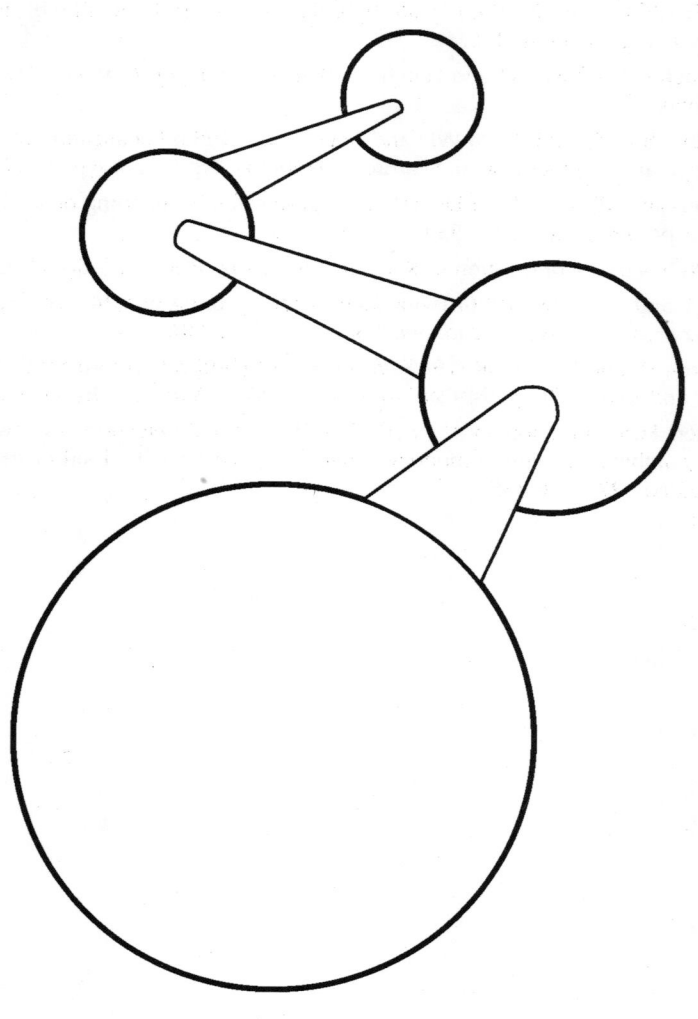

14

Fundamentals of Recombinant DNA Technology

I INTRODUCTION

The most dramatic developments in molecular biological began in the early 1970s and will continue to have significant effects on biology far into the future. One of the revolutionary changes that have occurred in the biological sciences over the past few decades can be directly attributed to the ability to manipulate DNA in defined ways. This chapter revises briefly the principal steps that are involved in recombinant DNA technology.

Composite molecules in which foreign DNA has been inserted into a vector molecule are sometimes called DNA *chimeras* because of their analogy with the mythological Chimaera – a creature with the head of a lion, body of a goat and the tail of a serpent. However, a much better analogy that represents the composite form created by a recombination of biological systems from quite diverse origins with little DNA homology is available in Hindu mythology. Lord Ganesha, also called Vigneshwara (Figure 14.1), is a classic and mythological example of a recombinant form (distantly related body parts grafted together). Ganesha was created by combining the head of an elephant and the body of a human by the power of three Hindu gods **Shiva** (destroyer, in molecular term means, **digestion**), **Vishnu** (conservator, in molecular term means, **ligation**) and **Bramha** (creator, in molecular term means, **construction**). Digestion, ligation and creation are the main molecular events involved in the generation of various biological molecules that constitute life forms and biological activities. A novel composite creation, Lord Ganesha, represents the power of wisdom. In the same way, many people are hopeful that recombinant DNA technology will have the potential to generate novel products that may meet humanity's multiple needs.

The construction of such composite or artificial recombinant molecules has also been termed **genetic engineering** or **gene manipulation** (also called **molecular cloning** or **gene cloning**) because of the potential for creating novel genetic combinations by biochemical means. This field of study encompasses a number of methodologies that enable new combination of genetic material to be artificially constructed in the laboratory and then extended to the field. This emerging new technology led to the construction of DNA molecules composed of nucleotide sequences taken from different sources. Genetic recombination that results from the breaking and rejoining of DNA molecules is subjected to strong taxonomic constraints *in vivo*. However,

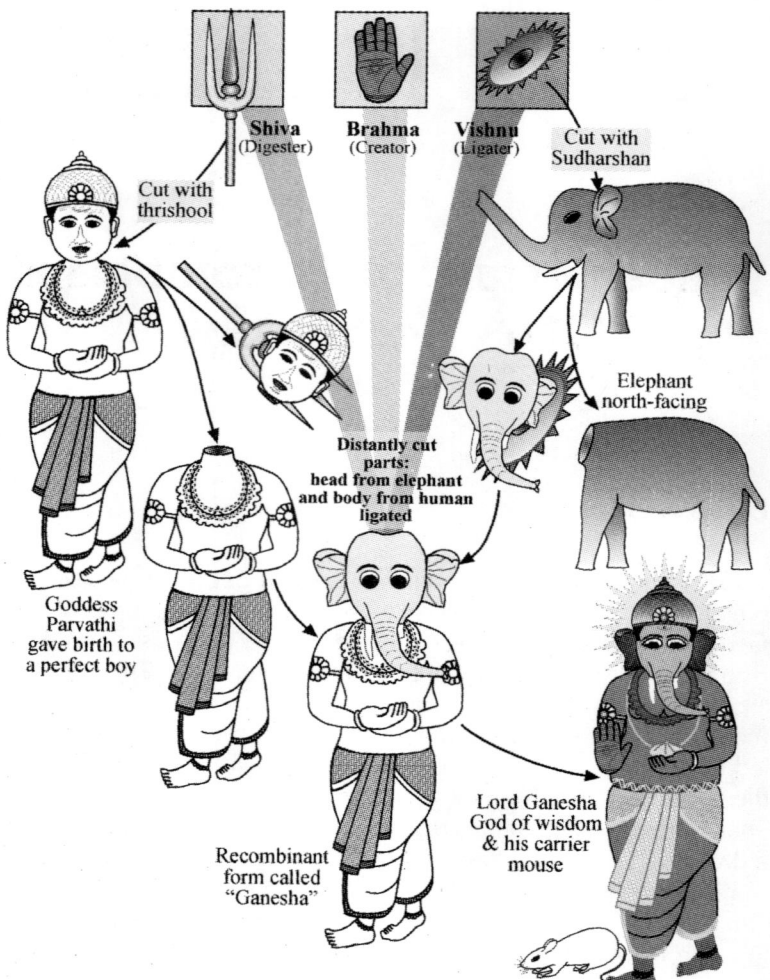

Figure 14.1 *Hindu mythological example for distant recombinant form*

recombination between DNA molecules *in vitro* is not subjected to these taxonomic restrictions. Consequently, genetic material from organisms of quite diverse origins (across the kingdoms) with little DNA sequence homology can now be introduced into host cells for propagation and expression.

The products of these innovations, recombinant DNA molecules, opened exciting new avenues of investigation in molecular biology and genetics. Genetic engineering is the application of recombinant technology to study the manipulated genes, and which enables amplification of any particular DNA segment, regardless of source, within bacterial or eukaryotic host cells. Vectors and other physical means (e.g. micro-injection, electroporation, gene-gun, etc.), carry these new sequences into host cells where they can be propagated and amplified many times over. This allows multiple copies of a piece of genetic material to be isolated and sequenced, and also, in various cases, to be transcribed into mRNA and translated into protein for various uses.

II BASICS OF RECOMBINANT DNA TECHNOLOGY

Recombinant DNA is generated *in vitro* by covalently joining DNA molecules from different origins. Recombinant DNA technology mainly involves the construction and application of recombinant DNA molecules, in all organisms and for various purposes. The basis of recombinant DNA technology was the discovery of a set of enzymes and techniques, which allowed DNA to be manipulated and modified precisely. The major enzymes used and their functions are discussed in detail in Chapter 15. The most fundamental techniques and steps involved include: (1) Cutting DNA with sequence-specific bacterial endonucleases (**restriction endonucleases**) to generate defined DNA fragments. (2) Joining of the restriction enzyme digested fragments by using DNA ligase enzyme. (3) Separating the nucleic acids on the basis of size and structure by gel electrophoresis. (4) Detecting of specific sequences in a complex mixture by nucleic acid hybridization and other techniques. (5) Introducing DNA into host cells for further propagation (6) Amplification of specific DNA molecules, either by molecular cloning or the polymerase chain reaction. (7) Screening recombinant molecules by various methods, such as the various blotting techniques, the hybridization techniques and the immunotechniques.

In cell-based *molecular cloning*, the DNA fragment of interest is amplified *in vivo* in a population of proliferating host cells, unless a DNA molecule is inserted into host cell chromosomes, will not replicate. Because it requires a sequence origin of replication, which is provided by a **cloning vector**, a genetic element derived from a plasmid or virus or artificial chromosome which has been modified to carry extra DNA (**donor, foreign, insert** or **passenger DNA**). Plasmids are widely used as vectors for transformation and cloning of different size of DNA fragments. The donor and vector DNAs are covalently joined, producing a **recombinant vector** (see Chapter 15, **Gene constructs and cloning vectors**). The recombinant molecule is introduced into host cells, and replicates as the cells proliferate, generating a number of copies (clones) of the donor DNA sequence. Since the recombinant vector is episomal and can therefore be separated from chromosomal DNA on the basis of its unique physico-chemical properties (see Chapter 7), and this facilitates the recovery of large amounts of cloned donor DNA from bulk cultures. Alternatively, DNA can be amplified *in vitro* using PCR (refer to Chapter 12), which is quicker and easier than molecular cloning and is extremely sensitive and robust. However, only relatively short sequences can be amplified and the cost of amplification is very high compared to growing bacteria containing recombinant plasmids. The enzymes used in PCR tend to be less accurate than those found in cloning hosts, leading to some heterogeneity in PCR products. Furthermore, no previous knowledge of the donor DNA sequence is required for molecular cloning, whereas PCR requires primers designed to anneal at sites flanking a specific DNA target.

The DNA inserted into a vector for cloning arises from either a primary or secondary source. A **secondary source** is a previously isolated DNA clone – the donor DNA isolated from within that clone for further manipulation, in a procedure known as **subcloning**. On the other hand, **primary cloning** is the isolation of donor DNA from its original or **primary source**, which is either whole genomic DNA or a population of cDNAs. Although it is occasionally possible to isolate a particular cDNA or genomic DNA fragment directly from a source of low complexity. Primary cloning usually requires a DNA library – a representative collection of all the DNA fragments from a given source cloned in vectors (refer to Chapter 17). The desired clone must be isolated from this library by exploiting some unique property of the donor DNA, e.g. its sequence, structure or functional properties of the protein it encodes (refer to Chapter 19).

Once a desired DNA clone has been isolated, it may be exploited in a great number of ways. The research applications may fall into one or more of the following broad categories, which are discussed in more detail in this and other chapters: (1) Characterization of gene and genome structure, and of gene expression; (2) Physical gene mapping and positional cloning (see Chapter 11); (3) Expression-cloning; (4) Functional analysis of genes and their products and regulatory elements (see Chapters 20–22); (5) *In vitro* mutagenesis; (6) Gene transfer and transgenesis (see Chapters 18 and 20–22).

Principles of molecular cloning

Molecular cloning is an *in vitro—in vivo* technique for producing large quantities of a particular DNA fragment. This technology generally involves several steps. Briefly, the major steps involved in gene manipulation are as shown in Figure 14.2 and further discussed below:

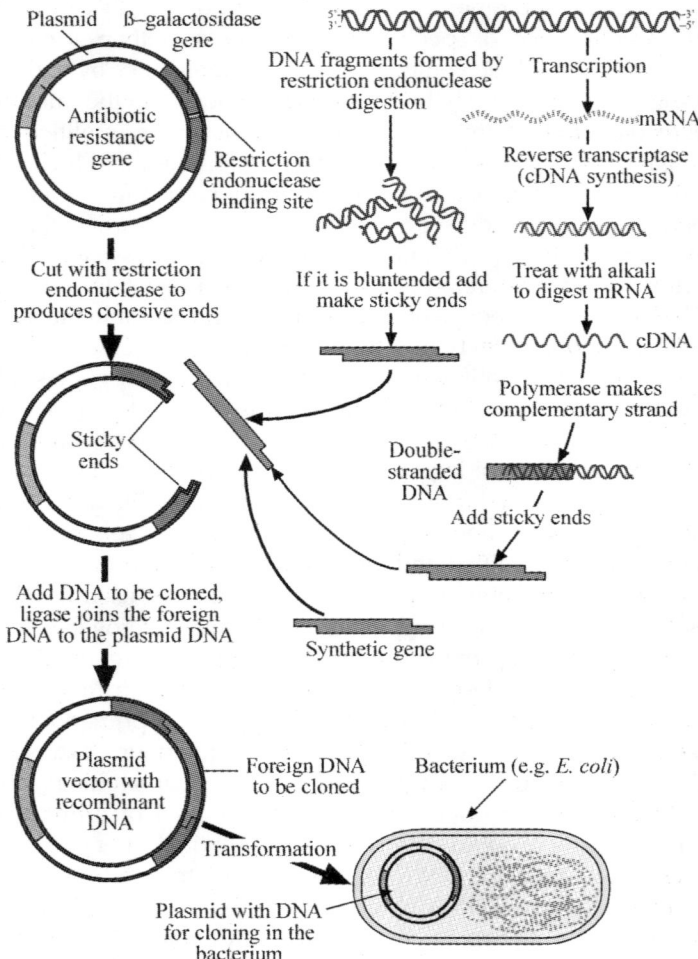

Figure 14.2 *Genetic engineering. Different steps involved in a procedure designed to clone a gene*

1. Isolation and purification of DNA fragments coding for proteins of interest, or which can be synthesized chemically (**Source of DNA**).
2. Isolation and purification of plasmid DNA (**vector**, *a vehicle for DNA transfer and multiplication*).
3. Generation of DNA fragments for genetic manipulation (***generating DNA fragments for cloning: restriction endonuclease digestion***).
4. Splicing of the DNA fragments into a composite molecule (creating ***recombinant DNA*** molecules).
5. Transfer of the recombinant vector into a host cell for replication (***transformation***).
6. Selection of cells that carry the desired recombinant DNA molecules (***replication as clones***).
7. Application of recombinant DNA technology to different systems and in diverse areas (**use in academics and for commercial purposes**).

III ISOLATION OF GENES FROM CELLULAR CHROMOSOMES

Because a single gene is only a very small part of any chromosome, isolating a DNA fragment containing a particular gene often requires two procedures. First, a DNA library is constructed that contains many thousands of DNA fragments derived from a genome. Second, the DNA fragment containing the gene of interest is identified by taking advantage of the key property that distinguishes it from other DNA fragments – its sequence. Chromosomal location of DNA sequences of a gene is a critical step for the isolation of genes. On the basis of phenotypic effect and organization in the chromosomes, genes (nucleotide sequences) can be broadly subdivided into the following groups:

1. **Simple genes**: Their product is a single polypeptide or single RNA (e.g., ribosomal RNA, tRNA, histone, insulin, or monomeric enzymes).
2. **Complex genes**: Many multimeric proteins are aggregates of several non-identical polypeptide subunits. All of these genes coding for the various parts of the functional enzyme aggregate may not be clustered together on the chromosome. For example, RNA polymerase holoenzyme, tryphtophan synthetase, fatty acid synthetase, hydroxylases etc.
3. **Operons**: Many genes specifying degradative and biosynthetic pathways are clustered adjacent to each other on the chromosome, and these gene clusters are called operons. Many operons are polycistronic and transcribe a single RNA molecule that codes for more than one protein. Examples are *lac* operon and histidine operon.
4. **Regulon**: Certain genes specifying a biosynthetic function such as arginine biosynthesis are organized in several mini-operons widely scattered on chromosomes. However, all such genes are regulated coordinately and are called regulons.
5. **Multiple regulons**: Many products, e.g. antibiotics, are assembled from precursors derived from several different pathways. For example, L-cysteine, L-valine, and L-α-aminoadipate are involved in the biosynthesis of β-lactam antibiotics. Multiple regulons participating in the synthesis and regulation of the biosynthesis of such complex molecules are probably scattered all over the chromosome. Isolation of all genes involved in antibiotic synthesis can therefore be a difficult task.

Genes are composed of stretches of nucleotide sequences in a DNA. To isolate genes one must isolate DNA. Isolation of specific genes from the total DNA of eukaryotes is not easy and many properties of genes are used to isolate the genes. Most genes in eukaryotic genomes are single copies. Very few genes exist as multi-copies. To increase the probability of success in cloning a particular gene from such a highly complex genome, DNA sequences have to be enriched before cloning into a vector. A combination of several methods could yield an enrichment of several thousand folds. Genes located on plasmids or phages are greatly enriched by purifying the plasmid or phage DNA. Genes located on eukaryotic DNA has to be enriched by many techniques such as RFLP, PCR, genomic and cDNA libraries etc. After preparations that have been enriched for a desired gene, recombinant DNA technology is used to isolate these genes.

A. THE BASIS OF SELECTION OF GENES

1. Function

Many genes can be selected by their function if mutants of a host organism lacking that function are available. For example, the genes determining resistance to antibiotics are located on plasmids. These genes can be enriched by purifying plasmid DNA from the bacteria that harbor the plasmid containing antibiotic resistance genes on nutrient agar plates supplemented with antibiotics. Plasmids containing different drug resistance genes can be combined *in vitro* and transformed into drug-sensitive bacteria. The transformant clones can be selected by directly plating cells on nutrient agar plates supplemented with antibiotics. On such media, only cells with acquired recombinant plasmids survive since they carry the antibiotic resistance genes. However, such experiments do not tell about the activity of the genes they associated with, the base sequences of regulatory elements (promoters) or the level of gene expression.

(a) DNA libraries: Techniques for the construction and screening of a complete library of clones for any genome is a practical starting point for amplifying specific genes. A complete library of genome fragments is a set of independent clones that, statistically, contain the entire genome among the recombinant DNA molecules. Maintenance and screening of DNA libraries are discussed in Chapter 15.

(b) *In situ* hybridization: Colonies or phage plaques with the desired gene can be identified *in situ* by hybridization to a RNA/DNA labeled (radioactively or fluorochrome) probe complementary to the desired gene. This procedure is more useful when the desired clones cannot be assayed for the function or where cells can be fixed on slides and stained (eukaryotic cells). Under such circumstances, DNA sequences for the gene of interest are detected by taking advantage of the complimentarity of the RNA or DNA probe (more details in Chapter 17).

(c) *In situ* immunoassay: Radioisotopes such as ^{125}I have been used for many years as labels for monitoring the distribution of reagents in immunological assay systems. Non-radioactive labels such as bacteriophages, enzymes, stable free radicals, and fluorescent and chemiluminescent groups have also been used. The method depends on the presence of the gene product and requires expression of the cloned gene in the host. Both viral and phage vectors can be used, since the technique is applicable to phage plaques as well as colonies. The combination of radio-immunoassays with gel electrophoresis can be used to characterize immunoreactive proteins. Enzymes involved in the biosynthesis of peptide antibiotics have been purified to homogeneity.

(d) Sib selection: This method can be used to isolate clones from a genome library. The whole genome library is divided into pools, each containing many different clones. The pool is tested by the pertinent assay procedure for the presence of a desired gene. The pool containing the gene of interest is subdivided and re-tested until the individual clone is identified.

2. Physical differences

(a) Density: If the genes differ significantly in base composition from the total cellular DNA and is present in multiple copies per genome, the repeated sequences band in a CsCl gradient as a satellite in addition to the main chromosomal band. Ribosomal RNA genes of *Xenopus leavis* were isolated on the basis of density differences and were the first eukaryotic genes to be cloned in *E. coli*. Many genes with multiple copies in different organisms have been enriched by CsCl and sucrose gradient methods.

(b) Size: Many enrichment procedures utilize information about the size of the DNA fragment required for the desired gene. Restriction enzyme-generated fragments are separated according to size by gel electrophoresis. For organisms with a high efficiency of transformation, the bands of endonuclease-digested DNA can be isolated after electrophoresis and correlated with specific genes after transformation. Gel electrophoresis is routinely used for determining the fragment composition of recombinant DNA molecules constructed by ligation of cohesive ends generated by restriction enzymes. As little as 10μg of DNA per band can be visualized by staining the gel with ethidium bromide. DNA can be extracted from these resolved bands and used for further analysis (restriction digestion, nucleotide sequencing, transformation etc.). Standards of λ and FX174 DNA already cleaved with restriction enzymes are commercially available and can be used as markers for determining size of unknown DNA fragments.

(c) Electron microscopy: The most accurate and reliable method for determining the size of DNA is electron microscopy. Heteroduplex analysis is used to build the map of λ DNA digested with *Eco*RI. Such maps of plasmid and phage genomes as well as other DNA segments are routine in these days and can be confirmed by comparing the electrophoretic pattern of DNA fragments obtained by restriction cleavage of recombinant molecules differing by deletion or by the addition of DNA from other genomes. New methods for visualizing RNA–DNA hybrids by electron microscopy permit precise localization of individual genes.

(d) Genetic methods: DNA coding sequences for yeast tyrosine transfer RNA have been located on eight different *Eco*RI restriction fragments. There is significant natural variation among various yeast strains in the sizes of *Eco*RI and *Hind*III fragments that hybridize with tyrosine tRNA. Restriction fragments of various sizes hybridizing with the same tRNA genes occupy genetically homologous chromosomal sites. These size variants of *Eco*RI-*Hind*III restriction fragment behave as Mendelian traits. Now, cloned DNA sequences can be correlated with the corresponding genetic loci by transforming into appropriate strains of yeast and testing for functional expression of the suppressor. Restriction fragment size variants have been used to analyze inheritance of plant and animal mitochondrial DNA, recombinant DNA (rDNA), and 2 μm plasmid sequences of yeast, and *Drosophila* mitochondrial DNA. Widespread occurrence of restriction fragment size variants can have an impact on molecular genetics similar to that of electrophoretic variants on proteins and may speed up the genetics of eukaryotes.

3. Complementary RNA

(a) cDNA: Many genes in eukaryotic DNA are split genes. The mRNA coding sequences (exons) are interrupted by intervening sequences (introns) of unknown function. These split genes are transcribed into an RNA precursor that matures into mRNA by a splicing mechanism. Since split genes are not found in bacteria, they are not expressed in *E. coli*. Therefore, cDNA made *in vitro* has been used to get an expression of several mammalian proteins in *E. coli*. Several methods for isolating biologically active RNA from various sources have been described. Total RNA containing poly (A) tails is sometimes translated in a cell free system and the polysomes precipitated by the purified antibody. The enriched RNA is further purified by affinity chromatography on an oligo(dT) column. RNA-dependent DNA polymerase (reverse transcriptase) from avian myeloblastosis virus is used for the synthesis of complementary DNA (cDNA) strand from RNA. The cDNA can be cloned into a vector by homopolymer tailing, blunt-end ligation, or utilization of linkers.

(b) Complementary RNA as a probe: Development of gene-specific probes allows the selection of the recombinants containing the sequences of interest. A gene-specific probe functions by specific hybridization. Ribosomal RNA has been iodinated (^{125}I) and successfully used as a probe to detect ribosomal gene sequences in mouse and bacteria, *Euglena gracilis*. 3'-poly(A) tail of most eukaryotic mRNAs can be hybridized to column-bound poly(U) or poly(T). Specific mRNAs can be isolated by hybridizing total mRNA to the complementary DNA sequences obtained by restriction digestion and immobilized on nitrocellulose filters after fractionations on agarose gels. The RNA species that are so selected by Southern hybridization are eluted and their products identified by translation in a cell-free system.

(c) Hybridization arrest of complementary RNA translation: *In vitro* cell-free protein-synthesizing system using wheat germ embryos can be used to translate mRNA into proteins. Proteins can be identified by urea polyacrylamide gel electrophoresis and specific antibody reactions. A bank of clones containing cDNA is generated. Their recombinant plasmid DNA is sonicated, denatured, and hybridized to the enriched mRNA mixture which is to be translated. A recombinant DNA that will arrest the translation of one of the mRNAs into the desired protein is further used as a probe for purifying more mRNA or DNA. When a specific mRNA cannot be purified sufficiently for it to be used as a probe, hybridization arrest of RNA translation can be employed to select specific DNA sequences.

(d) R-loop hybridization: RNA can hybridize to double-stranded DNA in the presence of formamide by displacing the identical DNA strand. The three-stranded complex so formed is called an R-loop and can be separated by CsCl density gradient centrifugation. In order to obtain an efficient rate of RNA–DNA hybridization, the reaction conditions must be near to or above the melting temperature of the DNA sequences that will hybridize with RNA.

4. Synthetic nucleotide primers

When poly(A)-RNA isolated from rat insulinoma is translated in a wheat germ cell-free system, about 25% of the synthesized protein is immunoprecipitated with anti-insulin sera. However, restriction endonuclease *Hae*III analysis of the dscDNA represents insulin-coding sequences. To overcome the difficulties of obtaining bacterial clones containing full-length insulin cDNA inserts, Chan et al. (1979) constructed a highly specific cDNA probe corresponding in sequence to the 5'-region of rat insulin mRNA. This was achieved by using a specific deoxydecanucleotide as a primer for the reverse transcription of total rat insulinoma poly(A)-RNA. The synthetic decanucleotide primer d(C-T-C-C-A-G-C-A-G) corresponded in sequence to the region coding for amino acids 11–13 and the first nucleotide of the codon for residues 14 of the insulin B chain as documented by Ullrich et al. (1977). Colony hybridization with decanucleotide-primed cDNA was used to identify seven clones coding for rat insulin I mRNA and nine for rat insulin II mRNA.

B. ISOLATION OF GENES CODING FOR KNOWN SPECIFIC PROTEINS

Isolation of genes that are coding for a known protein can be isolated by different methods. A standard method generally followed is as explained in Figure 14.3. First, the protein product of the known gene must be isolated. Using these proteins as an antigen, antibodies can be raised in an appropriate organism. Later these antibodies can be used for identification of proteins, which are still under synthesis (attached to ribosomes) in polysomes. The antibody precipitates with specific protein in the polysomes, which can be separated out and mRNA from this complex can be isolated. Using this mRNA, cDNA is synthesized and then cloned into the appropriate vector. This cDNA can be used as specific DNA probe for identification of genes from DNA libraries (genomic- and cDNA-library).

C. ISOLATION OF GENES WITH KNOWN OR UNKNOWN GENE PRODUCTS

If molecular probes, either DNA or RNA are available, they can be used for isolation of a specific gene. These probes can be obtained either from another plant species or be artificially synthesized using a part of the amino acid sequence of the protein product of the gene in question. The probes obtained from one plant species and used for another plant species are called **heterologous probes**. Heterologous probes have been very effective in identifying gene clones during **colony hybridization** or **plaque hybridization** or **Southern blots**. By this method the gene for **chalcone synthase** (**CHS**) has been isolated from *Antirrhinum majus* and *Petunia hybrida* using heterologous cDNA probe from parsley. Some other examples are *Antirrhinum* cDNA probe which was used for the isolation of CHS gene from barley. Heterologous probes from maize were used for the isolation of barley genes waxy gene (*Wx*) and aleurone gene (*A1*). Heterologous probes are generally used with cDNA library.

Figure 14.4 describes steps involved in the use of cDNA or synthetic probes for the isolation of a specific gene. Once we purify the protein using standard techniques of protein purification, it can be sequenced to 5–15 consecutive amino acids. This sequence data can be used for the synthesis of corresponding oligonucleotides (See Chapter 4). These oligonucleotides may then be directly utilized for screening of cDNA or genomic libraries for isolation of specific genes.

D. ISOLATION OF GENES CODING FROM AN UNKNOWN GENE PRODUCT

In some situations we know the phenotypic effect but we do not have any information about the gene. To isolate such genes we often use **reverse genetics**. Reverse genetics is an approach to discovering gene function, which begins from the DNA sequence (gene) and protein and

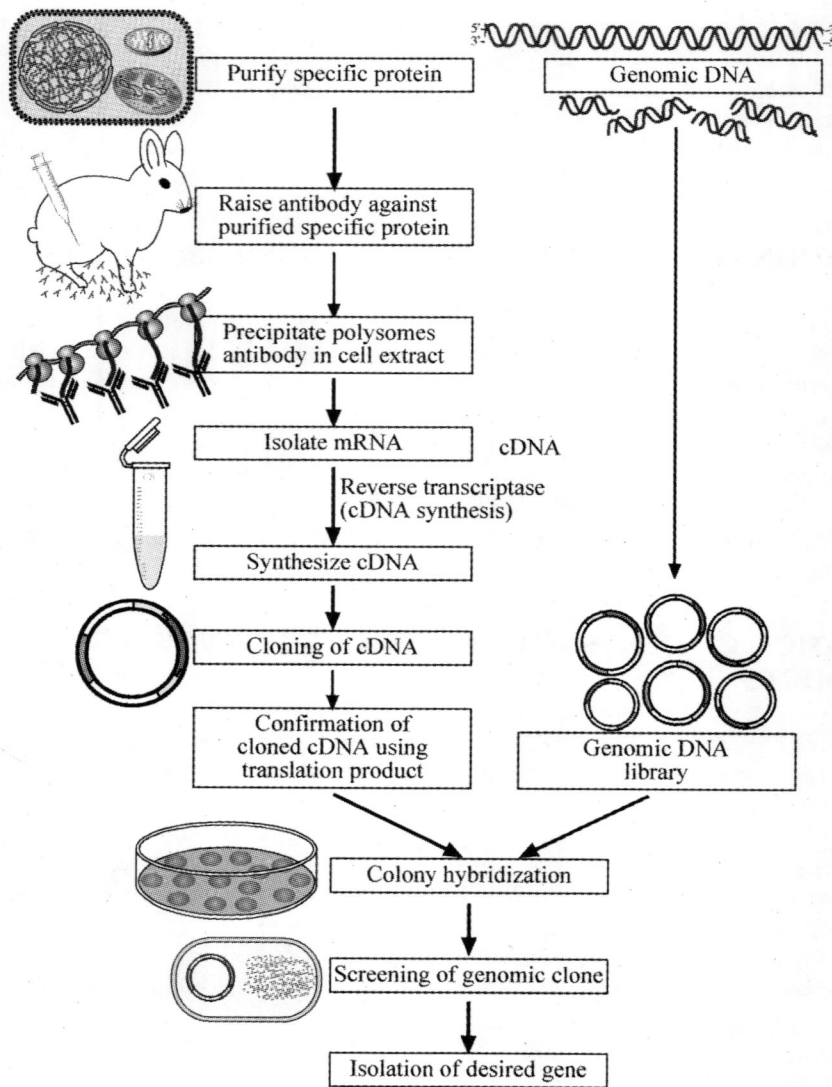

Figure 14.3 *Diagrammatic illustration of steps involved in the isolation of a gene coding for a specific protein*

then creates mutants to characterize the gene's function. Some important techniques of this type are differential display and substractive hybridization.

Differential display technique for isolation of genes: In this technique mRNA is prepared from contrasting tissues of plants, for example, (i) tissues/plants exposed to different environmental conditions (e.g. normal and drought situations), (ii) tissues those differ in function(s), (iii) plants at different developmental stages, etc. In all these cases, two contrasting samples of mRNAs are isolated and subjected for differential screening (Figure 14.5). First, mRNA from the two samples are extracted separately and used for synthesis of cDNA by using reverse transcriptase.

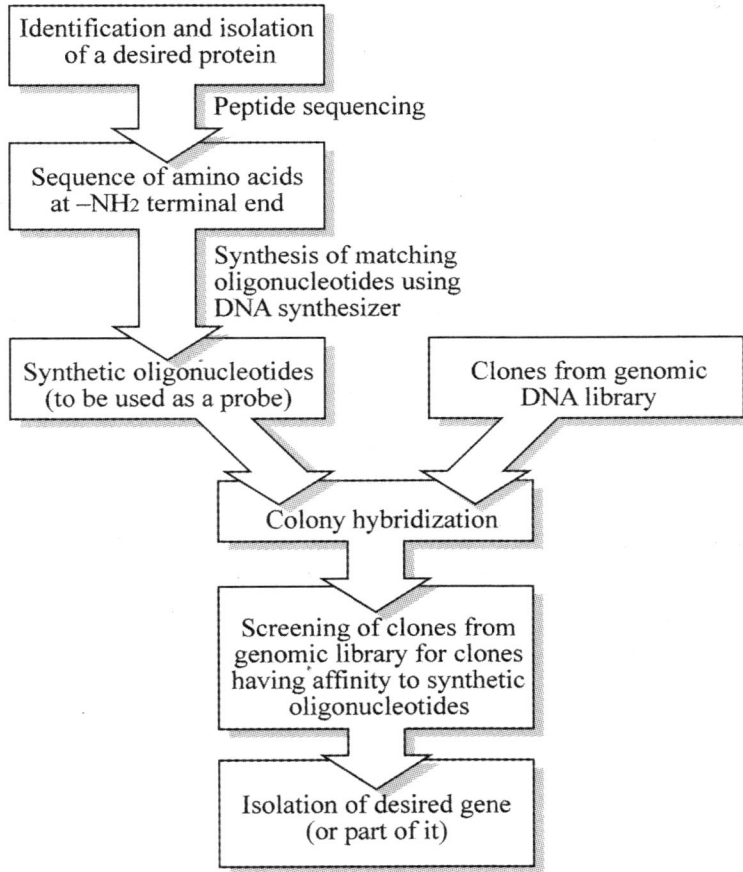

Figure 14.4 *Flow chart for steps involved in the isolation of a gene using the synthetic probe synthesized on the basis of amino acid sequence of a portion of a specific protein*

The cDNAs thus obtained from both samples of mRNA are labeled and used for sequential probing of a genomic DNA library. The clones that are more strongly hybridized by the cDNA from one mRNA sample than from another contain genes that are differentially expressed in the samples.

Substractive hybridization for gene isolation: Substractive hybridization may work better in isolation of a gene for which corresponding deletion mutants are available. In this method, the genomic DNA of a wild type may be digested and hybridized by an excess of sheared and biotin labeled DNA from the deletion mutant. The DNA from a normal plant, that does not hybridize, can be separated by passing the DNA through a column with avidin coated beads so that the biotin labeled and hybridized DNA will be bound on the beads and the unlabeled unbound DNA will pass out from the column. This process may be repeated through several cycles. The eluted DNA is amplified using PCR and then cloned (Figure 14.6). The clones are then used for hybridization on the Southern blots of the mutants and wild type to select clones that hybridize only to the wild type and not to the mutant.

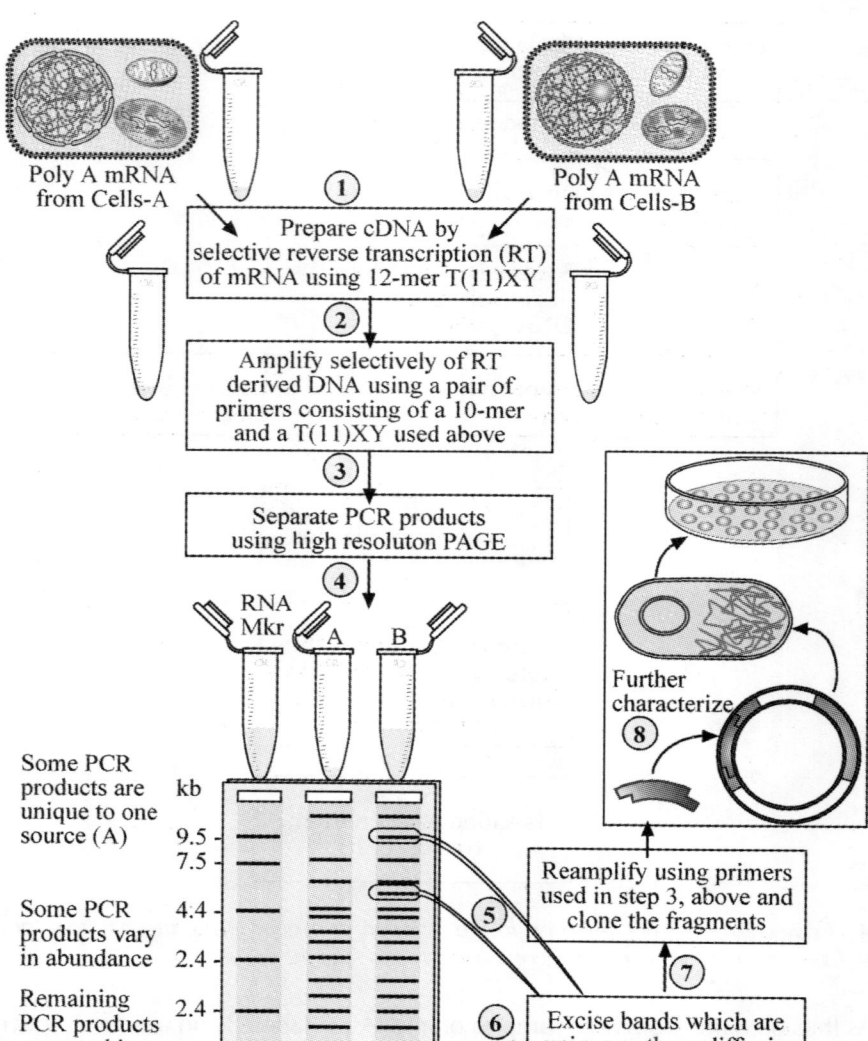

Figure 14.5 *Schematic illustration of steps involved in the isolation of a gene using differential display technique*

E. CHEMICAL SYNTHESIS OF GENES

Work on the chemical-enzymatic synthesis of helical DNA was begun by H. Gobind Khorana in the mid 1960s as part of an attempt to decipher the genetic code. Khorana and co-workers continued to refine their techniques for about a decade, as a result of which they succeeded in synthesizing a complete bacterial tyrosine tRNA gene, including the non-transcribed promoter region. The gene, about 126 bp, was put together from over 20 segments, each of which was

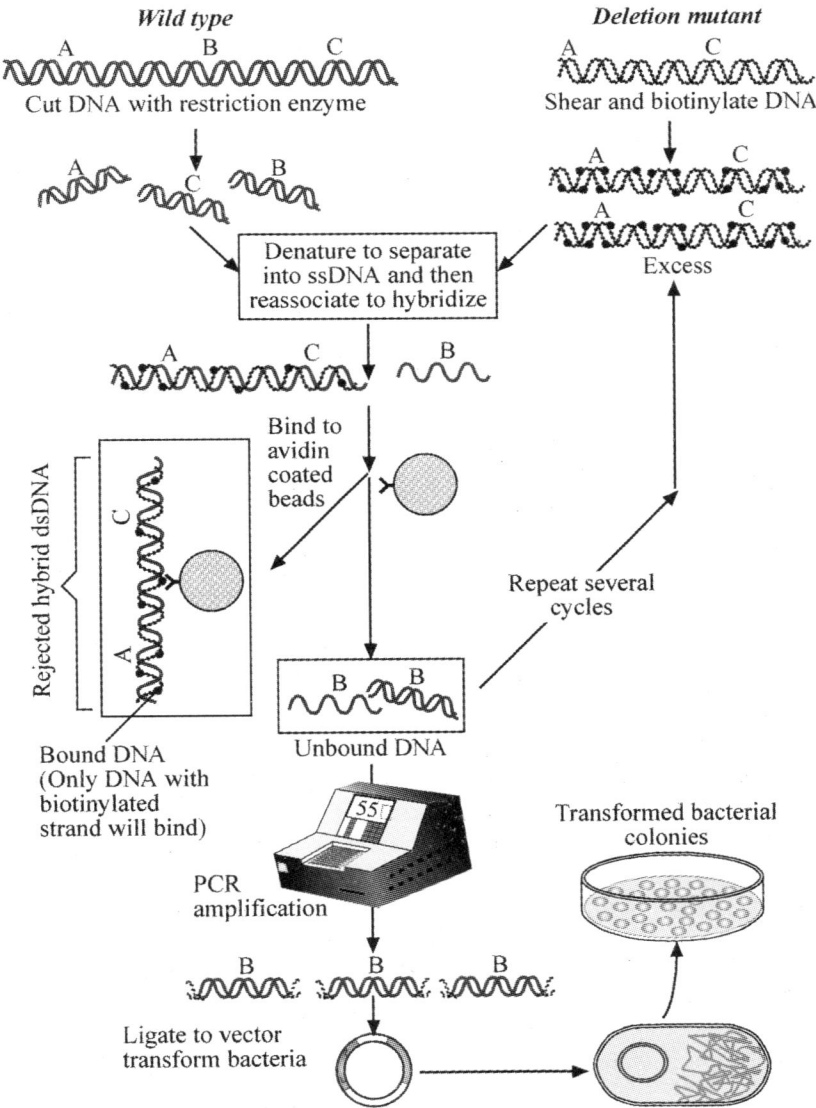

Figure 14.6 Different steps involved in the isolation of a gene using "substractive hybridization"

individually synthesized and later joined enzymatically. The total synthesis of the biologically functional tyrosine suppressor tRNA gene of *E. coli* is an achievement of unparalleled importance in the history of biology and chemistry (Khorana et al., 1976). The total synthesis of the 207 base pair-long DNA included the control elements, as well as the *Eco*RI restriction endonuclease-specific sequences at both ends. This synthetic gene had amber suppressor activity *in vivo* when introduced into a suitable *E. coli* host after being inserted into a double amber mutant of bacteriophage lambda.

The second landmark in the field of gene synthesis came in 1977 when a gene encoding the small hypothalmic hormone somatostatin was synthesized by Keiichi Itakura and co-workers at the Hope Medical Center. The mammalian hormone somatostatin, which could be useful in treating acute pancreatitis and insulin-dependent diabetes, consists of 14 amino acids. Somatostatin is produced in the hypothalamus at the base of the brain and inhibits the secretion of several hormones including growth hormone, insulin, and glucagon. Somatostatin was originally obtained in milligram quantities by extracting the brain tissue of a half million sheep. The synthesis of milligram quantities of somatostatin from a few litres of *E. coli* implanted with synthetic gene was the first demonstration of the utility of recombinant DNA technology for producing human peptides in *E. coli*. This synthetic somatostatin was inserted at the carboxy-terminal end of the β-galactosidase gene of a plasmid vector. The carrier β-galactosidase was purified from *E. coli* cells carrying the synthetic somatostatin gene. The tertiary structure of β-galactosidase-somatostatin-fused protein prevented proteolytic degradation of the somatostatin.

Microbial synthesis of a polypeptide with the size and immunological properties characteristic of mature human growth hormone was achieved by the recombinant DNA technology. Human growth hormone (HGH) consists of 191 amino acids. A precursor consisting of a signal peptide attached to the N-terminus of a growth hormone is produced in the anterior lobe of the pituitary. Genetech scientists used the knowledge of the restriction endonuclease map and sequence of cDNA of HGH to tailor a bacterial plasmid that directs synthesis of mature HGH in an *E. coli* cell. This synthetic–natural hybrid gene was transcribed in *E. coli* from a *lac* promoter, yielding 2.4 mg/litre of HGH that was stable and could be recovered by breaking the cells open.

A gene of average-sized protein, human interferon, was synthesized in 1981, an effort that required the production of a single duplex of 514 bp containing initiation and termination signals recognized by the bacterial RNA polymerase. Several fused proteins such as β-lactamase, rat proinsulin, β-galactosidase-chicken ovalbumin have been expressed by cloning cDNA.

During the last decade, Narang and his associates have developed the triester method for the unambiguous chemical synthesis of short deoxyribo-oligonucleotides with defined sequences of bases. The establishment of the chemical procedures required to assemble nucleotides has led to the development of automated DNA-synthesizing machines that can synthesize polynucleotides of any desired sequence approximately 75–100 nucleotides at a time from the 3'—>5' end of the segment (refer to Chapter 4 for more details). Once such short DNA fragments are synthesized, they can be covalently joined to one another to generate synthetic DNA molecules of considerable length. The simplicity, efficiency, and speed of this method have already enabled the synthesis of many biologically useful nucleotide sequences.

F. ENZYMATIC AMPLIFICATION OF DNA SEQUENCES BY PCR

Desired gene sequences can be selectively amplified to large number of copies and can be cloned into vectors for further studies. In 1983, a new technique was conceived by Kary Mullis of Cetus Corporation that has become widely used to amplify specific DNA fragments without the need of cloning in bacterial cells. This technique is known as **polymerase chain reaction (PCR)**. There are many different PCR protocols used for a multitude of different applications (see Chapter 12 for more details on the protocol). RT–PCR (enzyme reverse transcriptase used in PCR to synthesize cDNA from mRNA) has been extensively used to create cDNA library and to isolate genes of those expressed in a specific environment. In addition to its use in the amplification of specific DNA fragments, PCR can generate large amounts of DNA from minuscule starting samples, such as that in a single cell.

G. SITE-DIRECTED MUTAGENESIS

Variously altered DNA sequences for the study of its diverse roles can be constructed by using a recently developed technique called "**site-directed mutagenesis**". Isolation and characterization of naturally occurring mutants in various organisms has played an enormously important role in determining the function of genes and their products. But natural mutations are rare events and obtaining natural mutants for desirable characters and having mutations in a particular amino acid is very difficult. However, using *in vitro* molecular techniques, small sections of a DNA sequence can be deleted, additional nucleotides inserted, or one specific base changed to another one. This versatile *in vitro* procedure was developed by Michael Smith, who termed it as **site-directed mutagenesis** (**SDM**). SDM is generally accomplished by first synthesizing a DNA oligonucleotide containing the desired change, allowing this oligonucleotide to hybridize to a single-stranded preparation of the normal DNA (generally on a vector DNA), and then using the oligonucleotide as a primer for DNA polymerase. The polymerase extends the 3' end of the primer by adding nucleotides that are complementary to the normal DNA. The modified DNA can then be cloned and the effect of the genetic alteration determined by introducing the DNA into an appropriate host cell. For example, if the site under investigation is part of a regulatory region, then the effect of the alteration on the rate of gene expression can be monitored. If the site is located within the coding region of a gene, the resulting alteration in amino acid sequence provides insights into the role of that site in the overall structure and function of the protein. The altered gene may manifest at the phenotypic level also.

IV MAKING RECOMBINANT DNA MOLECULES *IN VITRO*

A. CUTTING DNA MOLECULES

DNA can be cleaved randomly by a number of mechanical, chemical or enzymatic methods. The extent of the treatment governing the average size of the fragment produced. Even though such methods may be useful for generating random, overlapping fragments of genomic DNA, the only way to generate precise and defined fragments is to use restriction enzymes (see Chapter 15). Most class II restriction enzymes recognize restriction sites with dyad symmetry, and cleave phosphodiester bonds at the same position in the half recognition sequence on each DNA strand. A few enzymes cleave at the axis of symmetry and produce blunt-ended fragments, but most cleavage sites are displaced from the axis of symmetry, so that the fragments produced have complementary single-stranded overhangs (cohesive ends). By hydrogen bonding, these can associate with the ends of other DNA fragments generated by the same enzyme. The base pairing between the overhanging ends holds the terminal residues of each fragment in adjacent positions allowing them to be efficiently joined by DNA ligase.

However, cutting the donor and vector DNA with the same restriction enzymes and joining them by ligase allows the vector to reclose without an insert, generating a high background of non-recombinant vectors. There are two procedures which prevent vector self-ligation. (1) The open vector can be dephosphorylated using the enzyme alkaline phosphatase (see Chapter 15) to remove 5' phosphate groups at each end of the linearized vector. Thus the vector cannot reclose unless 5' phosphate groups are provided by a bridge of donor DNA. (2) The vector and donor can be prepared using a pair of restriction enzymes which generate incompatible ends, so the vector cannot reclose unless the gap is bridged by an insert prepared using the same enzymes. The second strategy is advantageous since the orientation of the insert can be predicted (**directional cloning**).

B. MODIFYING DNA MOLECULES

In many cases, compatible ends are not available for ligation (e.g. where donor and vector must be cut with incompatible enzymes, where the donor DNA is randomly sheared and thus has ragged ends, or where the donor is cDNA). Under these circumstances, there are alternative joining strategies, which involve DNA end-modification. One common method is to make all the fragments blunt-ended by filling or trimming overhangs. Bacteriophage T4 DNA polymerase performs both functions. Blunt-end ligation is less efficient than sticky-end ligation and is non directional. Another strategy is to add linkers or adaptors to the ends of the donor and/or vector DNA.

1. Homopolymer tailing

Khorana and co-workers (1970) discovered that DNA ligase produced by *E. coli* virus T4 catalyses end-to-end linkage of DNA. T4 ligase link short synthetic sequences of nucleotides with single-strand projections into longer DNA pieces. DNA of various sources can also be treated with lambda exonuclease to remove a small number of nucleotides from the 5'-end. Terminal deoxynucleotidyl transferase is then used for stepwise addition of a series of identical nucleotides at 3'-OH ends of a DNA strand. A block of identical nucleotides is added to one population of DNA and a block of complementary nucleotides is added to DNA from another source. The preparations of DNA are then annealed by hydrogen bonding and covalently linked by the combined action of exonuclease III, DNA polymerase I, and DNA ligase. If the homopolymer tails are longer than 40 nucleotides, the joint is stable, so the hybrid molecule can be transformed without *in vitro* ligation. This method will join any two species of DNA irrespective of their size and base sequences of their ends, and has been successfully used for cloning DNA.

2. DNA linkers

(a) Blunt-end linkers: Restriction site DNA linkers for manipulating DNAs that lack cohesive ends (synthetic DNA, cDNA, sheared DNA) are commercially available. The decanucleotide d(C-C-G-G-A-T-C-C-G-G) contains recognition sequences of *Bam*HI, *Hpa*II, and *Hpa*II. When joined to blunt-ended DNA by polynucleotide ligase, this linker adds the recognition sequences of these restriction enzymes to both ends of the DNA fragment, which can be inserted into a vehicle by using the restriction enzyme cohesive termini. If the DNA fragment is not flush-ended, it can be first digested with the nuclease S1 to produce even ends, since this one linker can be used with several restriction enzymes and gives great stability due to the presence of a high number of G-C base pairs.

(b) Preformed adapters: Such linkers do not require cleavage with restriction enzymes for generating cohesive ends and have a blunt end and a single-stranded cohesive end corresponding to one of the restriction enzyme recognition sites. The adapter will join to DNA only at the blunt end if the cohesive end lacks 5'–phosphate. The DNA joined to the adapter is phosphorylated before ligating to the vector.

(c) Conversion adapter: When a DNA fragment obtained by digestion with one restriction enzyme is inserted into a different restriction site in the vector, conversion adapters are used. For example, using such an adaptor, an *Eco*RI-cleaved piece of DNA can be inserted into the *Bam*HI site of the vector. There are two methods of preparing a conversion adapter: (1) blunt-end ligation of two different types of preformed adapters and (2) synthesis of two different single-stranded decadeoxynucleotides, each containing a sequence of four nucleotides at the 5'–end corresponding to the central part of the recognition sequence of two restriction enzymes, and a six-nucleotide-long d(C-C-C-G-G-G) self-complementary sequence at the 3'–end. Annealing these two decadeoxynucleotides causes a conversion linker to form with a *Bam*HI sticky terminus on one side and an *Eco*RI, 5'–sticky terminus on the other. This adapter can be used to clone a DNA segment that has *Eco*RI cohesive ends.

(d) Single-stranded adapter: Single-stranded adapters can be used to clone a DNA fragment with 3'–protruding ends (*Pst*I–digested DNA into the *Bam*HI, or *Eco*RI–digested vector with 5'–protruding ends). These linkers are single-stranded decadeoxynucleotides containing a five nucleotide sequence complementary to the 3'–protruding end and the other pentanucleotide sequence complementary to the 5'–cohesive end. Two such linkers, the *Eco*RI–*Pst*I (5'–A-A-T-T-C-C-T-G-C-A–3') and the *Eco*RI–*Hae*II (5'-A-A-T-T-C-G-G-C-G-C–3') adapters, have been synthesized by the triester method.

Linker molecules have to be so designed that, after linking, triplets are introduced on either side of the DNA to be cloned. This would leave the reading frame for protein synthesis unchanged. A variety of linkers are available commercially. These may offer promoters, operators, new restriction sites, ribosome-binding sites, chain initiation, and termination sites flanked by restriction enzyme recognition sequences.

C. CHOICE OF CLONING VECTORS

Molecular cloning involves the amplification of donor DNA by replication in a host cell. However, since donor DNA generally lacks an origin of replication, it must be joined to a suitable replicon to facilitate cloning. Such a replicon is termed a **cloning vector**, and is a derivative of a plasmid, virus or chromosome. Refer Chapter 16 for details on cloning vectors.

D. CONSTRUCTING AND ANALYZING RECOMBINANT DNA

This chapter describes a number of procedures commonly used to construct, select, and characterize recombinant DNA. They could, for example, be used to obtain a recombinant DNA clone that contains a particular gene. The task of isolating a desired recombinant from a population of bacteria or phage depends very much upon the cloning strategy that has been adopted. For instance, when a cDNA derived from an abundant mRNA is to be cloned, the task is relatively simple – only a small number of clones need to be screened. Isolating a particular single-copy gene sequence from a complete eukaryotic genomic library requires techniques in which thousands of recombinants need to be screened. In Chapter 15 an overview of the general principles employed in recombinant selection and screening procedures are described.

Joining the ends of DNA Molecules: Joining two DNA ends together is the key operation when forming *in vitro* recombinants. There are basically three strategies for joining DNA fragments. (1) Ligation of sticky ends generated by restriction endonuclease digestion. (2) Ligation of blunt ends generated by, for example, shear breaking and either digestion with the single-strand specific nuclease, S1, or treatment with *E. coli* DNA polymerase I. (3) Addition of adapters and then joining the adapters together either by ligation or annealing to form stable joints. Despite this variety of strategies for joining molecules, only a few enzymatic reactions are involved – In this S1 digestion, ligation (blunt and sticky), lambda exonuclease digestion, and homopolymer tail formation (See Chapter 15).

E. RECOMBINANT DNA TRANSFER TO CLONING HOST

Once a recombinant vector has been constructed *in vitro*, it must be introduced into host cells for cloning. *E. coli* is the major host for general cloning purposes, but this state of competence can be brought about by chemical treatments. In addition to this general method, many transformation methods, both vector-mediated and non-vector based methods have been developed, which can be used for both prokaryotes and eukaryotes. Many of these techniques are common for both prokaryotes and eukaryotes. For more details refer to Chapter 16.

F. VECTOR AND RECOMBINATION SELECTION

Neither DNA manipulation nor gene transfer procedures are 100% efficient. Thus, at the start of any cloning experiment, there will be a large population of cells lacking the vector, and of those containing the vector, a moderate proportion will contain non-recombinant vectors. The problem is that the non-recombinant cells may proliferate at the expense of the recombinant population, so it is desirable to identify and preferably eliminate such cells. For more details refer to Chapter 19.

G. RECOVERY OF CLONED DNA

After transfer of recombinant DNA to host cells, the cells are cultured for a short time to allow recovery and then plated out to form colonies or plaques under appropriate selective regime, in case of bacterial or phage infected colonies. For selection for other organisms, various methods have been developed (refer to Chapter 19 for more details).

H. SCREENING STRATEGIES

The desired clone must be isolated from cultures and libraries, by exploiting some unique property of the donor DNA, e.g. its sequence, or a structural or functional property of the protein it encodes. This process is known as **screening**. Classical screening strategies require some knowledge of the biochemistry of the gene, either some sequence information, or an exploitable property of the product, which will allow the gene to be isolated from an expression library. These approaches are explained in detail in Chapter 19.

I. APPLICATION OF RECOMBINANT DNA TECHNOLOGY

Recombinant DNA technology has been applied in almost all systems and for myriad uses. However, the significant contribution of recombinant technology and its commercial applications can be seen in microorganisms, animals and plant systems (refer to Chapters 20–22 for more details).

V SUGGESTED READING

Abeles RH, Frey PA and Jencks WP (1992) *Biochemistry*, Jones and Bartlett Publishers, Inc. Boston, London.

Brown TA (ed.) *DNA Cloning: A Practical Approach,* 2nd Edn (4 volumes), IRL Press, Oxford.

Brown TA (ed.) *Essential Molecular Biology: A Practical Approach* (2 volumes), IRL Press, Oxford.

Garrett RH and Grisham CM (1999) *Biochemistry*, Saunders College Publishing, Fort Worth, Philadelphia, New York.

Gupta PK (1999) *Cell and Molecular Biology*, Rostogi Publications, Meerut.

Old RW and Primrose SB (1996) *Principles of Gene Manipulation: An Introduction to Genetic Engineering*, Blackwell Science, Oxford.

Sambrock, J., Fritsch, E.F. and Maniatis, T. (1989) *Molecular Cloning, A Laboratory Manual*. Second edition, Cold Spring Harbor Laboratory Press, USA.

Walker JM and Gingold EB (1993) *Molecular Biology and Biotechnology*, Panima Education Book Agency.

Watson JD, Gilman M, Witkowski J and Zoller M (1992) *Recombinant DNA*, 2nd Edn. Scientific American Books, New York.

15

Enzymes in Molecular Cloning

I INTRODUCTION

The discovery and characterization of a number of key enzymes have played an important role in the development of various techniques for the analysis and manipulation of DNA. A DNA sample can be characterized for individual identity according to its chemistry or sequence information. Cloning and manipulating genes requires the ability to cut, modify, and join genetic material, and to check parameters, such as size of the molecules that are being manipulated. The major tools for this type of manipulation are the enzymes that catalyze specific reactions on DNA molecules. Over the years, the number and applications of these enzymes have increased as new molecules have been discovered (Table 15.1). In this chapter, some of the properties and uses of these important components in molecular cloning procedures will be briefly discussed.

Table 15.1 *Enzymes used to manipulate DNA*

Enzymes and source	Functions in genetic engineering
Nucleases	
Restriction endonucleases (many bacteria)	Recognize specific sequences of four to six nucleotides in dsDNA, and cleave both strands once within the recognition sequence. Over 1000 enzymes are commercially available. Used for clone mapping and preparing DNA fragments.
Exonuclease *Bal*31 (*Alteromonas espejiana*)	Removes nucleotides from both strands of dsDNA.
Exonuclease HI (*E. coli*)	Removes nucleotides from the end of just one strand of dsDNA.
DNase I (Pancreas)	Non-specific endonuclease that cleaves NA into oligonucleotides. Used to introduce nicks into DNA and to remove contaminating DNA from RNA.

Continues...

...Continued

Enzymes and source	Functions in genetic engineering
RNase A	General RNA endonuclease, used to remove contaminating RNA from DNA. RNases are used in mapping studies
RNaseH	Used in second strand cDNA synthesis, it degrades the RNA strand of RNA: DNA hybrid molecule
Spleen phosphodiesterase	Works on both RNA and DNA, exonuclease at 5'-OH end, produces 3'-P mononucleotides.
S1 nucleases (*Aspergillus*)	ssDNA specific endonuclease, produces 5'-P mono nucleotides. Used for mapping DNA–RNA hybrids.
Micrococcal nuclease	ssDNA specific endonuclease, produces 3'-P mononucleotides.
DNA polymerases	
DNA polymerase I (*E. coli*)	Synthesizes a new strand of DNA complementary to an existing DNA template. Also has 5'→3' and 3'→5' exonuclease activity that provides an editing function.
Taq polymerase (*Thermus aquaticus*)	Thermostable polymerase enzyme used in PCR to amplify DNA sequences *in vitro* and also to introduce restriction sites and point mutations into a DNA molecule.
Klenow fragment (*E. coli*)	Derived from DNA pol I, it has only the polymerase activity and lacks 5'→3' exonuclease activity. Used for general DNA synthesis purposes, e.g., labeling by random priming, end-filling, second strand cDNA synthesis activity, primer extension.
T4 DNA pol (phage T4)	Active 3'→5' exonuclease activity. Used for replacement labeling, generating blunt-ended DNA.
Tth DNA pol	Thermostable DNA polymerase with RT activity
T7 DNA pol (phage 7)	Rapid and highly processive. Modified version, **sequenase**, lacks 5'→3' exonuclease activity and is used in DNA sequencing
RNA polymerases	
T7, T3, SP6 RNA pol	DNA-dependent RNA polymerase with strong promoter-specificity used for *in vitro* transcription, RNA labeling.
QB replicase	RNA-dependent RNA polymerase, used in RNA amplification

Continues...

... *Continued*

Enzymes and source	Functions in genetic engineering
Reverse transcriptase	
AMV-RT (from several RNA tumor viruses)	Synthesizes DNA from an RNA template, used in cDNA synthesis, RNA sequencing.
Ligases	
T4 DNA ligase (phage T4)	Links 5'-phosphate and 3'-hydroxyl ends via a phosphodiester bond. Used for ligation of cohesive and blunt termini. It uses ATP.
DNA ligase (*E. coli*)	Seals ss nicks between adjacent nucleotides and OH-terminus, prevents unwanted ligation of fragments of DNA while cloning but requires NAD^+.
Nucleic acid-modifying enzymes	
Alkaline phosphatase / Calf intestinal phosphatase (CIP)	Removes the phosphate group at the 5' terminus of a DNA and RNA molecule. Used to block self-ligation and for 5' end-labeling
T4 Polynucleotide kinase (phage T4)	Adds phosphate groups to free 5' termini of DNA and RNA. Used for 5' end-labeling.
Terminal deoxynucleotidyl transferase (TdT) (mammalian thymus)	Adds one or more nucleotides to the 3' terminus of a DNA molecule. Used for 3' end-labeling, tailing e.g. addition of homopolymer tail.
DNA methylases (many sources)	These enzymes introduce methyl groups on nucleotides.
DNA-binding proteins (many cell types)	Single-stranded DNA-binding protein (SSB): SSBs bind cooperatively to ssDNA, but not to dsDNA.
RecA Protein (*E. coli*)	This protein has several useful activities; It binds ssDNA, promotes uptake of homologous ssDNA into dsDNA, and is a DNA-dependent ATPase.
Topoisomerase I (Prokaryotes)	Topoisomerase I catalyzes the removal of superhelical turns from covalently closed circular dsDNA by transient breakage and rejoining of phosphodiester bonds.
Gyrase	Type II topoisomerase catalyzes conversion of relaxed DNA to a superhelical form; requires ATP.
Glycosylase	Hydrolysis of nucleoside bases–glycosidic bonds, resulting in production of an apurinic/apyrimidinic (AP) site in DNA and a free nucleoside base.

II CUTTING

Enzymes that break down nucleic acids are called nucleases. Those that break down RNA are called ribonucleases (RNases), and those that break down DNA are called deoxyribonucleases, or DNases. There are two ways of breaking down a linear nucleic acid molecule: by dismantling it bit by bit from the ends (exonucleolytic activity), or by breaking it into pieces by cutting within the molecule (endonucleolytic activity).

A. RESTRICTION ENDONUCLEASES

1. Characteristics of restriction enzymes

The discovery of restriction **enzymes** that cleave double-stranded DNA (dsDNA) at discrete nucleotide sequences was probably a breakthrough in molecular biology. Different species of bacteria contain their own sets of endonucleases and corresponding methylases. More than 900 (some sequence-specific and others not) restriction enzymes and 130 different nucleotide recognition sequences have been identified from 230 different bacterial strains. The term "**restriction**" refers to the function of these enzymes in restricting the host range of bacteriophage infection. For prokaryotic cells, they function in nature as restriction–modification systems and will cleave any foreign DNA that enters the bacterial cell (e.g., bacteriophage) but not the host DNA that has been "protected" or modified by methylation. Based on mechanistic differences between several types of bacterial restriction systems, three classes (Type I, II, and III) of restriction enzymes have been described. Type I and Type III enzymes are **bifunctional** and carry out both modification (**methylation**) and ATP-dependent restriction (**cleavage**) activities in the same protein. The state of methylation at the restriction site determines the subsequent enzyme action. Type III enzymes cut DNA at the recognition site and then dissociate from the substrate. However, Type I enzymes bind to the recognition sequence but cleave at random sites when DNA loops back to the bound enzyme. These enzymes require ATP, magnesium, and S adenosylmethionine as cofactors for activity.

Type II systems are required for most biotechnology applications. Type II restriction systems are binary systems that carry the endonuclease and methylase activities on separate proteins and are useful in molecular cloning. A large number of Type II restriction enzymes have been isolated (Appendix 11, Table 2), many of which are routinely used in molecular cloning. The vast majority of Type II restriction enzymes recognize specific sequences that are 4–8 nucleotides (mostly **4-base** and **6-base** cutters) in length and display twofold symmetry (i.e., **palindromic**). A few enzymes, however, recognize longer sequences. The location of cleavage sites within the axis of dyad symmetry differs from enzyme to enzyme: Some cleave strands exactly at the axis of symmetry, generating fragments of DNA that carry **blunt ends**; others cleave each strand at similar locations on opposite sides of the axis of symmetry, creating fragments of DNA that carry protruding single-stranded termini called **sticky ends** (Figure 15.1). Most Type II restriction endonucleases require only magnesium for their activity. These enzymes are used for DNA sequencing, isolation, physical mapping, cloning, and structural analysis of highly repetitive satellite DNAs and eukaryotic genomes. These enzymes have also been used to study site-specific interaction of DNA with protein molecules such as RNA and DNA polymerase, transcriptional repressors, and activators. Several enzymes cleave both double- and single-stranded DNA.

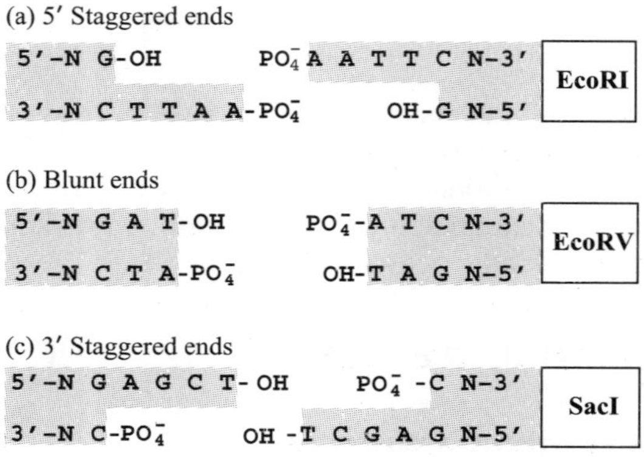

(a) 5' Staggered ends

```
5'-N G-OH           PO₄⁻A A T T C N-3'
3'-N C T T A A-PO₄⁻       OH-G N-5'
```
EcoRI

(b) Blunt ends

```
5'-N G A T-OH       PO₄⁻-A T C N-3'
3'-N C T A-PO₄⁻       OH-T A G N-5'
```
EcoRV

(c) 3' Staggered ends

```
5'-N G A G C T-OH   PO₄⁻ -C N-3'
3'-N C-PO₄⁻          OH -T C G A G N-5'
```
SacI

Figure 15.1 *Restriction enzymes cleave palindromic DNA sequences to produce double-stranded breaks. Restriction enzyme cleavage results in the formation of 5' PO and 3' OH termini with (a) 5' staggered ends, (b) blunt ends, or (c) 3' staggered ends*

2. Restriction enzymes created termini

Cohesive termini

(a) Compatible cohesive termini: If the restriction enzyme cleaves each strand of the substrate DNA on the 5' side of the axis of dyad symmetry, the resulting staggered break yields fragments of DNA that carry protruding cohesive 5' termini. Enzymes that make staggered cuts are especially useful. The enzyme DNA ligase is used to covalently link the 3'-OH terminus of one strand to the 5'-phosphate end of a second strand. Thus the ligase can repair single-strand breaks such as those present in hydrogen-bonded sticky ends of two DNA preparations that have been cleaved with the same restriction enzyme. For example, the enzyme *Eco*RI recognizes the sequence 5'——GAATTC—— 3' in double-stranded DNA and cleaves it as follows:

```
      ↓
5'-NNNNNNGAATTCNNNNN-3'
   |||||||||||||||||
3'-NNNNNNCTTAAGNNNNN-5'
              ↑
```

The hydrogen bonding of the four base pairs between the sites of cleavage is not favored under the conditions used for digestion. Therefore, the original segment of DNA separates into two fragments.

```
5'-NNNNNNG_OH              pAATTCNNNNNN-3'
   |||||||                    ||||||
3'-NNNNNNCTTAA_p            _OHGNNNNNN-5'
```

If fragments bearing compatible protruding termini are incubated under conditions that favor the formation of base pairs, they can anneal with one another, and the cleaved phosphodiester bonds can then be released with a DNA ligase. Because all DNA fragments created by cleavage with *Eco*RI carry the same protruding 5' termini, they can be joined in novel combinations. In fact, *Eco*RI was the first enzyme used to create recombinant DNA molecules.

If the restriction enzyme cleaves each strand of the DNA on the 3' side of the axis of dyad symmetry, a staggered break is generated that yields fragments of DNA with protruding cohesive 3' termini. For example, the restriction enzyme *Pst*I recognizes the sequence 5'—CTGCAG—3', in double-stranded DNA and cleaves it as follows:

```
         ↓
5'-NNNNNNCTGCAGNNNNN-3'
   |||||||||||||||||
3'-NNNNNNGACGTCNNNNN-5'
              ↑
```

The four base pairs of cleavage separate into two fragments.

```
5'-NNNNNNCTGCA_OH              pGNNNNNN-3'
   ||||||                      ||||||
3'-NNNNNNG_p                 OHACGTCNNNNNN-5'
```

Any fragment carrying a protruding terminus generated by cleavage with *Pst*I can be joined to any other fragment carrying the same terminus.

(b) Incompatible cohesive termini: The joining of DNA fragments with protruding 5' termini that are ordinarily not compatible can be accomplished by partial filling of the recessed 3' termini using the Klenow fragment of *E. coli* DNA polymerase I. The partial filling also eliminates the ability of compatible termini to pair with each other, thus reducing the frequency of self-joining catalyzed by T4 DNA ligase (Figure 15.2). For example, the restriction enzymes *Xba*I and *Hind*III generate non-compatible cohesive termini. However, if the termini generated by each enzyme are partially filled with the appropriate two nucleotides, the resulting complementary cohesive termini can be joined efficiently by T4 DNA ligase. Because the partially filled termini can no longer be joined to themselves, there is no competing self-ligation reaction, and the altered termini can therefore be ligated efficiently to one another.

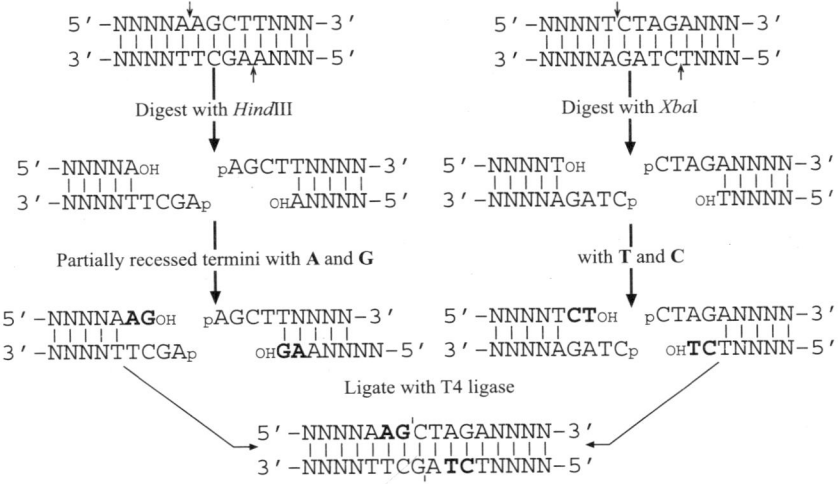

The inability of certain partially filled termini to self-ligate is useful in such applications as the construction of genomic DNA libraries – take for example, genomic DNA partially digested

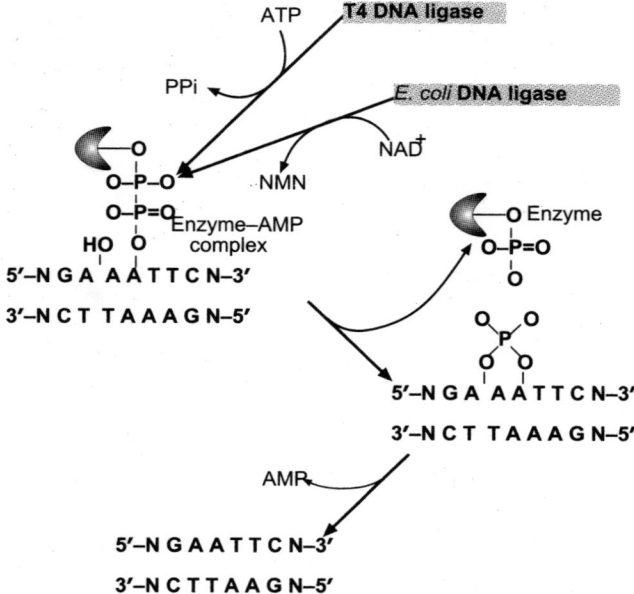

Figure 15.2 *DNA ligases catalyze the formation of a phosphodiester bond in nicked double-strand DNA. An enzyme–AMP complex binds to a nick bearing 3'–OH and 5'–P groups. The AMP reacts with the phosphate group. Attacks by the 3'–OH group on this moiety generates a new phosphodiester bond which seals the nick*

with *Sau*3AI (IGATC) into a bacteriophage lambda vector that had been cleaved with *Sal*I (GITCGAC). The cohesive termini generated by *Sau*3AI and *Sal*I are not normally compatible, but partial repair of the two different tetranucleotide 5' extensions with two nucleotides generates complementary termini and destroys the ability of the original 5' termini to self-anneal. Thus, neither the vector nor the genomic DNA fragments can join to themselves, although they can ligate efficiently to join one another. However, the resulting hybrid joints often cannot be cleaved by known restriction enzymes, unless they generate restriction sites upon joining, but this problem can be overcome by cloning into vectors that carry polycloning sites.

(c) Blunt end ligation: T4 ligase enzyme can join (blunt-ended) fully base-paired DNA duplexes. Even though blunt-end joining occurs at a much slower rate than sealing of single-stranded nicks, it is of considerable importance for joining DNA molecules that lack cohesive ends. Any DNA segments obtained by shear, by using two different restriction enzymes, or by using restriction enzymes that produce blunt ends, can be joined without adding homopolymer tails. Chemically synthesized DNA sequences can also be linked by this procedure, where it is not necessary to add additional bases.

Cleavage of both strands of DNA at the axis of dyad symmetry produces blunt-ended fragments. For example, the enzyme *Hae*III recognizes the sequence **GGCC** in double-stranded DNA and cleaves it as follows:

```
5' -NNNNNNGGCCNNNNNN- 3'
            ↓
    |||||||||||||||
3' -NNNNNNCCGGNNNNNN- 5'
            ↑
```

to yield fragments that carry blunt ends digest with *Hae*III

```
5'-NNNNNNGG_OH      pCCNNNNNN-3'
   ||||||||         ||||||||
3'-NNNNNNCC_p      _OHGGNNNNNN-5'
```

Ligating blunt ends by T4 DNA ligase, we get

```
5'-NNNNNNGGCCNNNNNN-3'
   ||||||||||||||||
3'-NNNNNNCCGGNNNNNN-5'
```

Thus, blunt ends generated in this manner can be joined using bacteriophage T4 DNA ligase, although the efficiency of the reaction is somewhat lower than for ligation of cohesive termini. Nevertheless, joining of blunt-ended DNA molecules is extremely useful because of its universal application.

3. Isoschizomers

In general, different restriction enzymes recognize different sequences. However, there are many examples of enzymes isolated from taxonomically different organisms that cleave within the same target sequences. These are known as **isoschizomeres**. In addition, many enzymes recognize tetranucleotide sequences, but in some cases, these tetranucleotides occur within hexanucleotide target sequences of other enzymes. For example, MboI and Sau3AI recognize the sequence

```
5'-NNNNNNNGATCNNNNNN-3'
   |||||||||||||||||
3'-NNNNNNNCTAGNNNNNN-5'
```

whereas *Bam*HI recognizes

```
5'-NNNNNGGATCCNNNNN-3'
   |||||||||||||||
3'-NNNNNCCTAGGNNNNN-5'
```

In each case, however, an additional protruding tetranucleotide terminus is produced. In a few cases, ligation of fragments generated by one restriction enzyme to those generated by a second enzyme results in hybrids that are recognized by neither of the parental enzymes. For example, when fragments generated by *Sal*I (G^TCGAC) are ligated to fragments generated by *Xho*I (C^TCGAG), the resulting hybrid target sites are cleaved by neither *Sal*I nor *Xho*I.

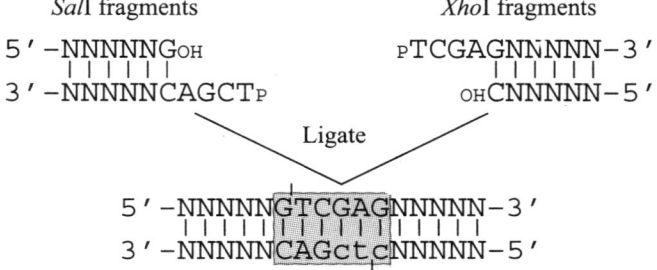

This site is not recognized by *Sal*I or *Xho*I enzymes

B. NUCLEASES

When nucleic acids need to be cut randomly, or sometimes partially digested into approximate sizes, endonucleases are not useful. There are many nucleases with specific digesting properties that are found *in vivo* and also isolated for *in vitro* investigations.

1. Nuclease BAL 31

BAL 31 is predominantly a 3' exonuclease that removes mononucleotides from both 3' termini of the two strands of linear DNA. BAL 31 is also an endonuclease; thus, the single-stranded DNA generated by the 3'-exonuclease activity is degraded by the endonuclease. The degradation is absolutely dependent on the presence of calcium, and the reaction can therefore be stopped at different stages by the addition of the chelating agent EGTA. Because degradation occurs relatively uniformly from the termini of DNA, digestion with BAL 31 can be used to map restriction sites in small fragments of DNA. BAL 31 can also be used to remove unwanted sequences from the termini of DNAs before cloning.

2. Nuclease S1

Nuclease S1 degrades single-stranded DNA or RNA to yield 5'–phosphate mono- or oligonucleotides. Double-stranded DNA, double-stranded RNA, and DNA–RNA hybrids are relatively resistant to the enzyme. However, double-stranded nucleic acids are digested completely by nuclease S1 if they are exposed to the enzyme at very high concentrations. This enzyme is mainly used for analyzing the structure of DNA–RNA hybrids, for removing single-stranded tails from DNA fragments to produce blunt ends, and for opening the hairpin loop generated during synthesis of double-stranded cDNA.

Protocol for making blunt ends by S1 digestion

DNA ends generated by shear or by restriction digestion are often poor substrates for subsequent joining reactions. The shear-generated ends are a mixture of 5' and 3' protruding single strands and blunt ends. The ends generated by particular restriction enzymes are either 5' or 3' single strands or blunt ends. In order to make more uniform substrates, it is frequently necessary to digest them with the single-strand specific nuclease (S1) of *Aspergillus oryzae*. Digestion with S1 will make most ends blunt or almost blunt. Such ends are fairly good substrates for blunt-end ligation and terminal transferase tailing.

Procedure

1. Resuspend ethanol-precipitated DNA in a 1.5 ml polypropylene microfuge tube to achieve a final concentration of 10 µg/ml in 200 mM NaCl, 30 mM Na^+–acetate, 5 mM $ZnSO_{14}$ at pH 4.5.
2. Incubate at 20°C for 5 min., then add 2000 units/ml of S1 and incubate for 30 min.
3. After the 30-min. incubation, chill the reaction on ice and extract twice with an equal volume of water-saturated phenol at 4°C. Then extract twice with chloroform and re-extract the phenol and chloroform with a small volume of TE pH 8.0. Combine the extracts and ethanol-precipitate the DNA.

3. Mung-bean nuclease

Mung-bean nuclease degrades single-stranded DNA to mono- or oligonucleotides with phosphate groups at their 5' termini. Double-stranded DNA, double-stranded RNA, and DNA–RNA hybrids are relatively resistant to the enzyme. Although mung-bean nuclease and nuclease S1 are similar to each other in their physical and catalytic properties, mung-bean nuclease may be less severe in its action than the S1 nuclease. This enzyme is primarily used to convert protruding termini of DNA to blunt ends.

4. Ribonucleases

(a) Ribonuclease A: RNase A is an endoribonuclease that specifically attacks single-stranded RNA 3' to form pyrimidine residues and cleaves the phosphate linkage to the adjacent nucleotide. The end products are pyrimidine 3' phosphates. RNase is primarily used for removing unhybridized regions of RNA from DNA–RNA hybrids and for mapping single-base mutations in DNA or RNA. In this method, single-base mismatches in RNA–DNA or RNA–RNA hybrids are recognized and cleaved by RNase A.

(b) Ribonuclease T1: RNase T1 is an endoribonuclease that specifically attacks the 3'-phosphate groups of guanine nucleotides and cleaves the 5'phosphate linkage to the adjacent nucleotide. The end products are guanosine 3' phosphates and oligonucleotides with terminal guanosine-3'–phosphate groups.

5' pApGp'Gp'CpCpGp'ApApGp'UpGp'CpApGp'C3'

Reaction with RNase T1 gives

5' pApGp + Gp + CpCpGp + ApApGp + UpGp + CpApGp + C3'

Ribonuclease T1 is also primarily used for removing unhybridized regions of RNA from DNA–RNA hybrids.

5. Deoxyribonuclease 1

DNase is an endonuclease that hydrolyses ss- or dsDNA preferentially at sites adjacent to pyrimidine nucleotides. Products are complex mixtures of 5'-phosphate mono- and oligonucleotides. In the presence of Mg^{2+}, DNase 1 attacks each strand of DNA independently, and the sites of cleavage are distributed in a statistically random fashion. DNase 1 is used for (i) introducing random nicks into dsDNAs in preparation for radiolabeling by nick translation; (ii) introducing a single nick into closed circular DNAs in preparation for resection prior to bisulfate-mediated mutagenesis; (iii) generating random clones for sequencing in bacteriophage M13 vectors; (iv) analysis of protein–DNA complexes (DNase foot-printing).

6. Exonuclease III

Exonuclease III catalyses the stepwise removal of 5' mononucleotides only from the 3'–hydroxyl termini of dsDNA. Linear double-stranded and circular DNA containing nicks or gaps are the substrate. The activity of the enzyme results in the formation of long single-stranded regions in dsDNA. The enzyme also carries out three other activities: an endonuclease specific for apurinic DNA, an RNase H activity, and a 3' phosphatase activity, which removes 3'-phosphate termini. The uses are: (i) generating partially resected DNAs that can be used as substrates for the Klenow fragment of E. coli DNA polymerase I; (ii) generating nested sets of deletions of the terminal sequences of linear dsDNAs. This reaction is generally carried out in conjunction with mung-bean nuclease or nuclease S1 and is an alternative to using BAL 31.

Some methods of site-specific mutagenesis use thiophosphate derivatives of the dNTPs for second-strand synthesis primed by the mutagenic primer. The parental template strand can be preferentially degraded with exonuclease III, increasing the frequency of mutants obtained upon transformation of E. coli, since exonuclease III will not cleave thioester bonds.

Protocol for lambda exonuclease digestion to yield free 3' ends

The technique of adding homopolymer tails to DNA for cloning purposes works considerably better if all the DNA strands to which tails are to be added extend beyond their complementary strands. This is because the tail-forming enzyme, terminal transferase, adds much more rapidly to extended ends and will give a very heterogeneous population of tail lengths if it is adding tails to a mixed substrate population of extended and unextended ends. Some reaction conditions for the addition of homopolymer tails have been devised that decrease the specificity for adding to extended ends, but they simultaneously introduce another problem, namely, of tails being added to nicks in the DNA. Greater success can be achieved when overhanging 3' ends are generated on all molecules by performing a limited digestion with lambda exonuclease. The general strategy for this reaction is to saturate the DNA ends with enzyme and then to digest them at a low rate so as to partially degrade all 5' ends. The reaction is run at low temperature because the enzyme digests rapidly at 37°C. Also, the reaction is run at enzyme excess because the enzyme digests progressively. The conditions described below will remove approximately 30 bases from 5' ends that are base paired or are extending only a few bases beyond the 3' ends of a duplex.

Procedure

1. To a DNA solution that is 3 mM in DNA ends and 10 mM Tris–HCl pH 8.0; 1 mM EDTA; add 1 M K^+–glycinate pH 9.5 to 67 mM 0.1 M $MgCl_2$, to 5 mM and 1 mg/ml bovine albumin to 50 µg/ml.
2. Bring the reaction mixture to 0°C in an ice water bath in a cold room. Also chill all pipettes to 4°C because the temperature coefficient of lamda exonuclease activity is high.
3. Add 8 units (about 2 pmol) of enzyme for every picomole of DNA ends.
4. Halt the reaction after a 45-min. incubation at 0°C by adding an equal volume of water-saturated phenol, 0°C. Extract with phenol twice and chloroform twice and re-extract the phenol and chloroform with an equal volume of TE pH 8. Combine the aqueous phases from the extraction and re-extraction and ethanol-precipitate the DNA. All operations are to be done in microfuge tubes. For subsequent terminal transferase reactions, resuspend the DNA in 10 mM Tris–HCl pH 8; 0.1 mM EDTA at a concentration of 20 nM in ends.

C. Physical stress

In addition to enzymatic means, physical shearing to cleave DNA at random has also been used. This can be accomplished in several ways; for example, by simply stirring the solution, by forcing it through a narrow opening, such as a syringe needle or a pipette tip, or by sonication. In practice, sonication is the preferred method, since it is the easiest to control and is more reproducible.

III DNA LIGASES

The enzymes used to join two pieces of DNA are called DNA ligases. They generate phosphodiester bonds between neighboring 3' hydroxyl ends and 5' phosphate ends of dsDNA chains. Ligases are part of the routine battery of enzymes required by a cell for the maintenance of its DNA. A deficiency of one of the ligases found in human cells is responsible for Bloom's Syndrome, in which replication is slowed and the cell becomes highly sensitive to DNA damaging agents. Ligases come from two sources, *E. coli* or cells that have been infected by viruses. The ligase used most often in cloning is encoded by bacteriophage T4. Bacteriophage T4 DNA ligase, a polypeptide, catalyzes the formation of phosphodiester bonds between adjacent 3'-hydroxyl and 5'-phosphate termini in DNA. This enzyme can be used to join DNA molecules both with cohesive and blunt-end termini. However, the rate of blunt-end ligation is much slower than the cohesive termini ligation.

Bacteriophage T4 RNA ligase is capable of covalently joining single-stranded RNA (or DNA) molecules containing 5'-phosphate and 3'-hydroxyl termini. This enzyme, however, is primarily used for 3' end-labeling of RNA, by incorporating ^{32}P-labeled mononucleoside 3',5'-bisphosphate (pNp), which is added to the 3'-hydroxyl termini of RNA. Synthesis of oligodeoxynucleotides can also be done.

The DNA ligases isolated from *E. coli* and T4 differ in their coenzyme requirement and also in substrate specificities. However, one of the most remarkable properties of T4 ligase is its ability to accomplish blunt-end ligation of dsDNA molecules. The catalytic activity of ligases is highly influenced by temperature. The optimal ligation temperature has been found to be 37°C. The concentration of DNA in the reaction solution also plays an important role on the product.

Method–1: Ligation with T4 DNA ligase

A wide variety of conditions can be used to ligate the ends of DNA. The following general procedures work in most cases.

Requirements

Ligase buffer (10X): 300 mM Tris–HCl, pH 8.0; 70 mM MgCl$_2$; 12 mM EDTA; 2 mM ATP; 100 mM dithiothreitol; 500 µg/ml bovine albumin

Reaction mixture: 1.5 µl 10X ligase buffer; 0.1 µg DNA; 5 kb in length in TE pH 8.0; Sterile distilled water to 15 µl total volume.

Procedure

1. Make the reaction mixture in a 1.5 ml polypropylene microfuge tube on ice.
2. Add 0.01 units of T4 DNA ligase for cohesive-end ligation and 0.1 units for blunt-end ligation and incubate at 4°C for 18–24 hr.

IV ENZYMES INVOLVED IN MODIFICATION

A. DNA METHYLASES

Most strains of E. coli contain two enzymes that methylate DNA – the **dam** methylase and the **dem** methylase.

dam methylase: This enzyme introduces methyl groups at the N6 position of adenine in the sequence 5' GATC 3'. The recognition sites of many restriction enzymes contain this sequence.

dem methylase: This enzyme introduces methyl groups at the C6 position of the internal cytosine in the sequence 5' CCAGG 3' or 5' CCTGG 3'. One enzyme affected by dem methylation is EcoRII.

For many of the Type II restriction enzymes, a corresponding methylase has been isolated that modifies the cognate recognition sequence and renders it resistant to cleavage. These methylases are very useful in a number of tasks in molecular cloning. For example, in the construction of genomic DNA libraries, random fragments of genomic DNA generated by partial cleavage with the restriction enzymes AluI and HaeIII can be treated with EcoRI methylase prior to the addition of synthetic EcoRI linkers. When the linkers attached to genomic DNA are subsequently digested with EcoRI, the natural restriction sites within the genomic DNA are protected from cleavage. Methylase can also be used to alter the apparent cleavage specificity of certain restriction enzymes. These alterations are accomplished *in vitro* by methylation of a subset of the sequences recognized by certain restriction enzymes. Only the methylated subsets will be resistant to cleavage. Sensitivity to site-specific DNA methylation is clearly not limited to restriction enzymes but is a property of DNA-binding proteins in general. ^{m4}C, ^{m5}C, ^{hm5}C, and ^{m6}A site-specific modification at non-canonical sites will block several type II methylases.

B. DNA POLYMERASES

Many steps in molecular cloning involve the synthesis of DNA by *in vitro* reactions catalyzed with DNA polymerases. Most of these enzymes require a template (single-stranded DNA) and synthesize a product whose sequence is complementary to that of the template. DNA synthesis require a pre-existing DNA or RNA primer with a 3' hydroxyl, whereas RNA synthesis can initiate synthesis *de novo*. The properties of different polymerases are briefly described below.

1. DNA dependent DNA polymerases

(a) DNA polymerase I (Pol-I)

DNA polymerase I consists of a single polypeptide chain that can function as a 5'→3' DNA polymerase, 5'→3' and 3'→5' exonuclease, and has an inherent RNase H activity. The RNase H activity is essential for cell viability in E. coli but has not been used in molecular cloning. The 5'→3' exonuclease activity of E. coli DNA polymerase I is often troublesome because it degrades the 5' terminus of primers that are bound to DNA templates and removes 5' phosphates from the termini of DNA fragments that are to be used as substrates for ligation. The 5'→3' exonuclease activity can be removed proteolytically without affecting either the polymerase activity or the 3'→5' exonuclease activity. The uses of DNA Pol-I holoenzyme and Klenow fragment are as listed below:

1. Labeling DNA by nick translation. Of all the polymerases, it is the DNA pol-I that can carry out this reaction, since it alone has a 5'→3' exonuclease activity that can remove nucleotides from the DNA strand ahead of the advancing enzyme.
2. Synthesis of the second strand of cDNA in cDNA cloning. But this has since been suppressed by reverse transcriptase or the Klenow fragment of DNA pol-I.
3. Synthesis of double-stranded DNA from single-stranded templates during *in vitro* mutagenesis.
4. Sequencing of DNA using Sanger dideoxy-mediated chain termination method.
5. End-labeling of DNA molecules with protruding 3' tails.
6. Filling the recessed 3' termini created in DNA with restriction enzymes.
7. Labeling the termini of DNA fragments by using [^{32}P]dNTPs to fill recessed 3' termini (end-labeling).

(b) Bacteriophage T4 DNA polymerase

Bacteriophage T4 DNA polymerase and the Klenow fragment of *E. coli* DNA Pol-I are similar in that each possesses a 5'→3' polymerase activity and a 5'→3' exonuclease activity that is more active on ssDNA than on dsDNA. However, the exonuclease activity of bacteriophage T4 DNA polymerase is more than 200 times that of a Klenow fragment. Since it does not displace oligonucleotide primers from ssDNA templates, the T4 DNA polymerase works more efficiently than the Klenow fragment in mutagenesis reactions *in vitro*.

(c) Bacteriophage T7 DNA polymerase

The DNA polymerase induced after infection of *E. coli* by bacteriophage T7 is a complex of two tightly bound proteins, the bacteriophage T7 gene 5 protein and the host protein thioredoxin. This complex is the most processive of all known DNA polymerases and a single molecule of T7 DNA polymerase can synthesize much greater than DNAs synthesized by other DNA polymerases. This higher processivity of T7 DNA polymerase has considerable advantages in certain circumstances, as for example, when sequencing DNA by the Sanger dideoxy chain-termination method and it has 5'→3' exonuclease activity, which is 1000 times more potent than other DNA polymerases. Bacteriophage T7 DNA polymerase does not have 5'→3' exonuclease activity. This polymerase enzyme can be used for primer-extension reactions that require the copying of long stretches of template and rapid end-labeling by either filling or exchange (replacement) reactions such as those described for T4 DNA polymerase.

(d) Taq DNA polymerase and amplitaq

Taq DNA polymerase is a thermostable DNA-dependent DNA polymerase originally purified from the extreme thermophile *Thermus aquaticus* and now available in a genetically engineered form (AmpliTaq; Perkin Elmer Cetus). These enzymes have a 5'→3' polymerization-dependent exonuclease activity. The characteristic feature of this enzyme is that it incorporates nucleotides best at 75–80°C, depending on the target sequence. *Taq* DNA polymerase is used for DNA sequencing and to amplify specific sequences of DNA *in vitro* by the polymerase chain reaction (Explained in detail in Chapter 12).

2. RNA-dependent DNA polymerases

These enzymes, also known as **reverse transcriptases**, and often abbreviated to RTases, are encoded by retroviruses, which have an RNA genome that has to be turned into DNA as part of their life cycle. Two forms of reverse transcriptase (RT) are commercially available and differ from one another in a number of respects. But both the enzymes lack a 5'→3' exonuclease activity on DNA and catalyze the reactions, and are therefore prone to error. The Km of RT for its dNTP substrates is very high. RT has 5'→3' polymerase activity, uses RNA or DNA as template, and has 5'→3' polymerase and 5'→3' RNase H activity. RT can be used to synthesize single-stranded copies of DNA templates using oligonucleotide primers. However, both double-stranded and single-stranded cDNAs are generated from RNA templates. Self-primed synthesis is much less efficient than synthesis from the added oligonucleotide primers.

3. DNA-dependent RNA polymerases

(a) Bacteriophage SP6

Bacteriophage SP6 synthesizes a DNA-dependent RNA polymerase that recognizes and initiates synthesis of RNA on double-stranded DNA templates that carry the appropriate bacteriophage-specific promoter. The polymerase is used *in vitro* to generate large quantities of RNA complementary to one strand of foreign DNA that is immediately downstream from the promoter in plasmids specifically designed for this purpose.

(b) Bacteriophages T7 and T3

Bacteriophage T7 and T3 also synthesize DNA-dependent RNA polymerases that recognize and initiate synthesis of RNA on double-stranded DNA templates that carry the appropriate bacteriophage specific promoters. These polymerases are used *in vitro* just like SP6 RNA polymerase.

These polymerases are used in molecular cloning for synthesis of single-stranded RNA for use in hybridization probes, functional mRNAs for *in vitro* translation systems, or substrates for *in vitro* splicing reactions. The bacteriophage T7 transcription system has been used to express cloned genes in bacteria.

4. Template Independent DNA polymerases

Terminal transferase (**Terminal Deoxynucleotidyl Transferase**) is an unusual DNA polymerase found only in prelymphocytes and in the early stages of lymphoid differentiation. In the presence of a divalent cation (Mg^{2+} or Co^{2+}), the enzyme catalyses the addition of dNTPs to the 3'-hydroxyl termini of DNA molecules. The minimum chain length of the acceptor DNA is three dNTPs, and as many as several thousand dNTPs can be incorporated if the ratio of acceptor to nucleotide is adjusted correctly. The enzyme strongly prefers to use DNAs with protruding 3' termini of DNA as acceptors. However, blunt or recessed 3' termini are also used, albeit less efficiently. Terminal Transferase is used for addition of complementary homopolymeric tails to vectors and cDNA and can be used to label the 3' termini of DNA fragments with labeled dNTP.

Protocol for addition of homopolymer tails

Requirements

2X TT solution: This double-strength reaction solution is made by adding compounds to water in the order listed below. Adjust the pH, and then bring the solution to the final volume: 0.2 M cacodylic acid, adjust to pH 7.0 with KOH; 0.016 M $MgCl_2$; 0.2 mM dithiothreitol; 0.015 M KH_2PO_4; 1.0 mM $CoCl_2$; 300 µg/ml bovine albumin. This solution may be stored at −20°C for months to years without loss of effectiveness. When thawed, there will be a precipitate that includes the lavender $CoCl_2$. The precipitate should be resuspended by vortexing before the solution is used.

Procedure

Addition of (dT)n tails:
1. Vacuum-dry an appropriate amount of 3H–labelled dTTP in a 1.5 ml polypropylene microfuge tube. Usually 25 µCi will allow sufficient resolution in detecting addition of nucleotides, but go through the calculations for your particular conditions.
2. Resuspend labeled triphosphates by adding 100 µl 2X TT, 2 µl 1.0 mM dTTP and 100 µl DNA. The reaction mixture is usually about 10 nM in 3′ DNA termini. The DNA may be in TE pH 8, but try to minimize the EDTA concentration in the reaction by using a more concentrated DNA stock and bringing the reaction to volume with distilled water.
3. For unknown reasons, the reaction rate seems to be somewhat different in different reaction mixes. Therefore, when the length of tails has to be controlled within narrow limits, it is wise to run a test reaction with a 15 µl aliquot of the above reaction mixture. This will give a reliable prediction of the incorporation rate for the subsequent main reaction.
4. Incubate the reaction mixture for 5 min. at 37°C and then add 1 unit of calf thymus terminal deoxynucleotidyl transferase for each picomole of 3′ termini. Incubate at 37°C. If necessary, dilute the enzyme in 2X TT. Take 5 µl samples of the reaction at 15-min. intervals and assay incorporation by the TCA precipitation method. Under the conditions described, an average of 100 bases will be added per terminus in 30–60 min. At the time that the reaction should be complete, put it on ice and assay the time samples. The reaction rate will be slowed by several orders of magnitude when the temperature is lowered to 4°C. After a period of 1–2 hours at 0–4°C, the reaction can be reinitiated at the former rate if it is returned to 37°C. More enzymes are usually necessary if the reaction is kept cold for a longer period.
5. Terminate the reaction by extracting twice an equal volume of water-saturated phenol at 4°C. Then extract twice with chloroform and re-extract the phenol and chloroform with 100 µl TE pH 8. Combine the extracts and ethanol-precipitate the DNA.

Note:

1. A simple calculation of average number of bases added to 3′ end is as following:

$$\text{Bases/end} = \frac{\text{nM triphosphates in reaction}}{\text{nM DNA ends in reaction}} \times \frac{\text{cpm precipitable in reaction sample}}{\text{cpm total in reaction sample}}$$

2. An average of 100 bases/end is adequate for most cloning purposes.
3. Most problems with this reaction are due to impurities in the transferase enzyme precipitation (e.g. nicking enzyme).
4. Poly–dA and poly–dT tails at about 1 nM may be efficiently annealed within a few hours at 42°C in 0.1 M ACl, TE pH 8.

C. KINASES

Bacteriophage T4 polynucleotide kinase catalyzes the transfer of the γ-phosphate of ATP to a 5′ terminus of DNA or RNA (Figure 15.3). This enzyme is used for radiolabeling 5′ termini in DNA for sequencing by the Maxam–Gilbert technique, for nuclease-S1 analysis, and for other uses requiring terminally labeled DNA. It is also used for posphorylating synthetic linkers and other fragments of DNA that lack terminal 5′ phosphates in preparation for ligation.

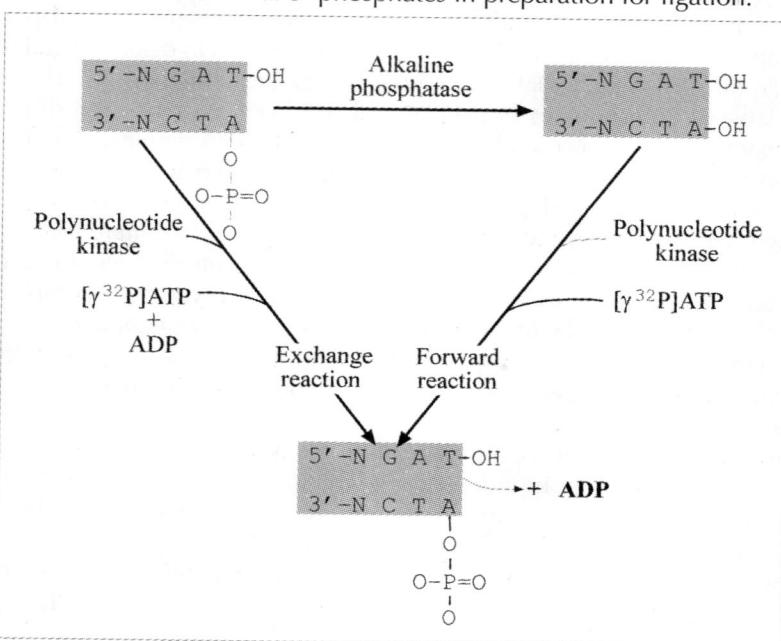

Figure 15.3 T4 polynucleotide kinase can be used to label radioactively the 5′ ends of DNA in a reaction using [$\gamma^{32}P$]ATP. DNA termini with 5′ phosphates can be dephosphorylated by the enzyme bacterial alkaline phosphatase to produce a 5′ hydroxyl that is the optimal substrate for polynucleotide kinase. In addition to the forward reaction, phosphorylated DNA can also be labeled with T4 DNA kinase in a two-step exchange reaction utilizing unlabeled ADP as an intermediate

D. PHOSPHATASES

Both bacterial alkaline phosphatase (BAP) and calf intestinal alkaline phosphatase (CIP) catalyze the removal of 5′-phosphate residues from DNA, RNA, rNTPS, and dNTPs. Thus, these enzymes are used for removing 5′ phosphates from DNA or RNA prior to labeling 5′ termini with 32P and also for removing 5′ phosphates from fragments of DNA to prevent self-ligation.

Alkaline phosphatases: Removes phosphate from 5' ends of double- or single-stranded DNA or RNA.

V DNA-BINDING PROTEINS

A. SINGLE-STRANDED DNA-BINDING PROTEIN (SSB)

SSBs bind cooperatively to ssDNA, but not to dsDNA. Their usefulness as reagents in molecular cloning derives from their ability to destabilize intrastrand secondary structures; they accelerate reanealing of complementary polynucleotides and increase the processivity of DNA polymerases by removing intrastrand secondary structures that form barriers to the progression of these enzymes. Because of this property, SSBs have become useful reagents in DNA sequencing. The primary uses of SSBs are: (i) visualizing DNA by electron microscopy; (ii) site-directed mutagenesis involving D-loops; (iii) increasing the chain length of the product in any reaction catalyzed by specific DNA polymerases used in DNA sequencing reactions.

B. *RECA* PROTEIN

This protein is encoded by a recA gene of *E. coli*. It has several useful activities associated with its role in recombination in the cell. It binds ssDNA, promotes uptake of homologous ssDNA into dsDNA, and is a DNA-dependent ATPase. The uses are (i) analysis of DNA structure by electron microscopy; (ii) *in vitro* mutagenesis using D-looping; (iii) selection of specific sequences from libraries of dsDNA *via* the formation of a D-loop.

C. TOPOISOMERASES

Topoisomerase I catalyzes the removal of superhelical turns from covalently closed circular dsDNA by transient breakage and rejoining of phosphodiester bonds. Topoisomerase I is relatively insensitive to the superhelical density of the DNA and is active in solutions containing EDTA. In contrast to the prokaryotic type I topoisomerases, the calf thymus enzyme relaxes both positive and negatively supercoiled DNA to completion.

VI SUGGESTED READING

Ausubel FM, Brent R, Kingston RE, Moore DD, Seidman JG, Smith JA, and Struhl K, (Eds)., (1999) *Short Protocols in Molecular Biology* (4th Edn), John Wiley and Sons, Inc.

Chirikjian JG (ed) (1999) *Biotechnology Theory and Techniques*, Vol I and II. Jones and Bartlett Publishers.

Davis CG, Dibner MD, and Battery JF (1986) *Methods in Molecular Biology*, ElsevierScience Publishing Co., Inc.

Docherty K (ed.)(1996) *Gene Transcription: Analysis: Essential Techniques*, John Wiley and Sons,Chichester.

Freifelder D (2000) *Molecular Biology (2nd Edn)*, Narosa Publishing House, New Delhi, Bombay.

Garret RH and Grisham CM (1999) *Biochemistry*, Saunders College Publishing.

Glick BR and Pasternak JJ (1996) *Molecular Biotechnology*, Panima Publishing Corporation, Bangalore, New Delhi.

Lewin B (2000) *Genes VII*, Oxford University Press and Cell Press.

Miesfeld RL (1999) *Applied Molecular Genetics*, Wiley–Liss, John Wiley and Sons, Inc, NY.

Old RW and Primrose SB (19) *Principles of Gene Manipulation, An introduction to Genetic Engineering.*

Sambrook J, Fritsch EF and Maniatias T (eds.) (1989) *Molecular Cloning: a Laboratory Manual (2nd Edn)*. Cold Spring Harbor Laboratory Press. Cold Spring Harbor, NY.

Steinberg M, Guyden J, Calhoun D, Staiano–Coico L and Coice R (1993) *Recombinant DNA Technology: Concepts and Biomedical Applications*, PTR Prentice Hall, Inc.

Watson JD, Hopkins NH, Roberts JF, Steitz JA, and Weiner AM (1987) *Molecular Biology of the Gene (4th Edn)*, The Benjamin/Cummings Publishing Company, Inc.

Weaver RF (1999) Molecular Biology. WCB/McGraw-Hill.

Wilson K and Walker J (2000) (Ed). *Practical Biochemistry: Principles and Techniques (5th Edn)*. Cambridge University Press.

Winnacker E-L (2003) *From Genes to Clones*. Panima Publishing Corporation.

16

Gene Constructs and Cloning Vectors

I INTRODUCTION

The goal in cloning is to multiply cells that contain a single vector molecule that has exogenous DNA. A vector is the agent used to replicate or multiply the exogenous DNA. Generally, vectors are segments of DNA with an **origin of replication** so that the vector is replicated after it is introduced into the host cell. Cloning vehicles can be either plasmids, bacteriophages, viruses, or hybrid structures called cosmids, phagemids, and artificial chromosomes. Vectors are mainly used to transfer DNA molecules from a test tube into a living cell. The main objective of using a number of cloning vectors all described in previous sections is to clone a gene or DNA segment. Cloned genes in these vectors need not express themselves either at the transcription level or at the translation level. But, when the cloned gene is used for transformation to develop transgenics (microorganisms, plants, or animals), then the cloned genes must express themselves. Often high levels of expression of cloned genes are required and the expression should be either constitutive or regulated to tissue or the stage of development. This objective can be achieved through the use of promoters and expression cassettes.

II GENE STRUCTURE AND CONSTRUCTION

A. TYPICAL PLANT GENE

The protein coding genes in both prokaryotes and eukaryotes have many DNA sequences with similar functional organization (Figure 16.1). But the regulatory sequences in eukaryotes are more complex. In prokaryotes most of the protein coding genes are polycistronic (more than one gene under the control of the same promoter) in nature and normally free from intron sequences (Figure 16.1a). A typical plant gene has the following regions starting from a 5′–end: (i) promoter (for transcription initiation); (ii) upstream activating sequence (UAS)/enhancer/silencer (concerned with regulation of gene activity); (iii) transcriptional start site/cap site; (iv) an untranslated region or leader sequence; (v) initiation codon; (vi) exons; (vii) introns; (viii) region of termination (stop codon); (ix) a second untranslated region and (x) a poly (A) signal site (Figure 16.1b).

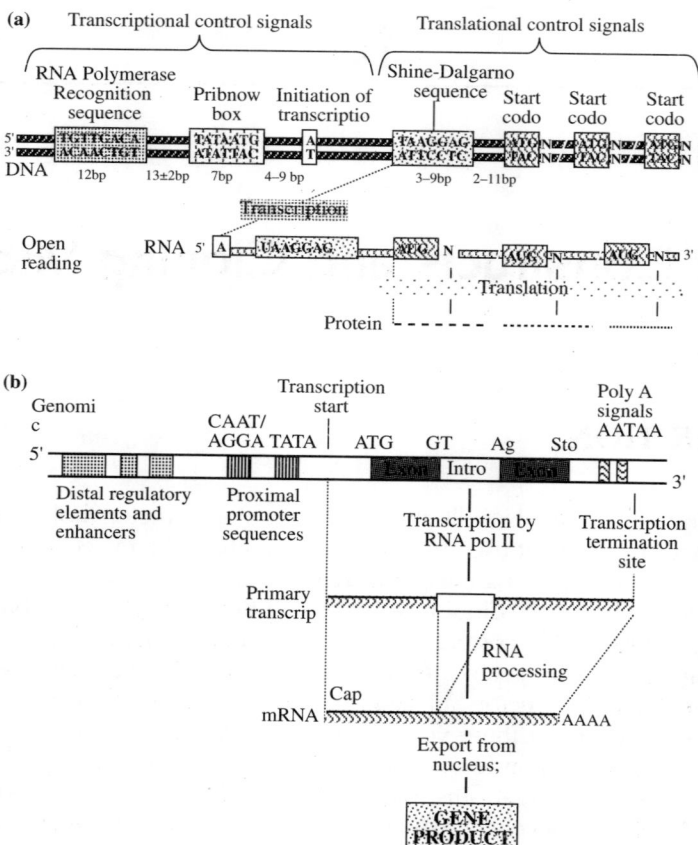

Figure 16.1 *Idealized representation of a polypeptide-encoding genes and signals involved in the regulation of gene expression (a) Prokaryotic polycistronic gene and (b) Eukaryotic nuclear gene*

1. Promoters

While cloning, it is essential to match the promoter to be used with the type of host cell expression system. In principle, any promoter should be satisfactory to control the expression of a gene, but in practice, each host system has a short list of favorite promoters. These are chosen to provide two main functions: (i) constitutive/inducible expression and (ii) strong/weak expression. Unfortunately, the factors that control the levels of expression are as yet so ill-understood that it is not feasible to predict which promoter will turn out to be the most effective in a particular system. The only sensible approach is to plan at the outset to try a number of different promoters in a number of different host cell strains.

The ***promoter*** provides the site for binding of RNA polymerase, and hence is involved in transcription initiation. The **CAAT** and **TATA** boxes represent a conserved (***consensus***) sequence within promoters. The enhancer/suppressor sequences on the other hand, make the regulation of gene action tissue/developmental stimulus-specific. ATG (in mRNA) is the initiation codon for mRNA translation, and hence marks the beginning of the coding sequence of the gene. The sequence lying between the **cap site** and ATG is not translated, and forms the **5′–leader sequence** of mRNA. The end of the gene coding sequence is specified by the chain termination TAG (in mRNA, UAG) or one of the other two stop codons (UGA, UAA). The stop codon is followed by a stretch of **non-translated region** and at the end is present the **polyadenylation signal site**, which denotes the end of transcription. Proteins have diverse functions. Many proteins either move across membranes or become inserted into them. Genes encoding such proteins have at their 5′ ends (immediately after the start codon AUG) a coding sequence, which encodes a **signal peptide**. The signal peptides are cleaved, after the protein has been translocated to the desired location, to generate mature functional proteins.

For an efficient expression in plant cells, particularly foreign genes, the gene sequence must have an appropriate promoter, 5′ leader, and 3′ terminator sequences. A suitable enhancer sequence will also be needed if the gene is required to be expressed either in a specific tissue, during a specific developmental stage or in response to a specific stimulus. A variety of promoter sequences have been used to drive genes in plant cells (Table 16.1). The 35S promoter is the most commonly used constitutive promoter both in dicots and monocots. The maize alcohol dehydrogenase 1 (*Adh*1) promoter shows anaerobic induction, i.e. expression in roots. The *Adh*1 intron 1 and maize *shrunken* 1 locus intron 1, enhance the efficiency of the 35S promoter 10-fold when they are placed in between the promoter and the gene. Indeed modifications of some of the promoters, e.g., the 35S RNA promoter, have resulted in increased promoter efficiency. The 35S promoter consists of several domains which confer on it developmental tissue-specific expression patterns.

Promoters from other organisms express poorly in plants. Therefore, to obtain correct expression, genes transferred to plants should be linked to plant-specific promoters (Figure 16.1). These promoters can also be interchanged so as to confer on the transferred gene, a specific pattern of expression. Some examples of promoters are (Table 16.1) discussed below.

Nopaline synthase (*nos*) promoter from T-DNA: This promoter is 200 bp long and contains several DNA sequence motifs, which direct the expression of the linked gene. The upstream element from −97 to −130, when duplicated increases the expression threefold, suggesting that this may be an enhancer sequence. The *nos* promoter is most active in the basal regions of plants, its activity slowing down in the vegetative parts at the onset of flowering, even though in the flower, its activity is increased.

Dual promoter of mannopine synthase (*mas*) genes 1 and 2: These promoters are present in T-DNA fragments 4667 bp long. The expression seems to be developmentally regulated and the linked genes seem to be most active in the basal region of the plant. Expression is also induced on wounding. Since they are closely linked, if a marker gene is linked to one of these two promoters and a foreign gene is linked to the other promoter, selection for high expression of marker gene will lead to a high level of expression of the foreign gene.

35S RNA promoter of CaMV: This is the most extensively used promoter in a wide variety of plants. The promoter is 343 bp long and contains a strong transcriptional enhancer, which, on duplication, results in a 10 fold increase in expression, even at a distance of 2 kbp. High expression is mainly observed in leaf tissue.

Polyhedron promoter from baculovirus: Baculovirus can synthesize polyhedron protein in infected cells, so that the polyhedron makes up to 50% of the total protein in the cell, giving up to 1 g of protein per litre of inset cell culture. In view of this, the promoter has been used to construct expression vectors to allow a high level of expression of any gene under the influence of this promoter. This is being used for developing biopesticides and even for the production of specific chemicals in industry.

Table 16.1 *Some common promoters used for gene expression studies in transgenic plants*

Promoter	Source	Relative activity
nos	*Agrobacterium* nopaline synthase gene	Moderate activity; constitutive
ocs	*Agrobacterium* octopine synthase gene	Moderate activity; constitutive
mas	*Agrobacterium* mannopine synthase gene	Moderate activity; constitutive
35S	CaMV 35S RNA gene	Constitutive, high activity
35S+*Adh*1-I 1	35S promoter + first intron of maize *Adh* 1 gene	Enhanced promoter activity; constitutive
35S+*sh* 1-I 1	35S promoter + first intron of maize shrunken 1 gene	Better than 35S + Adh1-I 1 in monocots; Constitutive
19S	CaMV 19S RNA gene	Constitutive; moderate activity
Gene VI	CaMV gene VI encoding the matrix protein	Constitutive
*Adh*1	Promoter of alcohol dehydrogenase gene of maize	Moderate activity in cereals; anaerobic expression
Emu	Modified from *Adh* 1 promoter and its first intron	Moderate activity in cereals; anaerobic expression
*Act*1+*Act*-I 1	Rice actin gene + its first intron	Moderate activity in cereals; anaerobic expression
*Ubi*1+*Ubi*1-I 1 (or 16)	Maize *Ubiquitin*1 gene promoter plus its first (or sixth) intron	High activity in cereals; constitutive
PHA-L	*Phytohaemagglutinin* gene of *P. vulgaris*	Strong, seed-specific activity
Vicilin promoter	Pea vicilin storage protein gene	Seed-specific promoter
Extensin promoter	Carrot extensin gene long transcript promoter	

2. Enhancers/Regulatory elements

The expression of many genes is confined to specific tissues and/or is induced by specific stimuli. Such genes are called tissue-specific genes or responsive elements. Such specificity in gene expression is due to the presence of certain DNA sequences called enhancers/silencers (Figure 16.2), which may be present within or at a considerable distance (several kilobases) away from the promoters they affect. An **enhancer** may be defined as a DNA sequence, which increases the activity of the promoter of a gene, while those DNA sequences which suppress promoter activity are called **silencers**. Thus enhancer/silencer sequences regulate the activity of promoters but are not by themselves involved in the promoter activity *per se*. These sequences act on promoters located both upstream or downstream of the gene, vary considerably in size, and can exert positive or negative effects on promoter action.

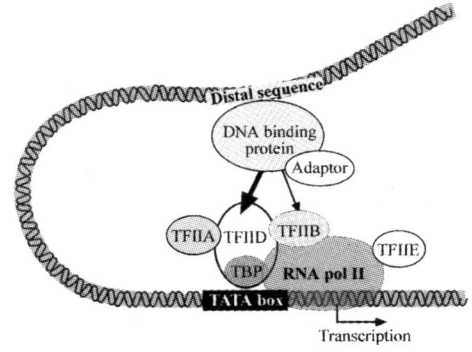

Figure 16.2 *Simplified model for the interaction of a trans-acting factor, bound to its specific recognition sequence (the cis-acting sequence), with the transcription initiation complex*

Regulatory sequences located upstream of many tissue-specific genes have been analyzed by producing suitable transgenic plants. One such nuclear gene (*rbc*S) encodes the smaller subunit of enzyme RuBISCO (ribulose, 1, 5–bisphosphate carboxylase/oxygenase), which is the key enzyme in CO_2 fixation during photosynthesis. The expression of this gene is induced by light. It contains a sequence called **light responsive element** (**LRE**) the upstream (−149 to −166 bp) of the promoter and it also regulates the tissue specific expression (in green tissues). Another sequence (−410 to −166 bp) upstream of LRE regulates the high expression (enhance) of *rbc*S in mature leaves. An example of a negative regulating element, i.e., silencer, is the sequence associated with the gene *Adh*1 of maize. This gene is expressed only under anaerobic conditions, and hence, only in the roots. The *Adh*1 enhancer sequence seems to suppress *Adh*1 expression under aerobic conditions. Anaerobic conditions do not seem to induce *Adh*1 expression, which is expressed simply because the enhancer is unable to suppress its expression in the absence of the O_2 stimulus. A number of other tissue-specific or stimulus responding regulating sequences have been identified (Table 16.2).

Table 16.2 *Selected regulatory sequences determining tissue- and/stimulus-specific gene expression*

Regulatory sequences	Source	Site of expression	Stimulus
Ribulose 1,5–bisphosphate carboxylase/oxygenase small subunit (*rbc*s)	Pea, soybean, tobacco	Leaves (high) and stem (low)	Light
Chlorophyl a/b binding protein (cab)	Pea, wheat, tobacco	Leaves and stem	Light
Chalcone synthase	Antirrhinum majus	–	Light
ST–LS 1	Potato	Green tissue	Light
Heat shock protein (*Hsp*)	Maize, soybean, Drosophila	–	Heat shock

Continues...

... *Continued*

Regulatory sequences	Source	Site of expression	Stimulus
Protease inhibitor	Potato	–	Wounding
Potatin	Potato	Tuber	–
Leghemoglobin	Soybean	Nodule	–
Nodulin	Soybean	Nodule	–
β–Phaseolin	French bean	Seed	–
Lectin	Soybean	Seed	–
Glutenin	Wheat	Seed	–
Hordein	Barley	Seed	–
5-Enolpyruvil shikimate–3–phosphate synthase (ESPS)	Petunia	Pollen	–
Sucrose synthase 1	Maize	Phloem cells	–

III CLONING VECTORS

The central component of a gene cloning experiment is the **vector** or **vehicle**, which transports (carries) the gene of interest into a host cell and is responsible for its replication. By cloning, it is possible not only to store a copy of a particular fragment of DNA but also to produce unlimited amounts of it (ability to replicate). Most segments of DNA do not have an inherent capacity for self-replication (**origin of replication**, the site where DNA replication begins) and have to be linked to DNA sequences that are self-replicating. Genetic signals required for replication are host-specific. An antibiotic resistant marker gene must be inserted to select a gene-inserted vector. An efficient translation initiation region (TIR) must be present. If foreign DNA is to be replicated in an organism, it should be inserted into a host system or ligated to a suitable DNA fragment (**vector**) that can replicate in that organism. The recombinant DNA molecules consisting of the desired foreign DNA and the vector can be introduced into the host, and the cells that have received the DNA chimeras can be identified. The yield and purity of cloned genes and efficiency of ligation and transformation are greater with small vectors. However, origin of replication, gene insertion sites, and genes useful for selection in transformation determine the minimum size of the vector. For easy recovery of the recombinants, an insertion site in a vector gene whose function is easily identifiable is a desirable attribute. Then clones containing a plasmid with inserted DNA that inactivates the gene can be selected easily from clones with wild plasmid phenotype. Another desirable property of the vector is its role as a promoter site for initiating transcription across the inserted fragment. Foreign genes can be transcribed from this vector promoter if their normal promoters are not recognized by the host transcription machinery. Promoters can be followed by the ribosome binding sites, initiation codon, and then a restriction site to insert the DNA for obtaining expression.

A. CLONING

To **clone** means to make **identical copies**. Clone, a complex term, not only refers to populations of cells or organisms with identical hereditary components, but also to populations of identical recombinant DNA molecules. In classical biology, a clone is a population of identical organisms derived from a single parental organism. For example, the members of a colony or bacterial cells that arise from a single cell on a Petri plate are clones. Molecular biology borrowed the term "clone" to mean a collection of molecules or cells all identical to an original molecule or cell. So, if the original cell on the Petri plate harbored a recombinant DNA molecule in the form of a plasmid, the plasmids within the millions of cells in the bacterial colony represent clones of the original DNA molecule. The process of making multiple identical copies of a molecule or a cell or an animal is called cloning. Thus the growth of biotechnology relies directly on the development of techniques that allow the cloning of biologically active molecules. In fact, many scientists today consider molecular cloning the single most important development in the growth of biotechnology research.

DNA cloning involves separating a specific gene or DNA segment from a larger chromosome, attaching it to a small molecule of carrier DNA, and then replicating this modified DNA thousands or millions of times through both the increase in cell number and the creation of multiple copies of the cloned DNA in each cell (Figure 16.3). The result is selective amplification of a particular gene or DNA segment. Cloning of DNA from any organism entails five general procedures: (1) cutting of DNA at precise locations; (2) joining two DNA fragments covalently; (3) selecting a small molecule of DNA capable of self-replication (various types of **cloning vectors**, a vector is a delivery agent); (4) moving the recombinant DNA from the test-tube to a host cell that will provide the enzymatic machinery for DNA replication; (5) selecting or identifying host cells that contain the recombinant DNA. The methods used to accomplish these and related tasks are collectively referred to as recombinant DNA technology, or more informally, as genetic engineering.

B. CHOICE OF CLONING VECTORS

The main objective of recombinant DNA technology is the cloning of desired DNA segments isolated from various sources (randomly fragmented DNA segments, cDNA, specific genes, synthesized genes, etc.). The cloning of desired DNA segments is possible only with the help of another DNA molecule that can self-replicate in a host cell. This autonomously replicating DNA molecule which can harbor DNA fragments from different origins is called a **vector**. A useful feature of any cloning vector is the amount of DNA it may accept or have inserted before it becomes unviable. The vector must be capable of undergoing replication to produce multiple copies by itself. The vector could be a plasmid, a bacteriophage, a derived cosmid or phagemid, a transposon or even a virus (Table 16.3). Vectors generally contain different antibiotic resistance genes, selection markers, and cloning sites with many restriction sites (polylinkers).

IV BACTERIAL CLONING VECTORS

Popular cloning vectors commonly used in experiments with *E. coli* are **plasmids**, **bacteriophages**, and **bacterial artificial chromosomes** (**BACs**).

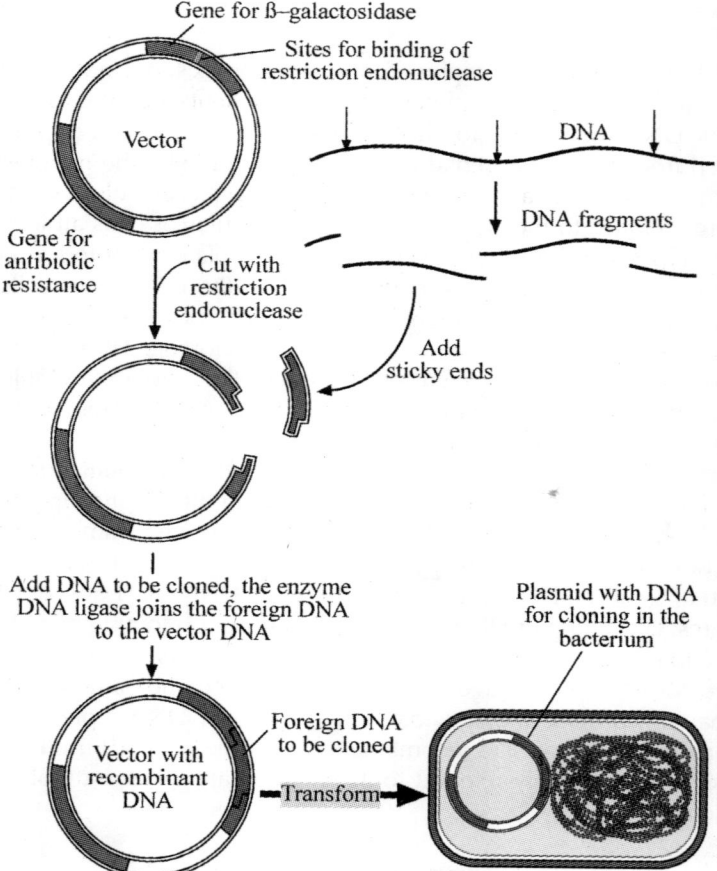

Figure 16.3 *Summary of a procedure designed to clone a gene. The vector acts as a carrier for the foreign DNA which may be a new gene. The vector will replicate inside the bacterium, making many identical copies of itself and the new gene (clones)*

Table 16.3 *Some generally available vectors for cloning DNA fragments*

Vector	Host cell	Vector structure	Insert range (kb)
General purpose cloning vectors			
Plasmid	E. coli	Circular plasmid	1–5
Phagemid vectors, M13	E. coli	Circular virus	1–4
Phage λ	E. coli	Linear virus	2–25
Cosmids	E. coli	Circular plasmid	35–45
High capacity cloning vectors			
Bacterial artificial chromosomes (BACs)	E. coli	Circular plasmid	50–300
P1 vectors and P1 artificial chromosomes (PACs)	E. coli	Circular plasmid	~100
Yeast artificial chromosomes (YACs)	S. cerevisiae	Linear chromosome	100–2000

A. PLASMID VECTORS

Bacterial plasmids are naturally occurring, double-stranded, covalently closed, circular, extra-chromosomal DNA molecules. The genome of plasmids range in size from 1 kb to more than 200 kb. They are found in a variety of bacterial species. Nevertheless, they depend on enzymes and proteins encoded by the host for their replication and transcription. Natural strains of *E. coli* and many other bacterial species isolated from different sources carry diverse plasmids (e.g. plasmids classified into **Fertility (F)** plasmids, **Resistance (R)** plasmids, **Col** plasmids, **Degradative** plasmids, and **Virulence** plasmids, e.g. Ti plasmids). Some plasmids are cryptic (confer no detectable phenotype), but others confer phenotypic characters. Often plasmids carry genes specifying novel metabolic activities that are advantageous to the host bacterium. Among the metabolic activities conferred by the plasmids are resistance to antibiotics; production of antibiotics; degradation of complex organic compounds; and production of colicins, enterotoxins, and restriction and modification enzymes. These plasmids are classified on the basis of their DNA replication origin (replicon) and the encoded antibiotic-resistance gene(s). The most common plasmids used today contain the ColE1 *ori*, which has a relaxed mode of replication. Plasmids can perpetuate themselves and inherit independently of the bacterial chromosome. Thus, they have become molecular vehicles (vectors) of recombinant DNA technology. Since restriction endonuclease digestion of plasmids can generate fragments with overlapping (sticky) ends, artificial plasmids can be constructed by ligating different fragments together. As long as they carry a site signaling for DNA replication (origin of replication, *ori*), these recombinant molecules can be autonomously replicated, propagated in a suitable bacterial cell, and be isolated with a large amount of plasmid copies for further genetic analysis.

The nomenclature of plasmid cloning vectors, for example, the name 'pBR322' conforms with the standard rules for vector nomenclature. 'p' denotes a plasmid; 'BR' identifies the laboratory in which the vector was originally constructed (BR stands for Bolivar and Rodriguez, the two researchers who developed pBR322); '322' distinguishes this plasmid from others developed in the same laboratory (e.g. pBR325, pBR327, pBR328, etc.).

(a) **Replication of plasmids**: Replication of plasmid DNA is carried out by subsets of enzymes used to duplicate the bacterial chromosome. However, different plasmids replicate to different extents in their host cells ranging from one to as high as 700 copies per host cell. Plasmids may be grouped into two broad categories on the basis of number of copies, relative to the chromosome of host cells. In plasmids, which are under stringent control, initiation of new rounds of replication is tied to initiation of chromosomal replication, and the copy number is low (<10 copies per cell). Typical stringently controlled plasmids are the F factor, certain large R factors, and the plasmid pSC101. By contrast, those such as Col E1, are not tied to chromosomal initiation, occur in copy numbers of 30–60, and may even run as high as 200. Initiation of replication of the chromosome and stringent plasmids is dependent upon new protein synthesis whereas relaxed plasmids can replicate in the absence of protein synthesis. The control of plasmid copy number resides in the origin of replication, which is usually one per molecule along with cis-acting control elements (the whole genetic unit is called "replicon"). Most vectors in current use carry a replicon derived from the plasmid pMB1, which was originally isolated from a clinical specimen. By testing the ability of different plasmids to coexist in the same cell, it is possible to assign them to incompatibility groups. Under natural conditions, plasmids are transmitted to new hosts by a process of bacterial conjugation. The mobilization of plasmids through a conjugation process is regulated by the mobilization gene (mob). Nevertheless, some plasmid vectors (e.g. pUC322) can be mobilized by a conjugative plasmid if a third plasmid (ColK) is present in the cell.

(b) Selectable markers: In the laboratory, plasmid DNA can be introduced into bacteria by the artificial process of transformation. However, even under the best conditions, plasmids become stably established in only a small minority of the bacterial population. To identify these transformants, selectable markers encoded by the plasmids are used. These markers confer a new phenotype that allows the bacteria transformed to be selected. The most commonly used selectable markers are genes that confer resistance to antibiotics such as ampicillin, tetracycline, chloramphenicol, and kanamycin (neomycin). The mode of action of each of these antibiotics is different (Appendix 2). Virtually all plasmid vectors in use carry one or more of the antibiotic resistance genes described above. However, other selectable markers are occasionally used for more specialized purposes.

(c) Construction of chimeric plasmids: Cohen et al. (1973) were the first to construct a small plasmid cloning vector (pSC101). It carried the DNA sequences essential for replication in E. coli, a selectable gene conferring tetracycline resistance and a single site for EcoRI that was outside regions determining replication of the vector and a selectable gene of tetracycline. Now many derivatives of plasmid ColE1 (pBM9, pBR322) have been developed as vectors for cloning foreign DNA into E. coli. These plasmid vectors are present as 30–50 copies per E. coli cell under normal conditions of growth. Their DNA can be amplified to about 45% of the total cellular DNA if the E. coli is grown in the presence of chloramphenicol. Consequently, very high yields of both plasmid DNA and any DNA segment inserted into the plasmid can be obtained.

The incorporation of DNA fragments into plasmid vectors not only allows foreign DNA to be replicated, but it can also be designed so that cells transcribe and translate the insert into protein. The creation of chimeric plasmids requires joining the ends of the foreign DNA insert to the ends of a linearized plasmid. This ligation is better facilitated if the ends of the plasmid and the insert have complementary, single-stranded overhangs. Then these complementary ends can base pair with one another, annealing the two molecules together. An alternative method for joining different DNAs is by blunt-end ligation. This method depends on the ability of phage T4 DNA ligase to covalently join the ends of any two DNA molecules.

Maps of widely used plasmid vectors that contain many of the useful features of these vectors are presented in the following sections. Each of these vectors has been completely sequenced.

1. pSC101

This plasmid vector is the earliest used vector and contains the replication module (ori) for replication in E. coli, tet^r gene for resistance to tetracycline and single recognition sites for restriction endonucleases EcoRI, HindIII, BamHI and SalI. Insertion of gene into the EcoRI site leaves the tet^r gene intact and functional. As a result, E. coli cells transformed by pSC101 become tetracycline resistant but such cells may either have a non-recombinant pSC101 (without the DNA insert) or a recombinant one (having the DNA insert). On the other hand, insertion of DNA fragment into the other sites disrupts the tet^r gene and makes it non-functional. Thus these transformed cells become sensitive to tetracycline. However, non-transformed cells are tet^s too. Hence they cannot be separated from those having the recombinant pSC101, clearly indicating that, pSC101 does not permit a direct selection of cells containing the recombinant vector. At the same time, it contains unwanted DNA, causing problems for replication. Therefore it is not commonly used.

2. Plasmid pBR322 and its derivatives

The plasmid pBR322 is probably the most widely used plasmid vector. Naturally occurring plasmids may not possess all the essential properties for cloning. Therefore, one may have to reconstruct them by inserting genes for antibiotic resistance, relaxed replication, and cloning sites. The pBR322 is a 4,362 bp dsDNA plasmid cloning vehicle designated to allow simple and rapid preparation of cloned recombinant DNA fragments. Generally, it contains two antibiotic resistance genes (tetracycline and ampicillin), an origin of replication, and a variety of useful restriction sites for cloning or subcloning restriction fragments (Figure 16.4). pBR327 is

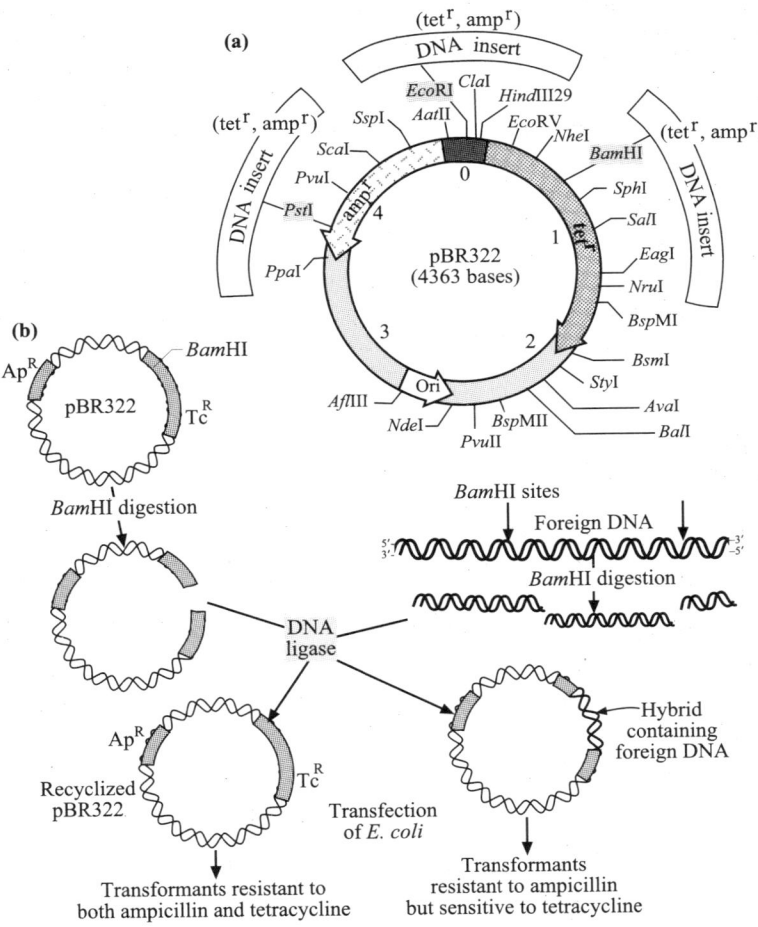

Figure 16.4 (a) The structure of plasmid pBR322, a typical cloning vector. Several restriction enzyme sites are shown that could be used as insertion sites for foreign DNA, depending on the cloning strategy. (b) The use of plasmid pBR322 as a cloning vector, showing how insertion of foreign DNA causes inactivation of the tetracycline resistance gene, permitting easy identification of transformants

identical to pBR322, except that it lacks all the nucleotides between 1,427 and 2,512. These sequences are deleted to reduce the size of the cloning vector and to eliminate sequences that were shown to interfere with the expression of cloned DNA in eukaryotic cells. pBR322 and its derivative plasmids are very common plasmid vectors. The two antibiotic resistance genes possess unique cloning sites for insertion of DNA fragments. Opening the plasmid with RE and inserting a compatible DNA segment with DNA ligase inactivates this resistance gene. However, selection can be achieved by growing and plating the transformed cells in the presence of the other antibiotic.

3. The pUC family of plasmids

Another series of plasmids that are derivatives of pBR322, and are used as cloning vectors, belong to the pUC (plasmids from the University of California) series. These plasmids are 2,700 bp long and possess, the ampicillin resistance gene, the origin of replication derived from pBR322 and the *lac Z* gene derived from *E. coli*. Within the *lac* region, a polylinker sequence with unique restriction sites is inserted. When DNA segments are cloned in this region of the pUC, the *lac* gene is inactivated. These plasmids when transformed into an appropriate *E. coli* strain, which is lac– (e.g. JM103, JM109), and grown in the presence of **IPTG** (isopropyl thiogalactoside) and **X-gal** (substrate for the enzyme), will give rise to **white** or **clear colonies**. On the other hand, pUC having no inserts and transformed into bacteria will have an active *lac Z* gene and will therefore produce **blue colonies**, thus permitting identification of colonies having the pUC vector with cloned DNA segments. Vectors belonging to the pUC family are available in pairs with reversed orders of restriction sites relative to the *lac Z* promoter. For example, pUC8 and pUC9 (Figure 16.5), pUC12 and pUC13, pUC18 and pUC19 (Figure 16.6) are some pairs available.

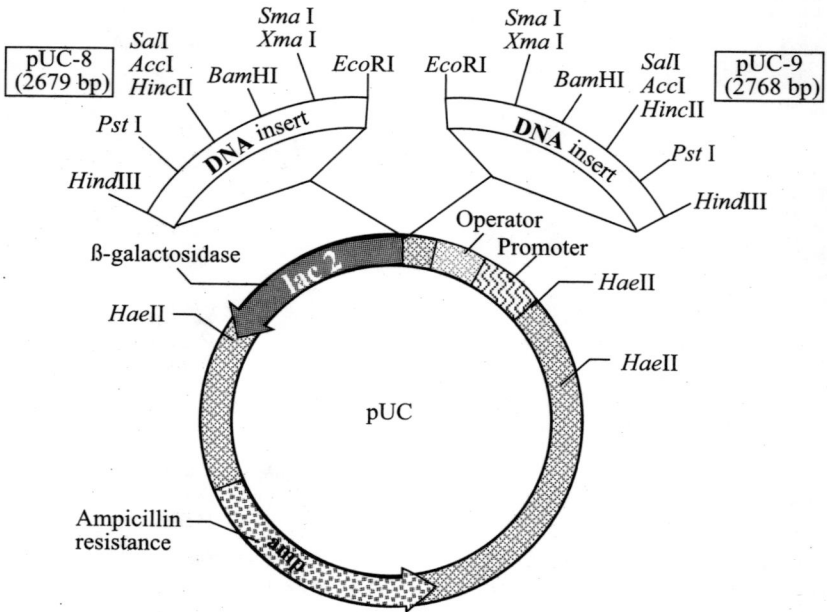

Figure 16.5 *Diagram of pUC plasmids, indicating restriction maps of polylinker regions of pUC 8 and 9*

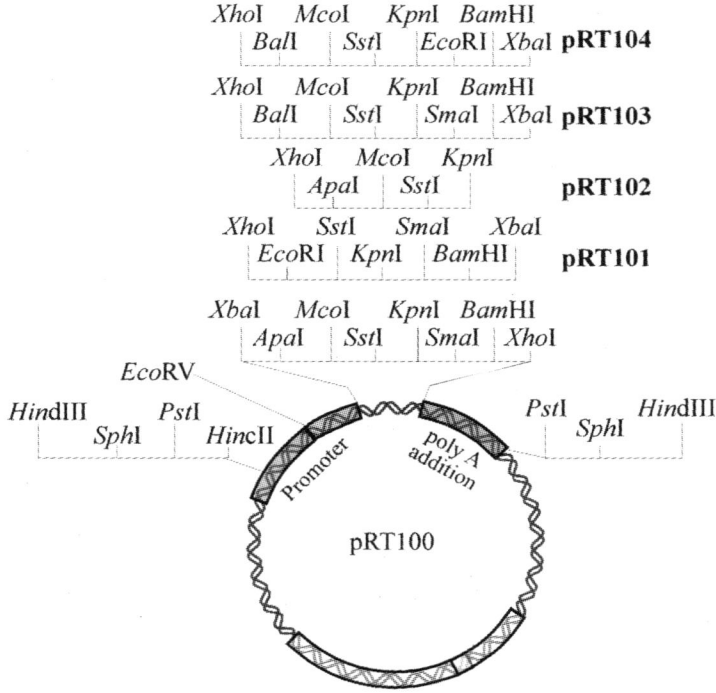

Figure 16.6 *pRT series of expression cassettes based on pUC 18/19*

The DNA of interest is inserted into a site located in one of the two genes for resistance against antibiotics, so that it will inactivate one of the two resistance genes. The insert bearing plasmid can be selected by its ability to grow in a medium containing only one of the two antibiotics and by its failure to grow in a medium containing both the antibiotics. The plasmids carrying no insert on the other hand, will be able to grow in media containing one or both the antibiotics.

4. The pEMBL family of plasmids

These plasmids (pEMBL) can replicate either as double- or single-stranded molecules. This arrangements obviates the need for subcloning target fragments from one plasmid to another. Construction of the pEMBL plasmids was based on the observation that when a segment of DNA containing the origin of replication and cis-acting replication elements from the genome of a single-stranded phage f1 was inserted into pBR322, then it could be induced to replicate as a single-stranded molecule when super-infected with phage f1. The f1 fragment is inserted into a unique *NarI* site common to all pUC plasmids. The new pEMBL plasmids thus obtained contain the *bla* and *lac* genes of the pUC plasmids and have the single-stranded cloning capabilities of the associated M13 vectors without the need for sub-cloning. Recently, the yeast genes *LEU2*, *URA3*, and *TRP1*, and in some cases, the 2 μ circle origin of replication were introduced into the pEMBL plasmid to make both insertional and episomal yeast vectors. These vectors are known as pEMBLY plasmids.

B. BACTERIOPHAGE VECTORS

Bacteriophages (or phages) are bacterial viruses that infect their hosts by attaching themselves to specific receptor proteins on the outside of the cell. Phage particles also differ in their physical structures from species to species and often certain features of their life cycles are correlated with their structure. The three basic structures are (i) Icosahedral tailless, (ii) Icosahedral tailed, and (iii) Filamentous structure. The most common type of nucleic acid in phages is linear dsDNA but other types of nucleic acids are also found. Phages generally gain entry into the host cell by injecting their DNA directly into the cell or by being internalized by host cell processes. Their ability to transfer DNA from the phage genome to specific bacterial hosts during the process of bacterial infection gave molecular geneticists the idea that specially designed phage vectors would be useful tools for certain types of gene cloning applications. Two of the most common phage-based methods are the use of gene cloning vectors constructed from components of the temperate bacteriophage λ (lambda) and the production of single-strand DNA using cells infected with vectors based on the filamentous bacteriophage M13.

1. Bacteriophage lambda

Biology of bacteriophage lambda (λ): The lambda phage is genetically complex, but it has been well characterized physically, genetically, and chemically. The λ phage particle contains two major structural components called a head and a tail. The genome of bacteriophage λ is a 48.5 kbp DNA molecule that is packaged into the phage head. The DNA is carried in the bacteriophage particles as a linear double-stranded molecule with single-stranded complementary termini 12 nucleotides in length (cohesive termini, 5' GGGCGGCGACCT). Soon after entering a host bacterium, the cohesive termini associate by base pairing to form a circular molecule with two staggered nicks 12 nucleotides apart. These nicks are rapidly sealed (form a *cos* site) to generate a closed circular DNA molecule that serves as the template for transcription during the early phase of infection. The phage chooses one of the following two replication pathways: lysogenic or lytic growth. During **lysogenic** growth, the infecting bacteriophage genome becomes integrated into the bacterial host DNA as a prophage by site-specific recombination and is subsequently replicated and transmitted to progeny bacteria like any other chromosomal gene (dormant). During **lysogeny**, only one bacteriophage gene "cI" (repressor) expresses. During lytic growth, the circular viral DNA is replicated manifold, a large number of bacteriophage gene products are synthesized, progeny bacteriophage particles are assembled, and the cell eventually lyses, releasing its many new virus particles.

Lambda cloning vectors (λgt10, λgt11, **EMBL3, EMBL4, Charon**): Studies of λ phage mutants showed that as much as 18 kb of the genome was dispensable for lytic growth, primarily by eliminating genes required specifically for lysogeny. This meant that portions (40%) of the λ genome could be replaced with foreign DNA using *in vitro* recombinant DNA methods. The utility of λ vectors for gene cloning was significantly enhanced following the development of two major improvements in the general strategy. First, the improvement which overcame the low efficiency of $CaCl_2$-mediated DNA transformation, as compared to phage infection. A second improvement in λ cloning systems led to an increase in the ratio of recombinant to non-recombinant phage in the pool of infectious particles. With this improvement, several genetic strategies were devised to select genetically against non-recombinant λ genomes. Two of these selection strategies are illustrated in Figure 16.7. One involves small DNA insertions into the λ cI gene (Figure 16.7a and b), which controls the lysogenic pathway, and the other (Figure 16.7c and d) relies on replacement of the vector "stuffer" DNA encoding two λ genes that promote lysogeny (red, gam), with large fragments of foreign DNA (genomic DNA 15–20 kb in size).

Engineered vectors of lambda are of two major types. **Insertion vectors** have a single target site at which foreign DNA can be inserted, while **replacement vectors** have a pair

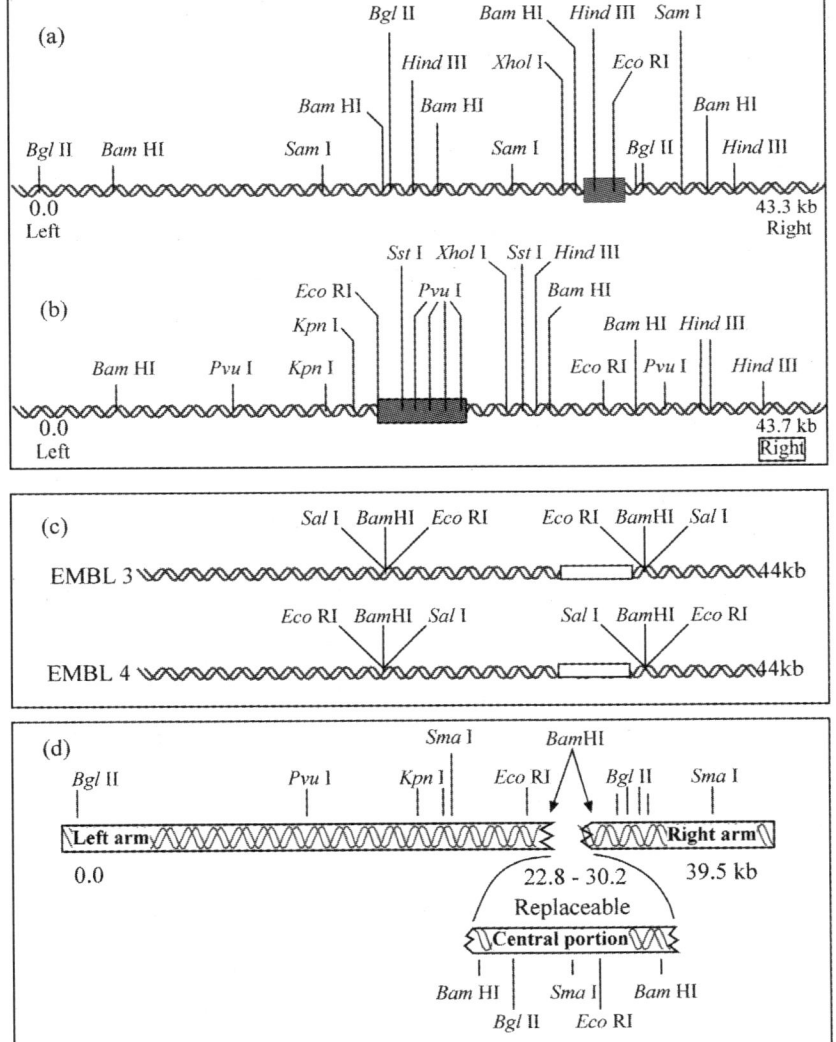

Figure 16.7 Diagram of a) λgt10, b) λgt11, c) EMBL3 and EMBL4, and d) Charon 28 #2221 vectors and showing major RE sites in multiple cloning regions

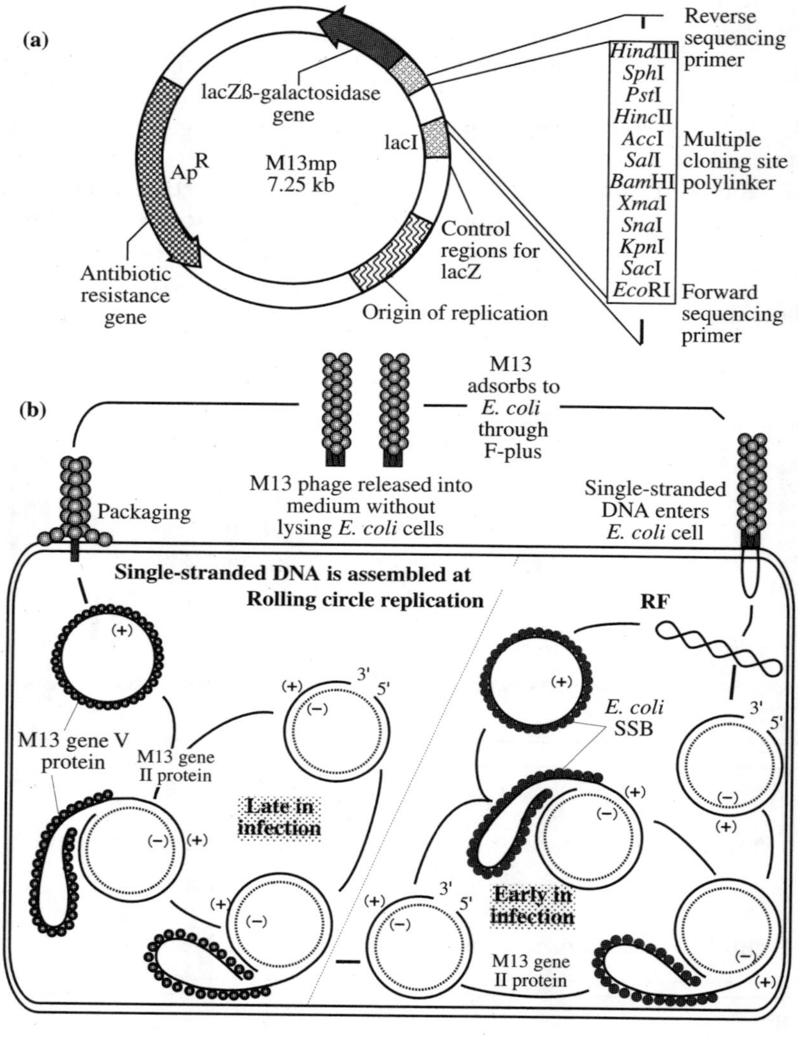

Figure 16.8 (a) Genetic map and important features of bacteriophage vector M13 (b) Life cycle of bacteriophage M13

of sites defining a fragment that can be removed and replaced by foreign DNA. Once exogenous DNA has been inserted into the vector lambda, this molecule can be multiplied by inserting it into host *E. coli* cells in one of two ways. Naked lambda DNA molecules lacking a protein coat can be introduced into *E. coli* cells in a process called **transfection** or by *in vitro* **packaging**. *In vitro* packaging requires the DNA molecules to be packaged with high concentrations of phage head precursors, proteins that participate in the packaging process and phage tails. These packaging ingredients are obtained by combining a very concentrated mixture of the lysate from two different lambda strains that are lysogenic.

2. Filamentous single-stranded (M13) bacteriophage

The availability of single-stranded DNA is of major importance for a variety of applications in recombinant technology. For example, in DNA sequencing, site-directed mutagenesis, construction of strand-specific cDNA libraries and the generation of highly radioactive probes. The development of M13 vectors has been a major breakthrough in recombinant DNA technology. M13 is a filamentous bacteriophage (Figure 16.8) that infects *E. coli* through the bacterial sex pili normally used for bacterial conjugation. M13 has the combined properties of phage and plasmids, termed phagemids. Since the length of virus particles is determined by the length of viral DNA, these viruses can accommodate almost any size of inserted DNA. Such recombinant viruses may or may not be defective. The filamentous bacteriophages specific to *E. coli*, which include a closely related M13, f1 and fd, each contain a single-stranded closed circular DNA molecule about 6.4 kbp long, packaged into long, thin tubes constructed from thousands of protein monomers. When used as cloning vectors, these bacteriophages have the unique advantage of generating large quantities of DNA molecules that carry the sequence of one strand of the foreign DNA. The morphogenesis of filamentous bacteriophages is extraordinary. Because the viral genome is not inserted into a preformed structure, there is no strict limit to the size of the single-stranded DNA that can be packaged. Rather, the length of the filamentous particle varies according to the amount of DNA it contains. Inserts of foreign DNA seven times longer than the wild type viral genome have been cloned and propagated in M13. The filamentous bacteriophages replicate with that of the host bacterium, and infected cells are not lysed, but continue to grow while producing several hundred virus particles per cell per generation. But they subsequently produce single-stranded virions for infection of further bacterial cells (lytic growth). All genes in a phage are essential, and vectors must therefore carry the full complement of coding sequences. The foreign DNA can be inserted into the region located between genes II and IV, but it is not dispensable, since it contains a cis-acting signal for the packaging and orientation of the DNA within particles, sites for initiation and termination of synthesis of +ve and −ve strand DNA, and a signal for p-independent termination of transcription. The replicative form of double-stranded DNA can easily be purified from infected cells, manipulated, and reintroduced into cells. The double-stranded DNA then re-enters the replication cycle, eventually generating progeny particles that contain only one of the two strands of the DNA (+ strand).

C. COSMID VECTORS

Another vector designed (**engineered vectors**) especially for cloning large DNA fragments is called a cosmid. Cosmids are plasmids usable as gene-cloning vectors in an *in vitro* packaging system by coliphage λ. These plasmids carry a copy of the λ bacteriophage *cos* site DNA sequences (***cos*** sequences) required for packaging into Pseudobacteriophage λ particles (Figure 16.9). In their plasmid mode, cosmids can replicate autonomously and behave like plasmids in every respect. However, the *cos* sites permit the cosmid to become packaged into a lambda capsule and transferred to a new host by transduction. In order for packaging to take place, several requirements must be met. (1) The inserted target DNA must increase the total cosmid size to 37–52 kb (i.e. 75–105% of the lambda genome), (2) The ligations must be carried out under high DNA concentrations so that linear concatemers will be formed, (3) The *cos* sites must be in the same orientation.

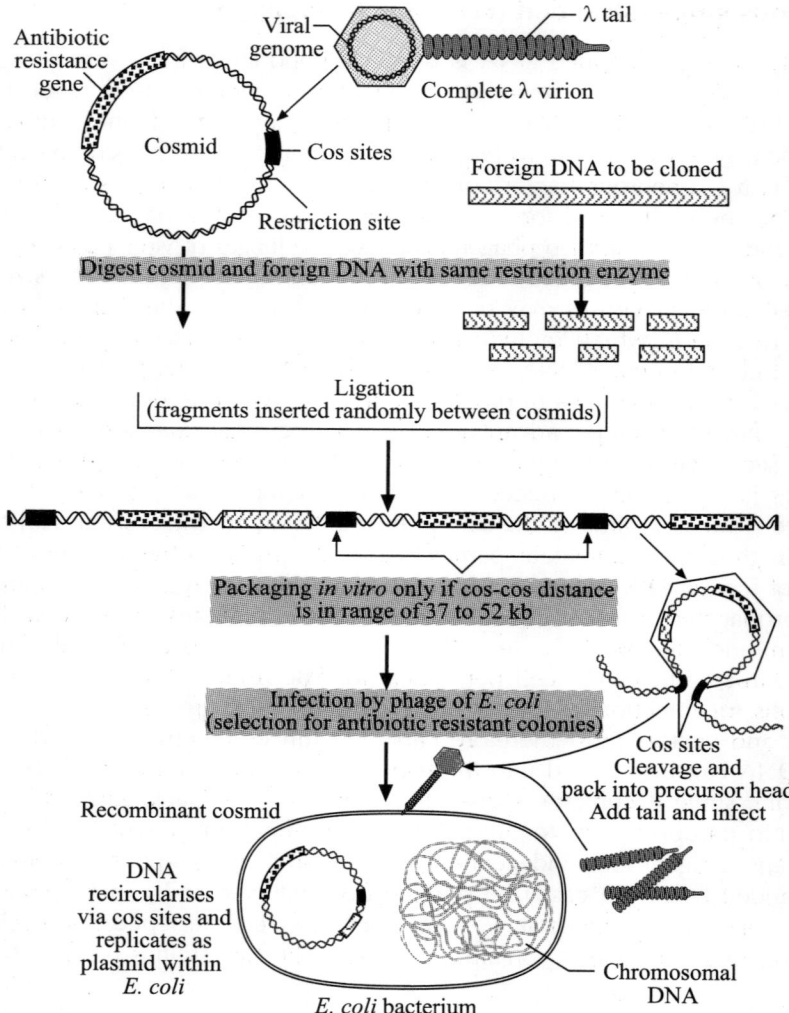

Figure 16.9 *Scheme for cloning foreign DNA fragments in cosmid vectors in vitro*

Cosmid particles adsorb to recipient *E. coli*, and introduce the large recombinant DNA into the host (up to 45kb). The DNA incorporated into phage heads by bacteriophage λ packaging systems must meet only limited specifications. It must possess a 14-bp sequence known as **cos** (which stands for cohesive end site) at each of its ends, and these **cos** sequences must be separated by no fewer than 36 kbp and no more than 51 kbp of DNA. Essentially, any DNA satisfying these minimal requirements can be packaged and assembled into an infective phage particle. Other cloning features such as an **ori**, selectable **markers**, and a **polylinker** are joined to the **cos** sequence so that the cloned DNA can be propagated and selected in host cells. These features have been achieved by placing **cos** sequences on either side of cloning sites in plasmids to create cosmid vectors that are capable of carrying DNA inserts of about 40 kbp in size. Since cosmids lack essential phage genes, they reproduce in host bacteria as plasmids.

In addition to the basic features discussed above that are common to all cosmid vectors, a variety of additional modifications have been made either to alleviate problems or to confer special advantages on individual vectors: (i) Incorporation of two *cos* sites into a single cosmid vector greatly simplifying directional cloning in cosmid vectors; (ii) the construction of cosmid vectors that lack homology with commonly used plasmid vectors so that they are unlikely to recombine with one another when they are carried within the same recombination-competent cell; (iii) the presence of promoters for bacteriophage-encoded DNA-dependent RNA polymerases that enable RNA probes specific to either end of the cloned insert to be generated *in vitro*; (iv) incorporation of selectable markers, to simplify the identification of eukaryotic cells, that have been transected with recombinant cosmids; (v) the incorporation of eukaryotic origins of replication that allows transected cosmids to replicate autonomously in mammalian cells; (vi) the addition of a range of prokaryotic markers to facilitate rescue in bacteria of cosmid sequences transected into mammalian cells; (vii) the use of bacteriophage origins of replication rather than plasmid origins of replication. No single cosmid vector contains all of these improvements, and it is therefore necessary to chose a particular vector that is most suitable for the task at hand.

Commonly used cosmid vectors include the pJB8 and the pcosEMBL family. pJB8 is probably the most useful for constructing genomic libraries. Because this vector is small, fragments of up to 45 kb can be inserted into it and packaged into phage heads. The pcosEMBL family was designed to simplify isolation of specific recombinants from cosmid libraries and to speed up isolation of large regions of complex genomes in an ordered array of overlapping clones. The vectors in this family differ by having different cloning sites and different numbers of *cos* sites.

D. PHAGEMIDS

Another class of vectors with single-stranded capability has also been developed. Several phagemids have been developed, that combine the desirable features of both plasmids and filamentous bacteriophages. These vectors are plasmids with EoIE1 origin of replication and a selectable antibiotic resistant marker that also carries a copy of the major intergenic region of a filamentous bacteriophage. This region contains all the sequences required in cis for initiation and termination of viral DNA synthesis and for morphogenesis of bacteriophage particles. Segments of foreign DNA cloned in these vectors can be propagated as plasmids in the conventional way. A number of phagemids have been developed using intergenic regions derived from various filamentous bacteriophages and inserted into different plasmids. Phagemids have several attractive features. They provide the stability and high yields of double-stranded DNA that are characteristic of conventional plasmids. They circumvent the tedious and time-consuming process of subcloning DNA fragments from plasmids to filamentous bacteriophage vectors. They are sufficiently small so segments of foreign DNA up to 10 kb in length can be obtained in single-stranded form. In summary, phagemids show great potential as dual-purpose vectors. However, many phagemids suffer from a major disadvantage – they sometimes give poor and generally irreproducible yields of single-stranded DNA that are only obtained after infection with helper bacteriophages.

One popular variety of phagemids (Figure 16.10) goes by the trade name pBluescript (pBS). Like the pUC vectors, pBS has a multiple cloning site inserted into the *lac* Z gene, so

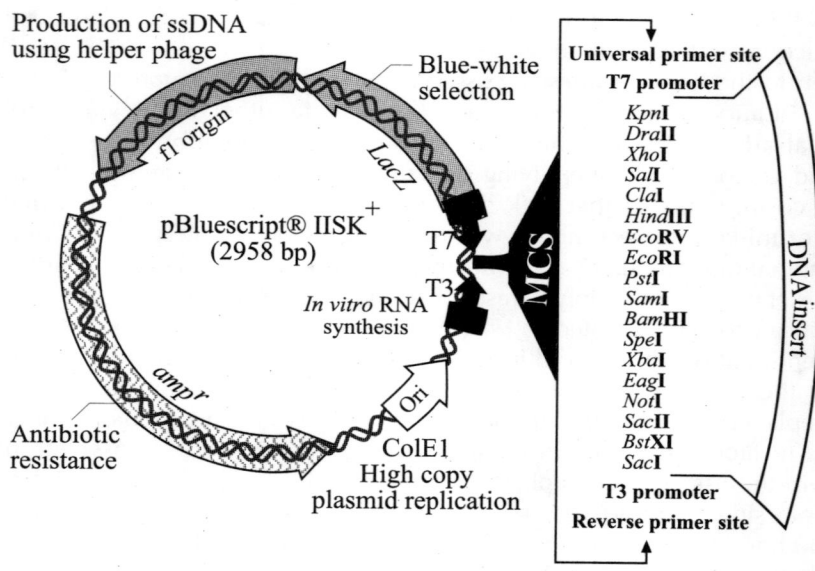

Figure 16.10 *Physical and genetic map of phagemid pBluescript II SK$^+$ cloning vector. Arrow indicate the direction of transcription for bacteriophage promoters. Ori, origin of replication; LacZ, a-peptide of b-galactosidase*

clones with inserts can be distinguished by white versus blue staining with X-gal. This vector also has the origin of replication of the phage f1. This means that a cell harboring a recombinant phagemid, if infected by the f1 helper phage, will produce and package the single-stranded phagemid DNA. Another useful feature of this class of vectors is that the multiple cloning site is flanked by two different phage promoters. For example, pBS has a T3 promoter on one side and a T7 promoter on the other. This allows us to isolate the double-stranded phagemid DNA and transcribe it *in vitro* with either of the phage polymerases to produce pure RNA transcripts corresponding to either strand.

Many of the commercially available vectors are phagemids and these may eventually supersede M13 cloning systems. Foreign DNA is cloned into pUC118 and propagated as dsDNA. There is a 476-bp fragment of M13 in these vectors, as well as a gene for ampicillin resistance. If cells carrying the phagemid are infected with the helper virus M13K07, then phage particles containing ssDNA are produced.

V EUKARYOTIC VECTORS

Genetic engineering of eukaryotes began about 25 years ago with the isolation of genes from the yeast *Saccharomyces cerevisiae*. At that time, vector systems were not available for the transformation and recovery of plasmids in yeast and molecular manipulation depended on the technology developed for prokaryotes. Later, the number and variety of vector systems as well as the host range expanded enormously. Specialized vectors for production of medically and commercially important proteins, production of chemicals, the engineering of plants and animals as well as vectors for human and animal therapy have been developed. There is always a temptation to construct the "best" vector system and organism or to try to develop a single vector system, which can be used in every instance. However, it is unlikely that any particular organism will have universal utility. The best vectors combine certain desirable properties, which include high stability, high transformation efficiency, larger size of foreign DNA acceptance, easy recovery of vector, high copy number, broad variable host range and the ability to express in host cells.

The different eukaryotic plasmid vectors can be divided into two types according to their mode of replication. **Integrating vectors** are replicated as part of the host chromosomal DNA while episomal vectors replicate **autonomously**. Some vectors, however, have properties of both types. For example, some vectors will integrate in some hosts and replicate as episomal plasmids in others. It is also possible to convert some integrating vectors into episomal vectors and *vice versa*.

1. Integrating vectors

Vectors that insert into and become part of the host chromosomal DNA are referred to as integrating vectors. The replication and maintenance of these types of vectors depend on chromosomal replication functions. The main advantages of integrating vectors is the high mitotic stability, so the vector lost from the host cell is low. The copy number is usually one per cell, but the number can sometimes be increased by gene amplification. There are three classes of integration events, type I, type II, and type III (Figure 16.11). In **type I integration**, a single crossover event between a DNA segment on the plasmid and the corresponding homologous region on the chromosome results in the insertion of the entire vector into the host DNA. The homology is usually between a host gene on the plasmid and the corresponding gene on the chromosome. If two regions contained within the vector have homology with chromosomal sequences, integration can occur at either of the two chromosomal locations. **Type II integration** results in integration at random locations throughout the host chromosome. The entire vector is inserted into the host DNA, and the integration is with limited homology between sequences on the vector and the host chromosome. **Type III integration** is due to a double crossover event. Only the portion of the vector between the crossovers is inserted. This type of integration results in a gene replacement or gene conversion. There are many types of integrating vectors.

2. Episomal vectors

Episomal plasmids can replicate independently of the chromosomal DNA. The main advantages of these types of plasmids are that the copy number and transformation frequency are usually high and the vectors can be easily isolated. The most useful vectors are those that replicate in both the host cell and *E. coli* (called **shuttle vectors**). DNA manipulation on these "shuttle vectors" can therefore be done in *E. coli*. Then the vectors can be transformed into the desired host. One important application of this type of vector is in the cloning of host genes: gene banks containing DNA of desired character can be constructed in *E. coli*, and then the bank can be transformed into the host where transformants with desired properties can be identified (complementation effect). The simplest episomal plasmids to develop are those that use the entire genomes of self-replicating endogenous plasmids or viruses. For example,

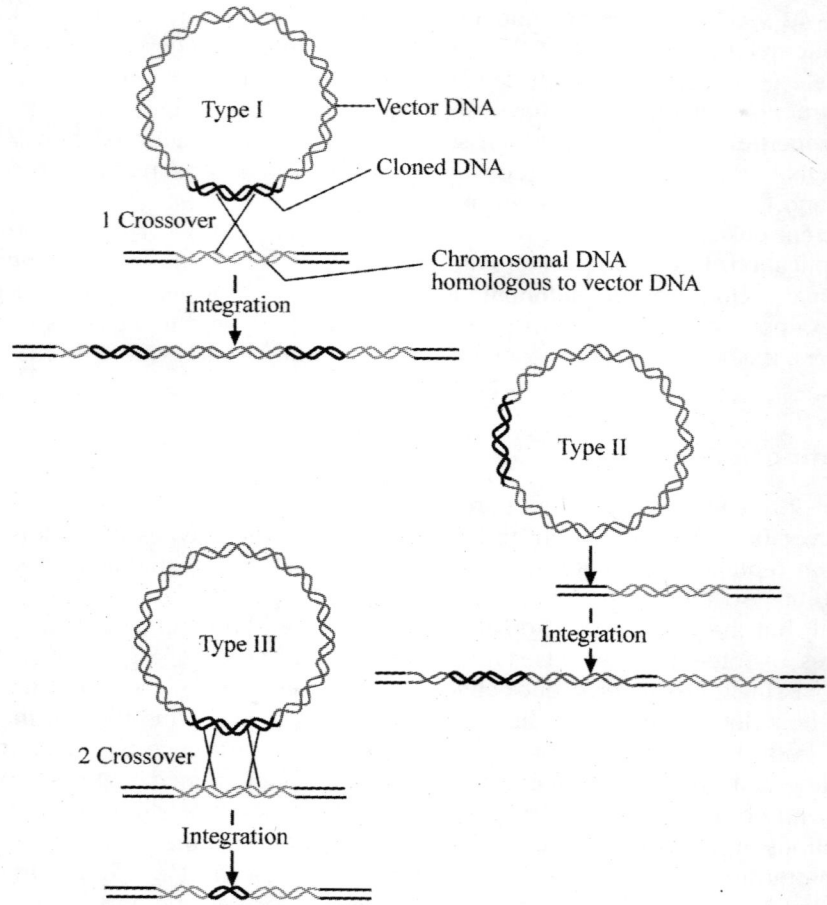

Figure 16.11 *Integration of vector DNA into the host chromosome*

the Bovine Papilloma virus, the Epstein Barr virus, and the SV40 virus have been used to develop mammalian episomal vectors, and the yeast 2μ plasmid has been used for yeast episomal vectors. A second type of episomal vector contains various chromosomal elements that allow autonomous replication. The simplest vectors of this type contain a chromosomal origin of replication and a selectable marker. Additional elements, such as centromeres, can be added to ensure mitotic stability.

The advantage of these eukaryotic vectors is that they accept fragments of DNA larger than phage λ or cosmids, which means that fewer clones need to be screened when one is searching for the foreign DNA of interest. Recent developments have allowed the production of large-insert capacity vectors based on bacterial (BACs), mammalian (MACs), and virus P1 (PACs), artificial chromosomes. However, perhaps the most significant development is vectors based on yeast artificial chromosomes or YACs. Several other useful vectors have been designed for cloning genes into eukaryotic cells. In this chapter only a few of them are discussed.

A. YEAST CLONING VECTORS

The main problem with the cloning vectors we have discussed so far in this chapter is that they do not hold enough DNA for large-scale physical mapping. Geneticists interested in a given region of the human genome aim to assemble a set of clones called a **contig**, which contains contiguous (actually overlapping) DNAs spanning long distances. Fortunately, we now have high-capacity vectors that will accommodate hundreds of kilobases each (Figure 16.12). These are called **yeast artificial chromosomes**, or **YACs**. YAC is digested with restriction enzymes at the SUP4 site (a suppressor tRNA gene marker) and BamHI sites separating the telomere sequences. This produces two arms, and the foreign genomic DNA is ligated on these arms to produce a functional YAC construct. The main advantage of YAC-based vectors is their ability to clone very large fragments of DNA. DNA molecules several hundred kilobase pairs in length have been successfully propagated in yeast by YACs (Figure 16.13). For these large DNAs to be replicated in the yeast cell, YAC constructs must include not only an origin of replication (known in yeast terminology as an autonomously replicating sequence or ARS) but also a centromere and telomeres. Recall that centromeres provide the site for attachment of the chromosome to the spindle during mitosis and meiosis. Telomeres are nucleotide sequences defining the ends of chromosomes. Telomeres are essential for proper replication of the chromosome. The centromere is placed adjacent to the left telomere, and a huge piece of human (or any other) DNA can be placed in between the centromere and the right telomere, as shown in Figure 16.13. The large

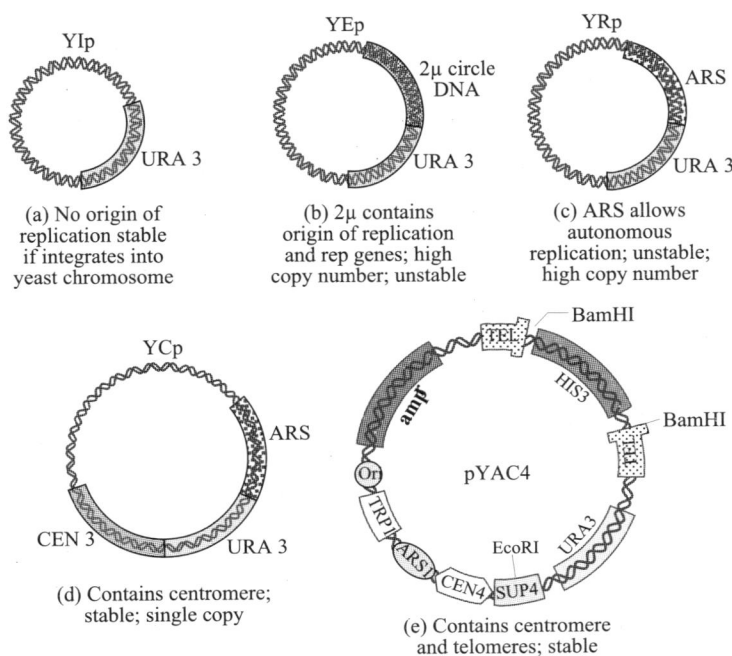

Figure 16.12 *Five types of yeast plasmid vectors. CEN, centromere sequence; ARS, autonomous replicating sequences; TEL, telemere sequence*

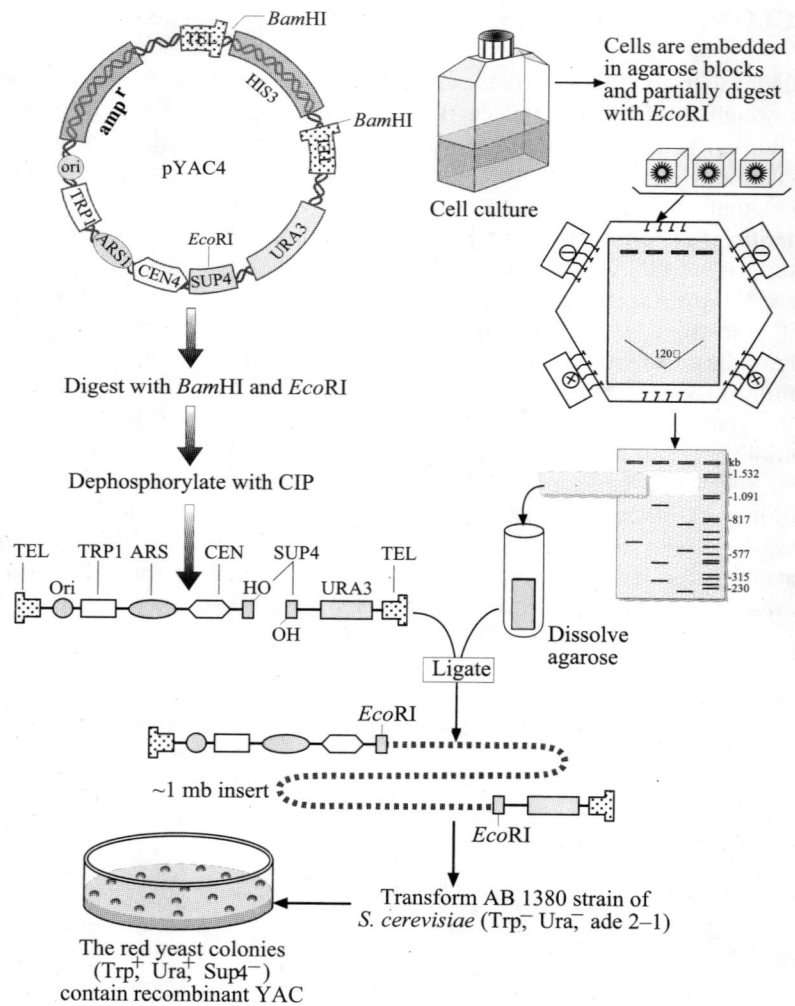

Figure 16.13 *Scheme for cloning large fragments of DNA into YAC cloning vectors contain functional elements for chromosome maintenance in S. cerevisiae*

DNA inserts are prepared by slightly digesting long pieces of human DNA with a restriction enzyme. The YACs, with their human DNA inserts, can then be introduced into yeast cells, where they will replicate just as if they were normal yeast chromosomes.

Examples of some commonly used yeast plasmid vectors are: (i) **Yip/YTp** or yeast integrative plasmids (which contain *E. coli* plasmid origin), which allow transformation (low frequency) by crossing over (maintained episomally) and have no replication origin. It gives stable transformations and is useful for surrogate genetics; (ii) **YEp** or yeast episomal plasmids, which carry 2μ DNA sequence including the origin of replication and *rep* genes. It gives high transformation frequency but has unstable maintenance and is used for functional analysis by complementation, or stable integration; (iii) **YRp** or yeast replicating plasmids, which carry autonomously replicating sequence (yeast chromosomal ARS element). This sequence is very common with many yeast genes so that stable transformation can be achieved by crossing over. It is used for functional analysis by complementation or stable integration; (iv) **YCp** or yeast chromosomal ARS and centromere (CEN) containing plasmids, which function as true chromosomes and segregate during mitosis and meiosis. It is used in functional analysis, especially if gene dosage effects are deleterious; (v) **pYAC** or yeast artificial chromosome which carries both centromere and telomere sequences and, therefore, can be maintained stably and used to obtain artificial chromosomes. It is the highest capacity cloning vector available, and is useful for generating libraries of large eukaryote genomes; (vi) **Ty** or **Ty retrotransposon**, Disarmed Ty vectors in YEps are strong, expression constructs. Autonomous Ty vectors integrate into the host chromosome. They are used as a high yield expression vectors.

The above vectors have been extensively utilized for the study of the yeast genome and are also used as shuttle vectors. Shuttle vectors are plasmids capable of propagating and transferring ("shuttling") genes between two different organisms, one of which is typically a prokaryote (*E. coli*) and the other a eukaryote (yeast). Shuttle vectors must have unique origins of replication for each cell type as well as different markers for selection of transformed host cells harboring the vector (Figure 16.14). Shuttle vectors have the advantage that eukaryotic genes can be cloned in bacterial hosts, yet the expression of these genes can be analyzed in appropriate eukaryotic backgrounds.

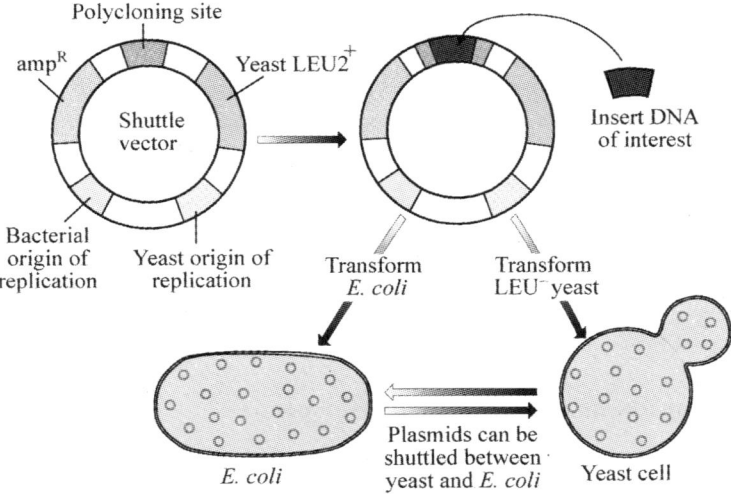

Figure 16.14 *A shuttle vector having both yeast and bacterial origins of replications*

B. FILAMENTOUS FUNGI VECTORS

Filamentous fungi have been used commercially particularly for the production of metabolites and extracellular proteins. Because of their commercial importance, considerable efforts have been devoted to develop transformation/expression systems. Although, *Neurospora crassa* and *Aspergillus nidulans* have been well studied, these organisms are not known for high-level production of extracellular proteins. The related organisms, A. niger, A. awamori, and A. oryzae do produce large quantities of extracellular proteins, but their genetic systems are less advanced. Transformation systems for filamentous fungi have been developed by analogy to those in yeast. The procedure is to treat cells with wall degrading enzymes and then to stimulate the uptake of DNA by the addition of polyethylene glycol and $CaCl_2$.

Vectors in fungi often use genes that complement nutritional markers for plasmid selection. For example, the QA2 (dehydroquinase) and AM^+ (glutamate dehydrogenease) genes are used in Neurospora and ARGB, AMDS, TRPC, and PYR4 genes are used in A. nidulans. It is not easy to use nutritional markers for selection in filamentous fungi. One reason is that host mutations are not always available. Use of an antibiotic resistance gene (e.g. G418) avoids the problems with nutritional selection. Transformation in filamentous fungi is almost always carried out via stable integration. The frequency is about 10–20 transformants per μg of vector DNA and occurs randomly. Considerable efforts have been made to develop autonomously replicating shuttle vectors. Filamentous fungi do not appear to harbor endogenous plasmids, so that such plasmid DNA is not available for vector development. In Neurospora, there are some mitochondrial plasmids, but use of these for developing episomal vectors has not been successful. Attempts have been made to isolate ARS elements from Aspergillus by using the S. cerevisiae selection system for ARS activity. Several strategies have been developed for the cloning and isolation of genes in filamentous fungi. Very recently there was a report on using Agrobacterium transformation of fungi.

C. PLANT VECTORS

The technology to manipulate the genome of plants has been a desirable goal for years. Geneticists have dreamed of improving crop yields, decreasing production costs, and developing plants with novel characteristics (e.g. herbicide and pest resistance, organic chemical production, high nutritive value, vaccine production, etc.). To achieve these goals, it is necessary to engineer the genome of plants with vectors that can be stably maintained during regeneration of plants from single cells and through successive generations.

The best studied system for introducing foreign genes into plants is the naturally occurring crown gall tumor system. This system is caused by a soil pathogen *Agrobacterium tumefaciens*, which carries a plasmid that contains tumor inducing (Ti) genes in plants.

1. Ti and Ri plasmids as vectors for higher plants

(a) Molecular genetics of *Agrobacterium*: Bacteria of the genus *Agrobacterium* are gram-negative rods that belong to the bacterial family of Rhizobiaceae. Well-known A. tumefaciens strains such as Ach5, A6, B6, C58, T37, Bo542 and 15955 all belong to biotype 1. They can proliferate at temperatures up to 37°C (28 to 29°C is optimum). Classic E. coli media such as LB, NB, and TY can be used for growth, although incubation in a defined minimal medium such as MM or YMB also permits growth. Laboratory strains of A. rhizogenes include the biotype 2 strains NCPPB 1855, ACTCC 15834, NCIB 8196, and A4. The characteristic of biotype 2 is the production of ketolactose from lactose. The biotype 2 strains proliferate poorly at temperatures higher than 30°C, and therefore they are grown at 28 or 29°C as are the biotype 1 strains. They grow well on media such as TY+Ca, YMB or in a defined minimal (RMM). Their ability to grow on erythritol as a carbon source is a phenotype by which they can be distinguished from other *Agrobacteria*.

(b) Ti plasmid vector systems: *Agrobacterium* cells naturally harbor only one sort of Ti plasmid (Figure 16.15). *Agrobacterium* lacking Ti plasmids do not induce gall disease or synthesis of opines in infected plants. The southern blot analysis of crown gall showed that copies of the T-DNA segment (about 20 kbp) were found covalently integrated into the DNA of tumor cells. The T-DNA segments integrate at various, seemingly random places in the host chromosomes. The only DNA that moves from the Ti plasmid to the host plant cell chromosomes is the T-DNA. The enzyme(s) responsible for inserting T-DNA into plant cell chromosomes must be coded either by genes on other segments of the Ti plasmid or by host chromosomal genes. Two properties of the T-DNAs of Ti plasmids make them virtually ideal vectors for introducing foreign genes into plants. First, the host range of *Agrobacterium* is very broad; they are capable of transforming cells of all dicotyledonous plants. Recently, the range has been broadened to include monocotyledons also. Second, the integrated T DNA is inherited in a Mendelian way, and its genes have their own promoters to which foreign genes can be coupled and expressed or they can even be expressed under foreign gene promoters.

Among higher plants, Ti plasmid of *Agrobacterium tumefaciens* or Ri plasmid of *A. rhizogenes* are the best known vectors. T-DNA, from Ti or Ri plasmid of **Agrobacterium**, are potential vectors for cloning and transformation of higher plants. The use of wild type pTi as a vector presents the following three problems: (i) presence of oncogenes in T-DNA, which causes disorganized growth and a loss of regeneration potential of the cells having T-DNA in their genomes; (ii) wild Ti plasmids being large in size makes the handling

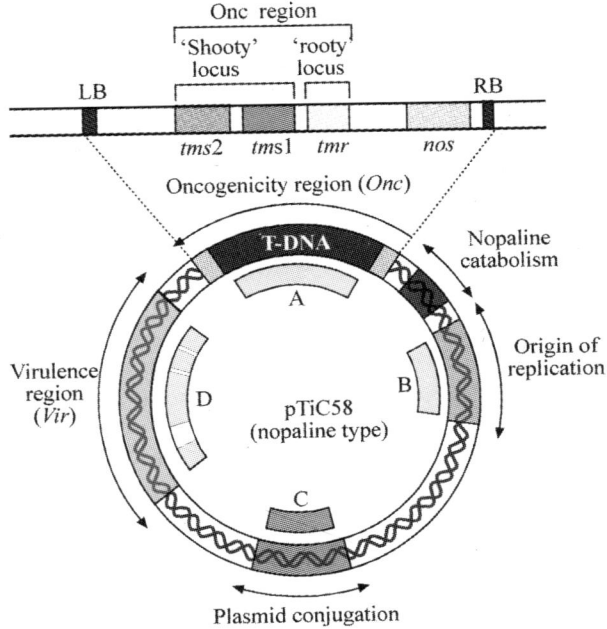

Figure 16.15 *General features of Ti plasmids and structure of T-DNA*

procedure during cloning tedious; (iii) native types do not have unique cloning sites within the T-DNA, which are needed for the insertion of DNA segments to be cloned. These problems have been resolved by deleting the oncogenes from the T-DNA (**disarming**) and by developing intermediate vectors (IV); to yield co-integrate vectors) and binary vectors to facilitate gene cloning procedures.

(c) **Disarming**: The removal of genes controlling auxin and cytokinin production (oncogenes) from T-DNA of Ti plasmid is known as **disarming**. It was discovered when oncogenes of a wild type pTi (pTiC58) were replaced by pBR322 sequences, retaining only the *nos* gene and the left and right borders (LB and RB), and the Agrobacterium containing this disarmed plasmid still transformed the modified T-DNA into plant cells. However, the transformed cells were non-tumorous and produced only nopaline and regenerated into plantlets. These results revealed that only the LB and RB sequences of T-DNA are necessary for the transfer of any DNA insert placed between them.

(d) **Co-integrate pTi vectors**: The genes of interest to be transformed into plants are initially cloned in *E. coli* for obvious reasons of ease in the cloning procedures. pTi cannot be used for this purpose due to its large size and the non-availability of unique cloning sites within the T-DNA. This difficulty is resolved by using a suitably modified *E. coli* plasmid for the initial cloning of genes. The subsequent gene transfer into plants may utilize one of the two strategies represented by co-integrate (Figure 16.16) and binary vectors. A co-integrate vector is produced by integrating the modified *E. coli* plasmid (used for cloning and containing the gene construct to be transformed) into a disarmed pTi. The co-integration of the two plasmids is achieved within *Agrobacterium* by homologous recombination. Therefore, the *E. coli* plasmid (e.g. pBR322) and the disarmed pTi must have some sequence common to both for recombination to occur. The pBR322 is suitably modified to produce an intermediate vector (IV). The IV must contain the origin for replication in *E. coli*, the pBR322 sequence present in the T-region of the disarmed pTi, the T-DNA (without the borders) from pTi and appropriate selectable markers, e.g., neo gene for selection of plant cells containing the recombinant T-DNA (T-DNA containing the DNA insert) and kan^r (kanamycin resistance) for the selection of the co-integrate vector in *Agrobacterium*. The IV must contain the origin for replication in *Agrobacterium*. Thus the IV can be propagated in *E. coli* and the DNA inserts can be readily and conveniently inserted with the T-DNA present in it to yield a recombinant IV.

(e) **Binary vector**: The *vir* region of the Ti plasmid need not be present in the same plasmid for an efficient transfer of T-DNA. The T-DNA present in one plasmid (*vir* region is absent) is readily transferred into plant cells in response to *vir* genes present in another plasmid contained in the same bacterial cell (Figure 16.17). This property has been exploited to construct binary vectors of pTi. A binary vector consists of a pair of plasmids of which one plasmid contains disarmed T-DNA sequences, while the other contains *vir* region, and ordinarily lacks the entire T-DNA including the border. The plasmid containing disarmed T-DNA is called **mini-Ti** or **micro-Ti**, e.g. Bin19, and has the origins for replication in both *E. coli* and *Agrobacterium* and also contains polylinkers and the α peptide gene of *E. coli* lac Z locus. The DNA inserted is integrated within the T-region of mini-Ti, and the recombinant mini-Ti is cloned in *E. coli*. The transfer of recombinant mini-Ti from *E. coli* into *Agrobacterium* is achieved either by a three-way cross or by direct transformation of an *Agrobacterium* strain containing the helper Ti plasmid.

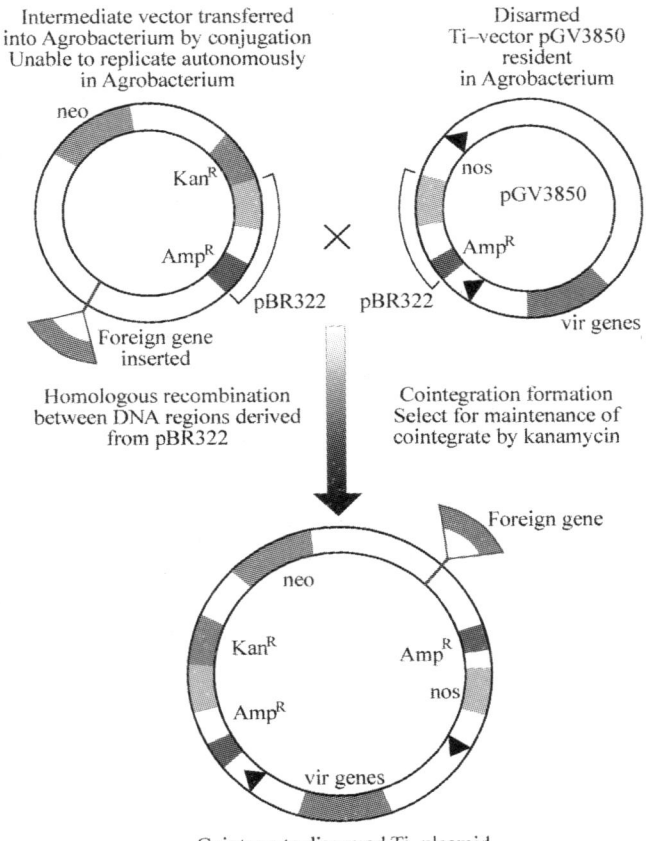

Figure 16.16 *Production of recombinant disarmed Ti–plasmid by cointegrate formation*

(f) Binary and shuttle vectors: Transformation of plants can be carried out by a number of binary vectors. These vectors are based on the pPCV (plant cloning vector) series of plasmids. Binary vectors contain a conditional mini-RK2 replicon, which is maintained and mobilized by *trans*-acting functions derived from the plasmid RK2 replicon (origins-ori, and oriT). The plasmid RK2 can be introduced into both *E. coli* as well as *Agrobacterium* (GV 3101, which contains a Ti plasmid lacking T-DNA) to facilitate replication of binary vectors in both these hosts, so that the vector can be maintained and shuttled between both. Hence the name shuttle vector.

D. ANIMAL VECTORS

Expression in animal cells has become an area of considerable interest to the scientific community because of the potential for engineering animal and human genomes. There is also interest in the production of human pharmaceuticals, which require expression in mammalian cell lines for proper biological activity of the product. For this purpose, a number of

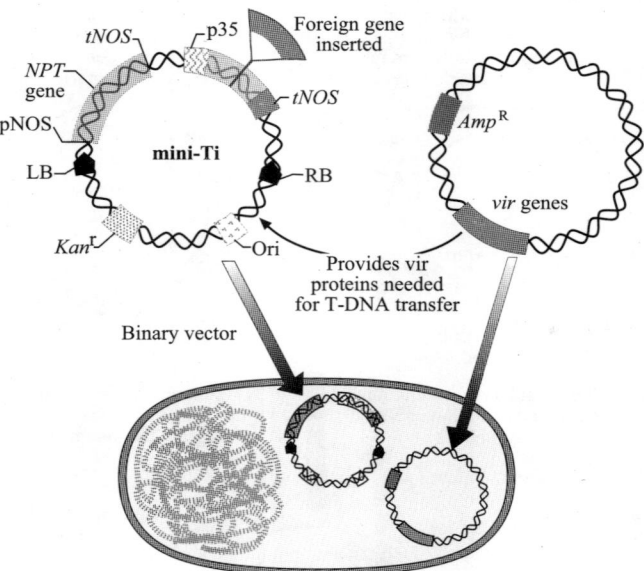

Figure 16.17 *A schematic representation of the binary vector system*

different expression systems have been explored. Description of all the vectors is beyond the scope of this chapter. Instead, a few select systems that are commonly used, and have novel features or show greater potential, will be discussed.

1. SV 40 vectors

Almost all the mammalian expression vectors are based on infectious viruses. The vector most commonly used in higher animals is the DNA tumor virus SV40 (sv, or the **simian vacuolating virus**, was first isolated from monkeys. However, it can transform normal mice, rabbit, and hamster cells also.). SV40 is an icosahedrral particle with a small (5,224 bp) chromosome, which is a circular, dsDNA molecule. SV40 virions allow foreign DNA to replace part of their DNA studies in one of two ways (Figure 16.18). SV 40 contains two promoters that express early, and late genes. The virus contains a replication origin which can be used in vectors for autonomous replication. Replication with this origin requires the large T antigen, which is one of the early gene products.

Construction of vectors is most easily accomplished in *E. coli*. Shuttle vectors containing *E. coli* sequences, the SV40 origin, and either the early or late promoters have been developed. SV40 has stringent packaging requirements. They can replicate and complete their life cycle with the help of non-recombinant viruses, or they can replicate in the host without making active virus particles by existing as circular plasmids in the cytoplasm or by integrating into the host's genome. To increase the amount of foreign DNA, vectors using this virus are of the replacement type. DNA inserts replace either the early or late coding region and are transformed into permissive host cells in the presence of a helper virus.

SV40 has become a valuable tool in mammalian genomic studies. For example, the rabbit β-globin gene was cloned in SV40, and enhancer sequences were discovered in SV40.

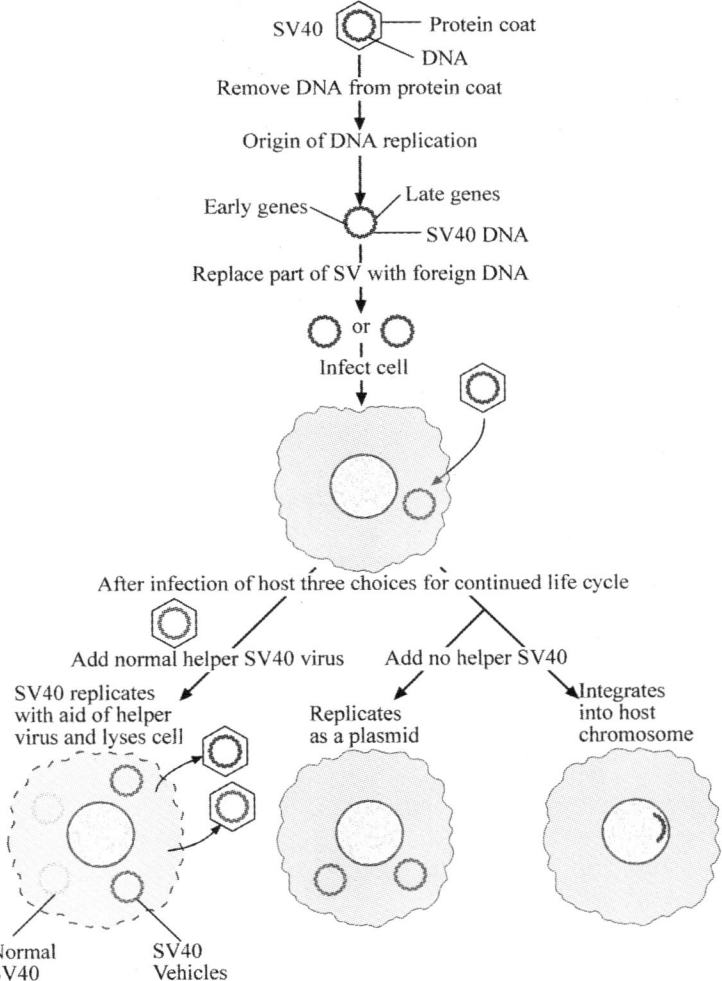

Figure 16.18 *SV40 virus can be used as a gene cloning vehicle*

DNA tumor viruses have also helped to understand the mechanism of transformation in eukaryotes (**oncogenesis**).

2. Bovine papilloma virus

Bovine papilloma virus (BPV) replicates as a circular plasmid (about 8 kb in size) at a copy number as high as several hundred per cell. All the ORFs are on the same strand of DNA and transcription is unidirectional. *E. coli* shuttle vectors have been developed using either the entire genome or a segment containing 69% of the genome. The advantages of this vector system are that the inherently high copy number results in a high expression level and shuttle vectors transform with high frequency.

Another vector that is similar in principle to the BPV-based vectors use the Epstein Barr virus (EBV). The addition of the replication origin and a DNA fragment containing a *trans* acting gene (EBNA–1) to vectors increases the transformation frequency 200-fold. These plasmids replicate episomally at a copy number of up to 60 per cell.

3. Retroviral vectors

Retroviruses have been frequently studied and used as vectors for a number of reasons. The host range is very broad and both individual cells and whole animals can be "infected". The transformation efficiency approaches 100%. Up to 10 kb of foreign DNA can be packaged, and the vectors integrate efficiently into the genome at a single or low copy number. The integrants are colinear with the genome, and rearrangements do not readily occur. Cells containing retroviruses can grow normally.

Retroviruses contain a single-stranded RNA genome that resembles an mRNA. Following the infection of a susceptible host, the genome is reverse-transcribed into a dsDNA provirus that subsequently integrates into the genome. The integrated virus contains *cis* acting sequences that are responsible for transcriptional initiation and termination in the 5' and 3' LTR, DNA replication, integration, and packaging (Figure 16.19). Splicing of the single RNA transcript in conjugation with proteolytic processing results in the expression of the three *trans* acting genes, *gag*, *pol*, and *env*. The RNA transcripts contain all the genetic information and can be packaged into virions that are released into the culture medium. These virions can then infect other cells.

In *in vitro* construction of the retroviral vector, the *cis*-acting sequences are retained while the *trans*-acting genes are replaced by foreign DNA. These vectors, however, require helper viruses for productive infections. *E. coli* shuttle vectors containing proviral DNA are used for DNA manipulations. The vector is then transformed into an appropriate mammalian packaging cell line where the virus is integrated into the genome. The packaging cell lines contain an integrated form of a helper virus that lacks the *cis*-acting packaging signal but is transcribed and produces the *trans*-acting elements required for replication and encapsidation. Because the helper virus lacks a packaging signal, it is not itself efficiently encapsidated and secreted. However, the vector DNA is transcribed, and its RNA is packaged and secreted. The culture supernatant is then used to infect the desired host. Only the transcribed RNA is packaged so that the vector is no longer capable of replication in *E. coli*. The packaged RNA lacks the *trans*-acting genes but retains the functions necessary for integration and expression. The host machinery will reverse transcribe the RNA into the proviral form and the DNA will integrate. The viral vector can infect only once because it lacks the necessary *trans*-acting genes. A number of different vector types have been constructed.

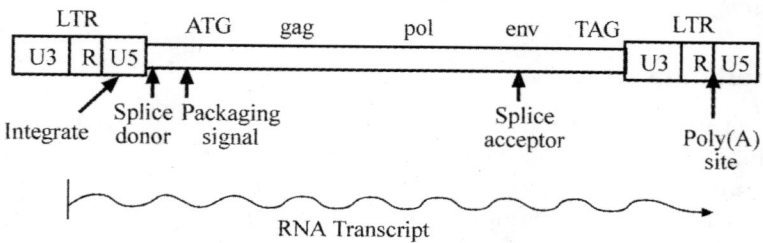

Figure 16.19 *Structure of Marine Leukemia virus*

One problem with retroviral vectors is that integration is random and the LTR can activate adjacent cells. To overcome this problem, self–inactivating (**SIN**) vectors have been developed. The correction of many other problems associated with retroviral vectors are in progress. There has been a lot of new vector type development and improvement and more advances can be expected in the future.

VI EXPRESSION CASSETTES

An area of particular interest in vector development is in the construction of vectors that exhibit high levels of expression of heterologous gene products. This is especially important in the biotechnology arena where the production of proteins and enzymes at high levels has an economic impact on the production process. Thus specialized vectors have been developed for the production of various pharmaceuticals, healthcare products, chemicals, and the like. In industrial microbiology, stain improvement takes on the added dimension of designing and constructing specialized genetic elements that are capable of "carrying" and over-expressing "foreign" (or cloned) genes in their new host cell environments. These specialized genetic elements (expression cassettes) are termed "**production level vectors or expression vectors**". Several points must to be taken into consideration when designing such vectors: (i) *Regulatory sequence*: the appropriate choice of regulatory sequences such as promoters, operators, and ribosomal binding sites; (ii) *the problem of codon choice*: selection and distribution of codons; (iii) *the final yield of recombinant product*: the copy numbers of recombinant molecules and stability of the plasmids; (iv) *the stability of final product*: nature and stability of translated proteins; (v) *host and vector incompatibility*: The problem of incompatible DNA sequences; (vi) the choice of host cells.

The expression cassette means gene constructs, which allow the insertion of foreign genes, either as transcriptional or translational fusions, behind specific promoters. Typical

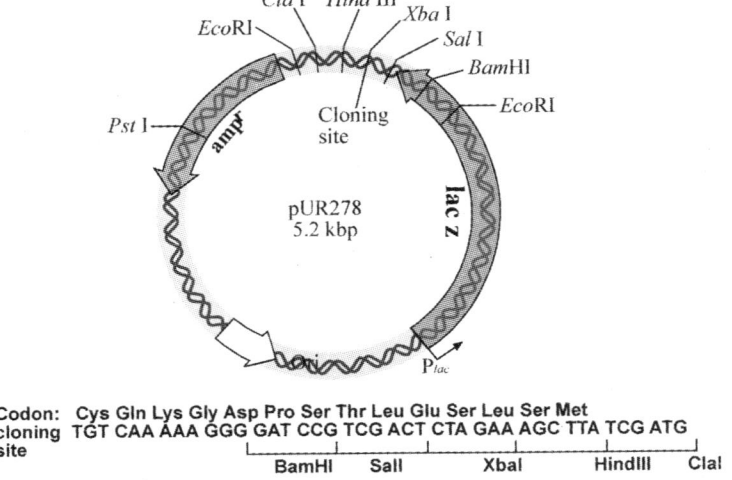

Figure 16.20 *A typical expression vector for the synthesis of a hybrid protein*

expression vectors, for example, pRT plasmids (e.g. pRT100, pRT101, pRT102, pRT103, pRT104), have been derived from pUC18/19. These vectors differ in polylinker sequences, each flanked by a 35S promoter of CaMV on one end and a sequence for poly A addition on the other end. A variety of marker genes (cat, nptII, gus, etc.) have been inserted into these cassettes and their expression studied, both in protoplasts and in stable transgenic tissues. Figure 16.20 shows a typical expression vector (pUR278) containing multiple restriction sites for cloning genes of interest.

Design of expression cassettes: Several strategies using regulatory sequences have been pursued to optimize the expression of genes. Since the genetic control of gene expression is different for different species, it is necessary to "link" the cloned gene to genetic control elements that are characteristic of the host species. The foreign gene is expressed and controlled through the normal biosynthetic and genetic "machinery" of the host cell. This is accomplished through the use of specialized extra-chromosomal elements termed production-level vectors that are specifically designed for overexpression of vector-borne genes. Modified plasmids are preferred for production vector constructions for ease of manipulation and high-level expression. The expression vectors often contain genetically-controlled, high-level promoters, efficient ribosome binding sites, multi-restriction insertion sites, and transcription terminators, translation, gene dosage, and product stability in order to control the timing and the level of expression of cloned genes. Conceptually, it is useful to make an artificial distinction between expression vectors and production vectors. **Expression vectors** are simply designed to express detectable levels of foreign proteins, usually at the "bench level". **Production level vectors**, on the other hand, are specifically designed for stable large-scale production of gene products at economically significant levels. Usually this means production level vectors have additional genetic elements such as par sequences, transcription terminators downstream from the cloned gene, etc.

Heterologous expression vectors: Two strategies are in use for the heterologous expression of gene products. These are **direct expression** and **secretion**. With direct expression, promoter sequences are linked directly to the ATG translation initiation codon followed by the coding region of the heterologous gene. The protein product accumulates in the cytoplasm of cells. This method has two drawbacks. In many cases, the initiator methionine is not removed. If the authenticate amino acid is not a methionine this may not be desirable. Second, many proteins contain disulfide bonds that must be formed to permit the protein to exhibit biological activity. The cytoplasm is not conducive to correct disulfide bond formation, and proteins must be purified and folded *in vitro*. Secretion vectors contain one additional element—a signal peptide. Secretory proteins contain at their amino terminus a 15–30 amino acid hydrophobic leader. This signal peptide directs the protein into the secretory pathway followed by eventual arrival at the cell surface. The signal peptide is generally removed during secretion by a signal peptidase and disulfide bond formation also occurs during this process. Another advantage of this system is post-transcriptional modification, i.e. addition of carbohydrate (glycosylation) during secretory organelles.

A number of variables in the design of the vector can affect the expression level, the promoter element perhaps the most. Usually strong homologous promoters are desirable. An alternative strategy is to use "promoter finders". Genes encoding enzymes, which have easily assayable properties, are introduced into vectors lacking promoter elements. In some cases, it is possible to use heterologous promoters for expression. For example, certain promoters from *Neurospora* function in *Aspergillus*. Often heterologous promoters are found to be inactive, particularly in distantly related organisms. The 5', and to a lesser extent, the 3' untranslated regions of mRNA, can have a dramatic effect on the expression level. Therefore, replacement of the untranslated regions of heterologous genes with analogous untranslated regions from host genes can increase the expression level. Codon bias is another important factor in the case of highly expressed genes. Heterologous expression can in some cases show toxic effects, particularly at high expression levels.

A. BACULOVIRUS VECTORS FOR EXPRESSING FOREIGN GENE IN INSECT CELLS

E. coli cells infected with plasmid or phage expression vectors have been used to express foreign genes. However, it is often difficult to obtain complex polypeptides derived from eukaryotes in a biologically active form from prokaryotes. This is because post-translational changes (mainly **glycosylation** and **phosphorylation**) are often brought about in proteins of eukaryotes that cannot be performed in *E. coli*. Consequently, eukaryotic expression vectors have been developed for use in a variety of eukaryotic cells, including yeast and insect cells.

The most effective expression vectors for use in insect cells have been engineered from baculoviruses (O'Reilly *et al.* 1992). Baculoviruses are viruses with double-stranded, circular DNA genomes contained within a rod-shaped protein coat. The Baculoviridae are divided into three subgroups – **nuclear poly-hedrosis viruses** (**NPV**), **granulosis viruses**, and **nonoccluded viruses**. Most NPVs primarily infect lepidopterans, where they produce nuclear inclusion bodies in which progeny virus particles are embedded. Polyhedrin is the protein component of the crystalline matrix that protects the viral particles when they are outside their insect host.

Several factors have contributed to this popularity—it is an eukaryotic expression system, thus no special modifications are required for expression in eukaryotic cells. It produces a large amount of recombinant proteins without the need of helper viruses for their multiplication. It also accommodates large segments of foreign DNA. Very important, is the fact that it is non-infectious to vertebrates. Currently, the most widely used baculovirus expression system utilizes a lytic virus known as *Autographa californica* nuclear polyhedrosis virus (AcMNPV). This virus is the prototype of the family *Baculoviridae*. The baculovirus expression system takes advantage of some unique features of the viral life cycle (Figure 16.21). It has a relatively broad host range, infecting over 28 different lepidopteran species. The life cycle of the wild-type baculovirus begins when caterpillars eat the protein matrix (polyhedrin) and release the virus particles. Two types of viral progeny are produced during the life cycle of the virus: extracellular virus particles (nonoccluded viruses) during the late phase and polyhedra-derived virus particles (occluded viruses) during the very late phase of infection. Polyhedrin production begins only when the caterpillar is near death, and the virus resumes polyhedrin production until approximately 20% of the insect host cell proteins consist of polyhedrin. The polyhedrin protein of occluded viruses serves to sequester and thereby protect hundreds of virus particles from proteolytic inactivation during host lysis.

The baculovirus expression system takes advantage of several facts about the polyhedrin protein: it is expressed at very high levels in infected cells, constituting more than half of the total cellular protein late in the infectious cycle, and is non-essential for infection or replication of the virus; also viruses lacking the polyhedrin gene have a plaque

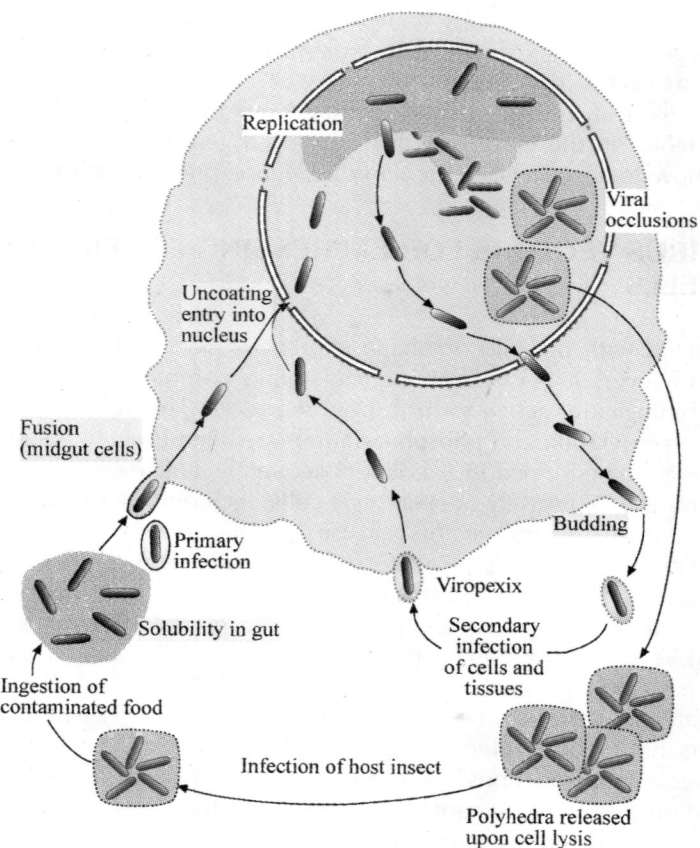

Figure 16.21 *A schematic diagram representing the biphasic life cycle of a typical baculovirus*

morphology that is distinct from that of viruses containing the gene. Recombinant baculoviruses are generated by replacing the polyhedrin gene with a foreign gene through homologous recombination. In this system, the distinctive plaque morphology provides a simple visual screen for identifying the recombinants. To produce a recombinant virus that expresses the gene of interest, the gene is first cloned into a transfer vector. Most baculovirus transfer vectors contain the polyhedrin promoter followed by one or more restriction sites for foreign gene insertion. Once cloned into the expression vector, the gene is flanked on both 5' and 3' by viral-specific sequences. Next, the recombinant vector is transferred along with wild-type viral DNA into insect cells. In a homologous recombination event, the foreign gene is inserted into the viral genome and the polyhedrin gene is excised. Recombinant viruses lack the polyhedrin gene and in its place contain the inserted gene, whose expression is under the control of the polyhedrin promoter (Figure 16.22). Linearization of the wild-type baculovirus DNA before cotransfection with plasmid DNA increases the proportion of recombinant virus to ~30%.

Biotechnology companies are developing proteins expressed in insect cells as commercial products. Companies are conducting research to learn how to scale up the culture of

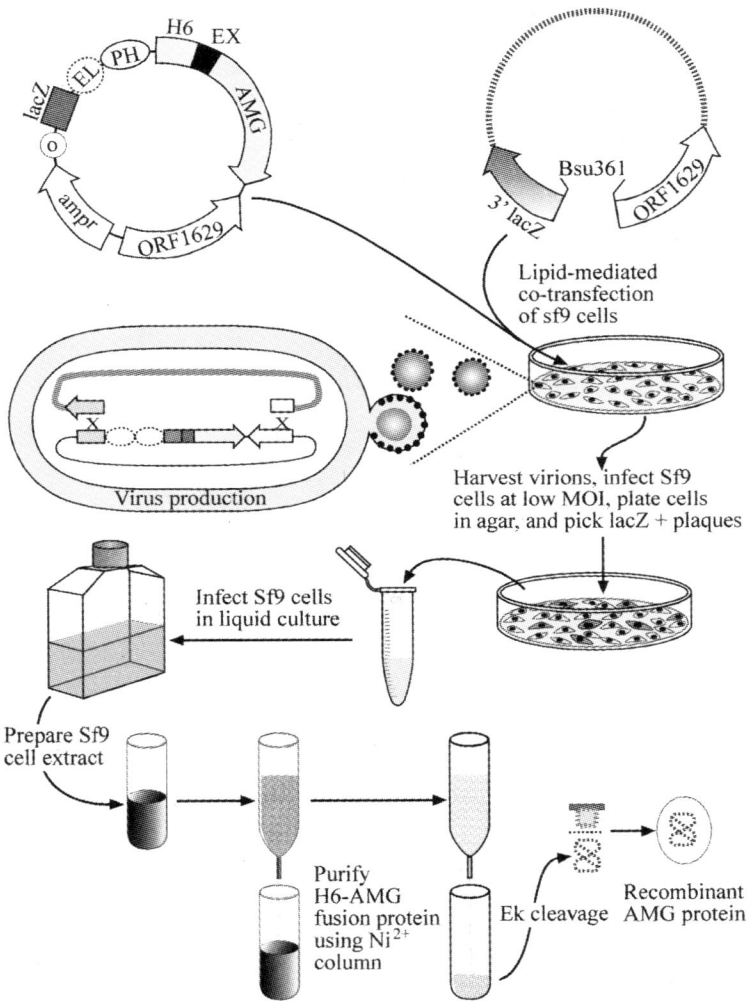

Figure 16.22 *Baculovirus expression systems are used to produce large quantities of recombinant proteins that can be purified using affinity chromatography*

insect cells in commercial-sized fermenter systems. However, many problems remain before these products can be widely used.

VII TRANSPOSONS AS VECTORS

Transposons of higher plants (Ds, Ac or Mu i of maize): Maize transposons, **Ac** and **Dc** (Figure 16.23a) are popularly known transposons. Each represents a transposon with short terminal repeats enclosing a long DNA segment, which measures more than 4500 bp in Ac

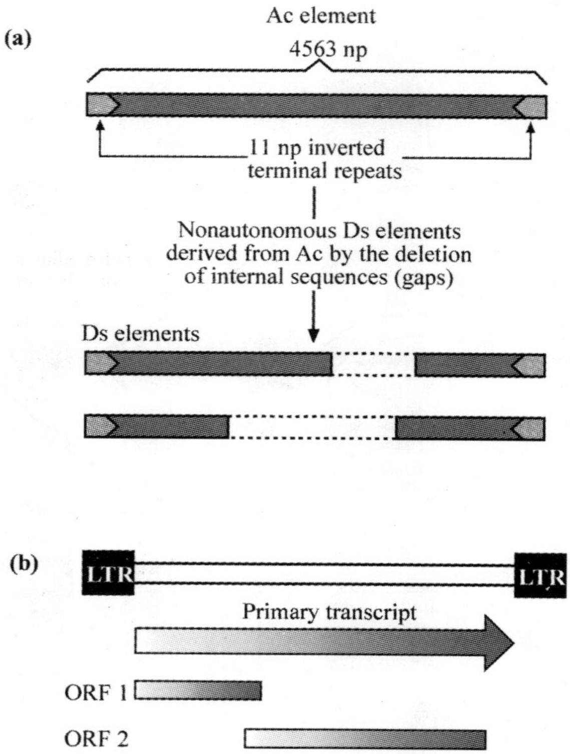

Figure 16.23 *(a) Structural organization of autonomous and nonautonomous members of the Ac/Ds family of transposable elements in maize. (b) Retrovirus-like vectors. Structure of a typical Ty element. ORF 1 and ORF 2 represent the two open reading frames. Long terminal repeats are indicated by LTR*

and about 400 bp in Ds. Each possesses genes including the gene for the transposase enzyme responsible for translocation. Part of this region can be deleted and the transposon can be used for cloning of the foreign DNA segment in much the same way as in other cases of gene cloning. Figure 16.23b shows the structure of a typical transposon of yeast (Ty element), which has been used as a retrovirus-like vector.

VIII VIRUSES

A number of plant and animal viruses have been used as vectors for both introduction of foreign genes into cells, gene amplification, and expression in host cells. Some of the attractive features of virus vectors are (i) viruses infect cells of adult plants, (ii) they produce a large number of copies per cell, leading to gene amplification, and production of the recombinant protein in large quantities and (iii) some viruses are systemic in that they spread throughout the plant. Most of the plant viruses do not integrate their genome into plants. As

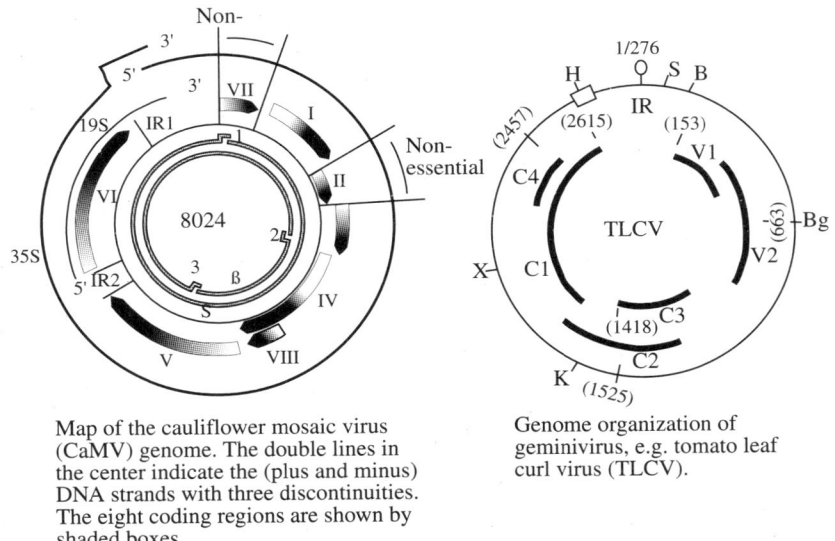

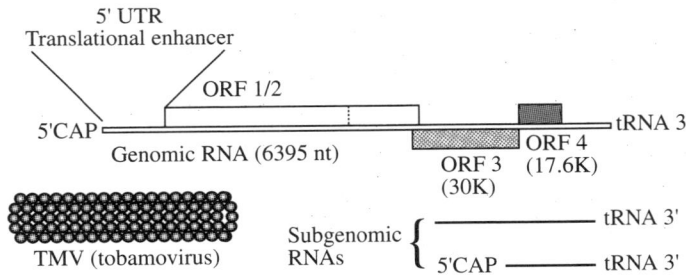

Tobacco mosaic virus structure and genome organization.

Figure 16.24 *Genomic organization of different representative plant viruses that have been used as vectors for plant transformation*

a result, they cannot be used to produce stable and heritable transformations. But they can be used to express transgene with a view to either improving the phenotypic performance of host plants or producing large quantities of valuable proteins. Some of the plant viruses that have been commonly used as vectors (Figure 16.24) are the Cauliflower mosaic virus (CaMV), Gemini viruses, Tobacco mosaic virus (TMV) and Brome mosaic virus (BMV).

IX SUGGESTED READING

Brown TA (Ed) (1991) *Essential Molecular Biology: A Practical Approach*, Vols. I and II, IRL Press, Oxford.
Glover DM, Ed. (1985) DNA cloning, vol I. A Practical Approach, IRL Press, Oxford.

Jarvis DL, Fleming JG, Kovacs GR, Summers MD and Guarino LA (1990) Use of early baculovirus promoters for continuous expression and efficient processing of foreign gene products in stably transformed lepidopteran cells. *Bio/Technology* **8**:950–955.

Luckow VA and Summers MD (1988) Trends in the development of baculovirus expression vectors, *Bio/Technology* **6**: 47–55.

Old RW and Primrose SB (1989) *Principles of Gene Manipulation: An Introduction to Genetic Engineering*, (4th Ed), Blackwell Scientific, Oxford.

O'Reilly DR, Miller LK and Luckow VA (1992) *Baculovirus Expression Vectors: A Laboratory Manual*, Freeman, New York.

Pouwels PH (1991) Survey of cloning vectors for *Escherichia coli*. Pp. 179–239. In: *Essential Molecular Biology: A Practical Approach*, vol. I. (Ed) Brown T A, Ed. IRL Press, Oxford.

Sambrock J, Fritsch E F and Maniatis T (1989) *Molecular Cloning, a Laboratory Manual (2nd Edn)*, Cold Spring Harbor Laboratory Press, USA.

Turner PC, McLennan AG, Bates AD, and White MRH (1998) *Instant Notes in Molecular Biology*, Bios Scientific Publishers Limited.

Watson JD, Hopkins NH, Roberts JF, Steitz JA, and Weiner AM (1987) *Molecular biology of the gene (4th Edn)*, The Benjamin/Cummings Publishing Company, Inc.

Weaver RF (1999) *Molecular Biology*, WCB/McGraw-Hill.

Williams JG and Patient RK (1988) *Genetic Engineering*, IRL Press, Oxford.

Wilson K and Walker J (2000) (Ed) *Practical Biochemistry Principles and Techniques (5th Edn)*, Cambridge University Press.

Winnacker E, L (2003) *From Genes to Clones*, Panima Publishing Corporation.

17

DNA Libraries

I INTRODUCTION

An essential early step in gene cloning is to isolate the gene of interest from all the other DNA of the organism. Researchers are interested in cloning individual genes so that those genes can be studied in detail. Many protocols have been developed to isolate DNA from the cells of an organism and to precisely cut genomic DNA into smaller fragments for cloning. The construction of a DNA library is a procedure that involves the insertion of manageably small pieces of DNA into a convenient vector. In most molecular biology laboratories, cloning experiments involve the production of DNA libraries. A DNA library is considered *"**representative**"* if the combined set of DNA segments contained within the library corresponds to all possible DNA sequences present in the original DNA (source DNA). The relative abundance of DNA segments should be maintained when converting the source DNA into a library. There are two sources of DNA for gene libraries: the genomic DNA and the mRNA/cDNA of a differentiated cell. For example, it should be possible to isolate DNA segments from a representative genomic DNA library that together cover an entire chromosome(s). When the library covers all the fragments of the whole genome, then it is called a **genomic library** and if the fragments represent only individual chromosomes, it is called a **chromosome library**. *In a cDNA library*, the relative abundance of each cDNA clone reflects the steady-state level of RNA transcripts in the original sample used to synthesize the cDNA. A critical parameter to consider when constructing a representative library is how redundant are the DNA segments with regard to containing overlapping regions of a particular gene. The main principles of constructing genomic and cDNA libraries are shown in Figure 17.1.

All DNA libraries are constructed using four basic steps: 1) selection and preparation of donor genetic material; 2) construction of donor DNA into recombinant molecules; 3) insertion of the recombinant molecules into the host cell; and 4) growth of host cells and selection of transformed cells. Subdividing the DNA source into clonable fragments using restriction enzymes is the first step in library construction. However, the choice of enzyme and the reaction conditions used for the digestion determine what type of cloning vector is most suitable for the intended libraries. Similarly, depending on how a cDNA library is going to be screened, for example, either by nucleic acid hybridization or protein expression, it is important to choose the most suitable cloning vector prior to library construction (refer to Chapter 16).

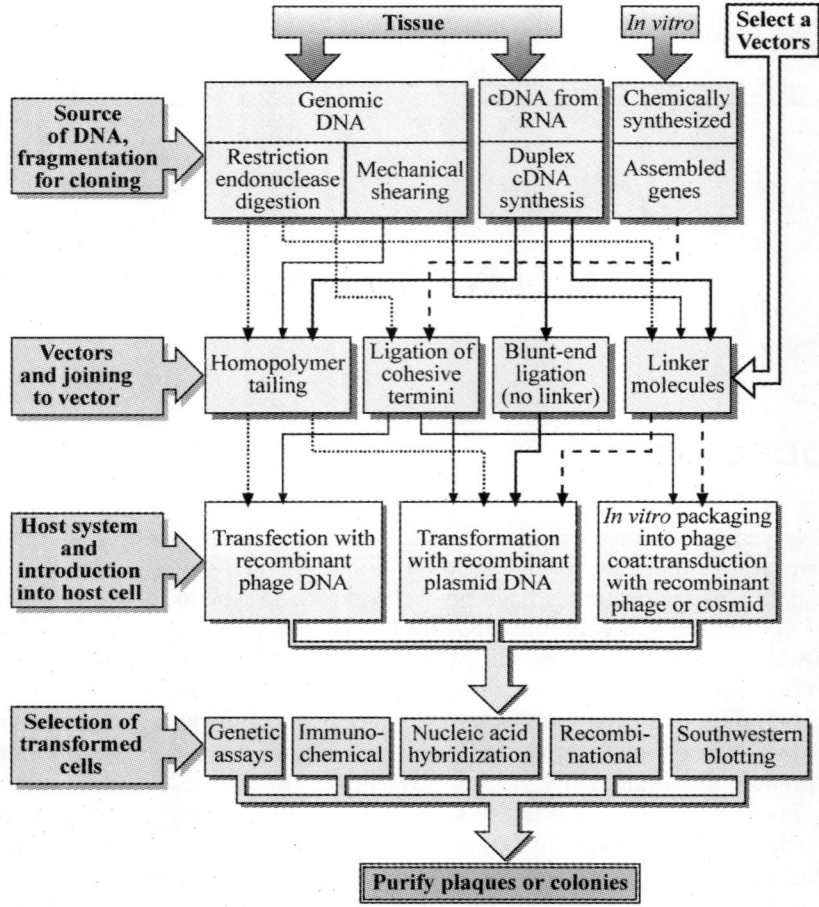

Figure 17.1 *Generalized scheme for construction of DNA libraries and screening libraries*

II CONSTRUCTION AND SCREENING OF GENOMIC LIBRARIES

A *genomic library* is a collection of clones that hopefully, contains at least one copy of every DNA sequence in the genome. Genomic DNA libraries may be required for a number of purposes. First, to isolate genes from tissue or stages where **tissue RNA is difficult** to obtain. Second, to produce a cDNA clone for a **gene of interest** which has already been isolated. Genomic clones may contain sequences missing from incomplete cDNAs, and a genomic fragment may provide clues as to the regulation of expression of the gene from its genomic environment. The **intron–exon** structure will also be apparent from the genomic sequence, with attendant implications for intra-protein domain boundaries and possible insights into gene evolution.

Any particular gene constitutes only a small part of an organism's genome. For example, if the organism is a mammal whose entire genome encompasses some 10^6 kbp and the gene is 10 kbp, then the gene represents only 0.001% of the total nuclear DNA. It is impractical to attempt to recover such rare sequences directly from isolated nuclear DNA because of the overwhelming amount of extraneous sequences. Instead, a genomic library is prepared by isolating the total DNA from the organism, digesting it into fragments of suitable size, and cloning the fragments into an appropriate vector. This approach is called **shotgun cloning** since the strategy has no way of targeting a particular gene but instead seeks to clone all the genes of the organism at one time (Figure 17.2). The intention is that at least one recombinant clone will contain at least part of the gene of interest. Usually, the isolated DNA is only partially digested by the chosen restriction endonuclease so that not every restriction site is cleaved in every DNA molecule. Then, even if the gene of

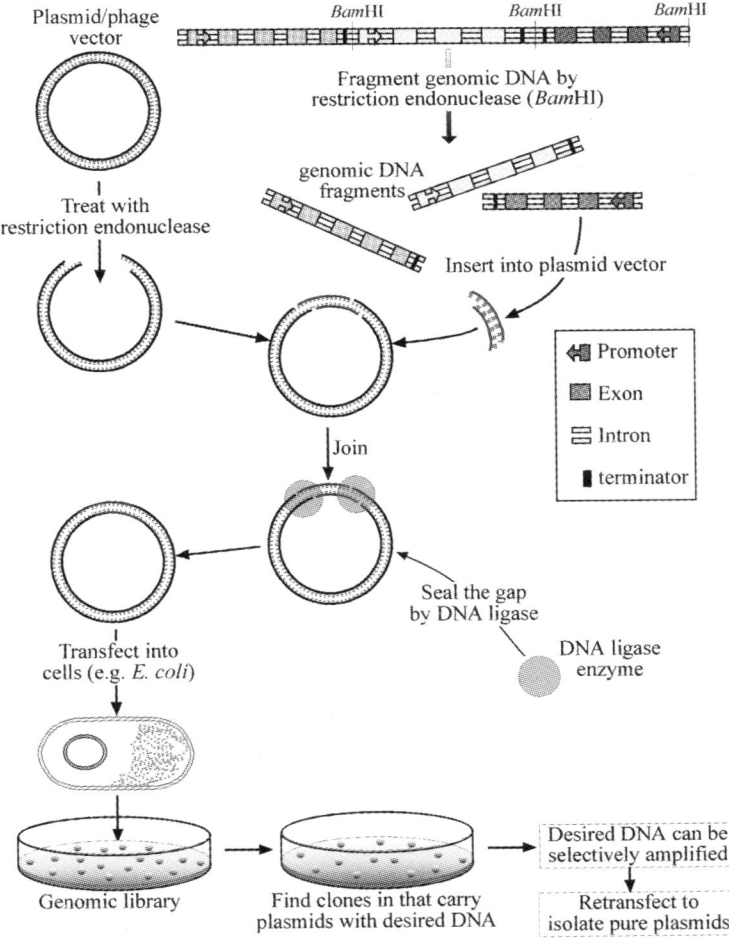

Figure 17.2 *Genomic DNA libraries contain linear segments of the genome. The genomic DNA insert is identical to the corresponding segment of DNA present in the chromosome*

interest contains a susceptible restriction site, some intact genes might still be found in the digest. Genomic libraries have been prepared for hundreds of different species.

Many clones must be created if one is to be confident that the genomic library contains the gene of interest. The probability, P, that some number of clones, N, contains a particular fragment representing a fraction, f, of the genome is

$$P = 1-(1-f)^N$$

Thus, $N = \dfrac{\ln(1-P)}{\ln(1-f)}$

For example, if the library consists of 10 kbp fragments of the *E. coli* genome (4720 kbp total), over 2000 individual clones have to be screened to have a 99% probability ($P = 0.99$) of finding a particular fragment (Since $f = 1/n = 10/4720 = 0.0021$ and $P = 0.99$, $N = 2193$). For a 99% probability of finding a particular sequence within the 3×10^6 kbp human genome, N would equal almost 1.4 million if the cloned fragments averaged 10 kbp in size. The need for cloning vectors capable of carrying very large DNA inserts becomes evident from these numbers.

Genomic libraries are generally constructed from large DNA fragments generated by partial digestion with frequent-cutting restriction enzymes (usually four-base recognition), although random cleavage by DNase I in the presence of Mn^{2+} ions, or sonication, can be used. For expression cloning, smaller fragments suitable for λgt10 or λgt11 are usually prepared by random cleavage (DNase I or sonication) to a mean size of 1 to 1.5 kb before insertion (Figure 17.3). The larger fragments require the use of either lambda replacement vectors such as EMBL 3 and EMBL 4, or cosmid vectors (refer to Chapter 16). Cosmid vectors are packaged *in vitro* like the lambda phage, but replicate in the host as a plasmid. They can accommodate up to 45 kb of DNA.

The improvement in *in vitro* packaging systems and the development of *in situ* hybridization techniques soon allowed much larger libraries to be constructed from unfractionated populations of restriction fragments (Figure 17.4), thus eliminating the need to enrich the starting DNA for the sequences of interest. The inherent limitations of genomic DNA libraries constructed from complete digests of genomic DNA with a restriction enzyme such as *Eco*RI then quickly became apparent. These limitations include the following: (1) Libraries are of large size: Because of their large size, libraries of complete digests are laborious both to create and to screen; (2) A sequence can contain more recognition sites. If the sequence of interest contains one or more recognition sites for a particular restriction enzyme, it will be cloned in two or more non-overlapping recombinants; (3) because of the quasi-random distribution of restriction sites in eukaryotic DNA, the sequence of interest may by chance be located on a fragment of genomic DNA that is too large or too small for the vector to accept.

A. VECTORS USED TO CONSTRUCT EUKARYOTIC GENOMIC DNA LIBRARIES

Many cloning vectors have been developed for the construction of DNA libraries (Table 16.3). However, two kinds of vectors commonly used to construct genomic DNA libraries

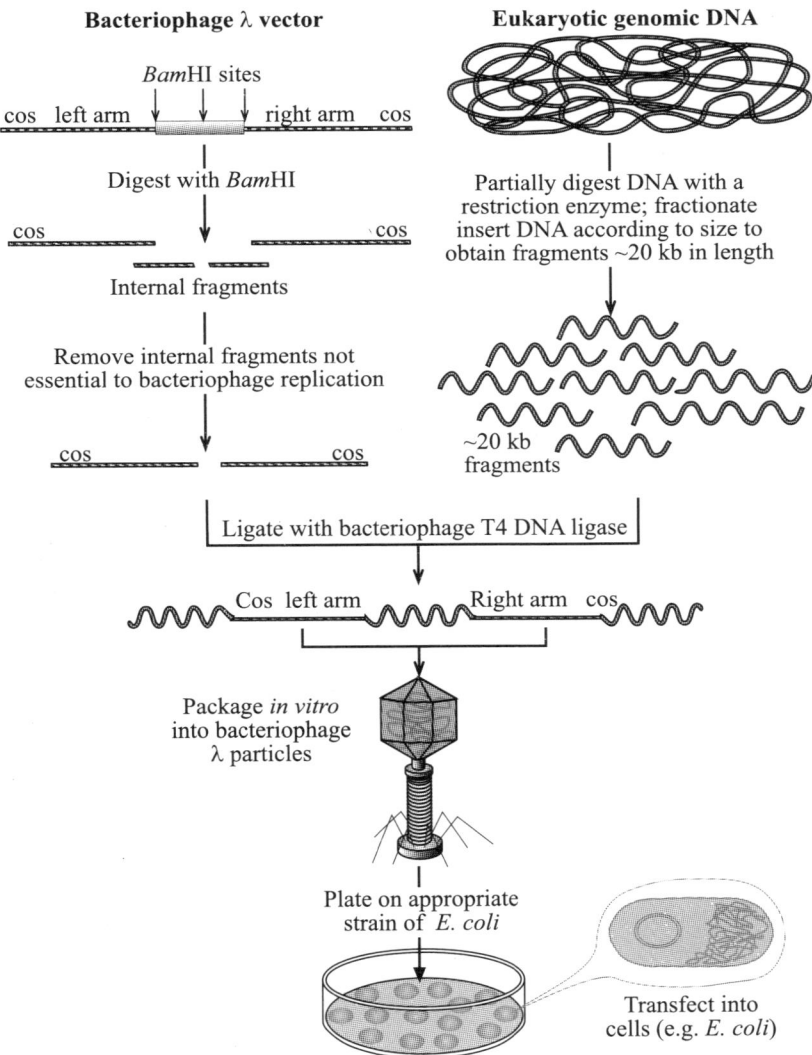

Figure 17.3 *A strategy used to construct libraries of random fragments of eukaryotic DNA*

are **bacteriophage lambda** and **cosmid** vectors (refer to Chapter 16). In both cases, large segments of eukaryotic DNA, generated by quasi-random fragmentation, are ligated with vector DNA to form concatemers that can be packaged into bacteriophage lambda particles. Library-constructed phage vectors are stored and propagated in the form of such recombinant bacteriophage particles. In cosmid cloning, however, these particles serve merely as vehicles to introduce the recombinant genomes into bacteria, where they are propagated as large plasmids.

Although vectors based on the bacteriophage lambda have been extremely powerful tools for the isolation of both cDNA and genomic versions of many eukaryotic genes, they can

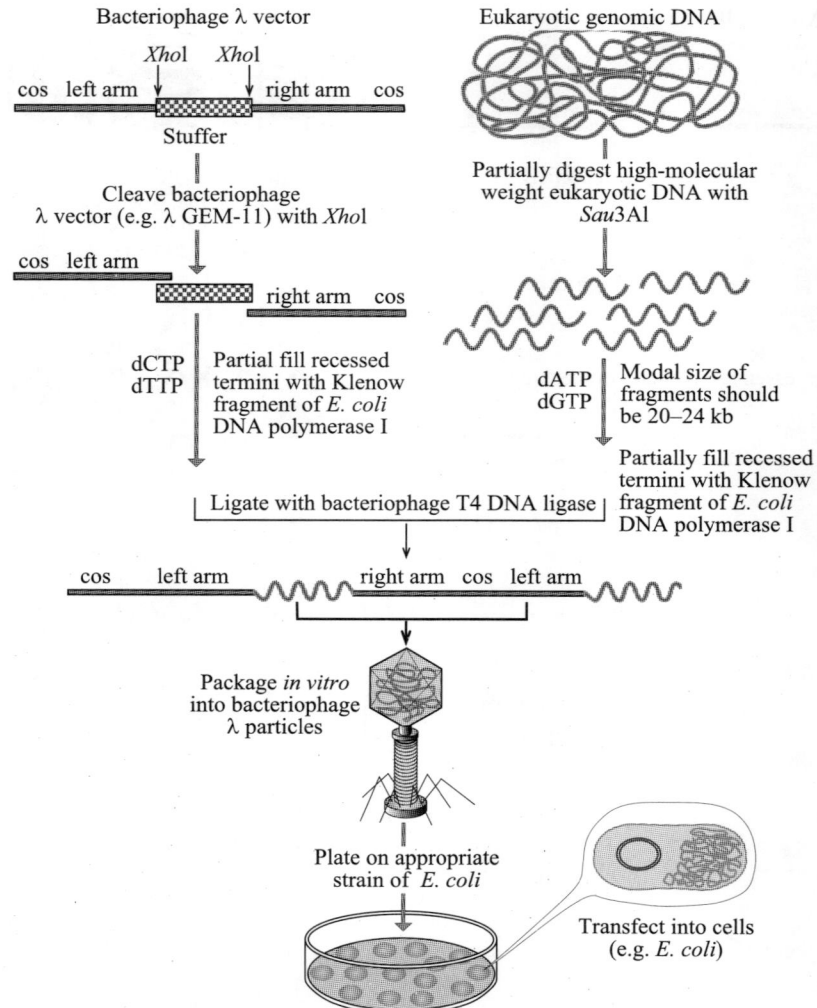

Figure 17.4 *A strategy used to construct libraries of unfractionated genomic DNA*

only accommodate inserts that fall within a defined size range. The need to clone large segments of DNA has been at least partly addressed by the development of yeast artificial chromosomes (YACs). The development of this cloning system was made possible by several other events: (1) the identification and characterization of cis-acting elements required for chromosomal stability in the yeast; (2) the development of a system to transform yeast with high efficiency; and (3) the ability to resolve very large DNA fragments by pulsed-field gel electrophoresis. The principle of the YAC system is similar to that used in conventional cloning of genomic DNA. Each of the arms of the YAC vector carries a selectable marker as well as appropriately oriented DNA sequences that function as telomeres in the yeast. In addition, one of the two arms carries two small DNA fragments that function as a centromere and as an origin of replication (also called an ARS element – autonomously replicating sequences).

B. PREPARATION OF GENOMIC DNA FOR INSERTION INTO EMBL 3

Isolation of high molecular weight DNA from different organisms has been described in Chapter 7. Low molecular weight contaminants are removed by dialysis. The only method by which DNA can be fragmented in a truly random fashion, irrespective of its base composition and sequence, is mechanical shearing. However, DNA prepared in this way requires several additional enzymatic manipulations to generate cohesive ends compatible with those of the vectors used to generate genomic DNA libraries. On the other hand, partial digestion with restriction enzymes that recognize frequently occurring tetranucleotide sequences within eukaryotic DNA yields a population of fragments that is close to random and yet can be cloned directly.

Genomic DNA (gDNA) is cleaved with *Sau*3AI, a four base cutter (^GATC), which gives sticky ends compatible with *Bam*HI (G^GATCC) sticky ends in the vector DNA. The insert is phosphatased with calf intestinal phosphatase to prevent re-ligation of non-contiguous fragments.

Materials

High molecular weight gDNA (>100 kb); *Sau*3AI restriction enzyme and buffer; Calf intestinal phosphatase (CIP) at 1 U/μl in buffer, 1 M $ZnCl_2$; Agarose, TAE buffer, EtBr, Mini-gel equipment, Marker DNA; Phenol; chloroform; 3 M Sodium acetate; 96% Ethanol; Microfuge; Water baths at 37°C and 65°C.

Procedure

1. Take 5 μg of gDNA in 100 μl 1x *Sau*3A I buffer on ice. Transfer the gDNA with a wide bore (cut off) pipette tip to avoid shearing. Perform all mixings very gently.
2. Prepare four tubes with 75 μl 1x *Sau*3A I buffer on ice, add 25 μl of the gDNA to each. Add 1 U *Sau*3A I to the first tube and mix gently.
3. Transfer 10 μl of this to the second tube and mix. Transfer 10 μl from the second tube to the third, mix, and transfer 10 μl from the third to the fourth tube. Take 10 μl from the fourth tube and discard.
4. Place the four tubes in a 37°C water bath for 15 min. Transfer to 65°C for 5 min.
5. Analyze by electrophoresis on a 0.4% agarose gel run in TAE buffer. Include an undigested DNA as control (1 μg) and as markers undigested 1 DNA and Hind III digested λ DNA. Stain the gel with EtBr and view under UV transillumination. Look for the digestion conditions which give a modal fragment size of 24 kb (just above the largest lambda Hind III marker).
6. Check these conditions, or perform additional control digestions to obtain the optimal digestion. Using these optimal conditions, cut μg of DNA. Heat to 65°C for 5 min. and place on ice.
7. Add $ZnCl_2$ 10 mM and 1 U of CIP per 10 μg DNA. Incubate at 37°C for 30 min. Heat to 65°C for 5 min. Extract with Phenol, phenol–chloroform and chloroform very gently before precipitating with sodium acetate and ethanol.

C. CONSTRUCTION OF AN EMBL 3 LIBRARY

The EMBL 3 and 4 vectors use spi selection to ensure that only recombinant phage will grow on the permissive host bacteria. A central 13.2 kb stuffer fragment contains genes which render the wild type phage sensitive to phage P2 interference (spi), and these will not grow in the NM539 host cells, which are lysogenic for phage P2. To clone in EMBL 3, the central fragment is excised with *Bam*HI and re-ligation inhibited by further cutting with an enzyme (*Sal*I) with sites in the stuffer fragment but not in the two 'arms'. The arms, 19.9 kb and 8.8 kb in length, are then ligated to the *Sau*3AI partial cut genomic DNA, and the recombinants packaged *in vitro*.

Materials

EMBL3 DNA (NBL, Boehringer); *Bam*HI and *Sal*I enzymes; 10x One-For-All Buffer: 100 mM Tris-acetate, pH 7.5; 500 mM Potassium acetate; 100 mM Magnesium acetate; Phenol, Chloroform, 3 M Sodium acetate; 96% Ethanol; T4 DNA ligase;

10x ligase buffer: 0.5 M Tris.Cl pH 7.6, 100 mM $MgCl_2$, 100 mM DTT, 5 mM ATP.

E. coli NM539 and C600 made competent for phage infection. Packaging extracts, e.g. Gigapack Gold (Stratagene).

Procedure

1. Ligate the cos ends of the EMBL3 vector by incubating 20 µg of DNA with 5 U ligase in a 100 µl reaction at 16°C for 2 hours. Heat to 65°C for 5 min., extract with phenol/chloroform and chloroform (very gently; the DNA is very long and sensitive to shearing) and precipitate with sodium acetate and ethanol. Re-suspend the dried pellet in 40 µl water.
2. Cut the ligated EMBL3 DNA with 10 U of *Bam*HI in a 50 µl reaction (37°C for 2 hr). Halt the reaction by heating to 65°C for 5 min. Add 10 U of *Sal*I and 1 µl buffer to a volume of 60 µl and return to 37°C for 2 hours.
3. Check that the DNA is digested by analyzing 0.2 µg on a 0.6% agarose gel.
4. Extract with phenol/chloroform and chloroform, then precipitate with sodium acetate and ethanol.
5. Mix 1 µg of EMBL 3 DNA with 0.1 to 1.2 µg of insert in 8 µl. Heat to 65°C for 4–6 hours. Include as control a reaction containing cut vector but no insert. Stop by heating to 65°C for 5 min.
6. Package the ligated DNA in 0.25 µg vector equivalent lots. As control, also package 0.1 µg uncut vector and 0.1 µg cut vector, and the re-ligated vector from Step 5. Stop the packaging reaction by adding 500 µl of SM.
7. Plate out 1 µl of a 1:1000 dilution of the library onto competent *E. coli* NM539 cells. Similarly, plate out the vector packaging and the library packaging onto C600 cells to check non-recombinant titre.

Notes

1. This ligation reaction covalently joins the cohesive 12 bp overhangs of the vector DNA (*cos* ends). On ligation to the insert, this will form concatemers of recombinant phage DNA. These concatemers are better substrates for the *in vitro* packaging reaction later.

2. For inserts of 20–24 kb, these figures give approximate molar ratios of arms to insert of 8:1 to 0.2:1. Ideally, a series of ligations spanning these ratios should be tried for each batch of insert and vector.

D. DNASE I CLEAVAGE OF GENOMIC DNA (gDNA) FOR EXPRESSION LIBRARIES

This method uses DNase I in the presence of manganese ions to randomly cut the DNA. Under these conditions, the enzyme produces blunt-ended fragments, so no fill-in reaction is necessary.

Materials

DNase I at 1 µg/ml; High molecular weight gDNA; Agarose; TAE buffer; EtBr; Mini-gel apparatus; Marker DNAs; 0.5 M EDTA pH 8.0; Reagents for cloning by linker addition into λgt11.

10X Mn–DNase I buffer: 500 mM Tris.Cl pH 7.2, 10 mM $MnCl_2$

Procedure

1. To determine the digestion time needed to produce fragments of the required size (0.5–4 kb), a series of analytical digests, using 1 µg of DNA in a reaction volume of 10 µl, should be done as follows: 0.1 ng DNase for 1 min.; 0.01 ng DNase for 1, 5, and 10 min. All incubations are carried out at 37°C in 1x Mn-DNase buffer.
2. Each digest is stopped by adding EDTA to 5 mM and heating at 65°C for 10 min. Digests are then analyzed on a mini-gel, run with suitable size standards, and the correct amount of enzyme and length of incubation determined. The correct digest can then be scaled-up and used to digest 5 µg of DNA. A small quantity of this should be tested on a gel, before proceeding.
3. Extract DNA with Phenol:chloroform and chloroform precipitated with sodium acetate and ethanol.
4. The vacuum dried pellet is redissolved in 10 µl of water.
5. The insert DNA has to be proceeded with as if it were a cDNA in preparation for ligation into λgt11 and packaging as explained earlier.

E. SCREENING GENOMIC LIBRARIES

The basic reason for constructing a DNA library is that it provides a pool of gene sequences that can be accessed using biochemical methods. A common method of screening plasmid-based genomic libraries is to carry out a colony hybridization experiment, using a radioactive DNA probe in a membrane hybridization reaction to physically locate plaques or bacterial colonies that contain complementary DNA segments. The protocol is similar for phage-based libraries except that bacteriophage plaques, and not bacterial colonies, are screened. In a typical experiment, host bacteria containing either a plasmid-based or bacteriophage-based library are plated out on a Petri dish and allowed to grow overnight to form colonies (or in the case of phage libraries, plaques). A replica of the bacterial colonies (or plaques) is then obtained by overlaying the plate with a nitrocellulose disc. The disc is removed, treated with alkali to dissociate bound DNA duplexes into ssDNA, dried, and placed in a sealed bag with a labeled probe. If the probe is dsDNA, then it must be heat-denatured by heating at 70°C. The probe and target DNA complementary sequences must both be in a single-stranded form if they are to hybridize with one another. Any DNA sequence the complementary to probe DNA will be revealed by autoradiography of the nitrocellulose disc. Bacterial colonies (phage plaques) containing clones bearing target DNA are identified on the film and can be recovered from the master plate (Figure 17.5).

F. RESTRICTION ENZYME MAPPING AND SUBCLONING

Restriction enzymes are endonucleases that bind to specific DNA sequences and create double-stranded DNA breaks by enzymatic cleavage (see Chapter 15 for more details).

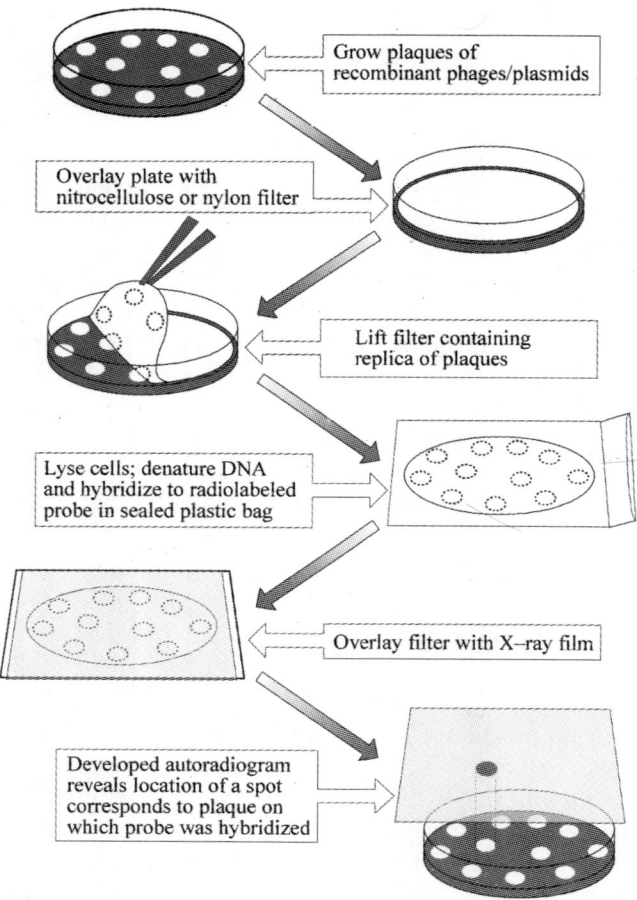

Figure 17.5 *Screening a genomic library by colony hybridization*

Because restriction enzyme recognition sites are highly specific, it is possible to use commercially available enzyme preparations to map cleavage sites within cloned DNA using agarose gel electrophoresis. The mapping strategy involves several rounds of DNA digestion with infrequent cutting enzymes (recognition site of 6–8 bp), followed by agarose gel electrophoresis to determine fragment lengths. Using this information, a tentative restriction map can be constructed as the basis for designing a second round of experiments using multiple enzyme combinations.

The first task when constructing a restriction map is to identify one or more unique restriction sites in the cloning vector to anchor a relative map position (5' or 3' of the insert). Although it is possible to construct a more detailed restriction map using additional enzyme digestions in various combinations, and by performing partial enzyme digests with end-labeled DNA fragments, it is now more efficient to use subcloning and automated DNA sequencing to locate restriction enzyme sites precisely. In order to better characterize a large genomic DNA sequence, it must be subdivided into fragments small enough to be cloned into a plasmid vector suitable for functional analysis and DNA sequencing. Figures 17.6 and 17.7 show a common strategy for subcloning DNA inserts into a plasmid vector. Generally, two methods are followed for subcloning: (1) a directed cloning strategy based on a previously determined restriction enzyme map; and (2) random subcloning using frequent cutting restriction enzymes or sheared DNA. The latter approach is called "shotgun" cloning and is often the first step in high-throughput DNA sequencing using automated instrumentation. DNA sequence information obtained from the shotgun cloning method can be analyzed by a computer-algorithm. To compile aligned sequences that can be used to find all known restriction enzyme sites within that DNA segment, computer-algorithms have been efficiently used.

III CHROMOSOME LIBRARIES

Many organisms (including humans) have genomes so large that many thousands of clones are needed to represent the entire genome. This makes searching the library for a gene of interest very time-consuming. One approach for reducing the screening time of large genomes is to make libraries of the individual chromosomes in the genome. In humans, for example, this gives 24 different libraries, one each for the 22 autosomes, and two each for the X and Y chromosomes. Individual chromosomes of an organism can be separated if their morphologies and sizes are distinct enough, as is the case for human chromosomes. One procedure currently used to isolate large chromosomes individually is **flow cytometry**. In this procedure, chromosomes from cells in the mitotic phase of the cell cycle are stained with a fluorescent dye. Chromosomes released from the cells are passed through a laser beam connected to a light detector. This system sorts and fractionates the chromosomes based on their differences in dye binding and resulting light scattering. Once the chromosomes have been fractionated, a library of each chromosomal DNA with restriction enzymes is heated and the fragment inserted into a cloning vector. As a result of the application of these procedures, libraries of DNA prepared from all human chromosomes are now available to researchers.

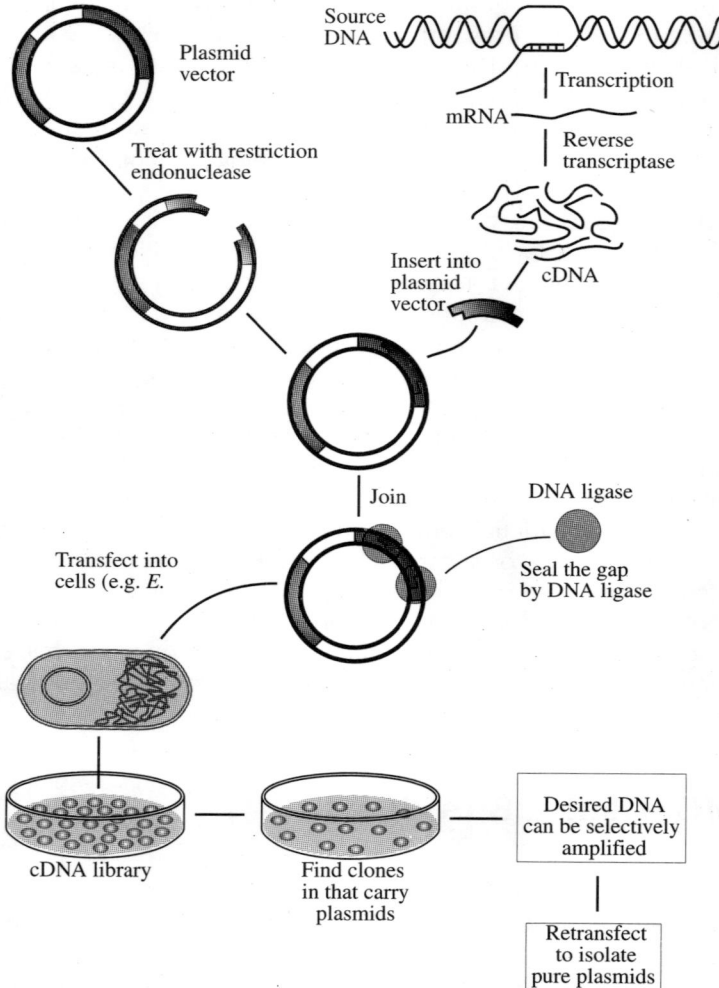

Figure 17.6 *Direct cloning a gene by cDNA library*

IV C-DNA LIBRARY AND SCREENING

The construction of cDNA (short form for complementary DNA or copy DNA) libraries is now well within the range of any competent laboratory. Comprehensive cDNA libraries can be routinely established from small quantities of mRNA, and a variety of reliable methods are available to identify cDNA clones corresponding to extremely rare species of mRNA. The cDNA library can be a set of clones representing as many as possible of the mRNAs in a given cell type at a given time. At other times, we may want to make one particular cDNA – a clone containing a DNA copy of just one mRNA. The analysis of that cloned cDNA molecule can then provide information about the gene that encoded the mRNA. Since

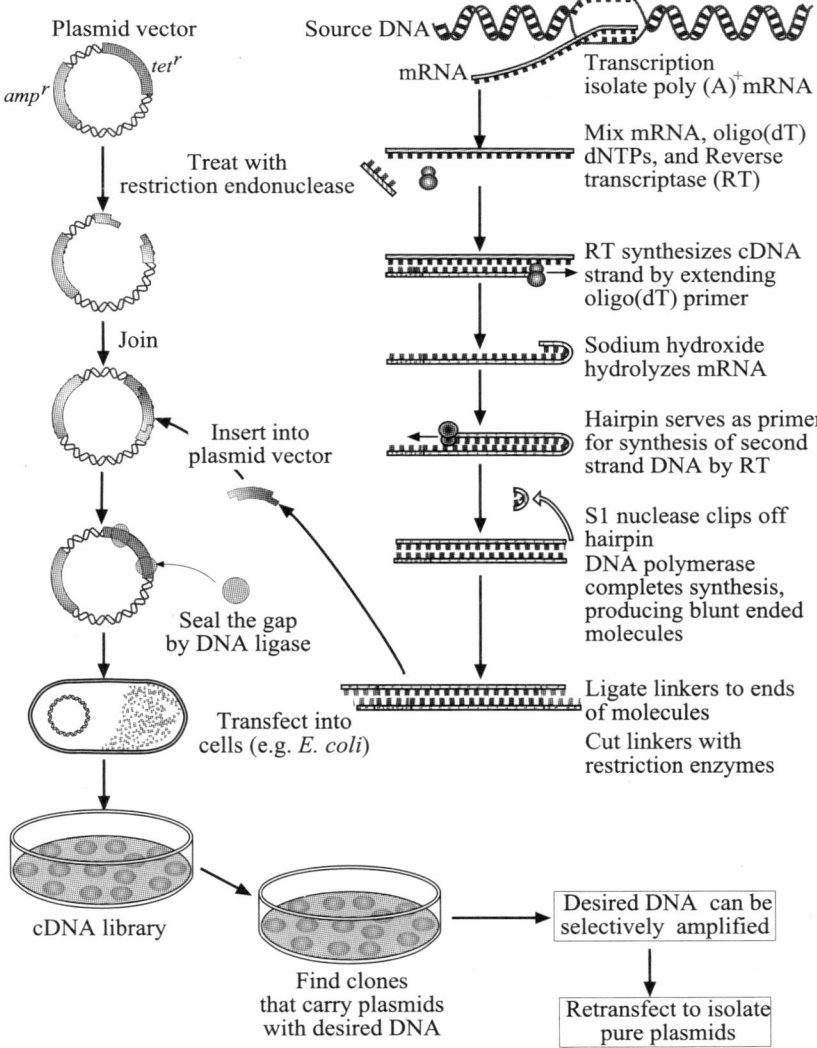

Figure 17.7 *Cloning a gene by cDNA library by hairpin amplification method*

the cDNA library reflects the gene activity of the cell type at the time the mRNAs are isolated, the construction and analysis of cDNA libraries is useful for comparing gene activities in different cell types of the same organism. Because there would be similarities and differences in the clones represented in the cDNA libraries of each cell type, the technique we use depends in part on which of these goals we wish to achieve. As the enzymatic reactions used to synthesize cDNA have improved, the sizes of cloned cDNAs correspond to large mRNAs. Before embarking on the synthesis and cloning of cDNA, however, it is essential to consider carefully which methods, vectors, and screening procedures offer the best chance of success.

The cDNA libraries are constructed by synthesizing cDNA from purified cellular mRNA. These libraries present an alternative strategy for gene isolation, especially eukaryotic genes. Because different cell types in eukaryotic organisms express selected subsets of genes, RNA preparations from cells or tissues where genes of interest are selectively transcribed will be enriched for the desired mRNAs. Since most eukaryotic mRNAs carry 3'-poly(A) tails, mRNA can be selectively isolated from preparations of total cellular RNA by oligo(dT)-cellulose chromatography. DNA copies of the purified mRNAs are synthesized by first annealing short oligo(dT) chains to the poly(A) tails. These oligo(dT) chains serve as primers for reverse transcriptage-driven synthesis of DNA. DNA polymerase is then used to copy the DNA strand and form a double-stranded (duplex DNA) molecule. Linkers are then added to the DNA duplexes rendered from the mRNA templates, and the cDNA is cloned into a suitable vector. Once a cDNA derived from a particular gene has been identified, the cDNA becomes an effective probe for screening genomic libraries for isolation of the gene itself.

A. STRATEGIES FOR cDNA CLONING

1. Preparation of mRNA for cDNA cloning

(a) **Source of the mRNA**: Clearly, the higher the concentration of the sequences of interest in the starting mRNA, the easier the task of isolating relevant cDNA clones. This can be achieved, for example, (1) by using immunoprecipitation to measure the amount of the protein of interest that is synthesized in cell-free systems by mRNAs prepared from different cell lines or tissues; 2) by increasing the concentration of the relevant mRNA by using drugs to select cell lines that overexpress particular proteins. 3) Some workers have taken advantage of the fact that treatment of infected cells with inhibitors of protein synthesis causes extended transcription of the early genes of mammalian DNA viruses. Whenever possible, estimates should be obtained of the frequency with which the mRNA of interest occurs in the starting preparation.

(b) **Integrity of the mRNA**: Since the cDNA library cannot be better than the mRNA from which it is derived, it is important to check the integrity of the preparation of mRNA before it is used as the template for synthesis of the first strand of cDNA. The following tests are commonly used: (a) The ability of mRNA to direct the synthesis of high-molecular-weight proteins in cell-free translation systems derived from reticulocytes; (b) the ability of mRNA to direct the synthesis of the polypeptide of interest in a cell-free system derived from reticulocytes; (c) the size of the mRNA; (d) the ability of the bulk mRNA preparation to direct the synthesis of long molecules of first-strand cDNA.

(c) **Percentage of mRNAs present**: mRNAs that represent less than 0.5% of the total mRNA population of the cell are classified as "low-abundance" or "rare" mRNAs. The isolation of cDNA clones for mRNAs of this type presents two major problems: a good chance of being represented in the cDNA library and identification and isolation of the clone(s) of interest.

(d) **Methods of enrichment**: A typical mammalian cell contains between 10,000 and 30,000 different mRNA sequences. But not all these sequences are represented equally in the steady-state population of mRNA molecules. Instead, the proportional representation of each sequence depends on its rate of synthesis and half-life. The simplest method to enrich preparations of mRNA for sequences of interest is to *fractionate* them according to size. An alternative method of enrichment involves the use of antibodies to purify polysomes that synthesize the polypeptide of interest (***Immunological purification of polysomes***).

2. Synthesis of cDNA molecules

Messenger RNA is unstable and not amenable to prolonged manipulation. It can, however, be faithfully copied by a retroviral enzyme, ***reverse transcriptase***, which catalyses the production of a cDNA strand from a single-stranded mRNA substrate. The process of enzymatic conversion of poly(A)$^+$ mRNA to dsDNA and the insertion of this DNA into prokaryotic vectors has become a fundamental tool of eukaryotic molecular biology. The synthesis and cloning of cDNA are still not easy, but as a consequence of a wide range of technical and theoretical advances, cDNA cloning is now well within the range of any competent laboratory.

3. First-strand cDNA synthesis

All methods for synthesis of the first strand of cDNA use the enzyme RNA-dependent DNA polymerase (***reverse transcriptase***) to catalyze the reaction (Figure 17.8). Two different forms of reverse transcriptase are available commercially: avian reverse transcriptase and murine reverse transcriptase. The avian enzyme available from most manufacturers is adequate for routine construction of cDNA libraries. However, the high level of RNase H activity of the avian enzyme tends both to suppress the yield of cDNA and to restrict its length by cleaving the template near the 3' end of the growing strand. Thus, the murine enzyme is a safer choice when attempting to obtain full-length cDNA copies of mRNAs (> 2–3kb). For cloning of cDNAs, the most frequently used primer is oligo(dT), 12–18 nucleotides in length, which binds to the poly(A) tract at the 3' terminus. The primer is added to the reaction mixture in a large molar excess so that each molecule of mRNA binds several molecules of oligo(dT)12–18.

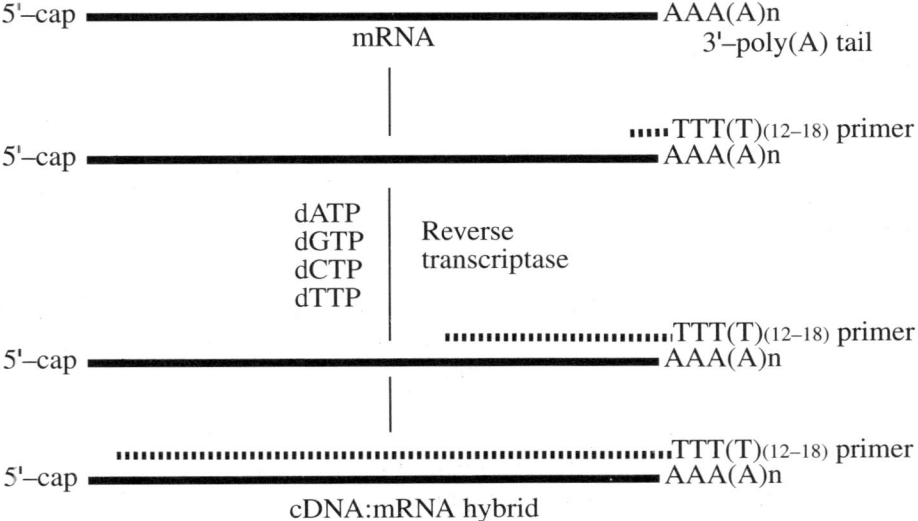

Figure 17.8 *Synthesis of the first strand of cDNA using an oligo (dT) primer and reverse transcriptase*

Before carrying out the first-strand reaction, many workers treat the mRNA with methylmercuric hydroxide to denature any regions of secondary structure that could impede the progress of the reverse transcriptase. To minimize the deleterious effects of RNases that may contaminate the solutions or reagents, a potent protein inhibitor (e.g. human placental) of RNase, is usually included in the reaction.

Materials

Poly(A)$^+$ RNA; Oligo-dT12–18, 1 mg/ml, kept at –20°C; Reverse transcriptase; RNasin (human placental RNase inhibitor). First-strand buffer: 100 µl of 1 M Tris-HCl pH 8.3 at 44°C, 140 µl of 1 M KCl, 10 µl of 1 M MgCl$_2$, 10 µl of 1 M DTT, 50 µl each 20 mM dNTPs. α-^{32}PdCTP; 0.5 M EDTA; Phenol and chloroform; 4 M ammonium acetate; 95% Ethanol kept at –20°C.

Procedure

1. Resuspend 2–5 µg of ethanol-washed poly(A)$^+$ RNA in sterile water at 1 mg/µl.
2. Heat to 65°C for 5 min. to remove any secondary structure.
3. Add the RNA to 10µCi of α-^{32}PdCTP contained in 25 µl of first-strand buffer.
4. Add 5 µl oligo–dT12–18, 1 µl RNasin (RNase inhibitor) and water to 35 µl.
5. Add 10 U of reverse transcriptase per µg RNA.
6. Incubate at 44°C for 45 min.
7. Stop reaction with 2 µl of 0.5 M EDTA.
8. Extract with phenol/chloroform.
9. Add an equal volume of 4 M ammonium acetate and two volumes of –20°C ethanol.
10. Incubate for 30 min. in dry ice-ethanol bath and microfuge for 30 min. at 4°C.
11. Re-suspend the pellet in 50 ml of water and repeat steps 9 and 10.
12. Re-suspend in 50 µl water and pass the products down a G50 Sehadex spun column and vacuum dry the second precipitate ready for the second-strand synthesis. Store at –20°C.

4. Second-strand cDNA synthesis

Careful consideration should be given to the method used to synthesize the second strand of cDNA, since it can determine the choice of vector and can dictate the means used to link the cDNA to the vector. The available options are discussed in the following sections.

Method-1: Replacement synthesis of second-strand cDNA

In this method, the product of first-strand synthesis – a cDNA–mRNA hybrid – is used as a template for a nick-translation reaction (Figure 17.9). RNase H produces nicks and gaps in the mRNA strand of the hybrid, creating a series of RNA primers that are used by *E. coli* DNA polymerase I during the synthesis of the second strand of cDNA. The reaction has three main virtues: it is very efficient, it can be carried out directly using the products of the first-strand reaction, which need no further treatment or purification, and it eliminates the need to use nuclease S1 to cleave the single-stranded hairpin loop in the ds-cDNA – a reaction that is difficult to control and frequently results in a great loss of cDNA. Most cDNA libraries are now constructed using a replacement reaction to synthesize the second strand of cDNA.

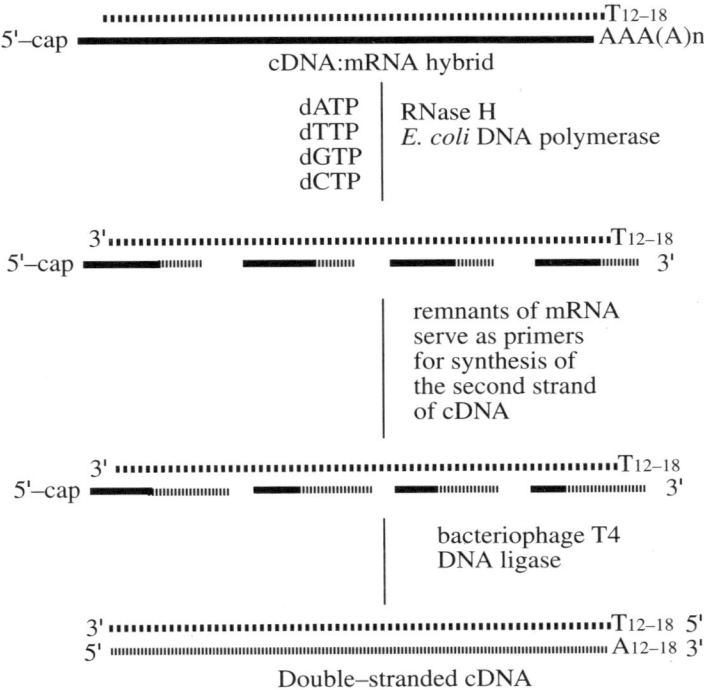

Figure 17.9 *Replacement synthesis of double-stranded cDNA*

Materials

First-strand product; α–^{32}PdCTP (dry down under vacuum); *E. coli* DNA polymerase I (holoenzyme); RNase H; *E. coli* DNA ligase; 0.5 M EDTA, pH 8.0; Phenol and chloroform; 4 M Ammonium acetate; 95% Ethanol kept at –20°C. Second-strand buffer: 20 µl of 1 M Tris–HCl pH 7.5, 100 µl of 1M KCl, 5 µl of 1 M MgCl$_2$, 5 µl each of 10 mM dNTPs, 355 µl water.

Procedure

1. Redissolve the first-strand product in 50 µl water.
2. Add 20 µCi of α–^{32}PdCTP diluted in 50 µl of second-strand buffer.
3. Add 25 U of DNA polymerase I, 1 U of RNase H and 10 U of DNA ligase.
4. Incubate at 12°C for 1 hour and then at 22°C for 1 hour.
5. Stop the reaction with 2 µl of 0.5 M EDTA.
6. Extract with 1 volume of phenol/chloroform.
7. Add an equal volume of 4 M ammonium acetate and 2 volumes of –20°C ethanol.
8. Incubate for 30 min. in dry ice-ethanol bath and spin down precipitate in microfuge for 30 min. at 4°C. Re-suspend the pellet in 50 µl of water; repeat steps 7 and 8, and vacuum dry. Re-suspend the pellet in 20 µl water and store at –20°C.

Method-2: Second-strand cDNA synthesis by self-priming

For reasons that are not fully understood, the 3' termini of single-stranded cDNAs (sscDNA) are capable of forming hairpin structures that can be used to prime the synthesis of the second strand of cDNA by the Klenow fragment of *E. coli* DNA polymerase I or reverse transcriptase (Figure 17.10). To allow these structures to form, it is necessary to denature the DNA–RNA hybrid molecules that are the products of the first-strand reaction. The product of the self-primed synthesis of the second strand is a double stranded cDNA (dscDNA) molecule closed at the terminus corresponding to the 5' terminus of the mRNA by a hairpin loop. This loop is then digested with the single-strand-specific nuclease S1 to give the dscDNA molecule. The self-primed synthesis of the second strand of cDNA is at best a poorly controlled reaction. The conditions used to achieve synthesis of full-length second-strand cDNA depend on the particular polymerase used.

Method-3: The heteroduplex formed in the first-strand reaction

This method is very simple and the full length of dsDNA can be synthesized. RNase H makes nicks in the RNA strand of an RNA–DNA hybrid, and these nicks can provide starting points (RNA primers) for DNA polymerase-catalyzed nick translation. *E. coli* DNA ligase joins the nicks left in the second DNA strand.

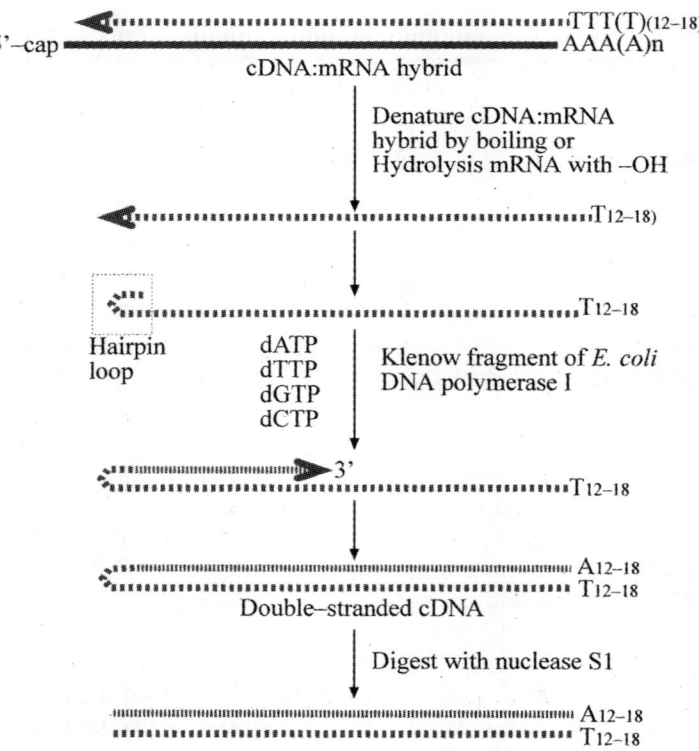

Figure 17.10 *Synthesis of double-stranded cDNA by the self-priming method*

Method-4: Primed synthesis of the second-strand of cDNA

To clone the 5'-terminal sequences of eukaryotic mRNA with high efficiency, several more complicated methods are available. For example, after completion of first-strand synthesis, terminal transferase can be used to add homopolymeric tails of dC residues to free 3'–hydroxyl groups, which serve as a primer for synthesis of a full-length second strand of cDNA. This process can be improved further by using synthetic primer–adapters that carry both homopolymeric tails for priming the synthesis of the first and second strands of cDNA and restriction sites for cloning into plasmids and bacteriophage lambda vectors. The use of primer-adapters has three advantages. (i) The number of steps involved in the synthesis and cloning of cDNA is reduced. For example, the cDNA–mRNA hybrid is attached to a bacteriophage lambda vector by synthetic adapters before synthesis of the second-strand of cDNA. This eliminates several steps that are required when dscDNA is cloned by the addition of synthetic linkers. The increased efficiency of priming of second-strand synthesis yields libraries that contain a comparatively high proportion of full-length cDNA molecules. This method can be adapted to allow the amplification of first-strand cDNA by the PCR. However, there are also some disadvantages. All the clones generated by homopolymeric priming of second-strand synthesis carry a tract of dG:dC residues immediately upstream of the sequences corresponding to the sequences of the mRNA template. The presence of these additional sequences may inhibit transcription of DNA both *in vivo* and *in vitro*. Furthermore, they may form a barrier to the Klenow fragment of *E. coli* DNA polymerase I during DNA sequencing by the Sanger's dideoxy chain-termination method, requiring that reverse transcriptase is used instead and the conditions adjusted accordingly.

5. Preparation of cDNA for ligation into a vector

In this protocol, three steps must be carried out before the insert cDNA can be ligated into a cloning vector such as phage λgt11. First, any internal *Eco*RI sites in the insert must be methylated to prevent their cleavage when the *Eco*RI linkers are cut. Second, any ragged ends on the cDNA must be blunt-ended by filling them in, using T4 DNA polymerase. Third, *Eco*RI linkers must be added to the insert by blunt-end ligation using T4 DNA ligase. These preparatory steps are applicable to any DNA fragments prior to their insertion into λgt11 by linker addition.

The bacteriophage lambda vectors commonly used for construction of cDNA libraries are λgt10 and λgt11. λgt10 is used to construct libraries that are to be screened only by nucleic acid probes; λgt11 is an expression vector that is used to construct libraries that are to be screened with immunological probes to isolate DNA sequences that code for specific antigens.

Method-1: Methylation reaction

Materials

*Eco*RI methylase; SAM: S–adenosyl methionine, a donor of methyl groups.

2X Methylase buffer: 40 µM SAM, 200 mM Tris pH 8.0, 5 mM EDTA, 100 mM $MgCl_2$;

10X One-For-All buffer: 100 mM Tris–acetate pH 7.6, 100 mM Mg acetate, 500 mM K acetate.

Procedure

1. The amount of methylase required can be tested by using DNA with a known number of EcoRI sites e.g. wild type lambda. A suggested series of test reactions are outlined below:

DNA (1 µg/µl)	0.5	0.5	0.5	0.5	µl
EcoRI methylase (20 U/µl)	0	0.1	0.5	1	µl
2X methylase buffer	5	5	5	6	µl
Water	4.5	4.4	4	3.5	µl

   Incubate the reactions at 37°C for 1 hour.
2. Stop the reactions by heating to 65°C for 10 min.
3. Extract the bulk cDNA reaction once with phenol/chloroform and once with chloroform. Add 0.1 volume of 3 M sodium acetate (pH 5.2) and two volumes of ethanol. Recover the precipitated cDNA by centrifugation at 12,000g for 15 min. at 4°C. Add 200 µl of 70% ethanol and re-centrifuge. Remove the ethanol by gentle aspiration, dry the pellet in the air, and then redissolve the DNA in 8 µl of TE (pH 7.6).
4. Samples must now be digested with EcoR I:

DNA	8 µl
10X one–For–All	3 µl
EcoRI	5 µl

   Incubate for 1 hour at 37°C. Heat-inactivate and run on a mini-gel. Use uncut DNA as control, and find the minimum units of methylase required to prevent restriction site (so giving the profile of uncut control). Calculate the number of units required per site per µg, and use this to methylate the insert.
5. Re-suspend the cDNA in, or dilute it to 10 µl, and add 10 µl 2X methylase buffer. Add the required number of units of methylase, and incubate at 37°C for 1 hour.
6. Heat-inactivate the methylase at 70°C for 10 min.

Fill-in reaction

Materials

T4 polymerase buffer: 100 mM of $MgCl_2$, 150 mM of DTT, 250 mM of KCl, 1 mM each of dNTPs.

T4 DNA polymerase; 0.5 M EDTA pH 8.0.

Procedure

1. Mix: 20 µl of cDNA (from methylase reaction), 10 µl of T4 polymerase buffer, 10 U of T4 polymerase.
2. Incubate at 37°C for 10 min.
3. Add 65 µl water and 5 µl 0.5 M EDTA.
4. Extract with phenol/chloroform.
5. Ethanol-precipitate and re-suspend in 5 µl water. Store at –20°C.

Method-2: Addition of synthetic DNA linkers

Synthetic linkers containing one or more restriction sites provide an alternative method for joining ds-cDNA to both plasmid and bacteriophage lambda vectors. The ds-cDNA, generated by one of the methods described earlier, is treated with bacteriophage T4 DNA polymerase or *E. coli* DNA polymerase I – enzymes that remove protruding single-stranded 3′ termini with their 3′→5′ exonucleolytic activities and fill in recessed 3′–hydroxyl termini with their polymerizing activities. The combination of these activities therefore generates blunt-ended cDNA molecules, which are then incubated with a very large molar excess of linker molecules in the presence of bacteriophage T4-DNA ligase, an enzyme that catalyzes the ligation of blunt-ended DNA molecules. The products of the reaction are cDNA molecules carrying polylinker sequences at their termini. These molecules are cleaved at a restriction site in the linker, purified, and ligated to a vector that has been cleaved with a restriction enzyme that generates cohesive termini compatible with those of the linker (Figures 17.11 and 17.12).

6. Checking expression libraries for recombinants

Phage λgt11, for example, has a single *Eco*RI site in the *lac* Z gene into which one can introduce foreign DNA. The inserted DNA results in the loss of the enzymatic activity of β-galactosidase and recombinants can be identified by their inability to convert a chromogenic substrate 5-bromo-4-chloro-3-indolyl-β-D-galactopyranoside (X-Gal). This test is carried out in the presence of isopropylthiogalactoside (IPTG), which induces β-galactosidase by binding the repressor (*lac* I gene product) away from the *lac* Z operon, allowing the *lac* Z gene to be transcribed. Non-recombinants expressing functional enzyme produce blue plaques under these conditions, whereas recombinants, under the same conditions, produce clear or colorless plaques. On this basis, the proportion of recombinants in an expression library can be calculated.

Materials

    E. coli Y1088 host cells; Ampicillin (250 mg/ml); Luria–Bertani (LB) culture medium, and 20% maltose; 90 mm LB-ampicillin plates, top agar; 10 mM $MgSO_4$; SM (phage storage buffer); 1 M IPTG (dissolve 1 g in 4.2 ml water); 2% (w/v) X-gal in dimethylformamide.

Procedure

1. Prepare phage-competent cells. Pick one colony from a fresh Y1088 stock plate and transfer to 10 ml LB medium plus 100 µg/ml ampicillin. Culture overnight at 37°C on an orbital shaker (200 rev/min).
2. Transfer 2 ml of the overnight culture to 200 ml LB plus 0.2% maltose. Culture at 30°C on the orbital shaker until OD_{600} nm reaches 0.6 to 0.9.
3. Spin cells at 4,000 rpm for 10 min. at 4°C. Re-suspend pellet in 1/20 original volume 10 mM $MgSO_4$. Store at 4°C.
4. For each 90 mm plate mix 100 µl of Y1088 competent cells with various dilutions of phage, e.g. 10^{-2} to 10^{-5}. Mix in a 1.5 ml microfuge tube and incubate at 37°C for 20 min.
5. Aliquot 4 ml lots of top agar into sterile tubes in a water bath (48°C) or constant temperature block. Add 10 µl IPTG and 35 µl of X–gal to each.

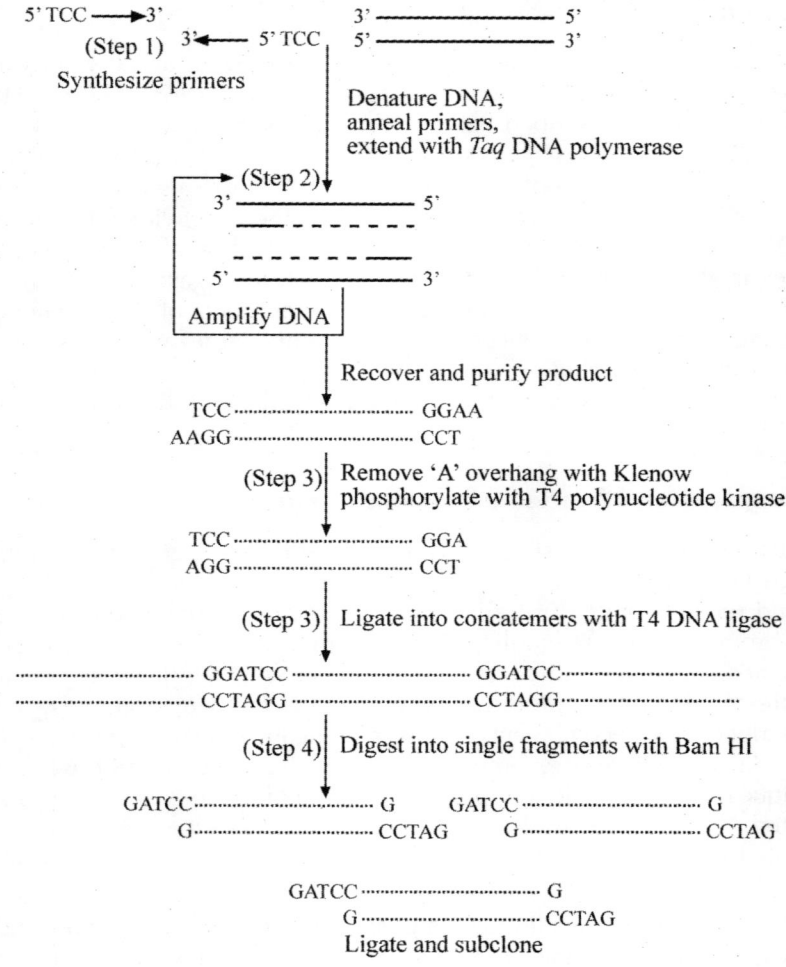

Figure 17.11 *Generation of restriction enzyme half-sites. Fragments generated by PCR using primers containing a three 3′ nucleotides of a six base recognition site at their 5′ ends*

6. Allow to cool briefly. Add cells and mix by swirling. Pour onto pre-warmed LB plates and allow to set for 10 min. at room temperature.
7. Invert and incubate overnight at 42°C.
8. Divide the plates into sections, count the clear plaques (=recombinants), and blue plaques (=wild type) on each plate.

7. Amplification of bacteriophage lambda and its derivatives

Amplification of bacteriophage λgt11 and its derivatives and λZAPII libraries: cDNA libraries constructed in expression vectors such as λgt11, λgt18, λgt20 and λgt22 should be

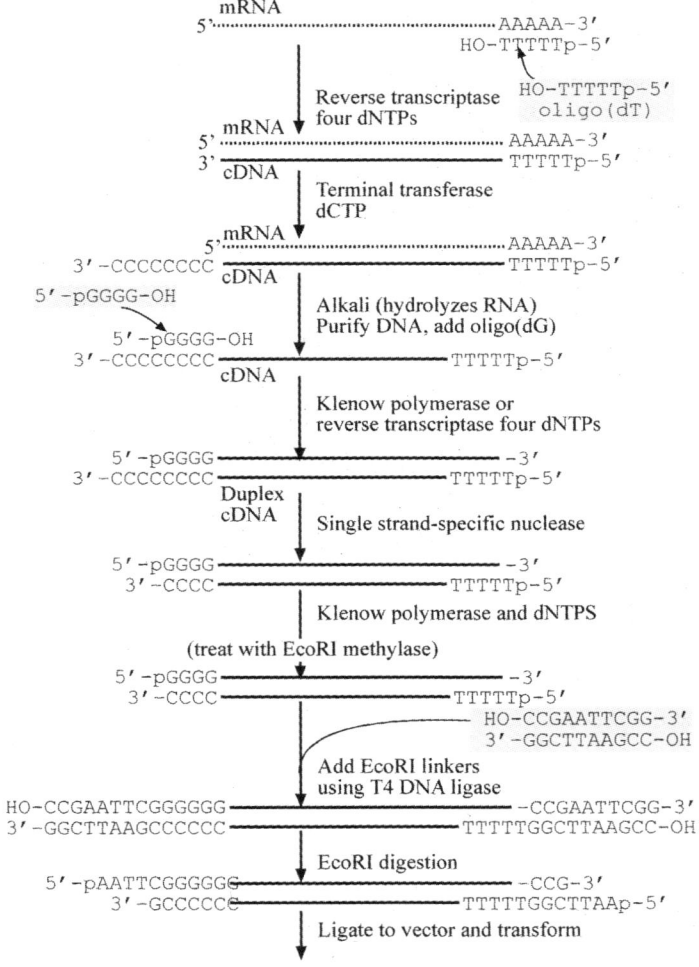

Figure 17.12 cDNA cloning. First and second strand synthesis and end preparation by adding linker to duplex cDNA

amplified on *E. coli* Y1090*hsd*R and λZAP vectors on *E. coli* strain BB4. These strains are defective for host-controlled restriction but carry an active methylation system.

1. Mix 10^5 bacteriophages with 600 µl of plating bacteria of *E. coli* strain Y1090*hsd*R (or strain BB4 for λZAP). After incubating for 20 min. at 37°C, plate the infected culture on a 150-mm Petri dish containing NZCYM agar. Incubate for 12 hours at 42°C.
2. Prepare and store the library as described before.
3. Titrate the λgt library on strain Y1090*hsd*R (or the λZAP library on strain BB4) on LB agar plates using 3 ml of top agarose containing 40 µl of a stock solution of X-gal (20 mg/ml in DMSO) and 4 µl of a solution of IPTG (200 mg/ml). The ratio of blue to colorless plaques is a measure of the proportion of the library that consists of nonrecombinant bacteriophages.

8. Problems commonly encountered with cDNA cloning

1. The first strand of cDNA is too short or the yield is too low. Although these problems can result from any of the components used, more frequently they reflect the poor quality of the mRNA template. Re-check the integrity of the mRNA and other components.
2. The library contains shorter inserts of cDNA than desired. This problem is generally caused by (a) inefficiency in the synthesis of the second strand of cDNA, resulting in ds–cDNA that contain gaps or traces of mRNA, (b) small pieces of DNA contaminating the final preparation of cDNA. The best course of action is to digest with appropriate restriction enzyme or re-fractionate.
3. The cDNA inserts contain more sites for *Not*I, *Sal*I, or *Eco*RI than expected. If these sites occur more frequently than expected, it is likely that tandem arrays of cDNA molecules have been cloned into the vector. This problem can be avoided by decreasing the ratio of inserts to arms in the ligation mixtures.
4. The frequency of *Eco*RI sites within the cDNA inserts is less than 1 site per 4 kb. In this case, it is usually necessary to check the methylation reaction in a series of pilot reactions and then repeat the methylation afresh.
5. No cDNA inserts can be excised from the bacteriophage lambda vector by digestion with the appropriate restriction enzyme. This problem is almost always caused by inefficiency while adding synthetic linkers to the ends of the ds-cDNA. Re-treat the ends of the ds-cDNA with bacteriophage T4 DNA polymerase, and repeat the large-scale ligation reaction with a mixture of labeled and unlabeled linkers. Analyze an aliquot of the reaction by polyacrylamide gel electrophoresis to check that the linkers have formed a series of ligation products that exhibit a ladder-like distribution.

V PCR VERSUS GENOMIC AND cDNA LIBRARIES

In this chapter so far, the construction of genomic and cDNA libraries have been described. Efficient methods exist for ensuring that large, representative libraries can be constructed. However, there are alternative routes to obtaining specific genomic or cDNA libraries, which are based upon PCR. For many applications, a PCR-based approach is quicker and simpler than library construction and screening. If a suitable library already exists, a PCR-based approach may be attractive (For more details refer to Chapter 12).

VI SCREENING GENE LIBRARIES

Once a cDNA or genomic library has been prepared, the next task is the identification of the specific fragment of interest. In many cases, this may be more problematic than the library construction itself, since many hundreds of thousands of clones may be in the library. One clone containing the desired fragment needs to be isolated from the library and therefore a number of techniques have been developed.

A. METHODS OF SCREENING

Three methods are generally used to screen DNA libraries for clones of interest: (1) Nucleic acid hybridization; (2) Immunological detection of specific antigens; (3) Sib selection either by hybrid selection and translation of mRNA or by production of biologically active molecules. Most cloning projects today are aimed at isolating cDNA corresponding to rare mRNAs and therefore require screening of large numbers of recombinant clones. This can be carried out effectively with only two types of reagents: antibodies and nucleic acid probes. Nucleic acid probes are the most preferred because they can be used under a variety of stringencies that minimize the chance of undesirable cross-reactions and detect all the cDNA clones.

1. Nucleic acid hybridization

This is the most commonly used and most reliable method of screening cDNA libraries for clones of interest. Screening by nucleic acid hybridization allows extremely large numbers of clones to be analyzed simultaneously and rapidly, does not require cDNA clones be full-length, nor does it require that an antigenically or biologically active product be synthesized in the host cell.

Homologous probes: Homologous probes contain at least part of the exact nucleic acid sequence of the desired cDNA clone. They are used in a variety of circumstances, as for example, when a partial clone of an existing cDNA is used to isolate a full-length clone from a cDNA library. Usually, a fragment derived from one end or the other of the existing clone is isolated, radiolabeled *in vitro*, and used to probe a library.

Partially homologous probes: Partially homologous probes are used to detect cDNA clones that are related, but not identical, to the probe sequences. If neither antibody nor nucleic acid probes are available, a number of alternative strategies can be considered. For example, if the same gene has already been cloned from another species or if a related gene has been cloned from the same species, it would be worthwhile carrying out a series of trial experiments to determine whether there is sufficient conservation of nucleic acid sequence to allow the screening of a cDNA library by hybridization. This is most easily accomplished by performing a series of Southern and Northern hybridizations at different stringencies.

Total cDNA probes: Total cDNA probes are prepared by uniform incorporation of radiolabeled nucleotides with reverse transcriptase or end-labeling of total or fractionated poly(A)$^+$ mRNA. They can be used to screen libraries of cDNA for specific clones if the cDNA clones of interest correspond to screen libraries of cDNA species present in the initial population at a frequency of at least 1 in 200. It is not possible to detect cDNA clones homologous to species that are represented rarely in the mRNA preparation.

Subtracted cDNA probes: Subtracted cDNA probes are often used to probe cDNA libraries for clones that correspond to mRNAs which are depleted of sequences present in a second type of mRNA by subtractive hybridization. Typically, the cDNA is hybridized two or three times in succession to a 20-fold excess of the second mRNA, and the cDNA–mRNA hybrids are this time recovered by chromatography on hydroxyapatite. The unhybridized cDNA is then annealed to a 100-fold excess of the mRNA preparation from which it was originally synthesized, and the resulting cDNA–mRNA hybrids are this time recovered by alkaline hydrolysis. The cDNA, which is highly enriched for sequences specific to the original mRNA, is used to probe a cDNA library for clones homologous to these sequences.

Subtracted cDNA probes are particularly valuable when there are very few differences between the two starting mRNA preparations (<2%).

A slightly different approach is used when two preparations of mRNA share sequences that are present at different concentrations. Examples of such sib pairs might be mRNAs extracted from control cells and cells that have been exposed to heat shock, drugs, or hormones. cDNAs corresponding to mRNAs whose expression is altered by such treatments can often be detected by **differential hybridization**.

Synthetic oligonucleotide probes: Synthetic oligonucleotide probes are tracts of dNTPs of defined sequence that have been synthesized *in vitro*. The sequences of these probes are deduced using the genetic code, from short regions of the known amino acid sequence of the protein of interest. Because of the degeneracy of the genetic code, it is very unlikely that a given sequence of amino acids will be specified by a predictable single oligonucleotide of defined sequence. Three solutions have been found to this problem: 1) a family of oligonucleotides can be synthesized containing all possible sequences that can code for a given sequence of amino acids; 2) a longer (40–60 base) oligonucleotide of unique sequence can be synthesized using the most commonly used codon for each amino acid; 3) an oligonucleotide can be synthesized that contains a base such as inosine at positions of high potential degeneracy. Inosine can pair with all four conventional bases without seriously compromising the stability of the resulting hybrid.

2. Colony and plaque hybridization

Colony hybridization is one method used to identify a particular DNA fragment from a plasmid gene library. A large number of clones are grown to form colonies on one or more plates, and these are then replica plated onto a membrane placed on solid agar medium. Nutrients diffuse through the membranes and allow colonies to grow on them. The colonies are then lysed, and the liberated DNA is denatured and bound to the membrane, so that the pattern of colonies is replaced by an identical pattern of bound DNA. The membrane is then incubated with a pre hybridization mix containing non-labeled non-specific DNA such as salmon sperm DNA to block non-specific sites. Following this, a denatured labeled gene probe is added. Under specific hybridization conditions, the probe will bind to cloned fragments containing at least part of its corresponding gene. The membrane is then washed to remove any unbound probe and the binding detected by autoradiography of the membrane. By comparing the patterns on the autoradiograph with the original plates of colonies, those colonies that contain the desired sequence can be identified and isolated for further analysis. A similar type of procedure is used to identify the desired genes cloned into bacteriophage vectors. In this case, the process is termed plaque hybridization. It is the DNA contained in the bacteriophage particles found in each plaque that is immobilized onto the nylon membrane. This is then probed with an appropriately labeled complementary gene probe and the detection undertaken as for colony hybridization (for more details refer to Chapter 19).

3. Immunological detection of specific antigens

The cDNA libraries constructed in expression vectors can be screened with an antibody directed against the protein of interest (Figure 17.13). Antibodies that specifically cross-react with a particular protein of interest are often available. If so, these antibodies can be used to screen a cDNA expression library to identify and isolate cDNA clones encoding the

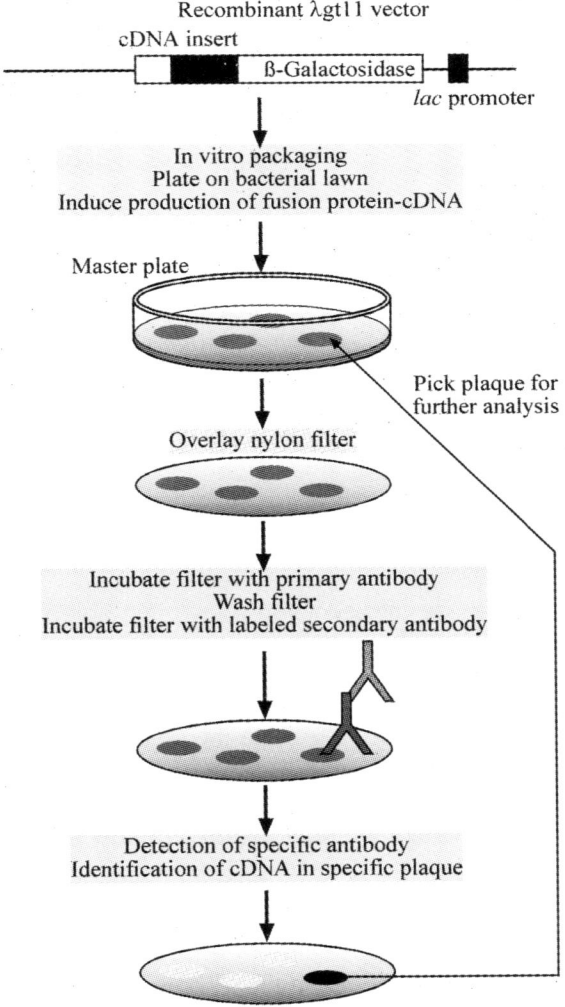

Figure 17.13 *Screening of cDNA libraries in expression vector λgt11*

protein. The cDNA library is introduced into the host bacteria, which are plated out and grown overnight, as in the colony hybridization scheme previously described. Nitrocellulose filters are placed on the plates to obtain a replica of the bacterial colonies. The filter is then incubated under conditions that induce protein synthesis from the cloned cDNA inserts, and the cells are treated to release the synthesized protein. The synthesized protein binds tightly to the filter, which can then be incubated with the specific antibody. Binding the antibody to its unique protein product reveals the position of any cDNA clones expressing the protein, and these clones can be recovered from the original plate. The key to success with this method lies in the quality of the antibody. It is essential that the antibody efficiently recognizes the denatured protein.

4. Sib selection of cDNA clones

Two methods of screening are based on the concept of dividing a large cDNA library into a manageable number of pools, each consisting of between 10 and 100 colonies. These pools are then tested for the sequence of interest. After a pool is identified that scores positively, it is subdivided into successively smaller and smaller pools, each of which is re-tested until the cDNA clone of interest is isolated.

Hybrid selection: In this method, cDNA clones carrying sequences complementary to specific mRNAs are denatured, immobilized on a solid matrix, and hybridized to preparations of mRNA. The mRNA–cDNA hybrids are then heated to release the mRNA, which is then translated in a cell-free protein-synthesizing system. The translation products are identified by immunoprecipitation and/or by SDS-polyacrylamide gel electrophoresis.

Production of biologically active molecules: A few groups have used sib selection and analysis to screen cDNA libraries for the production of biologically active protein in cells, e.g., human lymphokines colony-stimulating factor and interleukin-3). This approach is usually undertaken when no other methods of screening are available and when the protein product is small enough to give reasonable assurance that the cDNA library will contain full-length clones.

5. Identifying specific DNA sequences in libraries using heterologous probes

It is also possible to identify specific genes in a genomic library by using clones of equivalent genes from other organisms as probes. Such probes are called heterologous probes and their effectiveness depends upon a good degree of homology between the probes and the genes. The greatest success with this approach has come by using either highly conserved genes or probes from a species closely related to the organism from which a particular gene is to be isolated.

6. Identifying genes in library by complementation of mutants

For those organisms in which there are well-defined mutations, it is possible to clone genes by complementation of those mutations, as for example, cloning a yeast gene by complementation from the wild type yeast strain in a replicative shuttle vector such as YEp24. The library is used to transform a host yeast strain carrying a mutation to enable transformants to be selected. Let us consider the cloning of the ARG1 gene, the wild type gene for an enzyme needed for arginine biosynthesis (Figure 17.14) by complementation of an arg1 mutation. A yeast strain carrying the arg1 mutation would have an inactive enzyme for arginine biosynthesis and, hence, a growth requirement for arginine. A genomic library is made using DNA isolated from a wild type (ARG1) yeast strain. When a population of ura3 arg1 yeast cells is transformed with the YEp24 genomic library, some cells will receive plasmids containing the normal (ARG1) gene for the arginine biosynthesis enzyme. The plasmid's ARG1 gene will be transcribed, and the resultant mRNA will be translated to produce a normal, functional enzyme for arginine biosynthesis. The cell will be able to grow on a minimal medium (absence of arginine) despite the presence of a defective arg1 gene on the cell's chromosomes. The ARG1 gene is said to overcome the functional defect of the arg1 mutation by complementation of that mutation. The plasmid is then isolated from the cells and the cloned gene is characterized.

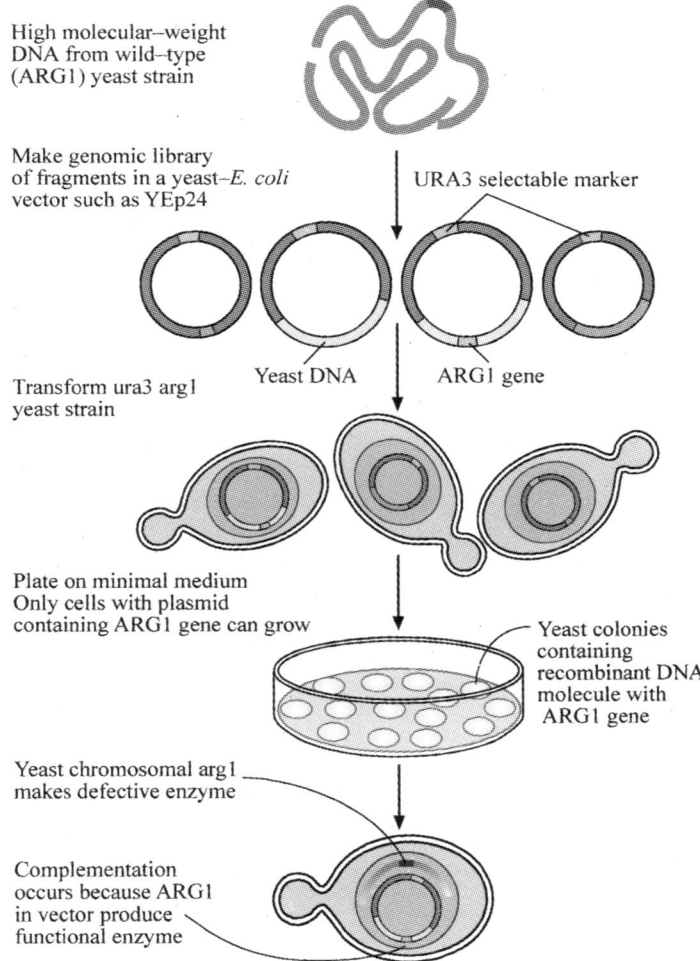

Figure 17.14 *Cloning a gene by complementation of mutations: the cloning of the yeast ARG1 gene. (Redrawn from Genetics by Russell PJ, Benjamin/Cummings, 1998)*

7. Identifying genes or cDNAs in libraries using oligonucleotide probes

This method requires that at least some of the amino acid sequences must be known for the protein encoded by the gene. In that case, it may be possible for consensus sequence to be determined from previously cloned versions of the gene that are available in *GenBank*. Then since the genetic code is universal, oligonucleotides about 20 nucleotides long can be designed which, if translated, would give the known amino acid sequence. Because of the degeneracy of the genetic code – up to six different codons can specify a given amino acid, all of which could encode the targeted amino acid sequence. These mixed oligonucleotides are labeled and used as probes to search the libraries with the hope that at least one of the oligonucleotides will detect the gene or cDNA of interest.

VII CONSTRUCTION OF EXPRESSION LIBRARIES

Libraries of gene sequences cloned into vectors in *E. coli* which permit expression of the products encoded by the inserted sequences, provide a primary source from which tissue specific proteins can be identified with antibodies and selected for analysis. The two sources of genetic material for expression libraries are cell/tissue specific RNA, which will reflect only genes activated in the stage and tissue from which the RNA is taken, and genomic DNA, which include both non-coding and repetitive DNA.

A. EXPRESSION VECTORS

Expression vectors are engineered so that any cloned insert can be transcribed into RNA, and in many instances, even translated into protein. The cDNA expression libraries can be constructed in specially designed vectors derived from either plasmids or bacteriophage lambda (refer to Chapter 16). Proteins encoded by the various cDNA clones within such expression libraries can be synthesized in the host cells, and if suitable assays are available to identify a particular protein uniquely, its corresponding cDNA clone can be identified and isolated. Expression vectors designed for RNA expression or for protein expression, or for both, are available. However, if we are interested in high expression of our cloned gene, specialized expression vectors are needed. Bacterial expression vectors typically have two elements that are required for active gene expression; a **strong promoter** and a **ribosome-binding** site near an initiating ATG codon.

B. CONSTRUCTION OF GENOMIC EXPRESSION LIBRARIES

The cDNA strands are generally short enough to be inserted into either phage or plasmid vectors without further treatment. Genomic DNA, however, must be cleaved to a suitable length, by physical disruption such as sonication, or by enzyme digestion. In a library made from DNA fragments with identical restriction sites at each end and cloned into a single site in the vector, only one in six of the copies of a particular gene will be in frame. This must be taken into account when calculating how many recombinants need to be screen based on a measure of the abundance of the DNA or RNA species of interest. A gDNA expression library, however, is designed to give an equal probability to all the genome to be expressed, and a significantly greater number of recombinants will have to be screened to obtain the clone of interest. This is especially true of organisms with a large genome size ($>5 \times 10^8$ bp). Genomic libraries, however, are not biased by the stage-specificity of expression of genes, and are thus a source of expression clones for an antigen from a stage intractable to cDNA cloning.

C. RNA EXPRESSION

A vector for *in vitro* expression of DNA inserts as RNA transcripts can be constructed by putting a highly efficient promoter adjacent to a versatile cloning site. Linearized recombinant vector DNA is transcribed *in vitro* using SP6 RNA polymerase. Large amounts of RNA product can be obtained in this manner, if radioactive ribonucleotides are used as substrates, and labeled RNA molecules useful as probes are made.

D. PROTEIN EXPRESSION

The cDNA of eukaryotic genes are uninterrupted copies of the exons (lack introns) of expressed genes. It is feasible to express these cDNA versions in prokaryotic hosts that otherwise lack the capacity to process the complex primary transcripts of eukaryotic genes. To express an eukaryotic protein in *E. coli*, the eukaryotic cDNA must be cloned in an expression vector that contains regulatory signals for both transcription and translation. Accordingly, a promoter, where RNA polymerase will initiate transcription as well as a ribosome binding site to facilitate translation are engineered into the vector just upstream from the restriction site for inserting foreign DNA. The AUG initiation codon that specifies the first amino acid in the protein (the translation start site) is contributed by the insert.

The main function of an expression vector is to yield the product of a gene in high amounts. Therefore, expression vectors are ordinarily equipped with very strong promoters to get more RNA and protein product. Strong promoters can drive the synthesis of foreign proteins to levels equal to 30% or more of total *E. coli* cellular protein. An example is the hybrid promoter, *Ptac*, which was created by fusing part of the promoter for the *E. coli* genes encoding the enzymes of lactose metabolism (the *lac* promoter) with part of the promoter for the genes encoding the enzymes of tryptophan biosynthesis (the *trp* promoter). In cells carrying *Ptac* expression vectors, the *Ptac* promoter is not induced to drive transcription of the foreign insert until the cells are exposed to inducers that lead to its activation. Analogs of lactose (a β–galactoside) such as IPTG, are excellent inducers of *Ptac*. Thus, expression of the foreign protein is easily controlled. The bacterial production of valuable eukaryotic proteins represents one of the most important uses of recombinant DNA technology.

The analogous system for expression of foreign genes in eukaryotic cells include vectors carrying promoter elements derived from mammalian viruses, such as simian virus 40 (SV40), the Epstein–Barr virus, and the human cytomegalovirus (CMV). A system gaining widespread use for high level expression of foreign genes uses insect cells infected with the baculovirus expression vector. In engineered baculovirus vectors, the foreign gene is cloned downstream of the promoter for polyhedrin, a major viral-encoded structural protein, and the recombinant vector is incorporated into insect cells grown in culture. Expression from the polyhedrin promoter can lead to accumulation of the foreign gene product to levels as high as 500 mg/L.

E. IDENTIFYING A SPECIFIC CLONE BY A PCR

A relatively new technique called polymerase chain reaction (**PCR**) can also yield a DNA fragment for cloning, and can also be very useful for detection of DNA fragments in a clone (for more details refer to Chapter 12).

VIII SUGGESTED READING

Allen JB, Walberg MW, Edwards MC and Elledge SJ (1995) Finding prospective partners in the library—the 2-hybrid system and phage display find a match, *Trends Biochem. Sci.* **20**:511–516.

Anderson S (1981) Shotgun DNA sequencing using cloned DNase I generated fragments, *Nucl. Acids Res.*, **9**:3015–3027.

Berger SL and Kimmel AR (Eds) (1987) *Guide to Molecular Cloning Techniques* (Methods Enzymol. vol. 152). Academic Press, San Diego.

Brown TA (Ed.) (1991) *DNA Cloning: Practical Approach* (2nd Edn) (4 volumes). IRL Press, Oxford.

Frischauf AM, Lehrach H, Poutska A and Murray N (1983) Lambda replacement vectors containing polylinker sequences. *J Mol Biol* **170**:827–842.

Goman M, Langsley G, Hyde JE, Yankovsky NK, Zolg JW and Scaife JG (1982) The establishment of genomic DNA libraries for the hyman malaria parasite *Plasmodium falciparum* and identification of individual clones by hybridization, *Mol Biochem parasitol*, **5**:391–400.

Kimmel AR and Berger SL (1987) Preparation of cDNA and the generation of cDNA libraries: Overview, *Methods Enzymol.* **152**:307–316.

Russell RJ (1998) *Genetics*, (5th Edn), The Benjamin/Cummings Publishing Company, Inc. Menlo Park, CA; New York; Sydney; Madrid; Amsterdam.

Sambrock J, Fritsch EF and Maniatis T (1989) *Molecular Cloning, a Laboratory Manual*, (2nd Edn), Cold Spring Harbor Laboratory Press, USA.

Watson JD, Gilman M, Witkowski J and Zoller M (1992) *Recombinant DNA*, (2nd Edn), Scientific American Books, New York.

Williams JG and Patient RK (1988) *Genetic Engineering*, IRL Press, Oxford.

Zhang JZ and Chai JH (1995) Methods for finding new genes in positional cloning, *Prog. Biochem. Biophys.* **22**:126–132.

18

Molecular Biology of Gene Transfer Systems

I INTRODUCTION

The best way to characterize a functionally cloned gene is to put it into live cells and determine what effects the encoded gene product has on defined cellular processes. Similarly, to identify DNA elements required for the transcriptional control of a cloned gene, an appropriate reporter gene must be constructed and tested *in vivo*. Both these objectives require an efficient means of transfecting experimental DNA into cells under conditions where the expression vector or reporter gene will be functional. Three basic DNA transfection strategies have been developed to deliver cloned DNA to cells. These three methods are (1) infection by vectors, (2) attack by force, and (3) stealth strategy of DNA transfection. The transfection method of DNA is the *infection strategy*, which uses recombinant eukaryotic viruses to deliver DNA to host cells. **Agrobacterium-mediated** DNA transfection in plants and two of the most efficient mammalian cell viral vectors are derived from **retroviruses** and **adenoviruses**. The *attack strategy* uses physical methods to literally force experimental DNA into cells. Three attack DNA transfection methods have been developed: **electroporation, biolistics** (gene gun), and **microinjection**. The overall efficiency of these transfection methods is relatively independent of cell type. The *stealth strategy* is based on the use of positively charged carrier molecules that are mixed with the experimental DNA *in vitro* and then applied directly to the cell culture media. These carrier DNA complexes attach themselves to cell membranes and stimulate the uptake of the "stealth" DNA molecules. The three most commonly used stealth transfection methods are **calcium phosphate precipitation, DEAE dextran-mediated** gene transfer, and **liposome-mediated** gene transfer (Figure 18.1).

The presence of the cell wall and the enormous diversity that exists in the plant species and cell types, require the development of a variety of techniques suitable to individual conditions. In this chapter, the different methods of gene transfer followed for various cell types (particularly, plant cells) are briefly discussed.

Target cells for transformation: The first step in gene transfer technology, will generally be the selection of target cells that are capable of division and also differentiation into complete organisms, for e.g. regeneration of the plant cell into a whole plant. In case of plants,

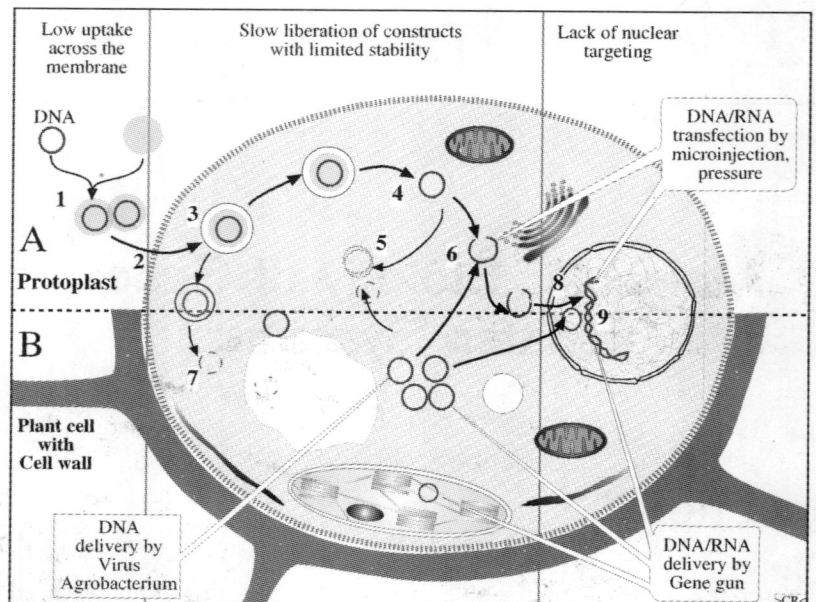

Figure 18.1 *Schematic illustration of DNA delivery pathways and the major barriers. A. A cell without intact cell wall (protoplast). B. Plant cell with cell wall, different steps involved in DNA entry into a cell and finally into a nucleus. (1) DNA-complex formation. (2) Uptake. (3) Endocytosis (endosome). (4) Escape from endosome. (5) Degradation (endosome). (6) Intracellular release. (7) Degradation (cytosol). (8) Nuclear targetting. (9) Nuclear entry and expression*

transformation without regeneration has a very limited value. In many plants, selecting the right cell type (tissue) is very difficult. Cultured cells or protoplasts are often used for transformation, since plant cells are **totipotent** and can be regenerated into whole plants *in vitro*. However, *in vitro* plant regeneration of unorganized cells imposes a degree of 'genomic stress'. This may lead to chromosomal or genetic abnormalities in regenerated plants (called **somaclonal variation**). In contrast to this, gene transfer into organized cells/tissues (meristem, anthers, pollen grains, ovary/ovule, buds, etc.,) may allow the development of a normal transgenic plant (see Chapters 32 and 33).

Methods for introducing DNA into cells (transformation): The type of plant cells employed for transformation will largely depend on the objectives of the study. In plant cells, stable transformations may be either non-integrative, or integrative. In **non-integrative stable transformation**, the transgene is maintained stably in an extra-chromosomal state. But such transformations will not pass on to the next generation. On the other hand, **integrative stable transformation** results when the transgene becomes integrated into the plant genome; these integrations are heritable.

Requirements of plant transformation technology: Three essential elements of plant transformation technology are: (i) A defined/cloned gene along with suitable elements such as promoters, enhancers, and targeting sequences for its regulated expression; (ii) a reliable method for delivery and stable integration of DNA into cells; (iii) a method to recover complete plants from transformed cells.

II VECTOR-MEDIATED DNA TRANSFORMATION

Recent advances in recombinant DNA technology, bacterial genetics, and tissue culture techniques, have made it feasible to isolate specific genes, manipulate them *in vitro*, and introduce them into various cell types. This not only opens up the exciting possibility of genetically manipulating cells, but also provides a powerful tool for studying regulation of plant gene expression and, possibly, for molecular cloning so as to regulate a specific gene expression and molecular cloning of selectable plant genes from gene libraries. Transformation, in this chapter, refers to the stable introduction of foreign genetic material into cells. There are several potential transformation techniques and DNA vehicles for the efficient introduction and replication of foreign genes in cells. These vehicles include vectors and physical methods.

A. PLANT VIRUSES AS POTENTIAL TRANSFORMATION VECTORS

Application of viruses as a natural means of gene transfer is well established for animal cells, so it is not unexpected that plant viruses are considered as candidates for plant gene vectors. In fact, a simple but economically significant example of plant genetic engineering using a viral vector already exists. One phenotype that is usually conferred on a plant by viral infection is called **cross-protection** – a plant infected by one virus usually cannot be super-infected by a second strain of a related virus. Attenuation of the virus must be a stable characteristic, so that reversion to a severe strain does not occur.

The development of plant viral vectors has proved to be very useful in plant virology and biotechnology. Over 300 plant viruses have been identified and classified into 25 different groups. Biological characteristics, which affect the choice of a potential viral-vector include host range, virulence, ease of mechanical transmission, rate of seed transmission, and the potential to carry additional genetic information. In plants, many viruses or their isolated genomes are capable of infecting intact plant tissue and spreading systematically. Consequently, relatively large amounts of plant viruses can be produced in transgenic plants. This has made them suitable for use as plant transformation vectors. But since viral genomes cannot integrate with plant genome, they can be used only for a study of transient expression of transferred genes. All the three kinds of viruses including caulimoviruses (double-stranded DNA), gemini viruses (single-stranded DNA) and tobacco mosaic virus (positive single-stranded RNA) are capable of delivering genes into intact plant tissues, where they are expressed (refer to Chapter 16, Figure 16.22). However, virus vectors cannot be used for the production of stably transformed plants, popularly described as transgenic plants.

Plant virus vectors can be used in two distinct ways. One would be a massive infection which is deleterious to the plant and would perhaps ultimately kill it, but which meanwhile leads to the expression of a foreign gene product. The alternative would be a less harmful systemic infection. This leads to more-or-less healthy plants in which the foreign protein is expressed either to alter the plant to make it a better crop plant, or to express a foreign protein which is a desirable product when purified from the plant tissues. Crop plants can produce biomass cheaply without the use of high technology. Ultimately the production of proteins for therapeutic [or other] uses may be undertaken not in expensive fermenters with microorganisms such as yeast or bacteria as the host, but in fields of genetically engineered plants. These plants could be engineered by gene-transfer systems, or by using virus vectors. At present, virus vectors have not been developed to the stage where they are widely used as vectors, but progress has been made with the only two groups of plant viruses which have DNA genomes: the caulimoviruses and geminiviruses. Most plant viruses have RNA genomes. Plant RNA viruses, which have great potential as prototype vectors are the Brome mosaic virus and the tobacco mosaic virus (TMV).

1. Cauliflower mosaic virus (CaMV)

Among all the viral vectors, CaMV is the best studied of the caulimovirus group. This is a group of spherical viruses which contain a circular double-stranded DNA genome of about 8 kb. The caulimoviruses are responsible for a number of economically important diseases of cultivated crops. A unique feature of CaMV, which makes it attractive as a vector, is the ability to infect plants and then move systematically through the host, which does away with the need for cell culture. In laboratory experiments, virus particles or DNA infect a plant when rubbed on the surface of a leaf in the presence of a small amount of abrasive compound. In infected cells, refractile, round-inclusions form, which consist of many virus particles embedded in a protein matrix. The matrix protein is virus-encoded and can account for up to 5% of the protein in the infected cells. In nature, CaMV is transmitted by aphids and the transmission depends upon a transmission factor which is also a virus-encoded protein present in infected cells as a separate protein.

CaMV genome: CaMV lends itself easily to the manipulations involved in recombinant DNA technology. Plants can be inoculated by rubbing the leaves with engineered CaMV. Several different isolates of CaMV have been sequenced. The DNA of CaMV has an unusual structure. It is about 8 kbp long and several varieties have been sequenced. The DNA exists in linear, open, circular, and twisted or knotted forms. There are three discontinuities in the duplex two-in-one strand (non-coding or non-transcribed) and one in the other (coding or transcribed). These are regions of sequence overlap. The DNA has a single *Sal*I site and *Sal*I–linearized DNA is infectious even in the absence of re-ligation *in vitro*. Recircularization occurs in the plant cell. CaMV DNA has been cloned into an *E. coli* vector and propagated in *E. coli*. When the CaMV was released from the vector as a linear DNA molecule it was found to be infectious, despite the lack of the single-strand discontinuities in the inoculating DNA. Another unusual feature of CaMV DNA is the presence of ribonucleotides covalently attached to the 5′–termini of the discontinuities. These and other observations have led to the conclusion, that CaMV replication involves reverse transcription with an RNA genomic intermediate. The replication cycle resembles that of retroviruses and the hepatitis B virus. The sequence of CaMV DNA reveals eight closely-packed reading frames. There are only two small intergenic regions, and the only non-essential genes are the two small genes II and VII. Thus, most attempts to alter the DNA by insertion or deletion have caused loss of infectivity. The CaMV gene II has been shown to be dispensable for viral replication. When this gene is replicated with a bacterial gene the resulting engineered CaMV is able to systematically infect and express the bacterial gene in inoculated plants.

CaMV as a vector: The CaMV virus as a vector can help researchers to introduce genes into plants in a way that would encourage their spread and expression. However, the CaMV envelop does not allow the DNA which it encapsidates to be substantially larger than normal. This was evident when foreign DNA was inserted at the unique *Xho*I site, which lies in the non-essential gene II. If DNA longer than a few hundred nucleotides was inserted, the infectivity was destroyed. This packaging limitation and the absence of long non-essential sequences, which can be deleted in the genome without effecting the infectivity of the virus, severely limit the capacity for foreign DNA insertion. Knowledge of CaMV has been of value in providing very strong promoters eg., (the 35S promoter and the promoter of gene VI) for driving expression of other genes in plants.

Brisson et al (1a) replaced the non-essential open reading frame II of CaMV with the dihydrofolate reductase gene and used the construct to transform turnip plants. However, the vector had a restricted host range and limited space for inserting DNA, and it was necessary to eliminate the 5' and 3' non-coding sequences. In order to circumvent these defects, pKR612B1 was constructed, containing the NPTII (neomycin phosphotransferase) gene under the control of the CaMV gene VI promoter. This construct was similar to but more versatile than pABDI and was used to transform protoplasts of *Brassica campestris* var. rapa by PEG-induced plasmid uptake. However, transformation was achieved only when the hybrid gene was supplied to protoplasts with wild type viral DNA.

2. Agroinfection with a geminivirus

Geminiviruses are attractive for vector development because members of this group infect a wide range of monocot and dicot crop plants. The virus particles are twinned (geminate) and contain one or two, circular single-stranded DNA molecules, which replicate via double-stranded intermediates in the plant nucleus. Many copies of the replicative form of a gemimivirus genome accumulate inside the nuclei of infected cells (channarayappa et al. 1992). All monocot infecting gemimiviruses such as the maize streak virus (MSV) are transmitted by leaf-hoppers. Their genomes are very small, 2.7 kb. The dicot–infecting geminiviruses, such as the cassava latent virus (CLV) and TLCV, are transmitted by the whitefly *Bemisia tabaci*, and have a two-component genome consisting of two circular single-stranded DNAs, which have only a small region of about 200 nucleotides in common.

One advantage of this group of viruses is that they replicate via a double-stranded intermediate, which would make *in vivo* manipulation in bacterial plasmids more convenient. Replacement-type vector derivatives of geminiviruses have been investigated. The coat protein gene of CLV is required for insect transmission, but is not essential for replication and infectivity of the viral DNA. It has a strong promoter. Coat protein replacement constructs have been produced, but have yet to find application. Three bacterial genes that have been inserted into the wheat dwarf virus genome have been successfully replicated and expressed after transfer into cultured cells of the cereals. This illustrates the potential of geminiviruses for serving as the basis of autonomously replicating expression units in plants.

3. Brome mosaic virus

Since a majority of plant viruses have RNA genomes, the range of potential vectors is very great. In addition, many of these RNA viruses have filamentous morphology. So it is expected that the length of a viral nucleic acid determines the size of a virus particle. Thus there should be no strict size limitation on the RNA to be packaged. Another common feature of plant viruses is that they may be multi-component. An example of this is the Brome mosaic virus. Bromoviruses are a group of plant single-stranded RNA viruses that belong to the *Bromoviridae*. Their polyhedral particles are 26 nm in diameter and have the icosahedral T–3 surface lattice symmetry, with 180 identical polypeptides. The virus infects a number of *gramineae* including barley. There are three separate genomic RNA components, each of which is packaged into a separate particle. Each of these RNAs is also mRNA and during infection, a fourth mRNA is produced which is a subgenomic derivative of RNA 3. This RNA 4 is the mRNA encoding the very abundant viral coat protein. The cloned cDNAs corresponding to RNAs 1–3 can be transcribed *in vitro* to produce transcripts, which are infectious when mixed and introduced into barely protoplasts. Only RNAs 1 and 2 are necessary for the replication and expression of the genome, so that the coat protein sequence on RNA 3 is available for manipulation, although in the absence of the coat protein, no virus particles are produced.

4. Tobacco mosaic virus

The TMV genome is a single component RNA, which is also a messenger RNA. The genome encodes at least four polypeptides. The 130 kDa and 180 kDa proteins are translated directly from the same initiation codon on the genomic RNA. The other two proteins, 30 kDa and coat protein, are translated from processed subgenomic RNAs. The 130 kDa and 180 kDa proteins are involved in viral replication, whereas the 30 kDa protein is necessary for cell-to-cell movement of virus. These three proteins are therefore probably essential for TMV propagation in whole plants. The coat protein is not essential for viral multiplication, but is necessary for long-distance spread of infection in the plant. Since the coat protein is synthesized in large amounts, and can expresses a foreign gene. Takamatsu *et al.* (1987) have modified a full-length TMV cDNA clone from which infectious TMV RNA can be transcribed *in vitro*. A bacterial CAT gene sequence was placed just downstream of the initiation codon of the coat protein gene. When *in vitro* transcripts of the recombinant TMV cDNA were inoculated into tobacco plants, CAT activity was observed in the inoculated leaves, although the infection was unable to spread systemically throughout the plant.

Viral vectors, have very little potential for production of transgenic cereals although they can be easily used for amplification of genes and gene products. The discovery that RNA virus genomes could be reverse–transcribed to yield cDNA clones that again were infective, opened up the possibility of using genetic engineering of a larger group of RNA viruses. Although it would be very difficult to use viral vectors for integrative transformation, they have invaluable potential for gene amplification and systemic spread within individual plants. Because viral vectors are limited to the host species, which they naturally infect and cannot be sexually transmitted, a more general mechanism of DNA incorporation into plant cells is needed.

5. Viroids

The smallest pathogenic agents known which have the potential to be used as a plant gene vector are viroids. They are small, 300–400 bases long, circular, single-stranded and consist of naked RNA. These non-protein coding viroids replicate in the host using host enzymes, probably host RNA polymerase II. They are mechanically transmissible, able to move throughout the sap and infect other parts of the plant. Some viroids may also be transmitted through the seed. They replicate in the nucleus and cause disease in a wide variety of tropical plants.

B. *AGROBACTERIUM*-MEDIATED GENE TRANSFER

Members of the genus *Agrobacterium* are soil-borne phytopathogenic bacteria. Four distinct species are recognized in *Agrobacterium*: *A. tumefaciens*, *A. rubi*, *A. rhizogenes*, and *A. radiobacter*.

Agrobacterium tumefaciens naturally infects many dicotyledonous and gymnosperm plants predisposed by some form of wounding. *Agrobacterium* infection causes tumorous plant growth, commonly called crown galls, by introducing DNA into the plant cells at the wound site. Crown gall is a plant tumor, a lump of undifferentiated tissue. The tumor-forming ability of *Agrobacterium* depends on the presence of a **large plasmid** (150–250 kbp) called the tumor-inducing (Ti) plasmid. Molecular analysis has revealed a small and defined region of the Ti plasmid, referred to as the T-DNA, which is transferred into the plant cell and covalently integrates into a plant chromosome by means of a process thought to be analogous to bacterial conjugation. The integration of T-DNA into plant cells is a random process, requiring the *cis*-acting **T-DNA border sequences** and *trans*-acting **virulence** (*vir*) functions encoded on both the plasmid and bacterial chromosome. T-DNA itself contains genes referred to as oncogenes, which are responsible for tumor induction. The oncogenes code for the production of indole acetic acid and zeatin riboxide, natural plant hormones (Figure 18.2). Overproduction of these hormones results in

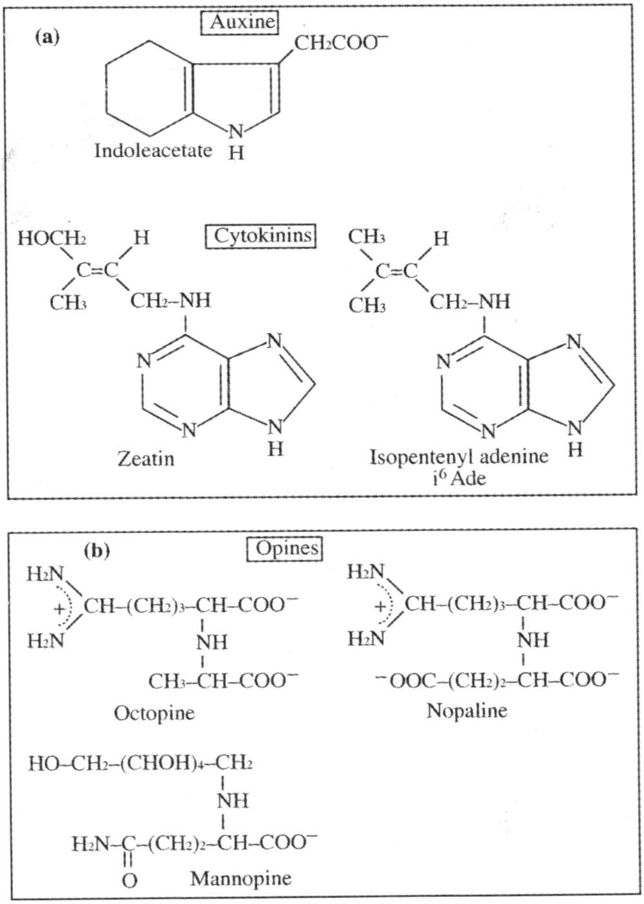

Figure 18.2 *Metabolites produced in Agrobacterium-infected plant cells. (a) Natural plant hormones, auxins and cytokinins are plant growth hormones. (b) Opines are generally derived from amino acid precursors*

the tumorous growth of plant cells. The T-DNA also encodes several genes responsible for the synthesis of compounds called opines (Figure 18.2), which are metabolic substrates for the bacteria. By transforming the opine genes to the plant cell genome, bacteria are able to subvert the plant's own metabolism.

Crown gall cells, unlike normal plant cells, fail to regenerate intact plants. But the removal of sequences in the T-DNA required for tumor promotion does not interfere with transfer of the T-DNA to the plant genome, and their replacement with chimeric genes constitutes the basis for *Agrobacterium*-mediated transformation of plants. Although the nature of T-DNA transfer is not completely understood, molecular characterization of the T-DNA sequence reveals flanking direct repeats at the boundary regions of the T-DNA (25 nucleotide imperfect border sequences). Genes that are to be introduced into plant cells must be inserted between the borders or adjacent to at least one border of the T-DNA. By cloning foreign DNA into the T-DNA of the Ti plasmid, plant molecular biologists are able to exploit the natural ability of *Agrobacterium* to transfer new DNA into the plant genome (Figure 18.3).

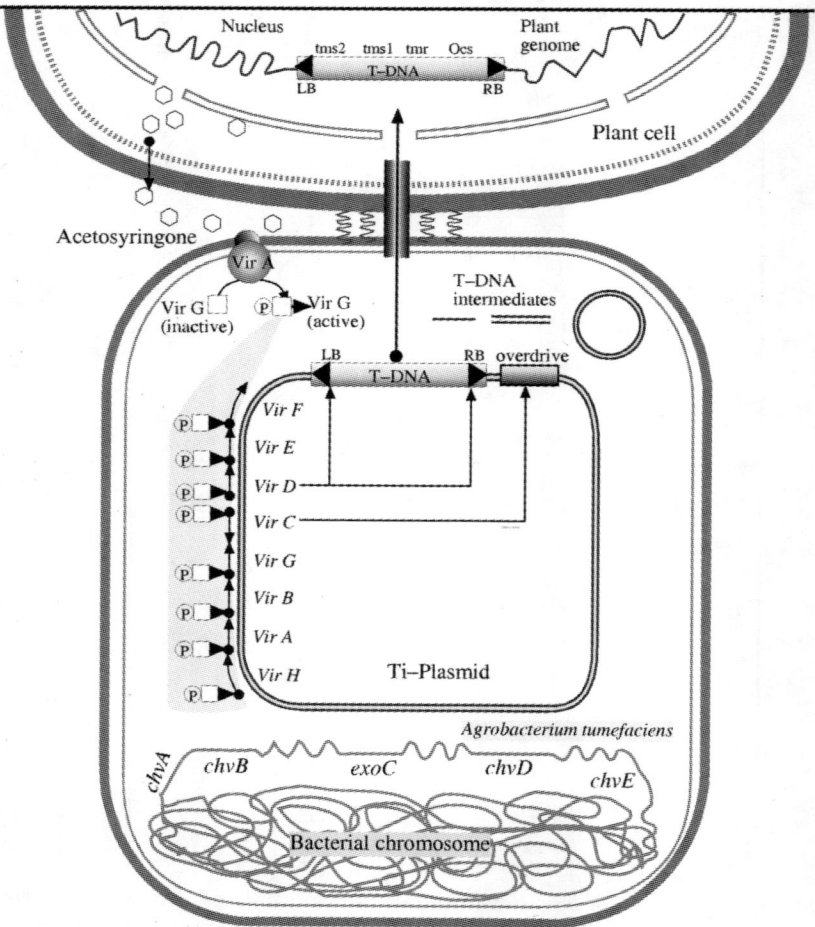

Figure 18.3 *Schematic illustration of the interaction between Agrobacterium and plant cell*

Agrobacterium-mediated transformation is simple and widely applicable, provided with a suitable protocol for regeneration of complete plants from transformed cells. However as it is a biological process, a number of factors such as strain of bacterium, host genotype, physiological conditions of host cells, environmental factors etc., affect transformation frequency.

1. Introduction of foreign genes into *agrobacterium*

The DNA that is placed between the 25-bp border sequences in *Agrobacterium* retaining a full complement of *vir* functions can be efficiently transferred and stably integrated into the plant genome. This property provides the basis for producing Ti plasmid vector systems for plant cell transformation. Since the Ti plasmid is about 200 kb in size, it is very difficult to introduce foreign genes into the T-DNA of the Ti plasmid directly by using the general recombinant DNA technique. Therefore, two methods – binary vector systems and co-integrate vector systems – have been developed for introducing gene constructs into *Agrobacterium*.

2. Co-integrative vector systems

These are derivatives of a wild-type Ti plasmid from which the T-DNA-encoded tumor genes have been replaced by the common cloning vector pBR322. In this system, both the T-DNA and the *vir* region are present on the Ti plasmid (refer to Chapter 16, Figure 16.15). Insertion of foreign DNA between T-DNA border sequences depends on a co-integration event between homologous sequences present in the cloning vector and the Ti plasmid T-DNA. For example, the pGV23850 vector comprises a Ti plasmid containing the T-DNA from which all of the oncogenic genes have been deleted and replaced by the cloning vector pBR322. Therefore, foreign DNA cloned in pBR322 derivatives can be integrated into *Agrobacterium*, harboring pGV3850 and inserted between the T-DNA border sequences by a single-homologous recombination. Since pBR322-derived vectors cannot replicate in *Agrobacterium*, exconjugants possessing the co-integrative vectors can be easily isolated by selection for antibiotic-resistant markers present in pBR322 derivatives.

Co-integrative vectors have the disadvantage that the co-integration efficiency is not high, because it relies on the frequencies of both conjugation and homologous recombination. Moreover, it is impractical to isolate the large-sized Ti plasmid, and Southern blot hybridization for the total *Agrobacterium* DNA has to be carried out to confirm that the expected co-integration has occurred.

3. Binary vector systems

The binary vector system is based on the fact that the *vir* region and the T-DNA need not be located on the same replicon for the T-DNA transfer (see Figure 18.4). **Binary vectors** consist of origins of replication activity in both *E. coli* and *A. tumefaciens* and selectable markers also functional in both bacteria; hence, they can be selected in both bacteria. These vectors can be introduced into *A. tumefaciens* from *E. coli* by conjugation, using a helper plasmid for mobilization (Tri-parental mating) or introduced directly by either transformation or electroporation. Furthermore, binary vectors contain multiple restriction sites between the 25 bp border repeats for cloning of foreign DNA and a dominant selectable marker gene fused with a plant promoter that is functional in plant cells. The host of *Agrobacterium* for plant transformation contains a helper Ti plasmid deleted of its T-DNA and provides the necessary *vir* functions to transfer the

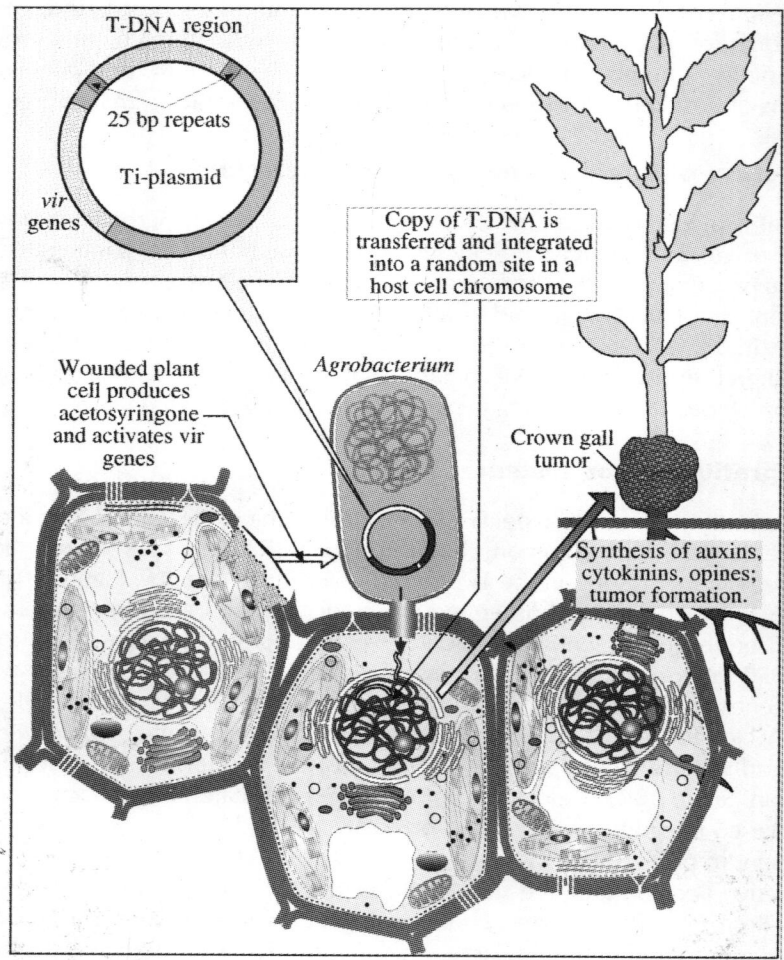

Figure 18.4 *Agrobacterium transfers T-DNA into a plant cell and induces tumor*

DNA sequences between 25 bp border repeats in the separate replicon. However, the helper Ti plasmid need not be disarmed, as the co-transformed plants have been regenerated with a fully virulent plasmid as a helper. The advantage of binary vectors is that the exconjugants are obtained with a frequency of 10^{-1} to 10^{-2}, while cointegration vectors are stabilized with a frequency of only 10^{-5} to 10^{-6}.

An important point in the use of binary vectors is the order of the border sequences, foreign DNA, and a selectable marker gene. Since the right border sequence can determine the orientation of the T-DNA transfer, it is desirable that foreign DNA is inserted between the right border sequences. A selectable marker gene; otherwise, a transgenic plant containing only a selectable marker gene have frequently been obtained because of incomplete integration of the inserted DNA between the border sequences. The major disadvantages of the binary vectors now available are that most of them are still rather large and carry a limited number of cloning sites. In addition, the stability of these plasmids is not always optimal in *Agrobacterium*, and require selection to confirm their maintenance.

4. Introduction of plasmids into *agrobacterium*

Plasmids can be introduced into *Agrobacterium* either by conjugation or by transformation (***co-transfection***). The latter technique can be applied to a few strains (e.g. C58) and is not very efficient since 1 µg of DNA may give only up to 1000 transformants. Therefore, in general, the conjugation process is used to introduce plasmids into *Agrobacterium*. As far as is known, only plasmids of the *E. coli inc* groups P, Q, and W can be stably maintained by *Agrobacterium*. *Agrobacterium* may, in addition, naturally harbor plasmids of more than five different *inc* groups called Rh-1 to *inc* Rh-5. In order to be able to shuttle plasmids between different hosts, it is necessary to have good selectable markers in the strains used.

5. Methods for transformation of plant cell by *agrobacterium* vector

The *Agrobacterium*-mediated transformation system is becoming a general method for introducing foreign DNA into plants. Improvement of tissue culture techniques for the regeneration of plants from explants or protoplasts has begun extending this system to several plant species, including economically important crops, such as rice. Gene transfer through *Agrobacterium* is achieved in the following two ways: (i) co-culture with tissue explants and (ii) *in planta* transformation. Some of the useful methods for the transformation of plant cells are described below.

(a) Co-culture with tissue explants

The appropriate gene construct is inserted within the T-region of a disarmed Ti plasmid; either a co-integrate or a binary vector. The recombinant vector, which contains a **chimeric gene** (since it contains sequences from several genes), is placed in the *Agrobacterium*, which is co-cultured with the plant cells or tissues to be transformed for about 2 days.

(i) Leaf-disk technique

Because of its simplicity and efficiency, the leaf-disk technique is now becoming the most useful method for the production of transgenic plants. This technique involves the incubation of leaf disks with a suspension of *Agrobacterium* (co-culture) followed by culturing on a medium that allows the induction of the callus, subsequently forming the shoots when placed on the appropriate medium (Figure 18.5). During the leaf disc-*Agrobacterium* co-culture, the acetosyringone released by the plant cells induces the *vir* genes, which bring about the transfer of recombinant T-DNA into many of the plant cells. The T-DNA become integrated into the plant genome, and the transgene is expressed. As a result, the transformed plant cells become resistant to kanamycin. After two days, the leaf disks are transferred onto a regeneration medium containing appropriate concentrations of kanamycin and carbencillin. Kanamycin allows only transformed plant cells to divide to form the callus and later regenerate to form shoots in about 3–4 weeks, while carbencillin kills the *Agrobacterium* cells. The shoots are separated and rooted. Mature plants can be obtained when the shoots are subcultured to form roots. The leaf-disk technique can be easily modified for use with explants other than leaves, such as stems, roots, and epidermal strips derived from flowering branches. Transgenic plants can be

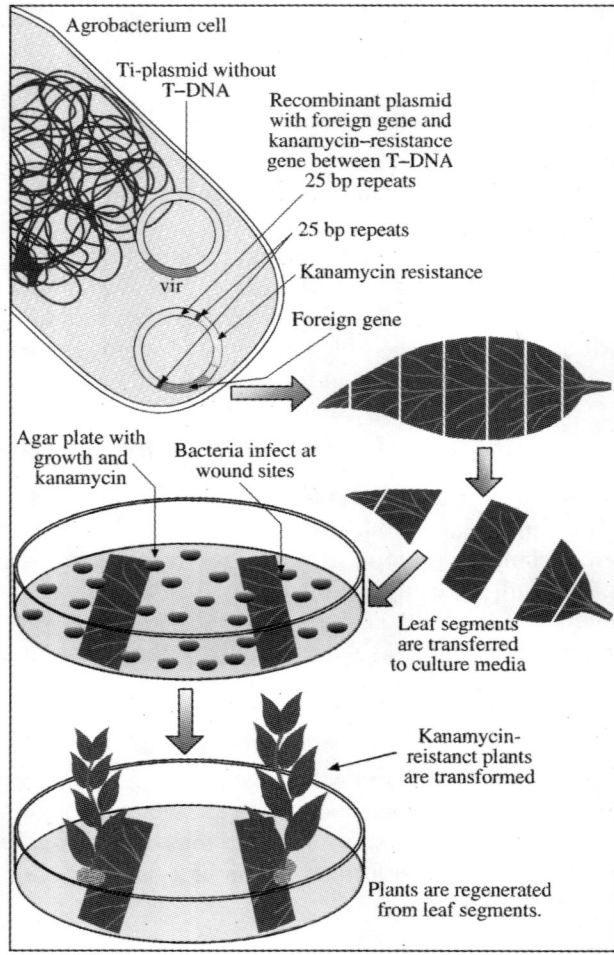

Figure 18.5 *A two-plasmid strategy to create a transgenic plant*

regenerated more rapidly when plant tissues are transformed directly without protoplast isolation. The use of leaf disk for co-culture is better than that of protoplasts or cultured cells since cultured cells are likely to show somaclonal variation. When using *A. rhizogenes*, explants inoculated with the bacteria form hairy roots at the site of infection, and these can be induced to regenerate mature plants.

(ii) Protoplast co-cultivation technique

Large populations of isolated cells can be transformed by the protoplast co-cultivation technique (Figure 18.6). In this technique, isolated protoplasts, which have been allowed to partially regenerate their cell wall, are co-cultured with *Agrobacterium*. Following the co-cultivation period, the cells derived from the protoplasts are cultured further to form the callus in media supplemented with appropriate antibiotics to prevent the growth of bacteria and allow

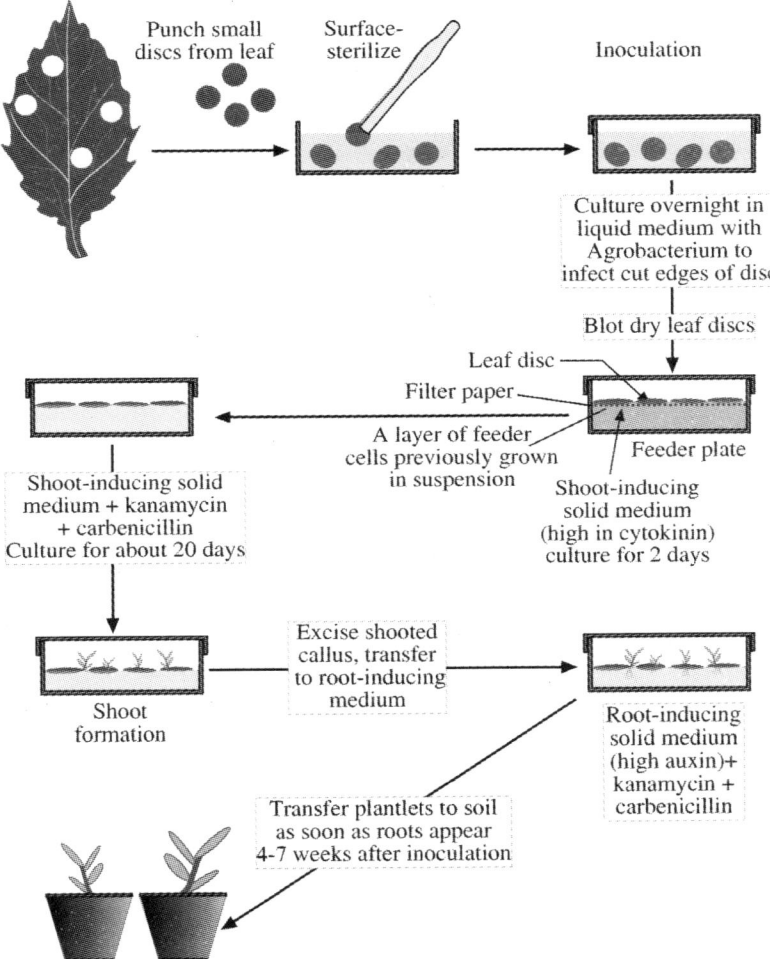

Figure 18.6 *Leaf-disc transformation by co-cultivation of leaf with Ti-plasmid vectors*

selection of transformed plant cells. Shoots can be regenerated by subculturing the callus, depending on their totipotency. Although many individual transformed plants can be regenerated by this technique, it should be noted that the resulting plants are subjected to somaclonal variation.

(iii) Suspension cell transformation

The suspension cell procedure is very efficient and has been reliably demonstrated in the *Nicotiana tabacum* suspension cell line. The transformation efficiency depends on the cell lines of *N. tabacum*, and the lack of released plant signal molecules for inducing the *vir* expression in some cell lines. Thus, pre-induction of the *Agrobacterium vir* genes with acetosyringone and initial culture of the transformed tissues under non-selective conditions may improve the transformation efficiency.

(b) In planta *transformation*

In the seed-inoculation technique, germinating seeds are inoculated with *Agrobacterium*, the resulting plants are selfed, and the progeny are germinated in the presence of appropriate antibiotics to obtain transgenic plants. Although there are many factors that might affect transformation efficiency, several T-DNA-tagged mutants have been successfully isolated in Arabidopsis. It appears that Agrobacterium cells enter the seedlings during germination, are retained within the plants, and when flowers develop they transform either the zygotes or the cells that give rise to zygotes.

Transformation of *Arabidopsis* plants is also achieved by immersing the flowers in a fresh culture of *Agrobacterium*, and applying partial vacuum to facilitate entry of bacterial cells into the plants. Later the plants are grown, selfed, and the progeny screened for identification of transformants.

(c) *Transformation of monocot plants by* Agrobacterium

Agrobacterium also infects some monocot plant species and forms crown galls, (e.g. Asparagus, *Dioscorea bubbifera* and *Allium cepa*). However, the efficiency of transformation is rather low. Efficient transformation of monocot cells can be obtained by providing acetosyringone during the co-culture of plant cells with *Agrobacterium*. Using this approach, successful transformation has been obtained in barley, wheat, maize and rice.

Method-1: *Introduction of plasmids into Agrobacterium*

Procedure

1. Prepare fresh plates with bacteria.
2. Inoculate the bacteria in a liquid medium and grow them overnight in a shaker at 29°C.
3. Dilute the culture to an OD_{666} of 0.02 by adding a fresh medium and incubate for 4 hours at 29°C with shaking.
4. Centrifuge the cells at 5,000 rpm for 5 min., wash the pellet with 10 mM Tris–HCl (pH 7.4–7.6), and re-pellet the cells by centrifugation.
5. Concentrate the cells 10-fold by re-suspending in 1/10th volume of rich medium.
6. Mix 20 µl of this bacterial suspension with 10 µl plasmid DNA (1–5 µg). Incubate this mixture for 5 min. at –70°C (dry ice-ethanol).
7. Bring the mix to 37°C and incubate for 25 min. at this temperature.
8. Dilute the suspension with 70 µl fresh medium, incubate at 29°C for 1 hour (for phenotypic expression) and plate the suspension onto a selective medium.

Competent cell preparation:

1. Prepare fresh plates with bacteria.
2. Inoculate the bacteria in a small volume (2 ml) of LB medium and grow in a shaker at 29°C for 6 hours.
3. Inoculate 100 ml of LB medium plus 0.1% glucose with 100 µl of the pre-culture and grow overnight at 29°C to an OD_{660} of 1–1.5.
4. Chill the culture in ice for 15 min., spin down the cells (20 min., 5,000 rpm), and wash the pellet three times with 10 ml of 1 mM HEPES (pH 7.0) and once with 10 ml of 10% glycerol.
5. Suspend the pellet in 500–750 µl of 10% glycerol (cell density of $1-5 \times 10^{11}$ bacterium/ml).

6. Distribute the bacterial suspension in 45 µl aliquots in microfuge tubes, freeze in liquid N_2 and store at –70°C.

Electroporation:

Materials

SOC medium: 2% tryptone, 0.5% yeast extract, 10 mM NaCl, 2.5 mM KCl, 10 mM $MgSO_4$, 10 mM $MgCl_2$, 20 mM glucose pH to 7.0.

1. Thaw the cells in ice.
2. Add 1–10 µl of alkaline lysis, small scale-isolated plasmid DNA suspended in distilled sterile water and mix.
3. Transfer the mixture to a pre-chilled electroporation cuvette (0.2 cm gap).
4. Apply electric pulse: 5 milli seconds at a field-strength of 12.5 kV/cm.
5. Dilute the cells immediately with 1 ml of SOC medium, incubate at 29°C (optional shaking) for 1–1.5 hours and plate onto selective medium.

Checking presence of plasmids in Agrobacterium:

Materials

Solution I: 50 mM glucose, 10 mM EDTA, 25 mM Tris.Cl (pH 8), 4 mg/ml lysosome (freshly added).

Solution II: 1% SDS, 0.2 N NaOH

Alkaline phenol: 2 volumes of 0.2 N NaOH, 1 volume Tris-saturated phenol.

Phenol–chloroform: 1 volume Tris-saturated phenol, 1 volume of chloroform/isoamyl alcohol (24:1).

Procedure

1. Incubate fresh bacteria in 2 ml of LB medium and grow overnight at 29°C.
2. Spin down 0.5 ml of the bacterial culture and discard supernatant.
3. Suspend the pellet by vortexing in 100 µl of solution I. Incubate for 10 min. at room temperature.
4. Add 200 µl of freshly prepared solution II, mix by inverting the tube four times, and incubate for 10 min. at room temperature.
5. Add 30 µl of alkaline phenol and mix by brief vortexing.
6. Immediately add 150 µl of 0.3 M sodium acetate (pH 4.8), mix by inversion, and incubate at –20°C for 20 min. Centrifuge for 3 min. and quickly collect supernatant.
7. Add 400 µl of phenol-chloroform and mix by brief vortexing. Centrifuge for 3 min. and collect the upper phase.
8. Add 800 µl of ice-cold 96% ethanol and mix by inversion. Spin down for 10 min. and discard supernatant.
9. Wash the pellet with 250 µl of ice-cold 70% ethanol, and dry briefly in a speed vacuum dessicator. Suspend the pellet in 25 µl of distilled sterile water or TE.
10. Use 1 µl of the DNA suspension for checking plasmid presence in a 0.6% agarose gel, 5 µl of the DNA suspension for restriction analysis, and 1–10 µl for the transformation of *Agrobacterium*, e.g. by electroporation.

C. AGROINFECTION

Agroinfection is a technique in which a virus infects a host as a part of T-DNA of Ti plasmid carried by the *Agrobacterium*. Introduction of a viral genome into plant cells can be carried out by placing it within the T-DNA of a Ti plasmid, and T-DNA can be delivered using the *Agrobacterium* containing this recombinant plasmid for co-culture with the plant cells. After infection, the viral DNA is released to form a functional virus that replicates and spreads systematically. Agroinfection may also lead to the integration of the viral DNA so that a transgenic plant containing integrated viral DNA can be produced. Agroinfection has great potential for studies in virus biology, because it can transfer deletion mutations or even single viral genes. This strategy is used with viruses, that have to be transmitted by an insect vector for successful infection, e.g., geminiviruses. Agroinfection has been shown for at least two geminiviruses—maize streak virus (MSV) and wheat dwarf virus (WDV).

III DIRECT GENE TRANSFER

Introduction of DNA into plant cells without the involvement of a biological agent, e.g., *Agrobacterium*, and which lead to stable transformation is known as ***direct gene transfer*** (Figure 18.7). The spontaneous uptake of DNA by plant cells is very low. Therefore, different chemical and physical methods are employed to facilitate the entry of DNA into plant cells. The physical delivery methods can be performed on almost all tissue types and where there is a problem of tissue regeneration. Low spontaneous uptake of DNA by the cells and unavailability of safe biological agents for gene transfer limits the gene transfer technology. These two limitations fueled the search for alternative methods for gene transfer. The various methods of direct gene transfer are as follows: (i) chemical methods, (ii) electroporation, (iii) particle bom-

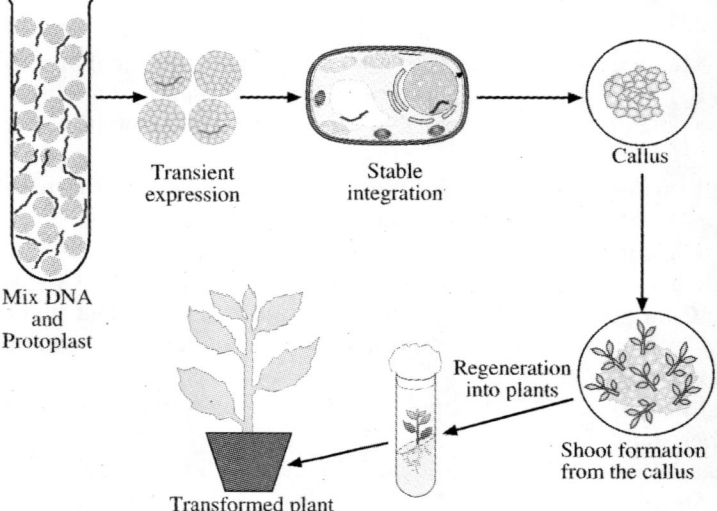

Figure 18.7 *Direct gene transfer into protoplasts. DNA is introduced directly into protoplasts, and the gene products can be assayed by transient expression or for permanent integration after selection*

bardment, (iv) lipoinfection, (v) microinjection, (vi) macroinjection, (vii) pollen transformation, (viii) delivery *via* growing pollen tubes, (ix) laser-induced, (x) fiber-mediated gene transfer, (xi) transformation by ultrasonication, and (xii) Calcium phosphate precipitation.

A. PROTOPLAST TRANSFORMATION BY CHEMICAL METHOD

Several methods have been published for transformation of protoplasts without the use of electroporation. The majority of these methods result in a transformation frequency lower than the electroporation. However, these methods easily produce enough transformants for the majority of cases. Certain chemicals, for example, PEG (polyethylene glycol), polyvinyl alcohol, and calcium phosphate, enhance the uptake of DNA by plant protoplasts. A generalized approach for PEG-mediated DNA delivery is as follows. Plant protoplasts are suspended in a transformation medium rich in Mg^{2+} ions in place of Ca^{2+} ions. Linearized plasmid DNA containing the gene construct is added to the protoplast suspension, following which PEG (up to 20% concentration) is added and pH adjusted to about 8. PEG at high concentration (15–20%) will precipitate ionic macromolecules such as DNA and stimulate their uptake by endocytosis without any gross damage to the cells. The protoplasts may be given a 5-minute heat-shock at 45°C followed by transfer to ice just before the addition of DNA, since this increases the frequency of transformation by several orders of magnitude. After a period of incubation, the PEG concentration is reduced while that of Ca^{2+} is enhanced. The procedure presented below is a simple method for transformation of tobacco protoplasts. PEG can also improve the efficiency of liposome uptake and electroporation.

Materials

Solutions:

- 0.4 M mannitol containing 6 mM $MgCl_2$, 0.5% MES pH 5.6 with KOH; 300 mM $MgCl_2$ solution;

- W5 solution: 145 mM Nacl, 125 mM $CaCl_2$, 5 mM KCl, 5 mM glucose, pH 5.6–6.0.

- DNA solution: 0.2 mg/ml linearized plasmid DNA and 1 mg/ml of calf thymus DNA in double-distilled water; The DNA is sterilized by precipitation in 70% ethanol, a wash in 70% ethanol and drying in a sterile flow hood.

- K3/H (1:1 mix) medium containing 0.6% Sea Plaque agarose (agarose sterilized dry, the K3 medium added and melted followed by H medium).

- PEG solution: 24% w/v PEG 6000 in 0.4 M mannitol, pH 5.6. This PEG solution prepared with sufficient magnesium chloride added (ca. 30 mM) to bring the resistance when measured in the chamber of the electroporator into the region of 1.2 k&Ω.

Procedure

1. Prepare the protoplasts and wash two times by re-suspension in W5 solution and centrifugation at 600 rpm for 5 min., followed by counting.
2. Re-suspend the protoplasts at a density of 1.6×10^6/ml in the mannitol/magnesium solution. Heat shock for 5 min. at 45°C followed by cooling to room temperature and distribute 0.3 ml aliquots into 5 ml sterile plastic tubes.

3. Add 30 µl of the DNA solution and mix, followed by 300 µl of the PEG solution. Incubate for 25–30 min. at room temperature with occasional shaking.
4. Gradually add 10 ml of W5 solution over about 10 min. and then centrifuge for 10 min. at 600 rpm.
5. Re-suspend the protoplasts in 1 ml of K3 medium, transfer to a 9 cm Petri dish and add 7 ml of a 1:1 mixture of K3 and H media containing 0.6% w/v Sea Plaque agarose. Mix the protoplasts well but gently with the agarose medium, and allow this to set. Do not disturb the dishes until the medium is solid, as this will cause damage to the protoplasts.

Note

1. Synchronized tobacco protoplasts when transformed during the S or mitotic phase yield 3% transformed colonies, while non-synchronized protoplasts give only 1.5%.

B. TRANSFECTION OF PLANT CELLS BY ELECTROPORATION

When a cell is exposed to a very high electric field, its cell membrane becomes permeabilized and allows external molecules to diffuse in. This phenomenon is called **electroporation** (Figure 18.8). Because the cell membrane must be re-sealed later to allow the cell to recover, the external electric field can only be applied in a transient manner, that is, in a pulsed form. Samples are pulsed with high-voltage pulses in the solutions containing cells and DNA. DNA enters the cells through the reversible pores, which are created in the membrane of the cells by the action of short electrical pulses. Electroporation has been used as a gene transfer method in many different cell types, including bacteria, yeast, and plant and animal cells. Electroporation has a number of advantages over other conventional methods of loading exogenous substances. (1) It requires less training and not very expensive equipment; (2) millions of cells can be transformed at one time; (3) this method has very few biological or chemical side-effects; (4) since it is a physical method, it is less cell-type-dependent.

The mechanism for induction of the high-permeability-state of cells is not fully understood. The reversible permeabilization of the plasma membranes of whole plant cells to charged molecules, which are otherwise unable to enter the cell, can be achieved by application of electrical fields. Since the plant cell wall is a barrier to DNA uptake by intact cells, its removal is essential for efficient uptake of DNA.

1. Factors influencing electroporation

Electroporation has been standardized in different species to introduce DNA into a large proportion of protoplasts while maintaining reasonably sufficient cell viability.

(a) **Electrical parameters:** Strong direct electric fields reduce cell viability, but weak direct current/electric fields, applied over extended periods of several days, have been shown to enhance plant regeneration from the callus, probably by increase in the polar transport of auxin. After charging the selected capacitor (100–1000 V), it is discharged over a chamber/cuvette holding the protoplasts. Pulse duration varies with the capacitance of the medium in the electroporation chamber. Rectangular (square) pulse machines include electronic components for the modification of electric pulse to square waves. In these machines, the pulse duration can usually be selected in the range of 10 to 1000 µs and the voltage from 1000 to 2000. In

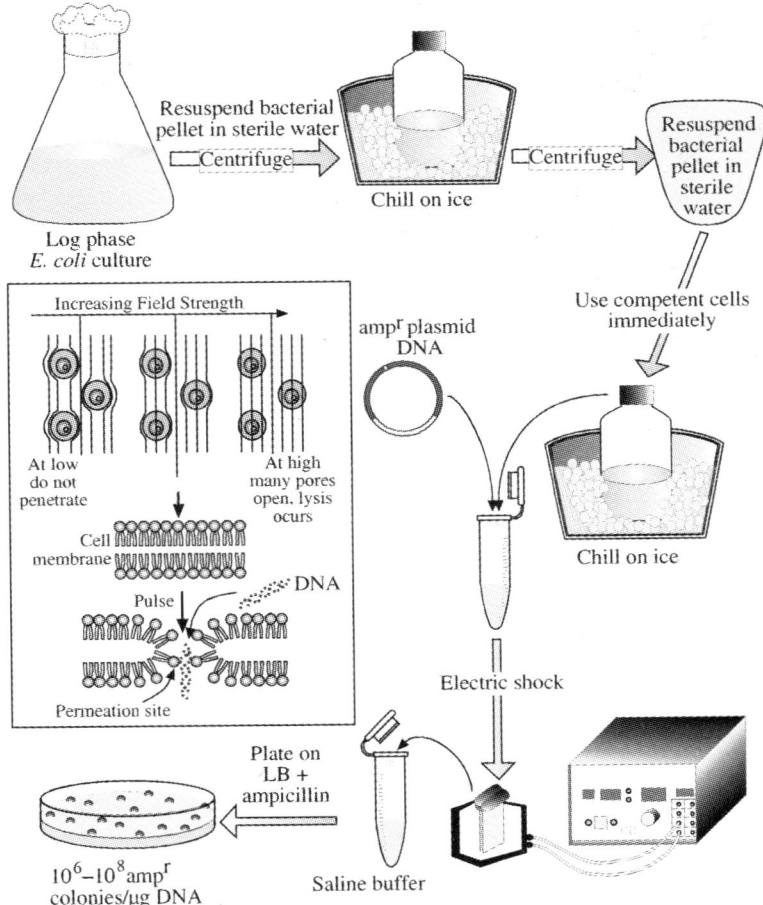

Figure 18.8 *DNA transformation by electroporation is done by delivering a short electric pulse to cells that are suspended in low ionic strength solution in a special cuvette. In the insert, the principle of electroporation is shown*

order to induce the uptake of DNA through permeabilization of the plasma membrane, the electric field strength must exceed a certain minimum value (critical field strength) to induce the uptake of DNA. Alternatively, a low voltage (350 V) pulse of long duration can also be used (device can be made easily at home) for electroporation.

(b) Size of plasmid: A recent trend which has improved transformation efficiency is reduction in the size of the vector DNA. It has been shown that the frequency of stable transformation of protoplasts is influenced by the concentration and type (linear or super coiled or carrier DNA supplied with the plasmid) of exogenous DNA. Increasing the concentration of DNA in the electroporation medium also increases the transformation frequency. A linear correlation between the transient expression in electroporated cells and the plasmid DNA concentration up to 40 µg/ml has been reported for many crops. Addition of carrier DNA has been used in some studies in order to protect the plasmid DNA from degradation.

(c) Electroporation media and incubation temperature: In addition to plasmid size, electrical parameters and electroporation media are also interrelated. Several workers have used different media for electroporation. The pulse strength decreases with increasing conductivity of the medium when exponentially decaying pulses are used. Addition of Ca^{2+} to the electroporation medium is advantageous due to its membrane–stabilizing properties. In some studies the protoplasts are incubated on an ice-bath some minutes before and immediately after electroporation in order to enhance DNA uptake by prolonging the duration of the high permeability state of the membranes.

(d) Effect of physiology of protoplasts and incubation period: The ability to be transiently transformed varies enormously with species and also among varieties/genotypes within one species. Also with respect to stable transformation, large differences are found in the competence of protoplasts. Due to very low regeneration of most monocotyledonous protoplasts, intact cells and other cell types have been used for electroporation. Pectin appears to play a significant role in electroporation in plants. It may inhibit DNA transfer, either as a physiological barrier or as a chemical.

(e) Influence of cell cycle: The cell cycle of recipient protoplasts influences transformation. Many studies show that synchronization of protoplasts at the S or M phase by culturing them in the presence of various cell division inhibiting (aphidicolin, 2,6–dichlorobenzonitrile, irradiation etc.) compounds has enhanced transformation more than twofold.

Method-1: DNA transformation of protoplasts by electroporation

Protoplast preparation

See Chapter 33 for procedure for protoplast preparation

Stable transformation of tobacco protoplasts with electroporation

Samples of protoplasts are pulsed with high-voltage pulses in the chamber/cuvette. The pulse is applied by discharge of a capacitor across the cell. The decay constant of the pulse (time taken to decay to 1/e of the initial voltage) is of the order of 10 µs with a capacitor of 10 nF and a chamber resistance of 1 k&Ω. The resistance across the chamber is measured with an alternating current multimeter operating at 1 kHz. The protocol given is for mesophyll protoplasts of *N. tabacum*. These have an average diameter of 42 µm. The field strength required is inversely proportional to the diameter of the protoplasts being treated, and may vary a little from species to species in the strength required for a given size of protoplast.

Materials

Equipments:

Electroporation apparatus (shock chamber); High voltage power supply; etc.

Solutions:

0.4 M mannitol containing 6 mM $MgCl_2$, 0.5% MES pH 5.6 with KOH; 300 mM $MgCl_2$ solution;

DNA solution: 0.2 mg/ml linearized plasmid DNA and 1 mg/ml of calf thymus DNA in double-distilled water; The DNA is sterilized by precipitation in 70% ethanol followed by a wash in 70% ethanol and drying in a sterile flow hood.

K3/H (1:1 mix) medium containing 0.6% Sea Plaque agarose (agarose sterilized dry, the K3 medium added and melted followed by H medium).

Procedure

1. Transfer a 0.35 ml aliquot of protoplasts to the chamber of the electroporator, with a Gilson pipette, and measure the resistance. This should be in the region of 1–4 k&Ω. Add the appropriate amount of $MgCl_2$ solution to the protoplast suspension to adjust the resistance to a value of 1–1.1 k&Ω.
2. Heat shock the protoplasts by treatment for 5 min. at 45°C in a water bath, followed by cooling to room temperature on ice.
3. Dispense aliquots of 0.25 ml of protoplast suspension into 5 ml capped plastic tubes.
4. Add 20 µl DNA solution followed by 0.125 ml of PEG solution. Ten minutes after addition of the DNA and PEG, transfer samples to the chamber of the electroporator and pulse three times at 10-s intervals with pulses of an initial field strength of 1.4 kV/cm.
5. Return each sample to a 6–cm diameter Petri dish and wait for 10 min.
6. Add 3 ml of a 1:1 mixture of K3 and H media containing 0.6% w/v Sea Plaque agarose. Mix the protoplasts well but gently with the agarose medium, and allow this to set. Do not disturb the dishes until the medium is solid, as this will cause damage to the protoplasts.

C. DIRECT DNA TRANSFER INTO INTACT PLANT CELLS VIA GENEGUN

To overcome the limitations of protoplast regeneration, high velocity microprojectiles are being used to deliver nucleic acids directly into intact plant cells or tissues. **Biolistics** is the delivery of microprojectiles, usually of tungsten or gold coated with DNA and propelled into the target cells by acceleration which enables their entry into plant cell/nuclei. The particle acceleration can be produced by an explosive charge (cordite explosion) using gun powder, gases (such as helium, CO_2), or by shock waves initiated by an electric discharge (velocity around 1,4000 ft per sec.). The design of two particle-guns used for acceleration of microprojectiles are shown in Figure 18.9. This technique can introduce DNA into virtually any tissue (e.g. meristems, leaf blades, immature and mature embryos, mature pollen, root and shoot sections etc.) from any cultivar (Figure 18.10). However, success depends critically upon the ability of the target tissue to proliferate and give rise to a fertile plant after transformation. In addition to the above mentioned factors, the efficiency of gene transfer by particle bombardment is highly dependent on various physical parameters that influence the particles' momentum and distribution upon the target tissue. The size and shape of the particle will also influence its potential for cell penetration. The number of particles per unit area that impact the target tissue is controlled to a large degree by the method of acceleration. Another important parameter controlling the efficiency of the process is the method used to adsorb DNA to the particles. Biological factors such as the size of the target cells, their turgor pressure, and stage of development are also of importance in particle bombardment.

1. Requirements for stable transformation

Minor alterations in the standard biolistic protocol have yielded tremendous improvements. Some of these include: (a) proper pre-culture of the explant material, (b) use of baffling screens,

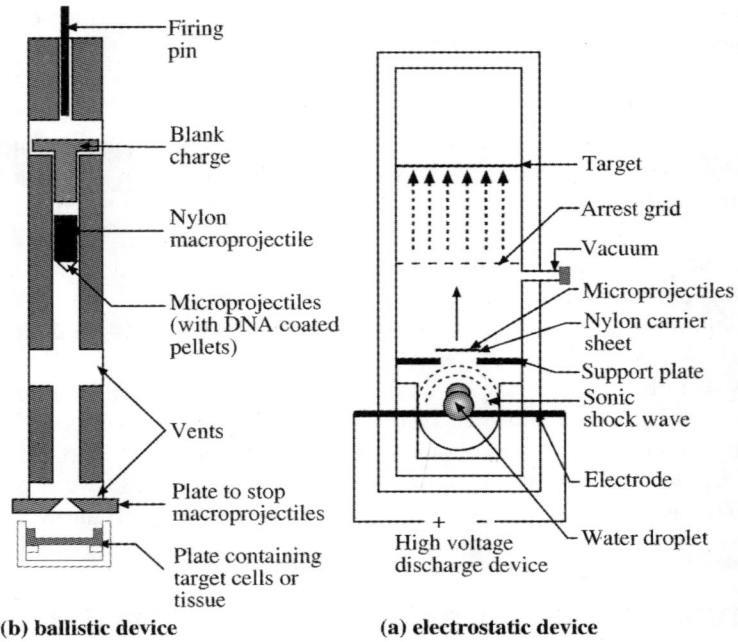

Figure 18.9 *Microprojectile acceleration devices. Microprojectiles coated with DNA and delivered into cells are tissues by these devices*

(c) use of small-size microprojectiles (0.5–3 µm in diameter) and (d) subjecting the tissue to an osmotic pre-treatment by either partial drying in a laminar flow hood or culturing in a medium containing an osmotic agent. The use of the gene gun requires cells that are capable of cell division (single cells or organ), that are healthy and can be enriched by selection with antibiotics or herbicides that correspond to the transforming DNA. Tungsten particles of 1.2 µm diameter have been used most often with the biolistic apparatus while gold particles have been used with the electric discharge device. The most important parameters for increasing the frequency of both transient and stable transformation is the coating of the microparticle with DNA by precipitation and the number of bombardments (2 to 3). Several studies have investigated the general conditions for efficient binding of DNA to microprojectiles, which include the amount of DNA, precipitating agents such as $CaCl_2$ or $CaHPO_4 \cdot 2H_2O$, polyamines, spermine, and amount of particles.

The number of transformed cells is relatively low compared to the mass of the tissue explant treated. Enrichment for transformed cells requires both high expression levels in transformed cells and efficient selective conditions. Efficient production of stable transformants in many species will require the use of an efficient selectable marker. The target tissue will more likely determine the probabilities of recovery of stable transformed plants with or without selection rather than the method used to deliver particles into individual cells. Molecular analysis of plants obtained by biolistic transformation generally reveals a complex pattern of transgene integration. The biolistic technique has been used to produce stable gene transfers in cotton, maize, rice, sorghum, soybean, sugarcane, tobacco, wheat, papaya, etc.

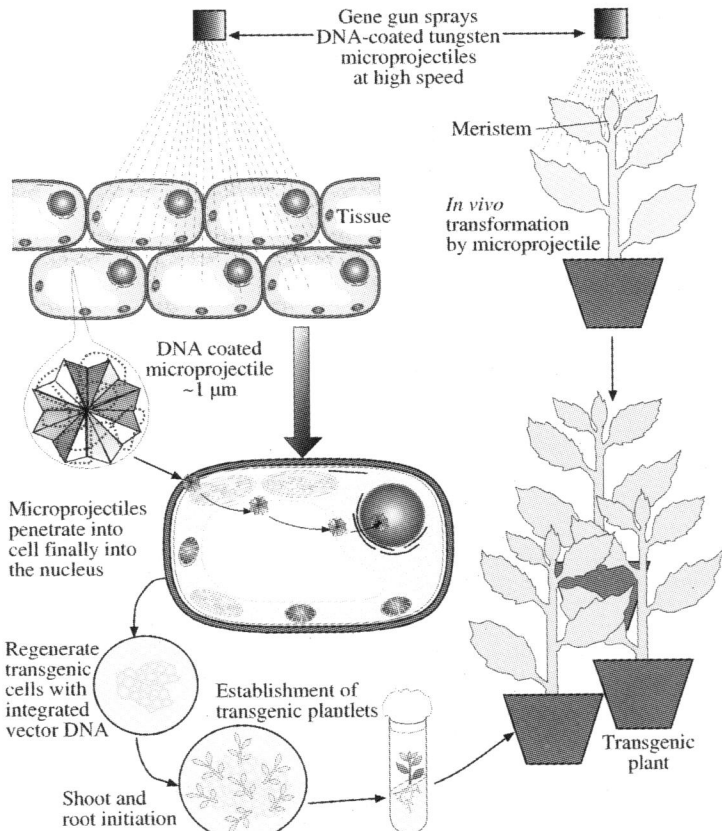

Figure 18.10 *Illustration of microprojectile bombardment for plant transformation: a particle gun accelerates DNA-coated tungsten microprojectiles at high speed into cells such as embryogenic callus or directly into meristem*

Method-1: Preparation of DNA (microprojectiles) for particle bombardment

Procedure

1. Sterilize 100 mg tungsten, size 1.2 μm microprojectiles by suspending them in 2 ml of 95% EtOH in a 15 ml centrifuge tube.
2. Sonicate on ice with a continuous pulse for 10 min. using a 20% duty cycle at level 2 output.
3. Following sonication, pipette microprojectiles into a 2 ml Eppendorf tube. Centrifuge at 9000 rpm to pellet microprojectiles for 1 min. Withdraw supernatant solution and replace with one ml sterile deionized water.
4. Centrifuge to re-pellet microprojectiles at 9000 rpm for 2 min. Withdraw supernatant solution and replace with 2 ml sterile deionized water.
5. Centrifuge to re-pellet microprojectiles at 9000 rpm for 2 min. Withdraw supernatant solution and make final replacement with 2 ml sterile deionized water.

6. Sonicate 2 ml vial of microprojectiles prior to aliquoting 25 µl samples into 1.5 ml Eppendorf tubes. Re-sonicate after every 2 aliquots to ensure uniform bead concentration per aliquot.
7. Add 10 µg of DNA to each aliquot of microprojectiles and mix with the pipettor. Standard DNA concentration for this preparation is 1 µg/µl.
8. Add 25 µl of 2.5 M $CaCl_2$ to DNA/microprojectile mixture. Mix with the pipettor.
9. Add 10 µl of 0.1 M spermine to DNA/microprojectile mixture. Finger-vortex and allow particles to settle for at least 10 min.
10. Centrifuge DNA/microprojectile mixture at 9000 rpm for 2 min. Withdraw supernatant solution to a final volume of 30 µl.
11. Sonicate DNA/microprojectile mixtures and pipette 2.0 µl onto a sterile microprojectile. Pipette 2–3 aliquots before re-sonicating. Six aliquots per tube can be taken.
12. Place macroprojectile with DNA in gun barrel, place power level 1 blank in chamber. Insert the stopping and tissue sample; attach the detonator and draw vacuum. When vacuum reaches 68–71 cm Hg, fire the gun.

Method-2: Transformation of tobacco leaves

Materials

Preparation of media:

MS inorganic salts (mg/l)		MS micronutrients (mg/l)		MS vitamins (mg/l)	
KNO_3	1900	Na_2EDTA	37.3	Thiamine-HCl	20
KH_2OPO_4	190	H_3BO_3	6.2	Pyridoxine-HCl	100
NH_4NO_3	1650	$MnSO_4.H_2O$	19.2	Nicotinic acid	100
$CaCl_2.H_2O$	440	$ZnSO_4.7H_2O$	8.6	Glycine	400
$MgSO_4.7H_2O$	370	KI	0.83		
$FeSO_4.7H_2O$	28.8	$Na_2MoO_4.2H_2O$	0.25		
		$CuSO_4.5H_2O$	0.025		
		$CoCl_2.5H_2O$	0.025		
B–5 vitamins (mg/l)					
Inositol	20000	Thiamine-HCl	2000	Pyridoxine-HCl	200
Nicotinic acid	200				

Preparation of **media A, B, C, D, E**, and **F** (per litre)

Chemical	Med. A	Med. B	Med. C	Med. D	Med. E	Med. F
Inorganic salts	4.3 g	4.3 g	4.3 g	4.3 g	4.3 g	4.3 g
MS vitamins	2.5 ml	2.5 ml	2.5 ml	–	–	–
B–5 vitamins	–	–	–	5 ml	5 ml	5 ml
Myo-inositol	100 mg	100 mg	100 mg	–	–	–
Sucrose	40 g	40 g	40 g	30 g	30 g	–
Kanamycin sulfate	–	200 mg	50 mg	100 mg	100 mg	–
Gel-rite	2 g	3 g	1.5 g	–	–	–
Phytagar	–	–	–	8 g	8 g	–
Hormones						
NAA	–	–	–	–	2 ml	
BAP					0.5 ml	0.5 ml
PCPA						0.01 ml
KinetinpCPA						4 ml
Final pH	5.6	5.6	5.6	5.6	5.6	5.6

Autoclave all the above ingredients with the exception of kanamycin sulfate. Add the kanamycin sulfate when medium has cooled to 60°C in a water bath.

Procedure

1. Surface-sterilize seeds in 10% Chlorox plus one drop of Tween 20 per 100 ml of solution for 15 min., then rinse three times in sterile water. Seed can be air dried in a laminar flow hood and sealed in Petri plates for use at a later time. Place 16 surface sterilized seeds in a Petri dish with germination medium (Medium A). Incubate seeds on culture medium for 14 days at 28°C with 16th day length at 100 $\mu E\ m^{-1}s^{-1}$. After 14 days, remove cotyledons by placing seedlings in a Petri dish containing 5 ml of liquid medium for tobacco (Medium F), then excise cotyledons with a scalpel.
2. Place excised cotyledons or leaf pieces 1 cm × 0.5 cm on Whatman filter paper in 60 mm × 20 mm Petri dishes containing 15 ml callus medium with 100 mg/l kanamycin (Medium D). Concentrate ten leaf pieces in the center of each Petri dish in order to maximize exposure to bombardment.
3. Sonicate DNA/microprojectile mixture and pipette 1.5 µl onto a sterile macroprojectile. Pipette 2–3 aliquots before re-sonicating. Six aliquots per tube can be taken.
4. Place macroprojectile with DNA in the gun barrel; place the power level 1 blank in chamber. Insert the stopping plate and tissue sample, attach the detector and draw vacuum. When vacuum reaches 68–71 cm of Hg, fire the gun.
5. Following bombardment, maintain leaf strips in a growth room at 28°C with 16 h day length, at 100 $\mu E\ m^{-1}S^{-1}$. Maintain leaf strips on callus medium with 100 mg/l kanamycin for 14 days (Medium D). Callus colony development should be noticeable on the cut or bombarded surfaces of the leaf strips after 14 days.
6. After 14 days on the callus medium, transfer all leaf strips to a regeneration medium containing 100 mg/l kanamycin (Medium E). Transfer to fresh regeneration medium every 14 days.
7. As plantlets develop from the callus, they should be cut at the base of their attachment to the callus and placed in sterile Flow boxes containing 50 ml of root formation medium with 50 mg/l kanamycin (Medium C). Wrap flow boxes with Parafilm to prevent contamination. Plantlets should remain in culture boxes until root formation and shoot growth is sufficient to allow transfer to the greenhouse.

D. LIPOSOME–MEDIATED DNA DELIVERY

Introduction of DNA into cells *via* liposomes is known as *lipoinfection*. Lipoinfection is the method of choice for DNA delivery into animal cells. It was also one of the first methods of DNA delivery into plant protoplasts. The system is currently receiving the most attention and is considered to hold the most promise. Liposomes are small artificial **lipid vesicles** (bags), in which large numbers of DNA (plasmids) are enclosed (Figure 18.11). Liposomes can be formulated by different methods (Figure 18.12) or prepared from phosphatidylcholine and stearylamine by a process known as reverse-phase evaporation. Nucleic acid entrapped in such liposomes renders them highly tolerant to attack by nucleases. Techniques for fusing these liposomes to plant cell protoplasts have been evolved. They can be induced to fuse with protoplasts using devices like PEG and other agents. This technique offers the following advantages: (i) protection of DNA/

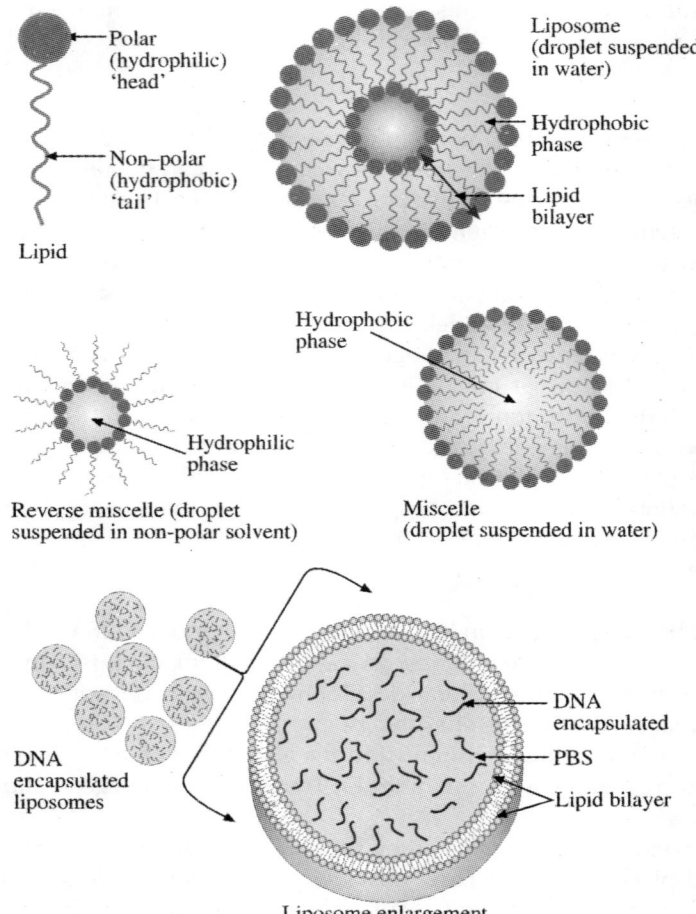

Figure 18.11 *Liposome structures with multiple layers*

RNA from nuclease digestion, (ii) low cell toxicity, (iii) stability and storage of nucleic acids, (iv) high degree of reproducibility, and (v) applicability to a wide range of cell types. The mechanism of lipoinfection is (Figure 18.13), DNA enters the protoplasts due to endocytosis of liposomes, involving the following steps: (a) adhesion of the liposomes to the protoplast surface, (b) fusion of liposomes at the site of adhesion and (c) release of plasmids inside the cell. Transformation frequencies of 4×10^{-5} have been reported. The integrated DNA apparently does not undergo re-arrangements, but multiple copies may become integrated in tandem. In some cases, plasmid DNAs for 9 kb and larger have been integrated intact. However, higher transformation frequencies with PEG and electroporation make them more attractive. In the experiments with tobacco protoplasts, TMV RNA was encapsulated in a variety of different liposome preparations. Incubation of tobacco protoplasts was done with negatively charged, phosphatidyl serine liposomes containing TMV RNA. Virus production could also be detected following incubations with neutral or positively charged liposome preparations.

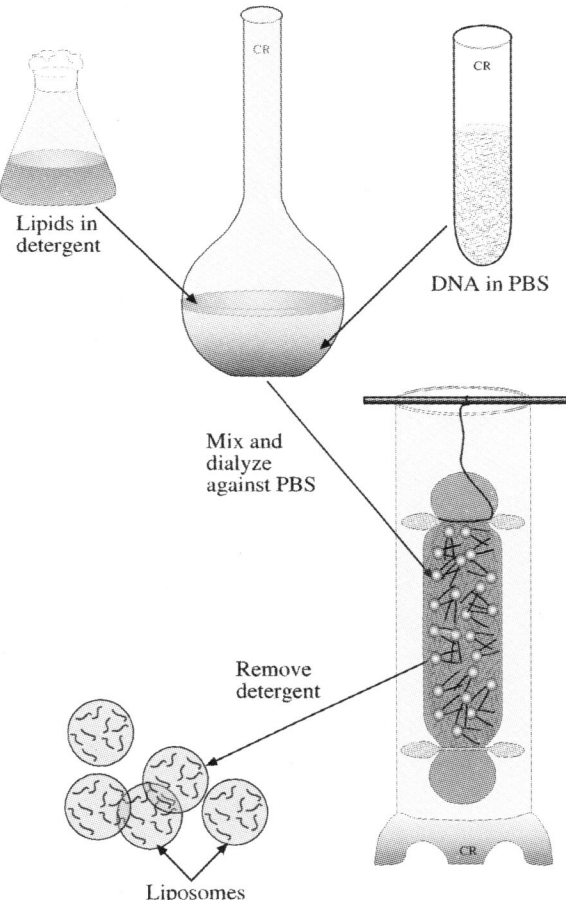

Figure 18.12 *Liposome formation using the detergent extraction method*

E. MICROINJECTION

In case of microinjection, the DNA solution is injected directly inside the cell using capillary glass micropipettes (Figure 18.14) with the help of micromanipulators. In plants, microinjection can be done to protoplasts and many immature structures (immature embryos, meristems, immature pollen, germinating pollen, isolated ovules, embryogenic suspension cultured cells, etc.,) conveniently. The process of microinjection is technically demanding and time-consuming; a maximum of 40–50 protoplasts can be microinjected in one hour. In plants transformation, a frequency of up to 14 to 66% has been achieved. It is most successful with densely cytoplasmic, non-vacuolated embryonic cells. The most promising approach appears to be the use of young embryos for microinjection. About 80% of these embryos produced plants of which about 50% were stably transformed. But the main drawback of this technique is the production of chimeric plants with only a part of the plant transformed. However, from the chimeric plant itself, transformed plants of single cell origin can be subsequently obtained. Regeneration of

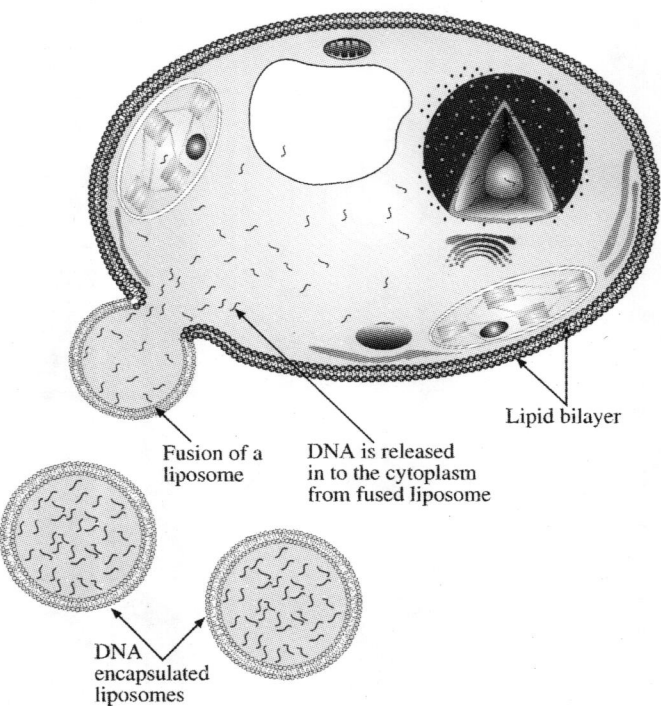

Figure 18.13 *Fusion of DNA (or plasmids) filled liposomes with isolated protoplasts as a method of DNA uptake*

transformed cells from chimeric tissues can produce stably transformed plants and can also be subjected to secondary embryogenesis.

When cells (or protoplasts) are used as targets in the technique of microinjection, glass micropipettes with 0.5–10 μm diameter tip are used for transfer of macromolecules into the cytoplasm or the nucleus of a recipient cell. The recipient cells are usually immobilized in agarose or on glass coated with polylysine or by being held under suction by a micropipette. Often a specially designed micromanipulator is employed for microinjecting the DNA.

F. SILICON CARBIDE FIBER-MEDIATED DNA DELIVERY

In this method, DNA is delivered into the cell cytoplasm and nucleus by a silicon carbide fiber of 0.6 μm diameter and 10–80 μm length. In one study, suspension culture cells were mixed with plasmid DNA having *gus* gene and silicon carbide fibers, all suspended in the culture medium. The mixture was vortexed, and the cells were assayed for transient *gus* expression; the frequency of *gus* positive cells was 10^{-4}. It was shown that the fibers mediated the delivery of DNA into the cytoplasm and nuclei of cells in a manner similar to microinjection. The method was successful with both maize and tobacco suspension culture cells. However, it is not known if the transformation resulted in stable integration of the transgene.

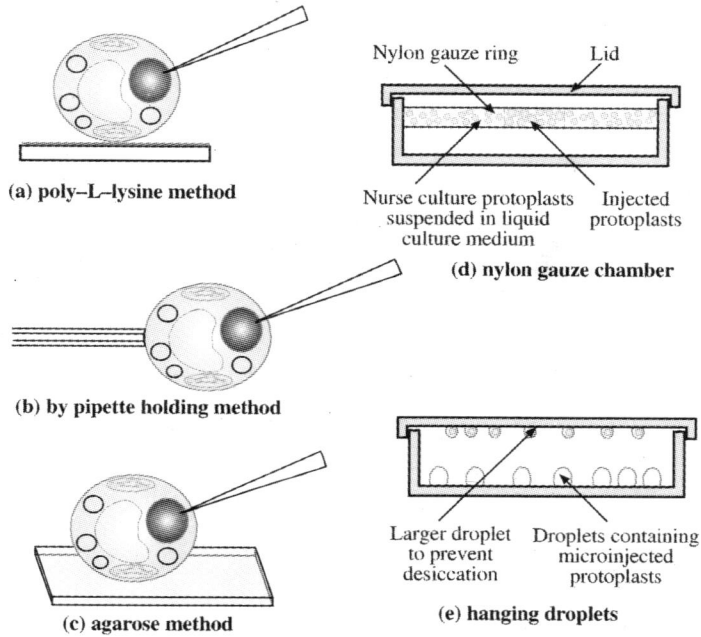

Figure 18.14 *Methods for the microinjection and culture of plant protoplasts*

The fiber-mediated DNA delivery appears to be widely applicable, and is the most rapid and inexpensive method of DNA delivery provided stable integrations are achieved. However, the hazardous nature (carcinogenic) of silicon fibers may have hampered their routine use.

G. LASER-INDUCED DNA DELIVERY

The use of focused, high-energy laser in surgery is well established. In a similar way, a laser is used to bore micropores into cells, which are thus induced to take up DNA from the surrounding medium. Lasers have been successfully used for high frequency (10^{-3}) transfection of animal cells. They puncture transient holes in the cell membrane through which DNA may enter into the cell cytoplasm. Lasers have been used to deliver DNA into plant cells, but there is no information on transient expression or stable integration.

H. POLLEN TRANSFORMATION

Some reports have claimed gene transfer by the soaking of pollen grains in DNA solutions prior to their use for pollination. It is hoped that DNA can be taken up by the germinating pollen and can either integrate into sperm nuclei or reach the zygote through the pollen tube pathway. However, other workers using cloned genes (when marker genes were used) could not substantiate these reports. Factors that contributed to failure include the presence of cell wall, nucleases, heterochromatic state of acceptor DNA, callose plugs in pollen tube, etc. Though the method is attractive in view of its simplicity and general applicability, there is so far no definite evidence for a transgene being transferred by pollen soaked in the DNA solution.

Sperm as vector of DNA. The use of sperm as carriers of exogenous DNA has been evaluated for honeybees and the Australian sheep blowfly *Lucilia cuprina*. If sperms are used to transfer genes, the exogenous DNA should bind to the sperms so that it can be carried into the nucleus of the egg. This approach to gene transfer would probably be limited to species such as the honeybee, for which semen can be collected and used for artificial insemination.

I. DNA DELIVERY VIA GROWING POLLEN TUBES

In this approach, the stigma of a flower is cut off some time after pollination, and a DNA solution is applied onto the cut surface. The time of stigma excision would depend on the rate of pollen tube growth and may range from 5–20 minutes to 2–3 hours. In the case of rice, plasmid DNA containing *npt*II gene was applied to the cut surface of stigma; up to 20% of the seeds so produced contained the *npt*II gene in copy numbers ranging from 1–300. But in a similar work with barley, transformation frequency was only 10^{-3} to 10^{-4} of the seedlings so obtained; the expression level of *npt*II was low, and the mature plants and their progeny failed to show any *npt*II expression. In any case, this method is simple, easy, and very promising provided consistent results and stable-transformations are achieved. Its use would necessitate a much better understanding of the mechanism of DNA transfer into the zygotes and the factors affecting it.

J. MACROINJECTION

The injection of plasmid DNA into the lumen of developing influorescence using a hypodermic syringe is called **macroinjection**. It is hypothesized that the DNA is taken up by microspores during some specific stage of their development. In one such study, the DNA was injected into the developing influorescence of rye, and a low frequency (0.07%) of transformed plants was recovered in the progeny. But in subsequent experiments with barley and wheat, stable transformation was not obtained. Even though the approach is simple and easy, the main problem is the low frequency and consistency of stable transformants obtained.

K. DIRECT DNA UPTAKE BY IMBIBATION OF SEEDS, EMBRYOS, TISSUES, OR CELLS

Incubation of dry seeds, embryos, tissues, or cells in known DNA solution has been tried in many cases and expression of defined genes has been observed. When dry isolated embryos of wheat, barley, rye, pea, and bean (isolated by grinding the seeds in carbon tetrachloride or cyclohexane) are imbibed in a DNA solution, they take up the DNA and show the expression of a marker gene. Dry seeds, whose seed coats have been removed by grinding, also take up DNA when imbibed in a DNA solution. The imbibed seeds/embryos are germinated on appropriate selective media to isolate the transformed embryos. In case of isolated wheat embryos, amplification of the marker gene inserted in the wheat dwarf virus genome was observed, confirming the uptake of DNA by embryos. However, much work is needed to refine the technique and to understand the process of DNA uptake to enable its application for the production of transgenic plants. It is thought that the DNA is taken up by embryos through the cells injured during their isolation. The DNA then moves through the plasmodesmata to other cells of the embryos.

L. CALCIUM PHOSPHATE PRECIPITATION METHOD

Foreign DNA can also be introduced into protoplasts with the Ca^{2+} ions. Calcium ions can carry DNA by precipitating with DNA in the form of calcium phosphate and releasing it into the cell. Even though the efficiency of this method is low in plant cells, in the past it was the standard method for gene transfer in plants.

M. TRANSFORMATION BY ULTRASONICATION

The work conducted at the Biotechnology Research Center at Beijing, China, showed the use of ultrasound for DNA transformation of wheat, tobacco, and sugarbeet. In this method, explants, after being cultured for a few days, were sonicated with plasmid DNA (carrying marker genes like *gus*, *cat*, and *npt*II). The sonicated calli were transferred to a selective medium and shoots were obtained. In tobacco, up to 22% of transformed plants were obtained.

N. SCRAPEFECTION

This method was initially described as a general method for introducing functional macromolecules into adherent cell-lines. It has since been demonstrated to function well in DNA transfection experiments. The protocol involves incubating a washed monolayer of cells with a buffered DNA solution (1–50 µg cm^{-3}) prior to scraping with a rubber policeman. The scraped cells are then distributed to fresh culture plates and analyzed for transient expression of the transfected DNA 1 to 5 days later. Up to 80% of the scraped cells were shown to express the DNA and cell viability of 70% was obtained.

O. RETROVIRAL MEDIATED GENE TRANSFER

Retroviruses are ssRNA viruses. Upon transfection, viral RNA is transcribed into dsDNA, which integrates into the host genome. This proviral DNA integrates, replicates, and stably inherits along with the host chromosome. Cellular RNA polymerase II can transcribe the DNA to make viral RNAs which, in turn, can be packaged into virions which are capable of infecting other cells. The retroviral genome consists of the group-specific antigen gene (*gag*) for the synthesis of core proteins, the reverse transcriptase (*pol*) gene for synthesis of RNA-directed DNA polymerase, the envelope (*env*) gene for synthesis of envelope glycoproteins and a number of cis-acting elements for the replication and encapsidation of the viral genome into virions. It is composed of two parts: the genetic material of interest and the virion, which infects into the target cell. Retroviral infection is a multi-step process (see Chapter 16).
The use of retroviruses as vectors for gene transfer involves the following main steps.
1. *In vitro* construction of a hybrid DNA containing *cis*-acting retroviral elements and the gene of interest.

2. Transfer of hybrid DNA into target cells by calcium phosphate precipitation.
3. Expression in the transfected host cells of transcripts of the recombinant provirus by host RNA polymerase.
4. Packaging of viral RNA into a viral particle by supplying the *gag*, *pol*, and *env* gene products in *trans*, either by superinfection of transfected cells with a replication competent helper virus or by co–transformation of the same cells with plasmids containing *gag*, *pol*, and *env* genes.
5. Harvesting of this virus and the reintroduction of the hybrid RNA into cells by infection rather than transfection.
6. Once integrated, the provirus behaves as a transcription unit, which expresses the foreign gene.

Retroviral vectors have been successively used to introduce genes into hematopoietic progenitors in several *in vitro* hematopoietic progenitor cells. The Moloney murine leukemia virus (MoMuLV) based vector N2 has been used to introduce the bacterial neo^r gene into hematopoietic cells of mice, humans, and dogs. The expression of this drug resistance gene has been used to determine the efficiency of retroviral-mediated gene transfer, and to evaluate means by which the gene transfer efficiency could be improved in the case of human bone marrow cells. The simplified protocol for bone marrow gene transfer using a retroviruses vector is as follows (Minhas BS and Voelkel SA, 1989).

Procedure

1. Bone marrow aspirated into a heparin containing sterile tube.
2. Nucleated bone marrow cells prepared over ficoll density gradient centrifugation.
3. About 5×10^6 nucleated bone marrow cells added to flasks confluent with producer cells and co-cultivated at 37°C for 18–21 hours (co-cultivation procedure). Another option is the supernatant procedure in which 1×10^7 nucleated bone marrow cells are mixed with 40 ml of viral-containing supernatant (e"10^5 cfu/ml). This infection procedure requires about 2 hours.
4. Non-adherent bone marrow cells are recovered by gently pipetting medium over the adherent producer cells and concentrated by centrifugation.
5. The recovered bone marrow cells can be evaluated by *in vitro* colony assays and/or *in vivo* reconstitution of a lethally irradiated animal.
6. Colony forming units (CFU) granulocyte/macrophage those expressing resistance to G418 following transformation with the N2 vector and are estimated.

There are a number of advantages for the use of retroviral vectors, the major one being that they can infect a wide variety of cell types with extremely high efficiency. Technically, it is quite straightforward to transfect embryos at various developmental stages (Figure 18.15). It has been found to be easier to isolate the host sequences flanking a proviral insert than those flanking an insert derived from pronuclear injection.

IV BACTERIAL TRANSFORMATION

Most methods for bacterial transformation are based on the observations of Mandel and Higa (1970), who showed that bacteria treated with ice-cold solutions of $CaCl_2$ and then briefly heated could be transfected with bacteriophage lambda DNA (Figure 18.16). Apparently, the treatment induces a transient state of "competence" in the recipient bacteria, during which they are able

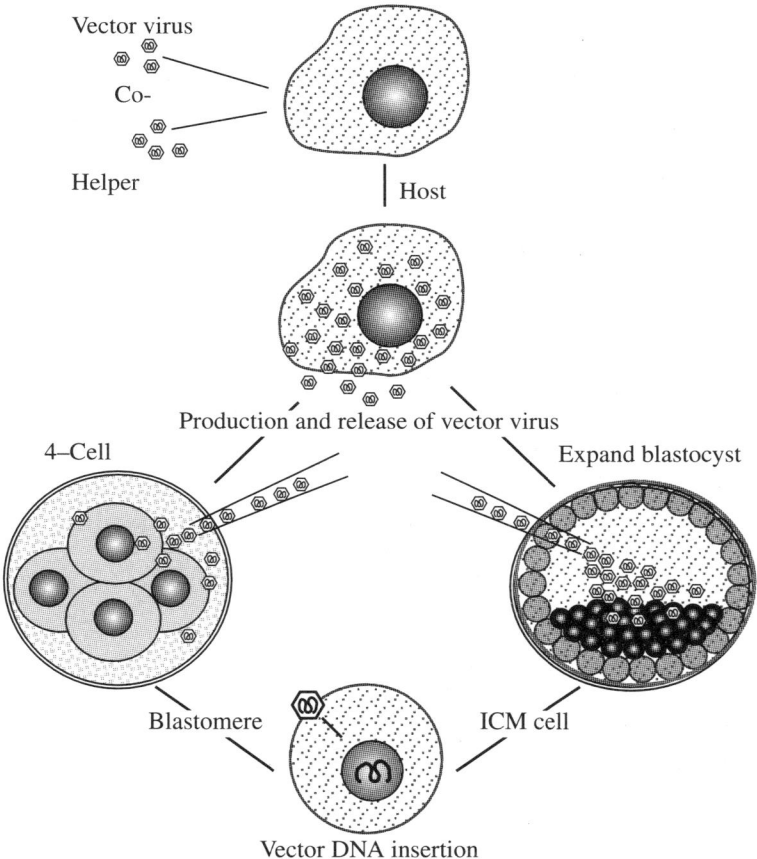

Figure 18.15 *Embryo transformation by infection with replication defective retroviral vector*

to take up DNAs derived from a variety of sources. Many variations of this basic technique have since been described, all directed toward optimizing the efficiency of transformation of different bacterial strains by plasmids. Bacteria treated according to the original protocol of Mandel and Higa yield 10^5–10^6 transformed colonies/µg of super coiled plasmid DNA. This efficiency can be increased 100- to 1000-fold by exposing improved strains of *E. coli* to combinations of divalent cations for longer periods of time and treating the bacteria with DMSO, reducing agents, and hexammine cobalt chloride.

Method-1: E. coli Transformation with plasmid DNA

The procedure described below which works well with most strains of *E. coli*, is quick, more reproducible, and yields 5×10^6 to 2×10^7 transformed colonies per microgram of super coiled plasmid DNA. Competent cells made by this procedure may be preserved at −70°C, although there may be some deterioration in the efficiency of transformation during prolonged storage. The method described here is based on a procedure developed by Mandel and Higa (1970).

506 MOLECULAR BIOTECHNOLOGY: PRINCIPLES AND PRACTICES

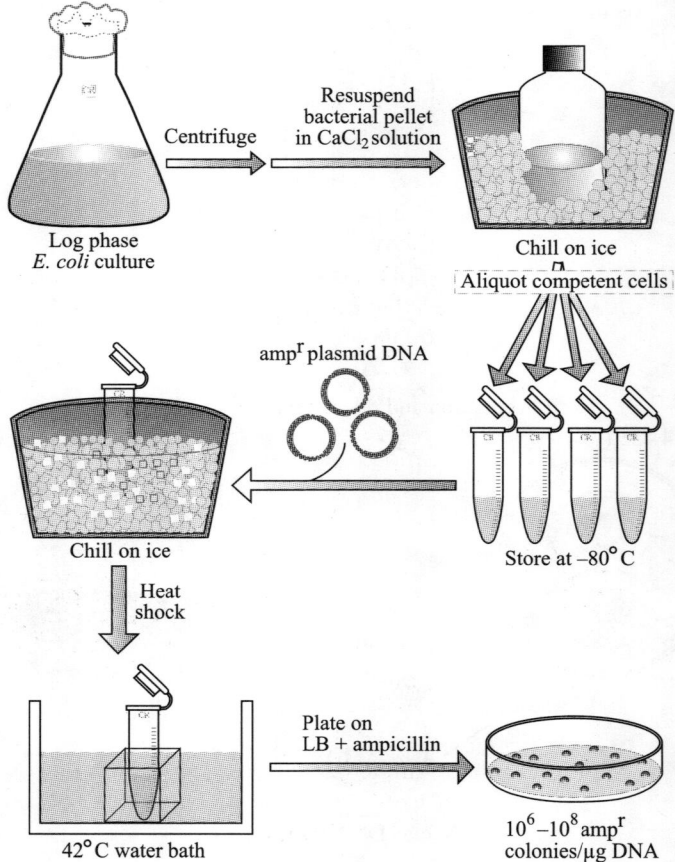

Figure 18.16 *DNA transformation of E. coli by $CaCl_2$ precipitation and heat stimulation*

Materials

Single colony of *E. coli* cells; LB medium; ice cold $CaCl_2$ solution; LB plates containing ampicillin; Plasmid DNA; Beckman JS–5.2 rotor or equivalent.

Procedure

Preparation of fresh competent E. coli using calcium chloride
1. Grow an overnight culture of the HB101 strain *E. coli* K-12 in 10 ml of LB broth at 37°C.
2. Dilute the fresh overnight culture at least 20-fold into LB broth. A 50 ml culture in a 250 ml flask is convenient for most purposes. Grow at 37°C with vigorous shaking, that is, a setting of 7 to 8 on a New Brunswick G76 shaker bath.
3. When the cell density reaches 5×10^8 cells/ml, centrifuge 40 ml in a sterile 50 ml polypropylene tube at 7,000 rpm, for 5 min., at 4°C.
4. Discard the supernatant and re-suspend the cell pellet by vortexing in 20 ml of sterile 0.05 M $CaCl_2$, 0°C. Leave the tube in ice for 15 min. and then re-centrifuge as before.

5. Discard the supernatant and re-suspend the cells by vortexing in 4 ml of 0.05 M $CaCl_2$, 0°C. This cell suspension can be stored with little or no loss of transformation competence by adding sterile glycerol to 15% (v/v), freezing aliquots in 1.5 ml polypropylene microfuge tubes with a dry ice-ethanol bath, and storing at –70°C.

Transformation of E. coli

1. Thaw the competent cells in ice water.
2. Add 0.2 ml of the cell suspension to a DNA sample tube that is in an ice water bath. This should be done shortly after the cells have been re-suspended or thawed. Each DNA sample tube has 0.1 ml of 10 mM Tris–HCl pH 7.0; 10 mM $CaCl_2$, and 10 mM $MgCl_2$. Generally, higher the concentration, lower the efficiency of transformation.
3. Shake the rack of sample tubes in the ice bath to give a good mixing of the solutions. Incubate the tubes for 25 min. in ice water, then 2 min. at 37°C and 10 min. on the lab bench.
4. Add 1 ml of LB broth and shake for 30 min. at 37°C. Then either spread 0.2 ml directly onto selective plates or plated in top agar. The former method allows more reliable replica plating because the colonies are of more uniform size. The latter method is easier if many plates must be made. But since cells in the agar will grow with poor aeration until they reach the agar surface, they will take longer to become visible. Wait 2 days for the poured plates to develop all of their colonies.

Notes

1. Expect $3-5 \times 10^6$ transformants/μg of intact pMB9 or pBR322 DNA. This efficiency may be increased by modifications such as prolonged incubation at 0°C in step 4 or the substitution of RuC_{l2} for $CaCl_2$.
2. 10 μg/ml of tetracyclin to select transformants gives high efficiency. Higher concentrations, however, lower the transformation efficiency.
3. E. coli strains vary in their ability to be transformed. Most C600 strains are transformed efficiently.

Method-2: High-efficiency transformation by electroporation

Electroporation with high voltage is currently the most efficient method for transforming E. coli with plasmid DNA. The procedure described may be used to transform freshly prepared cells that have been previously grown and frozen (See next section for more details).

Materials

Single colony of E. coli cells. LB medium (Appendix 1). Ice-cold water. Ice-cold 10% (v/v) glycerol. SOC medium (Appendix 1). LB plates containing antibiotics. Beckman JS–4.2 or equivalent and adapters for 50 ml narrow-bottom tubes. Electroporation apparatus with a pulse controller. Chilled electroporation cuvettes, 0.2 cm electrode gap.

Procedure

1. Incubate a single colony of E. coli cells into 5 ml LB medium. Grow 5 hours to overnight at 37°C with moderate shaking.
2. Inoculate 2.5 ml of the culture into 500 ml LB medium in a 2 litre flask. Grow at 37°C with shaking (300 rpm) to an OD_{600} of 0.5 to 0.6.

3. Chill cells in an ice-water bath 10 to 15 min. and transfer to a 1 litre prechilled centrifuge bottle. Centrifuge for 20 min. at 5000g, 2°C. Re-suspend pellet in 5 ml ice-cold water.
4. Add 500 ml ice-cold water, mix, and centrifuge as in step 3. Repeat once. Pour off supernatant immediately and re-suspend pellet by swirling in remaining liquid.
5. *For fresh cells*: Place suspension in pre-chilled, narrow-bottom, 50 ml polypropylene tube, and centrifuge 10 min. at 500g, 2°C. Estimate pellet volume (~500 µl/500 ml culture), add an equal volume of ice-cold water to re-suspend cells (2×10^{11} cells/ml), and aliquot 50 to 300 µl cells into chilled microcentrifuge tubes (fresh cells work better).
For frozen cells: Add 40 ml ice-cold 10% glycerol, mix, and centrifuge as described above. Estimate pellet volume, add an equal volume of ice-cold 10% glycerol to re-suspend cells, and aliquot 50 to 300 µl into pre-chilled microcentrifuge tubes. Freeze on dry ice and store at –80°C.
6. Set electroporation apparatus to 2.5 KV, 25 µF and pulse controller to 200 or 400 Ohms.
7. Add 1 µl plasmid DNA (5 pg to 0.5 µg) to tubes containing fresh or thawed cells (on ice) and mix.
8. Transfer mixture to pre-chilled cuvettes, dry outside of cuvette, and place into sample chamber.
9. Apply pulse. Remove cuvette, and immediately add 1 ml SOC medium and transfer to a sterile culture tube with a Pasteur pipette. Incubate 30 to 60 min. with moderate shaking at 37°C.
10. Plate aliquots on LB plates containing antibiotics (transformation efficiency $>10^9$/gDNA).

V THE FATE OF DNA TRANSFERRED TO EUKARYOTIC CELLS

DNA transferred to higher eukaryotic cells is rarely maintained episomally, but is integrated into the genome. Exogenous DNA can interact with the genome in three ways: (1) it can integrate randomly by illegitimate recombination; (2) it can integrate at a specific site by single cross-over homologous recombination, and (3) it can replace a fragment of the genome by double cross-over homologous recombination or gene conversion (**transplacement**). In yeast, homologous recombination is extremely efficient and most genome integration events occur either by single-cross-over integration or transplacement. In mammals, illegitimate recombination occurs 10^5 times more frequently than homologous recombination. A vector designed specifically to introduce DNA into the genome is described as a gene *delivery vector* or *suicide vector*. Homologous interactions between the host genome and vector occur only if there is a shared region of homology (enhanced by linearization).

VI INTEGRATION OF THE TRANSGENES

The success of transformation depends on integration of introduced DNA into recipient cells. The major part of DNA introduced into plant cells by direct DNA delivery methods is degraded. Only a small portion of it will be integrated into the recipient cell genome. The introduced DNA fragments (or plasmids) may become rearranged, deleted, or become joined in tandem repeats. The perfect transformant would contain a single copy of the transgene that would segregate as a Mendalian trait, with uniform expression from one generation to the next. As gene integrations are essentially random in the genome, variability is often observed from one transgenic to another, a phenomenon ascribed to 'position effect variation'. Some of the main features of transgene integration from direct DNA delivery into the plant cell genome are as follows: (i) DNA integration is random; (ii) the efficiency of integration of linearized DNA is much more than the circular DNA; (iii) introduced DNA fragments usually become linked and show co-transformation (>50%); (iv) multicopy integrations usually occur in tandem; (v) higher DNA concentrations increase the integration frequency, and the number of integrations also increase (5–10); (vi) use of carrier DNA in some cases enhances the stable integration of directly delivered DNA. Most of the features mentioned above will also be applicable to *Agrobacterium* mediated gene transfer. In addition, gene integration by homologous recombination has also been reported; in some transformed cells, restoration of defective gene activity was observed.

Genomic characterization of many T-DNA insertions convincingly showed the occurrence of many different plant/T-DNA junctions. *In situ* hybridization and genetic analysis demonstrates that T-DNA insertions can be found on every chromosome. Furthermore, sequence analysis of different left and right T-DNA/plant junction sequences proved that all are different and that no preference for a unique site or particular sequence motif can be demonstrated.

VII INHERITANCE OF TRANSGENES

Transgenic plants can presently be obtained by *Agrobacterium* transformation or by alternative methods. The inheritance of the genes introduced in these plants can be studied by the techniques of classical genetics. Segregation data allow the determination of the number and the stability of the introduced loci in the transformed clones. In most cases, stably integrated transgenes inherit in Mendelian fashion, and usually show dominance. But in many cases, in subsequent generations, some instability may occur probably due to gene modifications, which includes methylation, rearrangements of the T-DNA region. Homologous recombination between transgenes can also create instability. Gene silencing by various mechanisms has also been found to contribute to the gene inactivation.

VIII VARIATION OF GENE EXPRESSION

The level of expression of a particular gene varies in different transgenic plants. This has been studied in detail for T-DNA insertions but the same principles probably also hold for inserts obtained from other DNA transformations. The variation of expression can have several reasons: inserted DNA copy number in transgenic plants, *cis*-acting elements such as **silencers** and **enhancers** in the target sites, transcriptional interference of inserted DNA and target-expressing units, and the general chromatin structure produced by nearby heterochromatic DNA. This variation in expression between different transgenic plants can be screened for position effects, which increase the expression of particular linked genes.

IX SUGGESTED READING

Brown TA (1991) (Eds) *Essential Molecular Biology: A Practical Approach* (2 volumes), IRL Press, Oxford.

Channarayappa (1992) Resistance mechanism of *Lycopersicon* species to the vector *Bemisia tabaci* and ultrastructural changes in leaf curl virus infected tomato plants. Ph.D. Thesis Submitted to the University of Agricultural Sciences, GKVK, Bangalore, pp151.

Channarayappa, Muniyappa V, Schwegler–Berry D and Shivashankar G (1992) Ultrastructural changes in tomato infected with tomato leaf curl virus, a whitefly–transmitted geminivirus. *Canadian Journal of Botany*, **70**:1747–1753.

Chung CT, Niemeia SL and Miller RH (1989) One step preparation of competent *E. coli:* Transformation and storage of bacterial cells in the same solution, *Proc Natl Acad Sci* USA **86**:2172–2175.

Hanahan D (1985) Techniques for the transformation of *E. coli*, In: *DNA Cloning, A Practical Approach* (Ed Glover DM) **1**:109–135.

Minhas BS and Voelkel SA (1989) Transgenic Animals. In: *Biotechnology: A comprehensive Treatise in 8 Volumes*, (Eds) Rehm HJ and Reed G, (vol 7b) *Gene Technology*, (Eds) GR Jacobson and Jolly SO.

Old RW and Primorose SB (1996) *Principles of Gene Manipulation: An Introduction to Genetic Engineering*, Blackwell, Science, Oxford.

Sambrook J, Fritsch EF, and Maniatis T (1989) *Molecular Cloning: A Laboratory Manual* (3 volumes), Cold Spring Harbor Press, Cold Spring Habor, NY.

Saul MW, Shillito RD and Negrutiu I (1988) Direct DNA transfer to protoplasts with and without electroporation, *Plant Molecular Biology Manual*, Kluwer Academic Publishers.

Tomes DT, Ross M, Higgens R, Rao AG, Staebell M and Howard J (1990) Direct DNA transfer into intact plant cells and recovery of transgenic plants via microprojectile bombardment. *Plant Molecular Biology Manual*, **A13**:1–22.

Watson JD, Gilman M, Witkowski J and Zoller M (1992) *Recombinant DNA*. (2nd Edn) Scientific American Books, NY.

19

Selection and Screening of Recombinant Molecules

I INTRODUCTION

Generally, the first step of a cloning experiment is the identification of recombinant phages, clones of bacteria, or cells. This chapter describes a number of basic and applied procedures commonly used to select, and characterize recombinant DNA products. These procedures could, for example, be used to obtain a recombinant DNA clone that contains a particular gene. The task of isolating a desired recombinant from a population of bacteria or phages depends very much upon the cloning strategy that has been adopted. For instance, when a cDNA derived from an abundant mRNA is to be cloned, the task is relatively simple – only a small number of clones need to be screened. Isolating a particular single-copy gene sequence from a complete eukaryotic genomic library requires techniques in which thousands of recombinants need to be screened. In this chapter, an overview of some basic and applied techniques employed in recombinant selection and screening procedures are described.

As mentioned before, isolating a desired recombinant from a population of transformed bacteria depends very much on the cloning strategy that has been adopted; for example, if a synthetic gene has been cloned, no selection is necessary because every transformed cell will contain the correct sequence. A number of different methods have been devised to facilitate screening of recombinants. These include genetic methods, immunological methods, and methods based on nucleic acid hybridization.

II BASIC TECHNIQUES USED IN RECOMBINANT DNA TECHNOLOGY

A. TECHNIQUES THAT USE RADIOACTIVE TRACERS

Radioactive tracers have been available for decades, and they are easy to use and detect. Radioactive tracers allow an enormous increase in sensitivity, such that very minute quantities of a substance may be detected. In molecular biology, most of the substances used in the experiments are present in very tiny amounts. In this section, some important techniques which molecular biologists use to detect radioactive tracers such as autoradiography, phosphorimaging, and liquid scintillation counting are briefly discussed.

1. The nature of radioactivity

An atom is composed of a positively charged nucleus that is surrounded by a cloud of negatively charged electrons. The mass of an atom is concentrated in the nucleus and is made up of two particles, protons and neutrons. Protons are positively charged particles. The number of orbital electrons in an atom must be equal to the number of protons present in the nucleus. This number is known as the atomic number (Z). Neutrons are uncharged particles with a mass approximately equal to that of a proton. The sum of protons and neutrons in a given nucleus is the mass number (A). Since the number of neutrons in a nucleus is not related to the atomic number, it does not affect the chemical properties of the atom. Atoms of a given element with different mass numbers (different numbers of neutrons) are called isotopes. In general, the ratio of neutrons to protons in the nucleus will determine whether an isotope of a given element is stable enough to exist in nature (Figure 19.1). Unstable isotopes or radioisotopes are often pro-

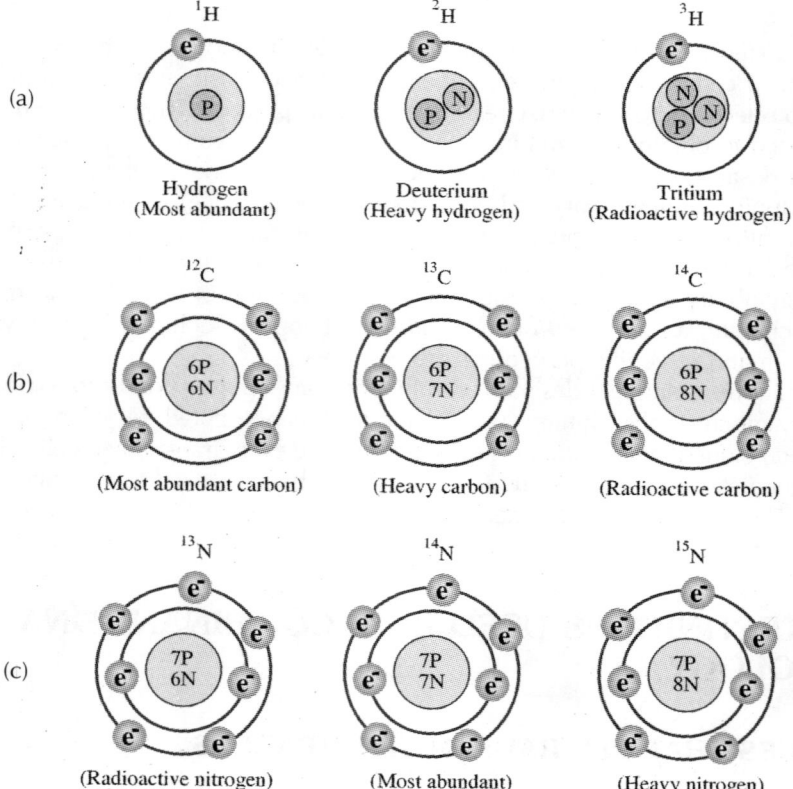

Figure 19.1 *Selected examples of atomic structures and their isotopes: (a) Hydrogen, (b) Carbon, and (c) Nitrogen*

duced artificially. Radioisotopes emit particles and/or electromagnetic radiation as a result of changes in the composition of the atomic nucleus.

2. Applications of radioisotopes in the biological sciences

Isotope tracers are very powerful tools for identifying the metabolic fate of precursors and for following metabolic turnover. Numerous examples of the application of radiolabeling may be found in literature. Some of the applications are in: (1) Investigating aspects of metabolism, (2) Tracing metabolic pathways, (3) Following Metabolic turnover times, (4) Studies of absorption, accumulation and translation, (5) Pharmacological studies, (6) Analytical applications, (7) Enzyme and ligand binding studies, (8) Isotope dilution analysis, (9) Radioimmunoassay, (10) Radiodating, (11) Molecular biology techniques, (12) Clinical diagnosis, (13) Ecological studies, (14) Sterilization of food and equipment, (15) Mutagenesis studies, (16) Other applications.

3. Methods based upon exposure of photographic emulsions

Ionizing radiation acts upon a photographic emulsion to produce a latent image much as visible light does. For a photograph, a radiation source, an object to be imaged, and a photographic emulsion are required. For an autoradiograph, a radiation source emanating from within the material to be imaged is also required, along with a sensitive emulsion. The emulsion consists of a large number of silver halide crystals embedded in a solid phase such as gelatin. As energy from the radioactive material is dissipated in the emulsion, the silver halide becomes negatively charged and is reduced to metallic silver, thus forming a particulate latent image. This process, which is known as autoradiography, is very sensitive and has been used in a wide variety of biological experiments.

(a) Autoradiography

Autoradiography is a means of detecting radioactive compounds with a photographic emulsion. A variety of emulsions are available with different packing densities of the silver halide crystals. The most preferred form of emulsion is a piece of x-ray film. As shown in the Figure 19.2, the DNA fragments are radioactively labeled and resolved by gel electrophoresis. Then the gel is placed in contact with the x-ray film (direct autoradiography) and left in the dark for a few hours or days. The time of exposure depends upon the isotope, sample type, level of activity, film type, and purpose of the experiment. The radioactive emissions from the gel bands of DNA will expose the film and dark bands corresponding to the DNA bands will form on the film. This is called *autoradiography*, since the DNA bands in the gel take a picture of themselves.

If we want to enhance the sensitivity of the autoradiography process, we can use fluorography or an *intensifying screen*. In case of fluorography, fluor (e.g. 2,5-diphenyloxazole (DPO) or sodium silicate) is infiltrated into the gel, and the gel dried and then placed in contact with a pre-flashed film. The sensitivity can be increased by several orders of magnitude. This is because the negatrons emitted from the isotope will cause the fluor to become excited and emit light, which will react with the film. In case of intensifying screens, the screen is coated with a compound (solid phosphor) that fluoresces when it is excited by b-rays at low temperature. Thus, we can put our radioactive material (e.g. gel) on one side of a film and the intensifying screen on the other. Some b-rays expose the film directly, but others pass right through the film and would be lost if we did not use a screen. When these high-energy electrons strike the screen, they cause fluorescence, which is detected by the film.

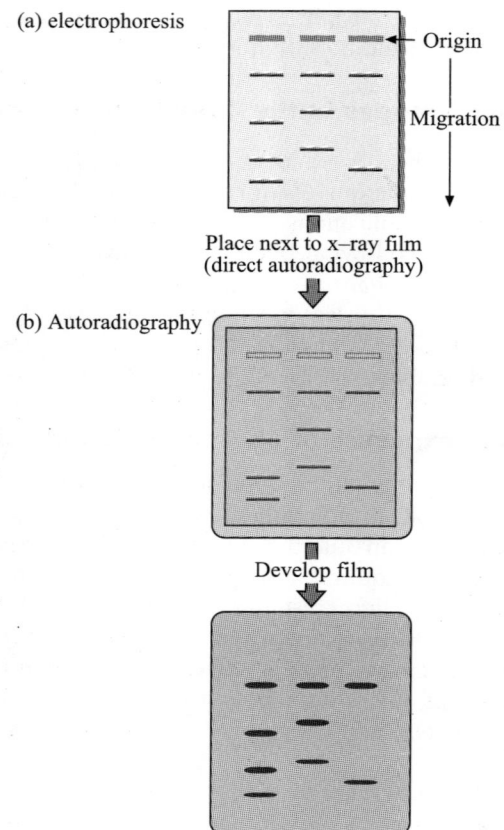

Figure 19.2 *Diagram illustrating the autoradiography technique*

We can detect the presence of sequence of interest, the molecular weight, and a rough estimate of the amount of radioactivity by looking at the intensity of a band on an autoradiograph. A better estimate of radioactive intensity can be obtained by scanning the autoradiograph with a **densitometer**. There are many varieties of densitometers available and the choice made will depend on the purpose of the experiment. Quantitation is not reliable at low or high levels of exposure. However, more accurate readings can be obtained by scanning the gel with a phosphorimager, or by subjecting the sample to liquid scintillation counting.

(b) Phosphorimaging

The phosphorimaging technique has many advantages over standard autoradiography, but the most important is that it is much more accurate in quantifying the amount of radioactivity in a substance. With a standard autoradiography, a band with 50,000 radioactive disintegrations per minute (dpm) may look no darker than one with 10,000 dpm because the emulsion in the film is already saturated at 10,000 dpm. But the phosphorimager collects radioactive emissions and analyzes them electronically, so the difference between 10,000 and 50,000 dpm would be obvious. The working principle of the phosphorimager is that when we place a sample in contact with a phosphorimager plate, which absorbs β-rays, these rays excite molecules on the plate, and these molecules remain in an excited state until the phosphorimager scans the plate with a laser. At that point, the β-rays energy trapped by the plate is released and monitored by a computerized detector. The computer converts the energy it detects to an image which will be a false color image where different colors represent different degrees of radioactivity, from the lowest (yellow) to the highest (black).

(c) Liquid scintillation counting

The principle behind this method is to convert the radioactive emissions from a sample to photons of visible light that a photomultiplier tube can detect. This can be performed by placing the radioactive sample into a vial with **scintillation fluid**. This liquid contains flour, a compound that fluoresces when it is bombarded with radioactivity. The instrument counts the bursts of light or **scintillation**, and records them as counts per minute cpm. This is not the same as dpm because the scintillation counter is not 100% efficient. However, if radioisotope ^{32}P is used, the scintillation counter can directly count without the need of flour at lower efficiency.

B. NON-RADIOACTIVE TRACERS

Even though, radioactive tracers provide many advantages, non-radioactive detection methods, that have been recently developed, are now being widely used. Non-radioactive methods have a significant advantage because radioactive substances pose a health hazard and need special care in handling the material and the radioactive waste. Non-radioactive tracers gives comparatively higher sensitivity by using the multiplier effect of an enzyme. That is, if we can couple an enzyme with a probe that detects the molecule of interest, the enzyme will produce many molecules of the product, thus amplifying the signal. This works well if the product of the enzyme is chemiluminescent, because each molecule emits many photons, amplifying the signal again. Figure 19.3 shows the principle behind one such tracer method. The light can be detected by autoradiography with x-ray film, or by a phosphorimager. If we want to avoid the expense of a phosphorimager or an x-ray film, we can use enzyme substrates that change color instead of becoming chemiluminescent. These chromogenic substrates produce colored bands corresponding to the location of the enzyme, and therefore to the location of the molecule we are trying to detect. The intensity of the color is directly related to the amount of our molecule of interest, so this is also a quantitative method.

III GENETIC METHODS

A. IDENTIFYING RECOMBINANT PLASMIDS BY SIZE

A plasmid containing a fragment of introduced DNA is usually bigger than the original uncleaved plasmid. This increase in size can be used to detect plasmids with an insert by agarose gel electrophoresis. Single colonies of plasmid-containing cells can be lysed and directly examined by electrophoresis. In a single bacterial colony, enough plasmid DNA is visible as a single band moving far ahead of the chromosomal DNA. The plasmid DNA moves a distance related

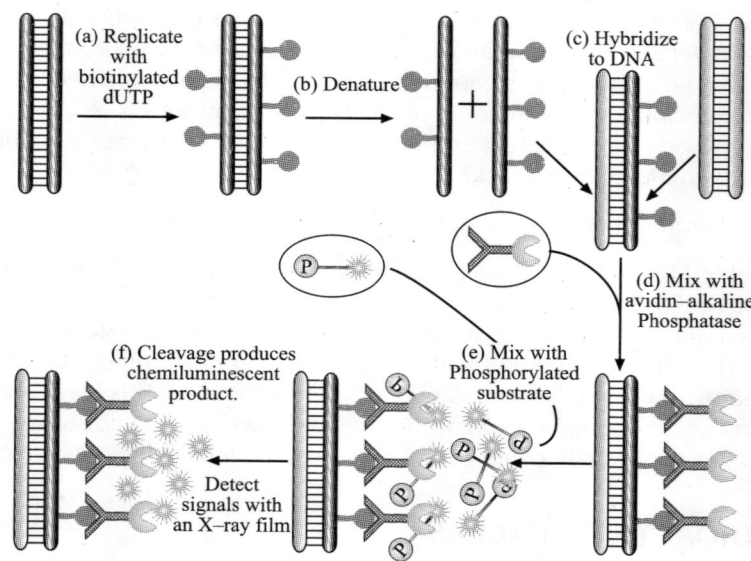

Figure 19.3 *Detecting nucleic acids with a non-radioactive probe*

to its molecular weight – the larger the DNA, the smaller the distance moved in a given time internal. Hence colonies containing larger plasmid DNA molecules, which contain donor DNA, move more slowly than those lacking donor DNA.

B. GENETIC COMPLEMENTATION TEST

A prerequisite for selection by complementation is the existence of auxotrophic or conditionally lethal mutations in the host organisms. A variety of mutants, in the metabolism of carbon source, amino acids, and nucleic acids are known for *E. coli*, and also for other organisms. Suppose a bacterial strain is available which has a mutation in a gene encoding an enzyme involved in the biosynthesis of amino acid histidine. Such a mutant will only grow in a medium supplemented with histidine. By cloning DNA from the normal strain, i.e. one that can synthesize its own histidine in the mutant strain and selecting those transformants which grow in the absence of histidine it is possible to isolate the gene of interest. This test can be used to select even yeast and mammalian genes. For example, a gene from the histidine biosynthetic pathway of the Archaea *Methanobacteria voltae* was identified by cloning random fragments of *M. voltae* into a plasmid, transforming the DNA into an *E. coli* his mutant and selecting those transformants which grow on plates with no histidine in the medium.

C. SELECTION BY MOUSE DIHYDROFOLATE REDUCTASE RESISTANCE ASSAY

In an analogous procedure, a clone carrying the mouse dihydrofolate reductase (DHFR) gene was selected from a population of recombinant plasmids containing cDNA derived from an unfractionated mouse cell mRNA preparation. In this instance, the premise was that mouse DHFR is more resistant to the inhibitor trimethoprim than *E. coli* DHFR. When transformed *E. coli* was grown on a medium containing trimethoprim, only those cells carrying the mouse DHFR gene survived.

D. SELECTION FOR PRESENCE OF VECTOR

Genetic selections, when combined with microbiological techniques, become a very powerful tool for screening large populations. Most laboratory-constructed vectors carry a selectable genetic marker or property, thus facilitating the selection procedure. Plasmid and cosmid vectors carry drug-resistance or nutritional markers and serve as selection pressure for screening. In the case of phage vectors, plaque formation is itself the selected property. Genetic selection for the presence of the vector is a prerequisite stage in obtaining the recombinant population. Insertional inactivation of a drug-resistance marker, or of a gene such as β-galactosidase for which there is a color test, are examples of this. With certain replacement-type lambda vectors, and with cosmid vectors, size selection by the phage particle selects recombinant formation.

E. SELECTION OF INSERTED SEQUENCES

If an inserted foreign gene in the desired recombinant is expressed, then genetic selection may provide the simplest method for isolating clones containing the gene. Cloned *E. coli* DNA fragments carrying biosynthetic genes can be identified by complementation of non-revertible auxotrophic mutations in the host *E. coli* strain. For example, Cameron et al., (1975) who cloned the *E. coli* DNA ligase gene in a phage λgt IB vector, exploited the inability of the λred⁻ phage (the vector is red⁻ by deletion of the C fragment) to form plaques on *E. coli* λigts at the permissive temperature, whereas λred–phage will form plaques on *E. coli* Lig⁺. Recombinant phage carrying the wild-type ligase function could therefore be selected simply by their ability to form plaques through complementation of the host deficiency when plated on *E. coli* λigts.

IV IMMUNOCHEMICAL METHODS

Immunochemical detection of clones synthesizing a foreign protein has also been successful in cases where the inserted gene sequence is expressed. A particular advantage of the method is that even genes that do not confer any selectable property on the host can be detected, but it does of course require that the specific antibody is available.

A. RADIOACTIVE ANTIBODY TEST

If the protein product of the gene of interest is made, two immunological assays allow the protein-producing colony to be identified. If the protein is excreted, a radioactive antibody test can be used (Figure 19.4). An autoradiograph will show which colony synthesized the gene product.

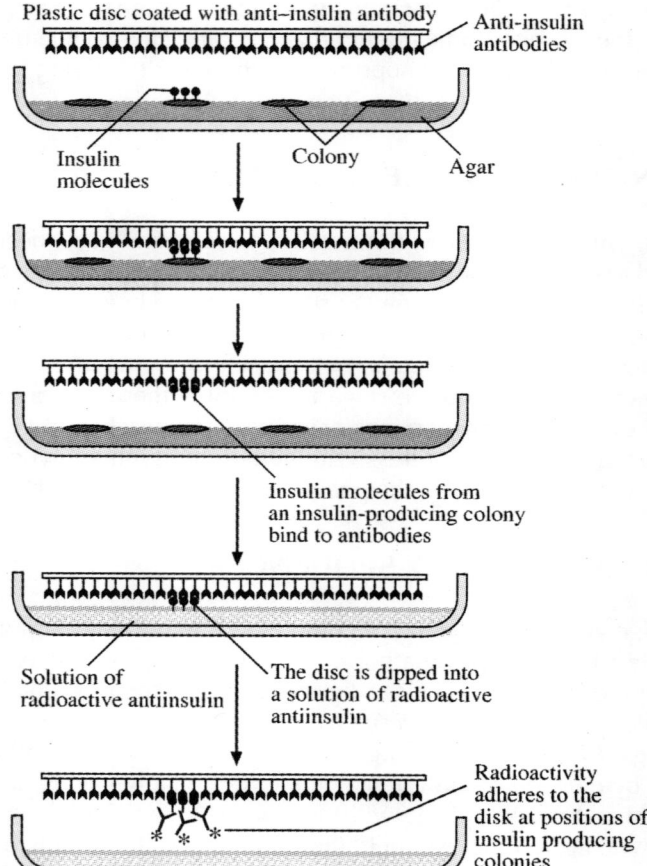

Figure 19.4 *Diagram illustrating the radioactive antibody test*

B. IMMUNOPRECIPITATION TEST

An immunoprecipitation test can also be used to identify a protein-producing colony by directly adding a specific antibody to the medium. If the protein is excreted, an antibody–antigen precipitate called **precipitin** forms a visible white ring around the colony producing the protein (Figure 19.5). Some slight modifications of this procedure allow one to detect a protein that is not excreted. In one modification, a host cell is used that is lysogenic for *lcI857*. A replica of the plate containing colonies to be tested is made by transforming the colonies onto agar containing antibody. As soon as the colonies on the antibody plate become visible, the temperature is raised to 42°C to inactivate the temperature-sensitive *cI857* repressor. This induces lysis of many of the cells, releasing the cell contents. The protein of interest then reacts with the antibody and forms a precipitin ring at the colony. In the second modification, soft agar containing antibody and the enzyme lysozyme is poured on the colonies, releasing the

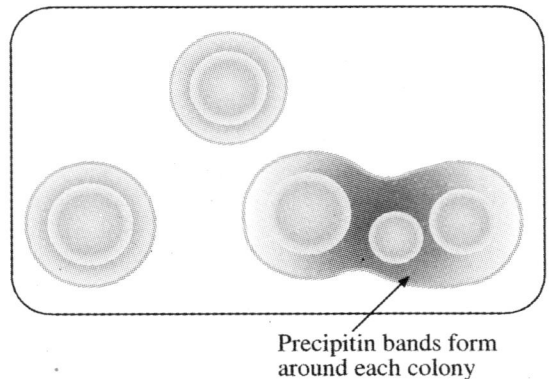

Figure 19.5 *Diagram illustrating the immunoprecipitation test. Colonies making β-galactosidase are formed on agar containing antibody to the enzyme*

intracellular proteins, resulting in the formation of a precipitin ring around the colonies that produced the appropriate protein.

A very efficient exploitation of the immunochemical detection method involves the phage λ expression vector λgt11. This vector carries the *E. coli lac* Z gene. A unique *Eco*RI site is located within the β-galactosidase coding region. Recombinant libraries can be constructed in which eukaryotic cDNA has been inserted, by means of linkers, into the *Eco*RI sites. In such recombinants the β-galactosidase is insertionally inactivated and, depending upon the translational phase at the fusion junction, hybrid proteins are expressed.

V NUCLEIC ACID HYBRIDIZATION METHODS

As with genetic selection, immunochemical identification requires that the target gene be expressed. However, expression of eukaryotic genes often requires so much effort that it is not feasible to begin expression work until the desired clone is definitely known to be in hand. Thus, it is essential to identify the target gene without expression. This can be accomplished by nucleic acid hybridization.

The colony or *in situ* hybridization assay allows detection of the presence of a particular gene. If an appropriate hybridization probe is available, it can be used to detect clones with the desired insertion. For example, if the DNA sequence of a homologous gene from another organism is known or if a partial amino acid sequence of the protein is known, a small oligonucleotide of the desired sequence can be made, radiolabeled, and used as a hybridization probe.

A. *IN SITU* HYBRIDIZATION

Grunstein and Hogness (1975) developed a screening procedure to detect DNA sequences in transformed colonies by hybridization *in situ* with a radioactive 'probe' RNA. A modification of the method allows screening of colonies plated at a very high density. The colonies to be screened are first replica plated onto a nitrocellulose filter disk that has been placed on the surface of an agar plate prior to inoculation (Figure 19.6). A reference set of these colonies on the master plate is retained. The filter bearing the colonies is removed and treated with alkali so that the bacterial colonies are lysed and their DNAs are denatured. The filter is then treated with proteinase K to remove protein and leave denatured DNA bound to the nitrocellulose, for which it has a high affinity, in the form of a 'DNA-print' of the colonies. The DNA is fixed firmly by baking the filter at 80°C. The defining, labeled RNA is hybridized to this DNA and the result of this hybridization is monitored by autoradiography. A colony whose DNA print gives a positive autoradiographic result can then be picked from the reference plate.

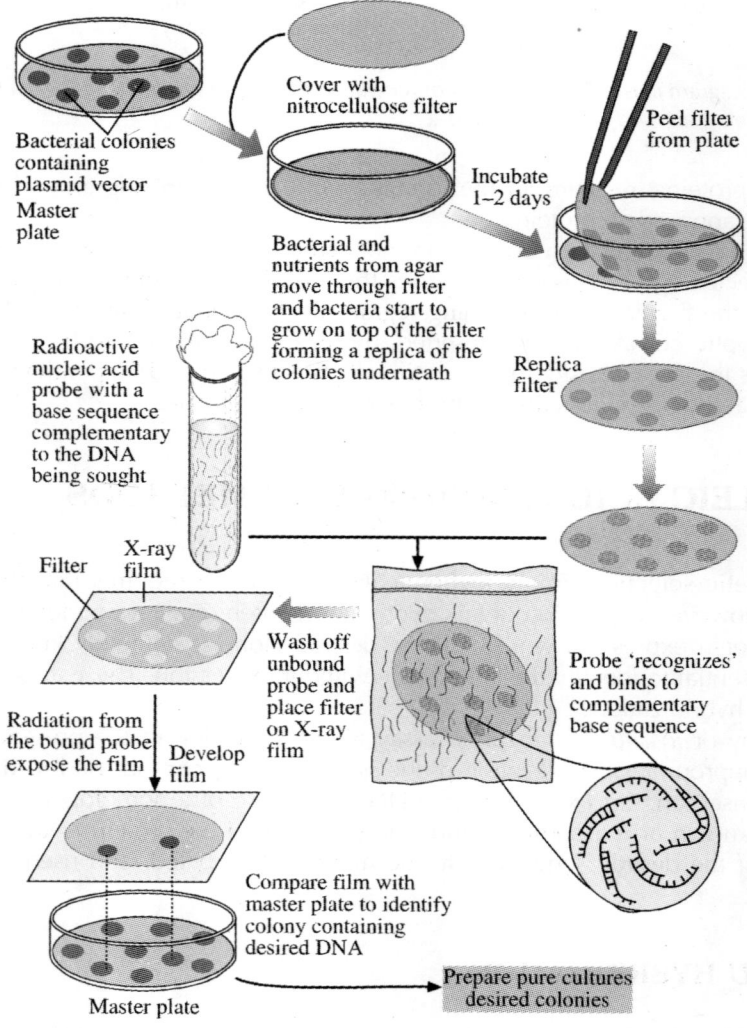

Figure 19.6 *Screening a library with a nucleic acid probe (a gene probe) to find a clone*

Variations of this procedure can be applied to phage plaques. Benton and Davis (1977) devised a method, in which the nitrocellulose filter is applied to the upper surface of agar plates, making direct contact between plaques and filter. The plaques contain phage particles as well as a considerable amount of unpackaged recombinant DNA. Both phage and unpackaged DNA bind to the original 'plaque-lift'. This procedure has the advantage that several identical DNA-prints can easily be made from a single-phage plate. This allows the screening to be performed in duplicate, and hence with increased reliability, and also allows a single set of recombinants to be screened with two or more probes (Figure 19.7). This is the prototype for a variety of plaque-lift screening procedures. The great advantage of the hybridization methods is generality. They do not require expression of the inserted sequences and can be applied to any sequence, provided a suitable probe is available.

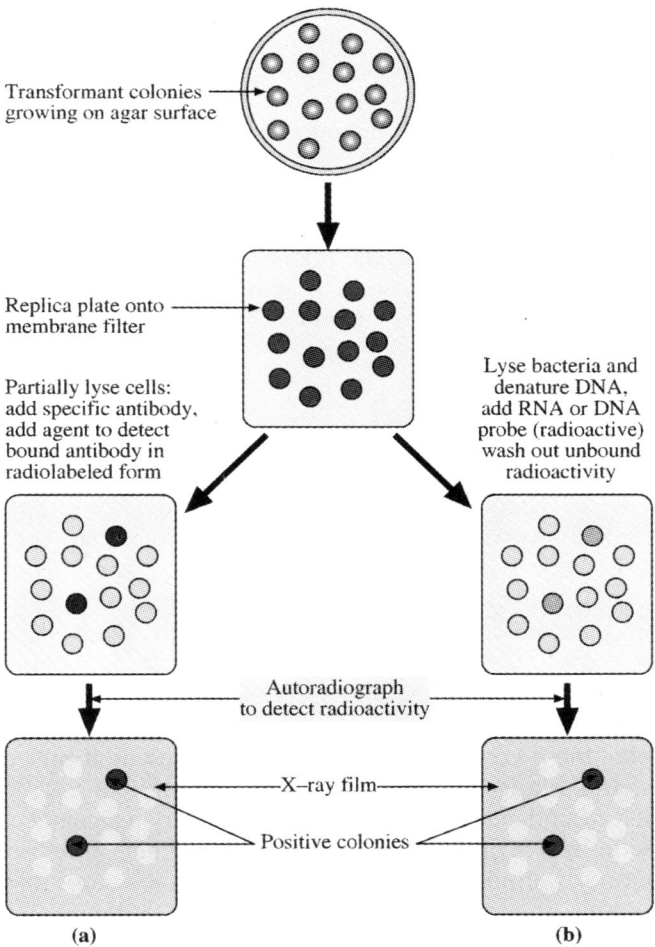

Figure 19.7 *Finding the right clone (a) Technique for detecting production of protein by use of specific antibody. (b) Method for detection of recombinant clones by colony hybridization with a radioactive nucleic acid probe*

The probe problem

Screening procedures, which rely on nucleic acid hybridization, as we have seen, are general in application and powerful. Using these procedures it is now possible to easily isolate any gene sequence from virtually any organism, if a probe is available. The problem of gene isolation, then, is the problem of obtaining a suitable probe. There are several solutions to this.

In certain specialized cell types or tissues, particular mRNAs are abundant. Thus, the corresponding cDNA clones can be isolated by screening small numbers of recombinants directly by sequencing. These cDNA probes can then be used to isolate genomic sequences.

Nucleic acid sequences encoding certain proteins have been sufficiently conserved in evolution such that cross-species nucleic acid hybridization is possible. This means that the particular biological advantages of one experimental system can be exploited to isolate a gene sequence, which may then provide a probe for corresponding genes in other organisms.

An oligonucleotide, which needs to be only about 14–20 nucleotides long, and which corresponds to a part of the sequence encoding the protein in question, can be synthesized chemically. What length of oligonucleotide is required for reliable hybridization? Even though 11-mers can be adequate for southern blot hybridization, longer probes are necessary for good colony hybridization. Mixed probes of 14 or 16 nucleotide lengths have been successful.

In vitro radiolabeling of DNA and RNA: Most of the techniques used to analyze and identify nucleic acids require some kind of hybridization reaction that is assayed by following radioactively labeled RNA or DNA. By far the highest specific radioactivities in nucleic acids are achieved by *in vitro* methods and these high activities lead to the most sensitive assays. Different methods of probe labeling are explained in Chapter 13.

B. DIFFERENTIAL SCREENING

This is a variant of the nucleic acid hybridization method that is particularly suitable for isolating tissue-specific or developmentally-regulated cDNA sequences or clones derived from mRNAs that are induced by particular treatments. Let us consider, for example, the isolation of cDNAs derived from mRNAs which are abundant in the gastrula embryo of the frog *Xenopus* but which are absent, or present in low abundance, in the egg. A cDNA clone library is prepared from gastrula mRNA. Replica filters carrying identical sets of recombinant clones are then prepared (Figure 19.8). One of these filters is then probed with ^{32}P-labelled mRNA (or cDNA) from gastrula embryos, and one with ^{32}P-labelled mRNA (or cDNA) from the egg. Some colonies will give a positive signal with both probes; these represent cDNAs derived from mRNA types that are abundant at both stages of development. Some colonies will not give a positive signal with either probe, these correspond to mRNA types present at undetectably low abundance in both tissues. This is a feature of using probes derived from mRNA populations: only abundant or moderately abundant sequences in the probe carry a significant proportion of the label and are effective in hybridization. Importantly, some colonies give a positive signal with the gastrula probe, but not with the egg probe. These should, therefore, correspond to the required sequences.

Subtractive cDNA cloning and differential screening

As we have seen, differential screening is applicable in many biological situations where it is desirable to isolate cDNAs derived from mRNAs that are induced by a particular treatment. In

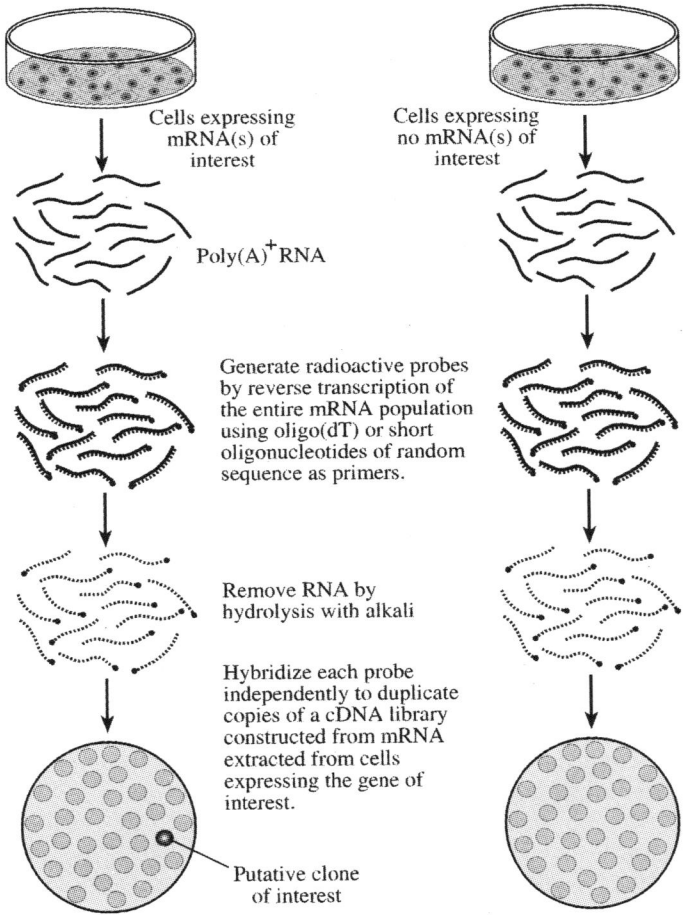

Figure 19.8 *Differential screening*

order to increase the efficiency of the procedure, it is beneficial to be able to create a cDNA library that is enriched in the desired sequences. This can be achieved by subtractive hybridization, the essence of which is to remove those cDNA sequences that are ubiquitous or not induced.

C. SCREENING OF PHAGE PLAQUES

The recombinant phages are plated and allowed to develop until the plaques are almost confluent. Some of the phages from each plaque are then transferred to filter paper and the DNA in them is denatured and immobilized on the paper. This DNA is then allowed to hybridize with a [^{32}P]-labeled nucleic acid and, after washing, the location of hybridization is detected by autoradiography. Phages from the positive plaques can be recovered from the plaques on the original plate.

Method-1: Recombinant phage screening

Procedure

1. Grow 50 ml of an appropriate bacterial host in LB broth plus 0.2% maltose overnight with vigorous shaking. The A_{600} should reach approximately 2.
2. Pellet 40 ml of the cells in a sterile 50 ml polypropylene tube for 10 min. at 5000 rpm in the SS34 Sorvall rotor. Re-suspend the cells in 20 ml of 10 mM $MgCl_2$ at 4°C.
3. When screening large numbers of phages, it is most convenient to screen them at the highest manageable concentration. Adsorb the phage to cells at 37°C for 15 min. in the proportion of 5×10^4 phage/1.75 ml of cells. Dilute 1.75 ml of this with 30 ml of 0.7% (wt/vol) agarose in LB broth at 47°C and pour onto 3 LB-agar petri plates of 140 mm diameter or onto 10 standard 90 mm diameter petri plates. If only 10^4 or fewer phage must be screened it is more convenient to adsorb at the same concentration and then spread at a greater dilution so that the plaques will be well separated. This can be done by diluting with a much larger volume of LB-agarose and then spreading on more plates.
4. Let the agarose harden for 10 min. at room temperature and then incubate the plates at 37°C, face up, until the plaques are nearly confluent. On high-density plates, this usually requires about 6 hours and on low-density plates 8–12 hours.
5. Incubate the plates for at least 1 hour at 4°C and then fit the nitrocellulose filter paper onto the agarose surface of a plate so that no air bubbles are trapped between the agarose and the paper. Do this by holding the circle with tweezers at two points 180°C apart and allow the middle of the circle to curve down onto the agarose. Do not allow the plates to warm up before transfer. If you have to work at room temperature, then use the plates immediately after removing them from 4°C. The filters should be marked for orientation by dipping a 22-gauge needle into India ink and making an asymmetric set of dots on the circumference of the circle, poking holes through the surface of the filter and into the agarose. This will simultaneously label the plate and the filter and will orient the filter relative to the plaques on the plate. The nitrocellulose paper will shrink upon heating. For reasons that will become obvious, the filters should be pre-shrunk before they are placed on the phage plates. Do this by boiling them in water for 5 min. and blotting to damp dryness or by autoclaving for 5 min. Leave the paper on the agarose for 2 min. and then lift it with tweezers. For a second paper replica of the plate, apply another paper for 3 min.; for a third, use 4 min. As with Southern transfers, handle the nitrocellulose paper with tweezers and wear gloves so that you can grab the paper should an emergency arise.
6. At room temperature, immerse the replica filters in 0.1 N NaOH; 1.5 M NaCl for 2 min., then 0.2 M Tris–HCl pH 7.5; 2X SSC for 2 min., and then twice for 15 min. in 2X SSC. Blot-dry with Whatman 3 MM paper. Bake and, if the filter-bound DNA is to be hybridized to radioactive DNA, soak in rehybridization solution as described in the Southern transfer procedure.
7. Hybridization is done at the aqueous (non-formamide) conditions described for Southern transfers. The washing is also essentially the same as for Southern transfer filters. Wash the filters twice in 2X SSC for a few min. at room temperature. Then wash for 3–6 hours at 65°C in 0.5X SSC, 0.5% SDS. Dry the filters thoroughly and mark the orientation spots with a radioactive compound. Autoradiography is done as described for Southern transfers.
8. Filter paper replicas may be reused by repeating steps 6 and 7; however, the paper may become more brittle and cracks with frequent use. Also, the hybridization signals diminish with successive uses probably because DNA falls off the filters.

9. To be sure that a radioactive spot is the result of specific hybridization, hybridize and autoradiograph two filter transfers from each phage plate. If both autoradiograms give a signal in the same spot, then specific hybridization has usually occurred.

Phage recovery procedure:

1. Place the developed X-ray film on a light box and align the phage plate on top of it. Remove the positive agar region by making a core of the plate surface with the mouth of a sterile Pasteur pipette. Dissolve the core in 1 ml of lambda dilution buffer (10 mM Tris–HCl pH 7.5; 10 mM $MgCl_2$).
2. Titer the phage concentration and then spread at approximately 200 pFU/plate and re-screen the paper replica in order to purify the positive phage. Two additional cycles of picking and screening are advisable when the original screen is done at high phage density. When picking phage plaques from low-density plates, use the tip of a Pasteur pipette. Always pick and screen one cycle after an easily identifiable and well-separated positive plaque has been found.

Growing phage stocks:

Stocks of the desired recombinant phage can be made either on plates or in liquid. The DNA can be purified by the methods described in Chapter 7. However, the following modification of the SDS–DNA extraction method is useful for a quick and dirty extraction of DNA that will be clean enough for restriction enzyme digests.

Procedure

1. Begin with 10 ml of a lysate or liquid lysate with phage at a concentration greater than 10^{10} particles/ml.
2. Add 2 ml of 0.25 M EDTA; 0.5 M Tris-HCl pH 9; 2.5% SDS and incubate at 65°C for 30 min.
3. Add 0.5 ml of 8 M K^+–acetate and chill on ice for 15 min. Centrifuge at 17,000 rpm in the SS34 Sorvall rotor for 15 min. Decant the supernatant into a clean tube. If any precipitate is carried over into the clean tube, repeat the centrifugation and decant the supernatant. Add 5.6 ml of 95% ethanol, mix and chill at –70°C for 20 min. or at –20°C for 30 min. Centrifuge at 17,000 rpm in a Sorvall SS34 rotor for 15 min., decant the supernatant, and remove any excess liquid with a Q-tip.
4. Dissolve the precipitate in 0.4 ml of 2 M NH_4^+–acetate, using a Pasteur pipette to break up the pellet. Transfer to a 1.5 ml microfuge tube and add 0.8 ml of 95% ethanol. Vortex and precipitate the DNA by the standard method for rapid precipitation.
5. This DNA may be digested with restriction enzymes by using about 1 µl of the DNA solution in 10 µl of reaction volume, adding sufficient restriction enzyme to digest 2 µg of DNA in 30 min., and incubating for 30 min. The DNA product is occasionally contaminated by a large amount of RNA. If this RNA obscures the restriction pattern on gels, then follow the restriction digestion with a digestion by 1 µg of RNase for 15 min. at 37°C.

Screening colonies:

The method described below is considerably simpler than the original Grunstein and Hogness (1975) method. It is also less expensive and produces filters that can be reused many times. The two methods appear to have about the same maximum sensitivity. However, to achieve this maximum with the method we describe, some attention must be paid to the rate of the hybridization reaction. For most applications, the rates can be largely ignored.

Procedure

1. Grow colonies to a diameter of 1–4 mm. Transfer them to Whatman 541 filter paper by placing the paper over the colonies and incubating at 37°C for 2 hours. When placing the paper over the colonies, do not trap air between the paper and the agar surface because an air bubble will prevent transfer.
2. If the plasmids have relaxed control of DNA replication, then amplify the plasmid copy number by transferring the paper, face down, onto LB agar, 250 µg/ml chloramphenicol, and incubating for 24 hours at 37°C.
3. Lyse the colonies and denature and immobilize their DNA on the paper by washing twice with agitation for 5 min. each in the following 3 solutions: 0.5 M NaOH; 0.5 M Tris-HCl pH 7.4; and 2X SSC pH 7. Wash briefly in 95% ethanol and dry in air. The filters have very high wet strength and the DNA is bound to the filters very rapidly. For these reasons, all filters can be treated as a single batch and no special care is necessary.
4. Label the filters with black ink "Sharpie" pens (Sanford's).
5. Hybridize in 50% formamide, 5X SSC pH 7.5; 250 µg/ml carrier tRNA or DNA in a volume of 75 µl/cm^2 of filter. In general, about 10^5 cpm of [^{32}P]–labeled RNA or DNA/ml of hybridization solution are hybridized with gentle agitation at 37°C for approximately 24 hours. The optimum length of hybridization time will vary with the complexity of the radiolabeled nucleic acid. With these filters it is best to minimize both the concentration of radiolabeled nucleic acid in the hybridization solution and the length of hybridization time, since the amount of nucleic acid that binds tightly and randomly to the 541 filter paper increases with time. This paper should be referred to if maximum sensitivity is required.
6. Following hybridization, wash the filters four times in a large volume of 2X SSC at room temperature. Filters hybridized to radiolabeled RNA should be digested with 10 µg/ml RNase A from bovine pancreas in 2X SSC for 30 min. at 25°C. Significantly higher digestion temperatures, such as 37°C, will remove the radiolabel from terminally labeled, hybridized RNA. If the filters are to be reused, residual RNase activity can be eliminated by incubation in 0.05% (v/l) diethylpyrocarbonate for 30 min. at 25°C.
7. Autoradiograph the washed air-dried filters using the same methods described for DNA gels. If the colony positions are not obvious on the autoradiogram, determine them by aligning the autoradiogram with the filters that have been stained with 1 µg/ml ethidium bromide and illuminated with UV light.
8. After use, the filters may be recycled by being passed them through the entire washing procedure described in step 3. This will remove more than 95% of the hybridized DNA and all of the hybridized RNA. The filters may be reused more than 5 times without a noticeable difference in screening sensitivity.

Method-2: Isolating DNA from a single colony

It is frequently necessary to screen 10–100 colonies in order to identify one that has a plasmid of a particular size or a particular set of restriction sites. For example, a particular *Eco*RI restriction fragment may have been cloned into a plasmid *Eco*RI site and a clone with one of the two possible insert orientations may be required. It would be useful to have a quick method for isolating, restricting, and electrophoresing DNA from 10 different transformed colonies so that colonies containing a plasmid with an insert in the required orientation could be identified. The method described below accomplishes the first step – rapid isolation of restrictable DNA. A single colony will yield DNA that can be seen in an ethidium bromide-stained gel (Barnes, 1977).

Procedure

1. Use the end of a toothpick to collect bacteria from a large colony or a streak of bacteria on a plate.
2. Re-suspend the cells in 25–50 µl of 50 mM EDTA; Tris pH 8.0; 5% Triton X-100; 8% sucrose. Do this in a 1.5 ml polypropylene microfuge tube. This is a crucial step. Make sure that all of the cells are re-suspended.
3. Add an equal volume of the same buffer containing 2 mg/ml of lysozyme. If you are working with only a few samples, it is possible to include the lysozyme in the re-suspension buffer. However, do not leave them in this buffer for more than 5 min. before you proceed to step 4.
4. Put the tubes in a 100°C oil bath for 3 min.
5. Cool the tubes on ice for 5 min.
6. Centrifuge for 10–15 min. in an Eppendorf microfuge.
7. Remove the supernatant with a mechanical micropipette, being careful to avoid the pellet. The pellet size and texture are variable. Deliver the supernatant to a new microfuge tube.
8. Add an equal volume of 2-propanol to the supernatant fraction and place at –20°C (not –80°C) for 30 min.
9. Centrifuge for 15 min. in an Eppendorf microfuge.
10. Decant the alcohol and drain dry. Do not vacuum desiccate or use any other method to completely dry the pellet.
11. Re-suspend the DNA in 25 µl of the desired restriction enzyme buffer. There is usually enough DNA to be seen in two different lanes of a gel.
12. After digestion with restriction enzyme(s), add RNase A to 20 µg/ml and incubate for 2 min. at room temperature before loading the gel. This last step is not always necessary. It removes a large amount of RNA from the low molecular weight (50–200 nucleotides) region of the gel.

D. HYBRID SELECT/ARREST TRANSLATION

The difficulty of characterizing clones and detecting a desired DNA fragment from a mixed library may be resolved by two useful techniques termed hybrid select (release) translation or hybrid arrest translation. After preparation of a cDNA library, the plasmid is extracted from part of each colony, and each preparation is then denatured and immobilized on a nylon membrane. The membrane is soaked in total mRNA, under stringent conditions, usually a temperature only a few degrees below T_m (at high stringently). Unhybridized mRNA is washed off. The bound mRNA is eluted, and can be later used for direct *in vitro* translation. By immunoprecipitation or electrophoresis of the *in vitro* translated protein, the mRNA coding for a particular protein can be detected, and the clone containing its corresponding cDNA is isolated. This technique is known as **hybrid release translation**. In a related method called **hybrid arrest translation**, a positive result is indicated by the absence of a particular translation product when the total mRNA is hybridized with excess cDNA. This is a consequence of the fact that mRNA cannot be translated when it is hybridized to another molecule.

E. PCR SCREENING OF GENE LIBRARIES

Now it is possible to use PCR to screen DNA libraries constructed in various types of vectors. This is usually undertaken with primers that anneal to the vector rather than the foreign DNA insert, since its sequence will not be known. The size of an amplified product may be used to characterize the cloned DNA and help in subsequent analysis. The main advantage of PCR over traditional hybridization-based screening is the rapidity of the technique: PCR screening can be undertaken in 3 to 4 hours whereas it may be several days before detection by hybridization is achieved. The PCR screening technique gives an indication of the size of the cloned inserts rather than their sequence; however, PCR primers that are specific for a foreign DNA insert may also be used if it is known. This allows a more rigorous characterization of clones from cDNA and genomic libraries. But the limitation of this method is that if the insert fragment is too big, then it may not amplify the fragment (for more details refer to Chapter 12).

VI TECHNIQUES USED FOR ANALYSIS OF NUCLEIC ACIDS AND PROTEINS

The phenomenon of hybridization – the ability of one single-stranded nucleic acid to form a double helix with another single strand of complementary base sequence – is one of the backbones of modern molecular biology. Indeed, hybridization is a very common part of the techniques of molecular biology. Different blotting techniques such as the **Southern blot** (identifying specific DNA fragments), the **Northern blot** (measuring gene activity) and the **Western blot** (locating proteins on gels) are explained in details in Chapters 9 and 25. **In situ hybridization** (locating genes in chromosomes), another useful hybridization technique is discussed in detail in Chapter 13. Different methods of **DNA sequencing** are explained in detail in Chapter 10. The **physical mapping** of genomic DNAs for characterization of useful traits and molecular markers which includes RFLP analysis, RAPD analysis etc., are discussed in Chapter 11.

Protein engineering with cloned genes (site directed mutagenesis): Cloned genes make investigation much more precise and allow microsurgery to be done on proteins. By changing specific bases in a gene amino acid can be changed at corresponding sites in the protein product. Then observations can be made on the effects of those changes on the protein function. This can be achieved by **site-directed mutagenesis**. This method has been explained in detail in Chapter 11.

A. MAPPING AND QUANTIFYING TRANSCRIPTS

One recurring theme in molecular biology has been mapping transcripts (locating their starting and stopping points) and quantifying them (measuring how much of a transcript exists at a certain time). Scientists use a variety of techniques to map and quantify transcripts. This is not only important for gene regulation studies but may also be used as a marker for certain clinical disorders. Other than the Northern analysis, three popular techniques are available for mapping the 5'-ends of transcripts, and one of these also locates the 3'-end. All of these methods rely on the power of nucleic acid hybridization to detect just one kind of RNA among thousands.

1. S1 mapping

The principle behind S1 mapping is to label a single-stranded DNA probe that can hybridize only to the transcript of interest, rather than labeling the transcript itself. The probe must span the sequence where the transcript starts or ends. After the probe is hybridized to the transcript, the **S1 nuclease** is applied, and this degrades only single-stranded DNA and RNA. Thus, the transcript protects part of the probe from degradation. The size of this part can be measured by gel electrophoresis, and the extent of protection tells us where the transcript starts or ends. Figure 19.9 shows in detail how S1 mapping can be used to find the transcription start site. First, label DNA probe at its 5'-end with ^{32}P-phosphate. The 5'-end of a DNA strand usually already contains a non-radioactive phosphate, so this phosphate has to be removed with an enzyme called alkaline phosphate before addition of the labeled phosphate. Then use the enzyme polynucleotide kinase to transfer the ^{32}P-phosphate group from [γ–^{32}P]-ATP to the

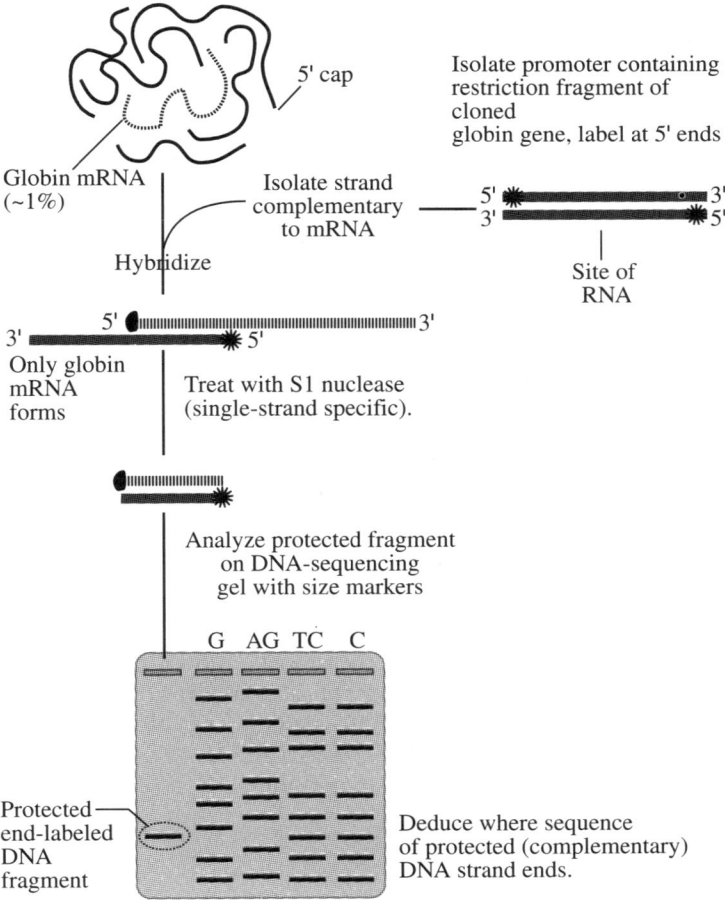

Figure 19.9 *S1 nuclease mapping for location of the 5' end of a particular transcript in a mixture of RNA molecules*

5'-hydroxyl group at the beginning of the DNA strand. We can also use S1 mapping to locate the 3'-end of a transcript. All that is it needed to do this is to prepare a 3'-end-labeled probe and hybridize it to the transcript. All other aspects of the assay are the same as for the 5'-end mapping.

B. PROTECTION ASSAYS

This section describes the uses of three extremely sensitive electrophoretic assay systems, 'footprinting' (DNase footprinting & DMS footprinting) and 'bandshift', in the analysis of sequence specific DNA-binding proteins.

1. Footprinting assays

Once we know that a protein binds to a given DNA, we can use footprinting to find out where the target site lies on the DNA. We can choose from several methods, but two are very popular: DNase and dimethylsulfate (DMS) footprinting.

(a) DNase footprinting

This method relies on the fact that a protein, by binding to DNA, covers the binding site and so protects it from attack by DNase. In this sense, it labels either strand, but only one per experiment (Figure 19.10). Next, the protein complex is bound with DNase I under mild conditions, so that an average of only one cut occurs per DNA molecule. Next, the protein is removed from the DNA, the DNA strands separated, and the resulting fragments electrophoresed on a high-resolution polyacrylamide gel alongside size markers. Of course, fragments will arise from the other end of the DNA as well, but they will not be detected because they are unlabeled. A control with DNA alone is always included and more than one protein concentration is used so that it can be seen by the gradual disappearance of the bands in the footprint region that protection of the DNA depends on the concentration of added protein. The footprint represents the region of DNA protected by the protein, and therefore tells us where the protein binds.

(b) DMS footprinting

This method also starts in the same way as in DNase footprinting, by end-labeling the DNA and binding the protein (Figure 19.11). Then the DNA–protein complex is methylated with dimethylsulfate (DMS), using a mild treatment so that on an average, only one methylation event occurs per DNA molecule. Next, the protein is dislodged, and the DNA treated with a reagent that removes methylated purines, creating apurinic sites on the DNA. Then the DNA is broken at these apurinic sites. Finally, the DNA fragments are electrophoresed and the gel autoradiographed to detect the labeled DNA bands. Each band ends next to a nucleotide that was methylated and thus was unprotected by the protein.

2. Bandshift

The basis of **bandshift** assay, often called the gel retardation assay, is that protein–DNA complexes are surprisingly stable to electrophoretic fractionation in gels and migrate as distinct bands more slowly than the free DNA fragments. The assay is simple and quick and the use of

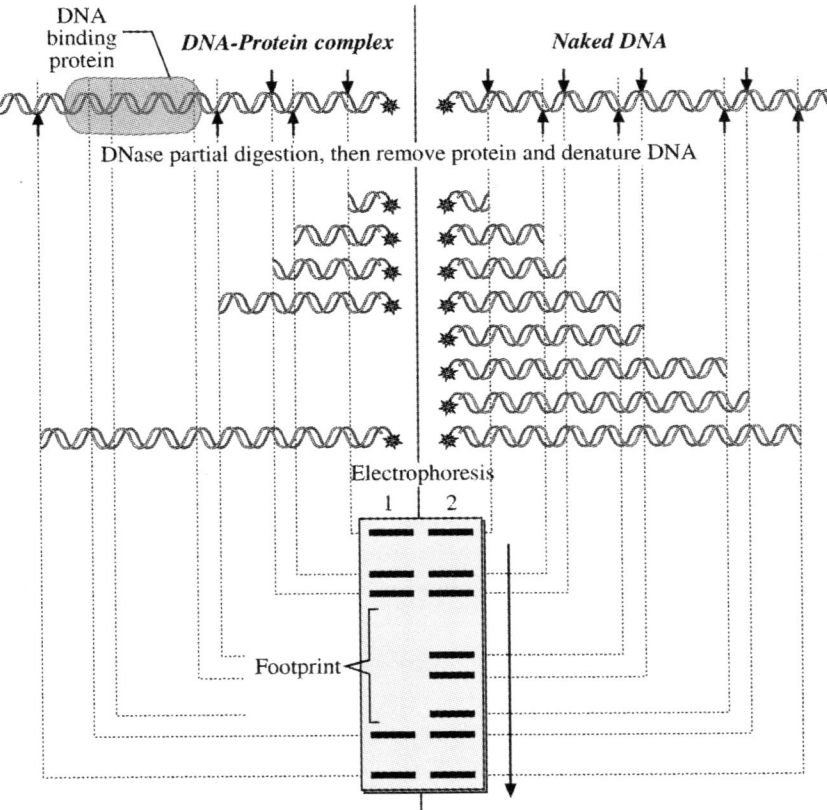

Figure 19.10 *Schematic representation of the footprinting assay. The right side of the illustration shows a naked DNA fragment and the left side the same DNA fragment containing bound protein. Arrows represents the points of cleavage by DNase*

radioactive binding-site DNA makes it highly sensitive. Its primary use is in the detection of DNA-binding proteins in crude cell extracts, but it can also be used to study the binding activity (complex formation) of purified DNA-binding proteins. Thus, the appearance of one or more slowly migrating bands in a gel may be taken as a sign that one or more proteins present in the extract bind to the binding-site on DNA fragment.

C. PRIMER EXTENSION

One drawback of the S1 method for mapping the 5'-ends of transcripts is that the S1 nuclease tends to "nibble" a bit on the ends of the RNA-DNA hybrid, or even within the hybrid where A-T-rich regions can melt transiently. On the other hand, sometimes the S1 nuclease will not digest the single-stranded regions completely, so the transcript appears to be a little longer than it really is. These can be serious problems, but another method called primer extension can tell us the 5'-end to the very nucleotide (Figure 19.12).

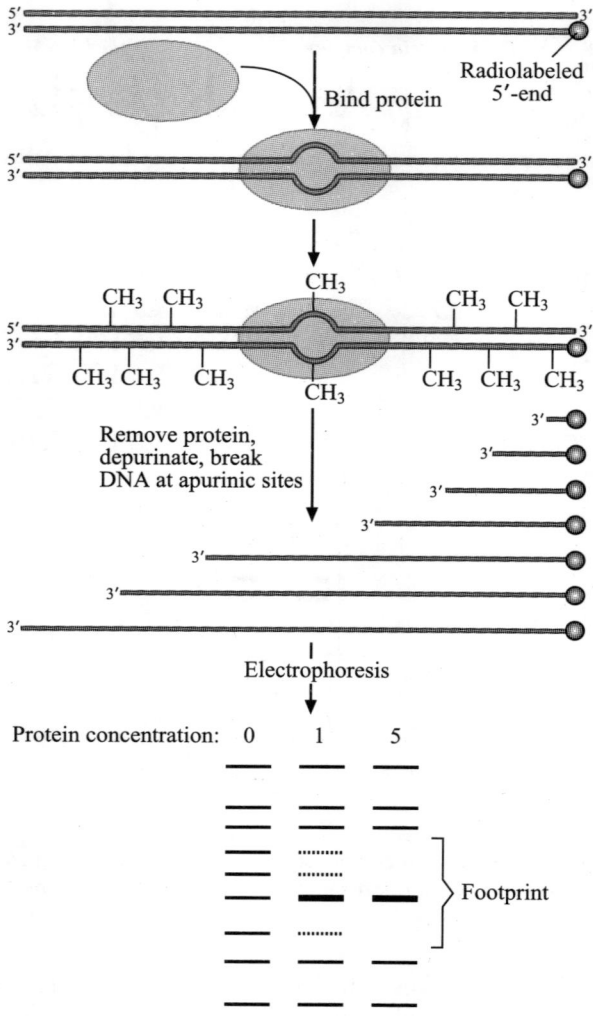

Figure 19.11 *DMS footprinting*

D. ASSAYING DNA-PROTEIN INTERACTIONS

Another of the recurring themes of molecular biology is the study of DNA–protein interactions. In this section, two methods for detecting protein-DNA binding and two examples of methods for showing which DNA bases interact with a protein are discussed.

1. Filter binding

Single-stranded DNA binds readily to nitrocellulose, but double-stranded DNA by itself does not. On the other hand, protein does bind, and if a protein is bound to dsDNA the protein-

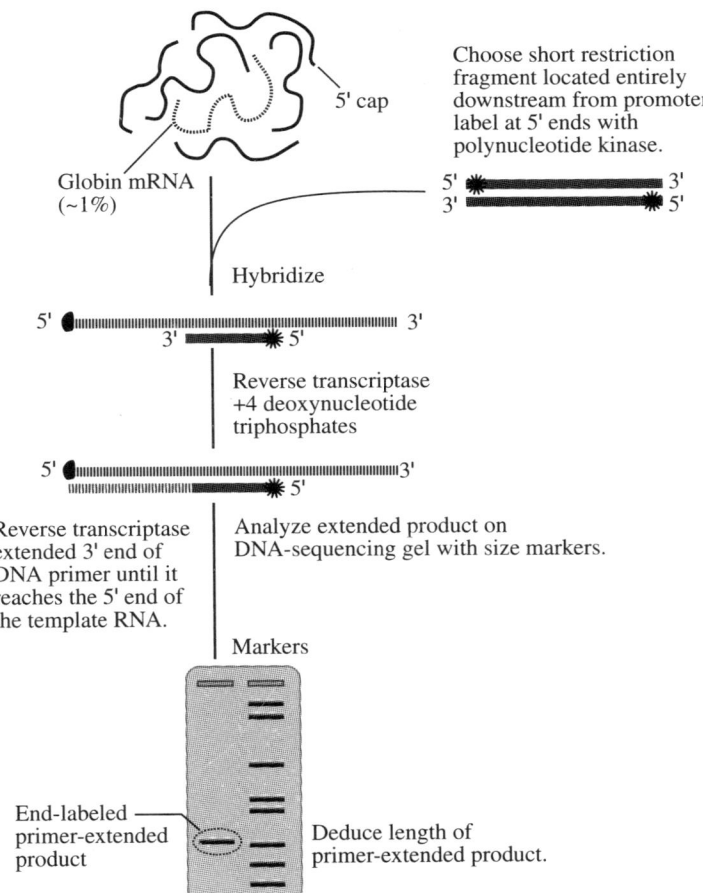

Figure 19.12 *Primer extension for location of the 5' end of a particular transcript in a mixture of RNA molecules*

1. Filter binding

Single-stranded DNA binds readily to nitrocellulose, but double-stranded DNA by itself does not. On the other hand, protein does bind, and if a protein is bound to dsDNA the protein-DNA complex will bind to nitrocellulose. This is the basis of the assay portrayed in Figure 19.13. Labeled dsDNA is poured through a nitrocellulose filter (Figure 19.13a). The label in the filtrate and the filter-bound material is measured. All the labeled material would have passed through the filter into the filtrate. This confirms that dsDNA does not bind to nitrocellulose. A solution of a labeled protein is then filtered (Figure 19.13b). All the protein is bound to the filter. This demonstrates that proteins bind by themselves to the filter. dsDNA is again labelled (Figure 19.13c), but this time mixed with a protein to which it binds. Since the protein binds to the filter, the protein-DNA complex will also bind, and the labeled material is bound to the filter, rather than found in the filtrate. Thus, filter binding is a direct measure of DNA-protein interaction.

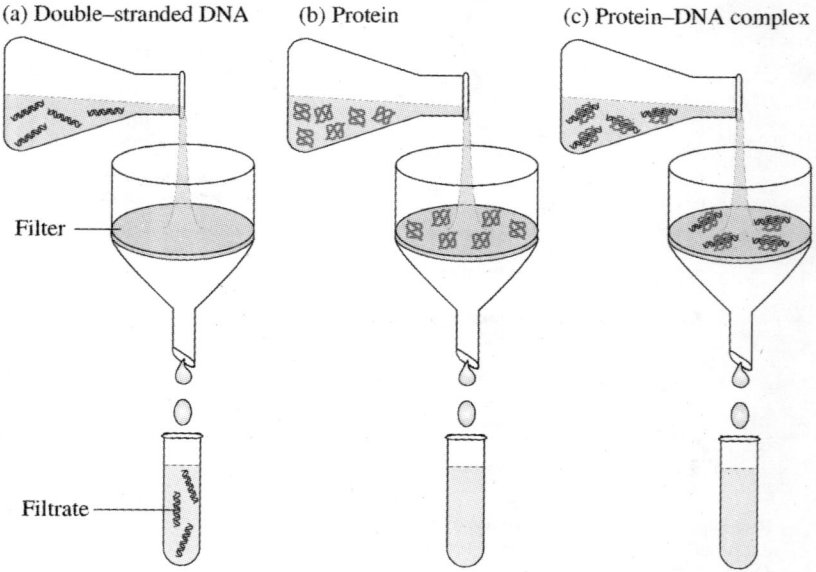

Figure 19.13 *Nitrocellulose filter binding essay*

2. Gel mobility shift

A second method for measuring DNA-protein interaction relies on the fact that a small DNA has much higher mobility in gel electrophoresis than the same DNA does when it is bound to a protein. Thus, we can label a short, dsDNA fragment, then mix it with a protein, and electrophorese the complex. Subject the gel to autoradiography to detect the labeled species. Figure 19.14 shows the electrophoretic mobilities of three different species. Lane 1 contains naked DNA, which has a very high mobility because of its small size. Lane 2 contains the same DNA

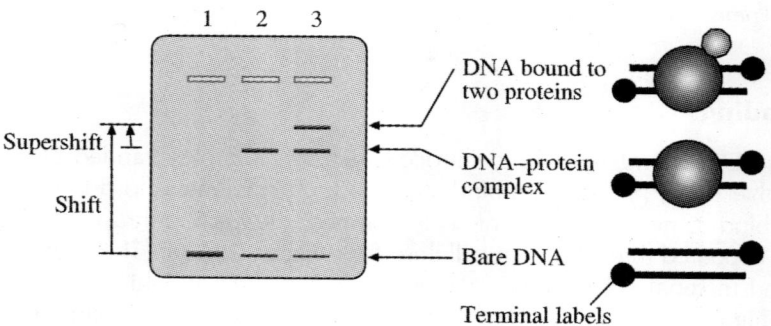

Figure 19.14 *Gel mobility shift assay. Pure labeled DNA or DNA-protein complexes electrophoresed, then autoradiographed the gel. Lane 1 shows the high mobility of bareDNA. Lane 2 shows the mobility shift that occurs upon binding a protein to the DNA. Lane 3 shows the supershift caused by binding a second protein to the DNA-protein complex*

bound to a protein, and its mobility is greatly retarded. This is the origin of the name for this technique: **gel mobility shift assay**. Sometimes it is also called a *"**bandshift assay**"*. The protein could be a DNA-binding protein, or a second protein that binds to the first. It can even be an antibody that specifically binds to the first protein. The small size of the DNA is important in this assay. We can prepare a short restriction fragment, or even a synthetic double-stranded oligonucleotide that will contain the target site for the protein, but little else.

E. KNOCKOUTS

The most frequently asked questions are what is the role of the gene being studied, what function does the gene play in the life of the organism? We can often answer this question best by deliberately mutating a particular sequence of the gene in a living organism. In many cases it is desirable to analyze the effect of certain genes and proteins in an organism rather than in the laboratory. Furthermore, the production of pharmaceutical products and therapeutic proteins is also desirable in a whole organism. This also has an important consequence for biotechnology and agricultural industries. The introduction of foreign genes into germ line cells and the production of an altered organism are termed transgenics. There are two broad strategies for transgenesis. The first is direct transgenesis in mammals, whereby recombinant DNA is injected directly into the male pronucleus of a recently fertilized egg. This is raised in a foster mother animal, resulting in offspring that are all transgenic. A second method is selective transgenesis, where the recombinant DNA is transfected into embryo stem cells (ES cells). The cells are then cultured in the laboratory and those expressing the desired gene are selected and incorporated into the inner cell mass of an early embryo. The resulting transgenic fetus develops in a foster mother, but in this case, the transgenic animal is mosaic or chimeric, since only a small proportion of the cells will express the protein. These two methods are random in nature for the gene integration. A refinement of this, however, is gene targeting, which involves the production of an altered gene in an intact cell, a form of *in vivo* mutagenesis.

Many techniques have recently been developed for targeted disruption of genes in mice. This is generally called knockout. Figure 19.15 explains one way to begin the process of creating a knockout mouse.

1. Select a cloned DNA containing the mouse gene we want to knock out.
2. Interrupt this gene with another gene that confers resistance to the antibiotic neomycin.
3. Introduce a thymidine kinase (tk) gene, elsewhere in the cloned DNA, outside the target gene. These extra genes enable the weeding out of those clones for which targeted disruption did not occur.
4. Mix the engineered mouse DNA with stem cells from an embryonic brown mouse. In a small percentage of these cells the interrupted gene will find its way into the nucleus and homologous recombination will occur between the altered gene and the resident, intact gene. This recombination places the altered gene into the mouse genome, and removes the *tk* gene.
5. Grow the cells in a medium containing the neomycin derivative G418. Cells without the homologous recombination will be removed.
6. Cells that experienced non-specific recombination will have incorporated the *tk* gene, along with the interrupted gene, into their genome, which can be killed with gangcyclovir, a drug that is lethal to *tk*$^+$ cells (The stem cells used are *tk*$^-$). These two treatments leave only the

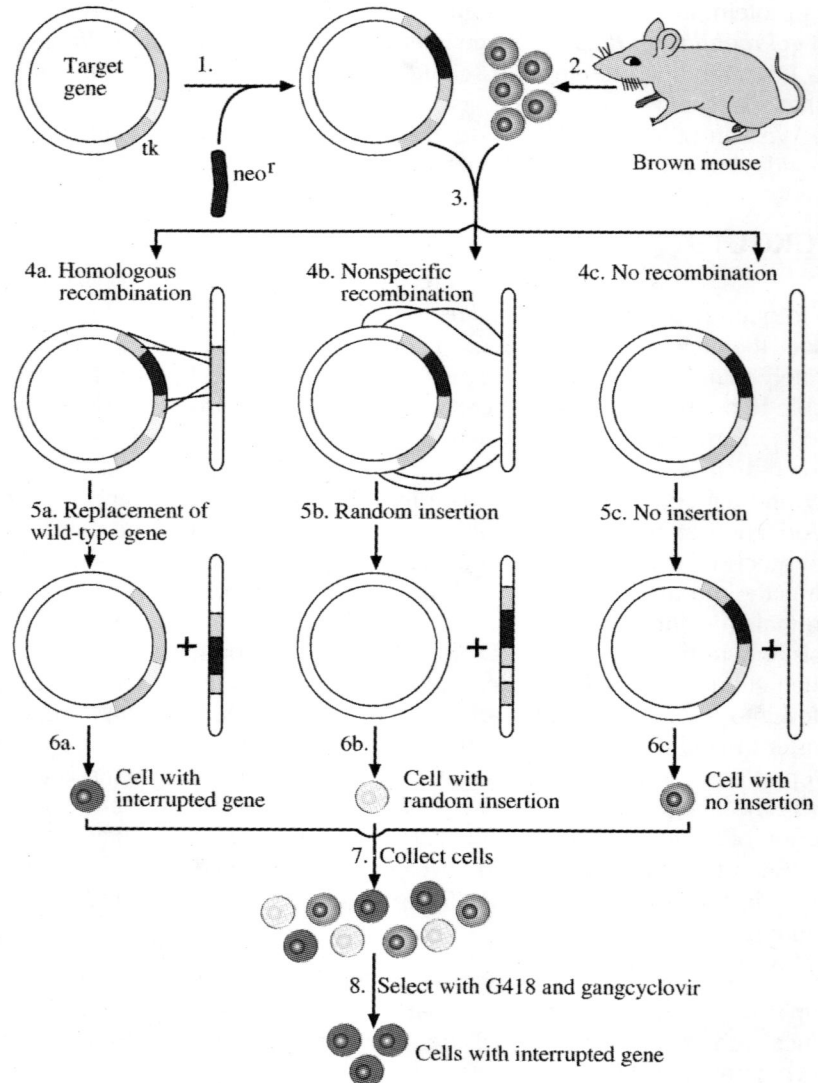

Figure 19.15 *Making a knockout mouse. Stage 1 creating stem cells with an interrupted gene*

engineered cells that have undergone homologous recombination and are therefore heterozygous for an interruption in the target gene.
7. Introduce these cells containing the interrupted gene into a whole mouse (Figure 19.16). Inject the engineered cells into a mouse blastocyst that is destined to develop into a black mouse.
8. Place the altered embryo into a surrogate mother, which eventually gives birth to a chimeric mouse.

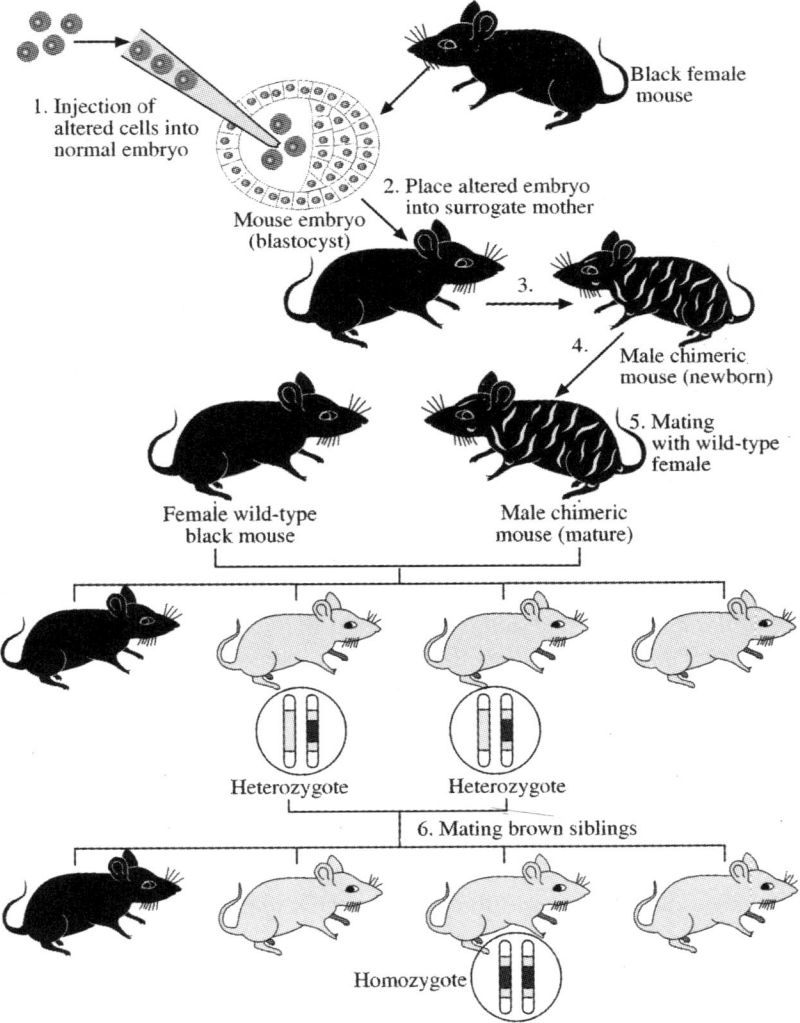

Figure 19.16 *Making a knockout mouse. Stage 2, placing the interrupted gene in the animal*

9. After maturation, mate the chimera with the black mouse. Since brown (agouti) is dominant, some of the progeny should be brown. Confirm that the brown mouse is carrying the interrupted gene by Southern blotting techniques.
10. To obtain homozygous knockouts, mate among the browns that carry interrupted genes.

F. ANALYSIS OF WHOLE GENOME

The most ambitious projects endeavored by molecular biologists in the last decade was the initiative to map and completely sequence a number of genomes from various organisms. At present initiatives are underway for the mapping and sequencing of certain organisms (bacteria, yeast, roundworm, fruit fly, puffer fish, and mouse), but the ultimate challenge, mapping the human genome, is currently in progress. The demands of such large-scale mapping and sequencing have provided the impetus for the development and refinement of many molecular biological techniques such as DNA sequencing. It has also led to new methods of identifying the important coding sequences that represent proteins and enzymes.

Physical genome mapping: In genome mapping, a physical map is the primary goal. Genetic linkage maps have also been produced by determining the recombination frequency between two particular loci. The development of vectors with large insert capacity has enabled the production of continuous, overlapping, cloned fragments that have been positioned relative to one another. From these maps, any cloned fragment may be identified and aligned to an area in one of the contig maps. Overlaps are created because of the use of partial digestion conditions with a particular restriction endonuclease when the libraries are constructed. This ensures that when each DNA fragment is cloned into a vector, it has overlapping ends that may be identified and the clones are positioned or ordered so that a physical map may be produced. In order to position the overlapping ends sequencing or fingerprinting or *in situ* hybridization methods are generally followed.

In order to define a common way for all research laboratories to order clones and connect physical maps together, an arbitrary molecular technique based on the PCR has been developed using sequence tagged sites (STS). An STS is a small unique sequence of length between 200 and 300 bp that is amplified by PCR. The uniqueness of the STS is defined by the PCR primers that flank the STS and act as a potential marker. An STS that occurs in two clones will overlap and may therefore be used to order the clones in a contig. Clones containing the STS are usually detected by Southern blotting, but a more rapid method of identifying STS within a clone is by the PCR-based screening method. STS elements may also be generated from variable regions of the genome to produce a polymorphic marker that may be traced through families along with other DNA markers and located on a genetic linkage map.

Gene discovery and localization: A number of disease loci have been identified and localized to certain chromosomes. This has been facilitated by the use of *in situ* mapping techniques such as FISH (refer to Chapter 13). This method of gene discovery is known as positional cloning. Recent investigations have found that certain active genes may be identified by the presence of so-called HTF (*Hpa*II tiny fragments) islands often found at the 5'-end of genes. These are C+G-rich sequences that are not methylated and form tiny fragments on digestion with the restriction enzyme *Hpa*II. A new gene discovery method that has been used extensively in the past few years is a PCR-based technique giving rise to a product termed as an expressed sequence tag (EST). This represents part of a putative gene for which a function has yet to be assigned. It is carried out on cDNA using primers that bind to an anchor sequence such as a poly(A) tail and primers that bind to sequences at the 5'-end of the gene. Such PCRs may subsequently be used to map the putative gene to a chromosomal region. A further gene isolation system that uses adapted vectors, **termed exon trapping** or **exon amplification**, may be used to identify exon sequences. Exon trapping requires the use of a specialized expression vector that will accept fragments of genomic DNA containing sequences for splicing reactions to take place. Following transfection of a eukaryotic cell line, a transcript is produced that may be detected by using specific primers (for splicing sequence) in an RT-PCR.

Human genome mapping: Mapping and sequencing of the human genome is one of the most ambitious projects ever undertaken and will certainly bring new insights into gene function and gene regulation. It also provides a means of identifying DNA mutations or lesions found in current genetic disorders. One interesting aspect is the sequencing of the whole genome in relation to the coding sequences. Much of the human genome appears to be non-coding and composed of repetitive sequences. However, about 90,000 genes appear to code for proteins, and mapping and sequencing them is exciting and has good prospects. The diversity of the human genome is also an area of great interest and is currently under study.

G. SOUTH–WESTERN SCREENING FOR DNA-BINDING PROTEINS

Blotting techniques has been used very successfully for screening and isolation of clones expressing foreign gene products. The method involves a nitrocellulose membrane 'plaque-lift' on which expressed fusion proteins are adsorbed. The procedure for performing the plaque-lift is the same as that just described. However, the screening is carried out by incubating the membrane with a radiolabeled duplex DNA oligonucleotide containing the sequence for which the DNA binding is minimized. Positively reacting plaques are identified by autoradiography of the filter. This technique therefore uses a radiolabeled DNA to detect polypeptide on the nitrocellulose, and has been called a 'South-western' procedure. It has been spectacularly successful in the isolation of clones expressing cDNA sequences corresponding to certain mammalian transcription factors. Clearly, the procedure can only be successful (a) where the binding activity is due to a single polypeptide chain, and (b) where it can be expressed in functional form when in a fusion polypeptide. It is also clear that the affinity of the polypeptide for the specific DNA sequence must be high. The procedure has been found most efficient when the oligonucleotide containing the binding sequence has been ligated into multimeric form. This may mean that a single DNA multimeric may be bound by more than one fusion polypeptide molecule on the filter, thus greatly increasing the average dissociation time.

VII CONFIRMATION OF GENE EXPRESSION IN TRANSGENIC PLANTS

Before analyzing the gene expression of a foreign DNA in transgenic plants, the structure and copy number of the DNA should be verified. Instead of Southern blot hybridization, which takes a long time, a rapid technique has been developed using the polymerase chain reaction (PCR) for the detection of the transferred DNA.

Some genes when transferred produce enzymes whose activities can be easily detected or used as a basis of selection for the transformed cells, e.g., genes for herbicide resistance. However, most genes need to be tagged with another gene called **reporter gene** to identify its presence. The reporter gene's expression is easily detected either through highly sensitive enzyme assays (**scorable reporter genes**) or through expression of resistance to a toxin (**selectable reporter genes**). Some commonly used easily detectable enzyme producing genes are *nos* (nopaline synthase, from *Agrobacterium*), *lux* (luciferase from bacteria or firefly), *cat* (chloramphenicol acetyltransferase from bacteria), and *gus* (β-glucuronidase from bacteria) (Table 19.1). The activities of the enzymes produced by scorable reporter genes are determined either *in situ* (in the transformed tissues) or by *in vitro* assays using plant tissue extracts. In addition, immunological methods may also be used to detect the protein products of marker genes either *in situ*, *in vitro*, through electrophoresis or by various blotting techniques (e.g. Western blotting).

Table 19.1 *Some common marker genes used for detection of transformed cells*

Marker gene	Source	Substrate, assay and identification
a) Scorable:		
ocs (Octopine synthase)	Agrobacterium	Arginine + pyruvate + NADH, electrophoresis; Octopine production
nos (Nopoline synthase)	Agrobacterium	Arginine + ketoglutaric acid + NADH electrophoresis; Nopaline production
nptII (Neomycin phosphotransferase)	Tn 5	Kanamycin + (35P) ATP (*in situ* assay) Radioactivity detection (also dot blots)
cat (Chloramphenicol acetyltransferase)	E. coli	(^{14}c) chloramphenicol + acetyl CoA; acetyl chloramphenicol by autoradiography
luc (Luciferase)	Bacteria, firefly	a) Decanal and FMNH2, b) ATP + O_2 + luciferin, Phosphorescence,(exposure to x-ray film)
lac Z (β-galactosidase)	E. coli	β-galctoside (X-gal), Blue color of colony
gus A (β-glucuronidase)	E. coli	Glucuronides (PNPG, x-GLUC, NAG, REG) Blue color, colorimetric, fluorometric fluorescence, histochemical
Diverse regulatory factors involved in anthocyanin biosynthesis	Maize	Enhanced anthocyanin pigments
gfp (green fluorescent protein)	Jellyfish	A reporter which, because it is autobioluminescent, has the distinct advantage that it can be used in living systems
b) Selectable:		**I. Antibiotics**
dhfr (dihydrofolate reductase)	Tn 7, mouse, plasmid R67	Methotrexate resistance
nptII (Neomycin phosphotransferase)	Tn 5	Kanamycin, neomycin, and G418 resistance
hptIV (Hygromycin phosphotransferase)	E. coli, Streptomyces hygroscopicus	Hygromycin B resistance
ble gene (enzyme unknown)	E. coli	Bleomycin
Gentamycin acetyltransferase	E. coli	Gentamycin
deh1 (Dehalogenase)	Pseudomonas putida	Dalapon resistance
– (Streptomycin and spectinomycin phosphotransferase)	E. coli	Resistance to streptomycin and spectinomycin
		II. Herbicides
als (mutant acetolacetate synthase)	Arabidopsis thaliana, tobacco	Resistance to sulfonyl ureas
bar (Phosphinothricin acetyl-transferase)	Pseudomonas hygroscopicus	L-phosphinothricin resistance (PPT; also known as bialaphos)
Bromoxynil nitrilase		Bromoxynil
Kanr [Aminoglycoside 3'-phosphotransferase H; APD (3') III	Bacteria	Resistance to kanamycin
aro A (5'-enolpyruvyl- shikimate–3-phosphate (EPSP) synthase		Glyphosate (Roundup)

A. GENETIC MARKERS IN PLANT CELLS

Regardless of the transformation strategy, genetic markers are required for the selection of transformed cells that have acquired the DNA of interest. Genetic markers include dominant selectable markers for the direct selection of transformed cells, assayable markers that allow the detailed analysis of gene expression, and negative selection markers that prevent the development of normal cells.

Dominant selectable markers: Although several dominant selectable markers have been developed, one of the most extensively used marker genes is the neomycin phosphotransferase II (NPTII) gene from the bacterial transposon T5, which confers resistance to kanamycin and G418. The hygromycin phosphotransferase gene from *E. coli* has also been used for selectable marker genes with several plant species. Other selectable markers, such as the bleomycin resistance gene and the dihydrofolate reductase gene, have also been used. New types of selectable markers are being developed that confer herbicide resistance, such as glyphosate resistance, and can be used for the selection of resistance cells in the presence of herbicide.

Negative selection markers: Negative selection markers have been developed that are lethal or interfere with the normal development of the plant. An example of this type of marker is the *tms2* gene on the T-DNA of the Ti plasmid of *A. tumefaciens*. The *tms2* gene encodes an enzyme that catalyzes the conversion of indole-3-acetamide (IAM) to indol-3-acetic acid (IAA), the natural auxin found in plants. Although auxins are normally present in low concentrations in plants and act as regulators of growth and development, they are lethal at elevated concentrations. Thus, the *tms2* gene product converts a substrate with lower toxicity, IAM or α-naphthaline acetamide (NAM), into a toxic product, IAA or α-naphthalene acetic acid (NAA) and can be suitable for negative selection when it is expressed at a high enough level and in the presence of high concentrations of IAM or NAM. Moreover, this system can be used as a conditional lethal marker, depending on individual plant promoters fused to the *tms2* gene. Another example of this type of marker that has been recently developed is the nitrate reductase (NR) gene. The NR gene catalyzes the first step of the nitrate assimilation pathway by converting nitrate and is known to confer chlorate toxicity by producing the toxic product, chlorite. The wild-type *Nia* gene, which encodes for the apoprotein of NR, is regulated by various factors, including the type of nitrogen sources. Very low levels of *Nia* transcript can be detected in the presence of ammonia compared with the growth on nitrate. Thus, the cauliflower mosaic virus (CaMV) 35S promoter-NR chimeric gene confers high levels of expression on both nitrate- and ammonium-supplemented media. On the basis of the quantitative difference in NR activity between the wild–type plants that contain the CaMV 35S-NR gene, on media supplemented with ammonium, the CaMV 35S-NR gene is used as a counter-selectable system in *N. plumbaginifolia* using chlorate selection combined with ammonium nutrition.

B. REPORTER GENES

Gene expression is controlled by a series of individual levels, including the initiation of transcription or translation and the processing, transport, or degradation of mRNA or protein. Thus, the use of gene fusions is required for delineating the contribution of transcriptional control by eliminating the gene-specific post-transcriptional controls and replacing them with easily assayable reporter genes. In addition, gene fusions enable one to study the expression of individual genes separate from the other members of multigene families. Reporter genes have to be well characterized both genetically and biochemically, and the enzymatic activity of most reporter genes is either not normally present in plant cells, or can it be easily distinguished from endogenous activities (Table 19.2).

To date, the most frequently used reporter gene in plant cells is the β-glucuronidase (GUS) gene. Since many plant species have high activity of β-galactosidase, one of the most commonly used reporter genes in other fields, the *lac* Z gene is of limited use in plant cells. The GUS gene has the great advantage that no detectable background activity is present in many plant species, although it has been noted that some plant species, such as rice and wheat, have significant background activity. Another advantage of this system is that it can be used to analyze the cell-specific expression using histochemical detection of enzymatic activity in cells and tissues of transformed plants.

The chloramphenicol acetyl transferase (CAT) gene and the luciferase gene from either the firefly or bacteria have been widely used as reporter genes in plant cells, especially when two different reporter genes have to be used. The neomycin phosphotransferase II (NPTII) gene, which is usually used as a dominant selectable marker gene, can be used for a reporter gene. The NPTII can tolerate NH_2-terminal fusions and is enzymatically active, making it useful for studying organelle transport in plant cells. However, plants have non-specific phospholyases that can phosphorylate kanamycin as a substrate. Therefore, it is necessary to use a gel assay system to distinguish the NPTII protein from plant phosphorylases.

Table 19.2 *A comparison of reporter genes*

Reporter genes (and their products)	Uses
lac Z (β-galactosidase) from *E. coli*	Commonly used reporter system. Induced by IPTG. The enzyme converts chromogenic substrate X-gal into blue precipitate for localization of gene expression. Converts ONPG into soluble yellow product for quantification.
luc (luciferase) from fireflies	Highly sensitive reporter which generates a bioluminescent product when exposed to substrate luciferin.
cat (Chloramphenicol acetyltransferae (CAT) from *E. coli* Tn9	Useful assay for *in vitro* analysis. CAT acetylates chloramphenicol and the extent of acetylation can be determined by thin layer chromatography in CAT assay
gfp (green fluorescent protein) from jellyfish	A non-invasive detection system, because it is autobioluminescent.
GUS (β-glucuronidase) from *E. coli*	Generally used reporter in plant systems, converts chromogenic substrate X-gluc into blue precipitate for localization of gene expression.

1. Selection of transformed cells

The mechanism by which any organism controls the production of proteins by the differential expression of its genes has fascinated many researchers since protein production and gene expression were first linked. In many instances, the detection of a polypeptide of interest, which may only be produced under certain conditions or in specific cell types within an organism, is rather complex or laborious since many of these gene products do not have easily detectable enzymatic or functional activities. The use of fusion between the regulatory or targeting signals of a gene of interest and a gene whose product is easily detected offers several advantages for the study and characterization of genetic systems. These latter genes have been named reporter genes since they supply information concerning the regulation or action of a different gene. The use of reporter genes to study gene expression simplifies the analysis, facilitates the comparison of different or altered regulatory sequences, and often enhances the sensitivity with which measurements of gene activity can be made. The use of *in vitro* generated gene fusions followed by transformation facilitates such an analysis.

The essential features of an ideal reporter gene are:
1. Lack of endogenous activity: The gene product must have an enzymatic activity that is not present in an easily distinguishable form of endogenous activities of the host;
2. The gene and gene products should be well characterized genetically and biochemically;
3. The gene products should be stable under different physiological conditions, i.e. organs or tissues, pHs or light conditions;
4. The enzymatic activity should be easily detectable and quantifiable;
5. The expression of the reporter gene should produce a selectable or visibly screenable change in the phenotype of the transformed cell.

There are several reporter molecules that are now widely used for plant cells. Some of those were originally developed for use in prokaryotic or animal cell systems. Some of the most commonly used systems are described in subsequent chapters. Each reporter gene has some advantages and disadvantages and none of them is ideally suited for all cell types.

2. Phenotypic analysis of regenerated plant (R) progeny from transgenic *N. tabacum*

Regenerated plants (R0) from *N. tabacum* are selfed and crossed as male and female onto control plants. Segregation ratios are tested by germination on kanamycin-containing medium.

Procedure

Maturation and pollination of regenerated plants in the greenhouse

1. Isolate regenerated plants (R0) from their tissue culture environment under water to avoid wilting. Transfer plants to 10 cm pots containing a porous soil mix of 1 part Peat-lite mix and 2 parts Perlite. Acclimatize plants by placing in a mist chamber set at 25°C day/20°C night with a 12 h day length and a light intensity of approximately 200 µE m–2s–1. Remove established plants, transfer to 3.8 litre pots, and mature in the greenhouse. Fertilize plants three times per week with 20–20–20 fertilizer and once weekly with blood meal.
2. Make buds from each plant self-pollinate and cross others as male or female onto control plants. Initiate test cross-pollinations first by removing the anthers from the designated female. Second, apply pollen from the designated male plant to the emasculated female by transferring the dehisced anther with forceps and touching it to the receptive stigma. In some instances, the anthers will remain intact and can be placed in a container of derite until the pollen has dehisced.

Seed germination *in vitro* and Chi-square analysis

3. Collect pods from selfed and crossed pollinations. Surface-sterilize as in description for growing plants for transformation.
4. Place sixteen seeds in 100 mm × 25 mm Petri dishes of germination medium containing 200 mg/l kanamycin (Medium B). In addition, place seeds on medium without kanamycin (Medium A) to check germination frequency. For medium A and B (see Chapter 18). Incubate the dishes for two weeks at 28°C in a growth room under approximately 100 µEm–2s–1 of light and a 16 h day length.
5. After 14–21 days, score the phenotype of germinated seedlings. Kanamycin-resistant plants have green cotyledons whereas susceptible plants have bleached or white cotyledons. Observe plants on control Medium A without kanamycin for overall germination frequency.
6. Genetic ratios can be determined by summarizing phenotypic data and testing best fit to various genetic ratios by chi-square analysis.

3. β-Galactosidase and blue or white selection

One version of these fusion protein expression vectors places the cloning site at the end of the coding region of the protein β-galactosidase, so that among other things, the fusion protein is attached to β-galactosidase and can be recovered by purifying the β-galactosidase activity. Alternatively, placing the cloning site within the β-galactosidase coding region means that cloned inserts disrupt the β-galactosidase amino acid sequence, inactivating its enzymatic activity. This property has been exploited in developing a visual screening protocol that distinguishes those clones in the library that bear inserts from those that lack them.

Cells that have been transformed with a plasmid-based β-galactosidase expression cDNA library (or infected with a similar library constructed in a bacteriophage lambda-based β-galactosidase fusion vector) are plated on media containing 5-bromo-4-cloro-3-indolyl-β-D-galactopyranoside, or X-gal. X-gal is a chromogenic substrate, a colorless substance that upon enzymatic reaction yields a colored product. Following induction with IPGT, bacterial colonies (or plaques) harboring vectors in which the β-galactosidase gene is intact (those vectors lacking inserts) will express an active β-galactosidase that cleaves X-gal, liberating the 5-bromo-4-chloro-indoxyl, which dimerizes to form an indigo blue product. These blue colonies (or plaques) represent clones that lack inserts. The β-galactosidase gene is inactivated in clones with inserts, so those colonies (or plaques) that remain "white" (actually, colorless) are recombinant clones.

4. Assay for β-glucuronidase activity in transgenic tobacco

A sensitive assay for β-glucuronidase (GUS) utilizes the fluorescence of 4-methyl-umbelliferone (4 MeU), the product of hydrolysis of the substrate 4-methyl-umbelliferyl-β-D-glucuronide (MUG) by the enzyme. The assay has been adapted to a microplate format and a facile method for quantitatively determining the amount of GUS in plant tissue extracts is included.

Materials

Lysis buffer: 50 mM sodium phosphate pH 7.0, 10 mM EDTA, 0.1% Triton X-100, 10 mM β-mercaptoethanol and 0.1% sarkosyl.

Substrate: 4-methyl-umbelliferyl-β-D-glucuronide (MUG). Prepare 10 mM stock solution in 0.1 M phosphate pH 6.5. Store frozen in 1 ml aliquots. Individual samples can be thawed and re-frozen 4–5 times. Further freezing-thawing causes an increase in background fluorescence.

Enzyme: β-glucuronidase (type VII from *E. coli*) from Sigma.

Protein standard: lyophilized BSA. Follow instructions for making solution. Store at 4°C.

Dye reagent concentrate from BioRad, store at 4°C; 0.2 M $NaCO_3$; Knotes tubes and pestles; Clear microplates for protein assay; Opaque microplates for fluorescence measurements; Titertek Fluoroskan II microplate fluorometer; Molecular Devices Vmax microplate reader.

Procedure

1. Grind 5 mg tissue in 50–200 μl of lysis buffer. Grind until completely homogenized. Pellet the cell debris in the microfuge, 15 min. at 10,000 rpm at 4°C. Transfer supernatant solution to a fresh 1.5 ml microfuge tube.
2. Perform a Bradford protein assay using BSA as standard, to estimate protein in tissue extract.
3. Determine background fluorescence in source tissue, resulting from non-specific hydrolysis of substrate.
4. Determine the maximum amount of protein from tissue extracts that can be used in the GUS assay. With tobacco leaf, tissue extracts up to 5 μg protein can be used without noticeable inhibition of the pure enzyme activity. A 10–20% inhibition is observed at protein amounts between 6 and 8 μg.
5. Determine amount of GUS expression in transformed tissue as follows:
    a. Add an appropriate volume of lysis buffer to all wells (final reaction volume is 50 μl).
    b. Add tissue extract corresponding to at least two different protein amounts, e.g. 2 and 4 μg.
    c. Initiate the reaction with 5 μl of MUG.
    d. Cover the plate and incubate at 37°C for 30 min.
    e. Stop the reaction with the addition of 150 μl of 0.2 M Na_2CO_3.
    f. Read the fluorescence and subtract the blank value from all experimental samples.
6. Calculate the amount of GUS as a function of total protein. To do this, first determine the relationship between enzyme concentration and fluorescence using the pure enzyme. For example, if this number is 115 fluorescence units ng–1 min–1, then:
    a. Divide the fluorescence of the experimental sample by the time of incubation (30 min). This gives fluorescence units/min.
    b. Divide above number by 115. This gives nanograms of GUS in experimental sample.
    c. Express this as a fraction of total protein used in the assay. For example, 0.1 ng of GUS in 2 μg of total protein would be equal to 0.0005% GUS.

5. Ketolactose test

Materials

Lactose medium: 1% lactose, 0.1% yeast extract, 2% Difco Bacto–agar.

Benedict's reagent: 256 gm Na_2CO_3-10 H_2O, 132 gm sodium citrate, 13.2 gm $CuSO_4.5H_2O$ per litre.

Procedure

1. Grow the strain to be tested on lactose medium plates.
2. After colonies or a streak of growth have appeared, flood the plates with Benedict's reagent. If the strains are positive, a yellow zone will form surrounding the area of growth.

6. NPTII assay to detect enzyme activity in transgenic plants

Assay allows greater sensitivity for detection of NPT, and it is possible to obtain quantitative data.

Materials

Soaking buffer (made fresh): 20 mM ATP plus 100 mM sodium pyrophosphate, 10 ml per 20 cm × 20 cm blot.

Wash buffer: 10 mM sodium phosphate buffer, pH 7.50.

Lysis buffer: 50 mM sodium phosphate, pH 7.00, 1 mM EDTA, 0.1% Triton X-100, 10 mM 2-mercaptoethanol, 0.1% sodium lauryl sarkosine, Filter-sterilize and store at 4°C.

Reaction buffer (5X stock stored at 4°C): 335 mM Tris, 210 mM magnesium chloride, 2 M ammonium chloride. Titrate to pH 7.10 with 1 M maleic acid.

Assay mixture: 4.936 ml 1X reaction buffer plus 5 µl of a 10 mM ATP solution (made fresh), 7 µl of a 22 mM neomycin sulfate solution (made fresh), 50 µl of a 1 M sodium fluoride solution, 15 µCi ^{32}P-ATP (10 µCi/µl, 1.5 µl), 250 mg BSA; mix well and centrifuge (600 Xg, 4°C, 10 min) to reduce foam.

Procedure

1. Soak Whatman P81 paper with the soaking buffer and allow to drip dry.
2. Add 50 µl cold lysis buffer to 5 mg tissue sample contained in a Knotes 'pellet-pestle' tube. Grind each sample with 20 strokes on ice. Centrifuge samples at 4°C for 15 min. at high speed using a microfuge to pellet debris.
3. Determine the protein concentration of the supernatant solution using the Bradford assay with BSA as a standard.
4. Add 20 µl of sample (containing 1–5 µg plant protein) to 20 µl of assay mixture contained in a chilled 1.5 ml Effendorf tube, mix well on ice, start the reaction in a 37°C water bath, and incubate for 30 min. To stop the reaction, place tubes on ice for at least 2 min. Spin the tubes in a microfuge briefly to collect liquid.
5. Spot a 30 µl aliquot of each reaction on the dry pre-soaked P81 paper in a grid pattern with gentle drying. Wash the dried blot in a 200 ml wash buffer, which has been pre-heated to 80°C, for two minutes with constant agitation. Remove blot and immerse in 200 ml of wash buffer at room temperature. Wash at room temperature for 10 min. with constant agitation. Wash twice more at room temperature. Hang blot to dry.
6. Analyze blot by scanning with AMBIS 2-D Beta-scanner for two hours. Alternatively, paper can be cut and sample radioactivity determined using a liquid scintillation counter.

VIII OTHER METHODS

The confirmation of transgene integration and analysis of its function is based on several approaches other than the methods mentioned above. These include detection of proteins by various assays including Western blot, Southern blotting, Northern blotting (refer to Chapter 9) and PCR amplification of integrated DNA segments (refer to Chapter 12). Some other methods, which are extensively used, but not described in this chapter are, marker inactivation technique, restriction enzyme cleavage patterns, chromosome walking, and *In vitro* translation techniques.

IX SUGGESTED READING

Ausubel FM, Brent R, Kingston RE, Moore DD, Seidman JG, Smith JA, and Struhl K (1999) (Eds) *Short Protocols in Molecular Biology* (4th Edn), John Wiley and Sons, Inc.

Becker JM, Caldwell GA and Zachgo EA (1991) *Biotechnology: A Laboratory Course*, Academic Press, Inc.

Davis CG, Dibner MD, and Battery JF (1986) *Methods in Molecular Biology*, ElsevierScience Publishing Co., Inc.

Gardner EJ, Simmons MJ and Snustad DP (1991) *Principles of Genetics* (8th Edn), John Wiley and Sons, Inc. New York, Toronto, Singapore.

Garret RH and Grisham CM (1999) *Biochemistry*, Saunders College Publishing.

Glick BR and Pasternak JJ (1996) *Molecular Biotechnology*, Panima Publishing Corporation, Bangalore, New Delhi.

Hartl DL and Jones EW (1998) *Genetics: Principles and Analysis* Jones and Bartlett Publishers, Inc.

Madigan MT, Martinko JM, and Parker J (1997) *Brock Biology of Microorganisms*, Prentice Hall International, Inc.

Miesfeld RL (1999) *Applied Molecular Genetics*, Wiley-Liss, John Wiley and Sons, Inc, Publication, NY.

Old RW and Primrose SB (1981) *Principles of Gene Manipulation, An Introduction to Genetic Engineering*, Oxford: Balckwell Scientific Publishers.

Peters P (1994) *Biotechnology: A guide to Genetic Engineering*, Wm. C. Brown Publishers.

Russell PJ (1998) *Genetics*, Benjamin/Cummings Publishing Company, Inc.

Sambrook J, Fritsch EF and Maniatias T (Eds) (1989) *Molecular Cloning: A Laboratory Manual* (2nd Edn), Cold Spring Harbor Laboratory Press. Cold Spring Harbor, NY.

Steinberg M, Guyden J, Calhoun D, Staiano–Coico L and Coice R (1993) *Recombinant DNA technology. Concepts and Biomedical Applications*, PTR Prentice Hall, Inc.

Watson JD, Hopkins NH, Roberts JF, Steitz JA, and Weiner AM (1987) *Molecular Biology of the Gene (4th Edn)*, The Benjamin/Cummings Publishing Company, Inc.

Weaver RF (1999) *Molecular Biology*, WCB/McGraw–Hill.

Wilson K and Walker J (2000) (Ed) *Practical Biochemistry: Principles and Techniques* (5th Edn), Cambridge University Press.

PART – V
Applications of Biotechnology

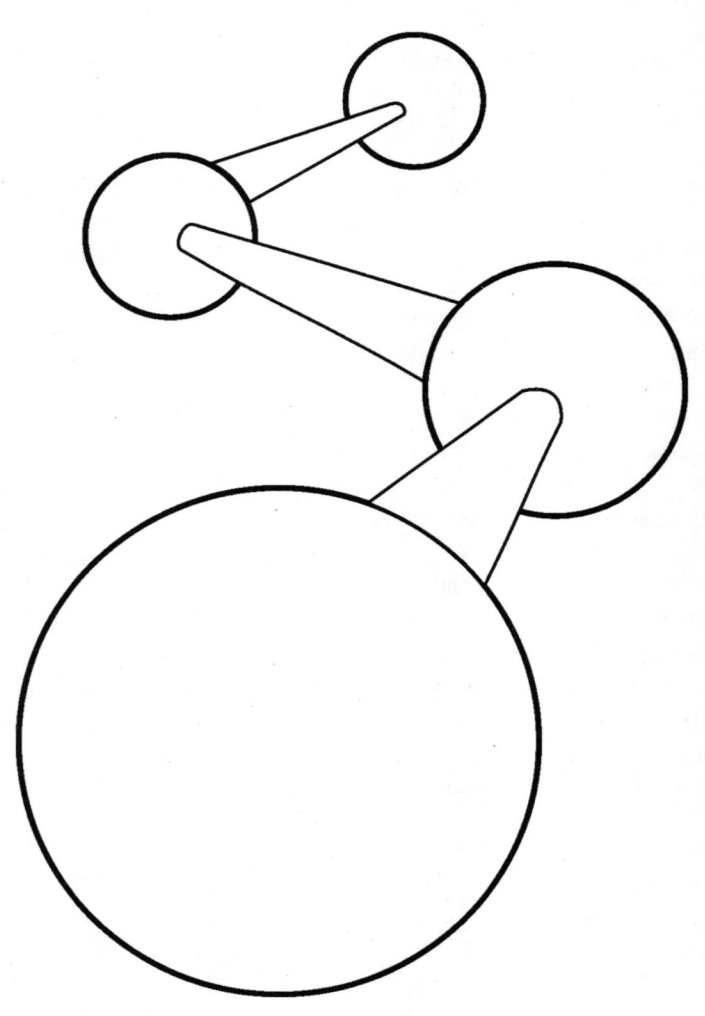

20

Genetic Engineering of Microorganisms

I INTRODUCTION

Since time immemorial, man has been engaged in the improvement of various activities such as the leavening of bread, wine making and brewing, food preservation and, several fermentation processes (e.g. soy sauce and vinegar). Although biotechnology operations started long ago, the recent advent of genetic engineering and recombinant DNA techniques have given a boost to biotechnological advancements exploited by academic and industrial needs (Figure 20.1). Microorganisms have certain properties that make them well suited for industrial processes. They possess a relatively high and rapid metabolic activity that permits a broad variety of enzymes to make an array of chemical conversions. They have a large surface area for quick absorption of nutrients, and release of end-products. Further, they usually multiply quickly, as evidenced by the 20-min. generation time for *E. coli* under specific conditions. In industrial processes, microorganisms act like chemical factories. They liberate a large amount of a single product that can be efficiently isolated and purified. It is also easy to maintain and cultivate. However, they should have genetic stability with infrequent mutations.

The understanding of the structure, organization, and function of genes of *E. coli* and its phages has progressed far beyond the knowledge about other organisms. This is because of the availability of techniques of genetic manipulations using F′ episomes, phage lambda, and many translocatable elements. Small segments of DNA can be moved to make merodiploids for any part of the chromosome, which allow the analysis of operons and their regulation by the study of interactions between different alleles of the same regulatory genes and controlling elements in the same cell. Induction with the gratuitous inducers and availability of non-inducing substrates (e.g. X-gal), have been used to distinguish constitutive from inducible strains.

The biochemical capabilities of microorganisms are vast, and a wide variety of new or unusual compounds may be produced from various microbial isolates. One of the major tasks of microbial geneticists is to develop procedures for obtaining new microbial metabolites. Some of the main approaches are: (1) Screening for the production of new metabolites with new isolates and/or new test methods. This is the only way to obtain completely new classes of substances; (2) chemical modification of known microbial substances; (3) bio-transformation, which results in changes in a chemical molecule by means of microbial or enzymatic reaction; (4) interspecific protoplast fusion, which is a means of recombining genetic information from

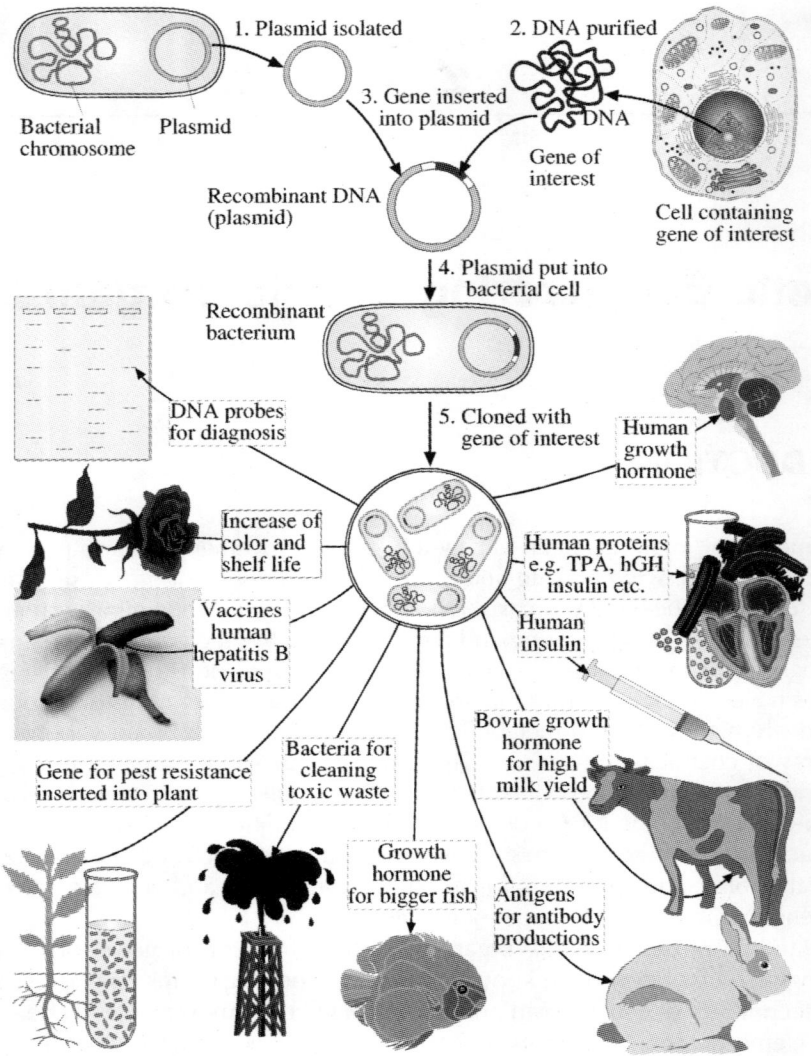

Figure 20.1 *Development of new products in bacteria by recombinant DNA technology*

rather closely related producer strains. New or hybrid substance (e.g. antibiotics) are expected; and (5) gene cloning, in which genes may be transferred between unrelated strains, which are producers of known substances.

II MICROBIAL PRODUCTION OF INDUSTRIAL CHEMICALS

The ultimate aim of biotechnology research is the development of commercial products. Consequently, unlike other scientific disciplines, molecular biotechnology is driven, to a great extent, by economics. The initial expenditure and the final expectation of financial gain are both responsible for the sustainability of biotechnology. In 1985, there were over 400 biotechnology companies in the US. Today, there are over 900 in the US and about 1200 worldwide. By the early 1990s, approximately 100 new drugs produced by molecular biotechnology companies were in the process of being tested in human trials and more than 400 other pharmaceutical products were in some stage of development. Product sales for the molecular biotechnology industry have increased from about $6 million in 1986 to over $3.4 billion in 1993. It has been estimated that by the year 2010, product sales for molecular biotechnology commodities will reach $ 60–100 billion per year. Biotechnology as a business segment for India has the potential of generating revenues to the tune of $ 5 billion and creating one million jobs by 2010 through products and services.

The major benefit from recombinant DNA technology today is the production of proteins and other compounds. Most recombinant products are currently produced microbially. In this case, the host microorganism is engineered to become a factory for the production of useful metabolites. Some of the important products produced through genetic engineering of microorganisms are briefly discussed here.

A. MICROBIAL GROWTH

Microorganisms are capable of growing on a wide range of substrates and can produce a remarkable spectrum of products in very little time and space. The recent advent of *in vitro* genetic manipulation has extended the range of products and has provided new methods for increasing the yields of existing ones. The growth of a microorganism may result in the production of a range of metabolites but to produce a particular metabolite the desired organism must be grown under precise cultural conditions at a particular growth rate. In defined nutrient concentrations and environmental conditions, microorganisms pass through a number of stages and the system is termed batch culture. There is initially a lag-phase (slow or no growth), followed by a log or exponential phase. In the log phase, cells grow at a constant and maximum rate, which may be described as $dx/dt = \mu x$, where x is the cell concentration (mg cm^{-3}), t is time of incubation (hr), and μ is the specific growth rate (h^{-1}). On integration, the equation gives $x_t = x_0 e^{\mu t}$, where x_0 is the cell concentration at time zero and x_t is the cell concentration after a time interval, t h. However, as the substrate (nutrient) is exhausted and toxic products accumulate, the growth rate of the cells deviates from the maximum. Growth eventually ceases and the culture enters the stationary phase. After a further period of time, the culture enters the death phase and the number of viable cells declines. However, if availability of medium components is the limiting factor, then growth may be extended by adding an aliquot of fresh medium. If the fresh medium is added continuously and at an appropriate rate, and the culture vessel is fitted with an overflow device, such that the culture is displaced by the incoming fresh medium, a continuous culture may be established. The flow of medium through the system is described by the term dilution rate, D, which is equal to the rate of addition of the medium divided by the working volume of the culture vessel. Hence, the growth rate of the organisms is controlled by the dilution rate, which is an experimental variable. Fed-batch culture is a system that may be considered to be intermediate between batch and continuous processes. The term fed-batch is used to describe batch cultures that are fed continuously, or sequentially, with a fresh medium without removal of the culture fluid. Thus, the volume of a fed-batch culture increases with time.

B. MICROBIAL FERMENTATION

Stanbury and Whitaker (1984) classified microbial fermentations into four major groups: (i) those that produce microbial cells (biomass) as the product; (ii) those that produce microbial enzymes; (iii) those that produce microbial metabolites; and (iv) those that modify a compound which is added to the fermentation and the transformation processes.

1. Production of whole cells (Biomass)

Microbial biomass is produced commercially as a single cell protein (SCP) for human food or animal feed and cells to be used in the industry.

(a) Inoculants

A successful example of inoculation of seeds with microorganisms for better crop yield is the *Rhizobium* treatment of seeds. Adding appropriate *Rhizobia* to the soil along with the seed, can minimize addition of nitrogen fertilizers. The Rhizobium species infects root hairs, leading to the beneficial nitrogen-fixing symbiosis, which makes legumes so agriculturally important. Due to climatic changes and crop rotation, inoculation is required in each planting season. Therefore, mass production of the rhizobium inoculant has industrial potential.

An organism whose commercial exploitation has begun only recently is *Pseudomonas syringe*. A surface component of this bacterium catalyses the formation of ice nuclei and is finding use in the formation of 'artificial' snow on ski slopes. The presence of this organism on the surface of leaves can promote frost damage to crops, so considerable benefit might be obtained if plants are sprayed with ice-nucleation defective mutants.

(b) Insecticides

Insecticidal toxin of *Bacillus thuringiensis*: An ideal microbial insecticide must be a microorganism that is relatively target specific and can act rapidly. It should be stable in the environment, easily dispensed, as well as inexpensive to produce. One of the most extensively used microbial-insecticides is the sporulating cells of a *Bacillus* species. This bacterium comprises a number of different strains, each of which produces a different toxin that can kill certain specific insects. *B. thuringiensis* produces toxic crystals (an alkaline protein) in older cells during the process of sporulation. When the *B. thuringiensis* spores are ingested by caterpillars, the insecticide dissolves substances in the caterpillar's gut that cement the cells of the gut wall together. As gut liquid diffuses between the cells, the larvae experience paralysis and bacterial invasion soon follows. This toxin appears to be harmless to plants and other animals. Toxins produced by harvesting the bacteria at the onset of sporulation and drying them into a commercially available dusting powder. Viruses also show promise as pest control devices partly because they are more selective in their activity than bacteria. Among the insects successfully controlled with viruses are the cotton bollworm, cabbage looper, and alfalfa caterpillar.

(c) Starter cultures

Addition of starter cultures to milk for the production of fermented products such as cheese, yogurt, sour cream, etc., is not a new technique. For thousands of years, man has depended upon the natural contaminating flora in milk for the synthesis of lactic acid and flavors and the formation of the desired product. However, the widespread thermal processing of milk to destroy pathogens such as the causative agents of tuberculosis and brucellosis alters the microbial flora of milk. Consequently, starter cultures have to be added and these are generally frozen pure cultures of bacteria. When these bacteria are added to milk, they cleave the lactose to glucose and galactose and these are metabolized to lactic acid. The lactic acid coagulates the milk proteins, principally casein, to form a continuous solid curd in which fat globules, water and water-soluble materials are entrapped.

2. Single-cell protein (SCP)

The use of microorganisms as a source of protein has been widely advocated. Genes directing the synthesis of proteins that are enriched in essential amino acids can be incorporated into organisms to improve the nutritional value of their protein. Such tailored organisms, grown on sewage and other garbage material, can be used as feed for poultry and cattle. Single cell protein (SCP) is the term that refers to a monoculture of microbial cells or the total protein extracted from pure cell cultures (microbial biomass), either of which can be used as a human or animal protein supplement. Using microbial biomass as a food source deserves serious consideration because of insufficient world food supply and the high protein content of most microorganisms (60–80%). Refer to Chapter 35 for more details on SCP.

3. Production of small biological molecules

In addition to the major products that we have surveyed, microorganisms provide a number of specialized materials. With recombinant DNA technology, it is possible to modify metabolic pathways (Figure 20.2) of organisms either by introducing new genes or by altering existing ones. **Gibberellins** (GA) are a series of plant hormones that promote growth by stimulating cell elongation in the stem. They have also been used for breaking seed dormancy and flowering. GA also increases the yield and size of fruit in grapes. They are also produced by the fungus **Gibberella fujikuroi**. A specialized material of microbial origin is **alginate**, a sticky substance used as a thickener in ice cream, soups, and other foods. Another commercial product produced by microorganisms is **perfume**. Musk oil, for example, is produced from ustilagic acid, a product of the mold **Ustilago zeae**, which ironically causes smut disease.

Microorganisms are useful in various fields. They are used in industry to produce a variety of organic compounds including acids, growth stimulants, and enzymes. In some cases, the production results from an apparent accident in nature where it produces thousand folds more than the amount necessary for its own metabolism.

Citric acid was one of the first organic acids to be made in bulk by microorganisms (the microorganisms most commonly used is the mold *Aspergillus niger*). Manufacturers use this compound in soft drinks, candies, inks, engraving materials, and in a variety of pharmaceuticals such as anti-coagulants, and effervescent tables.

Lactic acid is another important microbial product extensively produced. This compound is employed to preserve foods, finish fabrics, prepare hides for leather, and dissolve lacquers. Lactic acid is commonly produced by *Lactobacillus bulgaricus* in the fermentation from lactose.

Gluconic acid is an another valuable organic acid produced from carbohydrates by *A. niger* and a bacterium **Gluconobacter** cultivated in fermentation tanks. This product is useful in medi-

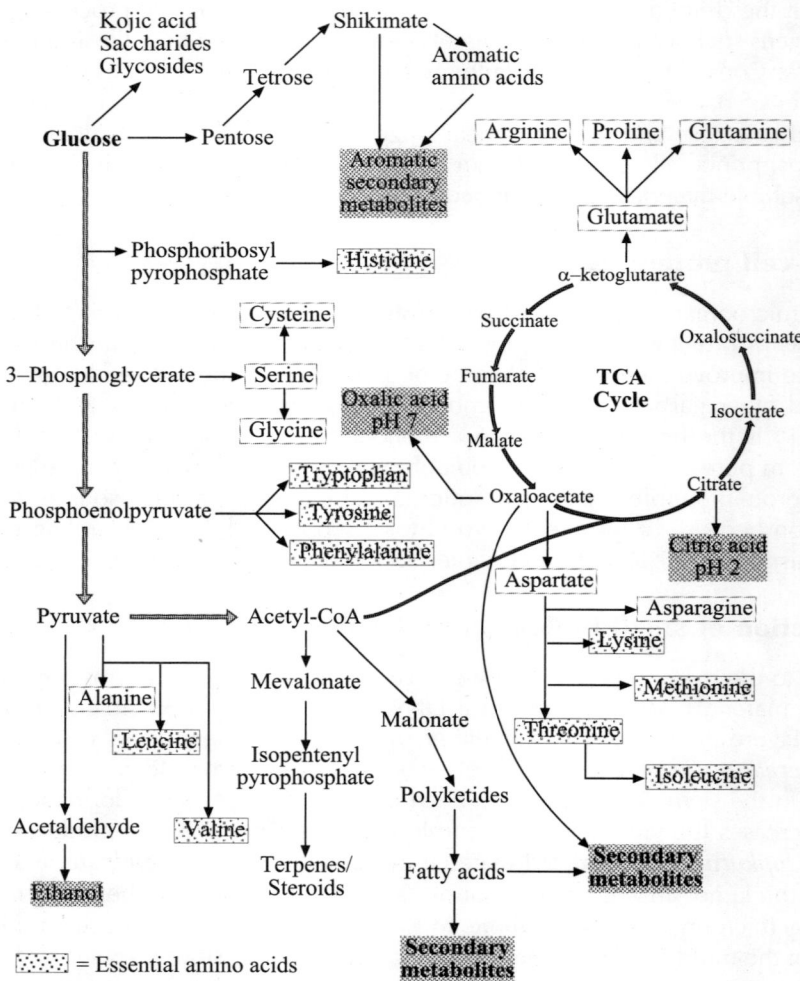

Figure 20.2 The interrelationships between primary and secondary metabolism (shaded text)

cines as a carrier for calcium, because gluconic acid is easily metabolized in the body, leaving a store of calcium for distribution. Calcium gluconate is also a good supplement for laying hens as it provides calcium that strengthens the eggshells.

Glutamic acid: Glutamic acid is a valuable food supplement for humans and animals, and its sodium salt, monosodium glutamate, is used in food preparations. Naturally, when the amount of amino acid produced by a microorganism exceeds the need, the remainder is excreted into the environment. The microorganism used to produce this acid is by *Micrococcus*, *Arthrobacter*, and *Brevibacterium*.

Lysine: Lysine is another amino acid used in breads, breakfast cereals, and other foods, and it is produced by two organisms. *E. coli* is first cultivated in a medium of glycerol, corn steep liquor, and other ingredients to produce diaminopimelic acid (DAP). Several days later,

Enterobacter aerogenes is added to the mixture. This produces an enzyme that removes the carboxyl group from DAP to produce the lysine.

Vitamins: Riboflavin (vit B2) and cyanocobalamin (vit B12) are two important vitamins that are also products of microbial growth. Riboflavin is produced by a mold (*Ashbya gossypii,* which produces 20,000 times more than required for its metabolism) and cyanocobalmin, which is produced by a selected species of Pseudomonas, *Propionibacterium,* and *Streptomyces.*

(a) Alcoholic Beverages

The fermentation of beer, wine, and other alcoholic beverages (brandy, rum, whiskey, and neutral spirits such as vodka, and gin) is among the most universal and extensively used biotechnology. Large-scale production of fermentation products is required for large quantities under optimal growth conditions, and the microorganisms should be genetically modified to efficiently synthesize the products. Therefore, a number of specific strains of microorganisms have been isolated and genetically modified to maximize the fermentation process.

(b) Antibiotics

Microorganisms are the source of various types of **antibiotics**. An estimated 100 to 200 new antibiotics are discovered each year, primarily through labor-intensive research programs in which many thousands of different microorganisms are screened to find those that produce unique antibiotics. Recombinant DNA technology can have a positive impact on this endeavor in two ways. First, this technology can be used to develop new, structurally unique antibiotics with increased activities against selected targets and decreased side effects. Second, genetic manipulation can be used to enhance yields and, as a consequence, lower the cost of production. **Penicillin** was the first antibiotic to be produced on an industrial scale. Penicillin is generally produced by *Penicillium notatum* and *P. chrysogenum*. Till today, over 5000 antibiotic substances have been described and about 100 such drugs are available to the medical practitioner. Although most antibiotics are produced by the species of *Streptomyces*, significant numbers are products of the *Penicillium* and *Bacillus* species. Antibiotic production is carried on in huge, aerated tanks of stainless steel similar to those used in brewing.

The biosynthesis of an antibiotic may include 10 to 30 separate enzyme catalyzed steps, so cloning all the genes for the synthesis of a particular antibiotic is not an easy task. One strategy for isolating the complete set of antibiotic synthesis genes consists of transforming one or more mutant strains to synthesize the antibiotic with DNA from a clone bank constructed from wildtype chromosomal DNA. Following the introduction of the clone bank into mutant cells, transformants are screened for their ability to produce the antibiotic. Then, the plasmid DNA from a single clone that complements a mutant strain is used as a DNA hybridization probe to screen another clone bank consisting of wildtype chromosomal DNA. In this way, DNA segments that are adjacent to the initial complementing DNA can be identified and cloned. This approach has been successfully used to isolate some of the genes for the biosynthesis of many antibiotics. In addition to cloning antibiotic biosynthesis genes by complementation, more direct strategies can be employed. One or more of the key enzymes in a biosynthetic pathway can be identified through either genetic or biochemical studies and then purified. The N-terminal amino acid sequence of the enzyme can then be determined, and based on this information, oligonucleotide probes for the gene can be prepared.

Eventually, new antibiotics with unique properties and specificities may be produced by the genetic manipulation of the genes involved in the biosynthesis of existing antibiotics. In addition to being a means for developing new antibiotics, genetic engineering can be used to enhance the yield and rate of production of known antibiotics. For example, the large-scale production of antibiotics by *Streptomyces* spp., is often limited by the amount of oxygen available to the cells. One strategy used to cope with oxygen-poor environments is to synthesize a hemoglobin-like molecule that can sequester oxygen from the medium and then deliver it to the cells. Following expression of the *Vitreoscilla* sp., hemoglobin gene in *S. coelicolor*, the *Vitreoscilla* sp., the hemoglobin represented approximately 0.1% of the total cellular protein.

4. Production of high molecular weight compounds

Microorganisms are used to produce two groups of high-molecular-weight compounds (macromolecules): polysaccharides, and proteins. The commercial production of both has increased dramatically in the last 10 years. Recombinant DNA technology has had an impact on proteins and polysaccharide production because it is possible to overproduce these high molecular compounds in microorganisms. Microorganisms can provide a constant and reliable supply of a polymer with relatively uniform chemical and physical properties.

(a) Biopolymers

Biopolymers are large, multi-unit macromolecules from living organisms. Some of these polymers have physical and chemical properties that are useful to the food processing, manufacturing, and pharmaceutical industries. Biopolymers can be derived from microorganisms, plants, and animals. The ability to genetically engineer organisms has stimulated researchers to design new biopolymers, replace synthetic polymers with biological equipments, modify existing biopolymers to enhance their physical and structural characteristics, and find ways to increase yields or decrease the costs of existing industrial processes. Polysaccharides are used commercially to produce gels, and to thicken and stabilize foods, medicines, and industrial products. They are also used as polymers in the fluids used to force oil to the surface in the tertiary oil recovery process. Although polysaccharides of plant origin have been used for many years, microbial polysaccharides (Table 20.2) have become widely used over the past several decades.

Table 20.2 Some important microbial polysaccharides with commercial applications

	Polysaccharide	Organism
1.	Xanthan	*Xanthomonas campestris*
2.	Alginate	*Pseudomonas aeruginosa, Azotobacter vinelandii*
3.	Gellan	*Pseudomonas elodea*
4.	Polytran	*Sclerotium glucanicum*
5.	Curdlan	*Alcaligenes*
6.	Scleroglucan	*Sclerotium glucanicum, S. delphinii, S. rolfsii*
7.	Pullulan	*Aureobasidum pullulans*
8.	Dextran	*Acetobacter* sp., *Leuconostoc dextranicum, Streptococcus mutans*
9.	Zanflo	*Erwinia tahitica*

Engineering *Xanthomonas campestris* for inexpensive xanthan gum production: *X. campestris* is a gram negative obligatory aerobic soil bacterium that produces the commercially important biopolymer xanthan gum, a high molecular weight exopolysaccharide, as a by-product of its metabolism. *X. campestris* was genetically engineered to make *Xanthomonas* grow on whey, a waste product of the cheese-making process. The *E. coli lac* ZY genes, which encode the enzymes b-galactosidase and lactose permease, were cloned onto a broad-host-range plasmid under the transcriptional control of a *X. campestris* bacteriophage promoter. This construct was introduced into *E. coli* and then transferred from *E. coli* to *X. campestris* by triparatite mating. Transformants maintained the plasmid, expressed the enzymes b-galactosidase and lactose permease at high levels, utilized lactose as a sole carbon source, and produced high levels of xanthan gum by using glucose (transformant produced 3.71 g/ml over 3.53 g/ml by wild-type), lactose (transformant produced 3.61 g/ml over 0.25 g/ml by wild-type), or whey (transformant produced 4.24 g/ml over 0.22 g/ml by wild-type) as a carbon source.

Microbial synthesis of a plant biopolymer: Natural rubber, cis-1.4-polyisoprene, is an extensively used biopolymer that is obtained from a large number of different plants. The biosynthesis of rubber starts from simple sugars and requires approximately 17 enzyme-catalyzed steps, with the final step being the polymerization of isopentenyl pyrophosphate onto an allylic pyrophosphate. The last step is catalyzed by the enzyme rubber polymerase. As an initial step toward genetically engineered microorganisms for rubber synthesis, the rubber polymerase gene from *Hevea brasilensis* has been successfully cloned and expressed in *E. coli*. This cDNA clone can now be used, possibly in concert with other genes in the rubber synthesis pathway, in an attempt to produce natural rubber in the microbial system. And it can also be used as a source of rubber polymerase to develop an *in vitro* catalytic system.

(b) Protein and other products

Prior to the development of recombinant DNA technology, most human protein pharmaceuticals were available in very limited quantities. They were costly, and often their biological mode of action was not well characterized. Since the development of recombinant DNA technology, over 300 different proteins that are potential human therapeutic agents have been cloned (Table 20.3).

Table 20.3 *Human proteins that have been produced by recombinant DNA technology*

Protein	Use
α_1-Antitrypsin	Treats emphysema
Calcitonin	Treats osteomalacia
Endorphins and enkephalins	Analgesic agent
Blood proteins	
Erythropoietin	Treats anemia
Factor VIII, Factor IX	Coagulation factor; treats hemophilia
Urokinase	Thrombolytic agent
Tissue plasminogen activator	Dissolves clots
Human hormones	
Andrenocorticotrophic hormone	Treats rheumatic diseases
Growth hormone, growth hormone releasing factor	Promotes growth
Insulin	Treats diabetes
Platelet-derived growth factor	Treats atherosclerosis

Continues...

... *Continued*

Protein	Use
Follicle stimulating hormone	Treats reproductive disorders
Lymphotoxin, Macrophage activator factor	Antitumor agents
Nerve growth factor	Promotes nerve damage repair
Somatomedin C	Promotes growth
Relaxin	Facilitates childbirth
Serum albumin	Supplements plasma
Epidermal growth factor	Promotes wound healing
Immune modulators	
Interleukins	Cancer therapy and immune disorders
Interferons (α, β, γ)	Anti-viral, anti-tumor, anti-cancer agent
Colony stimulating factor	Treatment of infections and cancer
Lysozyme	Anti-inflammatory
B-cell growth factors	Treats immune disorders
Tumor necrosis factor, Urogastrone	Anti-tumor agent
Vaccines	
Cytomegalovirus	Prevention of infection
Measles	Prevention of measles
Rabies	Prevention of rabies
Hepatitis B	Prevention of serum hepatitis

III MICROBIAL ENZYMES

Enzymes produced by microorganisms for commercial exploitation have been an important area of research in industry. The major commercial use of enzymes is in the food and beverage industries, although enzymes do have considerable application in clinical and analytical situations as well as for being used in washing powders. Enzymes can be produced from animals, plants, and microorganisms, but the production by microbial fermentation is the most economic and convenient method. Currently, over two dozen types of enzymes produced by microorganisms are in use, and the number is still growing. Some important microbial enzymes include *amylase* (produced by a mold **Aspergillus oryzae**), *pectinase* (a product of a **Clostridium** species), several *proteases* (produced by **Bacillus subtilis**, **Aspergillus oryzae**, and other microorganisms), and many recombinant restriction endonucleases. Some other microbial enzymes used in food and medicines are: **Invertase**, an enzyme from yeast, which is used to convert sucrose to glucose and fructose; **Streptokinase**, which is used to break down blood clots formed during a heart attack; and **Hyaluronidase**, which is used to facilitate the absorption of fluids injected under the skin. Numerous pharmaceutical products derived from the ergot poisons of the mold **Claviceps purpurea** are prescribed to induce labor, treat menstrual disorders, control migraine headaches, etc.

Furthermore, it is now possible to engineer microbial cells to produce animal or plant enzymes. Enzyme engineering deals with the development of useful products or processes based on the catalytic activity of enzymes; the enzymes are usually isolated from immobilized, non-growing cells. Probably the most widely used industrial example of immobilized enzyme application at present is that of glucose isomerase, used for producing HFCS. Several commercial forms of this enzyme are available in the market. These commercial preparations are made from diverse microorganisms such as *Streptomyces ribigenosus, S. wedmorensis, S. olivaceus,* and *Bacillus* coagulants. Industrial fermentation sometimes requires tailor-made enzymes to catalyze a reaction efficiently.

The power of genetic engineering to engineer natural products can be seen in many areas, as for example, in the detergent industry, where new products with increased efficiency of cleansing action are being developed. Recently, it has become very common to add proteolytic enzymes to remove protein stains from soiled clothing. **Subtilisin**, a serine protease produced by recombinant *E. coli*, is added to modern detergents but it cannot be used with bleach since the latter inactivates it. The methionine at position 222 in the enzyme has been found to be responsible for this inactivation. Genetic alteration (site-directed mutation) has enabled this methionine residue to be substituted by other amino acids. The alanine substituted enzyme showed 53% activity as compared to wild type subtilisin and was not sensitive to bleach.

IV TRANSFORMATION PROCESSES

In addition to the production of biomass and microbial products, microbial cells may be used to catalyze the conversion of a compound into a structurally similar but financially more valuable compound. For example, conversion of vinegar is the oldest and most well-established transformation process (the conversion of ethanol into acetic acid) and involves production of high-value compounds. Microbial processes have the advantage of specificity over the use of chemical reagents and of operating under relatively mild conditions. The main limitation however, is that a large biomass has to be produced to catalyze a single reaction. Nowadays, some processes have been streamlined by immobilizing either the cells themselves or the isolated enzymes, which catalyze the reactions. There are numerous ways in which biocatalysts can be confined to a restricted space leading to a heterogeneous reaction system. The key to successful operation is immobilization in such a way that the substrate solution can pass through this region easily and interact with the catalyst efficiently. The most common means of achieving this is to have the substrate and catalyst in different phases, the latter being retained in some type of reactor vessel.

The term biocatalyst covers a range of biological materials, which are capable of transforming many substrate molecules into products, without themselves being changed. Most immobilized processes are concerned with aggregating these active materials on support particles, which have larger dimensions than the free biocatalyst. A number of methods have been developed for immobilization of all classes of enzymes, microbial cells, algae, plant cells, and animal cells. Biocatalysts can be retained in reactors either by entrapment or by attachment to a fixed structure or matrix. Entrapment technique involves creating a barrier through which substrate and product molecules will pass freely but which is impermeable to the biocatalyst. Since most biocatalysts are significantly larger than the substrates on which they act, there are a number of ways in which barriers, which discriminate between materials on the basis of size, can be employed. Entrapment methods fall into two categories. In one, a single membrane or barrier encloses a defined area containing the catalyst. In the second case, the catalyst is dispersed within a three-dimensional gel.

The attachment methods of immobilization rely on binding the biocatalyst to a solid support. The means by which the catalyst is actually held exploit almost every aspect of chemical and physical chemistry. The attachment processes are broadly divided into those which involve a physical, often reversible interaction (adsorption), and those in which a covalent bond is formed (covalent binding).

The equipment in which an immobilized biocatalyst is used is often as important in determining the productivity of the system as the actual method of immobilization itself. The main function of the reactor is to retain the immobilized complex and to ensure efficient and controlled contact between the catalyst and substrate in order to maximize product formation. Reactors can be divided into those designs, which accept particulate biocatalysts and those that are intended to operate with fixed, specialist elements such as sheets, membranes, or tubes. Particle reactors are probably more common and are in general more versatile, since they will easily accommodate minor changes in shape and loading of material. The non-particle reactors generally have more clearly defined flow patterns but often require expensive capital equipment and can only accept a fixed volume of catalyst.

Method-1: Immobilization of cells and enzymes by gel entrapment method

Immobilization of cells or enzymes considers the physical separation during continuous operation of the biocatalyst from the solvent in such a way that the substrate and product molecules may be readily exchanged between phases. Of the different immobilization techniques (Figure 20.3), entrapping cells or enzymes in calcium alginate beads is simple and economical. In the presence of monovalent cations, such as Na^+, the alginate forms a viscous solution. However, in the presence of divalent cations (e.g. Ca^{2+}), gelation occurs. Thus, when sodium alginate mixed with a biocatalyst extrudes into the $CaCl_2$ solution, the solid matrix of calcium alginate beads with the entrapped biocatalyst can be obtained.

Materials

Sodium alginate, and $CaCl_2$, 10 ml syringe fitted with 1mm wide bore needle, magnetic stirrer, *E. coli* cells for cell immobilization and invertase enzyme solution, Acetate buffer (pH 4.7) for enzyme immobilization.

Procedure

Cell immobilization:

1. Grow *E. coli* cells in a shaker flask by bacterial culture method (refer to Chapter 27).
2. Centrifuge the fresh culture for 10 min. in cooling centrifuge at 7,000 rpm at 4°C.
3. Wash the cells once with distilled sterile water and re-suspend in distilled water. Adjust the cell number to obtain optical density of about 0.5 at 620 nm.
4. Prepare one litre of 0.2 M $CaCl_2$ solution.
5. Prepare 4% w/v of sodium alginate solution: The sodium alginate powder should be added to the distilled water while being stirred. It should be stored for an hour to dissolve the powder completely.
6. Mix 50 ml of cell suspension with 50 ml of 4% w/v sodium alginate solutions (alginate concentration reaches 2% w/v).

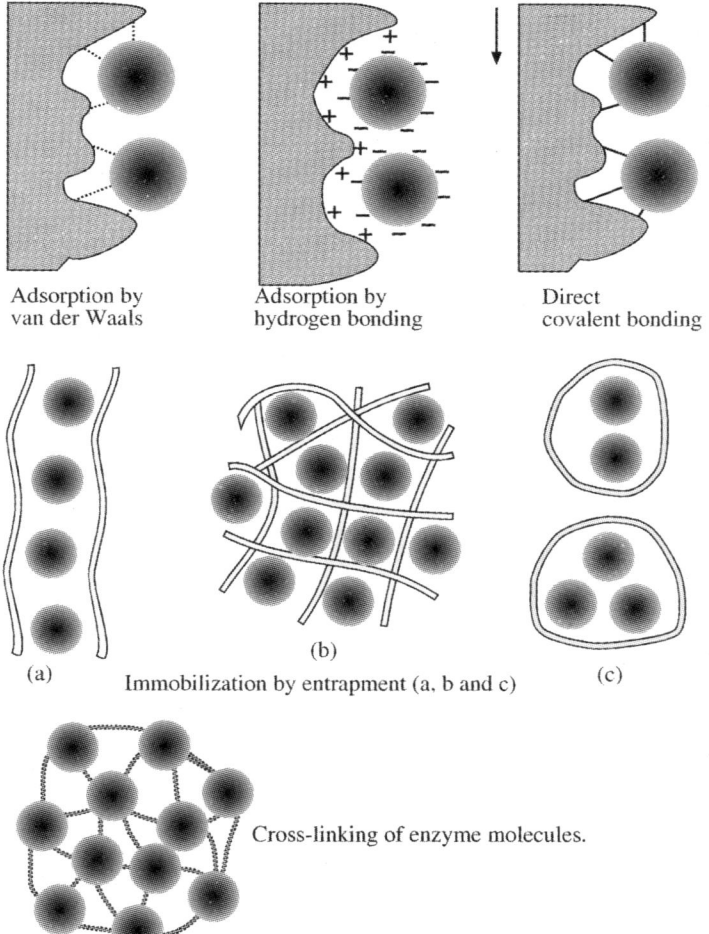

Figure 20.3 *Different methods of enzyme immobilization*

7. Extrude this solution drop-wise through a syringe fitted with a 1 mm bore needle from a height of about 20 cm into an excess of 0.2M $CaCl_2$ solution, i.e. one litre of $CaCl_2$ solution to extrude 100 ml of alginate solution, to form spherical beads.
8. Leave the beads of calcium alginate entrapped cells to harden in $CaCl_2$ solution for about an hour.
9. Filter the beads on a Buchner funnel and wash with distilled water.
10. Measure the mean diameter of the beads, (by randomly selecting a few beads) using vernier calipers and take the average.

Size and hardness of beads:

Extruding the solution through different bore sizes of needle will lead to different sizes of beads. Increasing the concentration of the sodium alginate solution will result in more tightly "cross-linked" gels, which will have more strength and stability but may cause mass transfer limitations.

Enzyme immobilization

The same procedure can be followed for enzyme immobilization in alginate beads. Instead of cell suspension, use 50 ml of enzyme solution prepared in 0.1M sodium acetate buffer with pH 4.7. Filter the beads and wash them in acetate buffer.

V THE DESIGN AND OPERATION OF A CONVENTIONAL FERMENTER

Humans have been exploiting microorganisms for many centuries for the production of fermented foods and beverages. The production of alcoholic beverages became in many countries the first truly industrial process. The large-scale culture of microorganisms in a submerged culture in industry is loosely defined as fermentation. The vessels used in the fermentation processes in the liquid medium are called fermenters and the technology associated with their design and operation is called fermentation technology. In this section, the design and operation of fermenters, the media used in industry fermentations, the basics of product recovery, the kinetics of growth in culture systems, how culture systems can be operated to maximize output, and how production organisms can be manipulated to maximize output will be briefly discussed.

A. OPERATION OF CULTURE SYSTEM

Design: Traditional fermenters used in brewing were open vats made of wood, slate, or stainless steel. Modern brewing vessels are closed cylindrical stainless steel vessels with a conical base and dished (domed) top (Figure 20.4). Enclosing such vessels allows CO_2 to be collected and reused and prevents contamination of the contents from air. External jackets allow temperature control. These tanks made from an appropriate grade of stainless steel and range from about 1000 litres to one million litres.

Sterilization: Most industrial fermentation processes operate pure cultures, and so the fermenter and its contents must be sterilized prior to inoculation of the production organism, and asepsis maintained thereafter. Sterilization *in situ* involves heating the whole fermenter to 120°C for 20 min. after it has been filled with the required volume of the medium by injecting pressurized steam into a jacket or coil around the fermenter wall. Alternatively, the empty fermenter is first sterilized by injecting it with pressurized steam, and then the medium is run into the fermenter through a sterilizing system.

Inoculation and sampling: Once the fermenter has been filled to the required level and sterilized, the medium is inoculated with the production organism. The inoculum is prepared from a stock culture of a carefully selected high-yielding production strain (volume of 5–10% of the working volume of the production fermenter). After inoculation, the fermenter is usually sampled regularly, so that the culture can be examined for signs of contamination and the concentration of product measured.

Aeration: Aeration of a fermentation medium has two functions: the supply of oxygen to the production organism and the removal of CO_2 formed as a result of aerobic respiration of the

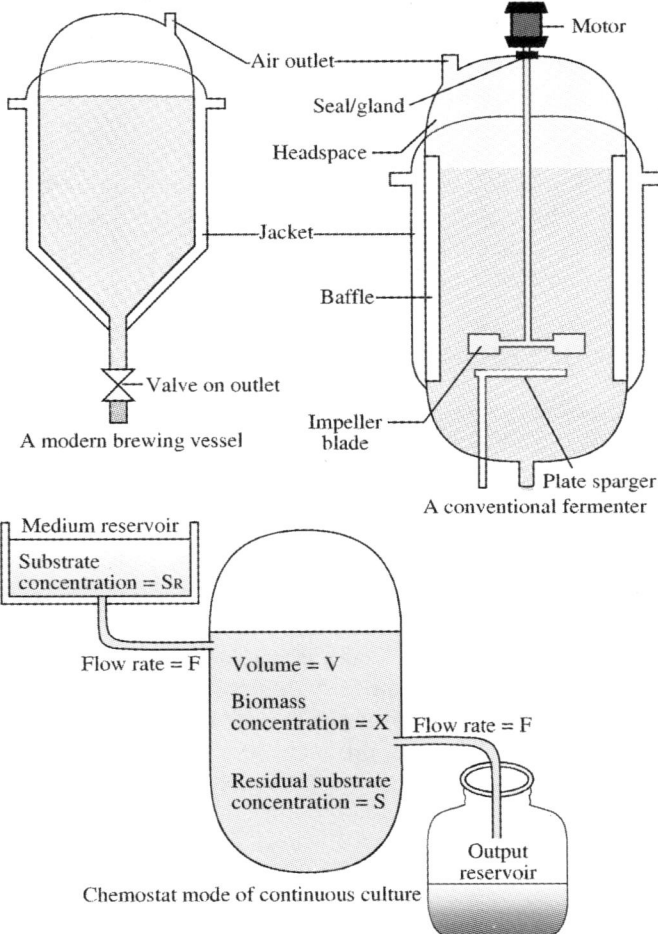

Figure 20.4 *The design of a modern brewing vessel and a conventional fermenters*

production organism. Oxygen is supplied from air compressed by a compressor and stored in tanks. It passes into the fermenter through a flow-meter/regulating valve system. In order to increase the aeration capacity of the system, the medium can be agitated (stirred) by paddles driven by a motor.

Control systems: During cultivation of microorganisms in a fermenter, parameters such as temperature, pH, dissolved oxygen concentration, and nutrient concentration will tend to vary. In order to maintain optimal conditions for growth and product formation, such variables should be monitored and controlled.

Cleaning: Once fermentation is complete, the fermenter is harvested. The fermenter is then prepared for the next fermentation. The time taken for the turnaround is called down time and it is essential to keep this as short as possible as it is non-productive. Because of this, automatic cleaning using high-pressure water jets is usually performed.

Variations in fermenter design: Not all fermenters are of the same design. Some are **spherical** with specialized stirrers and baffles to produce vigorous agitation. **Loop fermenters** are of the air-lift type, with one chamber up which air and the medium flow and a separate chamber down which the medium returns. **Tower fermenters** are tall cylindrical towers; those fermenters where air is simply bubbled up the column are referred to as bubble-column fermenters. A **pressure-cycle fermenter** is a tower/loop fermenter of large dimensions; the column of medium is high (>30 metres) so that high hydrostatic pressure is generated at the bottom of the column where air is sparged in. The high hydrostatic pressure increases the solubility of oxygen in the medium, increasing the aeration capacity. The advantage of such a fermenter is that it produces high aeration capacities without having any moving parts.

B. SOLID–SUBSTRATE FERMENTATION

Not all production processes of microorganisms involve a liquid medium; the organism can be grown on the surface of pieces of a solid substrate too. This solid–substrate fermentation (SSF) is an established traditional technology in many countries to produce commodities such as edible mushrooms, fungal-fermented foods, fungal-ripened cheeses, soy sauce, etc. Composting is regarded as a form of SSF. In practice, SSF is generally operated as a non-aseptic process, using pasteurization of the solid substrate to reduce the number of competitors, coupled with heavy inoculation to ensure rapid establishment of the production organism, and low moisture content to depress the growth of organisms surviving pasteurization. In traditional processes, shallow layers of inoculated substrate are placed in trays or baskets and incubated under conditions of controlled temperature and humidity. The major advantages of SSF compared to conventional fermentation in a liquid medium are that SSF uses solids, is low-tech, needs low-energy technology, requires less capital investment, produces less waste effluent, and can make product recovery easier.

C. CULTURE SYSTEMS

In fermentation processes, the microorganisms used are carefully selected to maximize yield of the product. The fermenters used are designed and engineered to provide optimal conditions for growth and product formation (Figure 20.5). There are four fundamental modes of culture that are applied in industrial fermentation processes. (i) **Batch culture**: Batch culture involves adding all nutrients to the fermenter prior to inoculation. This system is thus a closed system in terms of nutrition. (ii) **Fed-batch culture**: A fed–batch culture is a batch culture in which one or more nutrients are fed into the fermenter during the fermentation period. In this way, nutrients can be added at the same rate as they are used up. (iii) **Semi-continuous culture**: For products, which are growth-linked, such as ethanol or biomass, most of the product will be formed toward the end of the growth phase, when most growth occurs. If at the end of this phase a portion of the culture medium is removed from the fermenter and replaced by a fresh medium, then growth of the organism will continue. This process can then be repeated at appropriate intervals, and is known as semi-continuous culture. (iv) **Continuous culture**: Continuous culture involves removing the medium from the fermenter continuously and replacing it with a fresh medium, usually at the same rate. This is an open system.

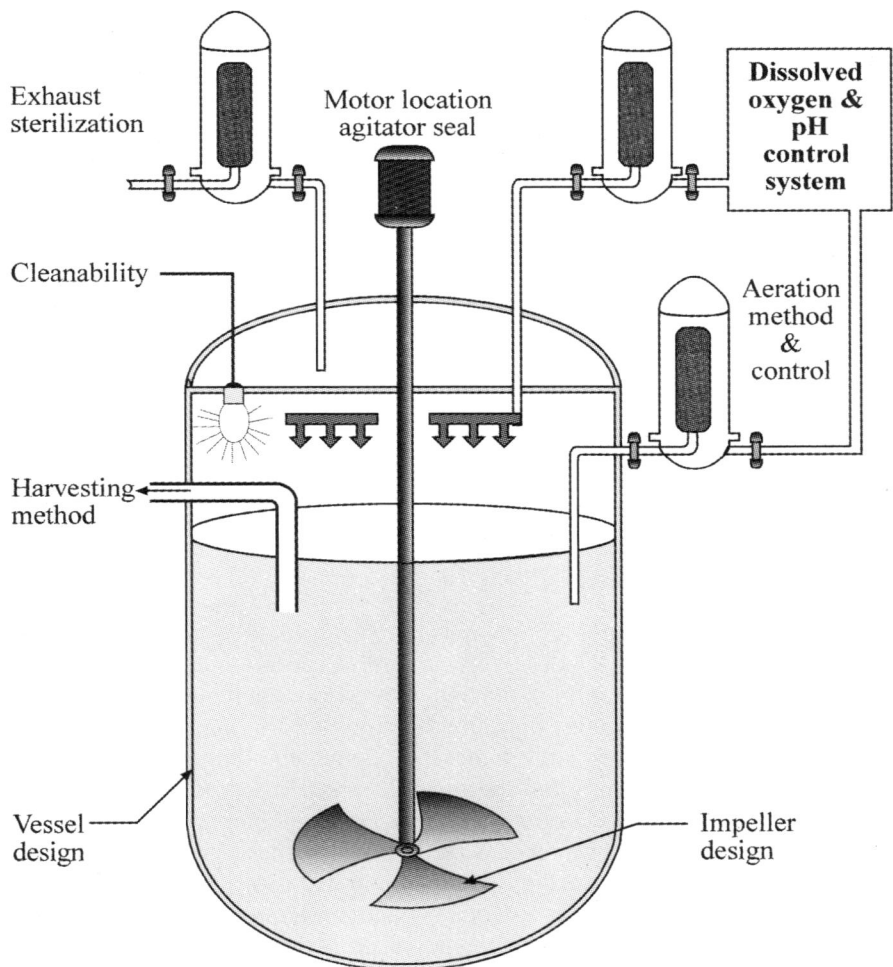

Figure 20.5 *A model bioreactor design and areas that must be controlled in selection and equipment specification*

VI GENETIC ENGINEERING OF MICROORGANISMS

Genetic engineering based on plasmid technology has been seen as the beginning of modern industrial microbiology. Advances in the use of recombinant DNA technology to produce pharmacologically interesting proteins, enzymes, or amino acids are proceeding at a rapid pace. The areas of applications in medical diagnostics, agriculture, and environmental protection as well as in basic research are also gaining importance. The significance of plasmid technology has been explained in detail in Chapter 16.

A. APPLICATION OF RECOMBINANT MICROORGANISMS IN BASIC RESEARCH

Genetic engineering has revolutionized research in basic biology. The clarification of control methods in prokaryotes and eukaryotes, embryological development and differentiation, and sequence determination of nucleic acids and proteins have all made enormous advances. The identification and characterization of antibiotic resistance genes, as well as regulatory and structural genes of biosynthetic pathways, the study of genetic instability and strain degeneration, and genome analysis are some other microbial applications.

1. Merodiploids construction and complementation test

Recombinant DNA technology can now be used to ligate small pieces of genomes of industrially important organisms to appropriate vectors that replicate in such organisms as well as in *E. coli*. This allows application of the complementation test for understanding regulatory mechanisms. Such vectors that replicate in *E. coli* as well as in some other organisms such as *B. subtilis* or yeast are already being used to shuttle genes between two different host organisms.

2. Operon fusions

Another application of this technology emerges because genetic engineering makes fusion of any two pieces of DNA possible. Genes of any organism can now be fused to the well-studied *E. coli* promoters with the expectation of efficient expression. Such gene fusions allow exploitation of a simple β-galactosidase gene controlled by promoters of other genes that are functional in industrial organisms. It has been shown that the operator binding site of the *lac* repressor is restricted to 59 N-terminal residues and the first 92 N-terminal residues are sufficient to bind the lambda repressor to the lambda operator. The type of genetic analysis that has been used for *lac* and phage lambda repressors can now be applied to other proteins. Negative dominance can be used as a tool to study the function of any protein that carries two different enzyme activities on one polypeptide chain such as threonine–deaminase/threonine kinase. Mutants lacking one of the activities can show which end of the polypeptide chain is responsible for each activity.

The knowledge, gained from repressor β-galactosidase fusions, that certain peptides that are degraded in *E. coli* can be stabilized by being fused to high-molecular-weight proteins has already been exploited to make somatostatin hormone in *E. coli*. Somatostatin was found to be stable in *E. coli* when it was fused to the β-galactosidase protein. Similarly, insulin is not destroyed when it is fused to the β-lactamase gene. Both insulin and somatostatin were fused to the carboxylterminal ends of protein and were transcribed from the promoters of the *E. coli* gene (Figure 20.6).

3. Directed secretion of proteins

To understand the mechanisms by which proteins are secreted from the cytoplasm to the cell membranes, genetic recombination was used to replace the amino-terminus of β-galactosidase with the amino-terminal sequence of membrane proteins such as the maltose-binding protein or phage lambda receptor lam B protein. In both cases, β-galactosidase activity became associated with membranes, which suggested that addition of a signal sequence to a

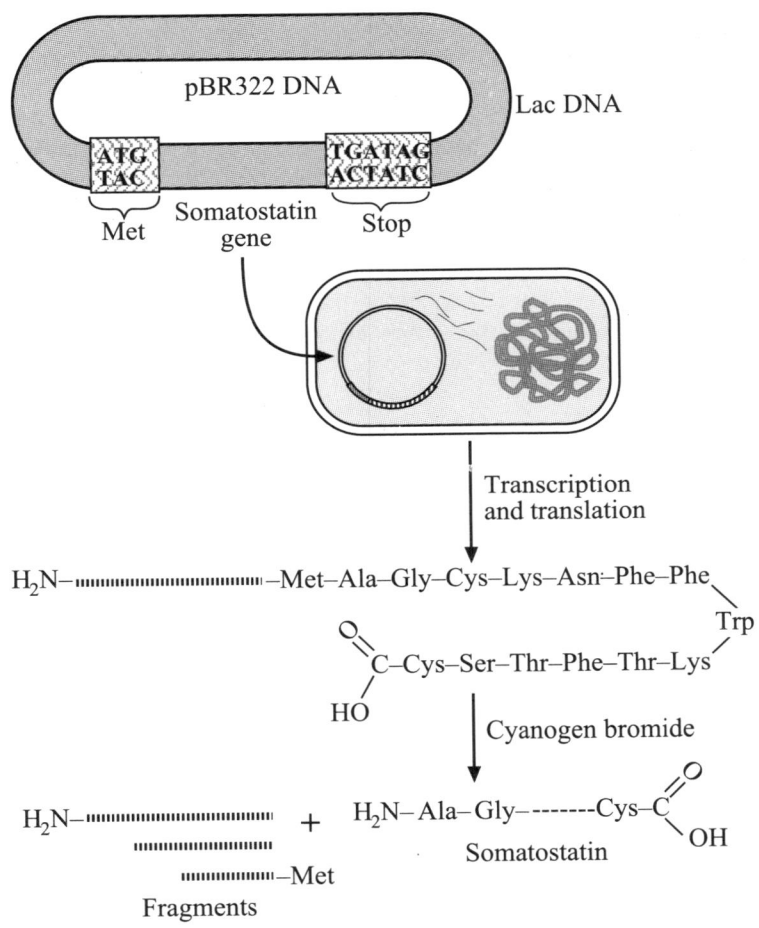

Figure 20.6 *Synthesis of somatostatin from a chemically synthesized gene joined to the plasmid pBR322 lac*

non-secreted protein would cause it to be secreted. On the basis of these results, Villa-Komaroff et al. (1978) fused *in vitro* the rat insulin gene cDNA to the β-lactamase gene of pBR322. The hybrid insulin β-lactamase protein found to be secreted, protects the insulin from proteolytic degradation by cellular enzymes. The insertion of a viral DNA sequence coding for surface antigen into a carrier protein gene could result in the secretion of a fused protein that could be used as a vaccine even though no correct virus product has been synthesized.

Various genes that determine the synthesis of enzymes secreted by industrial organisms might be fused to appropriate vectors. For example, the yeast invertase gene could first be inserted into a vector. Genes directing synthesis of peptides could then be inserted into this vector in such a manner that the amino-terminal end of the invertase facilitates the secretion of fused protein. Invertase would also be helpful to the yeast while growing on molasses, a cheap industrial carbon source.

4. *In vitro* localized mutagenesis

Every time a microbial cell divides, there is the probability of an inheritable change occurring. A strain exhibiting such a changed characteristic is termed as a mutant and the process giving rise to it, a mutation. Recombinant DNA technology can be used to isolate, amplify, and determine the structure of any DNA segment. The functional significance of a specific region of DNA can be explored by generating localized modification in the DNA molecule and then studying the resulting effects on the biological properties of the molecules *in vivo* or *in vitro*. *In vitro* mutagenesis offers the possibility of mutating isolated purified DNA sequences to obtain enzymes with altered properties. The genes that have been altered in pre-determined base sequences could also produce analogs of peptide hormones. DNA sequences of regulatory elements (operator, promoter, attenuator) can be subjected to *in vitro* localized mutagenesis, and mutated sequences can be used to study how corresponding proteins interact with them. Such studies have provided an insight into the types of interaction between the lac operator and repressor and the promoter and RNA polymerase.

Scherer and Davis (1979) developed a general method for replacing *Sacchromyces cerevisiae* chromosome segments with altered DNA sequences. Any isolated gene, deleted or altered by *in vitro* mutagenesis, can be propagated by attaching it to a vector with a selective chromosomal marker. Upon transformation, the composite DNA molecule, consisting of vector, selective marker, and altered DNA, recombinases with the homologous DNA sequences in the chromosome, yielding a direct non-tandem duplication separated by the vector and the marker DNA. The vector and marker DNA are lost owing to the second recombinational event on the other side of the altered sequence relative to the first. This replaces the altered sequence with the wild type DNA segment of the untransformed cell. Homologous sequences on both sides of the mutant DNA sequences must be present in the transforming DNA. The composite molecule does not have to replicate autonomously in the recipient host where altered DNA sequences are to be replaced. An internal deletion mutant of his-3 gene and a transposition of a galactose inducible region into chromosome XV are generated by using the ura-3 gene as a selective marker.

Scherer and Davis' method has general application and relies on the ability to transform cells to achieve integration by homologous recombination and on the availability of selective markers as cloned DNA fragments. The expected phenotype of the organism after insertion of a foreign DNA is not known in advance. The desired strain can be identified by restriction enzyme analysis of the clones with the selective marker. This allows deletion or alteration of genes with no known functions. If a good, non-reverting mutation for the selective marker is not available; such a mutation can be transplanted into the organism of interest after its creation *in vitro*. This technique could help the manipulation of the chromosomes of industrially important organisms, since it can provide well-characterized sequence variants of any genetic locus.

5. Genome organization

(a) Gene mapping

Restriction mapping is increasingly being used for gene mapping. The positions of about 350 genes on human chromosomes have been localized and the major part of the human gene map may be constructed in the next few decades. The availability of cloned and labeled probes will speed up chromosome mapping and add resolution to restriction mapping. Using labeled probes of complementary RNA prepared from cloned sea urchin genes, five human histone genes have been localized to the long arm of chromosome VII. Scott *et al.* (1979) and Haigh *et al.* (1980) have precisely localized the β-globin gene on the short arm of human chromosome XI. DNA was isolated from hybrid clones that contained Chinese hamster genome together with a section of human chromosome XI. The DNA was treated with restriction endonucleases and fragments resolved on agarose gels. The separated fragments were transferred onto nitrocellulose filters and hybridized to a radioactively labeled human β-globin DNA probe. The pattern of hybridization as revealed by autoradiography was used to conclude that the gene for β-globin was carried on a section of the short arm of human chromosome XI.

(b) Gene structure

Intervening sequences: The availability of homogeneous purified chromosomal DNA fragments has allowed the determination of their exact base sequence. This information has resulted in the unexpected discovery of intervening sequences (introns) within many genes of higher organisms (eukaryotes) and their cytoplasmic organelles (mitochondria). The diversity, complexity, and size of genes with introns were unexpected. Including its intervening sequences, the gene that codes for the enzyme dihydrofolate reductase in mammalian cells is about one million nucleotides long. About 10% of these nucleotides code for amino acids in the enzyme. The chick ovomucoid gene contains seven intervening sequences.

Overlapping genes: The total genomes of several small viruses have now been sequenced. Analysis of these sequences has confirmed that certain genes overlap and some pieces of DNA direct synthesis more than one protein. Overlapping genes have gone undetected for a long time, but DNA sequencing can now aid in their discovery.

Promoters in the middle of the gene: Promoters are sequences of nucleotides in a genome where RNA polymerase first binds before transcribing a gene into mRNA. In bacteria, such RNA polymerase binding sites are at the beginning of a gene. However, sequencing of the DNA of several higher organisms has shown that promoters might be in the middle of the gene.

Regulation by inversion of promoter and chromosomal rearrangement: *Salmonella* possess two distinct flagella antigens. In general, however, only one of the antigens is expressed at a given time. This phenomenon, called phase variation, has recently been understood through the application of recombinant DNA technology. Transition from expression of one antigen to another results from an inversion in a region containing a promoter-like DNA sequence that controls initiation of gene transcription. In one orientation, the transcription of genes that direct synthesise flagellin subunits is interrupted. The mechanism for regulating the inversion of the promoter DNA sequence is not known.

Transcription of structural genes as a developmental regulatory mechanism may be involved in the life cycle of homothallic bakers' yeast. Yeast haploid cells of one mating type (a or α) produce progeny of the opposite mating type during mitotic growth. A diploid phase (a/α) is established by conjugation between siblings of opposite mating types. Switches between mating types occur by transposition of a genetic element that carries genes for the mating type. By using a copy of the α mating type gene as a probe for Southern hybridization, the conclusion was reached that the mating type interconversion process in yeast involves genetic rearrangement. Silent copies of the mating type genes are located at the HML and HMR loci on chromosome III. Translocation of the replicas of these silent sequences to the mating type locus affects the switch between mating types.

Genome rearrangement, e.g., inversion and translocation, may be involved in the regulation of functions that are not essential for survival of the organism and are subjected to catabolite repression. Enzymes involved in antibiotic production by *Streptomyces* may be controlled by such a regulatory system.

6. Origin (Ori) of DNA replication

There have been plasmids constructed that contain DNA replication origin (ori) of *B. subtilis*, *E. coli*, phage G4, *Salmonella typhimurium*, yeast, and Drosophila mitochondrial DNA. A non-replicating *Eco*RI fragment conferring Km resistance was used for selecting two different *Eco*RI fragments containing the replication origin from the F factor and the *S. typhimurium* chromosome. Owing to incompatibility with the *E. coli* chromosome, *E. coli*, and *S. typhimurium* origin plasmids are not stably maintained in *E. coli* in the absence of selective pressure.

Recombinant DNA technology is being used for the analysis of plasmid structure and function. Segments of any plasmid can be inserted into *Coli*EI. The hybrid plasmids are used to transform Poly(A) *E. coli* that is deficient in DNA polymerase I for replication. Only those recombinant plasmids that have inserted a fragment with origin of replication inserted will replicate in Poly(A) *Escherichia*. This approach has been used to construct hybrid plasmid replicons that harbor the origin of replication of pSC101 and ColEI plasmids.

An autonomously replicating 859-base pair sequence has been isolated from yeast. Such replicator-containing plasmids are used for cloning foreign genes and for understanding DNA replication.

B. IMPROVEMENT OF INDUSTRIAL ORGANISMS

For commercial production of recombinant DNA products, some organisms may prove better than *E. coli*. Grooming of organisms already being used in the fermentation industry could prove highly beneficial. Even though experiments with industrial organisms have not received as much publicity as have the unraveling of the secrets of nature using *E. coli*, industrial organisms harbor metabolic pathways that are specific to them. For example, pathways that lead to the biosynthesis of antibiotics, gibberellins, and alkaloids exist in only a few microorganisms. The regulation of general metabolism in such organisms is not well coordinated. The lack of fine regulation of primary cellular metabolism leads to accumulation of many branch point intermediates and end products. These elevated levels of intermediates are the precursors of many commercial metabolites that are produced in significant quantities. For example, penicillin and glutamic acid are accumulated in concentrations of up to 50 and 120 gm/litre of growth medium, respectively. Efficient and economical processes for the industrial production of these microbial metabolites can now utilize the tools of recombinant DNA technology and make even further improvements.

1. Gene amplification

Many molecules such as insulin, interferon, and growth hormone are active in very minute amounts. Such molecules that are not needed in large amounts and can be sold at a good price may very well be commercially produced in *E. coli*. Many multicopy plasmid and phage vectors can be used to amplify genes in *E. coli*. Foreign genes on a multicopy plasmid are easy to isolate and purify in relatively large amounts. The products of such genes can be made in significant quantities if they are inserted in such a manner that their transcription is initiated by a strong promoter.

Gene cloning is an effective means of amplifying the number of copies of a specific gene. A pertinent demonstration of this is the construction of a hybrid bacteriophage lambda into which an *E. coli* gene coding for DNA ligase has been inserted. Upon infection of *E. coli* with this chimera, a 500-fold increase in the concentration of DNA ligase over that produced by uninfected *E. coli* is generated. DNA ligase accounts for about 5% of the total *E. coli* protein and is easy to purify in large amounts from such strains. DNA polymerase-I was amplified 100-fold following infection of E. coli with the lambda polA QS phage. Enhancement of yields of any protein could be achieved by fusing the respective gene to an efficient promoter and inserting it into a multicopy plasmid containing a heat-inducible lambda regulatory system. A copy number mutant of plasmid ColEI exists that produces 210 plasmid copies per *E. coli* chromosome. A tenfold amplification of the lactose carrier protein in *E. coli* has been achieved using a recombinant plasmid carrying the y gene of the *lac* operon.

Weisblum and co-workers (1979a,b) have isolated a high copy number mutant of the *S. aureus* plasmid pE194 that is replicated and maintained in *B. subtilis*. It is not known if this plasmid will be maintained in other gram-positive organisms such as *Streptomyces*. The mutant plasmid is present up to 100 copies per cell and determines resistance to macrolides, lincosamide, and streptogramin type B antibiotics. This plasmid could be used to clone and amplify genes in *B. subtilis*, which is a source of many commercial enzymes. Yeast and many streptomycetes are known to harbor plasmids. The size of these plasmids can be decreased by available techniques. Decreasing the size of the plasmid sometimes elevates the copy number.

2. Novel genes

With the increasing prices of media constituents for microbial fermentations, naturally abundant substrates such as cellulose might some day replace expensive carbon sources. Genes for the efficient utilization of cellulose and other substrates are not present in all organisms. Such genes might be isolated and cloned into organisms of commercial interest. Using such genetically tailored microbe that utilises cheap media constituents, more and more natural products might be made by fermentation, instead of by their expensive chemical synthesis.

Single genes that direct the synthesis of enzymes involved in the modification of functional groups may be isolated. Such genes (for example hydroxylation, chlorination, amination, acetylation, adenylation, phosphorylation, and glycosylation) can be introduced into industrial organisms to obtain altered metabolites. Genes determining the synthesis of a chlorinating enzyme, for example, could be isolated from chloramphenicol-producing Corynebacterium. Crynecin is analogous to chloramphenicol if it is resistant to excreted chloramphenicol and does not inactivate it.

3. Novel partial metabolic pathways

Recombinant DNA technology can be used to transfer genes using partial metabolic pathways to organisms that do not have genes for such pathways.

(a) **Catabolic pathways**: Many organisms possess catabolic pathways that are useful in microbial transformation programs. Sometimes mutants blocked in a catabolic pathway are used to accumulate valuable intermediates. Genes for a portion of a catabolic pathway that direct transformation of a substrate to the desired intermediate can now be isolated and introduced into a new organism. This could allow accumulation of the desired metabolite and avoid undesirable additional transformations. This strategy might have immediate application in the biotransformations of steroids.

(b) Anabolic pathways: Many secondary metabolites, e.g., antibiotics, alkaloids, and gibberellins, consist of subunits that are present in several unique biologically active compounds produced by unrelated organisms. Genes for some of the unusual moieties, e.g., sugars with side chains, could be isolated and introduced into a suitable organism to obtain hybrid molecules. Fusion of various macrolide and aminoglycoside antibiotics may now be possible and may yield a superior antibiotic. Fusing two natural products with no biological activity could also yield a superior antibiotic. Similarly, fusing two natural products with no biological activity could also yield an active metabolite. In *Penicillium chrysogenum*, a transacylase exchanges the L-α-amino adipyl side chain of isopenicillin N for non-polar-substituted acetic acid esters. This transacylase gene could be cloned and subjected to *in vitro* mutagenesis. A mutant gene may be obtained that will direct the synthesis of an enzyme that could exchange the D-α-aminoadipyl side chain of cephalosporins for non-polar-substituted acetic acid esters. If the suitable mutant gene is introduced and expressed in the appropriate β-lactam producing host, cephalosporins with non-polar-substituted acetic acid side chains may be produced. Such side chain precursors and their analogs have to be added to the growth medium of the tailored microbe.

Hybridization among β-lactam-producing streptomycetes does, indeed, offer many attractive possibilities. Expression of *Streptomyces* genes determining synthesis of hydroxylases and methylases in cephalosporin might allow efficient cephamycin production utilizing highly efficient strains and technology already in use in the fermentation industry. Benzylpenicillin acylase of *E. coli* hydrolyses benzyl penicillin to 6-aminopenicillanic acid and phenylacetic acid. At low pH, it catalyzes the synthesis of new penicillins from 6-aminopenicillanic acid and different side chain precursors. Several semi-synthetic penicillins that are sold commercially have already been prepared by this method.

Many enzymes may be used in the industrial hydrolysis of cephalosporin and penicillins and acylation of 7-aminocephalosporanic acid as well as 6-aminopenicillanic acid. The genes for these enzymes can be isolated and planted in appropriate organisms to make the process of preparing penicillin and cephalosporin analogs economical. Many streptomycetes produce several antibiotics of different chemical structures. These antibiotics are sometimes produced during various phases of growth. Self-cloning in streptomyces could result in increased levels of antibiotic production. Furthermore, clones that harbor amplified operators involved in antibiotic synthesis could neutralize the repressors, resulting in the constitutive production of antibiotics thus altering the relationship between various growth phases and antibiotic production. Alteration of this relationship will also result in production of a novel coupling of antibiotic moieties that were never linked in the parent culture because these various moieties were not produced during the same phase of growth. Some of the DNA sequences could be ligated to a vector whose replication is temperature-sensitive, and then amplified during any growth phase by manipulating growth temperature.

C. APPLICATION OF RECOMBINANT MICROORGANISMS IN MEDICINE

In medicine, the methods of gene technology have made important contributions in cancer research, immunology, circulatory, as well as in the nerve and endocrine systems. By the use of recombinant DNA technology, substantial quantities of proteins can now be produced. Previously these were only available in extremely small quantities. Such proteins can be used therapeutically in those patients deficient in these proteins (e.g. insulin, human growth hormone, factor VIII etc.). Another use is in the increase in an immune response, as for instance interferon. Over 100 proteins of human origin have been cloned and expressed in experimental cell systems. Several such products are now in clinical trails.

1. Human proteins

Advances in our understanding of physiology and cell biochemistry have made it possible to identify the precise protein deficiencies responsible for a variety of metabolic disorders, including diabetes, hemophilia, and dwarfism. Such deficiencies can be corrected by supplying the missing or under-produced proteins – insulin for diabetes, clotting factors VIII and IX for hemophilia, and human growth hormone for dwarfism. However, ensuring adequate supplies of such therapeutic proteins have proved difficult in the past. These proteins are produced in very limited quantities in the body and are thus laborious to isolate. For example, 8000 pints of blood are needed for purification of the clotting factor to treat a hemophiliac for one year, 7–10 pounds of pancreas from 70 pigs or 14 cows are needed to isolate enough insulin to treat a diabetic patient for one year, and extraction of the human growth factor to produce enough for one year's therapy requires the pituitary glands from at least 80 human cadavers. The risk of virus contamination is another important consideration in any therapeutic product purified from mammalian cells. Producing therapeutic proteins from cloned human genes inside bacterial hosts eliminates the risk of virus contamination and allergic sensitivity.

The only hope of obtaining proteins such as human interferon, human insulin, and human growth hormone in good amounts is recombinant DNA technology. The microbial production of some human proteins of importance to medicine has already been achieved using recombinant DNA technology. It has already been used to produce ovalbumin, human interferon, human insulin, human growth hormone, somatostatin, thymosin, and many other products in E. coli.

An outstanding example of the benefit to be derived from the application of the new technology would be the production of the highly desirable interferon. Fibroblast and leukocyte interferon genes have been cloned and expressed in E. coli by Genetech scientists. Induced leukocytes were used to make cDNA. The amino acid sequence of tryptic peptide of interferon was used to synthesize an mRNA probe. Transformants fell into six classes as analyzed by restriction mapping and nucleotide sequencing. One class that may be similar to interferon was independently cloned by C Weismann collaborating with the Biogen Laboratory. Two hundred and fifty million units of leukocyte interferon per litre of E. coli culture were produced. Non-glycosylated interferon appeared to be biologically active.

(a) Human insulin

Cloning of the human insulin (Humulin) gene in E. coli and the commitment to commercialize microbially produced human insulin by Eli Lilly and Company further supports the notion that industry is doing its best to ensure that the results of research are reaching the market place. Insulin is a small protein produced in patches of cells in the endocrine pancreas. The mature insulin molecule consists of two polypeptide chains – an A chain of 30 amino acids and a B chain of 21 amino acids – held together by disulfide linkages. However, this active insulin results from sequential modifications in two precursor molecules: preproinsulin and proinsulin. The insulin gene consists of two coding exons separated by a single intron. Following splicing of the pre-mRNA, a functional transcript is translated into a large polypeptide called preproinsulin. The molecule includes a 24-amino-acid signal peptide at its amino terminus, which is necessary for their proper transport through the cytoplasm. The signal peptide, which is the first part of the preproinsulin molecule produced, anchors the free-floating ribosome to the endoplasmic reticulum (ER) and is subsequently clipped off as the molecule passes through the ER membrane. The resultant 84 amino acid proinsulin, maintains its looped shape by cross-linking disulfide bonds. Proinsulin makes its way to the Golgi apparatus, where a converting enzyme removes 33 amino acids from the middle of the connecting loop (the C chain) – leaving the remaining A and B chains held together by the disulfide linkages. This yields active insulin, which is stored in a secretary granule for eventual release into the bloodstream (Figure 20.7).

The nucleotide sequences coding for the A and B chains of active insulin are chemically synthesized and cloned into separate expression plasmids. A bacterial strain containing each

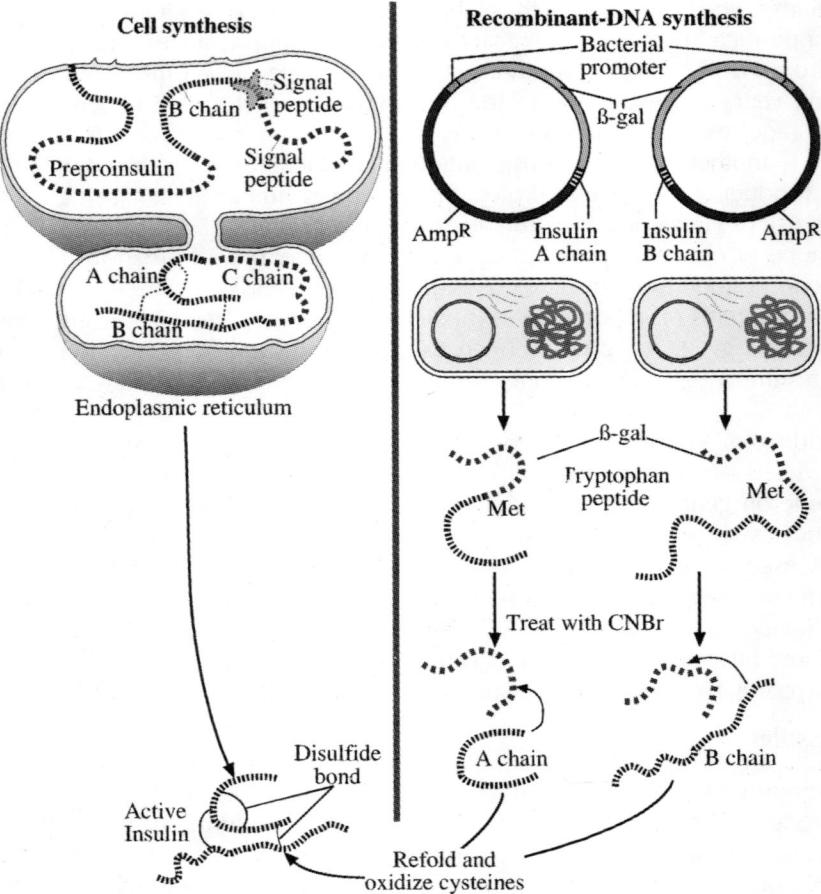

Figure 20.7 *Cellular and recombinant DNA synthesis of insulin*

expression vector produces a fusion bacterial/human polypeptide, which is harvested and subsequently treated with cyanogen bromide to remove the bacterial amino acids. Cyanogen bromide cleaves the fusion polypeptide at the methionine residue that begins the human sequence. This treatment, which also alters tryptophan, is only useful because, by happenstance, neither the A nor B insulin chain includes tryptophan or additional methionine residues. Purified A and B chains are then mixed in equal portions and incubated under conditions that form the disulfide linkages.

(b) Human growth hormone

The human growth hormone, a polypeptide of 191 amino acids, is also being produced in *E. coli* by the use of recombinant DNA techniques (Figure 20.8). The coding sequence for the first 24 amino acids of the expressed gene has been synthesized chemically, whereas amino

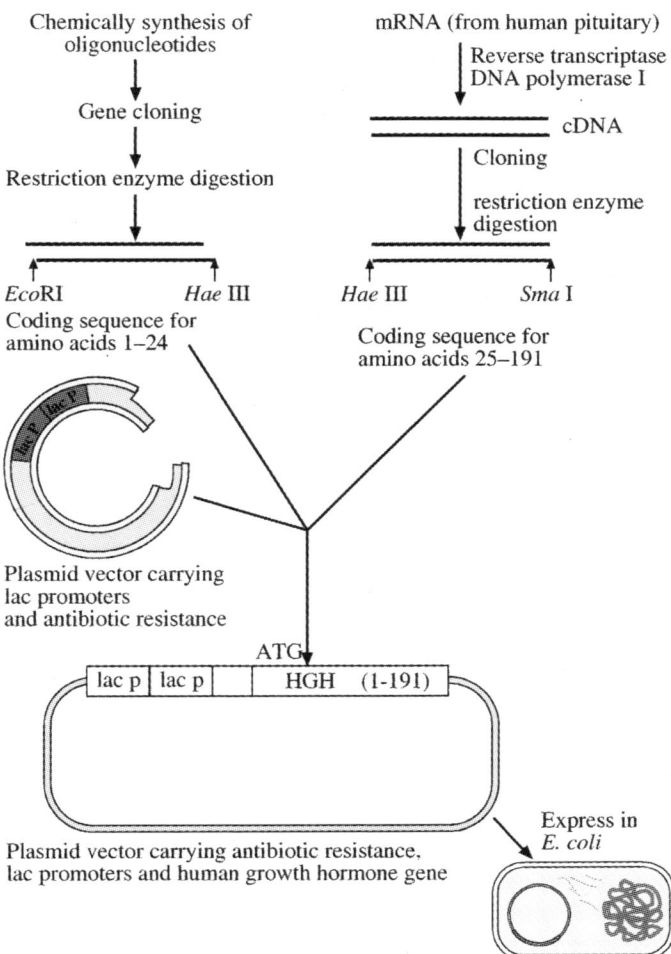

Figure 20.8 *Construction of an expression vector that directs the synthesis of human growth hormone (HGH) in E. coli*

acids 25 through 191 have been derived from a cDNA copy of the human growth hormone mRNA isolated from pituitary cells. The recombinant human growth hormone differs by one amino acid from the normal human growth hormone due to the fact that *E. coli* is unable to remove the initiator methionine residue that is removed post-translationally in human cells.

2. Recombinant vaccines

During immune response, clones of B lymphocytes secrete different antibodies that recognize many antigenic features on the surface of an invading virus. In theory, stimulation with only one or several such antigenic subunits should result in immunity to the live virus. Thus, modern vaccine construction depends on cloning the genes for viral surface antigens that elicit a strong immune response. The cloned gene can then be used to prepare several different types of vaccines. A subunit vaccine of the purified antigen can be prepared by expressing the cloned gene in *E. coli*. Immunization with this type of vaccine only stimulates B lymphocytes to produce antibodies. This is called a humoral response. Alternatively, one or more antigen genes can be cloned into the genome of a non-lethal carrier virus, such as vaccinia, which then expresses the subunits along with its own genes. Such a live/subunit vaccine gives a stronger and longer-lasting effect because relatively large amounts of the antigen molecule are expressed within the cells of the inoculated patient. This type of vaccine elicits both humoral (B-cell-mediated) and cellular (T-cell-mediated) immunity.

3. Human vaccines

The Hepatitis B virus infects man and can cause severe complications such as cirrhosis, hepatitis, and hepatocarcinoma. It has been estimated that there are one hundred and twenty million chronic carriers of hepatitis virus worldwide. This virus cannot propagate in cell cultures, and human serum is the only viral source. However, cloning of this virus in *E. coli* now allows the purification and determination of the complete nucleotide sequence of the hepatitis B virus genome. Since high titers of hepatitis B virus are not produced in tissue culture cells, Valenzuela *et al*. (1979) used recombinant DNA technology to obtain large quantities of the viral genome. Hepatitis B virus DNA was cloned in *E. coli* using the plasmid pBR322 as vector. Restriction endonuclease analysis of full-length hepatitis virus genome and the nucleotide sequence of an 892-base pair fragment encoding for hepatitis B surface antigen has been reported. The portion of the gene coding for surface antigen consists of 226 amino acids of MW 25.398. This information combined with the availability of a large amount of viral genome will help in the development of antiviral vaccines.

Segments of viral genome may be used to produce appropriate viral antigens in alternative hosts like *Saccharomyces cerevisiae* or *B. subtilis*. The core antigen gene is expressed in *E. coli* and, when injected into rabbits, the antigen produced by *E. coli* induces antibodies that react with the human serum core antigen. Recombinant DNA technology can be used to identify the genes that code for a surface protein of foot and mouth disease virus. Such genes can be introduced into the appropriate host to produce the protein that may be an effective vaccine. Eintage *et al*. (1980) inserted a DNA sequence from fowl plaque virus into an *E. coli* plasmid. The recombinant plasmid directs the synthesis of a protein that is 0.75% of total cell protein and reacts with antisera to FPV hemagglutinin. Transcription of the inserted fragment is from a tryptophan promoter.

D. APPLICATION OF RECOMBINANT MICROORGANISMS IN AGRICULTURE RESEARCH

In agriculture, genetic engineering microorganisms have been used for improvement of soil fertility, environmental stress tolerance (frost resistance), control of pathogenic microorganisms, plant growth promotion, and production of nutritional for humans and animals, etc.

1. Microorganisms in improvement of soil fertility

Under natural environmental conditions, successful plant development and high crop yields depend on the genetic constitution of the crop species, and the availability of nutrients. In general, microbial plant growth-promoting mechanisms that affect plants directly include fixing atmospheric nitrogen and supplying it to the plant; increasing the availability and/or sequestering iron and phosphorus from the soil and providing these minerals in a form that can be used by the plant. Recent genetic researches focused on creating microbial strains that augment the plant growth are in the following areas of study: the molecular basis of nitrogen fixation, root nodule formation by symbiotic bacteria, and the microbial synthesis of iron sequestering compounds (siderophores).

Soil microflora play a very important role in soil fertility. Many microbes could be improved so that they could easily establish themselves in plant rhizospheares. For example, most of the phosphorus in the soil is fixed in insoluble form. A number of phosphorus-solubilizing organisms such as *Bacillus megatherium* var *phosphaticum* could potentially be modified to enhance their activity, resulting in increased crop yield.

2. Plant growth enhancement

Under natural environmental conditions, successful plant development and high crop yields are also affected by the presence of certain beneficial microorganisms, and the absence of a pathogenic one in the surrounding soil. Beneficial indigenous soil bacteria and fungi include some species that act directly by providing a plant growth-enhancing product and other species that act indirectly. These latter beneficial organisms inhibit the growth of pathogenic soil microorganisms, thereby preventing them from deleteriously affecting plant growth. Microorganisms also synthesize phytohormones that trigger plant cell proliferation. The indirect stimulation of plant growth occurs when a beneficial microbial strain prevents the growth of a phytopathogenic soil microorganism that could interfere with normal growth and development. This action is called antibiosis and can be due either to the depletion of scarce resources by the beneficial microbial strain or the production and release of a compound by the beneficial strain that impedes the growth of the phytopathogenic strain. Most of the recent genetic research directed at creating microbial strains with augmented plant growth-promoting activity has focused on the pathways of microbial phytohormone production. These pathways are being studied in an effort to develop strains that will produce and release specified levels of selected phytohormones that could stimulate plant proliferation. A number of plant growth promoting rhizobacteria and several phytopathogenic microorganisms have been shown to be capable of synthesizing the plant hormones auxins and cytokinin. Although the auxin and cytokinin biosynthesis genes have not been isolated from any of the plant growth-promoting rhizobacteria.

3. Insecticides

Insects are the most abundant group of organisms on earth. They cause massive crop damage, and act as vectors of human, animal, and plant diseases. A number of chemical insecticides have been developed as a means of controlling the proliferation of noxious insect populations. The widespread use of agrochemicals allowed man to achieve unprecedented control over pests and has contributed to today's high agricultural productivity. However, researchers soon realized that chlorinated hydrocarbon insecticides and organophosphate insecticides had dramatic and immediate side effects and long-term and indirect effects on animals, ecosystems, and humans. Chlorinated hydrocarbons (e.g. DDT) were found to persist in the environment for 15–20 years and to accumulate in increasing concentrations through food chains. Also, the targeted insect pest population became increasingly resistant to treatment with many chemical insecticides. Since most insecticides lack specificity, beneficial insects were being killed along with those considered to be pests. Given all the drawbacks associated with the use of chemical insecticides, alternative means of controlling harmful insects have been sought over the past 20 years. Using insecticides that are produced naturally by either microorganisms or plants was an obvious choice. On the positive side, these compounds are usually highly specific for a target insect species, they are biodegradable, and slow to select for resistance. Consequently, interest is being turned to certain microorganisms which are natural pesticides. Although over 100 bacteria, fungi, and viruses that infect insects have been described, only a very few are in commercial production. Researchers can now manipulate genes that encode insect pathogenic agents to increase the effectiveness of microbial insecticides.

(a) Insecticidal toxin of Bacillus thuringiensis

One of the most extensively used microbial-insecticides is the sporulating cells of a *Bacillus* species. *B. thuringiensis* sub sp. *Kurstaki* contains an insecticidal protoxin gene. The procedure for isolating a protoxin-encoding DNA sequence is a standard one. Once the isolation and sequencing of a toxin gene was accomplished, the complete amino acid sequence was deduced. Under normal conditions, the *Bt* protoxin protein is only synthesized during the sporulation phase of growth. It is therefore advantageous to have the toxin gene transcribed and translated during vegetative growth, thereby relieving the synthesis of the toxin of its normal constraints. This system would also entail integration of this vegetatively expressed toxin gene into the chromosomal DNA of the sporulation-defective *B. thuringiensis* host. This manipulation would ensure that the insecticidal toxin gene is not lost due to plasmid instability during a continuous fermentation process.

Agricultural crops are attacked by many insect species. It would therefore be advantageous, if feasible, to create microbial insecticides that are each effective against a broad spectrum of target insects. Such a broad-specificity compound could be obtained by (1) transferring different toxin genes into a single host or by fusing portions of two different species-specific toxin genes to one another. The insecticidal specificity of Bt toxins from different sources was found to reside in the central region of each protein molecule. This segment of the toxin molecule has been called the hypervariable region, because it is distinct for each of the many insecticidal toxins that were examined. The remainder of the protein molecule in each case is largely invariant.

The Bt subsp. *Israelensis* insecticidal protein is highly toxic when ingested by mosquito larvae. However, the parasporal crystal of this species sinks rapidly after it is sprayed on water, dramatically decreasing its efficacy as a mosquitocide. To overcome this shortcoming, the insecticidal toxin gene could be introduced into an organism that is a common food source for mosquito larvae. Good candidate organisms for this purpose include *Synechocystis* and *Synechococcus* spp., photosynthetic cyanobacteria that are a food source for the larvae and proliferate near the water surface. The roots of many plants are attacked by insects that, because of their underground location, would normally be protected from Bt based insecticides. However, it is possible to introduce the toxin gene from a Bt strain into a bacterial species that colonizes the region adjacent to plant roots. The engineered bacteria could be introduced into the soil, where they would synthesize and release the insecticidal toxin into the area, thereby conferring protection against root-attacking insects.

(b) Baculoviruses as Biocontrol Agents

Baculoviruses are rod shaped, double-sided DNA viruses that can infect and kill a large number of different invertebrates. Subgroups of this viral family are pathogenic to several classes of insects. In nature, some of these baculoviruses are important for the control of certain pest insects, and several have been registered for use as biological insecticides. Baculoviruses are relatively slow in killing target insects. Depending on conditions, it can take from a few days to several weeks before the viral infection leads to the host's death. To remedy this ineffectiveness, several attempts have been made to enhance the virulence of baculoviruses by introducing foreign genes that will either severely impair or kill the target insect species. One approach has been to use a gene that will disrupt the normal life cycle of the insect when it is expressed within the host insect cells. During insect development, a reduction in the level of the juvenile hormone in the larvae initiates metamorphosis into pupae and leads to a cessation of larval feeding. The reduction in the juvenile hormone level is due to an increase in the amount of the juvenile hormone esterase. Therefore, researchers reasoned that an experimentally induced increase in the supply of juvenile hormone esterase should lower the endogenous level of the active juvenile hormone and cause a premature cessation of feeding. When *Trichoplusia ni* (cabbage looper) at the first larval instar stage was treated with the genetically modified baculovirus, the level of the juvenile hormone in the insect was reduced by the cloned juvenile hormone esterase, and larval feeding and growth were dramatically reduced relative to the control.

E. APPLICATION OF RECOMBINANT MICROORGANISMS IN ENVIRONMENTAL PROTECTION

Although it is still in the distant future, there is the possibility of genetically modified microorganisms (GMMs) being used for the elimination of toxic wastes. Many industrial products and byproducts are highly persistent in the environment and, because of their potential toxicity, need be eliminated. Genetic engineering can be used to develop new strains of microorganisms capable of breaking down these persistent chemicals, opening up the possibility that these new microorganisms could play a role in environmental protection. For more details on GMMs refer to Chapter 41.

VII SUGGESTED READING

Alcamo IE (1996) *Fundamentals of Microbiology*, Benjamin/Cummings Publishing Company.

Demain AL and Solomon NA (1986) *Manual of Industrial Microbiology and Biotechnology*, American Society for Microbiology, Washington, DC.

Glick BR and Pasternak JJ (1996) *Molecular Biotechnology*, Panima Publishing Corporation, Bangalore, New Delhi.

Madigan MT, Martinko JM, and Parker J (1997) *Brock Biology of Microorganisms*, Prentice Hall International, Inc.

Primrose SB (1999) *Molecular Biotechnology*, Panima Publishing Corporation, New Delhi/Bangalore.

Rehm HJ and Reed G (Eds) (1981–1988) Biotechnology. *A Comprehensive Treatise (Vol 1), Microbial fundamentals (Vol 2), Fundamentals of Biochemical Engineering (Vol 3), Biomass, Microorganisms for Special Applications, Microbial Products I (Vol 4), Microbial Products II (Vol 5), Food and Feed Production with Microorganisms (Vol 6a), Biotransformations (Vol 6b), Special Microbial Processes (Vol 7a), Enzyme Technology (Vol 7b), Gene Technology (Vol 8), Microbial Degradations*, VCH Publishers, Weinheim, Germany and Deerfield Beach, Florida.

Sambrock, J., Fritsch, E F and Maniatis, T (1989) *Molecular Cloning, a Laboratory Manual* (2nd Edn), Cold Spring Harbor Laboratory Press, USA.

Stanbury PF and Whitaker A (1984) *Principles of Fermentation Technology*, Pergamon Press, Oxford.

21

Genetic Engineering of Animals

I INTRODUCTION

The ability of living organisms to pass on their characters to their offspring is one of man's earliest scientific observations. This concept of heredity and the possibility of selective breeding enabled the Stone Age people of the Near East to develop some of our domestic animals and crop plants from wild relatives. Animal breeders have worked to improve the productivity of many animal traits such as growth, milk yield, and feeding efficiency since historical times. Genetic improvements through the application of quantitative genetics greatly increased the economic efficiency of many domestic animals. However, despite great improvements for some economically important traits, there are limitations to the traditional animal breeding. Molecular genome analysis of farm animals may help solve some of these problems. Complete genetic maps can be made and the entire genomes evaluated to dissect complex genetic traits into their individual entities. All this information can be channelized into animal improvement. Much of the increase in livestock production (mainly in hoofed animals—cattle, sheep, goats and pigs), witnessed over the last quarter century, is attributed to the fact that improved nutrition allows a pure-bred animal to better fulfill its genetic potential. Animal protein is an excellent source of dietary amino acids essential for human nutrition, but in addition to meat and hides, we are able to continually harvest milk and fiber from live animals, and of course use them for labor. Animal improvement is no longer a matter of interest; rather, it is a matter of necessity. The pressures of continued population growth along with shrinking resources force us to improve the quality and quantity of food and fiber. Only biotechnology can provide us with powerful tools with which to achieve this task.

In this chapter, we examine the four major types of transgenic organisms that have been engineered using standard molecular genetic techniques. The selected examples are (1) P element transformation of the laboratory fruit fly *Drosophila melanogaster*, (2) generation of transgenic mice as models of human disease, (3) production of transgenic livestock, and (4) many other transgenic methods often studied in many organisms (although not described here). The methods used to make transgenic organisms are similar to those for other organisms. Thus, principles and techniques used in other systems can also be cross-referred to fill in information uncovered in this chapter.

II TRANSGENIC ANIMALS

A. GENE TRANSFER TECHNIQUES IN ANIMALS

Livestock animals are a major source of food and fiber (wool), producing meat, milk, leather, as well as numerous by-products. Many limitations exist in the traditional methods of animal improvements and these have been counteracted through the use of recombinant DNA and gene transfer techniques. Gene transfer is an experimental process of modifying the heritable genotypic and phenotypic content of the cell by causing the cell to take up and express gene sequences of purified DNA from donor cells. Many gene transfer techniques have been developed. Some of these are very specific to animal transformation (see Chapter 18, Transformation techniques).

Gene transfer technology in animals: The transfer of foreign DNA to the genome of animals is emerging as a principal methodology in the study of animal genetics. A wealth of information concerning the structure and function of the animal genome is now being generated as a direct result of gene transfer research on animals.

B. MOLECULAR GENETICS OF *DROSOPHILA* DEVELOPMENT

The fruit fly *D. melanogaster* has been studied for more than a century as a laboratory model organism of embryonic development. It is an ideal system in which to study development because it is a relatively simple, segmented organism that progresses through distinct larval stages. Since the embryo develops externally, as opposed to *in utero*, changes can be readily observed. It completes its entire life cycle in two weeks. The discovery of a number of bizarre developmental mutations over the years has provided very useful information for the study of molecular events. These mutations are clearly mistakes in the spatial arrangement of segments within the overall body plan.

The process of embryonic development in *Drosophila* begins with the translation of maternally derived mRNAs that are present in the egg. These maternal gene products encode signaling proteins and transcription factors that induce a highly orchestrated set of molecular events at the onset of embryogenesis. Ultimately this transcription factor cascade leads to the regulated expression of structural proteins and intercellular signaling molecules, which together specify the framework of the *Drosophila* embryo. In this section, the *Drosophila* gene transfer technique that was worked out by Allan Spradling and Gerry Rubin to create fertile transgenic flies is discussed. This method, called **P element transformation**, has allowed *Drosophila* geneticists to demonstrate which *Drosophila* genes are responsible for the host of developmental mutants that have intrigued biologists for decades.

Method-1: P-element-mediated transformation

Transposable elements are movable genetic "info-bytes" that exist in the genomes of most organisms as self-sustaining entities. The process of integration and excision of transposable elements is called **transposition**. Transposons contain two functional regions: (1) DNA repeat sequences at the termini of the transposon that are required for genomic integration and excision and (2) an internal segment encoding a transposase enzyme that is required for transposition. Examples of transposons are the Ac/Ds elements of maize, the IS1 and Tn10 elements found in *E. coli*, and the *S. cerevisiae* transposon Ty1. Through a genetic analysis of a phenomenon known as hybrid dysgenesis, the presence of transposons in most *Drosophila* species have been discovered, and the best characterized of these is the P element transposon. The P element is a 2.9 kb DNA segment encoding an 87 kDa transposase enzyme flanked by an inverted repeat sequence of about 31 bp length. *Drosophila* in the wild contains about 30–50 genomic copies of randomly integrated P elements.

The trick Spradling and Rubin used was to replace the transposase coding sequence of a cloned P element with both a marker gene to identify transgenes and the experimental gene to be tested. By co-injecting *Drosophila* embryos with this modified P element transposition plasmid and a second expression plasmid encoding the P element transposase, Spradling and Rubin pioneered a reliable method to promote integration of cloned genes into the *Drosophila* germ line. Because the transposition plasmid lacks transposase and the co-injected transposase expression plasmid is not capable of integration, a stable transgenic *Drosophila* strain can be isolated following one round of mating. Figure 21.1 illustrates how P element transformation can be used to produce a transgenic fly that expresses the *lac Z* gene in specific cells known to be important for establishing pattern formation during *Drosophila* embryogenesis. These transgenic reporter gene experiments were invaluable in deciphering the transcriptional circuitry that underlies the early events in *Drosophila* development. The frequency of obtaining fertile transgenes by P element transformation is ~5% based on the number of embryos injected.

Method-2: P-element enhancer trap vectors

Walter Gehring and his colleagues designed a special P element cloning vector that contained the *lac Z* gene fused in-frame with the second exon of the transposase. This "enhancer-less" *lac Z* P element was used to identify cell-specific genes by identifying transgenic flies that express β-galactosidase in subsets of cells based on histocytochemical staining with X-gal. Because enhancer regulatory elements modulate transcriptional promoters in a manner independent of distance and orientation, this P element enhancer trap vector need only integrate in the vicinity of a gene regulatory region. Moreover, the weak promoter of the transposase–*lac Z* fusion gene is insufficient to allow significant transcription in the absence of a linked enhancer element. Enhancer trap strategies have been used to mark specific cells for fate analysis during different stages of development, as well as to clone novel genes based on enhancer activation of the *lac Z* marker gene.

A variation of the original enhancer trap strategy was developed by Andrea Band and Norbert Perrimon using P element insertions to identify cell-specific enhancers that selectively direct expression of the yeast Gal4 gene contained on the enhancer trap vector. By crossing one of these Gal4 transgenes with a fly containing an integrated copy of a Gal4-responsive *lac Z* expression vector, they showed that it was possible to obtain cell-specific expression of β-galactosidase activity. This elegant Gal4 enhancer trap strategy provides fly strains that can be used in a variety of experimental crosses to direct cell-specific expression of experimental target genes (Figure 21.2).

C. MAMMALS

The laboratory mouse has been the most widely used animal for the study of gene transfer experiments and then for the application of animal genetic research. The first transformation of mouse embryos with foreign DNA (SV40 DNA) was reported by Jaenisch and Mintz (1974). Subsequently, germ line transformation of foreign DNA was demonstrated by exposure of zona free 4- to 8-cell stage mouse embryos to the Moloney leukemia retrovirus (Jaenisch 1976). This

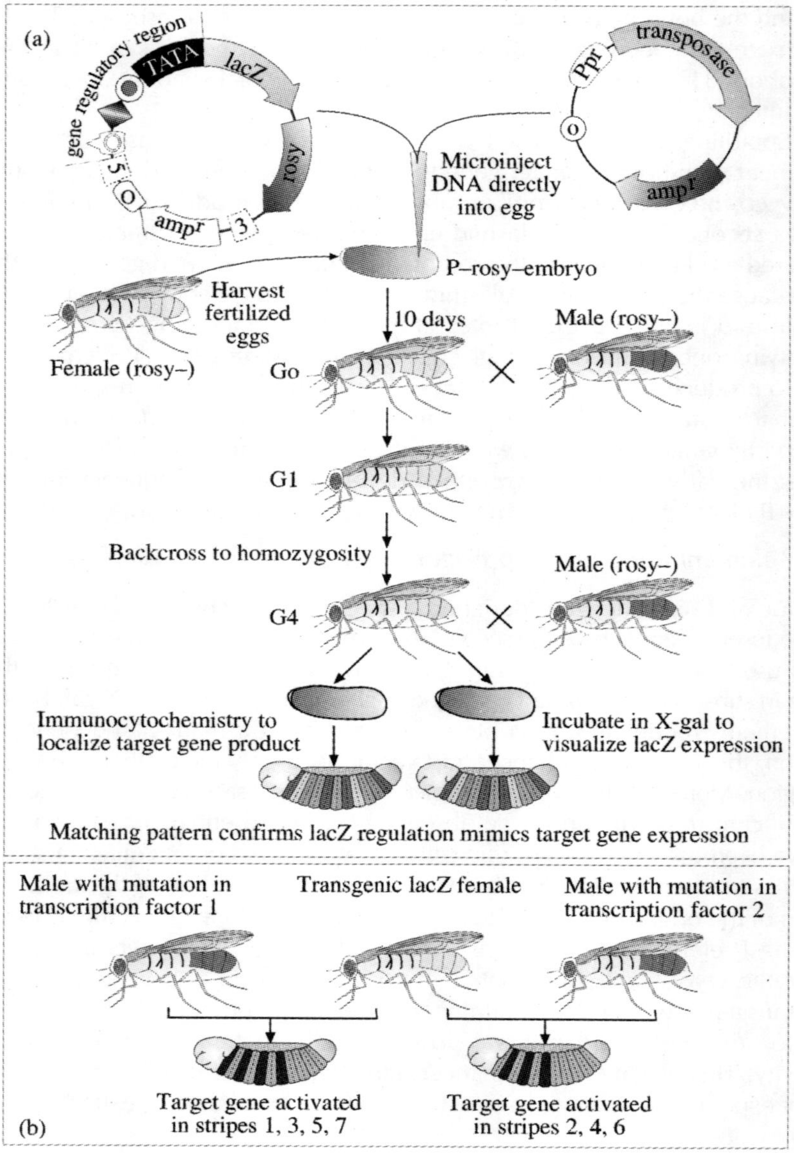

Figure 21.1 *P element-mediated gene transfer*

work demonstrated that it was possible to create new strains of animals by artificially introducing foreign DNA into the genome, which is heritable. The most significant of these early experiments with pronuclear injection was the ability to achieve germ line integration of the foreign DNA. As a direct result of having achieved these technical goals, a multitude of research ensued employing gene transfer as a research tool. The most recent efforts have been directed at the elucidation of gene expression regulation, immunity, oncogenesis, and cellular

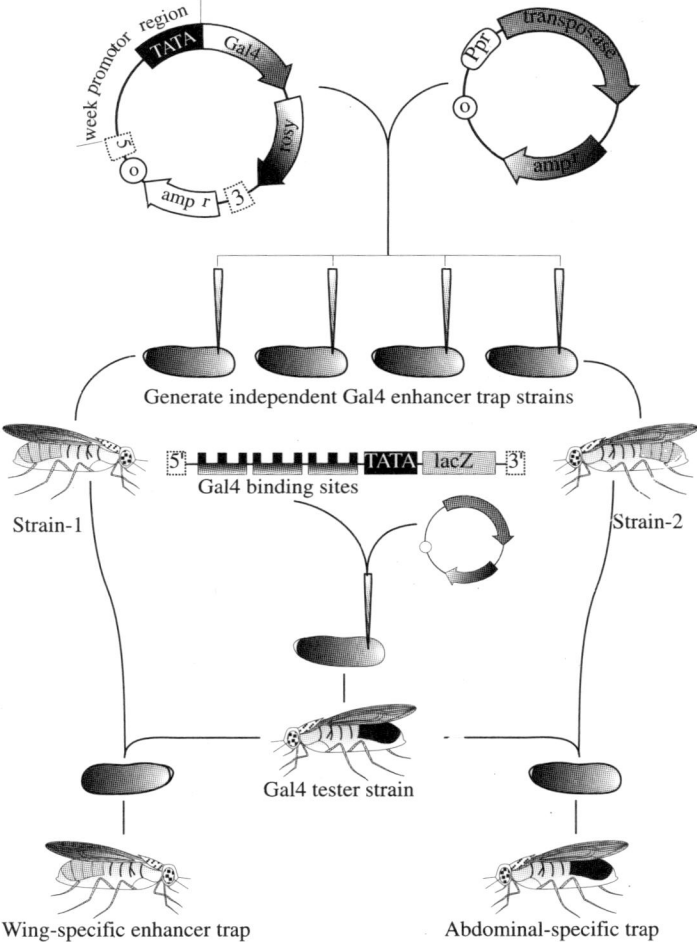

Figure 21.2 *The Gal4 enhancer trap strategy is used to identify cell-specific enhancers by screening for transgenic flies that display restricted expression of a Gal4-responsive lac Z gene*

differentiation. Furthermore, laboratory animals that have unique characteristics, important for research and unobtainable by conventional breeding and selection techniques can now be produced to enhance research addressing specific problems in genetics and cellular biology.

More recently, gene transfer is being centered around the development of transgenic animals that have unique commercial value. There have been two basic strategies employed in this regard. (1) Production of transgenic strains of domestic livestock, in this case, the primary goal is to transmit to the animal a high production capacity, which enhances the animal's value as a producer of a conventional animal product (e.g. meat, milk, leather, or wool); (2) producing transgenic animals that biosynthesizes a commercial product other than conventional animal products.

1. Transgenic mouse

The laboratory mouse, to molecular biologists, is the mammalian equivalent of *E. coli* because it provides the critical link between observational medical science and mechanistic experimental studies. Transgenic technology has been developed and perfected in the laboratory mouse. The recent development of gene-transfer techniques has made it possible to study cloned genes in whole organisms, such as mice. A "***transgenic***" animal is created by simply inserting a foreign gene into a single-cell embryo, which then ramifies throughout the somatic and germ line cells during development. In later generations, the foreign transgene becomes stably integrated into the genome and is inherited in Mendelian fashion by succeeding generations. Hundreds of different genes have been introduced into various mouse strains. These studies have contributed enormously to an understanding of gene regulation, tumor development, immunological specificity, molecular genetics of development, and other biological processes of fundamental interest.

Of the various gene transfer methods, retroviral vectors have the advantage of being an effective means of integrating the transgene into the genome of a recipient cell (Figure 21.3). However, they can only transfer small pieces (~8 kb) of DNA which may lack essential adjacent sequences for regulating the expression of the transgene. However, the emphasis in this section

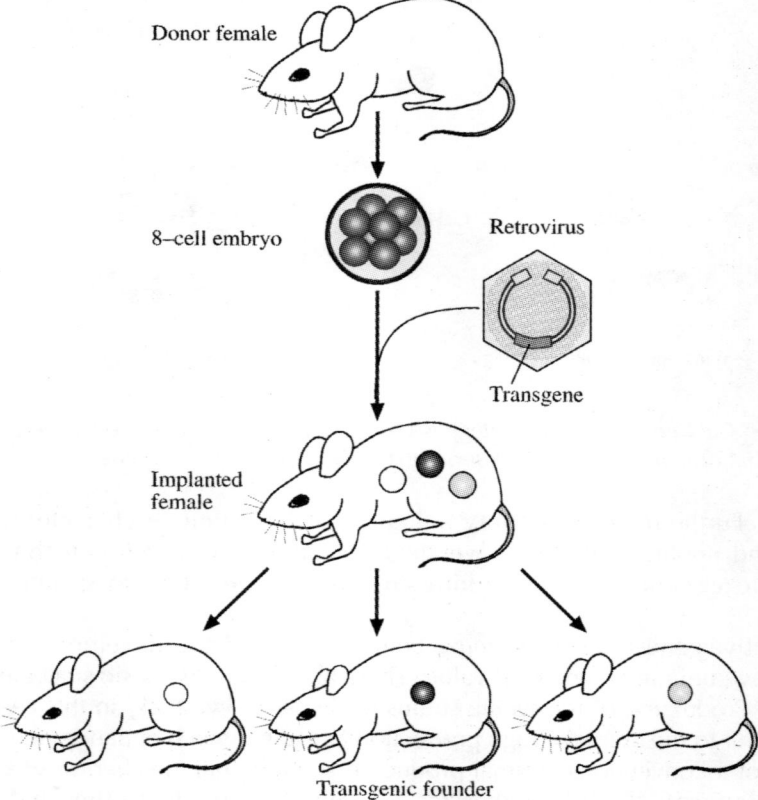

Figure 21.3 *Establishing transgenic mice with retroviral vectors*

is on describing the two other major approaches that molecular geneticists have taken to adopt transgenic mouse models to the study of complex human diseases. The first method involves **DNA microinjection** into single-cell embryos to produce transgenic mice containing one or more copies of randomly integrated expression vector DNA. A second transgenesis method generates **founder mouse lines** that contain specific genetic alterations as a result of homologous recombination events at defined genomic loci.

(a) Germ line transmission of mice

Method-1: Generating transgenic mice using fertilized egg cells

Germ line transmission of microinjected DNA in transgenic mice was accomplished by a number of independent researchers and the methods they refined were based on a working knowledge of early mouse development. DNA microinjection of single-cell mouse embryos is an attack DNA transfection strategy and is currently the preferred method for producing transgenic mice.

The general scheme for creating transgenic mice is described below; similar protocols have been developed for other laboratory and domestic animals.

1. The number of available fertilized eggs that are to be microinjected is increased by stimulating donor females to superovulate. This is done by initial injection of pregnant mare's serum and another injection about 48 hours later of human chorionic gonadotropin. A superovulated mouse produces about 35 eggs as compared to the normal 5–10 eggs.
2. These females are mated and then sacrificed, 12 hours after mating. The fertilized ova are collected from the oviduct of a female mouse. Embryos harvested at this time are still in the one-cell stage, prior to fusion of the male and female pronuclei.
3. The eggs are microinjected with cloned DNAs into either pronucleus using a capillary pipette whose tip has been drawn to a diameter of several microns. Microinjection is done to the male pronuclei (because the male pronucleus tend to be larger than the female pronucleus) of fertilized egg with about a picoliter of buffer containing several hundred molecules of linearized DNA as shown in Figure 21.4.
4. After inoculation, the surviving intact eggs – about 25–40 – are surgically implanted into the oviduct of a foster female mouse that is rendered "pseudo-pregnant" by mating with a vasectomized male.
5. The foster mother will deliver pups from the inoculated eggs about three weeks after implantation.
6. To identify the transgenic animals, three weeks after birth of the pups, genomic DNA from mouse tail snips are analyzed by PCR or Southern blot for the presence of the transgene DNA. Expression of the foreign gene in various tissues can be detected by Northern hybridization.
7. Mice that score positive by this test are mated with another mouse to determine if the transgene is in the germ line of the founder animal. Subsequently, the progeny can be bred to establish pure (homozygous) founder transgenic lines.
8. Microinjection is fairly efficient. Of 100 injected, approximately 70 survive microinjection, about 20 are carried to term, and approximately 5 offspring stably incorporate the transgene. Figure 21.5 outlines the general procedure for producing a transgenic mouse by embryo microinjection.

A significant breakthrough in mouse transgenesis, and the key factor in generating cell-specific gene knockouts, has been the use of experimental DNA expression cassettes that are based on tightly regulated cell-specific promoters. Cell-specific promoters provide a means of direct

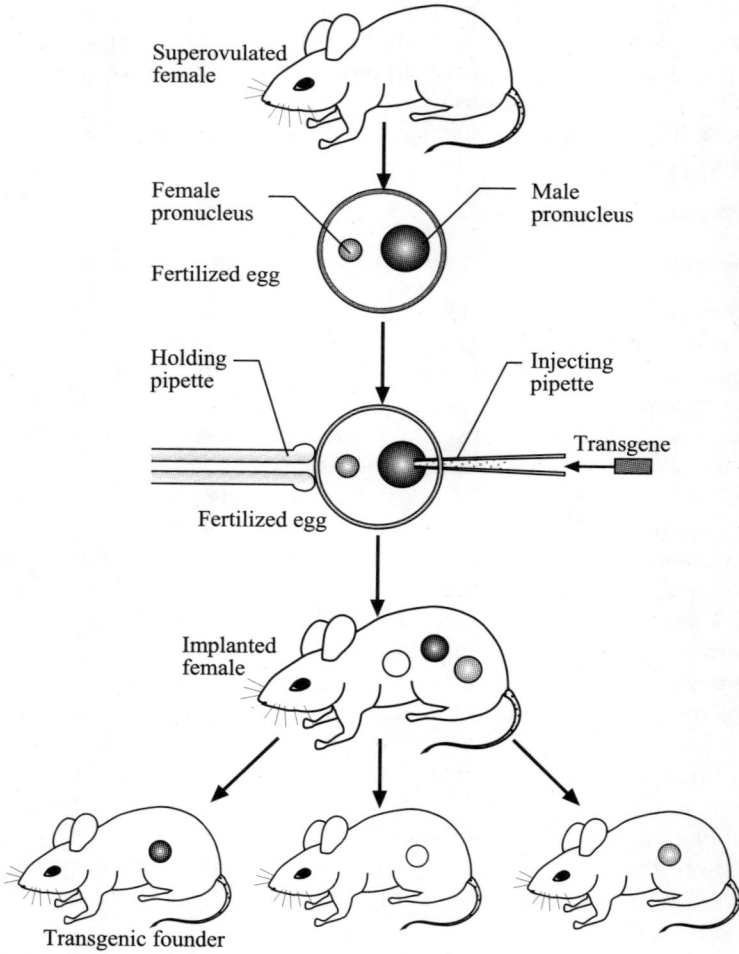

Figure 21.4 *Establishing transgenic mice. Purified DNA is directly injected into the male pronucleus of a fertilized mouse egg using a microinjection method*

expression to a limited number of cells owing to the restricted expression of cell-specific transcription factors. For example, founder mice obtained by microinjection of an experimental DNA construct containing the mouse lymphocyte-specific tyrosine kinase (lck) gene regulatory region linked to a CAT reporter gene display CAT activity only in T lymphocytes, even though the lck-CAT expression cassette is contained in every cell of the mouse. This can be a very powerful strategy because it permits the investigation of cell-specific phenotypes throughout mouse development.

(b) Engineered embryonic stem cell method

Cells from the blastocyst stage of a developing mouse embryo can proliferate in cell culture without losing its capacity of differentiating into all other cell types – including germ line cells – after they are reintroduced into another blastocyst embryo. Such cells are called pluripotent

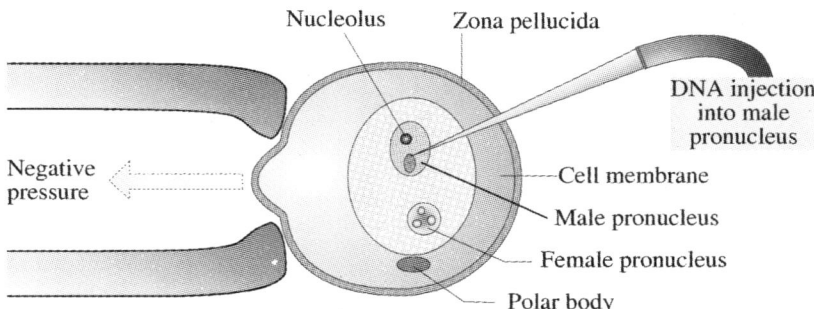

Figure 21.5 *A diagram illustrating the technique of microinjection.*

embryonic stem (ES) cells. When in culture, ES cells can be readily manipulated without their pluripotency being altered. With this system, for example, a functional transgene can be integrated at a specific site within a dispensable gene in the genome of ES cells. Then, the genetically engineered cells can be selected, grown, and used to generate transgenic animals (Figure 21.6). In this way, the randomness of integration that is inherent with the DNA microinjection and the retroviral vector system is avoided.

Creation of gene knockout mice: Two methodological breakthroughs were required before transgenic "knockout (KO) mice" could be created that contained germ line insertional loss-of-function mutations. The two key discoveries were (1) procedures to cultivate pluripotent mouse embryonic stem (ES) cells that could be combined with non-transgenic mouse blastocysts to produce a chimeric embryo, and (2) the development of DNA cloning vectors and cell line screening strategies that allow the identification of transfected cells that have undergone a homologous recombination event following transfection with experimental DNA. Each of these procedures is described and then several examples are given to illustrate how knockout mice have been used as experimental models of human genetic diseases.

Martin Evans and Gail Martin were the first to report methods to isolate pluripotent ES cells from the inner cell mass (ICM) of a 3-day-old mouse embryo. They showed that ES cells could be grown in culture under conditions that maintained their pluripotential functions. Within a few years, methods were developed to modify ES cells genetically, by either retroviral infection or stable DNA transfection, in a way that permitted ES cells to be used as embryonic material for generating transgenic mice. This feat was accomplished by exploiting microinjection techniques to deliver ~10 ES cells into freshly isolated 3-day-old embryos, which could then be reimplanted into pseudopregnant females to produce chimeric mice 17 days later. Because only a portion of the ES cells injected into the embryo give rise to gametic precursors, the offspring must be bred to non-transgenic mice to identify the founders. This screening procedure was facilitated by the use of ES cells and pre-implantation embryos from mice with different coat color markers to readily identify chimeric offspring. Figure 21.7 illustrates how stably transfected ES cells can be generated starting with freshly isolated blastocysts, and ES-derived transgenic founder mice are identified based on coat color.

The most commonly used method to create knockout (KO) transgenic mice is to insert the neo^r gene into the desired target gene by transfecting ES cells with DNA sequences that promote homologous recombination, while at the same time permitting genetic selection and/

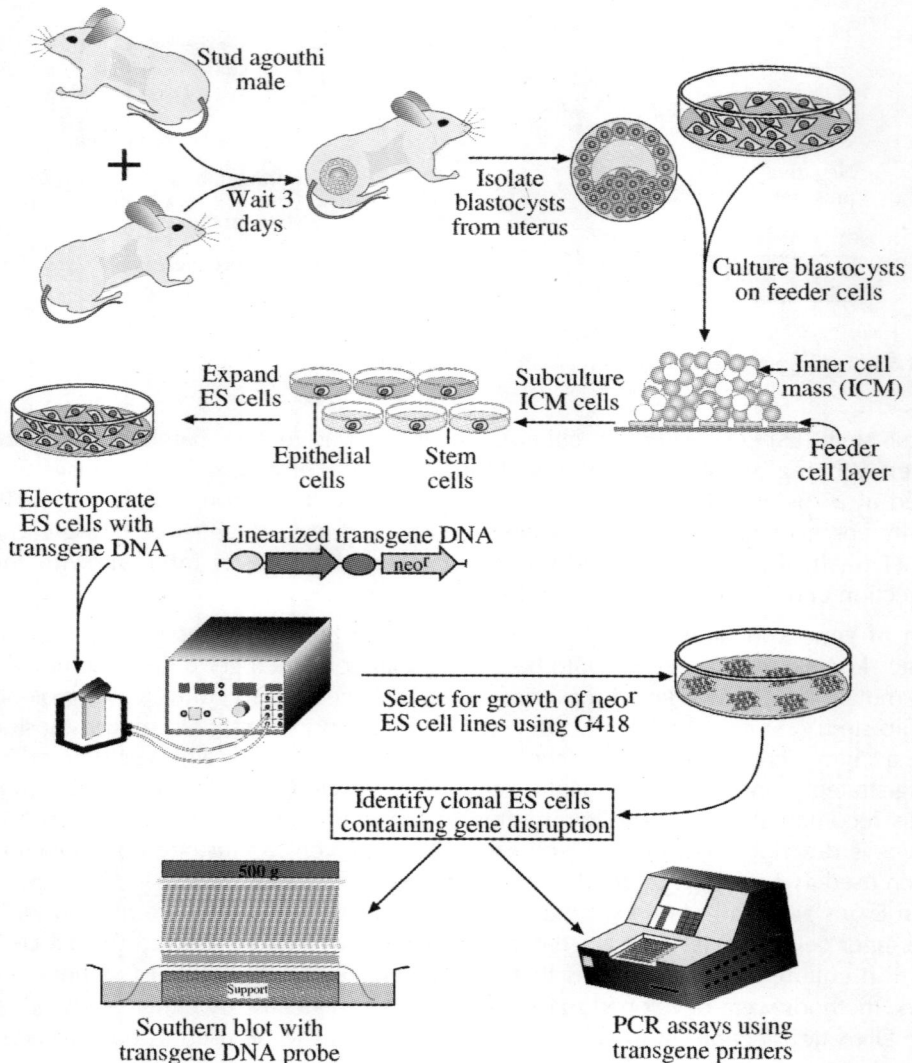

Figure 21.6 *Mouse embryonic stem (ES) cells can be manipulated in culture using standard DNA transfection methods*

or sensitive screening assays to detect rare targeted integration events. Numerous improvements in targeting vectors and screening strategies have led to two basic types of gene targeting strategies. The simplest type is based on a gene insertion process that requires a cross-over event between two ends of the targeting vector and the genomic sequence. The net result of gene targeting is by insertion of a selectable marker gene such as the neo^r gene (Figure 21.8a). A second type of targeting vector utilizes a gene replacement strategy. Gene replacement vectors contain two selectable markers that provide a convenient method to identify ES cells that have undergone a positionally defined double cross-over event in the target gene. As shown in

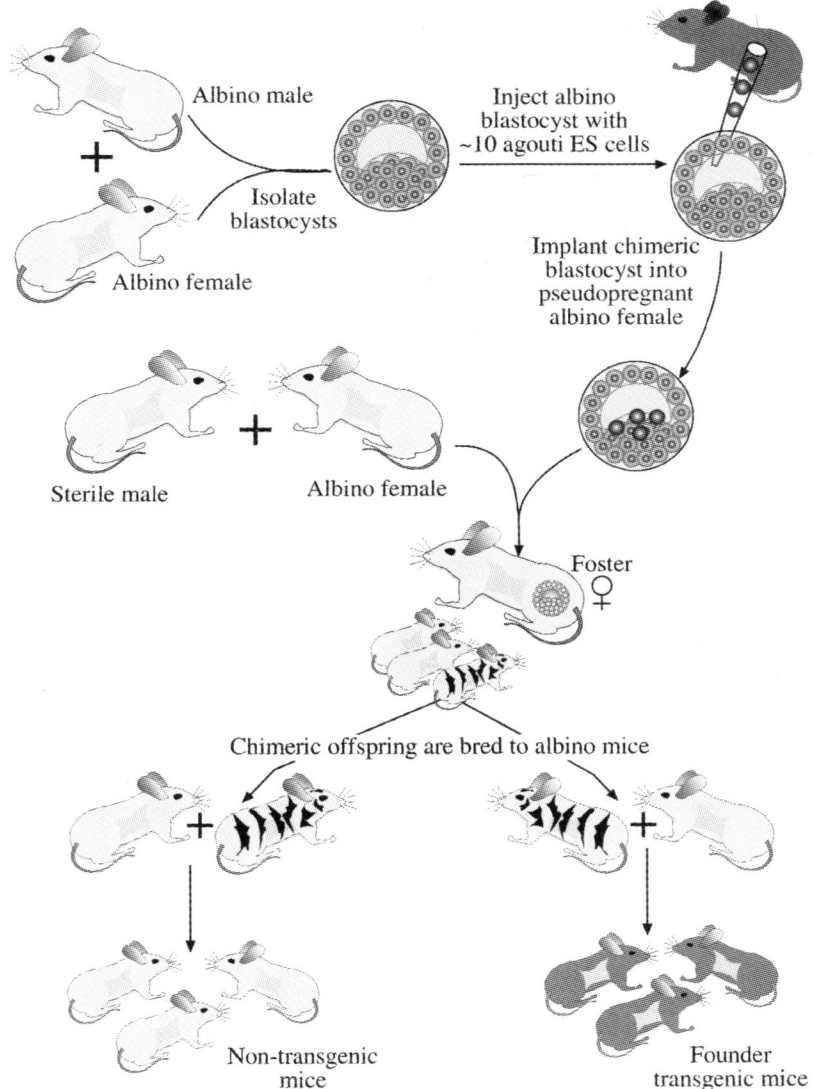

Figure 21.7 *Differences in the coat color of ES donor mice and blastocysts host mice are used to identify transgenic founder animals*

Figure 21.8b, homologous recombination events using a gene replacement vector produces ES cells that are resistant to both G418 and Ganciclovin (GC), where GC is a suicide nucleotide analog that is phosphorylated by the HSV-tk gene product, but is not a substrate for mammalian thymidine kinases.

After transfection of ES cells in culture with a DNA vector that is designated to integrate within a specific chromosomal location, some cells will have DNA integrated at non-target (spurious) sites; in other cells, integration will occur at the target (correct) site. To enrich the cells with

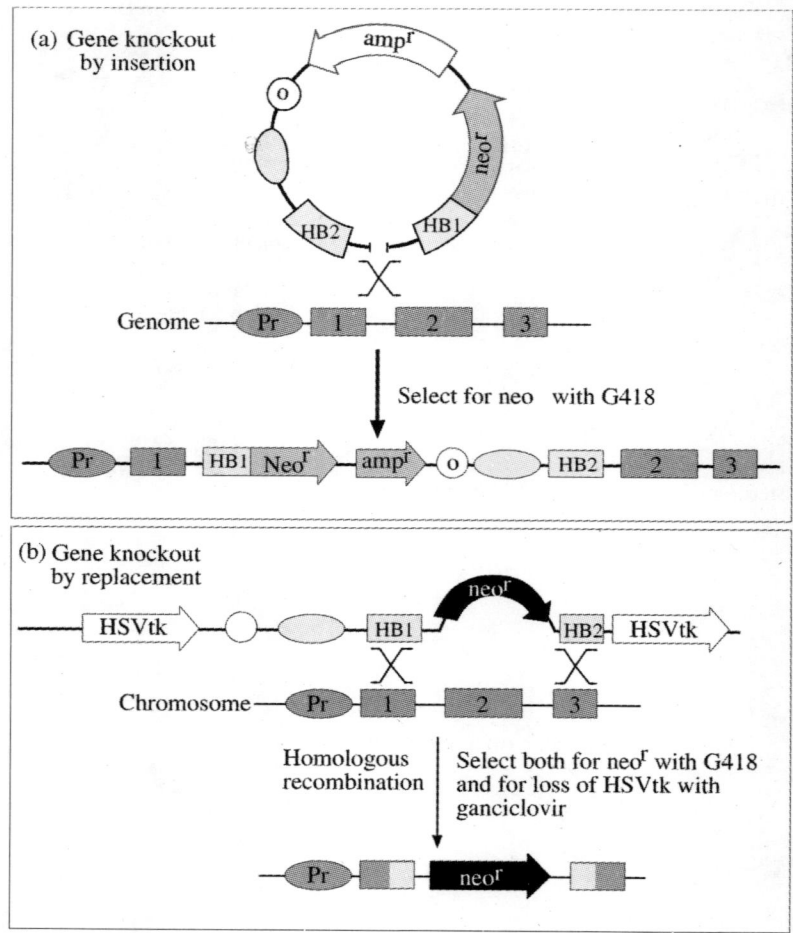

Figure 21.8 *Targeting vectors for gene KO strategies result in either (a) gene insertion or (b) gene replacement depending on the arrangement of targeting sequences and marker genes within the vector*

DNA integrated at the target site, a procedure called positive–negative selection is implemented. This strategy uses positive selection for cells that have vector DNA integrated anywhere in their genomes and negative selection against the vector DNA sequence that is integrated at non-target sites. A targeting DNA vector for the positive-negative selection procedure usually contains (1) two blocks of DNA sequences (HB1 and HB2) that are homologous to separate regions of the target site; (2) the transgene (TG), which will confer a new function on the recipient; (3) a DNA sequence that codes for resistance to the compound G418 (Neo^r); and (4) two different genes for thymidine kinase (tk1 and tk2) from herpes simplex virus type I and II (HSV-tk1 and HSV-tk2). Between the two blocks of DNA that are homologous to the target sites are the genes for the transgene and G418 resistance (Neo^r gene). Outside of each of the homologous blocks are the genes for HSV-tk1 and HSV-tk2. If integration occurs at a spurious site, then either one or both of the HSV-tk genes have a high probability of being integrated along with the other sequences. Alternatively, if the integration event is due to homologous recombination by a double crossover at the target site, then the HSV-tk genes will not be included and only the Neo^r gene will be incorporated into the genome. When transfected cells are grown in the presence of G418, all the cells that lack the Neo^r gene will be killed. Therefore, only those cells with integrated DNA will survive (positively selected). If the compound gancyclovir (ganciclovir) is added at the same time as G418, those cells that express thymidine kinase will be killed (negatively selected). Thymidine kinase converts gancyclovir to toxic compounds that kill cells. The cells most likely to survive this dual selection scheme will be those that have DNA integrated at the target site. However, the more direct way to detect ES cells that carry a transgene at a targeted chromosomal site is to use the PCR method.

Embryonic stem cells carrying an integrated transgene can be cultured and inserted into blastocyst stage embryos, and these embryos can then be implanted into pseudopregnant foster mothers. From those progeny in which the genetically engineered ES cells have contributed to the formation of germ line cells, transgenic lines can be established, by first mating founder transgenic mice to animals from the same strain and then crossing transgenic litter mates to create transgenic mice that are homozygous for the transgene.

The frequency of homologous recombination is ~1%, based on comparing the number of G418-resistant colonies to the number of G418- and ganciclovir-resistant colonies. This corresponds to an overall homologous recombination frequency of ~10^{-5}, when considering the total number of ES cells transfected. Factors influencing the frequency of homologous recombination are: (1) the length and arrangement of homologous sequences in the targeting construct; (2) the extent of homology between genomic sequences contained in the targeting vector and the target gene; and (3) the chromosomal location of the gene target. Because only one copy of the target gene is disrupted in the ES cell, it is necessary to perform a heterozygous cross between the first generation of transgenic mice to obtain a homozygous line, which, depending on the gene KO, could produce an embryonic lethal phenotype.

(c) Applications of transgenic mice

Transgenic mice can be used both as model systems for human diseases and as test cases to determine the value of a proposed production scheme. Whole animal models simulate both the onset and progress of a human disease and provide a system for testing potential therapeutic agents. Mouse models for human genetic diseases such as Alzheimer's disease, arthritis, muscular dystrophy, tumorigenesis, hypertension, neurogenerative disorders, endocrinological dysfunction, coronary disease, and many others have been developed. Transgenic mice have also been used as models for expression systems that are designated for secretion of the product of a transgene into milk.

2. Development of transgenic livestock

Since prehistoric times, farmers have relied on two methods to increase their animal stock—by buying additional animals from others or by breeding their existing animals. Many generations of selective matings are required to genetically improve livestock and other domesticated animals genetically for traits such as milk yield, wool quality, rate of weight gain, egg-laying frequency etc. This combination of mating and selection, although time-consuming and costly, has been exceptionally successful. However, once an effective genetic line has been established, it becomes difficult to introduce new genetic traits by selective breeding methods. These inherent limitations of traditional methods led the development of new technology that overcame these problems. Biotechnology has played a very important role in overcoming these limitations. Now animals can be injected with hormones exogenously, to induce superovulation (more than normal number of eggs). These eggs can then be fertilized *in vitro* and implanted into host mothers (surrogate mothers). Thus a single superior animal can give rise to eight or ten offsprings instead of the more traditional single offspring during the normal course of time. Also, the gene pool of animal populations can be greatly and specifically improved by genetic manipulation using embryo transplant technology. Depending on the nature of animal improvement, animal biotechnology discussions have been categorized into breeding and non-breeding strategies.

Very recently, gene transfer is being looked at as a means to generate transgenic animals with unique commercial value. Transgenic animals with commercial value arise from specific genotypic and phenotypic traits conveyed to the animals by their transformation with foreign genes.

(a) Non-breeding strategies

(i) BST and milk production

The action of the anterior pituitary growth hormone has been known for its strong galactopoietic action for many years. Its increased energetic efficiency of milk production in dairy cows has been recorded. The prohibitive cost of the hormones from the natural source made it difficult to commercially exploit these basic discoveries until gene cloning and recombinant DNA technology made possible the large-scale production of recombinant bovine growth hormone or bovine somatotropin (BST). Extensive trials have been done with BST, mainly in cattle but also in sheep and goats. Milk yields increase soon after treatment begins. The increase is 15% to 20% within a few weeks. Feed intake also begins to rise. No adverse effects have been found and the cows under treatment behaved like genetically superior dairy cattle.

The need for frequent BST injection has been overcome by slow-release implants and the effects are fully reversible. The technology therefore not only offers the possibility of leap-frogging 20 years of selective breeding but also helps to adjust the herd production to meet short-term market or quota requirements. The physiological mechanism of BST is thought to act by releasing an insulin-like growth factor (IGF-1) from other organs, such as the liver. IGF-1 has been found to increase milk by 30% when injected into the udder. However, the increase of IGF-1 in the circulation and milk following BST treatment do not exceed those occurring under natural conditions.

(ii) Growth and meat production

The physiology of growth of farm animals is still poorly understood despite the huge amount of information available on various metabolic processes of animals. The growth hormone has been found to direct metabolism toward protein synthesis in bone and muscle. Another equally important metabolic effect of the hormone in growing animals is its anti-lipogenic action, which reduces the adipose tissue in the carcass. Because of this, many attempts have been made to increase live weight gains and leanness by treatment with the recombinant growth hormone. The results have been variable. In pigs, growth-rate increases by 10–20% and improvements of feed conversion by 25–30%. In lambs, in favorable circumstances, a 36% enhancement of the growth rate could occur after two months' treatment with the bovine growth hormone associated with a 30% reduction of visceral fat. There are many reasons for supposing that IGF-

1 is a mediator of the protein anabolic action of the growth hormone. Confidence in the rational handling of all the above procedures will increase when we have a more comprehensive knowledge of the fundamental interplay of the various hormonal and other factors.

(iii) Ruminant nutrition

Microorganisms in the complex rumen affect digestion of cellulose in ingested herbage by cattle, sheep, and goats. In ruminants receiving normal diets, a significant proportion of dietary fiber remains undigested and an improved conversion of forage into meat, milk, and wool could be expected if cellulose and lingo-hemicellulose complexes were more efficiently degraded in the rumen. Recombinant DNA techniques and the introduction of genes encoding important degradative enzymes such as cellulose and hemicellulose offer means of modifying the function of the ruminal microflora and improving the efficiency of fiber digestion. Workers at Abraham and elsewhere have obtained encouraging results with cloned genes in ruminal organisms *in vitro*. Cellulase genes have been cloned and expressed in E. coli from the highly cellulolytic thermophile *Closteridium thermocellum* and from the rumen bacteria *Butyrivibrio fibrisolvens*, *Ruminococcus albus*, *R. flavefaciens,* and *Bacterioides succinogenes*. The genes and their encoded proteins have been extensively characterized and in some instances, primary sequences determined. Whether genetically manipulated organisms would survive extended periods in the rumen is uncertain.

(b) Breeding strategies

(i) Embryo transfer

Early developments were mainly centered on effective exploitation of the genetic potential of the male by improving techniques in artificial insemination. Recently, a comparable method was also developed for genetic exploitation of the female by introduction of the system for superovulating animals with gonadotrophic hormone, mating them, flushing out the fertilized ova, and transferring the embryos to the uteri of surrogate mothers. Culture methods have been developed to obtain enough fertilizable ova from oocytes harvested from ovaries at the slaughterhouse. Embryo culture under tight laboratory conditions is a prerequisite for many subsequent genetic manipulations. One of the major advantages of culturing oocytes is for *in vitro* fertilization to form zygotes. Later, these zygotes can be inserted non-surgically into dairy cows. Thus, high-quality beef animals can be obtained with substantial economic advantage.

Multiple ovulation with embryo transfer (MOET): In the reproductive cycle of an ovulating female, one ovarian follicle matures and ruptures, releasing its fertile egg, every 21 days (cow, horse) or 16 days (sheep, goat). However, some 20 or so ovarian follicles reach an early stage of maturity if oocyte maturation and ovulation can be triggered by a suitable dose of the gonadotrophic hormone. In the normal course of events, only one is so triggered by circulating hormones, but by increasing the hormone dosage at the appropriate time, a number of follicles can be induced to 'ripen' and ovulate. In practice, well-managed domestic cattle can yield 8–60 eggs when **superovulated**. Several protocols can be used to superovulate sheep and cattle including pregnant mare serum gonadotropin (PMSG) and various preparations of follicle stimulating hormone (FSH). Donor females are frequently injected with prostaglandin F2a (PGF2a) to induce a synchronized oestrous before treatment is started. Starting 10 days after oestrous, they are injected with FSH (follicle stimulating hormone) over four days, followed by PGF2a to induce oestrous. Oestrous should occur within two days and the superovulated donors are then mated within 24 hours. Since the purpose of MOET is to increase the number of genetically superior progeny, the donor is usually mated by artificial insemination. The fertilized embryos are recovered 6–8 days after insemination when they have developed to the stage of a compacted cell mass (morula) or early blastocyst or are still in the oviduct by non-surgically inserting a Foley catheter into the oviduct. A saline solution is allowed to drain into the sealed oviduct and the embryos are flushed out through the catheter into a collection bottle. After the embryos have been recovered, they are immediately transferred into synchronized recipients, using a procedure identical to artificial insemination (AI). Alternatively, they can be frozen for indefinite storage or they can be subjected to genetic manipulation.

In vitro **culture**: *In vitro* culture of embryos is very critical for extended periods and at present, it is inadequate. Often, early embryos of most mammalian species undergo a block to development *in vitro*. Because of these limitations, embryos are placed into an *in vivo* system, whether it is a temporary surrogate or the final recipient, as soon as possible, after microinjection of DNA to minimize the amount of time the embryos are held outside of the body. This complicates the logistics of the gene transfer procedure when working with large animals. Alternatively, *in vitro* embryology techniques, increase the number of zygotes available for use in gene transfer research and it is possible for the microinjected zygotes to be screened for viability prior to transfer to recipients. In Figure 21.9, a hypothetical situation is presented in which a number of advanced *in vitro* techniques are integrated into a gene-transfer program for mammalian farm animals.

Embryo splitting: The most efficient means of increasing pregnancy output rates is to produce identical twins by splitting embryos into two halves. In recent years, the procedure of embryo splitting has been refined. Embryos usually at the blastocyst stage are transferred for a few min. into a standard cell culture medium that contains two additional components: hypertonic sucrose and BSA. The medium penetrates the zona pellucida where its high osmotic strength causes the cells to contract. The BSA penetrates and adheres to the zona pellucida, giving it a net negative charge. The embryo is then transferred into the standard culture medium in a plastic Petri dish where it sinks to the bottom and sticks to it. The dish is placed on the stage of an inverted microscope, and a fine surgical blade controlled by a micromanipulator is used to orient the embryo and bisect it. At the blastocyst stage, the embryo comprises two different cell types: the trophectoderm, a single layer of trophoblast cells lining the inside of the zona (which becomes the fetal membrane of the placenta) surrounding a fluid-filled cavity (the blastocoel), and at one 'pole' of the blastocyst a mass of cells called the 'inner cell mass' (ICM) that will develop into the fetus. When an embryo is split to produce twin pregnancies, it is essential that the ICM is split into roughly equal halves (sometimes even into four). After splitting the hemispherical cell mass reform sphere, they are simply transferred into the oviducts of synchronized recipients as for normal embryo transfer.

(ii) Sex selection

In dairy and beef cattle, it is more desirable to use females than males. In part, this is because females are rate-limiting in reproduction and frequently have more desirable production characteristics. The sex of an animal is determined genetically. The gametes produced in the gonads of adult animals – sperm in the testes, ova in the ovaries – each contain just half of the chromosomes present in their somatic cells. With the recent development of large-scale dairy farms for milk production, many attempts have been made to sex spermatozoa in semen (sperm bearing X- and Y-chromosome) to eliminate unwanted bull calves in dairy cows. The sexed

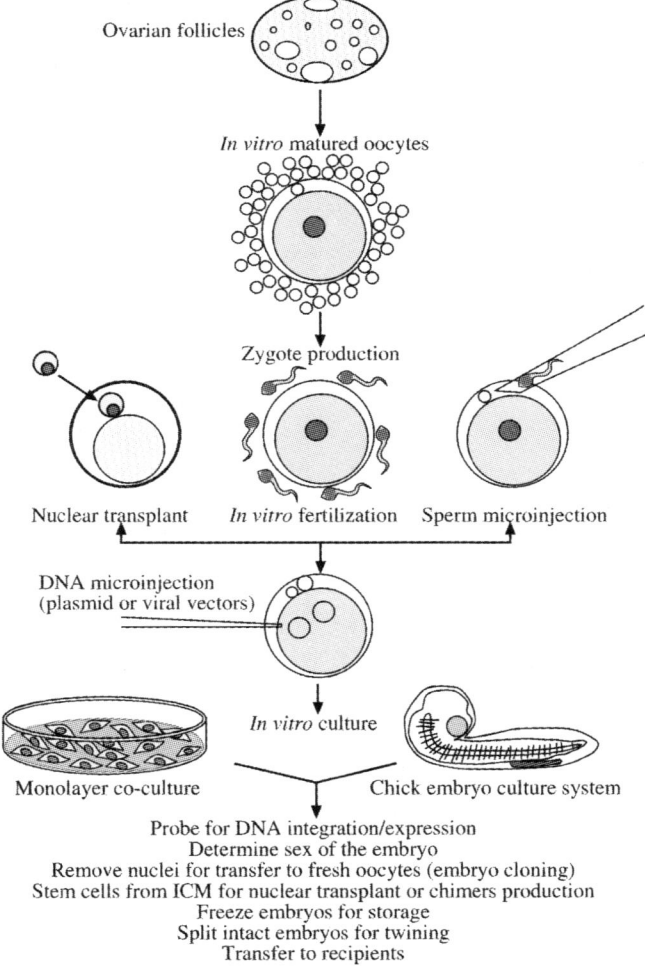

Figure 21.9 *Integration of in vitro embryology techniques to enhance gene transfer in farm animals (Redrawn from Minhas and Voelkel, 1989)*

sperm could be used either for direct insemination into cattle or perhaps more effectively for *in vitro* fertilization of oocytes for embryo production. An alternative strategy that could be more generally useful is embryo sexing. For this purpose, a method has been developed with a fluorescent probe for the H-Y antigen. The antigen could then be detected in the cells of the inner cell mass of expanded blastocysts.

(iii) Embryo cloning

The goal of embryo cloning is to multiply desirable genomes and produce transgenic animals. Pioneering work on embryo cloning in livestock was done by Willadsen at the Cambridge Building on Embryo-transfer Techniques. He developed a system for separating embryonic blastomeres by microdissection *in vitro* at the two-, four-, or eight-cell stages to form sets of genetically identical individuals from surrogate mothers. Blastomeres at the eight-cell stage rarely developed into viable embryos owing to insufficient cell mass and trophectoderm inadequacy, but this could be overcome by inserting cells from other embryos to form the placenta. It is possible to use Willadsen's embryo-splitting technique, say at the four-cell stage, and repeat the process *seriatum* until tens or hundreds of identical embryos had been produced in culture and stored in an embryo bank. The phenotype could be tested by transfer of some embryos to surrogate mothers for assessment of performance. However, most workers have chosen **nuclear transfer** as the method of forming cloned embryos. This technique involves separation of nuclei from blastomers from eight to 16-cell embryos and electrofusion with enucleated oocytes. The nuclear transfer will be affected by many factors, including age and source of the nuclei and recipient oocytes. Though cloned animals would contain no genetic benefits over their precursors, the methodology could provide means of upgrading genetically inferior herds, and incorporated into breeding programs, could accelerate the rate of genetic improvement.

(iv) Gene transfer in animals

After the success in gene transfer in *E. coli*, many attempts have been made to upgrade domestic animals by gene transfer. The field received enormous impetus from the spectacular success in producing giant mice by injecting a human growth hormone gene fused to a metallothionein promoter into the pronucleus of fertilized mouse eggs.

A modified mouse transgenesis DNA microinjection protocol has been developed for cattle. In brief, the essential steps entail: (i) Collection of oocytes from killed cows; (ii) *in vitro* maturation of oocytes; (iii) *in vitro* fertilization with bull semen; (iv) centrifugation of the fertilized eggs to concentrate the yolk, which in normal eggs prevents the male pronuclei from being readily seen under a dissecting microscope; (v) microinjection of input DNA into male pronuclei; (vi) *in vitro* development of embryos; (vii) nonsurgical implantation of one embryo into one recipient foster mother in natural oestrus, (viii) DNA screening of the offspring for the presence of the transgene.

3. Genetically engineered livestock

One of the goals of transgenesis of dairy cattle is to change the constituents of milk. The amount of cheese produced from milk is directly proportional to the k-casein content. Increasing k-casein production with an overexpressed k-casein transgene is reasonably likely. For a specific purpose, expression of a lactase transgene in the mammary gland could result in milk that is free of lactose. People who are intolerant to lactose, milk, and milk products would welcome such milk. Although transgenesis of cattle holds promise, progress toward producing large numbers of transgenic animals has been very slow.

For livestock in general, attempts have been made to create animals with inherited resistance to bacterial, viral, and parasitic diseases. For example, genetic resistance to bacterial diseases such as mastitis (mammary gland abscesses) in dairy cattle, neonatal scours (dysentery) in swine, and fowl cholera, occurs in some breeds. If the basis of this kind of resistance is a single gene, then it may be possible to create transgenic animals that are resistant to these bacterial infections.

(a) Characterizing and isolating animal genes

The best use of available genetic variation requires, first, the identification of genes that are responsible for economic traits and getting detailed information about them. There are several methods available to identify the desirable genes in an animal system, one of them being the use of markers in breeding selection instead of inferring the genotype from the phenotype of animals and their progeny. These markers facilitate effective gene tagging. In this method, the first step is to identify DNA sequences associated with economic trait loci. Then these genes must be incorporated into three maps: (i) A physical map of the chromosomes, in which the DNA sequences are assigned to specific sites on specific chromosomes; (ii) a linkage map, in which the linkage (co-inheritance) of different genes and markers is assigned (linked genes and markers are in close proximity on the same chromosome); (iii) a genetic map, in which the inheritance of economic traits is correlated with the inheritance of genes and markers.

(b) Homologues in human

In the initial stages of gene identification, in most cases, information available on other mammals including humans stored in the OMIM database (On–line database of Mendelian inheritance in Man) is used. The genomes of all mammals are remarkably conserved in both their genetic content and their arrangement of genes. For example, if a variant of a gene causes a heritable disease in one mammalian species, the homologous gene usually exists in similar variant forms in other species and causes a similar disease. This is known as the 'candidate gene' approach, where we recover a gene that is a probable candidate for causing the trait of interest. By this method, diseases like citrullinaemia, Pompe's disease, porcine stress syndrome, and many other genetic diseases have been identified. Bovine genes encoding milk proteins were isolated by means of their similarity with homologous rat genes, and genes encoding the growth hormone, many growth factors, and hormone receptors were isolated by homology with their human counterparts. The Human Genome Project is proving to be of immense value to livestock producers.

(c) Linked markers and linkage mapping

Not all genetic diseases have an unambiguous, identified homologue in humans or other animals. In complex cases, one could use an indirect approach to identify polymorphic DNA markers that are linked to variants of the unknown gene. Cattle have 30 pairs of chromosomes, each comparing an average of about 10^8 bp of DNA. Genes and other DNA sequences that are part of the same chromosome are said to be linked (or co-inherited). If we identify a specific (allele) of a DNA sequence that is known to be on the same chromosome as a defective gene allele, it is understood that the defective gene is also present since the variant sequence and the defective gene are linked and hence will be co-inherited. Linkage is defined by this co-inheritance. In the absence of any knowledge of a disease-causing gene, we can use an identified polymorphism in an 'anonymous' DNA sequence that is tightly lined to the gene variant to tell us whether that variant is present. Not more than 20% of the genome encodes genes. The remaining 80% is not subject to strong selective pressure and so accumulates a high number of mutations and DNA sequence variations (polymorphisms). The first type of anonymous DNA polymorphism that was identified is known as an RFLP followed by microsatellites, quantitative trait loci (QTLs), synteny, chromosome microdissection, etc.

(d) Direct identification of novel genes

The tools of molecular genetics also provide direct approaches to the isolation of unknown genes that are responsible for a genetic disease or a production trait. Information about expression of the gene allows us to use a variety of methods to identify it: (i) Identification from protein sequence, (ii) identification with antibodies, (iii) subtractive hybridization, (iv) differential display, and (v) livestock genome mapping projects.

4. Genetically engineered vaccines used with livestock

One of the most important advances in agriculture has been the development of vaccines to treat diseases such as hoof-and mouth disease, scours, trypanosomiasis, rabies, and such parasites as tapeworms and liver flukes. The recombinant subunit vaccine which is currently in use, is relatively inexpensive. Moreover, with it, yearly booster shots are not required.

In this section, we look at the use of transgenic animals as bioreactors for the production of recombinant proteins that have pharmaceutical applications. Currently, transgenic sheep, goats, pigs, and cattle are primarily produced by pronuclear DNA microinjection of fertilized egg cells, rather than by embryonic stem cell transfection, which has not been successfully applied to livestock animals. Although there have been numerous research reports demonstrating the feasibility of animal "***pharming***," a term referring to the use of farm animals to produce pharmaceutical products, to date relatively few pharmaceutical products derived from transgenic livestock are ready to be approved for human medical use. There are a number of reasons for the slow development of commercial transgenic livestock. First, unlike transgenic crop plants, large farm animals propagate at a comparatively slow rate and generate only small numbers of transgenic offspring. In addition, the inherent variability of gene expression levels, as a result of random DNA integration events arising from egg cell microinjection, leads to an unpredictable outcome. Second, there has been public concern that food products derived from transgenic animals, for example, low-fat meat products or humanized cows' milk, may not be widely accepted by consumers, owing to a perception that unknown risk factors could be associated with the production of cloned gene products. Third, animal rights activists have questioned the bioethics of animal pharming on the basis of its impact on decreased genetic diversity, and the animal safety of proposed large-scale production procedures.

Because of these formidable problems, established agricultural and pharmaceutical companies have not made significant investments in large animal transgenesis. However, this has begun to change as a result of two breakthroughs in this branch of the biotechnology industry. First, recent human clinical testing of milk-borne pharmaceuticals, such as antithrombin III and α-1-antitrypsin, have shown high biological activity of these products with no overt side-effects. Second, Ian Wilmut and colleges at the Roslin Institute in Edinburgh have reported that they were able to clone a sheep using nuclear transfer methods. Because of this renewed economic interest in the molecular genetic methodologies of milk-borne animal pharming, and in the application of nuclear transfer to animal cloning, we need to take a closer look at how these techniques are currently being performed.

Animal pharming using transgenic livestock

Several early attempts at producing genetically improved livestock animals by transgenesis indicated that regulated cell-specific promoters were going to be important components of protein expression vectors in animals. For example, high-level expression of the porcine growth hormone gene in transgenic pigs was accomplished by several groups that used constitutive-promoter to drive expression. Although it was found that these transgenic animals grew faster and had lower fat deposits than non-transgenic controls, the unregulated expression of the growth hormone led to serious health problems in the pigs. Because of this unexpected outcome, and the uncertainty that transgenesis could be used to improve livestock-derived food products, researchers instead began to develop methods to use animals as bioreactors to synthesize pharmaceutically important proteins. The most successful of these approaches has been to exploit the properties of milk as a renewable resource to produce functional enzymes, antibodies, and structural proteins using existing dairy manufacturing processes. The strategy has been to use the transcriptional promoters of mammary-specific genes to direct the expression of soluble transgenic proteins. Because milk-borne proteins do not cross into the blood or lymphatic fluid of the transgenic animal, there is less danger of producing undesirable side-effects in the animal owing to the species origin or biochemical function of the transgenic protein.

In addition to the use of transgenic animals to synthesize pharmaceutical proteins in milk, cell-specific mammary gland expression can also be exploited to produce nutrient-enriched dairy products. One such example has been to use transgenic cows to produce milk that contains human lactoferrin, an iron-transporting protein that is present in high levels in human breast milk. It should also be possible to produce specialized cheeses and cream products more efficiently using milk from transgenic animals that contains proteins normally added later in the manufacturing process. Table 21.1 lists some of the transgenic animals that have been engineered to produce human proteins in milk.

Table 21.1 *A few human proteins that have been expressed in the milk of transgenic animals*

Human gene product	Mammary gland-Pharmaceutical use	Transgenic specific promoter	animal
Factor IX	Blood clotting protein, treatment of hemophilia B	Sheep ß-lactoglobin	Sheep
α-1-Antitrypsin	Protease inhibitor, treatment of emphysema and cystic fibrosis	Sheep ß-lactoglobin	Sheep
Antithrombin III	Blood clotting protein, treatment of ATIII deficiency disease and use in open heart surgery	Cow casein	Goat
Tissue plasminogen activator	Dissolves blood clots, used as an acute treatment of heart attacks	Mouse whey acidic protein	Goat
Lactoferrin	Iron transport protein, infant formula additive	Cow α-S-casein	Cow
Protein C	Anticoagulant, treatment of hemophilia and used for surgery	Mouse whey acid protein	Pig

5. Genetic engineering of animals by nuclear transfer

It has been estimated that the cost of developing a single founder transgenic cow using pronuclear injection methodology can be as much as $500,000 owing to the associated cost of inefficient transgenesis (>97% failure rate) and variable gene expression levels. Moreover, extensive breeding of the founder line is required to produce a sufficient number of homozygous transgenic animals to constitute a cost-effective herd, and this could take up to 10 years.

To overcome these limitations, developmental biologists in agrobiotechnology have been trying for years to find conditions under which nuclear material from somatic diploid cells could be used as a pluripotent source of genetic information. In this way, founder animals could be quickly reproduced without the associated costs of breeding. Moreover, if the donor cell could be genetically manipulated in cell culture using standard stable DNA transfection protocols, or better yet, subjected to site-directed homologous recombination, then transgenic farm animals would be as common as transgenic mice. As a first step toward this goal, Wilmut and Cambell at the Edinburgh Biotechnology Company PPL Therapeutics, used a modified nuclear transfer method to produce a viable sheep using genetic material from a somatic cell that had been isolated from the mammary gland of a pregnant 6-year-old ewe. Using electroporation, the donor cell was fused to an unfertilized sheep egg cell that had been enucleated by micromanipulation. These reconstituted egg cells gave rise to 29 multicellular embryos that were implanted into 13 surrogate female sheep. A single lamb was born 148 days later that the Roslin Institute researchers named "Dolly," in honor of their favorite country singer, Dolly Parton. Subsequent molecular genotyping proved that Dolly's DNA was indeed derived from the donor cell nuclei. Dolly was later shown to be fertile, further substantiating that the regeneration process initiated by nuclear transfer was genetically complete. Figure 21.10 outlines Wilmut's animal cloning method, which is based on maintaining the nuclear donor cell in low concentrations of serum for several days prior to electroporation to induce cell cycle arrest in Go. However, Go quiescence may not be required for transgenesis in all large animals, because nuclear transfer experiments using bovine fetal firbroblasts have shown that actively dividing cells can also be used successfully.

The culmination of almost half a century of DNA-based biochemical, developmental, and cell biological research came in December 1997 when the Roslin Institute published its latest results describing the birth of two more transgenic lambs named "Molly" and "Polly". What made these transgenic farm animals special was that they were generated from fetal donor cells that had been stably transfected with a mammary-specific β-lactoglobulin expression vector containing the human Factor IX gene. Molly and Polly were not only genetically identical to each other with respect to sheep genes, but they also each possessed the potential to produce large quantities of the human blood clotting protein Factor IX in their milk.

Yanagimachi and his colleagues (1998) at the University of Hawaii reported another breakthrough. Figure 21.11 shows the procedure they pioneered called the "Honolulu technique," that was initially developed to clone laboratory mice. The most significant difference between the Honolulu technique, and the membrane electrofusion method developed by Wilmut is Yanagimachi's use of direct nuclear injection into enucleated eggs. The first transgenic animal made with this improved nuclear transfer technique was "Cumulina" the mouse, so named because of the adult cumulus cells used as the source of her genetic material.

One of the transgenic experiments that used pigs to produce human hemoglobin was performed as follows. A construct consisted of the regulatory region from the human β-globin gene joined to two human α1-globin genes and one human $β^A$-globin gene. Healthy transgenic pigs were used to produce human hemoglobin in their blood cells. Many clinical studies revealed that the human hemoglobin from the transgenic pigs had the same properties as natural human hemoglobin. Although preliminary, these results point to the possibility of replacing whole blood that is used for transfusions with human hemoglobin produced by transgenesis.

6. Transgenic sheep

Work is underway on introducing bacterial genes, which control cysteine production, into sheep. In sheep, the rate-limiting step in the production of wool is the availability of cysteine. Sheep

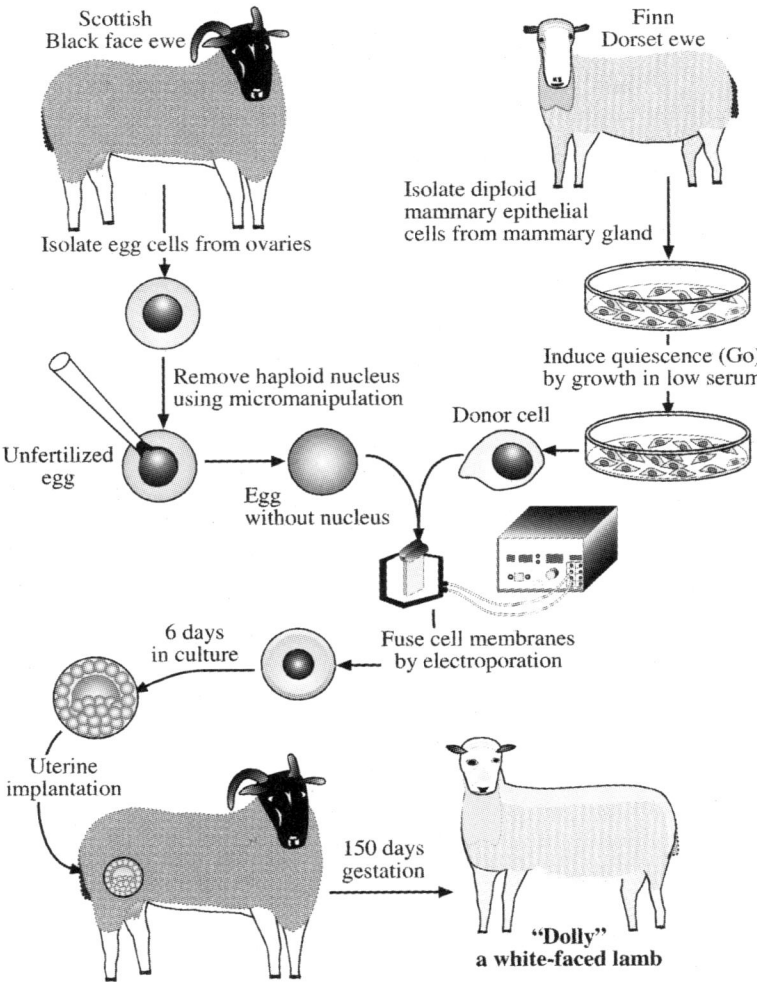

Figure 21.10 *Nuclear transfer can be used to clone livestock animals from enucleated egg cells that have been fused to a donor somatic cell by electroporation*

can be genetically transformed to synthesize their own cysteine by introducing two bacterial genes. When introduced into sheep, transgenic sheep have produced two times more wool than normal sheep. Another application of transgenic technology in sheep is the expression of insecticidal proteins in the wool to avoid the need for chemical spraying and to reduce the number of animals suffering as a result of fly strike. Chitinase genes, a gene found in plants, can be introduced into sheep. If sheep could be induced to secrete chitinase in the skin, insect larvae would be killed by ingesting the protein and die before causing damage to the sheep. This endogenous insecticide would eliminate the need for chemical insecticides. The chitinase gene may be expected to pose no risk when introduced into sheep because it is known to be non-toxic to mammals.

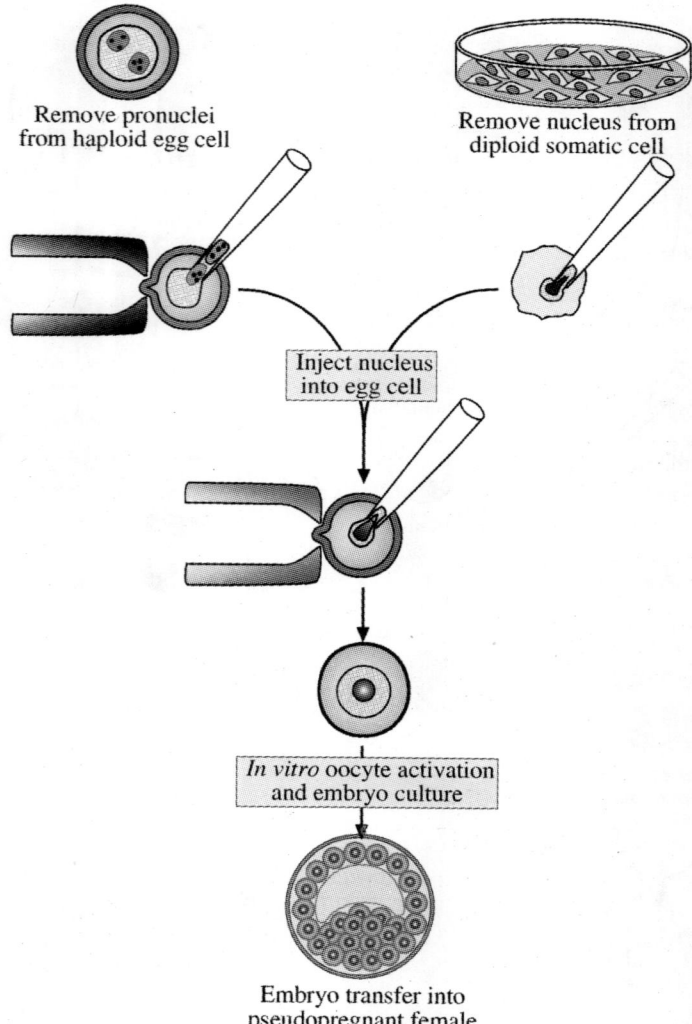

Figure 21.11 *The Honolulu nuclear transfer technique utilizes microinjection to generate diploid egg cells for animal cloning*

7. Transgenic goats

Transgenic goats, which produce heterologous proteins in their milk, have been produced. A research group from Tufts University, USA, and the Genzyme Corporation, USA, designed a construct coding for a variant of human tissue plasminogen activator (t-PA), which was linked to the mouse whey acid promoter. The modified t-PA (LAt-PA) was designed to be longer acting when used to treat myocardial infarction in humans. Two transgenic goats, male and a female, were identified from the 29 animals born. After mating and selection, a founder female produced LAt-PA in its milk during lactations. The LAt-PA was produced at a concentration of

3–6 g/ml. Higher LAt–PA expression levels have since been achieved in another transgenic goat after expressing the plasminogen under the goat β-casein promoter.

D. TRANSGENIC POULTRY

The application of gene transfer to poultry breeding is another area with remarkable potential for enhancing commercial meat and egg production. To transfer DNA to poultry eggs and embryos requires that special procedures be developed. One method of transforming poultry is through the use of irradiated spermatozoa. It should be noted that this procedure suffers from the problem of the inability to dictate which genes are transferred to the egg. Another potential means of transforming chicken embryos would be through isolation and propagation, through *in vitro* culture, or totipotent or pluripotent stem cells from embryonic tissues. These cells would be transformed by one of the various means of DNA insertion and transformed cells re-injected into developing embryos. This protocol is similar to that described for producing stem cell chimeras in other animals. The primary method now used for transforming poultry is with a retrovirus vector to transmit foreign DNA.

Microinjection of DNA into the fertilized eggs of birds to produce transgenic strains is limited because of several features that are unique to avian reproduction and development. For example, during fertilization in birds, several sperms can penetrate the ovum instead of one. Consequently, it is impossible to identify the male pronucleus that will fuse with the female pronucleus. Also, microinjected DNA into the cytoplasm does not become integrated into the genome of the fertilized egg. Finally, even if nuclear DNA microinjection were feasible, the technique would be difficult to implement because the avian ovum after fertilization becomes, in rapid succession, enveloped in a tough membrane, surrounded by large quantities of albumin, and enclosed in inner and outer shell membranes (Figure 21.12). By the time an avian egg outer shell membrane has hardened, the developing embryo (blastoderm stage) comprises two layers of 40,000–80,000 cells.

Another limitation is that till today, avian-specific embryonic stem cells have not been identified, so use of ES for genetic manipulation is not feasible in birds. However, a procedure using engineered cells from the embryo offers possibilities. Blastoderm cells can be removed from a donor chicken, transfected with cationic lipid (liposome) transgene DNA complexes (lipofection), and reintroduced into the subgerminal space of the embryos of freshly laid eggs (Figure 21.13). Some of the progeny may contain a mixture of cells, with some cells from the donor but most from the recipient. Such a genetic mixture is called a chimera. In some of the chimeras, cells that were descended from transfected cells may become part of the germ line tissue and form germ cells. Transgenic lines are then established from these chimeras by rounds of matings. The proportion of donor cells in chimeras can be increased to enhance the probability of obtaining germ line chimeras if the recipient embryos are irradiated with a dose of 540–660 rads for 1 hour prior to the introduction of the transfected cells. The radiation treatment destroys some but not all the blastoderm cells, thereby increasing the final ratio of transfected cells to recipient cells.

Transgenic chickens could be used to improve the genetic make up of existing strains with respect to built-in resistance to viral and coccidial diseases, better feed efficiency, lower fat and cholesterol levels in eggs, egg with high protein content, and better meat quality. A number of genes have been identified by poultry researchers as having potential value for improv-

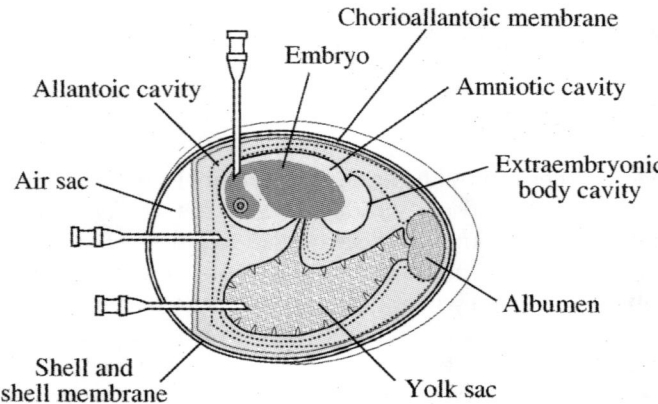

Figure 21.12 *Diagrammatic representation in sagittal section of the embryonated hen's egg (10 to 12 days old). The hypodermic needles show the routes of inoculation of the yolk sac, allantoic cavity, and embryo (head)*

ing disease resistance in commercial flocks. These include genes coding for light and heavy chain immunoglobulins, T-cell receptors, MHC, virus envelope antigens, and lymphokines.

E. GENETICALLY ALTERED FISH

As food supplement, from land grown agricultural crops fail to meet demand, worldwide food resources will depend more on aquaculture. Altering fish genetically using transgenesis is therefore a primary research objective. An advantage to the application of gene transfer techniques in fish as opposed to mammals is that fertilization and development naturally occur outside the body. Thousands of eggs can be produced from females under well controlled experimental conditions. The timing of fertilization can be predicted with accuracy, permitting the gene transfer to be performed at a prescribed time. The culture conditions are very simple. Typically, the eggs are incubated in a circulating bath of filtered, oxygenated water at a desired salinity and temperature.

In fish, the pronuclei are not readily seen under a microscope after fertilization, so linearized transgene DNA is microinjected into the cytoplasm of either fertilized eggs or embryos that have reached the 4-cell stage of development. The unique feature of fish egg development is that it is external; hence there is no need for an implantation procedure. Development can occur in temperature regulated holding tanks. Survival of fish embryos after DNA microinjection is high (35 to 80%), and the production of transgenic fish ranges from 10 to 70%.

In many animals, the rate of growth is controlled by the amount of growth hormone produced. It has been shown that the human growth hormone can increase the quantity and quality of meat, and the milk yield in cows and can make transgenic mice grow about twice as large as normal mice. Scientists at the Academy of Sciences in Wuhan, China, first inserted a gene sequence, which encodes the human growth hormone, into the eggs of goldfish. The engineered eggs then developed into mature fish. Analysis showed that approximately 50% of those fish had incorporated the human growth hormone gene and many of these grew four times

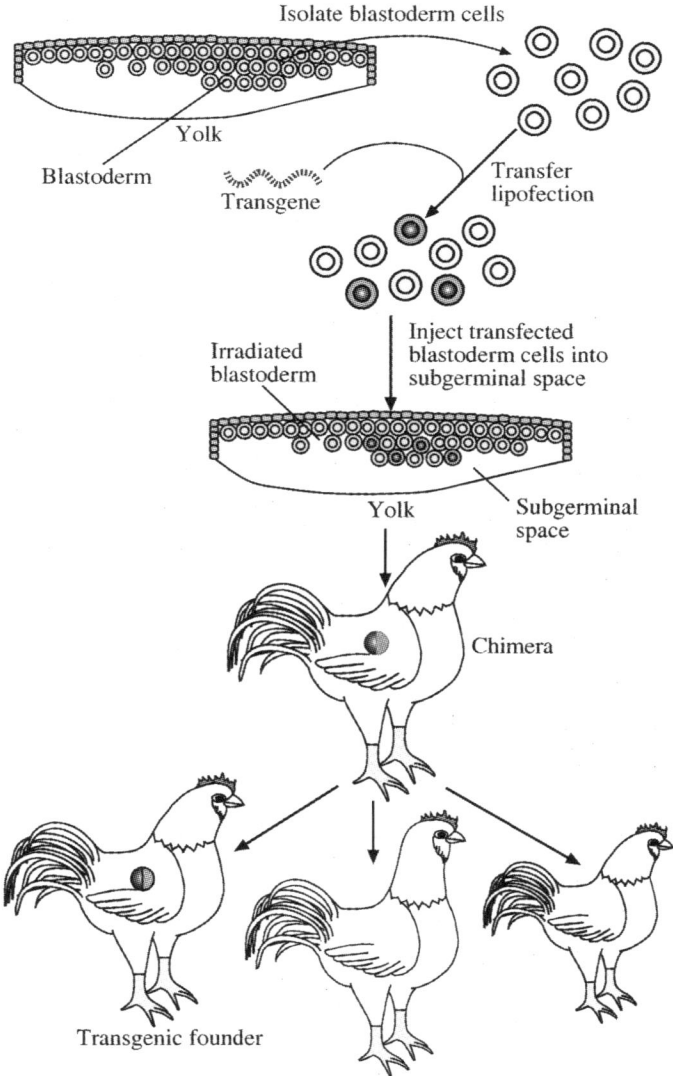

Figure 21.13 *Establishing transgenic chickens by transfection of isolated blastoderm cells*

bigger than the untransformed normal fish. The effect of the sockeye salmon growth-hormone gene driven by the regulatory region from a metallothionein gene in coho salmon has been well studied. The transgenic salmon averages 11 times the weight of the non-transgenic fish. Since then in many countries, similar attempts have been made and it has been extended to many other proteins, including the improvement in food value, resistance to frigid water conditions, better survival in polluted waters, disease resistance, resistance to low and high temperatures, etc. It is hoped that this type of research will lead to the provision of greater food reserves for a hungry world.

III REPLACEMENT THERAPY AND GENE THERAPY IN HUMAN

Over 3000 inherited human diseases have been catalogued to date out of which only a few such as phenylketonurea are currently treatable. Because genes have many different roles to play in the proper functioning of an organism, there is not a single biological consequence of a mutated gene. A mutation that alters the activity of an enzyme might cause either an accumulation of a substrate that is toxic or a deficiency of a compound that is vital for normal cellular operations. For many of these inherited diseases, the missing (or defective) gene-product cannot be supplied exogenously, for example, as insulin is supplied to diabetics. Most enzymes are unstable and cell membranes are not permeable to large macromolecules such as proteins; thus most enzymes must be synthesized within the cells in which their activity is required. Treatment of inherited diseases is largely restricted to those cases where the product of the defective enzyme is a small molecule that can be distributed to the appropriate tissues of the body through the circulatory system.

Replacement or supportive therapy: In this case, abnormal cells are replaced with normal cells as for example, in bone marrow transplantation. First, a patient with bone marrow cancer can be treated with radiation or chemotherapy to destroy all endogenous bone marrow cells. The bone marrow can then be reconstituted from cells preselected from the patient before radiation therapy, or with cells from a matched donor such as a close relative.

Somatic gene therapy, an alternative method: For most inherited diseases, *somatic cell gene therapy* appears to be the most promising approach. Somatic-cell gene therapy involves three sequential steps: (1) the removal of some of the patient's cells, (2) the introduction of normal, functional copies of the gene that is defective in the patient into these cells, and (3) the reintroduction of the transgenic "repaired" cells into the patient. Several inherited diseases are likely candidates for treatment by somatic cell gene therapy in the near future.

Gene therapy: Although the above therapies may benefit the patient, they often serve to alleviate the symptoms, rather than cure the disease. For some genetic diseases, neither drug nor replacement therapy is currently available. In addition, some genetic diseases (e.g. cancer) are thought to involve mechanisms regulating the expression of a number of genes (e.g. oncogenes), rather than the loss or malfunction of a single protein. The conventional treatments are ineffective in these cases. One recent and promising new approach to curing such diseases is **gene therapy**, or restitution of the normal gene *in vivo*. This technique seeks to repair the damage caused by a genetic deficiency through introduction of a functional version of the defective gene. Successful experiments with mice have shown that gene therapy is possible. To achieve this end, a cloned variant of the gene must be incorporated into the organism in such a manner that it is expressed only at the proper time and only in appropriate cell types. Many gene therapies have received approval for trials in human patients, including the introduction of gene constructs into patients. Among these are constructs designed to cure ADA–SCID (severe combined immunodeficiency due to adenosine deaminase [ADA] deficiency), neuroblastoma, or cystic fibrosis, or to treat cancer through expression of the E1A and p53 tumor suppressor genes.

Outline of procedure: A basic strategy in human gene therapy involves the incorporation of a functional gene into the target cell (generally, the somatic cells). The gene is typically in the form of an **expression cassette** consisting of a cDNA version of the gene downstream from a promoter that drives expression of the gene. A vector carrying such an expression cassette is introduced into the target cell, either *ex vivo* via gene transfer into cultured cells in the laboratory and administration of the modified cells to the patient, or *in vivo* via direct incorporation of the gene into the cell of the patient. Because retroviruses can transfer their genetic information directly into the genome of host cells, they provide one route to permanent modification of host cells *ex vivo*. A replication-deficient version of the *Maloney murine leukemia virus* can serve as a vector for an expression cassette up to 9 kb in size. Figure 21.14a describes a strategy for retrovirus vector-mediated gene delivery. In this strategy, it is hoped that the expression

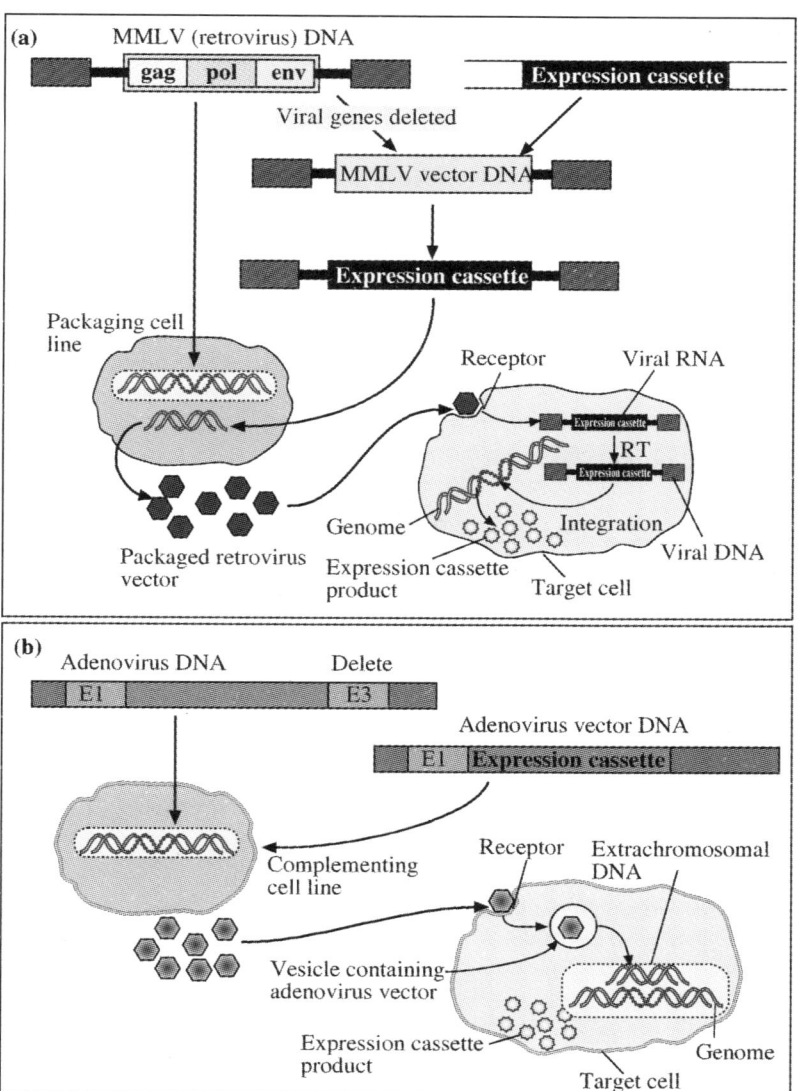

Figure 21.14 *Gene therapy (a) Retrovirus-mediated gene delivery ex vivo, (b) Adenovirus-mediated gene delivery in vivo*

cassette will become stably integrated into the DNA of the patient's own cells and will be expressed to produce the desired gene product. Alternatively, *adenovirus* vectors that can carry expression cassette up to 7.5 kb are a possible *in vivo* approach to human gene therapy (Figure 21.14b). Recombinant, replication-deficient adenoviruses enter target cells via specific receptors on the target cell surface; the transferred genetic information is expressed directly from the adenovirus recombinant DNA and is never incorporated into the host cell genome. Although many problems remain to be solved, human gene therapy as a clinical strategy is feasible.

Various strategies for implementing somatic cell gene therapy are beginning to emerge and can be grouped into three categories: **ex vivo gene therapy**, **in vivo gene therapy**, and **antisense therapy**. Usually, *ex vivo* gene therapy involves the collection of cells from an affected individual, correction of the genetic defect by gene transfer, selection and growing of the genetically corrected cells, and transfer of the corrected cells into the patient either by infusion or transplanting. However, *in vivo* gene therapy entails the direct delivery of a remedial gene into the cells of a particular tissue of a prospective patient. In contrast to the basic principles of *ex vivo* and *in vivo* gene therapies, antisense therapy is designed to prevent, or at least lower, the expression of a specific gene. Antisense therapy entails the introduction into a target cell of a nucleic acid sequence that is complementary to all or a portion of a specific mRNA. After the two RNA sequences (sense and antisense) are synthesized and hybridized (generally, RNA–RNA hybridize but DNA–RNA also hybridize), the amount of RNA for translation is drastically reduced. Consequently, a very small amount of target protein is synthesized.

IV SUGGESTED READING

Anderson WF (1992) Human Gene Therapy, *Science*, **256**:808–813.

Ausubel FM, Brent R, Kingston RE, Moore DD, Seidman JG, Smith JA, and Struhl K, Edits., (1999) *Short Protocols in Molecular Biology* (4th Edn), John Wiley and Sons, Inc.

Cross BA (1989) Animal Biotechnology, *Phil. Trans. R. Soc. Lond.*, B **324**:563–575.

Crystal RG (1995) Transfer of Genes to Humans. Early Lessons and Obstacles to Success, *Science*, **270**:404–410.

Garrett RH and Grsiham CM (1999) *Biochemistry*, Saunders College Publishing, Forth Worth, Phildelphia, New York.

Glick BR and Pasternak JJ (1996) *Molecular Biotechnology*, Panima Publishing Corporation, Bangalore, New Delhi.

Gordon JW (1989) Transgenic Animals, *Int. Rev. Cytol.*, **115**:171–230.

Gordon JW, Scangos GA, Plotkin DJ, Barbosa JA and Ruddle FH (1980) Genetic Transformation of Mouse Embryos by Microinjection of Purified DNA, *Proc. Natl. Acad. Sci. USA*, **77**:7380–7384.

Hartl DL and Jones EW (1998) *Genetics Principles and Analysis*, Jones and Bartlett Publishers, Inc.

Jahner D, Haase K, Mulligan R, and Jaenisch R (1985) Insertion of the Bacterial Gpt Gene into the Germ Line of Mice by Retroviral Infection, *Proc. Natl. Acad. Sci. USA*, **82**:6927–6931.

Lyon J and Gorner P (1995) Altered Fates, *Gene Therapy and the Retooling of Human Life*, New York. Norton.

Minhas BS and Voelkel SA (1989) Transgenic Animals, In: Biotechnology: A Comprehensive Treatise in 8 Volumes, (Vol. 7b) *Gene Technology* (Ed.), GK Jacobson and Jolly SO.

Monk M and Handyside (1988) Sexing of Preimplantation Mouse Embryos by Measurement of X–linked Gene Dosage in a Single Blastomere, *J. Reprod. Fertil.*, **82**:365–368.

Morgan RA and Anderson WF (1993) Human Gene Therapy, *Annual Review of Biochemistry,* **62**:191–217.

Starr C and Taggart R (1995) *Biology, The Unity and Diversity of Life*, Wadsworth Publishing Company.

Steinberg M, Guyden J, Calhoun D, Staiano-Coico L and Coice R (1993) *Recombinant DNA technology: Concepts and Biomedical Applications,* PTR Prentice Hall, Inc.

Watson JD, Hopkins NH, Roberts JF, Steitz JA, and Weiner AM (1987) *Molecular Biology of the Gene*, (4th Edn), The Benjamin/Cummings Publishing Company, Inc.

Wilmut I, Schnieke AE, McWhir J, Kind AJ, Cambell KHS (1997) Viable Offspring Derived from Fetal and Adult Mammalian Cells, *Nature*, **385**:810–813.

Wilson K and Walker J (Ed) (2000) *Practical Biochemistry Principles and Techniques* (5th Edn), Cambridge University Press.

22

Genetic Engineering in Plants

I INTRODUCTION

Agriculture is both the oldest and the largest of the world's industries, and it is likely to remain so in the foreseeable future. Since the earliest times, humans have nurtured plants and animals that are good to eat or are useful for work and clothing. The domestication of plants and animals is perhaps the single greatest civilizing factor in human history. Considerable effort has gone into developing varieties of plants that produce increased yields and have enhanced nutritional value. Biotechnology is a sophisticated extension of this age-old penchant for adapting living things to serve our own purposes and improving them further for their superior traits. The introduction of recombinant DNA (rDNA) into plants has enormous implications for agriculture, making it possible to alter and characterize the genetic make-up of plants. There are three major reasons for developing genetically modified plants (transgenic plants). First, the addition of a gene(s) often improves the agricultural value of a crop plant. Second, transgenic plants can act as living bioreactors for the inexpensive production of economically important metabolites. Third, plant genetic transformation provides a powerful means for studying the action of genes during development and other biological processes. Some of the genetically determined traits that can be introduced into plants by genetic engineering are insecticidal activity, protection against viral infection, resistance to herbicides, delay of senescence, tolerance of environmental stresses, altered flower pigmentation, improved nutritional quality of seed proteins, and self incompatibility. To date, numerous transgenic plants have been generated and in the future, biotechnology will have an enormous impact on traditional plant breeding programs.

Many plant proteins are deficient in amino acids that are essential for human health. Genes determining the synthesis of a superabundant protein that is rich in essential amino acids could be introduced to improve the quality of plant protein (alteration of nutritional profile). It also might be possible to move genes for resistance to pests, herbicides, microbial toxins, and nitrogen fixation. Fertile plants of some species may be generated from a single transformed cell, enabling an introduced gene to be transmitted to progeny through the seed.

Resistance to plant pests and diseases is often conferred by a single gene, which is known as a major gene. Resistance based on a single major gene or a few such genes becomes ineffective against new races of pathogens and pests after a few years of its introduction into crop plants. Therefore, novel strategies are needed for obtaining crops resistant to pests and pathogens. Breeding of resistant varieties is one solution to crop diseases. An integrated approach involving resistant cultivars, pesticides, fungicides, herbicides, biological control, and agronomical methods is usually the most effective.

The term genetic engineering is a label adopted by the popular media to describe the method of transferring human genes into bacteria using applied molecular genetic techniques. In this chapter, we examine the few major types of transgenic crops that have been engineered using standard molecular genetic techniques.

II APPLICATIONS OF BIOTECHNOLOGY IN PLANT IMPROVEMENT

Plant biotechnology mainly deals with the generation of useful products or services from plant cells, tissues, and often small organs. Such genetically modified plant material is either continuously maintained *in vitro* or it passes through a variable phase to enable regeneration of whole plants which can eventually be grown in the field. Therefore, plant tissue culture forms an integral part of any plant biotechnology work.

A. CONSTRUCTION OF TRANSGENIC CROP PLANTS

In contrast to animal cells, plant cells exhibit a variety of characteristics, which distinguish them from other cells. Plant cells have a unique property of regenerating into a whole plant from a single cell (**totipotency**). This property of plant cells allows a researcher to manipulate just a single cell, see and that cell develop into a complete mature plant (for more details refer to Chapters 29–37), and examine the whole spectrum of physical and growth effects of the original manipulations.

Two approaches have traditionally been taken to increase the quality and quantity of crop yields. First, plant breeders have incorporated desirable traits into seed stocks using standard genetic manipulations (i.e. by **conventional breeding methods**). However, because not all desirable traits are found within a single plant type, it has not been possible to breed a "super" plant crop with this approach. A second way to increase crop yields is to inhibit the deleterious effects of pests (both diseases and insects) and weeds by spraying crops with selective pesticides and herbicides. The overuse of these agents on a long-term basis may cause a negative impact on the environment. High cost is also involved in buying and applying them. One way to circumvent these limitations of plant breeding and chemical treatment is to use molecular genetic strategies to engineer transgenic crop plants that are of higher quality owing to increased resistance to drought, salinity, herbicides, and insects.

Plants have several unique characteristics that influence their stability for transgenesis. Most importantly, many types of dicots can be manipulated in tissue culture by hormones and nutrients to permit whole plant regeneration from a dedifferentiated tissue mass called a callus. Plants also cannot move, so they undergo self-fertilization, and thus produce a large number of progeny that are easy to harvest. Genetically, however, plants can present some problems with regard to gene copy number and clonality. First, some plant genomes are polyploid, meaning that they contain many copies of the same genome, which can make it difficult to interpret genetic crosses. Second, although plant tissue culture is a powerful means to regenerate whole plants, the resulting material can display somaclonal variation, which indicates that genomic alterations occur in somatic cells. Nevertheless, researchers have been able to define conditions that provide a "window of opportunity" for gene transfer through molecular genetic intervention. This allows tissue regeneration and genetic back-crossing to produce true-breeding transgenic plants.

B. IMPROVEMENT OF PHOTOSYNTHETIC EFFICIENCY OF PLANTS

Because of the fundamental and well characterized role of carbon fixation in plant metabolism, and the abundance of the protein ribulose bisphosphate carboxylase-oxygenase (RuBISCO), photosynthesis and photorespiration have attracted a great deal of attention as processes amenable to manipulation by genetic engineering techniques. The RuBISCO is probably the most abundant protein in nature composed of eight identical large and eight identical small subunits, encoded in the chloroplast and nucleus respectively, and the genes, which comprise a multigene family. This highly soluble protein occurs in the stromal compartment of the chloroplast and represents up to 65% of the total soluble protein in leaf extracts. The function of RuBISCO is twofold: first, in the fixation of carbon dioxide, providing carbon skeletons for intermediary metabolism (the dark reactions of photosynthesis), and second, in the oxygenation of ribulose bisphosphate to form phosphoglycolate, in the presence of light (photorespiration). The improvement of carbohydrate yield can therefore be attempted by increasing the relative rate of CO_2 fixation and/or decreasing photorespiration.

To improve the specific rate of carboxylase activity, it is necessary to consider those factors which may limit the process and which are amenable to genetic manipulation. RuBISCO from tetraploid varieties of rye grass has a higher affinity for carbon dioxide than does the same enzyme from diploid varieties. Diploid leaves have a higher photorespiration rate than do tetraploid leaves. Because of the tremendous importance of decreasing photorespiration in agriculture, many wild plants should be screened for this enzyme with the desired properties. Naturally occurring genes coding for efficient RuBISCO could be isolated and introduced into crop plants of agronomic values. Cloning of the large subunit gene from chlamydomonas and maize and the small subunit gene from pea have already been achieved. *In vitro* mutated versions of the cloned gene can be put back into plant cells when such technology becomes available.

Biomass could be used as a source of energy if photosynthetic efficiency could be increased, as the basic processes of photosynthesis are fully understood. They can be mimicked to provide energy-like hydrogen, fixed carbon, and nitrogen. The red photosynthetic bacterium *Thiocapsa roseopersicina* contains a thermostable hydrogenase that has a temperature optimum of 75°C and is stable in air. These hydrogenases can be used in biophotolytic systems that give off hydrogen gas. Application of recombinant DNA technology can speed up the genetic analysis of such useful organisms. A promiscuous sex factor has already been used to mobilize segments of the *Rhodopseudomonus capsulata* chromosome that codes for photopigment synthesis, reaction centers, and tryptophan synthetase. These R– prime plasmids can now be isolated and the photo pigment genes can then be cloned for use in the construction of gene structure, function, and regulation.

C. RESISTANCE TO DISEASES

Another area of agriculture that has been affected by the development of genetic engineering technology for plants is the control of plant diseases. The genetic engineering of disease resistance in plants is and will be a challenging task. Over the years, many agricultural methods have been employed to effect disease control in crop plants. Approaches that have been taken include plant quarantines, isolation of fields, control of weeds that may harbor pathogens, eradication of vectors, and the use of pathogen-free seed or vegetative stocks. These approaches have been variably effective depending on the environment in which the crops were grown. Moreover, the use of pesticides and other chemicals may not always be effective or safe and could be very expensive too. Therefore, new means to control plant diseases are needed. rDNA technology is now being used to address this issue. Since single genes govern many naturally resistant traits, these traits appear to be approachable targets.

Most plant pathogens belong to five groups: bacteria, fungi, viruses, nematodes, and insect vectors. To a significantly lesser degree, algae and protozoans also cause disease. However, these will not be individually discussed here.

1. Developing virus-resistant plants

Plant viruses often cause considerable crop damage and significantly reduce yields. For example, the tobacco mosaic virus (TMV) causes losses of over $50 million per annum globally to tomato plants. Unfortunately, little is known about the genetics of virus resistance and no chemical pesticide that controls virus multiplication has been developed so far. Therefore, in the absence of effective chemical treatments, agriculturists have, over the years, inoculated plants with a mild strain of virus or viroid in an attempt to induce resistance to more virulent strains (**cross-protection**). Although the mechanism of co-protection is unknown, infection by the inoculated virus suppresses or delays the onset of symptoms caused by the second (super-infecting) virus.

(a) Genetic engineering crops for resistance to viruses

Genetic engineering has made possible breakthrough in controlling viruses. Dissection of molecular details of viral infection and spread have been central to this success. Engineering virus resistance into crop plants is an important area in crop biotechnology. Several strategies have been developed to introduce virus resistance into desired crop plants. Natural virus resistance can be achieved in different ways: viral transmission can be blocked; establishment of the virus can be blocked; or viral symptoms can be bypassed or resisted. Recently, researchers have used the technique of genetic engineering to develop non-conventional types of virus-resistant transgenic plants. These methods used "vaccination" with viral coat protein genes, other viral genes, or viral gene antisense sequences to confer resistance. Generally, resistance to viruses can be achieved by two ways, (i) host encoded resistance and (ii) resistance using virus-encoded genes.

(i) **Host-encoded resistance**: A number of plant encoded virus-resistance genes have been studied by genetic analysis. Recent molecular biology techniques make it possible to isolate these resistant genes from resistant plants and introduce them into susceptible crop plants. Many pathogenesis-related (PR) proteins, specific or broad range, have been isolated from plants hypersensitive to the viruses.

(ii) Resistance using virus-encoded genes: One of the methods is to use virus-encoded genes to generate transgenic plants which produce compounds of the virus that confer cross-protection without causing viral disease. The strategies, which have been shown to be effective, include the deployment of the following genes. (i) coat protein, (ii) nucleo–capsid proteins, (iii) replicase, (iv) movement proteins, (v) satellite RNA, (vi) antisense RNA, (vii) aberrant RNA, (viii) RNase specific to double-stranded RNA, and (ix) Cosuppression.

(iii) Coat protein mediated transgenics: Some earlier studies have indicated that the tobacco mosaic virus (TMV) coat protein severely inhibited the translation of the TMV RNA *in vitro*. Taking a cue from this, a team of scientists used *Agrobacterium* transformation to induce cross-protection against the TMV. The gene encoding the TMV coat protein was fused to the cauliflower mosaic virus promoter (CaMV 35S) and nopaline synthase polyA signal. This fusion gene was inserted into a Ti plasmid and then introduced into tobacco plants using *Agrobacterium*. Transgenic tobacco plants created by infection with the recombinant Ti plasmid expressed the TMV coat protein. The coat protein gene was stably integrated into the genome of the transformed plants and was inherited in a Mendelian fashion by their progeny. The transgenic offspring showed delayed onset of symptoms following inoculation with live TMV, and up to 60% of the transgenic plants showed no symptoms at all. Coat protein-mediated protection reduced infection frequency, as well as local and systemic spread of infection. It should be noted that not every coat-protein positive plant within a line was equally resistant to virus infection. Subsequent experiments showed that the TMV coat protein gene also protects tomatoes and potatoes from TMV infection. Most surprising, the "vaccine" is multivalent – it also confers resistance to the tomato mosaic virus.

Following this remarkable success, coat protein-mediated protection has been successfully engineered for over 30 different viruses (e.g. the alfalfa mosaic virus, ALMV; the potato virus X, PVX; the potato virus Y, PVY; the tobacco streak virus TSV and the tobacco rattle virus). By using similar approaches, coat protein-mediated protection was obtained against a variety of viruses possessing different structural morphologies. For example, resistance to the alfalfa mosaic virus (AIMV, a tripartite, bacilliform virus) was achieved by expressing the AIMV CP gene from either the CaMV 35S or the 19S promoter in transgenic tobacco plants. In spite of the remarkable success of coat protein-mediated protection, the molecular basis of it has not yet been clearly understood.

(iv) Replicase: The expression of non-structural 54 kD protein, representing a part of the replicase enzyme of TMV, conferred a very high level of resistance in tobacco against TMV. The transgenic plants engineered with the 54 kD protein gene make the transcripts but do not make detectable levels of the protein. It is not clear at this stage whether the protection seen in the transgenic plants is due to the 54 kD protein or the transcript.

(v) Movement protein: Intracellular and intercellular (systemic spread) translocation of viruses are an important process of virus spread and pathogenesis. The systemic infection of virus results from rapid cell-to-cell spread and long distance spread through vascular tissues. A 30 kD movement protein (MP) of TMV binds to TMV-RNA and some host proteins and enables the ribonucleoprotein complex to pass through the plasmodesmata thus mediating cell to cell spread of virus. In contrast, transgenic tobacco plants expressing a mutated 30 kD MP showed a reduction in the final yield of infective TMV particles in infected plants. It was proposed that the defective 30 kD MP of the transgenic plants competed with the wild-type TMV-encoded 30 kD movement protein and thus reduced the spread of the virus. In another study, the 32 kD movement protein of the brome mosaic virus (BMV) expressed in transgenic tobacco effectively lowered the spread of TMV. This protein probably binds only to the movement-related host protein or to TMV-RNA but not to both. But for effective movement of the viral spread, binding of MP to both the host protein and viral RNA may be necessary. This strategy also appears to work for the single-stranded DNA viruses (gemini virus) as well. A recombinant MP having parts from the tomato golden mosaic virus and the African cassava mosaic virus has been shown to severely interfere with the spread of both the viruses.

(vi) **Transmission protein**: Most viruses require some kind of transmission agent (called vectors) for plant-to-plant spread. But there is a high level of specificity in the interaction between plant viruses and the insect vectors that spread them from one host to other. For example, the leaf curl virus in tomatoes spreads only through the vector *Bemissia tabaci*, In the case of aphid-transmitted CaMV, a bi-functional virus-encoded protein recognizes the viral coat protein and also a 'helper component' of the aphid, referred to as the 'aphid–transmission factor'. *In vitro* mixing of a defective (mutated) helper component of the aphid with the virus severely interfered with the transmission of the virus by the aphid. The defective helper component apparently competed with the aphid protein in binding to the bi-functional insect spread protein of the virus. By this mechanism it may be possible to inhibit viral spread through insect-transmitted viruses by engineering crops to express a defective virus-transmitted protein.

(vii) **Expression of satellite RNA**: Satellite RNAs infect plants only with the help of helper viruses by encapsulating them along with the respective helper viruses. These satellite RNAs either increase the severity of the symptoms or at times attenuate the symptoms caused by the helper virus. The latter property of satellite RNAs has been used in biologic control of the spread of certain viruses. For example, when the cucumber mosaic virus (CMV) infects pepper plants, severe symptoms appear. However, when CMV is co-inoculated with a satellite RNA, the disease symptoms are attenuated and the yield loss is decreased. However, the approach does have the drawback that the first infection of the satellite RNA (mild disease) causes a decrease in yield. An alternative approach to avoid the need to use a helper virus is to introduce the DNA sequence corresponding to satellite RNA into the plants. The approach of minimizing the effects of plant virus infection was developed by expression of the satellite RNA of tobacco ring spot virus and such plants were phenotypically resistant to infection with the tobacco ring spot virus itself. In this approach, a portion of viral genomes (SAT RNAs) other than the CP genes have been expressed in plants. Researchers have exploited this phenomenon and have expressed genes encoding the SAT RNAs associated with CMV and the tobacco ring spot virus in transgenic tobacco plants. The mechanisms responsible for these effects are still not clear.

(viii) **Defective interfering DNA**: Defective interfering (DI) DNAs and RNAs are very rare in plants. Similar to satellite RNAs, DI nucleic acids can intensify or ameliorate the symptoms of their respective parental viruses. This method has been successfully tested in the case of ACMV, a Gemini virus with two ssDNA genomes (A and B). A subgenomic B-component of ACMV was engineered into tobacco plants, which were susceptible to ACMV. The DI of ACMV interfered with the replication of both A- and B-DNAs in the transgenic tobacco plant and ameliorated the symptoms of virus infection.

(ix) **Antisense RNA**: In both eukaryotes and prokaryotes, the expression of an RNA molecule that is complementary to a normal gene transcript (mRNA), is called ***antisense RNA***. Antisense RNA can reduce the expression of the normal gene. The presence of antisense RNA (reverse oriented, i.e. $3' \rightarrow 5'$ as opposed to $5' \rightarrow 3'$) can decrease the synthesis of the gene product by forming a duplex molecule with the normal sense mRNA, thereby preventing it from being translated. Antisense RNA technology has various potential applications. (1) In various systems, mutants with leaky mutations can be produced; (2) promoter specificity, transcriptional and translational control can be studied; (3) various steps in metabolic pathways can be studied; (d) fruits with delayed ripening and longer shelf life can be achieved; (e) pigmentation in flowers can be altered; (f) seed oil components can be altered; (g) viral and fungal infections can be checked; (h) tumor growth in animal cell can be inhibited.

The antisense RNA–mRNA duplex also rapidly degrades, a response that diminishes the amount of that particular mRNA in the cell. Theoretically, this technique should be able to prevent plant viruses from replicating and, subsequently, damaging plant tissues by creating transgenic plants that synthesize antisense RNA that is complementary to virus coat protein mRNA. Several groups of scientists have constructed transgenic plants that synthesize antisense RNA copies of virus coat protein genes (e.g. PVX, CMV, and TMV) and tested these plants by challenging them with viruses. In all instances, it was observed that the plants were protected against the invading virus only when low concentrations of the virus were used. Although it may be premature to give up on using the antisense RNA approach as a means of creating virus-resistant plants, significant improvements are required before it will gain acceptance.

(x) Ribozyme-mediated virus protection: Recently it has been reported that certain naturally occurring RNA molecules possess the property of self-catalyzed cleavage. Viroid RNAs and SAT RNAs (e.g. avocado sunblotch viroid and SAT viroids of TobRV and the lucerne transient streak virus (sLTSV), possess this characteristic *in vitro*. The cleavage reactions are intermolecular and presumably occur when the RNA molecule assumes the configuration that allows reactive groups to interact. Furthermore, the sites of catalysis are specific and are associated with conserved sequence domains. It has been reported that synthetic oligonucleotides can possess highly sequence-specific endoribonucleolytic activities. It may be possible to produce transgenic plants that express ribozymes to specific viral transcripts based upon their nucleotide sequence specificity.

(b) Resistance genes from animals

Recently there have been many reports on introgression of resistance genes from animal and other organisms into plants. Genes for mouse monoclonal antibodies engineered into plants were expressed well and functional antibodies were shown to assemble within plant cells. The biological significance of such antibodies (called ***plantibodies***) has been demonstrated using monoclonal antibodies against phytochrome. Phytochrome–related functions were inhibited in plants expressing phytochrome plantibodies. After these experiments, attempts are being made to introduce genes for plantibodies against important viral proteins such as coat proteins and replicase so that multiplication of viruses within the plant cell can be limited. The usefulness of plantibodies (i.e. use of animal immune mechanisms) in the control of viral diseases is currently in the testing stage.

2. Fungal resistant transgenics

Plants resist fungal infections in a number of ways. In a passive resistance mechanism, plant cell walls can be strengthened to provide a physical barrier to limit penetration by fungal hyphae. However, in the active resistance called 'hypersensitive reaction', a resistant plant synthesizes toxic antimicrobial compounds called 'phytoalexins'. The hypersensitive reaction also leads to the induction and accumulation of many defense–related metabolites, for example, proteinase inhibitors and lytic enzymes such as chitinase and β-1,3-glucanase. Hypersensitive reaction is a specific process that occurs as a result of interaction between a product of a dominant resistance-gene (R) of the host and a product of a dominant avirulence-gene (Avr) of the fungal pathogen. This kind of plant–microbe interaction is known as an incompatibility reaction, and it leads to resistance in the host. In spite of extensive research, the precise mechanism that activates a hypersensitive reaction is not known.

Because of the greater complexities of fungal–plant interactions, there has been less research on developing molecular approaches to confer host resistance to fungal than to other plant pathogens. However, recently, a significant amount of progress has been made in developing resistance to plant pathogenic fungi. For example, expression of a bean chitinase can protect tobacco and oilseed rape from post-emergent damping off caused by *Rhizoctonia solani*. A hybrid chitinase gene was constructed by replacing the 5' regulatory sequences of a bean chitinase gene with the CaMV 35S promoter to achieve constitutive expression. After introduction of the hybrid gene into tobacco plants, the bean chitinase was expressed in the transgenic plants and the precursor enzyme was correctly processed post-translationally. Tobacco plants that expressed bean chitinase showed marginally higher resistance to the fungus as compared to controls. *Brassica napus* plants engineered with bean chitinase showed significant resistance to *R. solani*. Overexpression of chitinase confers resistance only against fungi that have chitin in their cell wall.

Tomato plants expressing the grapevine genes for stilbene synthase demonstrated increased resistance to infection by *Botrytis cinerea*. Similarly, tomato plants expressing the bacterial gene tabtoxin detoxification protected plants against *Pseudomonas syringae* infection. It is well known that certain plant varieties are more resistant to disease than others. Some progress has been made in identifying the genes responsible for such resistance. For example, in maize, resistance to the fungus *Cochliobolus carbonum* is mediated by a reductase, which inactivates the phytopathogenic toxin. RFLP mapping also has identified a tomato gene that confers resistance to the fungus *Fusarium oxysporum* and the construction of a tomato YAC library has led to the identification of genes conferring resistance to the tobacco mosaic virus and *Pseudomonas syringae*.

3. Resistance to bacterial pathogens

Many transgenic plants resistant to pathogenic bacteria have been developed. For example, the peptide fragment (cercopin) from a giant silkworm haemolymph protein, which is thought to act as an ionophore against a broad spectrum of bacterial species, has been introduced and expressed constitutively in transgenic potato plants. Field tests have indicated that expression of cercopin gave significant resistance to the bacterial diseases soft and **black-leg** (stem rot). Similarly, lysozome, an enzyme, which degrades the peptidoglycan bacterial cell wall has been expressed in transgenic plants.

D. RESISTANCE TO INSECTS

Pests are a major cause of yield loss and hence the use of pesticides to obtain good harvests is almost universal. Pesticides and insecticides are chemical agents that kill various types of insects and other pests while leaving plants and most animals relatively unharmed. A global survey in 1988 revealed that the annual cost of use of chemical pesticides was in the range of $6 billion. Unfortunately, many pesticides are non-selective: they kill insects which are beneficial to plants and the environment along with those which are not. In addition, pesticides are toxic to humans and other non-target animals and they contaminate soil and water supplies. The environmental contamination may then lead to a second level of adverse effects on both plant and animal life. Also, prolonged use of insecticides can lead to the development of resistant insect strains, which are no longer susceptible to the toxic effects of those chemicals. Because of these problems, a major goal of biotechnologists has been to provide plants with a natural endogenous resistance (intrinsically tolerant) to a variety of pests through the technique of genetic engineering.

Many genetic sources of insect resistance have been identified and targeted for developing pest resistant transgenic crop cultivars. These include *Bacillus thuringiensis* (Bt) bacterial toxin, protease inhibitors, snow drop-lectin, hevein, and wheat germ agglutinin. Two main strategies have successfully been used to confer resistance against insect predators. The first involves a gene for an insecticidal protoxin produced by one of several subspecies of the bacterium *Bacillus thuringiensis*. The second strategy uses genes for protease inhibitors that have been shown to be effective against a wide variety of insects; their presence restricts the ability of insects to digest food by interfering with the hydrolysis of plant proteins by insects. Tremendous progress has taken place in the genetic engineering of plants using both these approaches.

1. Endotoxin gene of *Bacillus thuringiensis*

Bacillus thuringiensis, a gram–positive soil bacterium, possesses endogenous insecticidal activity directed against a variety of insects including mosquitoes. Bt produces crystalline (cry) proteins (endotoxins) on sporulation. The proteins are harmless to humans, domestic animals, most insects, and bees. However, the cry protein products produced by different Bt strains are toxic to specific classes of insects. For example, cryI is effective against *Lepidoptera* larvae, cryII is toxic to *Lepidopteran* and *Diptera* insects, cry III to the Colorado potato beetle and other *Coleoptrans*, cryII and IV to *Diptera* larvae, and cryV and cryVI against nematode worms. To exploit this lethal characteristic, *Bt* spores are combined with water to form a mixture, that can be sprayed over an insect-infested area. When insect larvae ingest the bacterial spores, the pores release a chemical toxin which, in turn, causes the death of the insect larvae. However, the insecticidal effects are transient due to the limited field survival of the spores, so long-term insecticidal activity requires repeated applications of spores to the treated area. Despite their remarkable safety and efficacy, they have not been a commercial success, basically due to the high cost of production and the low persistence and stability of the product on the surface of plants.

Recombinant DNA technology attempts to circumvent the need for repeated spore treatments and the Bt gene has been the transgene of choice in a number of insect resistance experiments. The Bt toxin, believed to be an environmentally safe insecticide, is active against a number of caterpillars, including the tobacco hornworm and gypsy moth. The strategy has been to link the toxin gene to a constitutive promoter that will express the toxin in other organisms, which are better suited for survival in the field. The *B. thuringiensis* sub sp., the *Kurstaki* toxin protein gene is not particularly well-expressed in plants. In an effort to raise the level of the expressed protein, researchers truncated the gene (minimum sequence required for toxin activity) so that only the N-terminal portion of the insecticidal toxin was produced (Figure 22.1). Examination of the bacterial gene indicated that it differed significantly from plant genes in a number of ways. For example, localized regions of AT richness resembling plant introns, potential plant polyadenylation signal sequences, ATTTA sequences which could destabilize mRNA and rare

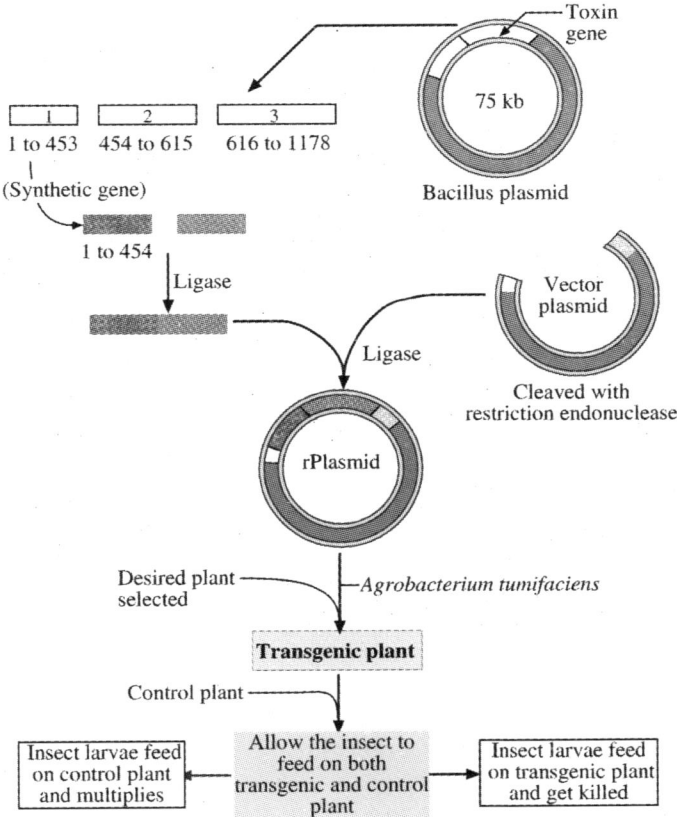

Figure 22.1 *Construction of a transgenic plant and expression of Bt toxin gene to control insect larvae*

plant codons were all found. Elimination of many of these undesirable sequences resulted in greatly enhanced expression of the insecticidal toxin and good insect resistance of the transgenic plants in field tests. A second approach involved linking the *Bt* toxin gene to a constitutively expressed promoter and introducing it directly into the cells of the plant, which is to be protected. T-DNA transformation has been used to move the gene into tobacco and tomato plants, where it appears to express itself strongly enough to kill a large percentage of caterpillars. The Bt gene is inherited in Mendelian fashion, indicating that it has been stably integrated into the plant genome.

In an effort to dramatically increase the level of expression, two other unique approaches have been tried. In the first attempt, an isolated insecticidal toxin gene was modified by site-directed mutagenesis to change any DNA sequences that might inhibit efficient transcription or translation in a plant host. This partially modified gene had a nucleotide sequence that was 96.5% unchanged from the wild type gene and encoded the identical insecticidal toxin protein. The transgenic plants with the modified toxin gene produced a tenfold higher level of insecticidal toxin protein. In the second attempt, a fully modified version of the insecticidal toxin gene was designated and chemically synthesized. This fully modified gene contained codons more commonly used by plants. It was also modified to eliminate any potential mRNA secondary structure or chance of plant polyadenylation sequences, which might decrease the gene expression. It had a G+C content of 49% and a nucleotide sequence that was only 78.9% identical to the wild type gene. This fully modified gene containing the transgenic plant had approximately a 100-fold higher level of toxin protein.

Another strategy that is designed to increase the effectiveness of relatively low levels of Bt insecticidal toxin activity entails combining the toxin with a serine **protease inhibitor**. These two genes, fusion protein produced an insecticidal activity that is 20-fold that of the Bt toxin alone.

An indirect approach to pest management bypasses the problem of plant transformation altogether. It involves inserting a toxin gene into the genome of a leaf- or root-colonizing bacterium, which synthesizes and secretes the pesticide *in situ* on the leaf surface. Examples of organisms that have been transformed in this manner include *E. coli* and the bacterium *Pseudomonas fluorescens*. *P. fluorescens* lives on the roots of many different types of plants. Tests show that when *P. fluorescens* transformed with a cloning vector containing the genes for the *B. thuringiensis* toxin is sprayed on corn plants, the transformed bacterial cells colonize the root area. As a result, the *B. thuringiensis* toxin is synthesized at the site of the plant itself, thus endowing the corn plant with certain endogenous insecticidal abilities.

2. Protease inhibitors

Most plant species, including crop plants, have evolved a natural mechanism of defense against herbivorous insects by accumulating proteinase inhibitors (compounds that inhibit the activity of protease enzymes) at concentrations that will cause metabolic inhibition upon ingestion by insects. Proteinase inhibitors (PI) have a broader spectrum of metabolic inhibition and so can be used to control many different types of insects. Thus, there have been constant attempts to introduce PIs into crop plants that normally do not make these proteins.

Proteinase inhibitors are categorized into four classes: serine-, thiol-, metallo-, and aspartyl proteinase inhibitors. Serine-protenase inhibitors and thiol-proteinase inhibitors are important where pest control is concerned since different types of insects use either serine proteinases or thiol proteinases or in some cases both. Proteinase inhibitors offer several advantages as insect control agents: PIs can be alternative to Bt or they can be used with Bt to reinforce the resistance already achieved through Bt. PIs become inactivated by cooking and their main limitation is that they require high concentrations of protein expression in order to be effective. Transgenic tobacco plants expressing high levels of the trypsin-inhibitor protein encoded by a gene of an African variety of cowpea showed decreased insect damage by *Heliothis virescences*, the tobacco bud worm, a serious pest of tobacco. It was also reported that the cowpea trypsin inhibitor killed a wide variety of *Lepidopteran* and *coleopateran* insects including the armyworm (*Spodopteran littoraalis*), the corn earworm (*Heliothis zea*) and the corn root worm (*Diabrotica undecimpunctats*).

3. Baculovirus

Similar types of studies that make use of the endogenous insecticidal activities of a type of virus known as **baculovirus** have been done. Like *B. thuringiensis*, the baculovirus particles can be combined with water and then sprayed onto foliage. When damage-causing insects eat the virus-treated foliage, they are subject to a viral infection, which leads to their death. Large-scale field applications of both *B. thuringiensis* and baculovirus-derived insecticides await further studies.

E. HERBICIDE RESISTANCE

Early transgenic work in plants was mainly concentrated on engineering herbicide resistance due to the following reasons: (1) The mode of action of herbicides was well known and many genes for resistance had been identified; (2) there was a clearly defined market for such crops and profits on investments were almost assured; (3) it was quite easy to screen and select transgenics expressing high levels of tolerance. Hence, a number of crops were engineered for tolerance to herbicides, mainly by private companies who also patented and traded these herbicides.

Approximately 10% of global crop production is lost through weed infestation every year, despite the expense of $10 billion spent on more than 100 different chemical herbicides. Herbicides are chemical agents that, when applied to various types of plants, result in the death of the plant. Herbicides are used primarily to kill non-desirable plants that compete with a desired plant for space, water, and nutrients. Research in herbicides is mainly concentrated on making the herbicides act only on non-desirable plants. Herbicides typically disable target enzymes in metabolic pathways unique to plants, such as those involved in photosynthesis or biosynthesis of essential amino acids (Table 22.1). These processes are shared by both crops and weeds. Thus developing herbicides which are selective for weeds is very difficult. An alternative approach is to modify crop plants so that they become resistant to broad-spectrum herbicides. Two approaches have been adopted to engineer herbicide resistance. In the first of these, the target molecule in the cell is either rendered insensitive or is over-produced. In the second approach, a pathway that degrades or detoxifies the herbicide is introduced into the plant.

Table 22.1 *Mode of action of herbicides and method of engineering herbicide resistant plants*

Herbicide	Pathway inhibited	Target enzyme	Engineered resistance
Glyphosate	Aromatic amino acid biosynthesis	5-enol-pyruvyl shikimate-3-phosphate (EPSP) synthase	Overexpression of the plant EPSP gene or introduction of bacterial glyphosate-resistant *aro*A gene
Sulphonylurea	Branched-chain amino acid biosynthesis	Acetolactate synthase (ALS)	Introduction of resistant ALS gene
Imidazolinones	Branched-chain amino acid biosynthesis	Acetolactate synthase (ALS)	Introduction of mutant ALS gene
Phosphinothricin	Glutamine biosynthesis	Glutamine synthetase	Over-expression of glutamine synthetase or introduction of the bar gene, which detoxifies the herbicide
Atrazine	Photosystem II	Q_b protein	Introduction of mutant gene for QB protein or introduction of gene for glutathione-S-transferase, which can detoxify atrazines
Bromoxynil	Photosynthesis		Introduction of nitrilase (*bxn*) gene which detoxifies bromoxynil
2,4–D	Growth regulation	–	*tfd*A (bacteria)

Many broad-spectrum herbicides are often toxic to crop plants, as well as to the weeds they are intended to kill. Herbicides act by specifically affecting some key metabolic pathways of plants. Therefore, the major thrust has to be on identifying and transferring herbicide resistance genes into major crop plants. The following four strategies are used for attaining herbicide resistance: (1) Inhibit the uptake of the herbicide; (2) stimulate over expression of the target protein of the herbicide, so that enough of it will escape disablement by the herbicide; (3) insert a genetically altered form of the target protein that is less sensitive to the herbicide; and (4) endow plants with the capability to metabolically inactivate the herbicide (Cloning of herbicide resistance—degrading genes). Of these possibilities, the last three have been implemented to produce herbicide-resistant transgenic plants.

The first engineered recombinant plant of commercial value was the herbicide resistant transgenic plant. An economically important herbicide glyphosate (Roundup) is a non-selective herbicide that inhibits an enzyme in a metabolic pathway found only in plants and microorganisms. The pathway is the *shikimic acid pathway*, and glyphosate is a competitive inhibitor of the enzyme *EPSP synthase (5-enolpyruvylshikimic acid-3-phosphate synthase)*. *EPSP synthase* is an enzyme necessary for the biosynthesis of aromatic amino acids in the chloroplast (Figure 22.2). Resistance plants developed for the herbicide glyphosate are "environmentally friendly", because glyphosate is readily degraded to non-toxic compounds in the soil. Two approaches have been used to produce plants that can overcome the toxic effect of glyphosate. In the first method a resistant petunia, in which the EPSP gene is highly amplified, was isolated by growing cultured cells in increasing levels of the herbicide. Petunias transformed with a fusion gene linking the viral promoter to a wild-type EPSP gene also show resistance. In the second approach a mutant EPSP gene (*aroA*) was isolated from *E. coli* which was 6000 times less sensitive to glyphosate than the wild-type enzyme, but lacks the plant "transit peptide" that is needed to transport EPSP into the chloroplast. To overcome this problem, a fusion gene was constructed, linking the *aroA* coding region with a dicot transit peptide sequence. Petunias and tobacco plants transformed with this fusion gene express the *aroA* protein in the chloroplast, in addition to their own EPSP. The third approach to glyphosate resistance has yet to be achieved. However, metabolic pathways have been defined in species of *Pseudomonas* and other bacteria that allow the use of glyphosate as a sole phosphate source. Efforts are under way to clone the genes that encode the glyphosate-degrading enzymes.

Resistance due to herbicide inactivation has been developed for bromoxynil (3,5-dibromo-4-hydroxybenzonitrile), a herbicide that acts by inhibiting photosynthesis. In this case, resistant plants were created by the introduction of a bacterial gene that encodes the enzyme nitrilase, which can inactivate bromoxynil before the herbicide can act. The gene for the enzyme nitrilase was isolated from the soil bacterium *Klebsiella ozaenae* and placed under the control of the light-regulated promoter from the small subunit of the enzyme ribulose bisphosphate carboxylase before it was transferred to the tobacco plants. The transgenic plants expressed nitrilase activity and were resistant to the toxicity of the herbicide.

Phosphinothricin (PPT) is an irreversible inhibitor of glutamine synthetase in plants and bacteria. Bialaphos, produced by *Streptomyces hygroscopicus*, consists of PPT and two alanine residues. When these residues are removed by peptidases, the herbicidal component PPT is released. To prevent self-inhibition of growth, bialaphos-producing strains of *S. hygroscopicus* produce an acetyltransferase that inactivates PPT by acetylation. The *bar* gene that encodes the acetylase has been introduced into potato, tobacco, and tomato cells using *Agrobacterium*-mediated transformation. The resultant plants were resistant to commercial formations of PPT and bialophos

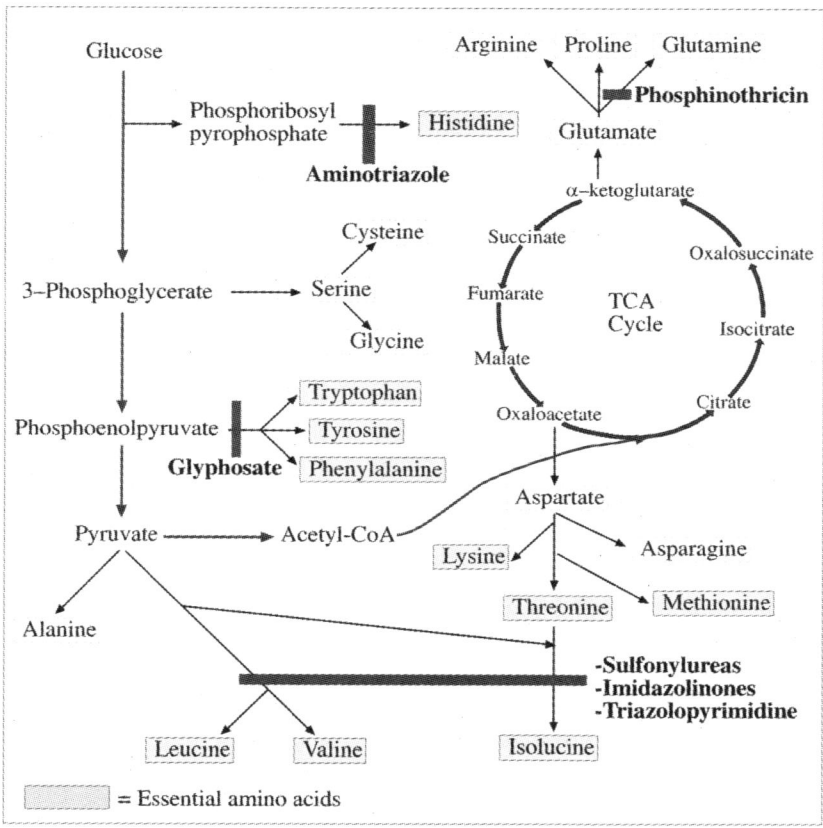

Figure 22.2 *Metabolic pathway for amino acid biosynthesis and the target sites for herbicide action*

in the laboratory and in the field. Surprisingly, rice plants after treatment with bialaphos herbicide, bialaphos–resistant transgenic rice plants, showed resistance to bialophos. The herbicide was also toxic to fungi and protected the plants from fungal sheath blight disease.

F. IMPROVING PLANT NUTRITIONAL VALUE

1. Improvement of quality of seed storage proteins

Seeds that are used as sources of both carbon and nitrogen in the human diet and animal feed contain a limited number of amino acids, which are organized into repeating units. Often these proteins are nutritionally deficient because they lack one or more of the amino acids (generally, lysine or methionine) that are essential for human health. The amino acid composition of the seed storage proteins can be altered to a limited extent by breeding programs, but recently, genetic engineering strategies have been used.

Protein molecules are composed of varying arrays of twenty different amino acids, of which the human body can synthesize twelve. The remaining eight amino acids, called essential amino acids, must be provided to the body by ingestion. All of the eight essential amino acids are present in a wide variety of animal products, including red meat, poultry, and milk. In contrast, no source of plant food contains adequate supplies of all eight of the amino acids. Cereals are the staple food and major source of protein for a large percentage of the earth's population. Protein comprises 10% of the dry weight of grain. However, grains lack one or more essential amino acids and thus offer only incomplete nutrition. Therefore, the effort to engineer missing amino acids into cereal protein will potentially have the greatest impact on worldwide nutrition of any agricultural improvement.

Consider, for example, the lowly bean. While beans have more than enough of the essential amino acid **lysine**, they lack the amino acid **methionine**. Wheat and rice, on the other hand, both contain suboptimal levels of lysine while containing a sufficient amount of methionine. Scientists are currently using the knowledge gained through biotechnology to try and alter the genes of a variety of plant proteins. For example, perhaps the genes of the bean could be altered so as to encode a protein that contains sufficient quantities of methionine. If the gene for **phaseolin** – the primary protein molecule of a broad bean – could be altered to contain codons, which specify the amino acid methionine without altering the overall structure or growth pattern of the plant, phaseolin could become a complete protein. In transgenic plants, the altered proteins, unlike unaltered proteins, are not always properly compartmentalized into protein bodies in the seeds. However, recent studies have shown that if the desired amino acids are inserted near the hypervariable region of the C-terminal end of the seed storage protein molecule, the overall structure of the protein is not perturbed. Moreover, the product of these transgenes can be properly packaged during the development of the seed.

Much attention has been focused on storage proteins, which are engineered to provide nourishment to the embryo in the germinating seed. Storage proteins, such as zein in corn, constitute approximately half the total protein content of the grain seed. Because zein lacks one amino acid – lysine – we, in theory, need only to employ site-directed mutagenesis to insert nucleotides containing the lysine codon into the protein-coding region of the zein gene. However, this is not an easy task. First, the additional codon must be inserted in such a way that the reading frame of the zein gene and construction of the protein are not disrupted. Second, the altered zein gene must be introduced into the corn genome and the transgenic plants recovered. The lack of a reproducible means for transforming and regenerating corn plants makes this a serious obstacle. Finally, accurate tissue-specific expression of the transgene must be achieved.

2. Manipulation of starch biosynthesis

Starch biosynthesis is the major determiner of yield in many cereal (e.g. corn and rice) and tuber crops (e.g. potato and cassava). The recombinant DNA technique has been used to manipulate both the quality and quantity of starch in potato. Starch is composed of two types of polysaccharides—amylose and amylopectin. The first attempt at manipulation of starch biosynthesis was directed toward reducing the amylose synthesis. This was achieved by using antisense RNA against the starch synthase gene (responsible for amylose synthesis) using the CaMV 35S promoter. The tubers of transgenic potato plants were devoid of amylose and contained only amylopectin.

An increase in the starch content of potato has been achieved recently. Starch biosynthesis in plants is regulated at a step catalyzed by ADP-glucose pyrophosphorylase, which is allosterically regulated by several metabolites, including phosphate. *E. coli* ADP-glucose phyrophorylase was mutated to alter its allosteric properties and was expressed in potato plants with a transit peptide sequence that would permit targeting of the E. coli enzyme to the stroma of amyloplasts. These transgenic plants displayed increased starch level and a higher dry matter content in the tubers (25% more).

3. Genetic engineering of plant oils

Vegetable oils provide a vast renewable source of oils. Almost two-thirds of the vegetable oil produced are used as food and the remaining for a variety of industrial purposes. Plant oil production in the world amounts to 60 million tonnes, with a value of about 20 billion US dollars. Considering the high value of oil producing crops and the vast market for them, enormous efforts are being made to increase the yield of these crops and to alter or improve the quality of the oils by structural modification, both by conventional plant breeding and by recombinant DNA technology. A number of genetic engineering programs have been developed throughout the world to improve oil quality. They involve (i) To decrease of linoleic acid content in soybean oil for increased stability and less tendency to spoil; (ii) development of a temperate-zone, annual crop as source of lauric acid. This helps to grow lauric acid crop in temperate zones and provides stable supplies at lower prices for the soap and detergent industry; (iii) development of high palmitic lines of rape seed to provide plant oils at low price to industries; (iv) development of high palmitic lines of rape seed for increased use of canola oil in margarines; (v) decrease of levels of unsaturated fatty acids in cocoa seeds to prevent melting of chocolates in warm weather; (vi) increase of oil yield to sell at low price and to substitute it for diesel. Oils are synthesized in various compartments of the plant cells and stored in lipid bodies. The cell compartments involved in oil biosynthesis are the proplastid, cytoplasm, and endoplasmic reticulum (Figure 22.3).

Modification of fatty acid composition by conventional plant breeding: For the first time, the modification of the fatty acid composition by plant breeding was reported by Canadian plant breeders in rape seed (*Brassica rapa*) variety with reduced levels of erusic acid. This helped to market rape-seed oil as edible "Canola oil". Conventional breeding also resulted in improving sunflower with a high oleic acid content (90% of seed oil). This is used for industrial purposes. Since many desirable traits such as high erusic acid content (>90%) and synthesis of medium chain fatty acids are found in plants that cannot be crossed with rape-seed, the next logical step is to overcome the limitation of conventional plant breeding by manipulation of plant oil biosynthesis by genetic engineering.

Modification of plant oil biosynthesis by genetic engineering: Because of the complexity of plant oil biosynthesis, genetic engineering of plant oils necessitates careful planning for gene manipulation. One should remember that all plant cells need lipids as constituents of membranes. As a result, any disturbance in fatty acid composition or concentration in non-seed cells may drastically affect the normal physiology of these cells. The modified gene should always be expressed only in the seed comportment, where oils are synthesized and stored. Medium chain fatty acids (C8 to C14) such as lauric acid (C12) are in high demand in the detergent industry. The natural sources of these fatty acids are the oils of palm kernel and coconut. A gene responsible for the synthesis of medium-chain fatty acids has been isolated from an unconventional oil plant, California bay (*Umbellularia californica*), which accumulates high levels of caproate (C10) and laurate (C12). The gene for the C-12 acyl carrier protein thioesterase (BTE) also coded for an N-terminal transit peptide sequence for plastid targeting. For seed comport-

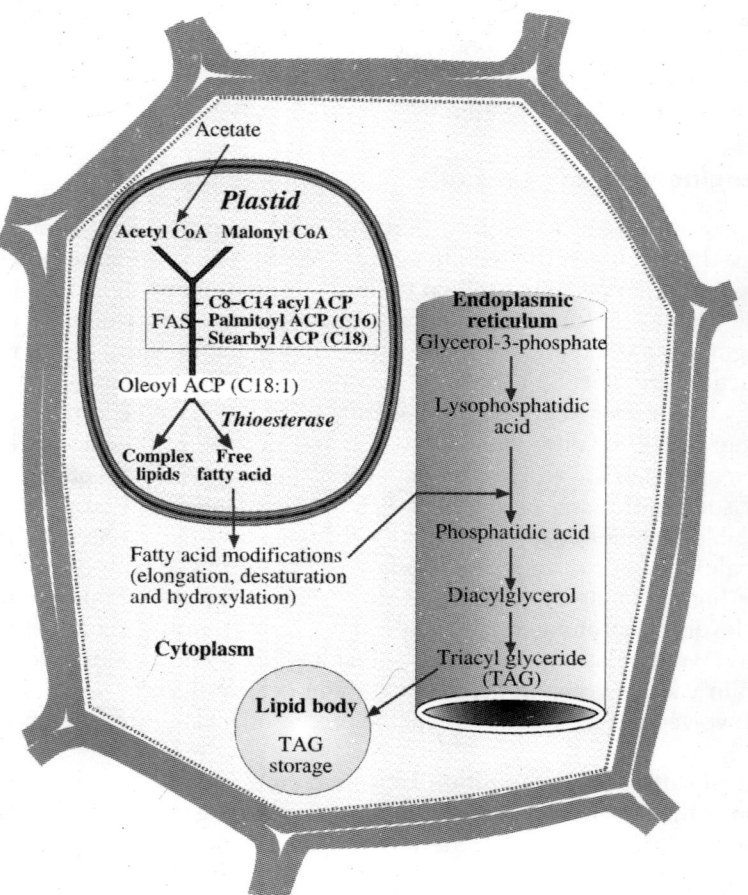

Figure 22.3 Biosynthesis and storage of triacylglycerides in plants (FAS=fatty acyl synthase; ACP=acyl carrier protein, TAG=triacylglyceride

ment specific expression, the DNA construct containing the BTE gene was fused to a seed storage protein (napin) promoter of *B. rapa* and introduced into a model plant *Arabidopsis* using *Agrobacterium* vectors. The transgenic plants showed expression of BTE activity and accumulated high levels of medium chain fatty acids. These results show the feasibility of altering the fatty acid chain length.

A second approach has been reported recently, in which a qualitative change in fatty acid synthesis in *B. rapa* and *B. napus* was achieved by blocking a step in the fatty acid biosynthesis pathway using the antisense RNA approach. The main objective in this case was to reduce the conversion of stearate (C18:0) to unsaturated fatty acids such as oleic acid (C18:1), linoleic acid (C18:2) and linolenic acid (C18:3). The first desaturation step, conversion from stearoyl-ACP to oleoyl-ACP, is catalyzed by the enzyme stearoyl-ACP desaturate. A gene for stearoyl-ACP denaturase was isolated from *B. rapa* and the complementary sequence cloned in reverse orientation under the control of two seed-specific promoters, namely napin and acyl carrier protein (ACP) of *Brassica*. The two seed-specific promoters were used to enhance antisense RNA transcription and thereby ensure the presence of antisense RNA throughout seed development. The antisense RNA transcribed in the seeds is thought to hybridize with the sense RNA coding for stearoyl-ACP desaturase thereby inhibiting RNA translation and synthesis of protein. Consequently, antisense RNA genes in *B. napus* expressed about 20 to 25% stearic acid, when compared to 1.3% in control plants, with a concomitant decrease in the activity of the stearoyl-ACP desaturase.

Genetic manipulation of lipids has also been shown to influence very complex traits such as the 'chilling sensitivity' of plants. It had previously been postulated that plants that tolerate chilling had high levels of *cis*-unsaturated fatty acids in the chloroplast-localized phosphatidylglycerol. The chloroplast enzyme glycerol–3–phosphate acyltransferase was considered to be important in promoting phosphatidyl glycerol fatty acid unsaturation. Introduction of this gene into chill-sensitive tobacco plants exhibited high levels of chilling tolerance, suggesting that a single enzyme such as glycerol-3-phosphate acyltransferase could control a complex trait such as chilling tolerance.

G. POST-HARVEST QUALITY IMPROVEMENT

A significant proportion of crop harvested (10–30%) is spoiled during transit and storage. This happens particularly for highly perishable commodities such as vegetables, fruits, and flowers. The introduction of antisense constructs of pigmentation genes into a fully-colored Petunia has been used to develop variants with reduced or erratic pigmentation or white flowers. The same antisense constructs were also effective in *Solanum* and *Nicotiana* sp. Antisense RNA constructs have been used to block the replication of tomato golden mosaic virus in tobacco plants. The resulting transgenic plants were much more resistant to infection by the virus than were the controls.

During transcription of a gene, only one strand of DNA (sense strand) is copied into mRNA. The location of the promoter 5' to the gene ensures that the mRNA is produced from the correct strand. If the gene is positioned in the opposite orientation relative to the promoter, the 'antisense' strand is transcribed and RNA is made that is complementary to the mRNA is made (Figure 22.4). This molecule is termed antisense RNA. Although antisense RNA has been shown to work in many cases, it is not clear how exactly it functions. Some of the mechanisms proposed are: (i) duplex formation with the DNA template to block transcription; (ii) blocking intron splicing; (iii) failure of the antisense mRNA duplex to be transported to the cytoplasm; (iv) promoting rapid RNA degradation; and (v) blocking initiation of translation. In practice, more than one of these mechanisms may be operating simultaneously.

1. Genetic engineering for extended shelf-life of fruits

Delaying fruit ripening has become an important commercial aspect of crop improvement by plant genetic engineering. Delay of fruit ripening extends shelf-life and the keeping quality of fruits, thus enabling long distance shipping, with minimum loss. Gene manipulation for delay in ripening has been successfully exploited in tomatoes, since a great deal of physiological, biochemical, and molecular biology work has already been done.

The plant growth regulator, ethylene, is involved in fruit ripening and flower senescence. The ethylene biosynthetic pathway in plants is well understood and key genes coding for

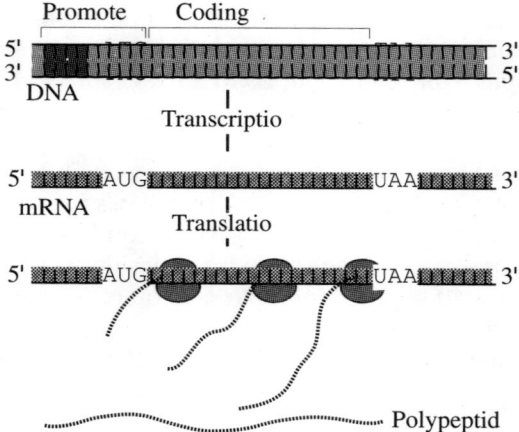
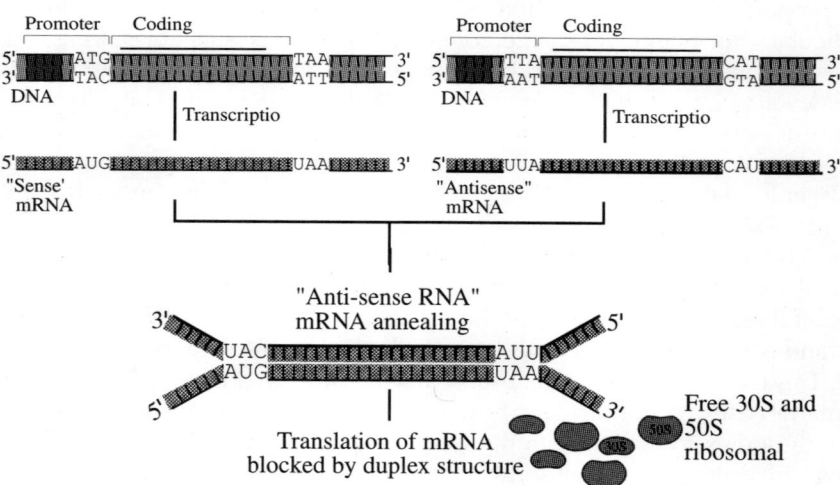

Figure 22.4 *Schematic diagram of the use of the "anti-sense RNA" technique to block the expression of a specific gene. (a) A normal cell showing the production of the polypeptide gene product by transcription and translation. (b) A cell containing the homologous anti-sense gene in addition to the normal sense gene*

enzymes such as ACC (1-aminocyclopropane-1-carboxylic acid) synthase and ACC-oxidase have been cloned from several plants. By using antisense technology, the production of ethylene is drastically reduced. This leads to delayed ripening or senescence. Since ethylene plays an important role in plant development, the suppression of ethylene has to be restricted to the desired organ by using tissue specific promoters. For example, Oeller *et al.* (1991) engineered tomato for delayed ripening through antisense suppression of both ACC oxidase and ACC synthase genes. These tomatoes remain fresh after harvest and ripen slowly, thereby allowing easy transport and ensuring long shelf-life. This technology has also been extended to other crops such as apple, banana, melon, peach, and pineapple. Other strategies to achieve similar effects include diverting ethylene precursors to other pathways using bacterial ACC-deaminase or viral-S-adenosyl methionine hydrolase.

Fruit ripening is a complex phenomenon. It also involves modification of cellular structures and carbohydrate and sugar compositions. The enzyme polygalacturonase, which breaks down plant cell walls, is primarily responsible for fruit softening. Two research groups have found that transformation of an "antisense" copy of the polygalcturonase gene in tomato plants decreases expression of the softening protein by 90%. To make the antisense gene, a cDNA clone of the polygacturonase gene was fused, *in reverse orientation*, to a constitutive promoter. This reverse, or antisense, gene was linked to T-DNA and introduced into tomato plants. Each transformed plant thus carried an antisense gene, as well as a normal copy of the polygalcturonase gene. During gene transcription, antisense messenger RNA (mRNA) molecules are produced that are complementary to normal mRNA molecules. Presumably, the antisense mRNAs bind to a proportion of normal RNAs, making them unavailable for translation into protein. The first commercial transgenic crop variety the FLAVR SAVR™ tomato was engineered for antisense suppression of polygalacturonase. Thus by combining the above two strategies, long shelf life has been engineered into melons. Other side benefits of this approach were that the fruits could be left longer on the plants, allowing more photosynthates to be translocated. This increased fruit solid content. This technique also helped to reduce the rotting of the fruits.

Long vase-life flowers have been produced in a similar way by co-suppression (both native and transgenes are suppressed due to gene silencing) or antisense suppression of ethylene production. Delaying senescence is also being attempted by increasing cytokinin levels in a tissue specific manner.

2. Genetic manipulation of flower pigmentation

Plants are widely used for ornamental purposes. The flower industry is a billion-dollar business, which is continually attempting to improve flower appearance and shelf life. Largely out of a desire to understand the basis of colors produced in a variety of plant parts, anthocyanin chemistry and genetics have been studied for many decades. Pigmentation of flowers (creamy yellow to bright red, from pale pink to blue and violet) is mainly due to three classes of compound: the flavonoids, the carotenoids and the betalains. Of these, the flavonoids are the best characterized with much information now available concerning chemistry, biochemistry, and molecular genetics. Uniquely colored flowers can be developed by manipulation of genes for enzymes in the flavonoid biosynthesis pathway. Although earlier the emphasis was on genetic engineering of anthocyanins per se, the focus is now broader, to include the engineering of any factor that can impact flower color produced by anthocyanins. For example, the factors determining the flower color produced with anthocyanins include the anthocyanin base, its subsequent chemical modification, interaction with other flavonoids and interaction with ions, particularly protons.

Why use genetic engineering to modulate flower color when such variety already exists? The reasons are two-fold: (1) to introduce into specific ornamental crops a color type not available naturally, and (2) to modify flower color in a specific cultivar which has a commercially valuable combination of agronomic and consumer characteristics. To understand the methods available to plant molecular geneticists for manipulation of flower color, it is necessary to understand the biochemistry and genetics of pathways for anthocyanin biosynthesis as obtained in several plants.

Biosynthesis pathway of anthocyanin: The principal reactions of anthocyanin biosynthesis, is elucidated in *Petunia hybrida, Antrrhinum majus,* and *Zea mays,* with reference to the production of one anthocyanin, ***pelargonidin-3-glucoside***. In this example, the pathway is considered as three distinct blocks which can produce overlapping sets of specialized products. The first four steps in the pathway, block 1, are common to a range of different secondary metabolites, and they comprise the core phenylpropanoid biosynthetic pathway. Chalcone synthase is the first step in flavonoid biosynthesis; it initiates the second block of the overall pathway. This portion of the pathway terminates with the synthesis of dihydrokaempferol. A range of side reactions and flavonols can interact with anthocyanins to modify flower color. The final portion of the pathway specific to anthocyanin biosynthesis, consists of a series of four reactions, and terminates with the addition of glucose to the 3-hydroxyl of an anthocyanidin. With the addition of glucose to the 3-hydroxyl group, the first anthocyanin is produced, as the glycosylation of anthocynidins is essential for stable expression of color. It should be noted that the anthocyanins accumulate in vacuoles, although most biosynthetic reactions occur in the cytosol.

For determining the final color expressed by an anthocyanin, three additional sets of reactions are important. First, additional hydroxyl groups can be added to the 3' or 5' positions of the anthocyanidin B-ring. The addition of hydroxyl groups causes a shift in the maximum amount of absorption, from 520 nm for pelargonidin (monohydroxylated) to 535 nm for cyanidin (3'4'-hydroxylated) to 546 nm for delphinidin (3',4'5'-trihydroxylated). Flavonoids can also be synthesized with the same set of different hydroxylations, resulting in kaempferol, quercetin, or myricetin, generally at the 3 and/or 5 positions. The glycosylations and acylations can alter color as well, as exemplified by the difference in flower color observed in the rt and gf mutants of *Petunia hybrida*.

A number of the genes encoding enzymes of the biosynthetic pathway, or encoding proteins which regulate the expression of these structural genes, have been identified now. Creating plants that produce flowers of reduced color intensity requires a disruption in a step of the biosynthetic pathway. If the goal is to obtain varieties producing a range of flower colors, then the disrupted step could be virtually any of the steps of the anthocyanin biosynthetic pathway which are general with respect to late modifications. These steps are chalcone synthase (CHS), flavanone-3-hydroxylase (F3H), and dihydroflavonol reductase (DFR).

Antisense technology has been used to reduce the expression of endogenous genes with a sequence related to the introduced gene. In one series of experiments, petunias were transformed with antisense RNA-producing versions of one or more of the genes of the pathway. A number of plants were created with unique colors and color patterns. Van der Krol et al. (1988) reported that the introduction of a full-length antisense CHS gene construct resulted in suppression of anthocyanin biosynthesis. A range of phenotypes was obtained, including flowers with chaotic white sectors, flowers with uniformly reduced pigmentation, and flowers with nearly white color throughout the corolla. In these experiments, expression of the full length coding sequence was programmed for expression in the antisense orientation by the 35S promoter from cauliflower mosaic virus, a strong constitutive promoter. The alteration in flower pigmentation was attributed to a reduction in chalcone synthase gene expression. The CHS protein and mRNA were reduced in white corolla tissue, which lacked all flavonoids normally present. This demonstrates that pigment synthesis could be restored by supplementing the product of the CHS reaction with naringenin-chalcone.

H. FERTILIZER MANAGEMENT

1. Nitrogen fixation

All living things, including plants, require a source of nitrogen to synthesize essential carbon-containing molecules, such as amino acids, nucleic acids, proteins, and vitamins. Despite the fact that the atmosphere around us consists of approximately 70% nitrogen, this requirement presents a big problem: nitrogen in its atmospheric form is not usable by animals, plants, fungi, or the vast majority of bacterial strains. Before it can be used in essential growth processes, atmospheric nitrogen must be converted into appropriate nitrogen containing compounds, usually by the addition of either hydrogen or oxygen. These modified nitrogen compounds can, in turn, be used in various biosynthetic processes. This process of conversion called nitrogen fixation – is carried out only by a limited number of bacterial species, which then share fixed nitrogen with nitrogen-deficient.

The bacterial species that can fix nitrogen include **Rhizobium**, often associated with the roots of legume plants such as beans and peas, **Azospirullum**, associated with the roots of grasses such as rye and wheat and **Frankia**, often associated with the roots of some shrubs and trees. It is interesting to note that despite the absolute requirement of nitrogen fixation for successful growth, plants depend on associated bacteria to carry out this task. Although other bacteria also fix nitrogen, *Rhizobium* shares fixed nitrogen with its associated plants most efficiently.

In most cases, the nitrogen fixing bacteria enter root cells of a target plant and multiply. Eventually, a lump or a **nodule** forms at the site of bacterial entry. Symbiosis begins with attachment of rhizobia in the soil to the root hair of a germinating legume. Because each species of *Rhizobium* has a narrow host range, this initial attachment is thought to depend on a cell recognition event between surface molecules on the bacteria and the plant. Following attachment, the bacterium is encapsulated within a pocket of the plant cell wall and carried deep inside the root by formation of an infection thread consisting primarily of cellulose. Cortical cells on the outside of the root dedifferentiate to form a growing meristem that bulges out into a nodule. Proliferating infection threads then release the bacteria into the cytoplasm of the host plant cells, where the bacteria known as **bacteroids** are covered with a special membrane synthesized by the host. Within this extremely protective environment, the bacteria finally differentiate into the nitrogen-fixing state by derepressing the genes for nitrogenase and bacteroid-specific cytochromes. At this point, the bacteria make use of a special set of enzymes known as the **nitrogenase complex**. This complex involves six proteins and the activities of two distinct enzymes, one of which is simply called nitrogenase while the other is called **nitrogenase reductase**. These proteins and enzymes work together to satisfy the nitrogen requirements of the growing plant by carrying out the complex process of nitrogen fixation within the root nodules of the plant.

In the rhizosphere of many crops, despite the presence of nitrogen fixing bacteria, the growth of crops such as corn, wheat, and rice quickly depletes the useful supply of nitrogenous compounds available in the soil, thus limiting the growth of the plants themselves. In order to continue to grow these crops, an exogenous source of nitrogen compounds must be added to the soil, usually by adding generous amounts of nitrogen containing fertilizers. Although this practice allows continued crop growth, it also create problems. Having to add nitrogen fertilizer to the costs, and these expenses cannot always be met, especially in poor and developing countries. The application of chemical based fertilizers to growth areas can contaminate water supplies as some of the fertilizer gets washed away during irrigation. Thus adversely affect supplies of drinking water and decrease aquatic animal populations as the growth of algae and other plants is encouraged by the presence of excess fertilizer derived nitrogen in lakes and rivers.

For these and other reasons, genetic engineers would like to increase supplies of fixed nitrogen, making them available to a wide variety of plants. This type of research takes many routes and is more promising than other types. In one avenue of examination, scientists would like to modify the factors, which cause host-restriction in the relationship between *Rhizobium* and legume plants. This may take place either by altering the genes of the *Rhizobium* itself or by altering the genes of the associated plant. Although research continues in these areas, it is hampered by the extreme complexity of the genetic systems involved in the symbiotic relationship between the plant and the bacteria, which provide fixed nitrogen. In a second avenue of examination, researchers are attempting to determine why some bacteria, while capable of fixing nitrogen, do not share this valuable product with their associated plants. If this characteristic results from a specific gene function, genetic engineers should be able to identify, locate, and modify this gene or gene family and its controlling mechanisms. Although these and other research questions are being examined in the area of nitrogen fixation, answers and solutions to the problems depend on the ability of researchers to unravel the extremely complex processes of nitrogen fixation and on the ability of any genetic modifications to survive and produce in the field.

Cloning the symbiosis genes of rhizobia: The bacterial genes required for formation of functional symbiotic nitrogen-fixing nodules are known as **sym** genes. The *sym* genes have been provisionally divided into three classes, depending on which step in the nodulation process appears to be blocked by mutation. The **nod** genes are required for initiation and the early steps of nodule formation. The **fix** genes are required for later steps in the construction of a nodule capable of nitrogen fixation. And finally, the **nif** genes encode the three highly conserved subunits of the *nitrogenase* (which reduces molecular nitrogen to ammonia) as well as auxiliary proteins required for the nitrogenase to function. The *sym* genes can be identified and cloned using the technique of **transposon tagging**.

I. ENGINEERING MODIFIED REPRODUCTIVE SYSTEMS

1. Male sterility with suicide gene

Another practical application of genetic engineering is in the production of male sterile plants needed for the production of F1 hybrids. Hybrid plants are usually superior in numerous respects, including higher yield and increased disease resistance. Cytoplasmic male sterility is widely used in corn breeding, but analogous mutations are not available in many crop plants. A genetically engineered system of male sterility and fertility restoration has been introduced into a number of plant species, including the oilseed rape, *Brassica napus*. The genetic engineering of male sterility and fertility restoration makes use of two engineered genes. The basis of the sterility is an extracellular RNA nuclease called **barnase**, which is produced by the bacterium *Bacillus amyloliquefaciens*. The *barnase* nuclease is an extremely potent cellular toxin. In the use of *barnase* to produce male sterility, the coding sequence of the bacterial gene is fused with the regulatory sequence of a gene, *TA29*, that has a tissue-specific expression in the tapetal cell layer that surrounds the pollen sacs in the anther (Figure 22.5). When the artificial

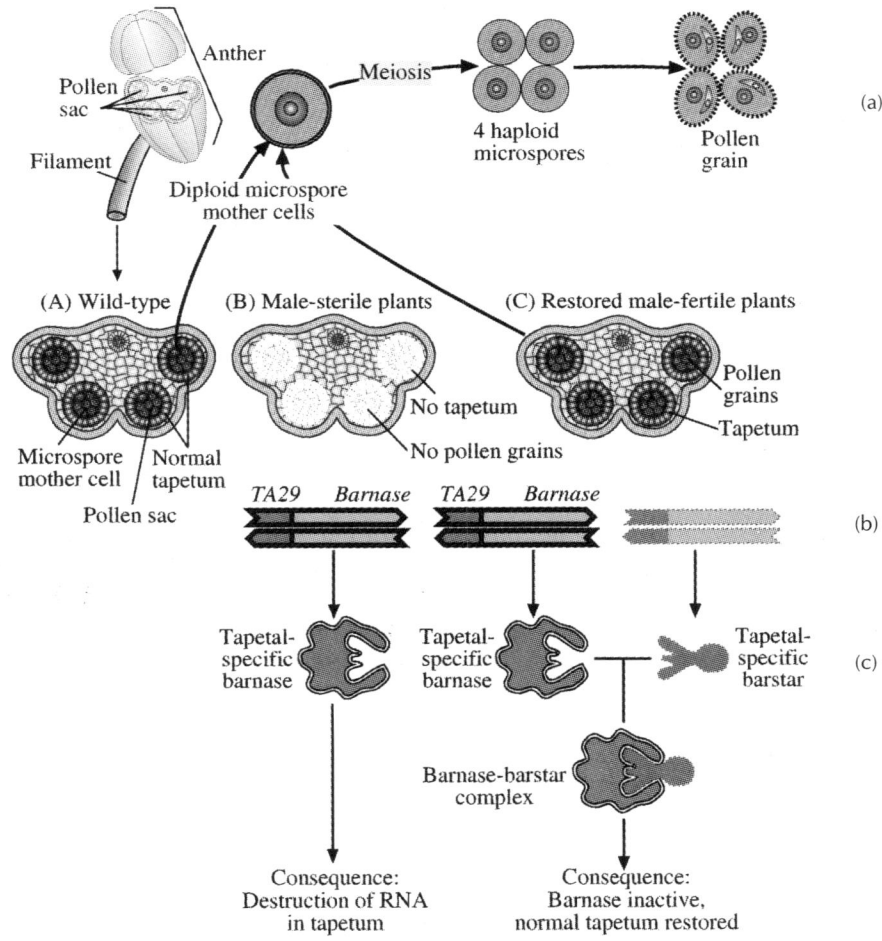

Figure 22.5 *Engineered genetic male sterility and fertility restoration (a) Pollen grains develop within a thin layer of tapetal cells, (b) Expression of the barnase RNA nuclease under the control of the tapetum-specific regulatory region of gene 'T'A29 destroys the tapetum and renders the plants male-sterile, (c) Expression of barstar (a barnase RNA inhibitor) in tapetal cells inactivates burnase and restores fertility*

TA29-barnase gene is transformed into the genome, the resulting plants are male-sterile because of the destruction of tapetal cell RNA by the barnase enzyme and the resulting lethality of cells in the tapetum. Fertility restoration makes use of another protein, called *barstar*, also produced by *B. amyloliquefaciens*. Barstar is an intracellular protein that offers protection against the lethal effects of barnase by forming a stable, enzymatically inactive complex in the cytoplasm. Plants transformed with an artificial *TA29-barstar* gene are healthy and fertile; they merely produce *barstar* in the tapetum because of the tapetum-specific expression of the TA29 regulatory region. However, when the male-sterile *TA29-barnase* plants are crossed with those carrying *TA29-barstar*, the resulting genotype that combines *TA-barnase* with *TA-barstar* is male-fertile (Figure 22.5). The reason for the restoration of fertility is that the *barstar* protein combines with the *barnase* nuclease and renders it ineffective. In the seed production program, a herbicide resistant gene *bar* conferring resistance to phosphinothricin was linked to the *barnase* gene. This allowed selective elimination, through herbicide spray, of male fertile progeny (50%) resulting from pollination of hemizygous male-sterile lines (*barnase*) with non-transgenic maintainer lines (–/–). The remaining (50%) male sterile lines could be then used for F1 hybrid seed production. This system is currently being commercialized by Plant Genetic Systems (PGS).

Similar approaches to selective cell ablation to achieve male sterility have been reported. These include expression of the diphtheria toxin gene, plant growth regulator gene *rol*B from *Agrobacterium*, endoglucanase gene, bacterial gene for breaking down non-toxic chemicals into phyto-toxic compounds and antisense inhibition of flavonoid biosynthesis.

2. Inducing parthenocarpy

Seedlessness is a desirable feature in many vegetable and fruit crops. Traditionally, such varieties have been developed through cytogenetic manipulations or by using natural genetic variations. Seedlessness can also be induced (for example, in grapes) by spraying gibberrellic acid on young fruits. Fruit development in many cases is triggered by auxin produced by the growing embryos. To achieve fruit development, in the absence of seed development (i.e. without pollination) the bacterial gene *iaa*M from *Pseudomonas* for auxin synthesis was expressed in developing ovules. This permitted fruit development to proceed normally without pollination. However, upon pollination, the normal seed set could be obtained. Thus by combining male sterility and parthenocarpy, seedless fruit and vegetable varieties can be developed.

J. DEVELOPING STRESS- AND SENESCENCE-TOLERANT PLANTS

Plants have evolved physiological strategies to cope with adverse environmental conditions such as intense light, herbicides, ozone, heat, chilling, freezing, drought, salinity, flooding, heavy metals, etc. Many genes, which involve in biotic and abiotic stress-tolerance have been identified. Transgenic plants, which overexpress for some of these abiotic stresses, have already been tested for stress tolerance (Table 22.2).

At the molecular level, one of the undesirable consequences of physiological stress is the production of oxygen radicals. Many, if not all forms of environmental stress (abiotic and biotic) result in oxidative stress. This may occur mostly due to exposure to ozone pollution or ionizing radiation, pathogen attack, water deficit stress, etc. The oxidative stress arises from the production of free radicals and the subsequent ensuing cascade of reactions. Common types of potentially damaging oxygen radicals (reactive oxygen species, ROS) are the superoxide anion, hydrogen peroxide and the hydroxyl-radical species. The ROS cause damage as a result of their reactions with cellular macromolecules (membranes, proteins DNA, etc.). Plants contain a number of enzymes that catalyze the cascade of ROS and convert them to less reactive products. Biologically, the enzyme superoxide dismutase detoxifies superoxide anion by converting it to hydrogen peroxide, which, in turn, is broken down to water by any of a number of cellular peroxidases or catalases.

Two general strategies have been adapted for engineering tolerance to oxidative stresses: one, increase the level of enzymes that remove ROS or increase the level of antioxidant compounds that react with ROS. In one experiment, tobacco plants transgenic for superoxide dismutase gene was produced that under the control of the 35S promoter from cauliflower mosaic virus expressed the enzyme superoxide dismutase and also acquired tolerance to oxygen radical damage. An additional potential benefit of increasing the superoxide dismutase level in plants is an increased resistance to the herbicide methyl viologen (paraquat) and a higher tolerance to light stress. A superoxide dismutase gene might also be used to maintain the quality of cut flowers during shipping. If one could increase endogenous oxygen-scavenging capacity in cut flowers then their shelf life might increase as well. Alternatively, expression of three enzymes – ascorbate peroxidase, glutathione peroxidase and glutathione reductase – in *Arabidopsis* and tobacco plants have been shown to have some effect on tolerance to various abiotic stresses.

Table 22.2 *Transgenics produced for improved drought and salt tolerance*

Transgenic Over-expressing	Crop plants	Tolerance to
Mannitol	Tobacco	Tolerance to salinity
Sorbitol	Tobacco, persimmon	Oxidative stress, salt
Trehalose	Tobacco, potato	Drought
Glycine-betaine	Tobacco chloroplast	Marker for osmoprotectant
Proline	Rice, tobacco, barley, soybean	Stress protein, drought resistance
Fructans	Tobacco, sugar beet	Resistance to drought stress
Osmotin	Tobacco	Drought, salt stress
LEA	Rice	Tolerance to water deficit and salinity
Glutamine	Rice	Salt, chilling
D-Ononital	Tobacco	Drought, salt stress

Different abiotic stresses create a water deficit

Water stress, in plants are created by various environmental conditions (heat, chilling, freezing, drought and high salt concentrations). Water deficit is a particular problem to plant cells because it inhibits photosynthesis, increases the concentration of toxic ions and causes the loss of the protective hydration 'shell' around vulnerable molecules. Plant cells respond to osmotic stress by producing non-toxic compounds called 'compatible solutes' ('osmolytes' or 'osmoprotectants') to reduce the osmotic potential. These osmolytes compounds can accumulate intracellularly in higher concentrations without disrupting the hydration shell around proteins and membranes (Figure 22.6). Plants produce a range of compatible solutes, which fall into two categories: sugars and sugar alcohols and zwitterionic compounds. Some of the sugar alcohols are mannitol, sorbitol, pinitol and D-ononitol, and oligosaccharides such as trehalose and fructose. The zwitterionic compounds include amino acids such as proline, and quaternary ammonium compounds such as glycine betaine. Basic strategies for engineering resistance to water-deficit stress have therefore focused on the production of osmoprotectants as a mechanism for overcoming the osmotic stress. These strategies require the determination of the biosynthetic pathways for various osmoprotectants, isolation of the relevant genes and appropriate engineering of constructs to target gene expression and protein destination.

A strategy used by plants to tackle salt stress is avoiding the accumulation of sodium ions in the cytoplasm. Thus improvement of transportation of ions out of the cytoplasm into the vacuole may provide plants resistance to salt stress. The vacuolar Na/H antiport protein AtNHX1 of

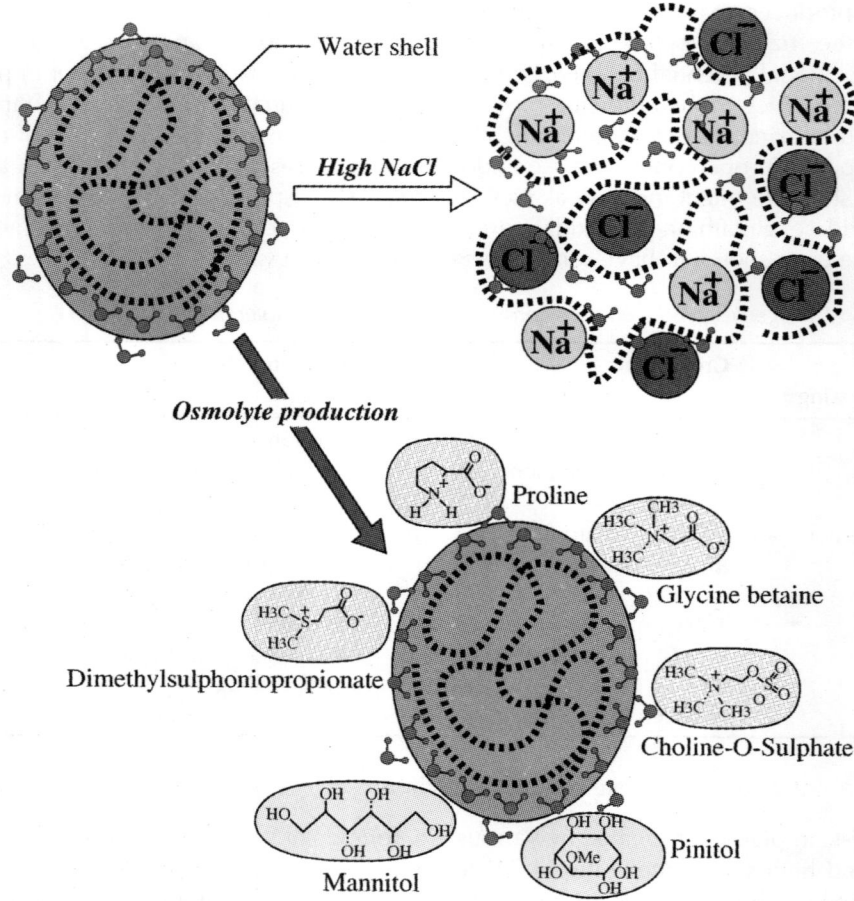

Figure 22.6 *The water shell around macromolecules. The water shell around proteins is disrupted when the concentration of ions (salts) is increased, which leads to the protein denaturation. An equivalent concentration of compatible solutes synthesized during the stress do not disrupt the water shell and the protein is not denatured*

Arabidopsis, has been extensively studied and is known to be coupled to proton pumps such as AVP1, a vacuolar H^+-translocating pyrophosphatase. To increase the ion traffic through the membrane, one could therefore either increase the number of channels, or provide more energy for the existing channels to spin faster. Tomato and canola plants after transformation with the *Arabidopsis* AtNHX1 antiport protein gene became resistance to high-salt environment. Alternatively, overexpression of the gene AVP1, initially in *Arabidopsis*, substantially improved not only the salt tolerance but also the drought tolerance.

Many plants differ enormously in their ability to withstand cold and freezing temperatures (less than −30°C). There are some examples of the expression of cold-induced proteins in transgenic plants. For example, constitutive expression of the small hydrophilic, chloroplast-targeted COR protein COR15a in *Arabidopsis* improved the freezing tolerance. It is well-known that in response to heat stress, many organisms induce a specific set of 'heat shock' proteins (HSPs: HSP100, HSP90, HSP70 and HSP60). These HSPs appear to be involved in counteracting the effects of heat stress by protecting or refolding denatured proteins. When the fusion protein of AtHSF1 and *gus*A reporter gene overexpressed in *Arabidopsis* constitutively, transgenic plants demonstrated enhanced thermotolerance.

Decrease of fruit senescence by genetic engineering

As previously described, the major problem in fruit marketing is premature ripening and softening during transport. These changes are part of the natural aging (senescence) process of the fruit. Some of the genes that are induced during ripening encode the enzyme cellulase and polygalacturonase. It has been hypothesized that by interfering with the expression of one or more of these genes, the ripening process might be delayed. This interference could be achieved by creating transgenic plants with antisense RNA-producing gene for polygalacturonase. In the transgenic tomatoes both polygalacturonase mRNA and enzymatic activity were reduced by 90%. Thus, lowering polygalcturonase production inhibits fruit ripening in tomatoes.

K. MODIFYING THE PROPERTIES OF PLANT PRODUCTS

Over the years, plant breeders have been extremely successful in optimizing the useful properties (e.g. protein or oil content) and increasing the productivity (yield) of a large number of crop plants. However, traditional breeding approaches to crop improvement are both difficult and few, and they are intrinsically limited by existing genetic information of cross-breeding strains. In contrast, the use of genetic engineering techniques allows scientists both to dramatically speed up the process of developing plants with improved characteristics and to introduce traits that would otherwise be impossible to develop by conventional methods. For example, on a laboratory scale, genetic engineering has been used to improve (1) the nutritional quality of several different plants (e.g. corn and pea by modification of the amino acid content of some of their seed storage proteins); (2) the fatty acid composition of both edible and non-edible oil producing crops; and (3) the taste of fruits and vegetables by the introduction of monellin, a sweet tasting protein.

Even though a fruit or vegetable may have high nutritional value, if it is not tasty, people usually will not eat it. Although palatability of food can be achieved by adding many ingredients during preparation, it would be advantageous to the food industry if certain foods could be made intrinsically more appetizing. Monellin, a protein that is found in the fruit of an African plant (*Dioscorephyllum cumminsii* Diels), is approximately 100,000 times sweeter than sucrose on a molar basis. This feature makes monellin a candidate as a sugar substitutive, with the added bonus that, because it is a protein, it would not have the same metabolite impact as sugar. Monellin is a dimer with A-chain of 45 amino acid residues and B-chain of 50 residues, held together by weak non-covalent bonds. Unfortunately, monellin is composed of two separate polypeptide chains. This limits its usefulness as a sweetener because it is readily dissociated – and as a consequence, it loses its sweetness (by heating or exposure to acids). Also the need to clone and express in a coordinated manner two separate genes complicates efforts to produce the protein in either transgenic plants or microorganisms. To overcome this problem, a fusion gene of both A and B chains was chemically synthesized and expressed in tomato and lettuce plants. In the experiment with tomatoes, expression was directed by the tomato fruit ripening-specific promoter E8 and for the lettuce 35S promoter. Each construct used the transcription termination –polyadenylation site from Ti plasmid nopaline synthase gene. Monellin was detected in ripe and partially ripe, but not green tomatoes, and in lettuce leaves. The monellin level in tomatoes could also be elevated by a burst of the plant hormone ethylene. However, comprehensive state tests have not been reported for these genetically sweetened foods.

L. PLANTS AS PRODUCERS OF SPECIALITY CHEMICALS

The advantages of plant genetic engineering is that the crop plants are being modified not only for their traditional economic products but also being considered for elaborating novel compounds. Cultivation of plants is easy and does not require special equipment or chemicals (self-sustainable). Where materials accumulate to high levels in plants, their production costs very little (e.g. production of starch from maize costs about one cent per kilogram). Genetic manipulation in plants is largely free from ethical or moral considerations. Thus there is great interest in using plants as 'factories' for materials, which are normally expensive to produce. Plants have been used to produce bacterial (e.g. bacterial polyhydroxybutyrate, for biodegradable plastic) and mammalian proteins (e.g. interferon, enkephalins, and human serum albumin). Several compounds of pharmaceutical value are now targeted for production in plants. In particular, plants have been shown to be capable of assembling complex antibodies with proper biological activities.

1. Immunotherapeutic drugs

Plantibodies

Antibodies are proteins produced by the vertebrate immune system (refer to Chapter 26 for more details). Transgenic plants are also being looked upon as a source of antibodies which can provide passive immunization by direct application. Hiatt *et al.* (1989) first demonstrated that biologically active antibodies can be produced in plant cells by expressing corresponding antibody genes from animals. Subsequently, a number of reports have shown that a range of whole antibodies, chimeric or antibody fragments can be elaborated in plants (Table 22.3). In many cases, the levels of these proteins in the plants have been surprising (e.g. transgenic plants expressing immunoglobulin genes produced antibody light or heavy chains at levels of more than 1% of soluble leaf protein). Immunotherapy is not employed currently due to lack of adequate supply of antibodies. Therefore, high levels of production in plants may make immunotherapy cost effective in times to come. Ma *et al.* (1995) successfully produced multimeric secretary IgA (SIgA) molecules in plants, which represent the predominant form of immunoglobulin in mucosal secretions (Figure 22.7). A hybrid monoclonal antibody (IgA/G) having constant regions of IgG and IgA fused has been used successfully against human dental caries caused by the bacterium, *Streptococcus mutans*. Interestingly the hybrid antibody exhibited similar affinity to surface adhesion protein of *S. mutans*, but SIgA/G survived longer in the oral cavity than the IgG.

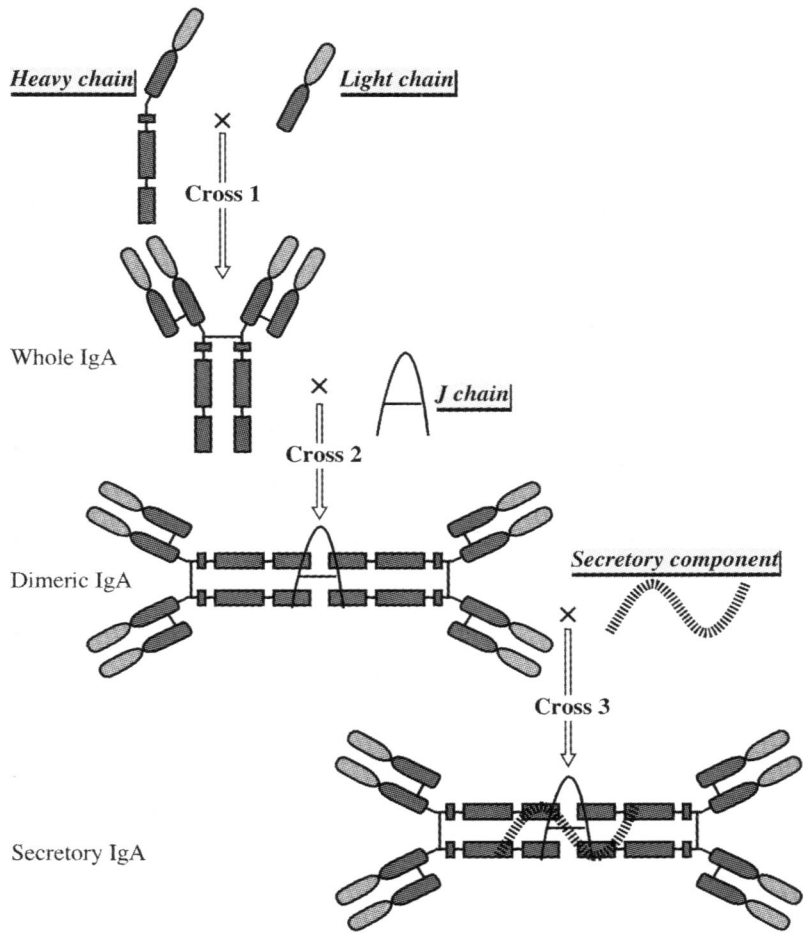

Figure 22.7 *Synthesis of plantibody, a secretory IgA (sIgA) molecule in transgenic plants. Four separate parental plant lines are constructed, each expressing a separate piece of the antibody: the H, L, or J chain, or the secretory component*

2. Edible vaccines

The fine epitope mapping of viruses has led to the identification of short amino acid sequences that are potentially highly immunogenic. Interest in using such sequences as the basis for vaccine development has intensified, and several approaches directed at the mass production of useful epitopes are being investigated. These include chemical synthesis and coupling to carriers, expression in both prokaryotic and eukaryotic (animal and plants) cells as well as subcellular organelles, and insertion into live viruses. Construction of tropical and semi-tropical food plant-based oral immunization methods for microbial enteropathogens could be an inexpensive alternative to conventional fermentation systems for vaccine production. Plants produce large quantities of vaccine, but its purification may prove costly. Therefore attention has to be paid to the production of such antigens that stimulate mucosal immune system for producing secretary IgA (S–IgA) at the mucosal surface.

Table 22.3 *Antigens produced in transgenic plants*

Antibody	Antigen	Plant
IgG (k)	Transition stage analogy	Tobacco
IgM (?)	NP (4-hydroxy-3-nitrophenyl) acetyl hapten	Tobacco
Single domain (dAb)	Substance P	Tobacco
Single chain Fv	Phytochrome	Tobacco
IgG	Glycoprotein B of herpes simplex virus	Soybean
Fab:IgG (k)	Fungal cutinase	Tobacco
IgG (k) and SIgG/A hybrid	*Streptococcus mutans*	Tobacco
Single chain Fv	Abscisic acid	Tobacco

Source: Dubey (1993)

Vaccination involves introduction of antigenic substances into the body or blood stream to stimulate production of antibodies that enable the body to fight infectious agents in subsequent infections. Although there are many examples of oral vaccines (e.g. polio), they are less commonly employed because higher quantities of antigen are required. If antigens can be produced in plant cells (particularly in edible parts) most of the problems of production, storage, and transportation could be minimized, thereby expanding the vaccination program to the poor. Vaccines from plants have many other advantages. Often vaccines purified from animal cells are contaminated with deadly viral pathogens. Since they are of animal origin, there is a high probability of cross-contamination during vaccination. Plant-based vaccines on the other hand, are safe and easy to administer. Plants can be grown in a natural way without any special conditions (indigenous technology). Plant based vaccines can be produced *in vitro* (tissue culture) or in the field itself as a normal field crop. Mason et al. (1992) first reported the production of Hepatitis B surface (HBsAg) antigen in tobacco. These plant-derived antigens resembled the commercially available Recombivax (R) vaccine produced from yeast and were capable of eliciting an immunogenic response. Similarly, antigens of cholera-toxin B subunit of Vibrio cholera and heat-labile enterotoxin B subunit of *E. coli* have been produced in plants (Table 22.4). Also, an antigen for another major agent of gastroenteritis, the Norwalk virus, has been expressed in tobacco and potato. They showed that by producing antigens in edible parts (e.g. potato tubers) and administering it directly as food, immunogenic response could be elicited in mice. Thus plants are now considered choice candidates for production of pharmaceutical compounds under what is called *'biopharming'* technology.

3. Edible interferons

Scientists at ICGEB have successfully produced transgenic tobacco and maize plants that secrete the human interferon (IFN–I). It was produced by transformation of nucleus and chloroplast through the particle bombardment method. It has been found that interferon was 10–15 fold greater in chloroplast-transformed plants than nucleus-transformed plants.

Table 22.4 *Antigens produced in transgenic plants*

Proteins	Plants
Surface protein gene spaA	Tobacco
Hepatitis B surface antigen	Tobacco
Rabies virus glycoprotein	Tomato
Malarial epitope	Tobacco
Norwalk virus capsid protein	Tobacco
E. coli heat-labile enterotoxin ß-subunit	Potato
Cholera toxin ß-subunit	Potato, tobacco
Mouse glutamate dehydrogenase	Potato
VP1 protein of Foot and mouth disease virus	*Arabdiospsis*
Insulin	Potato
Rhinovirus 14 epitope	Cow pea
HIV-1 epitope	Cow pea
Glycoprotein of swine–transmissible gastroenteritis coronavirus	*Arabdiopsis*

III SUGGESTED READING

Anderson EJ, Stark DM, Nelson RS, Tumer NE and Beachy RN (1998) Transgenic Plants that Express the Coat Protein Gene of TMV or AlMV Interfere with Disease Development of Non–related Viruses, *Phytopathology*, **12**:1284–1290.

Arakawa T, Chong DKX and Langridge WHR (1998) Efficacy of Food Plant-based Oral Cholera Toxin B Subunit Vaccine, *Nature Biotech*, **16**:292–298.

Ausubel FM (1984) Developmental Genetics of the Rhizobium/Legume Symbiosis In *Microbial Development*, (Ed) R., Losick and L., Shapiro, Cold Spring Harbor, NY, Cold Spring Harbor Laboratory, pp 275–298.

Brown TA (2001) *Gene Cloning and DNA Analysis* (4th Edn), Blackwell Science Ltd.

Chopra VL (Editor) (2000) *Plant Breeding Theory and Practices*, Oxford & IBH Publishing Co. Pvt. Ltd.

Dubey RC (1993) *A Text Book of Biotechnology*, S Chand & Company Ltd.

Glick BR and Pasternak JJ (1996) *Molecular Biotechnology*, Panima Publishing Corporation, Bangalore, New Delhi.

Grierson D (Ed.) (1992) *Plant Genetic Engineering*, Blackie, Glasgow.

Grierson D and Covey S (1988) *Plant Molecular Biology* (2nd Edn), Blackie, London.

Hiatt AC Cafferkey R and Bowdish K (1989) Production of Antibodies in Transgenic Plants, *Nature*, **342**:76–78.

Hooykaas PJJ and Schilperoort RA (1992) *Plant Molecular Biology*, **19**:15–38.

Khanna-Chopra R and Sinha SK (1998) Prospects of Success of Biotechnological Approaches for Improving Tolerance to Drought Stress in Crop Plants, *Current Science*, **74**:25–34.

Kumria R, Verma R and Rajam MV (1998) Potential Applications of Antisense RNA Technology in Plants, *Current Science*, **74**:35–41.

Lindsey K and Jones MGK (1990) *Plant Biotechnology in Agriculture*, Prentice Hall, New Jersey.

Ma JKC, Hiatt A, Hein M, Vine ND, Wang F, Stabila P, van Dolleweerd C, Mostov K and Lehnew T (1995) Generation and Assembly of Secretary Antibodies in Plants, *Science*, **268**:716–719.

Mason HS and Arntzen CJ (1995) Transgenic Plants as Vaccine Production Systems, *Trends Biotechnol.*, **13**:388–392.

Mulligan JT and SR Long (1985) "Induction of *Rhizobium meliloti* nodC Expression by Plant Exudate Requires nodD," *Proc. Nat. Acad. Sci.*, **82**:6609–6613.

Oeller PW, Min-Wang L, Taylor LP, Pike DA and Theologis A (1991) Reversible Inhibition of Tomato Senescence by Antisense RNA, *Science,* **254**:437–439.

Wilmut I, Schnieke AE, McWhir J, Kind AJ, Cambell KHS (1997) Viable Offspring Derived from Fetal and Adult Mammalian Cells, *Nature,* **385**:810–813.

PART – VI
Working with Proteins

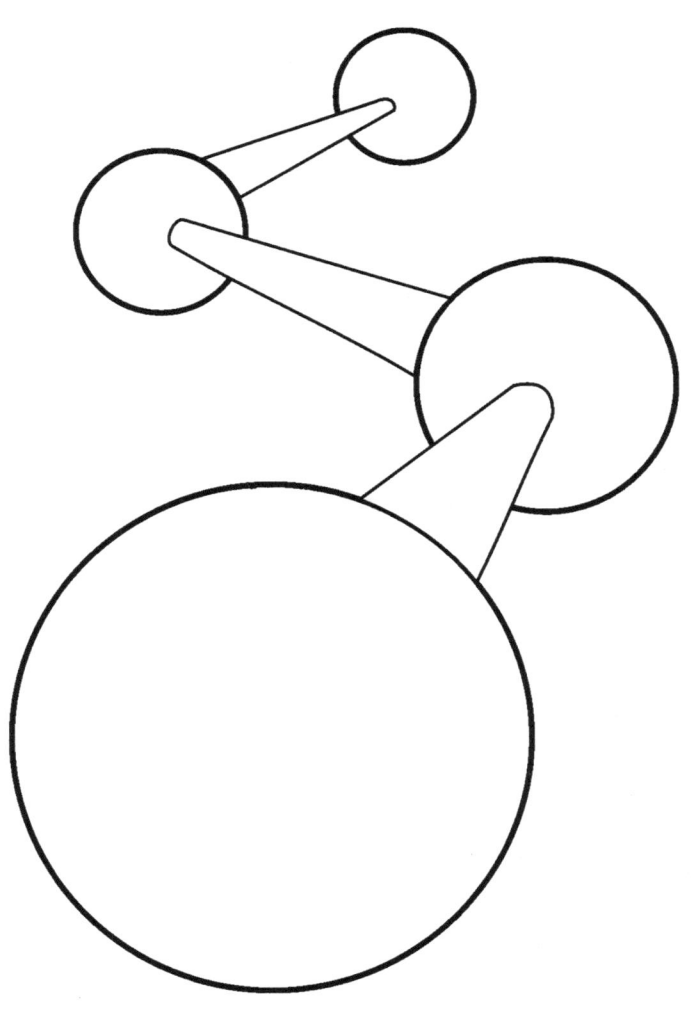

23

Protein Purification Techniques

I INTRODUCTION

Our understanding of protein structure and its function has been derived from the study of many individual proteins. Most biological applications involve either the use or production of enzymes or proteins. To study a protein in detail it must be separated from all other proteins. The development of techniques and procedures for the separation and purification of biological macromolecules such as proteins has had a significant impact on the advances in bioscience over the last three decades. Enzymes are increasingly been employed as reagents in clinical chemistry, as therapeutic agents in chemotherapy, and as catalysts in industrial processes. Now it has been realized that downstream processing costs constitute a very large percentage of overall production costs. The processing cost can range between 50–80% or could, at times be, even higher. Therefore the emphasis now is more on the development of highly efficient and easily scalable methods. The development of new generations of chromatographic media and a wide variety of chromatographic column packaging materials facilitated increased efficiency and selectivity in protein purification. The development of new electrophoretic techniques for rapid analysis of protein composition and purity has also had an impact on protein separation and characterization.

The isolation and analysis of proteins is integral to designing oligodeoxynucleotide probes and gene cloning, confirming DNA sequence data, and synthesizing peptides for eliciting antipeptide antibodies. Some biological materials themselves constitutes clear or nearly clear protein solutions suitable for direct application to chromatography columns after centrifugation or filtration (e.g. blood serum, urine, milk, extracellular medium after cultivation of microorganisms, and cell suspension cultures). Proteins are remarkably labile structures, sensitive to pH, monovalent and divalent ions, temperature, and protein concentration. In most cases, however, one has to extract the desired activity from a tissue or a cell lysate. This means that a considerable number of contaminating molecular species will be set free and that proteolytic activity will make preparation work more difficult. Thus it is important to pay close attention to buffers when working with proteins. The remarkable diversity in properties and sizes of proteins complicates purification procedures. Various methods are available for the homogenization of cells or tissues. Major problems in preparing a protein are in general denaturation, proteolysis, and contamination with pyrogens, nucleic acids, bacteria, and viruses. These can be limited by the proper choice of the extraction medium and procedures.

II STRATEGIES FOR PROTEIN PURIFICATION

A. CLASSIFICATION OF PROTEINS

Proteins can be classified in several ways, such as by structure, by function, or by physicochemical characteristics. Each protein species consists of identical molecules of exactly the same size, amino acid sequences, and three-dimensional shape. In this way, a solution of a mixture of proteins differs from a solution of synthetic polymers or sheared DNA, both of which contain a complete spectrum of varying sizes centered around the average. The protein mixture has only discrete sizes of molecules corresponding to each type of protein present. A useful structural classification takes into consideration both shape and oligomeric structure. Structure reflects biological location and origin. Simple, fairly rigid protein molecules occur in the extracellular environment while more complex and readily deactivated molecules are found intracellularly and hydrophobic proteins are associated with membranes. Classification by function is even more relevant. Proteins can simply be stores of amino acids, be structural, or have specific binding functions. The most "functional" proteins are enzymes, which have both binding and catalytic roles.

1. A list of proteins classified on the basis of structural and functional characteristics

A. Based on structure

1) Monomeric: Single polypeptide proteins usually extracellular, often have disulfide bonds. Examples are lysozome, growth hormones, etc.
2) Oligomeric subunits:
   (a) Oligomeric identical subunits: Mostly intracellular enzymes; rarely have disulfide bonds. For example, Glyceraldehyde-3-phosphate dehydrogenase, catalase, alcohol dehydrogenase, hexokinase, etc.
   (b) Oligomeric mixed subunits: Allosteric enzymes; different subunits have separate functions. For example, aspartate carbamoyltransferase and pertusis toxin.
3) Membrane-bound proteins:
   (a) Peripheral: Readily solubilized by detergents. For example, Mitochondrial ATPase and alkaline phosphatase.
   (b) Integral: Require lipid for stability. For example, Porins, cytochromes P450 and insulin receptor.
   (c) Conjugated: Many extracellular proteins contain carbohydrates. For example, Glycoproteins, lipoproteins, nucleoproteins, etc.

B. Based on function

1) Amino acid storage: Seed proteins. For example, glutens, gliadins, etc.
2) Structural:
   a) Inert: Collagen, keratin, etc.
   b) With activity: Actin, myosin, tubulin, etc.

3) Binding:
   a) Soluble: Albumin, hemoglobin, hormones, etc.
   b) Insoluble: Surface receptors (e.g., insulin receptor, etc.), antigens (e.g., viral coat proteins, etc.).
   c) With activity: Enzymes, membrane transporters, for example, amino acid uptake systems, ion pumps etc.

B. PROTEIN PURIFICATION STAGES

The time required for protein purification is relatively very high compared to purification of other biological macromolecules (DNA, RNA etc.). There is no simple and universally applicable procedure for protein purification. Before the commencement of protein purification, it is imperative to have information regarding the availability of amount of material, quantity of material required, nature of protein (native configuration or denatured), purity, activity (enzymatic, regulatory, antibody etc.) and location (nucleolus, membrane systems, cytoplasm, cell organelles etc.) of protein in the material. There are other real limitations—not only the cost and availability of facilities, but the physical limitations in the techniques themselves. The techniques employed should be as gentle as possible. Possibly the most important preliminary step is to develop appropriate assays. The success of purification is often most dependent on the sensitivity of proteins. Finally, the protein assays have to be simple reproducible, specific, and reliable.

Protein purification flow charts are presented to give a broad outline of the methods used for different types of proteins. In most cases, the first stage is to obtain a solution containing the desired protein, after which it can be dealt with by the many separation techniques described in the following chapters. Purification procedures are commonly divided into three stages. (1) The primary steps, which deal with crude mixtures of proteins and other molecules present in the raw material; (2) secondary processing, which generates a product near to homogeneity; and (3) the polishing steps, which remove minor contaminants, a process that is especially important for therapeutic proteins.

III PREPARATION OF EXTRACTS FROM PLANTS

A. SOURCE AND CHOICE OF MATERIAL

High-value proteins, other than the products of genetic engineering, are also isolated from different starting sources. For example, blood/plasma and human tissues; plant and animal tissues; bioreactors; fermentation and cell culture; monoclonal antibodies; and chemical synthesis (peptides). Levels of expression vary significantly in the single–cell vehicles. Bacteria can express >20% of their mass as recombinant proteins. Yeast typically expresses 3–5%, whereas the mammalian cells secrete proteins only in mg/L amounts.

A wide variety of plant tissues are the source of proteins, such as enzymes. These proteins must be isolated from the crude source by a combination of chemical and mechanical processes. For preparation from cytoplasmic, nuclear, or mitochondrial parts, a tissue devoid of chloroplasts is desirable. Commonly used sources are roots from dark-sprouted peas, soya, wheat, maize, or other seeds. Other useful sources are cauliflower influorescence and carrot roots. Wheat embryo is widely used because it is very active for many enzymes, is convenient to work with, and is available in large amounts. For purification of chloroplast activities, green leaves free of fibers and low in tannins are best. An ideal source is spinach, which is available in bulk from markets. The freshest possible material should always be used. The use of genetically characterized lines or cultivars is recommended especially when protein isolation is a prelude to isolation of the corresponding gene.

B. PREPARATION OF EXTRACTS

Several questions need to be answered during preliminary trials. First, what plant species and tissues are most favorable for purification of the desired activity? Second, if the activity is organellar, can it and should it be purified from isolated organelles, or from a mixed preparation of cytoplasm and lysed organelles? Third, what cofactors, metal ions, and reductants, if any, must be present during purification to yield an active enzyme? These considerations are specific to each enzyme and will not be discussed here. Fourth, what substances present in the crude extract inhibit enzyme activity, and what protective agents which are compatible with the necessary "activators" mentioned above need to be added?

C. TYPE OF EXTRACT

Consideration must be given to the subcellular location of the activity being sought, and to the trade-offs between speed, yield, and purity. The presence of chloroplasts makes it extremely difficult if not impossible to prepare from plant leaves a cytoplasmic extract, completely free of chloroplast contents. Even the most careful breakage of cell walls will damage some chloroplasts. If enzymes with the desired activity are found both in cytoplasm or nucleoplasm and in chloroplast, the extract must be made from a non-green tissue, or the purification scheme must be designed to separate cytosolic from chloroplastic activities. Similarly, purification of chloroplast enzymes may start with isolated chloroplasts or with a total cell homogenate in which the chloroplasts have already been lysed. Plastid purification itself accomplishes substantial purification of enzyme activity and in some instances removes interfering material.

D. HOMOGENIZATION BUFFER

The buffer used for homogenization of plant tissues will typically include high concentrations of reductant, polyphenol inactivators, and of covalent protease inhibitors. Labile compounds are added to the buffer immediately before tissue homogenization. Covalent inhibitors should be present only during extract preparation; while competitive inhibitors should be present at all stages of the purification.

1. Reducing agents

Many plant enzymes, for example those of the chloroplast stroma, require a reducing environment for activity. But when released into the buffer they are exposed to a more oxidizing environment. Since most proteins contain a number of free sulphydral groups, these can undergo oxidation to give inter- and intramolecular disulphide bridges. To prevent this, reducing agents are added. Typically, the preferred reductants are dithiothreitol (DTT) at 2 to 5 mM, or 2–ME at 14 mM (a 1:1000 dilution of the ~14 M stock solution). For isolation of chloroplasts, commonly used reductants are the more physiological compounds ascorbate and reduced glutathione. Some chloroplast enzymes are regulated by dithiol reduction, and can be activated only by DTT.

1. DTT: Prepare a 1 M stock in water; store at –20°C.
2. Ascorbate: Prepare a 0.5 M stock solution of sodium ascorbate in 50 mM HEPES–NaOH, pH 7.6; store at –20°C.
3. Glutathione: Prepare a 0.1 M stock of reduced glutathione plus 5 mM DTT in water; store at –20°C.

All these reagents are prone to oxidation. Stocks should be stored in small aliquots and discarded if there is any question about their effectiveness.

2. Polyphenols

Plant tissues contain considerable amounts of phenolic compounds that can bind to enzymes and other proteins by non-covalent forces, causing protein precipitation. These phenolic compounds are also easily oxidized, predominantly by endogenous phenol oxidases, to form quinones, which are highly reactive and can combine with reactive groups in proteins causing cross-linking, and further aggregation and precipitation. Polyphenols are inactivated as follows: (1) by complexing with 1.5% (w/v) insoluble poly(vinylpolypyrrolidone). Solid insoluble polyvinylpyrrolidone (PVP) is added, to the buffer 2 to 24 hours before use to hydrate fully. Insoluble PVP, which mimics the polypeptide backbone, is therefore added to adsorb the phenolic compounds, which can be removed by centrifugation. (2) By maintaining a strong reducing environment to counteract the effect of phenol oxidases, DTT or 2–ME is added as above. Alternatively, 20 mM sodium diethyl dithiocarbamate or 10 mM sodium metabisulfite is added. (3) Using buffers containing borate (which binds *cis-diols*, including many polyphenols) is reported to inhibit polyphenol oxidation. (4) By using buffer exchange chromatography on BioGel P-6DG or Sephadex G-50, inhibitory material is effectively removed but this is feasible only when the extract is prepared on a small scale.

Preparation of active extracts from plant species with high levels of tannins and other phenolic compounds has been examined for the purpose of routine assay for isozyme markers. The most important consideration is the inclusion of PVPP or PVP (2 to 4% soluble PVP, Mr 400,000) in the extraction buffer.

3. Protease inhibitors

Whenever protein molecules are extracted from living organisms, there is always the problem that the desired material will be degraded by proteases released during the extraction process. To slow down unwanted proteolysis, extraction and purification steps are carried out at 4°C, and in addition a range of protease inhibitors is included in the buffer. Each inhibitor is specific for a particular type of protease. Below (Table 23.1) are outlined the properties of a range of protease inhibitors all easily obtained from Sigma and a protocol for a basic cocktail. Some inhibitors are only soluble in organic solvents. Therefore, two stock solutions should be made, and added to solutions as required.

Table 23.1 *Protease inhibitors*

Substance	Mol. Wt. (Daltons)	Working concentration	Enzymes inhibited
α1 Antitrypsin	54,000	0.1 μM	Serine, Thiol proteases
Aprotinin	6,511	0.3 μM	Serine proteases
E–64	357.4	2 μM	Thiol proteases
EDTA	292.2	1 mM	Metalloproteases
EGTA	380.3	1 mM	Ca/Mg dependent
pHMB	360.0	1 mM	Thiol proteases
Leupeptin	475.0	0.1 μM	Serine, Thiol proteases
NEM	125.1	1 mM	Thiol proteases
Pepstatin A	685.9	0.1 μM	Aspartyl proteases
1, 10–P	198.2	1 mM	Metalloproteases
PMSF	174.2	1 mM	Serine proteases
SBTI	20,100	10 g/ml	Serine, Thiol proteases
TLCK	369.3	0.2 mM	Trypsin, Thiol proteases
TPCK	351.9	0.1 mM	Chymotrypsin, Thiol proteases

The protocols given below are stock solutions that produce the working concentrations of inhibitors to cover the range of likely proteases.

(a) Serine protease inhibitors: *Phenylmethylsulfonyl fluoride* (PMSF): A 100 mM (100X) stock in dry 2-propanol or "anhydrous" ethanol (95% ethanol, 5% 2-propanol) is prepared. The stock is stored at room temperature and is stable for at least 6 months. Immediately before use, the buffer is brought to 1 mM PMSF. The stock solution is added slowly with constant stirring, from a plastic disposable pipette whose tip is submerged in the buffer. *Benzamide*: A 0.1 M stock is prepared in ethanol. It is convenient to dissolve benzamide together with PMSF (in 2-propanol or anhydrous ethanol). *Benzamidine-HCL*: An aqueous 0.1 M stock is prepared; used at 1 mM. *e-Amino-n-caproic acid*. A 0.5 M stock is prepared in water; used at 5 mM final concentration. Benzamidine and caproic acid are dissolved together to give a 100X stock.

(b) Cysteine (thiol) protease inhibitors: *Sodium p-hydroxymercuribenzoate* [PHMB; the sodium salt of p-chloromercuribenzoic acid (PCMB)]: A 100 mM stock is prepared in water; adjusted to pH 8 with NaOH or solid HEPES as necessary. This is added to homogenization buffer to a final 1 mM. The competitive inhibitory proteins *antipain* and *leupeptin* (Sigma): These can be added to 1 μg/ml from 0.1 to 1 mg/ml stocks in distilled water.

(c) Aspartate (acidic) protease inhibitors: Working at pH 7 or above is probably a reasonable defense against aspartate protease. Further protection can be afforded by the following: *Pepstatin* (Sigma), a weak competitive inhibitor: It is added to 0.1 mg/ml from a 10 mg/ml stock in MeOH. *Diazoacetylnorleucine methyl ester* (DAN): In the presence of Cu(II) this is a covalent inhibitor. The requirement for copper ion precludes the simultaneous use of this reagent with chelators used to inhibit metalloproteases. A 100 mM stock of DAN is prepared in methanol; cupric [Cu(II)] acetate is dissolved in water at 100 mM. Each solution is added to 5 mM final concentration.

(d) Metalloprotease inhibitors: EGTA [ethylene glycol bis(β–aminoethyl ether) N,N,N', N'-tetraacetic acid]: An efficient chelator of divalent metal cations other than Mg^{2+}. To prepare a 0.5 M stock, 0.05 mol is suspended in 80 ml water, titrated to pH 8 with concentrated or solid NaOH, and diluted to 100 ml. This is added to buffers at 10 mM final concentration. Mg^{2+} may be added to these buffers at normal levels if necessary. The chelator 1, 10–phenanthroline: A replacement for EGTA. A 100 mM stock is dissolved in 95% EtOH and used at 5 mM.

(e) Stock solutions that produce the working concentrations of six inhibitors to cover the range of likely proteases.

Procedure

1. Make up each of the following stock solutions:

EDTA (pH 8.0)	2.92 g/10 ml water	(10 ml of 1 M)	Inhibits metallo
EGTA (pH 8.0)	3.80 g/10 ml water	(10 ml of 1 M)	Metallo
NEM	1.25 g/10 ml water	(10 ml of 1 M)	Thiol
Pepstatin	6.85 mg/10 ml ethanol	(10 ml of 1 mM)	Aspartyl
PMSF	581 mg/10 ml ethanol	(10 ml of 0.33 M)	Serine
TPCK	352 mg/10 ml ethanol	(10 ml of 0.1 M)	Serine/Thiol

    Store each stock solution in 1 ml aliquots at –20°C.

2. Mix 1 ml aliquots of the water soluble inhibitors and make up to 5 ml with water. This gives a 200x solution, designated PI-A (aqueous). Store in aliquots at –20°C.
3. Mix 3 ml of 0.33 M PMSF with 1 ml of 0.1 M TPCK and add 1 ml of 1 mM Pepstatin. Warm in 65°C water bath to obtain complete dissociation. This gives a 200x solution, designated PI-B (organic). Store in aliquots at –20°C.
4. Add 5 µl each of PI-A and PI-B to every 1 ml of sample extract.

    Abbreviations: pHMB= p-Hydroxymercuribenzoic acid; NEM= N-Ethylmaleimide; 1,10–p= 1, 10–Phenanthroline; SBTI= Soybean trypsin inhibitor; TLCK= N-α-p-Tosyl-L-lysine chloromethyl ketone HCl; TPCK= N-Tosylamide-L-phenylalanine chloromethyl ketone.

4. Enzyme substrate and cofactors

Low levels of substrate are often included in the extraction buffer when an enzyme is purified, since binding of substrate to the enzyme active site can stabilize the enzyme during purification processes.

5. EDTA

Ethylene diamine tetra acetate (EDTA) can be present to remove divalent metal ions that can react with thiol groups in proteins giving **mercaptids**. $R-SH+Me^{2+} \rightarrow R-S-Me^{+}+H^{+}$.

6. Sodium azide

For buffers that are going to be stored for long periods of time, antibacterial and/or antifungal agents are sometimes added at low concentrations. Sodium azide is frequently used as a bacteriostatic agent.

7. Membrane proteins

Membrane-bound proteins require special conditions for extraction as they are not released by simple cell disruption procedures alone. Two classes of membrane proteins are identified. Extrinsic membrane proteins are bound only to the surface of the cell, normally via electrostatic and hydrogen bonds. These are predominantly hydrophilic in nature and are relatively easily extracted by raising the ionic concentration of the extraction buffer. Intrinsic membrane proteins are those that are embedded in the membrane. These invariably have significant regions of hydrophobic amino acids and have low solubility in aqueous buffer systems. These proteins can be extracted with a buffer containing detergents. The choice of detergent is mainly one of trial and error but can include ionic detergents such as sodium SDS, sodium deoxycholate, cetyl trimethylammonium bromide (CTAB) and CHAPS, and non-ionic detergents such as Triton X-100 and Nonidet P-40.

E. PRE-TREATMENT

Washing and imbibing: The leaves are washed in tap water, rinsed for 15 seconds in a solution of 0.1% sodium hypochlorite (1:50 dilution of household bleach) plus 0.05% non-ionic detergent (Brij 35 or Nonidet P–40) in deionized water. The leaves are patted thoroughly; dry and large ribs and veins are removed. Dry seeds or beans are surface sterilized by soaking for 30 sec in 0.1% sodium hypochlorite and rinsing thoroughly with deionized (distilled quality) water. The seeds get swollen by soaking overnight at room temperature in a sterile water.

Organic floatation: Floatation is used to separate embryo from bran and endosperm (starch-containing vacuoles) in ground seeds or in raw wheat germ preparations. Wheat embryo is pale yellow; bran is brown and starch is white. Embryos purified by floatation have lower levels of non-specific nucleases. An initial trial should be performed to determine whether enzyme stability or yield during purification is affected by this treatment. The floatation is usually omitted in a large–scale preparation. In a fume hood, prepare a solution containing 250 ml CCl_4 plus 70 ml cyclohexane. Add 125 g raw wheat germ and stir gently with a glass rod. Stop stirring for a min. or so to allow the suspension to separate into floating and non-floating material. Collect material in the upper 20% of the solvent. Spread the floating wheat germ on a thick filter paper, dry 15 min. in the fume hood, and finish drying overnight at 4°C. Store the dry germ at –70°C.

F. CELL DISRUPTION

Unless one is isolating proteins from extracellular fluids such as blood, protein purification procedures necessarily start with the disruption of cells or tissue to release the protein content of the cells into an appropriate buffer. This initial extract is therefore the starting point for protein purification. The different methods available for disrupting the cells depends on the nature of the cell wall/membrane being disrupted. There are three main methods for the release of intracellular proteins from different cell types: enzymic, chemical, or physical. Not all of the techniques available are suitable for use on a large scale.

1. Enzymic methods of cell disruption

Lysozyme, an enzyme produced commercially from hen egg white, hydrolyses β–1,4–glycosidic bonds in the cell walls. Other enzymes are infrequently used, but microbial glucanases and lysostaphin may be used.

2. Chemical methods of cell lysis

Alkali: Treatment with alkali has met with considerable success in small and large-scale extractions. The success of this method relies on the alkali stability of the desired product. The high pH level may inactivate proteases, and the method is of value for the combined inactivation and lysis of microorganisms.

Detergents: Detergents, either ionic, for example sodium lauryl sulfate, sodium cholate (anionic), and cetyl trimethyl ammonium bromide (cationic), or non-ionic, for example Triton X-100 or X-450 or Tween-20, have been used to aid cell lysis. The presence of detergents can affect subsequent steps, in particular salt precipitation.

3. Physical methods of cell lysis

Osmotic shock: Osmotic shock has been used for the release of hydrolytic enzymes and binding proteins from the periplasmic space of a number of bacteria. The method involves washing the cells in buffer solution to free them from the growth medium, and then suspending them in 20% buffered sucrose. After being allowed to equilibrate, the cells are harvested and rapidly re-suspended in water at about 4°C.

Grinding with Abrasives: This technique involves the grinding of cell pastes in a mortar with an abrasive powder, such as glass, alumina, or kieselguhr. Many factors influence the rates of cell breakage, such as the size and concentration of the glass beads, the type, concentration, and age of the cells, the agitator speed, the temperature, buffer, etc.

Liquid shear: Liquid shear is the choice for the large-scale disruption of microbial cells, and it finds widespread application in both industrial processes and in research. Large-scale work usually requires a homogenizer of the type developed for emulsification in the diary industry.

Materials

Waring Blender (1 litre); Miracloth; Cheesecloth (autoclave moist); Kitchen spatula; Plastic beakers; Large plastic funnel to fit beaker.

Homogenization Buffers:

50 mM Tris.HCl, pH 8 at 25°C; 5% glycerol; Monovalent cation (0.1 M KCl or NH_4Cl, or 0.05–0.2 M $(NH_4)_2SO_4$); 1X Inhibitor mix. Buffer is equilibrated to 0°C before use, and should be used as a slurry prepared by allowing buffer to freeze partially until ice starts to form, or by freezing solid in a household ice-cube tray and crushing in a household blender ice-crusher attachment. Mix crushed frozen buffer with cold liquid buffer to obtain pourable slurry.

Inhibitor mix:

Protease inhibitors: 1 mM PMSF, 1 mM benzamide, 1 mM benzamidine, 10 mM EGTA, 1 μg/ml antipain, 1 μg/ml leupeptin, 0.1 mg/ml pepstain.

Reductants: 5 mM DTT; 20 mM sodium diethyl dithiocarbamate.
Antiphenolic: 1.5% PVPP (insoluble PVP).
For isolation of phosphorylated enzymes, add KF to 50 mM.

Preparation of sand:
Stir (e.g.) 50 g quartz sand into ca., 100 ml chromic acid for 5 min. Allow to settle down, decant the acid, and wash the sand exhaustively with tap water. Rinse the sand with distilled water and collect on a Buchner funnel. The sand is oven dried, saturated with 5% dichlorodimethylsilane in CCl_4, and washed thoroughly in 95% ethanol. Alternatively, the damp sand may be treated with an equal volume of aqueous siliconizing reagent and rinsed in distilled water. Either treatment is completed by baking at 150°C.

Small-scale extracts:
Small amounts of plant material such as wheat germ, leaves, or wet seeds may be homogenized by grinding with acid-washed siliconized sand, if necessary, in a mortar and pestle.

Procedure

Wheat germ or non-leafy plant material:
1. A 5 g sample is placed in a mortar on ice and ground to a powder with an equal volume of acid-Homogenization buffer is added and grinding is continued until a smooth paste is formed.
2. Leaves and other soft material, and all fibrous tissues are first frozen in liquid nitrogen, then placed into a mortar on dry ice and ground to a fine powder without sand. Homogenization buffer is then added. For either type of extract, the paste is diluted to a pourable consistency with homogenization buffer, transferred to 15- to 50-ml centrifuge tubes, and centrifuged for 20 min. at 12,000 g. The supernatant is decanted into clean tubes and spun 45 min. at 40,000 g. This clarified S40 extract is divided into aliquots, quick frozen, and stored at −70°C.

Cytoplasmic extracts of non-green tissue, wheat germ:
All procedures are carried out in the cold room. Wheat germ (50–500 g) is processed at a ratio of 400 ml homogenization buffer/100 g wheat germ. Suspend 500 g raw wheat germ in 2 litres homogenization buffer (ice cold or a partially frozen slurry) in a 1-gal blender pre-cooled to −20°C. Large amounts are processed in 500 g batches. Smaller amounts are scaled down proportionately, and a blender jar is chosen which will be about two-thirds to three-quarters filled. Grind at high speed for four bursts of 15 seconds, with 30–second rests. Be careful to avoid foaming. The use of a foam arrester attachment may prove helpful. Check the pH with pH paper at this point and add solid Tris-base or 2 M Tris-base if necessary to attain pH 7.0–7.5 at 4°C. Decant into 500 ml leak proof centrifuge bottles. Rinse the blender jar with 500 ml cold homogenization buffer and add to the extract. Remove debris by centrifugation (at 2°C) for 30 min at 12,000 g (8500 rpm) in the Sorvall GS–3 rotor). Decant the supernatant through one layer of sterile Miracloth into a plastic beaker on ice.

Fleshy tissue (Cualiflower inflorescence, carrot root):
Proceed as for wheat germ, but use 1 litre homogenization buffer/kg chopped plant material. Filter the homogenate through four to six layers of cheesecloth and one layer of

Miracloth prior to centrifugation. When preparing extracts from ripe fruits or vegetables containing pectin, EGTA must be present to prevent gelation of the extract.

Fleshly beans or seeds:

Dry seeds (peas or soybeans) are washed and imbibed as described above. The beans, at a ratio of 1 litre buffer/kg initial dry weight, are ground five times at low speed in a Waring blender for 30 seconds each time with 15- to 30-sec rests. The homogenate will contain substantial debris that can be separated in a commercial juice extract. Filter the homogenate through four layers of cheesecloth and one layer of Miracloth. If a juice extractor is unavailable, use six to eight layers of Miracloth, or six to eight layers of cheesecloth followed by two layers of Miracloth.

Whole–cell extracts from leaves:

Soft, non-fibrous leaves and similar tissue may be ground wet in the homogenization buffer. Fibrous leaves, and all leaves, which are high in polyphenolic contents, are first quick frozen in liquid nitrogen and ground to a powder prior to addition of the homogenization buffer. This procedure allows more efficient cell lysis, and subjects enzymes to less contact with vacuolar contents and other compartmentalized inhibitory compounds. Freezing plant tissue at $-20°C$ is not recommended because during slow freezing, intracellular compartments are ruptured with the release of deleterious substances.

Non-fibrous leaves:

1. Wash, dry and de-rib fresh leaves as described earlier. Pack the leaves into a Waring blender of appropriate capacity and add ice-cold homogenization buffer at 1.5 litres/1 kg leaves. Grind for a few 5-second bursts, packing intermittently with a kitchen spatula to reduce the volume of leaves; then grind for 1.5 to 3 min. in 30-second bursts. Less span of time may be used for soft leaves like those of spinach. Frozen, crushed homogenization buffer may be added to maintain the temperature at 0–4°C.
2. Filter the homogenate through six to eight layers of cheesecloth, then through two layers of Miracloth. Remove debris by centrifugation for 15 min. at 1,200 g. When large volumes are being processed, the crude filtrate may be used directly as starting material for ammonium sulfate or acetone precipitation.

Fibrous or phenol-rich leaves:

Start with washed, de-ribbed leaves. Pack the leaves into a thick-walled styrofoam box until half full and carefully add liquid nitrogen (LN_2) to just cover the leaves. After 30–60 seconds, pour off the LN_2 into another box of leaves. Transfer the frozen leaves plus a minimal amount of LN_2 to a stainless steel Waring blender container (previously chilled to $-20°$ or $-70°C$). Grind the leaves for 1.5 to 3 min. at high speed, or until they are reduced to a fine powder. (It might be necessary, with some samples, to complete the process manually with a pestle and a mortar placed on dry ice.) The LN_2 should evaporate by the end of the grinding. Add the homogenization buffer, 200 ml/100 g initial dry weight of leaves, and mix for 30 seconds at high speed. Continue with filtration and centrifugation as for non-fibrous leaves above. Large amounts of buffer may be required for maximal yield and activity from fibrous, viscous, or tannin-rich homogenates.

Chloroplast extracts:

For a small-scale preparation, or when initiating a purification trial, it is best to use plastids that are intact and have been purified on Percoll density step gradients. Percoll is preferable to sucrose unless specifically contraindicated. Later, when a chloroplast location has been demonstrated, and if there is minimal interference from cytoplasmic activity, large-scale preparations can be done with bulk chloroplasts, which have been washed twice to remove cytoplasmic contamination. If desired, a preliminary low-speed spin can be added to remove most nuclei. Mitochondria remain largely in the post-chloroplast supernatant if the centrifugation is not prolonged.

Preparation of chloroplast lysate:

(a) *Hypotonic lysis*:
1. Wipe the walls of the tube or bottle with KimWipes. Re-suspend the chloroplast pellet in a small amount of hypotonic lysis buffer.
2. Measure the choloroplast volume as the difference between the re-suspended volume and the volume of added buffer.
3. Dilute the suspension to a final concentration of 5–7 ml lysis buffer/1 ml packed chloroplasts (ca. 7 ml buffer/100 g leaves). The ratio may vary seasonally or with plant maturity. To further hasten plastid rupture, transfer the suspension to a Dounce–type homogenizer vessel and homogenize with several strokes every few min.

(b) *Detergent lysis*:
1. Alternatively, chloroplasts may be lysed with non-ionic detergent. The washed plastid pellet is re-suspended in several volumes of GR buffer and diluted with an equal volume of hypotonic lysis buffer containing an inhibitor mix and 0.2–1% peroxide–free Triton X–100. Stir on ice 15 min., then spin 15 min. at 20,000–30,000 g.

Method-1: Thylakoid extracts

A preparation of thylakoids that are intact is obtained after vigorous homogenization of leaves in an iso-osmotic buffer. Chloroplast envelopes are largely disrupted and thylakoids are isolated by differential centrifugation. The thylakoids remain as judged by their content of plastocyanin, a lumenal protein. The thylakoids are extensively washed to remove ribuloase-1, 5-bisphosphate carboxylase/oxygenease, the major stromal protein. To obtain unstacked thylakoids, perform these washes in a buffer containing 10 mM NaCl. This protocol gives active chloroplast ATP synthase in good yield without protease inhibitors or reducing agents.

Materials

Reagents and solutions:

Thylakoid homogenization buffer: 0.4 M sucrose (or sorbitol), 20 mM Tricine-NaOH (pH 8), NaCl to 10 mM (to prepare unstacked thylakoids) or 100 mM (to prepare stacked thylakoids).

Thylakoid wash buffer: 20 mM Tricine-NaOH (pH 8), 10 or 100 mM NaCl (as homogenization buffer).

Procedure

1. Wash 1–1.5 kg de-veined spinach leaves and pack tightly into a 1 litre Waring blender.
2. Add thylakoid homogenization buffer at a ratio of 1.4 litres/kg leaves.

3. Homogenize the mixture at high speed for 15 seconds after maximum speed is reached. A few initial 1 second bursts alternated with packing may be necessary to achieve a uniform slurry.
4. Filter the homogenate first through two layers of cheesecloth into a plastic beaker on ice. This filtrate is then passed through six layers of cheesecloth and two layers of Miracloth and collected in a beaker in ice.
5. Remove the cell debris first by centrifugation of the filtrate for 10 seconds at 2000 g (or 2 min. at 300 g); transfer the supernatant to fresh centrifuge bottles and pellet the thylakoids for 10 to 15 min. at 3000 to 6000 g.
6. Wash the thylakoid pellets re-suspension in a final total volume of 2 litres ice-cold thylakoid wash buffer and centrifuge as before. Repeat this wash twice. Prolong centrifugation times to 30 min. for the second wash and 40 min. for the third.
7. Re-suspend the final pellet in wash buffer. Thylakoids may be lysed by re-suspension in hypotonic lysis buffer or hypotonic buffer containing 0.1 to 1% Triton X-100. The lysate is clarified by centrifugation (15 min., 20,000 g).

IV   ISOLATION OF ENZYMES

An enzyme may consist of a single polypeptide chain, may be composed of multiple subunits, or may form part of a multi-enzyme complex. In general, enzyme-catalyzed reactions occur at moderate pressures, temperatures and pH values. Furthermore, enzymes usually show a remarkable specificity, for the reactants and for the reaction that is catalyzed, including the ability to distinguish between optical isomers. Therefore, side-reactions are kept to a minimum. This is important for an organism, as there is no opportunity to purify the product of one reaction before it is utilized by the next enzyme in a metabolic pathway, as might be done in a chemical laboratory. Enzyme activity is also tightly regulated in an organism as part of its homeostatic mechanisms. Nearly 2500 different enzyme-catalyzed reactions have been catalogued by the Enzyme Commission of the International Union of Biochemistry and there are often several different enzymes that promote a given reaction.

A. ISOLATION OF RESTRICTION ENZYMES

Method-1: *Isolation of restriction enzymes*

Sequence-specific, or Type II, endonucleases are commonly described as restriction enzymes (see Chapter 16). Type II endonucleases generate reproducible nucleotide fragments from specific DNAs. They cleave double-stranded DNA by hydrolyzing one phosphodiester bond in both the strands, within defined nucleotide sequences. More than 1,500 enzymes have been discovered since the first one reported by H. O. Smith and collaborators, with a variety of bacterial cells as their source. The name of a restriction enzyme is derived from the genus and species of bacterium from which it is isolated. The first letter of the genus name and first two letters of the species are combined to form the enzyme name. This is followed by a strain designation if needed. If a bacterium contains more than one restriction enzyme, then each enzyme is assigned a Roman numeral.

Structurally, most restriction enzymes reported in purified form are composed of two equal subunits with molecular weights of 20,000 to 25,000 or single polypeptides with molecular weights of 30,000 to 35,000. Enzyme activities can be differentiated from each other by the characteristic patterns of digestion of small viral or bacteriophage, or plasmid DNAs. A given sequence in DNA is recognized and can be cleaved by more than one restriction enzyme. The term "isoschizomers" describes a group of enzymes that recognize the same sequence in DNA. The sequence recognized by these enzymes are for the most part centrosymmetric "**palindromic**" sequences that are usually hexamers, pentamers, or tetramers.

Enzymes in this family are amenable to purification by conventional isolation procedures. Phosphocellulose at nearly neutral pH is a widely used separation matrix for extracts that have been freed of cellular nucleic acids. At this stage of purification, short-term assays often make it possible to visualize fractions that contain enzymes in the presence of non-specific exonulceases. A variety of enzymes have been fractionated with affinity chromatography. This method takes advantage of biospecific interactions not offered by conventional fractionation methods. Enzymes fractionated by such procedures can be concentrated selectively and eluted by gradients containing substrates or salt. The advantages of affinity chromatography are speed of purification and protection against denaturation during fractionation. A general procedure to purify restriction enzymes utilizes Cibacron blue. Cibacron blue F3GA is a blue dye that appears to have biospecific affinity for nucleotide–requiring enzymes. Before fractionation of enzymes on blue-CNBr-separose unbound ligand should be removed by salt washes and extraction done with a hydrophobic solvent such as dioxane.

Several reports have described apparent changes in the specificity of restriction endonucleases in association with an altered reaction environment. Conditions that alter specificity have included changes in the ionic concentration, in pH of the reaction buffer, and the introduction of glycerol into the reaction mixture (Figure 23.1).

Materials

Cell-seed cultures for one or more (Bam HI, Pst I and Xba I)

DEAE-Cellulose; Media-Tryptone, yeast extract; Cybacron blue matrix; Glycerol and/or DMSO; Lambda DNA; Electrophoresis buffer; Agarose.

Reagents for the experiment are available in kit form from EDVOTEK Inc.

Cibacron blue sepaharose can be prepared by using CNBr-activated sepharose or by activating sepharose with cyanogen bromide. CNBr-activated Sepharose can also be purchased commercially from Pharmacia Inc.

In order to activate the sepharose 4B (Pharmacia), 100 g is suspended in 100 ml of distilled water and 100 mg of finely divided solid cyanogen bromide (CNBr) is added over 5 min. The pH is adjusted and manipulated at 11 by addition of 8 N NaOH with staining. After 10 min. at 20°C, 200 ml of cold distilled water is added. The suspension is washed and activated sepharose is used. This has to be done in a well-ventilated hood and should not be attempted by students.

CNBr-activated sepharose, either purchased or prepared, is washed with excess 1 mM HCl, and Cibacron blue F3GA is coupled at pH 8.3 in 0.1 M $NaHCO_3$ and 0.5 M NaCl at 4°C for 16 hours. If necessary, pH is maintained at 8.3 with 0.1N NaOH. The ratio of starting

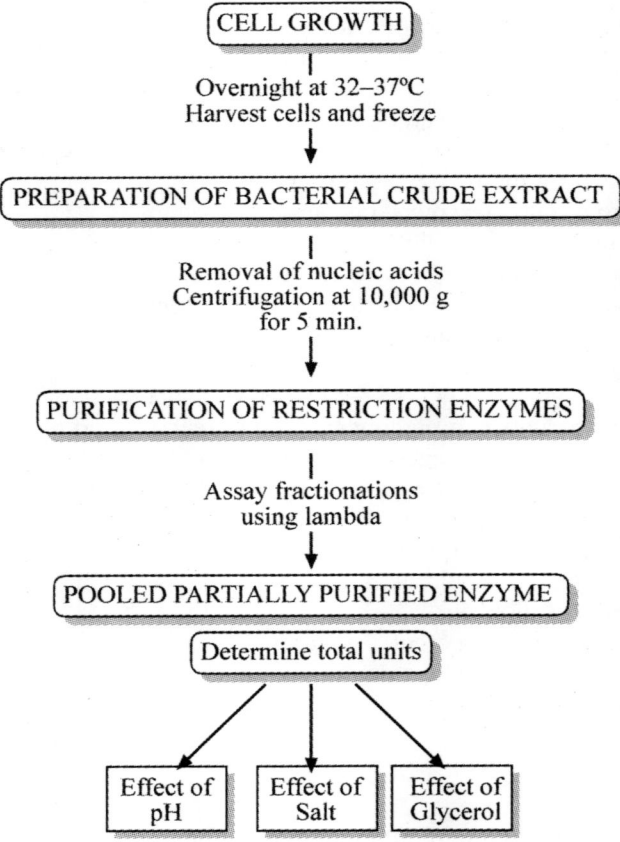

Figure 23.1 *Experimental outline for isolation of enzymes*

blue dye to swelled activated sepharose is 1 g to 10 ml. To block the remaining active groups, 1 M Tris.HCl pH 9 is added and incubated for 1 hour at 22°C. The matrix can be stored at 4°C in the presence of high NaCl and a bacteriocide and will remain stable for several months.

Procedure

1. Cells can be grown by inoculating the specific medium with a 2% overnight culture.
2. After growing for 16 hours, cells are harvested by centrifugation at 8000 g, washed with 0.9% saline and frozen at −90°C.
3. Growth media and temperature conditions for the different strains are as follows: *Bacillus amyloliquefaciens. Providencia stuarti* 164 is grown on tryptone (11 g/l), yeast extract (22 g/l), glycerol (4 ml), K_2HPO_4 (8.9 g) or KH_2PO_4 (2.14 g) per litre and NaCl (5 µg/l) at 37°C; *Xanthomonas badrii* is grown on nutrient broth (8 g/l) at 32°C.

B. PREPARATION OF BACTERIAL CRUDE EXTRACTS

A typical preparation utilizes 15 g of pre-washed bacterial cells.
1. Cells are suspended in 35 ml of buffer A (20 mM Tris-HCl pH 7.2, 1 mM Na2EDTA, 7 mM β-ME and 10% glycerol).
2. Cells are at least 80% lysed by sonication using 30 second pulses for 8–10 min. with intermittent on-off cycles.
3. The temperature of the slurry is maintained at 10°C through the lysing, and all procedures to follow are carried out at 4°C unless otherwise specified.
4. The cell extract is then separated by two sequential centrifugation steps: at 10,000 g for 1 hour each. Only supernatant (approximately 40 ml), referred to as crude extract, is recovered and pellets discarded.
5. Nucleic acids are then removed by one of several procedures, e.g., batch elution from DEAE-cellulose under conditions where the enzyme is recovered, or by precipitation with streptomycin sulfate and PEG-6000. Since nucleic acids will bind to the column reducing the capacity of the matrix, acid was found to be a prerequisite.
6. The supernatant after the removal of nucleic acid is submitted to fractionation by Cibacron blue sepharose as described in the legends.

C. PURIFICATION OF RESTRICTION ENDONUCLEASES

Cibacron blue sepharose is effective in the separation of restriction endonucleases from nucleic acid free bacterial extracts. For this experiment, three enzymes *Bam* HI, *Pst* I and *Xba* I, from different genera can be selected. For the initial experiments, a 1.5 ml column of blue-CNBr-sepharose should be equilibrated with buffer A. Columns must be developed with a 20 ml salt gradient 0–0.8 M in Buffer A. Salt concentration at the peak of enzyme activity for these and other enzymes tested is summarized below and in most preparations yields can be expected between 2,000 and 10,000 units.

Organism	Enzyme	Salt concentration (Molar) (approximate)
Bacillus amyloliquefaciens	*Bam* HI	0.45 M
Providencia stuartii 164	*Pst* I	0.30 M
Xanthomanus badrii	*Xba* I	0.25 M

D. ANALYSIS OF DNA ENDONUCLEASE ACTIVITIES BY GEL ELECTROPHORESIS

Assays are performed in 50 μl volumes in either optimal buffers (details for each specific enzyme are indicated in Table 23.1).

Procedure

1. Selection of the lambda DNA is based on obtaining a characteristic and recognizable fragmentation pattern.

2. Incubations (10 µl) of column fractions and assays to locate activity at 37°C are done for 1 hour or overnight.
3. Reactions are stopped by the addition of 5 µl of reaction stop solution (for routine assays, gels at a concentration of 0.8% agarose are prepared and electrophoresed in Tris (48.4 g/l), sodium acetate (16.4 g/l) and EDTA (7.4 g/l); adjusted to pH 7.8 with acetic acid and run for 3 hours at 6 volts/cm).
4. Gels are stained with methylene blue Plus™ or ethidium bromide, and DNA is visualized with short wavelength UV light source.
5. Photographs are obtained through the use of a red filter (Kodak 23A) and Polaroid type 57 film.

Unit determination

Unit determination is performed with serial dilution of enzyme using a fixed amount of substrate. A unit of enzyme is defined as the minimum amount of enzyme, which will totally digest 1 µg of Lambda DNA in 60 min. at 37°C.

V SUGGESTED READING

Birk Y (1976) Protease Inhibitors, *Methods Enzymol.,* **45**:701–703.

Creighton TE (1997) *Protein Function, A Practical Approach* (2nd Edn), Oxford University Press.

Gupta MN and Mattiasson B (1994) in *Highly Selective Separations in Biotechnology* (Ed) Street, G, Blackie Academic & Professional, Glasgow, UK, pp 7–33.

Malaikan A (1994) in *Highly Selective Separations in Biotechnology (Ed)* Street. G, Chapman & Hall, Glasgow, UK, pp 34–54.

Roy I and Gupta MN (2000) Current Trends in Affinity–based Separations of Proteins/Enzymes, *Current Science,* **78**:587–591.

Walker JM and Gingold EB (1999) *Molecular Biology and Biotechnology*, Panmia Publishing Corporation. New Delhi, Bangalore.

Wilson K and Walker J (2000) *Practical Biochemistry, Principles and Techniques* (5th Edn), Cambridge University Press.

24

Protein Detection and Estimation

I INTRODUCTION

The need to determine the protein concentration in a solution is a routine requirement during protein purification. The only accurate method for determining protein concentration is to acid hydrolyze a protein of the sample and then carry out amino acid analysis on the hydrolysate. However, this is relatively time-consuming, particularly if multiple samples are to be analyzed. Fortunately, other quicker methods that give a reasonably accurate assessment of protein concentrations of solution are also acceptable (Figure 24.1). The amount of protein contained in a solution can be conveniently determined by various methods (Table 24.1). The appropriate choice of method depends on five major criteria: The amount of protein available to assay, the concentration of the protein, the specificity of the assay, the presence of chemical compounds which may interfere with the assay, and the ease and reliability with which the assay can be performed. An approximation of sensitivity is given for each assay. Only those assays that are easy to perform, require simple instrumentation, and are highly sensitive will be discussed, although there are many other excellent methods available (Table 24.2). Three general techniques are given here for the determination of the concentration of a known protein by UV absorbance (**Method 1**) or unknown proteins or mixtures by dye-binding (**Method 2, 3, 4**) or copper interaction (**Method 5**).

Table 24.1 *General methods for the detection of peptides*

Method	Applicability in solution	On paper/ thin layers
UV absorbance (206–230)	+	−
Reaction with ninhydrin	+	+
Reaction with fluorescamine	+	+
Reaction with o-phthaldehyde/mercaptoethanol	+	+
Alkaline hydrolysis, followed by reaction with ninhydrin, fluorescamine or o-phthalaldehde	+	−

Continues...

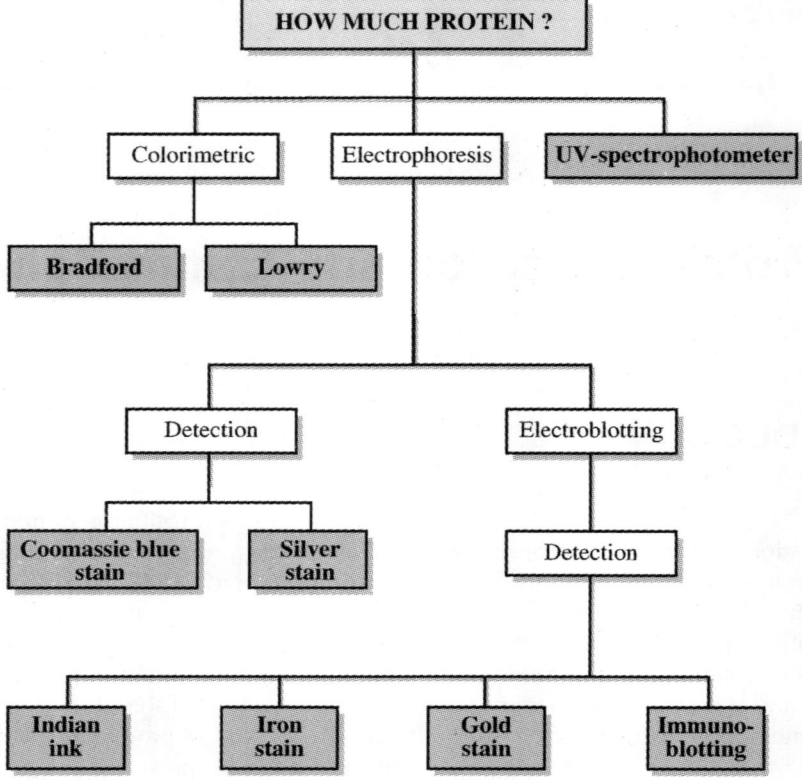

Figure 24.1 *Techniques for quantitation of protein concentrations*

... Continued

Method	Applicability in solution	On paper/ thin layers
Acid hydrolysis and reaction with ninhydrin	−	+
Acid hydrolysis and amino acid analysis	+	−
Acid hydrolysis and thin-layer electrophoresis at pH 2	+	−
Dansylation, followed by hydrolysis and thin-layer chromatography	+	−
Acid hydrolysis, dansylation and thin-layer chromatography	+	−
Transfer to thin-layers for spot tests, chromatography or electrophoresis	+	−
SDS-gel electrophoresis, acid/urea gel electrophoresis or isoelectric focussing, followed by staining with Coomassie blue	+	−
Chlorination, followed by reaction with o-tolidine or starch/KI	−	+

Table 24.2 *Methods for the detection of peptides containing particular residues*

Method	Amino acid residues detected	In solution	On paper
Uv absorbance at 280 nm	Trp, Tyr, some modified residues	+	−
Fluorescence	Trp (weak), oxidized Trp, some modified residues	+	+
Liquid scintillation counting	Radio-labeled residues. E.g. S-[^{14}C]carboxymethyl cysteine. [^{32}P]phosphoserine	+	+
Radioautography	Radio-labeled residues	−	+
Gamma-counting	Some radio-labeled residues (e.g. [^{125}I]iodoTyr)	+	+
Reduction and reaction with DTNB	Cystine	+	+
Phenol H$_2$SO$_4$	Glycopeptides	+	−
Iodoplatinate	Met, Cys	−	+
Phenanthrenequinone	Arg	−	+
p-Dimethylaminobenzaldehyde	Trp	−	+
Pauly test	His	−	+
Nitrosonaphthol	Tyr	−	+
Isatin	N-terminal Pro	+	+

Photometry

The principle of photometry depends on the phenomenon of light absorption by molecules in solution. The specificity of a compound to absorb light at a particular wavelength (monochromatic light) is exploited in the laboratory for quantitative measurements. Colorimeter and spectrophotometer are the laboratory instruments used for this purpose. These instruments work on the principle that when a light at a particular wavelength is passed through a solution (incident light), some amount of it is absorbed and, therefore, the light that comes out (transmitted light) is diminished. The nature of light absorption of protein in a solution is governed by the "**Beer–Lambert law**". Beer's law states that the amount of transmitted light decreases exponentially with an increase in the concentration of the absorbing material. According to Lambert's law, the transmitted light decreases exponentially with an increase in the thickness of the absorbing molecules (i.e. concentration of material in the medium). By combining the two laws, the following mathematical formula can be obtained.

$$I = I_o^{\varepsilon ct}$$

where I=Intensity of the transmitted light, I_o=Intensity of the incident light, ε=Molar extinction coefficient of substance being investigated, c=Concentration of the absorbing substance (moles/l or g/dl) and t=Thickness of medium through which light passes. The ratio of transmitted light (I) to that of incidence light (I_o) is referred to as transmittance (T).

$$T = \frac{I}{I_o}$$

Absorbance (A) or optical density (OD) is very commonly used in laboratories. The relationship between absorbance and transmittance is expressed by the following equation.

$$A = 2 - \log_{10} T = 2 - \log \% T$$

Colorimeter

A colorimeter is used for the measurement of colored substances. This instrument is operative in the visible range (400–800 nm) of the electromagnetic spectrum of light. The colorimeter in general consists of light source, filter, sample holder and detector with display (meter or digital). However, all colorimetric assays have the disadvantage that different proteins produce different absorbance and thus, unless the same protein as that being assayed is used as the standard, the protein values obtained are relative rather than actual values. Protein standards should be assayed at the same time and in the same solution as the unknown to take into account interactions of the buffer with the assay's reagents, reagent decomposition, time differences, temperature changes, etc.

Spectrophotometer

The spectrophotometer primarily differs from the colorimeter by covering the UV-region (200–400 nm) of the electromagnetic spectrum. Further, the spectrophotometer is more sophisticated with several additional devices that ultimately increase the sensitivity of its operation several-fold when compared to a colorimeter. A precisely selected wavelength (234 nm or 610 nm) is used for measurements. The spectrophotometer has similar basic components as described for a colorimeter and its operation is also based on the Beer–Lambert law.

II ULTRAVIOLET ABSORPTION METHOD (OPTICAL DENSITY)

Ultraviolet light (UV light) absorption methods for protein quantitation have several advantages: (i) They can be performed directly and non-destructively on the sample without the addition of any reagent; (ii) they can be performed very rapidly since no incubations are required; and (iii) the relationship between protein concentration and absorbance is linear. The aromatic amino acid residue tyrosine and tryptophan in a protein exhibit an absorption maximum at a wavelength of 280 nm. Since the proportions of these aromatic amino acids in proteins vary, so too does the extinction coefficient for individual proteins. However, for most proteins, the Isoelectric focussing (pI) lies in the range 0.4 to 1.5. So for a complex mixture of proteins, it is a fair approximation to say that a solution with an absorbance at 280 nm (A_{280}) of 1.0, using a 1 cm path length, has a protein concentration of approximately 1 mg cm^{-3}.

The majority of the proteins absorb light in the UV region with two maxima, a peak at 280 and 200 nm. Absorption spectroscopy involves the absorption of a photon by an electron. Only those photons with a certain energy level can be absorbed as defined by the difference in energy between the orbital of the unexcited electron and a higher orbital. This is why there are absorption maxima. Photons with higher energy have shorter wavelengths. Thus, electrons that are excited at 280 nm have absorbed less energy than those at 200 nm. Less energy is required for the electrons which absorb at 280 nm because these electrons lie within aromatic rings which stabilize the excited state due to resonance. Nearly all proteins characteristically absorb

UV light with a peak at 278–280 nm, due to the presence of aromatic rings in tryptophan and tyrosine residues. The absorbance of an individual protein is therefore a function of the relative proportion of these residues. If this absorbance value is known, this is the most accurate way of determining concentration. Measurement of absorbance (optical density, OD_{280}) requires quartz cuvettes as glass absorbs light at this wavelength.

The extinction coefficient gives the absorbance at 280 nm of a 1 cm path length of a 1% (10 mg/ml) solution of a given protein. These can vary widely and it is not valid to extrapolate from BSA or any other standard protein. Some examples are: Bovine serum albumin (6.67), Human serum albumin (5.31), Lysozyme, Chicken (29.0), Immunoglobulin (14.0), Ovalbumin (7.35).

Materials

Solutions

Protein solution to be measured; Spectrophotometric standard protein solution; Buffer in which protein is dissolved.

Equipments

Spectrophotometer with UV lamp; Quartz cells-1 cm. Light path

Method-1: Absorbance at 280 nm (range: 20–3,000 µg)

Procedure

1. Turn on the UV lamp of the spectrometer and warm up the instrument for 15 min. at least. Adjust to the required wavelength needed (280 nm).
2. Zero the spectrometer using the buffer in which the protein is dissolved as a blank.
3. Measure the absorbance (A) of the protein solution. If the value obtained is >2.0 at 280 nm, dilute an aliquot of the sample, until reading falls between 0.1 and 2.0.
4. For unknown or protein mixture, use the following formula for range estimate (For a protein whose absorption coefficient is known).

$$\text{Concentration (mg/ml)} = \text{Absorption of protein at 280 nm}$$

$$\text{Concentration (mg/ml)} = \text{Absorbance}/A_{1cm}^{1mg/ml}$$

$$\text{Concentration (\%)} = \text{Absorbance}/A_{1cm}^{1\%}$$

$$\text{Concentration (M)} = \frac{\text{Absorbance}}{E_m} \quad (E_m = \text{molar extinction coefficient})$$

Percentage protein and molarity can easily be converted to milligrams per millilitre protein by the following formulas:

$$\text{Concentration (mg/ml)} = \frac{\text{Percentage protein}}{10} = \frac{\text{molarity}}{\text{protein molecular weight}}$$

To correct for nucleic acid content, perform the following steps in addition to those outlined above: Zero the cuvette containing the buffer at 260 nm. Next, place the solution containing the protein in the cuvette and read the absorbance. To determine the protein concentration of the solution, use the following formula:

$$\text{Protein concentration (mg/ml)} = 1.55\, A_{280} - 0.76\, A_{260}$$

Method-2: Calibration curve method (range: 10–3,000 µg)

Requirements

Standard protein (BSA)
Standard solution = 10 mg/100 ml (= 100 µg/ml)

Procedure

1. Prepare different known concentration tubes of standard protein with distilled water.
2. Take absorbance at 280 nm against water as blank.
3. Prepare calibration curve of known concentration of proteins (mg/ml) vs absorbance.
4. Prepare different concentrations of unknown protein and take absorbance at 280 nm against water as blank.
5. Find out the unknown protein concentration from this calibration curve on the graph sheet.

Method-3: Absorbance at 280 and 260 nm (range: 10–3,000 µg)

Although proteins exhibit an absorption maxima at 280 nm due to the amino acids tyrosine and tryptophan, the third aromatic amino acid phenylalanine exhibits an absorption maxima at 255 nm. Thus, it is sometimes advisable to take absorbance of a protein solution at 260 nm and 280 nm. Moreover, nucleic acids due to their constituent purines and pyrimidines, absorb at 260 nm, the dual wavelength absorbance measurement gives an indication of the concentration of the protein solution by nucleic acid.

Procedure

1. Set the wavelength of the spectrophotometer at 280 nm and zero the absorbance with the buffer in the cell.
2. Read absorbance of the sample in the same cell.
3. Repeat steps 1 and 2 at 260 nm.
4. Calculate the ratio of 260:280 nm. This should be below 0.6, higher ratios indicate that the protein is contaminated with interfering substances notably nucleic acids.

$$\text{Concentration of the sample} = \frac{\text{Absorbance at 260 nm}}{\text{Extinction coefficient at 280 nm}} \times 10\, \text{mg/ml}$$

For mixtures of proteins or any proteins with unknown extinction coefficient:

$$\text{Protein concentration (mg/ml)} = 1.55 \times \text{absorbance at 280 nm} - 0.77 \times \text{absorbance at 260 nm}$$

Notes

1. Remember to calculate the absorbance for the original solution in a 1 cm light path if it is diluted or measured in thinner cells.
2. If a buffer or protein solution is cold, the outside of the cuvette may need to be wiped between each reading with a lint-free wiper and the readings made quickly.
3. Crude cell extractions containing RNA and/or DNA will produce erroneously high protein estimates. To determine the protein concentration of the solution use the following formula:

    Protein concentration (mg/ml) = $1.55 \, A_{250} - 0.76 \, A_{260}$

4. If the absorbance of the protein solution is greater than 2, dilute the protein sample in the same buffer as the original solution.
5. Particles of lipid droplets cannot be present in the solution since they have a strong tendency to scatter light at short wavelengths.

III COLORIMETRIC ASSAYS

Method-1: Biuret method (0.2–1.0 mg)

The Biuret reaction for protein determination is one of the first colorimetric protein assays developed and is still used widely. It is mostly used in applications requiring fast but not highly accurate, determinations. The Biuret reaction occurs with all compounds that contain two or more peptide bonds. The reagent consists of a solution of dilute copper sulfate in strong alkali. The purple-blue color produced is attributed to the formation of a coordination complex between the Cu^{2+} and four nitrogen atoms, two from each of two adjacent peptide chains. The name of the reaction is derived from the organic compound Biuret. The failure of ammonium sulfate to interfere with color formation makes it advantageous for determinations during the early steps of purifying a protein. Substrates interfering with this assay include buffers which are amino acids or peptide in nature, such as Tris and Good's buffer because they give positive color formation. The Cu^{2+} ions are also susceptible to reduction. This can be detected in practice by the appearance of a reddish blue color in the reaction mixture.

Materials

Solutions

Preparation of Biuret reagents:

1. Place a 1.50 g $CuSO_4$–$5H_2O$, 6.00 g sodium potassium tartarate $(NaKC_4O_6–4H_2O)_4$, and a stirring bar in a 1 litre volumetric flask.
2. Add 500 ml glass-distilled water to the flask and dissolve the above solids.
3. While stirring the contents of the flask vigorously, add 300 ml 10% (w/v) NaOH.
4. Remove stirring bar from the flask and bring the volume of liquid to 1 litre with glass distilled water.
5. Mix the contents thoroughly and transfer them to a plastic bottle.

6. The Biuret reagents just prepared should be a deep royal blue. This solution may be stored indefinitely. If a black precipitate is observed in the storage container, discard the solution and prepare a fresh one.

Chemicals

BSA; $CuSO_4$-$5H_2O$; $NaKC_4O_6$-$4H_2O$; NaOH.

Equipment

Hot water bath; Stirrer with magnetic stirring bar; Glass test tubes (16×150 mm); Spectrophotometer.

Procedure

1. Take 10 clean glass test tubes (16×150 mm). Into each test tube, carefully pipette one of the following volumes of a 10 mg/ml solution of BSA: 0, 0.1, 0.2, 0.3, 0.4, 0.5, 0.6, 0.7, 0.8, and 1.0 ml.
2. Bring the total volume of liquid in all tubes to 1.0 ml by adding an appropriate amount of glass-distilled water.
3. Add 4.0 ml Biuret reagent to each tube and vortex thoroughly.
4. Incubate the tubes for 20 min. at 37°C.
5. Determine the O.D. at 510 nm of each sample. Color is stable for 1–2 hours but slowly increases over a period of several hours. Therefore, read the samples as soon as possible.
6. Plot the data obtained at step-5. Assay unknown samples in the same way and determine their protein content from the standard curve just prepared.

Method-2: Lowry method (range: 2–100 µg)

The lowry (Folin–Ciocalteau) method (Lowry et al. 1951) is another colorimetric method for protein estimation. It is nearly a 100-fold more sensitive than the Biuret method and is an extension of the Biuret method. In the initial step, the copper sulfate reacts with the peptide bond in an alkaline medium and forms a complex, as in the Biuret method. Then this complex—an aromatic amino acid e.g. tyrosine and tryptophan—reduces the phenol reagent i.e. phosphomolybdophosphotungustic acid, a blue colored complex. The intensity of the blue is measured colorimetrically. The method is more time-consuming and is sensitive to interfering ions, mercaptans, and a number of compounds. For estimating proteins in membranous fractions, a modified Lowery's method is used. It involves SDS in extraction medium as well as in reagents C.

Materials

Preparation of Lowry's reagents:

1. 2% Na_2CO_3 in 0.1 NaOH (reagent 'A')
2. 0.5% $CuSO_4$-$5H_2O$ in 1% sodium potassium tartrate (reagent 'B')-
3. Alkaline copper solution: Mix 50 µl of A and 1 ml of B prior to use (reagent 'C').
4. 0.2 N Reagent: Mix 10 ml 2N Folin (Folin-Ciocalteu) "phenol reagent" with 90 ml water. This solution is stable for several months at room temperature if stored in amber bottles. (reagent 'D').
   Na-molybdate with orthophosphoric acid available commercially.

5. Protein standard: 1 mg/ml.

For modified Lowry's method:

1. SDS 10 g in 100 ml water.
2. NaOH 4 g in 100 ml water to make 1 M solution.
3. Mix 3 parts of copper reagent with 1 part of SDS and 1 part NaOH. This reagent may be stored for 2 to 3 weeks.

Procedure

1. Pipette out 0.2, 0.4, 0.6, 0.8, and 1 ml of standard into a series of test tubes.
2. Pipette out 0.1 ml and 0.2 ml of the sample extract into other test tubes
3. Make up the volume to 1 ml in all the test tubes. A test tube with 1 ml of water serves as the blank.
4. Add 5 ml of reagent C to each tube including the blank. Mix well and allow to stand for 10 min.
5. Then add 0.5 ml of reagent D. Mix well and incubate at room temperature in the dark for 30 min. Blue color develops.
6. Take the reading at 660 nm.
7. Draw a standard graph and calculate the amount of protein in the sample.

Notes

1. For good results, vortex the samples immediately after the addition of the Folin reagent.
2. Mix tubes thoroughly. There should not be any yellow gradient from top to bottom—the solution should appear homogeneous.
3. Heating the protein with the alkaline copper solution to 100°C for 20 min. is reported to equalize the reaction for different proteins.
4. The assay can be accelerated with the addition 10 µl of 20 mM DTT 3 min. after the addition of the Folin reagent.
5. Many compounds such as the Tris buffer, sucrose, ammonium sulfate, sulfhydryl compounds, urea, etc., also interfere with this assay.

Method-3: Coomassie blue (bradford) protein assay (0.2–20 µg)

The determination of protein concentrations by dye-binding is generally preferable to Lowry on account of speed, convenience, stability of reagents and few interfering substances. The basis of the assay is the shift in peak absorption by Coomassie Brilliant Blue from 465 nm to 595 nm when the dye is bound by protein. The dye principally reacts with arginine residues and to a lesser extent with histidine, lysine, tyrosine, tryptophan, and phenylalanine residues. Few substrates interfere with this reaction, except detergents, which will also bind dye and give spurious positive results (Bradford 1976). In addition, many proteins will not dissolve properly in an acidic reaction medium.

Materials

Bradford reagent:

1. Dissolve 100 mg Coomassie brilliant blue (final concentration 0.01%) in 50 ml of 95% ethanol (final concentration 4.7%) and 100 ml 85% Phosphoric acid (final concentration 8.5%).

2. Make up the volume to 1 litre with distilled water.

Alternatively the reagent can be substituted with 3% perichloric acid for 8.5% phosphoric acid.

Bradford Reagent, can be purchased from BioRad (Cat. No. 500–0006).

Standard protein solution: BSA 100 ng/ml).

Procedure

1. Warm up spectrophotometer 15 min. before use.
2. To 100 µl of protein sample (containing 1–100 µg of protein), add 0.9 ml (for 1–20 µg) or 5 ml (for 20–100 µg) of reagent. Mix well by vortex or inversion. Set up duplicates of the sample to be determined. Include a range (1 µg/ml to 1 mg/ml) of a reference protein (e.g. BSA) to construct a standard line, and include two tubes with reagent alone.
3. Read after two minutes and before 60 min. at 595 nm, measuring differences between the sample and the blank. Draw a graph of protein standards and read off unknown concentrations by OD value.

Notes

1. Protein-dye complex binds to quartz cuvettes but not to glass or plastic. Therefore, polystyrene cuvettes are recommended.
2. Bound dye has a broad absorption peak. Readings may be taken up to 620 nm.
3. Reference protein concentrations should be determined by OD at 280 nm. For BSA, the extinction coefficient for a 1 mg/ml solution is 0.66.
4. The Serva Blue G has a greater dye content than Coomassie Blue G and produces less difference in color yield from protein to protein.

Method-4: Kjeldahl analysis

Kjeldahl analysis is a general chemical method for determining the nitrogen content of compounds. This method is not normally used for the analysis of purified proteins or for monitoring column fractions but it is frequently used for analyzing complex solid samples and microbiological samples for protein content. The sample is digested by boiling with concentrated sulphuric acid in the presence of sodium sulphate (to raise the boiling point) and a copper and/or selenium catalyst. The digestion converts all the organic nitrogen to ammonia, which is trapped as ammonium sulphate. Completion of the digestion stage is generally recognized by the formation of a clear solution. The ammonia is released by the addition of excess sodium hydroxide and removed by steam distillation in a Markham still. Ammonia is collected in boric acid and titrated with standard hydrochloric acid using methyl red–methylene blue as indicator. It is possible to carry out the analysis automatically in an autokjeldahl apparatus. Alternatively, a selective ammonium ion electrode may be used to directly determine the content of ammonium ion in the digest. Although the Kjeldahl analysis is a relatively accurate method for the determination of nitrogen, the determination of the protein content of the original sample is complicated by the variation of the nitrogen content of individual proteins and by the presence of nitrogen in contaminants such as DNA. In practice, the nitrogen content of proteins is generally assumed to be 16% by weight.

IV GENERAL INSTRUCTIONS

A. CUVETTES

Glass cuvettes may be used for wavelengths above 320 nm. Although quartz cuvettes may be used for the entire ultraviolet to visible spectrum, they are generally not used for visible wavelength readings because of their cost and also because the Coomassie Blue dye sticks to quartz more readily than glass. Disposable styrene cuvettes may be used at wavelengths above 340 nm and acrylic cuvettes may be used at 275 to 350 nm. Cuvettes are fragile and easily scratched, so non-scratching pipettes that will fit into standard cuvettes should be used. Cuvettes should be washed immediately after use to prevent proteins from adhering to the inside surface of the cuvette. The cuvettes have to always be handled by their sides to prevent fingerprints from increasing the absorbance of the sample. It should be insured that the cuvettes in place are properly adjusted i.e. the cuvettes should not wobble in their holders. The cuvette has to be left in place and the solutions removed and replaced with a pipette.

B. PROTEIN STANDARDS

Most methods of protein quantitation compare an unknown quantity to a standard curve of a "known" protein amount. Absolute protein quantitation can be tricky for several reasons. First, different proteins will react to varying extents in different protein assays. For example, 10 µg of gelatin will produce only 69% as much absorbance as 10 µg BSA in the Lowry assays. In preparing standards, salts and moisture can add to the weight of a protein powder. Different pH and ionic strength, can alter the UV absorbance of proteins. For these reasons, protein values in many publications are actually relative values and thus it is very important that information be provided about the standard protein and assay used.

Commonly, BSA is used as a protein standard. This is because it is extensively studied and makes present and future protein determinations readily comparable. It is also easy to handle and inexpensive. Standards can be prepared as follows: The powder should be stored at –20°C in a desiccator. The jar should be allowed to warm up to room temperature before being weighed. Enough protein should be weighed out for accurate measurement. It should be dissolved in water to make 2 mg/ml solution, aliquoted into 0.5 ml fractions, and stored in a freezer at –20°C. It should be noted that not all proteins are soluble in water.

C. REMOVAL OF INTERFERING SUBSTITUTES

a) *Neutralization:* Strong acids, bases, or buffers should be neutralized or adjusted to the optimal pH of the assay system by the addition of HCl or NaOH. EDTA, which interferes with the copper-based assays by chelating the copper, may be counteracted with the addition of equimolar $CaCl_2$. The $CaCl_2$ precipitates in alkaline solutions and must be removed by centrifugation before absorbance readings are taken. Thiols, such as DTT, mercaptoethanol, and cysteine, can be neutralized by the addition of 8 Ml 2 M iodoacetic acid dissolved in 2.5 M NaOH.

b) *Precipitation*: Buffers, lipids, reducing agents, and certain detergents can be removed by precipitating the protein using TCA, PCA, or acetone and re-suspending the protein in a suitable solution for the cohesen assay.

c) *Dialysis and ultrafiltration:* The time required for dialysis is inversely dependent on pore sizes. Dialysis can take just a few hours. Ultrafiltration is still faster and can take as little as 30 min. to perform.

D. INSTRUMENTATION

The choice of the spectrophotometer purchased depends on the types of protein assays planned. The most versatile instrument is one with both a visible and ultraviolet source. The spectronic 20 spectrometer is convenient for routine assays using light in the visible range. A relatively new convenient innovation is the microwell plate reader, which automates spectrophotometric readings. All the colorimetric procedures described above have been modified for those plate readers. However, the microwell plate reader doesn't have UV light capabilities.

V DETECTION OF PROTEINS IN GELS

The location of a protein in a gel can be determined by either Coomassie blue staining or silver staining. The former is easier and more rapid; however, silver staining methods are considerably more sensitive and thus can be used to detect smaller amounts of proteins.

A. GEL RECOVERY AND FIXATION

Once the electrophoresis is complete, the gel is removed carefully from the apparatus for localization of the separated polypeptides. Gloves should be worn throughout this procedure, as contact with the skin can lead to the production of artifacts during staining. Slab gels are simply dismantled by removing the spacers and carefully lifting the glass plates apart with a spatula. Gels cast in capillary tubes, such as those used for the first IEF dimension of 2-D PAGE are best removed by hydrostatic pressure. This is achieved by attaching a syringe filled with water to the glass tube with silicone rubber tubing. As a last resort, the glass tube is carefully broken, using a hammer, to release the gel.

Proteins can be visualized directly as unfixed proteins within gels, but these are usually applied only if it is wished to recover the separated components (e.g. purification of enzymes). For most visualization techniques, it is essential to precipitate and immobilize the separated proteins within the gel and to remove any non-protein components which might interfere with subsequent staining. The best general fixative is 20% (w/v) trichloroacetic acid (TCA), whilst sulfosalicylic acid (1–20%, w/v) or mixtures of TCA and sulfosalicylic acid (0% each) are less efficient. Methanolic solutions of acetic acid (e.g. methanol, distilled water, acetic acid, 9:9:2, v/v) are very popular for gel fixation, but the limitation is that the low molecular species, basic proteins, and glycoproteins may not be adequately fixed. Aqueous solutions of formaldehyde (5% w/v) or 2% (w/v) gluteraldehyde can be used to cross-link proteins covalently to the gel matrix.

B. GENERAL PROTEIN STAINS

Organic dyes such as Bromophenol blue and Amido black 10B are popular protein staining reagents (sensitivity range from 0.2–0.5 µg of protein per band). The level of sensitivity has been further increased with the development of autoradiographic and fluorographic methods for radio-labeled proteins. But the introduction of fluorescent (10 ng of protein per band) and silver (0.1 ng of protein per band) staining techniques has further enhanced the resolution.

Method-1: Coomassie blue staining

The most popular protein staining procedures following electrophoresis are based on the use of no-popar, sulfated triphenyl-methane Coomassie stains. This is the standard method of protein detection. Detection of protein bands in a gel by Coomassie blue staining depends on non-specific binding of the dye to the proteins and the thickness of the gel (10% polyacrylamide gel of 0.5 mm thickness should be stained for about 2 h). In practice, immersion of the gel in staining solution overnight is often convenient. In order to visualize the protein bands, it is necessary to destain the gel by gentle agitation in the same acid-methanol solution without dye. The detection limit is 0.3–1 µg/protein band. Staining methods using CBB are simple and convenient, but suffer from the disadvantage that they are less sensitive (0.2–0.5 µg of protein per band).

Materials

Staining solution: 0.1% Coomassie Brilliant Blue R-250 (w/v) in 40% methanol (v/v), 10% acetic acid (v/v). Filter the staining solution through a Whatman No. 1 filter after the dye has dissolved. The staining solution is re-usable. Store at room temperature.

Fixing solution: 40% methanol (v/v), 10% trichloroactic acid (w/v).

Destaining solution: 40% methanol (v/v), 10% acetic acid (v/v).

Polyacrylamide gel 7% (v/v) aqueous acetic acid; Cellophane membrane backing sheets; Whatman 3MM filter paper (optional); Deionized, distilled water.

Procedure

1. Place polyacrylamide gel in a plastic container and cover with 3 to 5 gel volumes of fixing solution. Agitate slowly for 2 hours on an orbital shaker.
2. Pour out fixing solution, cover gel with Coomassie staining solution, and agitate slowly for 4 hours.
3. Pour out staining solution. Rinse gel briefly with ~50 ml fixing solution.
4. Pour out fixing solution. Cover the gel with de-staining solution and agitate slowly for 2 hours. Pour out de-staining solution. Add fresh de-staining solution and continue destaining until blue bands and a clear background are obtained. Store the gel in 7% acetic acid or water.
5. If desired, photograph gel. Use a yellow-orange filter, such as a Wratten #8 or #9 filter series A with a fine-grained panchromatic halftone film, such as Kodak T-Max 100, for photography.
6. Alternatively, place gel on two sheets of Watman 2MM filter paper and cover top with plastic wrap. Dry in a conventional gel dryer for 1 to 2 hours at 80°C.

Note

1. For long-term storage, keep gels in a covered tray of 5% methanol, 7.5% acetic acid, or dry gel on transparent cellophane or filter paper.

Method-2: Fluorescent staining methods

An increased sensitivity of detection of proteins in gels can be achieved using fluorescent compounds. Two fluorescent techniques are commonly used. In the first approach, proteins are coupled with a fluorescent dye before electrophoresis and fluorescent protein bands are detected after electrophoresis by scanning. For example, dansyl chloride and fluorescamine have sensitivity as low as 10 ng and 5 ng protein per band, respectively. The main disadvantage of pre-electrophoretic staining procedures is that they can cause protein charge modifications. Consequently they alter protein mobility during electrophoresis in buffers, which do not contain SDS.

The second approach, which also overcomes the problem of protein charge modifications, is to label the proteins with fluorescent molecules such as aniline-8-naphthalene sulfonate (ANS) and 0-phthalaldehyde (OPA) after the electrophoretic separation has been completed. However, these methods are not sensitive (can detect 0.5 µg protein per band) and need special scanning and visualization equipments.

Method-3: Silver staining

Silver staining of proteins following electrophoresis was introduced by Swizter et al. (1979). This method, developed by Merril and co-workers, can be as much as 100 times more sensitive than dye staining. Bands containing 10–100 ng of protein can be easily seen. Detection of protein bands in a gel by silver staining depends on binding of silver to various chemical groups (e.g. sulfhydryl and carboxyl moieties) in proteins. The detection limit is 0.1 to 5 ng/protein bands. The high sensitivity of silver staining methods makes them ideal for the detection of trace components within a protein sample or for the analysis of protein samples available in only limited quantity. However, high background staining, surface 'mirror' staining, poor reproducibility, high cost, and the facts that it is slow, labor-intensive and that certain proteins stain poorly, negatively or not at all are major disadvantages of this method.

Materials

BioRad Silver Stain Kit (161–0443), containing oxidizer, silver reagent and developer. Dilute oxidiser and silver reagents as required; make up stock bottle of developer (115 g makes 3.6 L) and store at room temperature.

Methanol, ethanol, and acetic acid.

Procedure

1. Fix the gel immediately after electrophoresis in about 400 ml of 40% methanol, 10% acetic acid (v/v) (or 40% methanol, 10% trichloroacetic acid) for 30 min. to overnight.
2. Fix twice in 400 ml 10% ethanol, 5% acetic acid (v/v) for 15 min. twice, to remove SDS.
3. Soak the gel for 3–10 min. in 200 ml of fresh 10% oxidizer solution (0.0034 M potassium dichromate, 0.0032 N nitric acid).

4. Wash the gel three or four times for 5–10 min. in 400 ml deionised water, until the yellow color has been washed out.
5. Soak the gel in 200 ml fresh 10% silver reagent (0.012 M silver nitrate) for 15–30 min.
6. Wash the gel with 400 ml deionised water for 1–2 min.
7. Wash the gel for about 1 min. in the developer (0.28 M sodium carbonate, 1.85% paraformadehyde).
8. Replace the developer with fresh solution and incubate for 5 min.
9. Replace the developer a second time and allow development to continue until satisfactory staining has been obtained. Reaction time can vary with gel thickness.
10. Stop development in 400 ml of 5% acetic acid (v/v) for 15 min. and transfer to water.
11. Pour off solution. Wash several times in water.
12. Photograph gel. Dry gel or store in sealable plastic bag after soaking in 0.03% sodium carbonate for 10 min. (6 to 12 months). Use a blue-green filter such as a Wratten #58 filter with Kodak T-Max 100. Photograph gels as soon as possible because there may be slight changes in color intensity and increases in nonspecific background.

Notes

1. Vertical streaks and sample-independent bands in the 50- to 70-kDa region are sometimes seen in silver-stained gels. These artifacts have been attributed to reduction of contaminants inadvertently introduced into the samples. They can be eliminated by adding excess iodoacetamide to samples after treatment with SDS-reducing buffer.
2. If bands do not appear after the initial silver staining, the gel should be soaked in deionized water for 30 min. and the process repeated from step 5. The developing time can be adjusted to prevent over-development and high background.

Method-4: Radioactive detection methods

Protein samples to be analyzed by electrophoresis can be radio-labeled synthetically by the incorporation of radioactive amino acids. This approach is generally used in tissue culture systems where the culture medium should be depleted of the amino acid used for labeling. Techniques are also available for synthetic labeling of specific proteins, for example, $^{32}PO_4^{3-}$ for phosphoproteins and [^{3}H]glucosamine for glycoproteins. Proteins can also be radio-labeled post-synthetically, prior to electrophoresis, using a variety of approaches such as iodination with ^{125}I or reductive methylation with [^{3}H]sodium borohydride.

For techniques in which the gel is to be placed in direct contact with X-ray film (i.e. autoradiography and fluorography), it is necessary that the gel is first dried. Thin and ultra-thin gels cast on glass or plastic supports can, after equilibration in 3% (w/v) glycerol, be dried simply in air or in an oven at 40–50°C. Unsupported gels can be dried by equilibrating gel in 3% (w/v) glycerol to prevent cracking and then placed between two cellophane sheets supported in a plastic frame. The most popular method for drying gels is by heating under a vacuum. Suitable apparatus is available from several manufacturers.

VI SUGGESTED READING

Bradford MM (1976) A Rapid and Sensitive Method for the Quantitation of Microgram Quantities of Protein Utilizing the Principle of Protein-dye Binding, *Anal Biochem,* **72**:248–254.

Deutscher M (1990) *Guide to Protein Purification*, Academic Press Inc., London and New York.

Doonan S (1996) *Protein Purification Protocols*, Human Press, Totowa, NJ.

Dunn MJ (1993) *Gel Electrophoresis: Proteins*, Bios Scientific Publishers.

Lowry OH, Rosebrough NJ, Farr AI and Randall RJ (1951) Protein Measurement with the Folin-phenol Reagent, *J Biol. Cem.,* **193**:265–268.

Scopes RK (1993) *Protein Purification–Principles and Practice*, Springer-Verlag, Berlin.

Stoscheck CM (1990) Quantitation of Protein, *Methods in Enzymology,* **182**:50–68.

Wilson K and Walker J (2000) *Practical Biochemistry, Principles and Techniques* (5th Edn), Cambridge University Press.

25

Protein Fractionation Techniques

I INTRODUCTION

Most research in molecular biology uses proteins at every turn. Frequently it is necessary to purify these proteins. Proteins are remarkably labile to pH, monovalent and divalent ions, specific ions and specific metals, temperature, and protein concentration. Thus, it is important to pay close attention to buffers when working with proteins. The removal of low-molecular weight solutes and concentration of protein solutes are the most frequently performed operations in protein purification and analysis. The highly specific characteristics of an individual protein and the requirements of the experiment must be taken into consideration. In most cases concentrating and de-salting of a protein solution can be achieved without substantial losses or effects on the biological activity of the protein. The objectives of the procedure may be as diverse as concentrating and de-salting bulk quantities of proteins at the initial steps of a purification or the preparation of microgram quantities of labile enzyme for physical, chemical, or biological analysis. Often the composition of protein samples has to be adjusted for a preparative or analytical separation. Electrophoretic techniques, ion-exchange chromatography and protein sequencing are particularly sensitive to the presence of salts. Additionally, the techniques require concentrated protein samples as is the case in size-exclusion chromatography and many other advanced techniques such as high-performance liquid chromatography (HPLC), mass spectrometry, and protein microsequencing techniques (Figure 25.1).

II IONIC PROPERTIES OF AMINO ACIDS AND PROTEINS

Twenty amino acids varying in size, shape, charge, and chemical reactivity are found in proteins. Since they possess both an amino group and a carboxyl group, amino acids are ionized at all pH values, i.e. a neutral species represented by the general formula does not exist in solution irrespective of the pH. This can be seen in Figure 25.2. Thus, at low pH values, an amino acid exists as a cation and at high pH values, as an anion. At a particular intermediate pH an amino acid carries no net charge, although it is still ionized. This ion is called a *zwitterion*. This ionization confers upon its physical properties characteristic of ionic compounds, i.e.,

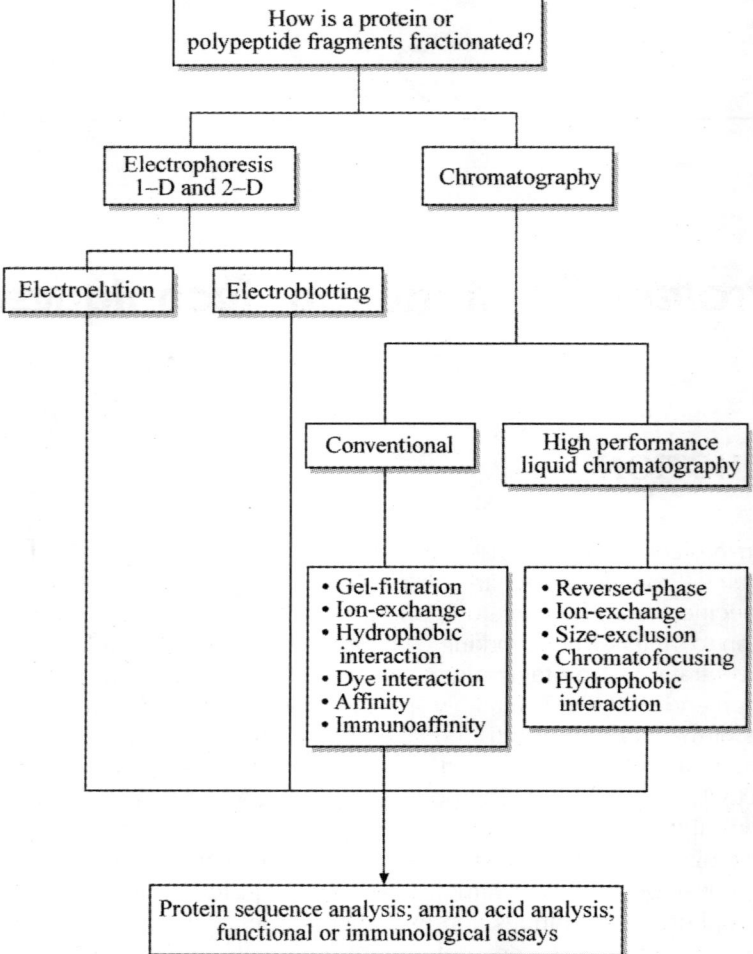

Figure 25.1 *Fractionation of proteins and polypeptide fragments*

high melting point and boiling point, water solubility and low solubility in organic solvents such as ether and chloroform. The pH at which the zwitterion predominates in an aqueous solution is referred to as **isoionic point** because it is the pH at which the number of negative charges on the molecule produced by ionization of the carboxyl group is equal to the number of positive charges acquired by proton acceptance by the amino group. In the case of amino acids, this is equal to the **isoelectric point** (**PI**), since the molecule carries no net charge and is the therefore electrophoretically immobile. As an alternative to possessing a second amino or carboxyl group, an amino acid side chain may contain in the R group of the general formula quite a different chemical group that is also capable of ionizing at a characteristic pH.

Proteins are formed by the condensation of the α-amino group of one amino acid with the α-carboxyl of the adjacent amino acid. With the exception of the two terminal amino acids,

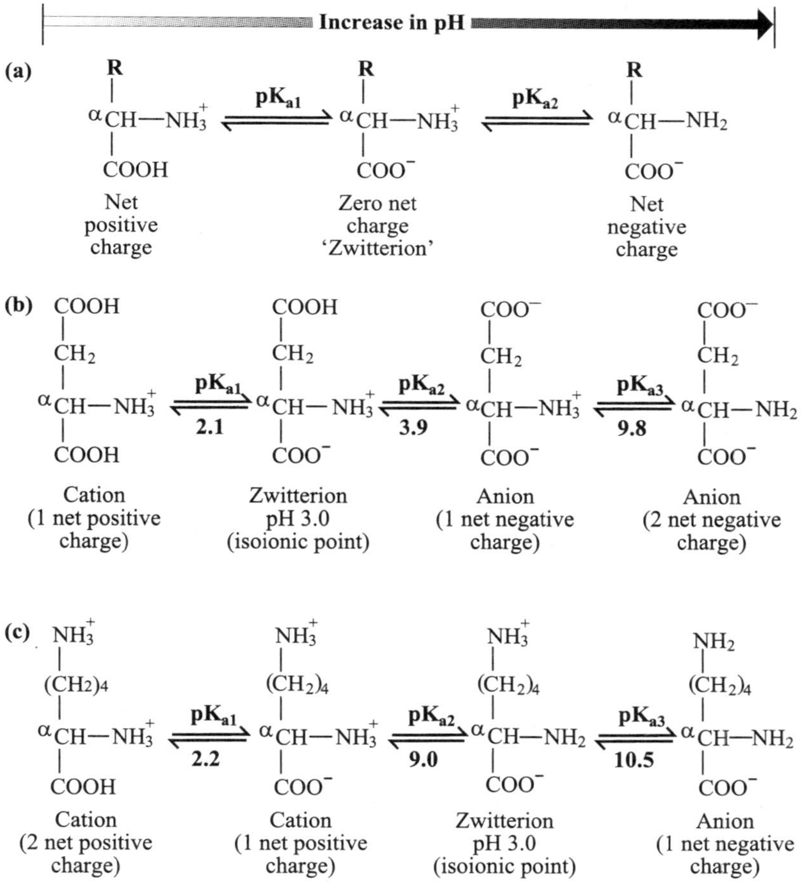

(a)

$$\underset{\substack{\text{Net}\\\text{positive}\\\text{charge}}}{\overset{R}{\underset{\text{COOH}}{\alpha\text{CH}-\overset{+}{\text{NH}_3}}}} \underset{}{\overset{pK_{a1}}{\rightleftharpoons}} \underset{\substack{\text{Zero net}\\\text{charge}\\\text{'Zwitterion'}}}{\overset{R}{\underset{\text{COO}^-}{\alpha\text{CH}-\overset{+}{\text{NH}_3}}}} \underset{}{\overset{pK_{a2}}{\rightleftharpoons}} \underset{\substack{\text{Net}\\\text{negative}\\\text{charge}}}{\overset{R}{\underset{\text{COO}^-}{\alpha\text{CH}-\text{NH}_2}}}$$

Increase in pH →

(b)

$$\underset{\substack{\text{Cation}\\(1\text{ net positive}\\\text{charge})}}{\overset{\text{COOH}}{\underset{\text{COOH}}{\overset{|}{\underset{|}{\text{CH}_2}}}\overset{|}{\underset{|}{\alpha\text{CH}-\overset{+}{\text{NH}_3}}}}} \overset{pK_{a1}}{\underset{2.1}{\rightleftharpoons}} \underset{\substack{\text{Zwitterion}\\\text{pH 3.0}\\(\text{isoionic point})}}{\overset{\text{COOH}}{\underset{\text{COO}^-}{\text{...}}}} \overset{pK_{a2}}{\underset{3.9}{\rightleftharpoons}} \underset{\substack{\text{Anion}\\(1\text{ net negative}\\\text{charge})}}{\text{...}} \overset{pK_{a3}}{\underset{9.8}{\rightleftharpoons}} \underset{\substack{\text{Anion}\\(2\text{ net negative}\\\text{charge})}}{\text{...}}$$

(c) Lysine ionization: pKa1 = 2.2, pKa2 = 9.0, pKa3 = 10.5

- Cation (2 net positive charge)
- Cation (1 net positive charge)
- Zwitterion, pH 3.0 (isoionic point)
- Anion (1 net negative charge)

Figure 25.2 (a) Amino acids are ionized at all pH values and neutral species represented by the general formula does not exist in solution irrespective of the pH, (b) For acidic amino acids such as aspartic acid, the ionization pattern is different. In this case, the zwitterion will predominate in aqueous solution at a pH determined by pKa1 and pKa2, and the isoelectric point is the mean of pKa1 and pKa2, (c) In the case of lysine, which is a basic amino acid, the ionization pattern is different and its isoionic point is the mean of pKa2 and pKa3

therefore, the α-amino acid and carboxyl groups are all involved in peptide bonds and are no longer ionisable in the protein. Amino, carboxyl, imidazolyl, guanidino, phenolic, and the sulphydryl groups in the side chains are, however, free to ionize and of course there may be many of these groups. Proteins fold in such a manner that the majority of these ionisable groups are on the outside of the molecule, where they can interact with the surrounding aqueous medium. The relative numbers of positive and negative groups in a protein molecule influence aspects of its physical behavior, such as solubility and electrophoretic mobility. The isoionic point of a protein and its isoelectric point, unlike that of an amino acid, are generally not identical. In practice, protein molecules are always studied in buffered solutions, so it is the isoelectric point that is important. It is the pH at which, for example, the protein has minimum solubility.

III FRACTIONATION METHODS

A. PRELIMINARY PURIFICATION STEPS

The initial extract (or homogenate) produced by the disruption of cells and tissue will invariably contain insoluble matter. This is most easily removed by filtering through a double layer cheesecloth or by low speed (5000g) centrifugation. If the fat is floating on this surface, it can be removed by filtration through glass wool or cheesecloth. The solution will still be cloudy with organelles and membrane fragments, and these can be removed during the salt fractionation or precipitation with Celite, or by a cell debris remover (CDR) a cellulose-based absorber, or any number of flocculants such as starch, gums, tannins, or polyamines. The cell extract may also contain other molecules such as nucleic acids, carbohydrates, and lipids as well as any number of small molecular weight metabolites. Small molecules tend to be removed later on during **dialysis** steps or steps that involve fractionation based on size (e.g. **gel filtration**) and are therefore of little concern. Some workers include DNase I in the extraction buffer to reduce viscosity, the small DNA fragments generated being removed at the later dialysis/gel filtration steps. Likewise, RNA can be removed by treatment with RNase. The addition of a solution of protamine sulphate to the extract precipitates most of the DNA and RNA, which can subsequently be removed by centrifugation. The protein can best be removed by totally precipitating the protein with ammonium sulphate leaving the gummy material in solution. The protein can then be recovered by centrifugation and re-dissolving it in the buffer. The clarified extract is now ready for the protein fractionation steps that are to be carried out.

1. General properties of proteins

Most purification methods are based on exploiting those properties by which proteins differ from one another. These different properties, and the techniques that exploit these differences, are:

(a) **Stability**: Denaturation fractionation exploits the differences in the heat sensitivity of proteins. The native structure of proteins is maintained by a number of forces, mainly hydrophobic interactions, hydrogen bonds, and sometimes disulphide bridges. If a protein is denatured, these bonds will be disrupted and the protein chain unfolds to give the insoluble, 'denatured' protein. However, different proteins will denature at different temperatures, depending on their different thermal stabilities. In a similar way, proteins differ in the ease with which they are denatured at extremes of pH (<3 and >10). More sensitive proteins will precipitate and can be removed by centrifugation.

(b) **Solubility**: Proteins differ in the balance of charged, polar, and hydrophobic amino acids that they display on their surfaces. Charged and polar groups on the surface are solvated by water molecules, thus making the protein molecule soluble whereas, hydrophobic residues are masked by water molecules found adjacent to these regions. Since solubility is a consequence of solvation of charged and polar groups on the surfaces of the protein, under particular set of conditions, proteins will differ in their solubilities. In particular, one exploits the fact that proteins precipitate differentially from the solution on the addition of species such as neutral slats or organic solvents. However, the importance of these methods is that they precipitate native protein that has become insoluble by aggregation and the protein will not be denatured. Salt fractionation is frequently carried out using ammonium sulphate. As the salt concentration increases, freely available water molecules that can solvate the ions become scarce. At this stage, those water molecules that have been forced into contact with hydrophobic groups on the surface of the protein are the next most freely available water molecules and these are therefore removed to solvate the salt molecules, thus leaving the hydrophobic patches exposed. As the ammonium sulphate concentration increases, the hydrophobic surfaces on the protein are progressively exposed. Thus these hydrophobic patches cause proteins to aggregate by hydrophobic interaction, resulting in precipitation.

(c) **Charge**: Proteins differ from one another in the proportions of the charged amino acids that they contain. Hence proteins will differ in net charge at a particular pH. This difference is exploited in ***ion-exchange chromatography***, where the protein of interest is bound onto a solid support material bearing charged groups of the opposite sign (ion-exchange resin). Proteins with the same charge as the resin pass through the column to waste. After this, bound proteins, containing the protein of interest, are selectively released from the column by gradually increasing the strength of the salt ions in the buffer passing through the column or by gradually changing the pH of the eluting buffer. These ions compete with the protein for binding to the resin, the more weakly charged protein being eluted at the lower salt strength and the more strongly charged protein being eluted at higher salt strength. Another feature of the different charged groups found in proteins is the fact that most proteins will differ in their isoelectric points (pI), i.e. they will differ in the pH value at which they have zero overall charge. This difference in pI can be exploited using chromatofocusing.

(d) **Affinity**: Certain proteins bind strongly to specific small molecules. This property can be exploited by developing an affinity chromatography system where the small molecule (ligand) is bound to an insoluble support. When a crude mixture of proteins containing the protein of interest is passed through the column, the ligand binds the protein to the matrix while all other proteins pass through the column. The bound protein can then be eluted from the column by changing the pH, increasing the salt strength, or passing it through a high concentration of unbound free ligand.

(e) **Hydrophobicity**: Proteins differ in the amount of hydrophobic amino acids that are present on their surface. This difference can be exploited in salt fractionation but can also be used in a higher resolution method using ***hydrophobic interaction chromatography*** (**HIC**). A typical column material would be phenyl-Sepharose where phenyl groups are bonded to the insoluble support Sepharose. The protein mixture is loaded on the column in a high salt solution, to ensure that the hydrophobic patches are exposed allowing hydrophobic interactions to occur between the phenyl groups on the resin and hydrophobic regions on the proteins. Proteins are then eluted by applying a decreasing salt gradient to the column causing the proteins to emerge from the column in order of increasing hydrophobicity.

2. Engineering protein for purification

The advances in recombinant technology have enhanced the ability to clone and over-express genes for proteins. This technology has also facilitated the purification process by manipulation of the gene of interest prior to expression. These manipulations are carried out either to ensure secretion of the proteins from the cell or to aid protein purification.

Secretion of proteins: Recombinant genes that are being expressed in microbial or eukaryotic cells have an advantage that they can manipulate cells to secrete proteins. (i) They facilitate purification, (ii) prevent intracellular degradation of the cloned genes, (iii) reduce intracellular concentration of toxic proteins, and (iv) facilitate post–translational modification of proteins.

Fusion proteins to aid purification: In this approach, an additional gene sequence (tag) is attached to the gene of interest such that the protein is produced as a fusion protein. This fusion protein will be selectively removed from the cell extract and can then be cleaved to release the protein of interest from the tag protein.

B. CONCENTRATION AND DE-SALTING

1. General considerations

The first parameter to be considered in a concentration or de-salting experiment is the initial volume of the protein sample and the volume that is desired after the procedure. Detailed information has to be had about the physical and chemical properties of the protein to be concentrated and/or de-salted facilitates the choice of the best method. Its molecular weight predetermines the exclusion limit of membranes used for dialysis and ultrafiltration, although the three-dimensional structure of the macromolecules strongly influences the behavior of the individual protein in the techniques. The choice of a suitable buffer system is dictated by the solubility of the protein and possible effects of the solution on its stability. A bad choice of buffer may result in the protein being denatured and/or precipitated. The composition of the desired solution used to concentrate also has to be carefully selected. Appropriate ionic strength, pH, and volume have to be adjusted in order to maintain the solubility of the protein. Particular attention has to be paid to the need for essential metal ions and cofactors that may be required to retain the native structure and biological activity of the protein.

To avoid significant losses of protein during concentration or de-salting, the tendency of the protein to adsorb to surfaces should be investigated. Many proteins containing charged or hydrophobic domains show a high affinity to various surfaces that may lead to irreversible binding or denaturation. An important decision to be taken is whether glass or plastic containers should be used since the adsorptive properties of proteins differ widely on these materials. Those involving the use of semi-permeable membranes are commonly more prone to adsorptive losses. Care must be taken to avoid chemical destruction of the protein during concentrating and/or de-salting. Oxidative and reductive interaction of reactive chemical compounds may cause considerable modification and damage to sensitive amino acid side-chain groups. The properties of the solutes are of critical importance for maintaining the stability of the protein. Some commonly used buffers or solutes, such as bicarbonate and urea, may contain or generate reactive, mostly electrophilic, compounds that modify proteins. The use of high-purity solvents and chemicals is therefore recommended in any concentrating and de-salting procedure. Another important factor that should be considered is the time required for concentration and de-salting of the protein solution. Techniques such as dialysis and lyophilization may take days to be completed. The destructive influence of chemical and enzymatic activities present in the protein sample or introduced as impurities due to the solvent and buffer used may become significant during such long periods of purification. Not only the loss of protein, but also the modification of the protein and microbial contamination are some of the other factors to be considered.

2. De-salting methods

Dialysis techniques for protein concentration

Dialysis is one of the oldest techniques for removal of low-molecular-weight solutes or exchange of the buffer. The method is based on the properties of a semi-permeable membrane separating the protein solution from the dialysis buffer. It allows free passage of molecules below a certain molecular weight, the so-called "molecular weight cutoff," while macromolecules cannot penetrate the pores of the membrane. The process of dialysis is driven by the difference in concentration of the solutes on the two sides of the membrane. As the equilibrium concentration is approached, the diffusion of solutes becomes equal in both directions. Further reduction in the solute concentration in the protein solution can only be achieved by changing the dialysis buffer. Thus, the number of buffer changes is more important than the total volume of the dialysis buffer. Dialysis is markedly accelerated by increasing the ratio of membrane area to the volume of the solution. Unfortunately, losses of protein due to adsorption to the membrane increase concomitantly. Hence, a compromise has to be made in the practical case between required dialysis time and optimum protein recovery.

The diffusion of a solute is dependent on the temperature and viscosity of the solution. Although a high temperature increases the rate of diffusion, in most cases the stability of the protein requires dialysis to be conducted at 4–8°C, normally in a cold room. A protein solution with too high a viscosity usually has to be diluted prior to dialysis because the rate of diffusion is drastically reduced with the increasing viscosity of the solution.

Considerable progress has been achieved in the development of dialysis membranes. Recent dialysis membranes possess a more precise molecular weight cutoff, and show less variation between batches in terms of average pore size and content of impurities. Membranes are available in a variety of sizes, in the form of tubing or as membrane sheets for special applications. They are also available in a number of thickness and pore sizes. Thicker membranes are tougher, but restrict solute flow and reach equilibrium more slowly. Pore size is defined by "molecular weight cutoff" (MWCO), i.e., the size of the smallest particle that cannot penetrate the membrane. A membrane with a pore size that is much smaller than the macromolecule of interest should be used; MWCO ranges from 1000 up to 50,000, so it is possible to select the optimal membrane for the individual protein.

Most dialysis membranes are made of derivatives of cellulose. They come in a wide variety of MWCO values, and also vary in cleanliness, sterility, and cost. Dry dialysis membranes contain significant amounts of sulfurous compounds and heavy metals. Protein dialysis should be done with clean membranes. In some cases, wetting of the membrane and washing with the intended buffer solution may be sufficient as a pre-treatment. However, the following procedure to remove impurities is recommended.

(a) Selection and preparation of dialysis membrane/tubing

Materials

    10 mM NaHCO$_3$; 10 mM Na$_2$EDTA pH 8.0; 20% to 50% (v/v) ethanol.

Procedure-1

1. Wear gloves and carefully transfer the entire roll of dry dialysis tubing as supplied to a 4 litre beaker containing 2 litres of a 10 mM NaHCO$_3$, and 10 mM Na$_2$EDTA solution

adjusted to pH 7.0. Alternatively, remove the membrane from the roll and cut into usable lengths (usually 8 to 12 in.) and wet membrane in a large excess of 10 mM $NaHCO_3$, 10 mM Na_2EDTA solution.
2. The beaker is placed in a shaking water bath, covered, and the temperature is brought to 60°C. Gentle agitation is continued for 2 hours.
3. The incubation is repeated with fresh solution.
4. The cleaning solution is replaced by 2 litres of double-distilled water and the dialysis tubing is washed for 1 hour. This step is repeated several times until the solution appears clear.
5. After slowly cooling to 4°C, the tubing is stored in a fresh volume of double-distilled water including 1 ml chloroform/litre as a preservative. It is also stored at 4°C in 20% to 50% ethanol to prevent growth of cellulytic microorganisms.

Procedure-2

1. Cut the tubing into pieces of convenient length (10–20 cm).
2. Boil for 10 min. in a large volume of 2% (w/v) sodium bicarbonate ($NaHCO_3$) and 1 mM EDTA (pH 8.0).
3. Rinse the tubing thoroughly in distilled water.
4. Boil for 10 min. in 1 mM EDTA pH 8.0 (Note 2).
5. Allow the tubing to cool, and then store it at 4°C. Be sure that the tubing is always submerged. From this point onward, always handle the tubing with gloves.
6. Before use, wash the tubing inside and out with distilled water.
7. Store the tubing as in step 5 of previous method.

Notes

1. It is recommended that solutions be pre-warmed to avoid sudden temperature changes that may cause alterations of the pore size of the membrane. Dialysis tubing should be handled with gloves to avoid contamination with proteases.
2. Instead of boiling for 10 min. in 1 mM EDTA pH 8.0, the tubing can be autoclaved at 20 lb/sq. in. for 10 min. of liquid cycle in a loosely capped jar filled with water.

(b) Basic protocol for large-volume dialysis

Materials

Macromolecule-containing sample to be dialyzed, Appropriate dialysis buffer, Dialysis membrane, Clamps.

Procedure

1. Wear gloves, remove dialysis membrane from ethanol storage solution, and rinse with distilled water. Secure clamp to one end of the membrane or knot one end.
2. Fill membrane with water or buffer, hold the unclamped end closed, and squeeze membrane. A fine spray of liquid indicates a pinhole in the membrane; discard and try a new membrane.
3. Replace the water or buffer in the dialysis membrane with the macromolecule-containing sample and clamp the open end. Again, squeeze to check the integrity of the membrane and clamps.

4. Immerse dialysis membrane in a beaker or flask containing a large volume of the desired buffer. Dialyze for several hours at the desired temperature, while gently stirring the buffer.
5. Change the dialysis buffer two or more times as necessary.
6. Remove dialysis membrane from the buffer. Hold the membrane vertically and remove excess buffer trapped at the end of membrane outside the upper clamp. Release the upper clamp and remove the sample with a Pasteur pipette.

(c) Simplest method of dialysis

For the simplest method of dialysis, the protein solution is transferred to a length of dialysis tube with the aid of a small funnel, a pipette, or syringe. The tubing should be sufficiently long to accommodate the volume of the solution, to allow for a possible increase in volume during dialysis, and to permit tight closure at both ends. Leak-proof closure of the dialysis tubing can be achieved by tying knots at both ends of the tubing or, more conveniently, by using plastic clamps. The filled tubing is then submerged in the dialysis buffer contained in a beaker, Erlenmeyer flask, or most commonly, a cylinder (Figure 25.3). The dialysis buffer is mixed with a magnetic stirrer. Care must be taken to avoid contact of the tubing with the stirring bar, as this can easily damage the tubing. Dialysis can be accelerated by mixing both the protein solution and the dialysis buffer, since dissipating the solutes throughout the entire volume prevents the formation of concentration gradients. For this purpose, a variety of devices have been developed.

For practical purposes, the use of dialysis tubing is limited to volumes of 1–100 ml. For dialysis of smaller volumes, (0.01–1 ml) several "microdialyzer" devices are available commercially.

Gel Filtration

Gel filtration permits the preparative separation of molecules by their native molecular weight, because of differential penetration into the gel matrix by molecules of different sizes. In contrast to PAGE, large molecules move more rapidly than small ones as they diffuse over a smaller volume while traveling through a column (Figure 25.4). This principle is in cases where proteins are to be separated from low-molecular weight molecules. They can be rapidly de-salted by gel filtration. The choice of the gel matrix and the conditions of the column chromatography, are dictated by the molecular weight range of the proteins involved. Gel filtration is a rapid means of removing low-molecular weight salts, reagents, or reaction products. This procedure is suitable where unwanted contaminants are very much smaller than the macromolecules, where sufficient material is available to measure optical density (OD) or is radio-labeled, and where the dilution effect of gel filtration is of little consequence.

Materials

Sephadex G-25 medium beads; Buffer in which de-salted protein is required.

Procedure

1. If not using a pre-packed column, swell beads in buffer and pour into a 2 ml syringe, using glass wool or a circle of filter paper to support the beads.
2. Wash well in buffer, allow surplus to drain away without letting the bed run dry.
3. Add sample and buffer to the column, rinsing down walls of the column.

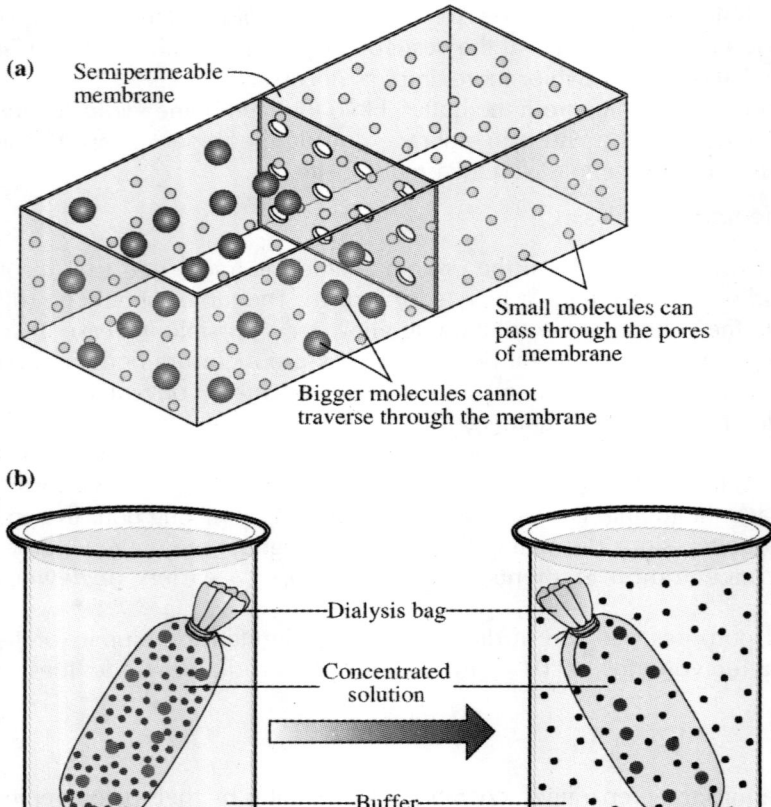

Figure 25.3 *Separation of molecules on the basis of size by dialysis: (a) principle, (b) the actual dialysis. Protein molecules (bigger molecules) are retained within the dialysis bag, whereas small molecules diffuse into the surrounding medium*

4. Collect fractions of 10% of the column volume and measure each fraction for protein (OD or TCA-precipitable cpm) and for tracer (OD or TCA-soluble cpm).

Notes

1. Some substances (e.g. lectins) may react with the dextran of Sephadex; BioRad P6 polyacrylamide beads are an alternative matrix for de-salting.
2. Inclusion of detergent (0.5% Triton X-100) reduces loss of protein due to non-specific adherence to the column.

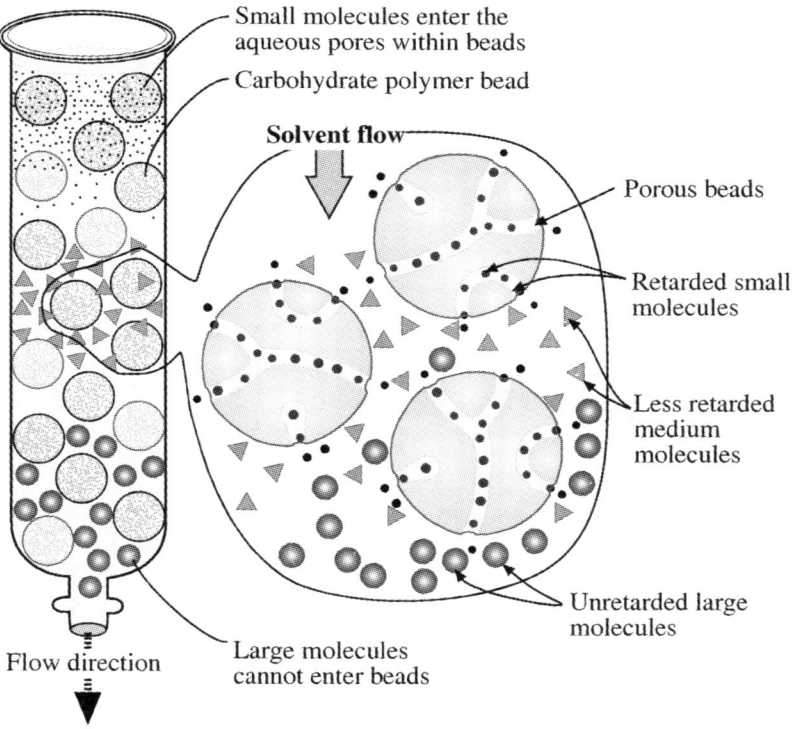

Figure 25.4 *Separation of three different sizes of proteins by gel-filtration chromatography. Large proteins comes down sooner than small ones because they cannot enter the internal volume of the beads*

3. Concentration methods

Protocol-1: Concentration using solid-phase or slurry absorbents

A cheap, low-tech alternative to dialysis and ultrafiltration is the use of solid-phase or slurry absorbents. In this method, the sample is placed in a dialysis membrane and surrounded with dry matrix or immersed in absorbent slurry, whereupon water and small solute molecules are drawn out of the sample. This technique is slow and requires constant monitoring; otherwise, complete drying can occur to the potential detriment of the sample. Common absorbents are polyethylene glycol (PEG) and carboxymethyl cellulose, Sephadex beads, and polyacrylamide derivatives.

Protocol-2: Lyophilization

Lyophilizaion is a means of concentrating protein solutions by drying the materials in the frozen state with the solvent being removed by sublimation. The technique has a high capacity and is very easy to perform. The system required for lyophilization (lyophilizer) consists of a sample manifold with standard taper joints, a cooled condensation trap, and a high-capacity vacuum pump. The protein solution is frozen in a suitable container that is able to withstand the external pressure, the stress of freezing, and has little tendency to adsorb protein. To enable rapid removal of the solvent and to prevent additional stress to the container due to expansion during freezing, the container should be filled to only one-quarter of its volume. Rapid freezing is important for the stability of the proteins which is to be lyophilized. A device for the concentration of a relatively small sample volume is the Speed Vac concentrator (Savant, Farmingdale, NY). A low speed centrifuge is used to keep the protein solution at the bottom of the container at all times. Therefore, the samples generally need not be frozen to avoid foaming, boiling, or squirting. Concentration proceeds by connecting the centrifuge to the vacuum system, which maintains the rotor chamber at 45°C.

C. PURIFICATION BY PROTEIN PRECIPITATION: BULK METHODS

1. Ammonium sulfate precipitation of proteins

Crude extracts can be roughly fractionated by ammonium sulfate precipitation. As already indicated, ammonium sulfate is the precipitant used most frequently in the salting out of proteins. Typically, about a 5-fold increase in protein purity can be obtained and extraneous DNA or tRNA is often removed as well. Its major advantages are (1) at saturation, it is of sufficiently high molarity so as to cause the precipitation of most proteins; (2) it does not have a large heat of solution, so that the heat generated is easily dissipated; (3) even its saturated solution has a density (1.235 g cm^{-3}) that is not so large that it interferes with the sedimentation of most precipitated proteins by centrifugation; (4) its concentrated solution prevents or limits most bacterial growth, and (5) in solution it protects most proteins from denaturation. Because of this last property, one often preserves purified proteins as suspensions in concentrated solutions of ammonium sulfate. A limitation in the use of ammonium sulfate for fine fractionation of a protein is that, in going from one step to the next as one increases the concentration of the salt, the purification achieved is usually only 2 to 5 times over the previous fraction.

Precipitation of antiserum with ammonium sulfate to purify IgG

The IgG fraction can be partially purified by selective precipitation with high concentrations of salts such as ammonium sulfate (40% saturated). Two procedures that have been widely used are as follows. The first procedure yields a higher degree of purity while the second procedure is simpler and faster; the second procedure is the one usually used prior to further purification by ion exchange, affinity chromatography, etc.

Materials

Preparation of saturated ammonium sulfate (SAS) pH 7.4.

Ammonium sulfate, Ammonium hydroxide, Bunsen burner or hot plate, Flask or beaker (4 litres), Buchner funnel, Filter paper, Vacuum flask, 4 litres, pH meter.

Precipitation of immunoglobulins:

Antiserum, SAS, High speed centrifuge, Magnetic stirrer and bar.

Reconstitution of precipitated immunoglobulins:

NaCl, M_r = 58.4; Appropriate buffer; Dialysis equipment.

Procedure-1

Preparation of SAS (pH 7.4)

1. Add about 900g of ammonium sulfate to 1.0 litre of water. Heat the mixture until the ammonium sulfate dissolves completely, and quickly filter the solution while it is hot. Let the solution cool to room temperature (crystals will form and should always be present in the bottle).
2. Adjust to pH 7.4 with ammonium hydroxide.

Precipitation of Immunoglobulins

1. Clarify antiserum by centrifugation (10,000 g, for 5 min.).
2. Add per ml of antiserum, 1 ml of de-mineralized water and 1.3 ml cold saturated ammonium sulfate, pH 7.0–7.2, drop-wise while stirring (final ammonium sulfate concentration 40%).
3. Mix for 30 min. and refrigerate overnight.

Reconstitution of precipitated immunoglobulins

1. Re-suspend the 40% SAS slurry and centrifuge an aliquot at 8,000g for 10 min. and collect precipitate.
2. Re-suspend precipitate gently in 1 ml of 0.01 M phosphate buffer, pH 7.2, containing 0.15 M NaCl (PBS).
3. Add 1 ml water and 1.02 ml saturated ammonium sulfate per ml of re-suspended precipitate (final sulfate concentration 33%).
4. Mix for 30 min. and centrifuge as above.
5. Re-suspend pellet in PBS at one-half the original volume of antiserum.
6. Dialyze precipitation against several changes of PBS over 24–36 hr.
7. Filter through 0.2 µm filter, add sodium azide to 0.025%, and store long-term at –20°C or short-term at 4°C.

Procedure-2

1. To 1 ml antiserum add 9 ml distilled water.
2. Add 10 ml saturated ammonium sulfate solution.
3. Leave at room temperature for 30–60 min.
4. Centrifuge for 5 min. at 8000g and collect precipitate.
5. Re-suspend pellet in 2 ml half strength (1/2X) PBS.
6. Dialyze 3 times against 500 ml of 1/2X PBS.

Notes

1. The solubility of ammonium sulfate is 103.4 g/100 g water at 100°C, and 70.6 g/100 g water at 0°C.
2. Note that precipitation in 40–45% SAS is satisfactory either for storage, as a preliminary step to further IgG purification (e.g. DEAE chromatography), or for concentrating purified solutions of IgG.

2. Precipitation of proteins with polyethylene glycol

The use of non-ionic, water-soluble polymers, in particular polyethylene glycol (PEG), for fractional precipitation of proteins was introduced in 1964 by Polson et al. This method was improved later and the approach is applicable to any complex mixture. The advantages of PEG as a fractional precipitating agent stem primarily from its well-known benign chemical properties. Unlike organic solvents, PEG has little tendency to denature or otherwise interact with proteins even when present at high concentrations and elevated temperatures. The low heat of the solution and the relative insensitivity of PEG precipitation curves to minor variations in temperature eliminate the need for controlling temperature during reagent additions. Another advantage of PEG is the shorter time required for the precipitated proteins to equilibrate and achieve a physical state suitable for centrifugation.

Careful measurements with a variety of purified proteins indicate that their solubilities decrease exponentially with the increasing concentration of PEG. The precipitation of proteins with PEG is relatively insensitive to pH and ionic strength, but markedly dependent on the size of the PEG up to about 6000 Da. Most workers use material with a nominal average molecular weight in the 4000–6000 range. Polymers larger than this offer no advantage, since their solutions are more viscous and the precipitation curves are not very different from PEG 6000. Decreasing the molecular weight below 4000 spreads the precipitation of the mixture over a broader range of PEG concentrations. PEG can be removed from proteins by different methods including ion-exchange or affinity columns. Alternative approaches to removing PEG include ultrafiltration and salt-induced phase separation. For many research purposes, it is probably unnecessary to remove all traces of polymer from the final product, since it is optically transparent and helps prevent loss of protein by absorption of glass.

D. PURIFICATION PROCEDURES: CHROMATOGRAPHIC METHODS

The most powerful methods for fractionating proteins is to make use of **chromatographic** techniques. The term chromatography refers to a group of separation techniques, which are characterized by a distribution of the molecules to be separated between two phases, one stationary (solid, gel, liquid or a solid/liquid mixture) and the other mobile (liquid or gaseous). The choice of stationary and mobile phases is made so that the compounds to be separated have different distribution coefficients. The basis of all forms of chromatography is the partition or distribution coefficient (K_d), which describes the way in which a compound distributes itself between two immiscible phases (e.g. phases A and B).

$$K_d = \frac{\text{concentration in phase A}}{\text{concentration in phase B}}$$

Chromatography separations may be achieved by two basic techniques: **Column chromatography** and **thin-layer or planar chromatography**. Molecules with a high tendency to stay in the stationary phase will move through the system at a lower velocity than will those which favor the mobile phase. The most common physical configuration is column chromatography in which the stationary phase is packed into a tube, a column, through which the mobile phase, the eluent, is pumped (Figure 25.5). The sample to be separated compound is introduced into

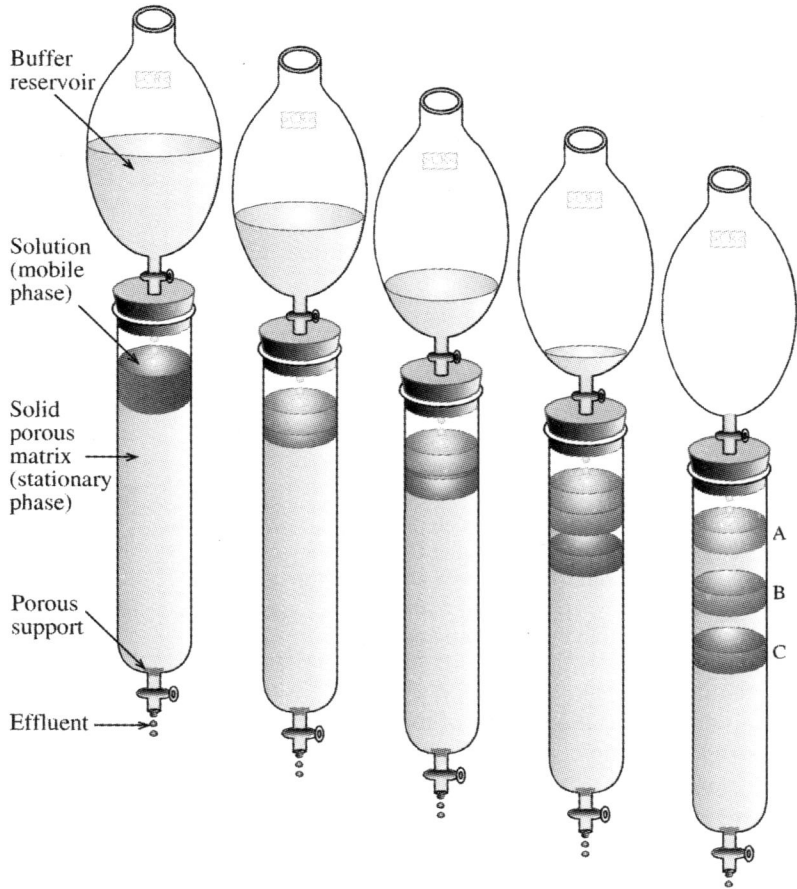

Figure 25.5 *Column chromatography and the standard elements of a chromatographic column. The protein solution to be separated is layered on top of the column and allowed to percolate into the solid matrix. Additional solution is added on top (A, B, C are protein fractions separated after known time)*

one end of the column. The various sample components travel with different velocities through the column and are subsequently detected and collected at the other end. For protein separation, several versions of liquid chromatography are used, differing mainly in the types of stationary phase (Table 25.1). One of these, gel filtration chromatography is based on quite different principles than are other versions of liquid chromatography.

1. Chromatographic techniques

Modern column chromatography utilizes sophisticated equipment to obtain high-resolution separations. However, for some applications this might not be necessary. The most simplest way to carry out an adsorption experiment is batch-wise, by simply stirring the adsorbent with the protein sample choosing proper conditions for adsorption and subsequent desorption.

Column chromatography can be performed at low pressure (standard), medium pressure, or High Pressure Liquid Chromatography (HPLC). For de-salting experiments, adsorption tests, etc., simple column chromatography equipment is often sufficient. High resolution, fast techniques require small diameter beads and equipment able to withstand the often high pressures necessary to force the buffer or solvent through the column. The most crucial point in column chromatography is to achieve good column packing.

Table 25.1 Versions of protein liquid chromatography

Separation principles	Type of chromatography	Resolution
Size and shape	Gel filtration (Size exclusion)	Low
Net charge	Ion exchange chromatography	Medium
Isoelectric point	Chromatofocusing	High–medium
Hydrophobicity	Hydrophobic interaction chromatography	Medium
	Reversed phase chromatography	Medium
Biological function	Affinity chromatography	High
Antigenicity	Immunoadsorption	High
Carbohydrate content	Lectin affinity chromatography	High
Content of free –SH	Chemisorption (covalent chromatography)	High
Metal binding	Immobilized metal ion affinity chromatography	High
Miscellaneous	Hydroxyapatite chromatography	Medium
	Dye affinity chromatography	High

The packaging techniques used differ depending on the type of gel matrix. The most important discriminating parameter is the rigidity of the gel matrix. It is thus convenient to distinguish between soft, semi-rigid, and rigid gel matrices. Particle shape, diameter, and size distributions are also important parameters to consider in column packing. The first step is to mount the column with its extension tube on a steady stand and to ensure that the column tube is perfectly vertical. The stationary phase slurry is de-gassed to remove all trapped air. Rigid gel materials such as silica with particle diameters in the range 5–15 microns are preferably slurry packed in dry acetone or chloroform (10% slurry concentrations and 300 kg/cm^{-2} packing pressure). Semi-rigid gels such as Sephacryl$^{(R)}$HR, Sepharose$^{(R)}$FF and Superose$^{(R)}$ are preferably packed in two steps. Every column should be tested for packing quality. A zone of acetone (0.5% Vt) at 30 cm h^{-1} is suitable for this purpose. The packing buffer composition for semi-rigid gels does not seem to be critical. The column bottom frit or filter mesh should be wetted and all air removed. Ethanol of 20% is recommended for this purpose. A critical point in the column packing is the application of the adaptor on the packed bed. As a general recommendation one should allow the adaptor to compress the upper part of the bed for approximately 5 min. For short columns (5–15 cm in length) normally used in various types of adsorption chromatography, packing quality is less critical. Soft gel matrices are in principle, packed in the same way as the semi-rigid gel materials.

Method-1: Size-exclusion chromatography or gel filtration

Among the chromatographic techniques employed for protein purification, size-exclusion chromatography (also called gel filtration) is unique in that fractionation is based on the relative size of the protein molecules. Many models have been proposed to explain the separation mechanisms in gel filtration. In contrast to conventional filtration, none of the proteins are retained by the gel filtration column. The basic components of gel filtration chromatography are the matrix, column, and the elution buffer. The function of the matrix is to perform the separation. Separation is accomplished by pore sizes and internal channels available in the gel comprising the matrix (porous beads). The gels are made of cross-linked polymers. Gel filtration is performed using porous beads as the chromatographic support. A column constructed from such beads will have two measurable liquid volumes—the external volume, consisting of the liquid between the beads, and the internal volume, consisting of the liquid within the pores of the beads. Large molecules equilibrate only with the external volume while small molecules will equilibrate with both the external and internal volumes. A mixture of proteins is applied in a discrete volume or zone at the top of a gel filtration column and allowed to percolate through the column. The large protein molecules are excluded from the internal volume and therefore emerge first from the column while the smaller protein molecules, which can access the internal volume, emerge later.

The dimensions important to gel filtration are the diameter of the pores that access the internal volume and the hydrodynamic diameter of the protein molecules. Proteins whose hydrodynomic diameter is small relative to the average pore diameter of the beads will access all of the internal volume and are described as being included in the gel matrix. Proteins whose hydrodynamic diameter is large relative to the average pore diameter will be unable to access the internal volume and are described as being excluded (Figure 25.6). Thus, the complex mixtures of macromolecules may be separated by their size differences and their respective molecular weights estimated by the relative migration of marker proteins.

The separation of molecular weights that the matrix is capable of separating is called the fractionation range, and it lies between 1000 to 100,000 Dalton. Molecular weights of 1000 Dalton or less will penetrate the beads completely. These molecules take the maximum volume, which is equal to one bed volume. Molecules having molecular weights in the range of 1000 to 100,000 Dalton will enter the beads with varying efficiencies and be partially or completely separated from one another. Molecules with molecular weight greater than 100,000 Dalton will not enter the beads. They will be eluted in the void volume, since they will not sieved by the matrix. Partially or completely separated molecules eluted from the column are called peaks. Peaks consist of increasing and decreasing concentration gradients of molecules.

The column is a tube with a frit fitted at the bottom. This frit is a membrane or porous disk that supports and retains the matrix in the column, but allows water and dissolved solutes to pass through. The length of the column affects the resolutions of separation. Column lengths of 20 to 30 centimeters are frequently used. The diameter of the column as well as its length influences the kind of resolutions one may obtain during the chromatographic process. The elution buffer (mobile phase) flows through the matrix and out of the column. The function of the elution buffer is to provide a means for developing the matrix with the applied sample contained in the column. This means that molecules in the sample are carried by the flow of a buffer into the matrix where they are gradually separated. The shape of the molecule is also a critical factor in the separation. Thus, larger molecules tend to spend time flowing around and between the beads of the matrix, while smaller molecules tend to spend more time in the pores and maze of the channels of the matrix's beads. Consequently, the larger, higher molecules take the faster and more direct path that involves less time in the column.

One can determine, graphically, the void volume (V_o) and effluent volume (V_e) for the separation process. The void volume is the volume of the space surrounding and outside the particles of gel. It is usually determined by measuring the volume necessary to elute a solute that is excluded from the pores of the gel. A commonly used solute is blue dextran dye. This is a

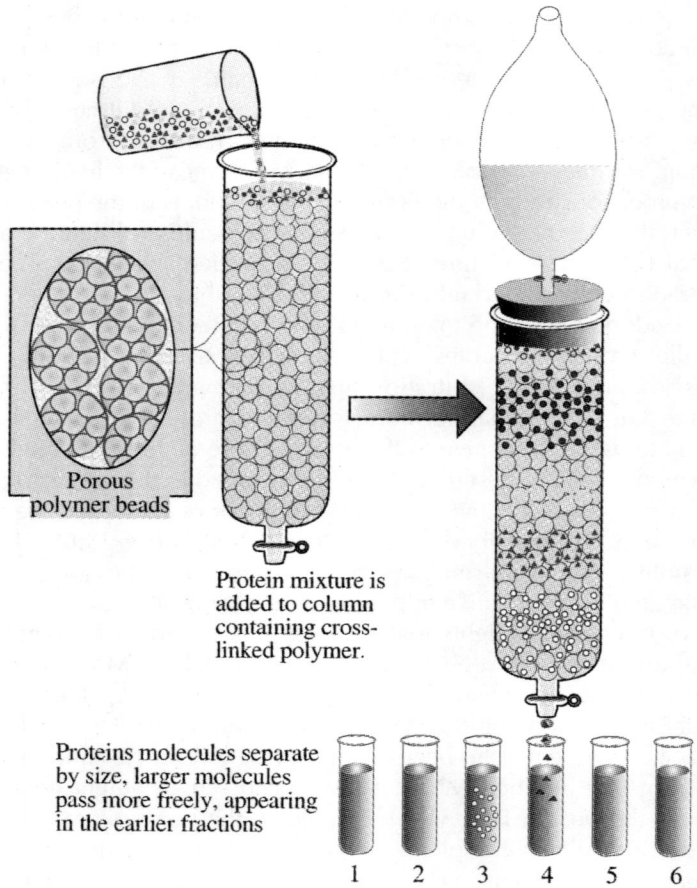

Figure 25.6 *Size-exclusion chromatography used in protein purification*

high molecular weight polysaccharide dye complex. The elution volume is the volume of solvent necessary to elute a solute from the time the solute enters the gel bed to the time it begins to emerge at the bottom of the column. The V_e is measured as the volume of solvent that has flowed through the column when the leading side of the solute peak is extrapolated to the base line of an elution profile. The total volume (V_t) is the total volume occupied by the packed gel bed. This is obtained most easily by water calibration of the column prior to packing it.

Applications

1) De-salting of proteins by gel filtration has been used for almost thirty years since the separation is simple and the requirements on the gel and the chromatographic system are modest.
2) Fractionation of protein mixtures: The most favorable situation is when it is possible to use a gel that will exclude the protein of interest and include the contaminants or vice versa (i.e. a filtration process). Contaminants that show strong affinity for the gel matrix may be

removed by filtering the sample solution through a small bed of the gel before applying the sample on the column.

3) Analytical gel filtration: In analytical applications the resolution is of utmost interest. Gel filtration may be used in the analytical mode to monitor a purification process selectively. Interactions between the matrix and the solute should be absent in analytical gel filtration. Analytical gel filtration is frequently used for the assay of molecular weights or molecular weight distributions of hydrophilic macromolecules or polymers.

4) Determination of pore size: In gel filtration the elution volume of a solute will be affected by the relationship between the solute size and the pore size of the matrix. Thus, by eluting molecules of known sizes and applying the established theory for gel filtration, approximations of the pore size can readily be achieved. By establishing the entire calibration curve with standards, estimations of the pore size, pore volume, and surface area may be obtained.

5) Mixed mode separations: In mixed mode separations, the elution of a solute is not due to size parameters only but is affected by adsorption or affinity to or exclusion from the matrix or layers of solvents or solutes adsorbed to the surface of the matrix. These types of solute-matrix interactions provide the basis for many of the separation principles developed. Thus, this information is very valuable when selecting materials and conditions for which separations based on gel filtration alone are desired.

Method-2: Fractionation of antigens

Materials

Gel matrix selected according to the molecular weights of the material to be fractionated, as indicated in the Table 25.2 below.

Fraction collector, UV monitor, and recorder as available from Pharmacia-LKB and other manufacturers.

Table 25.2 *Gel filtration matrices*

Matrix		Purpose	Molecular weight range
Sephadex	G–25	For general purposes (e.g. de-salting)	1,000 – 5,000
	G–50		1,500 – 30,000
	G–75		3,000 – 80,000
	G–100		4,000 – 150,000
	G–200		5,000 – 600,000
Sephacryl	S–100 HR	High resolution	1,000 – 100,000
	S–200 HR		5,000 – 250,000
	S–300 HR		10,000 – 1,500,000
	S–400 HR		20,000 – 8,000,000
Sepharose	6B	For very large macromolecules	10,000 – 4,000,000
	4B		60,000 – 20,000,000
	2B		70,000 – 40,000,000

The above are all available from Pharmacia-LKB. The range given applies to globular proteins; above the stated range, molecules are completely excluded from the gel beads, below it, they are completely included. Many other matrices are available from other suppliers, e.g. BioGel P6 (BioRad).

Procedure

1. Select a column with a volume approximately 100 times that of the sample to be fractionated.
2. Pour slurry of swollen beads (Note 1), equilibrated with the running buffer, into the column, connect buffer reservoir to the column and outlet to fraction collector.
3. Load sample (<1% of column volume), collecting required volume in each tube.
4. Measure each sample for cpm (for radio-labeled preparations), OD (for tubes which the UV monitor has shown to contain protein) and/or by any other appropriate assay (e.g. IRMA).

Notes

1. Sephadex and BioGel beads require swelling, while Sepharose and Sephacryl are supplied pre-swollen. Swelling times vary with bead type: follow instructions provided with the beads.
2. An even, uninterrupted flow rate is essential to sharp separation of peaks in gel filtration, as is sample loading without disturbing the column bed. The rate of flow is determined by the height of the reservoir above the outlet, or may be controlled by a peristaltic pump.
3. If the sample volumes exceed 2% of the column volume, peaks will begin to broaden and may therefore overlap.
4. Store columns when not in use in 0.1% sodium azide.
5. Extended column runs or separation of bioactive molecules should be performed in cold (4°C) rooms or in cabinets.

E. ION EXCHANGE CHROMATOGRAPHY

Ion exchange chromatography (IEC) exploits differences in the sign and magnitude of the net electric charges of proteins at a given pH. Most biological compounds are positively or negatively charged when exposed to pHs in the range of 2–10. When the pH is varied, the net charge can change from zero to the opposite charge. Adsorption chromatography of protein depends on several types of protein ligand interactions. The first of these to be successfully employed were the ionic interactions used in IEC. It is by far the most utilized chromatographic technique, included in about 75% of purification protocols followed by affinity chromatography (60%) and gel filtration (50.5%). The reason for the popularity of IEC is its versatility, its high resolving power, its high capacity, and its straightforward basic principle. The basis for the IEC process is the competitive binding of ions of one kind (e.g. proteins), for ions of another kind (e.g. other proteins or salt ions of the same charge), to an oppositely charged chromatographic medium, the ion exchanger (Figure 25.7). The interaction between the proteins and the ion exchanger depends on several factors. Net charge and surface charge distribution of the protein; the ionic strength and the nature of the particular ions in the solvent; pH, or strictly speaking the proton activity; and other additives to the solvent, such as organic solvents, etc.

IEC, the solid support (adsorbent) contains either a permanent positive or negative charge. They are called cation and anion exchanger respectively. The permanent charge on the exchanger is attracted to the opposite charge on the molecules. The separation of compounds is based on an equilibrium between the molecules-exchanger and elution solvent. This equilibrium can be shifted gradually by changing the ionic strength or pH of the eluting buffer, thereby weakening the electrostatic forces and de-adsorbing the molecules from the exchanger. This allows the separation of molecules with small differences in net charges. The solid support is

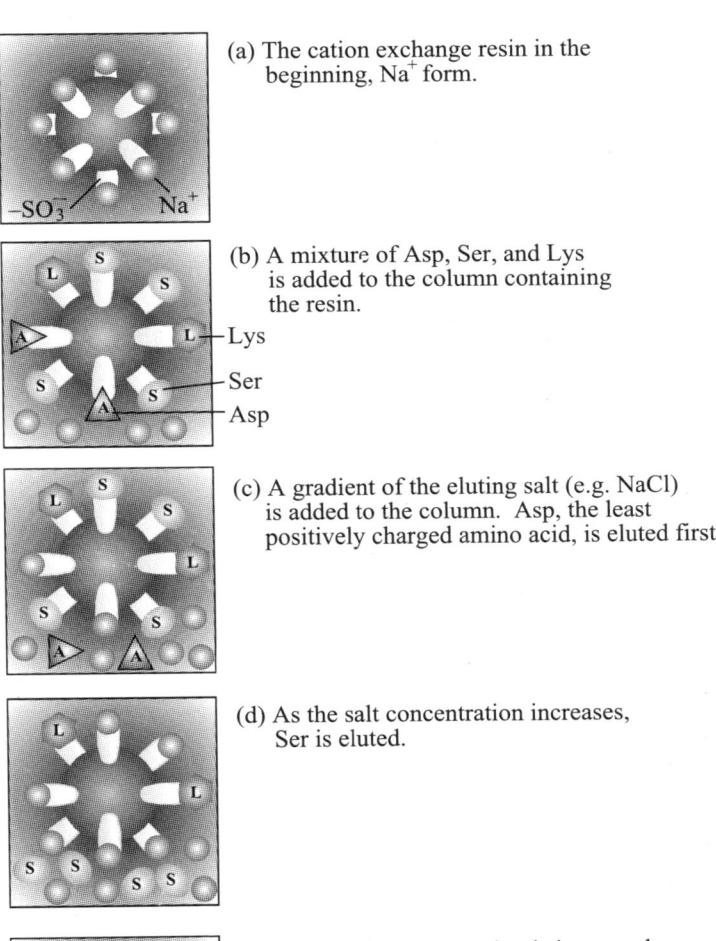

(a) The cation exchange resin in the beginning, Na⁺ form.

(b) A mixture of Asp, Ser, and Lys is added to the column containing the resin.

(c) A gradient of the eluting salt (e.g. NaCl) is added to the column. Asp, the least positively charged amino acid, is eluted first.

(d) As the salt concentration increases, Ser is eluted.

(e) As the salt concentration is increased further, Lys, the most positively charged of the three amino acids is eluted last.

Figure 25.7 *Principle steps of a cation exchange column, separating a mixture of Asp, Ser, and Lys*

usually a synthetic resin (cross-linked polystyrene) or cellulose derivative covalently bonded to the desired functional group to create a weak or strong exchanger. A weak cation exchanger's functional group is a carboxylic acid and strong exchanger is sulfonate, whereas the anion exchangers are derivatives of either secondary or tertiary amines.

Carboxymethylcellulose (CM-cellulose) has the $-CH_2OH$ groups of cellulose derivatized to $-CH_2OCH_2COOH$, and the corresponding cation exchanger is substituted with –

$CH_2OCH_2CH_2N(CH_2CH_3)_2$ (DEAE–cellulose). The exchanger is supplied from the manufacturer with a counter ion, which can be Na^+, H^+, Cl^- etc. The capacity of the exchanger is determined by a number of meq/ml of a standard material that can be adsorbed. Highly cross-linked and large capacity resins can be used for small molecules. The resin acts as a molecular sieve, which can block large molecules from entering the interior of the resin beads. In the case of cellulose, there is some limit to the number of substitutions that can be made per unit of cellulose. Celluloses are the preferred supports for large biologically active proteins because they do not denature (de-activate) the protein as readily as resins.

The adsorption and separation are based on the differences between electrostatic interaction of the molecules and the support. The following example demonstrates the exchange principle. Ion exchange chromatography can be used to separate both small molecules, such as amino acids and large ones like proteins, RNA, and DNA (Figure 25.8).

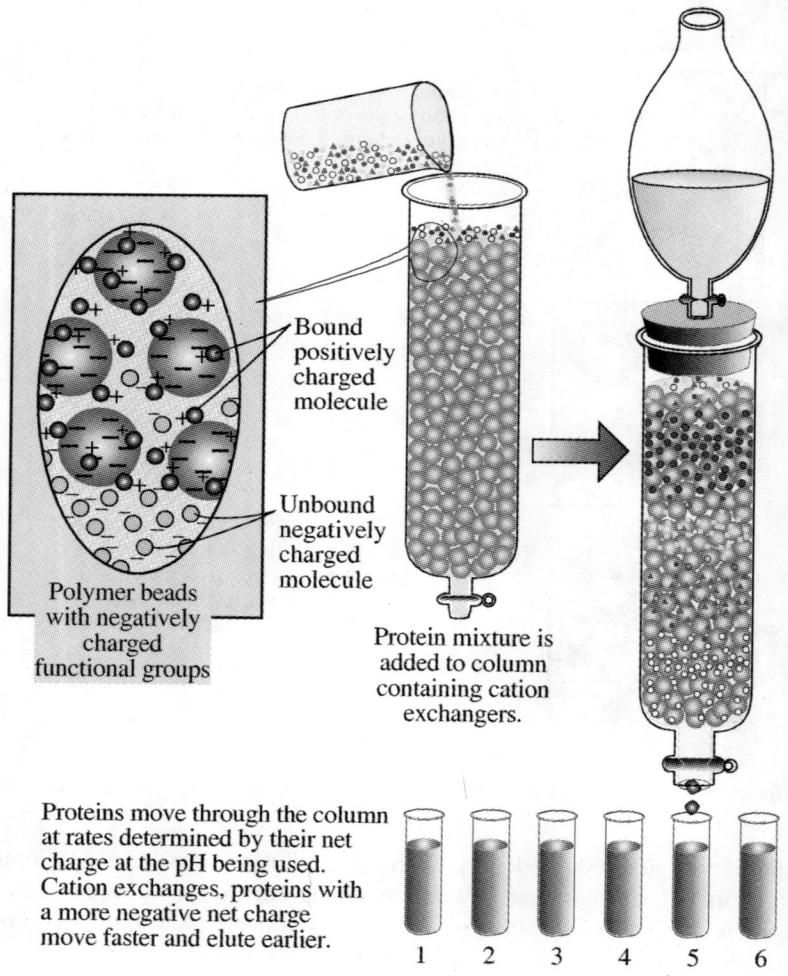

Figure 25.8 *Ion-exchange chromatography used in protein purification*

Protocol-1: *Ion exchange chromatography using CM–Sephadex column*

Materials

CM-Sephadex (G-50) prepared in 0.1 M potassium acetate buffer pH 6.0. CM-Sephadex is a weak anion exchanger (RCOO$^-$). When equilibrated in buffer at pH 6, it becomes RCOO$^-$ K$^+$, and RCOOH when equilibrated in 0.1 M HCl. Blue dextran is a large non-ionic polysaccharide, molecular weight >500,000, with a blue color. Cytochrome c's pI=10.7 (pH where the net charge = 0), and has a net positive charge (cation) at the pHs of the experimental conditions. Cytochrome c has a molecular weight of 12,400. DNP-glycine is in the anionic form above pH 3.0, has a molecular weight of 241 and a yellow color.

0.1 M Potassium acetate buffer, pH 6.0 (Buffer A).

1.0 M Potassium acetate buffer, pH 6.0 (Buffer B).

Dye mixture: Blue dextran (1 mg/ml), cytochrome c (2 mg/ml), and DNP–glycine (1 mg/ml).

10 ml chromatography column; Pasteur pipette with dropper bulb can substitute for the chromatography column; rubber tube and pinch clamp.

Preparations

Suspend the CM-Sephadex G–50 (30 g) in 500 ml of H$_2$O and vacuum filter. Re-suspend the CM-Sephadex in 0.1 N HCl (83 ml conc. HCl into 920 ml H$_2$O), vacuum filter and rinse with 3–4 volume of H$_2$O (vacuum filter). Re-suspend in 2 volumes of 1.0 M potassium acetate, pH 6.0, and vacuum filter. Repeat this process a second time (final cake). Re-suspend the CM-Sephadex cake in 2 volumes of the 1.0 M potassium acetate buffer for column packing. Repeat (or take half of final cake) and re-suspend in 2 volumes of 0.1 M potassium acetate, pH 6.0. Vacuum filter and repeat the process a second time. Re-suspend the cake in 1 volume 0.1 M potassium acetate.

1.0 M potassium acetate, pH 6.0: Dissolve 98 g of potassium acetate in 950 ml of H$_2$O in a beaker and adjust to pH 6.0 with glacial acetic acid.

0.1 M potassium acetate pH 6.0. Take 100 ml of 1.0 M potassium acetate buffer and dilute to 1000 ml.

Sample: In a sample vial, add 1 mg of blue dextran, 2 mg cytochrome c, and 2 mg of DNP-glycine and dilute to approximately 5 ml.

Procedure

1. Set up two columns. Place a piece of glass wool in the restriction and label each column.
2. Attach a piece of rubber tube and clamp to each column. Close the clamp.
3. Fill a column with each of the CM-Sephadex buffer solutions (buffer A or B). Add the slurries using a pipette and bulb. Open the pinch clamp and allow the slurry to settle until resin bed is 4–5 inches. Do not allow the column to run dry. Add the proper buffer (as used for packing columns) dropwise to maintain the liquid level just above the resin bed. Close the clamps.
4. Add approximately 0.2 ml of the solution to be separated. Form a narrow band at the top of the resin. If any of the solution adheres to the side of the column above the resin, rinse

with a few drops of the buffer on to the resin (gradually open the clamp to allow the solution to move onto the resin bed).
5. Place a 10 ml graduated cylinder under each column.
6. Open the clamp, and fill the column with buffer.
7. Allow the buffer to flow through the columns until a clear separation of at least an inch has taken place. Measure the distance bands have travelled in centimeters from the top of the resin. Measure the amount of buffer collected from each column.
8. Relate the distance traveled by the band to the amount of buffer collected.
9. Measure the volume of buffer required to elute each band (collect the bands together) (Table 25.3).

Table 25.3 *For results after one-inch separation (approximate)*

	0.1 M (Buffer A)			1.0 M (Buffer B)		
	A	B	C	A	B	C
Vol buffer	/	/	/	/	/	/
Distance (cm)	/	/	/	/	/	/
R*f	/	/	/	/	/	/
Rf	/	/	/	/	/	/

$$Rf = \frac{Distance\ (cm)}{Volume\ collected} \quad \text{or} \quad Rf^* = \frac{Distance\ (cm)}{Total\ volume\ of\ elution^*}$$

*Total elution volume to remove the equipment
A=Dextran, B=DNP-glycine, C=cytochrome c

Method-2: Ion exchange chromatography using DEAE-cellulose resin column

The information contained herein should be used in conjunction with the essential information contained in the manufacturer's literature. Here, the discussion is only on commonly used ion exchange resins, DEAE-cellulose, TEAE-cellulose, and phosphocellulose. Many other ion exchangers, such as the basic, acidic, and mixed-bed ion exchange resins and affinity columns as well as hydroxyapatite, are also used to separate proteins. They may be used with essentially the same methods as described in this chapter.

Procedure

Preparing ion exchange resins

1. Wash dry forms of DEAE as slurry in 0.5 M HCl, in water until the pH reaches 4, in 0.5 M NaOH, in water until the pH reaches 9, and finally wash with buffer.
2. Repeat a wash more than once if orange or brown color continues to be eluted from the resin.
3. After washing, suspend the resin in a 10-fold larger volume of the desired buffer and let it settle for about 10 min.
4. Suck off the supernatant liquid that contains the fine particles. Repeat this removal of fine particles twice or until the resin particles settle uniformly. DE-52, a washed, pre-swollen

microgranular DEAE-cellulose, can be used without washing with acid and base, but fine particles do have to be removed. Whatman P-11 and other cellulose phosphates are prepared like DEAE except that the base wash precedes the acid wash.

5. The pH of ion exchange resins is crucial for reproducible chromatography. Phosphocellulose is particularly troublesome because it titrates very slowly. For this reason, before pouring the resin in a column, it is usually best to adjust its pH with a concentrated buffer and then equilibrate it with the required low-ionic strength buffer. The pH ranges that are usually used for DEAE and phosphocellulose are between 6.5 and 9. Above pH 9.5, the DEAE groups become uncharged, and the resin loses its ion exchange capacity. Therefore, for pH above 9.5 use TEAE-cellulose, a quaternary amino ion exchanger with properties similar to DEAE. It is best to use a buffer that is not adsorbed by the resin: for example, use a Tris buffer for DEAE and phosphate buffers for phosphocellulose. Divalent cations should not be present in the phosphocellulose buffer since they will bind tightly to the resin.

Pouring columns

1. Before pouring a column, de-gas the resin by aspirating for a few minutes. This is especially important if the resin has been stored at 4°C and the column is being prepared at room temperature.
2. Select a column of appropriate size – a length-to-width ratio of about 10:1 is often satisfactory. Close off the bottom and add a few millilitres of buffer to wet the bottom.
3. Prepare slurry of the equilibrated resin in about 3 volumes of the desired column buffer. Pour the slurry into the column and let it settle at least several minutes before starting the flow. After this settling, pour the remainder of the resin bed with the column buffer draining slowly through the resin. It is best to pour the column in one pass, usually by attaching an extension to the top of the column.
4. It is important to avoid the formation of an interface between a packed resin surface and the new packing slurry.
5. After pouring, check that the pH and conductivity of the eluent are the same as those of the column buffer. Usually, it is wise to pass several column volumes of buffer through the column before the column is loaded with a sample.
6. Be sure to choose a column that has sufficient exchange capacity. About 50 mg of crude *E. coli* proteins per ml of packed resin will not overload a DE-52 column. At least 10 times as much can be applied to phosphocellulose. The theoretical protein-binding capacity of phosphocellulose is 10 times that of DEAE, but the adjacency of phosphate groups may limit this. Only 10% of a mixture of crude *E. coli* proteins bind to phosphocellulose whereas 95% bind to DEAE; hence, in theory one could apply 100 times as much crude protein to a phosphocellulose column as to a DEAE column of the same size. However, it is best to use about 10% of these amounts, i.e., 5 mg/ml for DEAE and 50 mg/ml for phosphocellulose.

Loading and eluting columns

1. Apply the sample to the column without disturbing the column surface. DEAE may be loaded under pressure because adsorption is so rapid that it occurs in minutes. Phosphocellulose appears to bind proteins more slowly than DEAE. Rinse the loaded column with sample loading buffer until no protein is found in the eluent. This usually takes about three column volumes.

2. For maximal separation between proteins, it is best to elute them from the resin with a continuous salt gradient. Step changes in salt concentration applied to a column are usually most valuable for concentrating protein solution or for removing bulk impurities such as nucleic acids. The gradient can be about 10 column volumes. Arrange the gradient apparatus so that the buffer volume at the top of the column is small but sufficient to prevent the following liquid from disturbing the surface of the resin. DEAE and phosphocellulose columns can be operated under pressure. A convenient flow rate can usually be obtained with a hydrostatic pressure of 0.5–1 m. In general, the slower the flow rate, the better the resolution.
3. To elute proteins from DEAE, KCl gradients are usually used. Of a mixture of *E. coli* proteins, approximately 5–10% flow through DEAE at 0.5 M KCl, and 5–10% still bind at 0.30 M KCl. To elute proteins bound to phosphocellulose, phosphate gradients are frequently used. Phosphate gradients have the advantage of providing increasing buffering power at higher ionic strengths. This may be important because the resin gives off protons as the ionic strength increases and the pH has a significant effect on a protein's adsorption to the resin. Most of the proton release occurs in the range below 0.15 M phosphate.
4. After running the column, it is informative to learn the salt concentration in the various fractions. This is particularly useful for smaller columns in which it is difficult to control development of the salt gradient. Since conductivity is nearly proportional to ionic strength and is easily quantitated, it is the measurement of choice. A conductivity meter is calibrated with samples of known salt concentration in the column buffer.
5. The elution properties of protein may change in the presence of nucleic acids. Nucleic acids generally bind more tightly to DEAE than do proteins and hence proteins tightly stuck to the DNA, for instance, may elute from a DEAE column only at very high ionic strengths. On phosphocellulose, the opposite effect may occur, that is, proteins bound to nucleic acids may flow through the column under conditions where they might otherwise bind. Proteins can be separated from nucleic acids by dextran-polyethylene glycol phase partition, protamine sulfate.
6. After the gradient is finished, the column can be re-used if it is washed in a highly concentrated salt solution, for example, 1.0 M KCl or phosphate. If a crude extract labeled with ^{32}P is applied to DEAE, some counts still stick after a wash; hence, these washes do not totally regenerate the columns. The columns may be reused for at least 10–20 times if proteins that are fairly free of nucleic acid are applied to columns.

F. AFFINITY CHROMATOGRAPHY

Generally, biochemists' choice of techniques are ammonium sulfate based purification protocols. The protein obtained from this method is good enough for structural and mechanistic studies. However, in cases where the enzyme is to be used for bioconversion, synthesis, or analysis on a large-scale, one has to seriously think about the cost and time involved. One smart move is to bring affinity-based separations early into the downstream processing strategy. Many alternative formats for affinity-based separations that have emerged over the years are given in Table 25.4.

Table 25.4 *Some alternative affinity-based separation techniques for downstream processing of proteins or enzymes*

Membrane based Affinity cross-flow ultrafiltration	High affinity ligand is attached to the target protein. It creates an 'affinity escort' to a polymer that shows minimum non-specific adsorption and the membrane retains the protein ligand complex while the contaminants pass through the membrane pores.
Membrane affinity filtration	Many pre-activated membranes and affinity membranes for specific target proteins are used.
Affinity adsorbant based Continuous affinity recycle extraction	In this method, a continuous operation is achieved by recirculation of the adsorbent particles between two or more vessels. The feed containing the crude extract is fed into the vessel containing the adsorbent. The contaminants are washed off while the target molecule-bound adsorbent is passed on to a vessel containing the eluent.
Smart polymer-based Affinity precipitation	This technique exploits the affinity interactions in free solution by combining a macroaffinity ligand with a target protein. The key element involved here is a reversibly soluble–insoluble polymer (also known as **smart polymers**). These polymers can exist both in solution and suspension (precipitation) form depending on their environment. Changing simple parameters like the pH, temperature, ionic strength, etc., or adding of metal ion can bring down the polymer from the solution to the precipitate form.
Polymer-based Aqueous two-phase affinity partitioning	The use of affinity ligands to phase-forming polymers offers selectivity and is known as 'affinity partitioning'.
Detergent-based Reversed micellar extraction	Reverse micelles are thermodynamically stable aggregates formed by self-aggregation of surfactant molecules in organic solvents. The partitioning of proteins between the bulk aqueous phase and the reverse micelles depends on different parameters such as ionic strength, pH of the aqueous solution, interaction potential between surfactant, protein molecules and ion, etc. Thus, the efficiency of this method depends on the ease of back-extraction of protein into the aqueous phase.
Electrophoresis Affinity electrophoresis	In this method, two substances with the same charge but different mobilities, are separated, with the faster moving away from the slower one on gels. However, when two substances having opposite charges are applied at different positions and electrophoresed, they will cross each other during electrophoresis. In both the cases, the electrophoretic pattern will change at the crossing or passing point.

All biological processes depend on specific interactions between molecules. Examples can be found from all areas of structural and physiological biochemistry such as in multimolecular assemblies, effector–receptor interactions, DNA–protein interactions, and antigen–antibody binding (Table 25.5). Affinity chromatography owes its name to the exploitation of these various biological affinities for adsorption to a solid phase. In affinity chromatography, the material to be separated or purified (S) binds to a ligand (L) immobilized on an insoluble matrix (M) such as agarose. The technique represents a powerful and efficient method for separating proteins and nucleic acids (Figure 25.9). The general protocol is to immobilize the ligand by covalently binding it to a suitable matrix. The substance of interest is then bound with high specificity to the ligand to form a complex while the impurities pass through the column. The substance S is finally eluted from the column by an eluent, which has an even higher binding capacity for S than the immobilized ligand, L.

Table 25.5 *Examples of biological interactions used in affinity chromatography*

Ligand	Counter-ligand
Antibody	Antigen, virus, cell
Enzyme	Substrate analog, inhibitor, co-factor
Lectin	Polysaccharide, glycoprotein, cell surface receptor, membrane protein, cell
Nucleic acid	Nucleic acid-binding protein (enzyme or histone)
Hormone, vitamin	Receptor, carrier protein
Sugar	Lectin, enzyme or other sugar binding protein

Choice of matrix: As in all adsorption chromatography, an adsorbent with a large surface area is desirable to maximize the capacity of the affinity adsorbent. Hydrophilic gels with a high surface-to-volume ratio are very suitable as matrices. For affinity chromatography applications the ideal gel material should meet the following characteristics: 1) Be macroporus to accommodate the free interaction of large molecular weight proteins with ligands which can themselves be proteins or other macromolecules; 2) be hydrophilic and neutral to prevent the proteins from interacting non-specifically with the gel matrix itself; 3) contain functional groups to allow derivatization by a wide variety of chemical reactions; 4) be chemically stable to withstand harsh conditions during derivitization, regeneration, and maintenance. 5) be physically stable to withstand hydrophobic stress in packed beds and, when applicable, sterilization by autoclaving; 6) be readily available at low cost to facilitate industrial applications.

Properties of ligand: For the preparation of the affinity adsorbent, the ligand should be compatible with the solvents used during the coupling procedure. It should possess at least one functional group by which it can be immobilized to the matrix. Commonly used groups are –NH_2 (amino), –COOH (carboxyl), –CHO (aldehyde), –SH (thiol), –OH (hydroxyl). It should also possess a functional group for coupling which is non-essential for its binding properties, i.e. the binding properties of the ligand should be not adversely affected as a result of its immobilization. Ligands of high-molecular weight type (e.g. proteins), with a large number of suitable functional groups, can normally be immobilized without adversely influencing the structure or function. If the affinity interaction decreases, it might be necessary to chemically modify the ligand, to provide it with an appropriate functional group for immobilization. The functional group used should permit the formation of a stable covalent bond so that the ligand is not released from the matrix. For proteins "multi-point attachment" between ligand and matrix is rather common. It is, of course, also essential that the ligand remains intact during the immobilization procedure and that it is sufficiently stable to allow the planned affinity chromatography to be carried out. It is essential that the ligand reagent is as pure as possible and in particular

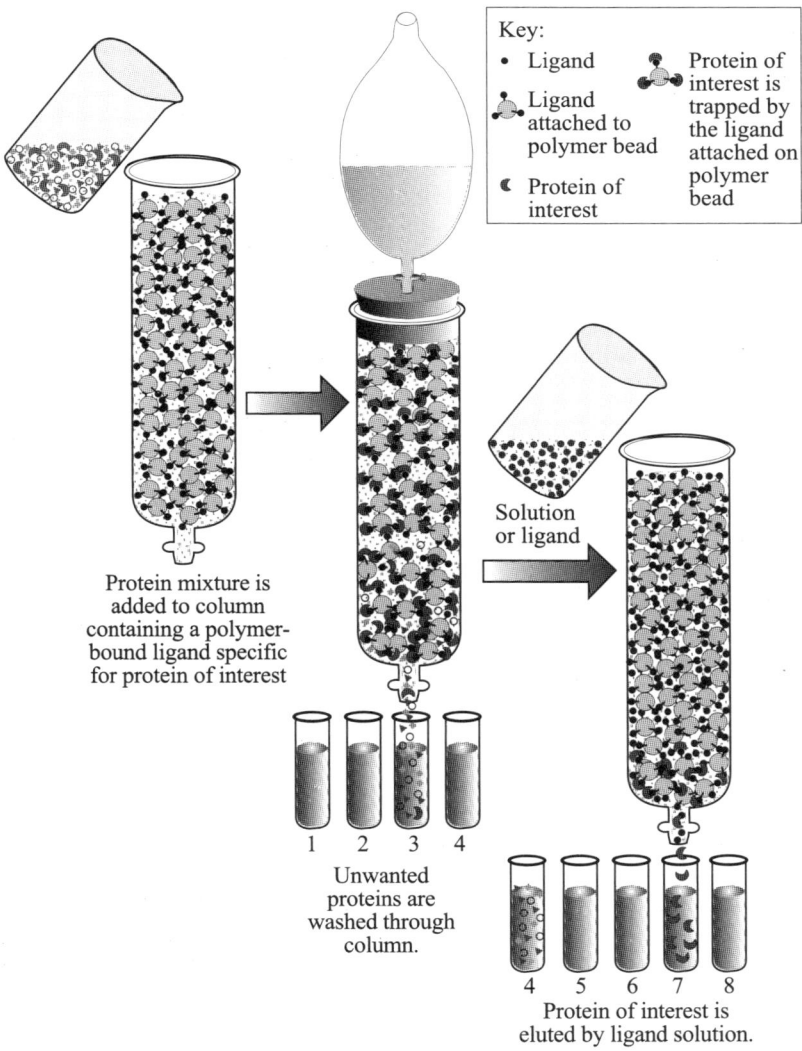

Figure 25.9 *Affinity chromatography used in protein purification*

does not contain substances with functional groups that can react competitively in the immobilization.

Immobilization techniques: In general, the immobilization procedure consists of three steps: (1) Activation of the matrix to make it reactive toward the functional group of the ligand; (2) coupling of the ligand; (3) de-activation or blocking of residual active groups by a large excess of a suitable low-molecular weight substance such as ethanolamine. Several reactions are used to couple the ligand to the solid matrix depending on the nature of the ligand and the type of matrix. In the present application known as immobilized metal affinity chromatography (IMAC), a chelating ligand, iminodiacetate (IDA) is coupled to epoxy-activated agarose. A transition metal such as Cu^{2+} is then bound to the imminodiacetate. Other metal ions used are Zn^{2+}, Fe^{3+} and Ni^{2+} and chelating ligands include Tris (carboxyethyl) ethylenediamine (TED). The IDA ligand, tridentate, occupies three coordinating positions around the central metal ion thus leaving three other coordinating positions for electron donors like proteins and nucleic acids containing N, O, or S atoms on their surfaces.

When a mixture of (e.g., serum) protein is added to the column, some proteins, especially those containing histidin residues, are preferentially bound to the metal. The degree of binding depends on the type of protein and the pH and ionic strength of the surroundings. The pH should be such that the proteins are not too highly protonated. Thus buffers with pH close to neutral and are weakly coordinating are frequently used during the adsorption stage. Acetate (pH 5–6) and phosphate (pH 8–9) are common choices.

Unbound proteins pass through the column fairly rapidly and so a partial separation is effected. The bound proteins can then either be eluted as a group or individually by a proper choice of eluting conditions. The most obvious choice is to lower the pH so that the protein is protonated and therefore released from the metal–protein complex. However, care must be taken not to denature the protein at the low pH. Another common method of desorption is to use a ligand in the eluent that has a stronger affinity for the metal than the protein. Tris (hydroxymethyl) aminomethane (Tris) buffers and imidazole are good choices.

A 'model' protein system has been chosen consisting of a mixture of rabbit serum IgG and rabbit serum albumin (RSA). The affinity column is IDA-agarose with bound Cu^{2+}. The experiment will serve to demonstrate the usefulness of the technique in separating the two proteins. The effect of the metal ion on the efficiency of separation could also be examined.

Protocol-1: Affinity chromatography

Materials

5 ml rabbit albumin (RSA) solution at 1 mg/ml; 2 ml of IgG solution at 0.5 mg/ml;

Protein mixture: 1 ml of a mixture of rabbit serum IgG and rabbit serum albumin (0.5 ml each of above solutions).

100 ml of 50 mM sodium acetate and 100 mM NaCl, pH 5.5 (buffer A). 100 ml of 100 mM Tris-Cl buffer pH 8.0 (buffer B). 10 ml of 50 mM Cu^{2+} solution (as copper (II) sulfate in water). 100 ml of 50 mM EDTA and 0.5 M NaCl solution, pH 7.0; 2 prepacked 0.8 × 5 cm IDA-agarose columns (Sigma IDA–5); UV-VIS spectrophotometer; 1 80-tube fraction collector (optional); 1 Peristaltic pump (optional); 6 Pasteur pipettes and bulb; Electrophoresis or Immunoelectrophoresis apparatus (optional).

Procedure

1. Clamp the pre-packed IDA-agarose column in a ring stand.
2. Drain the solution from the column to a level just slightly above the top of the gel. (Do not allow column to go dry).
3. Wash the column twice with 2 ml portions of Buffer A. Discard the wash.
4. Add 1 ml of the 50 mM Cu^{2+} solution to the column with a Pasteur pipette and allow it to enter by slowly opening the outlet.
5. Fill the column with buffer A (or pump it) and continue to wash the column until all excess Cu^{2+} has been eluted (about 15 ml).

6. Attach column outlet to a fraction collector or alternatively collect fractions by hand.
7. Add 1 ml of the protein mixture (RSA and IgG mixture) to the column and allow it to enter the gel.
8. Fill the column with buffer A (or pump it) and begin to elute. Collect 2 ml fractions.
9. Measure the absorbance at 280 nm of your fractions from the start of the elution process and continue to do so until the absorbance returns to zero (about 10 ml, use buffer A as reference).
10. Now fill the column with buffer B and elute as you collect 2 ml fractions again.
11. Monitor the absorbance at 280 nm of your new set of fractions until the absorbance returns to zero (about 20 ml; use buffer B as reference).
12. Pool all the fractions eluted with buffer A that contain protein and likewise pool all the fractions eluted with buffer B that contain protein.
13. Identify the protein in each sample by PAGE electrophoresis or immunoelectrophoresis.
14. Record the data and plot on 2 separate graphs with A_{280} as the ordinate and fraction # as the abscissa.

Method-2: High-performance liquid chromatography

The principles exploited in high-performance liquid chromatography (HPLC) are the same as those used in common chromatographic methods. Very high-resolution separations can be achieved quickly and with high sensitivity in HPLC using automated instrumentation. Another version of HPLC, Reverse-phase HPLC, is a widely used chromatographic procedure for the separation of nonpolar solutes. In reverse-phase HPLC, a solution of nonpolar solutes is chromatographed on a column having a nonpolar liquid immobilized on an inert matrix; this nonpolar liquid serves as the stationary phase. A more polar liquid that serves as the mobile phase is passed over the matrix, and solute molecules are eluted in proportion to their solubility in this more polar liquid.

Method-3: Thin-layer (planar) chromatography

In a thin layer chromatography, a thin layer of the stationary phase is formed on a suitable flat surface. The movement of the mobile phase across the layer, generally by simple capillary action, is rapid, there being little resistance to flow. As the mobile phase moves across the layer from one edge to the opposite, it transfers any analytes placed on the layer at a rate determined by their distribution coefficients, K_d, between the stationary and mobile phases. Analyte movement ceases either when the mobile phase reaches the end of the layer and capillary action flow ceases or when the plate is removed from the mobile phase reservoir (Figure 25.10). The movement of the analyte is expressed by the retardaton factor, R_F such that

$$R_F = \frac{\text{Distance moved by analyte from origin}}{\text{Distance moved by solvent front from origin}}$$

Procedure

1. **Preparation of thin layer**: A slurry of the stationary phase, generally in water, is applied to a 20 cm square plate, as a uniform thin layer by means of a plate spreader starting at one end of the plate and moving progressively to the other. The thickness of the slurry layer

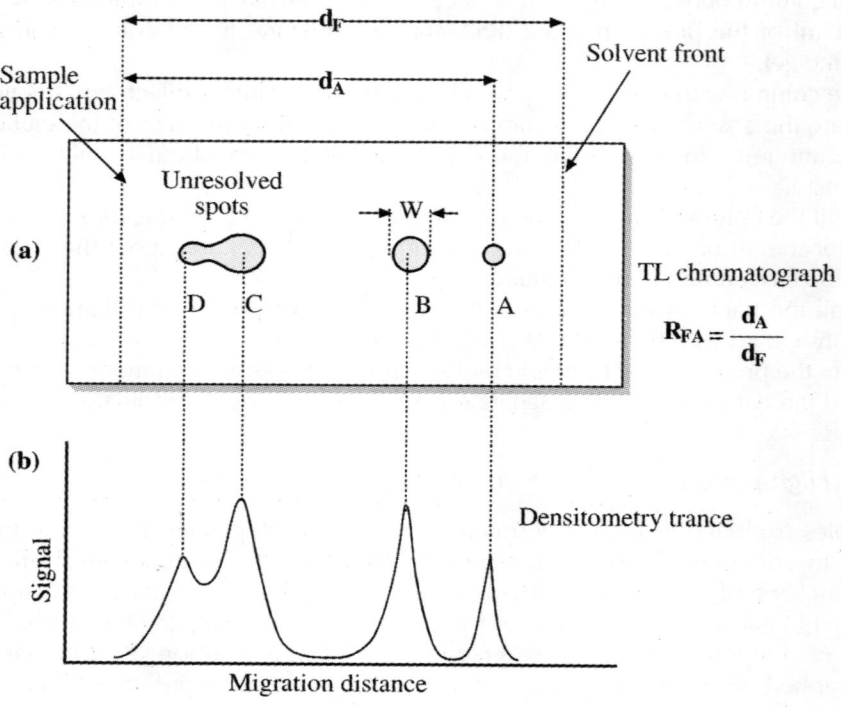

Figure 25.10 *Thin layer chromatograph of a mixture of compounds A to D (a) and the corresponding densitometer trance (b) from which quantitative data can be calculated*

used is dictated by the nature of the chromatographic separation (range 0.25 to 2 mm thick). Once the slurry layer has been prepared, the plates are dried to leave the coating of the stationary phase. A range of prepared plates is available commercially. So-called polyamide layer sheets, which consist of poly-ε-caprolactam coated onto both sides of a solvent-resistant polyester sheet are used. They can also be re-used if cleaned immediately with ammonia–acetone, and are widely used in protein sequencing studies by the Dansyl–Edman method.

2. *Sample application*: The sample is applied to the plate 2.0 to 2.5 cm from the edge by means of a micropipette or microsyringe. The solvent may be removed from the spot by gentle heating or by use of an air blower, care being taken in the case of volatile or thermolabile compounds. It is then possible to apply more samples to the spot if necessary. For preparative thin layer chromatography, the sample is applied as a band across the plate rather than a single spot.

3. *Plate development*: Separation most commonly takes place in a glass tank that contains the developing solvent (mobile phase) to a depth of about 1.5 cm. This is allowed to stand for at least 1 hour with a lid over the top of the tank to ensure that the atmosphere within the tank becomes saturated with solvent vapor. Unless this is done, irregular running of the solvent will occur as it ascends the plate by capillary action. After equilibration, the lid is

removed, and the thin layer plate is then placed vertically in the tank so that it stands in the solvent. The lid is replaced and separation of the compounds then occurs as the solvent travels up the plate. It is also possible to develop the plate in a horizontal plane by connecting the sample end of it to a reservoir of mobile phase by means of a suitable wick. One of the biggest advantages of TLC is the speed at which separation is achieved. This is commonly about 30 and will not exceed 90 min.

4. ***Analyte detection***: Several detection methods are available. Examination of the plate under ultraviolet light will show the position of UV absorbing or fluorescent compounds. Subjecting the plate to iodine vapor is useful if unsaturated compounds are being investigated. Spraying of plates with specific color reagents will stain certain compounds, for example, ninhydrin will locate amino acids and peptides. If the compounds are radio-labeled, the plates may be subjected to autoradiography, which will detect the spots as dark areas on x-ray film. A general, non-specific technique is to spray the plate with 50% (v/v) sulphuric acid or 25% (v/v) sulphuric acid in ethanol and heating it at 110°C, which will result in most compounds becoming charred and showing up as brown spots.

The amount of compound present in a given spot may be determined in a number of ways. On-plate quantification may be achieved by use of radiochromatograph scanning in the case of radio-labeled compounds or more generally by means of densitometry. Off-plate quantification may be carried out by scraping off the spot and the immediate surrounding stationary phase from the plate and eluting the compound with a suitable solvent. The amount of compound in solution can then be determined by standard methods, most commonly colorimetry or fluorimetry.

G. ELECTROPHORETIC METHODS

Another important technique for the separation of proteins is based on the migration of charged proteins in an electric field. This process is called **electrophoresis**. These procedures are not generally used to purify proteins in large amounts. They often adversely affect the structure and function of the proteins. Electrophoresis is, however, especially useful as an analytical method. Its advantage is that proteins can be visualized as well as separated, permitting a quick estimation of the number of different proteins in a mixture or the degree of purity of a particular protein preparation. Also, electrophoresis allows determination of crucial properties of a protein such as its isoelectric point (pI) and approximate molecular weight.

Fundamental to electrophoretic separations is the fact that proteins are electrically charged particles. The charges are derived from amino acids with ionogenic side groups. In addition, the proteins often have associated charged components of non-protein origin such as lipids or carbohydrates. Electrophoresis is used to separate complex mixtures of proteins, to investigate subunit compositions, and to verify homogeneity of protein samples. It can also serve to purify proteins for use in further applications. A great many electrophoretic systems have been developed and no attempt is made to summarize them here. The distinctions between the various "continuous" and "discontinuous" buffer systems, or alternative support matrices, or gradient gels are also not discussed. Only the most common (and most reliable) analytical SDS-PAGE procedure is described. In PAGE, the combination of gel pore size and protein charge, size, and shape determines the migration rate of the protein (Figure 25.11).

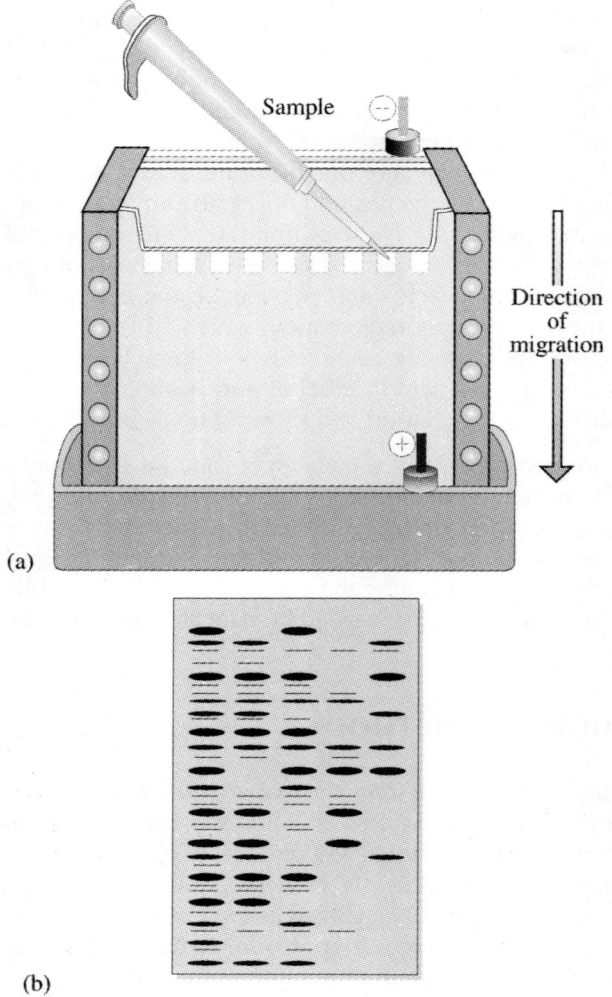

Figure 25.11 *Polyacrylamide gel electrophoresis of proteins. (a) Different samples are loaded in wells and the proteins move into the gel when an electric field is applied, (b) Proteins can be visualized after electrophoresis by treating the gel with a stain such as Coomassie blue*

1. Electrophoretic techniques

Equipment and mode of operation: Vertical electrophoresis with downward sample migration is the most common way to run PAGE. The sample is loaded on top of a gel, which has its ends in contact with an upper and lower electrolyte. Excellent equipment is available from a number of commercial sources. During recent years, horizontal electrophoresis has become more common. This trend has been accelerated by the introduction of automated equipment for horizontal electrophoresis. To a large extent, the choice between vertical and horizontal electrophoresis is a matter of taste.

Sample application and recovery: Obviously only protein molecules in solution can be analyzed by electrophoresis. If the sample is difficult to dissolve, agents such as detergents or urea may prove necessary. If the purified protein band has to be recovered, it can be easily accomplished by cutting out the band from the gel after localizing by any of the detection methods and then simply soaking the crushed gel in buffer. More efficient methods, however, take advantage of the principles of isotachophoresis.

Zone sharpening techniques: All protein zones tend to broaden with time due to diffusion. Fortunately, this effect can be counteracted by different methods. A very simple and straightforward zone sharpening technique is to make sure that the electrical conductance of the sample buffer is lower than that of the running buffer. This will increase the field strength in the sample zone. Proteins will then migrate quickly through the sample zone and into the running buffer where they will slow down and hence become concentrated in the lower field strength. A two to ten times dilution of the buffer will decrease the conductance proportionately with a concomitant proportional increase in field strength. Protein stacking and moving boundary electrophoresis is an application of isotachophoresis. Typical of this very efficient technique for sample concentration is casting of a special, concentrating or "stacking" gel on top of the separation gel. Since the technique utilizes discontinuities in buffer composition as well as in PAA concentration, it is frequently referred to as "disc-electrophoresis". The purpose of the stacking gel is simply to hold the buffer required for protein concentration before the protein zone enters the separating gel. The stacking gel is therefore prepared as a low porosity gel with minimal sieving properties.

Pore gradient gels: In this technique, the protein zones are concentrated during the separation. This is accomplished by using a gel with continuously changing acrylamide concentration. The sample is applied in the low concentration region and when proteins migrate into a denser gel network, the "friction" or sieving effect from the gel increases. The electrophoretic migration speed will thus decrease and asymptotically reach zero as the protein approaches the gel concentration of its "pore limit", primarily depending on the Stokes radius. Provided the pore limit is reached, gradient gel PAGE is a convenient way to estimate protein molecular weight.

2. Optimizing the separation

Choice of buffer pH and gel porosity: The electrophoretic mobility of a protein in a gel will be determined by the balance between the electrical force and the retarding "frictional" force. The electrical force is a function of the voltage applied and the protein charge. The retardation force is a function of gel concentration and protein size. The parameters which must be adjusted for optimization therefore are the buffer concentration, pH, and gel porosity. A conventional Ferguson plot analysis can be used as a powerful tool for establishing optimal conditions.

Choice of buffer composition and detergent: Electrophoresis under denaturing conditions in SDS is so efficient and well established that it is sometimes adopted too uncritically. In all cases where the biological function of the protein can be utilized to gain further relevant information, electrophoresis of the native protein, followed by specific detection of enzymatic or other activity etc., will normally be a better alternative or a powerful complement to SDS electrophoresis of denatured proteins. Ordinary globular proteins with good solubility in standard buffers can easily be separated into native conformation and with full activity. Many proteins, however, such as hydrophobic membrane proteins, filamentous proteins etc., with poor solubility in ordinary buffers, pose special problems in electrophoresis. In order to get these proteins in solution for electrophoresis, detergents such as SDS and reducing agents for cleavage of S-S bridges must be used.

Protocol-1: Electrophoretic methods

Method-1: *Electrophoresis in tris-tricine buffer systems*

Separation of peptides and proteins under 10 to 15 kDa is not possible in the traditional Laemmli discontinuous gel system because the co-migration of SDS and smaller proteins obscures the resolution. The Tris-tricine method uses a modified buffer to separate the SDS and peptides, thus improving the resolution. Several pre-cast gels are commercially available for use with tricine formulations.

Materials

Separating and stacking gel solutions (Table 25.6); 2X tricine sample buffer; Peptide molecular weight standards mixture (Table 25.7); Cathode buffer; Anode buffer; Coomassie blue G-250 staining solution; 10% (v/v) Acetic acid.

Procedure

1. Prepare and pour the separating and stacking gels, using Table 25.8.
2. Prepare the sample with 2X tricine sample buffer and treat the sample at 40°C for 30 to 60 min. prior to loading. Use the peptide molecular weight standards mixture for peptide separations (Table 25.7).
3. Load the gel and set up the electrophoresis apparatus using the tricine-containing cathode buffer or water to rise and fill wells. Fill the lower buffer chamber with the anode buffer, assemble the unit, and attach the upper buffer chamber. Fill the upper buffer chamber with the cathode buffer and load the samples.
4. Connect the power supply to the cell and run for 1 hour at 30 V followed by 4 to 5 hours at 150 V. Use a heat exchanger to keep the electrophoresis chamber at room temperature.
5. After the Coomassie blue tracking dye has reached the bottom of the separating gel, turn the power supply to zero and disconnect the power supply.
6. Disassemble the gel. Stain proteins in the gel for 1 to 2 hours in Coomassie blue staining solution. Follow by de-staining with 10% acetic acid, changing the solution every 30 min. until background is clear (3 to 5 changes). For higher sensitivity, use silver staining as a recommended alternative.

Preparation of separating gel

In a 25 ml side arm flask, mix 30% acrylamide/0.8% bisacrylamide solution, 4X Tris.Cl/SDS, pH 8.8, and water. De-gas under vacuum for ~5 min. Add 10% ammonium persulfate and TEMED. Swirl gently to mix. Use immediately.

Table 25.6 Recipes for polyacrylamide separating and staining gels

Stock solutions	Final acrylamide concentration in separating gel (%)									
	5	6	7	7.5	8	9	10	12	13	15
30% acrylamide/0.8% bisacrylamide	2.50	3.00	3.50	3.75	4.00	4.50	5.00	6.00	6.50	7.50
4X Tris.Cl/SDS, pH 8.8	3.75	3.75	3.75	3.75	3.75	3.75	3.75	3.75	3.75	3.75
Water	8.75	8.25	7.75	7.50	7.25	6.75	6.25	5.25	4.75	3.75
10% (w/v) ammonium persulfate	0.05	0.05	0.05	0.05	0.05	0.05	0.05	0.05	0.05	0.05
TEMED	0.01	0.01	0.01	0.01	0.01	0.01	0.01	0.01	0.01	0.01

Table 25.7 Molecular weights of protein standards for polyacrylamide gel electrophoresis

Protein	Molecular weight	Protein	Molecular weight
Cytochrome C	11,700	Lactate dehydrogenase (procine heart)	36,000
α–Lactalbumin	14,200	Aldolase	40,000
Lysozyme (hen egg white)	14,300	Ovalbumin	45,000
Myoglobin (sperm whale)	16,800	Catalase	57,000
β–Lactoglobulin	18,400	Bovine serum albumin	66,000
Trypsin inhibitor (soyabean)	20,100	Phosphorylase b (rabbit muscle)	97,400
Trypsinogen, PMSF treated	24,00	β–Galactosidase	116,000
Carbonic anhydrase (bovine erythrocytes)	29,00	RNA polymerase, E. coli	160,000
Glyceraldehyde–3–phosphate dehydrogenase (rabbit muscle)	36,000	Myosin, heavy chain (rabbit muscle)	205,000

Table 25.8 Recipes for Tricine peptide separation gels

Stock solutions	Separating gel	Stacking gel
30% acrylamide/0.8% bisacrylamide	9.80 ml	1.62 ml
Tris.Cl/SDS, pH 8.45	10.00 ml	3.10 ml
Water	7.03 ml	7.78 ml
Glycerol	4.00 g (3.17 ml)	–
10% (w/v) ammonium persulfate	50 μl	25 μl
TEMED	10 μl	5 μl

Method-2: Electrophoresis in single-concentration mini-gels

Separation of proteins in a small-gel format is becoming increasingly popular for applications that range from isolating material for peptide sequencing to performing routine protein separations. The unique combination of speed and high resolution is the foremost advantage of small gels. Small gels are easily adapted to single-concentration, gradient, and two-dimensional SDS-PAGE procedures.

Materials

Mini-gel vertical gel unit with glass plates, Clamps, and Buffer chambers; 0.75 mm Spacers; Single or multiple gel caster; Acrylic plate or polycarbonate separation sheet; 10- and 50-ml syringes; Combs; Long razor blade; Micropipette.

Procedure

1. Assemble each gel sandwich by stacking, in order, the notched or small rectangular plate, 0.75 m spacers, and the larger rectangular plate. Be sure to align the spacers properly with the ends flush with the top and bottom edge of the two plates when positioning the sandwiches in the multiple gel caster.
2. Fit the gel sandwiches tightly in the multiple gel caster. Use an acrylic plate or polycarbonate separation sheet to eliminate any slack in the chamber. Alternatively, gels can be cast singly with a stand-alone caster.
3. Place the front faceplate on the caster, clamp it in place against the silicone gasket, and verify alignment of the glass plates and spacers.
4. Prepare the separating gel solution as directed in Table 25.9. For five 0.75 mm-thick gels, prepare ~30 ml solution. Do not add TEMED and ammonium persulfate until just before use.
5. Fill a 50 ml syringe with the separating gel solution and slowly inject it into the caster until the gels are 6 cm high, allowing 1.5 cm for the stacking gel. Overlay each gel with 100 µl H_2O-saturated isobutyl alcohol. Allow the gels to polymerize for ~1 hour.
6. Remove the isobutyl alcohol and rinse with 1X Tris. Cl/SDS, pH 8.8.
7. Prepare the stacking gel solution (2 ml per gel) as directed in Table 25.9. Fill a 10 ml syringe with stacking gel solution and inject the solution into each gel sandwich. Insert combs, taking care not to trap bubbles. Allow gels to polymerize for ~1 hour.
8. Remove the faceplate. Carefully pull the gels out of the caster, using a long razor blade to separate the sandwiches. The gels can be stored tightly wrapped in plastic wrap with the combs left in place inside a sealable bag to prevent drying for ~1 week. Without the stacking gel, the separating gel can be stored for 2 to 3 weeks. Keep gels moist with 1X Tris.Cl SDS, pH 8.8, at 4°C.
9. Remove the combs and rinse the sample wells with 1X SDS electrophoresis buffer. Draw a line indicating the bottom of each well on the front glass with a marker.
10. Fill the upper and lower buffer chambers with 1X SDS electrophoresis buffer. The upper chamber should be filled to 1 to 2 cm over the notched plate.
11. Prepare the protein sample and protein standards mixture. Load the sample using a micropipette. Insert the pipette tip through the upper buffer and into the well. The mark on the glass plate will act as a guide. Dispense the sample into the well.
12. Electrophorese samples at 10 to 15 mA per 0.75 mm gel until the dye front reaches the bottom of the gel (~1 to 1.5 hours).
13. Disassemble the gel. Proceed with detection of proteins.

Method-3: Preparing multiple gradient mini-gels

Polyacrylamide gradients not only enhance the resolution of larger format gels but also greatly improve protein separation in the small format. Casting gradient mini-gels one at a time is not generally feasible because of the small volumes used, but multiple gel casters make it easy to cast several small gradient gels at one time. The gels are cast from the bottom in multiple casters, with the light acrylamide solution entering first.

Procedure

1. Assemble mini-gel sandwiches in the multiple gel caster as described for single-concentration min-gels (Method 1, steps 1–3).
2. Set up the 30 ml gradient maker, magnetic stirrer, peristaltic pump (optional), and Tygon tubing. Connect the outlet of the 30 ml gradient maker to the inlet at the base of the front faceplate of the caster.
3. Prepare light and heavy acrylamide gel solution (Table 25.9) for five 0.75 mm-thick min-gels. Keep heavy acrylamide solution on ice until used.
4. With the outlet and interconnecting valve closed, add the heavy solution to the reservoir chamber. Briefly open the interconnecting valve to let a small amount of heavy solution through to the mixing chamber, clearing the valve of air.
5. Fill the mixing chamber with light solution. Add ammonium persulfate and 4 µl TEMED per 12 ml acrylamide solution to each chamber and mix with a disposable pipette.
6. Turn on the magnetic stirrer. Open the interconnecting valves and allow the chambers to equilibrate. Then slowly open the outlet port to allow the solution to flow from the gradient maker to the multiple caster by gravity. Adjust the flow rate to 3 to 4 ml/min.
7. Close the outlet port as the last of the gradient solution leaves the mixing chamber, just before air enters the outlet tube. Fill the two chambers with plug solution and slowly open the outlet once again.
8. Allow the plug solution to push the acrylamide in the caster up into the plates. Close the outlet when the plug solution reaches the bottom of the plates.
9. Quickly add 100 µl H_2O-saturated isobutyl alcohol to each gel sandwich. Let the gels polymerize undisturbed for ~1 hr.
10. Prepare and pour the stacking gel (Method 1, step 7).
11. Disconnect the gradient maker, place the caster in a sink, and remove the front faceplate. The plug solution will drain out from the bottom of the caster. Remove the gels.

Method-4: Immunoelectrophoresis

The use of methods based on the specific reaction between an antigen (protein) and its antibody is more than a century old. The formation of specific immunoprecipitates allows several visualization assays not possible with ordinary protein electrophoresis methods. The specific reaction between an antigen and its corresponding antibody is known as the precipitin reaction. The precipitin reaction can be performed in free solution or in agarose gels where the antigen–antibody complexes can be seen as opaque precipitates (Figure 25.12). The precipitin reaction is temperature (4–37°C) and pH (7–9) dependent.

Method-5: High-resolution two-dimensional gel electrophoresis

Conventional PAGE or any other one-dimensional separation method will not always give the required resolution when the samples are complex mixtures of proteins. The development of two-dimensional (2D) PAGE by O'Farrell has revolutionized the field of protein analysis. The technique has proved to be one of the most powerful methods for protein mapping of cells or sub-cellular fractions. 2D-gel electrophoresis is the combination of two high-resolution electrophoretic procedures (IEF and SDS-PAGE) to provide much greater resolution than either procedure alone. The resulting 2D-gel contains numerous round or elliptical protein spots well separated from each other; depending on the sample, as many as 1500 protein spots may be detected by silver staining or autoradiography. The 2D-PAGE is now specifically applied to the separation of proteins in the first dimension according to their isoelectric points using IEF with carrier ampholytes after reduction of disulfide bonds, followed by separation in the second dimension according to their molecular weights using SDS-PAGE (Figures 25.13–15).

Table 25.9 *Light and heavy acrylamide gel solutions for gradient gels*

Stock solutions	Acrylamide concentration of light gel solution (%)									
	5	6	7	8	9	10	11	12	13	14
30% acrylamide/0.8% bisacrylamide	2.50	3.00	3.50	4.00	4.50	5.00	5.50	6.00	6.50	7.00
4x Tris.Cl/SDS, pH 8.8	3.75	3.75	3.75	3.75	3.75	3.75	3.75	3.75	3.75	3.75
Water	8.75	8.25	7.75	7.25	6.75	6.25	5.75	5.25	4.75	4.25
10% (w/v) ammonium persulfate	0.05	0.05	0.05	0.05	0.05	0.05	0.05	0.05	0.05	0.05

Stock solutions	Acrylamide concentration of heavy gel solution (%)									
	10	11	12	13	14	15	16	17	18	19
30% acrylamide/0.8% bisacrylamide	5.00	5.50	6.00	6.50	7.00	7.50	8.00	8.50	9.00	9.50
4x Tris.Cl/SDS, pH 8.8	3.75	3.75	3.75	3.75	3.75	3.75	3.75	3.75	3.75	3.75
Water	5.0	4.5	4.0	3.5	3.0	2.5	2.0	1.5	1.0	0.5
Sucrose (g)	2.25	2.25	2.25	2.25	2.25	2.25	2.25	2.25	2.25	2.25
10% (w/v) ammonium persulfate	0.05	0.05	0.05	0.05	0.05	0.05	0.05	0.05	0.05	0.05

Stacking gel (3.9% acrylamide)

In a 25 ml side-arm flask, mix 0.65 ml of 30% acrylamide/0.8% bisacrylamide, 1.25 ml of 4x Tris.Cl SDS, pH 6.8, and 3.05 ml water. Degas under vacuum 10 to 15 min. Add 25 µl of 10% ammonium persulfate and 5 µl TEMED. Swirl gently to mix. Use immediately.

The use of 2D-PAGE has become increasingly popular. It allows the resolution of a complex protein mixture into more discrete components than 1D-PAGE since it separates on the basis of protein charge in addition to molecular weight. The major advantages of large-scale 2D-PAGE is the improvement in reproducibility of protein patterns. This enables the researcher to directly compare the analysis of complex protein mixtures, whether the 2D-PAGE separations are conducted simultaneously or in different experiments. This feature makes 2D-PAGE a versatile and powerful tool in both basic and clinical research.

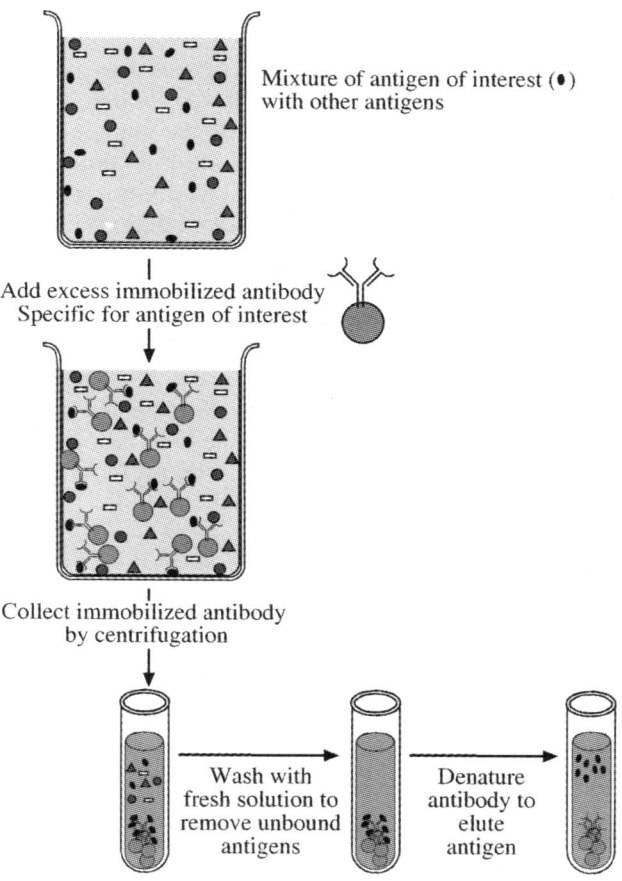

Figure 25.12 *Isolation of an antigen by immunoprecipitation. Immunoprecipitation can be used for purification, quantification and identification of an antigen*

3. Application of 2D-PAGE

The most common uses of 2D-PAGE are the analysis of complex mixtures of proteins and the analysis of the post-translational modification of proteins. 2D-PAGE can also provide valuable information about the molecular properties of proteins, including an estimate of the relative pI and molecular weights of proteins. However, it is generally inadequate to use this as the sole method for the precise determination of these parameters (e.g. the disulfide bonds of the proteins analyzed are usually reduced so the protein patterns may reflect subunit peptides). The pI and molecular weight values observed may therefore be different from those of the native proteins. Another common use of 2D-PAGE is to rapidly purify a specific protein, which can be cut from the gel and used directly to obtain an amino acid sequence or to purify antibodies. These antibodies can then be used for immunoaffinity purification of the original protein in quantities sufficient for detailed chemical characterization. They also provide an excellent method to analyze antibody specificity and to analyze carbohydrate or other epitopes. The use of 2D-PAGE with silver staining provides one of the best methods of estimating protein purity.

(a) Sample preparation and solubilization procedures

The preparation of the samples for 2D-PAGE analysis is the most critical step in guaranteeing excellent reproducible results. For instance, incomplete unfolding of a peptide chain can result in multiple spots in the SDS-PAGE direction while incomplete disaggregation of protein complexes might give multiple spots or streaking in both directions. Instead of analyzing whole cell extracts by 2D-PAGE, it is sometimes valuable to pre-fractionate the material into membranes and cytoplasmic components, or other subcellular fractions. Proteins present in low concentrations can be enriched to a level where it is possible to detect them by 2D-PAGE. All tissues and samples should be handled in the cold and stored at $-70°$. It is important that the ratio of solubilization buffer to protein concentration be optimized for each sample.

The reproducibility of 2D-PAGE is highly dependent upon the conditions used for solubilization and disaggregation of the proteins in the sample. It is essential to find conditions which: (1) solubilize the proteins in the sample quantitatively, (2) disaggregate protein complexes and unfold the peptide chains completely, and (3) are compatible with IEF. There are some proteins that resist solubilization even under reducing conditions in the presence of urea and a neutral detergent. The addition of small amounts of SDS to the sample buffer without decreasing the resolution in the IEF too much is possible.

Protein concentration in the sample must be high in order to detect minor components, to avoid diluting the sample buffer more than about 10%, and to keep the sample volume small. Soluble proteins often have to be concentrated. Lyophilization should be avoided, since charge modification due to oxidation and de-amidation can occur. High salt concentration in the sample must be avoided unless the ampholytes are covalently linked to the PA gel in the IEF. The presence of particulate or unsolubilized material, nucleic acids, lipids and some kind of proteins can adversely affect the resolution in 2D-PAGE.

Materials

SDS; Cyclohexylaminoethane (CHES); Glycerol; 2-Mercaptoethanol; 9 M Urea; Nonidet P-40; Ampholytes (pH 3.5–10: Bio-Rad); Deionized double-distilled water.

Solubilization buffers

1. SDS solubilization solution: 0.05 M CHES, 2% SDS, 10% glycerol, small amount of Bromophenol blue, pH 9.5. Add 2% 2–mercaptoethanol (2–ME) just before use. Samples should be suspended in an SDS solubilization buffer, placed in a tightly capped glass vial, and heated for 5–10 min. in a boiling water bath. It may be necessary to solubilize some samples at room temperature for 2–3 hours, with or without heating.
   Urea solubilization solution: 9 M urea, 4% Nonidet P-40. Add 2% 2-ME and 2% ampholytes to a small aliquot of solubilization buffer just prior to use. These reagents should be filtered to 0.2 µm with a syringe filter for best results. Samples should be suspended in the urea solubilization solution and incubated at room temperature for 2 hours.
2. Following the incubation, samples are centrifuged (100,000 to 200,000g, for 2 hr) to remove non-solubilized material and nucleic acids that may interfere with focusing or cause streaking in second-dimension protein patterns.

Method-1: First-dimension (isoelectric focusing) gels

(a) **Isoelectric focusing** (**IEF**) can be described as electrophoresis in a pH gradient set up between a cathode and an anode with the cathode at a higher pH than the anode (Figure 25.13).

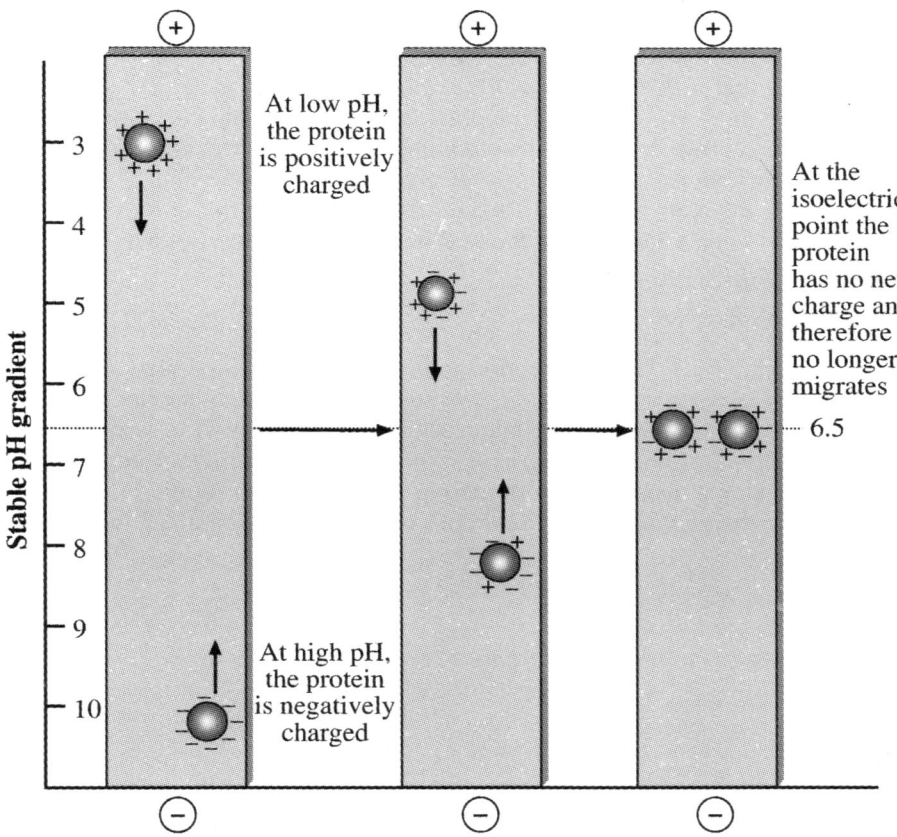

Figure 25.13 *Separation of protein molecules by Isoelectric focusing*

Proteins, being an amphoteric species will be positively charged at pH values below their pI and negatively charged above. This means that wherever a protein is in the pH gradient, it will migrate toward its pI. Under the influence of the electrical force, the pH gradient will be established by the carrier **ampholytes** (a mixture of low-molecular weight organic acids and bases), and the protein species focused at their pI. The focusing effect of the electrical force is contracted by diffusion, which is directly proportional to the protein concentration gradient in the zone. Eventually, a steady state is established where the electrokinetic transport of protein into the zone is exactly balanced by the diffusion out of the zone.

A large number of carrier ampholyte mixtures are available giving different pH gradients. Many can also be obtained in pre-cast gels ready for use. The optimal pH gradient will depend on the purpose of the experiment. For screening purposes, a broad range interval (pH 3–10 or similar) should be used. The exact gradient obtained depends on many factors such as choice of electrolyte solutions, the gradient medium, focusing time, etc. Despite the large number of pH intervals available, there may be occasions where none of them fits perfectly. In such cases, one can either choose to work with Immobiline or use "pH gradient engineering" in any of the following variants: A given pH interval should be extended by adding carrier ampholites covering the adjacent or a partly overlapping region. A certain pH area can be expanded by adding an amphoteric substance, "spacer", such as an amino acid. The spacer should be a "bad" ampholyte so that it does not focus too well. A certain pH range can also be extended by manipulating the thickness of the gel. The gradient will be shallower in areas with thinner gel. Manipulating the carrier ampholyte concentration will also affect the steepness of the final gradient. Areas with lower concentration will give shallower gradients. A mixture of three different pharmalyte intervals, which will maximize the number of carrier ampholytes in the region of interest has been found to give the best results. Generally, IEF gives a true representation of the isoelectric spectrum of the sample.

(b) Experimental techniques

Traditionally, preparative IEF has been performed in a vertical, coiled glass column stabilized toward heat convention by a density gradient of sucrose, glycerol, or similar substance (Figure 25.14). The many drawbacks of this technique can be circumvented by separation in a horizontal bed of Sephadex. Although being easy in principle, the technique requires some practice to get the right slurry density when preparing the bed, and in addition a new separation problem is created – the elution of the protein out of the gel. Analytical IEF is performed exclu-

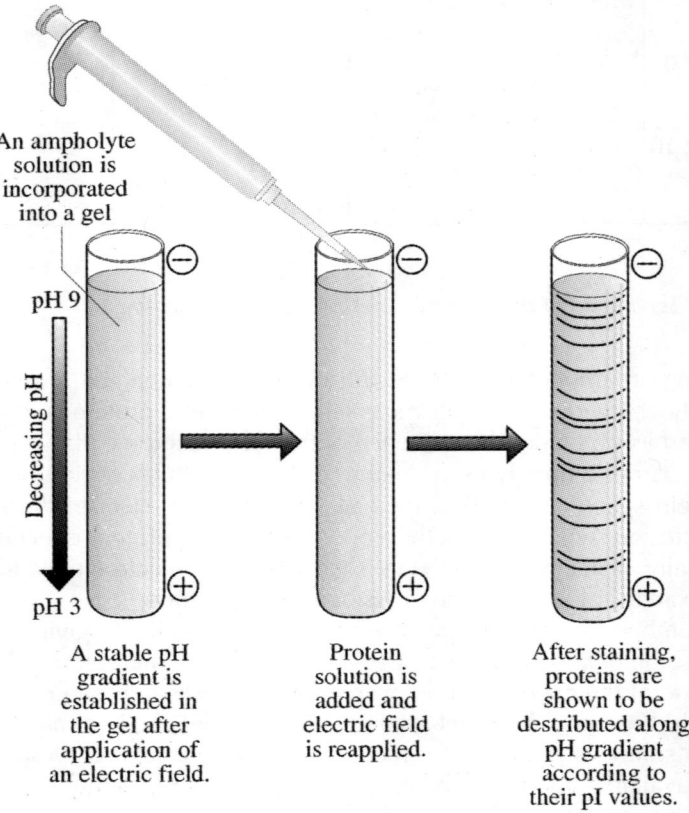

Figure 25.14 *Isoelectric focusing. This technique separates proteins according to their isoelectric points*

sively in gels of either PA or agarose. Horizontal slab gel IEF is the technique of choice when high resolution, high reproducibility, and high quality results are required. It is ideal for screening and comparison of many samples run side by side.

(c) Sample

The sample should have as low an ionic strength as possible. Too high a salt concentration in the sample will result in curved protein bands. Normally the sample is applied with the use of a plastic mask with cut-out holes or with the use of filter paper pieces soaked in the sample. High quality paper without unspecific adsorption of protein, or other undesired side effects, must be used. A good guide to the best sample application spot can be obtained by titration curve analysis. If possible, the protein should be applied from the side with the highest electrophoretic mobility. One advantage of IEF is the ability to apply very crude samples. Soluble proteins may be analyzed by applying pieces of tissue directly on the gel surface. Proteins will diffuse out of the tissue and into the gel where they will focus. In preparative IEF in Sephadex, the sample is usually mixed with the carrier ampholytes and slurry and applied as a sample zone with the aid of a special sample application tray. Within reasonable limits, the sample concentration is of less importance for the IEF process than the total amount of each protein and depends on a number of factors, mainly the purpose of the experiment.

(d) Optimizing the IEF system

Choosing IEF type: The optimal system and experimental conditions will of course depend on the separation problem, the type of sample, and the purpose of the separation. For preparative separations, flat bed IEF in Sephadex is generally recommended. To separate proteins differing by less than ca. 0.05–0.1 pH units, IEF in Immobiline gradients should be tried. In analytical IEF there are essentially three choices: (1) Conventional IEF in agarose, (2) Conventional IEF in PA, and (3) IEF in Immobiline (in PAA gel). If the protein to be analyzed has a high molecular weight (>150,000), agarose should be the first choice. In the majority of cases, however, IEF in PA is to be preferred.

Gel length and thickness: Intuitively, it seems advantageous to use long gels in order for the proteins to be spaced as far apart as possible. However, the use of long gels has a number of drawbacks: (1) the longer the gel, the longer the path for the proteins to migrate. Eventually some proteins may not be able to reach their pI during the lifetime of the pH gradient. (2) Very long gels cannot be run at the same field strengths as the shorter ones, and so the bands will be less sharp. (3) For a given pH span, the gradient will be shallower in the long gel. The bands will thus not focus as sharply as in the shorter ones.

In general, the effect of gel length on overall performance has little gain if the gel length is longer than ca. 10 cm. The trend in recent years has been to decrease the thickness of the gels used. Thin gels offer advantages such as more efficient cooling, apparently sharper bands, fast solvent penetration in staining and de-staining, and lower consumption of carrier ampholytes. The optimal gel thickness seems to be 0.2 to 1 mm for most applications.

Running conditions (voltage and power): The sharpest bands and best resolution are obtained with maximum voltage. However, the limits are set only by the ability of the system to control the heat produced. The following points should be considered for optimum performance: (1) An experimental set-up with good cooling capacity should be used, (2) carrier ampholytes with the most even conductivity profile should be used, (3) more carrier ampholytes and/or Immobiline than needed should not be used to stabilize the pH gradient, since excess carrier ampholytes will increase the current and the heat produced, making it impossible to use high voltage, (4) the experiment should be run at constant power.

Materials

Gel electrophoresis apparatus (any tube gel electrophoresis apparatus can be used if appropriate grommets or corks are prepared to fit small tubes).

1- to 3-mm-inner diameter glass gel tubes (0.2 ml disposable); 2.5- to 3-cm-inner-diameter gel casting glass tube; 50 µl, 1 ml, and 20 ml syringes; 0.2- or 0.45-µm filter capsule; Single-edge razor blade; Rubber grommets; Tube cell; 22-G hypodermic needle; 200 µl pipettor tip; 1-dram gel vials. Power pack, delivering up to 500 V.

Ampholytes, pH 3.5 to 10 or depending on needs of investigator.

Nonidet P-40; 10% (w/v) ammonium persulfate; TEMED.

PAGE chemicals: 28.38% Acrylamide/1.62% Bisacrylamide in nanopure water. Store in the dark for up to 2 weeks at 4°C.

IEF gel mixture: 5.50 g Urea (final 9.2 M), 2 ml Nanopure water, 1.33 ml Acrylamide stock (4%), 0.10 ml Ampholines pH 3.5–10, 0.40 ml, Ampholines pH 5–7 (2% total), 2 ml 10% NP40 (final 2%). Dissolve the urea at this point using a 37°C water bath and/or brief sonication if necessary. Then, immediately before pouring, add 7 µl TEMED and 10 µl 10% (w/v) Ammonium persulfate (fresh).

IEF electrophoresis buffer: Upper (–ve, cathodic) 0.05 M NaOH (2 g/L); Lower (+ve, anodic) 0.023 M phosphoric acid (1.44 ml/L)

Chromic acid cleaning solution: conc H_2SO_4, saturated with potassium dichromate.

Equilibration buffer: 10 ml glycerol (final 10%), 50 ml 10% SDS (final 5%), 6.25 ml 1 M (121 mg/ml) Tris pH 6.8 (final 62.5 mM), 1 ml 2% Bromophenol blue, 27.75 ml Nanopure water. Add conc HCl to pH 6.8. Add 5% 2ME before use.

Protein samples; Urea (ultrapure); Nonidet–P40, NaOH, and orthophosphoric acid.

Lysis buffer: 2.75 g Urea (final 9.5 M), 1 ml 10% NP40 (final 2%), 250 µl Ampholines pH 3.5 to 10 (final 2%), 40 mg DTT (0.78%). Add nanopure water to 5 ml. Aliquot and store at –70°C for up to 2 months. Do not refreeze aliquots.

4X lysis buffer: 180 µl NP40 (final 18%), 410 µl Ampholines pH 3.5–10 (final 41% total), 55 mg DTT (final 17%). Add nanopure water to 1 ml. Store in 50 µl aliquots at –70°C.

Immunoprecipitate elution solution: 250 ml 10% SDS solution (final 0.5%), 40 mg DTT (final 0.78%). Add nanopure water to 5 ml.

Ethanol–KOH bath solution: 10 g KOH in 500 ml 95% ethanol or methanol.

(e) Sample preparation:

1. Add 1 µl lysis buffer and 1 mg urea for every 1 µl of sample. For larger volumes, use 0.25 µl of 4X lysis buffer and 1 mg urea per 1 µl of sample. Include marker proteins with sample if desired.

2. For immunoprecipitates, boil precipitates in 50 µl of the immunoprecipitate elution solution. Cool, then add 12.5 µl of 4X lysis buffer and 12.5 mg urea. After loading, overlay with 10 µl 10% NP 40.

Procedure

1. Wash glass rods in water immediately after use. Immerse in chromic acid overnight, then wash before transferring to ethanol–KOH bath. Leave for at least one hour. Rinse in water then absolute alcohol just before pouring gels. Mark clean, dry 1.5-mm-i.d. gel tubes to indicate the desired height of the gel (usually 10–12 cm). Place a rubber band around the gel tubes so that they form a tight bundle (~12 tubes fit into a bundle). Hold the bundle vertically on a flat surface and push down on the tops of the tubes so that the bottoms are even.
2. Carefully seal one end of the 2.5- to 3-cm-i.d. gel-casting glass tube with three or four layers of Parafilm to form a strong, water-tight seal.
3. Place the bundle of gel tubes inside the gel casting tube and support the glass tube in a vertical position with a ring stand clamp to allow the sealed end of the glass tube to rest on a solid surface.
4. Pour the gels using a glass Pasteur pipette. Squirt the gel in vigorously at first to prevent air bubbles forming at the bottom of the rod. Tap rods on the bench to remove any small bubbles. The gels will polymerize in 15–30 min. Overlay the gel mixture with 20 µl of water. After 1–2 hours, remove the overlay and replace with 20 µl lysis buffer. Leave the gels for a further 2 hours before use, but not longer than 24 hours.
5. Gently run water down the outside of the gel tubes using a wash bottle. Add water until the level of the acrylamide solution inside the tubes reach the desired height.
6. Remove the Parafilm from the bottom of the gel casting tube and push the gel tubes containing the polymerized gel, out through the bottom. Cut across the gel-tube bottoms with a single-edge razor blade to remove excess acrylamide. Rinse the bottom of the gel tubes under running deionized water to remove residual acrylamide.
7. Place rubber grommet on the top of each tube, making sure that the top surface of the gel is visible below any unused holes with rubber stoppers.
8. Seat the tube and grommet assemblies in the holes of the upper buffer reservoir of the tube cell. Plug any unused holes with rubber stoppers.
9. Fill the lower reservoir with ~3 litres of 0.023 M phosphoric acid.
10. Place the upper reservoir into the lower reservoir and adjust lower buffer level to cover the entire gel.
11. Fill the upper buffer reservoir with 250 ml of 0.05 M NaOH. Fill the gel tubes to the top with 0.05 M NaOH using a 1 ml syringe equipped with a 22-G hypodermic needle. Be careful to eliminate any air bubbles in the gel tube.
12. Connect the tube cell to the power supply. The black (–) lead goes to the upper reservoir. Prefocus the gel for 1 hour at 200 V constant voltage. Disconnect the tube cell from the power supply.
13. Layer 10 to 30 µl of protein samples (100 to 150 µg) on top of the gels through the upper buffer with a 50 µl syringe. Fill the remainder of the tube with 0.02 M NaOH to eliminate any bubbles.
14. Place the lid on the upper reservoir and attach the electrical leads to a power supply. Turn on the power supply and adjust to the desired settings at constant voltage (700 to 800 V). Run the gel for 16 hours (11,000 to 13,000 V–hour).

15. Reduce the voltage settings to zero and turn off the power supply to end the run.
16. Extrude the gels from the tubes using equilibration buffer (without 2ME) pressure from a 1 ml syringe fitted with a 200 µl pipetter tip (cut off ~1 cm of the large end of the tip so it fits on the syringe). Extrude gels into 5 ml equilibration buffer. Add 5% 2ME and seal the tubes. Equilibration times can be varied from 15 min. to 1 hour. At this point, the gels can be loaded onto the second dimension or frozen in equilibration buffer at –70°C. The gels can be stored at –70°C for many weeks.
17. Soak gel tubes overnight in chromic acid cleaning solution, then rinse thoroughly under running deionized tap water for 15 min. Remove excess water from the gel tubes with suction and allow them to dry.

Note

1. Carbamyl ions form in urea solutions, which modify protein amino groups. Do not heat proteins above 37°C when urea is present. Equally, urea solutions should be freshly made up or deionized with mixed bed resin to remove carbamyl ions.

Method-2: Second-dimension gels

Combining isoelectric focussing and SDS electrophoresis sequentially in a process called two dimensional (2D) electrophoresis permits the resolution of complex mixtures of proteins (Figure 25.15). This is a more sensitive analytical method than either electrophoretic method alone. Two 2D-electrophoresis separates proteins of identical molecular weight that differ in pI, or proteins with similar pI values but different molecular weights. Second dimension gels are normal slab gels but with a flat stack. To ensure good contact between rod and slab gel, the top of the stack must not be wet.

Materials

30% Acrylamide/0.8% Bisacrylamide; Gel buffer; 10% (w/v) SDS; 10% (w/v) ammonium persulfate; TEMED; Isobutyl alcohol, water saturated; Stacking gel buffer (optional); First-dimension gel; Equilibrium buffer; Hot 0.5% and 1% (w/v) agarose (keep in boiling water); Protein molecular weight standards; SDS solubilization buffer; Reservoir buffer, prechilled to 10° to 20°C. Coolant; Gel plates, one long and one short; 1.5 mm spacers; casting stand; Gel identification tag (e.g. typed consecutive numbers on filtered paper); Nylon screen; 5 × 15 cm glass plate; Protean II electrophoresis cell (Bio Rad)

Procedure

1. Assemble the gel plates with 1.5 mm spacers. Position clamps on each side of the gel sandwich over the spacers and place on the casting stand. Be sure the plates and spacers are properly aligned, then tighten the clamps and cams to get a leak-proof seal. Make adjustments so that plates are leveled and vertical. Place the gel identification tag between the glass plates so that it rests in the lower right-hand corner.
2. Prepare the gel solution by combining 30% acrylamide/0.8% bisacrylamide, gel buffer, and water in a vacuum flask. De-aerate the solution by applying vacuum for 5 min.
3. Add 10% SDS and TEMED and swirl. Then add 10% ammonium persulfate and swirl.
4. Fill the gel sandwich to 5 mm below the top of the short plate and overlay with water-saturated isobutyl alcohol or water. Allow the gel to polymerize 1.5 hours.

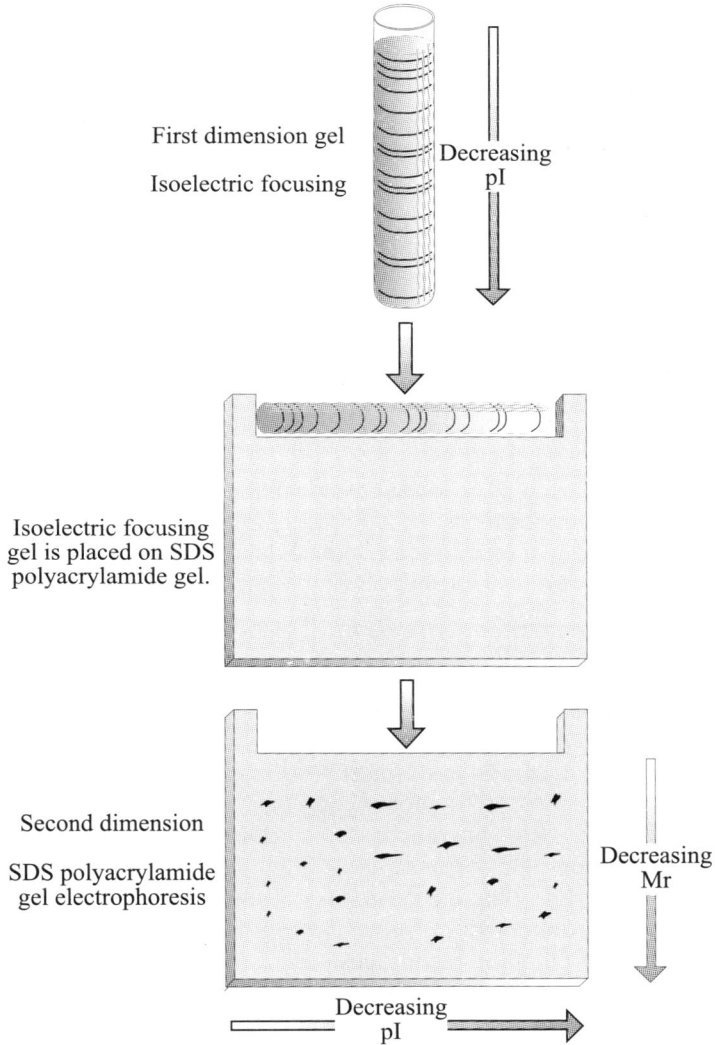

Figure 25.15 *Two-dimensional electrophoresis*

5. Add the equilibration buffer to completely cover the first-dimensional gel (thaw at room temperature, if necessary).
6. Pour the gel and equilibration buffer onto a nylon screen placed over a beaker and transfer the first-dimensional gel to a 5 x 15-cm glass plate. Using a spatula lay the gel out straight along one edge of the glass plate.
7. Pipette a very thin layer of hot 0.5% agarose on the top of the slab gel to be loaded.
8. Using a spatula, carefully slide the first-dimension gel off the glass plate and plate it across the top of the slab gel. Orient the first-dimension gel with the blue (basic) end to the right.

9. Pipette a thin layer of hot 0.5% agarose over the first-dimension gel to seal it in place. Allow the agarose to solidify.
10. Mount the gels on the electrophoresis cell. Fill the upper and lower reservoirs with prechilled reservoir buffer.
11. Attach tubing for coolant to the in and out ports and start the flow of coolant to maintain the temperature of the tank buffer at 10° to 20°C during the run to ensure that the gels are adequately cooled.
12. Attach the electrical leads to the power supply (the upper reservoir is connected to the negative lead). Electrophores at 15 to 20 mA/gel until the tracking dye reaches the end of the gel (or 3 to 5 mA/gel overnight).
13. Reduce the voltage setting to zero and turn off the power supply at the end of the run. Remove the gel from the electrophoresis unit and take off the clamps. Pry the glass plates apart with a spatula.
14. Stain the gel or process for immunoblotting or autoradiograph.

Note

1. Elution of spots excised from 2D gels may yield sufficient quantity of protein for immunization of animals to raise monospecific sera. Protein purity and quantity may also be adequate for amino acid sequencing or other microchemical analysis, generally conducted on samples transferred onto Immobilon (Millipore) polyvinylidene difluoride membranes.

Protein detection and quantitation methods in 2D-PAGE

2D-PAGE and electroblotting onto PVDF membranes have become widely used techniques for the characterization of proteins. In general, the same protein detection methods which are used for one-dimensional PAGE (Amido Black, Coomassie blue staining, silver staining, Colloidal gold, Ponceau S and radio-labeling) can be used for 2D-PAGE gels.

Identifications of protein spots: A 2D-protein map is of course only useful when the protein(s) of interest can be identified, or if differences in the 2D-patterns can be correlated to certain properties of the material from which the samples originated. There are several methods available for identification, based on different properties of the proteins: (1) Comparison with published 2D-maps, (2) coelectrophoresis, (3) specific intrinsic labeling, (4) genetic approaches, (5) physiological approaches, (6) blotting techniques.

Computer aided pattern analysis of 2D-protein maps: The amount of information in a 2D-PAGE of a complex protein sample is immense, but manual comparison is still possible, provided that only small series of experiments are performed. Advances in the methods for 2D-PAGE separation of proteins have been accompanied by the development of computer systems to analyze the resulting protein patterns, and to quantitate the individual protein components (Table 25.10). The 2D-PAGE may be digitized and analyzed by the computer to allow quantitative image analysis, automatic gel comparison, or specialized protein database searches. These data analysis systems range from simple programs for personal computers to more expensive systems (e.g. computer aided pattern analysis, CAPA), which allow the simultaneous analysis and comparison of complex protein patterns in 2D-PAGE gels. Before subjecting 2D-PAGE gels to computer analysis, there are several points to consider. First, computer quantitation of poor quality gels is of limited value. Therefore, the 2D-PAGE separation of components in a complex protein mixture must be optimized prior to analysis. Second, the information obtained from computer "quantitation" is relative to the method of protein detection used, and thus to the nature of the proteins themselves. Finally, useful information can be gathered from visual inspection of reproducible, high-quality 2D-PAGE gels without the assistance of computer programs. The lack of a computer system for analysis should not be a major factor in considering the use of 2D-PAGE.

Table 25.10 *Commercial computer software for analysis of 2-D gels*

System name	Derived from	Computer system	OS (operating system)	Supplier
Visage	-	SUN	UNIX	Millipore
PDQUEST	QUEST	SUN	UNIX	Protein and DNA
Kepler	TYCHO	VAX	VMS	Large Scale Biology Corporation
Biolog	HERMeS	SUN	UNIX	Biolog
Gemini	-	Custom	Unknown	Applied Imaging
IB-1000	-	IBM-AT	MS-DOS	Indiana Biotech
Microscan 1000	-	IBM-AT	MS-DOS	Technology Resources Inc
Phoretix-1	-	IBM-AT	MS-DOS	Biometra Ltd
QGEL/QBASE	-	IBM-AT	MS-DOS	Quanti-Gel Corporation

Method-3: Capillary electrophoretic separations

Electrophoretic separations in narrow bore plastic tubes, glass capillaries, and thin fluid films between parallel plates have received considerable attention within the last three decades. During this period, a number of instrumental approaches for capillary zone electrophoresis, capillary isotachophoresis, capillary isoelectric focusing, micellar electrokinetic capillary chromatography, continuous flow electrophoresis and electrical field flow fractionation or electropolarization chromatography have emerged. Capillary electrophoresis complements other existing capillary electrophoresis. Unfortunately, unlike chromatographic methods, classical electrophoresis, is rather slow, laborious, relatively irreproducible, difficult to quantify, and not really adaptable to automation. Thus, this technique is not discussed in this chapter.

IV PROTEIN RECOVERY

There are methods available for the recovery of proteins from preparative PAGE performed under native conditions by continuous elution during electrophoresis. The proteins are transported into an elution chamber, which is packed with some porous material and connected to the bottom of the column. The protein bands are then eluted from this chamber by using a peristaltic pump. Another recently introduced method, the Elfe™ system, uses a flow of liquid, which can be created by electro-osmosis, for elution. The amount of protein that can be purified in an analytical PAGE is eluted from slices of PA gels by rapid and efficient procedures. Proteins can be eluted from a homogenous gel by diffusion or by electrophoresis. The use of gels prepared with soluble cross-linking agents is also a possibility, but the reagents required for gel solubilization can be damaging to proteins.

A. ELUTION BY DIFFUSION

The gel piece containing the protein band of interest is cut into small pieces or homogenized, and placed in a test tube with a suitable buffer. The resulting gel slurry is then eluted, usually overnight at 4°C, with an appropriate buffer. For denatured proteins, efficiency of extraction can be increased by adding urea (4–8 M) or 0.1% (w/v) SDS to the elution buffer and also by eluting at elevated temperatures. The procedure is time-consuming, the yield for high molecular weight proteins is low and the reproducibility is not very good. Also, the protein preparation will contain polysaccharide contaminants. Thus, elution by diffusion can hardly be recommended for proteins.

B. ELUTION BY ELECTROPHORESIS

Many different devices have been constructed for electrophoretic elution of proteins from PA gels, and some of them are commercially available. The Elfe system can be used for electroelution of proteins from gel slices by casting only a stacking gel in the electrophoresis tube, and applying the gel slices on top of this stacking gel. Another possibility is to place the gel slice on top of a small column of Sephadex® and use displacement electrophoresis to force the proteins out into the Sephadex bed. The column is then connected to a peristaltic pump and eluted. A small syringe (2 ml), or a pre-packed PD-10 Sephadex G-25 column may also be used in combination with a displacement electrophoresis buffer system. In another device, the so-called Biotrap™ from Schleicher and Schuell, the gel slice is placed between two protein permeable membranes. The device is then placed in an apparatus for horizontal gel electrophoresis. The proteins are then eluted by electrophores into a second chamber, which is closed by a third membrane. A recovery of 80–97% depending upon Mr of the protein has been reported. A combination of a salt barrier and a sucrose gradient has been used for recovery of protein zones from PA gels.

C. UNSEPARATED PROTEINS CAN BE QUANTIFIED

To purify a protein, it is essential to have a way of detecting and quantifying that protein in the presence of many other proteins at each stage of the procedure. For proteins that are enzymes, the amount in a given solution or tissue extract can be measured or assayed in terms of the catalytic effect the enzyme produces (the rate of conversion of substrate into product). For this purpose, one must know (1) the overall equation of the reaction catalyzed, (2) an analytical procedure for determining the disappearance of the substrate or the appearance of a reaction product, (3) whether the enzyme requires cofactors or co-enzymes, (4) the dependence of the enzyme activity on substrate concentration, (5) the optimum pH, and (6) a temperature zone in which the enzyme is stable and has high activity. By international agreement, 1.0 unit of enzyme activity is defined as the amount of enzyme causing transformation of 1.0 µmol of substrate per minute at 25°C under optimal conditions of measurement. The **activity** refers to the total units of enzyme in a solution. The **specific activity** is the number of enzyme units per milligram of total protein.

After each purification step, the activity of the preparation (in units) is assayed, the total amount of protein is determined independently, and their ratio gives the specific activity. For proteins that are not enzymes, other quantification methods are required. Transport proteins can be assayed by their binding to the molecule they transport and hormones and toxins by the biological effect they produce. For example, growth hormones will stimulate the growth of certain cultured cells. Some structural proteins represent such a large fraction of a tissue mass that they can be readily extracted and purified without a functional assay. The approaches are as varied as the proteins themselves.

V STABILIZING PROTEINS

Proteins are fragile molecules that often require great care during purification to ensure that they remain intact and fully active. Although it is impossible to predict what steps will stabilize an uncharacterized protein, a number of procedures frequently help and are worth testing. Nowadays, many proteins are also purified in small amounts under denaturing conditions by various gel electrophoretic techniques, so that inactive proteins are obtained. But even here, it is usually advantageous to maintain the protein in an intact form. In the case of enzymes, and other proteins with assayable biological activities, maintenance of activity is generally of prime importance, both for following the protein during purification and for subsequent studies of its function. In this section, some of the major points to keep in mind with regard to maintaining the stability of proteins during purification and storage are described.

A. CAUSES OF PROTEIN INACTIVATION

Removal of proteins from the cellular environment subjects them to a variety of conditions and processes that can lead to loss of activity or alteration of structure. These include dilution, change in solution conditions, exposure to degradative enzymes, oxygen, heavy metals, and surfaces, and change in physical condition (e.g., freezing and thawing). If the protein of interest is lost or inactivated during the course of any procedure, determination of the reason for this loss will often suggest a simple solution. Thus, if possible, it should be examined whether the loss of activity is accompanied by loss of the protein or changes in its structure, or whether the protein remains but is now inactive. Distinguishing among these different possibilities might indicate what type of process is behind the problem and, thus, what an appropriate solution might be.

B. GENERAL HANDLING PROCEDURES

Obviously, to maintain the stability of proteins, treatments that denature it should be avoided. Thus, protein solutions should generally not be stirred vigorously or vortexed since this may lead to oxidation or surface denaturation. Protein solutions should not be exposed to extremes of pH, high temperatures, organic solvents, or any other condition that might promote denaturation. In an unfrozen state, bacterial and fungal growth can become a problem. It is best to make up all solutions that will come in contact with the protein with glass-distilled water, and to store the water in containers that do not have algal growth.

C. CONCENTRATION AND SOLVENT CONDITIONS

Extraction of proteins from cells inevitably leads to a change in their environment. Since proteins are generally stable *in vivo*, the theoretical goal is to try to reproduce the cellular milieu as closely as possible. This would mean very high protein concentrations (>1 mg/ml), close to a neutral pH (neutral buffer), and moderate ionic strengths (EDTA is usually added to chelate heavy metals, salts also added), reducing conditions (e.g., 2-mercaptoethanol or DTT)). Glycerol is added to maintain a certain ionic strength; on occasion, low levels of a detergent are added to prevent aggregation or the sticking of proteins to surfaces, such as glassware. Finally, it is good practice to include protease inhibitors, particularly in the early steps.

D. STORAGE CONDITIONS

One of the most important studies that can be performed during the course of new protein purification is a stability and storage study. It is highly recommended that after every step of the purification procedure, the stability and storage properties of the protein of interest should be determined. A different situation arises when one has completed a purification procedure and wants to store the purified protein for long periods of time by addition of high concentrations of glycerol, addition of stabilizing substrates, and even addition of extraneous protein such as serum albumin. The choice of storage conditions depends on what is effective for stabilization, and what the purified protein will be used for. Related to the question of storage is the problem of freezing and thawing solutions of purified proteins. One way to avoid repeated cycles of freezing and thawing is to store the purified protein in small portions and thaw individual samples once, as needed.

E. PROTEOLYSIS AND PROTEASE INHIBITORS

Proteolysis is a major problem for the purification of proteins. It is a partially insidious problem because in many cases the protein of interest is only partially degraded and retains biological activity. This results in erroneous conclusions about the size and structure of the protein. Proteolysis can be a problem at any stage of a purification procedure. As purification proceeds, even a small contamination with a protease could have a large effect because a larger fraction of the available protein substrate will be the one with which you are working. Cells contain a variety of different types of proteases. Fortunately, a number of protease inhibitors are available that can act on the various proteases.

VI SUGGESTED READING

Creighton TE (1997) *Protein Function* (2nd Edn), Oxford University Press.

Davis CG, Dibner MK, and Battey JF (1986) *Methods in Molecular Biology*, Elsevier Science Publishing Co., Inc.

Deutscher M (1990) *Guide to Protein Purification*, Academic Press Inc., London and New York.

Doonan S (1996) *Protein Purification Protocols*, Human Press, Totowa, NJ.

Gupta MN and Mattiasson B (1994) in *Highly Selective Separations in Biotechnology* (Ed) Street G, Blackie Academic & Professional, Glasgow, UK, pp7–33.

Jinno K (Ed) (1996) *Chromatographic Separations Based on Molecular Recognition*, Wiley VCH, New York.

Lowry OH, Rosebrough NJ, Farr Al and Randall RJ (1951) Protein Measurement with the Folin–phenol Reagent, *J Biol. Chem.*, **193**:265–268.

Malaikan A (1994) in *Highly Selective Separations in Biotechnology* (Ed) Street G, Chapman & Hall, Glasgow, UK, pp34–54.

Nelson DL and Cox MM (2000) *Lehninger Principles of Biochemistry* (3rd Edn), Worth Publishers.

Robards K, Haddad PR and Jackson PF (1994) *Principles and Practices of Modern Chromatographic Methods*, Academic Press, London.

Roy I and Gupta MN (2000) Current Trends in Affinity-based Separations of Proteins/Enzymes, *Current Science*, **78**:587–591.

Scopes RK (1993) *Protein Purification—Principles and Practice*, Springer-Verlag, Berlin, Heidelberg, New York.

Thormann W and Firestone MA (1989) Capillary Electrophoretic Separations, In *Protein Purification: Principles, High Resolution Methods, and Applications* (Eds) Janson JC and Ryden L, VCH Publishers, Inc.

Stoscheck CM (1990) Quantitation of Protein, *Methods in Enzymology*, **182**:50–68.

26

Immunochemical Techniques

I THE IMMUNE SYSTEM

All multicellular organisms need to defend themselves against invasion by potentially dangerous pathogens. Vertebrates have an immune system capable of distinguishing molecular "self" from "non-self" and then destroying those entities identified as non-self (foreign). In this way, the immune system eliminates viruses, bacteria, other pathogens, and molecules that may pose a threat to the organism. It combats tumors and neoplastic cells and can reject transfused cells and transplanted organs from genetically non-identical animals. Immune response can be classified as either **innate** or **acquired** (**adaptive immune response**). Invertebrates use innate response – relatively simple defense strategies that ingest and destroy invading parasites, infectious molecules, and microorganisms. Innate immunity does not require prior exposure to the foreign substance and is mediated mainly by cells of the monocytic lineage and polymorphonuclear leukocytes. Innate immunity is relatively non-specific, although it normally clearly distinguishes between self and non-self. It constitutes a potent, rapid-reacting, first-line defense against invasion and unwanted infection.

Vertebrates, too, depend on such innate **immune response** (i.e., pre-existing defenses of the body, e.g. barriers formed by skin and mucosa, toxins, and phagocytic cells) as a first defense. But, they can also mount much more sophisticated defenses, called **acquired** or **adaptive immune responses**. In vertebrates, the innate and adaptive immune systems work together. The innate immune response mounts very rapidly to an infection. It depends on pattern recognition receptors that recognize patterns of pathogen associated molecules (immunostimulants) that are not present in the host organism. Some of these receptors are present on the surface of professional phagocytic cells such as macrophages and neutrophils, where they mediate the uptake of pathogens, which are then delivered to lysosomes for destruction. Others are secreted and bind to the surface of pathogens, marking them for destruction by either phagocytes or the complement system. Some other cells are present on the surface of various types of host cells which activate intracellular signaling pathways in response to the binding of pathogen-associated immunostimulants. This leads to the production of extracellular signal molecules that promote inflammation and help activate adaptive immune responses (Figure 26.1).

Acquired immunity requires exposure ('priming') to the non-self material. It is primarily mediated by lymphocytes and may be further divided into cell-mediated and humoral immune

Immunochemical Techniques 735

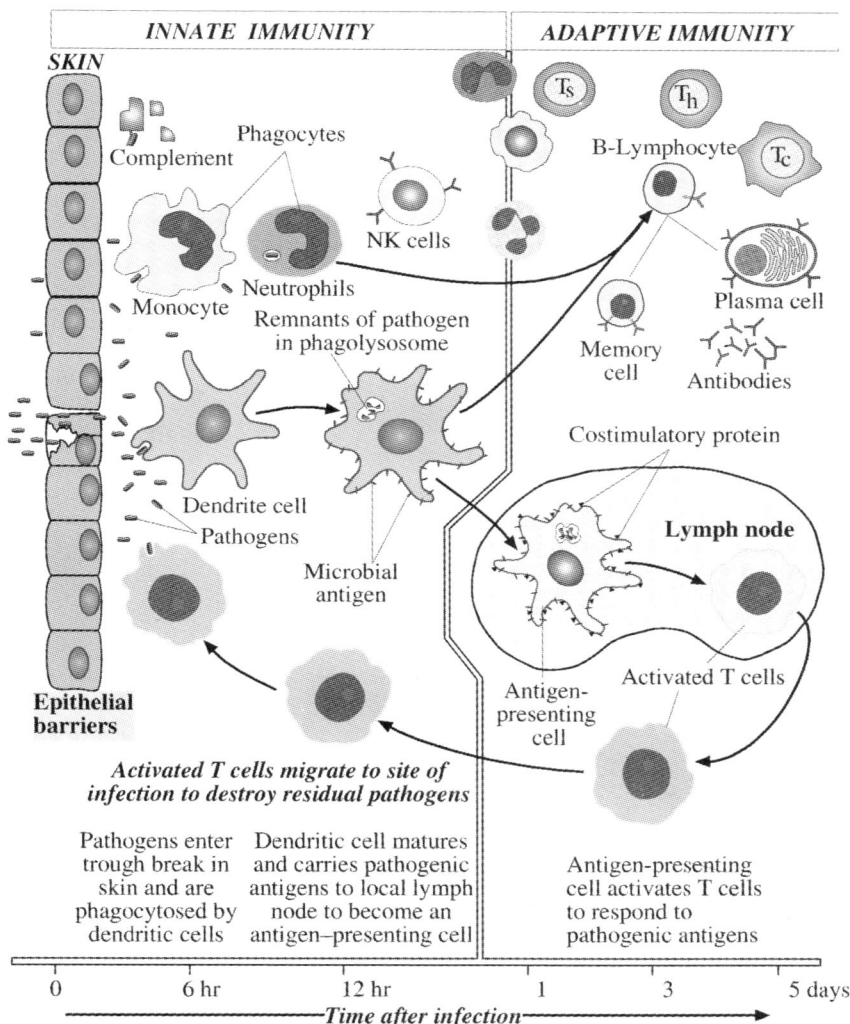

Figure 26.1 *Principle mechanisms of innate and acquired (adaptive) immunity. Innate immunity provides the initial defense against infections. The innate immune response stimulates the adaptive immune system*

response. The immune response features a specialized array of cells and proteins. Blood cells undergo progressive cell division and differentiation in the presence of mitogenic and differentiation-inducing growth factors from stem cells, originating in the bone marrow (Figure 26.2). The bone marrow contains lymphoid (Figure 26.3) and myeloid cells at various stages of differentiation. Peripheral blood, lymph nodes, the spleen, and the tonsils are classified as secondary lymphoid organs since they contain mature, fully functional lymphocytes. Functional myeloid cells include erythrocytes, granulocytes, monocytes, and platelets. The lymphoid cells include both B-lymphocytes (B cells) and different types of T-lymphocytes (T cells). The differentiated

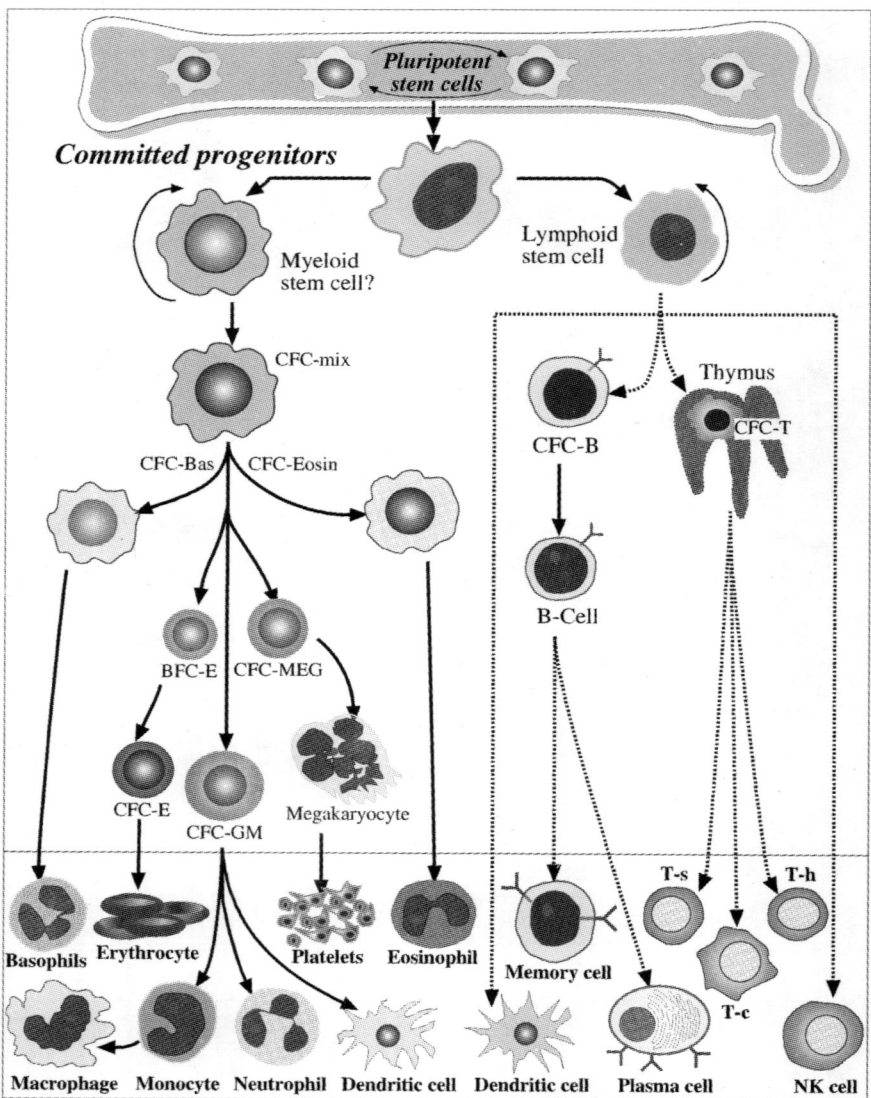

Figure 26.2 *An outline of lymphoid system in human being. Cell types within the box are terminally differentiated cells*

B-lymphocytes mature in the bone marrow and T-lymphocytes mature in the thymus; both of these lymphoid organs are classified as primary lymphoid organs. B-lymphocytes are the key cells in the production of antibodies following stimulation by an antigen.

Three different types of white blood cells play central roles in the immune response in vertebrates. These cells are (1) B-lymphocytes, (2) T-lymphocytes, and (3) macrophages. The immune response may be **humoral**, leading to antibody formation, or it may be **cellular**. When an

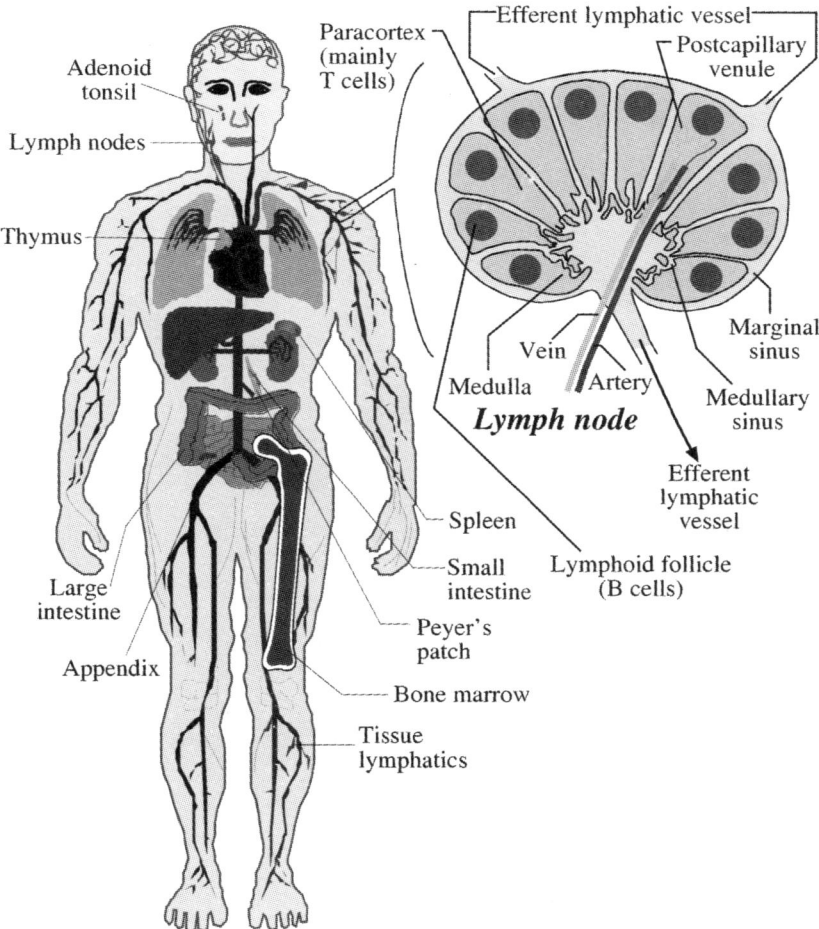

Figure 26.3 *An outline of lymphoid system in human being*

animal has been exposed to an antigen, **antigen-presenting cells** (APCs), T cells and B cells act in concert to stimulate the production of antigen-specific **antibodies** (**Ab**), or **immunoglobulins** (**Igs**). Unless these cells interact successfully, the B cells fail to produce antigen-specific antibodies. When the antigen is introduced into the animal body, APCs, such as macrophages or B cells, interact with it, ingest it, and break down or process it (Figure 26.4). The processed antigen is then returned to the surface of the APC where it is presented by the cell's **major histocompatibility molecules** (MHC). Subsequently, a T cell with a receptor specific for both the processed antigen and the MHC molecule on the APC interacts with the APC. A trimolecular complex is formed between the T cell antigen receptor, the processed antigen, and the MHC molecule on the APC. This complex triggers the APCs and T cells to produce a variety of soluble factors such as interleukin-1 (IL-1) and IL-2 (Figure 26.5), which stimulate T cell proliferation and differentiation. Properly activated T cells bind to antigen presenting B cells. These T cells then produce IL-4 and IL-5, which stimulate B cell multiplication and differentiation (Figure 26.5).

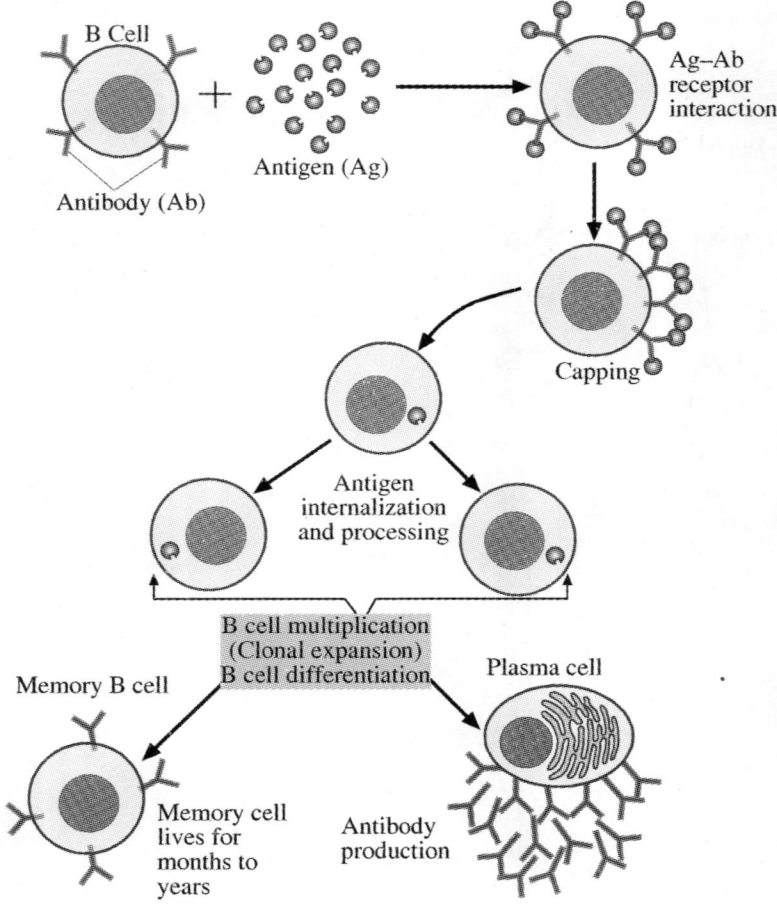

Figure 26.4 *Antigen (Ag) internalization and processing of B cell*

If properly stimulated, the B cell's immunoglobulin genome, which is responsible for the antibody's make-up, is spliced to yield the genetic information used to direct antibody formation. When an animal's immune system is further stimulated with the same antigen, that part of the immunoglobulin genome, which is responsible for the antigen-binding region of an antibody, undergoes a series of somatic mutations. These mutations lead to the production of antibodies with higher or lower affinity for the antigen. Those B cells producing antibodies with higher antigen affinity will attract more antigens and will be preferentially stimulated to grow, divide, and produce more of the high affinity antibody. T cell which secretes IL-5 along with other signals, allows B cells to differentiate into either short-lived (3–4 days) plasma cells which secrete large amounts of antibody or long-lived memory cells. Memory B cells are primed to respond to future antigen stimulation and rather than secrete, express the antibody on their surface.

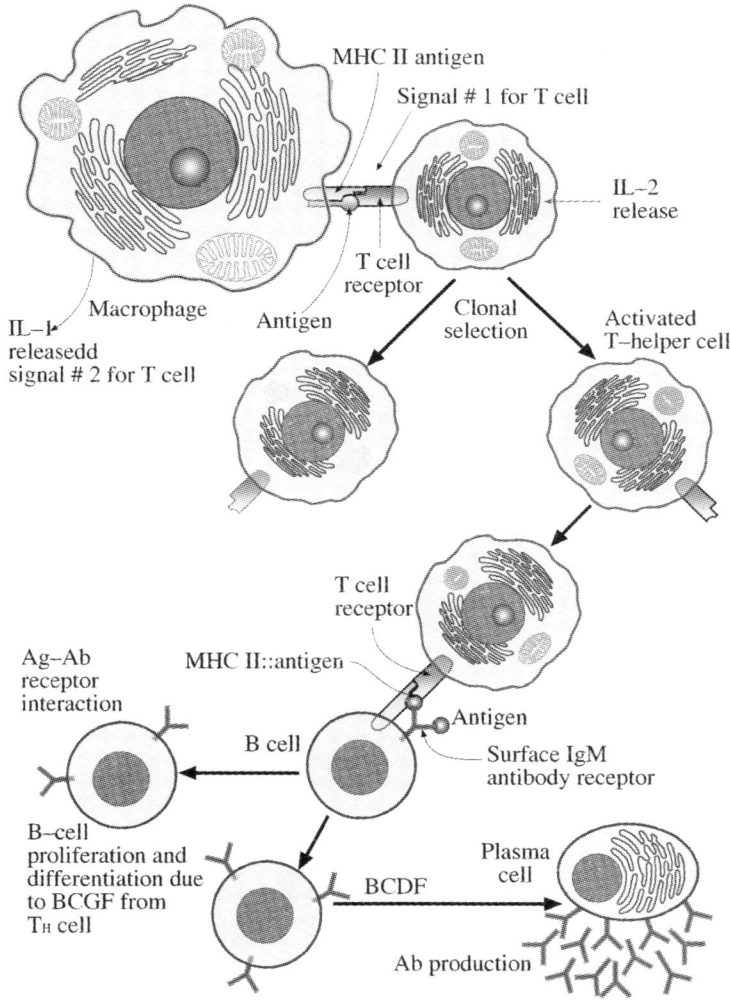

Figure 26.5 *Dependent antigen triggering of B cell. Schematic diagram of the events recurring in the interaction of macrophage, T-helper cells, and B cells that produce cell-mediated immunity*

A. STRUCTURES OF ANTIBODY

Antibodies consist of monomers or multimers of a basic antibody. Each basic antibody is a tetramer composed of four polypeptides, two identical light chains, and two identical heavy chains, joined by a disulfide bond. The light chains are about 220 amino acids long, and the heavy chains are about 440–450 amino acids long. Every chain, heavy or light, has an amino-terminal **variable region**, within which the amino acid sequence varies among antibodies specific for different antigens, and a carboxy-terminal **constant region**, within which the amino acid sequence is the same for all Abs of a given Ig class, regardless of antigen-binding specificity.

740 MOLECULAR BIOTECHNOLOGY: PRINCIPLES AND PRACTICES

The variable regions of antibody chains are about 110 amino acids long. Each antibody has two antigen-binding sites or domains, each of which is formed by the variable regions of one light chain and one heavy chain. In addition, the constant regions of the two heavy chains interact to form a third domain, called the **effecter function domain**, which is responsible for the proper interaction of the antibody with other components of the immune system. Antibodies directed to antigenically distinct regions associated with an Igs **paratope** or **hypervariable** region define an Igs idiotype. Each molecule has two or more antigen-binding sites termed **paratopes** (Figure 26.6). A paratope is capable of recognizing a single antigenic determinant known as an **epitope**. Antigenic determinants on variable regions are called **idiotypes** and the antibodies directed against these regions are termed **anti-idiotype**. Antigenic determinants that distinguish different types of constant regions (e.g. $C\mu$ from $C\gamma$, $C\lambda 2$) are called **isotypes**. Antigenic determinants that distinguish allelic genes are called **allotypes**. These antibodies do not recognize the antigenically distinct constant regions associated with either an Igs isotype or allotype. Anti-idiotope antibodies are characterized on the basis of their interactions with idiotopes on other antibodies from the same animal species or from different species; their interactions

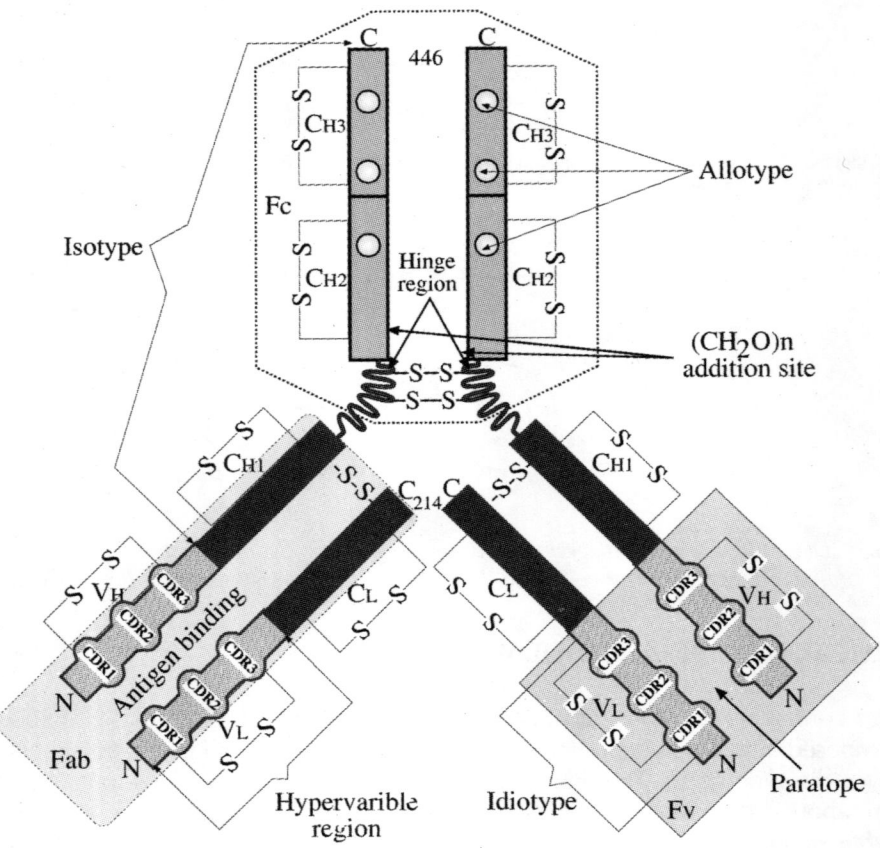

Figure 26.6 *Diagram of the organization of the IgG molecule*

with idiotype-presenting Ab in the presence of comparing antigens; and their *in vivo* biological activity. Antibodies on B cells recognize a wide variety of idiotopes on antibodies from intrastrain, interstrain, and interspecies animals. Shared idiotopes among antibodies are referred to as either **public idiotopes** (Idx) or **cross-reactive idiotopes** (CRIs). Those idiotopes which are not shared by all Igs are termed **private idiotopes** (IDI).

Although there are many types of Igs, five types have been distinguished which share the same type of combining site for antigen. They are designated as IgG, IgM, IgA, IgD, IgE, and eight subclasses (IgG1, IgG2, IgG3, IgG4, IgM1, IgM2, IgA, and IgA2) of antibody molecules. These classes serve specialized functions in the immune response and exhibit small structural differences. However, as mentioned before, each contains two types of polypeptide chains differing in size: a large one called the heavy (H) chain and a small one called the light (L) chain. IgG is the most abundant antibody class and has the simplest molecular structure. IgG, IgD, IgE, and serum IgA antibody molecules each consist of a pair of heavy and light polypeptide chains (Figure 26.7a). The chains are held together by disulfide bonds or, in the case of IgA2, by non-covalent linkages. IgM antibodies (Figure 26.7c) consist of five pairs of heavy and light chains while the secretary IgA (Figure 26.7b) consists of two to three such pairs. Pairs of heavy and light chains in IgM pentamers and IgA dimers are united by a J chain through disulfide bonds. In addition, secretary IgA also contains an Sc component. Digestion of IgG with papain yields three fragments (Figure 26.7d). Two of these fragments, each with a heavy and light chain, are equal in size and are antigen-binding (**Fab** fragments). The third fragment, consisting only of Ab heavy chains, does not bind antigen and crystallizes at 4°C (**Fc** fragment). The region between the Fab and Fc fragments is called the **hinge** region. This region gives the Ab molecule the flexibility needed for Fab segments to operate independently of one another. Digestion of IgG with pepsin yields two fragments [F(ab')$_2$] joined by disulfide bonds and **Fc'** fragments (Figure 26.7e).

The N-terminal ends of the heavy and light chains of Fab fragments, which make up the paratope, each contain three **complementary determining regions** (CDRs). The CDRs are oriented or anchored by framework (**Fw**) regions to the remainder of the Fab fragment. Essentially, the six CDRs found within the Abs paratope, and to some extent the Fw regions, are responsible for binding to or interacting with an antigen. Since the amino acid sequence for the CDRs varies among antibodies to different antigenic sites, they are also referred to as **hypervariable** regions. Regions outside those of the CDR and Fw vary less for a class or subclass of antibody and are referred to as **constant regions**. The Ab class and subclass can be immunologically identified with **anti-isotype** Abs directed to unique Ig heavy chain constant regions. Antibodies directed to other antigenically distinct constant regions on an Ig molecule can also be used to determine an Igs light chain (lambda or kappa) or allotypic characteristics. **Allotypic** characteristics are genetically inherited antigenic sites. These sites are present on a class of antibodies in some individuals and are absent from other individuals of the same species.

The first class of Abs produced by an antigenically stimulated animal during a primary immune response are IgM. Serum IgM concentration peaks approximately ten days after antigenic stimulation and then drops. If the antigen successfully elicits an appropriate T cell response, then IgG will be produced. Serum IgG concentration peaks at about fourteen days. If the same animal is antigenically re-stimulated to induce a secondary immune response, the time frame for the appearance of IgM and IgG antibodies will be similar to that seen during the primary immune response; however, the IgG concentration will generally be much greater during the secondary response. It must be remembered that if the MHC molecules of an animal are

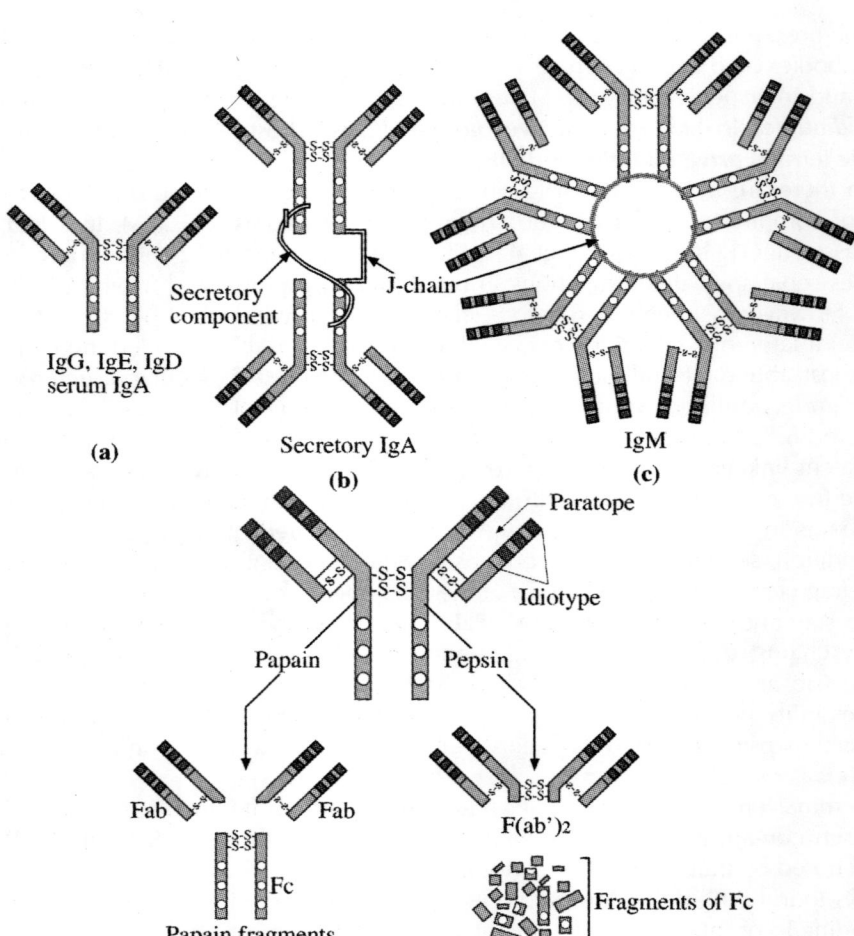

Figure 26.7 *Different forms of immunoglobulin molecules and creation of Fab, Fc and F(ab)2 fragments by specific proteolytic digestion of intact antibody molecules*

incompatible with the antigen, only IgM will be produced. Since IgA, IgD, and IgE are not as prevalent as IgG or IgM, they are seldom used as diagnostic immunological reagents.

B. ANTIGEN

An antigen is any agent that elicits an immune response in an animal circulatory system (either cellular or humoral). If an animal's B cells do not express antibodies on their surface capable of interacting with a substance, then an Ab to that substance will not be made. Although immunization is usually performed using fairly large complex structures, the antibodies recognize only a small portion of the complex. A low-molecular weight substance, such as a synthetic peptide

(10–15 amino acids), may be efficiently processed and presented by an APC's MHC molecule. However, it may be too small to be recognized simultaneously by both the MHC of the APC and a receptor of the T cell. A small polypeptide can be made more antigenic by chemically fixing it to a larger molecule exhibiting a ligand capable of interacting with a T cell receptor. In some cases, a large molecule may not elicit antibody formation in an animal because the animal's MHC molecules do not present or recognize the molecule.

A substance may exhibit one or more antigenic sites (epitopes) which are capable of eliciting and interacting with the Abs produced by an animal. Generally, the smallest peptide that demonstrates antigenic reactivity represents the epitope and is about 5–7 amino acids in size. Antibodies capable of interacting with random coil forms of a peptide define a **sequential** epitope. Antibodies that interact with a peptide exhibiting a specific conformation, define a **conformational** epitope. Antibodies to a conformational epitope will not react with the same peptide if it is unfolded (denatured). An expanse of amino acid residues constituting two or more groups of amino acids linked to one another is called a **continuous** or **contiguous** epitopes whereas, Abs that recognize a group of amino acids consisting of two joint, separate peptide chains or a folded single peptide chain is defined as discontinuous or discontiguous epitopes. Continuous or discontinuous epitopes, in essence, are conformational epitopes. Antibodies which interact with a denatured molecule, but which do not interact with the corresponding native molecule recognize epitopes known as **cryptotopes**. A **neotope** is an antigenic site on a monomeric subunit created by conformational changes brought about by the integration of the subunit with another molecule. An Ab, which interacts with an epitope on a substrate only when the substrate is bound to an enzyme, would define a **neotope**.

Generally, the degree of antigenicity of a molecule subunit is a reflection of its location or prominence on the molecule or its physical or chemical make-up. Hydrophilic amino acid residues, which tend to correspond to continuous epitopes usually appear on an antigen's exterior. Hydrophobic residues remain buried and are less accessible to the immune system unless the molecule is denatured. Epitopes that are mobile and capable of flexing to about 1–2 Å usually interact more strongly with an Ab than rigid molecules. The hydrophilic regions of a molecule that protrude and exhibit segmental mobility correspond to continuous epitopes. The amino and carboxy terminal ends of a molecule exhibit a high degree of mobility. These termini are usually close to one another, occupy sites on a molecule's surface, and typically constitute continuous epitopes. Sites on a molecule that exhibit little amino acid sequence conservation constitute epitope uniqueness useful for diagnostic distinctions. Apparently, these regions can tolerate changes brought about by DNA mutations since they are probably not involved in interactions responsible for the molecule's folding.

C. ANTIGEN–ANTIBODY INTERACTIONS

The interaction between an antigen and an Ab involves hydrogen bonds, salt bridges, electrostatic charges, and coulombic, hydrophobic, and van der Waal's forces. This interaction is reversible and the strength of the interaction (**affinity**) can be affected by temperature, pH, and solvent conditions. If these conditions are varied sufficiently, assays such as enzyme-linked immunosorbent assays (ELISAs) which employ the same antigen and Ab, can give divergent results. Stable semi-reversible immune complexes can be formed when multimeric interactions occur between Abs and antigens which exhibit either multiple copies of a single epitope or multiple copies recognized by different polyclonal Abs. Such multimeric interactions enable low-affinity Abs to bind tightly to antigens.

The ability to form stable immune complexes is known as **avidity**. IgM with ten paratopes are more multimeric and have greater antigen avidity than IgG with two paratopes. A paratope of an antibody may exhibit a low degree of affinity for an antigen. However, if this weak interaction is broken, the remaining paratopes of an antibody attached to nearby epitopes, can act to stabilize the antigen–antibody complex. A low-affinity IgM has nine other paratopes by which to stabilize this interaction, whereas a low-affinity IgG has only one. Consequently, IgMs which interact weakly with more than one antigen can appear to be less specific than IgGs, in general, simply because IgMs have greater capacity for stabilizing the antigen–antibody interaction. Therefore, a low-affinity IgG with the same antigenic specificity as an IgM may not remain bound to an antigen and may not be detected in an immunoassay while an IgM would be.

The hypervariable regions of the heavy and light chains, which make up the Abs paratope are arranged to form a pocket or groove. Typically, an antigen interacts with the CDRs found within or surrounding this groove. It is generally believed that these interactions assume a lock (antibody) and key (antigen) configuration. Structural changes apparently do occur in the antigen and/or antibody as a result of their interaction.

D. PRODUCTION OF ANTIBODIES

Virtually all immunochemical techniques rely on the use of Abs and their effectiveness is dependent on the quality of the Ab or antibodies employed. The nature of the antibody affects both the specificity of the methods and the sensitivity of the procedure. Proteins synthesized in heterologous systems can be detected either by assaying for a particular biological activity (enzymatic or other biological activity, such as ability to bind a specific ligand) or by employing assays that are independent of such activity. Even though many of these assays are very useful, they suffer from some practical limitations such as lack of sensitivity, endogenous expression of host genes of the same biological activity, and loss of specific activity of the protein due to mutations. For these and other reasons, it is essential to develop assays that are independent of biological activity and sensitive enough to measure very small amounts of the protein. The reagents of choice for these assays are Abs that react specifically with the foreign protein. Abs that react especially with the foreign protein fall into three general classes: (i) Abs that react with the foreign protein independently of its conformation, (ii) Abs that react only with epitope specific to the native of the target protein, and (iii) Abs that react only with denatured forms of the protein.

E. GENOME REARRANGEMENTS DURING B-LYMPHOCYTE DIFFERENTIATION

Vertebrate immune systems generate a nearly limitless variety of specific proteins. Antibody proteins are secreted by lymphocytes and circulate in the blood. Remarkably, however, complete genes encoding antibodies or immune receptor proteins do not occur in germ cells or in the early embryo. Instead, the genes are put together later in development, when lymphocytes are formed. The organization of immunoglobulin genes is extremely complicated, with a large array of variable region genes (V), diversity (D) segments in the case of VH, a small number of so-called J or joining segments and a cluster of constant region genes. In humans, these occur on the fourteenth chromosome but they are not immediately linked and recombination events are necessary to ensure synthesis of a particular immunoglobulin heavy chain. Each B-lymphocyte produces only a single type of antibody, that is, all the antibodies produced by a given B-lymphocyte have the same antigen-binding specificity. Each antibody chain is synthesized using information stored in several different "gene" or "gene segments" (Figure 26.8).

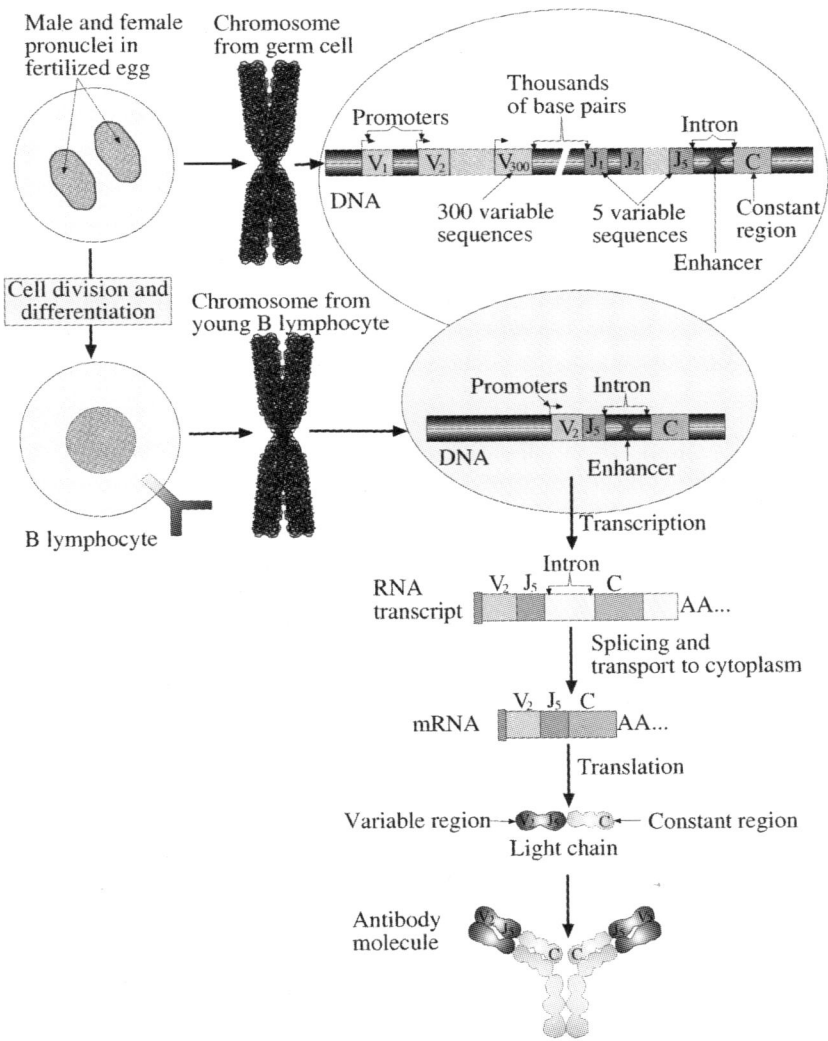

Figure 26.8 *Gene rearrangements that lead to the formation of a functional gene that encodes an immunoglobulin light (kappa or lamda) chain*

1. Kappa and lambda light chains

Synthesis of the kappa light chain is controlled by three different gene segments: (a) a Vk gene segment, code for the N-terminal 95 amino acids of the variable region, (b) a Jk gene segment (J for joining segment), coding for the last (constant region-proximal) 13 amino acids of the variable region, and (c) a Ck gene segment coding for the C-terminal constant region. A fourth gene segment, the Lk segment, codes for an N-terminal hydrophobic leader sequence of length 17–20 amino acids, which is essential for the transport of the antibody chain through the cell membrane. The leader sequence is cleaved off the chain as it passes through the membrane, and thus is not part of the final antibody. There are a large number (probably about 300) of Vk gene segments, each with a nearly equal number of Lk gene segments. On the other hand, there is only one Ck gene segment. Five Jk gene segments are located between the Vk gene and the Ck gene segments. During the development of a B-lymphocyte, the particular kappa light chain gene that will be expressed in that cell is assembled from one Lk–Vk segment, one Jk segment, and the single Ck segment by a process of somatic recombination. This process joins any one of the approximately 300 Lk–Vk segments with any one of the five Jk segments, with the deletion of all intervening DNA (Figure 26.8). The non-coding sequence between the Jk gene segment cluster and the Ck gene segment, and the Ck-proximal Jk segments, if any, remain between the fused VkJk segment and the Ck segment in the differentiated B-lymphocytes. This entire DNA sequence (Lk–VkJk–non-coding–Ck) is transcribed and the non-coding sequences are removed during RNA processing, just like the non-coding sequences or introns of any other eukaryotic gene.

Lambda light chain genes are also assembled from separate segments during B-lymphocyte development (Figure 26.9). The major difference is that each J-lambda gene segment comes with its own Ck-lambda gene segment. That is, the genome rearrangements required for lambda chain synthesis join L-lamb-V-lamb segments to J-lamb-C-lamb segments.

2. Heavy chains

The genetic information coding for antibody heavy chains is organized into LH–VH, JH, and CH gene segments analogous to those for kappa light chains; but there is one additional gene segment, called D for diversity, that codes for 2–13 amino acids of the variable region (Figure 26.10). The variable region of the heavy chain is thus encoded in three separate gene segments that must be joined during B-lymphocyte development. In addition, there are from one to four CH gene segments for each Ig class. In the mouse, there are a total of eight CH gene segments, all functional, arranged on the chromosome in the sequence CHμ, CH∂, CHγ3, CHγ1, CHγ2b, CHγ2a, CHε, CHα. CHμ, CH∂, CHε, and CHα code for the heavy chain constant regions of IgM, IgD, IgE, and IgA, respectively. Four gene segments, CHγ3, CHγ1, CHγ2b, and CHγ2a, code for the IgG heavy chain constant regions. In mouse germ line cells, there are about 300 LH–VH gene segments, about 10–50 D gene segments, 4 JH gene segments, and 8 CH gene segments, arranged on the chromosome in the preceding order. During the development of the B-lymphocyte from a stem cell, somatic recombination joins one LH–VH segment with one D gene segment and one JH gene segment, deleting the two intervening sequences of DNA, to form one continuous DNA sequence (VHDHJH) that codes for the entire heavy chain variable region.

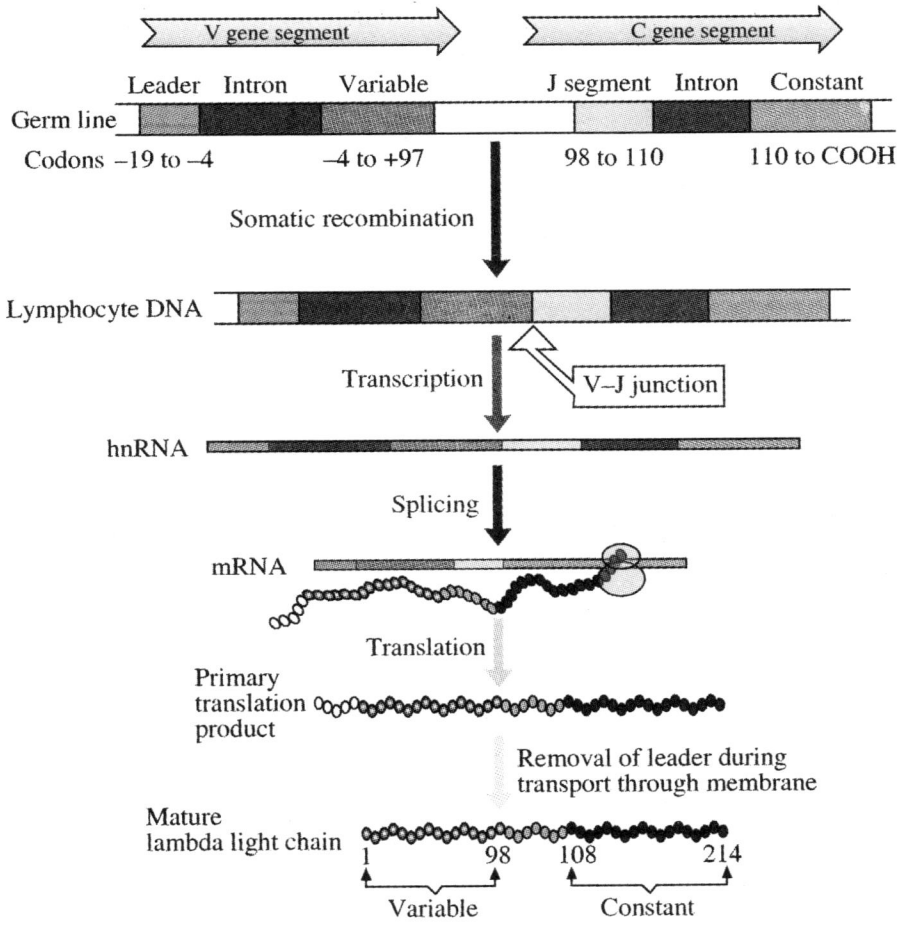

Figure 26.9 *Generation of functional lambda light chain by somatic recombination and splicing*

3. Class switching

At the time that antibody synthesis begins in the developing B-lymphocyte, all the CH gene segments are still present, separated from the newly formed LH–VHDJH gene segment by a short non-coding sequence. At this stage, all antibodies synthesized will have IgM heavy chains (CHμ gene products). If an antigen is recognized and bound to an antibody on the surface of a developing B-lymphocyte, however, that cell is stimulated to differentiate into a mature B-lymphocyte. During this differentiation, some B-lymphocytes will switch from producing antibodies of class IgM to producing antibodies of another class. This phenomenon called **class switching**, often involves further genome rearrangements during which the CH gene segments closest to the previously joined LH–VHDJH gene segments are deleted. The class of antibodies produced after class switching is determined by which gene is brought into the closest proximity with the LH–VHDJH gene segment.

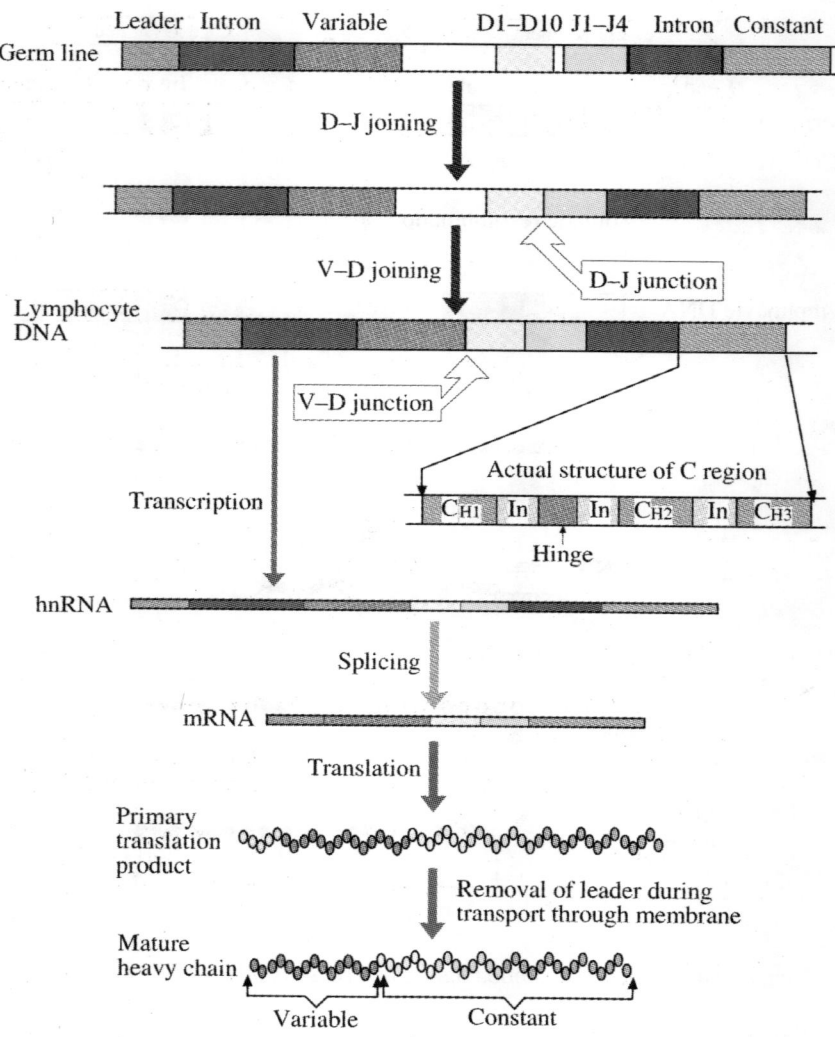

Figure 26.10 *Generation of functional heavy chain by sequential joining reactions*

4. Antibody diversity at the level of RNA splicing

A different type of class switching during B-lymphocyte differentiation occurs at the level of RNA processing (**splicing**). Certain mature B-lymphocytes produce both IgM and IgD antibodies, but these antibodies differ only in their effecter function domains. In these cells, a primary transcript that extends through both the CHμ and CH∂ gene segments is synthesized. During processing, the VHDJH transcript sequence may be spliced to either the CHμ sequence or the CH∂ sequence, such that both types of heavy chains are synthesized in the same cell. A further complexity observed in antibody synthesis is the sequential production of membrane-bound and secreted forms of a given antibody. The first antibodies to appear in developing B-lymphocytes are membrane-bound IgM molecules. Subsequently, these cells switch to the production of a secreted form of IgM. These two forms of IgM differ only in the C-terminal portions of their heavy chains. The heavy chain of the membrane-bound form is 21 amino acids longer than that of the secreted form. The membrane-bound heavy chain has a 41-amino acid long hydrophobic sequence at the C terminus that is probably responsible for anchoring it to the cell surface. This hydrophobic sequence is replaced by a 20-amino acid hydrophilic sequence in the secreted form.

The coding sequences (exons) of the CH gene segments are interrupted by non-coding sequences (introns). The CH gene segments contain four to six exons and three to five introns. In membrane-bound antibodies, the heavy chain constant regions are produced by splicing all six exons together (Figure 26.11). The last two exons code for the hydrophobic tails of the membrane-bound heavy chains. During synthesis of the membrane-bound form, the fifth CH exon is spliced to a site 20 codons from the end of the fourth exon, thus changing the amino acid sequence of this portion of the heavy chain constant region. In secreted antibodies, the heavy chain constant regions are therefore the product of the first four exons. Recent evidence suggests that similar alternative pathways of transcription and splicing are responsible for the production of the membrane-bound and secreted forms of the other classes of immunoglobulins as well.

5. Signal sequences control genome rearrangements

Several long segments of chromosomal DNA carrying clusters of V gene, D gene and J gene segments of both mice and humans have now been sequenced, and the resulting nucleotide-pair sequences suggests the presence of specific V–J, V–D, and D–J joining signals. The same

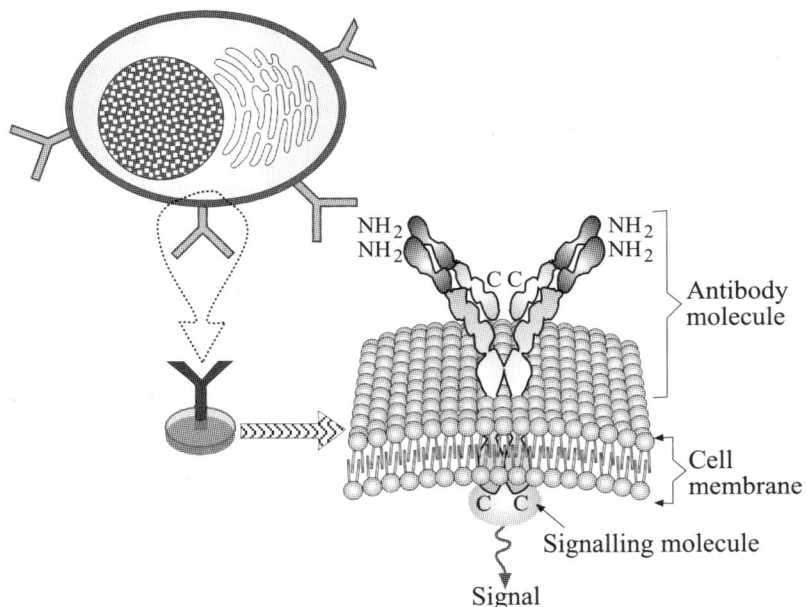

Figure 26.11 *An antibody bound in the cell membrane of a B-lymphocyte*

signal sequences are found adjacent to all V gene segments. Similarly, all J gene segments have identical signal sequences located adjacent to their coding sequences; however, their signal sequence is different from that adjacent to V gene segments. Likewise, D and C gene segments have their own adjacent signal sequences. The signal sequences controlling V–J, V–D, and D–J joining contain 7-base-pair (heptamer)- and 9-base-pair (nonamer)-long sequences separated by spacers of different, but specific lengths. For Vk–Jk joining, the spacers in the Vk signal sequence is 12 nucleotide-pairs long, whereas that in the Jk signal sequence is 22 nucleotide-pairs long. The heptamer and nonamer sequences located after the Vk gene segments are complementary to those preceding the Jk gene segments. These signal sequences have the potential to form "stem and loop" structures to bring the Vk and Jk gene segments into juxtaposition for joining. Apparently, joining will occur only when one signal sequence contains a 12-base pair spacer and the other contains a 22-base pair spacer. Very similar signal sequences appear to control VH–D and D–JH joining.

6. Antibody diversity due to variable joining sites and somatic mutations

A comparison of the diversity of amino acid sequences present in antibody molecules with that predicted from the sequences of gene segments that encode these antibodies revealed that there is more variation in amino acid sequences at the V–J junctions than is predicted by the nucleotide sequences. Subsequent studies show that much of this additional diversity could be explained by variation in the exact site of recombination during the V–J joining events. Thus, the use of alternate sites of recombination during the joining events that are involved in the assembly of mature antibody genes provides an additional mechanism for generating antibody diversity. Besides the vast array of antibody diversity produced by joining large families of V, D, and J gene segments and the use of alternate positions of recombination during the joining reactions, there is still another mechanism which causes the generation of antibody diversity. This has been established by comparing the nucleotide-pair sequences of expressed genes with the amino acid sequences of antibody chains. In essentially all variable regions, the changes have resulted from single nucleotide-pair substitutions. These nucleotide-pair substitutions are presumed to occur by some mechanism of somatic mutation that is restricted to the DNA sequences encoding the variable regions of antibody chains. Because these changes in the variable segments of antibody genes occur at such a high frequency, the process by which they occur is often called **somatic hypermutation**. The mechanism by which somatic hypermutation occurs is unknown. Somatic hypermutation of regions of antibody genes that encode antigen-binding sites may be of great value to the organism. Without this mechanism, the range of available antibody specificity would be fixed in terms of the sequences present in the genome at birth and the combinations that could be produced by the various levels of gene segment joining reactions.

7. Clonal selection

The question relating to how antibodies specific to antigens that it has not previously encountered are synthesized, can be explained by the **clonal selection theory**. As mentioned earlier, all the antibodies produced by a single B-lymphocyte have the same antigen-binding specificity. But different cells in a population of B-lymphocytes will have undergone different genome rearrangements leading to the production of antibodies with different specificities. Thus, the total population of B-lymphocytes in an organism will produce a very large variety of antibodies. The clonal selection theory states that the binding of a particular foreign antigen to an antibody on the surface of a B-lymphocyte stimulates that cell to divide, producing large numbers of this particular B-lymphocyte (a "clone" of identical cells) and thus large amounts of the particular antibody that recognizes the foreign antigen.

Clonal selection can explain immunological memory. When an animal first encounters an antigen, it produces a relatively slow primary immune response, because only a small number of its cells have membrane-bound antibody molecules or T cell receptors that can respond to the antigen. However, the next encounter with same antigen can enhance the production of anitbodies rapidly and there will be very high **secondary response**, because cells that can bind that particular antigen were stimulated to divide by the previous encounter (Figure 26.12). The first stimulation of B or T cells to a particular antigen lead not only to mature antibody-producing plasma cells or functional T cells, but also to expanded clones of long-lived, immature B and T cells called **memory cells**. Memory cells can respond well to the specific antigen at the next encounter, leading to the production of more functional B and T effector cells.

8. Allelic exclusion

One important point about the genetic control of antibody synthesis is that each B-lymphocyte makes only one type of antibody even though mammalian cells are diploid and carry two sets of genetic information coding for each of the antibody chains. Inspite of the diploid nature of the cell, only one productive genome rearrangement of light chain coding sequences and one productive genome rearrangement of heavy chain coding sequences occur in each B-lymphocyte! This phenomenon is called allelic exclusion because one of the "alleles" is excluded from being expressed. We still do not know how this mechanism is regulated. Clearly, there must be some feedback mechanism that arrests the recombination process(es) involved in these

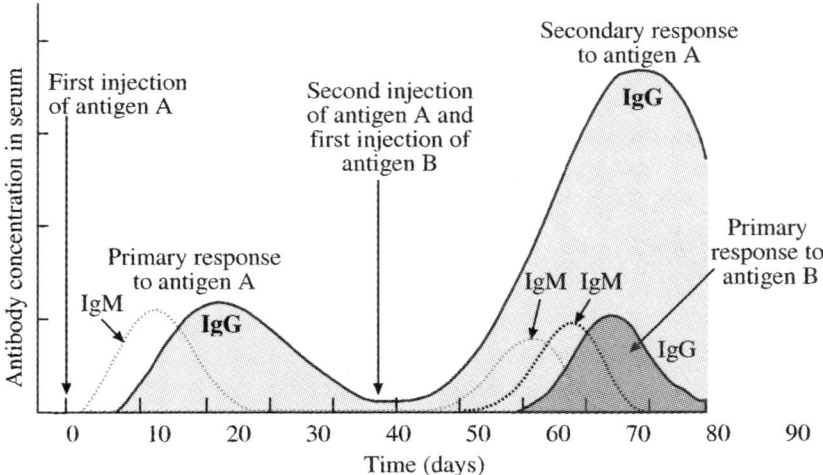

Figure 26.12 *Primary and secondary response to first and second dose of antigen. Note that the secondary response is faster and greater than the primary response and is specific for A, indicating that the immune system has specifically "remembered" encountering antigen A before B*

antibody gene rearrangements once a productive rearrangement has already occurred and the cell has started to synthesize a functional antibody.

9. The genetic basis of clonal restriction

The genetic basis of clonal restriction comes from: (1) the selection and translocation of a particular variable region sequence out of all those in the heavy chains variable region pool; (2) its association with particular D, J segments; and most importantly, (3) with a constant region gene segment. Analogous events occur in the case of light chains, but the diversity (D) segment lacks these chains. Kappa (κ) light chains, lambda light chains, and heavy chains occur on different chromosomes. In the general sense, the scheme underlying antibody production then would be that during development, the lymphocyte pool undergoes random translocation such that each lymphocyte can produce a particular light chain which has a complete variable and a constant region sequence. A particular heavy chain is characterized by one variable region hooked to a constant region, which in the precursor of the antibody secreting lymphocyte would be the μ-chain constant region. A cell would express only one surface receptor specifically as defined by the particular VH and VL structures exposed. If an antigen, which interacts with this structure with a certain binding energy (affinity) is introduced, the cell can be stimulated to become an antibody-producing cell. Some time after immunization and a secondary (booster) injection in mice and man, there is a switch to antibodies expressing the gamma chain constant region rather than the μ chain constant region. Eventually, the predominant antibody class in mammals is IgG. The first event in the process would be the translocation and fixation of a particular variable region by forming a contiguous association with D, J, and constant region gene segments. The switch from IgM to IgG or other classes consists of the translocation of this variable region to a contiguous association with constant regions distinct from μ. The increase in quality or capacity of the antibody to bind antigens reflects a switch to populations of variable regions, which bind more avidly to the particular antigen.

F. FACTORS AFFECTING THE IMMUNE RESPONSE

Antibodies may be used as versatile instruments for the analysis and purification of molecules contained within complex mixtures. The specificity of the immune response is determined by a number of factors and can influence the strength and specificity of the immune response. Thus, careful planning is needed to raise antisera in an animal. These include: (1) genetic differences between the species of origin of the antigen and the species of the recipient; (2) differences in responses between individuals within a species governed by both genetic and poorly understood non-genetic factors; (3) purity of antigen. It is often impossible to purify the antigen sufficiently. Even seemingly minor amounts of impurities may result in significant amounts of unwanted Abs; (4) cross-reactions between structurally similar molecules. For example, most of the Abs in the antisera against dinitrophenyl hapten will also react with trinitrophenyl, although a minor sub-population of Abs might distinguish between the two; (5) route of injection. Antigens stimulate the strongest immune response when they are injected directly into the popliteal lymph nodes. However, a much simpler route such as multiple intradermal or subcutaneous sites or into a single intramuscular site is more commonly used; (6) immunization schedule. It is difficult to make generalizations about immunization, because most of our information comes from empirical observations made in a variety of dissimilar situations. When an animal is to be immunized with bacteria (frequently killed) or mammalian cells (frequently living at the time of injection), good Ab titers can be obtained with inoculations repeated each week for a number of weeks. However, single or repeated injections with soluble antigens frequently do not produce the sustained, high-titer Ab responses that most investigators would like to see. The main reason for the inability of many antigens to produce the sustained response is that they are rapidly cleared from the animal's body. Hence, a number of procedures have to be used to prevent rapid clearance of the antigen.

II PRODUCTION OF POLYCLONAL ANTIBODY

Polyclonal antibody (**Antisera**) is defined as the total population of Abs present in animal serum. This complex population contains different Ab subclasses including IgG, IgM, IgA, and IgD. Each Ab represents the secretary product from a single stimulated lymphocyte and its clonal progeny. A complex antigen(s) such as a protein, glycolipid, etc., may contain many distinct antigenic determinants or epitopes, each of which is specifically recognized by antibodies from a single lymphocyte clone.

A large number of protocols developed for production of polyclonal Abs probably reflect the ability of the immune system to mount a response to many antigenic forms that arise in nearly any circumstances. Polyclonal Abs are easier to obtain and have higher specificity when made with highly purified antigen. Polyclonal antisera consist of heterologous populations of Abs with variable specificities. Different segments of the Ab population may be reactive when using different serological tests. The general protocol followed to generate high serum concentration of specific Abs is to use a moderate antigen dose, 100–500 μg for a rabbit or 10–100 μg for a mouse, administered with an adjuvant at least for the primary immunization, and to allow a one- to six-month resting period to generate long-lived immunological memory. Freund's adjuvant is mineral oil combined with (complete) or without (incomplete) killed mycobacteria. An emulsion of soluble protein in the oil provides a depot of antigen at the injection site, while the mycobacteria enhance the non-specific activation of macrophages participating in the response.

Antisera are generally tested against the target antigen by ELISA, immunoprecipitation, or western blotting. Where the antigen is in abundant supply, a simple test of lower sensitivity such as immunoelectrophoresis or even Ouchterlony immunodiffusion may be used to determine the level of response elicited. In many assays, two immunoglobulins binding proteins from Gram-positive bacteria have proved immensely valuable: these are *Staphylococcus aureus* Protein A and Group G *Streptococal* Protein G.

A. MAJOR ADVANTAGES AND LIMITATIONS OF POLYCLONAL ANTIBODIES

Major advantages
1. Multiple subclasses and high-affinity Abs are present in polyclonal antisera production.
2. Multiple specificities of Abs are likely to recognize sequential as well as conformational antigenic determinants.

3. These antibodies can recognize multiple determinants specific for a protein (important if screening gene expression libraries!).
4. Highly specific polyclonal antisera may easily be developed by immunization with proteins purified in a single step by 2D-PAGE.
5. Purification can be done by simple experiments involving immunoprecipitation, etc.

Major limitations
1. Immunogen must be highly purified and sufficient quantities are needed to obtain the desired specificity and to obtain good titer.
2. Individual domains of complex antigens are difficult to study because multiple antigenic determinants are recognized by the polyclonal antisera.
3. Quantities of Abs have to be produced every time and are limited to the life of the immunized animal.
4. Different bleedings have to be characterized individually due to changes in Ab affinity, specifically, and subclass.

B. ANTIGEN PURIFICATION AND PREPARATION

Antibodies and the cells that secrete them recognize an antigen primarily by conformational determinants on the surface of the molecule. The reaction between the antigen and the Ab involves not only steric factors (lock and key) but is dependent on a variety of interactions. Because of this, it is possible to envisage an almost infinite variety of antigens. One of the most interesting aspects of the immune system is that it appears to be capable of responding to an immense number of naturally occurring and completely synthetic antigens. However, the response of animals to antigens is not universally good. The reasons for this variability in response include intrinsic properties of the antigens (chemical and structural factors, size, and its conformation in solution) and the properties of the immune system of the responding animal.

The first and foremost thing in Ab production is obtaining sufficient quantities of the highly purified immunogen that will be used to immunize the animal. When protein preparations are used, even if the protein contains 1% contamination, the majority of Abs may recognize that contaminant if it is highly immunogenic. Immunogens purified by non-denaturing biochemical methods tend to elicit a strong response toward the conformational epitopes present in the molecule. Procedures that denature the immunogen tend to yield antisera reactive toward sequential determinants. In many instances, this is desirable since Abs made against denatured proteins are usually best for immunoblotting of SDS-PAGE separated proteins. However, denatured proteins are generally less immunogenic than the native proteins. Therefore, greater concentrations of denatured protein are necessary to enhance the immunogenicity.

C. CHOICE OF ANIMALS

Any hemothermic animal can serve as a source of the immune system. Rabbits, chickens, guinea pigs, rats, mice, sheep, and goats are commonly used for production of immune serum. However, though rats, mice, and guinea pigs are easy to maintain, they yield relatively small volumes of serum and there are risks of loss of animal when blood is collected by cardiac puncture. Chickens have high body temperature, which may prove detrimental to the integrity of some antigens. Larger volumes of serum may be obtained from larger animals such as sheep, goats, and horses, but more antigen is required for immunization and only one subclass of IgG binds to protein A, necessitating the use of a second Ab for detection. Rabbits are used most often to produce immune serum because they can be housed and cared for with little effort, are relatively easy to handle and inexpensive, and yield adequate volumes of high-titered serum in return for relatively small amounts of antigen used for immunization. Multiple subclasses of IgG bind protein A. The animals used for antisera production must be disease-free and quarantined. To produce antisera that reacts to a potential epitope(s), the species of animal should be phylogenetically different from the species from which the target protein was isolated. However, Ab producing animals should be genetically homogenous and their pedigree must be recorded. To avoid problems of variation in response between animals, at least two animals of the same species should be used for immunization.

D. IMMUNIZATION

The amount of immunogen required depends on the characteristics of the antigen, the means by which it prepared, and the method used for injection. Longer immunization schedules requiring more antigens normally result in a greater variety and quantity of Ab than shorter schedules, but they may produce Ab of lower specificity. Generally, a combination of injection methods is used to present immunogen to the Ab-producing sites in the animals. Both intramuscular (IM) and intravenous (IV) injections result in the most rapid distribution of antigen to many sites, but the antigen may be degraded quickly in the blood stream by the body's defense systems. Therefore, more than 8 i.v. injections may be required to obtain serum of usable antibody concentrations.

Requirements

Complete and incomplete Freund's adjuvant (Abbreviated as CFA and IFA respectively). Syringes 1 ml and 2 ml, either glass Leur-lock or plastic. The syringe needles should be 23 gauge and of the insulin type; Rabbits; Antigen solution.

1. Use of adjuvants: Antigens in buffer solutions are probably best prepared for intramuscular (IM) injections by emulsifying 1:1 (v/v) with Freund's adjuvant. The adjuvant consists of paraffin oil and an emulsifier, annide mono-oleate. Complete Freund's adjuvant consists of heat-killed Mycobacterium tuberculosis, or a similar acid-fast bacterium. The incomplete form contains no bacterium. The purpose of the adjuvant is to act as a depot for antigen within the muscle tissue, allowing for slow release, thereby stimulating more antibody production than would occur if the same amount of antigen were injected as an aqueous solution. Complete adjuvant may be more effective for enhancing the immune response than an incomplete adjuvant. In an IM injection series, complete adjuvant is usually administered in the first injection and the incomplete adjuvant given thereafter. Development of a complete, stable antigen–adjuvant emulsion before injection is important. The product of the IM injection should be a thick, creamy emulsion, which does not separate on standing.

Other adjuvant systems are available that are very effective in stimulating production of high titer serum. The Ribi Adjuvant System (RAS), from the Ribi Immunochemical Research lab, incorporates adjuvant and antigen into an oil phase instead of the antigen–aqueous phase of Freund's, thereby reducing the viscosity of the emulsion and permitting more rapid release of antigen into the animal tissue. Muramyl peptides, wax fractionations of purified cell walls of

Mycobacterium, have been used as adjuvants, and N-acetylmuramyl-L-alanyl-D-isoglutamine is described as the simplest active analog. Dimethyldioctadecyl ammonium bromide (DDA), a lipoidal quaternary ammonium compound has also been reported as an effective adjuvant. Several companies are increasing efforts to find better, purer, and more chemically defined adjuvants. A bacterium from genus Amycolate has shown adjuvant activity without producing side effects. Also, a glycoside extracted from tree bark (*Quillaja saponaria*), has also been found to be effective in microgram quantities which can boost vaccine activity and not produce side effects.

2. Injection methods: Animals can be immunized by injecting antigen (Ag) through different routes. These are as follows: (i) intravenous (IV), (ii) intramuscular (IM), (iii) intraperitoneal (IP), (iv) subcutaneous, (v) intradermal, and (vi) foot.

3. Injection schedule: Many factors influence the injection schedule – the animal, the antigen etc. General procedures for viral immunization, for example, that result in adequate antibody production in rabbits involve one IV and two IM injections of 1–2 mg each, given at weekly intervals. Peak titer usually occurs 5–7 weeks after the first exposure. The booster doses of antigen, however, are effective only after a significant drop from the maximum antibody titer is demonstrated. Antigens vary considerably in their ability to elicit antibodies, but an injection of excessive amounts of antigen rarely improves antibody titer. Rather, alteration of injection methods or antigen stabilization is usually more advantageous.

4. Preparation of antigen: Antigen should be prepared as described earlier in "1. Use of adjuvants". The antigen are of two types – soluble and particulate. The soluble antigen can be sterilized by passing through a millipore membrane of pore size 0.45 µM or 0.22 µM. Particulate antigen should be prepared under sterile conditions. Soluble proteins may be used in a wide range of concentrations from 1.0 µg to 1.0 mg/animal for immunization.

E. SERUM ACQUISITION AND TREATMENT

All animals used to produce antiserum should kept without food for 12–18 hr prior to obtaining blood samples, so as to eliminate the accumulation of lipids in the serum.

1. Blood collection: After immunization by antigen, the antisera can be collected from the animal by various bleeding pathways.

Requirements

Animal-holder. Hair clipper, Complete filtration assembly, glassware, non-heparinized capillary tubes, scalpel, spatula, scissors, cotton, syringe, and needle, Pasteur pipettes with bulbs, alcohol, xylene, sodium azide ether, anesthetic chamber.

1. **Bleeding from ear:**
    1. Rabbits can be bled from the marginal vein of the ear. The lateral margin of the ear is shaved and cleaned with 70% alcohol.
    2. The vein may be dilated either by rubbing the ear or by applying cotton swab moistened with xylene.
    3. With the help of a sharp blade, a diagonal incision is made across the vein and tanned blood is collected.
    4. If blood flow ceases, the clot can be removed by wiping with a dry cotton gauze. The bleeding can be stopped by pressing the vein at the site of the cut with dry cotton.

2. **Retro-orbital bleeding:**
    1. The mouse or rat is held in the left hand by the skin of the neck with the help of finger and thumb. With the right hand, the tip of a fine capillary is gently inserted into the inner angle of the eye and slid under the eye-ball to rupture the fragile venous capillaries.
    2. The blood starts flowing from the capillary, and is collected in a tube. After the collection of the blood, the capillary is removed and the blood around the eye-ball is wiped off with dry cotton wool. This method is good for mice or rats when only a small amount of blood is needed.
3. **Cardiac puncture:**
    1. Hairs on the chest must be shaved off and the shaved area wiped with 70% alcohol. The animal is laid in a supine position.
    2. The needle (20 or 21 G) is inserted through the gap between the last sterna rib or the left side of the midline.
    3. It should be penetrated downward at an angle of 30 till heart-beats can be felt.
    4. There should be a slight push forward and the plunger of the syringe should be retracted. If the blood does not enter the syringe, the needle may be beyond the heart, not in it.
    5. When blood begins to flow into the syringe, the plunger should be slowly withdrawn and the desired volume of blood collected. This is the best method of collecting blood from a guinea pig.
4. **From external jugular vein:**
    1. The animal is held by an assistant and the head is raised to expose the jugular vein.
    2. Hairs of the site are shaved and the area is cleaned with 70% alcohol.
    3. The venous blood returning to the vein can be blocked by pressing the jugular grove. A 18 G needle is inserted into the vein pointing towards the head and the plunger slowly retracted. When the blood begins to flow into the syringe, the plunger is slowly withdrawn and the desired volume of blood is collected. Horse, sheep, and goat can be bled by this method.

F. RAISING OF HYPER-IMMUNE SERUM

A variety of antigens can be used to raise polyclonal antiserum by immunizing an animal repeatedly at definite time intervals. Although monoclonal antibodies have unique advantages, the potential diversity of epitomes recognized by polyclonal antisera make these antisera useful – even a single molecular species of antigen can yield enormous benefits in immuno assays. Every individual has his own way of raising antisera. Antigens are more immunogenic when presented in an insoluble form or with the adjuvant. The commonly used adjuvants are Freund's complete and incomplete adjuvants besides alum, Spain, etc.

Method-1: Rabbit polyclonal antisera

Materials

Bovine serum albumin (BSA) fraction V

Complete and incomplete Freund's adjuvant (abbreviated as CFA and IFA respectively). Syringes 2 ml, either glass Leur-lock or plastic. Three-way Leur fitting stopcocks. One rabbit. NaCl, 0.85% (w/v) solution.

Procedure

1. Make an emulsion from 1 ml each of antigen solution (at 200 µg/ml) and CFA. Allow about 20 min. to emulsify by repeated passage under pressure between syringes.
2. Check that the stable water-in-oil emulsion will not disperse when a drop is placed on water in a beaker.
3. Inject 1.0 ml (100–200 µg) of emulsion intramuscularly into each of the hindquarters of rabbit (zero day).
4. After 2–3 weeks, repeat the immunization with Freund's incomplete adjuvant into 3–5 subcutaneous sites (second dose).
5. Two weeks after the second immunization, bleed and collect a 5 ml blood sample from the ear pinnae of the rabbit.
6. Allow the blood to clot for an hour at room temperature. The clot is released from the tube or beaker by rimming with a sterile applicator stick or glass rod. Heating the blood, after clotting, at 37°C for 30 min. helps to shrink the clot and increase the yield of serum.
7. After overnight refrigeration, decant the serum or aspirate it from the clot and centrifuge at 2,000 rpm to remove cell debris.
8. Filter the serum through a 0.2 µm millipore or comparable filter, and bring to 0.025% sodium azide. Aliquots of serum, mixed with equal volumes of glycerol, are stored at 20°C, or lyophilized. Avoid repeated freezing and thawing of raw serum in the absence of glycerol.
9. Test the antiserum for the presence of specific antibody response by appropriate immunoassay (interfacial ring test, immuno-electrophoresis, gel immuno diffusion, etc.).
10. Repeat bleeds, boosting if titer drops.

Notes

1. Emulsion may also be made by sonication for about 5 min.: but care should be taken not to overheat the antigen, as this may destroy the antigenicity.
2. If quantities of antigen are limited, direct injection of 10 µg of protein in IFA into the popliteal lymph node of a rabbit may efficiently prime the immune response.
3. Antibodies may be raised for insoluble proteins, particles, or re-suspended nitrocellulose blots mixed with the adjuvant. The nitrocellulose must be completely dry, and on mixing with a minimal volume of DMSO, it will dissolve. On addition of water, particles will form but these are small enough to be injected subcutaneously.
4. It is a good practice to obtain serum from the animal prior to immunization. Such a serum is called a pre-immune serum.

Method-2: Mouse polyclonal antisera

Procedure

1. Make emulsion as described above of equal volumes of antigen (1 mg/ml) and CFA. The emulsion will contain 500 µg/ml of antigen, or 100 µg in 0.2 ml.
2. Inject 0.2 ml of antigen in CFA, intraperitoneally into each of 4–5 mice.
3. After 4 weeks, inject 0.2 ml of antigen made up in IFA.
4. Bleed 4–5 days after challenge, from the tail vein. Collect serum from the blood as in method 1.
5. Boost after two weeks if required. Bleed out if required by cardiac puncture.

III MONOCLONAL ANTIBODIES AND HYBRIDOMA TECHNOLOGY

The problem of producing and characterizing antisera was radically altered when Kohler and Milstein (1975) demonstrated that antibodies producing hybrid cell lines could be generated by somatic cell hybridization. They introduced the novel approach of immortalizing specific antibody-forming cells by the fusion of spleen cells with myeoloma cells. This provided a method for the production of homogenous and biochemically defined immunological reagents, namely monoclonal antibodies (Mabs). Hybridization of antibody-forming lymphocyte cells with malignant myeloma cells results in a hybridoma, which combine the parental traits of specific antibody secretion and continuous growth. Cloning and further selection of hybrids allows the development of Mabs of identical specificity and desired activity directed against a unique epitope of the immunizing antigen. Mabs production has been adapted to almost all type of organisms.

When compared to conventional polyclonal antibody production, the ability to obtain practically unlimited quantities of the same antibody in a reproducible manner, and the ability to immortalize the production of such monospecific reagents by cryopreservation of the hybridomas for unlimited periods, are just two of the numerous advantages of the hybridomas-produced Mabs. One of the most striking advantages of the hybridoma approach is the ability to produce and select Mabs to almost any antigenic determinant, even when impure antigen or antigen mixtures are used as immunogen. In this chapter, readers are provided with details of the various techniques and the rationale used to develop the protocols and approaches.

A. MAJOR ADVANTAGES AND LIMITATIONS OF MONOCLONAL ANTIBODIES

Major advantages: Single homogeneous antibody to a defined antigen determinant can be produced. A specific antibody can be used to study the functional domain of the molecule. Large quantities of antibody can be obtained since theoretically, immortal cell lines can be developed. Antibodies with low-affinity binding can be selected during screening procedures (these antibodies are designed for immuno-affinity chromatography).

Major limitations: The procedure followed for monoclonal antibody production is expensive and time-consuming. Well-equipped tissue culture facilities are needed. The epitope recognized by the antibody may be shared among many different antigens not related to the antigen of interest. Hybridoma cell lines are frequently unstable due to chromosome loss or may be lost because of tissue culture contamination.

B. PRODUCTION OF MONOCLONAL ANTIBODIES BY HYBRIDOMA TECHNOLOGY

The generation of monoclonal antibodies involves a sequence of different techniques, many of which vary slightly between laboratories. Monoclonal antibodies can be produced in specialized cells through a technique now popularly known as **hybridoma technology** (Figure 26.13). The term hybridoma is applied to fused cells (antibody producing lymphocyte cell and

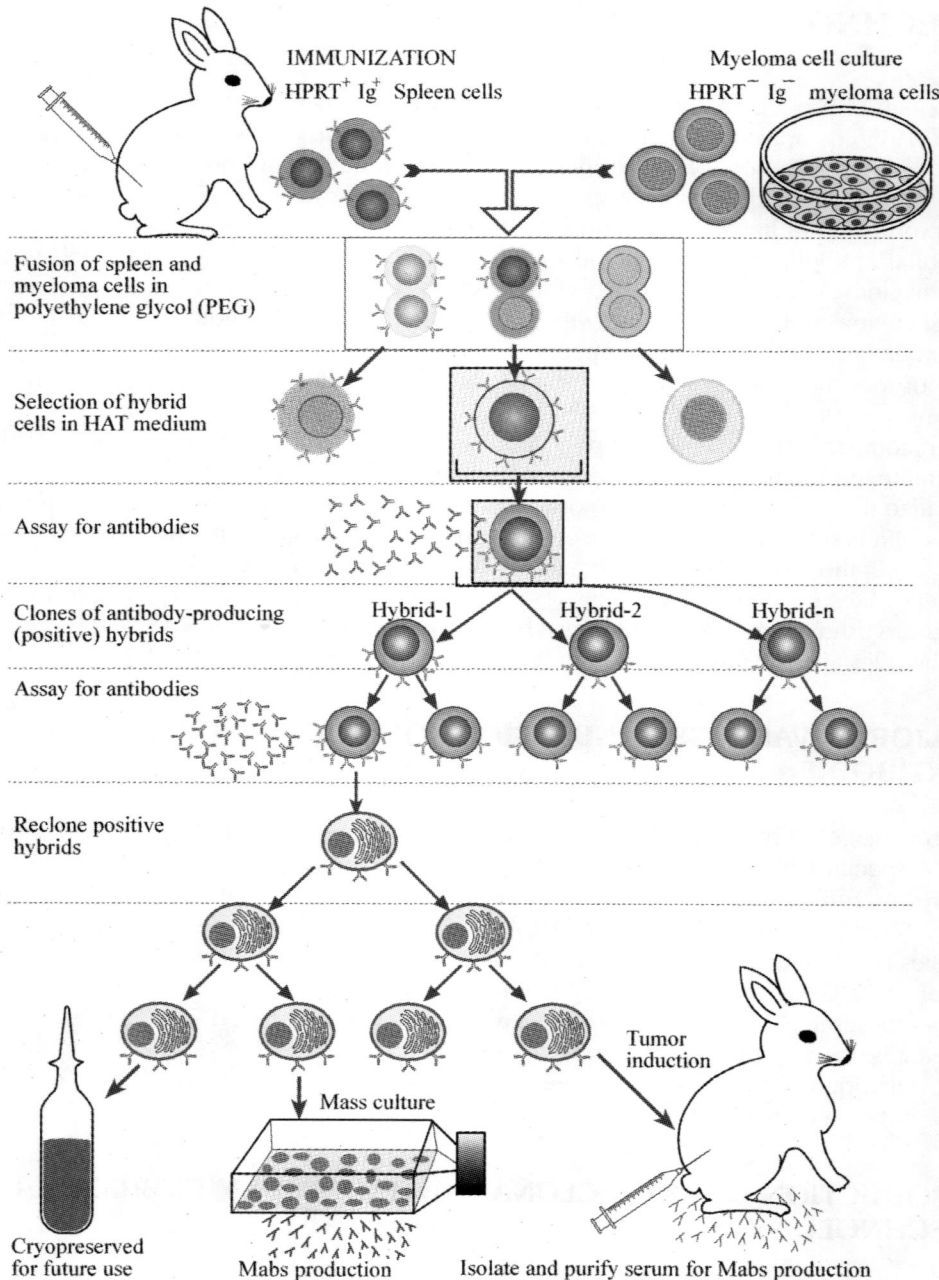

Figure 26.13 *Schematic diagram of the production of monoclonal antibodies (Mabs)*

immortal myeloma cell). Cultured myeloma cells (bone marrow tumor cell, which is capable of multiplying indefinitely, like a malignant cancer cell) are fused to spleen cells from an immunized donor (antibody producing lymphocyte). The standard mouse strain for monoclonal immunization is the BALB/c as this is the genotype of the available myeloma cell fusion partners. Hybrids between genetically different cells grow equally well *in vitro* but growth and asceitic fluid production *in vivo* requires histocompatible hybrid (F1) mice.

Immunization protocols vary infinitely, but the standard regimen is to inject 10–100 μg of antigen IP in Freund's complete adjuvant, followed at days 28, 29, and 30 by 1 μg soluble antigen IV. Spleen cells are taken for fusion two days after the final injection. Faster and more economical immunization procedures are to inject small quantities (20 μg) of antigen directly into the spleen 4 days pre-fresion, or to sensitize normal lymphocytes *in vitro*. The following steps are involved in the production of monoclonal antibodies using hybridoma technology. The procedure as a whole has been subdivided here into the following sections:

Materials

Equipment and plastic ware

Laminar flow hood; Humidified, temperature-controlled carbon dioxide incubator; Inverted microscope; Bench-top clinical centrifuge; Water bath, temperature controlled; –70°C freezer; Autoclave; Liquid nitrogen storage containers; Hemocytometer; Single and 8–12 channel variable-volume automatic pipettes and sterile tips; Periplastic pump refrigerators and freezers (–15°C); Spectrophotometer; ELISA reader (optional); Bright field, Inverted phase contrast and Fluorescent microscopes; Fluorescent-activated cell sorter (optional); Scintillation counter (optional); Animal holding facilities: Cages, bedding, feed and adequate temperature control, ventilation, etc.

Cell culture ware: 96-, 24-, and 6-well flat bottom sterile cell culture flasks; 1-, 5-, 10-, and 25-ml sterile plastic pipettes; ampules for liquid nitrogen preservation; 5, 10 and 50 ml sterile screw-cap centrifuge test tubes; Glass, cotton-plugged reusable pipettes; Disposable plastic or glass Pasteur pipettes, protective gloves; Petri dishes, Sterile filters, 1–50 ml syringes, 26–16 gauge needles, and surgical scissors.

In addition, investigators using these methods will require the laboratory equipment necessary for producing the antibody in usable amounts, analyzing antibody activity, and characterizing the antibody molecules produced by the hybrids. The exact equipment and materials required for each of these operations will depend on the nature of the antibodies being produced and the scale to which production is carried out.

Solutions

The establishment of hybridomas requires not only special attention, but also quality reagents, continuous monitoring, and nursing. All preparations are made with tissue-culture-grade, deionized, glass double-distilled water.

1. Media for cell culture

Three related culture media are involved in monoclonal production in order to induce the ability of aminopterin to block major pathways of nucleic acid synthesis. Normal cells possess HGPRT (hypoxanthine guanine phosphoribosyl transferase) and thymidine kinase, which provide bypass pathways for normal growth, but as the myeloma cell partner has been selected as

HGPRT–, it is killed in the presence of the aminopterin. Myeloma cells are therefore grown in basic tissue culture medium. Immediately after fusion, unfused myeloma cells are killed by culture in a medium containing hypoxanthine, aminopterin, and thymidine (HAT). Cells are grown in HAT for two weeks, followed by two weeks in HT (hypoxanthine, thymidine without aminopeterin) to ensure that no residual poisoning of hybrid cells occurs.

The most commonly used media for hybridoma production are DMEM and RPMI-1640 medium. The two media are very similar except that RPMI lacks pyruvate and DMEM lacks asparagine. Both media are usually buffered with carbonate/bicarbonate. Glutamine is unstable and is added to 2 mM of media, if medium preparation is older than 2 weeks. Some researchers also supplement the media with non-essential amino acids and other growth media. Prepared media should be stored (4°C) in the dark to prevent the production of highly toxic photoproducts.

(a) Antibiotics

The most common cell culture contaminants are bacteria, yeast, and fungi. The best control for these organisms is scrupulous use of sterile techniques. Most researchers add penicillin and streptomycin as a matter of routine (100 unit/ml) to minimize bacterial contamination. When necessary, or as an alternative, gentamycin (100 µg/ml) may be used as a broad-spectrum bacterial and mycoplasma antibiotic. Fungizone at 5–10 µg/ml is useful against fungal contamination.

(b) Sera

Fetal bovine serum (FBS) is used in nearly all hybridoma work. The quality of available sera for cell culture is now such that extensive prior testing of many different batches is usually unnecessary. The serum can be kept frozen at –20°C for at least 1–2 years. Most researchers heat the serum to inactivate the complement, then store it frozen.

(c) Serum substitutes, supplements, and serum-free medium

Iron-supplemented bovine calf serum and horse serum are sometimes substituted for FBS and are generally less expensive. However, these serums contain high levels of contaminating immunoglobulins, which can interfere with many immunoassays. Several companies offer defined or processed serum replacements (Mito +; CPSR–3) which reportedly have consistent chemical and performance characteristics and can be used in place of FBS. Serum free media that support the growth of several parent myeloma lines have been recently developed. These media are usually mixtures of RPMI and DMEM supplemented with hormones, transferrin, lipids, trace elements, and other factors. Unfortunately, none of these serum-free mediums are presently suitable for all hybridomas or myelomas.

(d) Preparation of media and working solutions

100x HT: 136.1 mg hypoxanthine and 38.8 mg thymidine are dissolved in 100 ml water at 60–70°C, 0.2 µm filter sterilized, and stored frozen at –20°C in 5 ml aliquots.

100x HAT: Prepare 100x HT as above and add 1.8 mg aminopterin and 0.5 ml 5N HCl and adjust to 100 ml before filter sterilizing. Store at –20°C as HT, protected from light.

Serum-free RPMI-1640: 50 ml 100 mM sodium pyruvate, 10 g sodium bicarbonate, 0.5 g gentamycin and a 5 litre package of powdered RPMI-1640 with glutamine (without bicarbonate) are thoroughly dissolved in 3600 ml of tissue-culture-grade water. The solution is dispensed by a peristaltic pump through a 0.2 µm filter into sterile 500 ml bottles. The medium is stored at 4–6°C until use.

Complete RPMI-1640: To 360 ml serum-free RPMI-1640 aseptically add 75 ml serum (15% serum), 50 ml myeoloma conditioned medium (10% 'conditioned medium'), 5 ml 100 mM glutamine (2 mM glutamine), 3 ml 1 M HEPES (6 mM HEPES), and 0.1 ml 0.1 M 2–mercaptoethanol (2×10^{-5} M 2MCE). Filter sterilize by milipore 0.2 µm.

Complete medium + HAT (or + HT): Add 5 ml 100x HAT (or 100x HT) to 500 ml complete medium.

Polyethylene glycol (PEG): At the time of fusion, a 45% solution is prepared by dissolving 2.0 g PEG (Mr 1000 to 4000 with 2.5 ml serum-free RPMI-1640 at 50–60°C, water bath). The slightly alkaline (pH 7.5–8.2; pink not orange or purple) solution is filter sterilized and held at 37°C until fusion.

Freeze medium: Aseptically add 5 ml serum (25%) and 4 ml DMSO (8%) to 40 ml complete medium + HT and store at 4–6°C.

2. Guidelines for antibody screening assays

Prior to the generation of monoclonal antibodies, it is essential to first establish a sensitive and rapid antibody-screening assay. Because hundreds of cell culture samples will usually have to be screened, adequate methods should be developed prior to setting up cell cultures. A variety of methods can be used to determine the presence, as well as the specificity, of the antibody. The immunoassay selected for screening antibody activity in the hybridoma culture supernatant fraction is critical to hybridoma production. In principle, any assay capable of detecting low antibody concentrations can be used. There are many ways to screen hybridomas. The screening method must be reliable, sensitive enough to detect all Mab-producing hybridomas of interest, simple, quick enough to test hundreds of samples at a time, and must be definitive.

Screening may be performed in two steps. The first step might involve a fast, non-specific assay that would reduce the number of hybrid cultures to be screened in the second more specific assay. A second factor in selecting assays is the nature of the antigen; e.g., purity, abundance, and physical form. Solid phase assays are suitable for most antigen preparations. Immunofluorescence and immunocytochemical techniques are suitable when screening for Mabs of cellular or tissue antigens. Mabs to functional antigens (e.g., enzymes) must be able to precipitate, modify, inhibit, or neutralize the activity mediated by the specific antigens.

Last, the screening method should be appropriate for the intended use of the antibody. Accordingly, the condition of the antigen in the screening assay should approximate that of the final assay. Mabs tend to be very assay-specific, useful in one assay, but not in another. This is due primarily to the strict specificity of Mabs for their respective epitopes and the "accessibility" of those epitopes in the assays. Proper selection of an assay can reduce the workload and ensure that the desired Mabs are identified.

3. Immunization schedule

The immune state of the animal from which the spleen is taken is very important in determining the success of a hybridoma experiment. The degree of immunization depends upon the choices of animal, immunogen, and the immunization schedule. Immunization protocols and schedules vary considerably. A protocol that works well for a membrane antigen may not necessarily be satisfactory with soluble protein. Hybrids between genetically different cells grow equally well *in vitro* but growth and ascitic fluid production *in vivo* requires histocompatible hybrid (F1) mice.

Procedure

1. Inject two-month-old mice or rats intraperitoneally with 10–100 µg antigen emulsified in 300 µl complete Freund's adjuvant or 1×10^7 cells in buffered saline or Freund's Incomplete adjuvant.
2. Inject 2 to 3 more times with the same dose (followed at days 28, 29, and 30) or inject 1 µg soluble antigen IV into animal body.
3. Take serum samples from individual animals 10 to 14 days later, and determine the titer of the relevant antibodies in the same assay(s) that will be used later for screening hybridomas.
4. Give an additional 1–5 week rest period (more than 3 weeks from the last injection) for animals showing the highest antibody titers prior to administering a final intraperitoneal boost of the same or higher dose in aqueous solution.
5. Three to five days later (after the final injection), remove the spleen and prepare for cell fusion.

4. Cell fusion between spleen and myeloma cells

The immediate event in somatic cell hybridization is the fusion of cell membranes, generating multinucleated (generally binucleated) cells or heterokaryons in which the cell membranes of the fusion partners surround a common cytoplasm with two or more nuclei. In a matter of days, synkaryons form when the nuclei fuse and are capable of synchronous mitosis; in the process, a variable number of chromosomes of both fusion partners are lost. With subsequent cell divisions, more chromosomes are lost, but the hybrid cell line eventually stabilizes.

A successful fusion protocol should bring spleen and myeloma cells together and allow the fusion to occur at a sufficiently rapid rate without causing more than minimal damage to the cells. Several procedures of the original PEG-induced fusion procedures have been described, and all variations seem to work well. The concentration (30–50%) and pH (maximum fusions at pH 7.8–8.2) of the PEG (1000–6000) mixture, the duration of exposure to PEG (2–10 min.), temperature (20–40°C), and the physical handling and processing of the cells during and after fusion are important factors. The mechanism of fusion is complex, involving cell agglutination, cell swelling, and membrane fusion and the optimal conditions for the three processes are often at odds.

Fusion protocol
1. Warm pre-packaged PEG 1500.
2. Set up a 37°C water bath or beaker in flow hood to accommodate fusion tube.
3. Prepare spleen and myeloma cell suspensions.
4. Mix cell suspensions containing cells from one spleen and myeloma cells (spleen to myeloma ratios of 1:1, 2:1, 5:1. 10:1 are common). Co-pellet in a 50 ml conical tube and remove all the supernatant liquid by careful aspiration.
5. Re-suspend pellet by gently tapping the tube and hold in the 37°C water bath for 1 min.

6. Using a 5 ml pipette, gently stir the cell pellet while adding 1.5 ml PEG drop-wise over 2 min., mixing continually but gently.
7. Slowly add 2 ml of serum-free medium drop-wise over 2 min. in the same manner.
8. Gradually accelerate the addition of serum-free medium to add a total of 20 ml over 10 min.
9. Spin down (400 g for 10 min.), re-suspend the pellet in HAT+ complete medium and check for viability and fusion by trypan-blue.
10. Transfer cells to a 75 cm^2 flask containing 100–180 ml complete medium + HAT (re-suspend in volume calculated to give a cell concentration of less than 1×10^5/ml).
11. Incubate flasks in incubator (37°C, 5–10% CO_2), standing up with loosened caps. Let them stand for at least 1 hour.
12. Carefully pour suspended fused cells to a new flask. Mix by gently pipetting up and down with a 25 ml pipette.
13. Then re-suspend fused cells in 200 ml HAT for distribution across ten 96-well plates (200 µl per well).
14. Plate out saved spleen and myeloma control cells in complete medium + HAT separately at cell densities similar to the fused cells, which serve as controls.
15. Place plates in 37°C incubator until feeding is required.

5. Selection and culture of hybrid cells (HAT selection)

Cell fusion is a random process and necessitates a means of selecting the desired hybrid cells. Fusion of a population of myeloma and immune spleen cells results in a mixture of fusion events (myeloma–myeloma, myeloma–immune spleen and immune spleen–immune spleen cells). Selection of myeloma–immune spleen cell hybrids is accomplished by culturing the fusion mixture in hypoxanthine-aminopterin-thymidine (HAT) medium.

Principle

Aminopterin (an analog of folic acid) blocks the *de novo* biosynthesis of purines and pyrimides. To survive in the presence of aminopterin (as in the HAT medium), cells must be able to synthesize these nucleotides by utilizing an exogenous source of hypoxanthine and thymidine (provided in the HAR medium). They do this via alternate nucleotide biosynthetic pathways aptly called the salvage pathways. (Aminopterin also blocks glycine synthesis, but the RPMI-1640 medium supplies enough exogenous glycine to meet this requirement). Myeloma cells are 8-azaguanine resistant and hence lack an enzyme, hypoxanthine-guaninephosphoribosyltransferase (HGPRTase), that is required in one of the salvage pathways of nucleotide biosynthesis. Myeloma and myeloma–myeloma fused cells are therefore not capable of growing when *de novo* nucleotide synthesis is blocked with the HAT medium. Should myeloma fuse with a normal, albeit antibody-producing spleen cell, the normal cell provides the fused partners with the required enzyme, HGPRTase. This allows the hybrid cell to utilize exogenous hypoxanthine and to grow in HAT medium. There is positive selection against the growth of infused normal spleen cells and spleen–spleen cell fusions in this scheme; hence it is called half-selection. Passive selection takes place because normal spleen cells have a limited growth potential in culture. By two weeks in culture, most spleen cells would have died. As a result of this half-selection, the desired myeloma–spleen cell hybrids are selectively grown.

Materials

Supplemented culture medium (see above), Thymidine, Hypoxanthine, Aminopterin, NaOH: 0.1 M (4 gm/litre).

Needle, 1½ in., 21 or 23 gauge, attached to tubing that is connected to an aspirator. Pasteur pipettes and bulb.

Preparation of 50X HT and HAT stock solutions and 1X HT and HAT media.

100X and 50X HT Stock solutions:
Prepare a 100X HT solution by dissolving 0.1361g hypoxanthine and 0.0388 g thymidine in 100 ml double-distilled water (ddH_2O) warmed to 70–80°C. The 100X HT stock solution is used in preparing the 50X HAT stock solution (see below). To prepare 50X HT stock solution, dilute the 100X solution to 50X with double-distilled water. Sterilize by membrane filtration and store in aliquots at –20°C.

1000X Aminopterin stock solution:
Dissolve 17.6 mg aminopterin in 80 ml ddH_2O. If the aminopterin does not dissolve readily, add several ml of 0.1 M NaOH. Bring volume up to 100 ml with ddH_2O. Store the stock solution in 10 ml aliquots at –20°C.

50X HAT stock solution:
Combine 100 ml of 100X HT stock, 10 ml of 1000X aminopterin stock, and 90 ml double-distilled water. Sterilize the solution by membrane filtration and store in aliquots at –20°C.

1X HT and 1X HAT media:
Add the 50X stock to an appropriate amount of supplemented culture medium. When thawing the aliquots of 50X stock, some material may come out of the solution. However, this material quickly dissolves when the 50X stocks are added to the culture medium.

HAT selection:
The following procedure describes a progressive HAT selection scheme. Two objectives are accomplished with this protocol: (a) selection for the growth of hybrid cells, and (b) dilution of immunoglobulin produced by spleen cells. The dilution eliminates some false positive test results in the subsequent assessment of antibody production by hybrid cells.

Procedure

1. On day 1 (i.e., the day after the fusion), add 0.1 ml of 1X HAT medium to each well. This must be done with sufficient accuracy by removing 100 µl from each well and replacing it with fresh HAT medium.
2. Repeat on day 5 and 10. Look for growing clones each day thereafter.
3. On day 14, take 100 µl from every well and replace with 100 µl of HT. Retain supernatants from wells containing hybrids for screening.
4. Continue feeding cultures with HT up to day 28, thenceforth with TCM.

Notes

1. On days 1, 2, and 3 the culture medium will appear quite acidic; thereafter, HAT selection will drastically deplete the cell numbers and the cultures will appear dead. With the aid of

an inverted phase-contrast microscope, cells approximately the size of the myeloma parent can usually be observed growing as colonies among the cellular debris by days 6 to 21.
2. At some point, cells that have grown in HAT medium are transferred to normal medium (RPMI-1640). However, before making this switch, it is necessary for them to grow in HT medium for about one week in order to dilute any remaining intracellular aminopterin.
3. As long as the cultures are fed regularly and not disturbed, the hybrid cells can remain in the 96-well plates for up to 4–6 weeks.

6. Screening to identify cultures producing relevant antibodies

A variety of techniques are available for screening hybrid cells for specific antibody production. The most important considerations are that any assay should provide quick results and be able to handle hundreds of samples at any one time. Between two and four weeks after cell fusion, the supernatants of cultures are harvested using individual pipettes for each culture plate well. The supernatants can be tested, undiluted or diluted. The type of assay used for antibody detection will depend on the goal of the investigator. It is possible to use cytotoxicity or lytic assays; however, these will only detect complement-fixing antibodies and will miss non-complement-fixing antibodies.

Detection is facilitated by preparing single cell colonies that will grow and can be used for screening for antibodies producing hybridomas; only one in several hundred cell hybrids will produce antibodies of the desired specificity. The hybridoma cell lines survive in culture to produce large amounts of an antibody that reacts with the immunizing antigen which was originally being produced by the spleen cell. Screening of specific antibody producing cells can be done by ELISA, immunoflorescence on intact cells, immunoprecipitation with labeled antigens (Figure 26.14), and western blotting techniques.

7. Subcloning hybridomas by limiting dilution

Positive wells in the primary screen may contain both positive cells and non-secreting cells, which will outgrow the secreting cells in time. For these reasons, cloning should be accomplished as soon as a positive result is obtained. Cloning may be undertaken either by counting cells and distributing less than one cell per well, or by serial dilution of a cell suspension. Two rounds of cloning, the first at a higher cell density, may be employed to guarantee monoclonality. At each stage of the subcloning procedure, some cells should be frozen and stored to ensure that a cell line is not lost.

Materials

Thymocytes, Culture medium, HT medium, for use only when cloning from master plate. Acride orange-ethidium bromide, for determining cell viability. 96-well tissue culture plates.

Procedure

The cloning medium generally includes 10^7 thymocytes per ml of 15% FCS in RPMI-1640. The thymocytes act as carrier cells in diluting the hybrid cells and also as feeder cells in culture. Again BALB/c thymocytes are used. The objective is to plate 36 wells of a 96-well tissue culture plate with an average of 5 cells/well, 36 wells with an average of 0.5 cells/well, and the

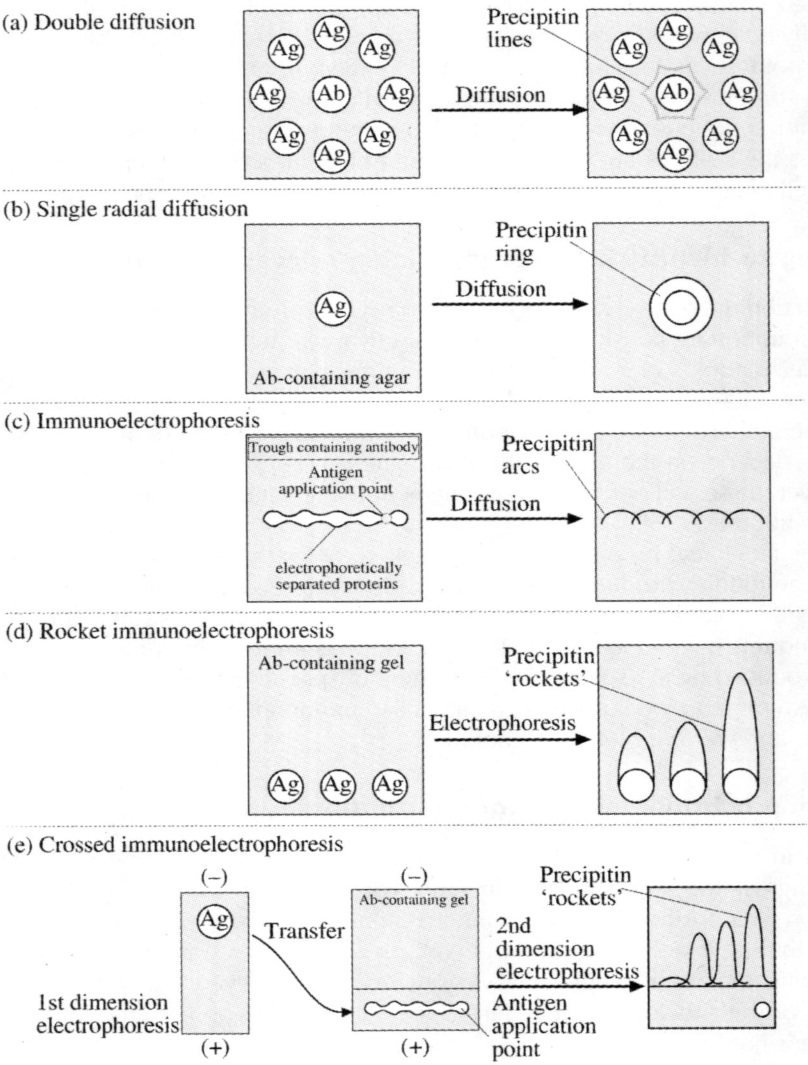

Figure 26.14 *Diagrammatic representation of immunoprecipitation techniques in agar*

remaining 24 wells with an average of 0.5 cells/well. One of these plating concentrations will yield wells with monoclonal growth. The dilutions are carried out as follows:

1. Harvest cells from each positive well when nearing confluency ($>10^5$ cells) with sterile tips and transfer to a plastic tube.
2. Determine the concentration of viable cells in a sample by staining with AO/EB.
3. Make dilutions of 50, 20, 5 and 2 cells/ml each in 5 ml volumes. Distribute 200 µl of each dilution among 24 wells of a 96 well plate containing feeder cells, preferably seeded the previous day.

4. Feed wells as above, taking supernatants for testing when wells are more than half confluent.
5. Select two or three positive wells, which are certain to be monoclonal by microscopic examination and dilution. Transfer these to 24-well plates and after a few days to 25 cm^2 flasks, then 75 cm^2 flasks. Always set up duplicate cultures in case of contamination.

8. Ascitic fluid production

Antibody production in culture supernatants ranges from 10–60 µg per ml; therefore, a litre of culture grown in roller bottles or Spinner flasks will yield 10 to 60 mg of antibody. Antibody production in mice can also be achieved by injecting healthy hybrid cells into appropriate H-2 compatible mice. Appearance of subcutaneous tumors means that the experiment is nearly successful and the mice can be continuously bled from the time the tumor first appears, which is anywhere from 10 to 30 days after injection. Analysis by agar or cellulose acetate electrophoresis of serum from the tumor-bearing mice shows a characteristic myeloma protein spike.

The hybridoma cells maintain many of the properties of the myeloma parent. They will grow continuously in culture, can be frozen and recovered, and form tumors when injected into animals. Since the hybrid arose through the fusion of a myeloma cell with a single antibody forming spleen cell, it produces a homogenous antibody with a single amino acid sequence. When the hybrid forms a tumor in a recipient animal, that animal accumulates as much as 1–10 mg/ml of that particular antibody in its serum or ascites fluid.

Procedure

1. Inject 0.5 ml pristane (2,6,10,14-tetramethylpentadecane:) IP into BALB/c mice or into F1 hybrids if non-BALB/c splenocytes used.
2. Transfer 1–4 x 10^6 hybridoma cells IP 7 to 21 days later. Days 10–14 are reported to be optimal.
3. From 7 days post passage, check mice daily for abdominal swelling. When swollen, tap fluid with 18 G needle, collecting into 10 ml tube containing 500 U heparin. It should be possible to collect further fluid at two-day intervals.
4. Centrifuge gently to pellet hybridoma cells, which can be passaged to a fresh recipient or cryopreserved. Note that passage of ascites cells usually produces a more rapid tumor than cells from culture. Keep a record of the passage number in case antibody secretion is lost.

9. Cryopreservation

Culture selected hybridoma cells have an added advantage for the production of Mabs in large quantities. These hybridoma cells may be frozen for future use and may also be injected in the body of an animal so that antibodies can be produced in the body and recovered later from the body fluid.

Procedure

1. Count cell suspension, and centrifuge.
2. Re-suspend in cold neat FCS, keep on ice and when ready to freeze, add an equal volume of cold 20% DMSO in TCM.
3. Aliquot into cryotubes at 10^6–10^7 viable cells per tube. Minimal volumes for the cryopreserved samples (0.2 ml) are preferable as thawing will be more rapid and cell survival increased.

4. Place in a slow freezer, after 4 to 24 hours. Quickly immerse tubes in liquid nitrogen to avoid warming up during transfer and store in liquid nitrogen.
5. When required, thaw rapidly by placing vial in water at 37°C. Add warm medium immediately and transfer vial contents to a 10 ml conical flask containing more warm medium. Centrifuge, re-suspend in medium, and perform dye exclusion viability test.

10. Isotype determination

It is very important to determine the class and subclass of any monoclonal antibody, an analysis best performed on culture supernatants where no host immunoglobulins are present. For example, those IgM subclasses which are not the subject of interest have to be eliminated from the cultures so as to reduce screening numbers. For this, characterizing subclasses of Mabs are essential. Because subtyping kits are now available commercially (Serotec Mouse Monoclonal Typing Kit, MMTOIK) this is easily done using the ELISA assay. Because some subtyping reagents are not specific for immunoglobulins from different strains of mice, it is helpful to use mice as spleen donors for antibody production, which are compatible with the antibody-subclassing reagents.

11. Improvement of optimal growth conditions and antibody yield in hybridomas

(a) Feeder layer cells: Single cells require feeder cells for effective growth. Therefore, whenever plating out cells immediately after fusion, or when cloning by limiting dilution, the cells should be placed in wells containing feeder cells. Peritoneal macrophages are the most effective feeder cells. It is often most convenient to use BALB/c cells, but allogenic (e.g. CBA) or even xenogenic (rat) cells are equally good. The most common feeder layers consist of (i) murine peritonial cells; (ii) macrophages derived from mouse, rat or guinea pigs; (iii) extra non-immunized spleen cells; (iv) human fibroblasts, human peripheral blood monocytes or thymus cells.

These feeder cells have some limitations like depletion of nutrients meant for hybridomas and contamination, but these problems can be overcome by using a purified or conditioned medium containing hybridoma growth factors (HGF) like interleukin-6 (IL-6) derived from human cells. For mice the procedure for adding feeder cells is given below:

Procedure

1. Kill mice by cervical dislocation and pin out. Douse in 70% ethanol and place on an open bench or, better, in a flow hood outside the culture room.
2. Carefully cut back the abdominal skin without puncturing the peritoneal membrane. Again douse in ethanol, and place ice on peritoneum.
3. Inject with 27 G (0.4 x 12 mm) needle, 5 ml of ice-cold serum-free RPMI-1640 containing 50 U/ml Heparin (Evans sodium heparin injection BP, 25,000 U/ml).
4. Withdraw needle and massage peritoneum. Use ice and alcohol liberally.
5. Insert 21 G (0.8 x 40 mm) needle and slowly withdraw fluid. Try to keep the bevel in sight through the peritoneal membrane. Take care not to block needle with peritoneal contents. Expect to recover about 4 ml fluid.
6. Pool the cell suspensions, wash (centrifuge 5 min. at no more than 400 g) and re-suspend in HAT or HT for fusion mixtures and clones respectively.

7. Expect to recover 1–3 × 10^6 cells per mouse, and plate out at 10^4 per well in 100 µl. The cells from one mouse, re-suspended in 20 ml, will cover two 96-well plates.

(b) Serum-free media: The second advancement in Mabs production by hybridoma technology is the use of different serum-free medium for bulk culture of hybridomas. The serum used to supplement the basal nutrient media is a highly complex and poorly defined mixture of components (like albumin, transferrin, lipoproteins, and various hormones/growth factors). Nevertheless, serum makes an essential component of media for culturing various animal cells. The use of serum, however, leads to difficulties in purification of antibodies. Further, it is an expensive technology for large-scale production of antibodies. In view of these difficulties, serum-free media are being increasingly used for culturing hybridomas.

12. Anti-idiotypic antibodies

If a homogenous antibody (e.g., a myeloma-produced antibody) is used as an antigen, certain portions of the molecule may be recognized as antigenic by the responding immunized host. The portion of the antibody molecule that recognizes its antigenic determinant is a unique site termed "idiotype". The sites are made up of particular amino acid sequences in the hypervariable portion of the variable region of the antibody. The antibodies produced by the host against these sites are therefore termed anti-idiotype. Anti-idiotypic antibodies have internal images of the original immunogen, and therefore are identified operationally as antibodies, which have activities that mimic those of the original immunogen. Anti-idiotypic antibodies that mimic such proteins and molecules as insulin and alprenolol have been described. While this technology is just becoming feasible, it has been demonstrated that the anti-idiotypic strategy is a potentially powerful approach for the preparation of monoclonal antibodies to receptors, which are difficult to isolate in quantities sufficient to be used as antigens.

IV PURIFICATION OF IMMUNOGLOBULINS AND THEIR FRAGMENTS

Mabs may need to be purified before they are used for a variety of purposes. Before final purification, the cultures (cell cultures or bacterial cultures with cloned genes) may be subjected to cell fractionation for enrichment of the antibody protein. Various methods are available for purification of antibodies from serum, ascites, or tissue culture supernatants, depending on the class(es) of antibody, the quantity and purity required, and the starting material involved. In *E. coli*, the antibodies may be secreted in the periplasm, which may be used for enrichment of the antibody, so that further purification is simplified. Alternatively, the antibodies may be purified from cell homogenate or cell debris obtained from the medium. Antibodies can be further purified by any one of the following techniques, such as precipitation with ammonium sulphate, ion-exchange chromatography, or antigen affinity chromatography.

Method-1: Partial purification by ammonium sulfate precipitation (refer to Chapter 25 for more details)

Method-2: Protein A-Sepharose Column Chromatography

It is frequently desirable to purify the antibodies synthesized by hybrid cell lines. The method described below provides rapid purification of immunoglobulins in a single step and utilizes protein A-Sephrose column chromatography. The Fc portion of IgG will bind to protein A and can therefore be used for affinity chromatography. The only restriction is that not all subclasses of IgG bind protein A. Elution is done by lowering the pH of the protein A column. The majority of mouse immunoglobulins bind to protein A at pH 8.6 and elute from the column at pH 4.3, or higher; thus harsh acidic elution can be avoided. A convenient form of protein A, bound to a chromatographic bead, is protein A-Sepharose CL-4B (Pharmacia).

Materials

Protein A-Sepharose CL 4B, Sodium azide (NaN_3), NaCl, NaOH, Tris concentrated HCl, Citric acid monohydrate, Trisodium citrate dihydrate, Na_2HPO_4, NaH_2PO_4 monohydrate, Sodium acetate trihydrate, Glacial acetic acid, Glycine-hydrochloride, Fraction collector, UV monitor (optional).

Buffers: The molarity of the buffering component may be varied. The important parameter is the pH.

0.05 M Tris, 0.5 M NaCl, 0.02% NaN_3, pH 8.6: for 1 litre of buffer, dissolve 6.06 gm of Tris and 8.76 gm of NaCl in 800 ml of distilled water. Add 10 M HCl to pH 8.6 and make up the volume to 1 litre with water.

0.05 M phosphate, 0.15 M NaCl, pH 7.0: For 1 litre of buffer, dissolve 4.34 gm of $NaHPO_4$, 2.70 gm of NaH_2PO_4 monohydrate, and 8.76 gm of NaCl in water to 1 litre; buffer pH should be 7.0.

0.05 M citrate, 0.15 M NaCl, pH 5.5: For 1 litre of buffer, dissolve 2.68 gm of citric acid monohydrate, 10.96 gm of trisodium citrate dihydrate, and 8.76 gm of NaCl in water to 1 litre; buffer pH should be 5.5.

0.05 M acetate, 0.15 M $CaCl_2$, pH 4.3: For 1 litre of buffer, dissolve 6.8 gm of sodium acetate, and 8.76 gm of NaCl in 800 ml of water. Add acetic acid to pH 4.3 and make up the volume to 1 litre with water.

0.05 M glycine-hydrochloride, 0.15 M NaCl, pH 2.3: For 1 litre of buffer, dissolve 5.6 gm of glycine-HCl and 8.76 gm of NaCl in 800 ml of water; add 10 M HCl to pH 2.3 and make up the volume to 1 litre with water.

Procedure

1. Swell 1.5 gm protein A-Sepharose CL 4B in Tris-buffered saline, pH 8.6. Pack resin in a suitable column (bed volume is 5–6 ml).
2. Harvest culture supernate and adjust to pH 8.6 by adding dilute NaOH.
3. Apply culture supernate to the protein A column. Wash column with Tris-buffer, pH 8.6. (One litre of culture supernate at pH 8.6 containing 10–60 mg of antibody is easily passed through the protein A column).
4. Carry out step elution with the buffered saline at pH 7.0, 5.5, 4.3, and 2.3 until the hybrid cell antibody is eluted, avoiding low pH buffers whenever possible. A UV monitor is useful in detecting antibody elution from the column.
5. Pool fractions containing antibody and dialyze using an appropriate buffer.
6. Regenerate column by washing with the glycine-HCl buffered saline, pH 2.3, and equilibrating with the Tris-buffered saline, pH 8.6 (including 0.02% NaN_3).

Method-3: Purification of IgG by ion-exchange chromatography

This procedure is relatively simple to perform and has been used extensively to purify IgG prior to conjugation with an enzyme for use in enzyme-linked immunosorbent assays. Purification on anion exchange cellulose, as described below, requires crude purification by ammonium sulfate precipitation prior to loading onto the column.

Procedure

1. Pre-condition DE23 cellulose (Whatman) by acid and alkali treatment according to manufacturer's instructions.
2. Equilibrate DE23 cellulose by stirring in 1/2X PBS at 6 ml/g of wet ion exchanger. Leave for 10 min. and decant. Repeat procedure until pH of cellulose is equal to buffer.
3. Add 7–10 ml equilibrated DE23 cellulose in 1/2X PBS at room temperature to a 12 x 120 mm column.
4. Wash column several times with 1/2X PBS, connect column eluant tube to UV monitor, and adjust absorbance at 280 nm to zero.
5. Add 2 ml ammonium sulfate-fractionated IgG to column.
6. Wash column with 1/2X PBS and monitor effluent at 280 nm.
7. Collect first peak from column.
8. Dilute effluent peak with 1/2X PBS to an absorbance of 1.4 (= 1 mg/ml). Store at –20°C in Eppendorf centrifuge tubes.

Note

1. IgG does not bind to the ion exchanger under the conditions described but rather passes through the column. The unwanted serum components bind to the column. A 14 x 80 mm column can thus be used most efficiently.

V IMMUNODETECTION OF ANTIGENS

Method-1: Immunoelectrophoresis (IEP)

Immunoelectrophoresis (IEP) separates antigens according to their native charge, and does not involve denaturation or dissociation of subunits. Small agarose gels are poured on microscope slides, and are run under straightforward conditions. Visualization of antigen, however, requires precipitation of complexes by an antiserum, and the method consumes relatively large amounts of antisera. The antigens are first electrophoresed and then exposed to the antibody along the whole length of the electrophoresis tank. Positive reactions appear as arcs, which are stable to prolonged washings. This stability makes the method useful for purifying individual antigens from Schistosoma mansont, for example. Excised arcs containing immune complexes have proved to be excellent immunogens for raising polyclonal antisera.

Materials

Agarose (ultrapure); TBE buffer; Bromophenol blue marker (0.25% in water); Flatbed electrophoresis tank; Double-width microscope slides, 50 x 76 mm.

Procedure

1. Make up 1% agarose in TBE, and boil to dissolve.
2. Slowly pour 10 ml over a 50 x 76 mm pre-washed glass slide.
3. Allow to set for 5 min. then cut the immunoelectrophoresis pattern on a template. Remove plugs to make wells, leaving troughs intact.
4. Load 5 µl of untreated antigen sample per well.
5. Add 1 µl of bromophenol blue to one of the wells.
6. Electrophorese at 20 mA for approximately 90 min. until the blue marker reaches gel end.
7. Remove troughs with a scalpel and fill with undiluted antiserum. Include control serum in one on the troughs.
8. Incubate overnight at room temperature in a humid box.
9. Examine against a dark background for arc formation.

Note

Because IEP depends upon formation of lattices of insoluble immune complexes, monoclonal antibodies reactive with a single epitope will not produce areas in this assay.

Method-2: Double immunodiffusion (DID)

This is the most commonly used technique of detection. It helps to compare different antigens (protein) and antisera directly. In this method (double diffusion, double dimensions), bound antigen and antibodies diffuse in the gel and form bands of precipitate layers at their zones of equivalence. The precipitin test gives a great deal of basic information but the method is quite time-consuming. Moreover, it does not allow a basis for quantitative analysis. However, semi-quantitative analysis can be done.

The technique is based upon the ability of antibodies to form precipitin lines specifically with the antigen. Free diffusion of both the antigen and the antibody takes place in agar gel and the resulting precipitates are normally visible to the naked eye. The disadvantage is that relatively large quantities of reagents, i.e. antibody and antigen, are needed. The local concentration of each reactant depends upon the absolute concentration of antigen and antibody in the wells besides their irrespective molecular weights. Multiple lines of precipitin will be present if the antigen and antibody contain several molecular species. Consequently, this technique has the advantage that several antigens or antisera can be compared around a single well of antibody or antigen.

Materials

Agarose, PBS pH 7.2 (0.05M), Normal saline, Glass slides, Humid chamber, Gel punch, Amino black B 10.

Procedure

1. Take glass slides, clean with alcohol and acetone, and place on a leveled surface.
2. Dissolve 1% agarose in PBS by boiling in a water bath.
3. To ensure good contact between the gel and glass plate, the glass plate may be pre-coated with a thin layer of agarose solution and dried.
4. Pour approximately 4 ml of 1% agarose solution on the glass slide, gently using a gel punch. Aspirate out the punched gel.

5. Load 20–40 µl of the antigen and antibody in separate wells and keep the slide in the humid chamber for 24 hours or more.
6. Note that the development of immunoprecipitation bands vary from time to time. When the precipitin lines are visible, slides may be stained with 1% (w/v) Amido back dye.

Method-3: Enzyme linked immunoabsorbant Assay (ELISA)

Enzyme-labeled antibodies can be used for rapid and simple immunoassay protocols, which do not depend upon expensive or sophisticated equipment. A range of enzyme labels and substrates for each enzyme may be selected to suit the particular application in mind. ELISA combines the specificity of antibodies or antigens coupled to an easily assayed enzyme, which also possesses a high turnover number. ELISA may be used for assaying antigens by either a competitive method or a double antibody method. Indirect ELISA also uses it for detecting antibodies by indirect ELISA.

a) Direct ELISA

This assay involves the immobilization of antigen onto the solid phase (e.g. micro well or a protein binding membrane such as nitrocellulose) followed by incubation with enzyme-labeled antibodies in the direct ELISA. The use of colorimetric chromogen/substrate combination can reveal either the extent of antibody binding if the colored product is soluble or the sites of binding, if the colored product is insoluble. The intensity of colored product is directly proportional to the amount of antigen or antibody in sample.

b) Indirect ELISA

This method may be used to measure both antigen and antibody. The test antiserum is reacted with specific antigen attached to a solid phase. Any specific antibody molecule bound to the antigen and all other material are washed away. Exposure to the enzyme-labeled anti-immunoglobulin antibody results in it being bound to any specific antibody molecules absorbed from the original serum. The complex is washed and the substrate is added, resulting in color formation which is proportional to the amount of specific antibody in the original serum.

c) Competitive ELISA

A mixture of a known amount of enzyme-labeled antigen and an unknown amount of unlabeled antigen is allowed to react with a specific antibody attached to a solid phase. After the complex has been washed, the enzyme substrate is added and the enzyme activity measured. The difference between this volume and that of a sample lacking unlabeled antigen is a measure of the concentration of unlabeled antigen. A major disadvantage of this method is that each antigen may require a different method to couple it to the enzyme. Alternatively, a known amount of standardized or purified antigen can be immobilized on the solid phase, and the antigen in the test sample incubated with the labeled antibody. In the absence of the antigen in the test sample, high amounts of labeled antibody can bind to the immobilized antigen. Increasing concentrations of antigen in the test sample lead to give decreasing of the antibodies.

d) Two-site ELISA assay for antigen detection (Sandwich ELISA)

The unknown antigen solution is reacted with a specific antibody attached to a solid phase, washed, and treated with the enzyme-labeled second antibody. After a further wash, the enzyme substrate is added. The amount of enzyme activity measured under standard conditions is directly proportional to the amount of antigen present. This assay is suitable when a monoclonal antibody is directed to the same antigen with more than one repeating epitope.

Method-4: Standard protocol for ELISA

This protocol combines horseradish peroxidase-coupled anti-immunoglobulin (HRP-anti-Ig) and an ABTS chromogen, which provides a strong visual signal. Alternative procedures are described in the following methods.

Materials

ELISA 96-well plates; ELISA reader; Incubator; Refrigerator; Rotary shaker.

Preparation of reagents

1. Coating buffer (0.06 M Carbonate buffer for antigen attachment):
   45.3 ml   1.0 M (8.4%) $NaHCO_3$
   18.2 ml   1.0 M (10.6%) Na_2CO_3
   Made up to 1 litre
2. Blocking solution:
   3% FCS; 0.5% skimmed milk powder; 2% BSA in distilled water
3. PBS with 0.05% Tween-20
4. HRP-conjugated anti-Ig of the appropriate species (e.g. anti-human and anti-mouse)
5. ABTS substrate (2,2'-azino-di-[3-ethyl-benzthiazoline sulphonate]) solution:
   Equal volumes of ABTS (solution A) and buffered hydrogen peroxide (solution B).

Procedure

1. Make up antigen to optimal dilution (e.g. 1 µg/ml) in 0.06 M carbonate buffer and add 200 µl per well. Incubate at 37°C for 3 hours or 4°C overnight or longer as required.
2. Remove antigen and add 200 µl per well of blocking solution for 30 min. at 37°C.
3. Wash three times in PBS-Tween, 3 min. per wash.
4. Add serum dilutions, 200 µl/well.
5. Incubate either for 30 min. at 37°C or 2 hours at room temperature.
6. Remove serum and wash three times in PBS-Tween as before.
7. Add 200 µl/well conjugate diluted (typically at 1/2000) in buffered PBS-Tween.
8. Incubate for 30 min. at 37°C or 2 hours at room temperature.
9. Make up fresh substrate solution: equal volumes of ABTS (solution A) and buffered hydrogen peroxide (solution B).
10. Remove conjugate, wash plate three times in PBS-Tween, rinse in water, and add substrate at 200 µl/well.
11. Stand plate at room temperature for up to 30 min. and read in ELISA reader between 398 and 415 nm.

Method-5: Immunoradiometric assay

An immunoradiometric assay (IRMA) is one in which a radio-labeled antibody is used to measure the presence of the target ligand. It differs from radioimmunoassay (RIA) in which a labelled antigen is used. The IRMA described here was first used with monoclonal antibodies in an assay for interferon (1) and subsequently applied to malarial sporozoites and to the phosphorylcholine epitope of *Onchoerca gibsont* (3), which is present on circulating antigen from all filarial species.

Like the two-site ELISA (sandwich ELISA), the IRMA test involves the binding of monoclonal antibody to a plastic surface, which is then saturated with a carrier protein (e.g. BSA). In a replica plate, the test serum is incubated with the same or different monoclonal antibody, which has been labeled with ^{125}I. This mixture is then added to the monoclonal-coated plate which, after further incubation, is washed and cut up so that each well can be counted in a gamma counter. This two-site assay therefore requires antigen to be bound at distant sites by each monoclonal antibody. In the case of repeating epitopes, the same monoclonal antibody may be used for both first and final layers.

Materials

PBS and Blocking Solution (PBS, 0.5% BSA, 0.1% sodium azide).

Diluent for labeled antibody (PBS, 5% NRS 0.05% Tween-20, 0.1% azide).

Washing solution (PBS, 0.05% Tween-20, 0.05% azide).

Monoclonal antibodies:
  (a) for coating plates: 1–10 µg/ml in PBS, as determined to be optimal. Prepare 5 ml per plate;
  (b) for iodinated monoclonal: Use at 2×10^6 cpm/ml in diluent. Prepare 3 ml per plate.

PVC 96-well plates.

Procedure

1. Coat the pipettes with Mabs in PBS by adding 50 µl to each well and incubate overnight at room temperature or longer at 4°C.
2. In a separate uncoated plate, add 30 µl per well of approximately diluted serum. Start with a 1:2.5 dilution in dilutent.
3. Add 30 µl of iodine-labeled Mab at 2×10^6 cpm/ml (60,000 cpm/well).
4. Incubate for at least 4 hours at 37°C, sealing plate.
5. Remove Mab from antibody-coated plate and add 60 µl of blocking solution. Incubate for at least 30 min. at room temperature, and wash three times in washing solution.
6. Transfer 50 µl of serum–Mab mixture from each well to the Mab-coated plate and incubate overnight, sealed, at room temperature.
7. Wash plates three times in washing solution, cut up plates, and count individual wells in gamma counter.

Method-6: Immunoblotting from polyacrylamide gels (Western blotting)

The transfer of protein bands from an acrylamide gel onto a more stable and immobilizing support is called protein blotting or Western blotting. This is an extremely sensitive method for visualizing a specific protein in a complex antigen mixture after separation on PAGE gels. Antibodies bound to bands on Western or immunoblots can be recovered as affinity purified ligand for micro-scale experiments.

As an alternative to immunoprecipitation, immunoblotting is often preferred because SDS dissociates protein complexes and allows precise visualization of antibody binding to target protein. Immunoprecipitation, however, can be performed on trace quantities of radiolabeled antigen for each track on a gel. A further factor is that many monoclonal antibodies bind well in immunoprecipitation but fail to react on blots. This is because two types of epitopes are destroyed during electrophoresis: (a) sites formed by the combination of or at the interface between subunits of different molecular weights and, (b) conformational epitopes abolished by exposure to SDS. The procedure given here reveals binding with an enzyme-linked reagent.

Materials

Semi-dry blotting apparatus and power supply capable of producing at least 200 mA and gel cassettes; Nitrocellulose membrane-pore size (0.2–0.45 µm); Whatman 3MM filter paper; Incubation trays – alternatively, a bag sealer and polythene bags, and a shaking table.

Specific rabbit antibodies against desired antigen; Goat anti-rabbit immunoglobulins conjugated with alkaline phosphatase.

BCIP/NBT substrate (Bromochloro indolyl phosphate/Nitro Blue Tetrazolium)

Preparation of reagents

1. Transfer buffer (Tris glycine buffer pH 8.3)
Tris-base	3.025 g	(25 mM)
Glycine	14.41 g	(192 mM)
Methanol	200 ml	(20%)

   The buffer is adjusted to pH 8.3 and made up to 1 litre. SDS to a final concentration of 0.1% may also be added to aid transfer of larger proteins. This can be stored at 4°C for 2–3 months and reused 3–4 times.
2. Washing buffer (PBST): add 0.05% of Tween-20 to PBS.
3. Quench buffer: 5% skimmed milk powder in PBS.
4. Blot solution A: 5% normal rabbit serum, fetal calf serum or skimmed milk powder pH 7.4 in PBS.
5. Blot solution B: 5% skimmed milk powder (or 1% NRS, FCS, etc.)
   Amido black stain (0.1% w/v):
Amido black	100 mg
Methanol	10 ml
Acetic acid	45 ml
Water	45 ml
6. De-staining solution: Methanol 45 ml, acetic acid 10 ml, and water 45 ml

Procedure

1. Run PAGE slab gel as explained earlier, loading 1–10 µg of protein, or up to 30 µg of complex antigen mixture per well.
2. Soak 12 pieces of 3MM paper cut to the size of the gel in transfer buffer. Soak eight pieces in transfer buffer and place on the lower (+) anode taking care to eliminate bubbles. Place wet NC paper on top, and then the gel on top of the NC paper. Again make sure no bubbles are trapped by gently rolling a pipette over each additional layer. Finally add the remaining four pieces of pre-soaked filter paper on top of the gel and position the upper (–) cathode on the sandwich. Put on the lid.

3. Transfer the assembled cassette in the buffer tank with the gel side facing the cathode and NCM side facing the anode. Top with more transfer buffer if necessary and place the power supply lid on the chamber. Connect the respective terminals of power supply. Transfer for 2 hours at 0.8 mA per cm^2 of gel. A standard 170 x 130 mm slab gel requires 180 mA.
4. Recover NCM paper. Before using blocking solution, remove strip containing marker proteins. Then quench the rest of the NCM in blot solution A for 2 hours at room temperature, or overnight at 4°C. The marker proteins can be visualized by staining for 10 min. in amido black, and de-staining in solution. Finally, rinse once in PBS-T.

Immunoblotting

1. Incubate in antibody diluted in blot solution B for 2 hours at room temperature or overnight at 4°C, with constant agitation, in incubation trays. Polyclonal antisera are typically diluted at 1:100 to 1:500.
2. Rinse three times in PBS-T, 10 min. each time.
3. Incubate the NCM in peroxidase-conjugated antibody, diluted to the manufacturer's recommendation (1:1000) in blot solution B, for 60 min. at 37°C or overnight at 4°C.
4. Rinse as in step 7 of method 5, with a final wash in PBS to remove detergent, which would otherwise inhibit the peroxidase enzyme.
5. Incubate the NCM in freshly prepared 3, 3'-diaminobenzidine substrate. To 100 ml of 50 mM Tris pH 7.6 add 50 mg of DAB and 10 µl of 30% hydrogen peroxide. Agitate gently for 30 min. or less if reaction proceeds rapidly.
6. Rinse in water, allow the NCM to air dry, and preserve the NCM.

Notes

1. Adding final concentrations of 0.03% cobalt chloride and nickel ammonium sulfate to substrate has been reported to raise sensitivity significantly.
2. Many variations are possible to reduce background levels of binding. TBS may give lower non-specific reactions than PBS, and performing all incubations at 4°C has been reported to improve signal–noise ratios.
3. Omit azide from all buffers when peroxidase reactions are to be used.

Method-7: Autoradiographic techniques

Although slower to yield a result, this autoradiographic technique has the advantage that the exposure may be varied to suit the level of binding. In addition, if labeled antigen is available, this can be run in one track and transferred to verify effective transfer of target molecules.

Materials

1. ^{125}I-labelled anti-immunoglobulin, Protein A other probe (e.g. lectins). X-ray film, cassettes, and chemicals. Add 0.1% sodium azide to buffers as preservative.

Procedure

1. Carry out transfer as in method 6.
2. Quench NCM paper in blot solution A.
3. Rinse and treat with test antibody in blot Solution B.
4. Rinse and incubate with 10^5–10^6 cpm/ml of radio-labeled probe in solution B. Rinse.

5. Ensure paper is completely dry and set up autoradiograph for 24 hours at –70°C; develop as in method 10.

Method-8: Direct staining of target cells

Materials

Cell non-toxic fetal calf serum (FCS); Fluorescent antibody in PBS (or some other physiological buffer), pre-titrated to determine the optimal concentration for staining.

MEM-FCS buffer: 5% FCS in minimum essential medium (MEM) plus 10 mM HEPES.

Phosphate-buffered saline, 3 ml conical glass centrifuge tubes.

Finely drawn out Pasteur pipettes.

Procedure

1. Load 2.5×10^6 target cells in MEM-FCS/3 ml tube.
2. Centrifuge at 300 g for 6 min. to pellet. Decant supernatant using drawn-out Pasteur pipettes.
3. Add 50–100 µl fluorescent labeled antibody to a 3 ml tube. Carefully re-suspend the cells, without forming bubbles, using drawn-out Pasteur pipettes. Incubate on ice for 15–20 min.
4. Add 1 ml FCS to tube, and mix with cell suspension.
5. Centrifuge at 400 g for 10 min., decant with drawn-out Pasteur pipettes. Re-suspend pellet in MEM-FCS, fill tube, and centrifuge.
6. Repeat MEM-FCS wash.
7. Prepare sample for microscopy.

Method-9: Microscopy

A. Preparation of slides

One can make wet mounts for immediate viewing or dry mounts for inspection later. Dry mounts will last for some months if they are refrigerated in the dark. However, they are found less satisfactory when the intensity of staining is low.

Materials

100% FCS; PBS; PBS containing 5% heat-inactivated (56°C, 30 min.) FCS; Glycerol-phosphate-buffered saline: 9 parts glycerol, 1 part PBS; 95% ethanol; Pasteur pipettes, Slides and coverslips (#1), Permount (or colorless nail polish)

Stock solution of 4% paraformaldehyde: Dissolve 4 g paraformaldehyde in 85 ml of 0.1 M phosphate buffer pH 7.3, by heating the solution to 70°C while stirring. Add 5 ml of 0.001 M $CaCl_2$, cool, and filter the solution through Whatman # 1 filter paper. Add 10 ml of 0.1 M phosphate buffer. The solution is stable for 1–2 weeks at 4°C. Just prior to use, dilute to 1:4 in PBS to obtain 1% fresh paraformaldehyde.

Procedure

Wet mounts
1. Pellet immunostained cells and re-suspend in a small volume of 100% FCS to a concentration of about $1–2 \times 10^7$ cells/ml. An ideal suspension should yield 20–30 cells/field at a magnification of 500X.

2. Place 1 small drop (5–10 µl) on a microscope slide, cover with a #1 cover slip, and seal with Permount or nail polish.

Dry mounts
1. Pellet cells and re-suspend in 100% FCS as above. Take up a small amount of the cell suspension into a Pasteur pipette by capillary action. Then streak 5–7 lines of the cell suspension onto a microscope slide.
2. Air-dry at least 30 min. at room temperature.
3. Fix for 20 seconds in 95% ethanol at room temperature. Air dry.
4. Add 1 small drop of glycerol-phosphate buffer, cover, and seal.

Fixation with Paraformaldehyde
An alternative method for fixing cells in dry mounts is by treatment with paraformaldehyde. This method of fixation can also be used to fix cells when staining two cell surface antigens to prevent movement of the first antigen during the staining of the second.

1. Wash cells free of serum.
2. Re-suspend pellet in 2 ml ice cold 1% fresh paraformaldehyde. Fix for 30 min. on ice.
3. Wash once in PBS without FCS, and once in PBS with 5% FCS.
4. Re-suspend pellet in FCS, streak, fix with ethanol, and mount as described above.

B. Viewing

Discount all cells in aggregates and cells at the edge of the field. Use the fine focus adjustment when judging fluorescence.

C. Microscope requirements

The following two systems of UV light source, filters, and objectives will provide incident-light excitation with a high yield and minimal quenching of fluorescence. They also permit visualization of specimens doubly stained with FITC and TRITC antibodies; the filter systems allow visualization of each fluorochrome while excluding the other.

Zeiss system (For example, or an equivalent)

UV light source:

50 W ultra-high-pressure Hg lamp (Osram) with 50 W DC power supply.

Filter system for FITC:

Excitation filters:
   No. 42–79–02 with KP 490 plus KP 500 interference filters
   510 LP dicroic beam splitter No. 46–63–04
   Barrier filter set No. 42–79–03 with KP 540 interference filter

Filter system for TRITC:

Excitation filters:
   No. 42–79–01 with two 546 BP interference filters
   580 LP dicroic beam splitter No. 46–63–05
   Barrier filter set No. 42–78–69 with LP 590 and LP 605 interference filters plus BG 18 red-attenuating filter No. 48–79–92.

Zeiss Objective:

Planapochromat 63/1.4 N.A. oil phase

Note

1. The properties of individual filters with a given designation vary such that each system must be checked for satisfactory operation.

Method-10: Flow cytometry and fluorescence-activated cell sorting

The tissue lineage, maturation stage, or activation status of a cell can often be determined by analyzing the cell surface or intracellular expression of different molecules. The flow cytometry is a specialized instrument that can detect fluorescence on individual cells in a suspension and thereby determine the number of cells expressing the molecule to which a fluorescent probe binds. A suspension of cells are incubated with fluorescently labeled probes, and the amount of probe bound by each cell in the population is measured by passing the cells one at a time through a fluorimeter with a laser-generated incident beam. The relative amount of a particular molecule on different cell populations can be compared by staining each population with the same probe and determining the amount of fluorescence emitted. For the preparation of flow cytometric analysis, cell suspensions are stained with the fluorescent probes of choice to characterize cells of interest. Most often, these probes are fluorochrome-labeled antibodies specific for a cell surface molecule (Figure 26.15). Alternatively, cytoplasmic molecules can be stained by temporarily permealizing cells and permitting the labeled antibodies to enter through the plasma membrane.

VI APPLICATIONS OF MONOCLONAL ANTIBODIES

Immunological applications are many and diverse. It is impossible to explain each and every application of antibodies. Here, a few applications are briefly mentioned.

The introduction of the Mabs has circumvented the two disadvantages of conventional antibodies: namely their heterogeneity and irreproducibility. In addition, the unique advantages of Mabs allow their use in ways which are not feasible for polyclonal reagents. Some of the areas in which Mabs have proved useful are: (i) immunological quantitation of other molecules, (ii) antigen purification, and (iii) biological and clinical applications.

As mentioned earlier, antibodies can be produced to protein or oligopeptide determinants, carbohydrates including individual monosaccharides, nucleic acids including individual nucleotides, as well as to macromolecules and to cells, all depending upon the way the immunization is carried out. The benefits of antibodies are enormous. Antibodies are essential tools in modern biomedical and agriculture research. Mabs are chemically defined reagents and in a single preparation, they react with a single epitope or antigenic determinant. Thus, Mabs bind very specifically and no background confusion arises due to the presence of antibodies of other specificities. Once a hybridoma producing a desirable antibody has been identified, hundreds of milligrams of that antibody can be generated. When the stock of the antibody is used up, the cells can be recovered from the freezer and the same exact antibody produced again and again in large amounts. The amounts of antibody, which can be generated and the ability to renew

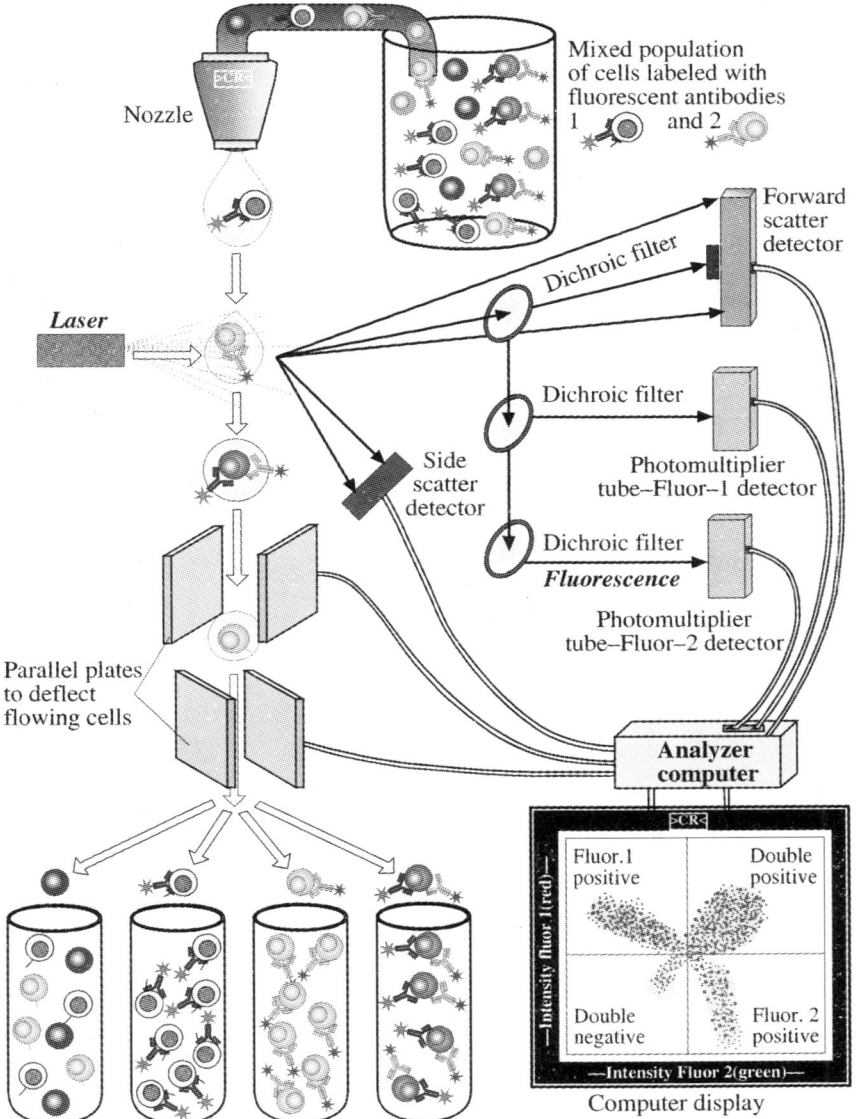

Figure 26.15 *Principle of flow cytometry and fluorescence-activated cell sorting*

the antibody whenever it is needed make it worthwhile to fully characterize its specificity, affinity, and effector functions. The identification of such Mabs is greatly facilitated by the nature of the techniques that essentially clone the individual antibodies, which make up the heterogeneous repertoire of antibodies in a conventional antiserum. It is thus possible to immunize with an impure antigen such as a whole cell and screen for a monoclonal antibody that reacts with one of the many determinants on the cell. The various applications of Mabs may be broadly grouped into the following three categories.

1. Diagnostic applications

   Mabs can be used:

   A. For classification of blood groups, e.g., ABO, Rh, etc.
   B. To detect pregnancy by the assaying of pregnancy specific hormones.
   C. To detect tumor antigens.
   D. To detect pathogens.
   E. For an accurate detection of specific chromosomes of a given species.

2. Taxonomy

3. Therapeutic applications

   Mabs can be used:

   As therapeutic agents
   For drug targeting
   For immunopurification
   As monoclonal antibodies for cytogenetic analysis

VII APPLICATIONS OF IMMUNOLOGY IN CONTROL OF HUMAN DISEASES

Vaccine development strategies

The birth of immunology as a science can be traced back to Edward Jenner's successful vaccination against smallpox in 1796. In recent days, the importance of prophylactic immunization against infectious diseases is best illustrated by the fact that worldwide programs of vaccination have led to the complete or nearly complete eradication of many deadly diseases (Table 26.1). Vaccines induce protection against various infections by stimulating the development of long-lived effector cells and memory cells. Most vaccines in routine use today work by inducing humoral immunity and attempts to stimulate cell-mediated immune responses by vaccination are going on. The success of active immunization in eradicating infectious diseases is dependent on many factors.

Table 26.1 *Vaccine approaches*

Type of vaccine	Examples
Live attenuated or killed bacteria	BCG, cholera
Live attenuated viruses	Polio, rabies
Subunit (antigen) vaccines	Tetanus toxoid, diptheria toxoid
Conjugate vaccines	Hemophilus influenzae, pneumococcus
Synthetic vaccines	Hepatitis (recombinant proteins)
Viral vectors	Clinical trials of HIV antigens in canarypox vector
DNA vaccines	Clinical trials ongoing for several infections

VIII SUGGESTED READING

Abbas AK and AH Lichtman (2000) *Cellular and Molecular Immunology* (5th Edn), Saunders an Imprint of Elsevier Science (USA).

Bergman Y (1999) Allelic exclusion in B and T lymphocytes, *Semin. Immunol.*, **11**:319–328.

Buck CA (1992) Immunoglobulin Superfamily: Structure, Function and Relationship to Other Receptor Molecules, *Semin. Cell Biol.*, 3:179–188.

Davies DR, Padlan EA and Segel DM (1975) Three-dimensional Structure of Immunoglobulins, *Ann. Rev. Biochem.*, **44**:639.

Garcia KC, Teyton L and Wilson IA (1999) Structural Basis of T Cell Recognition, *Annu. Rev. Immunol*, 17:369–397.

Lodish H, Baltimore D, Berk A, Zipursky SL, Matsudaira P and Darnell J (1995) *Molecular Cell Biology*, Scientific American Books, Inc.

Parham P (2001) *The Immune System*, Garland Publishing/Elsevier Science Ltd, New York, London.

Perham P, Androlewicz MJ, Brodsky FM, Holmes NJ and Ways JP (1982) Monoclonal Antibodies: Purification, Fragmentation and Application to Structural and Functional Studies of Class I MHC Antigens, *J. Immunol. Methods*, **53**:133–173.

Pieters J (2000) MHC Class-II Restricted Antigen Processing and Presentation, *Adv. Immunol.*, **75**:159–208.

Southern EM (1975) Detection of Specific Sequences Among DNA Fragments Separated by Gel Electrophoresis, *J. Mol. Biol.*, **98**:503–517.

Tonegawa S (1983) Somatic Generation of Antibody Diversity, *Nature*, **302**:575.

Towbin H, Staehelin T and Gordon J (1979) Electrophoretic Transfer of Proteins From Polyacrylamide Gels to Nitrocellulose Sheets: Procedure and Some Applications, *Proc. Natl. Acad. Sci. USA*, **76**:4350–4354.

Watson JD, Hopkins NH, Roberts JW, Steitz JA and Weiner AM (1987) *The Molecular Biology of the Gene*, Benjamin/Cummings Publishing Company, Inc.

Willerford DM, Swat W and Alt FW (1996) Developmental Regulation of V(D) Recombination and Lymphocyte Differentiation, *Curr. Opin. Genet. Dev.*, **6**:603–609.

PART – VII
Bacterial and Mammalian Cell Culture

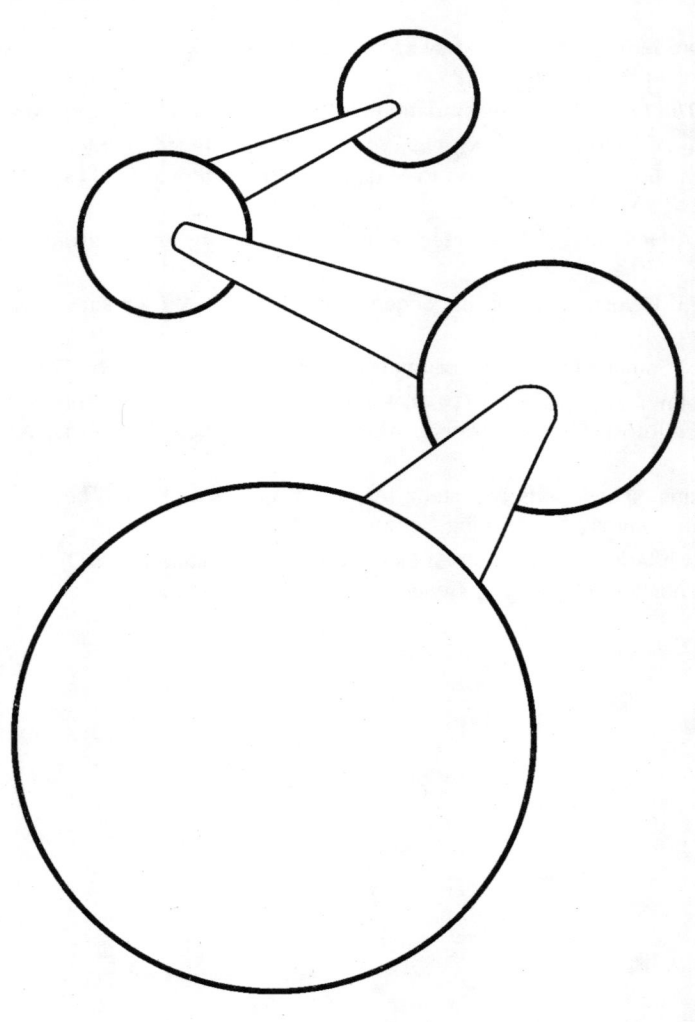

27

Biology of Bacteria

I INTRODUCTION

Before the rise of recombinant DNA technology, much of the research in molecular biology was carried out in bacterial systems, both as the object of experimental study and as the source of the experimental material. More recently, the diversity of organisms and problems being studied has greatly increased and many other organisms including viruses, prokaryotes, and eukaryotes are also being extensively used. *Escherichia coli* (*E. coli*) is often the most appropriate organism for the study of fundamental problems and the best source of DNA and proteins necessary for recombinant DNA research. *E. coli* is a rod shaped bacterium (Figure 27.1) with a circular chromosome about 3 million base pairs long. The following sections describe the use of *E. coli* as an object of experimental study and as a vehicle for other studies. Typically, there are three reasons for growing cells: (1) for the purification of subcellular structures or biomolecules; (2) for genetic manipulations; or (3) for measurements made on growing cells. In these instances, the variables of most concern are the strain, the growth medium, the temperature, etc.

Advantages and disadvantages of using microorganisms in biotechnology: There are many potential advantages of using microorganisms such as (1) They are easier to grow. They grow more quickly, more cheaply, and in unlimited quantity; (2) they are easier to manipulate in terms of their nutritional and environmental parameters; (3) they tend to be relatively simple at the whole cell level; (4) they exhibit unicellularity which makes them less demanding; (5) they are generally very versatile in their metabolic properties; (6) they allow populations of cells to be derived very quickly from a single cell, giving biological heterogeneity and thus predictability; (7) they do not pose the ethical problems which may arise from the use of animals; and (8) they have been used in genetic engineering for a long time. One of the main problems associated with the use of microorganisms is their biohazardous nature and the possibility of the development of new pathogenic strains.

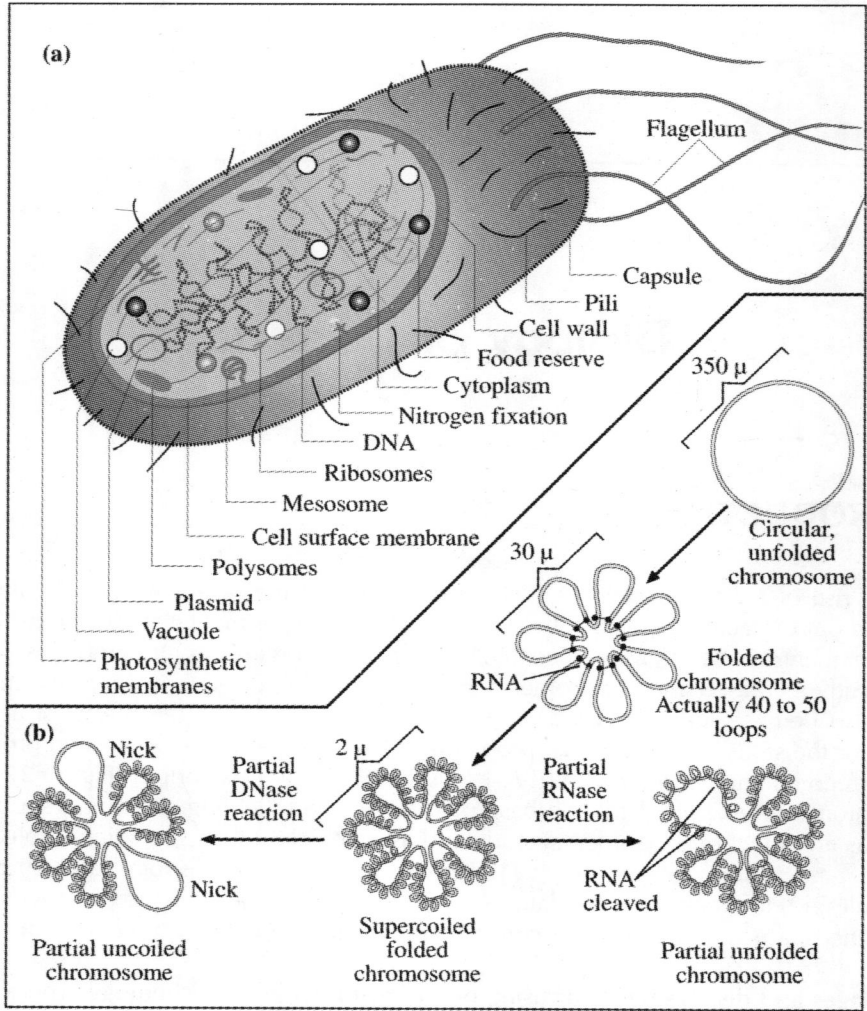

Figure 27.1 (a) Structure of a generalised rod-shaped bacterium (a typical prokaryotic cell), (b) Structure of the chromosome of E. coli

II CULTURE MEDIA

A. PREPARATION OF CULTURE MEDIA

The survival and continued growth of microorganisms depend on an adequate supply of nutrients and a favorable growth environment. The primary purpose of a growth medium is to enable or encourage one or more organisms to grow and divide. The first point to be considered when formulating the composition of media is which one and how much to put in. In addition to establishing the relative concentration of each element, care has to be taken while deciding in what form they should be added. From our studies to date, while we should have by now an idea of the basic components of all laboratory media – that is a mixture of simple mineral salts.

The culture medium used for cultivation of bacteria or yeast must provide certain basic nutritional requirements, including (1) a carbon source (CO_2, glucose, etc.), (2) water; (3) a nitrogen source (N_2, ammonium, nitrate salts, amino acids); (4) a phosphate source (phosphate salts); (5) a sulfate source (sulfur containing amino acids, sulfates, and elementary sulfur); (6) various mineral nutrients (Ca^{2+}, Zn^{2+}, Na^+, K^+, Cu^{2+}, Mn^{2+}, Mg^{2+}, Fe^{2+}, Fe^{3+}, etc.); (7) vitamins; and (8) energy (radiant energy, glucose, H_2S, and $NaNO_2$). *E. coli* and *S. cerevisiae* are capable of growth on a medium consisting of a single carbon source, such as the carbohydrate glucose; a simple nitrogen source, such as ammonium chloride or ammonium sulfate; and other inorganic salts to provide the phosphorus, sulfur, and minerals. Though *S. cerevisiae* also requires some vitamins. This kind of medium is termed as **defined** or **synthetic** because its exact chemical composition is known. For some kind of laboratory work, however, complex media in which the basic ingredients are provided by complex nutrients of which the exact composition is not known, are employed. Basically, all culture media are **liquid**, **semisolid**, or **solid**.

A **broth** media is one in which the components are simply dissolved in water without a solidifying agent. A broth medium supplemented with a solidifying agent (e.g. agar, a complex carbohydrate composed mainly of galactose, extracted from seaweed) results in a **solid** (1.5 to 1.8% agar) or **semisolid** (<1% of agar) medium. Solid agar melts at 90–100°C; liquid agar solidifies at about 42°C. Sterilization procedures eliminate all viable microorganisms. Media, glass, and plasticware (Figure 27.2) and all other equipment coming into contact with culture must be free from microbial contamination (for various sterilization techniques see Chapter 2). Because of the solidifying properties of agar, different organisms can be cultivated at about temperatures of 37.5°C without fear of their liquefying. A solid medium has the advantage of having a hardened surface on which microorganisms can be grown using specialized techniques for the isolation of discrete colonies. Each colony is a cluster of cells that originates from the multiplication of a single cell and represents the growth of a single species of microorganism. Such a defined and well-isolated colony is a **pure culture**. Another advantage of a solid medium is that, while in the liquefied state, solid media can be placed in test tubes, allowed to cool and harden in a slanted position, producing **agar slants**. If allowed to harden in the upright position, they are designated as **agar deep tubes**. Liquefied medium can be poured into Petri dishes, producing **agar plates**, which provide large surface areas for the isolation and study of microorganisms (Figure 27.3).

B. TYPES OF MEDIA

The main objective of growing microorganisms is almost always to isolate and maintain a pure or axenic culture. Such a culture contains only a single species of organism and may quite possibly be derived from a single cell. Pure cultures are artificial in that they rarely if ever occur in nature. The name of each of the media below gives a clue to its purpose. Examples of different types of media (also refer to Appendix 1) are:
1. **General medium**: It contains a balanced array of minerals together with glucose and many growth factors in the yeast extract.
2. **Enrichment (enriched) medium**: The yeast extract broth is an example of an enriched medium and is used for the cultivation of fastidious microorganisms – organisms that have

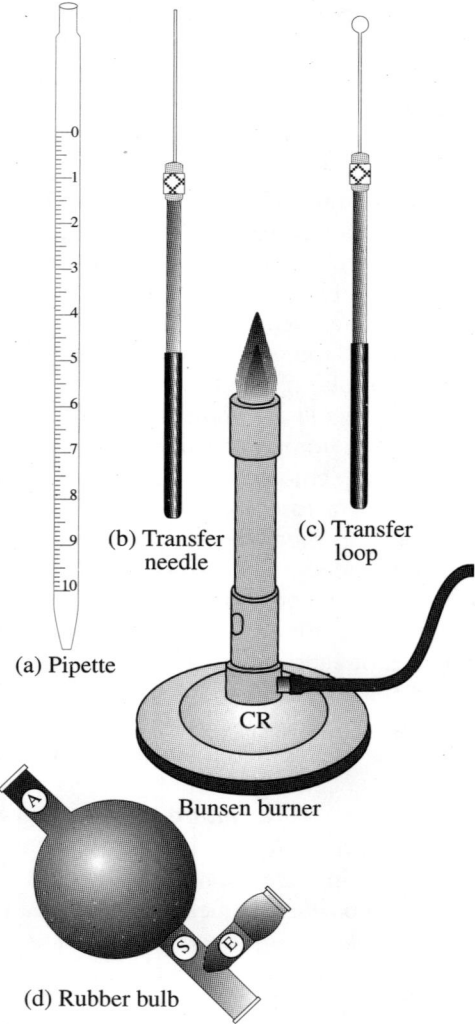

Figure 27.2 *Transfer instruments*

highly elaborate and specific nutritional needs. These do not grow or grow poorly on the basic artificial medium and require the addition of one or more growth-supporting substances – enrichments such as additional plant or animal extracts, vitamins, or blood.

3. **Selective medium**: This medium is used to select (isolate) specific groups of bacteria. They incorporate chemical substances that inhibit the growth of one type of bacteria while permitting growth of another, thus facilitating bacterial isolation. For example, staphylococci are cultivated on mannitol salt agar. This medium contains mannitol, an alcoholic carbohydrate fermented by *Staphylococci*, as well as a high salt that inhibits most other bacteria.

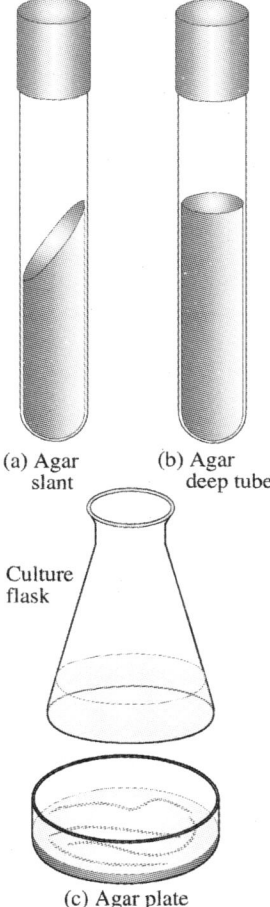

Figure 27.3 *Forms of solid (agar) media*

4. **Differential medium**: This medium makes it easy to distinguish colonies of one organism from colonies of other organisms on the same plate. MacConkey's agar is a selective and differential medium containing the following: peptone, lactose, bile salts, sodium chloride, neutral red indicator, crystal violet. The medium is one alternative for the isolation and identification of enteric bacteria – the bacteria living in the human gut.
5. **Diagnostic medium**: Many different diagnostic tests have been devised of which relatively few are in common use. Each of them tests one or more specific properties of a given cell. Such media often play little part in the primary isolation of organisms, rather they are important in distinguishing or identifying organisms at a later stage.
6. **Complex medium**: This is a complex mixture of nutrients, salts, and other solutes obtained from a natural source such as meat or milk, and therefore, essentially slightly variable in composition. Generally, this medium is cheaper than other media.
7. **Defined medium**: It is a mixture of known, purified constituents (specific organic and/or inorganic compounds) in known proportions and concentrations, which in theory should

be reproducible from batch to batch and from lab to lab. Its use requires knowledge of the organism's specific nutritional needs.
8. **Simple medium**: This medium consist of a single main constituent such as peptone, a mixture of amino acids obtained by hydrolysis of protein, without the additional supplements necessary for fastidious organisms.
9. **Minimal medium**: This is a basic mixture of essential salts with carbon and nitrogen and energy sources, and may support growth of organisms which have the ability to synthesize all their own organic constituents.
10. **Synthetic medium**: This is essentially the same as a defined medium, but the word is a little less precise, and reproducibility may not be total.
11. **Semi-synthetic medium**: This medium is hard to define and is also of limited scientific value; it usually consists of a defined or synthetic base with the addition of a complex supplement such as yeast extract, which negates the value of the defined base.

C. INTERMICROBIAL RELATIONSHIPS OF BACTERIA FOR NUTRITIONAL NEEDS

Closely allied to the nutritional needs of bacteria is their relationship with other organisms. The term **symbiosis** (dictionary meaning: living together) is applied to this relationship. Symbiosis implies a close and permanent association between two populations for the benefits of food, protection, support, or other life-sustaining factors. If a symbiotic relationship benefits both populations, the relationship is further defined as **mutualism** (e.g. *Rhizobium*). Another type of symbiosis is **commensalism**. This relationship occurs when one population benefits from the relationship while the other is neither benefited nor harmed (e.g. *E. coli* that lives in the human intestine). A third type of symbiosis is **synergism**. In this case, two populations or organisms live together and accomplish what neither population could accomplish alone. For example, at least two populations of bacteria must be present for infection of the oral cavity to occur.

When a symbiosis is beneficial to one organism but harmful to the other, the result is **parasitism**. In this situation, the organism that benefits is the parasite, and the one that suffers injuries is the host. For example, many diseases of humans, animals, and plants are due to parasitism.

D. METHODS AVAILABLE FOR PREPARATION OF MEDIA

Method-1: Preparation of YPD (Yeast extract, peptone, dextrose), a complex medium for yeast

Procedure

1. Add 500 ml of distilled water to a 1 litre Erlenmeyer flask. Weigh out 5 g of yeast extract, 10 g of peptone, and 10 g of dextrose with a top-loading balance; dissolve these in the water by mixing with a magnetic stirrer.
2. Divide the broth into two equal parts by adding 250 ml each into two 500 ml Erlenmeyer flasks.
   Flask A: Make no further additions.

Flask B: Add 5 g of agar; swirl to disperse the agar.
3. Dispense the broth media (Flask A) into 15 test tubes, adding 5 ml of broth to each tube.
4. Cap each test tube, but do not tighten excessively, and put aside for autoclaving later. Place Flask B and Flask A (with 175 ml of YEPD remaining) into the autoclave for autoclaving as well.
5. Load the autoclave with all the solutions and close the autoclave door. Most autoclaves have an automatic cycle (see Chapter 10 for liquid media sterilization).
6. After removal from the autoclave, allow the broth tubes and 175 ml of YEP (Flask A) to cool before tightening caps and storing for use in later experiments. After the medium in Flask B has been sterilized, place the flask in a 50°C water bath and equilibrate for 30 min. (the purpose of the cooling step is to prevent excessive water condensation on the Petri dish lid caused by evaporation from the agar)
7. To dispense the medium into sterile Petri plates, flame the mouth of the flask and, while carefully lifting the lid of a Petri plate, pour about 20 ml of agar from Flask B into a plate (enough medium to cover the bottom of the plate). Replace the lid and continue filling additional plates until all the medium has been dispensed with. Work quickly to minimize contamination, but carefully to prevent accidents.
8. Allow the agar plates to cool. After the agar has solidified, label the plates, and leave them at room temperature or incubate at 37°C overnight to allow for drying of the agar and detection of contaminated plates. Store the plates at room temperature in plastic bags or at 4°C for use in later experiments.

Method-2: Preparation of M9 medium, a defined medium for E. coli

Procedure

1. Add 450 ml of distilled water to a 1 litre beaker. Weigh out the following amounts of each component and dissolve them in the water: NH_4Cl 0.5 g, Na_2HPO_4 3.0 g, KH_2PO_4 1.5 g, NaCl 0.25 g.
2. Carefully adjust the pH to 7.4 with 1 N NaOH and bring the volume to 494 ml using a graduated cylinder. Transfer back to a 1 litre Erlenmeyer flask (Flask C) for autoclaving.
3. Make up 50 ml of the following three solutions in 100 ml screw cap bottles: 1 M $MgSO_4.7H_2O$, 20% (w/v) dextrose (glucose), 1 M $CaCl_2$.
4. Autoclave all solutions from steps 2 and 3 above.
5. At the time of the experiment, aseptically add 1 ml of $MgSO_4.7H_2O$ (1 M), 5 ml of glucose [20% (w/v)], and 50 µl of $CaCl_2$ (1 M) to Flask C to make up the M9 medium.

Method-3: Preparation of YNB medium, a defined medium for S. cerevisiae

Procedure

1. Add 6.7 g of yeast nitrogen base (YNB) to 100 ml of distilled water. Slight heating may be required to dissolve it. Filter sterilize for a 10X stock solution. Store in a refrigerator.
2. Prepare 100 ml stock solutions of 30 mg/ml lysine, tryptophan, and histidine. Also prepare 100 ml solutions of 2 mg/ml uracil and adenine sulfate. All solutions can be autoclaved except tryptophan and adenine, which should be filter sterilized.
3. Prepare a 100 ml 20%(w/v) dextrose solution, which should be autoclaved.
4. Prepare a 100 ml 4% agar solution. Autoclave. Place molten agar at 50°C to equilibrate for pouring of plates on addition of pre-warmed YNB.

5. Prepare minimal medium with YNB, supplements (amino acids and bases), and dextrose (Table 27.1). For agar plates, add 4% agar as indicated in Table 27.1. Be sure to equilibrate the medium components to 50°C before the addition of agar. Mix well, being careful not to introduce excess bubbles, and pour into plates.

Table 27.1 *Preparation of YNB medium*

Component	Medium Liquid	Plates
YNB (10X)	10 ml	10 ml
Agar [4% (w/v)]	–	50
Lysine (30 mg/ml)	0.1	0.1
Tryptophan (30 mg/ml)	0.1	0.1
Histidine (30 mg/ml)	0.1	0.1
Adenine sulfate (2 mg/ml)	1.5	1.5
Uracil (2 mg/ml)	1.5	1.5
Dextrose [20% (w/v)[	10	10
Distilled water	76.7	26.7

III HANDLING AND MAINTENANCE OF BACTERIAL CULTURES

A. ASEPTIC TECHNIQUE

Aseptic technique means using sterilized equipment and solutions and preventing their contamination while in use. Bacteria and fungal spores are abundant in most environments, including laboratories. The researcher uses a range of special techniques and apparatus, which are designed to prevent contamination of nutrient media. An aseptic technique is required to transfer pure cultures and to maintain the sterility of media and solutions. By aseptic techniques, biotechnology takes prudent precautions to prevent contamination of the culture or solutions by unwanted microbes. Many of the Petri dishes and tissue culture plates used for growing pure cultures of microorganisms are made of plastic and come pre-sterilized from the manufacturer. Filling those vessels with a sterile medium requires the use of an aseptic technique. A proper aseptic transfer technique also protects the biotechnologist from being contaminated by the culture, which should always be treated as a potential pathogen. An aseptic technique involves avoiding any contact of the pure culture, sterile medium, and sterile surfaces of the growth vessel with contaminating microorganisms. To accomplish this task, (1) the work area is cleaned with an antiseptic to reduce the numbers of potential contaminants, (2) the transfer instruments are sterilized, and (3) the work is accomplished quickly and efficiently to minimize the time of exposure during which contamination of the culture or laboratory worker can occur.

Typical steps for transferring a culture from one vessel to another are (1) flaming the transfer loop, (2) opening and flaming the mouths of the culture tubes, (3) picking up some of the culture growth and transferring it to the fresh medium, (4) flaming the mouths of the culture vessels and re-sealing them, and (5) flaming the inoculating loop. Essentially the same technique is used for inoculating Petri dishes and for transferring microorganisms from a culture vessel to a microscope slide. Developing a thorough understanding and knowledge of the aseptic technique and culture transfer procedures is a prerequisite for working with microbiological cultures.

B. STRAIN PURITY

Many of the bacterial strains often used in molecular biology are contaminated. Consequently, they vary in their growth requirements and other properties. Therefore, any strain which is to be used for the experimental purpose should be tested for its properties and, if possible, tested for the properties required by the particular experiment. This testing should usually be done on single colonies that have been obtained by streaking the strain on a YT plate.

C. STORAGE OF STRAINS

The shelf-life of strains varies widely. From time to time, every laboratory needs to acquire cultures whether for teaching, quality control or standardization purposes, or research work. For day-to-day manipulation of cultures in experiments, stocks will often need to be maintained for a few days by subculturing, usually on solid media. However, this may not be suitable for long-term storage. For long-term storage, samples of cultures should be either frozen or freeze-dried. Some strains remain viable for months when stored in a liquid medium in the refrigerator; others die within a few days under these conditions. For genetic purposes, many strains can be stored for years at room temperature in the dark in small, airtight vials half filled with slab agar. Probably the safest way to insure physiological constancy in a strain over a period of several years is to freeze the samples. A single colony can be grown, diluted into a medium containing 10 to 50% glycerol, distributed to a number of aliquot vials, and stored at –20° to –70°C. Such a vial provides viable inocula for about 30 cycles of freezing and thawing. For indefinite storage, freezing in liquid nitrogen can be used.

The strains containing plasmids, however, tend to be *cured* of (lose) plasmids when stored in or on nutrient agar. This is particularly true with some recombinant plasmids. Bacteria and their plasmids can be recovered after years of frozen storage and repeated freezing and thawing if the following procedure is used. Grow an overnight culture of the cells in LB broth and then add an equal volume of 2X freezing medium (see Appendix 1). Freeze and store at –70°C. Rapid freezing in liquid nitrogen and slow thawing at room temperature allow recovery of viable cells from a 100 µl culture after more than 15 cycles of freezing and thawing and after at least 5 years of storage.

D. PHAGE CONTAMINATION

Phage contamination of bacterial cultures can be a problem and may be particularly troublesome in laboratories where phages are grown in large quantities. For example, phage T1 presents some special problems because cells are often sickly. The problem is compounded because T1 is not killed by drying and therefore persists in a laboratory and spreads with dust. Phages

T2 and T4 are less problematic because they are killed by drying. RNA viruses like R17 are insidious forms of contamination because they frequently contaminate small cultures, preventing complete lysis. They also prevent the cultures from reaching the stationary phase. Several practices are commonly used to reduce problems of phage contamination. Using disposable culture dishes, properly sterilizing the contaminated or discarded cultures, using germicidal lamps at the work place and on laboratory apparatus, and following good laboratory practices are some of them.

IV ESTABLISHING A PURE CULTURE

For isolation of a particular organism, for whatever reason, basically two alternative sources are considered. First, isolation from a natural habitat or second, purchase from a recognized culture collection. Whatever the source of obtaining the microorganisms, the utilization of microbes in biotechnology depends on pure cultures, which consist of only a single species, and the maintenance of the purity of the isolates through subsequent manipulations. Most methods for obtaining pure cultures rely on some form of dilution technique. To isolate pure cultures from a mixture of bacteria in a suspension, **serial dilution** (**pour-plate**) methods or simple **streak-plate** and **spread-plate** methods are generally followed. The most useful and pragmatic approach is the streak-plate method in which a mixed culture is spread or streaked over the medium surface such that the individual cells become separated from one another. Each isolated cell grows into a colony and hence pure cultures because the cells are the progeny of the original single cell. Another plating method very commonly followed is the **replica-plating** technique. This is a very convenient method for characterization of the colonies by different genetic and biochemical tests.

Method-1: Serial dilution methods

In addition to isolation of individual colonies, studies involving the analysis of materials such as food, water, milk, and in some cases, air, requires quantitative enumeration of the microorganisms in the substances. Most of these experiments need diluted bacterial culture before plating. A saturated culture contains 10^9 live cells/ml. To be able to visualize several hundred isolated colonies on a plate we would have to dilute this culture 10 million-fold. One way to do this is by making a series of dilutions of either 1:10 or 1:100. A 10^{-1} dilution is achieved by pipetting 1 ml into 9 ml of dilution buffer, and a 10^{-2} dilution is achieved by adding 0.1 ml to 9.9 ml of the buffer (Figure 27.4). There is no particular need to use 10 ml, and often it is more practical to use larger volumes, which require fewer manipulations. The dilution buffer is either minimal salts (1 x A) or simply saline. A 10^{-6} dilution of a culture, which originally contained 2×10^{-9} cells/ml contains 2×10^{-3} cells/ml or 2000 cells/ml. If we pipette 0.1 ml of this, we are then delivering 200 cells. For most experiments, 200–300 colonies per plate are sufficient. Dilution tubes can be prepared directly before use according to exact needs. When a high degree of accuracy is required, dilution tubes should be prepared after autoclaving, since evaporation during sterilization is inevitable. Care should always be taken to wipe off the end of the pipettes before transferring to the next dilution tube. To minimize inaccuracies due to random error, important titers can be done with multiple determinations.

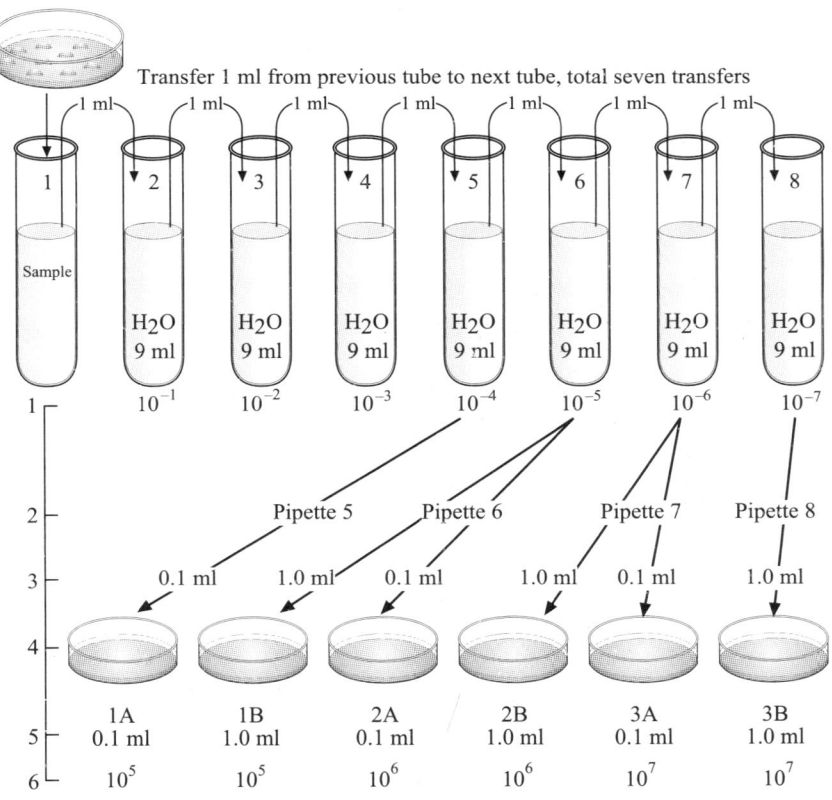

Figure 27.4 Serial dilution–agar plate procedure (1) Dilution, (2) Addition of sample, (3) Nutrient agar, 45°C mix by rotation of plate, (4) Dilution factor, (5) Incubation for 24 hr at 37°C and enumeration of number by colony counter

Materials

Culture: 24- to 48-hr nutrient broth culture of *E. coli*.

Media: 20 ml nutrient agar deep tubes and sterile 9 ml water blanks.

Equipment: Hot plate, water bath, thermometer, test tube rack, Bunsen burner, sterile 1 ml serological pipettes, sterile Petri dishes, colony counter, disinfectant solution, and glass marking pencil.

Procedure

1. Liquefy six agar deep tubes in an autoclave or by boiling. Cool the molten agar-tubes and maintain in a water bath at 45°C.
2. Label the *E. coli* culture tube as number 1 and the seven 9 ml water bath blanks as Numbers 2 through 8. Place the labeled tubes in a test-tube rack. Label the Petri dishes 1A, 1B, 2A, 2B, 3A, and 3B.

3. Mix the *E. coli* culture (tube #1) by rolling the tube between the palms of hands to ensure even dispersal of cells in the culture.
4. With a sterile pipette, aseptically transfer 1 ml from the bacterial suspension tube #1 to water blank tube #2. The culture has been diluted 10 times to 10^{-1}.
5. Mix tube #2 and with a fresh pipette, transfer 1 ml to tube #3. The culture has been diluted 100 times to 10^{-2}.
6. Mix tube #3 and with a fresh pipette, transfer 1 ml to tube #4. The culture has been diluted 1,000 times to 10^{-3}.
7. Mix tube #4 and with a fresh pipette, transfer 1 ml to tube #5. The culture has been diluted 10,000 times to 10^{-4}.
8. Mix tube #5 and with a fresh pipette, transfer 0.1 ml of this suspension to Plate 1A. Return the pipette to tube #5 and transfer 1 ml to tube #6. The culture has been diluted 100,000 times to 10^{-5}.
9. Mix tube #6 and with a fresh pipette, transfer 1 ml of this suspension to Plate 1B. Return the pipette to tube #6 and transfer 0.1 ml to plate 2A. Return the pipette to tube #6 and transfer 1 ml to tube #7. The culture has been diluted 1,000,000 times to 10^{-6}.
10. Mix tube #7 and with a fresh pipette, transfer 1 ml of this suspension to Plate 2B. Return the pipette to tube #7 and transfer 0.1 ml to plate 3A. Return the pipette to tube #7 and transfer 1 ml to tube #8. The culture has been diluted 10,000,000 times to 10^{-7}.
11. Mix tube # 8 and with a fresh pipette, transfer 1 ml of this suspension to Plate 3B. The dilution procedure is now complete.
12. Check the temperature of the molten agar medium to be sure the temperature is 45°C. Remove a tube from the water bath and wipe the outside surface dry with a paper towel. Using a sterile technique, pour the agar into Plate 1A and rotate the plate gently to ensure uniform distribution of the cells in the medium.
13. Repeat step 12 for the addition of molten nutrient agar to Plates 1B, 2A, 2B, 3A, and 3B.
14. Once the agar has solidified, incubate the plates in an inverted position for 24 hours at 37°C.
15. Identify the colony types and using a colony counter, observe all the colonies on the plate. Count only plates containing between 30 and 300 colonies. The number of organisms per ml of original culture is calculated by multiplying the number of colonies counted by the dilution factor.

Number of cells per ml = number of colonies × dilution factor

Method-2: Sterile plating and streaking techniques

There are many methods for obtaining a good streak plate, and each method requires some practice (Figure 27.5). To obtain single isolated colonies of a bacterial strain, either a wire loop or flat wooden toothpicks are used. As a general rule, a strain should go through at least two steps of single colony isolations before being stored. Strains which have been resuscitated from storage or have been received from another investigator should not be used unless they are re-purified and tested on single colonies.

Materials

Inoculating loops; LB agar plates; Bacterial culture; YEPD plates.

Bunsen burner; Incubator at 30°C; Incubator at 37°C.

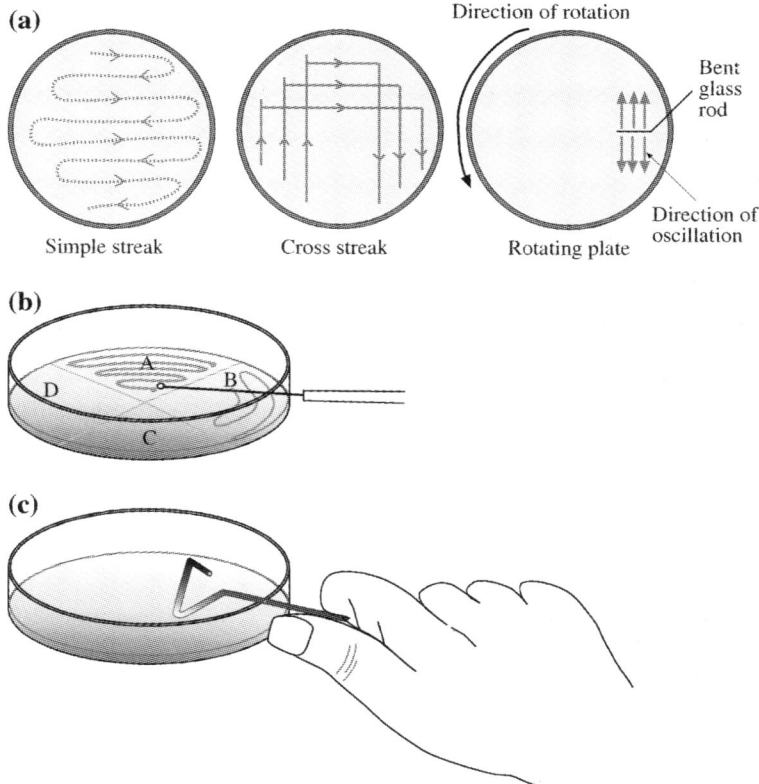

Figure 27.5 (a) Different types of streaking of cell suspension, (b) Quadrate type of streaking of a cell suspension for single colonies, and (c) Spreading of cell suspension over the surface of an agar medium in a plate

Procedure

1. Label LB and YEPD agar plates with name, date, and the name of the source culture used.
2. Use the quadrant streak method to prepare two streak plates (LB agar and YEPD agar). Repeat the procedure using the continuous streak method. Be sure to label each plate.

Quadrant streak method

a. Draw and number quadrants on the outside bottom of an agar Petri plate.
b. Flame the inoculating loop and be sure to allow the loop to cool before entering the broth culture.
c. Allow the loop to touch the surface of the agar lightly and slide gently over the surface in one quadrant in a continuous streaking motion. Use Petri dish cover to protect the agar surface and to prevent contaminants from falling onto the medium. Avoid digging the loop into the agar. Use reflected light to see where you have streaked and inoculated the plate. These areas will appear as faint scratch marks. This will allow you to better position your inoculations.

d. Flame the loop, allow to cool, cross-streak the previous area to produce a second quadrant.
e. Always sterilize the loop after inoculating each section of the plate; this will kill any cells adhering to the loop and prevent contamination of the next inoculation.
f. Repeat the procedure for each succeeding new area, until all four quadrants on the plate are inoculated.
g. Flame the loop when streaking is finished.

Continuous streak method

1. Take a small amount of culture inoculum on the loop and spread it in a single, continuous, back and forth motion over one-half of the plate.
2. Without flaming the inoculating loop, and using the same face of the loop, turn the plate 180° and continue the streaking procedure as you did in the initial area.
3. Prepare a control or sham inoculation in which only the sterilized inoculating loop is used without culture to prepare a streak plate by either streaking procedure. The control plate will be a good indicator of the aseptic technique, as nothing should grow on such a plate.
4. Invert the plates to prevent water condensation from spreading bacteria over the agar surface and place the plates in an incubator for 24–48 hours.
5. Observe well-isolated colonies and characterize them under microscopy for their cell morphology, etc.

Method-3: Spread-plate

The spread-plate technique requires that a previously diluted mixture of microorganisms be used. During inoculation, the cells are spread over the surface of a solid agar medium with a sterile, L-shaped bent rod (Figure 27.5).

Procedure

1. Dip the bent glass rod into the 95% ethyl alcohol to cover the lower, bent portion.
2. With a sterile loop, place a loopful of live culture in the center of the labeled nutrient agar plate and replace the cover.
3. Remove the glass rod from the beaker and pass it through the Bunsen burner flame, with the bent portion of the rod pointing downward. Allow the alcohol to burn off the rod completely. Cool the rod for 10 to 15 seconds.
4. While turning the Petri dish on a flat surface, lightly touch the sterile bent rod to the surface of the agar, and move it back and forth. This will spread the culture over the agar surface.
5. Replace the cover immediately and incubate.

Method-4: Replica plating

Replica plating is a convenient way to test colonies for their ability to grow under different conditions. Bacterial colonies are transferred from one plate to another in a way that maintains the original pattern of colonies (Figure 27.6).
1. Secure sterile velvet with a metal ring onto a replica block.
2. Press a master plate of well-spread colonies tightly onto the velvet.
3. Press new plates, oriented like the master plate, tightly onto the imprinted velvet to transfer colonies. Up to 10 plates/velvet can be replica plated.

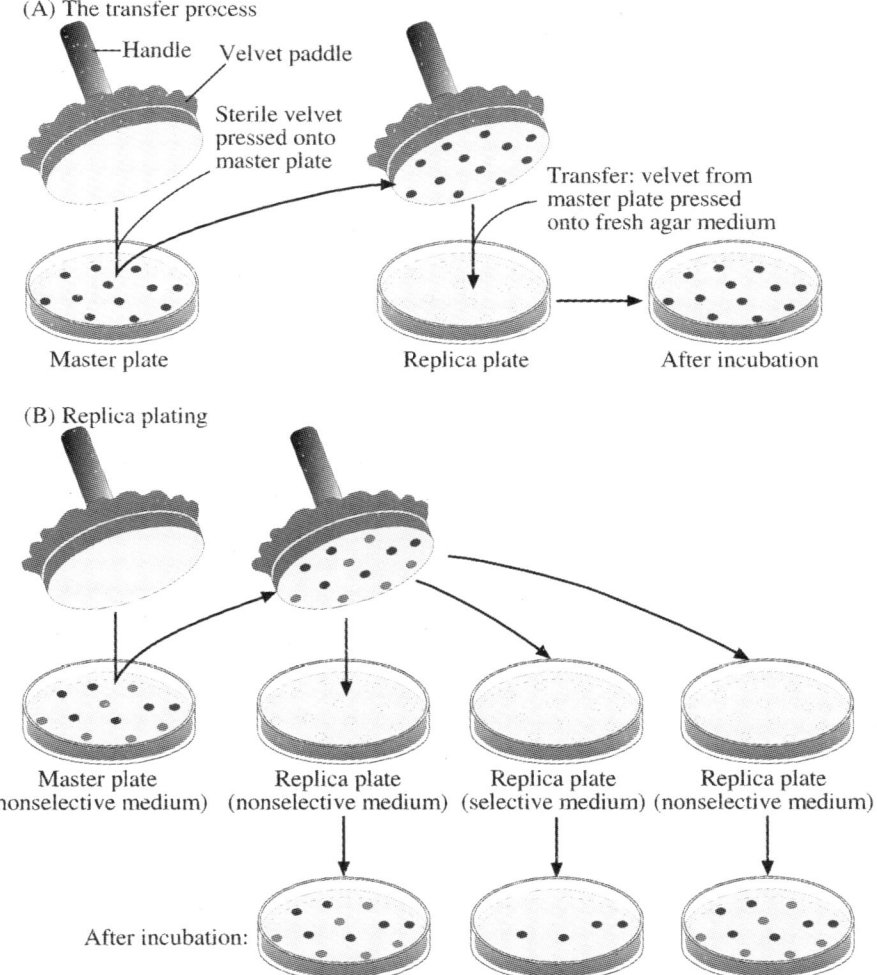

Figure 27.6 *Replica plating: (A) In the transfer process, a velvet-covered disk is pressed onto the surface of a master plate in order to transfer cells from colonies on that plate to a second medium, (B) For the detection of mutants, cells are transferred onto either non-selective (all form colonies similar to master plate) or selective medium (e.g. one spread with phages, mutant cells, T1–r)*

V CULTURE CHARACTERISTICS OF MICROORGANISMS

When microorganisms are grown on a variety of media, they exhibit differences in the macroscopic appearance of their growth. These differences, called **cultural characteristics**, are used as the basis for separating microorganisms into taxonomic groups. The cultural characteristics for all known microorganisms are contained in *Bergey's Manual of Systematic Bacteriology*.

802 MOLECULAR BIOTECHNOLOGY: PRINCIPLES AND PRACTICES

They are characterized by culturing the organisms on nutrient agar slants and plates, in nutrient broth, and in nutrient gelatin. The patterns of growth to be considered in each of these media are described below, and some are illustrated in Figure 27.7.

A. NUTRIENT

The solid medium may be liquefied by the enzymatic action of gelatinase. Liquefaction occurs in a variety of patterns (Figure 27.7a):
1. *Crateriform*: Liquefied surface area is saucer-shaped.

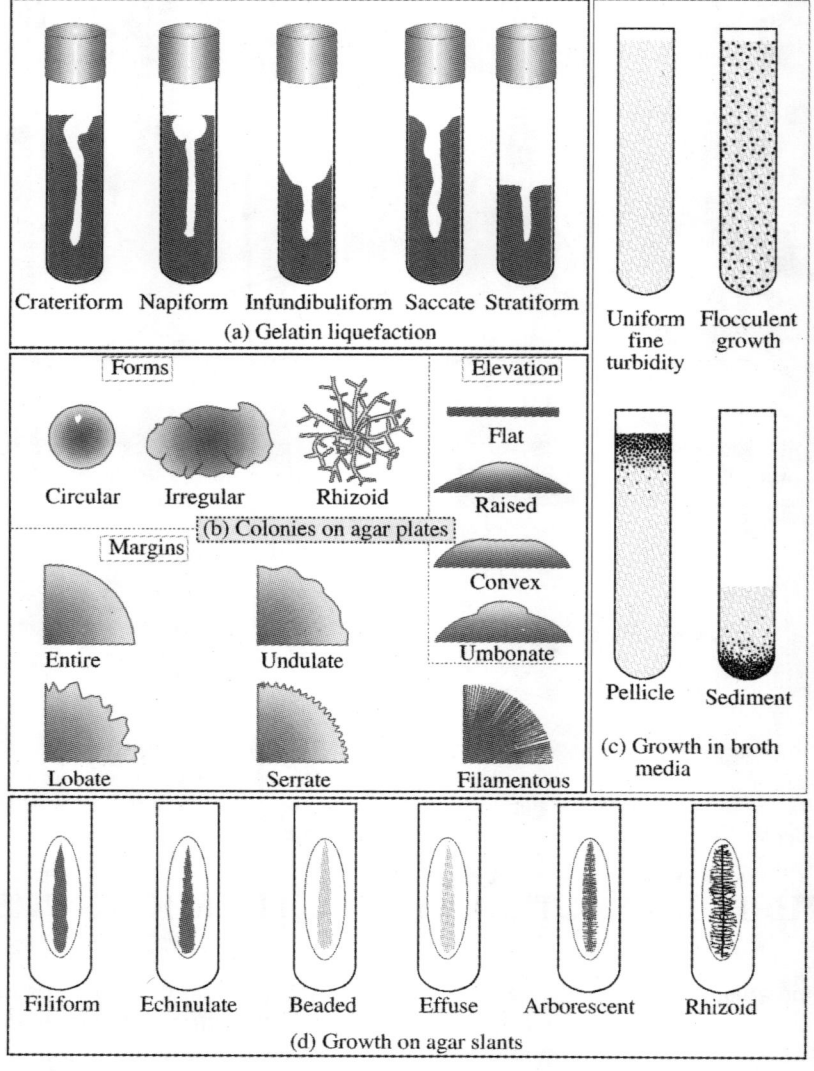

Figure 27.7 *Cultural characteristics of bacteria*

2. ***Napiform***: Bulbous-shaped liquefaction at surface.
3. ***Infundibuliform***: Funnel-shaped liquefaction.
4. ***Saccate***: Elongated, tubular liquefaction.
5. ***Stratiform***: Complete liquefaction of the upper half of the medium.

B. NUTRIENT AGAR PLATES

These demonstrate well-isolated colonies and are evaluated in the following manner (Figure 26.7b):
1. *Size*: Pinpoint, small, moderate, or large.
2. *Pigmentation*: Color of colony.
3. *Form*: The shape of the colony is described as follows:
    a. ***Circular***: Unbroken peripheral edge.
    b. ***Irregular***: Indented peripheral edge.
    c. ***Rhizoid***: Root-like spreading growth.
4. *Margin*: The appearance of the outer edge of the colony is described as follows:
    a. ***Entire***: Sharply defined, even.
    b. ***Lobate***: Marked indentations.
    c. ***Undulate***: Wavy indentations.
    d. ***Serrate***: Tooth-like appearance.
    e. ***Filamentous***: Thread-like, spreading edge.
5. *Elevation*: The degree to which colony growth is raised on the agar surface is described as follows:
    a. ***Flat***: Elevation not discernible.
    b. ***Raised***: Slightly elevated.
    c. ***Convex***: Dome-shaped elevation.
    d. ***Umbonate***: Raised, with elevated convex central region.

C. NUTRIENT BROTH CULTURES

These are evaluated as to the distribution and appearance of the growth (Figure 27.7c):
1. ***Uniform fine turbidity***: Finely dispersed growth throughout.
2. ***Flocculation***: Flaky aggregates dispersed throughout.
3. ***Pellicle***: Thick, pad-like growth on surface.
4. ***Sediment***: Concentration of growth at the bottom of broth culture, may be granular, flaky, or flocculent.

D. NUTRIENT AGAR SLANTS

These have a single straight line of inoculation on the surface and are evaluated in the following manner:
1. *Abundance of growth*: The extent of growth is designated as none, slight, moderate, or large.
2. *Pigmentation*: Chromogenic microorganisms may produce intracellular pigments that are responsible for the coloration of the organisms as seen in surface colonies. Other

organisms produce extracellular soluble pigments that are excreted into the medium and that also produce a color. Most organisms, however, are non-chromogenic and will appear white to gray.

3. *Optical characteristics*: Optical characteristics may be evaluated on the basis of the amount of light transmitted through the growth. These characteristics are described as **opaque, translucent**, or **transparent**.
4. *Form*: The appearance of the single-line streak of growth on the agar surface is designated as (Figure 27.7d):
   a. **Filiform**: Continuous, thread-like growth with smooth edges.
   b. **Echinulate**: Continuous, thread-like growth with irregular edges.
   c. **Beaded**: Non-confluent to semi-confluent colonies.
   d. **Effuse**: Thin, spreading growth.
   e. **Arborescent**: Bacterial colonies appear tree-like in size and form
   f. **Rhizoid**: Root-like growth.

Materials

Cultures: 24 hour nutrient broth cultures (e.g. *Pseudomonas aeruginosa*, *Bacillus cereus*, *Micrococcus luteus*, *Escherichia coli*, and *Staphylococcus aureus*).

Media: Nutrient agar plates, nutrient agar slants, nutrient broth tubes, and nutrient gelatin tubes.

Equipment: Bunsen burner, inoculating loop and needle, and glassware marking pencil.

Procedure

1. Using sterile techniques, inoculate each of the appropriately labeled media listed below in the following manner:
   a. **Nutrient agar plates**: With a sterile loop, prepare a streak-plate inoculation of each of the cultures for the isolation of discrete colonies.
   b. **Nutrient agar slants**: With a sterile needle, make a single-line streak of each of the cultures provided, starting at the butt and drawing the needle up the center of the slanted agar surface.
   c. **Nutrient broth**: Using a sterile loop, inoculate each organism into a tube of nutrient broth. Shake the loop a few times to dislodge the inoculum.
   d. **Nutrient gelatin**: Using a sterile needle, prepare a stab inoculation of each of the cultures provided.
2. Incubate all cultures at 37°C for 24 to 48 hours.

VI BACTERIAL GROWTH

Bacteria reproduce by an asexual process called **binary fission**. In this sequence of events, the genome (chromosome) replicates, the cell elongates, and the plasma membrane pinches inward at the center of the cell. When the nuclear material has been evenly distributed, the cell wall thickens and grows inward to separate the dividing cell. Once the cell division is complete, bacteria grow and develop the features that make each species unique. Bacteria can be grown on both a liquid and semisolid medium.

A. TYPES OF CULTURE

1. *Liquid batch cultures.* Cultures in liquid media to which no fresh nutrient is added during growth.
2. *Agar slope* (or slant) *cultures.* Test tubes or small bottles containing about 5 ml of solid medium dissolved and allowed to cool in a sloping position. The inoculum is either spread over the surface of the medium or applied in a thin streak using a wire loop.
3. *Stab cultures.* Tubes or bottles containing an agar or gelatin medium are allowed to solidify in the upright position. The medium is inoculated by plunging a long straight wire, charged with inoculum, vertically into the center of the tube.
4. *Semisolid culture.* Cultures grown in a medium containing sufficient agar (0.02–0.3%) to increase the viscosity of the medium, but insufficient to solidify the medium completely.
5. *Shake culture.* Test tubes or bottles containing a solid medium are used. The medium is dissolved, cooled to 45°C, inoculated, mixed well by rotating the tube between the hands, and allowed to solidify in the upright position.
6. *Plate cultures.*
    (a) Streak plates. Streak plates are used when well-isolated colonies are required either for the separation of mixed cultures or for the study of colonial form. In either case, the solid medium is covered with inoculum by a suitable streaking technique.
    (b) Pour plates. The inoculum is added to the tube of molten medium (at 45°C) as for shake tubes and mixed well before being poured into the plate. When isolated colonies are required by this method, as in the study of colonial form, it is advisable to inoculate and pour several tubes in a dilution series. When a quantitative technique is required, as in the estimation of viable numbers, a known dilution series is prepared in a suitable diluent and the inoculum is placed in the Petri dish and not in the tube of molten medium.

B. GROWTH IN LIQUID MEDIA

Method-1: Growing an overnight culture
1. Transfer 5 ml liquid medium into a sterile 16- or 18-mm culture tube.
2. Inoculate with a single bacterial colony on an inoculating loop, by dipping and shaking the loop in the medium.
3. Cap the tube and grow at 37°C to saturation (~6 hr) in a shaker or on a roller drum, 60 rpm. A freshly saturated culture can contain $1-2 \times 10^9$ cells/ml.

Method-2: Growing larger cultures
1. Dilute overnight cultures 1:100 in an Erlenmeyer or baffle flask that it is ≥20 times the volume of the culture.
2. Grow at 37°C with vigorous agitation, ~300 rpm.

VII BACTERIAL GROWTH CURVE

Bacterial population growth studies require inoculation of viable cells into a sterile broth medium and incubation of the culture under optimum temperature, pH, and gaseous conditions.

Under these conditions, the cells will reproduce rapidly and the dynamics of the microbial growth can be charted by means of a population growth curve. The growth curve can be used to explain the stages of the growth cycle. It also facilitates measurement of cell numbers and the rate of growth of a particular organism under standardized conditions as expressed by its *generation time* (the time required for a microbial population to double, *doubling time*).

A. BACTERIAL GROWTH CHARACTERISTICS

The development of a culture results in an increase in the bacterial numbers and in the total amount of protoplasm present. Bacteria transformed to a new and favorable medium grow, multiply and soon reach a maximal population. When an overgrown culture is diluted into a fresh medium, an initial **lag** phase is observed, giving way to the **log** phase. It ends with the **stationary** and **decline** phases. The cell sizes during the life cycle of the culture show that during these phases, they change in their size and shape. Initially, the cells will be smaller; during the lag phase they grow larger; they are the largest during the log phase; and then become smaller during the late-log and stationary phases. The interpretation of these observations is that bacteria pass through a series of stages in its life history. The life cycle of bacteria could be studied from birth to death in the same way that we might study the life cycle of a vertebrate organism. The cell sizes described above are not in balanced growth. In a fixed volume of nutrient media, the properties of cells will vary with the size and composition changing with time. When a cell is in balanced growth, the cell properties are constant and time-invariant.

Figure 27.8 shows the pattern of growth in a liquid medium in a closed environment. Note that two curves are shown – one for the **total number** of bacteria, including all those that have died. In practice, this is an easier number to determine. A more useful graph to study, though, is the number of living bacteria, known as the **viable count**. This curve has four distinct phases.

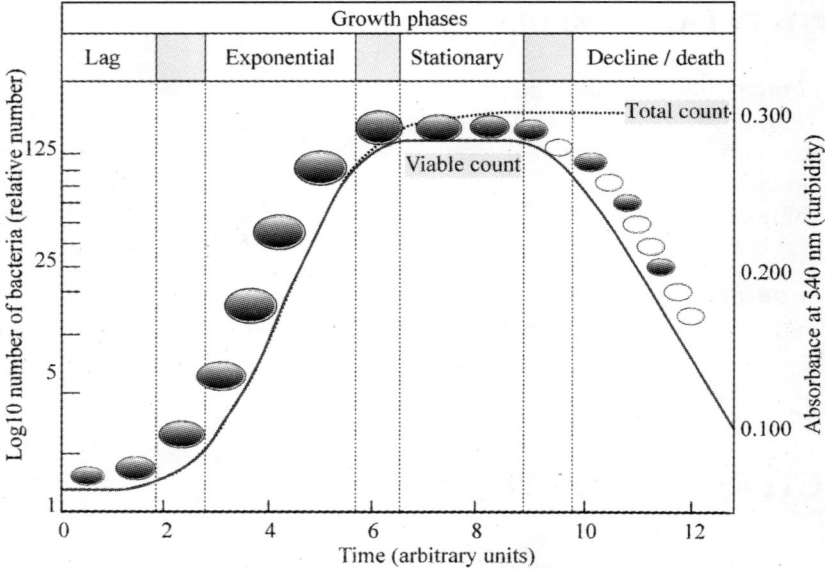

Figure 27.8 *Typical growth curve for a bacteria following inoculation*

1. **Lag phase**: This phase encompasses the first few hours of the curve. During this phase, the organisms adjust themselves to their new environment. The bacteria may, for example, be synthesizing new enzymes to digest the particular range of nutrients available in the new medium. Cellular metabolism accelerates, resulting in rapid biosynthesis of cellular macromolecules, in preparation for the next phase of the cycle. Although the cells are increasing in size, there is no cell division and therefore no increase in numbers.
2. **Logarithmic (log) phase**: When the bacteria have adjusted, they grow well and multiply rapidly by binary fission, entering the phase known as the ***log phase/exponential phase***. This phase is further subdivided as early-, middle-, and late-log phases, merely to describe harvested cells irrespective of their physiological state. During this phase, the doubling time is constant and at its shortest. The length of the log phase varies, depending on the organisms and the composition of the medium.
3. **Stationary phase**: As the numbers increase (1×10^8 to 1×10^{10} cells/ml), the quantities of toxic and waste products of metabolism increase and food supply decreases. Eventually, the growth of the colony begins to slow down and doubling time starts to increase. These factors combined together end the exponential phase and the rate of increase of organisms remains constant for a while. It then starts to enter the **stationary phase**. The rate of production of new cells is slower and may cease altogether. Any increase in the number of cells is offset by the death of other cells, so that the number of living cells remains constant. This phase is a result of several factors, including exhaustion of essential nutrients, accumulation of toxic waste products such as alcohol, and possibly, if the bacteria are aerobic, shortage of oxygen. Changes in pH sometimes occur, which also slow down growth.
4. **Decline or death phase**: Because of the continuing depletion of nutrients and the build-up of metabolic wastes, the microorganisms die at a rapid and uniform rate. When the medium is no longer conducive to growth, the culture enters the ***phase of decline (death phase)***. The death rate increases and becomes greater than the rate of multiplication. Eventually, the culture dies.

The time for these phases depends on the strain of an organism and the cultural environment (e.g. temperature, pH, aeration, nutrients, osmotic pressure, etc.). Construction of a complete bacterial growth curve requires that aliquots of a 24 hour shake-flask culture be measured for population period.

B. DETERMINATION OF EXPONENTIAL (LOGARITHMIC) GROWTH

When a bacterial culture growing in unlimited medium is kept below a given concentration by dilution at suitable intervals, the culture grows continuously and exponentially. Growing bacteria exhibit exponential or logarithmic growth kinetics until a point of saturation of the culture is reached. During this time, the increase in the number of bacteria (N) per unit time (t) is proportional to the number of bacteria present in the culture. If we plot the number of cells per volume against time, a straight line is produced on a semi-logarithmic paper. The growth curve is specified by:

$$\frac{dN}{dt} = kN \qquad (1)$$

where k is the growth constant; on integration this yields

$$\log(N) = \log(N_0) + Kt \qquad (2)$$

where $K=k/2.3$ and $N_0=$ the number of bacteria present when $t = 0$. If we plot log (N) or log (OD) versus the time (t), then the slope $= K$. The average properties of the culture are constant with time. The **extensive** properties of the culture (the amount and the number) increase. The **intensive** properties (the average cell size, the RNA per cell, the DNA per cell, and so forth) remain constant and invariant with time. In practice, a culture can be kept growing exponentially for many hours to ensure that it is in a constant state of exponential growth. Such an exponentially growing culture can be said to be in balanced growth.

Often we are interested in the generation time of the culture – the time required for the cells to double in number. In this case $N = 2N_0$. From Equation 2 we now have $\log 2 = Kt$, the generation time $= t = 0.3/K$. We can thus easily determine the generation time of a culture from a plot of the logarithm of the optical density versus the time (Figure 27.9).

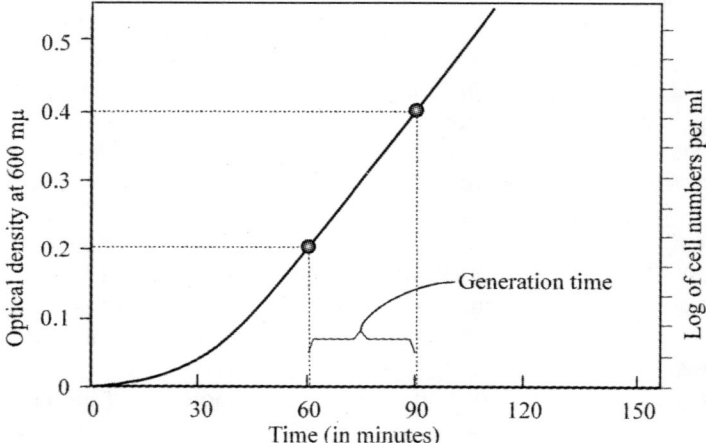

Figure 27.9 *Indirect method of determining generation time*

C. CULTURE CONDITIONS

Some of the incubation conditions that can be manipulated to our benefit are:
1. **Solid versus liquid media**: In the beginning of the experiment itself, we should decide the choice between solid and liquid media. A solid medium often has the same ingredients as a liquid medium except for the addition of agar. The major advantages of broth cultures is that virtually any volume of media may be prepared – from 1 ml to industrial fermenters containing thousand litres. The growth rate of microorganisms is usually quicker in liquid media. Solid media have the advantage of isolating a pure culture from a mixed culture. Organisms also remain viable on solid media for longer periods, making storage by refrigeration easy. A given species of microorganism always produces highly characteristic colonies on a particular growth medium under a specific set of growth conditions. This helps to tell whether a plated population is pure or a mixture and identification of a strain based on the colony characteristics is easy.

2. **Temperature**: Bacteria inhabit almost every environment of Earth because different species can tolerate the myriad conditions found on the planet. The growth of microorganisms can be considered as a series of chemical reactions and the rate at which the individual chemical reactions proceed is a function of temperature. It is important to remember that it is the effect of temperature on only growth and not on survival that is being considered. Many microorganisms will survive extreme temperatures – from subzero to above boiling temperature. The overall limits from growth and division of living cells are between about 110°C and –10°C. However, no single cell will grow over the whole range; most cells have individual optimum temperatures. Organisms are separated into three groups on the basis of their minimum, maximum and optimum temperatures – the so-called cardinal temperatures. The three groups are **thermophiles** (40°C to 110°C), **mesophiles** (10°C to 45°C), and **psychrophiles** (<0°C to 20°C). Most bacteria appear to be mesophiles. This is especially true of pathogenic bacteria growing on mammalian systems.
3. **Effect of pH on growth**: The majority of bacteria grow best under neutral pH conditions (the internal pH of bacteria is about 7.0). Bacteria can also tolerate acidic conditions as low as pH 6.5 (acidophiles) and alkaline conditions as high as pH 7.5. The growth of all microorganisms is affected by the initial pH, which in nature may vary from pH 1.0 in acid mine waters to pH 11.0 in ammonia-rich soils. One of the problems we must consider when preparing laboratory media is not only the initial pH but how to maintain a constant pH. Many organisms produce acidic end products and thus an unbuffered culture medium having an initial alkaline or neutral pH could quickly fall following microbial growth. The usual remedy is to buffer the medium.
4. **Light**: The provision of light is really only important for the growth of photosynthetic microorganisms. In order to obtain growth of the different types of phototrophic microorganisms, light of the correct wavelength has to be provided. In natural circumstances, photosynthetic organisms are subjected to cyclic day and night periods, but in the laboratory, the growth rate can be increased by providing a constant light source.
5. **Carbondioxide**: Atmospheric CO_2 concentration is adequate for the needs of heterotrophic microorganisms but autotrophs grow much more rapidly in higher concentrations of 1 to 5% CO_2. This can be provided simply in the laboratory by adding CO_2 to the air or, in the case of anaerobes, providing a nitrogen–carbon dioxide mixture.
6. **Oxygen**: Before we proceed, we have to differentiate clearly between the element oxygen and molecular oxygen (O_2). All living systems require elemental oxygen (about 20% dry weight) and this can be supplied in a variety of ways, for example as CO_2, organic compounds, mineral salts, and water. The relationship of microorganisms to molecular oxygen, is however, very different and allows us to distinguish between five major groups: (1) Obligate aerobes: organisms using molecular oxygen as a terminal electron acceptor (aerobic), they cannot ferment; (2) Facultative anaerobes: organisms which can respire aerobically or, in the absence of molecular oxygen, will grow by alternative means – generally by fermentation; (3) Obligate anaerobes: organisms which cannot grow in the presence of and may be poisoned by molecular oxygen; (4) Aerotolerant organisms: organisms which cannot use molecular oxygen but which are not directly affected by it; (5) Micro-aerophilic organisms: organisms which use molecular oxygen but can only tolerate the gas in low concentrations.

Method-1: Bacterial growth curves

In the following experiment, the growth rate of two *E. coli* strains will be determined. Both optical density reading and viable cell counts will be determined at different times from growing cultures. In addition to demonstrating the relationship between turbidity and number of viable cells, this experiment will also demonstrate the techniques of dilution and plating.

Materials

Culture: Overnight growth of bacterial culture (CSH51 and CSH61).

Media: 18, test tubes with either 0.5 ml broth each (for blank); all broth tubes should contain the identical broth as the growth medium; 16 LB plates.

Equipment: 36 dilution tubes; 42, 0.1 ml and 10, 1 ml pipettes for titering; 18, 1 ml pipettes for withdrawing samples; 125 ml or 250 ml Erlenmeyer flask; Shaking water bath at 37°C; 18 Pasteur pipettes; Spectrophotometer set at 550 mµ; Flasks containing side arms which conveniently fit into Klett reading devices.

Procedure

1. Subculture overnight each of the two strains to be used (e.g. CSH51 and CSH61). Dilute each fresh overnight culture 1:50 into pre-warmed rich broth (transfer 0.4 ml into 20 ml broth in a 250 ml Erlenmeyer flask). Shake vigorously in a 37°C room or in a shaking water bath.
2. Collect the culture samples by withdrawing 0.5 ml with a 1 ml pipette and immediately diluting 1:1 by delivering the sample into 0.5 ml chilled broth in a test tube. This should be done with as little disruption of the shaking of the cells as possible. The flask should be maintained at 37°C at all times throughout these manipulations. The 1 ml sample should be read at 550 nm in a spectrophotometer as soon as possible. If samples are to be further diluted for viable cell counts, 0.1 ml of this should be withdrawn directly before reading the optical density.
3. For viable cell counts, dilutions and plating should be done promptly. Both 10^{-5} and 10^{-6} dilutions should be plated for points after the first hour, but a 10^{-5} dilution is sufficient for points during the first hour.
4. Beginning 30 min. after the subculture is made, take points for optical density readings every 10 min. and time points for viable cell counts every 20 min. (thus every other optical density sample will also be diluted for plating).
5. Use as a blank an aliquot of the broth used for growth and for dilution.
6. When the optical density reading reaches 0.6, use 1:5 dilutions for the next samples, since readings much above this are less reliable due to the small amount of light transmitted. Continue to take points for 3 hours, or until the optical density has leveled off.
7. Construct a plot of optical density *vs.* time, and also the log of the cell density *vs.* time (on semi-log paper). Compute the generation time of each strain under these conditions. Incubate the plates at 37°C.
8. Count the colonies on each of the titer plates, and construct curves using these cell numbers. Compare the curves, and also the relationships between viable cell counts and optical densities at a given wavelength.

Method-2: Determination of generation time

Generation time (**cell doubling**) can be determined by indirect and direct methods by using the data on the growth curve. Indirect determination is made by simple extrapolation from the log phase as illustrated in Figure 27.9. Select two points on the optical density scale, such as 0.2 and 0.4 that represent a doubling of turbidity. Using a ruler, extrapolate by drawing a line between each of the selected optical densities on the ordinate and the plotted line of the growth curve. Then draw perpendicular lines from these end points on the plotted line of the growth curve to their respective time intervals on the abscissa. With this information, determine the generation time as follows:

$$GT = t(OD\,0.4) - t(OD\,0.2)$$

$$GT = 90\,min. - 60\,min. = 30\,min.$$

The direct method uses the log of cell number scale on the growth curve and the following formula:

$$GT = \frac{t \log 2}{\log b - \log B}$$

where GT = generation time;
 B = number of bacterial cells at some point during the log phase;
 b = number of bacterial cells at a second point of the log phase;
 t = time in hours or min. between B and b.

Materials

Cultures: 10- to 12 hour (log phase) brain–heart infusion broth culture of *E. coli*. Culture growth may be stopped at mid-log phase by immersion in an ice-water bath.

Media: 100 ml of brain–heart infusion in a 250 ml Erlenmeyer flask; 18 numbers of 99 ml sterile water blanks; and four 100 ml bottles of nutrient agar.

Equipments: 37°C water bath shaker incubator; Spectrophotometer; 13 x 100 mm cuvettes; Colony counter; 24 sterile Petri dishes, 1 ml and 10 ml sterile pipettes; Glassware marking pencil; Pipetting device, 1 litre beaker; and Bunsen burner.

Procedure

1. Separate the 18 numbers of 99 ml sterile water blanks into six groups each with three tubes and label each set as to time of inoculation (t_0, t_{30}, t_{60}, t_{90}, t_{120}, t_{150}) and the dilution to be effected in each water blank (10^{-2}, 10^{-4}, 10^{-6}).
2. Label six sets of four Petri dishes as to time of inoculation and dilution to be plated (10^{-4}, 10^{-5}, 10^{-6}, 10^{-7}).
3. Liquefy the four bottles of nutrient agar in a water bath. Cool and maintain at 45°C.
4. With a sterile pipette, add approximately 5 ml of the log phase *E. coli* culture to the flask containing 100 ml of brain–heart infusion broth. The approximate initial OD (t_0) should be 0.08 to 0.1 at 600 mµ.

5. After the t_0 OD has been determined, shake the culture flask and aseptically transfer 1 ml to the 99 ml water blank labeled t_0 10^{-2} and continue to dilute serially to 10^{-4} and 10^{-6}.
6. Place the culture flask in a water bath shaker set at 120 rpm at 37°C, and time for the required 30 min. intervals.
7. Shake the t_0 dilution bottle up and down. Plate the t_0 dilutions on the appropriately labeled t_0 plates. Aseptically pour 15 ml of the molten agar into each plate and mix by gentle rotation.
8. Thereafter, at each 30 min. interval, shake and aseptically transfer a 5 ml aliquot of the culture to a cuvette and determine its optical density. Also, aseptically transfer a 1 ml aliquot of the culture into the 10^{-2} water blank of the set labeled with the appropriate time, complete the serial dilution, and plate in the respectively labeled Petri dishes.
9. When the pour-plate cultures harden, incubate them in an inverted position for 24 hours at 37°C.
10. Perform cell counts on all plates. Cell counts are often referred to as colony-forming units (CFUs), because each single cell in the plate becomes visible as a colony which can then be counted.
11. Record the optical densities and corresponding cell counts in the chart.
12. On the semi-log paper, plot: optical densities on the ordinate and incubation times in the abscissa; log of the cell numbers on the ordinate and incubation times on the abscissa. On both graphs, connect the points with a ruler by drawing the best line between the plotted points. The log phase is represented by a straight-line portion.
13. Calculate the generation time for this culture by the direct method using the mathematical formula, and by the indirect method extrapolating from the OD scale on the plotted curve. Show calculations and record the generation time.

VIII CHEMICAL CONTROL OF MICROBIAL GROWTH

We have discussed in the earlier section (refer to Chapter 2 for more details), the ways in which microbial growth is inhibited by naturally occurring factors and how some of these may be intentionally applied to kill microorganisms (e.g. heat, irradiation, high solute concentration, etc). There are many instances where these methods of inhibiting microbial growth are unsuitable or too large. In such cases, chemical agents may be the answer. Antimicrobial agents can be either non-selective or selective in their action. If an agent selectively kills microbial cells without inhibiting host cells, it may become an important weapon for combating diseases. Antibiotics are antimicrobial agents produced by microorganisms, which can be tolerated by most larger eukaryotic organisms.

The antimicrobial properties of an antibiotic are related to its mode of action. Knowledge of how antibiotics inhibit or kill microbes is therefore important for establishing how the antibiotic should be used. Four different mechanisms can be distinguished in the way antibiotics affect microorganisms. (1) Inhibiting of cell wall synthesis; (2) influence on structure and permeability of the cell membrane; (3) inhibition of protein synthesis; (4) inhibition of nucleic acid synthesis (see Appendix 2).

IX ENZYME INDUCTION

Although bacteria possess a small and single chromosome, each cell is capable of synthesizing hundreds of different enzymes. Studies have shown that some enzymes, called **constitutive enzymes**, are synthesized at a constant rate regardless of the conditions in the cell's environment, whereas other enzymes, called **adaptive enzymes**, occur only when necessary, and are subject to regulatory mechanisms that are dependent on the environment. One such mechanism, **induction**, requires the presence of a substrate, the inducer, in the environment to initiate synthesis of its specific enzyme, called an inducible enzyme. An extensively studied inducible enzyme in *E. coli* is β*-galactosidase*, which acts on disaccharide lactose to yield the monosaccharides glucose and galactose. The gene for β-galactosidase is a member of a cluster of genes, called an **operon** that is involved in the metabolism of lactose (Figure 27.10). The member genes of the lactose operon function as a unit, all being transcribed only when the inducer, lactose, is present in the surrounding medium.

To illustrate gene β-galactosidase induction, two strains of *E. coli* are used: a prototrophic (wild type) strain (lactose-positive, *lac+*) and an auxotrophic (mutant) strain (lactose-negative, *lac–*), which carries a mutation in the gene for β-galactosidase as well as a mutation in the lactose operon regulatory gene. Both test strains will be grown in the following media: (1) Inorganic synthetic medium lacking an organic carbon and energy source that is required by the heterotrophic *E. coli*, (2) Inorganic synthetic medium plus glucose, which can be utilized by both strains as a carbon and energy source, and (3) Inorganic synthetic medium plus lactose, which can be utilized only by the prototrophic strain.

Orthonitrophenyl-β-D-galactoside (ONPG), a colorless analog of lactose, can serve as the substrate for the induction of β-galactosidase synthesis. As the inducer, it is hydrolyzed to galactose and a yellow nitrophenolate ion. Following a short incubation period, the growth in all the cultures will be determined by spectrophotometry. The induction of β-galactosidase synthesis and activity will be indicated by the appearance of a yellow color in the medium following the addition of ONPG, which occurs only in the presence of the nitrophenolate ion. The absence of this macroscopically visible color change indicates that enzyme induction in the lactose-negative strain did not occur.

Materials

Cultures: 25 ml inorganic synthetic broth suspensions of 12 hour nutrient agar cultures of a *lac+ E. coli* strain (ATCC e 23725) and a lactose-negative *E. coli* strain (ATCC e 23735) adjusted to an OD of 0.1 at 600 mµ.

Media: sterile 10% glucose, 10% lactose, and water.

Reagents: Toluene and orthonitrophenyl-β-D-galactoside (ONPG).

Equipment: 1 ml and 5 ml sterile pipettes, mechanical pipetting device, six sterile 13 × 100 mm test tubes, six sterile 25 ml Erlenmeyer flasks, Spectrophotometer, shaking water bath incubator, and glassware marking pencil.

Procedure

1. Label three test tubes and three sterile 25 ml Erlenmeyer flasks as "*lac+*" and the name of the substrate to be added (glucose, lactose, or water). Similarly label three sterile flasks "*lac–*" for each test organism.

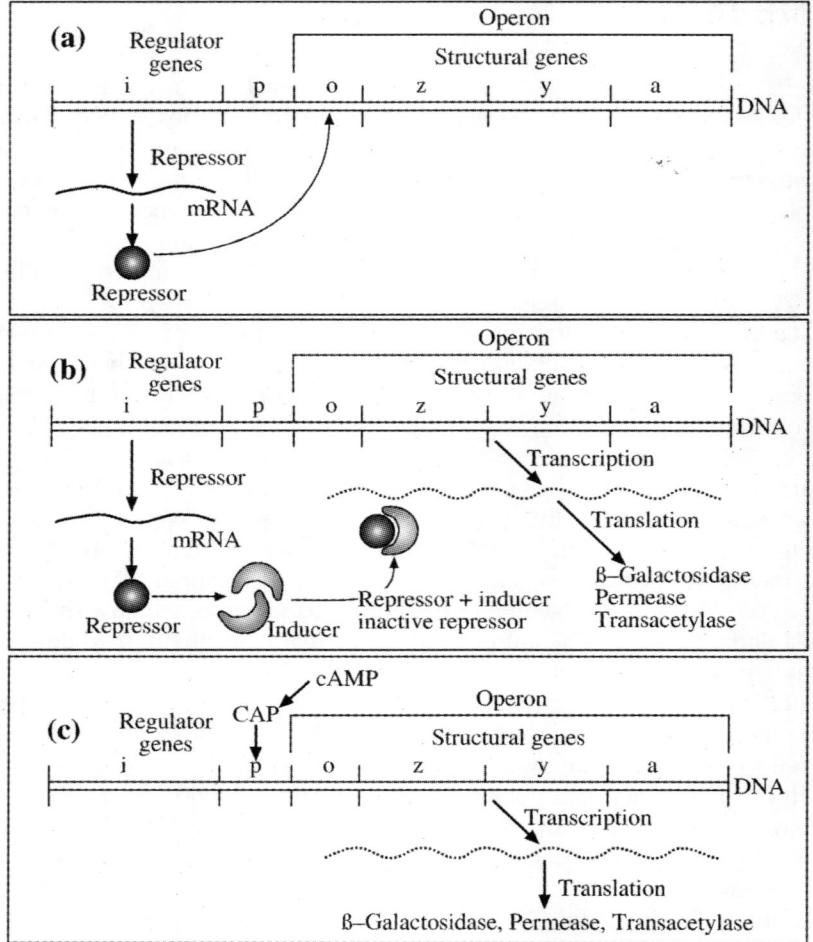

Figure 27.10 *The lactose (lac) operon. (a) Repressor of mRNA synthesis from lac operon: in the absence of inducer, the repressor binds to the operator (o) gene to prevent transcription. (b) Induction of mRNA synthesis: in the presence of inducer, the repressor binds to the inducer and become an inactive repressor, therefore transcription takes place. (c) Positive control of enzyme synthesis: Presence of cAMP activates the catabolite gene activator protein (CAP), which in turn activates transcription*

2. Using sterile 5 ml pipettes, aseptically transfer 5 ml of the *lac+* and *lac–* inorganic synthetic broth cultures to their respectively labeled test tubes.
3. Using a sterile 1 ml pipette, aseptically add 0.5 ml of the glucose and lactose solutions and 0.5 ml of sterile distilled water to the appropriately labeled tubes.
4. Determine the optical density (OD) of all cultures at a wavelength of 600 mµ.
5. Aseptically transfer each culture to its appropriately labeled flask. Incubate all flasks for 2 hours in a shaking water bath at 37°C and 100 strokes per min.
6. Following incubation, transfer all cultures back to their appropriately labeled test tubes.

7. Determine the OD for each culture at a wavelength of 600 mµ.
8. To each culture, add 5 drops of toluene and shake vigorously (toluene ruptures the cell, releasing intact enzymes).
9. To each culture, add 5 drops of ONPG solution. Incubate all cultures for 40 min. at 37°C. Observe all cultures for the development of yellow color.
10. Record the initial OD of each *lac+* and *lac− E. coli* broth culture at 600 mµ.

X STAINING THE BACTERIA

Although direct microscopic examination of living bacteria is possible using a phase contrast microscope, it is more common to kill and stain bacteria before examination. Stains are employed to (a) make organisms visible, (b) reveal their structure, (c) detect their chemical nature, and influence their growth and development. Dyes are synthetic chemical products of the aniline type. The chemical substance of bacterial cells that stains with the aniline dyes is almost exclusively protein. Whenever bacterial cells are stained with only one dye, the staining solution is termed a simple stain. Simple stains are usually employed because they are easily prepared, are less subject to deterioration, and give rapid results with only one application. Among the simple stains most frequently employed are Methylene blue, crystal violet, and dilute basic carbolfuchsin. Generally the entire cell is uniformly colored.

Table 27.2 *Types of bacterial staining techniques*

A. Simple staining: use of single stain
For visualization of morphological shape
cocci, bacilli, and spirilli
Arrangement
chains, clusters, pairs, and tetrads
B. Differential staining: Use of two contrasting stains
Separation into groups
Gram stain
Acid-fast stain
Visualization of structures
Flagella stain
Capsule stain
Spore stain
Nuclear stain

All microbiological staining procedures require preparation of smears prior to the execution of any of the specific staining techniques listed in Table 27.2. The different steps involved in preparation of microbial smears are as follows:

1. ***Preparation of glass slides:*** Clean slides are essential for the preparation of microbial smears. After cleaning the slides with detergent followed by 95% alcohol, dry the slides and place them on laboratory towels ready for use.

2. **Preparation of smears**: It is absolutely essential to avoid thick, dense smears. A good smear is one that, after it has dried, appears as a thin whitish layer or film. Those made from broth or solid cultures need variations in technique.
    a. **Broth culture**: One or two loopfuls of suspended cells should be applied directly to the glass slide with a sterile inoculating loop and spread evenly over an area of about 1.5 cm diameter.
    b. **Culture from a solid medium**: Organisms cultured in a solid medium produce thick, dense surface growth and are not amenable to direct transfer to the glass slide. These cultures must be diluted by placing a loopful of water on the slide, in which the cells will be emulsified. Suspension is accomplished by spreading the cells in a circular motion in the drop of water with the loop. The finished smear should occupy an area about 2 cm diameter and should appear as a semi-transparent, confluent, whitish film. Allow the smear to dry completely.
3. **Heat fixation**: Unless fixed on the glass slide, the bacterial smear will wash away during the staining procedure. This is avoided by heat fixation, during which the bacterial proteins are coagulated and fixed to the glass surface. Heat fixation is performed by the rapid passage of the air-dried smear two or three times over the flame of the Bunsen burner.

Method-1: Simple staining

In simple staining, the bacterial smear is stained with a single reagent. Basic stains with a positively charged chromogen are preferred, because bacterial nucleic acids and certain cell wall components carry a negative charge that strongly attracts and binds to the cationic chromogen. The purpose of simple staining generally is to elucidate the morphology and arrangement of bacterial cells. The most commonly used basic stains are methylene blue, crystal violet, and carbol fuchsin.

Materials

Cultures: 24-hour nutrient agar slant cultures of *E. coli* and *Bacillus cereus*, and a 24-hour nutrient broth culture of *Staphylococcus aureas*.

Reagents: Methylene blue, crystal violet, and carbol fuchsin.

Equipments: Bunsen burner, inoculating loop, staining tray, microscope, lens paper, bibulous paper, and glass slides.

Procedure

1. Prepare separate bacterial smears of the organisms following the procedure described earlier. All smears must be heat fixed prior to staining.
2. Place a slide on the staining tray and flood the smear with one of the indicated stains, using the appropriate exposure time for each: carbol fuchsin – 15 to 30 s; crystal violet – 20 to 60 s; methylene blue – 1 to 2 min.
3. Wash the smear with tap water to remove excess stain. During this step, hold the slide parallel to the stream of water.
4. Using bibulous paper, blot dry but do not wipe the slide.
5. Repeat this procedure for different strains and examine all stained slides under oil immersion.

Method-2: The gram stain

To bring out chemical differences existing within the cell or on its surface, special differential staining techniques have been introduced. Differential staining requires the use of at least three chemical reagents that are applied sequentially to a heat-fixed smear. The gram stain, usually employed for the classification of bacteria, is a differential stain having **primary** stain, **decolorizing** agent, and **counterstain** respectively). Gram positive bacteria possess a component, apparently in the cell wall, which complexes with crystal violet and iodine, the latter being relatively insoluble in alcohol. Gram negative bacteria either do not possess this component or the complex is more easily removed with alcohol, thus reflecting a difference in cell wall permeability. Before staining, all bacteria are colorless. After Gram staining, all positive bacteria are stained violet and Gram negative bacteria are stained red.

To demonstrate this special method of differentiating bacteria into two groups, namely, the Gram-positive (those retaining the blue color) and the Gram-negative (those which can be decolourized and counter-stained red) (Figure 27.11).

Materials

Cultures: 24 hr nutrient agar slant cultures of *E. coli, Staphylococcus aureus*, and *Bacillus cereus*.

Reagents: Primary (Basic) stain (crystal violet, 0.5% aqueous); Mordant (Lugol's iodine); Decoloriser (acetone-alcohol 50:50); Counterstain (Safranin 1% aqueous).

Equipment: Wire loop; Bunsen burner; Glass slides – scrupulously clean; Forceps; Staining rack set up over sink or dish; Distilled water; Blotting paper; Immersion oil and microscope with oil immersion lens.

Procedure

1. *Prepare a smear of bacteria on the slide*: Place a loopful or two of tap water on the center of a clean slide. Touch the wire loop lightly on a selected bacterial colony. Transfer the bacteria to the slide and gently mix with the water. Spread the bacteria over the slide, using the loop, to cover an area about 3 × 1 cm. It is important to achieve the correct thickness of the smear. Allow the smear to become perfectly dry in air.
2. *Fix the bacteria:* Holding the slide with forceps, pass it horizontally just over a yellow Bunsen flame three times. Fixing kills the bacteria by coagulating the cytoplasm and also makes them stick to the slide.
3. *Stain the bacteria*: Flood the slide with crystal violet stain. Leave for 30 seconds. This makes all bacteria violet.
4. Wash off with Lugol's iodine; flood with Lugol's iodine and leave for 30 seconds. Wash off the iodine with distilled water from a wash bottle. The iodine fixes the stain more permanently into the cells.
5. *Decolorizing*: Flood the slide with acetone-alcohol until no more color is seen to come off; immediately wash with water to prevent excessive decoloration. This decolorizes Gram negative bacteria. Gram positive bacteria stay violet.
6. *Counterstain*: Flood the slide with safranine and leave for 1 min. Wash off the stain with water. Gently dry the slide between sheets of clean blotting paper and finally air-dry. Safranine is described as a counterstain.

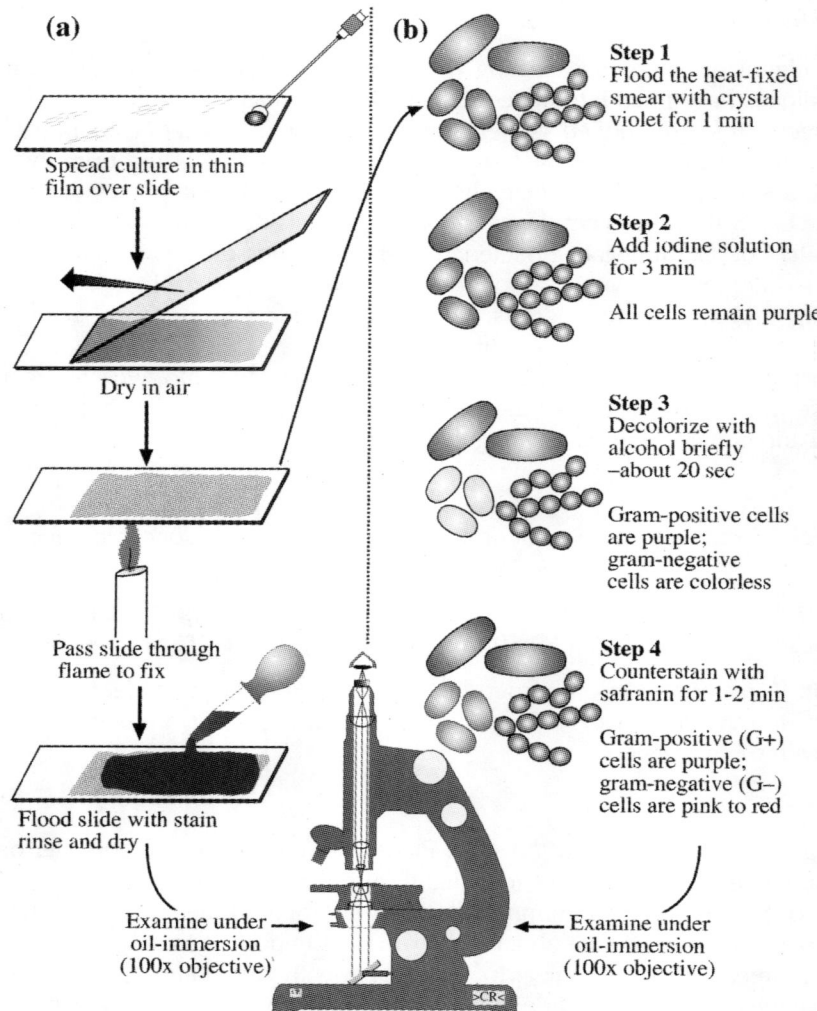

Figure 27.11 Bacterial staining procedures. (a) Steps in simple staining and (b) the Gram stain

7. Apply a drop of immersion oil and examine under the oil immersion lens.

Method-3: Negative staining

Negative staining requires the use of acidic stain such as India ink or nigrosin. The acidic stain with its negatively charged chromogen will not penetrate the cells because of the negative charge on the surface of the bacteria. Therefore, the unstained cells are easily discernible against the colored background (Figure 27.12). The practical application of negative staining is twofold: (1) since heat fixation is not required and the cells are not subjected to the distorting effects of chemicals and heat, their natural size and shape can be seen; (2) it is possible to observe bacteria that are difficult to stain, such as some spirilli.

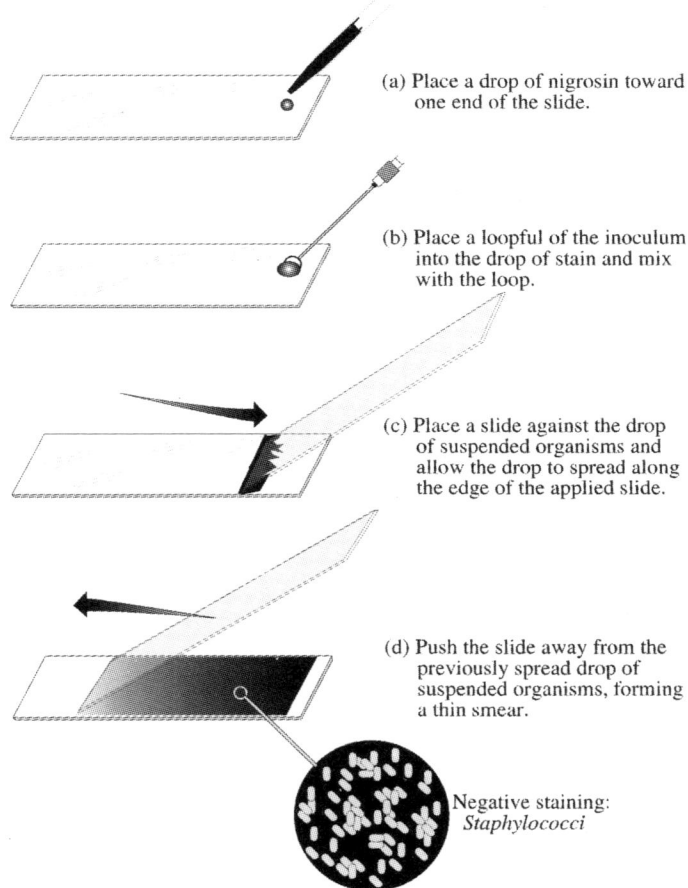

Figure 27.12 *Negative staining procedure*

Materials

Cultures: 24-hour nutrient agar slant cultures of *Micrococcus luteus, Bacillus cereus,* and *Aquaspirillum itersonii.*

Reagents: Nigrosin

Equipments: Inoculating loop; Bunsen burner; Glass slides – scrupulously clean; Forceps; Staining tray; Lens paper; Microscope.

Procedure

1. Place a small drop of nigrosin close to one end of a clean slide.
2. Using the sterile technique, place a loopful of inoculum from the *Micrococcus luteus* culture in the drop of nigrosin and mix.
3. With the edge of a second slide held at a 30° angle and placed in front of the bacterial suspension, push the mixture to form a thin smear.

4. Air dry. Do not heat fix the slide.
5. Repeat steps 1 to 4 for slide preparation of *Bacillus cereus* and *Aquaspirillum itersonii*.
6. Examine the slides under oil immersion.
7. Draw representative fields of microscopic observations.

Method-4: Staining of bacterial spores

Bacteria in the bacillus and Clostridium produce endospores, which are highly resistant to high temperature, lack of moisture and toxic chemicals. The endospores are also resistant to bacteriological stains. To stain them follow the procedure given below.

Procedure

1. Prepare a smear in the normal way, but heat fix very thoroughly by passing through a Bunsen flame 20 times.
2. Stain for 15 min. with a saturated aqueous solution of malachite green.
3. Wash gently with cold water for 10 seconds.
4. Counterstain with a 0.25% solution of safranine for 15 seconds.
5. Wash with water and blot dry.
6. Examine under the oil immersion objective.

XI MONITORING BACTERIAL CELL GROWTH

Two types of cell counts are possible, namely viable counts and total counts. The **viable count** is the total number of living cells only. The **total count** is the total number of cells, living and dead, and is often easier to measure. Different methods for measuring bacterial growth are summarized in Table 27.3.

Table 27.3 *Summary of methods for measuring bacterial growth*

Method	Applications	Expression of growth
Microscopic count	Enumeration of bacteria in vaccines and cultures	Number of cells per ml
Electronic enumeration	Same as for microscopic count	Number of cells per ml
Plate count	Enumeration of bacteria in milk, water, foods, soil, cultures, etc.	Colony-forming units per ml
Membrane filter	Same as plate count	Same as plate count
Turbidimetric measurement	Microbiological assay, estimation of cell crop in broth, cultures, or aqueous suspensions	Optical density (absorbance)
Nitrogen determination	Measurement of cell crop from heavy culture suspensions to be used for research in metabolism	Mg nitrogen per ml
Dry weight determination	Same as for nitrogen determination	Mg dry weight of cells per ml
Measurement of biochemical activity, e.g. acid production	Microbiological assays	Milliequivalents of acid per ml or per culture

A. TOTAL BACTERIAL CELL COUNT BY MEASURING TURBIDITY

More rapid and sensitive techniques of determining cell mass are based upon the fact that microbial cells scatter the light striking them. Because microbial cells in a population are of a roughly constant size, the amount of scattering is proportional to the concentration of cells present, within certain limits. Cell numbers can be measured with a spectrophotometer (Figure 27.13) since, up to an apparent optical density of about 1.5, the absorbence is proportional to cell density. The correspondence between absorbence and cell number can be measured for any particular strain in a particular medium.

This experiment will provide the basic knowledge needed to determine these parameters for a particular test strain of bacteria grown in a specific medium and temperature. If a suspension contains more cells, its **turbidity** or "cloudiness" will be greater. The cell mass is directly proportional to the optical density. The approach is to correlate the number of bacterial cells in a growing culture with the optical density (OD) at 450 nm of that culture and by following the changes in OD_{450} and cell number during growth, to determine the growth rate.

Materials

Exponential phase culture of bacterial strains, 50 ml of sterile LB medium in a 250 ml culture flask, LB agar plates, dilution tubes (sterile), sterile M9 salts for making dilutions (6 g/l

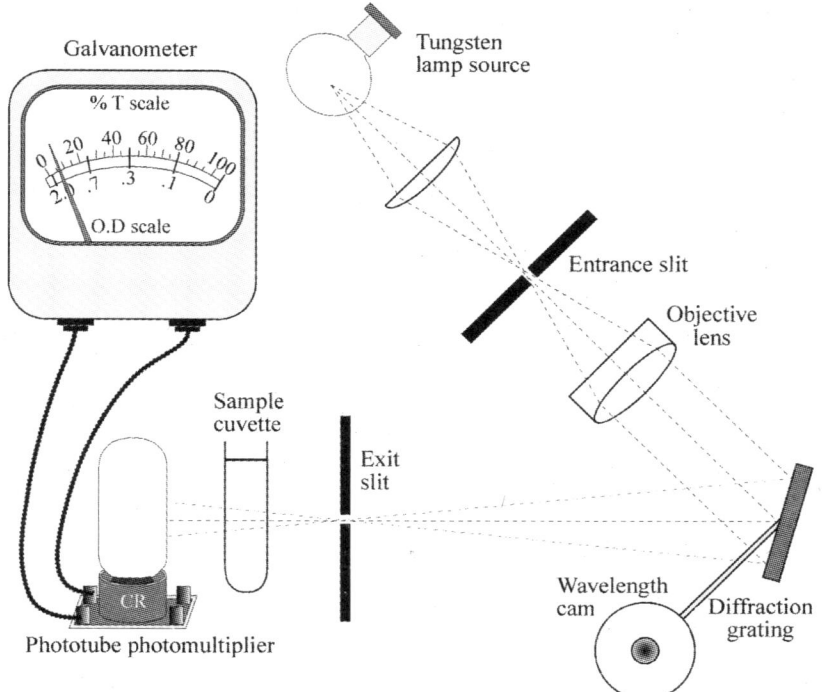

Figure 27.13 *Schematic diagram of a spectrophotometer*

Na$_2$HPO$_4$, 3 g/l KH$_2$PO$_4$, 0.5 g/l NaCl, 1 g/l NH$_4$Cl), glass Petri plates containing 97% ethanol, bent glass rods for spreading bacteria on plates, 37°C shaker incubator, spectrophotometer, 37°C air incubator.

Procedure

1. Add 1 ml of the exponential phase culture of *E. coli* strain to 50 ml sterile LB broth. Start timing the age of the culture at this point, and shake the culture at 37°C except when samples are removed for optical density (OD) measurements and viable cell counts.
2. Measure the OD at 450 nm of a 5 ml aliquot of the *E. coli* strain culture at 0, 30, 60, 90, and 120 min. to determine turbidity.
3. Plate dilute the culture to determine cell number. In plating dilutions, 0.1 ml of the dilution is spread on the agar plate. Thus, for a final dilution of 10^{-3}, 0.1 ml of the 10^{-2} dilution would be plated. Spread the bacteria around the agar plate using a sterilized bent glass rod.
4. Incubate the plates inverted at 37°C overnight or at room temperature for two days.

B. OTHER METHODS OF CELL COUNTING

1. Plate counts

Plate counts are essentially direct counts of numbers of colonies in, or on agar plates, each colony being assumed to represent the progeny of one organism. A concentrated microbial suspension can be inoculated in three main ways: spread plate, drop plate, and pour plate.

2. Limiting dilutions

This is an alternative to the plate count, which may in theory be used for organisms which do not grow as colonies and where a viable count is essential. The principle is to perform serial dilutions, and from each dilution to inoculate a series of replicates for liquid culture, perhaps in tubes. From the proportion of tubes showing growth at each dilution step, a value for the viable count can be estimated.

3. Visual counts

When the absolute number of microbial particles, dead or alive, must be known, the microscopic visual count is a useful method. It has the disadvantage that it is not very sensitive: a minimum of about 10^7 ml^{-1} of bacteria, or 2×10^6 ml^{-1} of larger cells are needed for statistically valid counting. Two types of counting chambers, haemocytometer and Thoma type, are commonly used.

4. Coulter counter

This is an alternative to the visual count, in that it counts all particles. The method is based on changes in electrical conductivity through or across a narrow orifice when cells in suspension are forced through it. Electric circuitry enables the very rapid counting of electrical pulses generated when particles pass through the orifice.

5. Dry weight

The measurement of cellular dry weight is one form of absolute determination of biomass, which is independent of all cellular variables such as viability, cell size, etc. Hence it may be useful, for example, in determination of biomass yields from different growth substrates.

XII GENETICS OF BACTERIA AND BACTERIOPHAGES

Bacteria stand at the forefront of genetic research for several reasons. In recent years, they have proved to be essential organisms in research into the structure and function of DNA, the universal genetic material. Their use is predicated on the following:
1. They have only a single chromosome. The haploid genetic state allows for the phenotypic, observable expression of a genetic trait in the presence of a single mutant gene.
2. Their rapid rate of growth, which permits observation of transmission of a trait through many generations.
3. The availability of large test populations that allows isolation of spontaneous mutants and their induction by chemical and physical mutagenic agents.
4. Their low cost of maintenance and propagation that makes it possible to perform a large number of experimental procedures.

A. BACTERIAL CHROMOSOME

The organization of a typical prokaryotic cell is characterized by the bacterial cell, particularly, *E. coli*. Most of the genetic information in a bacterial cell is contained within a single chromosome. The genetic material is located in a region that lacks clear confinement (lacks nuclear membrane) and is called the **nucleoid**. It occupies about 10% of the total volume of a cell and extended, it is about 1 millimeter long. The chromosome of *E. coli* has about 4000 genes. In addition to the chromosome, some microorganisms contain genetic material in closed loops of DNA called **plasmids**. Plasmids exist as independent units in the cytoplasm and multiply independently of the chromosome. Plasmids are not essential to the life of the cell, but they may confer selective advantages on those organisms that have them.

B. GENE TRANSFER

Genetic diversity, on which evolution depends, arises not only from mutations but also from a powerful additional mechanism, the recombination of genes from different individuals. In eukaryotes the evolution of sex made this process a regular feature of reproduction. However, more primitive forms of recombination evolved much earlier in prokaryotes as an occasional exception to clonal reproduction.

1. Bacterial recombination

Although a bacterial cell contains a major DNA molecule, the genetic exchange is usually between a chromosomal fragment from one cell and an intact chromosome from another cell.

Furthermore, a clear donor–recipient relationship exists: the donor cell is the source of a DNA fragment, which is transferred to the recipient cell by one of several mechanisms, and exchange of genetic material takes place in the recipient by means of a reciprocal recombination between homologous DNA sequences. Transfer of genetic material and its subsequent incorporation into the bacterial genome is also a source of genetic variation in some bacteria. Genetic variability is essential for the evolutionary success of all organisms. Three major types of genetic transfer (or bacterial genetic recombination) found in bacteria are:

1. **Transformation**: A genetic alteration in a cell resulting from the introduction of free DNA from the environment across the cell membrane (Refer to Chapter 18).
2. **Conjugation**: A mating process between "sexually" differentiated bacterial strains that allows one-directional transfer of genetic material in which donor DNA is transferred from one bacterial cell to another by direct contact.
3. **Transduction**: A bacteriophage-mediated transfer of genetic material from one cell to another.

Method-1: Bacterial transformation

Bacterial transformation is a process (Griffith, 1928; Avery, MacLeod and McCarty, 1944) in which recipient cells acquire genes from free DNA molecules in the surrounding medium. Transformation occurs naturally in many bacteria (e.g. *Bacillus*, *Streptomyces*, and *Haemophilus* sp.). The ability of a cell to be transformed depends on its **competence**, defined as the ability of a recipient bacterium to take up DNA from the environment. Factors affecting the cell surface are important to competence, particularly changes in membrane permeability or surface receptors. Transformation begins with the uptake of a DNA fragment from the surrounding medium by a recipient cell and terminates with one strand of donor DNA replacing the homologous segments in the recipient DNA (Figure 27.14). Generally, the binding and internalization of DNA is non-specific, although, for example, *H. influenza* takes up only DNA which contains a specific DNA uptake site, a sequence which is found with greater frequency. Many bacterial species have only a very limited ability to take up free DNA efficiently. Although they can be made competent to take up DNA, provided that the cells are subjected to an appropriate chemical treatment (Refer to Chapter 22). If the transforming DNA is a plasmid, it may be maintained in the recipient cell as an autonomous replicon. Linear chromosomal DNA may undergo recombination with the host chromosome, resulting in marker exchange. Both these events result in stable or permanent transformation. Occasionally, transforming DNA may integrate into the host chromosome by illegitimate end joining, although this is a more common occurrence when linear DNA is introduced in eukaryotic cells. Transformation is a convenient technique for gene mapping in some species.

Method-2: Bacterial conjugation

Conjugation is a mating process (Lederberg and Tatum, 1964) during which a one-direction transfer of genetic material occurs at physical contact between the donor cell and a recipient cell (two "sexually" differentiated cell types). The ability to transfer DNA by conjugation is conferred by a **conjugative plasmid**, a self-transmissible element, which encodes all the functions required to transfer a copy of itself to another cell by conjugation. If the conjugative plasmid can also facilitate the transfer of chromosomal genes, it is termed as a **sex factor**. This differentiation, or existence of different mating strains in some bacteria, is determined by a set of genes (a ***fertility factor***, or ***F factor***), within the cell. These genes are often present in a non-chromosomal, circular DNA molecule called a ***plasmid***. Cells that lack the F factor are recipients

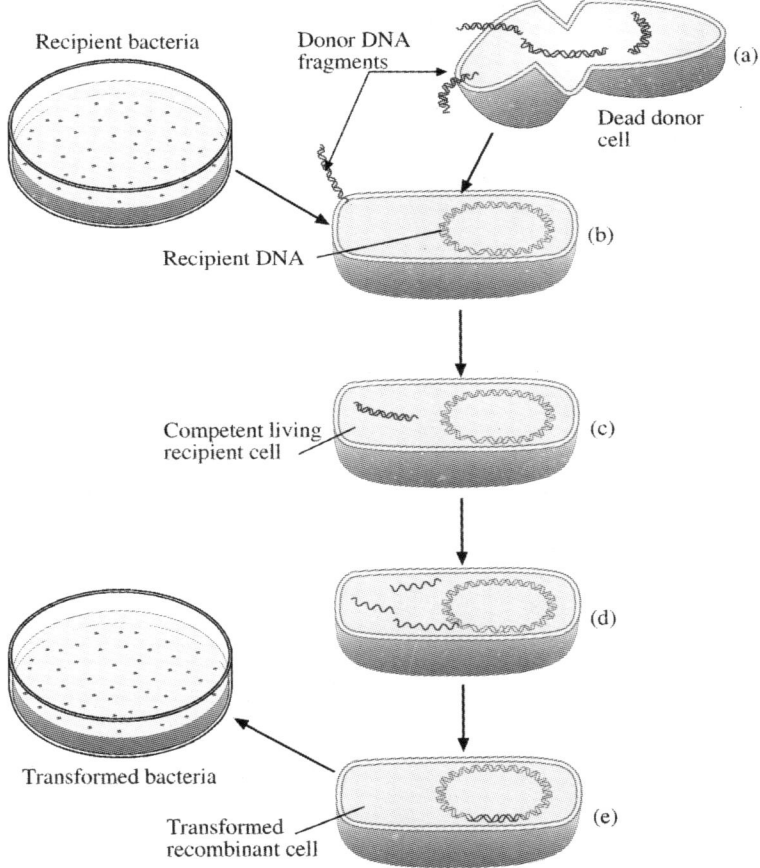

Figure 27.14 *Bacterial transformation (a) A fragment of donor DNA is released from a disintegrating bacteria (b) is taken up by a competent recipient cell. (c) Once the DNA segment cross through the cell wall, (d) one strand of DNA dissolves leaving one strand. (e) A strand of donor DNA replaces the strand of recipient cell DNA. The cell is now considered transformed*

(females) of the genetic material during conjugation and are designated as **F–**. Cells possessing the F factor have the ability to act as genetic donors (males) during mating. If this F factor is extra-chromosomal (a ***plasmid*** or ***episome***), the cells are designated as **F+**; most commonly, only the F factor is transferred during conjugation. If this factor becomes incorporated into the bacterial chromosome, there is a transfer of plasmid genes into chromosomal genes, although generally not involving the entire chromosome or the F factor. The resulting cells are designated **Hfr**, for ***high-frequency recombinants***. Conjugation begins with physical contact between a donor cell and a recipient cell. A tubular projection from the donor cell forms a passageway between the two cells (Figure 27.15).

In the following experiment preparation of mixed cultures represents a cross between an Hfr prototrophic (wild type) strain of *E. coli* that is streptomycin-sensitive (Str-s), and an F– auxotrophic (mutant) *E. coli* strain that requires threonine (thr–), leucine(leu–), and thiamine (thi–)

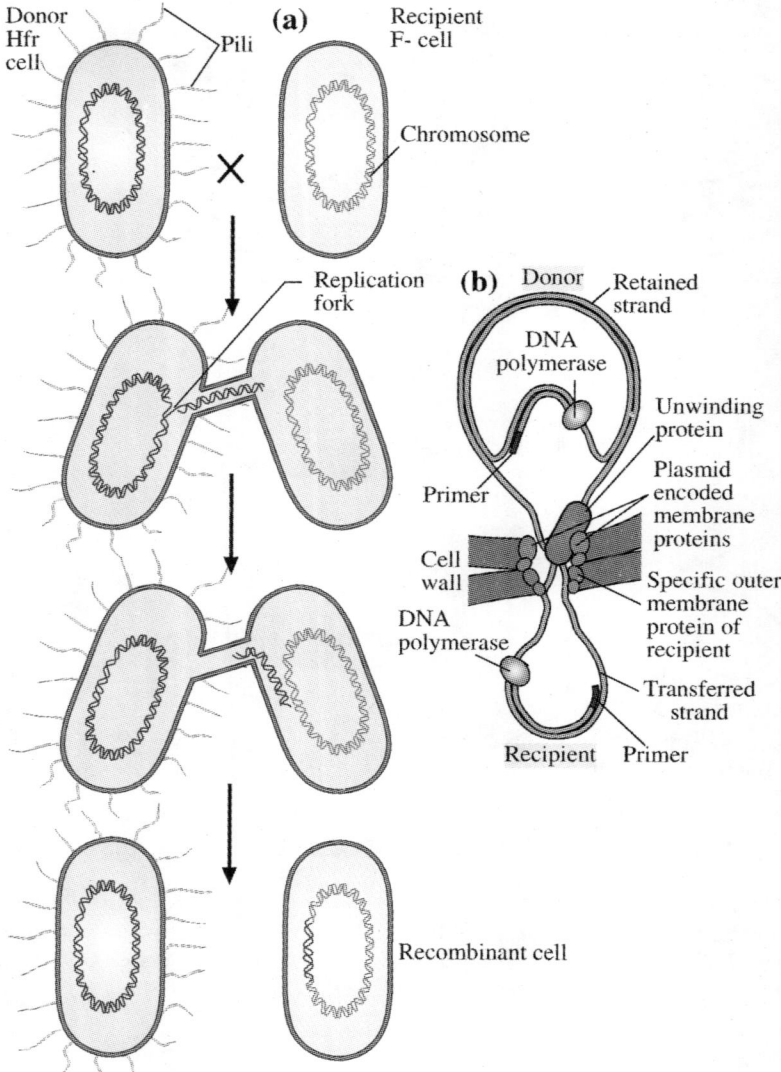

Figure 27.15 *Transfer of plasmid DNA by conjugation in bacteria (a) In this example, the F plasmid of an F+ cell is being transferred to an F− recipient cell (b) Details of the replication and transfer process*

and is streptomycin-resistant (Str-r). Following a short incubation period, the isolation of only the threonine and leucine recombinants will be performed by plating the mixed culture on a minimal medium containing streptomycin and thiamine. The streptomycin is incorporated into the medium to inhibit the growth of the wild type, Str-s parental Hfr cells. The thiamine is required as an essential growth factor for the thi− recombinant cells. Because of its distant location on the chromosome, this marker will not be transferred during the short mating period.

Materials

  Cultures: 12-hr nutrient broth cultures of F– *E. coli* strain thr–, leu–, thi– and Str–r (ATCC e 23724); and Hfr *E. coli* strain Str-s (ATCC e 23740).

  Media: Three plates of minimal medium plus streptomycin and thiamine.

  Equipment: Bunsen burner; beaker with 95% ethyl alcohol; L-shaped bent glass rod; 1 ml sterile pipettes; mechanical pipetting device; sterile 13 x 100 mm test tube; and glassware marking pencil.

Procedure

1. With separate sterile 1 ml pipettes, aseptically transfer 1 ml of the F– *E. coli* culture and 0.3 ml of the Hfr *E. coli* culture into the sterile 13 x 100 mm test tube.
2. Mix by gently rotating the culture between the palms of hands and incubate the culture for 30 min. at 37°C.
3. Prepare control plates of the parental Hfr and F– *E. coli* strains on two appropriately labeled minimal plus streptomycin and thiamine agar plates using the spread-plate technique.
   a. Aseptically add 0.1 ml of each *E. coli* strain to labeled agar plate.
   b. With a sterile glass rod, spread the inoculum over the entire surface of the agar plate.
4. Following incubation, vigorously agitate the mixed culture to terminate the genetic transfer.
5. Aseptically add 0.1 ml of the mixed culture to an appropriately labeled minimal plus streptomycin and thiamine plate and spread the inoculum over the entire surface with a sterile glass rod.
6. Incubate all plates in an inverted position for 48 hours at 37°C.

Method-3: Transduction

Bacterial recombination by the process of **transduction** was first reported in 1952 by Lederberg and Zinder. In the process of transduction, a bacterial DNA fragment is transferred from one bacterial cell to another by a phage particle containing the bacterial DNA (Figure 27.16). Such a particle is called a **transducing phage**. Two types of transducing phages are known – **generalized-** and **specialized-transduction** phages. A generalized transducing phage produces some particles that contain only DNA obtained from the host bacterium or plasmid, rather than the phage DNA. Since infection is a property conferred by the phage particle and not the nucleic acid it carries, this can be an efficient mechanism of gene transfer between cells, and any region of the chromosome can, in theory, be transduced. A specialized (or restricted) transducing phage produces particles that contain both phage and bacterial genes linked in a single phage DNA molecule, and the bacterial genes are obtained from a particular region of the bacterial chromosome. Virus genomes can be exploited as cloning vectors, and the transfer of cloned genes to the cloning host by first packaging the recombinant virus into its capsid can be regarded as artificial transduction; the use of recombinant bacteriophage λ vectors is an artificial specialized transduction, because the cloned DNA is covalently joined to the λ genome. Cloning using cosmid vectors is, however, more like generalized transduction because the λ genome is not used at all.

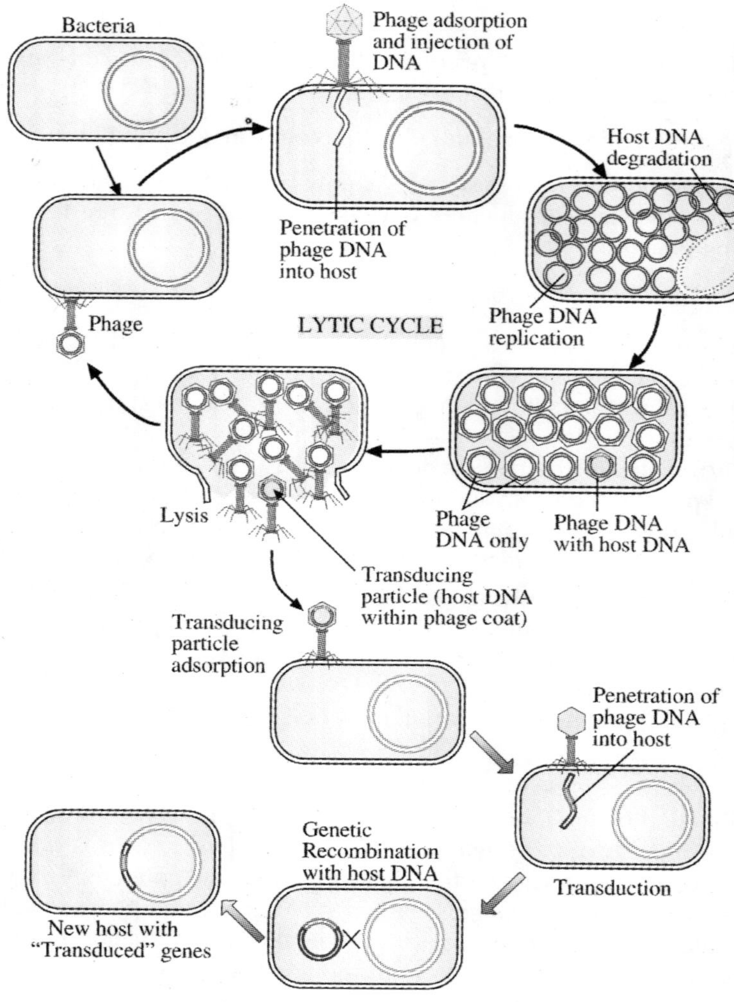

Figure 27.16 *The generalized transduction: one possible mechanism by which virus (phage) particles containing host DNA can be formed*

C. BACTERIOPHAGE GENETICS

Viruses are non-cellular biological entities composed solely of a single type of nucleic acid surrounded by a protein coat called the capsid. Much of our knowledge about the mechanism of animal viral infection and replication has been based on our understanding of infection in bacteria by bacterial viruses, called the **bacteriophages** or **phages** (*bacteriophage*, in Greek means to *eat bacteria*). A variety of different viruses or **bacteriophages** are capable of infecting and lysing *E. coli*. These phages, consisting of a nucleic acid molecule enclosed in a protein sheath (defined as **nucleoprotiens**) or "**head**", adsorb to receptor sites on the bacterial cell wall. Phage replication depends on the ability of the phage particle to infect a suitable bacterial host cell. Infection consists of the following sequential events: (i) Adsorption (particle binds to receptor sites on host cell); (ii) Penetration (infection); (iii) Replication (production of new phages); (iv) Maturation (phage assembly); and (v) Release (liberating from host cell). Infection begins when the phage DNA is injected into the cell. Viral genes are then expressed in ordered temporal sequences. Phage enzymes direct the synthesis of new phage DNA, the assembly of new virus particles, and the disruption of the cell wall (lysis), allowing the release of several hundred new phages (Figure 27.17). The process beginning with a single phage genome and ending with lysis and the release of new phage progeny is often termed the **lytic cycle**. Certain bacteriophages can undergo, in addition to the lytic response, a second process referred to as the **lysogenic cycle** (Figure 27.18). This occurs when the injected DNA is integrated into the bacterial chromosome and specific phage DNA and the expression of phage functions, which are required for the lytic cycle. Normally, the insertion of phage DNA is mediated by phage-encoded enzymes. These direct the integration of the phage genome at specific sites on the bacterial chromosome, referred to as **attachment sites**. The inserted phage genome, or prophage, is passively replicated as part of the bacterial chromosome. Bacterial strains carrying a prophage are termed **lysogens**. Phages which are capable of both the lytic and lysogenic response are referred to as **temperate** phages, while those which can undergo only the lytic response and cannot become prophages are termed intemperate or **virulent phages**.

1. Lysogenic phage – the lambdoid phage

At least two major groups of temperate phages can be distinguished on the basis of their ability to recombine with one another. One class includes the well-studied phage λ and its relatives, the λ phages. Included in this category are ø80, 434, and 21. Lambda phage particles consist of a double-stranded DNA molecule (MW 31 million Daltons) enclosed in an icosahedral protein capsule head about 0.054 microns in diameter. A 0.15 micron-long tube terminating in a fiber constitutes the tail. Lambdoid phages recombine with one another but do not share the same immunity specificity. A λ prophage directs the synthesis of a repressor molecule, which recognizes λ operators, represses the synthesis of λ proteins, and prevents the autonomous replication of λ DNA. A λ lysogen will thus prevent a second or **superinfecting** phage from entering the lytic cycle (Figure 27.18). A λ lysogen is therefore **immune** to superinfection by λ, but not by ø80 or 434 which are said to be heteroimmune with respect to λ.

The prophage enters the lytic cycle following the inactivation of the repressor. This process, termed **induction**, can be initiated by experimental means. Lambdoid phages are induced by exposing lysogens to UV irradiation. Also, many mutants of λ have been isolated which synthesize a heat-labile repressor, allowing the induction of the prophage carrying this mutation by simply heating a culture of a lysogen. Following inactivation of the repressor, phage-specific proteins direct the excision of λ from the host chromosome, the autonomous replication of phage DNA, and assembly of virus particles and lysis of the cell wall.

2. Transducing phage lines

The most important practical use of phages such as λ and ø80 is the ability of these viruses to incorporate bacterial genes into the phage genome and transduce recipient strains. These abnormal particles, or specialized transducing phages, give rise to a stable phage line, which

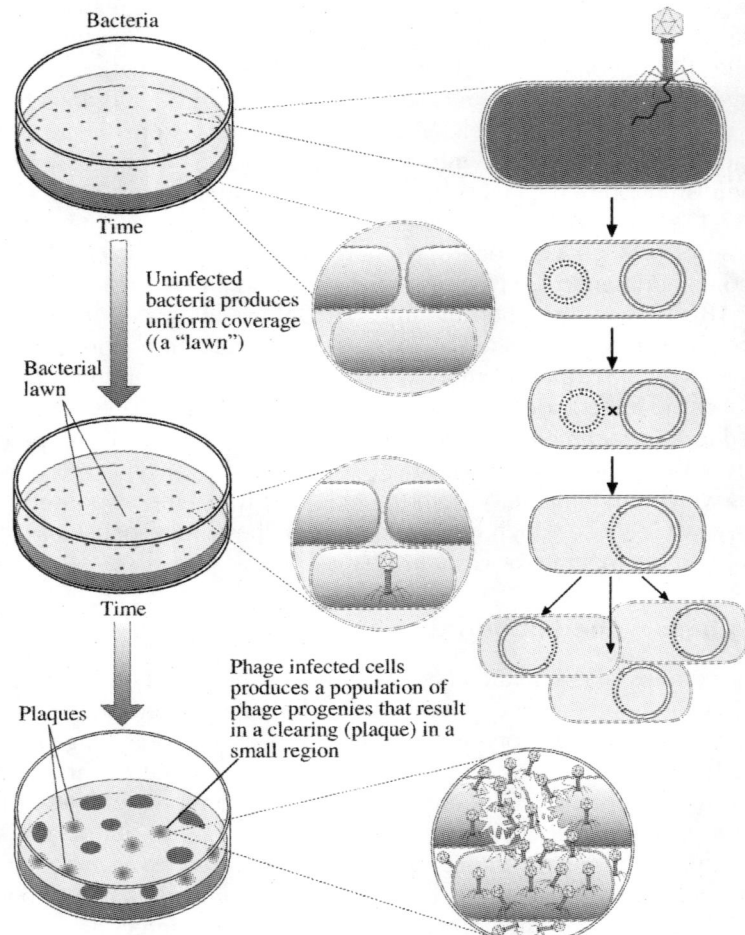

Figure 27.17 *Plaque formation by phages in bacteria. Large plaques in a lawn of E. coli formed by infection with a mutant of bacteriophage λ. Each plaque results from an initial infection by a single bacteriophage*

allows DNA enrichment for specific genes and makes possible the overproduction of specific enzymes after infection or heat induction. The isolation of specialized transducing phage lines is covered in detail elsewhere.

3. Other temperate phage

A second group of temperate phages includes P1 and P2. These phages do not recombine with λ phages and their prophages are not inducible by UV light. Lysates of P1 include defective particles which carry segments of the bacterial chromosome and which can transfer them to other cells. Because of this ability to mediate generalized transduction, P1 is widely used by bacterial geneticists. P1 has no specific attachment site on the *E. coli* chromosome and appar-

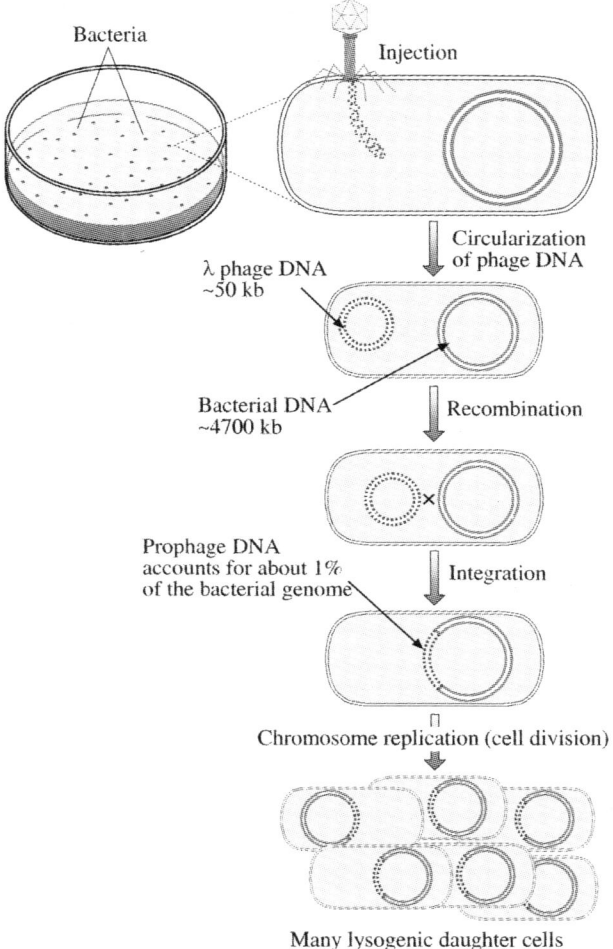

Figure 27.18 *The general mode of lysogenization by integration of phage DNA into the bacterial chromosome*

ently does not integrate into the host chromosome at all, but is instead passively replicated as an extra-chromosomal element. P2 differs from both P1 and the λ phages in that it has several different attachment sites on the *E. coli* chromosome.

4. Virulent phage

Certain bacteriophages cannot integrate into the host chromosome and are not subject to repression. These phages which never become prophages, and consequently form clear plaques on a sensitive host, are termed virulent phages. Virulent mutants of λ and ø80 which form clear plaques even on an immune host have been isolated (λv; ø80v). Among the virulent or intemperate phages, the T-phages (T1, T2, T3, T4, T5, T6, and T7) are a group of dsDNA viruses.

Also included in this class are a series of single-stranded RNA phages (R17, MS2, f2, M12, fr, and Qβ) and a group of ssDNA phages (fd, f1, and M13). These particular single-stranded viruses specifically infect strains of *E. coli* which carry the F factor. All of these phages form clear plaques on sensitive strains.

Method-1: Cultivation and enumeration of bacteriophages (plaque assay)

In order to understand the nature of viruses and virus replication, it is necessary to be able to quantify the number of virus particles. In general, viruses can only be quantified by measuring their effect on the host cells that they infect. When viable bacteria are spread on the surface of a nutrient agar plate, they form an even layer of cells after several hours of incubation. If a single virus particle is introduced, the phage adsorbs to and infects a single cell. After 15–60 min., the cell lysis, releasing several hundred new virus particles into the medium. These infect, in turn, the neighboring bacterial lawn (Figure 27.19). Virulent phages kill virtually all the infected bacteria and form clear plaques. Each plaque can be designated as a **plaque-forming unit** (**PFU**) and used to quantitate the number of infective phage particles in the culture. However, a percentage of the cells infected by a temperate phage becomes lysogenic and survives. The growth of lysogens results in the formation of turbid plaques.

5. Preparation of lysates

Lysates, stocks of particular phage lines, are prepared by infecting sensitive bacteria and then permitting additional growth to allow phage production and subsequent cell lysis. Chloroform is added to further disrupt the cell wall. Cell debris is then centrifuged, and the supernatant, which is the lysate, is stored in the cold. After infection, the cells are either plated in a layer of soft agar on nutrient plates (Figure 27.20), or else grown with aeration in liquid broth. In the former case, enough phages, usually 10^5, are applied to each plate to allow confluent lysis over the surface of the plate. This is a result of the overlapping of plaques. The soft agar layer is scraped into a test tube or centrifuge tube, and then treated as above.

The preparation of lysate in this manner gives high titers for virulent phages, but lower titers for temperate phages, which can form lysogens. An effective way of preparing lysate of temperate phages is to induce cultures of lysogens carrying these phages as prophages. Exposure to UV light results in the induction of many prophages, such as λ and ø80. The exact mechanism is not clear, but it is thought to involve the inactivation of the prophage repressor by accumulation of cell products after irradiation. In addition, certain mutant phages carry a mutation resulting in the production of a heat-labile repressor. Exposing cultures of the appropriate lysogens to high temperature is sufficient to induce the prophage in these strains. This is the most convenient method of preparing lysates.

Materials

Culture: 24-hour nutrient broth cultures of *E. coli* B and T2 coliphage.

Media: Five each of the following: tryptone agar plates and tryptone soft agar, 2 ml per tube, and 9 tryptone broth tubes, 9 ml per tube.

Equipments: Bunsen burner, water bath, thermometer, 1 ml sterile pipettes, sterile Pasteur pipettes, mechanical pipetting device, and glassware marking pencil.

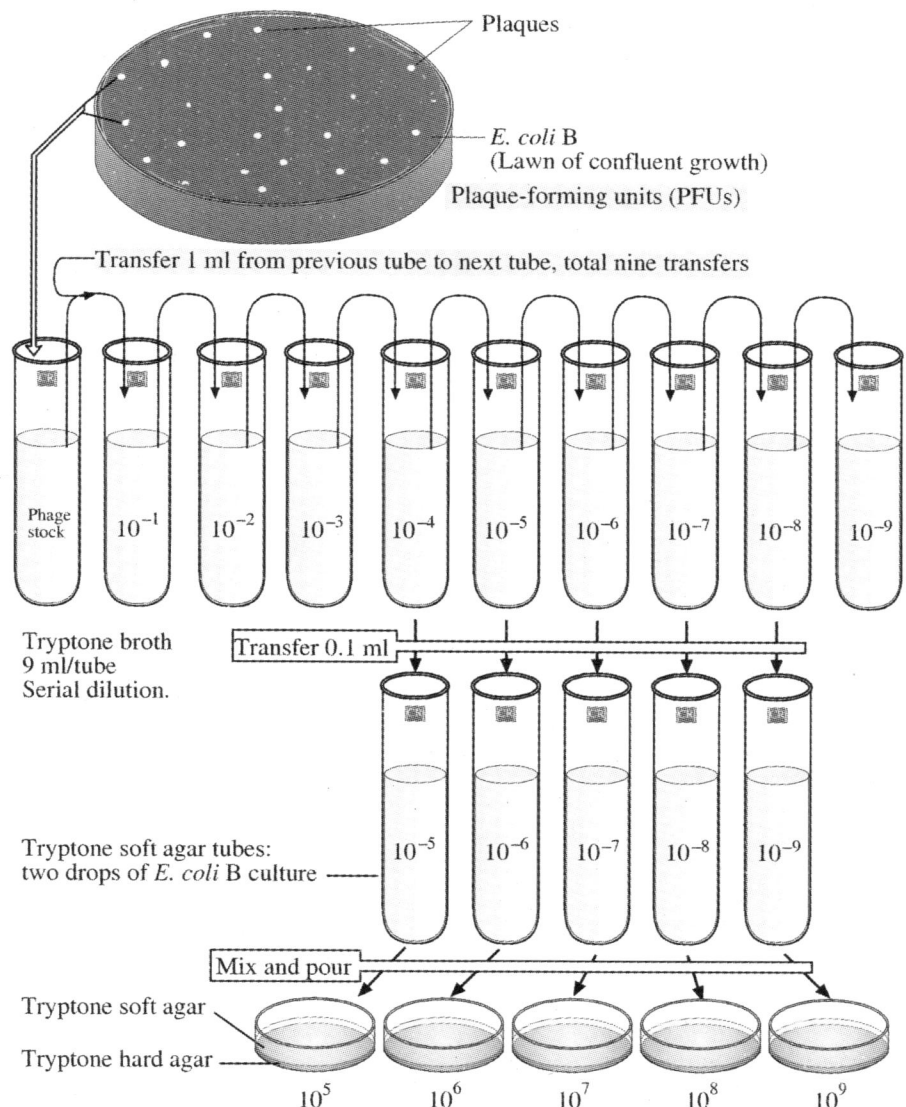

Figure 27.19 Dilution procedure for cultivation and enumeration of bacteriophages

Procedure

To perform the dilution procedure as illustrated in Figure 27.19:
1. Label all dilution tubes and media as follows:
    a. Five tryptone soft agar tubes: 10^{-5}, 10^{-6}, 10^{-7}, 10^{-8}, 10^{-9}.
    b. Five tryptone hard agar plates: 10^{-5}, 10^{-6}, 10^{-7}, 10^{-8}, 10^{-9}.
    c. Nine tryptone broth tubes: 10^{-1} through 10^{-9}.

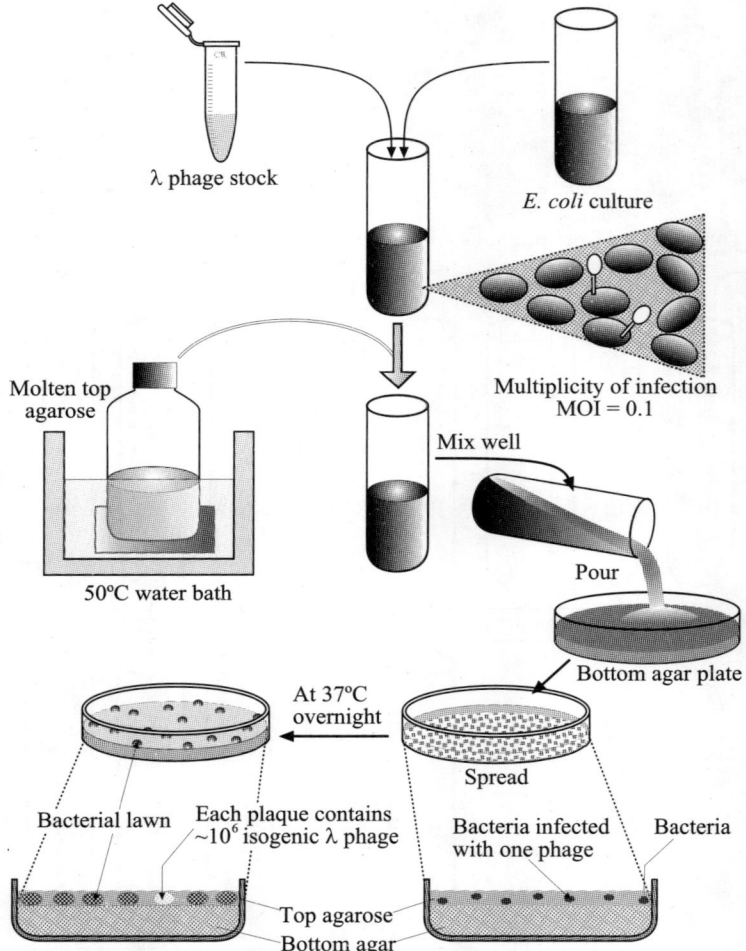

Figure 27.20 *The plaque formation assay can be used to isolate pure strains of λ phage by using a low multiplicity of infection (MOI)*

2. Place the five labeled soft tryptone agar tubes in a water bath. Water should be of a depth just slightly above that of the agar in the tubes. Bring the water bath to 100°C to melt the agar. Cool and maintain the melted agar at 45°C.
3. With 1 ml pipettes, aseptically perform a 10-fold serial dilution of the provided phage culture using the nine 9 ml tubes of tryptone.
4. To the tryptone soft agar tube labeled 10^{-5}, aseptically add two drops of the *E. coli* B culture with a Pasteur pipette and 0.1 ml of the 10^{-4} tryptone broth phage dilution. Rapidly mix by rotating the tube between the palms of your hands and pour the contents over the hard tryptone agar plate labeled 10^{-5}, thereby forming a double-layered plate culture preparation. Swirl the plate gently and allow to harden.

5. Using separate Pasteur pipettes and 1 ml sterile pipettes, repeat step 4 for the tryptone broth phage dilution tubes labeled 10^{-5} through 10^{-8} to effect the 10^{-6} through 10^{-9} tryptone soft-agar overlays.
6. Incubate all plate cultures in an inverted position for 24 hours at 37°C.

XIII GENETIC ENGINEERING OF BACTERIA

Experiments in bacterial recombination entered a new dimension in the late 1970s when it become possible to insert genes into bacterial DNA and thereby establish a cell line that would produce proteins according to the instructions of researchers. Genetic engineering refers to the development of organisms with genetic structure altered by biochemical manipulation. This kind of biochemical procedure is termed recombinant DNA technology and involves the use of plasmids as well as certain bacteriophages. In a nutshell, the technology consists of isolating, purifying, and identifying genetic material from one source; tailoring it for insertion into a new host; and isolating a colony of cells with the desired new genes (Refer to Chapters 14–20 for more details).

XIV SUGGESTED READING

Cappuccino JG and Sherman N (1999) *Microbiology: A Laboratory Manual*, The Benjamin/Cummings Publishing Company, Inc.

Cooper S (1991) *Bacterial Growth and Division*, Academic Press Inc.

Gibbs BM and Skinner FA (Eds) (1966) *Identification Methods for Microbiologists, Part A & B,* Academic Press, London, New York.

Harrigan WF (1998) *Laboratory Methods in Food Microbiology*, Academic Press Limited, CA, USA.

Madigan MT, Martinko JM and Parker J (1997) *Brock Biology of Microorganisms*, Prentice Hall International, Inc.

Malacinski GM (2003) *Essentials of Molecular Biology* (4th Edn), Jones and Bartlett Publishers.

Pelczar Jr MJ, Chan ECS and Krieg NR (1986) *Microbiology*, TATA McGraw-Hill Publishing Company Ltd.

Schleif RF and Wensink PC (1981) *Practical Methods in Molecular Biology*, Springer-Verlag.

Skinner FA and Lovelock DN (Eds) (1979) *Identification Methods for Microbiologists* (2nd Edn), Academic Press, London, New York.

28

Cultivation of Mammalian Cells *In Vitro*

I INTRODUCTION

Mammalian tissue culture refers to the culture of whole organs, tissue fragments, as well as dispersed cells on a suitable nutrient medium. Generally, It can be studied as an (i) organ culture or a (ii) cell culture, depending on whether the tissue organization is retained or not. **Cell cultures** are obtained from different sources including single cell populations (e.g. blood lymphocytes) and suspension cultures, either by enzymatic or mechanical dispersal of tissues into individual cells or by spontaneous migration of cells from an explant maintained as monolayers or as suspension cultures. In contrast, **organ culture** refers to whole embryonic organs or small tissue fragments cultured *in vitro* in such a manner that they retain their tissue architecture. Cultures of freshly isolated cells are called **primary cell cultures** and are usually heterogeneous and slow growing, but are more representative of the tissue of their origin. Once a primary culture is established to grow *in vitro* by subculturing, it transforms into a cell line, which may die after a few cell cycles of subculture or be immortalized to grow indefinitely.

A. HISTORICAL NOTE

Although animal tissue culture is said to be a very recent development, a similar technique finds mention in the Indian epic, the Mahabharat. The Adi Parva chapter of the Mahabharat describes a technique whereby complete human beings were created from a human embryonic tissue. The Kauravas, the 101 children of Queen Gandhari and King Dhritarashtra of Hastinapura, were said to have been regenerated from a fetal tissue (calcified tissue) by a technique that appears to be similar to the tissue culture technique. Sage Ved Vyasa is said to have used the technique to regenerate the Kauravas from an aborted fetus. As the story goes, Gandhari received the news that both Kunti and Madri, wives of Dhritarashtra's brother Pandu, had given birth to their babies before she had. In a fit of rage, Gandhari, who had probably been hoping that her child would be born first so that he would become heir to the throne, struck her womb a hard blow. Consequently, she brought forth a mass of flesh (called ***pinda***) as hard as an iron ball, which had remained within her womb for two years. When she saw the ***pinda***, she bitterly regretted what she had done and sought the help of Ved Vyasa, who had earlier granted her boon that she would have 101 children.

Vyasa immediately washed the 'pinda' with cold water (probably to prevent tissue from dehydration and slow down metabolic activities which would have caused metabolic deterioration). According to the epic, Vyasa divided the 'pinda' into a number of pieces (as big as the tip of his thumb) and covered them with cloth. He asked Gandhari to bring one hundred and one pots, filled with ghee (clarified butter). He put some herbs into the pots and then placed one piece of tissue into each pot. The pots were kept (incubated) in a concealed spot and watched carefully. Vyasa instructed Gandhari not to open the pots for two years. He then left for the Himalayas to perform penance. It was only after two years that the Kauravas were born (Figure 28.1). At the end of two years of incubation, 101 baby clones were born – one from each pot. Out of these, 100 were male baby clones and one was a female baby. The 100 males included Duryodhana and Dushyasana who played key roles in the Mahabharat. The one girl was called Dushule. These results indicate that the 'pinda' used for cloning was male. Genetically, it is possible to obtain male baby clones from male tissue since the Y chromosome essential for determining maleness, is present. But getting male baby clones from female tissue is not pos-

Figure 28.1 *Veda Vyasa could regenerate 101 human baby clones (Kauravas) from an aborted human tissue mass (called "pinda"). He fragmented the tissue mass into a number of pieces and cultured them in a pot containing a piece of tissue, ghee and herbs. The pots were incubated at a concealed place for two years. At the end of two years the Kauravas were born*

sible because of the absence of the Y chromosome in it. However, female clones can originate from a male tissue, if the developing tissue loses the Y chromosome from diploid cells. This could be the reason why 100 baby clones were male and only one baby was female. The resulting female, however, will have XO-monosomic condition for the sex chromosome. There are many reports available in literature about the existence of females with XO-monosomics in the human population. These XO-monosomic females can often grow normally to adulthood, but they may be infertile. The birth of Dushule is a typical example of somaclonal variation due to the loss of the Y chromosome, which is very common in *in vitro* tissue cultures (due to the various types of mutations, i.e. loss or gain of chromosomes or genes).

The procedure described in the 'Adi Parva' appears to be a scientific description of *in vitro* tissue culture and clearly indicates that the Kauravas were born from the fetal tissues. This ancient event raises many questions. Did Vyasa have enough knowledge of mammalian tissue culture to regenerate complete human beings from fetal tissue? What kind of medium did he use? Were the pots he used to incubate the tissue pieces very advanced incubators that could simulate all the conditions present in a womb, and so on and so forth. Even though it may not be easy to answer all these questions fully, it is clear that the procedure Vyasa used to regenerate human beings from tissue is comparable to some of the breakthroughs achieved recently in embryo cloning technology. On a logical basis, as earlier mentioned, the regeneration of more male clones and only one female baby clone from a male tissue has a genetic explanation. To answer the possibility of regeneration of embryonic tissues from somatic cells, there are several evidences now in literature. Recent experiments in animal embryo cloning have demonstrated the possibility of regeneration of complete adult animals from embryonic cells which have been transplanted with a somatic cell nucleus. These results provide evidence for conversion of the somatic cell nucleus into an embryonic genome that is both developmentally totipotent and endowed with the necessary epigenetic programming which will permit it to initiate the developmental pathway leading to the formation of a new organism. Recent studies also indicate that in mammals, the oocyte cytoplasm has the ability to modulate genome-imprinting information in the parental or introduced nuclear genome to convert it into an embryonic genome. This process involves the dedifferentiation of the somatic genome followed by redifferentiation into an embryonic genome. Surprisingly all the components required for this reprogramming are present in the oocyte cytoplasm itself. This has been well supported by many experiments that have shown the ability of the oocyte cytoplasm to terminate gene expression programs of transplanted nuclei and then to reactivate expression of appropriate stage-specific genes that support development. These results clearly suggest the possibility of regeneration of somatic tissue into embryonic cells, if cytoplasmic components of oocyte or equivalent conditions are provided. It has also been shown that the embryo can be cultured up to the stage (early gastrula), where embryonic cells are determined to lineage-commitment and which develop into specific organs in *in vitro* conditions.

But the next hurdle is to overcome the problem of growing embryonic cells into an adult organism, which requires the uterus for life supporting metabolites and conditions. At present, it is not possible to grow an embryo into a complete organism in *in vitro* conditions. The embryo either fertilized or genetically modified has to be transplanted into the uterus of a surrogate mother for complete development into an adult organism. If that is the case, then how did Vyasa grow tissue pieces in a pot filled with ghee and herbs into adult humans? Modern science can help to answer this question. Recent experiments in embryo cloning have enabled scientists to establish functional organ-specific tissues (brain, liver, neural, etc.) from embryo cloning *in vitro*, although further development is not possible at present. This is mainly due to the lack of a support system that is usually provided by the uterus in the female womb. In the development of embryonic cells into an adult organism, the womb plays multiple roles. It supplies essential metabolites and growth factors needed for embryo development, gives mechanical as well as immunological protection, facilitates gas exchange (e.g. oxygen and carbon dioxide exchange) and the removal of toxic substances from the developing embryo. It also provides comfortable and ambient conditions for normal development. Now the question is, is it possible to simulate all the functions and conditions of womb *in vitro*? Recent studies show that it is possible to reconstitute virtually a complete skin (both epidermis and dermis, called living skin equivalent, LSE), which can be grafted onto the burnt areas of a patient. The LSE then functions normally. Therefore, it is possible to establish a tissue layer that is similar to the uterine lining, which may function as an endometrium on which the embryo can be implanted. Also it may act as an extra-embryonic membrane such as the yolk sac in eggs with shells, which holds the nutritive yolk to support embryo development.

The origin of the Kauravas explains that all the conditions necessary for the development of the embryo into an adult organism, which is normally provided by the uterus, might have been provided by the magic pots (or incubators) prepared by Vyasa. With the available information from the text, it is difficult to tell whether the pots used by Vyasa were functioning as an egg with a shell, in which the embryo could complete its development without the need of a womb or whether they functioned like a womb in which the inner surface might have been covered with a tissue layer similar to the endometrium or whether they were, super-incubators that had the power to provide all the conditions necessary for the development of the embryo. Whatever mechanism used, the information available from recent *in vitro* embryo cloning experiments convincingly supports the possibility of Vyasa's claim of regenerating the Kauravas from embryonic tissue. Thus, we can be optimistic that a modern Vyasa(s) may soon be able to clone humans in laboratory jars (pots) like Vyasa did. Since the pros and cons of this technology are debatable, the discussion on this topic is limited to predicting the potential of tissue culture technology in future.

B. CONTEMPORARY MAMMALIAN TISSUE CULTURE

Contemporary mammalian tissue culture can be traced back to 1880 when Arnold showed that leucocytes could be divided outside the body. In the beginning of the present century, Jolly (1903) showed that cells could survive and divide in *in vitro* conditions. Ross Harrison (1907) developed a reproducible technique for tissue culture. He cultured the frog tadpole spinal chord in a lymph drop hanging from a coverslip into the cavity on a microslide. Later, Alexis Carrel (1912) used tissue and embryo extracts as culture media. Subsequently in 1913, Carrel developed a complicated methodology for maintaining cultures free from microbial contamination and demonstrated that animal cells could be grown indefinitely *in vitro*.

Tissue culture plays a significant role in many areas of biotechnology and in medical research. For example, for production of antiviral vaccines, cancer research, cell fusion techniques, for genetic manipulations, for *in vitro* production of monoclonal antibodies, for production of pharmaceuticals, for cytological analysis of normal and diseased cells/tissues, for bioassays to determine the toxicity and its effects of various compounds, for the study of the establishment of specific tissues committed for organ development (e.g. neural, liver, lung, etc.).

II ORGAN CULTURE

In vitro culture of animal organs or parts thereof in which their various tissue components (e.g. liver, lung, kidney, parenchyma, and stroma) are preserved both in terms of their structure and function so that the cultured organs closely resemble the concerned organs *in vivo* is called **organ culture**. In such cultures, new growth is in the form of differentiated structures (e.g. glandular structures in case of glands, small bronchi in the case of lung tissues; in tissues lined with one or the other type of epithelium, the epithelium differentiates in a pattern similar to that in the concerned organs *in vivo*). The *in vitro* cultured organs retain their physiological functions and the morphological architecture of the tissue. To achieve whole organ culture, it is essential that the tissue should never be disrupted or damaged and this requires careful handling. Media and laboratory conditions used for organ culture are generally the same as those used for cell culture, which will be discussed in the next section. The techniques of organ culture can be divided into (i) those cultured on a solid medium and (ii) those cultured in a liquid medium.

Culture of embryonic organs: Embryonic organs are easier to culture than organs from adult animals.

A. TECHNIQUES

The first attempt at organ culture was made by Loeb in 1897, who maintained adult rabbit liver, kidney, thyroid, and ovary on small plasma clots in test tubes and noted that these organs retained their normal histological features for three days. The technique of organ culture has since been considerably refined. These techniques utilize one of the following five approaches (refer to Freshney 1992, for more details).

(1) Plasma clot: In this approach, the explant is cultured on the surface of a clot consisting of three drops of chick plasma and one drop of chick embryo extract (50%) contained in a watch glass (also called watch glass technique) (Figure 28.2a). The watch glass may or may not be closed with a glass lid sealed with paraffin wax. The watch glass (one or more) is kept in a Petri dish lined with a moist filter paper or cotton wool to minimize evaporation of the clot. The Petri dish is usually incubated at 37.5°C. Fresh clots have to be provided every 2–3 days for avian tissues and every 3–4 days for mammalian tissues. In a modification of this approach, pieces of organs are placed on plasma clots kept on a cover slip, which is then inverted onto the cavity in a micro-concavity microscopic slide (Figure 28.2b); the coverslip is then sealed with paraffin wax. This method is inexpensive and permits light microscopic observation during culture.

(2) Raft methods: In this technique, the explant is placed onto a raft of lens paper or rayon acetate which is floated on serum in a watch glass. Four or more explants are usually placed on each raft. In a combination of raft and clot techniques, the explants are first placed on a suitable raft, which is then kept on a plasma clot. This modification makes media changes easy, and prevents the sinking of explants into the liquefied plasma.

(3) Agar gel: In this method, the culture medium (containing salts, serum, chick embryo extract or a mixture of some amino acids and vitamins) is gelled with 1% agar. This method avoids immersion of explants into the medium and permits the use of defined media. Additional mechanical support is also needed. The agar gels with explants are kept in embryological watch glasses (Figure 28.2c). The explants can be examined using a stereoscopic microscope.

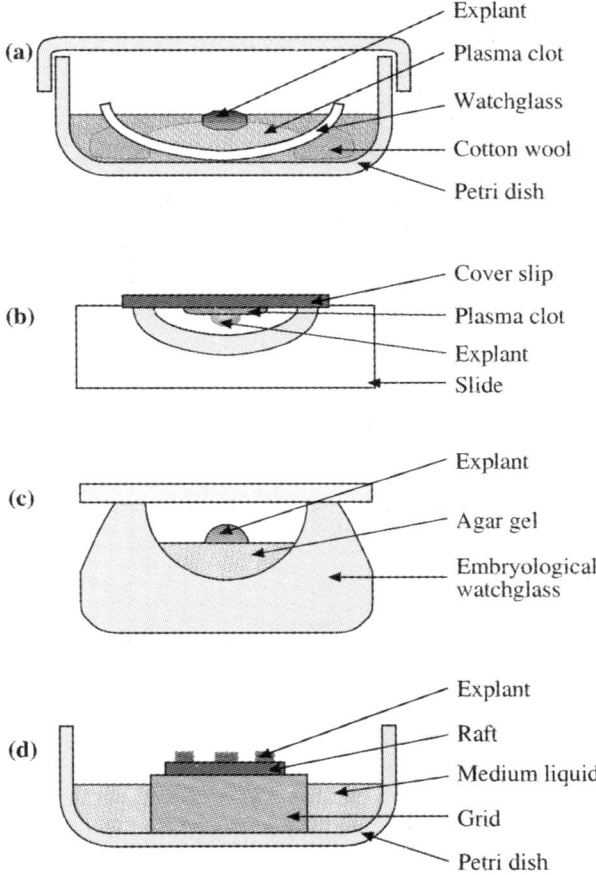

Figure 28.2 Techniques of organ culture. (a) Watchglass method of culture explants on a plasma clot; (b) Maximow's single slide technique; (c) Organ culture on agar gel; (d) The grid method uses a metallic wire mesh or a perforated sheet. Explants are placed on a raft of lens paper

(4) **Grid method**: This method was initially devised by Trowell in 1954. It utilizes a 25 mm^2 wire mesh or a perforated stainless steel sheet whose edges are bent to form four legs of about 4 mm height. Skeletal tissues are generally placed directly on the grid but softer tissues like glands or skin are first placed on rafts which are then kept on the grids (Figure 28.2d). The grids are placed in a culture chamber filled with fluid medium up to the grid and the chamber is supplied with a mixture of oxygen and carbon dioxide. A modification of the original grid method is widely used to study the growth and differentiation of adult and embryonic tissues.

(5) **Cyclic exposure to medium and gas phase**: In this technique, the explants are intermittently exposed to the fluid medium and the gas phase. The number of explants per dish varies from 2–18, depending on the organ cultured. The explants are attached to the bottom of a plastic culture dish and are covered with fluid medium. The dishes are enclosed in a chamber containing a suitable gas mixture and mounted on a rocker platform to increase gas phases.

B. APPLICATIONS OF ORGAN CULTURE

Organ cultures have been used for many purposes. (i) They have been used in studies on the pattern of growth, differentiation, and development of organ rudiments (e.g. fetal organs) and the influences of various growth factors; (ii) They are also used in experiments to conduct bioassays: the action of drugs, carcinogenic agents, etc., on animal organs is studied *in vitro* to asseses its effect on functional and developmental processes. (iii) The most important application of organ cultures is to produce tissues for implantation in patients. This is called tissue engineering. Human skin has been successfully produced *in vitro* and used for transplantation in people with severe burns, ulcers etc. The ultimate objective of tissue engineering is to reconstitute body parts *in vitro*. It is hoped that studies will permit the culturing and constitution of bones, liver, pancreas, and possibly the entire organism(?) in future.

Artificial skin production *in vitro* has been improved recently to permit reconstitution of virtually the complete skin (both epidermis and dermis), called the living skin equivalent (LSE). There are three biotech companies in USA working on artificial skin research and they have reached the stage of clinical trials. This technology employs a collagen matrix as a support for growth of tissue. The skin explants used for obtaining artificial skin may be either obtained from the patient concerned or from the foreskin (loose skin from the tip of the penis) of newborn babies.

III *IN VITRO* CELL CULTURE

Over the past several years, the spectrum of products from cell culture has rapidly expanded due to the advancement of cell culture techniques and the success in applying genetic engineering to eukaryotic cells, leading to increased productivity of the cell as well as to the development of new products. The increasing demand for these substances necessitates the application of recombinant DNA technology for the creation of new products and the establishment of large-scale bioreactors for animal cell culture. The first recorded attempt to grow animal cells in culture can be attributed to Ross Harrison in 1907, who was able to cultivate frog embryonic nerve cells using the hanging drop technique. This work was further improved by Barrows and others, who established the techniques for cultivation of a wide range of mammalian cells. The use of trypsin for cell disaggregation from tissue explants was a major innovation to obtain cell suspensions, which allowed single-cell cultures. The discovery of antibiotics in the late 1940s led to the development of improved cell culture techniques by reducing the risk of microbial contamination, which had been a serious problem. Ender's discovery in the late 1940s, that viruses could be propagated in cell cultures and used as vaccines, led to the development of large-scale animal cell culture. Polio vaccine production in the 1950s was the beginning of animal cell culture as a developing technology. Further milestones were the development of the first synthetically-based culture medium by Earle and co-workers.

Much of the molecular genetic work in higher eukaryotes, especially mammalian systems, has come from studies using immortalized cell lines. More than a century ago, biologists began to experiment with *in vitro* culture conditions that permitted dissected animal tissues to remain viable in culture for extended periods of time. Eventually, these studies gave rise to the development of various types of tissue culture media consisting of buffers, inorganic salts, trace elements, a source of carbon for energy production, amino acids, vitamins, and bovine serum.

Early molecular geneticists realized that cells derived from disaggregated animal tumors, or isolated from the blood of patients with leukemia, could be adapted to grow *in vitro* using the same media developed for tissue culture. It was soon discovered that many types of cancer cells have an unlimited capacity to proliferate in a serum-containing medium and are also able to immortalize many different primary cells. Clonogenic immortalized cells grown in culture are called **cell lines**. Table 28.1 lists some of the mammalian cell lines that are most commonly used in molecular genetic applications. Other types of eukaryotic cell lines have been established from immortalized insect, amphibian, avian, and fish cells. It is important to realize that immortalized cell lines are often dedifferentiated, and therefore have lost many of the cellular phenotypes associated with the normal differentiated state. Nevertheless, cell lines provide a convenient model system to study eukaryotic gene expression under defined conditions and have, in many cases, led to major breakthroughs in our understanding of human diseases.

Table 28.1 *Commonly used cell lines in molecular genetic studies*

Cell line	Species	Cell type of origin
293	Human	Kidney, transformed by virus
HeLa	Human	Cervical carcinoma
HEK-293	Human	Embryonic kidney
H1, H9	Human	Embryonic stem cells
LNCaP	Human	Prostate tumor
MCF7	Human	Mammary tumor
MDCK	Dog	Kidney
COS	Monkey	Kidney
CV-1	Monkey	Kidney
Rat-1	Rat	Embryo fibroblast
L6	Rat	Myoblast
PtK1	Rat kangaroo	Epithelial cell
PC12	Rat	Chromaffin cell
CHO	Hamster	Ovary epithelial
BHK21	Mouse	Fibroblast
3T3	Mouse	Embryo fibroblast
R1	Mouse	Embryonic stem cells
SP2	Mouse	Plasma cell
S2	*Drosophila*	Macrophage-like cells
BY2	Tobacco	Undifferentiated meristematic cells

IV ANIMAL CELLS AS BIOLOGICAL SUBSTRATES FOR INDUSTRIAL PROCESSES

Even though animal cells have many advantages when compared with systems based on bacteria or yeast, they do have many disadvantages as well (Table 28.2). This situation is well described by comparing the biomass productivity of a bacterium growing in a continuous culture with a doubling time of 20 min. with an animal cell continuous culture with a doubling time of 12 hours. However, the productivity of a system is not generally determined by biomass alone. The biological activity of the product and its ability to perform in an environment which reacts strongly and negatively to the presence of foreign bodies is a most relevant consideration. The presence of transforming DNA requires much testing to ensure safety. Also the presence of endotoxins, pyrogenic factors, or allergens will prevent products reaching the market place. Critically, it is in the area of post-translational processing that major and activity-determining events occur. Animal cells not only provide the appropriate and specific enzymes and cofactors but also a structured cytosol in which the proteins fold up and form the native disulfide bonds. This enable the molecule to express the unique three-dimensional structure which characterizes the enzymic, hormonal, and immunological reactivity of the molecule. Such factors have led to the enhancement of the productivity of materials, which can be made from animal cells. The method of choice to achieve this end is to engineer the animal cell genetically to 'over produce' a particularly desired material.

Table 28.2 *Comparison between animal cells and bacteria for biomass production in culture*

Parameter	Bacterial cell	Animal cell
Cell diameter (μm)	1	10
Surface area (cm^2)	3.1×10^{-8}	3.1×10^{-6}
Volume (cm^3)	5.2×10^{-13}	5.2×10^{-10}
SA/volume	59×10^3	5.9×10^3
Doubling time (h)	0.3	12
Cell concentration (ml^{-1})	5×10^9	4×10^6
Productivity (no. cells/L/h)	7.5×10^{12}	1.7×10^8
Productivity (gm cells/L/h)	3.75	0.085
Relative productivity	44	1
Relative SA/vol	10	1

V BASICS OF MAMMALIAN CELL CULTURE

A. LABORATORY FACILITIES FOR TISSUE CULTURE

The laboratory facilities for mammalian cell culture should provide a very special environment for sensitive cell culture operations, such as (i) sterile handling; (ii) incubation at regulated environment; (iii) preparation of glassware, media and tissue; (iv) wash-up; (v) sterilization, and (vi) storage. The complete working area has to be scrupulously clean. The equipments, media and other chemicals, and glassware have to be properly stored in the specified area. The introduction of laminar flow cabinets or safety hoods to provide a clean working atmosphere has greatly facilitated the maintenance of aseptic and sterile conditions, such that no separate room for sterile handling is needed. Most of the cell culture operations including cell preparations, media transfer, inoculation, and subculturing are done under safety hoods. Incubation of cultures under temperature, humidity, and CO_2- controlled atmosphere is done in incubators.

B. STERILE TECHNIQUE

Because cells or tissues must be grown in culture for days or weeks to obtain a sufficient numbers of cells for analysis, it is essential that the sterile technique (aseptic technique) be maintained. The aseptic technique involves a number of precautions to protect both the cultured cells and the laboratory worker from infection (also see Section 1). Protective apparel such as gloves, lab coats or aprons, and eye-wear should be worn when appropriate. Frequently, specimens received in the laboratory are not sterile and cultures prepared from these specimens may become contaminated with bacteria, fungus, or yeast. The presence of microorganisms can inhibit growth, kill cell cultures, or lead to inconsistencies in test results. The contaminants can grow faster than the cells and deplete nutrients in the medium and also produce substances that are toxic to cells. Antibiotics and fungicides may be added to the tissue culture medium to combat potential contaminants (Table 28.3). The antibiotic containing solutions can be used to wash specimens prior to culture and can be added to the medium used for tissue culture.

Table 28.3 Working concentrations of antibiotics and fungicides for mammalian cell culture

Additive	Final concentration
Antibiotics:	
Penicillin	50–100 U/ml
Streptomycin sulfate	50–100 µg/ml
Kanamycin	100 µg/ml
Gentamycin	50 µg/ml
Fungicides:	
Mycostatin	20 µg/ml
Amphotericin B	0.25 µg/ml

All materials that come into direct contact with cultures must be sterile. Sterile disposable dishes are preferable. If re-usable glassware is used, it must be washed thoroughly and autoclaved before using. With dry heat, glassware should be heated 90 min. to 2 hours at 160°C to ensure sterility. The media reagents, and other solutions that come into contact with the cultures must also be sterile, and they can be sterilized either by filter-sterilization or by autoclaving if they are heat-insensitive. Contamination can occur at any step in handling cultured cells. The neck bottles and flasks, as well as the tips of the pipettes should be flamed before the pipettes are introduced into the bottles. Although tissue culture work can be done on an open bench, some aseptic methods have to be strictly enforced. Tissue culture can also be performed in safety cabinets. All work surfaces both inside and outside of the hood should be kept clean and disinfected daily and after each use. Some safety cabinets are equipped with UV lights for decontamination of work surfaces. Cultures must be checked routinely for contamination. If contamination is confirmed with a microscope, infected cultures have to be discarded.

C. THE SUBSTRATES ON WHICH CELLS GROW

The majority of mammalian cells cultured *in vitro* have been grown as monolayers on an artificial substrate. Cell growth in suspension is restricted to a few cell types such as hemopoietic cell lines, rodent ascites tumor cell lines, and a few other selected cell lines. The majority of other cell types need supporting substrates in order to proliferate. Inadequate spreading on the substrate will inhibit cell proliferation. Those cells, which require attachment for growth, are said to be **anchorage dependent**. However, some cell types, which have undergone **transformation** (malignant cell types) frequently become anchorage independent and can grow in suspension. Whenever a substrate is needed, it may be adhesive (glass, plastic, palladium, etc. or a metallic surface; these surfaces may sometimes be variously modified) or non-adhesive (agar, agarose, methocel). Glass and plastic substrates are the most commonly used substrates, in the form of slides, flasks, Petri dishes, tubes, multi-well plates, etc.

Often substrate surfaces need to be variously treated to improve cell attachment and growth. The treatments with culture medium meant for another culture, or with purified fibronectin added to the medium or with collagen are commonly used. After coating with collagen, the substrate may be sterilized under UV. Sometimes, on the substrate surface, a monolayer of a special kind of cells is used, partly to supplement the culture medium for growth and partly for conditioning the substrate by cell products. These monolayers are called **feeder layers**, since they also feed the growing cultures.

D. GAS PHASE FOR TISSUE CULTURE

The animal cell and tissue cultures, depending on the type of cells require gas phases consisting of O_2 and CO_2, for their growth. CO_2 concentration, which is a complex factor, influences the pH and the HCO_3^- ion concentration of the culture medium.

E. CULTURE MEDIUM PREPARATION

The culture medium is the single most important factor for culturing cells and tissues. The choice of the tissue culture medium comes from experience. An individual laboratory must select the medium that best suits the type of cells being cultured. Chemically defined media are available in liquid or powdered form from a number of suppliers. The medium should be filter-sterilized and transferred to sterile bottles. The prepared medium can generally be stored ≤1 month in a 4°C refrigerator. Basic media such as Eagle minimal essential medium (MEM), Dulbeccos modified Eagle medium (DMEM), Glasgow modified Eagle medium (GMEM), and RPMI 1640 and Ham F 10 nutrient mixtures are composed of amino acids, glucose, salts, vitamins, and other nutrients. A basic medium is supplemented with the addition of L-glutamine, antibiotics, and usually serum to formulate a "complete medium." Where serum is added, the amount is indicated as a percentage of fetal bovine serum (FBS) or other serum. Some media are also supplemented with antimycotics, non-essential amino acids, various growth factors, and/or drugs that provide selective growth conditions.

The optimum pH for most mammalian cell cultures is 7.2 to 7.4. The pH of the medium has to be adjusted as necessary after all the supplements have been added. Buffers such as bicarbonate and HEPES are routinely used in the tissue culture medium to prevent fluctuations in pH that might adversely affect cell growth. HEPES is especially useful in solutions used for procedures that do not take place in a controlled CO_2 environment. Most cultured cells will tolerate a wide range of osmotic pressure and an osmolarity between 260 and 320 mOsm/kg is acceptable for most cells. Serum, fetal bovine serum (FBS or FCS are most commonly used serum), calf serum, horse serum, and human serum are also used; some cell lines are maintained in a serum-free medium. The complete medium is supplemented by 5 to 30% (v/v) serum, depending on the requirements of the particular cell type being cultured. All tissue culture media should be tested for sterility prior to use.

F. ESTABLISHMENT OF CELLS IN CULTURE

Provided appropriate conditions, most plant and animal cells can live, multiply, and even express differentiated properties in a cell culture dish. The first step in establishing cells in an *in vitro* culture medium is to dissociate organs or tissues into a single-cell suspension. The best yields of viable dissociated cells are usually obtained from fetal or neonatal tissues. Such a suspension can be used as a primary culture or as a starter culture to try and initiate a continuous, i.e., immortalized cell. The heterogeneous mixture of different cell types in the resulting single-cell suspensions is transferred into special culture dishes or flasks with a serum-containing medium providing near physiological conditions. Some cell types survive, attach themselves to the bottom of the culture dish, spread and proliferate until confluency. However, those cells which are transformed may be able to grow in multilayers or as a free suspension. For subcultivation, adherent cells have to be detached themselves from the culture vessel by short exposure to trypsin, diluted at a certain ratio with fresh medium and passaged into a new culture flask.

Method-1: Trypsinizing and subculturing cells from a monolayer

Most primary cultures or continuous cell lines grow as monolayers. These cultures need a periodic change of medium. The subculturing of monolayers involves the following steps (Figure 28.3).

Materials

Primary cell cultures; HBSS without Ca^{2+} and Mg^{2+}, 37°C; 0.25% (w/v) trypsin/0.2% EDTA solution, 37°C; Complete medium with serum: e.g., DMEM supplemented with 10% to 15% (v/v) fetal bovine serum; Pasteur pipettes; 37°C warming tray or incubator; Tissue culture plasticware or glassware including pipettes and 25-cm^2 flasks or 60 mm sterile Petri plates.

Procedure

1. Remove the medium completely from the primary culture with a sterile Pasteur pipette. Wash adhering cell monolayer once or twice with a small volume of 37°C HBSS without Ca^{2+} and Mg^{2+} to remove any residual FBS that may inhibit the action of trypsin.
2. Add enough 37°C trypsin/EDTA solution to the culture to cover adhering cell layer. Place plate on a 37°C warming tray for 1 to 2 min. Tap bottom of plate on the counter top to dislodge cells. Check cultures with an inverted microscope to be sure that cells have been rounded up and detached from the surface. If not, return plate to warming tray for an additional min. or two.
3. Add 2 ml of 37°C complete medium. Draw cell suspension into a Pasteur pipette, rinse cell layer twice or thrice to dissociate cells and to dislodge any remaining adherent cells. As

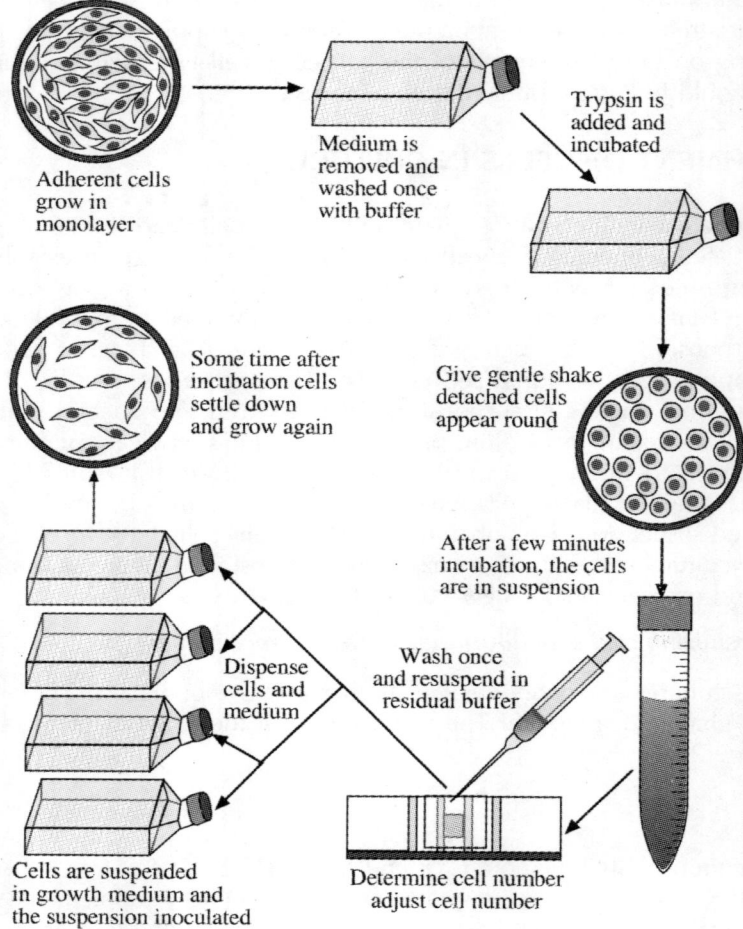

Figure 28.3 *Diagrammatic illustration of maintenance of adherent cell lines*

soon as the cells are detached, add serum or medium containing serum to inhibit further trypsin activity that might damage cells.

4. Add an equal volume of cell suspension to fresh 60 mm Petri plates or 25 cm^2 flasks that have been appropriately labeled. Be sure to label with date of subculture and passage number. Alternatively, cells can be counted using a hemacytometer or Coulter counter and diluted to the desired density (~5 × 10^4 cells/ml) so a specific number of cells can be added to each culture vessel.
5. Add 4 ml fresh medium to each new culture. Incubate in a humidified 35°C, 5% CO$_2$ incubator. If necessary, feed subconfluent cultures after 3 to 4 days, removing old medium and adding fresh 37°C medium.
6. Passage secondary culture when it becomes confluent by repeating steps 1 to 5, and continue to passage as necessary.

G. ESTABLISHMENT OF PRIMARY CELL LINES

Primary cell cultures are those that are composed of cells taken directly from a living normal animal. Primary cells normally contain a diploid set of chromosomes, have a limited life span, and undergo aging. In order to prepare a primary culture, an organ or tissue is removed from a freshly sacrificed animal and aseptically cut into small pieces and treated with enzymes such as trypsin, collagenase, pronase, or a combination of these. After the cells have been separated from the tissue, they are incubated in an appropriate cell culture medium in tissue culture flasks. The cells will adhere to the surface of the flask and replicate until they come into contact with each other. Thus, they attach and grow as a uniform layer of cells, or a monolayer, which is always one cell thick. These cells stop growing once the surface is filled up and contact follows. This is known as contact inhibition. Cells can be cultured as stationary monolayers usually inoculated with 2×10^5 cells in 5 ml of cell culture medium.

Some cell types can be grown in a suspended state in a culture medium. Primary suspension cultures are generally derived from bone marrow, peripheral blood lymphocytes, and spleen cells. The permanent cell lines of suspension cultures can be established directly from leukemic cells or by infecting normal lymphocytes with the Epstein-Barr virus. Suspension cell cultures will yield 5–10 times more cells per ml of medium than monolayers. In this chapter, protocols for establishing primary cell cultures of chicken embryo fibroblasts, bone marrow, and spleen, and peripheral blood lymphocyte culturing are described.

Method-1: Establishment of primary cell culture of chicken embryo fibroblasts

It is fairly easy to derive a primary culture of fast growing cells like fibroblasts, and as a result, most primary cultures consist of fibroblasts. Fibroblasts are connective tissue cells, which secrete the extra-cellular matrix of connective tissue. In this protocol, primary fibroblast cultures will be derived from chicken embryo. They will be made by dissociating the entire embryo from the proteolytic enzyme, trypsin. Embryo cells are somewhat easier to adapt to cells culture than adult cells, since embryo cells are less differentiated and more likely to divide rapidly. The culture that is first produced will contain a variety of cell types. It is possible, however, to eventually establish a homogeneous culture of fibroblasts, because the other cell types do not grow as fast and will be diluted out in subsequent passage subcultures.

Materials

Fertile chicken eggs, incubated 7–10 days; 0.25% trypsin solution; PBS; tissue culture flask 25 cm^2; Hank's minimum essential medium (HMEM) + supplemental 10% fetal bovine serum (heat inactivated); three 100 mm plastic Petri dishes; one 50 ml plastic beaker; 150 ml squeeze bottle with 70% alcohol; A sterile dissection kit containing sharp scissors, sharp scalpel and two pointed forceps; 95% ethanol; 5 ml plastic syringe; sterile gauze; two 125 ml flasks; trypsinizing flask; 15 ml centrifuge tubes; trypan blue (0.4%); 0.15 ml plastic microcentrifuge tube with an attached cap; penicillin-streptomycin 100X stock solutions; sterile 1 ml and 5 ml pipettes; CO$_2$ incubator.

Procedure

1. Keep 0.25% trypsin, HMEM and PBS in 37°C water bath.
2. Aseptically place 15 ml of PBS into each sterile Petri dish.

3. Place the egg in a 50 ml beaker with its blunt end up, and disinfect the entire egg shell with 70% ethanol.
4. With a sharp sterile forceps, puncture the top of the shell and remove the shell all the way up to the base of the air cell.
5. Locate the position of the embryo and carefully tear the membrane away from the embryo with sterile forceps.
6. Insert a sterile pair of forceps into the egg and grab the embryo firmly. Take it out of the egg and place it in the first dish of PBS.
7. Cut off the head and feet of embryo with a sterile pair of scissors. This procedure is done for a 10-day-old embryo, not for a younger one.
8. With a flamed sterile pair of forceps, transfer the remaining parts of the embryo to a Petri dish containing PBS and rinse thoroughly. Repeat the rinse until no color is found in the rinse solution in the dish.
9. Take a sterile 5 ml syringe and a sterile capped test tube. Remove the plunger from the syringe and rest it on sterile Kimwipes (paper towels). Use a sterile pair of forceps to transfer the remaining parts of the embryo from the Petri dish into the syringe. Remove the cap from the tip of the syringe without touching the tip with your fingers. Uncap the sterile test tube, place the cap on sterile paper towels, and keep the tube in a test tube rack. Position the tip of the syringe over the test tube. After inserting the plunger back in the syringe, push it all the way down with a steady and even pressure so that the embryo is pushed through the syringe into the test tube. The embryonic tissues are delicate and are homogenized to very fine pieces.
10. Transfer the tissue into a trypsinizing flask and add 10 ml of 0.25% pre-warmed trypsin solution.
11. Place the trypsinizing flask on the shaker in the 37°C water bath and allow the suspension to swirl on the shaker at low speed so that the tissues of the embryo barely graze the cutting edges of the trypsinizing flask. It is necessary to continue this action for 15–60 min., depending on the disaggregation process. This process has to be monitored closely. The flask is removed once the trypsin solution becomes cloudy with individual cells. Excessive trypsinization reduces the cell viability significantly.
12. Allow fragments to settle; the supernatant is a cloudy suspension of cells. Pour this cell suspension through the side spout into a sterile 125 ml flask containing 10 ml of pre-warmed HMEM with 10% FBS. The serum will inhibit further action of the trypsin and prevent further damage to cells.
13. Obtain another sterile 125 flask and place two or three layers of sterile gauze over the mouth of the flask and filter the cell suspension through this gauze to remove large cell clumps and other debris.
14. Take two clean sterile centrifuge tubes. Aseptically transfer the cell suspension into these tubes so that both tubes contain the sample quantity. Pellet the cells by centrifugation in a table top. Centrifuge for 5 min. at 1000 rpm. Discard the supernatant and re-suspend the cell pellets in 5 ml of HMEM containing 10% FBS. Combine the cell suspension in one tube and cap it.
15. Remove the cap of the tube containing the cell suspension and, using a sterile 10 ml pipette, aspirate the suspension by drawing it up into the pipette and then forcing it back into the tube. Repeat about six times. It is important to keep the cell suspension uniform and the clumps broken up as much as possible. Determine the viability and cell density.

16. With a sterile 1 ml pipette, aseptically transfer 0.1 ml of cell suspension into a 1.5 ml microcentrifuge tube and determine the cell density by the hemacytometer.
17. Adjust the cell density to 5×10^5 cells/ml by proper dilution. Aseptically transfer 5 ml of the diluted cell suspension into each of two 25 cm^2 tissue culture flasks.
18. Label each flask with group name, date, and place in an incubator at 37°C. Humidity can be provided by placing a pan of water in the incubator.
19. After 12–24 hours observe the cultures, pour off the old medium and replace it with 5 ml of fresh HMEM supplemented with 10% FBS. Change the medium every 2–3 days until the cells have become confluent. Notice how the culture changes over time into a homogeneous line of fibroblasts.

Method-2: Establishment of rodent bone marrow and spleen suspension cell cultures

These cells grow continuously in suspension. No enzyme treatment is required for sub-culturing. Subculturing can be done by diluting the cell density or by centrifugation followed by media replacement.

Materials

Animals
Disease-free, healthy, and genetically pure strains of animals are essential for good cell culture initiation. Sprague–Dawley male rats, 6–8 weeks old are used in this protocol. Immediately after receipt of the animals, they have to be acclimatized for a period of at least two weeks in the animal quarters or in some clean and disease-free area in the laboratory.

Equipments
Dissection box (scissors, forceps, needles, scalpel, etc.); 25 cm^2 Falcon culture flasks; 10 ml syringes with 20 gauge needles; 15 ml centrifuge tubes, 1-, 5- and 10-ml sterile pipettes; Spatula; Incubator with humidified atmosphere containing 5% CO_2; Hot water bath; Bench top centrifuge.

Solutions
Heat-inactivated fetal bovine serum (HI-FBS); RPMI 1640; penicillin and streptomycin; Phosphate buffered saline (PBS); F-12 medium; L-glutamine; 2-ME; lipopolysaccharide (Sigma); 70% alcohol; Pasteur pipettes with rubber bulb.

Bone marrow complete culture medium: F-12 nutrient mix, 20% FBS, 1% penicillin-streptomycin. Filter-sterilize and store at 4°C.

Spleen complete culture medium RPMI 1640, 20% HI-FBS, 1% penicillin-streptomycin, 1% 200 mM L-glutamine, 1×10^{-5} M 2-mercaptoethanol and 0.4 ml lipopolysaccharide.

Procedure

Isolation and culture of bone marrow cells
1. Sacrifice the animal in a closed chamber containing 30% CO_2. An alternative is to use cervical dislocation.
2. Wet the legs and body hairs with 70% ethanol. Cut away the skin, make an incision at the hip, and dislocate the leg at the hip joint, carefully remove the muscle with a pair of scissors and a sharp scalpel. Obtain the tibia in the same fashion. If several animals are prepared, store bones submerged in the growth medium.

3. Remove cartilage from both ends of the bone to expose marrow cavity. Fill a 5 ml syringe connected to a no 1 needle with the growth medium, insert needle into the bone cavity, and flush the bone marrow into the 15 ml centrifuge tube. Repeated flushing of the medium through the bone may be required to obtain a good yield of bone marrow cells.
4. Carefully decant the cell suspension into a fresh tube avoiding the tissue debris; make up the volume to 5 ml, and centrifuge at 285 g for 6 min.
5. Carefully remove the supernatant, wash the cell pellet once more in 6 ml of the growth medium.
6. Re-suspend the cell pellet with 1 ml complete culture medium.
7. Determine the cell concentration by using the hemacytometer after optimum dilution.
8. In a 25 cm^2 Falcon culture flask containing 6 ml of complete culture medium, culture 1.5×10^6 cells per flask.
9. Incubate culture flasks at 37°C in a humidified atmosphere containing 5% CO_2.
10. After a 20–24 hour incubation, observe mitotic cell division.

Isolation and culture of spleen cells

The preparation of lymphocytes from mouse spleen is a technique that readily provides a source of functional T cells, B cells, and macrophages. Unlike the thymus and bone marrow, which contain undifferentiated lymphoid and myeloid cells, spleen is classified as a secondary lymphoid organ along with lymph nodes and peripheral blood and contains active differentiated cells. The spleen is easily dissociated into a single cell suspension and provides a high yield (~$1-2 \times 10^8$ cells per spleen). Cells obtained from the spleen are an excellent starting material for primary cell cultures. Cells obtained from spleen may be fractionated into T-lymphocytes, B-lymphocytes, and macrophages and their activities studied using *in vitro* systems. Spleenic lymphocytes from immunized animals are key starting cells for the production of hybridomas and monoclonal antibodies.

Procedure

1. After removal of the hind leg bones, the same animal can be used for spleen separation. After the mouse skin has been wet with 70% ethanol, place the left side facing up on paper towels.
2. Make a cut through the loose skin by pulling gently upward and using blunt scissors on the skin flap. This will expose the peritoneal wall.
3. Pull gently in opposite directions on the two sides of the skin incision to expose a wider area of the peritoneal wall.
4. With a small pair of surgical scissors, make a cut over the spleen to expose it through the peritoneum.
5. Grasp the spleen with the forceps. Pull up gently and using the pair of surgical scissors, cut away attached connective tissue which appears white.
6. Transfer spleen to centrifuge tube containing 3 ml sterile PBS with 2% HI-FBS.
7. Mash the spleen with a sterile spatula by rubbing spleen tissue between spatula and bottom of the centrifuge tube.
8. With a sterile Pasteur pipette, bring the cell suspension up and down several times to dissociate large cell clumps. Allow the cells to settle for 5 min. to remove large debris.
9. Use the Pasteur pipette to transfer cell suspensions to fresh 15 ml sterile centrifuge tube.

10. Make up the volume to 6 ml with PBS+2% HI-PBS and centrifuge at 285 g for 7 min. Discard the supernatant and re-suspend the pellet in 1 ml complete spleen culture medium.
11. Determine the cell density and culture 3–4×10^6 cells per 10 ml complete spleen culture medium in a 25 cm^2 Falcon culture flask. Incubate the culture flasks at 37°C in a humidified atmosphere containing 5% CO_2.
12. 20–24 hours after incubation, the spleen cells start dividing.

Method-3: Isolation and establishment of human peripheral blood lymphocytes

The protocol described below works well for human peripheral blood lymphocytes. This method can also be followed for other mammalian blood lymphocytes. However, the mitogen should be standardized for individual blood types separately.

Materials

Equipment
Heparinised 15 ml vacutainer or centrifuge tubes; 20 ml sterile syringe with 22 gauge needle; 25 cm^2 Falcon culture flasks; 15 ml and 50 ml sterile centrifuge tubes; 1, 5, and 10 ml pipettes; bench top centrifuge.

Solutions
Heparin; RPMI 1640 with L-glutamine and 25 mM HEPES buffer; Hank's balanced salt solution (HBSS); L-glutamine; Heat-inactivated fetal bovine serum (HI-FBS); Phytohemagglutinin (PHA); Penicillin-streptomycin; Lymphocyte separation medium (LSM, Organon Teknika).

Wash medium: HBSS + 2% FBS

Complete culture medium: RPMI 1640 medium with L-glutamine and 25 mM HEPES buffer plus 20% HI-FBS, an additional 1 ml of 2 mM L-glutamine, 100 units penicillin and 100 μg streptomycin/ml, and 1.5% PHA. Store at –4°C until use. Just a few hours before use, pre-warm medium in a 37°C hot water bath.

Procedure

Isolation of peripheral blood lymphocytes (PBLs)
1. In a heparinized sterile syringe, collect blood by venipuncture from a healthy individual.
2. Isolate PBLs by using LSM or Ficoll gradient. Mix whole blood with HBSS, in a ratio of 1:3 and gently layer over LSM (final ratio of 1:1.5 blood:LSM dilution, respectively) in a centrifuge tube (centrifuge tube size varies depending on blood sample volume).
3. Centrifuge for 30 min. at 400 g. Carefully remove interface mononuclear cells (white band between LSM and sample solution) with a sterile Pasteur pipette into a fresh sterile centrifuge tube (Figure 28.4).
4. Dilute the PBLs with wash medium, top up the centrifuge tube with wash medium. Repeat washing three times.
5. After final wash, re-suspend the cell pellet in a culture medium and determine the cell number.
6. Culture the PBLs in a pre-warmed complete culture medium containing 0.5×10^6 cells/ml, 5–10 ml volumes in 25 cm^2 Falcon culture flasks.
7. Incubate the cultures at 37°C in a humidified atmosphere containing 5% CO_2.
8. PBLs start dividing mitotically 20 to 24 hours after culture initiation.

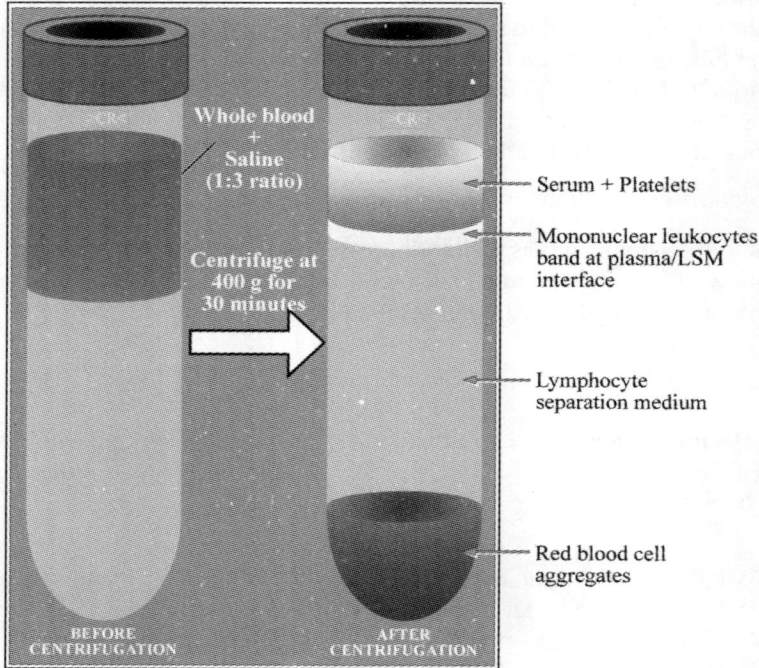

Figure 28.4 *Isolation of peripheral blood lymphocyte (PBLs). Collect fresh blood in a heparinized sterile syringe. Mix with physiological saline at a ratio of 1:3 and gently layer over lymphocyte separation medium (LSM) at room temperature. Centrifuge tube at 400g for 30 minutes. Mononuclear leukocytes band at plasma/LSM interface. Remove top layer of clear plasma then carefully remove the lymphocyte layer plus about half of LSM layer into a sterile centrifuge tube. Wash once or twice with saline supplemented with 2% bovine serum*

H. SYNCHRONIZATION AND ISOLATION OF SYNCHRONIZED CELLS

Cells grow in a metabolically regulated cycle of events. In mammalian cells, the stages of the cycle are quite distinct. They are: G_1, when the cell is making all the components for the new cell, except DNA; the S phase, when the DNA is replicated; G_2, when the cell organizes itself to divide; and the M phase (or mitosis), when the cell actually divides into two. The M phase is the phase during which the DNA is organized into macroscopic chromosomes, which can be seen under the microscope after staining. In a cell culture, cells are at a mixture of stages in the cell cycle. There are a range of chemical tricks to arrest them at one point (Figure 28.5), so that all the cells accumulate at, for example, the start of the D phase. After release, the cells march through the cell cycle together: this process is called *cell synchronization*, and the result is a synchronized cell population.

An asynchronous population of cells can be separated into fractions enriched for the G_1, S, and G_2/M phases by centrifugal elutriation. For mammalian cells in culture, the early G_1 cells are the smallest whereas cells in the G_2/M phase cells are the largest. Centrifugal elutriation can separate a large number of cells into populations of uniform size. Soon after mitosis, a mam-

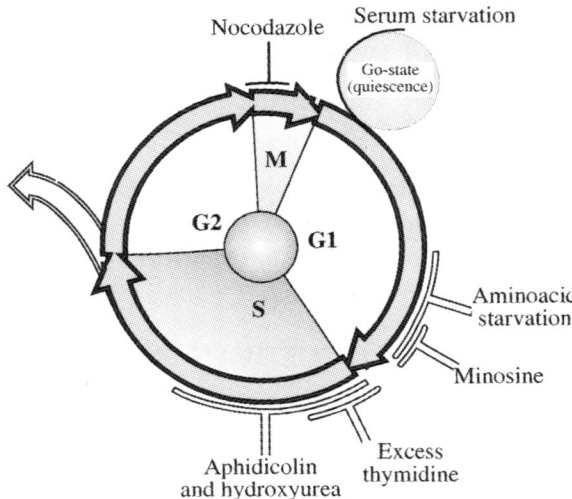

Figure 28.5 *Schematic illustration showing the point in the cell cycle that cells are arrested following applications of different synchronization methods. Note that not all the treatments outlined here are applicable for cell synchronization of every cell type*

malian cell will divide into two identical cells that are each half the size of the original mitotic cell. These early G_1 cells will continue to grow as they advance through each phase of the cell cycle so that S-phase cells are of intermediate size. Cells grown in suspension are easier to elutriate since they tend not to adhere to each other and have a rounded, uniform shape.

Centrifugal elutriation can be used to synchronize an exponentially growing population of cells in two different ways. A single fraction of the small, early G_1 cells can be collected and then re-inoculated into culture. Alternatively, cells can be elutriated into fractions enriched for early the G_1, S, or G_2/M phase (for details see Krek and Decaprio (1995).

VI MASS CULTIVATION OF CELLS IN BIOREACTOR SYSTEMS

In principle, cells can be cultivated under two basically different principles: As freely suspended cells or as an immobilized culture on solid phases or embedded in suitable materials. The choice of cultivation technique depends mainly on whether the cells are capable of growth in an adherent or anchorage-independent condition. In a suspension culture, the cells are prevented from clumping by medium agitation. In contrast, adherent cells are attached to and spread on a solid surface and are thus cultivated as an immobilized culture. Cells normally growing in free suspension may be induced to attach to and spread on a surface after suitable pre-treatment of the surface with attachment factors such as fibronectin or vitronectin. The mode of operation can be batch or continuous and variations of them such as repeated batch, fed-batch, or continuous culture with cell retention.

Integrated suspension culture: Reliability and reproducibility of the cell culture process are important preconditions to ensure quality and, thus, the activity and safety of the desired product. This requires the precise control of the biological and biochemical functions of all producer cells in the bioreactor and can be achieved only by optimal control of a fully defined cellular microenvironment affecting all physical and chemical, biochemical, and biological parameters (Figure 28.6). As each living cell permanently generates gradients due to its metabolism, an optimized process must be to keep the conditions in the reactor as homogeneous as possible. The only system that can ensure these conditions is a bioreactor with freely suspended cells in an agitated liquid medium of low viscosity. One of the most frequently used laboratory-scale vessels for suspension culture is a magnetically stirred spinner flask. This system has the advantage of simplicity and cheapness. It is also suitable for large-scale cultivation. Thus, a single-unit process in a well-adapted bioreactor is required.

A. IMMOBILIZED ANIMAL CELL-CULTIVATION SYSTEMS

The first step in immobilization of cells is to evaluate the material for cell adhesion.
(i) The most common materials used in microcarrier beads are a *charged surface*. DEAE sephadex is positively charged, whereas polystyrene is negatively charged. Gelatin beads can be slightly charged positive or negative, depending on the isoelectric point of the raw material and the method of preparation. For sufficient cell adhesion they must be coated with fibronectin. It seems that surface charge does not correlate with cell adhesion; rather, the important parameter for cell adhesion is the surface energy of the material. Most plastic materials in their native state show values of low surface energy of the material (30–40 ergs/cm^2). After surface treatment, the surface energy rises to 56 ergs/cm^2 or more, and cell adhesion becomes possible.

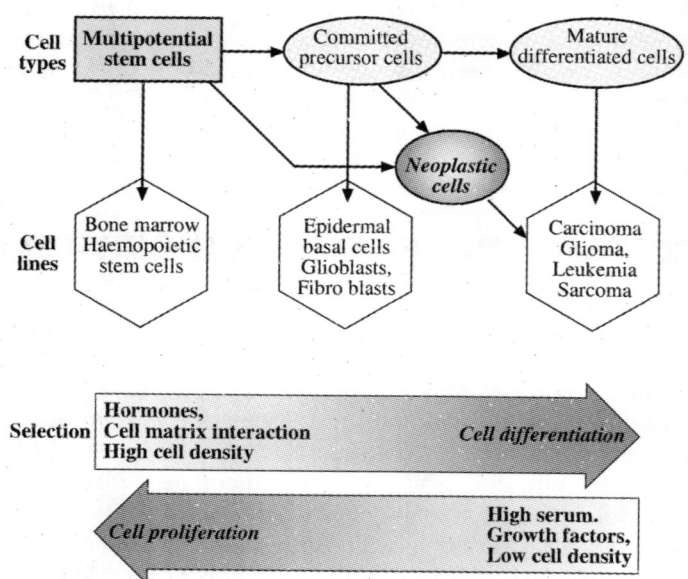

Figure 28.6 *Culture conditions affecting cell types and origin of cell lines (Redrawn from Dubey 1993, S. Chand & Company Ltd.)*

(ii) *Fixed immobilized beds*: Glass and ceramics, which have a naturally high surface energy, are well suited as cell carriers. In order to increase the surface area for mass cell cultivation, different materials for cell attachment are packed in different types of bioreactors (packed bed reactor). Among the various materials that have been developed and evaluated, glass of various shapes (spheres, rings, tubes, rods, and helices) has the advantage of stable packing with a regular channel size running through the bed. However, bed volumes over one litre cause severe problems due to nutrient depletion and waste product accumulation originating from unfavorable fluid dynamics in fixed bed reactor columns.

(iii) *Hollow fiber systems*: Conventional hollow fiber-filtration devices have been applied to higher density cell cultures. Thin hollow ultrafiltration fibers bundled together in a cylindrical cartridge provide a large surface area for perfusion of nutrients into the extracapillary space where the cells are maintained in densities of up to 10^8 per ml.

(iv) *In the opticell system, mammalian cells are immobilized by adsorption to a ceramic matrix*. The constructive principle originates from a car exhaust catalyzer. Many small square channels, each with a cross-sectional area of about 1.5 cm^2, run the length of a cylinder. They form a rigid cell support for adherent cells as well as for attachment of anchorage-independent cells. The matrix is built into a perfusion loop providing for the constant recirculation of the medium.

(v) *Fluidized bed systems*: In order to avoid concentration gradients within the reactor, porous microcarriers have been used as cell supports in suspension in fluidized bed systems. The density of the particles is adapted to that of the medium by incorporation of inert weights. As a consequence, the particles can be fluidized with a minimum of energy input by medium flow. A fluidized bed, as described, has the advantage over a fixed-bed system in that under optimal conditions, gradients are minimized to the scale of single beads. As growing cells do not block the porous structure, transport of nutrients into and cell products out of spheres is enhanced by medium flow resulting from the tumbling movements of spheres in the medium stream.

B. PRODUCT FORMATION

Product formation in animal cells involves two types of kinetics. Growth related production: an example of this is t-PA (tissue-type plasminogen activator) from CHO or melanoma cells, which express the desired protein mainly during the synthesis (S) phase of the cell cycle. In this case, the main objective of the process is to keep the maximum proliferation activity and high cell density states.

High production rates from cells resulting in the stationary phase: An example of this is antibody production by hybridoma cells. This often occurs in the state where cells are arrested in the late G_1 phase or when they are even in the G_0 phase due to unfavorable culture conditions. As many hybridoma cell lines are unstable with respect to antibody production, they must be periodically recloned to select for highly producing cells. This instability provides an incentive to design a process where cells, after having reached their maximal density, are maintained for a long time at high viability in a non-proliferative state.

C. LARGE-SCALE PURIFICATION OF PHARMACEUTICAL PRODUCTS

The downstream processing steps include all purification and handling steps of the product from the crude cell culture medium until storage as pure substance in a container. Highly controlled pharmaceutical procedures have to assure that a native substance reaches the final vial with a high lot-to-lot consistency. The final product quality demands are very high. For example, (i) there should be no structural changes brought about by unfavorable downstream conditions; (ii) the purity (extraneous protein, DNA, and endotoxin levels) must fulfill the specifications of the corresponding national administration guidelines; (iii) the structure and activity in the final container for application should be stable; (iv) the product should be free from high molecular weight aggregates that could reduce activity and induce harmful immunogenic side effects; (v) the product should be free from mycoplasmas and viruses.

The first crucial step in product purification is the rapid removal of the product from the unfavorable degradative conditions existing in the harvested culture medium. A continuous harvesting process with short product residence times, dealing with high medium flow rates from high-density reactors is favored. In order to minimize proteolytic degradation and aggregation, the medium is usually cooled down to 4°C as soon as harvested. Before purification, a 50- to 100-fold concentration of the product is performed by ultrafiltration with hollow fiber, plate, and frame, or a spiral cartridge system. Multiple ultrafiltered concentrates are pooled to obtain a larger lot size. After adjusting salt concentration and pH, purification is performed by means of chromatographic methods. All steps of processing are performed under the highest possible standards of hygiene.

VII CLONAL CULTURE AND COLONY SELECTION

Since hemopoietic populations are mixtures of cells of different lineages and at different stages of differentiation it is of course necessary to pre-fractionate the population to be cultured in order to determine the physical properties of a clonogenic hemopoietic cell. A number of methods have been standardized for pre-fractionation of mixed cell populations e.g. hemopoietic cells, peripheral lymphocytes, etc. It should be emphasized, however, that most hemopoietic cells rapidly die in culture in the absence of the relevant specific stimulating factor so that, for example, in cultures containing only GM-CSF, non-GM cells will rapidly disintegrate, leaving essentially pure populations of developing GM clones. For these and other reasons, there is rarely justification for elaborate prefractionation procedures simply on the grounds of removing irrelevant cells prior to culture. Where a prefractionation step has been used, it is essential to determine cell counts before and after fractionation and to record these in any published description of the data so that observed frequency of clonogenic cells can be related to the actual frequency in the starting population.

However, the situation is quite different when cultures are being used to detect colony stimulating or modulating factors. Precise assays of the concentration of a colony stimulating factor or determination of which cells are responding directly may well require the use of highly purified clonogenic target cells.

Clonogenic cultures are usually made semisolid by inclusion in the culture medium of either agar or methylcellulose. Each method has its own merits and demerits.

A. CLONOGENIC ASSAY IN SEMISOLID MEDIUM

1. Clonogenic cultures using semisolid agar medium

Most workers use disposable 35 mm plastic Petri dishes for hemopoietic cultures. The culture volume is usually 1 ml but these dishes will hold up to 2 or 3 ml. The principles involved in preparing semisolid cultures containing agar are to mix a double strength tissue culture medium and serum held at room temperature or 37°C with an equal volume of double strength agar (0.6% previously boiled for 2 min. to dissolve and sterilize the agar). The cells to be cultured are added and the cell suspension in the agar medium is pipetted into Petri dishes containing the stimulus required for the proliferation of the cells under study. After mixing, the cultures are allowed to gel and are then incubated.

A wide variety of cell culture media have been used successfully to support hemopoietic colony formation *in vitro*. Dulbecco's Modified Eagle's Medium (DMEM) or Iscove's Modified Dulbecco's Medium (IMDM) is routinely used. The availability of commercially prepared media in powder form greatly simplifies the preparation of the media. A double strength medium is most conveniently prepared by omitting the serum and can be stored at 4°C for up to 2 weeks.

Double strength DMEM: 10 g DMEM-HG 16 Instant tissue culture powder (GIBCO); 390 ml double distilled water; 3 ml L-asparagine (6.7 mg/ml; final concentration in medium 20 µg/ml); 1.5 ml DEAE Dextran (50 mg/ml final concentration in medium 75 µg/ml); 0.575 ml penicillin (2×10^5 IU/ml); 0.375 ml streptomycin (200 mg/ml); 4.9 g $NaHCO_3$. The DEAE dextran is included because of some evidence that it may inactivate toxic material in the agar, and additional L-asparagine is added because GMCF is highly sensitive to depletion of L-asparagine.

Bacto-agar (Difco) is the type of agar routinely used and has proved superior to more refined materials such as agarose. For most hemopoietic cultures, 20% fetal calf serum has been found to be optimal.

Preparation of agar cultures for eight one ml cultures
1. In a 15 ml plastic tube, add 3 ml double strength medium, then 2 ml of fetal calf serum, and finally, 5 ml of 0.6% agar.
2. Mix the medium thoroughly and add the cell suspension. The cell suspension gets thoroughly dispersed in the medium.
3. Transfer 1 ml of aliquots of the cell suspension in agar medium into the culture dishes, using a 5 or 10 ml pipette.

Procedure

1. Record batch numbers of medium, fetal calf serum, and other reagents to be used.
2. Number the empty culture dishes on the lids.
3. Collect and count the cells required for culture, adjusting cell concentrations where necessary.
4. Pipette the stimulating materials being used to the empty culture dishes. The usual volumes used for 1 ml cultures are 0.05–0.2 ml, and these will dry out if allowed to remain for too long before completion of preparation of the cultures.
5. Boil 0.6 g agar in 100 ml water for 2 min. and allow to cool at 37°C in a water bath.
6. Measure out the required volume of double strength medium and add the required volume of fetal calf serum. Mix well.

7. Add the required volume of agar. Mix well.
8. Add the required volume of cell suspension and mix thoroughly.
9. Pipette the 1 ml volumes of cell suspension in agar medium to each culture dish, using a pipette.
10. Mix the agar medium with the stimulus using a circular movement of the culture dish completing 2–4 circles, without splashing medium on the lid of the culture dish.
11. Allow cultures to gel on a bench that is cool and vibration free. Test that the cultures have in fact gelled, place an identifying label on the edge of the tray, then gently place the tray in a CO_2 incubator.
12. Rinse out all agar-containing glassware and syringes with hot tap water to remove visible agar and immerse glassware in a container with detergent.
13. Record the cultures in a daybook on the date due for examination and scoring.

Notes

1. The number of cells to be plated in a 1 ml culture depends on the type of experiment and the type of cells being cultured. In some instances, it may be at very low densities e.g. 25000 or even 50/ml, while in other instances as many as 10^6 cells/ml might be appropriate. In general, it is pointless attempting to grow dispersed colonies at more than 100/ml culture medium since colony crowding will prevent precise identification and counting of colonies.

2. Clonogenic cultures using semisolid methylcellulose-medium

The method of choice to clone hematopoietic cells and to isolate transformed clones is culture in semisolid medium (Figure 28.7). Methylcellulose (Methocel) is highly superior to soft agar since the colonies can be isolated and dispersed much more easily. Tested batches of methocel are commercially available.

Materials

Preparation of medium
1. Prepare the double strength Iscove's Modified Dulbecco's Medium (IMDM) by adding to 500 ml of double glass distilled water: 1 packet IMDM (GIBCO), 3.25 g $NaHCO_3$, 60 mg penicillin, 100 mg streptomycin, 5.9 µl 2-mercaptoethanol. Then mix this and pass through a millipore filter.
2. Add 20 g Methocel (65 HG, 4000m Pa.s,) to 500 ml boiling double glass distilled water. Cover the flask with aluminum foil and swirl the flask vigorously until no more lumps are visible. Boil for a further 5 min. vigorously shaking to sterilize.
3. Cool to room temperature and add to the 500 ml of sterile double strength IMDM. Hold overnight at 4°C stirring continuously.
4. Disperse in 50 ml aliquots into capped plastic centrifuge tubes. Shock-freeze and store at −15°C for 4 weeks.

Preparation of eight one ml cultures
1. Thaw a 50 ml tube of methylcellulose medium at room temperature for 2 hours (do not thaw on water bath, as this may create lumps in medium).
2. To prepare eight 1 ml cultures, place a series of eight 10 ml capped plastic centrifuge tubes in a rack, and add the growth medium (conditioned medium or growth factor) solution in a volume of 0.25 ml to each one.

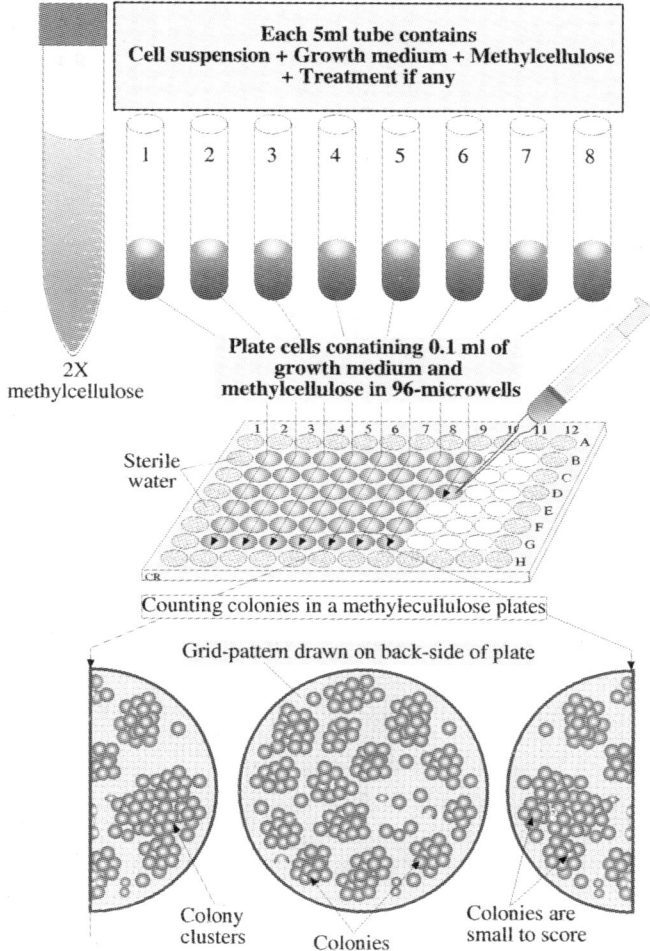

Figure 28.7 *A schematic representation of the clonogenic assay in methylcellulose*

3. Add to a 50 ml capped centrifuge tube 6 ml methylcellulose medium, using a 10 ml syringe and gauge 18 needle.
4. Add 2.5 ml FCS and the cell suspension in a volume of 1.5 ml.
5. Mix well by repeated inversion.
6. Using a 1 ml syringe and 18 gauge needle, dispense 1 ml of the cell suspension in methylcellulose medium to each tube containing the growth medium.
7. Mix methylcellulose growth medium and cell suspension using a Vortex agitator, and then dispense 1 ml to each Petri dish, using a 1 ml syringe and gauge 18 needle. The final culture will contain 1.2% methyl cellulose and 1X IMDM and growth medium.
8. Roll each culture dish to obtain uniform spreading of the medium over the bottom of the Petri dish.
9. Incubate for 2 days at 37°C in a fully humidified atmosphere of 5% CO_2 in air.

(a) Clonogenic assay in methyl cellulose plated in 96-microwell plates

Determination of plating efficiency (PE) and survival curves (dose response curves) are commonly used methods for suspension cell cultures. Clonogenic cells can be detected by their capacity to give rise to colonies in methylcellulose with growth medium. Drug-response curves (survival curves) in the methylcellulose can be obtained by adding increasing concentrations of agent to the wells, together with the cells, medium, and methylcellulose. The clonogenic assay in methylcellulose, plated in 96-wells have many advantages over other methods. First, it uses less cell culture medium (0.1 ml per microwell), colonies can be easily counted, and the individual colonies can be isolated and recloned. A large number of cultures can be set up in a single 96-microwell plate (60 out of the 96 wells can be used). There will not be any mix up between the treatments and the culture conditions can be maintained uniformly. Counting of colonies manually under an inverted microscope is also very fast.

Procedure

1. Repeat steps from 1 to 7 from the previous experiment. However, if immortal suspension cells are used, adjust the cell number 10^4 cells per 1 ml in a cell, medium, methylcellulose mix. For the unfractionated or partially fractionated primary cells, cell density should be higher, depending on their mitotic division.
2. Using a fine-tipped blue marking pen and plastic ruler, draw vertical and parallel lines on the reverse of a 96-microwell (ELISA) plate, in such a way that these markings form a grid pattern (~5 mm between the lines) on the outside-bottom surface of wells. The lids of the plates should not be opened during this process to ensure that the aseptic conditions of the culture wells are maintained. The marking of plates should be performed inside the safety hood to prevent any contamination. These markings not only help to locate the colonies but also facilitate easy and accurate counting of the colonies by scoring through the squares. Direct marking of lines on culture plates itself will facilitate bringing the markings and colonies to a single focal point.
3. Add about 0.3 ml of sterile distilled water to the border wells, to prevent drying off of the cell culture medium.
4. By using a 1 ml syringe and an 18 gauge needle, dispense 0.1 ml cell mix to each well for at least 4 wells (in a single row) per each treatment aseptically.
5. Label the plates and record all the operations including the number of treatments, number of replicates, date, and the initials of a worker.
6. Incubate the culture plates for 4–5 days at 37°C in a fully humidified atmosphere containing 5% CO_2.
7. Final colony counts can be done 4–5 days after plating under an inverted microscope.
8. For accurate counting, draw grid-lines on the reverse of the culture plate with the help of a fine marker pen and ruler as shown in Figure 28.7. Count the colonies between grid lines. This method helps avoid over or under counting of colonies.

(b) General procedure for scoring and analysis of colonies

Cultures of 1 ml of semisolid medium in 35 mm Petri dishes are three-dimensional structures that can contain up to several hundred objects ranging from 10 μm to 1–2 mm in diameter. Generally, an unstained culture can be scored using either an inverted microscope or a dissection microscope with a light source in the base plate. Counting colonies with an inverted microscope is intolerably slow because of the continuous re-focusing needed to inspect all depths of the culture and because of the slow action of most mechanical stages. Use of an inverted microscope for prolonged periods causes severe eyestrain. The variable size and shape of normal hemopoietic colonies and the likelihood that some colonies are superimposed in the three dimensional gels present formidable difficulties for the design and programming of automated counters. One or two instruments are available that are beginning to approach the necessary requirements but, in general, the scanning process is extremely slow and counts for colonies, other than those of very uniform and compact shape, can be unreliable. One approach to this problem has been the development of agar cultures in capillary tubes that can be scanned by an automatic counter. But this solution is impracticable for the average laboratory.

The scoring of virtually all types of hemopoietic colonies can only be performed comfortably and quickly using a dissection microscope with stereo-zoom optics and capable of adjustment to provide semi-indirect lighting. This instrument offers a continuous range of magnifications of up to 40X, and most colonies are best scored at 35X. Counting colonies in a 35 mm Petri dish is best performed by drawing parallel lines using a blue fine-tip marker pen on the outside bottom surface of the culture dish, to form a grid pattern (~5 mm between the lines). The direct marking of lines on the culture plates itself will help to bring markings and colonies to a single focal point. By this marking technique, it is possible to count most culture dishes without using a mechanical counter.

All culture experiments should include control cultures in which there is the absence or presence of the stable stimulating factor, or non-treated cultures in case of drug effect studies. Most cultures contain a wide variety of colony shapes and sizes. Scoring of cultures requires concentration, a commitment to accuracy, a preparedness to recheck when puzzling results occur, and an awareness that all cultures have the potential to reveal new phenomena, which may require thought and perhaps further experimentation.

(c) Criteria for scoring colonies

Colonies are conventionally defined as clones containing 50 or more cells. Most cultures contain in addition to colonies, larger numbers of clones of sub-colony size (clusters). With practice, it becomes quite easy to make a visual judgment as to which clones are large enough to be scored as colonies, and there is usually good agreement in colony counts made by two experienced observers. The beginner often has difficulty in deciding what an aggregate of 50 cells looks like, particularly if the cells are small and tightly packed. It is usual to underestimate the number of cells in colonies and therefore to underestimate colony numbers. The criteria used for scoring colonies ideally should be those used commonly by other workers in the field. The accurate enumeration of colonies requires that each individual clone be discrete and clearly separate from adjacent clones. Obvious problems will arise when too many colonies develop in a culture dish or where sedimentation in methylcellulose has led to superposition and merging of colonies. In most cases, this problem can be overcome by common sense but the occurrence of multicentric colonies can provide a formidable problem in deciding whether a collection of many discrete subcolonies really represents one multicentric colony.

(d) Picking off of colonies

This procedure is best performed using a finely drawn Pasteur pipette with a rubber bulb. The culture dish is held steady with the left hand and the pipette introduced into the culture under direct vision at an angle of 45°. Colonies are picked off using a very gentle but definite scooping action, scooping in the direction of the left hand. The colonies are not sucked into the pipette and the bulb is used essentially only to deliver the colony onto the slide. The volume of agar removed should be no more than that immediately surrounding the colony, and the agar droplets should not be allowed to be sucked far beyond the tip of the pipette. Colony removal should be confirmed by re-inspection of the area in the culture dish and the remainder of the colony removed if necessary. The transfer of clones is completed by placing the recipient culture under the microscope with the area of a reference circle in vision and by placing the agar droplet on the surface of the culture over the reference circle. The colonies from the methylcellulose can be directly transferred into the recipient medium in a 96-microwell.

VIII FROZEN STORAGE OF ANIMAL CELLS

To preserve cells, avoid senescence, reduce the risk of contamination, and minimize the effects of genetic drift, cell lines may be frozen for long-term storage, but a cryoprotective agent such as dimethylsulfoxide (DMSO) must be included.

Materials

Log-phase monolayer culture of cells in Petri plate; Complete medium; Freezing medium: complete medium supplemented with 10% to 20% (v/v) FBS and 5% to 10% (v/v) DMSO, 4°C. Bench top clinical centrifuge with 45°C fixed-angle or swinging-bucket rotor.

Procedure

1. Trypsinize cells in log-phase growth as explained earlier.
2. Transfer the cell suspension to a sterile centrifuge tube and add 2 ml complete medium with serum. Centrifuge for 5 min. at 300 to 350 g, at room temperature.
3. Remove supernatant and add 1 ml of 4°C freezing medium. Re-suspend pellet. Add 4 ml of 4°C freezing medium, mix cells thoroughly, and place on wet ice.
4. Pipette 1 ml aliquots of cell suspension into labeled 2 ml cryovials. Tighten caps on vials. Place vials 1 hour to overnight in a −70°C freezer, then transfer to liquid nitrogen storage freezer. Keep accurate records of the identity and location of cells stored in liquid nitrogen freezes.

Thawing and recovering cells

When cryopreserved cells are needed for study, they should be thawed rapidly and plated at high density to optimize recovery.

Materials

Cryopreserved cells stored in liquid nitrogen freezer; 70% (v/v) ethanol; complete medium containing 20% FBS, 37°C.

Procedure

1. Remove the vial from the liquid nitrogen freezer and immediately place it into a 37°C water bath. Agitate vial continuously until the medium has thawed (~60). Wipe top of vial with 70% ethanol before opening.

2. Transfer thawed cell suspension into a sterile centrifuge tube containing 2 ml warm complete medium containing 20% FBS. Centrifuge for 10 min. at 150 to 200 g, at room temperature. Discard supernatant. Wash cells with fresh medium to remove residual DMSO.
3. Gently re-suspend cell pellet in small amounts (~1 ml) of complete medium/20% FBS and transfer to properly labeled culture plate containing the appropriate amount of medium. Check cultures after ~24 hr to ensure that cells have attached themselves to the plate. Cultures are re-established at a higher cell density than that used for original cultures.
4. Change medium after 5 to 7 days or when pH indicator in medium changes color. Keep cultures in medium with 20% FBS until cell line is re-established.

Determining cell number and viability by hemacytometer

Determining the number of cells in culture is important in standardization of culture conditions and in performing accurate quantitation experiments. A hemacytometer is a thick glass slide with a central area designed as a counting chamber (Figure 28.8).

Materials

70% (v/v) ethanol; Cell suspension; 0.4% (w/v) trypan blue or 0.4% (w/v) nigrosin, prepared in HBSS; Hemacytometer with coverslip; hand-held counter.

Procedure

1. Clean surface of hemacytometer slide and coverslip with 70% alcohol.
2. Wet edge of coverslip slightly with tap water and press over grooves on hemacytometer. The coverslip should rest evenly over the silver counting area.
3. For cells grown in a monolayer culture, detach cells from surface of dish using trypsin. Dilute cells as needed to obtain a uniform suspension. Disperse any clumps.
4. Use a sterile Pasteur pipette to transfer cell suspension to the edge of hemacytometer counting chamber. Hold tip of pipette under the coverslip and dispense one drop of suspension, which will be drawn under the coverslip by capillary action. Fill second counting chamber.
5. Allow cells to settle for a few min. before beginning to count. Blot off excess liquid. View slide on microscope with 100X magnification.
6. Position slide to view the large central area of the grid; this area is bordered by a set of three parallel lines. The central area of the grid should almost fill the microscope field. Subdivisions within the large central area are bordered by these parallel lines and each subdivision is divided into sixteen smaller squares by single lines. Cells within this area should be evenly distributed without being clumped. If the cells are not evenly distributed, wash and reload hemacytometer.
7. Use a hand-held counter to count cells in each of the four corner and central squares. Repeat counts for the other counting chamber. Five squares (four corners and one center) are counted from each of the two counting chambers for a total of ten squares counted. Count cells touching the middle line of the triple line on the top and left of the squares. Do not count cells touching the middle line of the triple lines on the bottom or right side of the square.
8. Determine cells per ml by the following calculations:

$$\frac{\text{Cells}}{\text{ml}} = \text{average count per square} \times \text{dilution factor} \times 10^4$$

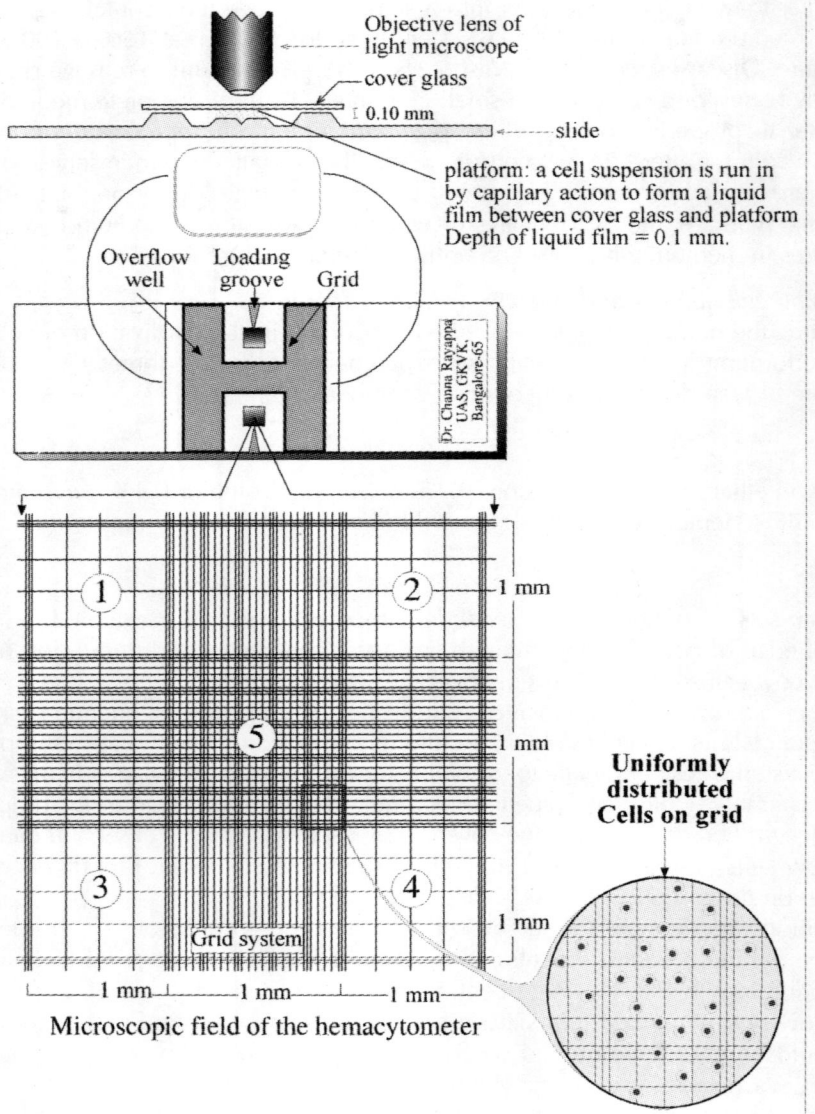

Figure 28.8 *Total and viable cell counts by haemacytometer*

$$\text{Total cells} = \frac{\text{cells}}{\text{ml}} \times \text{total original volume of cell suspension from which sample was taken.}$$

10^4 is the volume correction factor for the hemacytometer: each square is 1×1 min. and the depth is 0.1 mm.

9. Determine the number of viable cells by adding 0.5 ml of 0.4% trypan blue, 0.3 ml HBSS, and 0.1 ml cell suspension to a small tube. Mix thoroughly and let it stand for 5 min. before loading hemacytometer. Either 0.4% trypan blue or 0.4% nigrosin can be used to

determine the viable cell number. Non-viable cells will take up the dye, whereas live cells will be impermeable to it.
10. Count total number of cells and total number of viable cells. Calculate percentage of viable cells as follows:

$$\% \text{ Viable cells} = \frac{\text{Number of unsaturated cells}}{\text{Total number of cells}} \times 100$$

11. De-contaminate coverslip and hemacytometer by rinsing with 70% ethanol and then deionized water. Air-dry and store for future use.

IX GENETICALLY ENGINEERED ANIMAL CELLS

Since Enders discovered in 1949 that the Poliovirus could grow in primary green monkey kidney cells in roller tube cultures, animal cell cultures have formed the basis for the production of many virus vaccines. The step forward to genetically engineered animal cells began with the report that the fusion product of a human and a mouse cell could be viable, and produce antibody molecules. The era of "hybridoma" technology for the production of monoclonal antibodies (Refer to Chapter 26) had arrived. In addition to cell fusion products, there are a wide variety of alternative techniques, some of which have resulted in the production of new cell lines for proteins which are likely to be manufactured from some such cells in commercial areas.

There are many vector systems that can be used to transform animal cells (see Chapters 16 and 18). Both the SV40 and polyoma vectors have been used as lytic vectors in cells. They are harbored in constitutive fashion in viral genes, which complement for the genes, which have been removed from transducing vector. However, the calcium phosphate precipitation method is still a popular method of transformation. The selection of the cell, which has taken up the DNA, is a technology developing rapidly. A variety of markers are available which can be incorporated into an engineered vector or which can be added to the gene to be transferred before forming the calcium phosphate precipitate.

Many genes occur for positive regulations. The sequences, which respond to rapid heat shock, metal concentration, hormonal regulations, are all found upstream of the sites involved in the initiation of mRNA transcription. Such control sequences can be used in constructs to develop effective processes based on two-phase reaction systems. Enhancers, which are DNA sequences that can 'enhance' the expression of nearby genes, do not, however, have to be located upstream of the gene they are affecting and are not orientation specific. For maximum expression, the control elements have to be matched to the host cell in which the constructs are to be propagated. Enhancers normally increase gene expression in the absence of a particular stimulus and are of value in the development of one-stage systems with high constitutive levels of gene expression.

X SUGGESTED READING

Alberts B, Johnson A, Lewis J, Raff M, Roberts K and Walter P (2002) *Molecular Biology of The Cell* (4th Edn), Garland Science.

Balasubramanian D, Bryce CFA, Dharmalingam K, Green J and Jayaraman K (Eds) (1996) *Concepts in Biotechnology*, Universities Press (India) Ltd.

Channarayappa and E A McCulloch (1993) A Cell Culture Model for the Treatment of Acute Myeloblastic Leukemia with Fludarabine and Cytosine Arabinoside, *Leukemia*, 7:992–999.

Channarayappa, Nath J and Ong T (1990) Micronuclei Assay in Cytokinesis-blocked Binucleated and Conventional Mononucleated Methods in Human Peripheral Lymphocytes, *Teratogenesis, Carcinogenesis, and Mutagenesis*, **10**:273–279.

Krek W and Decaprio JA (1995) Cell Synchronization, *Methods in Enzymology*, **254**:114–124.

Lahtam KE (1999) Mechanisms and Control of Embryonic Genome Activation in Mammalian Embryos, *Inter. Rev. Cytl.*, **193**:71–124.

Metcalf D (1984) *Clonal Culture of Hemopoietic Cells: Techniques and Applications*, Elsevier.

Old RW and Primrose SB (1981) Cloning in Mammalian Cells, in *Principles of Gene Manipulation* (2nd Edn), Blackwell, pp121–137.

Xiao-Chun Shi, Sui-Guang Xing, He-Kan Bi, Ver Hochberg, Channarayappa, Gopalakrishna and Tong-Man Ong (1993) The Cytogenetic Effects of Benzene on Rat Bone Marrow and Spleen Cells, *Journal of Occupational Medicine and Toxicology*, **2**(1):53–63.

PART – VIII
In Vitro Plant Cell Culture and Crop Improvement

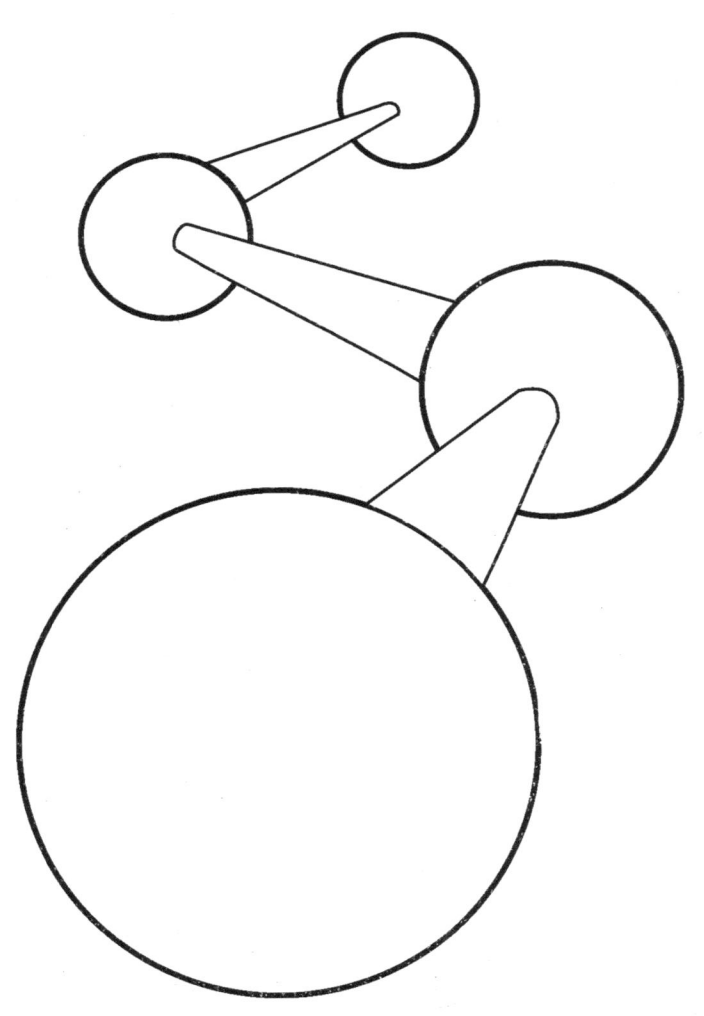

29

Plant Cell Culture Laboratory and Requirements

I INTRODUCTION

Plant tissue culture is a field of many facets (Figure 29.1) and has become a major biotechnological tool in agriculture, horticulture, forestry, and industry. It is used by the curious gardener multiplying plants in a home kitchen, to the renowned scientist working in a sophisticated laboratory for mass multiplication and industrial application. It is a rapid way for genetic engineers to grow material for identifying and manipulating genes or to transfer individual characteristics from one plant to another. **Plant tissue culture refers to the culture of cells and plant organs such as roots, shoot tips, and leaves**. In plants, most coordinated cell division takes place in the meristems and vascular cambial tissues. Embryonic tissues at early stages of development exist in an undetermined state which can rapidly proliferate (dedifferentiate) to produce **calli**. Calli can be grown in semisolid nutrient media whereas suspension cultures are maintained in liquid culture media. Under suitable cultural conditions, cell masses known as *calli* can form shoots and roots and eventually regenerate into whole plants. Morgan (1901) coined the term **'totipotency'** to denote this capacity of a cell to develop into an organism by regeneration.

Historical background: During the early 19th century, the concept of development of callus (a disorganized proliferated mass of actively dividing cells) from isolated stem fragments and root apices came into existence. The term 'tissue culture' can be applied to any multicellular culture growing on a solid medium containing many cells in protoplasmic continuity. In the early 1930s, RP White (US), Gautheret (France) and Nobercourt (France) independently cultured tissues excised from several plants on a defined nutrient media for a long period. In the middle of the 20th century, several important improvements and applications were made in the field of plant physiology and molecular biology. The role of growth hormones in rapid multiplication of totipotent cells was developed. Consequently, several chemicals were tested which stimulated callus and regeneration into whole plants. Miller (1953) developed a successful technique for the culture of single isolated cells which is commonly known as paper-raft nurse technique. Later on, attempts were also made for single cell culture by the hanging drop and agar plate method (Vasil and Hilderbrantt, 1965). In the 1960s, the role of enzymes, e.g. cellulase and pectinase, in the dissolution of the cell wall in a buffer solution at optimum pH, and

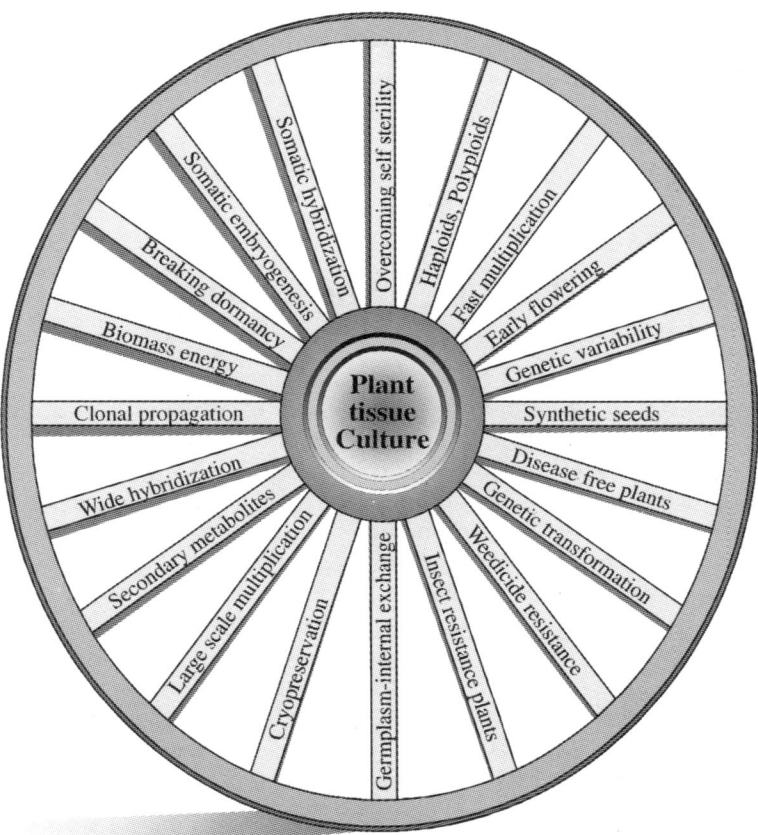

Figure 29.1 *Biotechnological applications of plant tissue culture*

isolation and culture of the protoplast was developed (Cocking, 1960). Guha and Maheshwari (1966) developed techniques for the production of vast numbers of embryos from cultures of pollens and sporogenous tissues of anther.

Significance: Regeneration of plants from cell culture offers a practical strategy for plant cloning, since all regenerants from a culture should be genetically identical. For commercially valuable plants that are difficult, costly or inefficient to propagate by cuttings or other asexual means, cell culture sometimes offers the only practical means of propagation. Most hybrid orchids, for example, are propagated today by the tissue culture method of "***meristemming***" or "***mericloning***". Plant cultures are also being investigated as sources of valuable plant products like drugs, flavors, and fragrances. Many strategies for genetic engineering of plants rely on plant tissue culture. Plant cells in culture can be genetically transformed by a number of techniques. Plant cell culture is central to these techniques in that cell culture allows transformants to proliferate and, sometimes, regenerate into genetically identical clones.

Cell culture laboratory: A cell culture laboratory mainly requires a laminar airflow to provide an aseptic environment for all culture manipulations because all tissue culture media can support the growth of many fast growing bacteria and fungi that will destroy the slow growing plant tissues. While many formulations for plant tissue culture media have been developed, all contain the same basic types of ingredients. Pre-made culture media are commercially available. However, plant growth regulators such as auxins and cytokinins tend to be omitted from the basic medium formulation because culture growth and differentiation are affected by these growth regulators. Synthetic auxins such as 2, 4-D and NAA can replace the naturally occurring auxin IAA. BAP (6-benzylaminopurine) and kinetin (6-furfurylaminopurine) are the most commonly used cytokinins.

II PLANT TISSUE CULTURE LABORATORY

Prior to the establishment of a tissue culture laboratory, one should have clear objectives such as purpose or goal, magnitude of the operation, resources available, and to some extent, the plants to be cultured. Plant tissue culture primarily involves **dedifferentiation** of the explant initially (freeing from inter-organ, inter-tissue and inter-cellular interactions), and then directing them to respond to the experimental conditions (Figure 29.2). To ensure the growth and development of an explant, a suitable nutrient medium and proper conditions for culturing have to be provided. These operations have to be carried out with special care and skill under aseptic conditions. The plant tissue culture laboratory has separate areas devoted to particular activities (Figure 29.3). The basic organization and facilities required to set up tissue culture laboratories are:

1. Location, design, and development of a tissue culture laboratory
2. Working space to carry out routine laboratory work, e.g. media preparation.
3. Washing sinks and drain racks. Ovens to dry the washed glassware.
4. Cabinets or storage shelves for the storage of cleaned glassware, chemicals, media preparations, and sterilized equipment.
5. Sterilization facility: autoclaves, steamers, and ovens for media sterilization, distilled water, glassware and instruments. UV sterilization can (sometimes) also be used.
6. Transfer/Inoculation room for aseptic transfers of plant tissues and cultures, where housing of laminar flow cabinets etc., are maintained cleanly and aseptically.
7. Culture room with incubator: Where cultures are maintained under controlled conditions. Incubators are used for the incubation of the cultures in a controlled environment with high accuracy.
8. Examination and analysis room: This is often a shared facility used only for microscopic, computer, and photomicrographic works.
9. General work laboratory: Used for class work and research work.
10. Service lines such as gas, water, electricity, vacuum, and compressed air.
11. Continuous supply of distilled and double-distilled water
12. Instrumentation room for keeping various instruments and equipment.

Though the setting up of a tissue culture laboratory appears very simple, it requires careful planning. Because the setting up of a facility and the subsequent activities are comparatively costly, careful planning is of paramount importance. Depending upon the availability of laboratory space and the tissue culture capacity, the laboratory space can be divided into separate rooms according to the operations and functions that are to be carried out during the course of work. Alternatively, due to severe space restrictions, some operations and functions can be carried

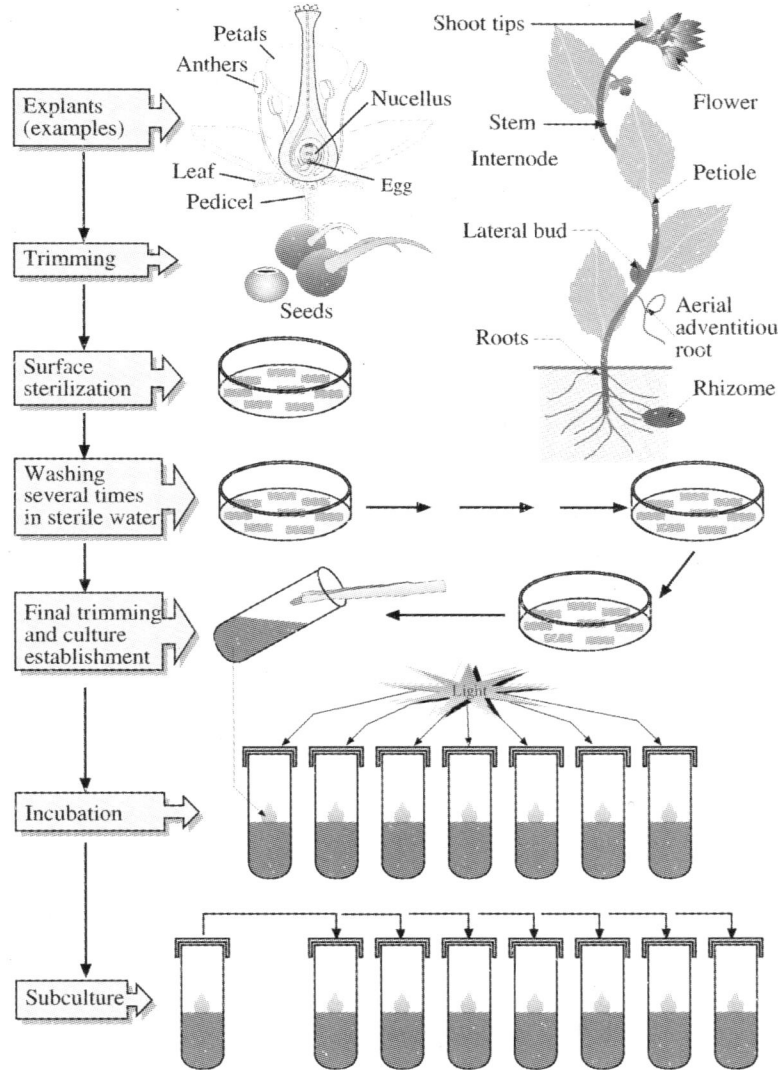

Figure 29.2 Basic procedure for establishing and maintaining a culture of plant tissue

out in a smaller area or in a common place rather than in separate working places. General guidelines for setting up a facility are to focus on a design that would provide a system analogous to a production line. The units and the activities need to be arranged in such a way as to make operational steps possible with the least amount of cross-traffic.

Research tissue culture laboratory: A well-constructed tissue culture laboratory with modern facilities is very important. The main emphasis in a tissue culture laboratory is on cleanliness and it being free from dust and microbial contamination. Thus, a plan for a tissue culture facility must be a design that allows for excluding dust and which can be cleaned readily and effectively. A tissue culture unit should provide distinct work areas as illustrated in Figure 29.3.

874 MOLECULAR BIOTECHNOLOGY: PRINCIPLES AND PRACTICES

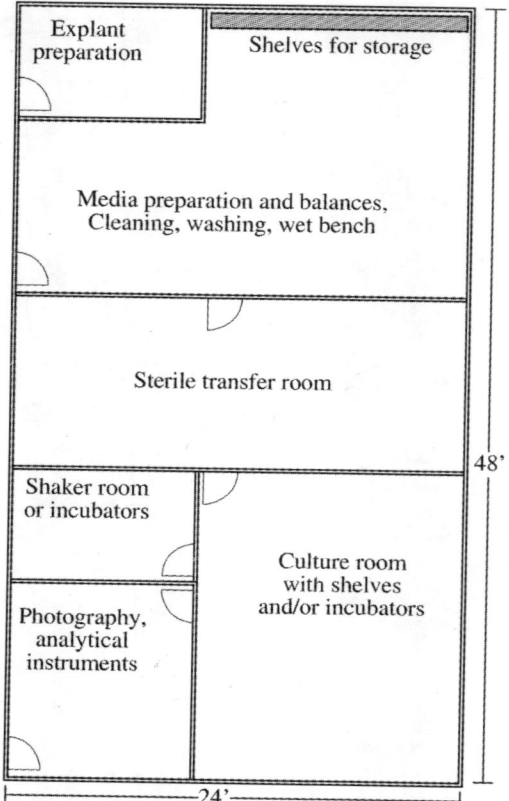

Figure 29.3 *An idealized design of a small to moderate-sized plant tissue culture laboratory*

The size of the laboratory depends upon the kind and the scale of tissue culture production. The location of the tissue culture laboratory is also very important for easy transportation. It has to be close to the glass house facilities, fields and other infrastructure facilities. A schematic representation of a tissue culture laboratory of about 74 m^2 is given in Figure 29.3. The research methods to be followed for a particular problem or plant can be developed here.

III PLANT TISSUE CULTURE LABORATORY DESIGN AND DEVELOPMENT

A. LABORATORY LOCATION AND DESIGN

Individual circumstances will, of course, determine the location of a proposed laboratory, but some general concerns need to be addressed in any location. In all aspects of planning, cleanliness should be considered one of the most important factors. An area that is relatively free from dust, smoke, mold, spores, and chemicals should be selected. A plant tissue culture laboratory consists of mainly three distinct areas: (1) a space for preparing media, (2) a chamber for transferring cultures, and (3) an area in which to grow the cultures. Usually these are located in three separate rooms, but in the face of other limiting circumstances, they may occupy the same room.

In all rooms, the walls, floors, and ceilings should be light colored and easily washable with detergent and standard disinfectants. Any outside air that is admitted into the 3 major rooms should be filtered, preferably by HEPA (high efficiency particulate air) filters, in order to reduce the particle count to an acceptable level (less than 10,000 particles per cubic metre of air). Rooms used for such functions as office and shipping does not require filtration of room air.

B. WASHING AND DRYING ROOM

In a tissue culture laboratory, all the culture facilities must be scrupulously clean. The washing room should be provided with the necessary service lines such as electricity, gas, water, compressed air, proper ventilation, light, etc. In one corner of the room, a washing sink can be fitted. It should also be furnished with brushes of various sizes and shapes, a washing machine*, and running hot and cold water. The drainage and drying racks could be placed adjacent to the sink for drying clean glassware. All the discarded cultures, as well as contaminated ones, should be autoclaved to liquefy the agar and kill any microorganisms that may be present. The culture vessels should be emptied, rinsed, and soaked in a detergent bath overnight. The glassware can then be scrubbed with a brush and rinsed three times with tap water followed by three rinses in distilled water (see Chapter 2). Even the new culture vessels should be washed thoroughly prior to use. The glassware should be stored in clean and closed cabinets after drying.

C. STERILIZATION ROOM

The sterilization room should preferably have two autoclaves. It should also be provided with a sink and all essential service lines. It should be well ventilated by fitting exhaust-fans. The oven for drying the cleaned vessels could also be accommodated in this room.

D. GENERAL LABORATORY

Tables of both standing and sitting height are necessary in this room for plugging tubes and wrapping the glassware before sterilization. This room should be provided with a microscope, refrigerator and deep-freeze for storing perishable compounds, stock solutions, etc. The working tables can be placed adjacent to the walls of the room to accommodate the appropriate equipment.

E. INOCULATION ROOM

The preparation and cutting of explants is generally done on a sterilized glass plate. However, it is good to carry it out in the inoculation room to avoid contaminations. In this room, the inoculation cabinet or laminar airflow cabinet are housed. A laminar airflow cabinet is one in which completely sterile air is blown over onto inoculation table. Since there is a continuous airflow through the inoculation cabinet, contamination from outside is completely prevented. It is recommended that the cabinet is serviced by the factory personnel once a year or earlier if needed.

The floor must be covered with linoleum or tiles to facilitate proper cleaning. The atmosphere in this room must be free from any contaminations and germicidal lamps should be fitted. The inoculation room should be preferably air-conditioned and should have double sliding doors. There should be a separate cubicle for hanging gloves, laboratory coats, etc.

F. TRANSFER ROOM

The transfer room contains the transfer chamber or hood, which is usually a bench with a partially enclosed area that is provided with sterile air. There are laminar horizontal flow sterile transfer cabinets, which are available in sizes of 8, 6, 4, and 3 feet. Cabinets for sterile transfers are available from many commercial sources. A stainless steel working surface is the most durable and easiest to keep clean. It is here that the explant preparation, processing and transfers take place. The transfer room and the culture growing room should be isolated as much as possible from outside doors and from significant foot traffic.

G. INCUBATION ROOM

After the aseptic transfers, the explants and tissues have to be incubated at a specified temperature – the most desirable temperature being around 25°±2°C, which can be maintained with an air-cooler or by window air-conditioners. Cultures can either be incubated in controlled environment chambers or on wooden or metal slotted angle racks. Cool daylight fluorescent tube lights (40 watts) can be fitted on the racks with a timing device to adjust the photoperiod required for the growth of cultures. Sometimes, cultures have to be incubated in continuous darkness, and there should be adequate provision for this. This is often done by using black curtains to cut off the light.

H. PLUGGING ROOM*

Tables of both standing and sitting heights are necessary in this room as described in the general laboratory. Sufficient space is required for storing the cleaned glassware and glassware plugged with non-absorbent cotton.

I. MEDIA PREPARATION ROOM

The media preparation area resembles a kitchen; there should be a sink with hot and cold running water, a refrigerator, a dishwasher, a stove or burner, and a pressure cooker or an autoclave for sterilization. This room should also be provided with pH meters, balances, a centrifuge, deep-freezers, magnetic stirrers, distilled water supply, vacuum lines, compressed air lines for filter-sterilization and transfer of solutions, etc. Benches of the height suitable to work while standing will be very convenient during media preparation.

J. MEDIA STORAGE ROOM*

A separate media storage room can be maintained when large-scale tissue culture work is needed. This room can be utilized for storing the media prior to use. This facility helps to keep media ready for timely use and can be verified by the different media and quantity required for the day's work.

K. STERILIZATION ROOM

This room should house different types and sizes of autoclaves. It should also contain an electric oven for drying the glassware. A sink should be fitted in one corner of the room and the room should be properly ventilated with the help of exhaust fans.

L. CONTROL ROOM*

The control room is mainly to regulate the power supply to various systems, to adjust the photoperiod by automatic timers, maneuvering the illumination, etc. The illumination of a particular rack can therefore be easily maneuvered from this room without having to enter the incubation room. The control room can also house a generator.

M. SHAKER ROOM*

Rotary shakers with variable speed controls ranging from 80 to 220 rpm are illuminated with fluorescent tube lights fitted about 60 cm above the shaker so that they can provide an intensity of 2000 lux to the cultures. The intensity of light in the inoculation and shaker room can be measured by the use of a lux-meter.

N. ACCLIMATIZATION ROOM

The acclimatization room is where the cultures are hardened after removal from the rooting media and transferred to pots. This room is provided with high intensity light (~3000 lux) and humidifiers (to achieve relative humidity of 90%).

O. ROOMS FOR OTHER PURPOSES*

Separate rooms for an office, library, meeting, and data-collections can be provided based on the availability of funds and space.

P. STAFF FOR TISSUE CULTURE LABORATORY

The staff for a tissue culture laboratory includes a qualified member who supervises the work, technicians, additional hands for washing and plugging the glassware, and laborers for low-skilled jobs like cleaning the labs, transportation, etc. The number of staff members required could, however, vary depending on the type of research work, production scale, etc.

IV TISSUE CULTURE LABORATORY WORKING PROCEDURE

The tissue culture laboratory is mainly devoted to manipulations of plant cells and tissues using *in vitro* procedures. Most of these operations require both expertise and effort. They also involve expensive chemicals and instrumentation. Highly aseptic conditions are a must for the growth and development of cells. Besides the plant material, nutrient media and other chemical substances used in the cultures are rich sources of nutrients for microorganisms. Thus, controlling microbial contamination in the tissue culture laboratory is one of the primary tasks. Some of the procedures used to reduce potential contamination of cultures are as follows.

A. PLANNING THE TISSUE CULTURE EXPERIMENTS IN ADVANCE

Before beginning tissue culture work, thorough planning of the experiment is very essential. The objectives of the experiment, the working protocol, materials required for the completion of the work, the schedule of operations, the final outcome expected, etc., should be planned well in time. Delay or deficiency in the operations may be disastrous to the outcome.

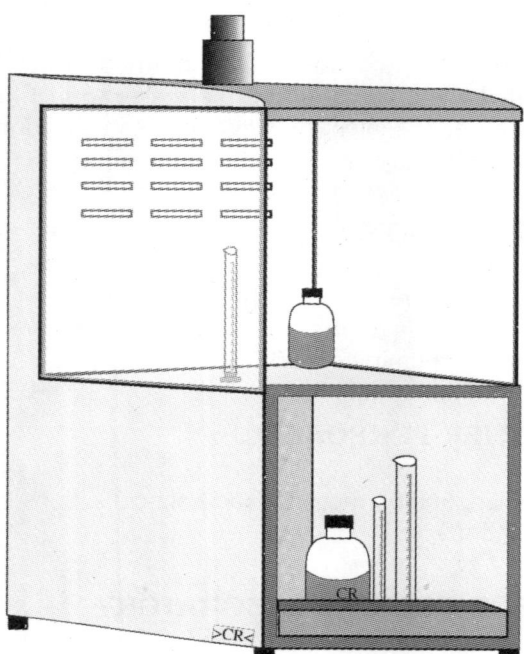

Figure 29.4 *Laminar air flow system*

B. STERILE TECHNIQUES

Most of the plant cell culture operations require very stringent sterile techniques. All operations must be carried out in laminar flow (Figure 29.4), sterile cabinets or equivalents. There are three categories of sterilization. 1) Preparations of sterile media and containers; 2) obtaining explants that are sterile or axenic (free of other organisms); and 3) maintenance of aseptic conditions of the cultures. Contamination also occurs most frequently during transfer of cultures.

(a) General aseptic techniques: Persons working in a tissue culture laboratory should have sound knowledge about working practices and aseptic techniques. The following are some of the points to be regularly checked.

i. Cultures should be regularly checked to remove contaminated cultures from the culture place to the disposal container.
ii. Only clean and essential items should be brought into the laboratory.
iii. The number of people entering the laboratory must be limited.
iv. Clean and uncontaminated lab coats and gloves should be worn.
v. Hair should be covered with a cap and it is recommended that a facemask be worn.
vi. Laboratory safety guidelines including fire safety, chemical disposal, decontamination procedure, etc., should be strictly followed.

(b) Use of aseptic practices in laminar flow cabinet

i. The laminar flow cabinet should not be used for storage of instruments, chemicals, or any other substances.
ii. A UV lamp should be on (preferably overnight) before the operation and it should be put off before the plant material is placed in the cabinet.
iii. Wipe the interior of the cabinet with 70% alcohol before switching on the blower.
iv. Thoroughly wash hands and arms with soap and water then swab with 70% alcohol before carrying out plant manipulations.
v. The work area must be restricted to within the cabinet so that working across or over containers and plant samples can be avoided.
vi. Any plant material and other substances which are accidentally dropped on the floor should be considered as contaminated and discarded.
vii. Immediately after the completion of the work, the area inside the cabinet should be thoroughly cleaned and disinfected. UV lamps should be turned on.
viii. The lights and blower should be switched off and the cabinet front panel should be closed or slid down after completion of work.

(c) Surface sterilants: Surface sterilization may be carried out with several different germicidal reagents (Table 29.1). Obviously, the best products are those which are low priced, non-toxic, and effective on a wide range of plant material.

V THE LABORATORY FACILITIES AND SUPPLIES

Any person who wants to establish a tissue culture laboratory will approach the cell culture project with a unique background, with specific objectives, and resources. Before planning a laboratory, one should determine the magnitude of the operation, the purpose, the plants to be cultured and the scale of cell culture production.

Table 29.1 *Effectiveness of some surface sterilizing agents*

Sterilizing agent	Concentration recommended	Duration (min)	Effectiveness
Calcium hypochlorite	9–10%	5–30	Very good
Sodium hypochlorite	2%	5–30	Very good
Hydrogen peroxide	10–12%	5–15	Good
Bromine water	1–2%	2–10	Very good
Silver nitrate	1%	5–30	Good
Mercuric chloride	0.1–1%	2–10	Satisfactory
Antibiotics	4–50 mg 1^{-1}	30–60	Fairly good

Equipments and apparatus

1. ***General needs***: The necessary equipments for setting up a tissue culture laboratory are listed below. For small-scale tissue culture work, simpler equipment can be used. Some items are not absolutely required, one can manage with an alternative arrangement*.
2. ***Service lines***: Good and continuous supply of gas, water, and electricity. Planning for electrical equipment and fire safety always deserves the best available professional advice. Electrical power failure and safety are other major concerns. It may be necessary to obtain a portable generator to keep the growth room at the correct temperature in the event of a power failure. A typical growth room uses cool-white fluorescent lamps above the cultures. In literature, light intensity in different units is often described, suggesting ranges of tissue cultures from 50 to 1200 micromoles per square metre per second, or from 20 to 100 microeinsteins per square metre per second. Light meters reading in foot-candles and lux are readily available and still serve the purpose for most tissue culture laboratories. The spectral curve of light from Gro-Lux lamps appears to coincide more accurately with the requirements of chlorophyll synthesis. However, for most the cultures, cool-white light is adequate.

    Culture growing room temperatures should generally be kept between 24 and 29°C, with 16 hours of light and 8 hours of darkness, although this will vary with the particular plants being grown. A centrally located thermostat controls heating and cooling equipment with blower. A 24-hour timer can be used to automatically control the light and help balance the heating and cooling demands of the laboratory. The timer should be set so that the 8-hour period of darkness occurs during the daytime and the 16 hours of light fall mostly during the night. By this arrangement, the heat given off by the lights at night helps keep the growing room warm and allows the room to be cooler in the daytime, to save cooling costs.
3. ***Different types of glassware***: Different types of vessels have been used to culture plant materials. The choice of type of culture vessels is dictated in some cases by the nature of the experiment. In others, it is guided mainly by convenience and availability. Culture dishes, Petri plates, test tubes, bottles, measuring cylinders, pipettes, syringes etc. are most commonly used. In many laboratories, the glass culture vials and other labware required for media preparation have been largely replaced by suitable plasticwares.
4. ***Hot plate and magnetic stirrer***: A hot plate with stirrer is very essential for quick solubilization and uniform mixing of compounds in a solution. For example, agar, a gelatinous

polysaccharide extracted from certain red algae, is required as a solidifying agent in many media formulas. Unless the mixture is constantly and effectively stirred until it boils (100°C) – the agar will settle and stick to the bottom of a flask and burn. The stirrer and hot plate features can be used at the same time or independently. An electric kitchen beater is an inexpensive alternative option for a stirrer and can be used for large-scale preparations.

5. **Balances**: A manual balance, preferably top loading for quick weighings and for analytical purposes, a single pan balance with a capacity of 100–200g and sensitivity 0.1 mg is used. An electronic balance, having top loading facility, with precision ± 0.005g, weighing range 0–200 g, with a digital read-out can also be used. Unless pre-mixed chemicals are to be used, a precision balance will be required to accurately measure the small amounts of chemicals required for tissue culture media, although the accuracy needed for analytical chemists is not usually required in plant tissue culture. The expensive, electronic one-pan balances are fast and precise. The dilution method is one alternative for weighing smaller amounts. In this method, if, for example, 10 mg of a compound is required per litre of medium, 100 mg (0.1 g) of the compounds can be placed in 100 millilitres (ml) of water. When weighing amounts of more than 10 g, use a triple-beam balance, saving the more sensitive balance for the smaller quantities. A good triple-beam balance costs less than the analytical balance.

6. **pH meter**. A digital pH meter with range of 0–14/0–1400 mv is generally used. The accuracy is I 0.1 pH/I 1 mv, with temperature compensation 0–100°C. Acidity or alkalinity (pH) of the media is crucial in tissue culture and is specific to the requirements of specific plants, just as it is in soils and potting mixers. A commercial laboratory should have a pH meter. A beginner, however, can get by simply with a pH indicator paper. Recommended pH ranges for test papers are 2.9–5.2, 4.9–6.9, and 5.5–8.0. These small ranges will measure adjustments of media to approximately pH requirements.

7. **Micro-wave oven***: For quick melting of agar media, etc.

8. **Electric oven**. Hot air oven (laboratory model) – double walled, thermostatically controlled, with a range 5°C above the ambient environment to 250° ± 2°C.

9. **Water purification equipment, Millipore filter system***: Water is the largest component of tissue culture media, so the quality of water used is of critical importance in establishing a successful operation. Tap water often contains dissolved minerals, particulates, and organic matter. These substances must be removed before the water is used in tissue culture media because they can upset the precise balance of nutrients in media formulas, or they can be toxic. The three most common methods used to remove dissolved chemicals from water are **distillation**, **deionization**, and **reverse osmosis**. Water purified by *distillation* has been the standard laboratory-grade water for many laboratories. The purity of the end product depends on the feed water and the efficiency of the still. *Deionization* is another effective way to remove dissolved chemicals from water. A single mixed resin bed tank, or cartridge, can provide adequate purity for normal tissue culture production. Various deionizing systems that use disposable cartridges are available from scientific supply companies. Reverse *osmosis*, a third method of water purification, is usually used in combination with a deionizer or still. It does not have the refinement of the other two methods, but it provides excellent pre-treatment.

10. **Media dispenser**: A small laboratory can easily forego high-priced equipment for dispensing media. A 10 ml polypropylene pipette can be used to dispense the media. A coffee urn can also sometimes be used. Alternatively, a reservoir for media that is gravity fed through

tubing can be used, with a pinch clamp to control the flow. An automatic pipetter to dispense media is a labor-saving device for large-scale production.

11. **Laboratory timer**
12. **Metal autoclave rack**s
13. **Material for vessel closure** (autoclave paper, aluminum foil, non-absorbent cotton etc.)
14. **Autoclaves**. Electrically operated, vertical or horizontal autoclaves provided with a safety-valve, pressure gauge, and steam-release cock, with varying sizes*. A good pressure cooker can be used for small-scale operations.
15. **Refrigerator**. Refrigeration for stock solutions and thermolabile materials. A household refrigerator is useful for storing perishable chemicals and stock solutions. A larger refrigerator is necessary if media of cultures are to be refrigerated as well. A refrigerator is a valuable means of treating or delaying the growth of cold-resistant cultures. Chilling is often called for if plantlets are ready too soon to be transferred to the greenhouse, or if stock cultures need to be slowed down. Cultures of some hardwood species benefit from cold treatment before acclimatization.
16. **Drying and draining racks***
17. **Growing room shelves**: Light shelves need to be constructed in the growing room to hold the test tubes or jars of cultures while they are growing. Shelves can be built using slotted steel angle supports for 4 8 ft shelving of particle board, plywood, or wire mesh of expanded metal, or 0.25- to 0.5-inch hardware cloth (Figure 29.5). Boards should be painted white to maximize available lighting.
18. **Microscopes**. A dissecting (low power) microscope for overall observation of tissue development and for isolating meristems is needed. A laboratory microscope should have facilities for bright field and phase contrast to examine the surface structure and structural composition of specimens. An inverted microscope, standard inverted microscope with provision for transmitted light examination of tissue cultures, protoplasts, etc.* can also be used. The decision to buy a microscope depends on the type of tissue culture work to be done, as well as on the interest and curiosity of the researcher. A stereo dissecting microscope is usually required for excising meristems. A compound microscope is not necessary for routine commercial production, but is essential for identifying contaminants.
19. **Low-speed centrifuge**. A centrifuge equipped with a continuously variable electronic speed control speed indicator, Amp meter, timer, dynamic break zero starting switch and safety fuse device is useful.
20. **Sterilizing equipment for media and transfer tools**: A modest laboratory can use a household pressure cooker for sterilizing test tubes or jars containing media. A wire basket to hold the test tubes must be built or purchased when using such a cooker. Autoclaves, a common method of sterilization for tissue culture, come in many sizes. Most are automated, more efficient, and accurate than cookers. A large operation should have an autoclave. Some organic chemicals, however, are heat labile, therefore, some media ingredients must be sterilized by filters. **Filter-sterilization equipment**. Syringe filter holder (2.5 cm) and pressure filter holder (4.7 cm) are usually used in tissue culture laboratory work.

In the past, tools used in the transfer hood primarily were sterilized by being dipped in alcohol (95% ethyl alcohol), followed by flaming using a Bunsen burner or alcohol lamp (spirit lamp). For sterilizing implements in the transfer hood, the preference today is the glass bead sterilizer, an insulated pot with a component that heats glass beads within it. A glass bead sterilizer can also be used for disinfecting the small transfer tools. Another choice used by a

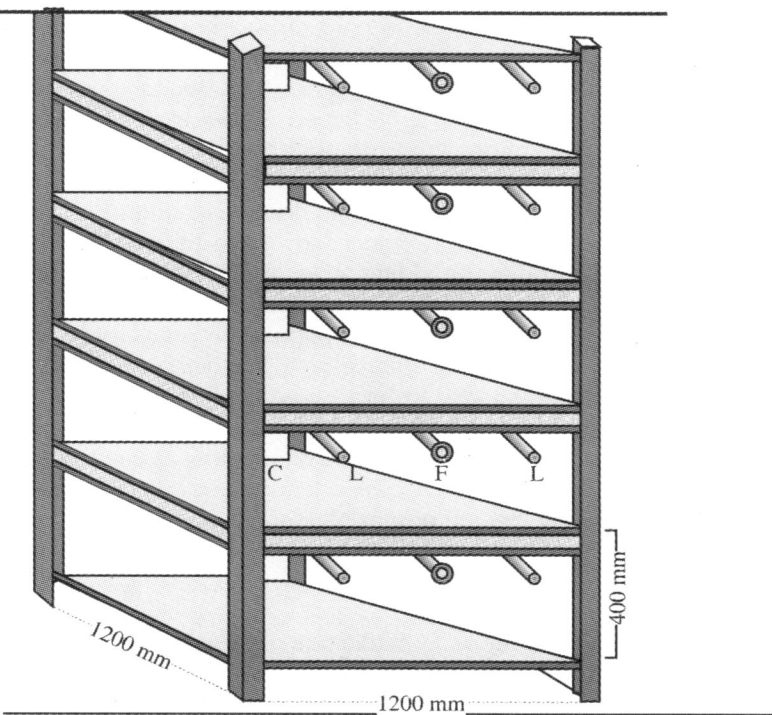

Figure 29.5 *Diagram of a plant cell culture shelving unit. C=control panel, F=fan, L=light source*

large number of laboratories today is the Bacti-Cinerator, an infrared sterilizer. Replacement heater elements may be needed here once or twice a year. The Bacti-Cinerator sterilizes instruments in 5 seconds at 870°C as the instruments are inserted into the red hot, hollow cone of the cylinder. Bleach solutions are satisfactory for sterilizing tools in some situations. Two concentrations of bleach solutions should be used with this method: a 10% solution to soak the instruments and a 1% solution to rinse them.

21. **Environmental growth cabinets**. These are required for the incubation of cultures under controlled temperature, light, and humidity. These are available in various sizes. The temperature range has to be 4–45°C with timer to regulate the photoperiod.
22. **Gyratory shakers**. The platform of the shaker is fitted with rubber discs to hold flasks from 50 to 1,000 ml capacities. The platform gives an orbital circular motion to the flasks. Shakers are available where the speeds can either be continually variable from 80 to 220 rpm or be kept constant.
23. **Laminar airflow cabinets**. To provide a constant flow of purified air across the width and height of a working area without turbulence. It maintains a continuous air current, which keeps this cabinet without the risk of contamination. Available in sizes of 0.6 m, 1.2 m, and 1.8 m capacities.
24. **Generator facility**. For emergency power supply, a fuel-based generator facility is very important, where frequent power failure problem is common. Otherwise the growth rates of cultures may be affected in addition to power problems for other operations.

25. **Dishwasher**: A conventional built-in household dishwasher will handle the dish-washing requirements of a small laboratory. Tubes in racks must be covered with wire mesh and turned upside down in the dishwasher. A commercial laboratory dishwasher of the same capacity costs at least 4 times as much as a kitchen dishwasher and yields few advantages.
26. **Transfer chamber or hood**: Laminar airflow transfer hoods are essential for commercial operations. They provide a sterile atmosphere in which to work with cultures. Air is forced through a HEPA filter, located at the back of the hood that strains out particles as small as 0.3 micrometers (µm). A gentle, scarcely detectable air stream flows through the filter, across the work area toward the worker, providing a sterile atmosphere in which the technician works. Many laboratory supply companies sell assorted designs of laminar flow hoods. Some larger units that are set on the floor have the option of air intake either at floor level or at the top of the unit. The unit that has air intake on top rather than at floor level is preferred because the air at floor level is likely to have more contaminants than the air nearer the ceiling.

    It is possible to construct, rather than buy, a satisfactory laminar flow hood, or have a local cabinet shop make one. Additional components necessary for building a transfer hood include pre-filters, a blower, plywood or plexiglass sides and top, a smooth bench top, and a fluorescent light. A still-air transfer chamber with a slanted glass front and partially enclosed hand access is an alternative for beginners.
27. **Additional supplies**

    Household items can often substitute for traditional glassware and tools used in the laboratory. Heat resistant Pyrex glass beakers are good vessels for cleaning explants. Aluminum utensils are undesirable because they can give off aluminum ions into the medium. An instrument holder is useful in the transfer hood, especially when using a Bacti-Cinerator sterilizer. A metal test tube holder standing on end makes a more convenient but also more expensive holder. Seedling containers are useful for holding test tubes in the growing room because they allow the test tubes to be slanted and time exposure to more light than if the test tubes are upright, as they are in the typical wire rack test tube holder.
28. **Miscellaneous items**: Air-conditioners*; Arrow heads; Deep-freeze; Fluorescent lamps/tubes; Heaters; Hot plate; Hot plates with magnetic stirrer; Metal trays and bowls for transport of cultures, tubes etc. Microtome*; Plastic carboys to store water and other solvents; Slide trays, metal or wooden; Slider containers with baskets; Spatula of different sizes to weigh the chemicals; Flaming instrument*; Dry sterilizers*; Stainless-steel dissecting knives; Stainless-steel sieves, 60–70 µm pore size; UV germicidal lamps; *Racks*: Wooden or iron racks with slotted angles fitted with lights for illumination of cultures; Wooden or metal racks to keep testtubes.; Filter paper; Petri dishes; Automatic dish washer*; Cleaning materials; Shaker*. Flasks (50, 100, 250, 500 ml and 1L, 2L, 5L); Volumetric flasks (500 ml, 1L, 2L, 3L); Measuring cylinders (25 ml, 50 ml, 100 ml, 1L); Graduated pipettes (1 mL, 2 mL, 5 mL, 10 mL); Pasteur pipettes; Culture vials (culture tubes, screw cap bottles of various sizes, Petri dishes, etc.) with suitable closure; Filter membranes to filter sterilize solutions; Hypodermic syringes, for filter-sterilization of solution; Trolley with suitable trays, to transport culture, media, and apparatus; Spirit-lamps or Bunsen burner, to flame instruments; Atomizer, to spray alcohol in the inoculation chamber; Instrument stand to support sterilized instruments during aseptic manipulations; Large forceps with blunt ends and pointed ends for inoculation and subculturing; Fine needles, for dissections; Scalpels for cutting plant material; Scissors; etc.

Chemicals

Sodium hypochlorite; 96% alcohol; Detergent; Chromic acid bath*

Note: * = Optional.

VI TYPES OF PLANT CELL CULTURE

The following are some basic plant cell culture methods that not only have the potential for plant regeneration but also have biotechnological applications in such areas as molecular biology, physiology, nutrition, disease and herbicide resistance, and in the production of secondary metabolite products. The requirements for these procedures are generally more sophisticated and demanding than for basic tissue culture. The tissue culture techniques are grouped into the following categories on the basis of the plant part used as explant and the type of development *in vitro*:

Tissue culture is commonly used as a collective term to describe all kinds of *in vitro* plant cultures. It is possible to culture both organized and unorganized tissues (Table 29.2) and morphogenesis (the development of form) can be inducible *in vitro*. The plant growth in *in vitro* conditions can be organized or unorganized.

Organized growth contributes to the establishment of defined structures. When plant organs such as the shoot or root meristems, leaf initials, flower buds or small fruits, are transferred to the culture, they continue to grow with their structure preserved. Coherently organized growth can also occur *in vitro* when organs are freshly developed from an organ or tissue placed in culture. The process of *de novo* organ formation is called organogenesis or morphogenesis.

Unorganized growth is very seldom seen in nature, but it is a very common feature among *in vitro* cultures. The unorganized tissues formed *in vitro* typically lack any recognizable structure and contain only a limited number of specialized and differentiated cells that are found in an intact plant. A differentiated cell is one which has developed a specialized form (morphology) and function. Until now, the formation of differentiated cell types could only be controlled to a limited extent in culture.

Table 29.2 *Some important plant tissue culture types*

1. Ovary and ovule culture	11. Anther and pollen culture
2. Nucellus culture	12. Endosperm culture
3. Seed and Seedling culture	13. Flower culture
4. Embryo culture	14. Somaclonal variation
5. Root culture	15. Somatic hybridization and transgenics
6. Stem culture	16. Synthetic seeds
7. Auxiliary and lateral bud culture	17. Salt or stress tolerance
8. Shoot–tip culture	18. Metabolism and its regulation
9. Leaf culture	19. Tissue culture for health-care products
10. Influorescence culture	20. Cell culture for germplasm conservation

Based on the type of explant, tissue culture methods are broadly grouped into organized and unorganized tissues. Some representative cell culture methods discussed in (Chapters 31–33) are:

Cultures of unorganized tissues: (1) Callus culture, (2) Suspension cell culture, (3) Protoplast culture, (4) Anther culture, and (5) Somaclonal variation.

Cultures of organized tissues: (1) Meristem culture, (2) Embryo culture, (3) Shoot tip culture, (4) Node culture, (5) Ovary culture, (6) Root culture.

VII SUGGESTED READING

Christou P (1992) *Genetic Engineering and In Vitro Culture of Crop Legumes*, Technomic Publishing Co. Inc.

George EF (1993) *Plant Propagation by Tissue Culture*, Exegetis Limited.

Khanna, V K (1998) *Plant Tissue Culture Practice*, Kalyani Publishers, Ludhiana, pp1–20.

Kumar HD (1998) *Modern Concepts of Biotechnology*, Vikas Publishing House Pvt. Ltd.

Kyte L and Kleyn J (1996) *Plants from Test Tubes: An Introduction to Micropropagation*, Timber Press.

Lindsey K (Ed) (1993) *Plant Tissue Culture Manual*, Kluwer Academic Publishers.

Reinert J and Yeoman MM (1983) *Plant Cell and Tissue Culture, A Laboratory Manual*, Narosa Publishing House, New Delhi, Springer-Verlag, Berlin Heidelberg, New York.

30

Plant Tissue Culture Media, Preparation, and Culture Initiation

I INTRODUCTION

Excised plant tissues (**Explants**) and organs will only grow *in vitro* on a suitable artificially prepared nutrient medium (known as **culture medium**) under a sterile and controlled environment. The vital activity of a cell is the absorption of nutrients through the cell membrane and the rapid proliferation into innumerable cells. Several techniques have been adopted for *in vitro* plant tissue culture, which includes preparation of nutrient media, sterilization, aseptic manipulation, and maintenance of culture. White (1934) observed the unlimited growth of isolated root tissues when provided with a nutrient medium containing inorganic salts, sucrose, vitamins, growth hormones, and a few amino acids. The composition of different culture media, preparation, and initiation of cell culture are briefly discussed in this chapter.

II PLANT TISSUE CULTURE MEDIA PREPARATION

The success in any technology employing plant cell, tissue, or organ culture is related significantly to the choice of nutritional components and growth regulators. Nutritional requirements for optimal growth of a tissue *in vitro* may vary with the species. Even tissues from different parts of a plant may have different requirements for optimal growth. As such, no single medium can be suggested as being entirely satisfactory for all types of plant tissues and organs. When starting with a new system, it is always advisable to work out a medium that would fulfil the species requirements. Some of the earliest plant tissue culture media, e.g. root culture medium of White (1943) and the callus culture media of Gautheret (1939) were developed from nutrient solutions previously used for whole plant culture while some calli (carrot tissue, tumor tissues, etc.) may grow on simple media containing only inorganic salts and unutilizable sugar. For most others, it is essential to supplement the medium with vitamins, amino acids, and growth substances in different qualitative and quantitative combinations. A medium containing only "chemically defined" compounds is referred as a '**synthetic medium**'.

A defined nutrient medium usually consists of inorganic salts, a carbon source, some vitamins, and growth regulators. Other components added for specific purposes include organic nitrogen compounds, tricarboxylic acid compounds, and plant extracts. Media compositions, which are frequently used, are listed in Table 30.1. The Murashige–Skoog (MS) or Linsmaier and Skoog (LS) salt compositions are the most widely used especially in plant regeneration procedures. The B5, N6, and Nitsch and Nitsch (NN), and derivatives of these media have had wide applications for different plant species and for different culture objectives (Appendix 13). An appreciation and knowledge of the nutritional requirements and the metabolic needs of the cultured cells and tissues is invaluable not only in a decision on the type of media to use but also in their preparation. Information about laboratory equipments and chemicals required for media preparation can be obtained from the previous chapter (Chapter 29). This section presents a few commonly used procedures for preparing media. It is important to consistently use caution, care, and common sense at all the steps.

A. MEDIA COMPOSITION

There is a large variety of plant cell culture media published in the literature. Included in the present chapter are a description of the composition and the preparation of media for special purposes. The choice of media is dictated by the purpose of the tissue culture technology which will be employed and the species or type of plants. The nutrient media for tissue culture generally contain the inorganic elements, carbon source, vitamins, growth regulators, and some organic supplements.

1. Inorganic salts

A relatively small number of mineral salts are used as components of media for plant tissue culture. For most purposes the medium should contain at least 30 mM of inorganic nitrogen and potassium. Inorganic salts supplied to plants in *in vitro* are grouped into macronutrients and micronutrients, depending upon the amount the cells utilize.

(a) Macronutrients

Macronutrients are compounds, which provide the major ions needed for the plants. Their addition to culture media is like applying fertilizers to naturally growing plants. The major macronutrients generally included are nitrate and ammonium (inorganic nitrogen source). Ammonium can be used at 2–20 mM. However, the effect of ammonium salts can vary from inhibitory to essential depending upon the tissue and the purpose of the culture. The use of ammonium salts of malate or citrate makes it possible to use ammonium as the sole nitrogen source. A concentration of 1 to 3 mM of calcium, sulfate, potassium, phosphorous, and magnesium is usually adequate. Sodium and chloride ions are also present in macronutrient compositions, even though it is not essential to supply these for most plants.

Ionic balance and concentration: Macronutrients ought to be defined according to the nutrient ions they contain, not according to the weights of compounds. It is generally not realized that a given set of balanced macronutrient ion concentrations can be prepared from several different combinations of salts. For example, the MS macronutrients can be made up from any of the three mixtures of salts shown in Table 30.1.

Table 30.1 *Alternative ways to prepare MS (1962) macronutrients*

Salt	Original MS mg/l	Alternative 1 mg/l	Alternative 2 mg/l
$NKNO_3$	1900	1308.7	1441.0
NH_4NO_3	1650	1650	1650
$MgSO_4.7H_2O$	370	370	370
KH_2PO_4	170	170	170
$CaCl_2.2H_2O$	440	9.2	106.0
$Ca(NO_3)_2.4H_2O$	0	690.9	535.7
KCl	0	435.9	338.4

Each of the three recipes in Table 30.1 gives the same concentration of nutrient ions, but from a casual glance the three sets of ingredients would appear as comparatively different preparations. Obviously other alternative MS salt mixtures could also be devised.

In whole plants, nutrients may be taken up into roots either passively or through active absorption (involves energy). Active uptake is less dependent on ionic concentration than passive (diffusive) uptake. For many types of *in vitro* culture other than root culture, it seems probable that passive uptake might be the more important mechanism. The concentration of macronutrient ions in solution would then maternally influence ion uptake as well as the period over which the medium can supply nutrients to the culture. Growth and morphogenesis can also be influenced by the total concentration of macronutrient salts, because macronutrients are usually present in relatively large amounts and so have an appreciable effect on the osmotic potential of a medium. The other component normally present in large amounts that has an effect on the osmotic potential of medium is sugars (e.g. sucrose).

(b) Micronutrients

These are compounds which supply the minor or microelements needed by plants, and include manganese, zinc, boron, copper, and molybdenum. Iron is required in small amounts. Thus, it can be included in the list of micronutrients even though it is often listed in macronutrients. The essential micronutrients and the amounts they are needed are as in the media composition shown in Table 1, Appendix 13.

Biochemical role of micronutrients: The essential micronutrient metals Fe, Mn, Zn, B, Cu, Co and Mo become components of plant cell proteins of metabolic and physiological importance. At least five of these elements are, for instance, necessary for chlorophyll synthesis and chloroplast function. Iron is required for the formation of amino laevulinic acid and protoporphyrinogen and is a component of ferredoxin proteins, which act as electron carriers in photosynthesis. Iron is also a component of many proteins, which regulate oxidation or reduction reactions. Manganese is necessary for the maintenance of chloroplast ultrastructure and the photosynthetic process. Molybdenum- and zinc-deficient plants have decreased chlorophyll content and poorly developed chloroplasts, while copper atoms occur in plastocyanin, another pigment that participates in electron transfer. Cobalt is the metal component of vitamin B12 analogues, which are concerned with nucleic acid synthesis. Micronutrients may have other roles in the functioning of the genetic apparatus. In bacteria, Zn and divalent Mg, Mn, or Co are required for DNA and RNA polymarase activities. A lack of boron can cause plant cell walls to be thicker or thinner than normal and also affect absorption of $H_2PO_4^-$ ions. Several micronutrients are also involved with the activity of growth substances. For example, Zn is a component of an enzyme concerned with the synthesis of the IAA precursor, tryptophan. Boron deficiency results in depressed cytokinin synthesis, but endogenous IAA levels increase. Mn ions are also known to act as one of the cofactors for the peroxidase enzymes, which lead to IAA oxidation in plant cells.

2. Carbon sources

Cultured plant organs, tissues, or cells may not be able to provide their own supply of carbohydrates through photosynthesis. It is thus common to add a sugar such as sucrose or glucose to the medium for this purpose. Generally, it will not be classified as a media ingredient. Sucrose and glucose are the standard sources of sugars. Fructose is utilized much less readily. All other sugars (cellobiose, galactose, D-Mannitol, D-Sorbitol, Hexitols etc.) are utilized only by a few variant cell lines. The addition of xylose, arabinose, and glucose had a favorable effect, and apparently could substitute for sucrose. Most media contain myoinositol, which is beneficial as a component of cell wall metabolism (Table 1, Appendix 13). However, an important point to take note is at a higher level, sucrose serves as an osmoticum and inhibits cell divisions.

3. Vitamins

Plants synthesize all vitamins, but cells in culture require thiamine. Vitamins also have enhancing effects on growth and development which are listed in Table 1, Appendix 13. Adenine ($AdSO_4$) is important to cells as a nuclear substance and has a weak cytokinin effect. It is used in culture media to promote shoot formation.

4. Growth substances and growth regulators

Some chemicals occurring endogenously in plant tissues have a regulatory, rather than a nutritional role in growth and development. Growth regulators (Figure 30.1) are compounds, which at very low concentrations are able to regulate plant growth and development. The five classes of substances with growth regulator activity are auxins, cytokinins, gibberellins, ethylene, and abscisic acid (Appendix 13). They are produced naturally in plants in low quantities and therefore they must be added selectively to culture media. The type of growth regulator and concentration used will vary according to the cell type and culture purpose. Auxins and cytokinins are by far the most important for regulating growth and morphogenesis in plant tissue and organ cultures. Auxins are required for the induction of cell division in cultured tissues. The indole compounds and NAA are also used to induce root formation. They are often used in combination with cytokinins. Cytokinins are adenine derivatives. They have an essential role in differentiation and plant regeneration of most species. Gibberellins are used in plant regeneration from meristems and after shoot primordia formation has occurred. The role of abscisic acid in tissue culture is not well defined. The compound has been implicated in somatic embryogenesis. No chemical alternatives to the natural gibberellins or abscisic acid are available, but some natural gibberellins are extracted from cultured fungi and are available for use as exogenous regulators.

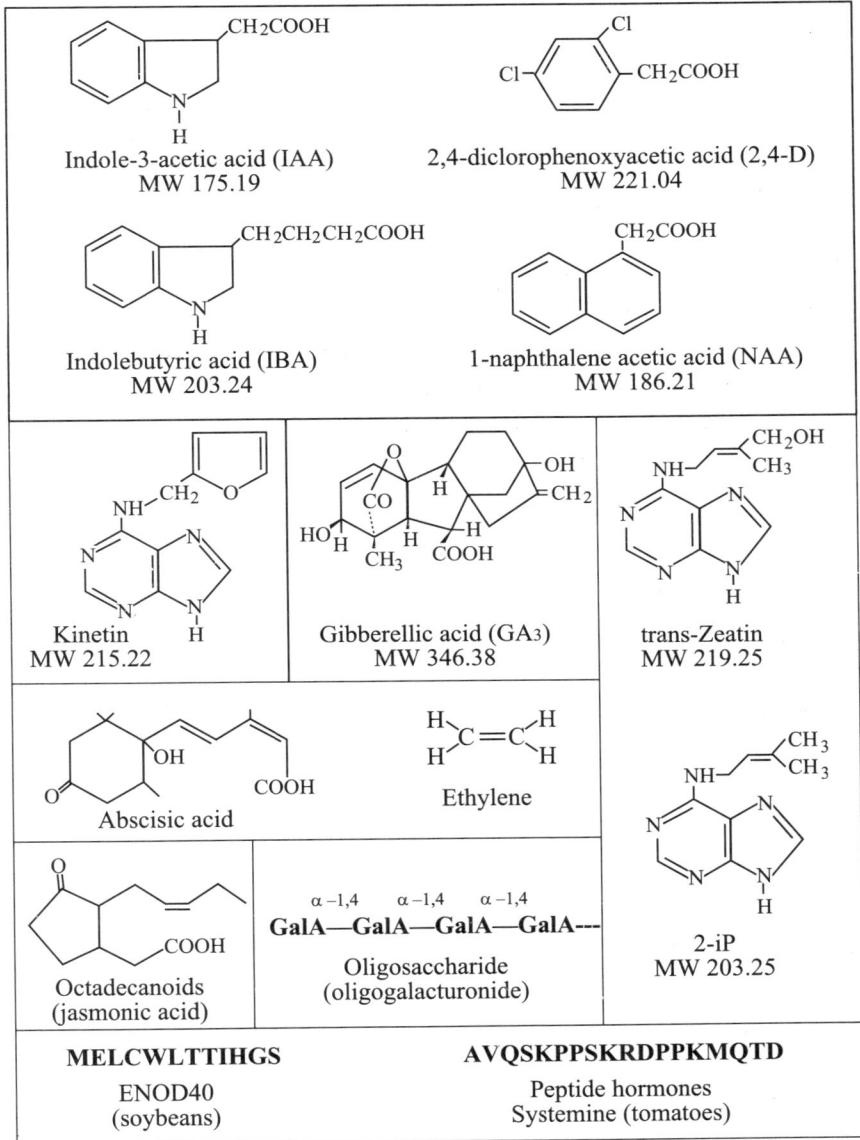

Figure 30.1 *Chemical structure of some important plant growth hormones*

5. Amino acids and amides

The addition of amino acids (as a readily-available source of reduced nitrogen) may enhance growth of cells and facilitate differentiation toward plant regeneration. Plant cells can grow on L-glutamine as the only nitrogen source. A special medium in which all inorganic nitrogen has been replaced by amino acids has been used effectively for different purposes.

6. Organic acids

Organic acids, salts of organic acids, and named buffers all have a temporary buffering effect on plant culture media. They are not common components and are only added for a particular purpose where stability of a medium is critically important. The addition of acids of the tricarboxylic acid cycle intermediates such as malate or citrate or fumaric is common in media for protoplasts. The compounds appear to alleviate any inhibitory effects of ammonium. Cells can tolerate up to 10 mM of the potassium salts.

7. Complex organics

Plant culture media are mainly composed of defined ingredients. However, often undefined supplements such as the juices of various fruits, plant extracts, and plant sap have been added. The most commonly used are protein hydrolysates and coconut milk. The enzymic hydrolyzed protein such as K-Z-amine Type A is preferable, since the amino acids are intact. Acid hydrolysis destroys several amino acids. Corn milk (corn endosperm), cornstarch, potato extract, yeast extract, antioxidants, etc have also been used in various types of cell cultures. Banana homogenate is another supplement still added by some people to media for orchid culture.

8. Gelling compounds

Agar, agarose, and gellan gum are the commonly used gelling compounds. The medium is solidified with 7.5 g/l of agar. The amounts of plant growth regulators added to the medium are variable, depending on whether the culture is to be maintained as callus or made to regenerate into whole plants. The medium used in this experiment contains Indole acetic acid (IAA, an auxin) at a concentration of 1.0 mg/l. No cytokinins are added. The effects of varying the concentrations of plant growth regulators on the cell culture will be investigated in another protocol.

9. Activated charcoal

Activated charcoal is often used in rooting media to adsorb root-inhibiting agents. Sometimes it is added to Stage I media to adsorb toxic phenolics, and also in Stage II media intermittently for curative purposes. It is generally added at 0.6 g per litre.

B. FACILITIES AND EQUIPMENT FOR MEDIA PREPARATION

The facilities for the use of tissue culture operations are best divided into distinct areas. One area should be designated for media preparation. This area would have top-pan balances, stirrers, a pH meter, and a microwave oven or equivalent. The area will also house stock solutions. In the same area or adjacently would be the cold room or refrigerator and freezer, as well as space, or a separate room, for analytical balances. Regularly used chemicals are also frequently stored in this area. Similarly clean glassware, containers or other supplies for the operations can be stored in the area. The other major and separate area is the room where the autoclaves, the water de-mineralizer, and distilling of water are located and where the dish washing is performed. After being autoclaved, the media can be stored in or close to this area.

C. MEDIA PREPARATION PROCEDURE

The easiest way of preparing media now-a-days is to use commercially available dry powdered media, containing inorganic salts, vitamins, and amino acids. The powder is dissolved in distilled water (10% less than the final volume of the medium), and after adding sugar, agar, and other desired supplements, the final volume is made up with distilled water. The pH is adjusted, and the medium autoclaved. Powdered media may be useful for routine purposes, such as micropropagation, etc. However, in experimental work where it is necessary to make major qualitative and quantitative changes in the organic and/or inorganic constituents of the medium, a suitable powdered medium may not be available. There are two possible ways of preparing the medium. One method is to weigh and dissolve the required quantities of the ingredients separately and mix them before the preparation of the medium. A more convenient and popular method is to prepare a series of concentrated stock solutions. Stock solutions are prepared as described in the following sections. The chemicals are dissolved in distilled or high purity water.

Stock solutions: The medium is distributed in culture vessels before autoclaving, or the vessels and medium may be autoclaved separately. The autoclaving conditions are generally 120°C for 15 to 20 min. Agar media are prepared in batches of different quantities (e.g. 500 ml). The agar is melted using the microwave oven prior to autoclaving and poured into sterile containers and cooled. The preferable temperature for media storage is about 10°C. Stock solutions are concentrated solutions of pre-mixed media chemicals that are prepared ahead of time and used to make several batches of media. They may be made in litre quantities of 10 or 100 times the concentration required in the final formula. Having stock solutions eliminates the need to weigh so many different chemicals every time a batch of medium is needed. Also, the quantities will be more accurate because they are weighed on a larger scale than would be required for a single batch of medium, and thus minor inaccuracies have less impact. Some of the ingredients will precipitate if mixed together in concentrated form, so each group is made up of chemicals that usually will not precipitate in the concentration of the stocks.

The salts must always be dissolved by being added one at a time. Precipitation is usually avoided by dissolving the inorganic nitrogen sources first. After the salts and other ingredients have been dissolved, the pH is adjusted by using 0.5 N HCl or 0.2 N NaOH or KOH. Stock solutions are prepared in 10- or 100-fold concentrations. There is no total agreement on the combinations of stock ingredients, nor on how or for how long they can be stored. After the ingredients are dissolved, the solution is distributed in sealable plastic containers or sterile glass bottles. Most stocks can be stored for a limited time without adverse reactions. If they have a shelf life, as do organics, then the chemicals will be stable for a longer time if they are stored in a refrigerator; hormone solutions tend to have a particularly short shelf life and so are generally made in small amounts. If the stock ingredients have a longer shelf life, like the inorganic salts, then they can be stored in a cupboard, but they run a greater risk of growing microbial contaminants because of the warmer temperature there.

1. Calculating amount of stock solutions per litre of medium

Although the calculations for the amount of stock solutions needed for making a particular medium can be directly done, it would be useful to know the method of figuring out how these amounts are determined. To determine the amount of stock solutions required for a medium, the easiest arithmetic to use is simple proportion. For example, a stock solution with 25 mg of BA in 250 ml water is available. Say, a particular medium requires only 0.4 mg of BA, how many millilitres of stock solution is to be used in order to obtain this amount of BA. We have to write the original milligram amount of BA (25 mg) over the original millilitre total of stock solution (250 ml). Next, we have to write the milligram amount needed (0.4 mg) over the unknown millilitre amount (? ml):

Then, we cross multiply:

$$\frac{25\,\text{mg}}{250\,\text{ml}} = \frac{0.4\,\text{mg}}{?\,\text{ml}}$$

$$25 \times ? = 250 \times 0.4$$

$$25 \times ? = 100$$

$$? = 4$$

The amount of BA stock solution needed is 4.0 ml.

2. Protocol for preparing plant cell culture media

A standard protocol for media preparation is described below and can be adapted to any medium preparation.

Procedure

1. Plan well in advance and establish a step-by-step routine for preparing a medium. This helps to maintain the identical composition of the medium each time and also makes it easier to avoid possible mistakes.
2. Check each component carefully when adding each time. Make sure each ingredient has completely dissolved before adding the next.
3. The containers for media preparation should be bigger than the final volume, but the amount of water taken should be smaller than the final volume. For example, if 1 litre of medium is required, start with about 400 ml water in a 1- or 2-litre beaker or conical flask on a magnetic stirrer.
4. Measurements of each component should be accurate. Use the correctly sized pipette or graduated cylinder and stay within the proper range of sensitive balances.
5. Once all the components have been added and have dissolved completely, make up the volume just below the final volume.
6. Adjust the pH of the medium to the required value by adding dropwise either 0.2 M KOH or 0.5 N HCl as needed, while stirring on the magnetic stirrer.
7. For solid media, add the known amount of gelling compound to medium taken in a bigger flask (e.g. for 1 litre medium taken in 1.5 or 2 litre flask), label the flask, and cap with aluminum foil.
8. Autoclave the media at 121°C at 105 kPa for 15 to 20 min. Allow sufficient air volume in the flask to prevent the medium from boiling over.

9. Cool the medium, and swirl the solution gently as it cools. Label the medium with name and date.
10. If filter-sterilized compounds are to be added to the medium, add to the autoclaved media when it is cooling but not yet solidified. Mix thoroughly by swirling the solution. Make sure the autoclaved medium and added filter-sterilized compounds sum up to the final volume.

D. TYPES OF CULTURE MEDIA

Plant material can be cultured either in a liquid medium or on a medium that has been partially solidified with a gelling agent. The method employed will depend on the type of culture and its objective.

Solidified media: Media which have had a gelling agent added to them so that they have become semisolid, are widely used for explant establishment. They are also employed for many routine cultures of callus or plant organs, and for the long-term maintenance of cultures. Agar is the most common solidifying agent, but a gellan gum is also widely used. Cultures grown on solid media are kept static. In the solid medium, only the lower surface of the explant, organ, or tissue is in contact with the medium. This means that as growth proceeds, there will be gradients in nutrients between the medium and the tissues.

Liquid media: Liquid media are essential for suspension cultures, and are preferred for critical experiments on the nutrition, growth, and cell differentiation in callus tissues. They are also used in some micropropagation work. Some kind of agitation is essential for suspension cultures to prevent cells and cell aggregates settling to the bottom of the flask. Other purposes served by agitation include increased aeration, the reduction of plant polarity, the uniform distribution of nutrients, and the dilution of toxic explant exudates.

E. CELL CULTURE MEDIA STORAGE AND HANDLING

Autoclaved media are dispensed in sterile culture vessels, which vary from Petri dishes and flasks to glass or plastic jars of different sizes. Before dispensing, the larger volumes of media must be thoroughly mixed. The media should be cooled as quickly as possible after autoclaving. The optimum storage temperatures are 4–10°C. Each vessel should be labeled to show the type of medium and date of preparation. Storage at room temperature should not be for more than 2–3 weeks before use. If the medium is dispensed in Petri dishes, turn the plates upside down to prevent moisture loss by evaporation.

F. PROBLEM SOLVING

1. Contamination in plant cell cultures

Contamination can occur in stock solutions of prepared media, which may be inadequately sterilized. Generally, yeast, fungus, and bacteria are the common contaminants detected. Any medium with actual or suspected contamination should be autoclaved as soon as possible after it is ensured that it is properly closed. Media can be stored for several weeks at 4°C.

2. Insoluble precipitation

The main causes of precipitation in the medium are the complexes formed between calcium, phosphate, and magnesium. The problem can be avoided if the compounds are added in the order: the nitrogen compounds, the magnesium compound, the calcium compound, and last, the phosphate compound. Each compound must be completely dissolved before the next one is added. The formation of precipitates can also be avoided if the calcium compound is dissolved separately and added in a solution form.

3. Quality indicator

A useful quality indicator is the color of the final medium. Consistent media preparation and sterilization procedures will result in a color which is reproducible. Incorrect media composition and improper autoclaving may change the color of the media. Improper mixing and too acidic a pH can cause problems in solidification of the gelled medium.

4. Problems of establishment

(a) Phenolic oxidation

Some kinds of plants, particularly tropical species, contain high concentrations of phenolic substances, which are oxidized when cells are wounded or senescent. Isolated tissue then becomes brown or black and fails to grow.

(b) Minimum inoculation density

Certain essential substances can pass out of plant cells by diffusion. Substances known to be released into the medium by this means include alkaloids, amino acids, enzymes, growth substances, and vitamins. The loss is of no consequence when there is a large cluster of cells growing in close proximity or where the ratio of plant material to medium is high. However, when cells are inoculated onto an ordinary growth medium at a low population density, the concentration of essential substances in the cells and in the medium can become inadequate for the survival of the culture. This means that there is a minimum size of explant or quality of separated cells per unit culture volume, for successful culture initiation. Large explants generally survive more frequently and grow more rapidly at the outset than very small ones. For commencing suspension cultures, it is commonly about $1–1.5 \times 10^4$ cells/ml.

The minimum cell density phenomenon is sometimes called a 'feeder effect' because deficiencies can often be made up by the presence of other cells growing nearby. Suspension cultures can be started from a low density of inoculum by 'conditioning' a freshly prepared medium. The use of conditioned media can reduce the critical initial cell density by a factor of about 10. It is also possible to overcome the deficiencies of plant cells at low starting densities by adding small amounts of known chemicals to a medium (growth regulators, organic acids, additional sugars, coconut milk, etc.).

G. SUBCULTURING

Once a particular kind of organized or unorganized growth has been started *in vitro*, it will usually continue if callus cultures, suspension cultures, or cultures of indeterminate organs are divided to provide new explants for culture initiation on fresh medium. Subculturing often becomes imperative when the density of cells, tissues, or organs becomes excessive, and when it is necessary to increase the volume of a culture or to increase the number of organs for micropropagation. The period from the initiation of a culture to the time of its transfer is sometimes called a *passage*. Suspensions regularly subcultured at the end of the period of exponential growth can often be propagated over many passages. The rate of plant propagation depends on the ability to subculture shoots from proliferating shoot or node cultures, callus, or suspension cultures. A further reason for transfer, or subculture, is to reduce the accumulation of toxic metabolites and the exhaustion of the medium.

III EXPLANT PREPARATION AND CELL CULTURE INITIATION

An **explant** is a piece of a plant from which a culture is initiated. The part of the plant from which explants are obtained depends on: (1) the kind of culture to be initiated, (2) the purpose of the proposed culture, and (3) the plant species to be used. Explants can therefore be of different kinds and a wide range of plant organs and tissues can be used as a source of explants for the initiation of callus culture. Explants range in size from a microscopic 10th of a millimeter to stem pieces several centimeters in length. Explants can be meristems, shoot tips, macerated stem pieces, nodes, buds, flowers, peduncle (flower stalk) pieces, anthers, petals, pieces of leaf or petiole, seeds, nucellus (the central part of an ovule) tissue, embryos, seedlings, hypocotyls, bulblets, bulb scales, cormels, radicals, stolons, rhizome tips, root pieces, or (though rarely) single cells or protoplasts (Figure 30.2).

Plants growing in the external environment are invariably contaminated with microorganisms and pests. These contaminants may be confined to outer surfaces of the plant or may be systemic within the tissues. Tissue culture is generally started from small explants and must be grown on nutritive media which are also favorable for the growth of microorganisms. Therefore, as far as possible, explants must be free from microbial contaminants when they are first placed on a nutrient medium. Meristems are usually free of virus infection because they are devoid of vascular tissue that transport viruses. They also have active mitotic division. While no precise instructions can guarantee successful explants, many protocols have been universally successful.

A. OBTAINING EXPLANT

Before collecting explants, stock plants should be moved to a clean greenhouse and maintained well with special care. The plants have to be washed with clean water, the foliage allowed to dry and thereafter they should be fed and watered only at the base. The new shoots that appear will provide relatively clean explants and, in addition, may provide some increased juvenility. Just as juvenility is a factor in selecting material for cuttings, so it is in selecting explants for

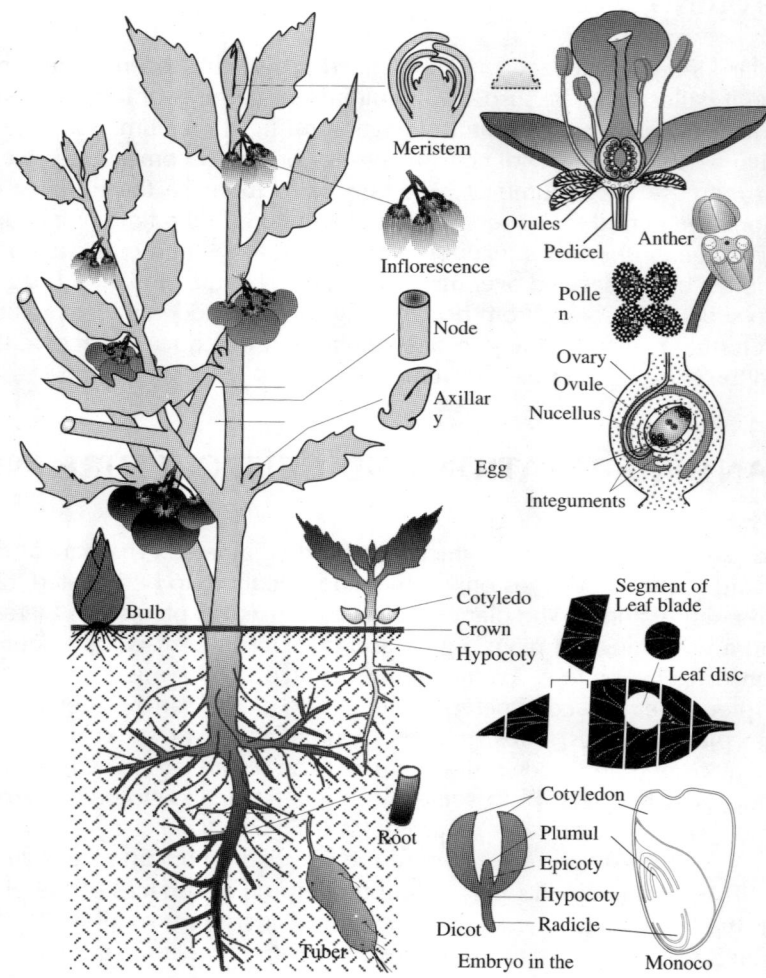

Figure 30.2 *Explant terminology. Explant source or portion of plant*

tissue culture. In general, the more juvenile the explant material, the greater the likelihood of success. When shoot tips, stems, buds, or flowers are taken for explants, they should be cut longer than the final size to facilitate processing. Occasionally, leaves or parts of leaves are used to start tissue cultures. Leaves, after cleaning, can be drilled with a cork borer and the discs placed in culture media. Leaf tissue is also a common source of protoplasts.

B. EXPLANT CLEANING AND TREATMENT

After explants are detached from the source plant, they must be submitted to cleaning treatments (Figure 30.3). There are no set rules for cleaning explants and the process selected also depends on how clean the explant must be and h ow delicate the explant is. The most common

Plant Tissue Culture Media, Preparation, and Culture Initiation

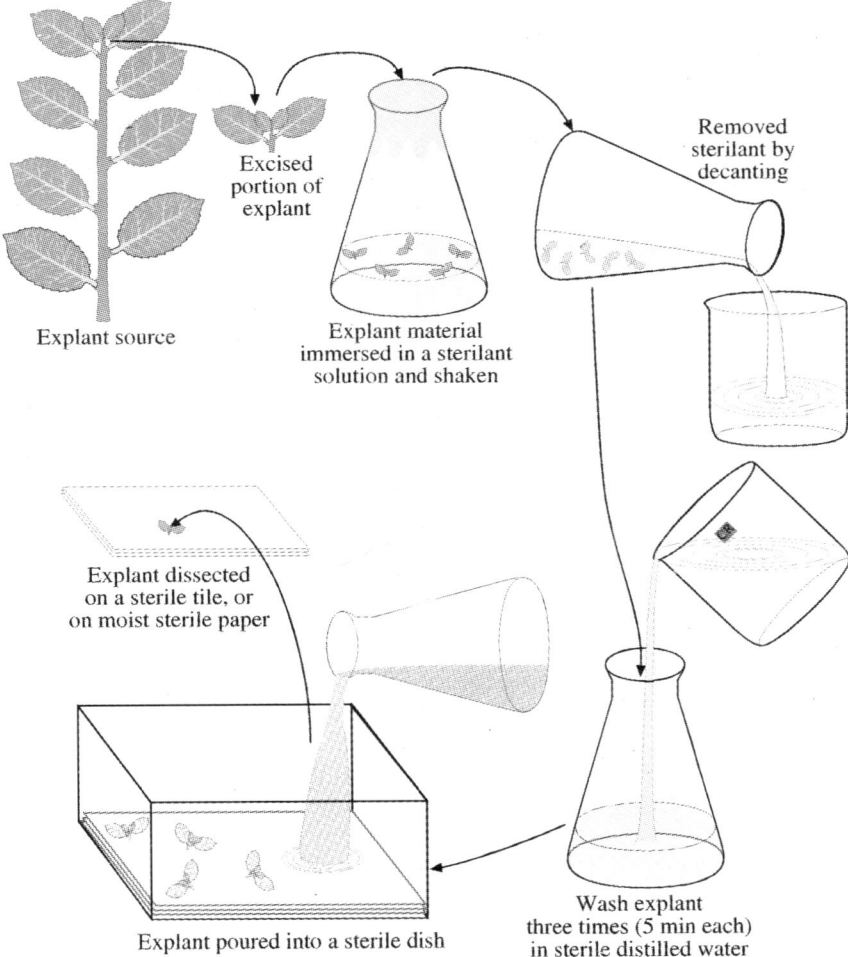

Figure 30.3 *A method of disinfecting the explants by detergent solution treatment followed by washing with sterile distilled water*

cleaning agent used to disinfect explants is bleach in various concentrations (5% sodium hypochlorite), used alone or in sequence with other disinfectants. Other disinfectants commonly used to clean explants include 70% ethyl alcohol or isopropyl alcohol; 5 to 10% calcium hypochlorite, 3% hydrogen peroxide and 0.1–0.2% mercuric chloride. The following is a general protocol for an explant cleaning treatment.

Materials

Five 100- to 250-ml beakers; 1 and 10% bleach; Dropper bottle containing Tween 20; Magnetic stir plate and stir bar; Explant material; Distilled water; 70% isopropyl alcohol; Forceps; Cutters or scalpel.

Procedure

1. Pour 100 ml of 1% bleach and 2 drops of Tween 20 into a sterile 150 ml beaker with the magnetic stir bar and place on the stir plate.
2. Add the explants to the beaker and stir for 10 min. Alternatively, the explants and the cleaning materials can be shaken by hand in a tightly closed jar. Pour off the bleach solution, and place the explants in a clean beaker.
3. Rinse the explants in 70% isopropyl alcohol (or 3% hydrogen peroxide) for 5 seconds. Again, pour out the alcohol and place the explants in a clean beaker.
4. Rinse the explants in distilled water and then place them in a clean beaker.
5. Pour 100 ml of 10% bleach and 2 drops of Tween 20 into a sterile 150 ml beaker with the magnetic stir bar and place on the stir plate.
6. Add the explants to the beaker and stir for 15 min. With the explants still in the beaker of 10% bleach solution, move them to the transfer chamber.

C. EXPLANT TRANSFER

Transfer of explant into the growth medium requires special care and preparation to prevent any contamination. Very often cultures are contaminated with different microorganisms including molds, bacteria, and fungal spores. Thus, plant transfers have to be made in the hood using sterile techniques.

D. STERILE TECHNIQUE

As we have emphasized, cleanliness is of primary importance in the transfer room, especially in the hood, where cultures are momentarily outside their sterile protective containers – a time when they are in the greatest danger of becoming contaminated. Microscopic organisms or contaminated particulates are literally everywhere, except on that which has been sterilized and subsequently protected. Air is sterilized by filtration. The air gently streaming through the transfer hood is filtered so as to open culture containers and make sterile transfers with reasonable assurance. The hood should be wiped down daily with a disinfectant such as 10% bleach, 70% alcohol, or Lysol or some other household disinfectant. All the transfer tools used must be sterilized. The more commonly used methods include alcohol, bleach, Bacti-Cinerators, and glass bead sterilizers. Deciding which is the best sterilization technique is a matter of individual preference.

Materials

10% bleach in a sterile plastic dish; 1% bleach in a sterile plastic dish; 100 ml sterile beakers containing sterile distilled water; 100 ml sterile beaker containing 1% bleach solution; Plastic gloves; Forceps; Stainless steel knife with disposable blade; Instrument holder; Pre-treated explants in beakers of 10% bleach; Sterile commercial-grade paper towels; Sterile test tubes with MS-based medium; Test tube racks for 10 and 40 tubes; Water basket, placed on the floor; Labeler; Planter trays or test tube racks; Parafilm/Food wrap.

Procedure

Sterile technique for explant transfer
1. Turn on the transfer hood blower for 10 min. before transfer.
2. Wipe or spray inside the hood with 10% bleach, 70% alcohol, or Lysol.
3. Rinse gloved hands in 10% bleach.
4. Using sterile forceps, rinse the treated explants in 1% bleach, then give 2 sterile distilled water rinses. Leave the explants in the sterile water rinse while performing steps 5 and 6.
5. Using forceps and knife, lay a sterile paper towel on the counter as far back in the hood as is practical to work.
6. Return forceps and knife to 10% bleach for one min., then rinse in 1% bleach.
7. Using forceps, place an explant from the sterile water onto the paper towel.
8. Pass the forceps to the left hand and pick up the knife with the right hand.
9. Using forceps and knife, trim the explant appropriately. For shoot tips, from 1 mm to several centimeters; 2 to 3 cm is usually a good length for stem. Petiole sections should more often be just a few millimeters.
10. Rinse the forceps with 1% bleach for 1 min. With the left hand, grasp a test tube containing the medium for the explant. Hold the test tube near its base. To prevent any microbes from falling into the test tube, hold it at about a 50° angle and facing right – parallel with the front of the hood so that it is facing neither the back filter nor the worker.
11. Take forceps in the right hand. While still holding the test tube and the forceps, grasp the test tube cap with the right-hand little finger, push and twist the cap slightly and remove the cap from the test tube.
12. Still holding the test tube, forceps, and cap, use the forceps to remove an explant from the sterile paper towel and place it firmly on the agar in the test tube.
13. Still holding the forceps, replace the cap on the test tube, seal it with parafilm, and place the tube in the test tube rack. Return forceps to 10% bleach.

Sterile technique for culture transfer
1. Place the test tubes containing cultures ready for transfer on a counter or cart outside the transfer hood.
2. Keep the fresh sterile test tubes containing agar medium ready to receive the transfers.
3. Follow steps 1 to 5 except step 4.
4. With the left hand, grasp a test tube containing the culture to be transferred. Check the label to confirm that the test tube contains the correct culture.
5. As described in steps 10 and 11 above, hold the test tube near the base and at about a 50° angle. While holding sterile forceps in the right hand, twist off the test tube cap with the right-hand little finger. Remove the cap and place it on the workbench.
6. With forceps remove the culture from the test tube and place it on the paper towel in the transfer hood.
7. Place the discarded test tube in a test tube rack, which can go directly into the dishwasher.
8. With the help of knife and forceps, cut, trim, and divide the culture as necessary. Trim away any brown or dead material.
9. Using forceps, place the culture in a test tube with the appropriate medium, following the procedures described in steps 10 and 12 above. Some shoot transfers should be laid on the surface of the agar to induce bud break if there are lateral nodes. Most shoot tips should

be inserted into the medium with their bases just deep enough in the agar to allow the shoot to be held upright. Transfers that are going into a rooting medium should be held upright.
10. Replace the cap on the tube, seal it with parafilm, and place the test tube in a rack ready for labeling.
11. After doing this for about 6 test tubes, replace the towel with a fresh sterile paper towel.
12. Label all the tubes properly with the date, culture name, and the workers initials.
13. Place the test tubes in planter trays or keep in test tube racks, and move the transferred cultures to the shelves in the culture growing room.

E. PHYSICAL ENVIRONMENT

1. Density

If the culture initiation is for protoplasts, then for successful culture of protoplasts, density is a very critical factor. Many studies show that the optimal density generally is between 10^4 and 10^5 protoplasts ml^{-1}, the exact value depends on the plant species, tissue employed, and the physiological conditions of the donor plant. Even though the exact cause for inability of protoplasts to grow at low densities is not known, the general belief is that at low densities protoplasts lose considerable amounts of vital substances to the medium. However, other effects cannot be ruled out entirely. In recent years, some success has been obtained in culturing protoplasts at low densities by the feeder-layer, nurse culture, and other techniques.

2. pH

A pH in the range of 5.5–5.8 is satisfactory for the culture of most protoplasts. However, pH values somewhat above 6.0 also markedly enhance cell divisions in protoplasts of pea, cowpea, and *Asparagus officinalis*. Generally, the pH of the medium is adjusted before autoclaving. But it is a common experience that autoclaving changes the pH of the medium. Recently, the use of a pH indicator, bromocresol purple, which can be added to the culture medium to monitor changes in pH, has been recommended.

3. Temperature

The temperature employed for culture of protoplasts has generally ranged between 22° and 28°C. However, at either extreme, sensitivity may be high and varies with different species used.

4. Light

Although light is very critical for culture initiation and differentiation, unfortunately, detailed investigations are still few on the effects of light and, from the current state of our knowledge, it is rather difficult to draw a generalization. There is some indication that light sensitivity may have a genetic basis.

F. HARDENING OFF

It is much easier to grow seedlings than it is to establish tissue-cultured plantlets in greenhouse conditions. If care is not exercised during this very critical stage, high losses can result. An abrupt change to the lower humidity and greater light of the greenhouse can be fatal to plantlets within a very short time. The quality of stage IV (*in vivo*) plantlets depends on the quality of the plantlets that come from the growth room. Often stage III (the rooting stage *in vitro*) is omitted and plantlets are rooted in soil directly from stage II. The success of this practice depends on the particular plant, its ease of rooting, and on the skill of the worker. Sometimes, the roots that develop in stage III are not functional in soil, in which case the plantlets may as well go to the soil from stage II. Root systems are also generally stronger if stage III is bypassed, except in some species.

Acclimatization can be started while plantlets are still *in vitro*, or it can wait until stage IV when the plantlets are moved from containers to the soil. If micropropagation is to be taken in stage II, then the hardening off will occur entirely in stage IV. A major problem faced in acclimatization is water loss. When a tissue cultured plantlet is transplanted into the soil, the stomata usually remain open until they are able to adjust to the lower humidity – tissue culture containers generally have high humidity – and greater light, or until new leaves are produced. This period varies with the particular plant, its conditions in culture, and its new environment. Leaves that develop in tissue culture often have an abnormally high number of stomata per given surface area. The unusual nature (chemical make-up and its appearance) of epicuticular wax on leaves and stems in culture has also been observed. The transpiration protection normally afforded by this waxy coat is depleted in cultures. Further water loss in tissue cultured plants results from guttation, the exudation of water from hydathodes.

In addition to these changes in morphology and transpiration, the entire photosynthetic process is confused in *in vitro* plantlets. In culture media, plantlets make little use of carbon dioxide because they draw on the sucrose in the medium for energy instead of depending on photosynthesis. In order to encourage photosynthesis in plantlets, CO_2 can be introduced into culture vessels or the greenhouse. This is done most readily in an acclimatization chamber rather than while the plantlets are still in culture. If CO_2 is to be introduced, there must be a corresponding increase of light for the CO_2 to be effective. Other characteristics exhibited by tissue cultured plantlets that are deviations from normal plants can include thinner roots and stems, thinner leaves, underdeveloped palisade layers, fewer trichomes, less collenchyma, reduced vascular tissue, and poorly distributed and less chlorophyll.

If plantlets are to be rooted *in vitro* in stage III, there are several ways of preparing for acclimatization. In some cases, small bags of desiccant (silica gel) can be hung in the containers to lower the humidity. Container covers can be placed to fit more loosely and allow more water vapor to escape, but if left for more than a few days contaminants may build up. Another effective method is to use lids with filters, which allow some desirable air exchanges and can be applied much earlier than when the lids are removed. Decreasing salt levels, especially nitrates, often help to induce rooting. The amount and type of cytokinin used in stage II can also affect rooting. Liriope, Schefflera, and Philodendron have been found to survive transplanting better by 80% when 2iP or kinetin are used instead of BA in stage II. In stage III, cytokinin is usually eliminated and the auxin level raised. The auxin level is very important – too much auxin can be worse than none at all. Lowering sucrose levels in stage III may encourage photosynthesis, but raising it may favorably affect the water potential. Phloroglucinol has proven beneficial in rooting of some fruit species. Although lighting requirements will vary widely, many cultures root better in increased light. On the other hand, some bulbous and rosaceous plantlets form root initials more readily in darkness.

The common procedure for transferring plantlets from the culture growing room to the greenhouse is to wash off the agar from the roots, plant the plantlets in artificial soil in undivided planter trays, and place them in high humidity in tunnels or tents on benches in a shaded greenhouse. Over a 2- to 4- week period, the sides of the tent should be gradually opened and the amount of mist gradually reduced to lower the humidity, thus allowing the existing leaves to adjust and/or assisting new leaves to grow. A fog system is an ideal way of maintaining high humidity without over-saturating the plants we are trying to harden off. Humidifiers are less expensive and will do the job well, especially in smaller areas. Greenhouse shade can be maintained at 50%. A growth room equipped with fluorescent lights over shelves, similar to the culture growing room, is a good option for establishing tissue cultured plantlets in soil. As plantlets are moved from stage II or stage III to soil mixes, it is desirable to wash off as much of the agar as possible because mold, yeast, bacteria, and insects thrive on the nutritive agar. There are probably about as many artificial soil mixture as there are growers. Soil media are not always defined in literature, which slights this very important aspect of hardening off.

IV FURTHER READING

Abo El-Nil MM, Hildebrandt AC (1976) Cell Wall Regeneration and Colony Formation from Isolated Single Geranium Protoplasts in Microculture, *Can. J. Bot.*, **54**:1530–1534.

Bidney DL, Shepard JF and Kaleikau E (1983) Regeneration of Plants from Mesophyll Protoplasts of *Brassica oleracea*, *Protoplasm*, **117**:89–92.

Gamborg OL and Phillips GC (1995) Media Preparation and Handling. In *Plant Cell, Tissue and Organ Culture* (Eds), Gamborg OL and Phillips GC, Narosa Publishing House.

Ignacimuthu S (1997) *Plant Biotechnology*, Oxford & IBH Publishing Co. Pvt. Ltd.

Khanna VK (1998) *Plant Tissue Culture Practice*, Kalyani Publishers, New Delhi.

Kite Land Kleyn J (1996) *Plants from Test Tubes: An Introduction to Micropropagation*, Timber Press.

Kumar De K (1997) *Plant Tissue Culture*, New Central Book Agency (P) Ltd, Kolkata.

Morrish F, Vasil V and Vasil K (1987) Developmental Morphogenesis and Genetic Manipulation in Tissue and Cell Cultures of the *Graminae*, In *Molecular Genetics of Development*, (Eds) Scandalis JG, *Advances in Genetics*, **24**:431–499.

Vasil IK (Ed) (1984) Cell Culture and Somatic Cell Genetics of Plants (Vol I) *Laboratory Procedures and Their Applications*, Academic, New York.

Wilson VM, Haq M, Evans PK (1985) Protoplast Isolation, Culture and Plant Regeneration in the Winged Bean, *Psophocarpus tetragonolobus* (L) DC, *Plant Sci*, **41**:61–68.

31

Micropropagation

I INTRODUCTION

Plants propagate and preserve naturally mainly through two developmental life cycles: the sexual (***seeds***) or asexual (***vegetative***). In the sexual cycle, new plants arise after fusion of the parental gametes, and develop into seeds or fruits. This method is also the most economical (large numbers are produced and can be stored for long periods), efficient (easily distributed and usually pest- and disease-free) and therefore, universally used for plant propagation. However, in most cases, the seedlings will be variable and each one will represent a new combination of genes, brought about during the formation of gametes (***meiosis***) and their sexual fusion; also, in several plants, seeds are either not formed or are produced in small quantities. Under such circumstances, vegetative propagation methods are followed (Figure 31.1). In the vegetative cycle, the unique characteristics of the individual plant selected for propagation (termed stock plant or ***ortet***) are perpetuated with high homogeneity because during normal cell division (***mitosis***), genes are typically copied exactly at each division. In most cases, each new plant (or ***ramet***) produced by this method may be considered to be an extension of the somatic cell line of one individual. A group of such asexually reproduced plants (ramets) is termed a ***clone***.

II CELLULAR TOTIPOTENCY

Unlike animals, where differentiation is generally irreversible, in plants even highly mature and differentiated cells (except sieve tube elements and xylem elements whose nuclei have disintegrated or fibers with cell walls thicker than 2 µm) retain the ability to regress to a meristematic state as long as they have an intact membrane system and a viable nucleus. When non-dividing, quiescent cells from differentiated tissues are grown on a nutrient medium that supports their proliferation, the cells first undergo certain changes to achieve the meristematic state. These include: (i) replacement of non-functional cellular components damaged by lysosomal activity during the process of cytoquiescence, (ii) reversing of mature cell to the meristematic state and forming undifferentiated callus tissue (termed ***dedifferentiation***), (ii) establishment of complete whole plant (multicellular cells with diverse functions) from unorganized callus (termed ***regeneration***, ***redifferentiation***).

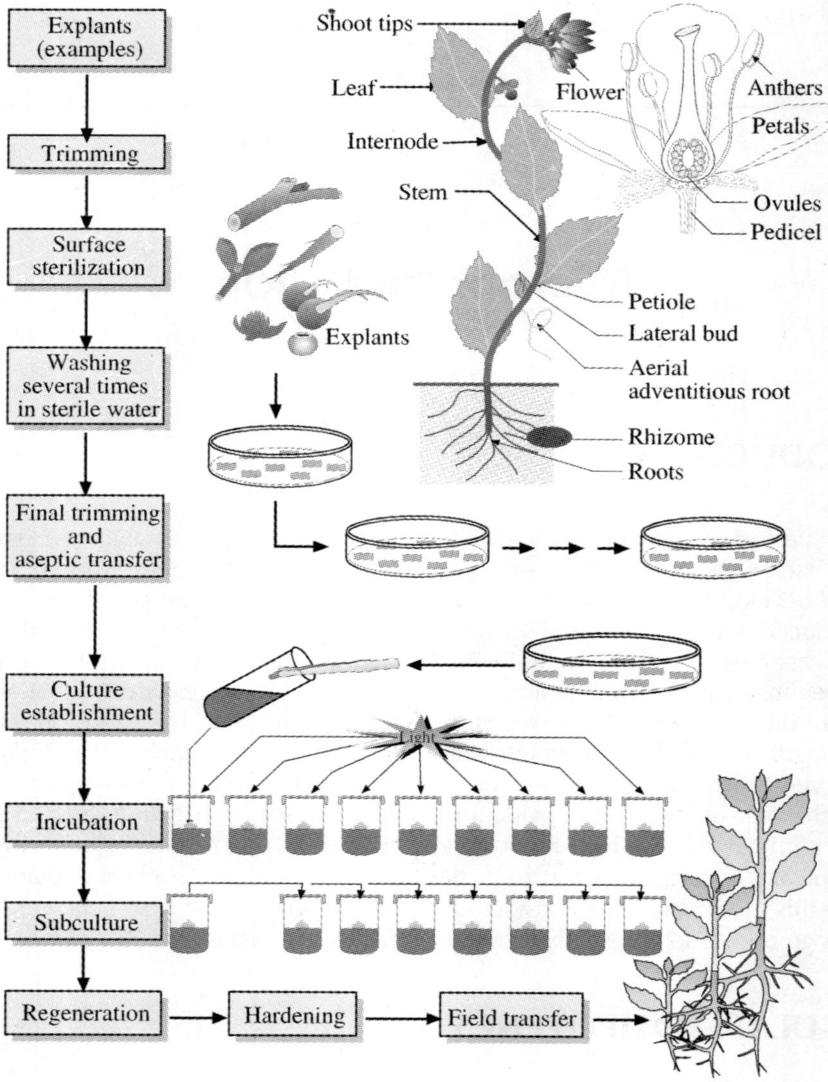

Figure 31.1 *Basic procedure for establishing and maintaining a culture of plant tissue*

The inherent capacity of a plant cell to give rise to a whole plant, a capacity which is often retained even after a cell has undergone final differentiation in the plant body, is described as **'cellular totipotency'**. In other words, totipotency is the cell characteristic in which the potential for forming all the cell types in the adult organism is retained. This morphogenic process offers not only an excellent opportunity to study the factors that elicit the totipotency of cells but also allows us to investigate the factors controlling cytological, histological, and organogenic differentiation.

III CYTODIFFERENTIATION

In an intact plant there are many kinds of cells all having different forms and functions. Meristematic cells, and thin-walled parenchymatous tissue, are said to be undifferentiated, while specialized cells are differentiated. The cells of the callus and suspension cultures are mainly undifferentiated, and it is not yet possible to induce them to become of just one differentiated type. The differentiated state is also difficult to preserve when cells are isolated from a plant. Differentiated cells are most effectively produced *in vitro* within organs such as shoots and roots. During cytodifferentiation *in vitro* and *in vivo* the main emphasis has been on vascular differentiation, particularly the xylem elements. The phloem has received less attention because of technological problems. Whereas tracheary elements can be easily stained and scored in macerated preparations of the tissue, differentiation of tracheary elements has become a model system for studying cytodifferentiation in cultured plants. The factors affecting vascular differentiation are: auxin, sucrose, cytokinin and gibberellin, and physical factors such as temperature and light.

The cell cycle: During division, a cell needs to pass through certain processes or phases which, because they are repeated sequentially from one division to the next, are jointly termed the cell cycle. According to the principal control point hypothesis, the initiation (continuance), or the cessation of plant cell division is primarily regulated by factors which operate during the G1 and G2 phases (Figure 31.2). Cells cease to divide and become arrested in either G1 or G2 when these factors are limiting. The differentiation of cells into defined types of tissue takes place in cells where division is arrested (also reported as arrested in a quiescent G0 phase). Much of the differentiation that a cell will ultimately show appear to be decided during the processes which led to its formation (preceding the cell cycle). Observations supporting this conclusion are that in the culture of some plants, the rate of xylem formation may exactly parallel the rate of cell proliferation, and that a preceding mitotic activity also seems to be a prerequisite. However, some changes in cell structure and function such as cell expansion and the induction of flowering are not dependent on cell division, and there appear to be circumstances where differentiation may occur without mitotic division. Cytodifferentiation appears to require prior DNA synthesis, either within the nucleus or within other cell organelles. Where there is previous cell division, this synthesis takes place during the S phase of the cell cycle. In non-dividing cells, DNA synthesis could occur by the processes of differentiation that are specific for the processes of differentiation that are to follow, or possibly by endopolyploidization. Endopolyploidy (or endomitosis, endoreduplication) is the duplication of the chromosome number (and DNA content) of cells, without the formation of a spindle and nuclear division. It seems to be associated with certain developmental processes in plants, and appears to occur frequently in cells forming tracheid elements.

Cell division is essential for differentiation: Whether cell division is a prerequisite for xylem differentiation or not is a subject of debate. Several workers have remarked that a cell must divide before the differentiation of xylary elements can occur. The chemical factors (auxin, cytokinin, sugars, etc.) reported to be involved in xylem differentiation are generally the same as those regulating cell division. Recent studies have, however, clearly established that cell division is not always a prerequisite for tracheary differentiation. A number of chemical and physical factors have been shown to have a profound effect (qualitative and quantitative) on cytodifferentiation. Chemical factors (such as auxin, cytokinin, gibberellin, and sugars), physi-

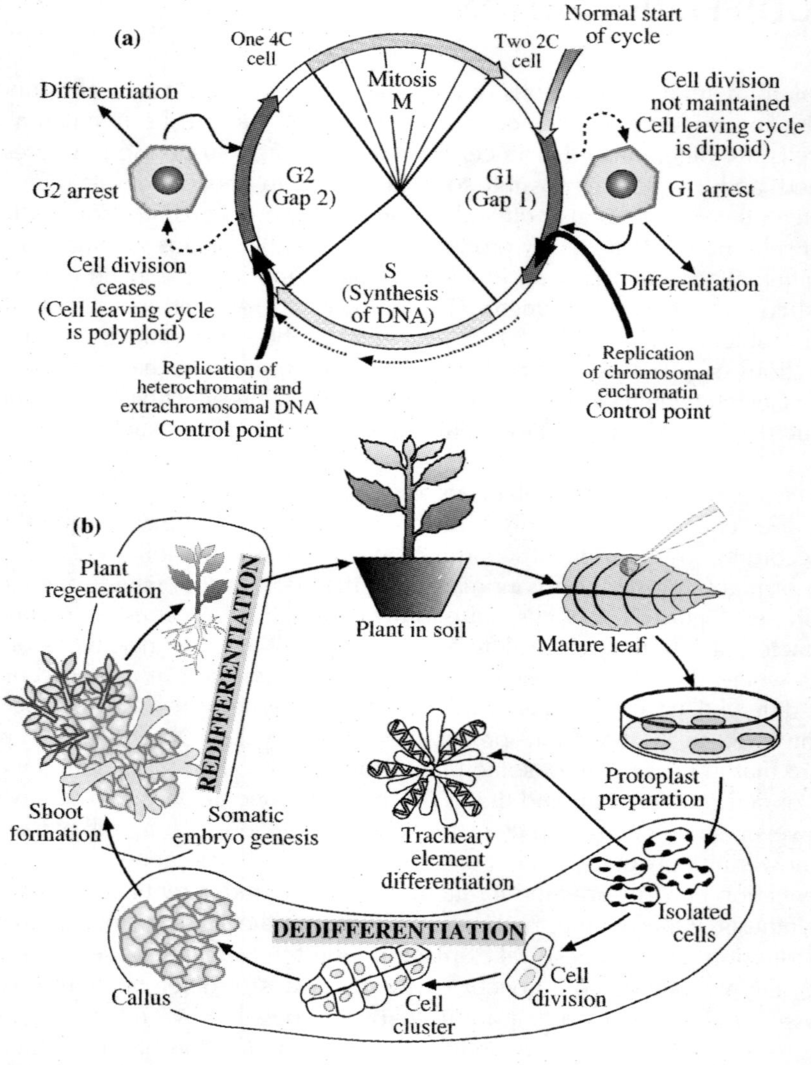

Figure 31.2 *(a) The cell cycle events (b) Scheme showing cytodifferentiation in plant cells*

cal factors (light, temperature, pressure, water stress), morphactins, methionine, and irradiation have been found to have a significant effect on cytodifferentiation.

IV MORPHOGENESIS

New organs such as shoots and roots can be induced to form on cultured plant tissues. Such freshly formed organs are said to be **adventive** or **adventitious**. The creation of new form and organization, where previously it was lacking, is termed **morphogenesis** of **organogenesis**. Tissues or organs, which have the capacity for morphogenesis/organogenesis are said to be **morphogenic** or **organogenic**. So far it has been possible to obtain the *de novo* (adventitious) formation of: (i) shoots (*caulogenesis*) and roots (*rhizogenesis*) separately, (ii) embryos that are structurally similar to the embryos found in true seeds, and (iii) flowers, flower initials, or perianth parts. Caulogenesis, rhizogenesis, and embryogenesis are important for plant multiplication (Figure 31.3). Plantlet regeneration is best achieved from adventitious shoots, which are afterward rooted, or from somatic embryos. Shoots, roots, and somatic embryos arise from single cells, or groups of cells, which become induced by the cultural conditions to become centers of active cell division (morphogenetic meristems), each capable of producing an organ of one kind. Morphogenetic meristems can theoretically occur in either of two distinct ways: (i) from the differentiated cells of a newly-transferred piece of whole-plant tissue, without the proliferation of undifferentiated tissue and (ii) from the unspecialized, unorganized, and dedifferentiated cells of callus tissues or suspension cultures. These two methods of morphogenesis are also called direct and indirect organogenesis respectively.

A. COMPETENCE

Cells that have retained the capacity for a particular kind of cellular differentiation or morphogenesis, or have acquired it in response to an appropriate stimulus, are said to be **competent**. Some consider that morphogenic competence usually conveys an ability to proceed toward only one particular developmental pathway. According to this hypothesis, a cell which is competent to undergo shoot morphogenesis may not be competent for root formation. Competence thus is the first step in the dedication of one or more undifferentiated cells (stage I of Figure 31.4) toward morphogenesis, or some other kind of specialized development (stage II). The second stage of dedication is then said to be the **induction** of determination in competent cells. Individual cells or groups of cells are said to be determined when they have become committed to follow a particular genetically-programmed developmental pathway. Determined cells differentiate to become the specialized competent tissues of mature plants.

B. DETERMINATION

During organogenesis, at some stage during the formation of new meristems from which organs will arise, the competent cells adopt a different inherent programming (determination) which decides their subsequent pattern of development. In the case of unorganized tissue cultures, the programming is apparently induced by the effect of growth regulators. Although growth regulators may also help to induce direct morphogenesis, cells in some parts of a plant appear to be partially pre-determined to a particular morphogenetic pathway. Even a slight change in environment induces the tissues of some explants to form a morphogenetic meristem instead of progressing to become a differentiated cell within the intact plant. Morphogenesis from cells, which are already committed to such development, is called *permissive*, while that from cells induced to become morphogenic by endogenous or exogenous growth regulators, is *inductive*. Permissive and inductive morphogenesis are thus almost synonymous with direct and indirect morphogenesis. It has been suggested that plant cells determined for subsequent differentiation are able to switch, by **transdetermination**, from one determined state to another. When mor-

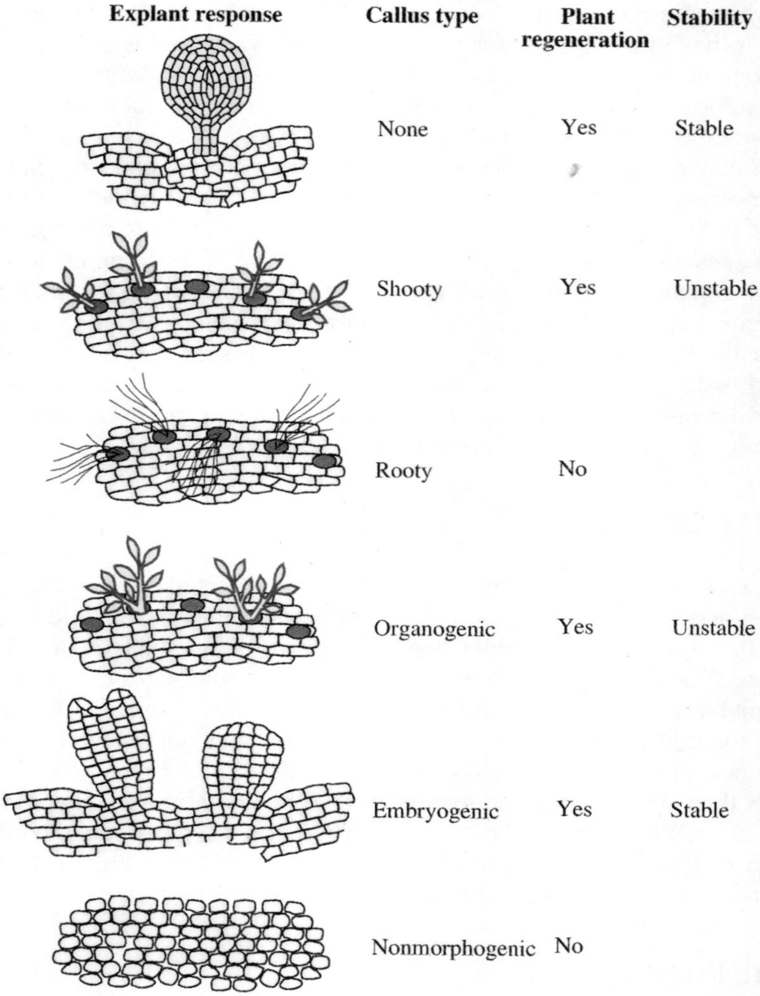

Explant response	Callus type	Plant regeneration	Stability
	None	Yes	Stable
	Shooty	Yes	Unstable
	Rooty	No	
	Organogenic	Yes	Unstable
	Embryogenic	Yes	Stable
	Nonmorphogenic	No	

Figure 31.3 *Explant response, plant regeneration capacity, and genetic stability in tissue cultures (Redrawn from Morrish et al. 1987)*

phogenic determination or cell differentiation becomes fixed, or irreversible, it said to have become **canalized**. Several kinds of morphogenetic determination are possible (root, shoot, leaf or floral meristems, etc.). Once canalization has occurred, progress through a single development pathway is usually inevitable unless the cells are capable of being stimulated to divide under conditions which promote dedifferentiated growth.

C. ORGANOGENIC DIFFERENTIATION

When relatively large pieces of intact plants are transferred to nutrient media, new shoots, roots, somatic embryos, and even flower initials are often formed without the prior growth of callus

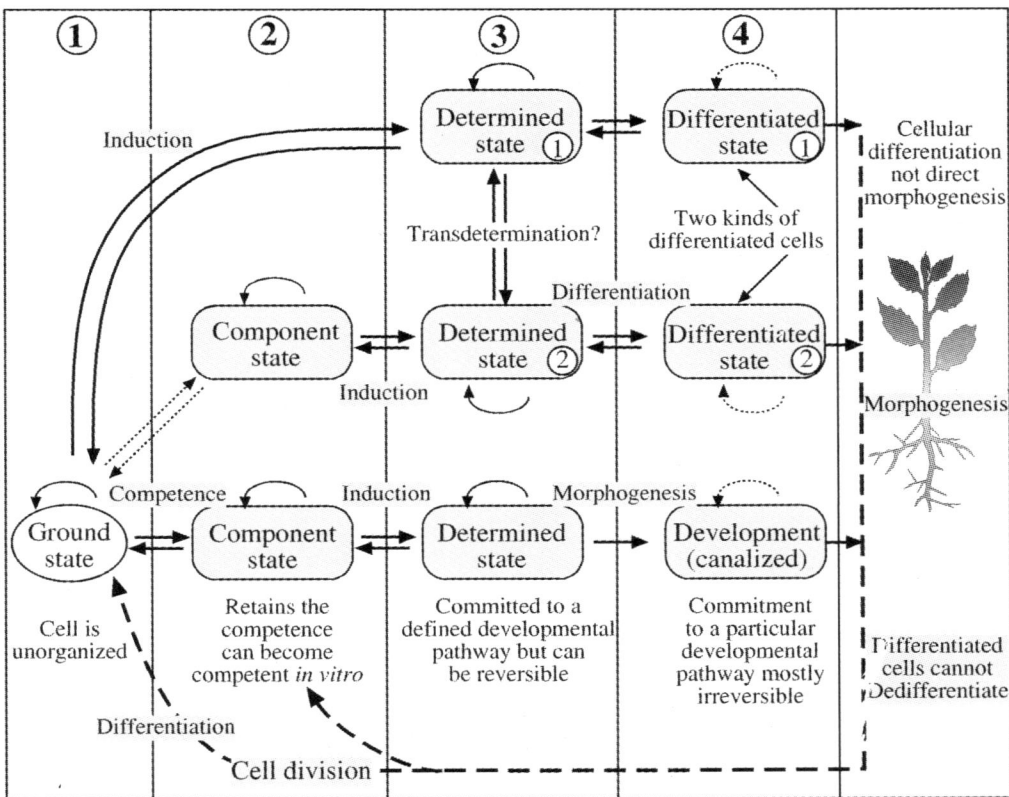

Figure 31.4 *Progressive steps in the capacity of a cell to become differentiated and/or morphogenic*

tissue. Small explants only rarely show organogenesis. The part of the original plant from which the explant is taken is important in influencing its morphogenetic potential. Whole plant regeneration from cultured cells may occur either through shoot bud differentiation or somatic embryogenesis. A shoot bud and an embryo are distinguishable on the basis of recognizable morphological differences between the two (Figure 31.5). The shoot bud is a monopolar structure and it develops procambial strands, which establish a connection with the pre-existing vascular tissue dispersed within the callus or the cultured explant (Figure 31.5 B,C,E). On the other hand, an embryo is a bipolar structure with a closed radicular end (Figure 31.5 A,D,F). It arises from a single cell and has no vascular connection with the maternal callus tissue or the cultured explant.

D. FACTORS AFFECTING SHOOT BUD DIFFERENTIATION

Many interacting factors influence organogenesis *in vitro*. The most widely studied of these are discussed in the following sections.

Chemical factors: A systematic approach to shoot/root induction *in vitro* started after Skoog and co-workers (1957) demonstrated that in tobacco the differentiation of the two organs can

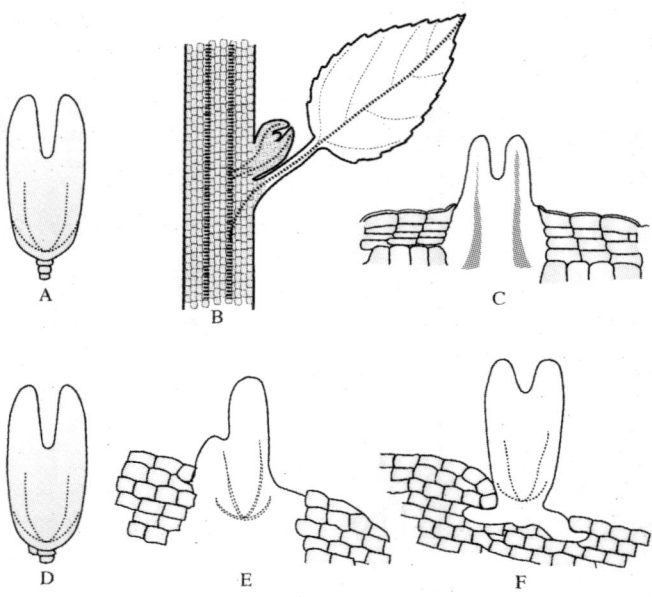

Figure 31.5 *Morphological differences in the basal ends of embryos (A, D, F) and shoot buds (B, C, E) under in vivo and in vitro conditions*

be induced by manipulation of the balance of IAA and adenine/kinetin in the medium. He also proposed that organ formation will be determined by quantitative interaction (ratios) rather than absolute concentrations, of substances participating in growth and development. IAA favors cell proliferation and root differentiation whereas relatively higher concentrations of adenine or kinetin promote bud differentiation. Thus, root–shoot differentiation is a function of quantitative interaction between IAA and kinetin. The promotion of bud formation by cytokinins occurs in several plant species. However, the requirement for exogenous auxin and cytokinin in the process varies with the tissue system, depending on the endogenous levels of the two hormones in the tissue. Besides kinetin, several other cytokinins, *viz.*, BAP, 2-ip, SD 8339 and zeatin, have been tested for shoot bud induction and generally 2-ip has proved most effective.

In most cereals, callus tissue exhibits organogenesis when it is transferred from a medium containing 2,4-D to a medium lacking it or having IAA or NAA in its place. Interestingly, the hormone ratio in the induction medium during the last four days is critical in determining the nature of the organ formed upon transfer to the regeneration medium. Gibberellin (GA3) inhibits shoot-bud differentiation in many crops (e.g. tobacco, begonia, and rice). Gibberellin is most effective at the stage of meristemoid formation. Once shoots have been formed, GA3 does not inhibit their further development. Complete inhibition by GA3 occurs only in the dark. However, in cells where endogenous gibberellin is low, it promotes bud differentiation. Despite the elegant demonstration of chemical control of organ formation in tobacco and its applicability to several other plants, a universal formula applicable to all plant species cannot be proposed.

Physical factors: White (1939) reported that in a solid medium, the tissue cultures of *N. glauca* x *N. langsdorffii* grew in a completely unorganized state, but in the liquid medium of identical composition, it formed leafy shoot buds. A striking alteration in the morphogenetic pattern with change in agar concentration in the medium occurs in thin tissue peels of tobacco. With 1% agar only flowers are formed, whereas, when the agar concentration is lowered, the frequency of flower formation drops and vegetative bud differentiation occurs. In a liquid medium, the tissue exhibits only callusing and vegetative bud formation. For shoot differentiation in callus cultures derived from mesophyll protoplasts of a diploid cultivar of potato it was essential to maintain the osmotic pressure between 200 and 400 millimoles by adding 0.2–0.3 M mannitol. High light intensity has been shown to be inhibitory for shoot bud formation in tobacco. Callus maintained under continuous light remains whitish and does not exhibit organogenesis. The quality of light also influences organogenic differentiation. Blue light promotes shoot bud differentiation whereas red light stimulates rooting in tobacco. The growth of the callus increases with arise in temperature up to 33°C. Other factors, which influence organogenetic differentiation in cultures are: genome, physiological state of the explant, cellular state of the explant, and endogenous hormone levels.

V CLONAL PROPAGATION *IN VITRO*

Recently, plant tissue culture has been extensively used for supplementing conventional methods and also for genetic manipulation of plants. Plant tissue culture is an experimental model system confined to discrete liquid or semisolid nutrient media, whereas micropropagation involves *in vitro* propagation of the selected genotype and ultimate establishment of the plant in the field or a glasshouse. Thus, **micropropagation** can be defined as "*in vitro* propagation of the selected *true-to-type* genotypes from organs, tissues, cells or protoplasts". True-to-type propagation has important benefits for highly heterozygous plants, where traditional plant breeding has failed to produce stable lines. Horticultural practice has traditionally used vegetative cloning of plant parts (shoots, leaves) for selective and rapid propagation of many plant species. Micropropagation has also been used for rapid multiplication during the release of new varieties prior to propagation by conventional methods (e.g., pineapple and strawberry). It also provides means of germplasm storage for maintenance of disease-free stock, both in controlled conditions, and in the long-term via cryopreservation. The chief disadvantages of *in vitro* methods, however, are the need of advanced skills and specialized and expensive production facilities.

Micropropagation is concerned with (1) mass multiplication of specific plants, (2) production of pathogen-free plants, (3) clonal propagation of parental stock for hybrid seed production, (4) year-round nursery production, and (5) germplasm preservation. It is often associated with the **organized part** of the plant, most often the bud, stem, root cuttings, etc. (Refer to Chapter 32). The culture process maintains this organization whilst directing subsequent growth and development toward multiplication and regeneration of whole new plants. This is distinct to some extent from cultures which involve the production of disorganized tissues such as callus at some stage in the process. Protocols, which have been developed for *in vitro* propagation of plant species can be divided into three types: (i) Callus culture followed by organogenesis. (ii) Proliferation of axillary buds and/or adventitious buds after repeated subculture on multiplication media containing cytokinin. (iii) Micro-cuttings from axillary buds of apically dominant shoots grown on hormone free or low cytokinin media.

The methods that are theoretically available for the propagation of plants *in vitro* are illustrated in Figure 31.6 and described in the following sections of this chapter. They are propagated either directly or indirectly: (i) by the multiplication of shoots from axillary buds, or (ii) by the formation of adventitious shoots, and/or adventitious somatic embryos. The most suitable and economic method for propagating plants of particular species could well change with time.

A. STAGES OF MICROPROPAGATION

In this chapter, a general description of stages 0–IV is provided. The requirements for the completion of each stage of micropropagation vary according to the method being utilized; the progress

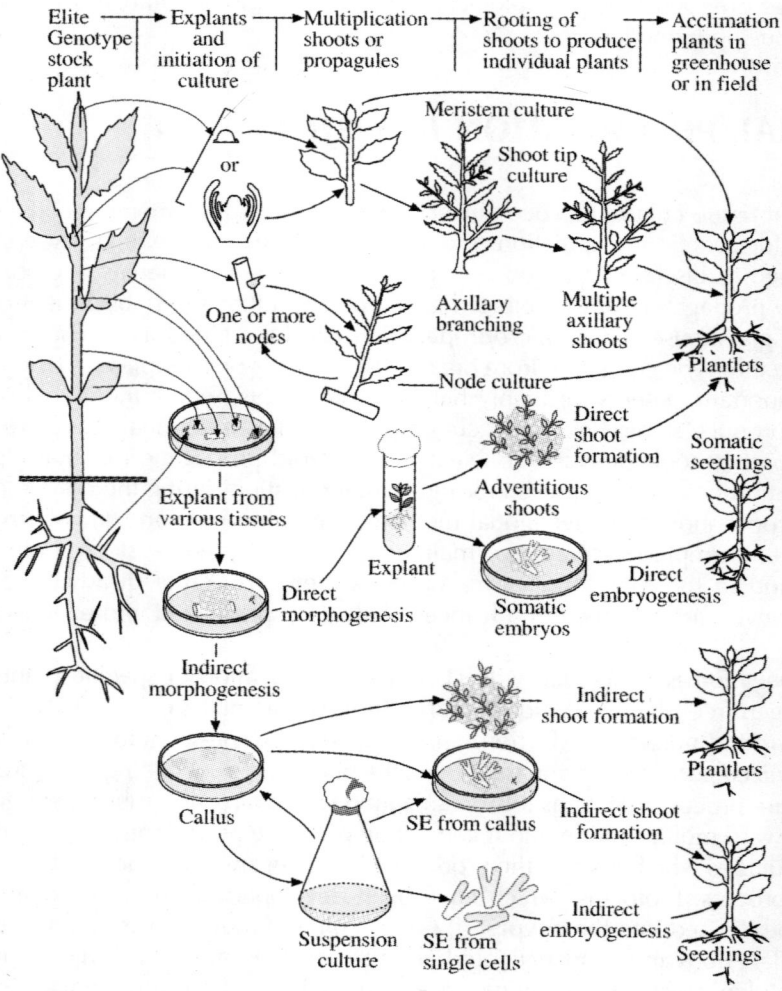

Figure 31.6 *The principal methods and stages of a typical micropropagation system*

of cultures will not always fit readily into neat compartments. Micropropagation involves some or all of the following stages, starting from the collection of the explant to development of new plants at the end.

1. Stage 0: Selection and preparation of explant

Before micropropagation commences, careful attention should be given to the selection of a stock plant or plants. The nature of the explant to be used for *in vitro* propagation is, to a certain extent, governed by the method of shoot multiplication to be adopted. For enhanced axillary branching only such explants are suitable which already carry a pre-formed vegetative bud. When the objective is to produce virus-free plants from an infected individual, it becomes obligatory to start with sub-millimeter shoot tips. Small shoot tip explants have a low survival rate and slow initial growth. A meristem tip culture may also result in the loss of certain horticultural characters which are controlled by the presence of a virus.

(a) **Stock plant selection**: The selection of a genetically distinct genotype for micropropagation is very important for obtaining superior plants. Explants should be collected from healthy and vigorous plants, which are more likely to produce successful cultures. Most commonly, it may be a bud or node but sometimes other parts of the plant depending on the species and requirement can also be selected. Small size explants are less likely to transmit endogenous infestations or introduce variability due to chimeras, but the smaller explants are more likely to be damaged during handling and have high initial culture failures. Shoot tips and most recent flushes of growth are the best parts of the plant for micropropagation. The explants have to be disease free. Apart from surface contamination, plant tissues may contain pathogenic organisms. Actively growing shoot tips tend to have less infestation due to lack of development of vascular bundles.

(b) **The plant age**: The age of the tissues in a developmental sense is distinct from chronological age. Mature tissues are produced after a number of growth cycles of the plant and are therefore usually some distance from the original seedling root system, even though they may be from a recent growth flush. Conversely, juvenile tissues are produced from the seedling part of the plant. Mature tissues have different physiological characteristics, which affect their culture requirements, than the juvenile tissues. Developing buds may be vegetative or generative (floral) depending on their position and the growth cycle of the plant. For most purposes, vegetative buds are preferable because they will produce a new shoot and therefore multiply the number of growing points. It is often recommended that the flowering period be avoided for collection of the vegetative buds, due to the different physiological state of the shoot tissues. During flowering, plants tissues go through cycles of growth activity and inactivity (dormancy) and these different states affect the response of the tissues to culture conditions.

(c) **Physiological conditions**: The objective of tissue culture is to control conditions under which the cultured explant is growing in order to manipulate its growth in the direction we require. The growth of any tissue or organ, whether in culture or as part of the intact plant, is ultimately determined by physiological conditions within the tissues. Changes in growing conditions will also affect the changes in physiological conditions in plant tissues. Thus, the resulting growth pattern is determined by the net physiological state as influenced by both internal and external conditions. This means that the precise conditions needed to elicit a particular growth response in culture will vary depending on the physiological state of the plant material. We can control some of these changes either indirectly by controlling the environment (e.g. temperature, light, water, nutrient), or more directly by applying plant growth regulators.

Many changes in the physiological state of the plant tissue are mediated by plant hormones. These chemicals occur in very low concentrations in the tissues and regulate the growth and development of the cells. Hormones in definition are produced in one part of the plant and move through the tissues to affect cell activities in another part. For example, cytokines produced in the roots may play a key role in stimulating cambial activity in the stem and the production of auxins at the shoot tips inhibit lateral bud development as the auxin passes down the stem. In this way, hormones provide communication between different parts of the plant and enable coordinated growth of the whole plant. By manipulating the hormone concentrations and proportions in the tissue culture medium, the pattern of growth can be regulated. Some broad generalizations about the response to plant hormones are: (i) Auxins usually promote root initiation and callus growth but inhibit root growth and lateral bud growth; (ii) cytokinins promote shoot proliferation and cell division but inhibit root initiation; (iii) gibberellins promote elongation and may overcome dormancy. But these are not reliable generalizations – there are even exceptions to these, so many other factors can influence the responses. Paramount among these is the physiological state of the explant when taken from the parent stock plant. We need to understand how the parent plant's physiological state is influenced by the conditions under which it is grown so that we may either compensate for this or preferably control it to our advantage.

(i) **Nutrient status**: The balance between the nutritional status of the parent stock plant and the nutrients available in the culture medium affect the overall growth of the explant. Plant tissues go through periods of net carbohydrate increase and decrease resulting in changes in the level of carbohydrate within the tissues, generally in association with the growth cycle. Shoots generally accumulate carbohydrates between periods of shoot or fruit growth and then consume these during the next growth period. Therefore, higher levels of carbohydrate would be expected at either the end or beginning of the growth season. Since most culture media contain a carbohydrate source (usually sucrose), the level of endogenous carbohydrate may not be critical. The explants should not be collected during the flowering or fruiting season. The carbohydrate balance of the stock plant can be manipulated by cultural practices such as cincturing or girdling and supplementary lighting or shading.

(ii) **Dormancy**: Plants do not grow continuously at the same rate. Growth of different parts of the plant is punctuated by periods of little or no growth, usually called dormancy. Dormancy may occur in specific growth centers of the plant including stems, roots, and leaves, as well as individual buds. Three types of dormancy can be discerned based on the origin of the growth inhibition. (i) Prevailing environmental conditions, e.g., extremes of temperature. (ii) Inhibition arising from other parts of the plant, e.g., an active terminal growing point may inhibit the growth of lateral buds below (apical dormancy) or a leaf inhibits the bud in its axil (correlative inhibition). (iii) Conditions, that occurred at an earlier time resulting in changes within the tissues which subsequently inhibit growth, e.g. winter dormancy of buds is due to conditions during the previous summer growth period (rest).

Since in tissue culture our aim is usually to encourage rapid growth and development from buds we need to avoid or overcome dormancy. Environmental dormancy can be avoided simply by providing favorable controlled environmental conditions. Apical dormancy can be overcome by the application of cytokinins and the correlative inhibition of a bud is overcome by isolating the bud or removing the terminal growing point and the leaves. However, the rest, by definition lies within the tissues themselves and cannot be removed by such direct treatment. The probable causes may include both physical and chemical changes within the tissues or surrounding structures of the organ. Some hypotheses assert that rest is caused by the accumulation of inhibitors (e.g., abscisic acid). Neutralization or application of hormones (e.g., cytokinins, gibberellins) is sometimes effective. Exposure to chilling temperatures or wounding of nearby tissues has also been effective. Rest may be avoided in cultures by taking explants from non-dormant parts of plants, i.e. by using actively growing shoot tips or buds just prior to the next growth flush.

(iii) **Light**: Light has diverse effects on plant growth apart from providing the energy source for photosynthesis. The exclusion of light from a plant or a particular tissue can affect its physiology. Carbohydrate levels may reduce and endogenous hormone levels or other physiological components can be affected by changes in light intensity, duration, or quality.

(iv) **Water stress**: Water stress can induce persistent physiological changes and these, in turn, may affect other physiological responses. Abscisic acid has been found to accumulate in some particularly in leaves under water stress. This may contribute to the induction of dormancy or rest. A period of sublethal water stress promotes the initiation of flowering in some cases.

2. Stage I: Establishment of aseptic culture

A customary step in the micropropagation process is to obtain an aseptic culture of the selection plant material. Success at this stage requires the following steps to be performed:

(a) Disinfection: Explants and cultures may be infested with a variety of microorganisms including fungi, bacteria, insects, or viruses. Many of them are not harmful to the host plant under normal conditions. However, the conditions *in vitro* which favor the explant, also favor the growth of microorganisms which often multiply and grow rapidly, smothering the explant. The initial explant is the major source of infestation but re-infestation is possible at any stage of the culture process.

Infestations may be on the surface of the plant, between the cells, or within the plant cells or endogenous – living within the plant tissues. Surface infestation can be dealt with by various washing and chemical treatments. The main limitations are to provide a sufficiently rigorous treatment to eliminate the infestation without damaging the explant tissue. Generally, the plant surface is covered with hairs or scales, so care must be taken to ensure penetration of the chemical since contact with the organism is necessary. Pre-treatment or management of the stock plants can greatly reduce the initial load of infestation thereby reducing the severity of disinfestation treatment needed and thus the resultant damage to tissues. Disinfection is generally achieved by adding detergent, by agitation, or by placing the submerged explants under reduced pressure to remove air bubbles, which could harbor microorganisms. Disinfecting endogenous infestations is more difficult. These may be controlled to some extent by spraying systemic pesticides applied to the stock plant prior to collection of explants or to the cultures themselves. Virus elimination, however, requires more special treatment.

Viruses usually reside within the cells of the plant tissue and are transferred to new cells during cell division. Therefore, they are transferred to the progeny during vegetative propagation. The main approach to elimination of a virus is the use of heat therapy. If the plant is grown at high temperatures it is possible to slow the replication rate of the virus so that the shoot apex can grow ahead of the infestation. The shoot apex can be removed and grown free of the virus. Heat treatment can be applied to shoots *in vitro* (e.g., 39°C for 7 days). Generally, meristematic tissues of shoot tips are virus-free because the viruses cannot move to this region due to the undevelopment of the vascular bundle cells. If one wants to produce a virus-free plant from a virus-infected individual, it is obligatory to start with a sub-millimeter shoot tip.

3. Stage II: Production of suitable propagules

The object of stage II is to bring about the production of new plant outgrowths or propagules, which, when separated from the culture, are capable of giving rise to intact plants. The multiplication *in vitro* can be brought about from newly-derived axillary or adventitious shoots, somatic embryos, or miniature storage or propagative organs. In some micropropagation methods, stage II will include the prior induction of meristematic centers from which adventitious organs may develop. Some of these propagules produced at this stage can also be used as the basis for further cycles of multiplication (subcultured) to increase their number.

(a) Shoot multiplication

Once aseptic cultures have been established, the next objective is to induce callus or shoot multiplication. In some species, explants may produce roots during this initial stage culture on a simple medium (i.e. they perform as micro-cuttings). This is not a desirable feature since the aim is to produce many plants, not single-rooted plants from each explant. In some other species, multiple shoots may form without further treatment. However, the need for more complex multiplication media will depend on the level of multiplication desired. The greatest potential of clonal multiplication is through initiation of the callus from an explant then to a complete plant. A high cytokinin/auxin ratio induces shoot formation and a high auxin/cytokinin ratio induces root formation. In many cases, a cytokinin alone is enough for optimal shoot multiplication.

(i) **Types of multiplication**: Shoot multiplication may be obtained in several ways: (i) The existing shoot tip or bud may elongate to give new nodes and internodes which can then be subdivided; (ii) lateral buds present on the explant may produce shoots which themselves have further buds along them. Often these lateral buds are hardly visible to the naked eye but most leaf axils contain numerous primordial buds; (iii) adventitious shoot development. In many species, plant organs, e.g. roots, shoots, or bulbs may be induced to form on tissues, which normally do not produce these organs. Such adventitious organogenesis has more potential than the induction of axillary buds for mass clonal propagation of plants. A single leaf, for example, may produce thousands of buds or shoots, each genetically identical to the explant; and (iv) somatic embryogenesis. The greatest potential of clonal multiplication is through somatic embryogenesis where a single cell can first dedifferentiate into the callus then redifferentiate to the embryo then to the complete plant. Somatic embryogenesis may occur in suspension cultures or occasionally in the callus. Induction of embryogenesis requires exposure to auxin, often 2,4-D followed by a reduction in auxin and also the source of nitrogen level.

(ii) **Multiplication controlling factors**: Axillary buds collected from plants may not grow in normal conditions because they might have been inhibited by the apical dominance (inhibition from the shoot apex or distal buds). Removal of the shoot tip or treatment with hormones can overcome apical dominance. The production of multiple shoots on cytokinin-rich media is often due to the release of existing buds from this apical dominance. In some plants, dormancy (rest) may be a factor which prevents buds from growing. Chilling treatments, gibberellin, or ethylene applications or long light periods and in some plants (apple trees) wounding the shoot distal to a bud can break rest.

Multiplying cultures may be repeatedly subdivided to produce many shoots. This is referred to as bulking-up. Sometimes, physiological or morphological anomalies (*Vitrification*) may occur when cultures are sub-divided through numerous cycles. Consequently, loss of vigor or increase in somatic mutations may occur in cultures. Because of this, it is advisable to maintain the mother stock culture, which is infrequently subcultured, and take sections of this for the mass production cycle. However, repeated sub-culture cycles may increase rootability.

(iii) Multiplication rate: The number of plants produced from each explant differs for different culture conditions and for different species. For example, in the case of strawberry, 1.5×10^7 plants can be produced in a year from a single explant. However, practically it may not be that simple to produce so many plants due to the involvement of many uncontrollable factors during repeated sub-culturing.

4. Stage III: Preparation for growth in the natural environment

Shoots or plantlets derived from stage II are small, and not yet capable of self-supporting growth in soil or compost. At this stage, steps are taken to grow individual plantlets, capable of carrying out photosynthesis, and survival without an artificial supply of carbohydrates. Some plantlets need to be specially treated at this stage so that they do not become stunted or dormant when taken out of the cultural environment.

(a) Shoot elongation: Once optimum shoot multiplication has been achieved, it may be necessary to provide particular conditions for shoot elongation to get shoots long enough to handle. Often transferring cultures to a hormone-free medium after the multiplication stage is sufficient to promote shoot growth. Instead of transplanting the shoots to a fresh medium, liquid media can be added to established cultures (double layer technique). The application of GA may induce shoot elongation by increasing cell size.

(b) Root formation: Rooting is a very important part of any *in vitro* propagation scheme. Adventitious and axillary shoots lack roots in the presence of cytokinin. Once a crop of shoots has been produced, root initiation may be carried out *in vitro* or the individual shoots harvested and treated as micro-cuttings under non-sterile conditions. For practical production of easy-to-root species, the rooting of the micro-cuttings out of culture is more economical. It avoids the preparation of the additional medium and the need for aseptic working routines. Where the species is more difficult to root, the extra labor of *in vitro* culture is often needed. Auxins such as NAA are usually required to induce rooting and activated charcoal can be added to the liquid medium to absorb any residual cytokinins.

High humidity conditions are necessary for the rooting of micro-cuttings to prevent desiccation of the soft shoots. Micro-cuttings may be treated with rooting hormones (auxin powder or liquid dip) just as in the case of conventional cuttings. Another advantage of rooting micro-cuttings out of culture may be the type of roots produced. Roots, which develop in agar or liquid medium, often have morphology adapted to water/nutrient uptake from culture mediums. These roots may be non-functional in soil and, therefore, have to be replaced if the plantlet is to survive. It could be quicker to root micro-cuttings directly in soil.

Some species form adventitious roots even during the multiplication stage while others produce roots when simply transferred to a cytokinin-free medium. However, many species require particular conditions, usually involving auxins, in the medium. Many other factors such as pH, light, nutrients (particularly calcium), and the hormone treatment also interact. Other than removing cytokinins from the medium, pre-conditioning (etiolation or chilling treatments) of the shoots before excision and transfer to the rooting medium may promote rooting. Whilst auxins may promote root initiation, they can inhibit subsequent root growth. Therefore, a transient exposure to the hormones may be more effective.

5. Stage IV: Transfer to the natural environment

The methods used to transfer plantlets from the *in vitro* to the *ex vitro* external environment (also called **transfer to a glasshouse or Planting out or Deflasking**) are extremely important. The micropropagated plantlets must be acclimatized to the environment of the glasshouse. Whether the shoots are being harvested as micro-cuttings or rooted plants are being transferred to soil, the plants are subject to a marked change in environment and are liable to be severely stressed unless adequate precautions are taken. This is often the critical stage in the overall tissue culture cycle where losses can be high. In *in vitro* conditions, plants are adapted to controlled environment which includes high humidity, freedom from pathogens, optimal nutrient supply, low light intensity, and a supply of sucrose plus a liquid or gel substrate. When exposed to the outside environment, the small plants must adapt and this generally occurs as new growth is produced rather than as a modification of existing organs. If the transition is too abrupt the plants will collapse.

Leaves produced under high humidity/low transpiration potential, tend to have thinner cuticular wax layers and a more open mesophyll tissue. Under low light intensity, they may have reduced chlorophyll levels. Gradual exposure to normal conditions leads to progressive morphological and physiological adaptation, i.e. a hardening off. This gradual effect can be achieved by modifying the culture conditions prior to transplanting to pre-condition the plant or by carefully controlling the environment for a period after transplanting. Acclimatization can proceed *in vitro* with bottom cooling – reducing the relative humidity in the headspace of the container. The culture vessels are uncapped and placed in the glasshouse several days prior to the removal of the plants from the culture medium. The plants are washed to remove agar because the agar will serve as a substrate for growth of disease-causing organisms. In the glasshouse, plants are given high humidity and low light intensity initially with a progressive reduction in humidity and an increase in light level over the following couple of weeks.

B. THE ADVANTAGES OF MICROPROPAGATION

Some of the main advantages of micropropagation are: (1) The speed of plant multiplication, (2) the quantity, (3) the uniformity of generated plants, (4) disease-free plants, and (5) germplasm storage.

VI METHODS AVAILABLE FOR MICROPROPAGATION

Micropropagation involves the vegetative propagation of plants through (1) establishment, (2) multiplication, (3) rooting, and (4) transplanting of clones *in vitro* starting with small explants and ending with a rooted plant established in a container or in the ground. Methods used for the *in vitro* propagation (micropropagation) of plants are listed in Table 31.1. In this chapter, only a few methods will be discussed; some other important regeneration methods commonly followed in micropropagation will be discussed in Chapters 32 and 33.

Table 31.1 *Techniques used to regenerate plants through tissue culture*

Structures formed	Regeneration methods	Explant source	General uses
Seedlings	Seed culture	Seeds	Orchid seed germination
	Embryo culture Embryo rescue Ovule culture Ovary culture	Embryos from seeds and fruits	Mature and immature embryos germinating in culture to form seedlings.
Plantlets	Meristem culture	Shoot tip (<1 mm)	Micropropagation and to get virus-free plants
	Shoot culture Axillary branching Nodal culture Pseudocorms Minitubers	Stem with one to four nodes, including leaves, shoot tip	Shoot cultures for micropropagation
	Organogenesis regeneration of –Diploid plants	Leaf, petioles bulb scales, stem internodes, roots, callus	Micropropagation
	–Haploid plant	Anther	Haploid breeding, to obtain somatic embryos
	Micrografting	Small scion, shoot tip	Virus elimination, micrografting
Callus	Callus cultures (Stationary)	Any vegetative tissue	In research, breeding, genetic transformation, and for biopharming.
	Callus cultures (suspension)	Callus from stationary culture	Uses are same as in previous case
	Protoplast cultures	Cells without cell walls	Study basic cell function, for protoplast fusion
Somatic embryos	Adventitious somatic embryogenesis Type 1	Nucellus or ovule.	Regenerating clonal copies of mother plant
	Type 2 (polyembryogenesis)	Embryogenic suspensor mass	Breeding and genetic transformation
	Type 3	Developing embryos or seedling parts.	Same as in previous case
	Induced somatic embryogenesis	Callus and cell suspension culture.	Same as above and used in synthetic seeds.

A. PROPAGATION OF PLANTS FROM AXILLARY SHOOTS

The production of plants from axillary shoots (Shoot culture and single or multiple node culture) has proved to be the most generally applicable shoots and the most reliable method of *in vitro* propagation. Basically this method depends on stimulating precious axillary shoot growth by overcoming the dominance of shoot apical meristems.

B. SHOOT (OR SHOOT TIP) CULTURE

The term shoot culture is now preferred for cultures started from explants bearing an intact shoot meristem, whose purpose is shoot multiplication by the repeated formation of axillary branches. In this technique, newly formed shoots serve as explants for repeated proliferation. This is the most widely used method of micropropagation.

1. Explant size

Shoots are generally collected from the apices of lateral or main shoots, up to 20 mm in length, dissected from actively-growing shoots or dormant buds. Larger explants are also sometimes used with advantage; they have better survival, commence growth more rapidly, and contain more axillary buds. Meristem tip or meristem cultures are used for virus elimination.

2. Regulating shoot proliferation

The growth and proliferation of axillary shoots in shoot cultures is usually promoted by incorporating growth regulators (usually cytokinins) into the growth medium, which removes apical dominance and promotes the formation of axillary shoots. In some plants, pinching out the main shoot axis is used as an alternative to the use of growth regulators for decreasing apical dominance and increasing shoot numbers. The most effective physical check to apical dominance can be achieved by pinching the tips, and/or placing shoot explant (2–3 nodes) horizontally on the medium.

3. The origin of shoots

Unfortunately not all the shoots arising in shoot cultures form axillary buds. Frequently, adventitious shoots also arise, either directly from cultured shoot material, or indirectly from the callus at the base of the subcultured shoot mass. The adventitious shoots, particularly those arising indirectly from the callus are not desirable, because they may be genetically deviant. The precise origin of shoots can sometimes only be determined from a careful anatomical examination. The formation of the callus on shoot tip explants and the subsequent development of adventitious shoots can be controlled by modifying the growth regulators in the medium. Fragmentation of a meristem tip, or its culture in a certain way, can lead to the formation of multiple adventitious shoots, which can be used for plant propagation.

Methods

In most herbaceous plants, shoot tip explants may be derived from either apical or lateral buds of an intact plant, and consist of the meristematic stem apex with a subtended rudimentary stem bearing several leaf initials. In the axils of the more developed leaf primordia there will be axillary bud meristems. Shoot tips from trees, or other woody perennials, can be difficult to decontaminate as they are more liable than those of herbaceous species to release undesirable phenolic substances when first placed onto a growth medium. Shoot tip or lateral bud explants are usually most readily induced to growth if taken from juvenile shoots such as those of seedlings or young plants.

Stage II subcultures are initiated from axillary shoots separated from primary shoot clusters. A high rate of shoot proliferation is often obtained from nodal explants or by subdivision of the basal shoot mass. To minimize the risk of genetic change in ramets, explants for subculture and shoots to be transformed to stage III should, as far as possible, be chosen from new shoots of axillary origin. It may be advisable to adjust the growth regulator content of the medium so that adventitious shoots are not formed, even though the rate of overall shoot multiplication is thereby reduced. Stage II cultures are typically without roots, and shoots need to be detached and treated as miniature cuttings which, when rooted, will provide the new plants, which are required.

C. MEDIA AND GROWTH REGULATORS

A major feature of shoot culture is the need for the high cytokinin levels at stage II to promote the growth of multiple axillary shoots. Cytokinin growth regulators are usually extremely effective in removing the apical dominance of shoots. Because of their nature, or the absence of an adequate method of culture, plants of some kinds fail to produce multiple shoots at stage II and retain their apical dominance.

Elongation. The length of the axillary shoots produced in shoot cultures varies considerably from one kind of plant to another. Species, which have an elongated shoot system *in vivo* will produce axillary shoots, which can be easily separated as micro-cuttings. At the other extreme are plants with a natural rosette habit of growth, which tend to produce shoot clusters in culture. This type can be propagated by dividing shoot clusters into pieces and re-culturing the fragments. Such shoot clusters can be induced to form roots when plants with a bushy habit are required. Otherwise it is necessary to specially elongate shoots before they are rooted. Shoot clusters are treated in such a way that axillary shoot formation is reduced, and shoot growth promoted.

Rooting and transfer. Single shoots or shoot clusters must be moved to a different medium for rooting *in vitro* before being transferred as plantlets to the external environment. An alternative strategy for some plants is to root the plant material *extra vitrum*. Treatment needs to be varied according to the type of growth, the nature of the shoot proliferation produced during stage II culture, and the plant habit required by the customer.

D. PROBLEM SOLVING

Contamination is one of the major problems in micropropagation, unless explants are collected from glasshouse-raised plants. Contaminants within the tissue are difficult to eliminate. Micropropagated plants may show morphological or genetic variation (***somaclonal variation***). Somaclonal variation may be heritable or epigenetic in basis. It may be desirable or undesirable. There is some evidence that the length of time in culture increases the frequency of somaclonal variation among regenerated plants. Cultural stress such as the use of improper media components or mutagens, certain growth regulator treatments, delayed subculture intervals leading to nutrient stress, or exposure to extreme or highly variable incubation conditions also have been implicated in the increased frequency of somaclonal variation. Explant source is a major determining factor. Explants consisting of meristems undergo orderly mitoses and produce a lower frequency of somaclonal variations than other sources.

Browning of medium: This kind of problem is commonly found during micropropagation of woody perennials. The browning could be due to the accumulation of inhibiting substances in the growth medium, specially during initiation of cultures. The phenolic substances produced may be toxic to tissues and inhibit their growth.

VII SUGGESTED READING

George EF (1993) *Plant Propagation by Tissue Culture Part 1* (Second Edition), Exegetics Limited.

Morrish F, Vasil V and Vasil K (1987) Developmental Morphogenesis and Genetic Manipulation in Tissue and Cell Cultures of the *Graminae*, In: *Molecular Genetics of Development* (Eds) Scandalis JG. Advances in Genetics, **24**:431–499.

Phillips GC and JF Hubstenberger (1995) Micropropagation by Proliferation and Axillary Buds, In: *Plant Cell, Tissue and Organ Culture: Fundamental Methods* (Eds), Gamborg OL and Phillips GC Narosa Publishing House.

Skoog F and Miller CO (1957) Chemical Regulation of Growth and Organ Formation in Plant Tissue Culture In Vitro, *Symposium on Society of Experimental Biology,* **11**:118–130.

White PR (1939) Potentially Unlimited Growth of Excised Plant Callus in an Artificial Medium, *American Journal of Botany,* **26**:59–64.

32

Cultures of Organized Tissues

I INTRODUCTION

Although all tissue cultures originate from organs or their sections, the progenitor organization need not always be retained during *in vitro* development. But an organ culture has as its aim the achievement of an organized structure, the morphology and physiology of which are identifiable with the specified organ. Organ culture is used as a general term for those types of culture in which an organized form of growth can be continuously maintained. It usually begins with a primordium explant. Differentiated plant organs can be grown in culture without loss of integrity. Thus organ culture is the choice of tissue culture method in many horticultural crops. Generally two types of organs are used in tissue culture. (1) **Determinate organs**: which are destined to have only a defined size and shape (e.g. leaves, flowers, and fruits). (2) **Indeterminate organs**: where growth is continuous (apical meristems of roots and non-flowering shoots). The first successful cultures of excised plant parts involved organ cultures, not cell or tissue cultures. An organ arises from a group of meristematic cells. In an indeterminate organ, such cells are theoretically able to continue in the same pattern of growth indefinitely. But in case of a determinate organ, as meristematic cells receive instructions on how to differentiate, their capacity for further division becomes limited. If the primordium of a determinate organ is excised and transferred to a culture, it will sometimes continue to grow to maturity.

Until recently, completely normal development was obtained in only a few cases, probably because of the use of suboptimal media composition. Plants cannot be propagated by culturing meristems already committed to producing determinate organs. In this chapter, vegetative organs (methods included meristem culture, root culture, and leaf culture) and reproductive organs (embryoculture, ovary culture, ovule culture, and flower bud culture) will be discussed with protocols and necessary illustrations.

II CULTURE OF INDETERMINATE ORGANS

A. MERISTEM CULTURE

Meristem culture (or micropropagation) is a method in which shoot apices containing the apical dome with a few primordial leaves are grown *in vitro*. This method is also known as apical-tip culture, shoot-tip culture, or culture of shoot apices. The application of the meristem culture technique to obtain plants free from systematically invasive viruses and other disease-causing pathogens was based on the initial observations that apical-tips of shoots and roots are generally free of viruses. With this information many virus-free plants have been developed by meristem culture in various crops, especially vegetatively propagated crops such as dahlias, potatoes, pineapple, carnation, chrysanthemum, strawberry, and sugarcane. This method can be used to isolate virus-free plants where chemotherapy and heat therapy fail. The size of the initial explant is, however, an important factor; the smaller size of the explant (smaller than 0.1 mm by 0.25 mm), the greater are the chances of virus elimination, but lower the chances of its survival. This tissue has no vascular system and is, therefore, less likely to be infected with systemic viruses.

For complete elimination of viruses it is often essential to pre-treat the explants. The parent plants or cuttings are grown in a controlled temperature cabinet at 30–40°C for 6–12 weeks (Figure 32.1) till the new shoots develop. Alternatively, viruses can be eliminated, culturing meristems at 30–40°C. Lower temperature treatment (5°C) was also shown to degrade virus particles following the blocking of viral-protein synthesis. When added to the culture medium, certain chemicals or tissue treatment with the chemical prior to culturing it prove effective in freeing the tissue of viruses. Virazole can eradicate the potato virus in cultured tobacco shoots and treatment with Odontoglossum Ring Spot Virus anti-serum (ORSV anti-serum) before culture eliminates ORSV from Cymbidium.

Plants obtained by meristem culture must be tested to ensure that they are virus-free after establishment in soil. The methods used for virus testing are grafting, electron microscopic examination of leaf and sap material, transmission of sap to susceptible hosts, serological test, serum-specific electron microscopy, and ELISA. Regenerated plantlets and virus-tested plants are generally grown as a foundation stock in glasshouses or in protected areas to prevent entry of insect vectors.

Case study: Meristem culture of sugarcane

Sugarcane production offers a continuing challenge to increase sucrose yields and develop disease- and pest-resistant clones. Techniques are now being studied that may have potential in manipulating plant systems at the cellular level for use in "asexual plant improvement". In addition to sexual breeding techniques, tissue and cell culture techniques provide many new methods for delivering the desired genetic variability. However, some species respond readily to one or more of these approaches, while other plants, including sugarcane, present more of a difficulty. The laboratory methods described here for meristem-tip culture of sugarcane is a standard technique, which has been used in various laboratories (Figure 32.2).

Media

The media generally used for the growth of sugarcane cell and tissue cultures are variations of the original White's medium or the Murashige and Skoog medium, with the addition of complex materials such as coconut milk, yeast extract, tomato juice, malt extract, etc. Generally, sugarcane clones in suspension culture grow better in a medium supplemented with yeast malt extract than in a medium containing coconut water supplement and sucrose and glucose are found superior to other carbohydrates. The best media for callus differentiation and plantlet development is MS or its modified media. The presence of an auxin, usually 2,4-D, but sometimes NAA is necessary for prevention of differentiation of plantlets from the callus. The use of

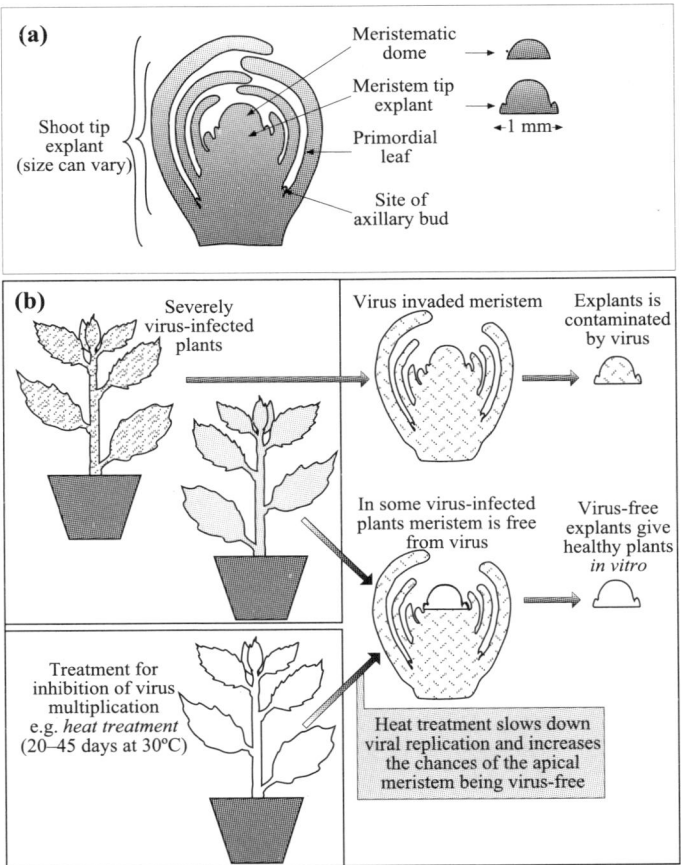

Figure 32.1 *Meristem tip culture (a) Section through a bud showing the locations and approximate relative sizes of meristematic dome; meristem tip and shoot tip explants (b) Strategies for obtaining virus-free plants by meristem culture*

IAA is not desirable, presumably because of the presence of a strong IAA oxidase system in sugarcane tissues. Although there are varietal differences, callus growth in sugarcane usually proceeds rapidly until there is an apparent exhaustion of some nutrients or until there is desiccation of the medium. For most rapid development, the transfer of the callus to fresh media is necessary every 3 to 4 weeks.

Establishing culture

Sugarcane cultures can be established almost from any portion of the plant. But the most rapid formation of the callus and the greatest totipotency can be obtained from young expanding leaf or young inflorescence tissue. However, one major problem often encountered during the establishment of callus cultures is the production and secretion of polyphenols into the medium, which often causes cultures to die, even if they initially become established. But this problem can be overcome by successive, rapid transfers to fresh media at intervals of a day or so.

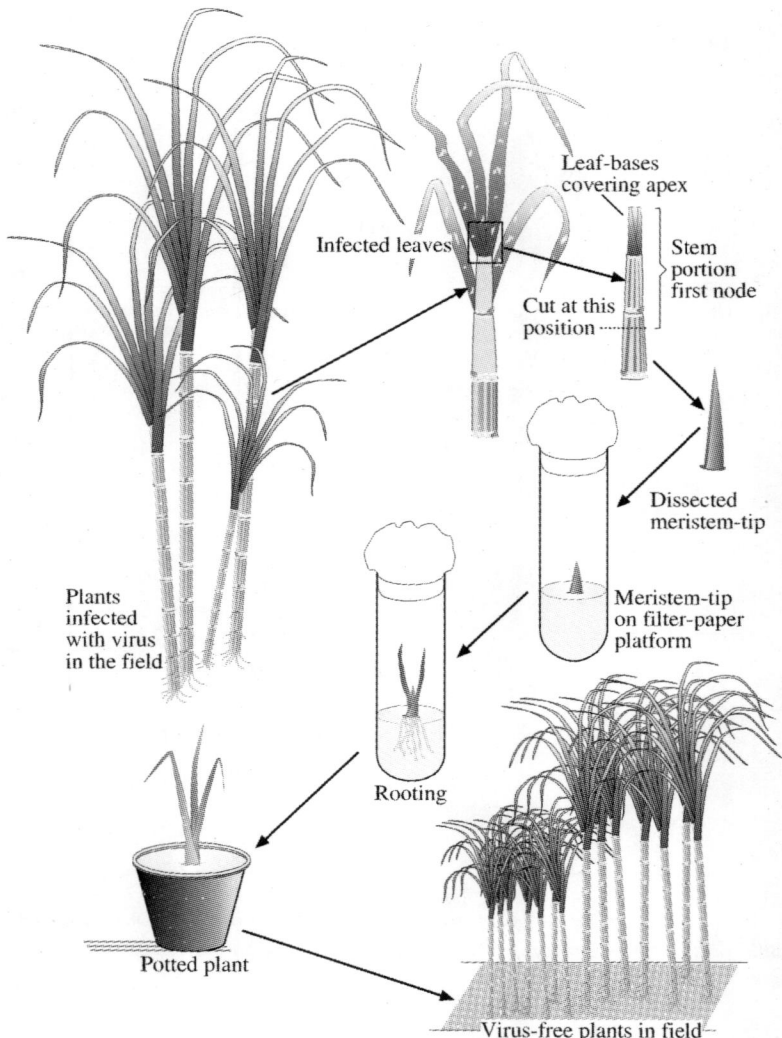

Figure 32.2 *Meristem-tip culture for sugarcane (Saccharum officinarum)*

Differentiation

Under proper conditions, the differentiation of either shoots or roots from the callus of most sugarcane varieties is easily obtained. However, obtaining roots from the differentiated shoots is more difficult (20 to 30% fail to form roots from callus-derived plants). The root production can be enhanced by some cultural modifications such as (i) separating the individual plantlets and transferring into fresh agar, (ii) after the shoots have developed to a few inches in height, transferring plantlets to aerated water, (iii) transferring plantlets to vermiculite after the initiation of small rootlets, (iv) transferring the rooted plants to soil under greenhouse conditions, and (v) transferring to the field. Regardless of variety or medium, root formation was highest at

15°C with almost no root formation at 10 and 24°C. Physiological reasons for this phenomenon have not yet been explored.

Nutrition

Sugarcane cell suspensions do not grow on nitrate, and grow poorly on an ammonium salt when these are used as sole sources of nitrogen. Other amino acids, particularly aspartic and glutamic acids, are important sources of nitrogen for sugarcane cell growth. In cell suspensions, the presence of arginine reverses the growth-depressing effects of a number of other amino acids. The cellular uptake of arginine by suspension cultures is rapid.

The commercial importance of carbohydrate deposition in sugarcane has helped to draw attention to accumulation, intracellular distribution, and storage physiology in stalk parenchyma tissue. After cells in suspension culture have passed through the phases of rapid division and entered a stationary phase, sucrose can accumulate to about 10% of the dry weight, presumably in the vacuole. A cell wall bound invertase with optimum pH of approximately 5 appears to control the uptake of sucrose. Sugarcane suspension cultures normally proliferate on galactose for only a limited period of time. Growth ceases by the 4th to 5th day. This pattern can be modified by prolonged incubation of cells on agar in 2% galactose instead of sucrose.

Protoplasts: Many methods for the isolation and culture of sugarcane protoplasts and protoplast fusions between different genotypes have been well established. Multiple divisions have been observed after fusion of the protoplasts.

Materials

Equipment

Laminar air flow hood; Manesty still; Glass distillation unit; Dissecting tools; Culture tubes 20 mm × 150 mm (Corning); Conical flasks 250 ml, and 500 ml; Petri dishes 80 mm diameter; Measuring cylinders 20 ml, 100 ml, and 250 ml; Pair of forceps; Steel scalpels; Needles, arrowhead; Cotton; Brown-bands; Autoclave; Dissecting microscope

Culture media, washing solutions, sterilizing agents

Detergent or wetting agent such as Tween-20; 70% ethanol; 0.1% w/v $HgCl_2$ solution; Glass distilled water for preparing media; Sucrose (Laboratory grade); IBA; GA3; Sterile water; Media are prepared in double distilled water and autoclaved at 15 psi for 20 min. Medium for shoot formation (Table 4, Appendix 13); Medium for root induction (Table 5, Appendix 13).

Solutions of IBA and GA3 are filter-sterilized (millipore 0.22 μm) and added aseptically.

Collection of sugarcane explant

Collect about twenty healthy disease- and pest-free canes from a sugarcane field, cut pieces (75 cm) from the top of the cane, and bring them to the laboratory.

Procedure

Sizing material for sterilization

1. Make a longitudinal incision on the upper portion of the cane, which is covered by leaf bases.

2. Remove the outer coarse green leaves.
3. Take care not to damage the apical region.
4. Locate the apex and first node and keeping only a few inner membrane's yellowish leaves around the tip, remove the unwanted portions above and below the tip to make 50–60 mm long pieces.
5. Transfer these pieces into a 250 ml flask containing distilled water (10 pieces per flask).

Surface sterilization

See Chapter 30 for more details.

Dissection of tips

The individual pieces after sterilization are dissected to expose the meristem-tip under a dissecting microscope placed in the laminar airflow cabinet.
1. Take out one piece of stem with the tip, and place it in a sterile Petri dish.
2. Hold the piece with a pair of forceps in the left hand.
3. With a scalpel in the right hand, make two superficial cuts – one longitudinally, and the second transversely (at the lower end of the leaf bases).
4. Slowly remove the inner membrane's yellowish leaves one by one with the help of an arrowhead, avoiding injury to the growing tip.
5. Carefully cut the tip to 2–3 mm by an incision at the base.

Inoculation

1. Place the tip vertically on the filter-paper support in the tube containing the nutrient medium (Appendix-12, Table 5) with the help of a pair of forceps.
2. Push down the filter-paper support carrying the shoot tip to make contact with the medium. The tip should be placed in the center of the tube – it should not touch the sides of the tube as this causes drying of the tip.

Incubation

1. Incubate the cultures at $28°\pm2°C$ with a 12 hour photoperiod at 800–1000 lux, supported by white fluorescent lamps.
2. Three to four days after inoculation, the meristem-tips begin to turn green and show visible elongation by the 8th day. When the phenolics start leaching out (after 8–10 days) coloring the medium, the growing tips have to be transferred to a fresh medium of the same composition (Appendix-12, Table 5) for further growth. When the shoots become 40–50 mm long (around 15–20 days), they are ready for root induction.

Rooting

1. When the shoots are 40–50 mm long, remove them from the shoot-forming medium (Appendix-12, Table 5) very carefully and place them in sterile Petri dishes.
2. Remove any dry leaves at the base with the forceps and scalpel.
3. Place the shoot in rooting medium (1 shoot/tube).
4. The basal end should dip slightly into the medium.
5. Incubate the culture at $28°\pm2°C$ with a photoperiod of 12 hours and light intensity 800–1000 lux. Roots start appearing within 7 to 8 days of transfer to the rooting medium. By 15

to 20 days, the root system is quite well developed and the plantlets are ready for transfer to soil.

Transfer to pots

A mixture of soil and vermiculture (1:1) is prepared and autoclaved at 15 psi for 60 min. Plantlets having a good root system (15–20 days on rooting medium) are transferred to the soil in pots according to the method given. There is 70–80% rooting and survival of plants in the field.

Hardening

1. Keep the plantlets in the small pots in a growth chamber at 25°±2°C with a 12 hour photoperiod at 1000–1200 lux for 8–10 days.
2. When new leaves appear, the pots can be transferred to a glasshouse.

Transfer to field

When the plants have attained height of 600–700 mm they are transferred from the glasshouse to the field for further growth (the survival rate in the field is 70–80%).

B. SHOOT CULTURE

Plant organ cultures are characterized by the maintenance of structural integrity, as a consequence of growth from defined meristems, and are thereby distinguished from callus and suspension cultures. Shoot cultures typically comprise rootless aerial parts (apical and/or axillary buds, stems, and leaves) growing on a defined medium supplemented with agar. Most commonly, shoot cultures are initiated from explant tissues containing meristems, such as apical buds, axillary buds or embryos, and are subcultured by the transplantation of excised buds onto a fresh medium (Figure 32.3). The precise composition of the medium may of course vary with the nutritional requirements of different species. The shoot culture of tobacco SR1 is an excellent source of leaf material. However, the age, health, and culture conditions of the shoot cultures are important factors in influencing transformation frequency and protoplast yield. By culturing shoots from the most apical cells only, it is possible to eliminate virus particles from infected plant material. A protocol for tobacco shoot culture has been discussed below.

Method-1: Initiation and establishment of tobacco shoot culture

Materials

Solutions

MS20 medium: 4.4. g/l MS basal medium (Sigma); 20 g/l sucrose; 8 g/l agar (not included in root culture medium); adjust pH to 5.8 with 1 M KOH.

Procedures

Surface Sterilization of tobacco seeds
1. Immerse seeds in 70% ethanol (to remove waxy substances and also to kill germs) for 20 seconds and agitate by gentle swirling in a beaker.

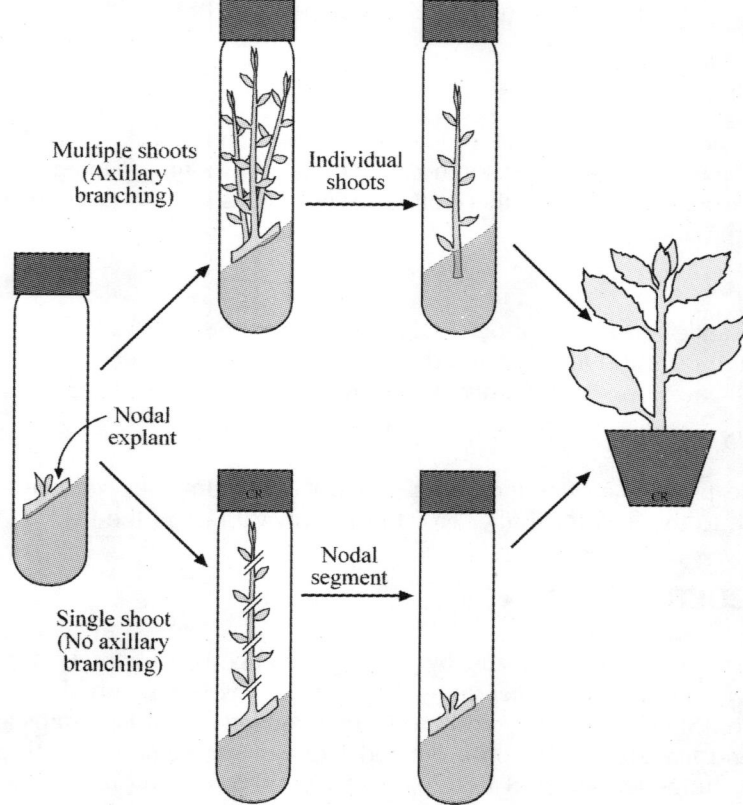

Figure 32.3 *Micropropagation by stimulation of axillary buds to develop into shoots*

2. Tobacco seeds tend to float, so remove the ethanol using a sterile pipette by keeping the pipette close to the bottom of the beaker.
3. Immerse seeds in a 5% (v/v) bleach (sodium or calcium hypochlorite) solution containing 0.05% (v/v) Tween-20 or 80, for 15 min. Agitate occasionally by swirling.
4. Remove the sterilant using a sterile pipette as described in step 2, and wash the seeds six times in sterile distilled water.

Germination of the seeds
5. Transfer sterilized seeds to 9 cm Petri dishes containing a solidified MS20 medium (in this medium the infected seeds can be identified early and removed). Due to the small size of tobacco seeds, this is most easily achieved by suspending seeds in a small volume of sterile distilled water and transferring them to the Petri dish with a pipette. Spread the seeds around the plate with a sterile glass rod, and remove excess water from the surface of the agar with a sterile pipette.
6. Seal the Petri dishes with Nescofilm or Parafilm, and culture at 24°C in the dark to induce germination. The first signs of germination typically occur within about 3–5 days.

Initiation and subculture of shoot cultures

1. When the germinated shoot is approx 1–2 weeks old, excise the shoot apex, cotyledons, and emerging first leaves at the hypocotyl and transfer to MS20, embedding the cut surface in the agar medium. Culture at 24°C in warm daylight with fluorescent tubes (50–100 μmol/m^2/s) under long day length conditions. Use culture vessels with loose or unsealed lids. After about 6 weeks, the shoots will be large enough to subculture.
2. To subculture the shoots, excise the apical buds together with the newest pair of leaves, or axillary buds and transfer to fresh MS20. Shoots should be subcultured routinely at intervals of approximately 4 weeks.

C. ROOT CULTURE

One of the pioneering events in the development of tissue culture techniques was the successful growing of clones of tomato roots reported by White in 1934. Root cultures are generally established by the transfer of aseptic root tips, isolated from germinated seeds or from root tips taken from primary or lateral roots of many plants into a liquid culture medium. Suitable explants are small sections of roots bearing a primary or lateral root meristem. If the small root meristems continue normal growth on a suitable medium, they produce a root system consisting only of primary and lateral roots (Figure 32.4). No organized shoot buds will be formed. Plants fall generally into three categories with regard to the case with which their roots can be cultured. There are some species such as *Datura*, tomato, clover, and *citrus*, where isolated roots can be grown for long periods of time, provided regular subcultures are made. In many woody species, roots have not been grown at all successfully. In other species such as pea, flax, and wheat, roots can be cultured for long periods but ultimately growth declines or insufficient lateral roots are produced. Isolated plant roots can usually be cultured on relatively simple media such as White media containing 2% sucrose. Liquid media are preferable, as growth in or on solid medium is slower. This is presumably because salts are less readily available to the roots from a solidified medium and oxygen availability may be restricted. Although roots will accept a mixed nitrate/ammonium source only, species and even varieties of strains, of plants, are found to differ in their requirement for growth regulators, particularly for auxins, in the root culture medium.

Isolated root cultures have been employed for a number of different research purposes. Cultured roots provide a useful experimental system to study aspects of root biochemistry and physiology such as ion uptake, root exudation, and the synthesis of root-specific secondary metabolites. They can also act as a source of protoplasts for the study of tissue-specific *trans*-acting factors in the regulation of root-specific gene expression. They have been particularly valuable in the study of **nematode infections**, **mycorrhizal fungi**, and to study the process of **root nodulation** with nitrogen fixing *Rhizobium* bacteria in leguminous plants. Root cultures are also useful in micropropagation. Unlike some other cultured tissues, root cultures exhibit a high degree of genetic stability. It has been suggested that root cultures could afford one means of storing the germplasm of certain species. For suitable species, root cultures can provide a convenient source of explant material for the micropropagation of plants, but they will only be useful in **micropropagation** if shoots can be regenerated from roots.

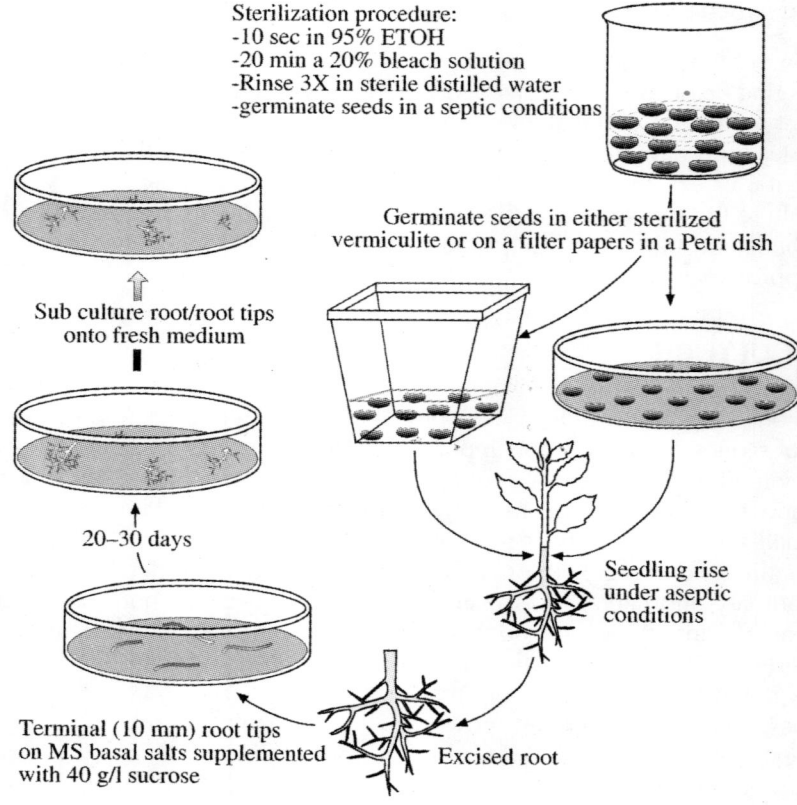

Figure 32.4 *Stages of dicotyledon seedling root culture*

Materials

Erlenmeyer flasks (250 ml); Sterile plastic Petri dishes (100 × 15 mm); Scalpels and 2 pairs of forceps; Ethanol; 1% aqueous solution of Tween-20; 20% chlorox solution; Sterile distilled water; About 50 dry, undamaged seeds.

Basic culture media: To MS salts and White's basal salts (Table 1, Appendix 13) add separately 15 g/l Sucrose, 0.5 mg/l Thiamin.Cl, and 100 mg/l Myoinositol.

MS20 medium: 4.4. g/l MS basal medium (Sigma); 20 g/l sucrose; 8 g/l agar (not included in root culture medium); adjust pH to 5.8 with 1 M KOH.

Procedure

Sterilization of seeds

1. Place the dry seeds in a 250 ml Erlenmeyer flask. Pour 200 ml 95% ethanol over the seeds and shake for 10 seconds, then pour off the ethanol. Pour 200 ml of Tween-20 solution over the seeds and shake the flask gently for 1 min. Examine the seeds individually and discard those that float or appear to have a translucent tissue beneath the seed coat. Steril-

ize the remaining seeds by pouring 200 ml of Clorox solution over them and disinfecting them for 20 min. under aseptic conditions. Following disinfection, wash the seeds three times in sterile distilled water.
2. Pour 20 ml of sterile distilled water into each of 10 Petri dishes and place a few (e.g. 2 pea seeds) seeds in each dish. Incubate the plates in the dark at 25°C for 48 hours.
3. Prepare a basic culture medium as described in the previous section and sterilize it. Under aseptic conditions, dispense 20 ml of medium into each of the 10 sterile Petri dishes. Repeat procedure for both media.
4. When the primary root of each pea seedling (step 2) reaches 20–25 mm, excise the terminal 10 mm and transfer the tip to a plate containing the basic medium. Inoculate two tips per plate. Repeat this procedure until all culture plates have been inoculated.
5. Incubate all cultures at 25°C in the dark.
6. Repeat steps 1 and 2 using tomato seed.
7. Supplement the basic culture medium with either 0.0 mg/litre 2,4-D (treatment A), 0.01 mg/litre 2,4-D (treatment B), or 1.0 mg/litre 2,4-D (treatment C). Under aseptic conditions, dispense 20 ml/plate of basic medium and supplemented media (A and B) into sterile Petri dishes. Prepare 5 replicates per treatment.
8. Remove primary root tips from tomato seedlings as described in step 4 and inoculate onto culture plates prepared in step 7. Incubate cultures at 25°C in the dark.
9. Subculture half of the cultures at 7-day intervals. Do not subculture the rest.

Initiation and subculture of root cultures from tobacco seeds
1. Germinate sterile seeds (as described in previous section).
2. When the germinated radicle is approximately 1 week old, excise the apical 1 cm and carefully transfer it to a liquid MS20 medium in a 9 cm Petri dish. Each dish should contain approximately 15 ml medium. A sterile microbiological loop is useful for the physical transfer of the root tips.
3. Seal the Petri dish with Parafilm, and incubate at 24°C without shaking in a dark incubator.
4. At intervals of 10–14 days, subculture the roots by excising the apical 1 cm and transferring it to a fresh liquid MS20 medium.

D. REPRODUCTIVE INDETERMINATE ORGANS

Method-1: Embryo culture

The embryo represents the beginning of the new sporophytic generation of the plant. Embryos of different developmental stages formed within the female gametophyte through the normal sexual processes, can be separated with relative ease from the bulk of the maternal tissues and cultured *in vitro* under aseptic conditions in media of known chemical composition. Embryo culture can be done *in vitro* either of an immature (polarized egg, zygote, pro-embryo) or mature zygotic embryo. Embryo culture is relatively simple in that the embryo is a miniature structure that can form into a complete plant and therefore no *de novo* differentiation of shoots or roots is needed. Embryos excised from the developing seed at or near the mature stage are completely autotrophic. They germinate and grow on a simple inorganic medium with a supplemental energy source. The possible practical applications of embryo culture used for a number of different purposes include:

1. **Embryo rescue to obtain rare hybrids**. In many interspecific and intergenic crosses, fertilization occurs normally; but poor or abnormal development of endosperm (incompatibility) causes premature death (embryo abortion) of hybrid embryos. Such embryos may be extracted while immature and before abortion and grown on culture *in vitro*. Perhaps the most commonly used technique is that of embryo rescue which produces an adult plant following intergenic crosses (e.g. wheat with barley, pearl millet with elephant grass, oats with wild oats, peanuts with wild *Arachis*, and soybean with wild glycine). All these crosses would be impossible by the normal seed production methods. By this method, various agronomic characteristics can be introgressed from wild relatives into the cultivating varieties.
2. **Speed up the breeding cycle of rare plants having slow developing embryo**. The seeds of some wild varieties of the cultivated plants and some germplasm collections do not germinate in nature. Seedlings can be obtained by culturing their excised embryos, e.g. seeds of *Musa bulbisian*, a wild relative of the commercial banana, and seeds of many orchid species which do not have an endosperm.
3. **To overcome seed dormancy of hard-to-germinate seeds**. The breeding work of deciduous trees is delayed due to the long dormancy periods of their seeds. Seed dormancy of many species is due to chemical inhibitors or mechanical resistance present in the structures covering the embryo. Excision of embryos from the testa and culturing them in nutrient media may by-pass such seed dormancies.
4. **To study morphogenesis and nutritional requirements**. It is an ideal technique for studying morphogenesis and nutritional requirements of the developing embryo.
5. **For rapid seed viability test**. *In vitro* manipulation of embryos helps to break seed dormancy. This technique is used for testing the viability of a particular seed batch rapidly.
6. **Haploid production.** Through elimination of chromosomes of one of the parents following distant hybridization, haploids can be produced. For example, in the cross of *Hordeum vulgare* X *H. bulbosum*, the chromosomes of *H. bulbosum* are preferentially lost resulting in haploid embryo plants.
7. **Enforced self-fertilization**. The production of viable plant is possible by embryo culture after enforced self-fertilization.
8. **Where mature seed cannot be recovered.** In those instances where mature seeds are not obtainable for a species because the seeds are shed and lost suddenly when the seed fruits are ripe or when developing seeds are highly susceptible to pests and diseases before they mature.

Development of Embryo and Endosperm

Embryo development: *In vivo*, the embryo follows a definite pattern of cell division and growth. In the early stage of embryo development, the first divisions usually lead to the production of a linear filamentous structure. From the funicular end of this structure, a globular embryo develops and the remainder forms the suspensor. In dicots, the globular embryo develops into a heart-shaped embryo and finally into the mature torpedo-shaped embryo with well-defined radicle and cotyledons. Depending upon the age of the embryo, the stimulus for its continued growth is thought to be present within its own cells or in the surrounding endosperm. The fertilized egg and pro-embryo develop on the nutritional resources of the endosperm. In general, full-grown embryos can be successfully cultured on a standard medium containing sugar, mineral salts, and vitamins. Younger embryos, in addition to the above, also require trace elements and other growth-promoting substances.

Endosperm development: The embryo depends on the endosperm for its nutritional requirements. Thus, the level of endosperm activity determines the status of the embryo's success. Embryo degeneration may commence when the endosperm is still healthy, or along with or after the disappearance of the endosperm. The extent to which disturbances in the endosperm in an inviable cross make it incapable of supporting the growth of the embryo remains elusive indeed. Although, a frequent correlation of embryo abortion with the onset of endosperm deterioration observed in many plants indicates that it is due, in part, to the activity of the latter the underlying causes and mechanisms of endosperm malfunction in hybrids remain obscure. Some reports have claimed that endosperm failure in certain crosses might result from abnormal behavior of the antipodals, which, presumably alter the nutrient supply to the endosperm. A point of view has also developed that the presence of a developing embryo weakens the endosperm and induces abnormalities in its development. Abnormalities are known to occur in the development of the somatic tissues of the ovule such as the integuments and nucellus of certain infertile hybrids, and some of these developmental anomalies have substantial interest because of their bearing on the problem of embryo abortion.

Basic nutrient requirements

The basic requirements for a successful culture of embryos is a well-balanced medium containing macro- and micronutrients and a carbon source. Supplementing the medium with organic nitrogen like amino acids or casein hydrolysate can enhance the growth of excised embryos. Sucrose has proved to be the best carbon source, though other sugars have also been used. The phytohormones like IAA; NAA; 2,4-D; Kn; GA3, etc., have a profound influence on the growth and development of the cultured embryos. Several other substances like vitamins, meso-inositol, malt extract, yeast extract, etc., induce growth of embryos when added as supplements to the basal medium. Coconut milk is the most common additive for an embryo culture medium. Alcohol-diffusates of some seeds have also been shown to promote the growth of excised embryos in cultures.

Environmental factors

Environmental factors such as osmotic pressure, light, temperature, and pH affect the growth of embryos *in vitro*. The nutritional requirements of embryos vary depending upon the age and size of the embryo. Globular embryos require high osmotic concentration. Low osmotic concentration results in precocious germination. The osmotic pressure of the medium can be manipulated by varying the concentration of salt or sugars, depending upon the species. Light and temperature play an important part in suppressing the precocious germination of embryos. Root elongation is inhibited in light. Optimum temperature for embryo growth is between 27° and 30°C for most of the crops. Growth of immature embryos is better at near neutral pH, but as the embryos mature an acidic medium is more suitable.

A case study: Embryo culture of papaya

The demonstration of the ability of embryos removed from non-viable seeds to grow successfully in culture encouraged attempts to grow embryos of interspecific crosses of horticultural varieties of deciduous trees which ripen early, but have a low yield of viable seeds. In breeding work with crop plants, the embryo culture method is useful in the production of hybrids endowed with desirable disease-resistant qualities. The cultivated papaya is severely affected by the papaya leaf-mosaic virus, and the ring-spot virus diseases. Using the embryo-culture technique (Figure 32.5) interspecific hybrids can be obtained from immature hybrid embryos of *Carica papaya* (a species susceptible to these diseases) X *C. cauliflower* (a wild resistant species).

Materials

Equipments

Laminar air-flow cabinet; Autoclave; Water distillation unit; Oven for dry sterilization; Environmental growth cabinet with temperature and light control; Millipore filter unit;

Bunsen burner; Pair of forceps 2; Scalpels 2; Stoppers or cotton-plugs wrapped in muslin-cloth; Refrigerator; Water-proof marking-pen; Erlenmeyer flasks 4 each of 250 and 500 ml; Petri dishes 4 (10 cm diameter); Culture tubes 20 mm × 150 mm (Corning) about 25; Measuring cylinders 20 ml, 100 ml, and 250 ml; Needles, arrowhead; Brown-bands; Dissecting microscope.

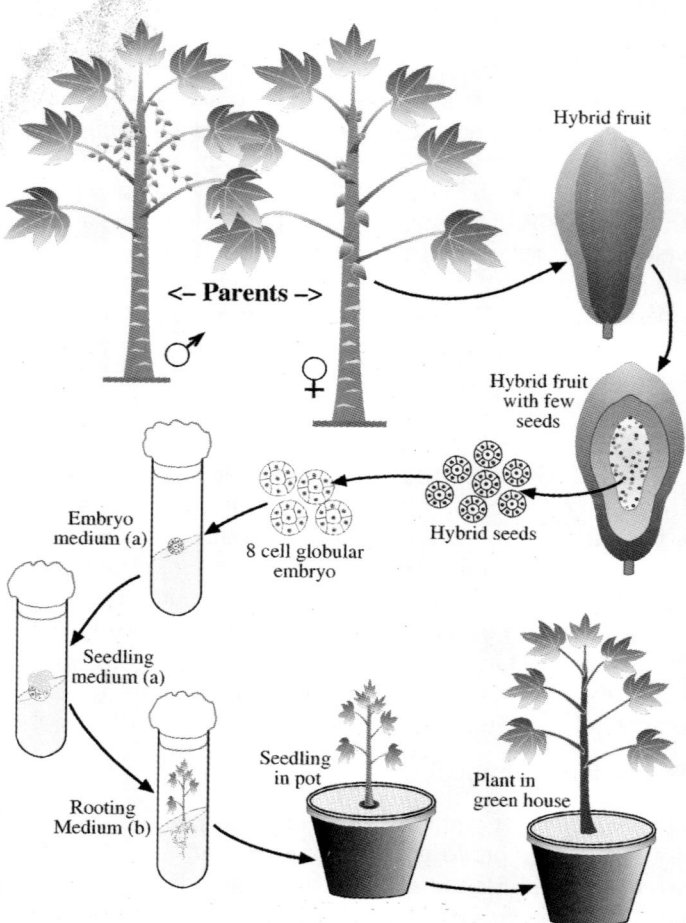

Figure 32.5 *Embryo culture of hybrid papaya (Carica papaya)*

Culture media, washing solutions, sterilizing agents

Detergent or wetting agent such as Tween-20; 70% ethanol; 0.1% w/v mercuric chloride ($HgCl_2$) solution; Glass distilled water for preparing media; Nutrient media (Table 6, Appendix 13); Media are prepared in double distilled water and autoclaved at 15 psi for 20 min.

Source of explant tissue

It is necessary to determine the right stage at which the hybrid embryo can be cultured.

Procedure

The source plant

1. Raise the F1 plant obtained from a cross between *C. papaya* var. 'Washington' (as the female parent) and *C. cauliflower* (as the male parent).
2. Before anthesis, cover the flowers of the female and male parents with butter-paper bags (8 cm × 12 cm).
3. When the female flowers fully open, collect the pollen from the male parent in a watch glass (between 9 and 11 am) and spray on to the stigma of the female flower with a fine brush. Repeat this step 2–3 times a day for 2–3 days. After pollination, cover the female flowers again with butter-paper bags to avoid contamination.
4. The fruits can be collected 89 days after fertilization and the seeds can be separated.
5. Surface sterilize the seeds (see Chapter 30), then transfer to a sterile Petri dish. Remove the testa carefully with a sterile forceps and scalpel.
6. Inoculate the globular-stage embryos on the culture medium (Table 6a, Appendix 13).
7. Incubate the cultures at $25° \pm 2°C$ in the dark for 30 days.
8. After this period, the culture can be transferred to a growth chamber with 16/8 hour photoperiod (2000 lux) at $25° \pm 2°C$. The embryos develop into seedlings of about 5 cm height within 3–4 weeks.
9. Then transfer the seedlings aseptically to another medium (Table 6b, Appendix 13). A healthy root and shoot system will develop on this medium in 15 days.
10. Remove the seedlings from the tubes and wash gently but thoroughly with sterile distilled water, and transfer to small pots (65 mm diameter) containing a mixture of sterile soil and vermiculite (1:1). Cover the pots with glass beakers to maintain humidity.
11. When a fresh set of leaves emerges after 3–4 weeks, the pots can be transferred to the glasshouse.
12. After 7 days the plants can be transferred to big pots (30 cm diameter) containing fertile soil.

III CULTURE OF DETERMINANT ORGANS

A. OVULE AND OVARY CULTURE

In angiosperms the female gametophyte, enclosing the egg (female gamete), is deep-seated in the ovarian cavity, well protected by the ovular tissues. The pollen grains are normally held at the stigma and there is no device for them to reach the egg. To effect fertilization the pollen grains germinate on the stigma by putting forth a tube (pollen tube) which grows through the stigma and style, finds its way into the ovule, and discharges two sperms in the vicinity of the egg. One of them fuses with the egg forming a zygote, while the other fuses with the polar nuclei forming the primary endosperm nucleus. The stigma receives a variety of pollen grains, but not all those that reach the stigma succeed in fertilization. The stigma and style are equipped with devices to allow pollen of only the right mating type to function normally; others are rejected. Some of the barriers to fertilization are:

1. Pre-fertilization barriers: (a) inability of pollen to germinate on foreign stigma; (b) failure of the pollen tube to reach the ovule due to excessive length of the style, or slow growth of the pollen tube which fails to reach the base of the style before the ovary abscises; and (c) bursting of the pollen tube in the style.

2. Post-fertilization barriers: Fertilization may occur normally, but the hybrid embryo fails to attain maturity due to embryo–endosperm incompatibility or poor development of the endosperm.

From time to time various techniques have been developed to circumvent the pre-zygotic barriers to fertility. These include bud pollination, stub pollination, heat treatment of the style, irradiation, mixed pollination, introduction of pollen grains directly into the ovary (intraovarian pollination) also called "test-tube fertilization", etc. The other promising and proven technique is *in vitro pollination*. Seed development following stigmatic pollination of cultured whole pistils has been referred to as ***in vitro pollination***.

The preliminary steps for in vitro pollination
1. Determination of the time of anthesis, dehiscence of anthers, pollination, entry of pollen tube into the ovules, and fertilization.
2. Emasculation and bagging of flower buds.
3. Collection of pollen grains.
4. Preparation of a suitable nutrient medium is required that will favor the germination of pollen grains and, more importantly, the development of fertilized ovules into mature seeds.

The pollen and ovule collection
1. *In vitro* pollination requires the maintenance of a reasonable sterility in pollen and ovule. To prevent chances of pollination of the buds to be used as the female partner, emasculate the buds before anthesis and bag.
2. One or two days after anthesis, bring the buds to the laboratory and prepare for aseptic culture. Remove the sepals and petals and the pistil along with pedicel. To surface sterilize, give a quick rinse in 70% alcohol; surface sterilize with a suitable detergent and finally wash thoroughly with sterile distilled water.
3. Remove the stigma and style and remove the ovary wall to expose the ovules. The whole placenta bearing the ovules and attached to a short pedicel is generally used for placental pollination. Alternatively, the placenta may be cut into two or more pieces each carrying a certain number of ovules and planted individually with their cut ends in contact with the medium.
4. For collecting the undehisced pollen under aseptic conditions, remove the anthers from the buds and keep in a sterile Petri dish until they dehisce. When the anthers are to be taken from open flowers they may be surface sterilized and left in sterile Petri plates containing a pre-sterilized filter paper until their dehiscence.

In vitro pollination
1. Collect the discharged pollens aseptically and deposit on the cultured ovules, or placenta, or stigma, as the case may be. It has been reported that pollen deposited on the ovules or placenta perform better than those spread on the medium around the ovules.
2. Incubate in aseptic condition until the seed formation is completed.

Factors affecting seed-set following In vitro pollination

Explant: In *Petunia axillaries*, *in vitro* pollinated ovules or a group of ovules attached to a piece of placenta did not form viable seeds. The pollen grains germinated normally, but the pollen tubes failed to enter the ovules. However, when intact placentae with undisturbed ovules were pollinated, normal events from pollen germination to the development of viable seeds occurred. In maize, the ovaries attached to cob tissue give better results than single ovaries, indicating the beneficial effect of parental tissue on ovary and ovule growth. The time of excising from pistils has a definite influence on the seed-set following *in vitro* pollination. The incidence of the seed-set is higher when the ovules are excised 1–2 days after anthesis (e.g., 3–4 days after silking in maize is optimal stage).

Culture medium: The method of *in vitro* pollination involves two major processes: (a) germination of pollen grains and pollen tube growth leading to fertilization, and (b) development of the fertilized ovules into mature seeds with a viable embryo. The success of the technique, however, will mainly depend on the composition of the medium, which would support both processes. The most important role of the culture medium is in supporting normal development of the fertilized ovules. It is, therefore, imperative that before attempting *in vitro* pollination the optimal nutritional and hormonal requirements are investigated. The salt mixture commonly used for *in vitro* pollination is the one developed by Nitsch (1951) for ovary culture. To this are added sucrose and vitamins. The composition of the modified Nitsch's medium widely employed for culturing *in vitro* pollinated ovules is given in (Table 8, Appendix 13). Almost invariably, sucrose has been used at a concentration of 4–5%. The information of the effect of various growth regulators and other supplements to the basal medium on seed development from cultured ovules is very meager. This seems to be the reason for the failure of seed-sets.

Incubation of cultures: Not much data is available on the precise effect of light on the response of *in vitro* pollinated ovules. However, it has been shown that in some systems temperature may influence seed-set.

Genotype: There is some evidence of genotypic variation in the response of *in vitro* pollinated ovaries of maize.

B. APPLICATIONS OF OVULE AND OVARY CULTURE

Ovule culture is important not only with reference to *in vitro* pollination, but also because it serves as an experimental system to study the *in vitro* response of zygotes and very young pro-embryos. Raising mature seeds by culturing ovules containing globular or older embryos is comparatively easy (Figure 32.6) and has been reported by several authors. Ovule culture holds a great potential for raising hybrids which normally fail due to the abortion of the embryo at a rather early stage when its excision and/or culture is either very tedious or impossible. The loss of a hybrid embryo due to premature abscission of fruits, as happens in many interspecific crosses in cotton, could be prevented by ovule culture.

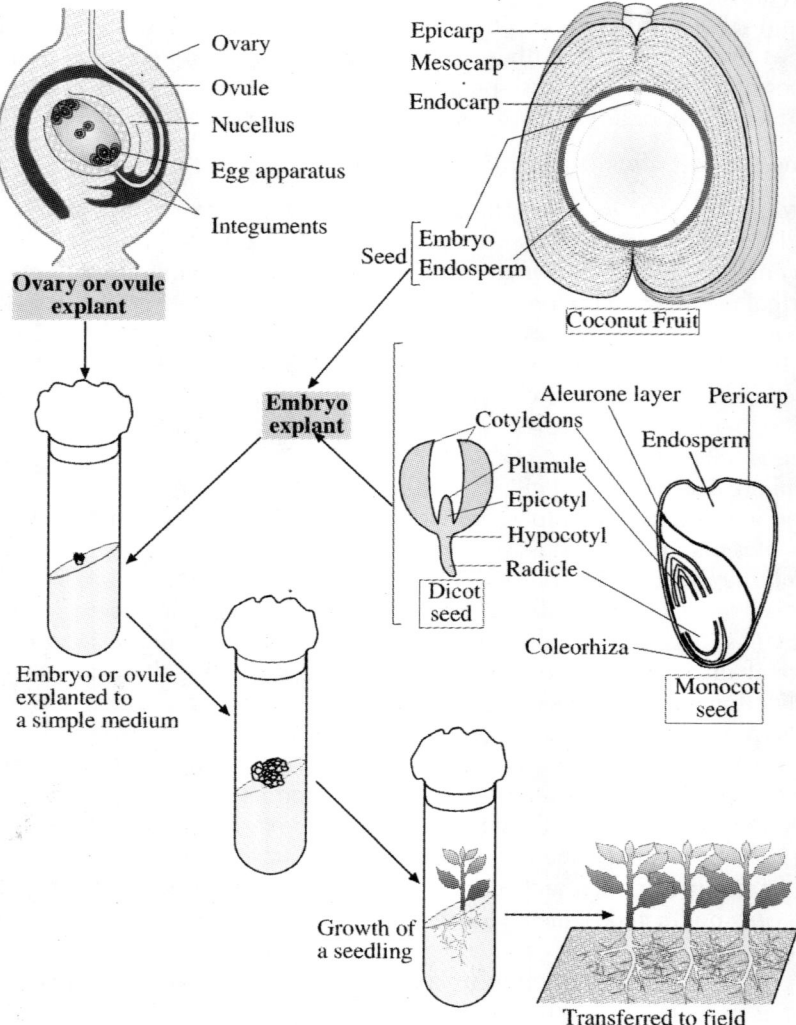

Figure 32.6 *Ovary (ovule) or embryo culture*

In vitro ovular and placental pollinations, where the stigmatic, stylar, and ovary wall tissues are almost completely removed from the path of the pollen tube, are potentially very useful in inbreeding and hybridization programs when the zone of incompatibility lies in the stigma, style, or ovary. *In vitro* pollination has proven useful in at least three different areas such as (a) overcoming self-incompatibility, (b) overcoming cross-incompatibility, and (c) haploid production through parthenogenesis. Potentially the most important application of *in vitro* placental pollination is in raising hybrids, which are unknown because of the pre-fertilization incompatibility barriers.

IV FLOWER BUD CULTURE

Flower bud cultures are particularly useful in examining the conditions that enable differentiation of such specific floral parts as stamens, pistils, petals, and so forth. They are also useful in exploring sex expression and other reproductive phenomena. Developing reproductively functional flowers has been achieved in only a few instances, for example, melon and *N. tabacum*. Usually, incomplete or non-functional flowers have resulted in many other crops such as *Aquilegia*, chrysanthemum, and *Kalanchoe*. The precise, nutritional needs of excised flower buds remain unestablished. The beneficial effects of exogenously supplied hormonal substances have been confirmed. The importance of association with developing leaves has been observed consistently. Blake (1966) noted that *Visceria* flower buds developed into mature flowers *in vitro* only when the explant had one pair of leaf primordia attached to the young explant. Without the leaf structures, the explants developed completely only if they had already initiated sepal primordia. However, it was found that kinetin could replace the leaves in *Nicotiana* flower primordium cultures.

Procedure (Hicks and Sussex, 1970)

1. Prepare a nutrient medium containing MS mineral salts and per litre: 0.4 mg thiamine. HCl; 100 myoinositol; 40,000 sucrose; and 0.1–1.0 mg kinetin.
2. Adjust pH of medium to 5.5 before autoclaving.
3. Disperse media in 25 × 150-mm culture tubes at 15 ml/tube or in 12.5 ml Erlenmeyer flasks in 45 ml aliquots. Cap or plug the culture vessels and autoclave at 121°C for 15 min.
4. Detach immature inflorescence from the tobacco plants.
5. Hold small portions of each inflorescence under a dissection microscope in a sterile transfer hood and expose individual flower primordia by removing the floral bract that subtends each primordia. Disinfesting the plant material is not usually necessary.
6. Using a fine surgeon's scalpel, excise the individual primordia and transfer to nutrient medium, one per culture vessel. Allow the primordia to float on the surface of the nutrient solution.
7. Incubate cultures at 27°C under Gro-Lux lamps (1,000 to 4,000 lux) for 16 hours daily.

V SUGGESTED READING

Blake J (1966) Flower Apices Cultured *in vitro, Nature,* **211**:990–991.

Heinz DJ, Krishnamurthi M, Nickell LG, and Maretzki A (1977) Cell, Tissue and Organ Culture in Sugarcane Improvement, In: *Plant cell, tissue, and organ culture* (Eds) Reinert J and Bajaj YPS, Norosa Publishing House.

Hicks GS and Sussex IM (1970) Development *in vitro* of Excised Flower Primordia of *Nicotiana tabacum.*, Canadian Journal of Botany **48**:133–139.

Iyer MS (1991) Meristem Culture, In: *Handbook of plant tissue culture*, Compiled. Mascarenhas AF. ICAR, New Delhi.

Thengane S (1991) Embryo Culture, In: *Handbook of plant tissue culture*, Compiled. Mascarenhas AF. ICAR, New Delhi.

White PR (1934) Potentially unlimited growth of excised tomato root tips in a liquid medium, Plant Physiology, Lancaster **9**:585–600.

33

Culture of Unorganized Tissues

I INTRODUCTION

Tissue culture is commonly used as a collective term to describe all kinds of *in vitro* plant cultures, but it strictly refers to the culture of unorganized aggregates of cells. Generally it includes (a) callus (or tissue) cultures, (b) suspension (or cell) cultures, (c) protoplast cultures, (d) anther cultures, and (e) somatic-embryo cultures.

II METHODS AVAILABLE

A. CALLUS CULTURE

When a tissue of a plant is injured, a wound repair response is induced. This response consists initially of the induction of division in the undamaged cells adjacent to the lesion, thus sealing off the wound. This is often followed by the hardening of this layer through deposition of lignin, suberin, wax, etc. However, if the damaged region is aseptically cultured on a defined medium, the initial cell division response can be stimulated and induced to grow indefinitely through the exogenous influence of the chemical constitution of the culture medium. The result is a continually dividing mass of generally poorly differentiated and disorganized plant cell aggregates termed a callus.

Callus in morphological terms varies extensively, ranging from being very compact (where the cells have extensive and strong cell to cell contact) to being 'friable' (where the callus consists of small, disintegrating aggregates of poorly associated cells and has a crumbly or creamy appearance). Friable callus is generally the most sought after as it is usually the fastest growing and most uniform type. It is best suited for the initiation of cell suspension cultures. Callus morphology is often explant-dependent but can usually be altered by the modification of the growth substance supplementation in the culture medium. Due to their size and nature, callus cultures have an inherent degree of heterogeneity. This heterogeneity is a disadvantage in the production of uniform biomass but it may also be an influential factor in plant regeneration.

Applications: Callus cultures have a wide range of applications. (i) They are used directly as a source of cell material for biochemical studies; (ii) callus is useful in secondary metabolite production; (iii) plant regeneration from callus is done to obtain somaclonal variants; (iv) It is used for screening of cells *in vitro* for types with useful characters; and (v) callus cultures are an intermediate step toward the initiation of somatic embryos, cell suspension cultures, or the production of plants from protoplasts.

1. Requirements for Callus Culture

(a) Laboratory equipment: Basically a plant tissue culture laboratory requires facilities for media preparation, sterilization of media and glassware, a sterile environment in which to carry out aseptic manipulations, and a constant environment room equipped for the control of light intensity, day length, and temperature.

(b) Plant material: The most commonly used starting materials are seeds; young, aseptically germinated seedlings and greenhouse-grown, young, healthy plants. In addition, if the plant material is at a very young stage there is a high potential for cell division within explants. Greenhouse-grown plants provide a larger source of explants material but sterilization can prove difficult both due to sensitivity to the chemicals used and the external morphology of the plant (e.g. presence of hairs and waxy layer). If such greenhouse-grown material is to be used for callus initiation it must be entirely free from infections, and insects and should be well maintained. Essentially all organs can be used as explant sources. However, the degree of success with different tissues can vary extensively and calluses with deferring morphologies are frequently obtained. For dicotyledonous plants, generally young leaves, petioles, stems, and hypocotyls are used. For monocots, meristematic regions are usually chosen (leaf bases, young inflorescences, etc.). Prior to culture, all material must be surface sterilized, and following sterilization the material must be washed thoroughly to remove all traces of the sterilant. The sterilized material is cut into suitably sized pieces (explant) and plated onto a solid medium. The choice of the medium is determined by the species to be used and the aim of the experiment (i.e. biomass production, plant regeneration, etc.). *In vitro* callus formation is often already visible after 4–7 days of culture although it can take much longer. Such a callus can grow indefinitely *in vitro* if it is provided with a constant supply of the appropriate nutrients and plant growth substances.

Method-1: The initiation and maintenance of Daucus carota (carrot) callus cultures

Plant regeneration by somatic embryogenesis from cultured cells was originally observed with carrot. The following protocol for the production of carrot *callus* from taproot material provides an excellent procedure required for **callus** initiation and maintenance. It also allows one to see how different tissues within the explant respond differently to the *in vitro* conditions. Callus formation should be visible within 7 days of establishment and rapidly-growing, friable callus cultures can be obtained within 5–8 weeks.

Materials

Plant material: Fresh, healthy and undamaged carrot roots;

Equipments: Sterile, sharp, stainless steel knife suitable for flaming; Sterile scalpel, long and short-handled forceps; 2 sterile 1000 ml (1L) beakers; 500 ml 2% (w/v) sodium hypochlorite solution in water; Sterile tap water and distilled water; Sterile filter paper, coarse grade;

Sterile 9 cm Petri dishes each containing 20 ml culture medium; Parafilm strips; and table top centrifuge.

Medium

MS (Ful strength)	4.71 g/l
Sucrose	30 g/l
2,4-D	1 mg/l
Agar	8 g/l
pH prior to autoclaving	5.8

Procedure

1. Collect the explant source – a healthy undamaged carrot taproot, 3–5 cm in diameter (explant).
2. Reject all diseased, damaged, or irregularly shaped individuals. Clean the explant thoroughly under running tap water and by scrubbing the carrots with a small nylon nailbrush.
3. Place the carrots in a 1,000 ml beaker, cover with the solution of sodium hypochlorite, and leave for approximately 30 min. Transfer the sterilized carrots to the sterile room.
4. Wash the carrots three times with sterile distilled water, agitating the beaker to completely remove the hypochlorite, and dry using sterile tissue paper.
5. Remove the uppermost 1 cm of the root (including all the remains of the shoot portion). Cut off all of the distal portion which is <2 cm thick and discard. Cut the remainder transversely into ca. 7 cm lengths.
6. Aseptically transfer one of the sterile segments onto a sheet of sterile filter paper.
7. Remove and discard 1 cm thick slices from both ends using a sterile sharp knife.
8. Cut as much as possible of the remainder into 1 mm thick transverse slices (Figure 33.1) and transfer these to a sterile Petri dish containing filter paper moistened with sterile distilled water, until 20–25 evenly-cut slices have been collected.
9. Place one slice on a suitable sterile cutting surface and using a sterile scalpel remove 2–3 (8 × 8 mm) explants so that each contains a segment of the cambial ring and the tissues on either side (phloem and xylem).
10. Aseptically, transfer these explants directly to culture dishes, 5 per dish and seal all dishes with Parafilm. It is preferable to retain the polarity of the explants by placing the root pole in contact with the culture medium. Incubate in the dark or under low intensity light (ca. 25 mmol/m^2/s) at 25°C.
11. Observe all dishes after a day, up to 14 days and discard any culture showing signs of contamination. Also monitor the growth patterns of cultures.
12. After 21 days, select the best-callusing explants, excise small pieces from callusing region into three evenly sized pieces and transfer to fresh medium (5 per dish). The callus pieces transferred should be ca. 5 × 5 × 2 mm. Seal all plates with Parafilm and incubate as above.
13. After 21 days, use the most friable callus for subculture as per step 12.
14. The cultures can hereafter be maintained by repeating step 13.

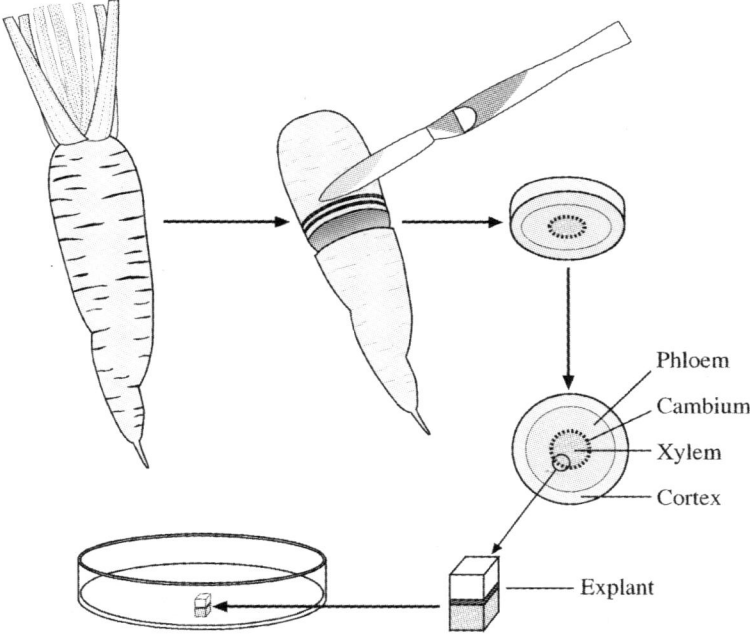

Figure 33.1 *Isolation of explants from carrot taproot*

Method-2: The initiation and maintenance of callus *cultures from tobacco pith tissue*

Pith is that tissue found in the center of the stem of most *Nicotiana* species. It consists entirely of parenchymatous cells – large, thin-walled, and highly vacuolate and represents one of the most uniform tissues which is readily available for *in vitro* culture. Indeed, the basal medium most commonly used for plant cell culture today, that of MS, was developed using tobacco pith callus as the experimental system (Murashige and Skoog, 1962). Callus formation from pith tissue is generally rapid – after 14 days the original explant will no longer be recognizable. The callus obtained is uniform and fast-growing and can be highly responsive to *in vitro* culture conditions, producing callus, shoots, leafy structures or roots according to the type, level, and balance of the growth substance supplements in the medium. As such, tobacco pith callus can be used as a reliable bioassay system for cytokinins and as an excellent system to demonstrate the influence of growth substances on plant morphogenesis.

Materials

Plant material and equipment

Healthy, greenhouse-grown tobacco plants ca. 0.5–1 m tall.

Sharp sterile scalpel, long- and short-handled forceps which, when not in use are stored with the ends in 70% EtOH. Before re-use, remove the alcohol by flaming; Non-sterile 9 cm Petri dishes containing moistened filter paper; Molten wax; 250 ml beaker containing ca. 150 ml 70% (v/v) EtOH in water; 2 sterile 250 ml beakers; 200 ml solution of 2%

(w/v) sodium hypochlorite in water; Sterile distilled water; Sterile 9 cm Petri dishes; 2 sterile 9 cm filter paper discs, coarse grade; Sterile 9 cm Petri dishes each containing 20 ml culture medium; Parafilm strips.

Medium

MS (Ful strength)	4.71 g/l
Sucrose	30 g/l
1-Naphthaleneacetic acid (NAA)	2 mg/l
Kinetin	0.25 mg/l
Agar	8 g/l
pH prior to autoclaving	5.8

Procedure

1. Use healthy, rapidly growing, non-flowering plants of *N. tabacum* varieties grown in the greenhouse, which are about 0.5–1 m tall.
2. Remove the upper 30 cm of the plant, cut off all the leaves, and chop up the remaining stem immediately above and below each node.
3. Collect all the internode segments which have a diameter >6 mm and cut them, if necessary, into 25 mm lengths. These should be stored in a Petri dish containing a disc of filter paper moistened with distilled water. Discard everything else.
4. Taking each internode in turn, dip both ends in molten paraffin wax (60–65°C) to a depth of ca. 1 mm. Allow the wax to harden fully.
5. Moving to the sterile work area, immerse the sealed internodes in 70% EtOH for 30 s and then rinse briefly (10 s) in distilled water.
6. Remove the foil from a sterile 250 ml beaker, pour in the well-mixed hypochlorite solution. Immerse the plant segments in this solution and leave for 25 min.
7. Using a pair of sterile, long-handled forceps transfer the internodes to a sterile 250 ml beaker and rinse with 100 ml sterile distilled water.
8. Wash the internodes three times with 200 ml sterile distilled water, first for 5 min., then 10 min. and finally for 15 min.
9. Transfer all the internodes to a sterile Petri dish containing a filter paper disc moistened with sterile distilled water.
10. Taking the internodes in turn, transfer onto a sterile cutting surface (Petri dish lid/base, thick filter paper, etc) and cut off and discard the waxed ends.
11. Place the internode on end and holding firmly with sterile forceps cut away all of the outer tissue by making 4 overlapping, longitudinal cuts to produce a rod of the central pith tissue ca. 10–12 mm long with a square cross-section.
12. Slice the pith transversely into 1.5–2 mm sections and transfer directly to the culture plates, 5 sections per plate.
13. Repeat for the other internodes until ca. 15 dishes have been prepared.
14. Seal all dishes with Parafilm and incubate in the dark at 25°C.
15. After 7 and 14 days examine all cultures and discard any dishes showing signs of contamination.

16. After 21 days, select the explants showing the best callusing response for subculture. Cut these up into 2 × 5 × 5 mm pieces and transfer onto fresh medium, 5 per plate.
17. Culture as per step 13. Cultures can hereafter be routinely maintained by repeating steps 15 and 16.

2. Troubleshooting

Even for experienced workers, the initiation of callus cultures from a new species or even a new genotype frequently does not go according to plan. The main problem generally is contamination and poor growth of the desirable callus type (**friable**). It is most important to maintain healthy source plants by checking them regularly and discarding immediately any which show signs of disease or damage. In addition to maintaining healthy source plants, inclusion of antibiotics in the culture medium can decrease the chances of contamination. The frequency and/or speed of callus formation can be enhanced by increasing the size of the explant or the surface area of the wound site. If there is little or no recovery of callus from cell suspensions, the amount of suspension cell mass placed on the agar should be increased. Alternatively, medium overlay techniques can be used to increase the frequency of callus colony formation.

B. SUSPENSION CULTURE

Suspension culture is essentially a product of *callus* culture, i.e. *callus* usually refers to a mass of undifferentiated cells. Once these are separated in liquid culture, it becomes a cell suspension. The culture of plant tissues in an agitated liquid medium eliminates many of the disadvantages ascribed to the culture of tissues on agar. Movement of the tissue in relation to the nutrient medium facilitates gaseous exchange, removes any polarity of the tissue due to gravity, and eliminates nutrient gradients within the medium and at the surface of the cells. Ideally, a plant cell suspension culture should consist of a population of single cells suspended, through continuous agitation, in a liquid nutrient medium, but in only the rarest of instances does such a culture exist. Increased cell dissociation means increased culture uniformity. However, some degree of cell aggregation generally has to be tolerated and fine-suspension cultures consist of micro- to sub-macroscopic colonies made up of ca. 5–200 cells. Furthermore, cultures consisting of larger aggregates (e.g. 0.5–1 mm in diameter) usually are more readily attainable, grow perfectly well, and, depending on the aim of the research, are often sufficient to meet all requirements (cell aggregation has been beneficial in retaining totipotent characters and enhanced yield of secondary metabolites). The most commonly used cell suspensions are of the closed type where the cells are grown in a fixed volume of liquid medium. These are routinely maintained through the transfer of a portion (ca. 10%) of fully-grown culture to a fresh medium at regular intervals.

Cells grown in a suspension culture generally resemble parenchymatous cells in having relatively large vacuoles, a thin layer of cytoplasm, and thin, rounded cell walls, but this is by no means always the case. The genotypes and medium composition used can influence *in vitro* cell morphology and different cell types can co-exist within a single culture. The balance can, however, usually be manipulated to favor one specific type through selective subculture. The growth curve of a cell suspension culture has a characteristic shape consisting of four essential stages: an initial lag phase, an exponential phase, a linear phase, and ultimately, a stationary phase. The duration of the different phases is dependent not only on the culture used but also on the medium and subculture regime. Inocula taken from a culture still in the linear growth phase will produce cultures with a shorter lag phase than those initiated using cells taken from a stationary phase culture. The lag phase is also shortened when relatively large inocula are used, although paradoxically, growth terminates earlier and overall biomass production is reduced. Conversely, a very small inoculum results in a greatly extended lag phase if indeed the culture grows at all.

1. Applications

Plant cell suspension cultures have a wide range of applications and are generally used in preference to callus cultures due to their more rapid growth rate, greater homogeneity, and the ease of experimental manipulation. Cell suspensions have been successfully used as a means of rapid and continuous production of uniform biomass for basic plant cell physiology studies, enzymological studies, and secondary metabolite production. Through the manipulation of the medium, cells can be induced to divide synchronously, thus providing a valuable tool for cell cycle studies. Cell suspensions have also proven to be excellent starting materials for the isolation of protoplasts to be used in a wide range of applications (cell fusion, embryogenesis, genetic manipulation, etc.).

Requirements: To a laboratory already equipped for growing callus cultures, essentially only three additions are required to enable the progression to suspension culture work; a supply of culture flasks, wide-mouthed glass bottles for subculturing, and a culture shaker. Most commonly, a rotating platform shaker is used consisting of a table onto which flasks of varying sizes can be firmly fixed and which rotates in a horizontal plane at selected speeds of 50–500 rpm with an amplitude of ca. 1 cm. This must be situated in a constant environment room or have an independently controlled culture chamber attached. Culture shakers must function 24 hours a day and around the year.

2. Types of Suspension Cultures

Basically there are two types of suspension cultures, viz., batch cultures and continuous cultures.

(a) **Batch culture**: Batch cultures are used for initiating single cell cultures and cell suspensions are grown in 100–250 ml flasks each containing 20–75 ml of culture medium. The cultures are continuously propagated by routinely taking a small aliquot of the suspension and transferring it to a fresh medium. During the incubation period the biomass of the suspension cultures increases due to cell division and cell enlargement. After some time, the growth stops due to the exhaustion of some factors or the accumulation of certain toxic metabolites in the culture medium. At this stage cells are subcultured in a fresh medium and with the transfer of small aliquots of cell suspension (ca. 5 X dilutions), the cell growth is revived. The cell growth (biomass) in batch cultures follows a fixed pattern, as shown in Figure 33.2. Initially the culture passes through a lag phase, followed by a brief exponential growth phase during which active cell divisions occur. After three to four cell generations, the growth declines and finally the culture enters the stationary phase. Cultures can be maintained continuously in the exponential phase by frequent (every 3–4 days) subculture of the suspensions. Prolonged maintenance of

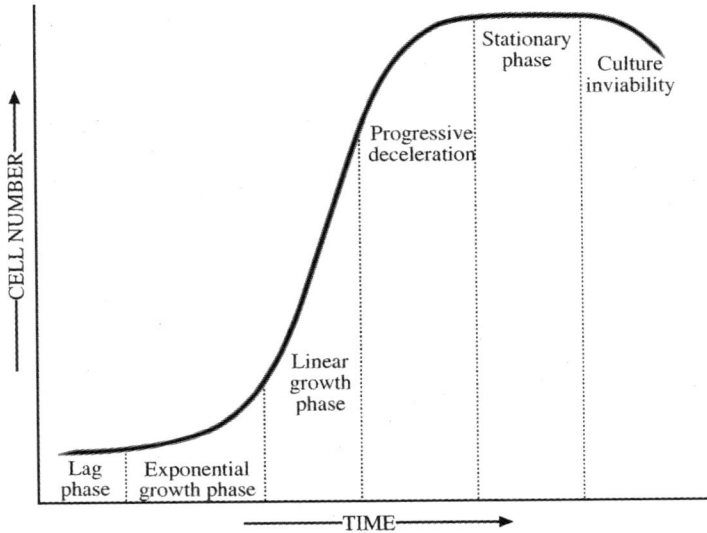

Figure 33.2 *Diagram showing the phases of growth in batch suspension cultures. Growth phases are labeled*

cultures in the stationary phase may result in extensive death and lysis of cells. The texture of a callus is genetically controlled and often it may be difficult to obtain a good dispersion of cells under any conditions. However, tissue dissociation can be improved by modification of media composition and subculture routine.

Owing to certain inherent drawbacks of the system, batch cultures are not ideal for studies of cell growth and metabolism. Batch cultures are characterized by a constant change in the pattern of cell growth and metabolism, and the composition of the nutrient medium. To a certain extent, these problems are overcome by continuous cultures.

b) Continuous cultures: A number of culture vessels have been designed to grow large-scale cultures under a steady state for long periods by adding fresh medium and draining out the used medium. Continuous cultures may be of the closed type or open type. In the closed type, the addition of fresh medium is balanced by outflow of the old medium. In contrast, in the "open continuous cultures", the inflow of the medium is accomplished by a balancing harvest of an equal volume of the culture. The rate of inflow of medium and culture-harvest are so adjusted that the cultures are indefinitely maintained at a constant, sub–maximal growth rate. Two major types of open continuous cultures are in use – 'chemostat' and 'turbidonstat'.

Method-1: Initiation of a carrot cell suspension culture

Carrot cell suspension cultures are one of the easiest to establish. Beginning with a friable callus, a suitably fine suspension culture can be achieved rapidly – within 3–4 weeks. The basic protocol involves the inoculation of the culture medium. Through the continuous agitation of the culture and the sustained division of the inoculum, single cells, and small cell aggregates are released into the medium (Figure 33.3). These are selectively subcultured and after 1–2 further cycles, the culture is established. Typically, this culture can be routinely maintained using a small inoculum at each subculture.

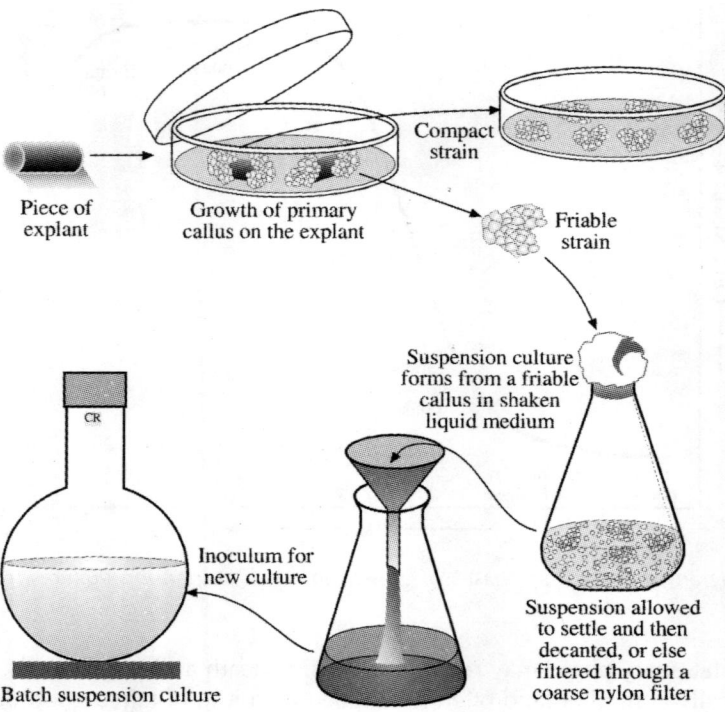

Figure 33.3 *Typical steps in the initiation of callus and suspension cultures*

Materials

Plant material: Established, healthy friable callus cultures.

Equipments: Autoclaved, 250 ml, wide-necked Erlenmeyer flasks containing 50 ml of culture medium sealed with a double layer of aluminum foil; a supply of sterile aluminum foil sheets 10 × 10 cm; a pair of sterile forceps and sterile spoon spatula; a series of sterile 250 µm and 100 µm nylon or stainless steel filters + glass filter funnels; sterile 100 ml Erlenmeyer flasks; sterile, wide-mouthed glass pipettes.

Medium

MS (Full strength)	4.71 g/l
Sucrose	30 g/l
1–Naphthaleneacetic acid (NAA)	2 mg/l
2,4–D	2.21 mg/l
Kinetin	2.15 mg/l
pH prior to autoclaving	5.8

Procedure

1. Use as the starting material friable, light-grown callus cultures (initial cell concentration $0.5 \times 2.5 \times 10^5$ cells /ml recommended), produced as described in the previous chapter, and which are still in their active growth phase (1–2 weeks following subculture).
2. Working in a sterile environment (laminar flow bench/UV-sterilized transfer room), quickly flame the neck of a culture flask and loosen but don't remove the aluminum foil cap.
3. Using forceps or a long-handled spoon spatula, aseptically remove the outer periphery of each clump of callus; remove the foil cap from the flask, flame the neck and transfer ca. 2 g of this peripheral cell material to the culture flask.
4. Repeat for each flask after which all can be firmly secured onto the rotary shaker under low intensity light (ca. 25 µmol/m^2/s).
5. After 10 days, filter each culture through a sterile 250 µm nylon/stainless steel mesh and collect the filtrate in a sterile 100 ml Erlenmeyer flask.
6. Wait 10 min. until the cells have settled and pipette off as much of the medium as possible. Use the remainder as inoculum for a fresh culture.
7. After 10 days, filter the culture through a sterile 100 µm nylon/stainless steel mesh. Repeat step 8 but use only one half of the remainder as the inoculum.
8. After 10 days and every subsequent 10 days thereafter, subculture as for an established culture, i.e. transfer directly, using a wide-mouthed pipette, 1 ml of the well-mixed culture per 10 ml fresh medium.

Note

For methods for the determination of suspension culture growth and development (Cell viability, cell number, measurement of media pH, packed cell volume, fresh and dry weights), refer to Chapter 35.

C. SINGLE CELL CLONES

Cultures can be initiated from a single cell, but only if a special environment is provided. Generally, suspension cultured cells are passed through a filter which removes coarse cell aggregates and allows only single cells to pass through. Upon culture on solidified medium, these cells form small cell clusters, which are assumed to have originated from a single cell. Cell lines originating from this single cell in this way are sometimes called **single cell clones** or **cell strains**. However, each cell clone has a minimum effective initial cell density or minimum inoculation density, below which it cannot be cultured. The minimum density varies according to the medium and growth regulators in which the cells are placed (generally about 10–15 cells/ml on standard media). The minimum inoculation density can therefore be lowered by adding to a standard medium either a filtered extract of a previously cell cultured medium (called conditioned medium), or special organic additives (said to be supplemented).

An insufficient number of single cells plated, which may not be enough initiation for spontaneous cell division, can initiate division by 'being **'nursed'** by tissue growing nearby. Single cells can be cultured using the following techniques, (1) filter paper raft-nurse tissue culture, (2) growing in Petri dish compartments, (3) microchamber technique, (4) microdrop method, (5) Bergman's plating technique, and (6) thin layer liquid medium.

Method-1: Filter paper raft-nurse tissue culture

1. Place an inoculum onto a sterile filter paper disc (a raft, 8 × 8 mm) or some other inert porous (Figure 33.4a), which is then put in contact with an established (several days in advance) callus culture of a similar species of plant (called **nurse cells or feeder layer**).
2. Incubate for a number of days under standard cell culture conditions.
3. When the single cells start dividing and form macroscopic colonies on the filters, isolate the colonies individually with care.
4. Culture each isolated colony in a standard medium to get a large number of cells for further establishment and study.

Method-2: Growing in Petri-plate compartments

Divide a Petri dish into compartments (Figure 33.4b). Nurse tissues cultured in some segments to assist the growth of cells or protoplasts in the other area.

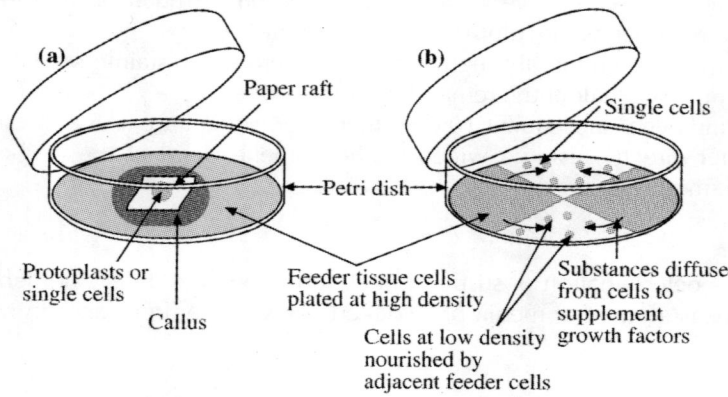

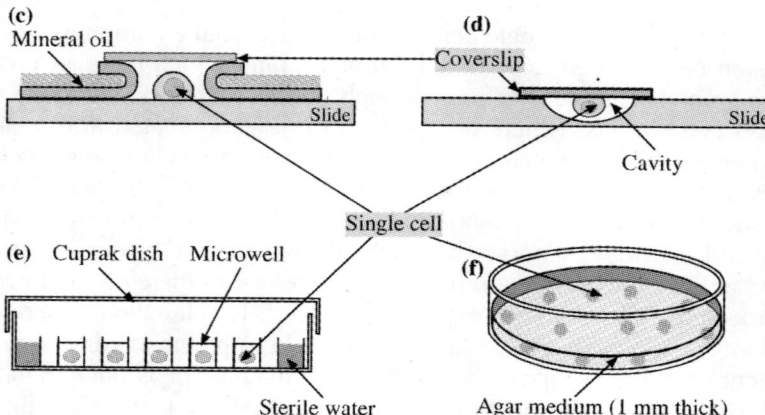

Figure 33.4 Techniques for culture of single cells (a) Filter paper raft-nurse tissue technique, (b) cell culture in Petri dish compartments, (c) and (d) microchamber technique, (e) microdrop technique, and (f) Bergamann's plating technique

Method-3: Microchamber technique

A microchamber can be constructed either by using a microscope slide and coverslips or by cavity slide (Figure 33.4c,d). The slide and coverslip can be held together by using mineral oil. Single cells are suspended in conditioned medium, and a drop of medium having a single cell in the microchamber. The microchamber is then covered by a coverslip. In case of a cavity slide, the drop is placed onto a coverslip which is then inverted into the slide cavity. The advantage of this method is that it allows microscopic observation.

Method-4: Microdrop method

A specially designed dish (Cuprak dish), having a smaller outer chamber for filling distilled water to avoid desiccation and a larger chamber containing a number of microwells is used. Microdrops of 0.25–0.5 ml are distributed in the microwells and the dish is sealed with Parafilm. The adjustment of medium is made in such a way that each microwell contains a single cell (Figure 33.4e).

Method-5: Bergmann's plating technique

This is a widely used method. The method consists of cells suspended in a liquid medium at a cell density that is twice the desired density in the plate. A sterilized agar (ca. 1%) medium is kept melted in a water bath at 35°C. Equal volumes of the liquid and agar media are mixed thoroughly and quickly spread in ca. 1 mm thick layer in a Petri dish. The cells remain embedded in the soft agar medium, are observable under microscope, and when macroscopic colonies develop, they are isolated and cultured separately (Figure 33.4f).

Method-5: Thin layer liquid medium culture

Cells can be plated in a thin layer of liquid medium to allow adequate aeration. Because cells are not fixed in position, it is not possible to follow up on an individual cell during culture. This method is common for protoplast cultures.

Note

1. Culture the single cells in the dark and make a limited number of microscopic observations.
2. Use either a conditioned medium or a suitably enriched medium, since standard tissue culture media are unsuitable.

D. PROTOPLAST CULTURE

The first requirement in any experimental procedure that involves protoplasts is to be able to isolate protoplasts of good quality that are viable and free from contaminating debris. **Protoplasts are plant cells whose cell walls have been removed**. The cell wall normally confines the cell, preventing it from bursting as a result of turgor pressure. For some biotechnology and genetic engineering protocols, however, the cell wall is a substantial barrier. Releasing plant cells from the confines of the cell wall allows them to be manipulated by microinjection, electroporation, and fusion. Many techniques for the production of protoplasts have been developed. Originally, tissues were bathed in a solution of high osmotic potential to shrink the

956 MOLECULAR BIOTECHNOLOGY: PRINCIPLES AND PRACTICES

cells. Then, the cell walls were broken mechanically by cutting or abrasion. Finally, by reducing the osmotic potential of the medium, the cells were made to swell and pop out of the broken cell walls. The yield of viable protoplasts from this type of protocol is generally low. Thus, except for some difficult tissues, this strategy is rarely followed. Most methods today depend on the enzymatic breakdown of cell walls. Protoplasts can be isolated from plant tissues or cultured cells by the enzymatic digestion of cell walls (Figure 33.5). The success of protoplast isolation depends especially on the condition of the tissue and the combination of enzymes being

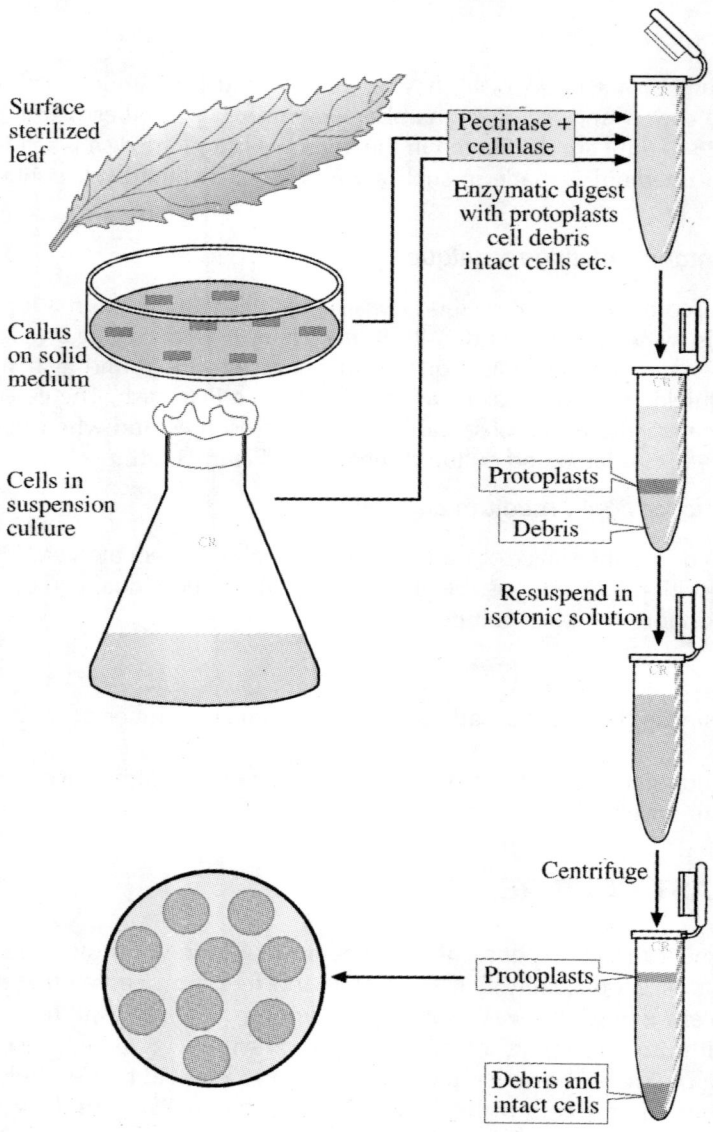

Figure 33.5 *Schematic outline of the enzymatic procedure for protoplast isolation*

used. Protoplast digestion mixtures generally include several enzymes (Table 33.1), the concerted action which efficiently releases cells from the cell wall. Because naked plant cells are fragile, and can burst in solutions of low osmotic potential, sorbitol or mannitol is included in the digestion mixture as an osmoticum. Since sugar alcohols are metabolically inert and infuse into protoplasts rather slowly, they have come into common use in recent years. Other ingredients included to improve the stability of the protoplasts are Ca^{2+} and Mg^{2+} ions. Another substance often included is potassium dextran sulfate, which helps in the adsorption of phenols due to its polyanionic nature.

Table 33.1 *Some commonly used enzymes for protoplast isolation*

Enzyme	Source
Cellulase R–10	*Trichoderma viride*
Meicelase–P	*Trichoderma viride*
Hemicellulase H–2125	*Rhizopus* sp.
Macerozyme R–10	*Rhizopus* sp.
Pectinase (purified)	*Aspergillus japonicus*
Pectolyase Y23	*Aspergillus japonicus*
Pectinol	*Aspergillus* sp.
Zymolyase	*Arthrobacter luteus*
Driselase	*Irpex lactes*

1. Isolation of viable protoplasts

Isolation of viable and culturable protoplasts in large quantities is affected by several **factors**, and optimum conditions for a system are established empirically. Factors affecting yield and viability of protoplasts are: (i) Source of material: leaf is very commonly used; (ii) Pre-enzyme treatments: surface sterilization, peeling off the lower epidermis to increase enzyme penetration, infiltration of leaf with enzyme solution and agitation of the incubation mixture; (iii) Enzyme treatment: the nature, concentration, combination of two or more enzymes, pH and temperature are important; (iv) Osmoticum (e.g., sorbitol and mannitol in the range of 450–800 mmol^{-1}).

Plant protoplasts have been in constant use for a long time and have become one of the most versatile analytical tools in plant biology. Depending upon the species and culture conditions, the protoplasts may have the potential to: (i) regenerate a cell wall (within 2–4 days in culture); (ii) dedifferentiate (division and callus formation); (iii) differentiate into shoots, roots, or embryos and produce a complete plant. They can be isolated in large quantities from a variety of tissues or cultured cells by enzymatic digestion to remove cell walls. The success of protoplast isolation depends especially on the condition of the tissue and the combination of enzymes being used. Freed from the cellulosic cell walls, the plasma membrane becomes accessible for investigation or purification. The relatively homogenous population of individualized wall-less cells can be subjected to various experimental treatments. These characteristics of plant protoplasts make them the material of choice for the following applications:

1. Protoplasts serve as recipient hosts for transformation by recombinant DNA. Transient gene expression experiments have considerable implications in (a) elucidating tissue-specific regulation and *cis*- and *trans*-regulatory interactions among foreign and/or endogenous genes, (b) analyzing translational processes independent of transcriptional events.

2. Molecular cytogenetics, with particular emphasis on *in situ* hybridization for gene localization on metaphase chromosomes (for example root meristem protoplasts).
3. Flow-cytometry experiments, comprising cell-cycle analysis and chromosome sorting.
4. Cloning of large DNA inserts (megabase cloning) in DNA preparations from agarose-embedded protoplasts.
5. DNA analysis and long-range mapping *via* pulsed-field gel electrophoresis.
6. Protoplasts are required in somatic hybridization by protoplast fusion. Somatic hybrids may be obtained by the fusion of two intact nuclear genomes from two different species.

One major requirement in performing these types of experiments is the production of high yields of robust protoplasts. There is no standard method for the isolation and culture of protoplasts. Recent advances in the isolation, culture, and regeneration of plants from protoplasts of a wide diversity of species have been reported. The protocols described here will be generally applicable to several plants of the dicotyledonous species including: tobacco, tomatoes, eggplants, potatoes, *Brassica*, *Medicago*, etc., but some modification may be necessary depending upon the species and the variety (Figure 33.6).

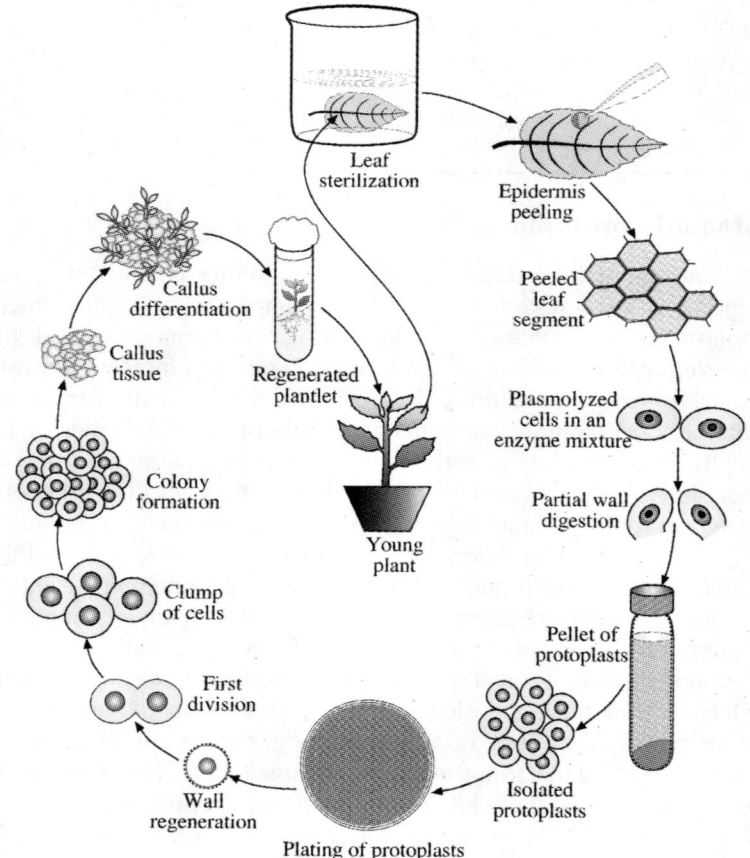

Figure 33.6 *Schematics sequence for the isolation, culture and regeneration of plants from leaf protoplasts*

Materials

Enzyme stock solutions

Each is prepared by dissolving in 0.4 M mannitol (72.8g/l) at pH 5.8, stored frozen at −20°C: Macerozyme R-10, 10% w/v, 0.1 g in 10 ml. Cellulase R-10, 10% w/v, 0.1g in 10 ml. When needed, the enzyme stocks are diluted to the desired concentration in 0.4 M mannitol and sterilized using 0.22 μm filters. *Enzyme solution 1 X*: Combine 1% (w/v) Macerozyme R-10 and 1% Cellulase R-10 in 0.4 M mannitol, pH 5.8, and filter sterilize.

Universal enzyme (ERG, You You Glob, Ukrainian Biotechnology Center for Agriculture, Kiev): 1% Driselase (Fluka), 0.3% Cellulose R-10, 0.2% Macerozyme R-10, 0.1% Cellulysin (Sigma); $CaCl_2.2H_2O$ 5 mM; Sucrose 0.5 M, pH≤5. A 50–70% strength of the enzyme mix is usually sufficient for a complete overnight digestion of leaf protoplasts.

Washing solutions:

W5 salt solution: 154 mM NaCl; 125 mM $CaCl_2$; $2H_2O$; 5 mM KCl; 5 mM glucose; pH 5.8–6.0.

Floating solution (ML 0.6): 0.6 M Sucrose, 15 mM $CaCl_2.2H_2O$, 0.1% MES, pH 5.6.

Equipment

Forceps, scalpels, and blades (sterile); Beaker 100 ml, sterile; Capped plastic centrifuge tubes; Rack for centrifuge tubes; Incubator; 500–100 μm sieve batteries with beaker; Counting chamber, type 'Thoma'; Parafilm; Sterile pipettes 1 ml, 5 ml, 10 ml; Petri dishes 9 cm, 6 cm.

Donor plant materials

Procedures are described for the isolation and culture of leaf tissue protoplasts of **Nicotiana tobaccum**: Haploid or diploid *in vitro* cuttings, grown on basal MS medium, obtainable, e.g. from Sigma with long days of maximum light intensity of 2500 lux, 25°C. Low protoplast yield and survival are observed when plants are grown under continuous light.

Plants can also be raised in fiber pots by growing seeds in a soil:sand:peat (2:1:1) mixture under a 16/8h day/night photoperiod, at 20/15°C. The plants are watered daily and receive weekly fertilizer with a 20:20:20 N:P:K liquid nutrient solution at 1 g/l.

Procedure

Preparation of explants

Isolate fully expanded healthy leaves, surface sterilize in 7% w/v of saturated calcium hypochlorite solution for 10 min., and rinse in sterile distilled water three times for 5 min. each. A sterile plastic Petri dish, 100 × 15 min., is opened and used as a sterile tray to cut the tissues from the leaves.

Enzymatic digestion to release protoplasts

1. Fully expanded leaves (7–8 days-old) are sliced in the enzyme solution (10 ml per Petri dish) and incubated overnight at 20°C. Care must be taken to ensure that over-digestion does not occur and that the protoplasts are not damaged by the enzyme treatment and the shaking. The addition of MES (2-N-morpholinoethanesulfonic acid) buffer at 1–5 mM, pH 5.8 helps to stabilize protoplasts during isolation, washing, and fusion procedures. Extreme

care must be taken to ensure that the protoplasts do not become contaminated with bacteria or fungal spores.
2. Next morning shake the dish gently to free the protoplasts and transfer the protoplasts, enzyme solution, and undigested tissues with a sterile Pasteur pipette onto a sterile 64 or 44 μm pore-size filter inside a sterile funnel or sieve them through sterile 50–100 μm sieves. Extreme care must be taken not to damage the protoplasts, which are very fragile at this stage. Vigorous pipetting, vigorous re-suspending of pellets during washing, or high-speed centrifugation are all examples of mishandling of protoplasts.
3. Transfer to 15 ml centrifuge tubes (round-bottom glass tubes preferable) and spin at 500g (or 700 rpm) for 5–10 min. Remove the enzyme solution carefully with a sterile Pasteur pipette.
4. Add 5 ml of 0.4 M sucrose to the pellet, and gently disperse and re-suspend the protoplasts in the solution.
5. When the protoplasts are completely dispersed, add carefully with a Pasteur pipette, by layering over the sucrose solution ML 0.6 and centrifuge at 500g for 6 min.
6. Remove the band of floating protoplasts with a pipette. The band will be easy to handle if sucked as a mixture of air and protoplast suspension. Dilute with 3–5 volumes of W5 salts solution.
7. Spin at 700 rpm for 5 min. Remove the supernatant with a pipette or vacuum pump. Shake the pellet gently and re-suspend it in 5–10 ml W5 salts solution. Count the cell density in a counting chamber (hemacytometer). Re-suspend the protoplasts to a density of 1×10^5 ml in culture medium. Measure the volume exactly.

2. Regeneration of protoplasts

Many investigations have been focused on the early events accompanying the formation of entire cells (called "**plastocytes**") from naked protoplasts and especially on those concerned with cell wall regeneration. Early investigations on cell formation relied largely on staining by **Calcofluor**, which binds to cellulose material comprising β 1–4 and β 1–3 linkages – perhaps through hydrogen bonds – and causes it to fluoresce. On occasions, other methods have also been used, e.g. polarization microscopy. Recently, investigations on microtubular organization have also been made employing both electronmicroscopy and indirect immunofluorescence. It does seem that microtubules are in some way involved in cell wall synthesis since they appear in rather large numbers at this time. Several investigators have focused attention on the problem of whether cell wall synthesis and nuclear division are related and whether these processes must proceed in a strict sequence. However, there is no conclusive correlation found between these events.

Cell wall formation is followed by an increase in the size of the cell and rearrangement of the chloroplasts, which become scant and yellow. The cell then divides equally or unequally. The various factors controlling growth and differentiation are discussed in more detail in the previous chapter.

Protoplast culture

The protoplasts may be cultured in one of three ways:
 – as thin liquid layers of 1–1.5 ml per 60 × 15 mm sterile Petri dish;
 – as microdroplets of 10–25 μl each, spotted in the Petri dishes;
 – as thin liquid layers of 1.5 ml over agarose underlayers.

1. Dilute with culture medium and distribute into Petri dishes at 5×10^4 protoplasts per ml. 10 ml per 9 cm dish. Incubate for 72 hours in the dark. Later the cultures are maintained in low light/dark cycles.
2. To embedded protoplasts in agarose, prepare fresh medium by mixing liquid K3M and autoclaved agarose powder (0.6 g per 100 ml to give 0.6%) by melting in a microwave oven. Surplus medium can be re-melted a couple of times for further use. Perform this 1 hour before use and keep the medium at 37°C until use. When embedding in agarose, make sure that the temperature of the medium is below 37°C.

3. Dilution and planting of dividing colonies

Steps in the procedure
1. After 7–8 days in culture, the SR1 protoplasts would have undergone 4–5 division cycles. They can be diluted to reduce the osmoticum concentration and to adjust the hormonal balance.
2. Dilute by a factor 10 with AG medium. The same conditions are used when selecting for resistance to antibiotics, herbicides, etc., in transformation experiments. In those cases in which agarose was included in K3M, sectors can be sliced in the original dish and transferred with a spatula to a new container. If more than 12 ml of the culture medium is to be used, shake the dilution mix for appropriate aeration.
3. After 2–3 weeks in culture, colonies visible to the naked eye are obtained. Refresh the medium by removing the old one. The osmotic pressure in the fresh AG medium should be 0.2 M. If agarose sectors undergo dilution steps, break them into smaller pieces to reduce local colony density.
4. Microcalluses of 2–3 mm can be picked manually or through wide-mouth pipettes and spread on top of agar plates containing MR1 medium. During two successive transfers onto MR1, a callus is produced and regeneration of multiple shoot structures should occur. If the regeneration frequency is too low, the BAP concentration be reduced to 0.25 mg/l.
5. Rooting and elongation of the shoots is obtained on half strength basal MS medium.
6. Rooted plantlets can be transferred to the greenhouse and/or maintained in culture cuttings.

4. Troubleshooting guide

Low protoplast yield: Check the quality of donor plant growth conditions. Uniformity of growth conditions, vigor, nutritional status, and other conditions that influence the physiological quality of materials will also influence the quantity and quality of protoplasts. Also, check the quality, concentrations, and/or digestion times for the enzyme sources being used. In some cases the osmoticum used in the enzyme and purification solutions can affect protoplast yield and quality. If many protoplasts burst, they are seen as cell "ghosts" sticking at the bottom of the Petri dish), and vacuolated cells may frequently exhibit low division rates. This problem can be reduced by the following method. Add <0.25% PEG to the protoplasts and/or pre-incubate at a high density ($2.5-5 \times 10^5$ protoplasts per ml) for 24–48 hours. Dilute afterward to the standard 5×10^4 protoplasts per ml. Furthermore, survival and division activities of the protoplasts are usually better when using a longer enzyme incubation time (overnight) and, thus, less concentrated enzyme solutions. The Ca^{2+} concentration during digestion is kept to a minimum (5 mM).

Adding NAA and BAP (1–5 mg/l each) to the enzyme can activate the protoplasts during the initial stages of culture.

Low plating efficiency from good quality protoplasts may be due to improper culture density or suboptimal medium. Osmoticum source and osmolality of the medium are critical for successful culture. The *method of culture system* can be a factor in culture success. Protoplasts are very sensitive to light. Excessive debris in the protoplast culture can result in inhibition of cell division due to the release or presence of toxic compounds. The purification procedure needs to be repeated or improved to remove the excess debris.

A. Stock solutions

MS salts Stock solutions (MS revised)

Macroelements MS (20 X)	**g/litre**
KNO_3	38.00
NH_4NO_3	33.00
$MgSO_4 \cdot 7H_2O$	7.40
$CaCl_2 \cdot 2H_2O$	8.80
KH_2PO_4	3.40

Microelements MS (1000 X)	**g/litre**
H_3BO_3	6.20
$MnSO_4 \cdot 4H_2O$	22.30
$ZnSO_4 \cdot 7H_2O$	8.60
$Na_2MoO_4 \cdot 2H_2O$	0.25
$CuSO_4 \cdot 5H_2O$	0.025
$CoCl_2 \cdot 6H_2O$	0.025

Morel-Vitamins Stock solution (500 X)	**mg/100 ml**
Thiamine . HCl	168.5
Pyridoxine . HCl	102.5
Nicotinic acid	61.50
Myoinositol	9000
Biotin	1.22
Ca pantothenate	238.0

K3M basal (protoplast culture)*	
Per litre:	
Macroelements K3 (10 X)	100 ml
Microelements K3 (1000 X)	1 ml
Fe-EDTA (100 X)	5 ml
Morel vitamins (500 X)	2 ml
Thiamine	100 mg
Glucose	0.45 M
Sea-Plaque agarose	0.6% (optional)[a]
pH	5.5

* Filter-sterilize and store frozen at –20°C; [a] Autoclave as dry powder

Macroelements K3	g/litre
KNO_3	1.9
NH_4NO_3	0.6
$MgSO_4 \cdot 7H_2O$	0.3
KH_2PO_4	0.17
$CaCl_2 \cdot 2H_2O$	0.6
KCl	0.3

AG medium (dilution of protoplast-derived colonies)
Per litre:

Macroelements (10 X)	100 ml
Microelements K3 (1000 X)	1 ml
Fe-EDTA (100 X)	5 ml
Morel vitamins (500 X)	2 ml
Sucrose	30 g
Mannitol	50 g
NAA	0.1 mg
BAP	1 mg
pH	5.7

Macroelements	g/litre
NH_4NO_3	1.01
$CaCl_2 \cdot 2H_2O$	0.44
$MgSO_4 \cdot 7H_2O$	0.74
KH_2PO_4	0.136
$(NH_4)_2$ Succinate	0.1

MR1 medium (callus formation and shoot regeneration)
Per litre:

Macroelements MS (20 X)	50 ml
Microelements MS (1000 X)	1 ml
Fe-EDTA (100 X)	10 ml
Morel vitamins (500 X)	2 ml
Thiamine-HCl	100 mg
Sucrose	30 g
BAP	(0.25-) 1 mg
Agar	0.6%
pH	5.8

E. ANTHER CULTURE

Haploids can be cultured in large numbers, screened for disease resistance, plant form, chemical content, and so forth, and then diploidized to produce homozygous. The primary requirements for successful anther culture are haploid cells that divide and organize, and a suitable culture environment. Methods used for production of haploids and diploidization may differ slightly according to the plant materials and laboratory objectives. Methods of haploid production were mainly classified into two types: (i) *in vivo*, including spontaneous occurrence of haploids related to polyembryony, pseudogamy, semigamy, and androgenesis (usually 0.001–0.01%) and (2) *in vitro*, including anthers (pollen) culture, unpollinated ovary culture, and chromosome elimination i.e. bulbosum method. Among these techniques *in vitro* induction methods, especially anther culture is simple and efficient. The formation of plants from pollen microspores in this way is sometimes called **androgenesis**. This technique has been variously modified and a great deal of attention has been paid to the production and utilization of haploids in fundamental and applied genetics.

Guha and Maheswari (1964) were able to regenerate haploid plants (gametic "**n**" number) from pollen of *Datura innoxia* by culturing intact anthers. Since then a great deal of research has been devoted to the subject. The occurrence of haploids was reported in 247 species of angiosperms. The basis of pollen and anther culture is that on an appropriate medium, the pollen microspores of some plant species can be induced to give to vegetative cells, instead of pollen grains. This change from a normal gametophytic pattern of development into a vegetative (sporophytic) pattern, appears to be initiated in an early phase of the cell cycle when transcription of genes for the gametophytic is blocked and genes for sporophytic genes are activated. Consequently, callus or pro-embryos are formed. In most of the cases, haploid plants are more readily regenerated by culturing microspores within anthers than by culturing isolated pollen. The presence of the anther wall provides a stimulus to sporophytic development. Species in which haploid plants can be regenerated reliably and at high frequency remain a comparatively small part of the total and they mainly comprise the *Solanaceous* species.

Haploid plants have many uses in genetics and plant breeding research. The protocols used for haploid production are many. From anther culture, an average of 1–3 embryos per anther (in *Brassica napus*) to 190 per anther in some other genotypes has been reported. When plants are regenerated from haploid cells, a haploid plant is produced. Haploid plants are sterile and can produce no seed. However, a spontaneous duplication of chromosomes often occurs within anther culture-derived callus cells, resulting in the production of fertile, doubled haploid plants. Because the two copies of genetic information within such plants are identical, the plants are fully homozygous and breed true. Colchicine can be used to induce polyploidization (diploidization) and offers the possibility of increasing the number of diploid plants produced. Anther culture provides a method for the production of homozygous lines over the course of a few months. The doubled haploid plants resulting from anther culture are homozygous and breed true. Also, because they harbor no hidden traits, the use of doubled haploids for breeding also improves the efficiency with which superior genotypes can be identified.

1. Factors affecting anther culture

Several factors can influence the formation of callus, embryoids, or shoots from anther cultures. The timing for obtaining anthers is critical because the stage of pollen development is crucial to success. In most species, anthers are best cultured when the pollen is unicellular and uninucleate, or just entering the first pollen mitosis (Figure 33.7). For example, a rule of thumb for tobacco plants is that the anthers are ready when the sepals and petals of the flower buds are of the same length. In tomato, this can vary from bud sizes of 3 mm to 6 mm, depending on the growth condition and genotype of the donor plant. Buds need to be pre-treated overnight at 46°F (8°C), cleaned in alcohol and bleach, and the anthers removed, and placed on an

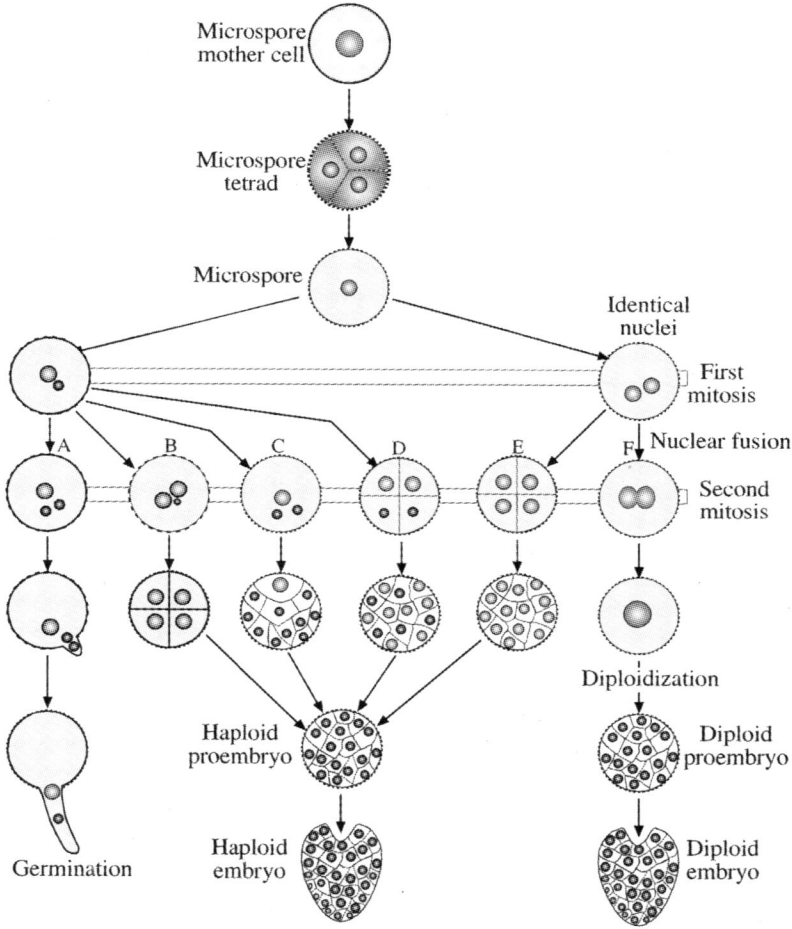

Figure 33.7 *Diagrammatic illustration of microsporogenesis, and various modes of development of pollen under in vivo and in vitro conditions*

appropriate growth medium under fluorescent light (18 hour photoperiod). Other factors are donor plant physiological conditions, basal medium composition, phytohormone balance, plating density, culture environment (e.g. light, photoperiod, temperature, gaseous atmosphere), and donor genotype.

As a result of meiosis in pollen mother cells *in vivo*, pollen tetrads are formed, which are eventually released in the form of microspores. By the first mitosis, a large and diffuse vegetative cell and a small dense generative cell are formed. The former remains quiescent while the latter divides to form sperms (Figure 33.7A). In culture, although androgenesis can be induced in anthers at the tetrad stage or at the binucleate pollen stage, microspores obtained just before, or at the time of the first mitosis, are most suitable for the induction of androgenesis. The early divisions in the responding pollen grains may occur in one of the following ways.

i. The uninucleate pollen divides unequally. The generative cell degenerates immediately or after undergoing one or two divisions. The callus/embryo originates due to successive divisions of the vegetative cell (Figure 33.7B).
ii. In some cases, the pollen embryos originate from the generative cell alone; the vegetative cell either does not divide or divides only to a limited extent forming a suspensor like structure (Figure 33.7C).
iii. For example, in the case of *Datura innoxia*, the uninucleate pollen grains divide unequally, producing generative and vegetative cells. But both these cells divide repeatedly to contribute to the developing embryo (Figure 33.7D).
iv. The uninucleate pollen grain may divide symmetrically to yield two equal daughter cells, both of which undergo further divisions (Figure 33.7E).
v. A homozygous diploid embryo is formed by the fusion of two similar nuclei of the pollen after the first mitosis (Figure 33.7F).

The responsive pollen grains become multicellular and ultimately burst open to release the cell mass. This cell mass may either assume the shape of a globular embryo and undergo the developmental stages of embryogeny, or it may develop into a callus depending on the plant species (Figure 33.8).

2. Applications of anther culture

Anther cultures are mainly developed for obtaining haploid plants. Haploid plants are very useful for the establishment of near isogenic lines (**breed true**) for genetic and plant breeding research. Anther cultures are of considerable value in obtaining homozygous diploids within a short period of time as compared to the several generations required using the conventional whole plant technique. The total number of promising strains developed in different crops exceeds 100 and represents improvements in yield, and/or in disease resistance, cold resistance, maturity duration, adaptability, etc. In general, anthers from the F1 hybrids from selected crosses are cultured to produce haploids. Haploids are sterile and can produce no seed. The chromosomes of these haploids are then doubled to obtain homozygous doubled haploid (**disomic**) plants. However, a spontaneous duplication of chromosomes often occurs within anther culture-derived callus cells, resulting in the production of fertile, doubled haploid plants. This approach is often referred to as **hybrid sorting**, since the heterozygous gene combination present in the hybrids is quickly resolved into homozygous combinations.

3. Protocol for tomato anther culture

Materials

Equipment: Dissecting and binocular microscopes; Laminar flow cabinet; Controlled growth chamber; Culture dishes; Media bottles; Forceps; Pasteur pipettes; Microscope slides, etc.

Solutions: Sodium hypochlorite; Ethanol; HCl; Distilled water sterile; MS medium; Sucrose; Phytohormones (NAA, BAP, and Kinetin); Agar; Sodium hydroxide; Colchicine.

Callus induction medium: A solid MS with NAA (1 mg l^{-1}) and K (1 mg l^{-1})

Shoot generation medium: A solid MS medium containing 2 mg l^{-1} kinetin or BAP should encourage formation of shoots *via* organogenesis.

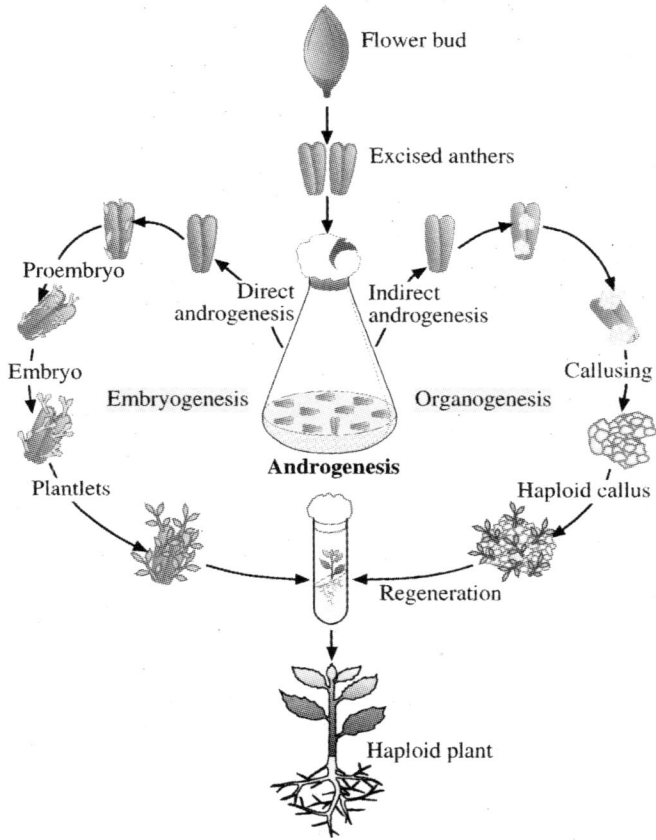

Figure 33.8 *The culture of excised anthers and the development of haploid plants directly by embryo formation, or through haploid callus*

Root induction medium: A solid MS medium with no or 0.1 mg l^{-1} IAA.

Procedure

1. Select buds which are at the correct stage (anthers should be taken from relatively young and healthy plants) and rinse in 70% ethanol for 1 min. Immerse in 1% sodium hypochlorite (with 1 drop of Tween-20) for 15 min. and rinse 3 times in sterile distilled water.
2. To culture anthers, make sure at all stages not to damage the anther wall as this may prevent growth response from the pollen and no callus induction from the anther wall which has the somatic cells.
3. First make an incision on one side of the flower bud and gently take out the stamens with a pair of fine forceps and collect in a sterile Petri dish. Carefully separate each anther from its filaments and plate the anthers from one flower on a callus induction medium. Damaged anthers should be discarded, as they often tend to produce callus from parts other than pollen.

4. The anthers are cultured on an agar-solidified medium in glass tubes or small Petri dishes. They can also be grown in liquid media in Erlenmeyer flasks and kept on a slow rotary shaker.
5. Seal the Petri dish with Parafilm and incubate at 24/32°C, in light about 2000 lux for 14 hour day length. The walls of responsive anthers turn brown and after 3–8 weeks they burst open due to the developing *callus* or embryos. Callus induction can be seen within 21 days. *Callus* originating from pollen is usually yellow or white, whereas callus from other tissues is usually green. This green tissue can be excised and discarded.
6. After 40–54 days on the callus induction medium, the small calli can be transferred to shoot regeneration medium. Depending upon the plant species, it takes about 3–8 weeks for pollen plantlets to emerge from the anthers.
7. When the seedlings (from embryos) or shoots (from callus) become 3–5 cm long, transfer them to autoclaved soil in the same way as other *in vitro* regenerated plantlets. To reduce sudden shock and to prevent desiccation, it is advisable to cover these plantlets with glass beakers, and keep them in a well-lit, humid greenhouse. After one week, remove the beakers, and after a further two weeks, transfer again to larger pots.

Shoot regeneration on callus:
Transfer the anther-derived callus from the induction medium to a shoot regeneration medium with high cytokinin and low/no auxin.

Root regeneration on shoots:
Once shoots have developed and attained about 10 mm, they can be transferred to a medium to encourage root induction. Excise shoots from the callus with a pair of fine forceps and a scalpel, and place cut end onto the solid root induction medium. Root regeneration should be performed in a jar or tube to allow room for growth of the shoot.

Identification of haploids by assessing chromosome number:
Regenerated plants may have the normal diploid chromosome number ($2n = 24$ for tomato), either due to spontaneous chromosome doubling, non-reduction of gametic tissues, or the fact that the plant arose from somatic tissue such as the anther wall or the filament. Therefore, it is necessary to discard those plants, that are not haploid. The best tips to use for chromosome counts are the root tips, which will divide rapidly, or the tips of young leaves. The cytological method used to determine chromosome number is discussed in Chapter 34.

4. Reconstitution of diploids from haploids by chromosome doubling

Several techniques can be used to obtain diploids from haploid plants. Diploidization can be done by two simple methods: (i) Colchicine treatment and (ii) Stem-segment culture.

Colchicine treatment
1. Treat the plantlets, while still attached to the anther, for 24–48 hours with a 0.5% colchicine solution, wash thoroughly, and replant.
   Or
   When plants are about 20 mm tall with well developed roots, carefully beak up the agar with a pair of forceps without damaging the roots.

2. Gently grip the base of the shoot with the forceps and slowly draw the roots from the agar. Wash off excess agar from the roots.
3. Dip the plantlets in a small tube containing 5 ml of 0.5% colchicine for 3 hours.
4. After colchicine treatment, the plantlets can be planted into pots.
   Or
   If plants mature, haploid plants are available. Then colchicines–lanolin paste (0.4%) may be applied to the axils of the leaves.

Diploidization by stem-segment culture
It is known that haploid callus cultures frequently undergo endomitosis to form diploid cells and this property can be exploited to obtain homozygous diploids. Grow a small segment of stem on an auxin–cytokinin medium to induce callus formation. During callus growth, diploid homozygous cells are produced by endomitosis and from these a large number of isogenic diploid plants can be differentiated.

Notes

1. The condition of the donor plant is critical. Use plants grown with optimal fertilization and other growth requirements.
2. Use only the first flush of flowers, because responsiveness generally declines with subsequent cycles of flowering.
3. Identify the optimal developmental stage of the immature pollen for culture by using anther staining procedures.
4. Optimal thermal shock conditions provided to anthers during pre-treatment may vary for the length of time as well as the temperature. It also depends on the species.
5. Medium composition can influence response.
6. Anthers often turn brown during culture. This is a typical response, and does not indicate that the anthers no longer have the potential to respond with callus formation or androgenesis.
7. A few diploids are recovered from anther culture. Colchicine treatments can be used to double the chromosome number in haploid plants. Often diploid plants may also be produced from anther wall or filament tissue and they may not be true breeding.

F. POLLEN CULTURE

Isolated pollen grains, when cultured *in vitro*, give rise to haploid embryos or callus; this approach is called **pollen culture** (Figure 33.9). Difficulties have been encountered in obtaining repeatable results, and anther culture generally gives higher yields in terms of number of plants recovered. About 50 anthers may be placed in 20 ml of medium and squeezed with a glass rod; the solution is filtered through a nylon mesh of suitable pore size (25–100 μm), and centrifuged at 500–800 rpm for 5 min. The pollen pellet is collected, washed twice, and suspended at a final density of 10^3–10^4 pollen/ml. In float culture, excised anthers are floated on a shallow liquid medium in Petri dishes; the anthers dehisce in a few days releasing their pollen grains into the medium. These anthers continue to shed pollen so that their serial subculture yields pollen samples in different stages of androgenesis. Anther cultures need crucial ingredients, some of which are glutamine, L-serine and inositol (for *Solanum* sp. aspargine and glutamine are needed). Androgenesis can be markedly improved by a pre-culture of anthers for 4–7 days before pollen isolation. Pollen culture offers many advantages over anther culture due to elimina-

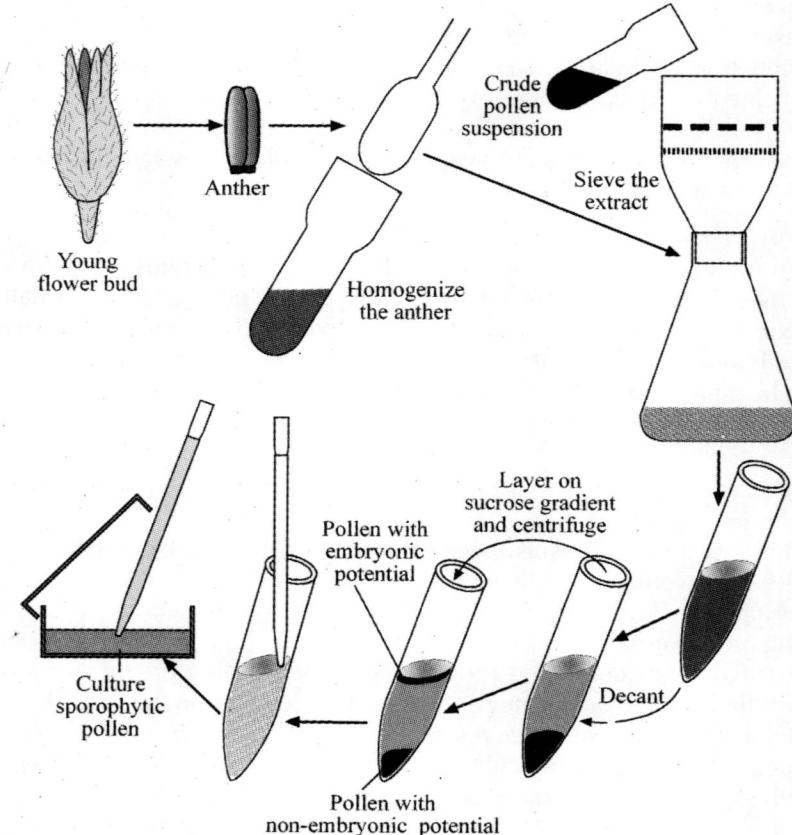

Figure 33.9 *Isolation of sporophytic pollen from gametophytic pollen by gradient centrifugation*

tion of the anther wall. (i) Studies on differentiation and development are easier and more precise, (ii) no callus formation can occur from wall tissue, and (iii) products from different pollen grains ordinarily do not get mixed up (this eliminates the risk of chimeras).

G. SOMATIC EMBRYOGENESIS

The formation of the zygote after fertilization triggers the egg-cell to divide and develop into an embryo (embryogenesis). However, to stimulate the egg to undergo embryogenesis, fertilization is not always essential. As happens in parthenogenesis, the pollination stimulus alone, or simply the application of some growth regulators may induce the egg to undergo embryogenic development. This can happen in any cell of the female gametophyte (embryo sac); even that of the sporophytic tissues around the embryo sac may give rise to an embryo. In nature there is no instance of *ex-ovulo* embryo development.

During the last two decades, considerable information has accumulated to establish the embryogenic potential of somatic plant cells. Virtually every plant organ has been shown to form embryos. It can be fairly inferred from the recent spectacular development in the culture of single cells of higher plants that any diploid cell, in which irreversible differentiation has proceeded too far, if placed in an appropriate medium, develops into an embryo-like structure and produces a complete plant.

1. Types of somatic embryos

Embryos formed in cultures have been variously designated as accessory embryos, adventive embryos, and supernumerary embryos. Kohlenbach (1978) has proposed the following classification of embryos.

i) Zygotic embryos: Those formed by the fertilized egg, or the zygote.
ii) Non-zygotic embryos: Those formed by cells other than the zygote.
   a) Somatic embryos – those formed by the sporophytic cells (except zygote), either *in vitro* or *in vivo*.
      Adventive embryos – somatic embryos arising directly from other embryos or organs, e.g., stem embryos in carrot and buttercup.
   b) Parthenogenetic embryos — those formed by the unfertilized egg.
   c) Androgenetic embryos — those formed by the male gametophyte (microspores, pollen grains).

In this chapter, embryos formed in culture have been referred to as somatic embryos or simply as embryos.

Somatic embryogenesis (SEs) is the development of embryos from cells and tissues within *in vitro* systems. SEs develop through stages similar to zygotic embryos. However, the final size of the cotyledons are usually reduced and there is no development of the endosperm or seed coat. Somatic embryos can arise from two different pathways: (a) adventitious embryogenesis, and (b) induced embryogenesis.

(a) Adventitious somatic embryogenesis. Somatic embryos develop directly from cells or callus which are associated with the explant. These cells have been embryonically predetermined prior to their excision as explants and can be referred to as **embryogenic**. This type can originate from three fundamentally different kinds of explants. **Type 1** includes nucellus or integuments of young ovules of some polyembryonic species including citrus, mango, as well as some monoembryonic species (rubber, apple, cacao, and grape). Type 1 reproduces the genotype of the mother plant as in apomixis. **Type 2** includes the **embryonal-suspensor mass** (**ESM**) at the earlier stage that precedes embryo development. Its appearance is white and mucilaginous, and it stains red with acetocarmine. **Type 3** includes the developing zygotic embryo at various stages of development or seedling tissue as the explant source. This is a common pattern of somatic embryogenesis in a wide variety of plants.

(b) Induced somatic embryogenesis: This type of embryogenesis results from callus and cell suspensions but only if the tissue is subjected to conditioning to induce embryogenic competence. Although embryogenic cell suspension cultures outwardly appear to be callus, on closer inspection, these cell masses are well organized as pro-embryogenic (Figure 33.10) masses (PEMs). PEMs continue to develop in suspension cultures until they are moved to a stationary medium to develop into mature somatic embryos. PEMs are often passed through sizing screens to get uniformity and synchrony of development.

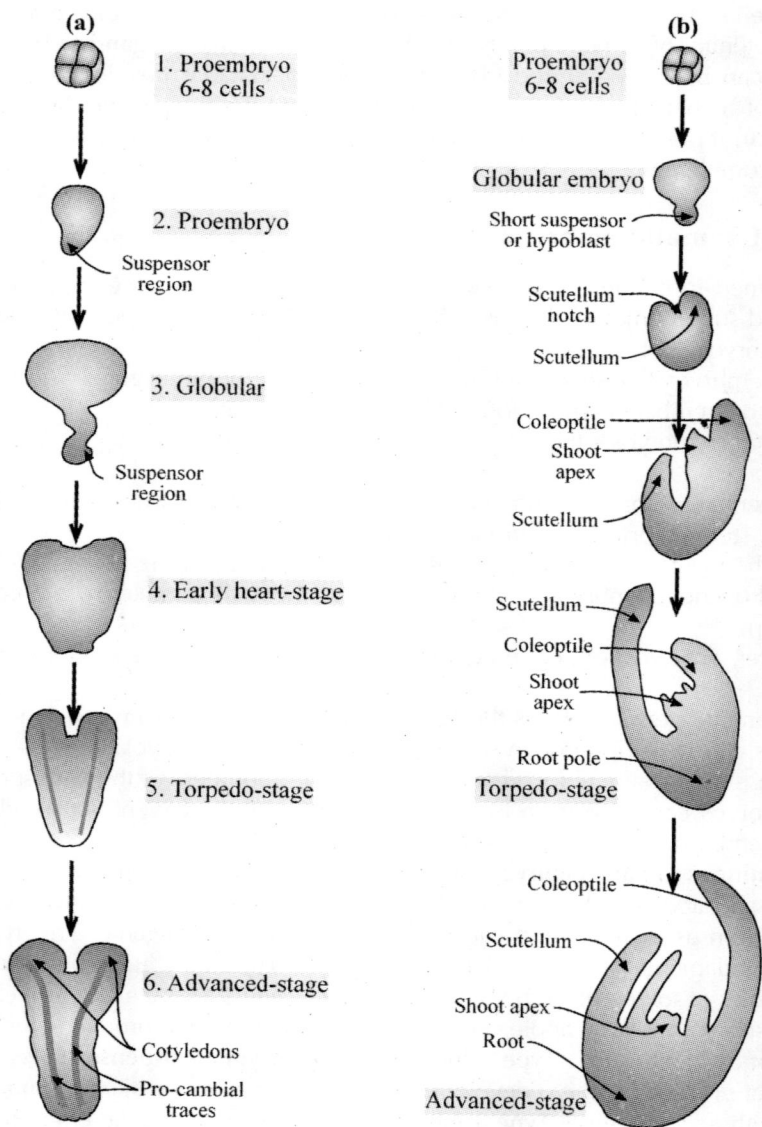

Figure 33.10 *Stages of somatic embryogenesis in (a) Dicotyledon and (b) Monocotyledons*

2. Factors affecting somatic embryogenesis

A number of factors have been shown to have a profound influence on somatic embryogenesis. The work to date has highlighted the importance of two media constituents in somatic embryogenesis, growth factors (e.g. auxin), source of nitrogen, and other factors.

(a) **Growth factors**: *In vitro* development of somatic embryos (SE) in carrot is a two-step process, each requiring a different medium. The callus is initiated and multiplied in an auxin-rich medium. The auxin generally used is 2,4-D in the range 0.5–1 mg^{-1}. On such a medium (proliferation medium), the callus differentiates into localized groups of meristematic cells called "embryogenic clumps" (EC). In repeated subcultures on the proliferation medium, the ECs continue to multiply without the appearance of mature embryos. However, if the ECs are transferred to a medium with a very low level (0.01–0.1 mg^{-1}) or no auxin at all (embryo development medium), they develop into mature embryos. The presence of an auxin in the proliferation medium seems essential for the tissue to develop embryos. Tissue maintained continuously in an auxin-free medium will not form embryos. In this respect, the proliferation medium could be regarded as the "induction medium" for SEs and each EC a disorganized embryo.

(b) **Nitrogen source**: Besides auxin, the form of nitrogen in the medium significantly affects *in vitro* embryogenesis. In the cultures of wild carrot raised from petiolar segments, embryo development occurred only if the medium contained some amount of reduced nitrogen. The calli initiated on a medium with KNO_3 as the sole source of nitrogen failed to form embryos upon removal of the auxin. However, the addition of a small amount (5 mmol^{-1}) of nitrogen in the form of NH_4Cl in the presence of 55 mmol^{-1} KNO_3 allowed embryo development. For the occurrence of embryogenesis in the cultured cells of carrot, the presence of a minimal amount of endogenous NH_4 is essential. This level of NH_4 within a cell is attained with a very low level of exogenous NH_4^+ (2.5 mmol l^{-1}).

(c) **Other factors**: Many other factors also affect embryogenesis. High potassium (20 mmol^{-1}) is necessary for embryogenesis in wild carrot. The amount of dissolved oxygen (DO) in the medium should be below the critical level of 1.5 mg l^{-1} to allow embryo development; higher levels of DO favor rooting. Embryogenesis in carrot can also be improved by activated charcoal.

Procedure

1. Selection and culture of appropriate explant material:
The selection of explant material is the most critical decision and may require a systemic analysis of the embryogenic potential of different explant sources within the plant. The first step is the production of the callus, cell suspension, or protoplast material by methods that have already been described.

2. Induction of embryogenic potential in the cell explants:
Induction is necessary for non-embryogenically determined cells and explants. Induction is achieved through the transfer of cells to a basal medium with a high concentration of auxin. The most effective auxins are 2,4-D or coconut milk plus low concentrations of NAA. After 1–2 weeks some small pro-embryos may be separated by different-sized screens for transferring to the differentiation medium. Smaller cells may be subcultured for continued production of somatic embryos.

3. Differentiation and maturation of somatic embryos:
After induction of the embryogenic potential on an auxin-containing medium, pro-embryo masses are shifted to an auxin-free basal medium, high in ammonium nitrogen. Somatic embryos arise from single cells in clumps or small masses, develop polarity, and follow a pattern mimicking normal zygotic embryogenesis. Development may be variable in rate and abnormal in appearance, with secondary embryos forming on primary embryos. Synchrony and normalized development may be obtained by separating PEMs into different sizes by screening or density gradients. Adding ABA to the medium has improved uniformity and promoted normal development. Dehydration may also induce embryo maturity similar to that produced by *in vivo* embryo development.

4. Plantlets formation:
Mature somatic embryos that have reached a "normal" size can be plated onto an agar medium devoid of any auxin but containing a low level of cytokinin.

5. Transplanting:
After the leaves and roots have formed, the plantlet can be transplanted to a medium in a container and handled just like any other seedling plant or plantlet.

3. Protocols for inducing somatic embryogenesis in cultures

Method-1: Somatic embryogenesis in carrot (Daucus carota) seeds

1. Surface sterilize the seeds in 10% calcium hypochlorite for 15 min. and wash three times in sterile distilled water.
2. Germinate the seeds on sterilized moistened filter paper in Petri dishes, in the dark, at 25°C.
3. Cut 1-cm long segments of roots from 7-day-old seedlings and culture them individually on a semisolid medium containing the inorganic salts of MS medium, organic constituents of White's medium, 100 mg l^{-1} my-inositol, 0.2 mg l^{-1} kinetin, 0.1 mg l^{-1} 2,4–D, 2% sucrose, and 1% bacto-agar. Incubate the cultures in the dark.
4. Six to eight weeks later, transfer pieces of root calli (0.2 g fresh weight) to fresh medium of the original composition and maintain the cultures in the light at 25°C. The tissues may be multiplied by subculturing every four weeks in a similar manner.
5. After the first passage, initiate suspension cultures by transferring ca. 0.2 g of callus tissue to a 200 ml Erlenmeyer flask containing 20–25 ml of liquid medium of the same composition as used for callus growth (without agar).
6. Incubate the flasks on a horizontal rotary shaker with 100 rpm, in the light at 25°C.
7. Subculture the suspensions every four weeks by transferring 5 ml to 65 ml of fresh medium (1:13).
8. To induce somatic embryogenesis, transfer callus pieces or portions of suspensions to 2,5-D-free medium of otherwise the same composition as used before.
9. After 3 to 4 weeks the cultures would contain numerous embryos in different stages of development.

Method-2: Protocol for carrot cells in suspension for embryo development

1. Decant the cell suspension medium and replace with a fresh MS liquid medium.
2. Continue agitation on the shaker.
3. In two weeks, decant the medium again.
4. Spread the cells on a half-strength MS medium with agar and without hormones in test tubes or Petri dishes. Continue from step 3 of previous protocol.

4. Isolation and fractionation of embryos

Embryogenically competent suspensions of carrot containing mixed populations of materials can be separated as follows.

Procedure

1. Flame the neck of the culture flask and cool for 30 seconds.

2. Pour the contents of the flask through the #20 screen into the beaker. The largest somatic embryos and plantlets are retained on this size sieve.
3. Wash the sieve with a rinse medium consisting of salts and vitamins and sucrose to flush out any smaller embryos remaining in the larger mass. Usually these larger embryos retained on the #20 screen are discarded.
4. After a few min., most of the embryos in the filtrate settle to the bottom of the beaker. Some of the upper filtrate can be decanted so that the volume of the filtrate does not become too large.
5. Repeat the same procedure of sieving by pouring the filtrate from the #20 sieve through a #40 sieve (380 µm) which also rests on a 250 ml beaker. The material retained on the #40 sieve is washed with rinse medium.
6. Collect the filtrate in the beaker, cover with foil, and set aside. Flip over the #40 sieve with retained embryos and place on a sterile 400 ml glass beaker.
7. Pour rinse medium through this sieve to rinse off these embryos into the 400 ml beaker.
8. Pour the embryos into a conical centrifuge tube and label. A small aliquot of these embryos can be subjected to observation under an inverted microscope.
9. Pass the filtrate in the beaker from the #40 sieve then through a #60 sieve in a similar manner and rinse with fresh medium.
10. Set filtrate aside again and flip over the sieve with retained embryos again. Collect the embryos in a 400 ml beaker, then pour into a centrifuge tube and label #40–60.
11. Repeat this procedure with #80 (190 µm) sieve and then a #100 (140 µm) sieve within embryos collected from each sieve, yielding #60–80 and #80–100 fraction.

The fraction collected from < #100 comprise small embryos at the heart stage, as well as smaller units and cells. This is done by centrifugation at about 300 rpm for 10 min. The fraction of the larger sizes generally settles quickly and completely, and thus does not require centrifugation. Re-suspend the embryos in a fresh medium without hormones or coconut water. The embryo concentration at this point can be quantified by counting the aliquots under the microscope and the concentration adjusted by adding a fresh medium.

5. Culture of embryos

The somatic embryos can be experimented with in various media and culture vessels, for biochemical analysis, etc. A typical experiment in a semisolid medium in small plastic disposable Petri dishes may be performed by mixing 6 ml of somatic embryos taken from a centrifuge tube with 24 ml of medium to which autoclaved agar-containing medium has been added. Other additions may be tested as desired.

6. Applications of somatic embryogenesis

(a) Mass propagation: Since embryos are produced from somatic tissue and have the genotype of the plant from which they were obtained, this creates the potential for mass clonal propagation, and (b) genetic improvement programs of plant cultivars: Somatic embryogenesis could be useful for isolating somaclonal genetic variation within populations of cells. Somatic embryos develop from a few cells (often single cells). This makes them attractive targets for genetic transformation.

H. SYNTHETIC SEEDS

True seeds are the products of fertilized ovules, consisting of a zygotic embryo enclosed by a protective coat. The zygotic embryo in the seed is an important part, which grows into a seedling by germination. The concept of synthetic or artificial seeds has been developed from somatic embryoids, which are formed adventitiously from *in vitro* cultured somatic tissue. In contrast to zygotic embryos, somatic embryos are not enclosed by seed coats. Therefore, scientists tried to develop a technique by which isolated somatic embryoids could be encapsulated by a protective gel-like substance so that embryoids could survive and would not desiccate even after being planted in the soil. Such encapsulated embryoids could be used as **artificial seeds**.

1. Methods for making synthetic seeds

Several steps are followed for making artificial seeds.
1. *Establishment of callus culture*: Refer to the previous sections for initiation of callus culture for carrots.
2. *Induction of somatic embryos*: Discussed in detail in previous section.
3. *Maturation of somatic embryos*: Maturation of somatic embryos means the completion of embryo development through some stages. Initially, embryos develop through the globular-shaped stage, then the heart-shaped stage and finally, the torpedo-shaped stage (Figure 33.10). In the final stage, the embryo attains maturity and develops the opposite poles for shoot and root development at the two extremities. Under optimal conditions, this embryo germinates to produce plantlets. However, in some plant species, such sequential development may not be followed and they may need some treatment (cold treatment, application of GA).
4. *Encapsulation of somatic embryos*: Water soluble hydrogels have been found suitable for making artificial seeds. Two standard methods have been used to coat somatic embryos: (i) gel complexation via a dropping procedure and (ii) molding. In the first method, isolated somatic embryos (Figure 33.11) are mixed with 0.5 to 5% (W/V) sodium alignate and dropped into 30–100 µM calcium nitrate solutions. Surface complexation begins immediately and the drops are gelled completely within 30 min. (Figure 33.12). In the second method, isolated somatic embryos are mixed in a temperature-dependent gel such as Gelrite™ and placed in the well of a microtiter plate. This forms gel when the temperature is cooled down.
5. *Test for embryoid to plant conversion*: After encapsulation, the effect of coating on somatic embryos is initially, very difficult to assess because the germination and continued development of the encapsulated embryos are sometimes very inconsistent after being planted in the field. So, to overcome this problem, embryo response in terms of embryo to plant development or conversion is tested under aseptic conditions. The embryo conversion frequency is the percent of somatic embryos that produce green-plants having a normal phenotype. Embryo to plant conversion includes the following steps. (1) Place the encapsulated embryos aseptically on simply agar medium with minimal nutrients. (2) Check somatic embryos for the uniform germination, growth, and development of root and shoot system.

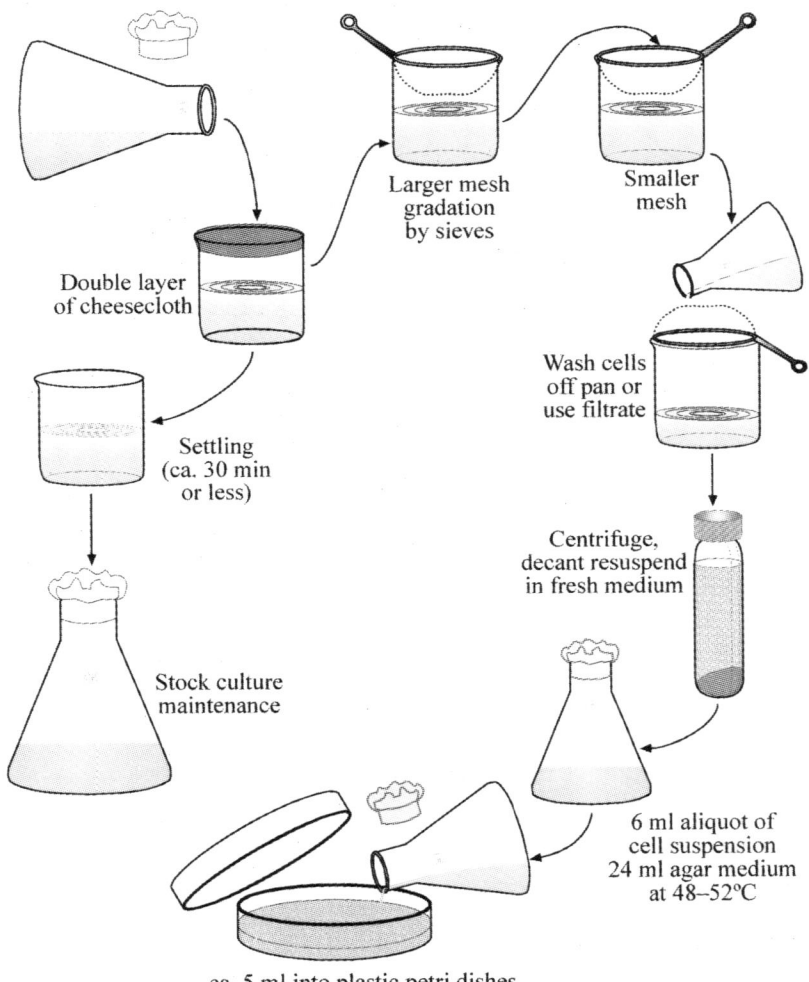

Figure 33.11 *Isolation and fractionation of embryos*

6. *Greenhouse and field planting*: The final assessment will be the greenhouse or field performance of artificial seed and their yield in comparison to plants derived from true seed.

2. Storage of somatic embryos

Storage of artificial seeds is a great limitation. When artificial seeds are stored at low temperatures the conversion percentage drops drastically. The limited storage time of artificial seeds is probably due to an anaerobic environment in the capsule. Since the somatic embryos are not developmentally arrested, they continue very active respiration. These problems are to some extent overcome by using small capsule volume, including a growth control agent in the encapsulation medium.

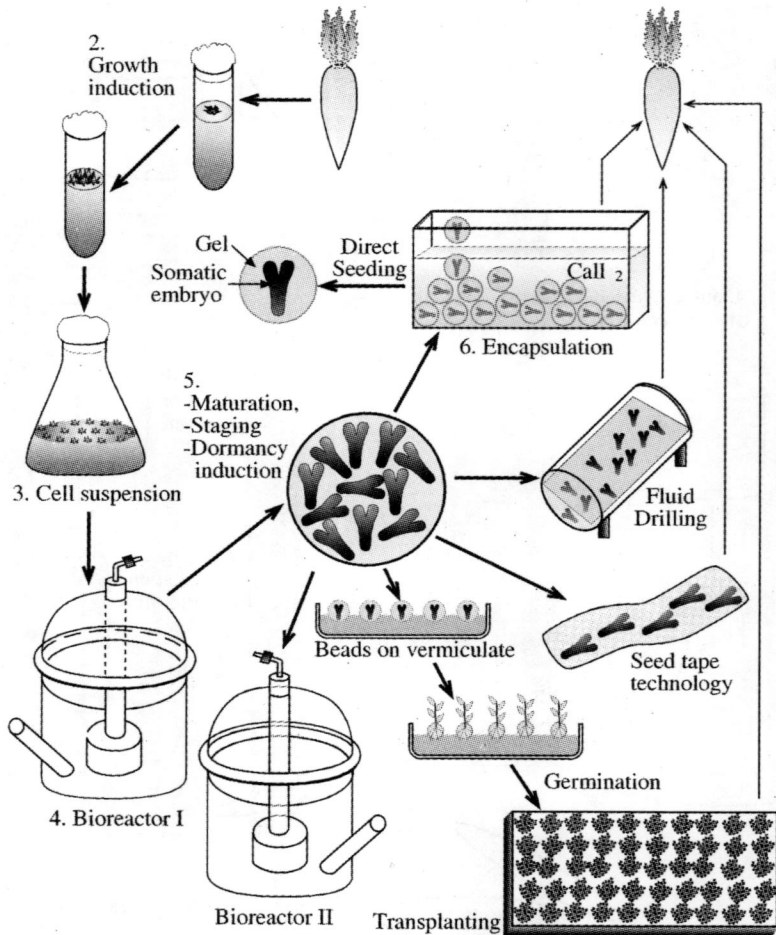

Figure 33.12 *Schemes for various propagation via embryogenesis and synthetic seed formation*

III FURTHER READING

Anonymous (1983) *Cell and Tissue Culture Techniques for Cereal Crop Improvement*, Science Press, Beijing, China, IRRI.Philippines.

Gamborg OL and Phillips GC (Eds) (1995) *Plant Cell, Tissue and Organ Culture*, Narosa Publishing House.

Guha S. and Maheshwari SC (1964) In Vitro Production of Embryos from Anthers of Datura, *Nature*, London **284**:497.

Jensen CJ (1974) Chromosome Doubling Techniques in Haploids. pp153–190. In KJ Kasha (Ed) *Haploids in higher plants: advances and potential; proceedings of the first international symposium*, Guelph, Ontario, Canada, June 10–14, 1974. University of Guelph, Guelph.

Kohlenbach SW (1978) Comparative Somatic Embryogenesis Frontiers of Plant Tissue Culture (Ed) Thorpe TA., International Association Plant Tissue Culture, Calgary pp.59–74.

Kumar De K (1997) *Plant Tissue Culture*, New Central book Agency (P) Ltd. Calcutta.

Murashige T and Skoog F (1962) A revised medium for rapid growth and bioassays with tobacco tissue cultures, *Plant Physiology*, Lancaster **15**:473–497.

Nitsch C (1977) Culture of Isolated Microspores. pp268–278, In Reinert and P Bajaj (Eds) *Plant cell, tissue and organ culture*, Springer Verlag, New York.

34

Cryopreservation and Distribution of Clonal Material

I INTRODUCTION

Wild species and their genetic variations constitute a pool of genetic diversity (germplasm or gene bank) which is invaluable for breeding programs. Due to the gradual disappearance of economic and rare plant species, the storage of genetic resources of plant species has became necessary. Conventionally, the cryobiology of plants was followed for the preservation of fruits, vegetables, and various plant products mainly, to increase the shelf-life and also for long-term storage. However, most of the information available today stems from the work on animals and their cell cultures including reports showing the revival of cells, larvae, and caterpillars. Cell cultures can be grown in synthetic media. The cells are not dependent on one another for their growth, unlike cells in an organized tissue or an organ which are interdependent. Plant cells, unlike animal cells, have the further advantage of totipotency and whole plants can be regenerated from them.

It has been estimated that the survival of more than 9,000 species of plants is in some way under threat and that the majority of these are from tropical regions. In addition, many new cultivars are replacing the primitive or conventionally used agriculture crops. Thus it is becoming important that displaced crops must be documented and conserved. Global climatic changes also affect the natural strategies. Plant material can be stored both for future consumption and as a resource for further propagation. *In situ conservation* is generally recommended. This method of conservation mainly aims at preservation of land races with wild relatives in which genetic diversity exists and/or in which the weedy/wild forms present hybridize with related cultivars. The *in situ* conservation of habitats has received high priority in the world conservation strategy programs. Institutional arrangements, especially in countries of the developing world, have been emphasized. The limitation of this method is the risk of material being lost due to environmental hazards. Further the cost of maintaining a large proportion of available genotypes in nurseries or fields may be extremely high.

Ex situ conservation is the chief mode of preservation of genetic resources, and may include both cultivated and wild material. Generally, seeds or *in vitro* maintained plant cells, tissues, and organs are preserved under appropriate conditions for long-term storage as gene banks.

This technique requires considerable knowledge of the genetic structure of populations, sampling techniques, methods of regeneration, and maintenance of varietal gene pools, particularly in cross-pollinated plants. Tissue culture techniques provide a novel way of storing the plant material needed for many purposes (Figure 34.1) connected with propagation and preservation (short- or long-term). The technology for the freeze preservation of plant cell cultures and the eventual regeneration of plants from them involves the following phases. (1) Raising the cell and tissue cultures aseptically; (2) adding cryoprotectants to avoid freeze-damage; (3) subjecting cell cultures initially to super-low temperatures and freezing them in liquid nitrogen for long-term storage; (4) thawing or rapid re-warming of cells and removal of cryoprotectant by repeated washing; (5) determination of viability; (6) reculturing of the retrieved cells; and (7) inducing of growth and regeneration of plants (Figure 34.2).

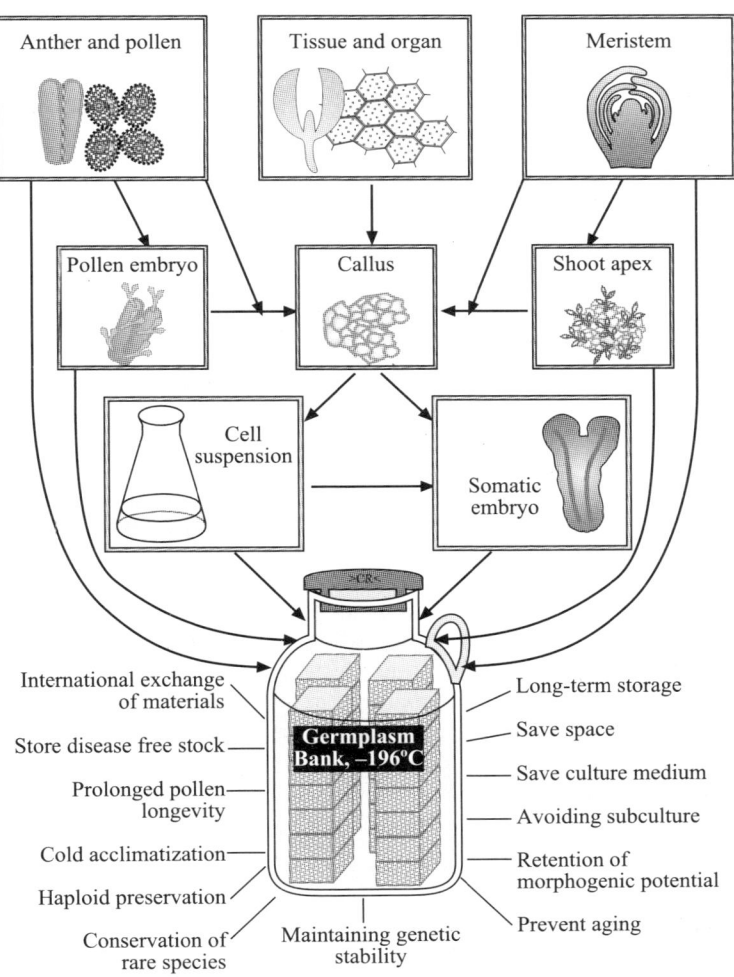

Figure 34.1 *Potentials and prospects of cryopreservation of plant material and establishment of 'Germplasm Bank'*

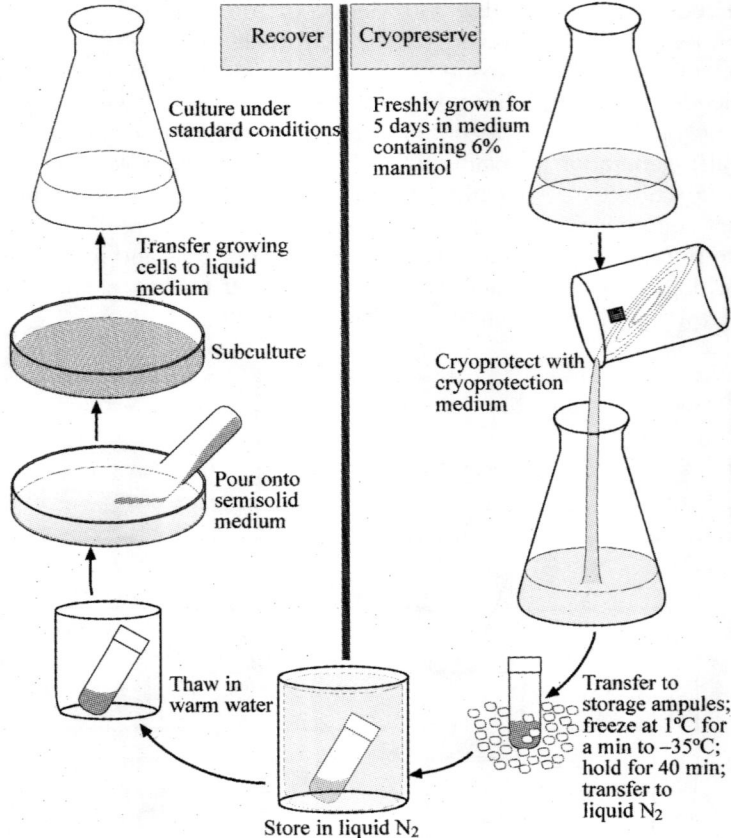

Figure 34.2 *A general procedure for cryopreservation and recovery of cell cultures*

In vitro methods are the most useful for conservation of vegetatively propagated plants, species with recalcitrant seeds, and genetically engineered materials. Some advantages of *in vitro* conservation include: (a) Little space is required for preservation of a large number of clonally multiplied plants, (b) maintenance of the material in an environment free of pests or pathogens, (c) there is protection against natural environmental hazards, (d) pure nucleus stock is available to propagate a large number of plants rapidly whenever necessary, and (e) the obstacles generally imposed by quarantine systems on the movement of live plants across national boundaries are minimized since they are raised and maintained in an aseptic environment.

II STORAGE TECHNIQUES

The most familiar procedure to initiate and maintain a culture involves introducing a piece of plant material into an aseptic culture, stimulating growth or development, and periodically transferring the culture to a fresh medium. For many cultures, the time between a transfer or subculturing can be divided into a lag phase, during which no increase in biomass is detected; a rapid exponential growth phase, followed by a stationary phase brought about by one or more limiting factors in the culture medium. Timely subculturing minimizes the stationary phase and helps avoid the risks of stress induced by prolonged exposure to suboptimal culture conditions. Under normal culture conditions, the subculturing interval may be as short as a few days or as long as several months, depending on the type of culture and species. Simply maintaining continuous growth may provide adequate short-term storage. However, this method cannot meet all requirements. It is necessary to seek means of either slowing growth or suspending it entirely in a storage medium at lower temepratures.

A. SHORT-TERM STORAGE

1. Desiccation of seed embryos

Since the 1980s, much research has been devoted to methods to sow somatic embryos directly into the soil. There has been limited success in sowing dried naked embryos or embryos surrounded by a protective capsule. Somatic embryos under storage should be able to survive as long as true seeds or for at least short-term storage, which will allow delays between production and sowing. A short term of dormancy can be induced in somatic embryos by desiccation. Attempts to induce dormancy in encapsulated somatic embryos have included the use of abscisic acid (ABA) and drying.

2. Short- to medium-term storage

Under special conditions, the genotype of plants propagated by node or shoot culture can be preserved without change. But this method is costly, because it involves frequent subcultures. Consequently, there is a high risk of losing material due to equipment failure and microbial contamination. By this method, cultures (single shoots, unrooted shoot clusters, somatic embryos, or rooted plantlets) can be stored generally between 6 and 24 months depending on the species.

3. Minimal growth techniques

There are different methods in which the life of cultured tissues can be extended.

(a) **Laissez-faire:** This is the simplest method, where cultures are left untouched on the shelf of the growth room. Their growth slows as the medium depletes. Shoots or tissues stay alive until they become desiccated or until the medium is exhausted or becomes toxic. Drying can be prevented or minimized by using appropriate closures. Tubes should be sealed with caps, or film, which allow gaseous exchange but prevent the escape of water vapor.

(b) **Control of temperature and light:** Intervals between transfers can often be greatly reduced by keeping cultures in weak light (or in the dark) and at a lower than optimal temperature. A cold room (close to 0°C) with low wattage lighting (a flux density of ca. 10–35 µmol $m^{-2}s^{-1}$; day length 12–16 hours or shorter) or darker for some species can be used for storage. Shoots, plantlets, and shoot-bearing cultures have been most frequently stored in this way.

(c) **Reduced oxygen tension:** Placing cultures in an environment with a low partial pressure of oxygen appears to have the potential for limiting their *in vitro* growth, and reduction in oxygen tension plays a role in the successful storage of cultures in tightly sealed vessels. Many methods have been developed to reduce oxygen by using overlays (overlay of mineral oil on callus cultures, using dissolving autoclaved silicone in water, or liquid paraffin oil). Reduced partial pressure of oxygen can be achieved by reduction in atmospheric pressure or by replacing some of the oxygen in the air at normal atmospheric pressure with inert gas (nitrogen gas).

However, one very simple way to reduce the supply of oxygen reaching tissues and organs is to place them in a long narrow vessel in which the distance between the culture and the point of gas exchange with the external atmosphere is the maximum. As the diffusion distance is increased, the rate of tissue growth decreases and this allows the frequency of subculturing to be reduced.

4. Altering the constituents of the culture medium

(a) Decreasing the carbohydrate/nutrient supply: The growth of a culture slows as the constituents of the medium are used up, but a spent medium which may contain toxic products and have an unfavorable pH may not be an ideal substrate on which to store shoots or plantlets for a considerable period. One containing a comparatively low level of sugar (low or no sucrose) could result in a reduced growth rate.

(b) Decreasing the supply of inorganic nutrients: The restricted growth of many plantlets (e.g. coffee, tomato) has been achieved by reduced supply of inorganic salts (3/4MS with 5% sucrose).

(c) Changing the osmotic potential: High sucrose levels can be used to maintain cultures in a dormant condition for long periods. This appears to be an osmotic effect. Sugars at high concentrations prevent the growth of somatic embryos.

5. The use of growth regulators

Growth regulators are normally added to culture media to promote and regulate plant growth *in vitro*. Withdrawal of these chemicals can assist in the storage of certain plants. The growth regulator abscisic acid (ABA, and others) can induce dormancy in plant meristems. The growth of somatic embryos can be arrested by 0.1 mM ABA. Somatic embryos can be stored for many weeks on ABA-containing media without germination or deterioration.

6. Treatments in combination

Minimal growth may be effectively induced by a combination of treatments. For example, shoot cultures of Pelargonium stock plants can be maintained at 12°C on a medium without growth regulators. Potato shoot cultures can survive at least 57 months when cultured at cool temperatures. This has been enhanced by the addition of ABA (5–10 mg/l), high sucrose (up to 8%) and/or mannitol (3–6%), to the medium. There may be alternative ways of storing cultures of any particular plant and combinations, which are equally effective.

7. Storage of autotrophic plantlets

It has been suggested that by using gas permeable vessels, it might be possible to grow plants *in vitro* that are capable of photosynthesis. This might be a convenient method for storing genotypes, provided the plants can be miniaturized and their growth slowed.

8. Applications

Minimal growth storage is clearly useful for the preservation of clones which are required as stocks for continued propagation *in vivo* or *in vitro*, or as parents in plant breeding programs. However, these techniques are not ideal for the long-term storage of genotypes.

B. LONG-TERM STORAGE (CRYOPRESERVATION)

Conventional dry seed storage is now widely used in germplasm repositories as an effective means of storing many plant genotypes. However, for plants with seeds which do not tolerate desiccation, and for plants, which are normally vegetatively propagated, it would clearly be desirable to have a reliable alternative method of storing important genotypes. Usually, the withdrawal of water should permit plant tissues other than seeds and zygotic embryos, to be stored.

Freezing offers an alternative method of making water unavailable to metabolic processes. By converting water into the solid or vitrified phase, at temperatures which are sufficiently low to create an absence of free energy available for molecular movement, biochemical reactions and diffusion are arrested. Therefore, in theory, living cells kept at ultra-low temperatures (**cryopreservation**) should remain indefinitely in suspended animation, without losing viability.

Conventional deep freezers are not cold enough to be suitable for long-term storage. In experimental cryogenics, biological specimens are usually stored in liquid nitrogen at $-196°C$ or, less frequently, in the vapor above liquid nitrogen (ca. $150°C$). Some seeds can be cryopreserved successfully, but it is more difficult to freeze and later retrieve without harm the hydrated plant cells found in somatic tissues. As success only results from freezing and thawing specimens evenly and rapidly, storage must be limited to cells samples, small pieces of tissue, or small organs. Compact and densely cytoplasmic cells with a low water content have a better survival potential than vacuolated ones. Aseptic techniques and sterile containers must be used in all procedures.

Cells taken from suspension cultures are suitable subjects for cryopreservation, but as has been explained earlier, they are prone to genetic change. The freeze storage of most cultures of this kind does not, therefore, provide a suitable method for germplasm preservation and needs to be treated with caution when the ultimate objective is to regenerate plants that are genetically identical to the plant from which the tissues or cells were originally derived. Problems with intrinsic genetic variability should be avoided by preserving shoot tip explants. Genetic variation arising from aberrant divisions of cell nuclei could occur during tissue culture before or after cryopreservation. During the storage phase, some genetic variation could be induced by the accumulation of mutations caused by background ionizing radiation.

1. Factors affecting survival

The accumulated information so far on storage conditions has revealed that the degree of success for the revival of plant cells subjected to ultra-low temperatures depends on a number of factors, the most critical being: (1) age, nature, and density of cells; (2) cryoprotective agent; (3) rate of cooling; (4) method of thawing; and (5) storage temperature.

(a) Age, nature, and density of the cells

The morphological and physiological conditions of the plant cell cultures considerably influence their reaction to cooling, and the extent of injury caused depends upon; (1) *age, nature, and density of cells*: periodically transferring and actively growing young suspensions of cells is much better than older cultures. Highly vacuolated cells are easily killed due to accumulation and bursting of ice crystals within the vacuoles. (2) *Water content*: higher the water content of a cell, more susceptible it is to freezing injury. (3) *Single cells vs clumps*: actively growing cell clumps or colonies of cells have shown better cell viability. (4) *Density of cells*: the density of the cells or biomass per ampoule is very critical. The ampoules containing thick and packed suspensions often show higher cell survival. (5) *Degree of cold hardiness*: the sensitivity of cells to the cold might also depend on the degree of frost hardiness of a plant.

(b) The Cryoprotectants

Dehydration of cells to an optimum level significantly promotes the chances of the material surviving freezing and thawing. In addition, it offers flexibility in the subsequent operation. When the water content of the cells has been adequately removed, the thawing rate is less critical. As a result of dehydration of cells before or during freezing, the intracellular concentration of the solutes increases before freezing of the protoplasm occurs. Various chemicals exhibit protective properties when biological samples are chilled to extreme temperatures. Such 'cryoprotectants' have little in common except a high solubility in water. Even though their mode of action is not entirely clear, it may depend on their ability both to reduce the concentration of salts within the cell, and to restrict the growth of damaging ice crystals. Usually it is advantageous to include one or more cryoprotectants in the cryopreservation protocol. An efficient cryoprotective agent should: (1) have low molecular weight, (2) be easily miscible with the solvent, (3) be non-toxic even at low concentrations, (4) be easily washable from the cells, and (5) permeate rapidly into the cells. Commonly used cryoprotectants are DMSO, glycerol, mannitol, sorbitol, and (poly)ethylene glycol, either alone or in combination. Proline (generally accumulates during stress conditions in plants), can give greater tolerance to poor thaw recovery conditions than other protectants. Sugars, amino acids, and methanol are also sometimes employed. The most appropriate cryoprotectant depends on the kind of plant being treated. So far, DMSO (5–15%) is the most popular. The total concentration of the protectants is usually 1–2 M.

(c) Cooling

The sensitivity of plant cells to low temperature varies considerably with the species. Therefore, no single method can be considered as being universally applicable. Generally, the material to be stored is suspended in the culture medium, treated with a suitable cryoprotectant and is transformed to sterile, polypropylene ampules with a screw cap and frozen by one of the following methods.

(i) Cold conditioning: When plants are subjected to low temperatures, compounds can accumulate, and changes in membrane structure occur that enable constituent cells to withstand subsequent freeze injury. Plants vary considerably in their liability to frost injury. In many species, conditioning mother plants by keeping them in the cold (e.g. 1–4°C) for several days is found to increase the probability of successful cryopreservation of callus cultures and shoot apical meristems.

(ii) Preventing injury during freezing: Nearly all the damage that may result to cells during transfer to a very low temperature or subsequent thawing, is a consequence of their water content. A possible exception is the damage caused by interference with biochemical or biophysical pathways during temperature reduction. Living cells are seriously harmed by the internal formation of ice crystals. Thus the main object of cryopreservation protocols must be to change the physical state of the water within the cells. The transformation of water from a liquid to a solid (glass-like) phase is called *vitrification*. Various precautions taken during freezing and thawing are as follows. (1) Slow freezing: One effective method of cryopreservation is to freeze material at ultra-low temperature at a slow and controlled rate to between –30°C and –60°C, before plunging it into liquid nitrogen. Slow cooling gives rise to cellular dehydration which progressively concentrates the cellular constituents and depresses their freezing point. Once cell solutions have been concentrated by slow cooling, the remaining water can usually be vitrified by freezing the plant material as rapidly as possible, thus avoiding further damage. In slow cooling protocols, material is cooled and frozen in a solution consisting of the medium plus cryoprotectants.

(iii) Step-wise cooling: As an alternative to gradual slow pre-cooling, specimens which have been treated with cryoprotectant chemicals are subjected to a series of sub-zero temperatures, before being plunged into liquid nitrogen (e.g. at 3 min. intervals at –15, –23, –30, –40, –50, and –70°C). But the stepwise freezing procedure has been found to vary depending upon the species. Cells which have dense cytoplasm and a relatively low water content can be preserved by rapid freezing. Small plant meristems have sometimes been stored in a viable condition when they have been cooled in the vapor above liquid nitrogen and then plunged into the liquid nitrogen.

(d) Storage

Maintaining the frozen material at the correct temperature is as important as proper freezing itself. Successful cryopreservation should enable valuable germplasm to be preserved for long periods. In an ideal protocol, frozen tissues should retain their viability over many years. But this varies from species to species. When storing large numbers of germplasm, it is vital to follow an efficient recording system. This not only facilitates what has been stored and for how long, but also reduces the time the other samples are exposed to ambient temperature when trying to remove a particular sample.

(e) Thawing

Reversal of the freezing process also has attendant hazards. The rates of warming considerably influence the LT50 of the cells. At around –130°C, ice structure can change from vitreous to crystalline. Ice crystals may damage if they appeared during freezing. At about –70°C and above, ice crystals may form and grow on the surface of smaller pre-existing crystals by a process of recrystallization. Slow thawing at room temperature has generally proved fatal. For these and other reasons (some reports contradict each other) very rapid thawing of frozen specimens is usually most successful in preserving viability. Ampoules are usually placed in water at 35–40°C until thawed and thereafter transferred to room temperature.

(f) Viability testing

A qualitative assessment of structural damage can be made by direct microscopic examination. Loss of osmotic property can be detected by treatment with hypertonic or hypotonic solutions. The true physiological condition is probably best assessed with the combined results from a wide range of tests and, ultimately, the ability of the cells or organ to grow and regenerate.

Fluorescein diacetate staining: This method is based on the fact that living cells stained with fluorescein diacetate show fluorescence under UV light, while the dead cells do not. This is a very simple staining method. First a drop of 0.1% fluorescein diacetate is mixed with a drop of cell suspension and allowed to stand for 5 min. Then this is observed under the tungsten light and the number of free cells or cell aggregates are counted in one field. The light source is then changed to UV and the percentage of cell survival is estimated by the number of cells showing fluorescence.

Triphenyl tetrazolium chloride method: In this method, cell survival is estimated by the amount of formazan produced as a result of reduction of triphenyl tetrazolium chloride (TTC), which gives a pink color.

Materials

Solutions: Buffer solution: 78% $Na_2HPO_4.2H_2O$ solution 0.05 M (8.9 g/l), 22% KH_2PO_4 solution 0.05 M (6.8 g/l). TTC solution: 0.6% TTC dissolved in buffer solution.

Procedure

1. In a 15 ml centrifuge tube, take about 150 mg of cell sample and add 3 ml of TTC solution; gently mix and incubate for 15 hours at 30°C.
2. Drain off the TTC solution and wash the cells with distilled water.
3. Centrifuge the cells and extract with 7 ml of ethanol (95%) in a water bath at 80°C for 5 min.
4. Cool the extract and make up the volume to 10 ml with 95% ethanol.
5. Record the absorbance (pink color) with a spectrophotometer at 530 nm. The amount of formazan produced by the frozen cells is expressed as a percentage (survival) of formazan produced by the control cell suspensions.

Growth measurements: By employing various parameters i.e. mitotic index, cell number, cell culture volume, dry and fresh weights and plating efficiency, the growth of frozen cultures can be compared with the control (refer to Chapter 35 for more details).

2. Re-culture

Routinely, before culturing, the thawed material is washed several times to remove the cryoprotectant which may otherwise be toxic to the cells. A gradual dilution of the cryoprotectant is desirable to avoid any de-plasmolysis injury to the cells. The plant material frozen at –196°C may exhibit some special requirement(s) for better survival when re-cultured. For example, frozen tomato seedling shoot tips developed directly into plantlets only if the medium was supplemented with GA3 but this treatment was not required by the controls (non-frozen). The overall survival of frozen and thawed plantlets of carrot was greatly enhanced by activated charcoal.

III COLD STORAGE

Germplasm can also be stored in cultures at non-freezing, low temperatures (1–9°C). At these temperatures the aging of the plant material is slowed down but not completely stopped as in cryopreservation. Consequently, subculture of the plant material is necessary though only very infrequently. Storing cultures for long periods at non-freezing low temperatures is not only very simple but also gives a high rate of survival of the plant material. The use of growth inhibitors, such as abscisic acid and high levels of sucrose (osmoticum) may help to further prolong the interval between two subcultures. The cold storage method of maintaining culture plants holds great promise in the nursery industry employing micropropagation techniques. During the periods of low demand for a particular genotype or species, the cultures may be simply shelved in a refrigerator (Figure 34.3).

Method-1: Freeze preservation of cells from suspension cultures

Procedure

1. Subculture the actively growing suspension culture (e.g. *Zea mays*) into a fresh medium supplemented with 10% proline.
2. After 3–4 days wash the cells in proline-free medium and suspend them in fresh medium.

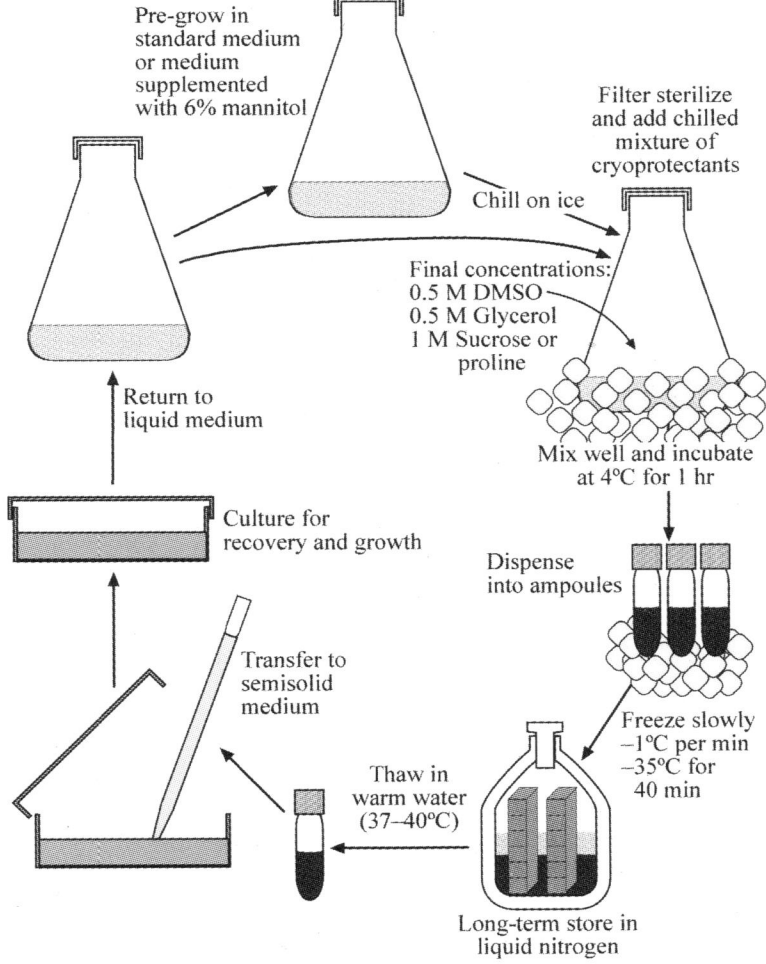

Figure 34.3 *Routine procedure for cryopreservation of cell cultures*

3. Add chilled culture medium containing 20% proline to an equal volume of the chilled cell suspension. The proline-containing medium should be added gradually, in four aliquots, over a period of 1 hour.
4. After maintaining the mixture on ice for 1 hour, transfer the cells to sterile polypropylene ampoules with screw caps (1 ml into each 2 ml ampoules).
5. Transfer the ampoules to a controlled freezing apparatus and cool to −30°C at a rate of about 1°C min^{-1}.
6. After holding at −30°C for 30–40 min. transfer the ampoules to the liquid nitrogen container or refrigerator.
7. To thaw, take out the ampoules from the storage container and agitate them in water at 40°C for 1–2 min.
8. Spread cells on agar medium along with the surrounding liquid medium for re-growth.

VI PROSPECTS

Progress in the development of plant cell, tissue and organ culture techniques for long-term germplasm conservation has been significant. The evidence available so far indicates that freeze-preservation is the most reliable approach to the long-term preservation of cell cultures, which possess the biosynthetic capacity for synthesis and accumulation of secondary metabolites. Since plant cells have totipotency and morphogenetic potential, the cryobiology of plant tissue cultures offers a wide range of prospects; some of the most obvious follow:

Conservation of genetic uniformity: Plant cells in cultures undergo a lot of genetic diversity and variability depending on the origin, the genetic constitution of the tissue, and the physical and growth conditions. Establishment of gene banks for the preservation of genetic uniformity of plant cells, especially for plants that can be propagated vegetatively by cell culture is highly desirable. This would ensure the genetic uniformity of the material.

Preservation of rare genomes: Somaclonal variations due to various kinds of genetic variabilities, even though undesirable for maintaining the uniformity and stability of clones, can generate a rare form. Those variables, which do not occur in nature, can be frozen and kept indefinitely. This technique has been well-exploited by plant breeders and geneticists.

Freeze-storage of cell cultures: Regular maintenance of cell cultures suggest that they have to be subcultured and transferred periodically and repeatedly over an extended period of time. This procedure requires lot of media, much space, and manpower. It can also create a room for contamination. To avoid periodical transfer of the cultures, freeze preservation would be an appropriate approach to suppress cell division and to avoid the need for periodical subculturing.

Maintenance of disease-free material: Disease-free stocks of rare plant materials could be frozen, revived, and propagated when needed. This method would be ideal for the storage for future use and also for international exchange of such materials.

Cold acclimation and frost resistance: Many novel mutants with resistance to frost have been isolated from the cultures subjected to super-low temperatures. Tissue culture will provide suitable material for the study on effects of temperature regimes for enhancing the hardiness of the callus to the cold.

Retention of morphological potential: In regular cell culture conditions, long-term tissue cultures are known to lose their ability to undergo morphogenesis. By freeze storage, the morphogenetic potential could be retained over an extended period of time.

Slow metabolism and aging: At ultra-low temperatures, there is a total arrest of growth and the cells are in a metabolic inactive state. This would prevent or virtually "stop" the aging process by avoiding accumulation of mutations associated with proliferation of cell cultures.

V   SUGGESTED READING

Bajaj YPS (1976) Regeneration of Plants from Cell Suspensions Frozen at –20°, –60°, and –196°C, *Physiol. Plant.* **37**:236–268.

Bajaj YPS (1980) Induction of Androgenesis in Rice Anthers Frozen at –196°C, *Cereal Res. Commun.* **8**:365–369.

Bajaj YPS and J Reinert (1977) Cryobiology of Plant Cell culture and Establishment of Gene-Banks. In: *Plant cell, tissue, and organ culture* (Eds) Reinert J and Bajaj YPS, Narosa Publishing House.

Bhojwani SS and Razdan MK (Ed) (1983) *Plant Tissue Culture: Therory and practice*, Elsevier.

Finkle BJ, Ulrich JM, Schaeffer GW and Sharpe F Jr (1982) Cryopreservation of Rice Cells, In: *International Rice Research Institute. Cell and Tissue Culture Techniques for Creal Improvement*, Los Banos, Philippines.

George EF (1993) *Plant Propagation by Tissue Culture Part I* (2nd Edn), Exegetics Limited.

Khanna, VK (1998) *Plant Tissue Culture Practice*, Kalyani publishers, Ludhiana, pp1–20.

Razdan MK (2000) *An Introduction to Plant Tissue Culture*, Oxford & IBH Publishing Co. Pvt. Ltd. New Delhi, Calcutta.

Reinert and Yeoman MM (1982) *Plant Cell and Tissue Culture*, Narosa Publishing House, New Delhi; Springer–Verlag, Berlin Heidekberg New York.

Withers LA and Ling PJ (1980) A Simple Freezing Unit and Routine Cryopreservation Method for Plant Cell Cultures, *Cryoletters* **1**:213–220.

35

Measurement of Plant Cell Growth and Cytological Analysis

I INTRODUCTION

Once a suspension culture with the desirable characteristics has been isolated, it is strongly advisable to carry out some basic growth analysis in order to chart the development of the culture and determine the time scales of the different phases. Growth analysis is best carried out by initiating sufficient cultures to enable whole flasks to be harvested (sacrificed) on each measurement day. Replicate flasks must therefore also be set up to gain an estimate of interculture variation. Alternatively, a smaller number of cultures can be set up from which samples are taken on each measurement day. The samples taken must be of just sufficient size to provide enough cells for accurate measurements. Measurements should be made every day for fast-growing cultures, which are routinely transferred at weekly intervals. For slower growing cultures, measurements should be made daily for the first six days and thereafter every second day.

II GROWTH PATTERNS IN SUSPENSION CULTURE

Under optimal light, temperature, aeration, and nutrient medium, the growth of suspension culture follows a predictable pattern or growth curve. The cell growth of the suspension culture can be easily monitored by a simple counting of the cell number per unit volume of culture and time. The data generated from such cell measurements can be used to prepare the growth curve. The growth curve for a typical higher plant suspension culture consists of the lag phase, logarithmic phase or exponential phase, linear phase, and stationary phase. Dry weight, total protein, DNA synthesis etc., can also be considered as other parameters for the preparation of identical growth curves.

A. MEASUREMENT OF CELL GROWTH

Method-1: Basic protocol for measurement of cell growth

1. Estimate the length the experiment is to run, number of harvest days, and number of replicates needed.

2. Prepare in advance a sufficient number of cultures to provide an excess (50%) of inoculum for the growth experiment.
3. Label all the experimental flasks 1, 2, 3, 4, 5, etc.
4. Inoculate all flasks with uniform samples taken from the stock culture.
5. Flame, re-seal, and secure all flasks firmly on the culture shaker.
6. Immediately harvest the flasks for the 0 day measurements beginning with # 1, 2, etc.
7. Carry out the measurements on the desired growth parameters.
8. Repeat 6 & 7 for each measurement day until the end of the experiment.
9. Repeat the experiment.

1. To make the measurements

Mix each culture gently but thoroughly and remove a representative 2 ml sample to determine cell viability and cell number.

Cell viability

a. Prepare a stock solution of fluorescein diacetate (FDA) in acetate (5 mg/ml).
b. To a 5 ml fresh culture medium, add drops of the FDA stock, mixing after each drop until the medium begins to show a slight turbidity.
c. On a microscope slide with sample well, combine one drop of well-mixed cell suspension with one drop of the diluted FDA stock.
d. Apply a coverslip and leave for 10 min. at room temperature.
e. For a randomly chosen field of view, count the total cell number under visible light.
f. Switch to UV light and count the fluorescent cells.
g. Repeat steps e and f until at least 500 viable (fluorescent) cells have been counted.
h. Calculate the production of the total cell number, which is viable and express the result as a percentage.

Cell number

When suspension is fine:

a. Mix the sample thoroughly by pumping it into and out of a Pasteur pipette 5–10 times.
b. Immediately remove a sample with the pipette and place in a hemacytometer slide (Figure 35.1).
c. Count the number of cells above the entire grid (=[A1]).
d. Repeat steps b and c 5 more times (=[A2]→[A6]).
e. Cell number (per ml) = $\dfrac{([A1] + \ldots + [A6])}{6 \times \text{vol. above grid (ml)}} \times (\text{dilution factor})$

When the aggregates are too large to count the constituent cells accurately:

a. Spin down the cells at 1000 X g for 5 min. and carefully pipette off the medium.
b. Re-suspend the cells in 2 ml 10% (v/v) HCl.
c. Add 2 ml 10% (w/v) chromium trioxide in water and incubate at room temperature in the dark for 1–5 days.
d. Mix the sample thoroughly by pumping it into and out of a Pasteur pipette 5–10 times.
e. Immediately remove a sample with the pipette and place in a hemacytometer slide.

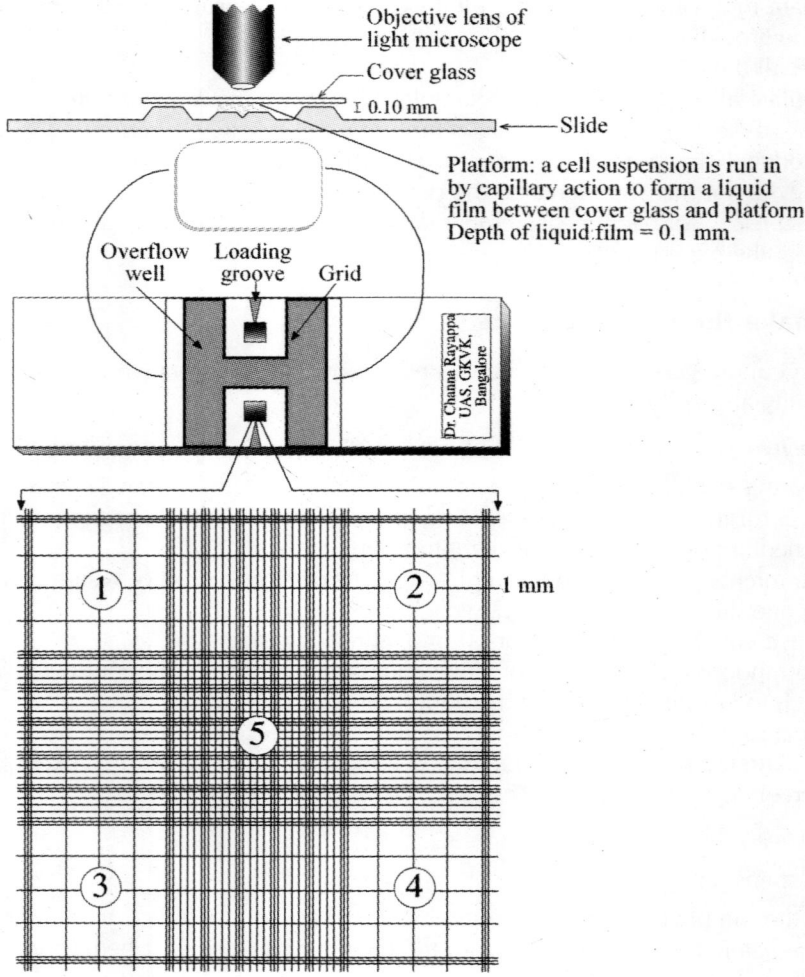

Figure 35.1 *Hemacytometer slide (Improved Neubauer) and coverslip*

f. Count the number of cells above the entire grid (=[A1]).
g. Repeat steps b and c 5 more times (=[A2]→[A6]).

h. Cell number (per ml) = $\dfrac{([A1] + \ldots + [A6])}{6 \times \text{vol. above grid (ml)}} \times \dfrac{\text{(Final sample volume)}}{\text{(Initial sample volume)}}$

2. Measurement of medium pH

Allow the cells in the remainder of the culture to settle down and measure the pH of the medium with a standard pH meter.

3. Determination of the packed cell volume (PCV)

PCV is a non-destructive determination of the amount of culture biomass. If necessary, it can be measured under sterile conditions and the cells reused. For fine cultures, a change in the PCV is the most accurate estimation of culture growth. Increased culture aggregation decreases accuracy.

a. Mix the culture thoroughly and quickly transfer (pour or use a wide-mouthed pipette) approximately 10 ml into a graduated conical centrifuge tube.
b. Centrifuge for 5 min. at 500 g using a swing-out rotor.
c. Record the precise volume of the sample taken and the volume of the cell pellet.

d. $PCV = \dfrac{\text{Volume of cell mass}}{\text{Total volume of sample}} \times 100 (\%)$

4. Determination of the 'fresh' and dry weights

Culture fresh weight (probably better-termed 'wet' weight) can only be determined following the separation of the cells from the medium. Herein lies a source of inaccuracy: over-drying small samples and under-drying large ones can sometimes cause problems. Consistent filtration is essential.

a. Combine the PCV sample with the remainder of the culture.
b. Record the weight of a disc of dry filter paper [A].
c. Place the filter paper in a Buchner funnel connected up to a side-arm flask, apply a vacuum, and thoroughly wet the filter with distilled water.
d. When the water stops dripping down, shut off the vacuum, remove and re-weigh the now moist filter paper [B].
e. Replace the filter paper and filter the culture, washing all the cells out of the flask with distilled water.
f. Weigh the cells + filter [C].
g. Weigh the dried cells + filter [D].

h. $\text{Fresh weight (culture)} = ([C] - [B]) \times \dfrac{\text{Total culture volume}}{\text{Culture volume} - 2\,ml}$

i. $\text{Dry weight (culture)} = ([D] - [A]) \times \dfrac{\text{Total culture volume}}{\text{Culture volume} - 2\,ml}$

5. Methods for further measurements of parameters such as DNA content, mitotic index, protein synthesis, etc. can be found in Street, 1977.

Notes

a. The acetone/FDA stock can be stored at 4°C in the dark in a tightly sealed flask for several months.
b. The combination of an intact plasmalemma and intracellular esterase activity in viable cells results in the production and accumulation of fluorescien, a dye which fluoresces yellow/green under UV light. FDA, in contrast, is colorless under UV light.

c. The chromium trioxide (corrosive) stock can be stored in the dark for several months.

III CYTOLOGICAL TECHNIQUES

Many kinds of rearrangements in the genetic material of a cell can occur in plant tissue or protoplast culture. These include changes in chromosome number, deletions and additions of chromosome parts, and rearrangement of the chromosomes such as translocations, inversions, etc. Many of the gross structural changes can be evaluated by conventional cytogenetic procedures that involve staining the chromosomes and evaluating the morphology at the metaphase of mitosis and their behavior in meiosis. Observation of mitotic chromosomes under the light microscope is a rapid and informative method of studying the genome in its entirety. Gross changes in chromosomes can be identified as changes in chromosome number and structure. If chromosomes are studied at the meiosis stage, more detailed information can be obtained on structural rearrangements and on aspects of the genetic system such as pairing and recombination. Now that techniques of protoplast fusion and transformation have been developed for direct genetic manipulation, such a simple method of looking at the genome should not be ignored. Most genetic manipulation approaches utilize tissue culture systems of plant regeneration from cultured cells. Although such systems are asexual, the regenerated plants may not be homogenous due to the occurrence of somaclonal variation during the culture phase. Genetically engineered plants may, therefore, contain uncontrolled genome modifications in addition to the changes engineered. The repercussions of genetic manipulation itself on the genetic system are also not yet known.

Basic cytological techniques, enabling accurate determination of chromosome number and structure, should be of standard use in plant tissue culture and genetic manipulation laboratories. The procedures outlined here include analysis of regenerated plants (mitotic and meiotic) and cell suspension and protoplast cultures. More advanced cytological procedures that combine molecular biology with cytogenetics, are discussed in the chapter *In situ* hybridization to chromosomes.

Cytological techniques involve four basic stages – collection, fixation, staining, and preparation of chromosome squashes. For analysis of mitotic chromosomes, a fourth stage of pre-treatment is usually added, to arrest mitosis at metaphase, so that the chromosomes can be visualized in their most condensed form. Although the techniques are simple on outline, chromosomes vary enormously among plant species, making it impossible to present a single method, which will account for all the problems that may be encountered. Some modifications of basic procedures are therefore to be expected and general guidelines highlighting points of importance are given in the following procedure.

A. METHODS AVAILABLE

1. Preparation of stains

Feulgen: 0.9 g basic fuchsin, 4.8 g sodium metabisulphite, 250 ml 0.15 M HCl. Mix in a conical flask, cover with foil, and shake for 24 hours. Add 5 g activated charcoal. Mix well. Filter (in fume hood). Repeat charcoal step until the filtrate is colorless. Store at 4°C in the dark.

Carbol fuchsin
   Solution A: 3 g basic fuchsin + 100 ml 70% ethanol
   Solution B: 10 ml sol. A + 90 ml 5% phenol.
   Solution C: 45 ml solution B + 6 ml acetic acid + 6 ml 37% formaldehyde.

Modified carbol fuchsin: 2–10 ml carbol fuchsin (solution C above), 90–98 ml 45% acetic acid, 1.8 g sorbitol.

Aceto-Orcein: Dissolve 2 g of orcein in 100 ml of 45% acetic acid by boiling gently. Shake and filter. Dilute 1:1 with 45% acetic acid when ready to use, for final concentration of 1% orcein. This can also be made in 45% propionic acid, which will stain chromosomes of alfalfa and *Brassica* more deeply than the acetic solution.

Aceto-Carmine: Aceto-carmine can be purchased in solution or can be made up as a 1% solution by dissolving 1 g carmine dissolved in 100 ml 45% acetic acid.

Alcoholic carmine: Gently boil 4 g of carmine in 15 ml of distilled water to which 1 ml of concentrated HCl has been added. After cooling, add 95 ml of 85% ethanol. Shake and filter.

Lacto-Propionic Orcein: Dissolve 2 g of orcein in 100 ml of a mixture of equal parts of lactic and propionic acids. Dilute 45 parts of the above mixture with 55 parts water.

Giemsa
   0.2N HCl: Make HCl stock solution, 84 ml concentrated HCl in 500 ml distilled water. Dilute this stock 1:9 with distilled water to make 0.2N HCl.

   Saturated $Ba(OH)_2$: Make saturated $Ba(OH)_2$ solution by dissolving $Ba(OH)_2 \cdot 8H_2O$ in distilled water so that there is a sediment. Prepare fresh each time. Stir for 30 min. and filter before use.

   2X SSC: 82.2g sodium citrate + 175.5g NaCl in 1000 ml of distilled water. Dilute this stock 1:9 with distilled water to make working solution. Stock can be stored for months.

   Phosphate buffer: Solution 1: 9.46g NCl in 1000 ml of distilled water. Solution 2: 9.07g KH_2PO_4 in 1000 ml distilled water. Stock solutions can be stored for several months. Working solution: 62 ml solution 1 + 38 ml solution 2, pH 6.8.

   Giemsa stain: 1–5 ml Giemsa in working solution of phosphate buffer to total 100 ml.

2. Staining schedules

Feulgen
1. Remove roots from fixative and wash thoroughly in water.
2. Immerse the roots in 1 N HCl at 60°C and hydrolyze for 10 min. (this step varies with the fixation time).
3. Wash roots in distilled water over a 5 min. period.
4. Place roots in Feulgen stain and allow to stain for approximately 1 hour.
   Feulgen specifically stains DNA. Feulgen staining is suitable for the root tips of several species. However, in many cases a second stain is required.

Carbol Fuchsin
1. Hydrolyze tissues of root tips or callus as for the Feulgen procedure to soften the tissue.
2. Stain in Feulgen and make squashes as above, including flattening under the coverslip.
3. Place a drop of the carbol fuchsin staining solution at the edge of the coverslip, and allow it to run under the coverslip. This can be hastened by placing a piece of paper towel on the opposite side of the coverslip to draw the stain across.
4. Flatten again. Carbol fuchsin is suitable for somatic cell preparations of cereals.

Carmine
Plant tissues can be placed directly into a few ml of carmine stain and incubated at room temperature. Squash the tissues in acetic acid or acetocarmine stain but not in alcoholic carmine. The incubation is usually for 15–30 min.

Anthers: Alcoholic carmine is suitable for the staining of meiocytes in anthers of cereal crops:

i. Place the immature inflorescence directly into the stain and allow to stain for several days.
ii. After staining, place individual anthers on a slide in a drop of acetic acid or aceto-carmine. Cut the anthers in half and squeeze the meiocytes out into the drop.
iii. Remove the anther walls and debris, and place a cover slip on the drop.
iv. Gently heat the slide to cause the cells to swell and the chromosomes to spread so that overlapping is minimized.

Orcein
Orcein stains are used very much like carmine stains. Hydrolyze the tissue in 1 N HCl for 5–10 min. Macerate the tissue in a drop of the stain. Add coverslip and heat gently. Squash. The orcein stain is more suitable than other stains for root tips and meiotic analysis of *Brassica* and *Medicago*.

Giemsa C-Banding
This protocol is suitable for cereal species.
1. Soften root tips in 45% acetic acid. Tease apart the tip in a drop of acetic acid, add coverslip, heat gently, and squash. Evaluate under microscope with phase contrast, or soften tissue in 1% aceto-carmine to determine if there are some good cells present.
2. Place slides on a block of dry ice, allow to freeze and prise off coverslip using a scalpel by placing the sharp blade under a corner.
3. Air-dry the slides for a short period, then incubate them for two and a half min. in 0.2 N HCl at 60°C.
4. Wash the slides briefly in distilled water, and incubate them for 7 min. in saturated barium hydroxide at room temperature.
5. Wash the slides carefully in distilled water to remove all the barium hydroxide, and incubate them for 1 hour in 2 × SSC at 60°C.
6. Rinse slides briefly in phosphate buffer and then place them in 1–5% Giemsa in the phosphate buffer. Begin with 1% Giemsa. The staining time varies from 10–45 min.
7. Rinse slides briefly in distilled water and air dry. Mount in Permount thinned with toluene.
8. Scan under microscope to find well-spread cells containing all of the chromosomes with few overlaps and with good Giemsa banding. Record karyotype.

Method-1: Root-tip squash preparations for mitotic chromosome analysis of regenerated plants

Procedure

Root tips
1. Place two layers of filter paper on a Petri dish and moisten filter paper with water. Place seed of cereal species on moistened filter paper. Cover and allow to germinate at room temperature, 23°C. This will require about two days.
2. Place the three seminal roots in cold water, using a separate vial for the roots from each seed or seedling.
3. Place in a refrigerator at 2°C for 24 hours. Pre-treatment to arrest cell division and accumulation of metaphase cells can be achieved by different methods as listed in Table 35.1.

Table 35.1 *Pre-treatment for arresting mitosis at metaphase*

Chemicals	Concentration	Time(hr)	Temp(°C)	Examples of suitability
8-hydroxyquinoline (in distilled water)	0.29 g in a litre (dissolve at 60°C)	3½–4 hrs	18–20°C	potato, oil seed, rape, sugar beet, tobacco
Colchicine (in distilled water)	0.05 g in 100 ml	4–6 hrs	18°C	*Vicia faba*, *Scilla*, Hyacinth
Ice-cold water	–	24 hrs	4°C	Cereals
α-bromonaphthalene (in distilled water)	saturated	3½–4 or 18 hrs	room temp. 4°C	Cereals Cereals
α-bromonaphthalene (in alcohol)	1 ml in 100 ml alcohol (stock) (Use 10 μl stock /10 ml water)	3½–4 or 18 hrs	room temp. 4°C	Cereals Cereals

Preparation of root squash:
1. Using a clean pair of forceps, collect healthy roots in a small vial containing distilled water (in water keep it less than 15 min.). Select healthy roots for good chromosome preparations. Roots can be collected from small plantlets in culture medium, vermiculite or soil, or from germinated seeds.
2. Transfer the roots to a suitable pre-treatment as quickly as possible and incubate for the appropriate time and temperature (range of treatments are given in Table 35.1).
3. Fix the roots by transferring them into 3:1 absolute alcohol: glacial acetic acid and incubate for at least 24 hours at 4°C. Roots in a fixative can be stored at 4°C for months. Fixations should be made for at least 24 hours, optimal after three days.
4. Hydrolyse the roots by incubation in 1 M HCl at 60°C for 4–9 min.
5. Wash the roots briefly in distilled water.

6. Transfer the roots to feulgen and leave to stain for 30 min.
7. Place a stained root on a clean glass slide. Remove the translucent root cap at the extreme tip, cut below the stained meristematic zone, and discard the unstained portion of the root.
8. Place the stained root-tip in a small drop of 45% acetic acid. Mix in a small drop of aceto-carmine (BDH). Aceto-carmine staining improves background staining, which colors cytoplasm and allows determining of the intactness of the cell and it will be very useful to locate chromosomes in plants with small DNA contents.
9. Tap the root-tip thoroughly in the drop with a flat-ended glass rod. Remove any remaining pieces with a needle. For a good chromosome squash, ensure that the root-tip cells are completely dispersed before adding the coverslip.
10. Gently lower a coverslip over the drop. Place a piece of filter paper over the coverslip and press gently to remove any excess stain.
11. Keeping the filter paper in place with the fingers of one hand on either side of the preparation, squash vertically downwards using the thumb or forefinger of your other hand. Do not rock your thumb or finger during the squash as this will roll the cells. Plants differ in the amount of pressure required to spread the chromosomes flat. It is essential to achieve a good spread for accurate counting, particularly in species with small chromosomes.
12. Examine quickly under the microscope and squash again if necessary.
13. Ring the coverslip with rubber solution and examine once the solution has dried. Chromosome counts should be made from a minimum of five cells, where the chromosomes are well spread, and from more than one root.

Method-2: Meiotic chromosome preparations

Procedure

1. Choose a suitable influorescence (Note 1). Peel away the outer leaves or bracts and place the influorescence into a tube containing Carnoys no. 2 fixative (6:3:1 alcohol:chloroform:acetic acid, to which are added a few drops of 10% (w/v) ferric chloride solution until a straw yellow color is obtained). Fix for at least one month, replacing with fresh Carnoys after 2–3 weeks. For immediate analysis of meiosis, anthers can be fixed in 3:1 alcohol:acetic acid and the chromosomes stained in feulgen (see root-tip preparations).
2. Transfer the inflorescence into a clean Petri dish containing 70% alcohol. Carefully dissect the anthers from one floret, noting from where they have been taken. Take one anther and place it on a clean slide.
3. Dab any excess alcohol with a piece of filter paper and add a drop of acetocarmine.
4. Cut the anther into two or more pieces depending on the size. Tap them thoroughly in the drop of aceto-carmine. Remove any remaining pieces of anther wall.
5. Carry out steps 10–12 of the procedure for root-tip squashes, but apply much less pressure and monitor carefully during the squash.
6. Identify the stage of meiosis and examine for abnormalities.

Notes

1. In cereals such as wheat and rye, meiosis occurs before complete emergence of the ear and it is relatively easy to select influorescence at the correct stage. You can feel how large the influorescence is with fingers and use the appearance of the flag leaf and the extent of

emergence as indicators of the stage. But in other plants it is difficult and trial and error may be the only way.

2. Warming of the slide before squashing is advantageous in many species. The slide should be warmed gently over a small burner and then allowed to cool a little before squashing. Do not overheat the slide.

Method-3: Chromosome preparations from cell suspension cultures

Procedure-1

1. Take a 50 ml aliquot of cell suspension 3–4 days after sub-culturing and transfer it to a clean conical flask. Add 50 ml of 0.02% colchicine and shake the mixture on an orbital shaker (100 rpm) for 1–2 hours at 25°C.
2. Transfer the contents to centrifuge tubes and centrifuge at 1100 rpm for 10 min. Discard the supernatant and re-suspend the pellet in 20 ml of 3:1 ethanol:acetic acid fixative. Refrigerate overnight at 4°C.
3. Centrifuge at 1100 rpm for 10 min., remove the supernatant, and re-suspend the pellet in 0.1 M sodium acetate (pH 4.5). Leave the mixture to settle for a few min. Centrifuge at 1100 rpm for 10 min., discard the supernatant, and transfer the pellet to a clean conical flask.
4. Re-suspend the pellet in 20 ml of an enzyme mixture containing 0.25 g Onozuka cellulase, 0.25 g of Macerozyme R10, and 49.5 ml 0.1 M sodium acetate buffer, pH 4.5. Incubate the mixture at 25°C for 2 hours.
5. Wash the cells in 0.1 M sodium acetate (pH 4.5) as described in step 3 but re-suspend the pellet in 10 ml of 45% acetic acid. Leave overnight in the fridge at 4°C.
6. Pipette 20 µl of the fixed suspension onto a clean slide and allow the fixative to evaporate. Add a few drops of a 1% solution of Macerozyme R10 and gently mix the cells into the drop. After a few min., tap the cells in the drop and continue tapping until the cells are dispersed (this may take up to 10 min. depending on the cell suspension).
7. Add a few drops of modified **carbol fuchsin**, mix the cells into the stain, and leave for 4–5 min. before adding a 22 × 50 mm coverslip.
8. Squash as for a root-tip preparation. If air bubbles appear, add a little more carbol fuchsin at the edge of the coverslip. Ring with rubber solution and examine.

Procedure-2

1. Transfer a 10 ml aliquot of cell suspension 3–4 days after sub-culturing into a glass vial.
2. Remove the culture fluid with a Pasteur pipette.
3. Make a stock solution of 1 ml α-bromonaphthalene in 100 ml absolute alcohol and shake vigorously. Add 10 µl of the stock solution to 10 ml of distilled water mix and add to the cell suspension in the glass vial. Cap the vial, shake to mix, and incubate at 4°C overnight.
4. Remove the pre-treatment with a Pasteur pipette.
5. Wash the cells with 3:1 absolute alcohol:acetic acid fixative by adding 5 ml to the vial, shaking the mixture, and then removing the fixative with a Pasteur pipette. Repeat the wash step once.
6. Add a final 5 ml of fixative and leave to incubate 24 hours at 4°C.
7. Make chromosome preparations as described above. You will need to squash several times to cover the large area of the coverslip.

Method-4: Chromosome preparations from protoplast cultures

Procedure

1. Transfer 5 ml of a protoplast culture 2–4 days after isolation into a small conical flask. Add 5 ml 0.02% colchicine solution and incubate for 6–10 hours at 25°C on an orbital shaker (100 rpm).
2. Transfer to centrifuge tubes and centrifuge at 1000 rpm for 10 min.
3. Remove the colchicine supernatant and re-suspend the pellet in 1 ml 3:1 ethanol:acetic acid fixative. Refrigerate at 4°C overnight.
5. Pipette 20 µl onto a clean slide and allow the fixative to evaporate.
6. Add a few drops of modified **carbol fuchsin** and allow to stain for 1–2 min. Cover with a 22 × 50 mm coverslip and squash gently.
7. Seal with rubber solution and examine.

IV SUGGESTED READING

George EF (1993) *Plant Propagation by Tissue Culture Part 1*, Exegetics Limited.

Karp A (1991) Cytological Techniques, In: *Plant tissue culture methods* (Ed) K Lindsey. Kluver., Academic Publishers.

Street HE (1977) *Cell Suspension Culture Techniques*, In: Plant Tissue and Cell Culture (Second Edition) (Ed) Street HE, Blackwell Scientific Publications, Oxford pp61–102.

Sharma AK and Sharma A (1980) *Chromosome Techniques Theory and Practice*, Butterworth and Co., London pp169–170.

36

Protoplast Fusion and Somaclonal Variation

I INTRODUCTION

Conventional plant breeding is often limited by pre- and/or post-zygotic incompatibility barriers and the fusion of somatic cells to generate somatic hybrid plants has been considered as a way to overcome such limitations. The transfer of genes using protoplast-based methods offers the opportunity to create diverse new plant types. Plant protoplasts represent the finest single cell system and offer exciting possibilities in the fields of somatic cell genetics and crop improvement. They also provide experimental material for many other fundamental and applied studies. Protoplasts from any two plants, regardless of the species, can be fused when they come into contact with each other. Production of hybrid plants through the fusion of protoplasts of two different plant species is called **somatic hybridization**. The lack of constraints to interspecific or intergenic protoplast fusion permits *hitherto* reproducibly isolated plant genomes to be combined at the protoplast (heterokaryon) level, thus, providing the basis for the generation of novel hybrids. Protoplast fusion also enables the genetic manipulation of vegetatively propagated crops, such as sterile or sub-fertile individuals, and those plants, including woody species, with naturally long life cycles. Somatic hybridization of highly heterozygous species also provides an element of predictability in relation to the hybrid, because meiotic recombination is avoided. Cytoplasmic factors, such as mitochondrial-based cytoplasmic male sterility may also be transferred from one species to another by protoplast fusion. The use of protoplasts for applied purposes is, however, critically dependent on protocols to enable plant regeneration.

Natural plant protoplast fusion

In natural systems, protoplast fusion has been known for a long time, but there is not much information available on the mechanism. Most commonly, fusion occurs as the initial step of zygote formation (fusion in the **sexual cycle**). The heterokaryotic state is established by fusion of either gametes or gametangiums. For this to happen, two types of barriers must be overcome: at least one cell wall and the plasma membranes. So far, little is known about the enzymatic functions that are needed for the removal of the wall material at the contact zones. It is likely that the fusogenic conditioning of the plasmalemma is established by changes of certain electric potentials of the surface and/or by alteration of the macromolecular structure of the membrane.

Fusion in development: Protoplasts of higher plants are united to a symplastic continuum by plasmodesmata. The formation of postgenital plasmodesmata has also been described, and this can be interpreted as protoplast fusion of a limited extent. Autoplastic connections are formed, for instance, during the formation of the false septum in the fruit of *Capsella bursa-pastoris*. Heteroplastic plasmodesmata occur in nature between plant parasites and the host. This can be well studied for e.g. in haustoria of *Cuscuta* in host tissue using the electron microscope.

Experimental protoplast fusion

Somatic hybridization (Protoplast fusion) experiments have been carried out in various plant cells. Cell fusion studies provide information on the physical properties of the plasma membranes by investigations on the induction and the process of fusion, to study fusion bodies and the development of fusion products under physiological and genetic aspects, and to obtain plants with new genetic combinations of basic and applied interests.

Cell fusion methods: During enzymatic degradation of cell walls some of the adjacent protoplasts fuse together forming homokaryons, each with 2–40 nuclei. This type of protoplast fusion is called "spontaneous fusion". So far as somatic hybridization is concerned, spontaneous fusion is of no value. It requires the fusion of protoplasts of different sources. To achieve induced fusion a suitable agent (fusogen) is generally necessary. During the last decade a variety of treatments have been tried to fuse plant protoplasts: $NaNO_3$, artificial sea-water, lysozyme, mechanically induced adhesion, virus, gelatin, high pH and high Ca^{2+}, polyethylene glycol (PEG), antibodies, plant lectin Concanavalin A, polyvinyl alcohol, and electric stimulation. Of these, only $NaNO_3$, high pH and high Ca^{2+}, and PEG treatments have been successfully used to produce somatic hybrid plants. Somatic protoplast fusion can either be between normal protoplasts of both parents (***symmetrical hybridization***) or the donor parent contributes only a few chromosomes, sub-chromosome fragment(s), or tiny pieces of the chromosome, which are transferred to the recipient partner (***asymmetrical hybridization***).

II INDUCED SOMATIC HYBRIDIZATION

The fusion of isolated protoplasts from different sources with the help of fusion inducing chemical or physical agents, is known as induced fusion. Normally, protoplasts isolated *in vitro* do not fuse because the outside surface of the plasma membrane of the isolated protoplasts carries a negative charge (–10 mV to –30 mM) on the outside surface of plasma membrane and thus, there is a strong tendency for them to repel each other. Therefore, a fusion-inducing agent is necessary to make the protoplasts fuse with each other.

Somatic hybridization involves six discrete, yet interrelated, stages. (1) Protoplast isolation; (2) induced protoplast fusion without loss of viability; (3) the development of a strategy for selection of viable somatic hybrid cells; (4) the confirmation of hybridity or cybridity and culture; (5) regeneration of hybrid plants; and (6) characterization of hybrid/cybrid plants.

A. PROTOPLAST ISOLATION

The ability to isolate protoplasts that, when cultured under defined conditions, divide mitotically and regenerate plants has now been established for many species. Protoplasts have been isolated from virtually all plant parts, but leaf mesophyll is the most preferred tissue (refer to Chapter 33 for more details). However, isolation of viable and contamination free protoplasts is not a trivial task, especially when working with new species or genotypes. It may require preliminary work to establish the correct physiological conditions for growing the source plant cultures, and the best combination of wall degrading enzymes, culture media, osmotic conditions, etc.

If protoplast fusion follows the regeneration of fused hybrid cells, it is always advisable to first establish a callus culture for one of the fusion partners. This not only helps to standardize the culture conditions but also to provide white color cells. The protoplasts isolated from fresh mesophyll leaf tissues are green. Fusing protoplasts from two different source plants (green protoplasts from leaf mesophyll and white protoplasts from the callus), may facilitate observation of the fusion and distinction of the dissimilar cells that are being fused. Protoplasts from two different plants can also be distinguished from one another microscopically by staining with different colored non-toxic vital stains.

Method-1: Isolation of protoplasts from leaf mesophyll cells

The leaf is the most convenient and poular source of plant protoplasts because it allows isolation of a large number of relatively uniform cells. Leaf mesophyll cells from a wide range of plants have been used as sources with success (Figure 36.1). Protoplast isolation from leaves involves five steps: (a) sterilization of leaves, (b) removal of the epidermal cell layer (if that is difficult, then the leaf can be cut into small pieces (ca 1 mm^2) and then combined with vacuum infiltration), (c) pre-enzyme treatment, (d) incubation in enzyme, and (e) isolation by filtration and centrifugation.

Materials

Plant material; Petri dishes or depression slides; Small diameter (60 mm) dishes or microscope slides with depressions or wells are the container of choice for digesting the plant tissue; small test tubes will also work; microscope; Pasteur pipettes.

Enzyme solution: 1.5% Cellulysin, 1.5% Macerase, 0.2% Pectinase, 0.4 M Mannitol, 2.0% Glycine. Filter the enzyme solution through a 0.45 µM filter before use. The solution can be stored frozen for several months without significant loss of activity.

Procedure

1. Collect leaves from young, fully expanded leaves and surface sterilize with bleach and wash thoroughly with distilled water.
2. Peel off the lower epidermis with a pair of forceps and cut the peeled areas with a razor blade or scalpel into small square bits of tissue (about 1 cm^2).
3. Place the tissue bits face down in an enzyme solution. Cover the container in which the tissue is being digested to prevent the enzyme solution from evaporating. Allow the enzyme to work for 1/2 to 1 hour. It will help if the mixture is occasionally agitated gently.
4. After one hour, remove with forceps any remaining bits of undigested tissue.
5. Remove a few drops of protoplasts on to a clean microscope slide. Protoplasts are spherical structures. Do not cover them with a coverslip. Check the slide on the microscope to check the presence of protoplasts. Cell number can be determined by using hemocytometer counting.

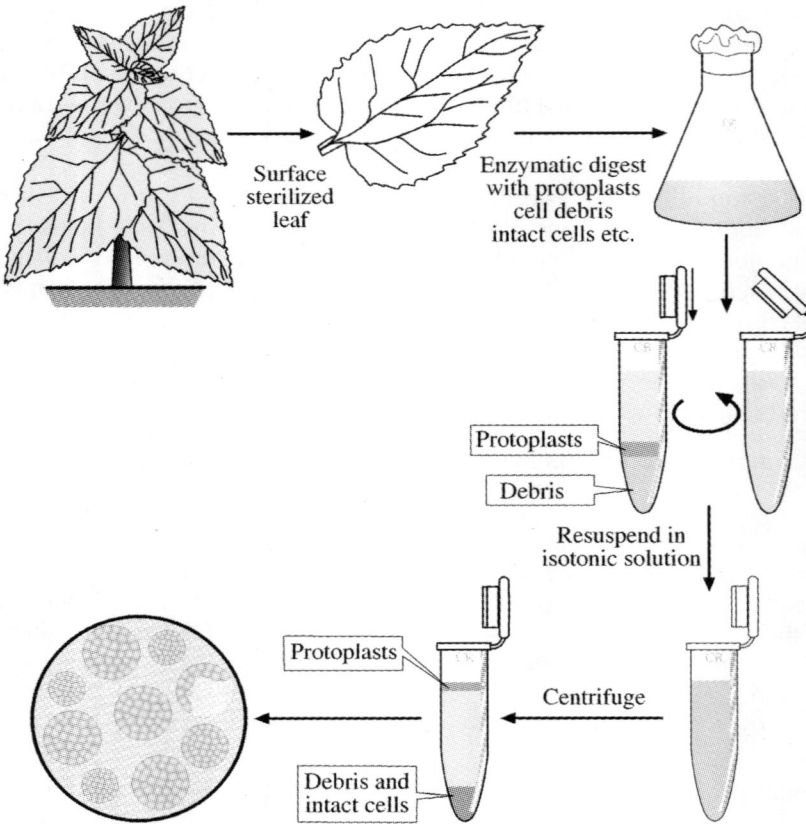

Figure 36.1 *Isolation of plant protoplasts from leaves*

6. Wash the protoplasts with a suitable washing medium in order to remove the enzymes and the debris.

Method-2: Isolation of Protoplasts from In vitro Starting Material

Starting material for protoplast cultures may be any rapidly growing tissue cultured callus or cell cultures. The protoplasts obtained from *in vitro* cell suspension cultures of callus are colored off-white.

Procedure

1. Pre-treat cells for 1 to 2 days in the dark.
2. Centrifuge cells in B5 medium and wash once with the same medium.
3. Incubate the cell with filter-sterilized enzyme solution for 1 to 2 hours.
4. Filter through a 60-micron screen and centrifuge.
5. Using microscope, observe for protoplast release.
6. Add to appropriate culture growth medium.

B. PROTOPLAST FUSION

Protoplast fusion as a tool for experimental hybridization was realized long ago. A number of strategies have been used to induce fusion between protoplasts of different strains/species in both somatic cell genetics and in biotechnology. These include spontaneous fusion, mechanical fusion, induced fusion, $NaNO_3$ treatment, high pH/Ca^{2+} treatment, PEG treatment, and electrofusion. Induced protoplast fusion can be achieved using chemical and electrical treatments. In both cases, fusion is a two-stage process. First, the protoplasts are brought into close membrane contact, the degree of plasma membrane adhesion depending on the parental protoplasts. Tight contact may occur only in localized regions between adhering protoplasts. Subsequently, the plasma membranes are stimulated to interact, for example, by modification of the electrical charges on the membrane resulting in protoplast fusion. Fusions generate products (**heterokaryons**) with two or more nuclei within a mixed cytoplasm containing organelles from the parental protoplasts (Figure 36.2). The cytoplasms derived from the respective parental protoplast mix at different rates within the heterokaryons, according to the protoplast types. Cell wall formation and nuclear fusion to produce hybrid cells occur early in culture. Nuclear fusion takes place either during the interphase by the formation of nuclear bridges, or at the first mitosis. The fate of the plastids in hybrid cells varies, and includes loss of one parental type or recombination between plastids of the two parents. Vacuoles in heterokaryons may fuse and microtubules integrate. However, the fate of other cell organelles is unclear.

The extent of protoplast fusion, heterokaryon formation, and survival of fusion products can be monitored using naturally occurring visual markers. Thus heterokaryons can be readily identified following the fusion of chlorophyll-containing leaf mesophyll protoplasts with suspension cell protoplasts lacking this pigment. Fluorescent dyes have also been used as visual markers to label protoplasts.

1. Chemically induced protoplast fusion

The plasma membranes of isolated plant protoplasts have a net negative electrical charge of approximately 10–35 mV, as a consequence of which adjacent protoplasts naturally repel each other. To induce the close membrane contact required for membrane fusion, the charges of the surfaces of protoplasts must be neutralized by exposure, for example, to polycations such as polyethylene glycol (PEG), or by the use of a high-pH solution. A number of protocols have been described for chemically induced protoplast fusion (Figure 36.3). The use of PEG coupled with solutions buffered at high pH in the presence of Ca^{2+} (high pH/Ca^{2+}) is the most commonly used method to induce protoplast fusion. Carbonyl-free PEG has been shown to improve protoplast fusion, to diminish the formation of large protoplast aggregates (more of heterokaryons), and to retain protoplast viability. PEG induces non-specific cell fusion, from entirely unrelated taxa, such as soybean–tobacco, animal cells–yeast protoplasts, and animal cells–plant protoplast.

Materials

>   CPW13M (mg/litre): KH_2PO_4 (27.2), KNO_3 (101), $CaCl_2.2H_2O$ (1480), KI (0.16), $MgSO_4.7H_2O$ (246), $CuSO_4.5H_2O$ (0.025), 13% (w/v) mannitol, pH 5.8, autoclaved.
>
>   CPW13M/Ca^{2+} (mg/litre): As above, but supplemented with 7.4 g $CaCl_2.2H_2O$ per litre.

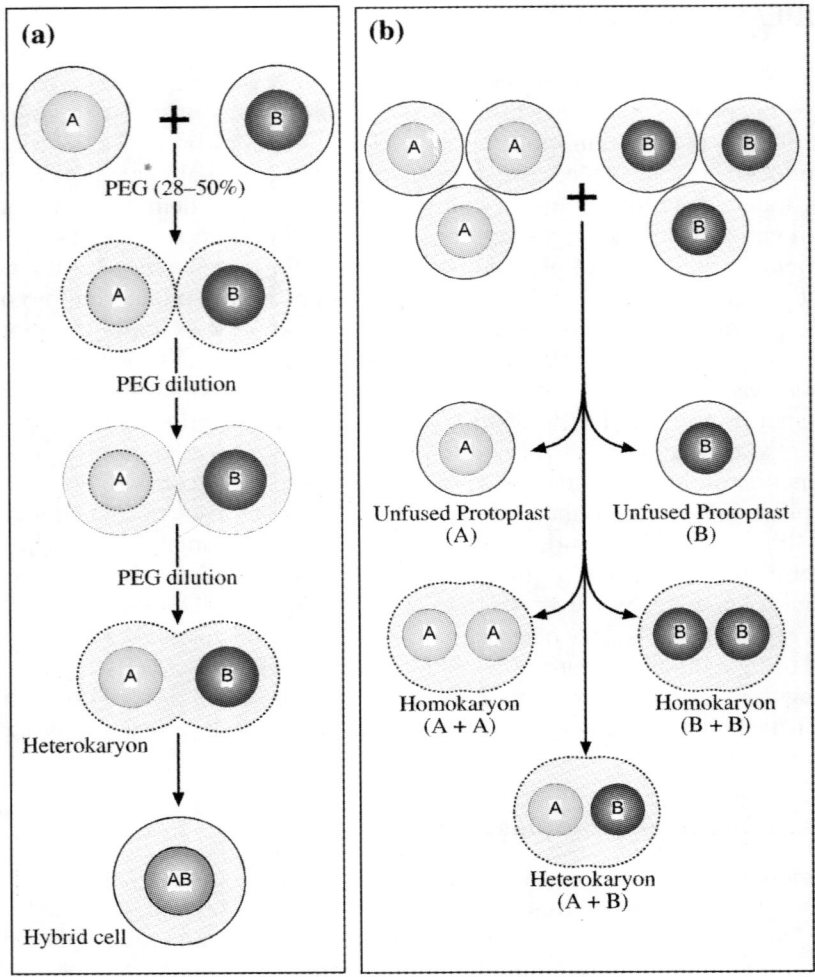

Figure 36.2 *An outline of protoplast fusion. (a) PEG induced protoplast fusion. Fusion occur during PEG dilution due to disturbances created in plasma membrane, (b) Different types of products recovered after fusogen treatment of protoplasts of two different species (A and B) mixed together (usually 1:1 ratio)*

CPW9M PEG: As CPW13M, but with 9% (w/v) mannitol

PEG: 30% (w/v) polyethylene glycol 6000, 4% (w/v) sucrose, 0.01 M $CaCl_2$ $2H_2O$, autoclaved.

High pH/Ca^{2+}: 0.05 M Glycine-NaOH buffer, 1.1% (w/v) $CaCl_2$ $2H_2O$, 10% (w/v) mannitol, pH 10.4, filter sterilized.

Purified PEG: PEG 1540 in N–2-hydroxyethylpiperazine–N'–2-ethanesulfonic acid (HEPES) buffer, pH 8.0 filter sterilized.

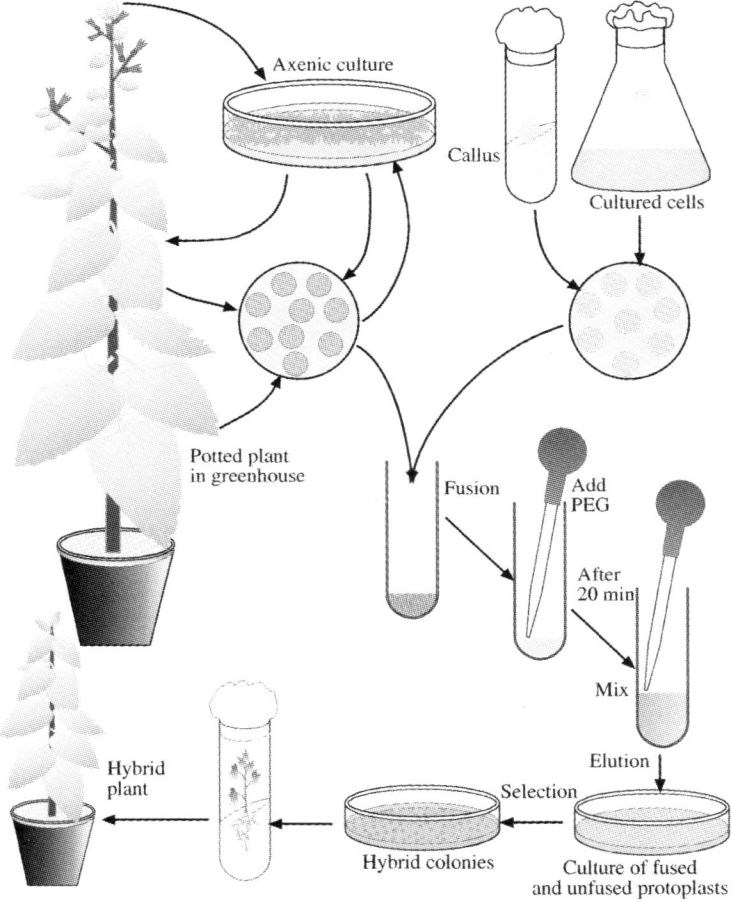

Figure 36.3 *Protoplast fusion between protoplasts of leaf tissue and callus tissue*

Procedure

PEG treatment

1. Mix 4 ml aliquots of freshly prepared protoplasts in a CPW13M solution, typically at a density of 2.0×10^5 ml^{-1}, in 1:1 ratio in 16 ml capacity screw-capped centrifuge tubes. Pass the suspension through a 62μm pore size filter and collect the filtrate in a centrifuge tube.
2. Pellet the protoplasts by centrifugation (60g; 10 min., 22°C).
3. Remove the supernatant and re-suspend the washed protoplasts in a CPW13M solution to make a suspension with 4–5% (v/v) protoplast ml^{-1}.
4. Add 2.0 ml of fusion mixture to the suspension drop by drop and gently re-suspend the protoplasts prior to incubation at room temperature (24°C) for 10–20 min.
5. Dilute the PEG solution, at 5-min. intervals, by the addition of 0.5, 1, 2, 3, and 4 ml aliquots of CPW9M solution.

6. Centrifuge the protoplasts (100g, 10 min., 22°C) and remove the supernatant.
7. Wash the protoplasts at least 2–3 times (5-min. intervals) in protoplast culture medium.
8. Subsequently, re-suspend them in an appropriate culture medium, before plating at a density of 5×10^4 ml^{-1}.

High–pH/Ca^{2+} treatment

1. Re-suspend the protoplasts in CPW13M solution, spun down as in step 1 of the previous section.
2. Remove the supernatant and add 8.0 ml of high–pH/Ca^{2+} fusion solution. Gently re-suspend the protoplasts, immediately centrifuge (60 g; 3 min., 22°C).
3. Place the centrifuge tubes in a water bath at 30°C for 15 min.
4. Add sterile distilled water (2.0 ml/tube) and gently mix with the fusion solution, leaving the protoplast pellet intact. Incubation is continued for a further 10 min. (30°C).
5. Remove the supernatant, wash the protoplasts once in CPW13M/Ca^{2+} solution. Re-suspend in the appropriate culture medium and culture.

PEG with high-pH/Ca^{2+} treatment

Fusion frequencies have been enhanced by the use of PEG in combination with high pH/Ca^{2+} (Kao and Saleem, 1986). The success of this latter modification probably relates to a combined effect of the two fusogens, which have separate modes of action. Primarily, PEG acts as a protoplast agglutinator, whereas high pH/Ca^{2+} modifies the surface charges of the plasma membrane.

1. The protoplasts are treated with PEG as described in steps 1 and 2 for the section on PEG treatment (above), but are diluted with 8.0 ml volumes of high-pH/Ca^{2+} solution per tube. The protoplasts are incubated at 22°C for 10 min.
2. The protoplasts are centrifuged (60 g; 3 min.) and treated as in step 3 of the previous section.

Purified PEG fusion treatment

PEG is known to reduce the viability of fusion products and this cytotoxic effect has been attributed to membrane dehydration and impurities in the polymer, such as α-tocopherol and phenolic derivatives. An improved procedure, using PEG preparations (MW 1540) with a low carbonyl content, has been developed for plant protoplasts. This method is applicable to a wide range of plant protoplast systems and results in a high frequency of heterokaryon survival compared with treatments using unpurified PEG.

1. Protoplasts of the species to be fused are suspended separately in 13% (w/v) mannitol solution at a density of 1.0×10^5 ml^{-1} and are allowed to stand for 5–10 min.
2. Equal volumes of the protoplasts suspensions are mixed and 1.0 to 1.5 ml aliquots dispersed into the wells of a 25-compartment 120 mm^2 grid dish.
3. An aliquot (0.5 ml) of the purified, low-carbonyl PEG solution is added and the mixture left for 15–20 min. at 22°C.
4. One ml of 5% (w/v) mannitol solution is added and the fused protoplasts are left for approximately 5 min. to become spherical.
5. The mixture of PEG and mannitol solution is removed and the protoplasts are washed in 13% (w/v) mannitol solution. The protoplasts are left in this concentration of mannitol solution for 30 min. before a final wash in 13% (w/v) mannitol solution and transferred to the culture medium.

C. CHARACTERISTICS OF PROTOPLAST FUSIONS

1. **Genetic peculiarities**: The fusion of two genetically different protoplasts has a number of implications: the choice of protoplast types for achieving a specific new constellation of genetic information.
2. **Characteristics of cell differentiation**: Different characteristics based on the developmental stages of the donor cells are useful for investigations on the formation and early development as well as for the purpose of mechanical selection of fusion bodies. It has been pointed out earlier that pairs of developmental markers are easily established in combinations of nearly any genetic constituent and hence, are highly useful in plant breeding programs.
3. **Artificially established specific protoplast properties**: Vital staining of protoplasts by fluorescent dyes is universally applicable for the detection of fusion bodies and also for automatic selection. The parental protoplasts can be labeled by different fluorescent dyes for easy identification. Staining may not affect the viability of the protoplasts. The metabolic inhibition of parental cells and subsequent complementation in fused cells can be used for early selection. Reversion of the membrane potential to positive charges of one of the protoplast types has been devised to obtain controlled heteroplastic fusion. Additionally, modifying protoplasts can be done by X-irradiation, which causes functional elimination of the nuclei.
4. **Subprotoplasts**: The genetic complexity of fusion-bodies reduces during subsequent development. Plastids and mitochondria segregate and the nuclei may be unequally distributed to the daughter cells, as for instance, when only one nucleus performs the mitotic cell cycle, while the other one is resting, and finally, chromosomes may be lost in mitoses. Consequently, various types of cell lines can be isolated.

D. SELECTION OF SOMATIC HYBRIDS

Induced protoplasts from two genetically different lines or species necessarily result in a variety of homokaryotic and heterokaryotic fusion products. The selection of the few true somatic hybrid colonies from a mixed population of regenerating protoplasts is a key step in successful somatic hybridization experiments. Despite efforts to increase protoplast fusion frequencies, the formation of viable, binucleate heterokaryons is typically restricted to less than 5% (0.5–10%) of the protoplast population. The protoplast suspension recovered after the treatment with the fusogen consists of various cell types (parental, homokaryones, heterokaryones, and a variety of other nuclear–cytoplasmic combinations). Therefore, it is necessary to select these fusion products against a background of other fusion bodies. Several selection methods have been described, but a universally applicable system has not been developed. Some commonly used selection systems are described below.

1. Genetic complementation

Complementation methods depend on the fusion of two protoplast systems, each of which carries different recessive selectable markers. The resulting somatic hybrid cells are functionally restored. A range of complementation systems has been used to recover somatic hybrid tissues and somatic hybrid plants. Examples are differential sensitivity of protoplasts to drugs (e.g. actinomycin D), and selection based on complementation of auxotrophic mutants.

2. Use of pigmented or chlorophyll-deficient mutants (albinos)

The fusion of protoplasts from two non-allelic chlorophyll-deficient lines results in somatic hybrid cells that are chlorophyll proficient, as in the case of the fusion of protoplasts from the albino cell lines of *Medicago sativa* and *Medicago borealis*. Selection can also be based on complementation between wild type and albino lines. Thus somatic hybrid cells between wild type mesophyll protoplasts of *Petunia parodii* and protoplasts from an albino line of *Petunia inflata* exhibit chlorophyll synthesis and sustained growth.

3. Use of light-sensitive mutants

The fusion of mesophyll protoplasts from a light-sensitive mutant of *Nicotiana lumbaginifolia* with wild-type mesophyll protoplasts of *Nicotiana gossei*, irradiated with 200 J kg^{-1} of ^{60}Coγ rays (0.066 J kg^{-1} s^{-1} dose rate) prior to fusion, has been used to select heterokaryon-derived green hybrid cell colonies. Regenerated plants have the morphology of *N. plumbaginifolia*, and normal green coloration.

4. Use of nitrate reductase-deficient lines

Following fusion, parental protoplasts from nitrate reductase-deficient (NR$^-$) cell lines are eliminated by their inability to utilize nitrates in the culture medium. This deficiency can be overcome in hybrid tissues through complementation of the other fusion partner. Thus, for hybrid cell/tissue selection, two selectable markers are required, with the result that protoplasts from NR$^-$ lines are combined with those carrying other selectable markers. For example, protoplasts from NR$^-$ *N. tabacum* fused with wild type *N. glutinosa* pollen-tetrad protoplasts, produced hybrid cells that utilize nitrate. Such cells regenerate to form green plants. Other types of autotropic plant mutants can be employed in somatic hybridization selection schemes, including amino acid autotrophic lines for the intraspecific fusion of *Datura innoxia* protoplasts.

5. Use of resistance markers

Dominant characteristics for traits such as resistance to herbicides and amino acid analogs are employed in selection. When protoplasts from two separate and mutually exclusive resistant lines are fused, the tolerance of each parental species is acquired by the somatic hybrid cells and the latter exhibit dual resistance. Unfused parental protoplasts and homokaryons are eliminated during selection.

6. Use of double mutants

Protoplasts of many potential fusion partners are of the wild type and, as a result, do not possess any markers suitable for selection. One method of overcoming this limitation is to construct a parental line carrying both negative and positive selectable markers, that is, an auxotrophic trait and a resistant trait. Only the heterologous fusion products with complemented auxotrophic-resistant traits will survive selection.

7. Use of transformed cell lines

Resistance markers used in somatic hybrid selection schemes can be introduced by transformation. Protoplasts from transformed lines of *S. tuberosum*, carrying kanamycin or hygromycin B resistance genes, are fused, resulting in a hybrid tissue that is resistant to both antibiotics. In forage legumes, kanamycin resistance combined with the use of the metabolic inhibitor sodium iodoacetate, are used to select somatic hybrids between *Lotus corniculatus* and *Lotus tenuis*.

8. Use of antimetabolites

Complementation selection systems can also be based on the use of irreversible biochemical inhibition, which blocks metabolic pathways when the parental protoplasts are treated prior to fusion. Inactivated parental lines cannot undergo cell division in their own right, but hybrid cells exhibit metabolic complementation and undergo sustained growth. The metabolic inhibitor sodium iodoacetate is used in combination with other markers, including lack of sustained cell division in one of the parental protoplast lines, to select somatic hybrid plants. An example of this selection system involves the fusion of sodium iodoacetate-inactivated *Oryza sativa* protoplasts with protoplasts from a range of wild *Oryza* species. Protoplasts of the wild species fail to divide in culture. Iodoacetate usage requires a careful determination of treatment levels, so as to minimize cross-toxicity from parental protoplasts.

9. Use of tumorous growth of F1 hybrids

To permit continued development of regenerated shoots from calli derived from the fusion of protoplasts of *N. langsdorffii* and *N. glauca*, the tissues are grafted onto plants of *N. glauca*. Tumor formation, a characteristic of the sexual F1 hybrid between these two *Nicotiana* species, is observed on the scion, thus providing a method for somatic hybrid selection.

10. Use of differential growth and plant regeneration

The differential response of parental protoplasts to culture conditions provides a method for selecting somatic hybrid tissues. Following the fusion of iodoacetamide-inactivated *O. sativa* protoplasts with those of *Echinochloa oryzicola*, the treated protoplasts are cultured in a medium that supports the growth of rice protoplast and somatic hybrid cells, but not protoplasts of *E. oryzicola*. The different mechanisms of plant regeneration also provide a method for somatic hybrid selection. Thus plant regeneration in *R. hirta* occurs through shoot formation, whereas shoot production in *R. laciniata* is via rhizogenesis. Somatic hybrids and plants of *R. laciniata* are regenerated through rhizogenesis. The somatic hybrids are identified by the presence of pigmented roots, a feature of *R. hirta*.

11. Use of electrical stimulation

Electrical pulse treatments have been shown to enhance the division of plant protoplast-derived cells, and to stimulate shoot formation from protoplast-derived cells of several plants, including woody species such as *Prunus avium* X *pseudocerasus*. This technology is applied successfully in somatic hybridization. Thus electroporation of parental protoplasts prior to electrofusion promotes the division of heterokaryons and facilitates the recovery of somatic hybrids between the two woody species, *Pyrus communis* var. *pyraster* and *Prunus avium* X *pseudocerasus*. Somatic hybrid tissues are not produced when parental protoplasts are not electrostimulated prior to fusion. Electrostimulation of protoplast division and plant regeneration may prove particularly useful in cases in which parental protoplasts respond to this treatment with increased growth and plant regeneration, especially if used in combination with other selection techniques.

12. Physical isolation of heterokaryons

Biochemical complementation/selection systems usually lead to preferential recovery of amphidiploid somatic hybrids. Asymmetric hybrids, such as those possessing genomes but only a few chromosomes of the other parent, are likely to be lost during selection due to an inability of the cells to survive the strong selection pressure, through incomplete complementation to growth proficiency. This, combined with the lack of suitable selectable markers for many parental species, makes physical identification, isolation, and culture of fusion products an important alternative. Heterokaryons can be identified by a dual-labeling system, such as red chlorophyll autofluorescence used in combination with the yellow-green fluorescence of fluorescein diacetate. Fluorescein diacetate labeling combined with the use of red fluorochromes such as rhodamine isothiocyanate has also been employed. Somatic hybrid tissues of the *Medicago* species and somatic hybrid plants of the *Solanum* species have been recovered from dual-labeled heterokaryons by using micromanipulation. However, micromanipulation is a laborious technique and the number of heterokaryons that can be selected with ease is limited. Flow cytometry is another procedure that permits the selection of larger numbers of labeled heterokaryons. Until recently, the range of somatic hybrid plants recovered from flow-sorted heterokaryons was limited to the genera *Nicotiana* and *Brassica*. However, sorting has been extended to fuse protoplasts from a wide combination of plant species, in some cases with somatic hybrid plant production.

13. Other criteria

Visual selection: (1) morphology of the tissue in culture, (2) morphology of leaf, (3) trichome characteristics, (4) chromosome number, (5) patterns of peroxidase isozymes, and (6) pigmentation or color of cells.

E. CONFORMATION OF HYBRIDITY

The first indication of the hybridity of cell lines/callus is their ability to survive the selection procedure. To eliminate potential problems such as reversion, cross-feeding, and residual leakiness from the selection system, additional confirmation is required at both the callus and plant levels. Verification of hybridity requires the demonstration of the presence and expression of genetic traits from both parents.

    1. **Morphological characteristics of regenerated plants**: Intermediate morphologies can be used to identify somatic hybrid material. Leaf shape and size and floral characteristics, including flower size, color, and number of ray florets, can be evaluated. Ideally, several independent characteristics should be considered. The more distant the taxonomic relationship between the parental species, the greater the number of morphological characteristics that are available for assessment. Morphological features, such as pigmentation and relative growth rates, can be

used to identify hybridity, even in protoplast-derived callus. In some cases, morphological analysis may be complicated by abnormalities arising from aneuploidy, somatic incompatibility, or somaclonal variation from the effects of the tissue culture procedure.

2. **The chromosome complement of hybrids:** The chromosome complements from actively dividing somatic cells, such as those from root tips, provide further evidence of hybridity and of ploidy levels. Hybrid plants are identified by their chromosome numbers, the structure, and size of somatic cell chromosomes, when compared with the karyotypes of parental species. In some cases, chromosome counts may be inaccurate due to doubling or elimination of chromosomes.

3. **Isoenzyme analysis:** The different electrophoretic mobilities of isoenzymes that catalyze basic cell functions can be used to identify hybrid tissue/plants, as in the case of somatic hybrids between wild pear and colt cherry, *Rudbeckia* species, and *Oryza* species. Hybrid tissue may possess isoenzyme band profiles characteristic of each parent, as well as additional bands. These additional bands may be regarded as possible artifacts or as hybrid molecules or genes present in parent cells that are expressed within the new genetic background.

4. **Molecular analysis**: The development of molecular techniques, such as restriction fragment analysis and DNA hybridization of nuclear and organelle DNAs, has permitted detailed analysis of the genetic constitution of somatic hybrids. Specific patterns of restricted DNA of both mitochondria and chloroplasts confirm hybridity, and elucidate organelle segregation and DNA recombination patterns. Species-specific DNA fragments are used to determine the relative parental contributions to somatic hybrids. Restriction fragment length polymorphism (RFLP) mapping permits a more detailed examination of the inheritance of nuclear and organelle genomes in somatic hybrids. Thus a variety of established methods are available that permit accurate determination of the presence of genetic material from both parents in somatic hybrids.

F. REGENERATION OF HYBRID PLANTS

Exploitation of protoplast fusion for crop improvements depends mainly on the success in regeneration from hybrid *calli*. Further, the hybrid plants must be at least partially fertile, in addition to having some useful traits. Even though a number of culture techniques have been refined to regenerate somatic hybrids, it is still not possible to recover hybrid plants from a number of somatic combinations; this phenomenon is called "***somatic incompatibility***'. Some somatic hybrid plants retain the full or nearly full somatic complements of the two parental species; these are called *symmetric hybrids*. Frequently, somatic hybrids (between distantly related species) are sterile. An approach to the improvement of apparently useless somatic hybrids is to fuse the protoplasts from the hybrid with those of one of the parental species (**somatic back hybridization**). Many somatic hybrids exhibit the full somatic complements of one species, while all or nearly all the chromosomes of the other species are lost during the preceding mitotic divisions; such hybrids are referred to as **asymmetric hybrids**.

In contrast to sexual hybrid cells (heterozygotes for nuclear genes and organellar genes only from female plants), somatic hybrid cells contain both nuclear as well as cytoplasmic complements from both the parental species. Somatic hybrids that contain the nucleus of one species but the cytoplasm from both parental species are called **cybrids** (cytoplasmic hybrids). The main objective of cybrid production is to combine the cytoplasmic genes of one species with the nuclear and cytoplasmic genes of another species. But the mitotic segregation of plasma genes leads to the recovery of plants having plasma genes of only one or the other species.

G. FATES OF NUCLEI AND CHLOROPLASTS IN FUSANTS

During the development of fusants, the heterokaryotic nature is lost rather early by degeneration or extrusion of nuclei of one type, by segregation particularly as a consequence of non-synchronized mitosis, or by the formation of hybrid nuclei via fusion of interphase nuclei or of mitotic figures.

1. Homo- and heterokaryons

Fusion bodies contain only one nucleus in case of subprotoplast fusion, two nuclei in most of the heterotypic fusion experiments, but frequently numerous nuclei. Multinucleate fusion bodies arise either from multiply-induced fusion or from the participation of spontaneous fusion products in induced fusion. Multinucleate protoplasts of various combinations are mostly deteriorates. The heterokaryon nature may be transmitted to the daughter cells. This is most reliably established when mitosis occurs unsynchronized, as it has been observed in the spontaneous fusants of soybean and in heterokaryocytes of *Vicia faba* X *Petunia hybrida*.

2. Cybrid formation

During normal sexual hybridization the plastome is contributed by only one of the parents (female), whereas in somatic hybridization the hybrid receives the plastomes from both the parents. Consequently, the cell fusion approach to crossing plants offers a unique opportunity to study the interaction of the cytoplasmic organelles from both parents. Such cytoplasmic hybrids are referred to as cybrids. Power et al (1975) demonstrated that following protoplast fusion and culture, cell lines can be isolated which carry the nucleus of one of the parents and the cytoplasm of both. Detailed characterization of the cytoplasm in presumptive somatic hybrids by different workers has revealed that often a selective elimination of the plastids of one of the parents occurs and all other possible combinations between the nucleus and organelles. This feature of somatic hybrid cell lines has been utilized in transferring cytoplasmic male sterility intra- and interspecifically.

In the experiments involving the fusion of full protoplasts, cybrids may arise through: (i) fusion of a normal protoplast with an enucleate protoplast, (ii) fusion between a normal protoplast and a protoplast containing a non-viable nucleus, (iii) elimination of one of the nuclei after heterokaryon formation, or (iv) selective elimination of chromosomes at a later stage. Enucleation of protoplasts can be achieved by centrifuging protoplasts at 20,000–40,000g for 45–90 min., in an iso-osmotic density gradient with 5–50% percoll.

3. Hybrids formation

Nuclei apparently composed of subunits repressing original parental nuclei have been found in fixed preparations. Nuclear material most frequently combines during mitosis of plastocytes. Fairly synchronized mitosis is demanded for this pathway. No clear information is available on the process of synchronization. Probably, close contact between the nuclei may have been involved in the fusion.

4. Fate of plastids

Plastid segregation and incompatibility in the early development of fusants have been discussed elsewhere. Unequal distribution of the different plastids to the daughter cells after the first cytokinesis has been found in protoplast-to-protoplast and in protoplasts-to-subprotoplast fusants. Their random location in the cytoplasm or incomplete mixture of the protoplasts might have been responsible for the unequal distribution.

5. Formation of genetic mosaics

Events during the development of fusants may lead very early to mosaics of variant cells. Naturally, the diversity is much lower in intraspecific than in interspecific fusants. Localization of certain types of cells and competition within the regenerants may decide on transmission or elimination of particular recombinants. It is, therefore, helpful in interspecific combinations to make use of selective pressure to a desired cell type or to separate the cells as early as possible. Segregation of nuclei, leading to cybrid cells or formation of hybrid nuclei are certainly completed after a few cell divisions. Segregation, loss and rearrangements of chromosomes, plastids, and mitochondria may be extended to later stages of development up to sexual progenies.

H. PRACTICAL APPLICATIONS OF SOMATIC HYBRIDIZATION AND CYBRIDIZATION

Protoplast fusion and somatic hybridization have been well exploited in crop improvement programs and genetic engineering studies. Some of the practical applications are:

1. **Means of genetic recombination in asexual or sterile plants**: Somatic cell fusion appears to be the only approach through which two different parental genomes can be recombined among plants that cannot reproduce sexually and those sexually sterile.

2. **Overcoming barriers of sexual incompatibility**: Sexual crossing at interspecific or intergenetic levels often fails to produce hybrids due to incompatibility barriers. This problem can be overcome by somatic cell fusion. For example, a fusion of the protoplasts of potato and tomato has resulted in production of a hybrid plant, the **pomato** (Figure 36.4).

3. **Cytoplasm transfer**: The genotype of cytoplasm codes for a number of practically important traits, such as the rate of photosynthesis, low or high temperature tolerance, male sterility, and resistance to diseases or herbicides.

III SOMACLONAL VARIATION

Clones of plants regenerated from callus often produce variants, which are sometimes called **calliclones** while those originating from protoplasts are called **protoclones**. Larkin and Scowcroft (1981) proposed the term **somaclone** to describe the plants originating from any type of tissue culture. The genetic variation found to occur between somaclones was then called **somaclonal variation**. In contrast to shoot tip propagation, the calli which were originally genetically homogenous yield plantlets that can show considerable variation in characteristics such as

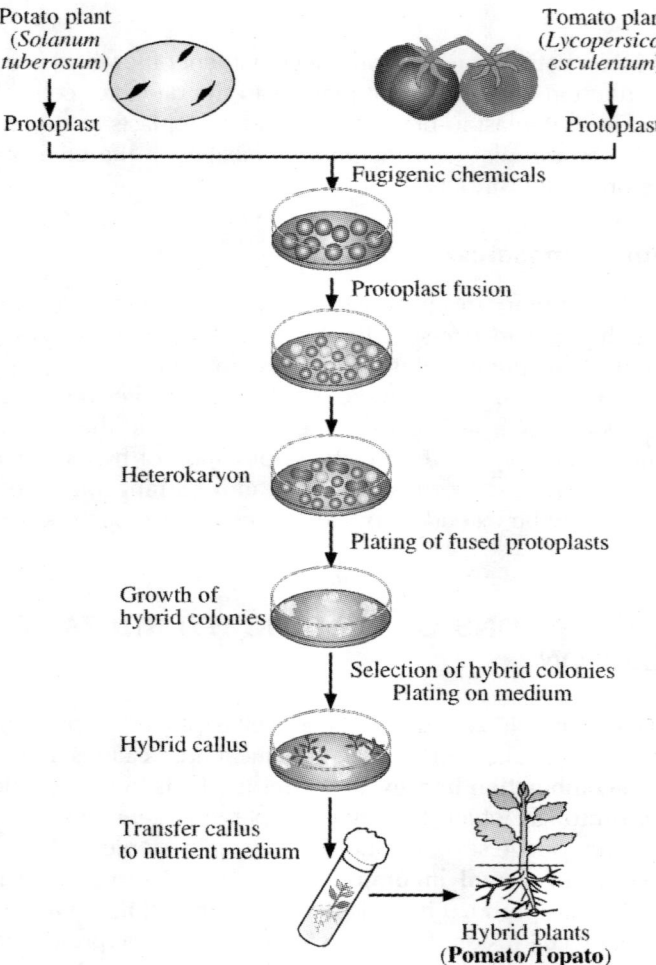

Figure 36.4 *Fusion of protoplasts of potato and tomato resulting in production of hybrid*

aneuploidy, sterility, and morphology. Progeny that differ significantly from the parent are called *'somatic variants'*. Somaclonal variation might be an appropriate technology for genetic manipulation of crops with ***polygenic traits***. It is an alternative tool to plant breeding for a generation of new varieties that could exhibit disease resistance and improvement in quality and yield. Introduction of variability in regenerated plants can be agriculturally useful. These **somatic variants** (some authors call them "protoclonal variation") have established new and important varieties such as the novel orange or nectarines. Potato and sugar cane 'sports' were found to be resistant to disease in contrast to the susceptible parent plants. Herbicide resistance, cold- and salt-tolerance are additional favorable traits that can be found in 'sports'.

A. MECHANISM OF SOMACLONAL VARIATION

Somaclonal variation may be attributed either to (i) pre-existing variations in the somatic cells of the explant (genetic) or (ii) variation generated during tissue culture (epigenetic). Often both factors may contribute to the variation. The original ploidy levels of the plant or plant organ from which the explant is taken may play an important role in somaclonal variation. When the cells of various genomic constitutions of the initial explants are induced to divide in culture, the cells may exhibit changes in chromosome number such as anueploids and polyploids. In contrast to pre-existing variations, when the callus or cell suspension is maintained for long periods, their morphogenetic ability is either lost or significantly reduced under the same culture conditions. Although factors affecting morphogenetic expression have not been precisely established, disorganized growth results in polyploidy, aneuploidy, chromosome breakage and reformation, gene deletions, gene translocations, gene amplifications and point mutations, and transposable elements (**chromosomal mosaicism**) are thought to lead to a change in or loss of morphogenetic capacity. DNA recombination or modification of extranuclear genes in mitochondria and chloroplasts can also account for some variation detected in regenerated plants. Transient variation in the regenerated plant may have an epigenetic cause where there is no change in genetic composition of the cell. These epigenetic changes are usually induced by culture conditions and are not useful for crop improvement.

The callus is produced in the intact plant as a natural wound response and only in a few instances are root and shoot initials produced from the callus itself in the intact plant. The callus proliferates with the use of growth regulators such as auxin and cytokinin. It is therefore not surprising that callus and cell suspension culture will have abnormal developmental pathways. Plant growth regulators, especially 2,4–D have been implicated as mutagens. Automutagenesis through self-synthesis of mutagenic compounds and pre-existing variation in the explant cells in normal plants also show a low level of mutation. Transient variation in the regenerated plant may have an epigenetic cause where there is no change in genetic composition of the cell. These epigenetic changes are usually induced by culture conditions and are not useful for crop improvement because they are not expressed in the progeny of regenerated plants. A selection screen is often applied to the tissue culture system (*in vitro* selection) to concentrate the types of mutant that are (e.g. salt, herbicides, fungal toxin, drought, resistance, pesticides, yields, anoxia, etc.). In some studies, the rate of somaclonal variation has been promoted by the use of ionizing irradiation or chemicals.

A summary of the most commonly accepted ways of decreasing or increasing the genetic variation in callus cultures and in the plants regenerated from them, is shown in Figure 36.5. For plant breeding purposes, the variability can be further increased by mutagenic treatments.

B. SOURCE MATERIAL AND CULTURE CONDITIONS

The evolution of new crop varieties through somaclonal variation is significantly influenced by genotype, explant source, duration of culture, and growth hormone effects. These factors can be readily varied with respect to some crops, while in others the constraints of regeneration may limit flexibility in manipulating one or more of these critical variables. The genotypes influence both the regeneration efficiency and the frequency of somaclonal variability. Potato plants regenerated from leaf discs yielded 12.3% phenotypic alterations in contrast to 50.3% variants from the callus of rachis and petiole.

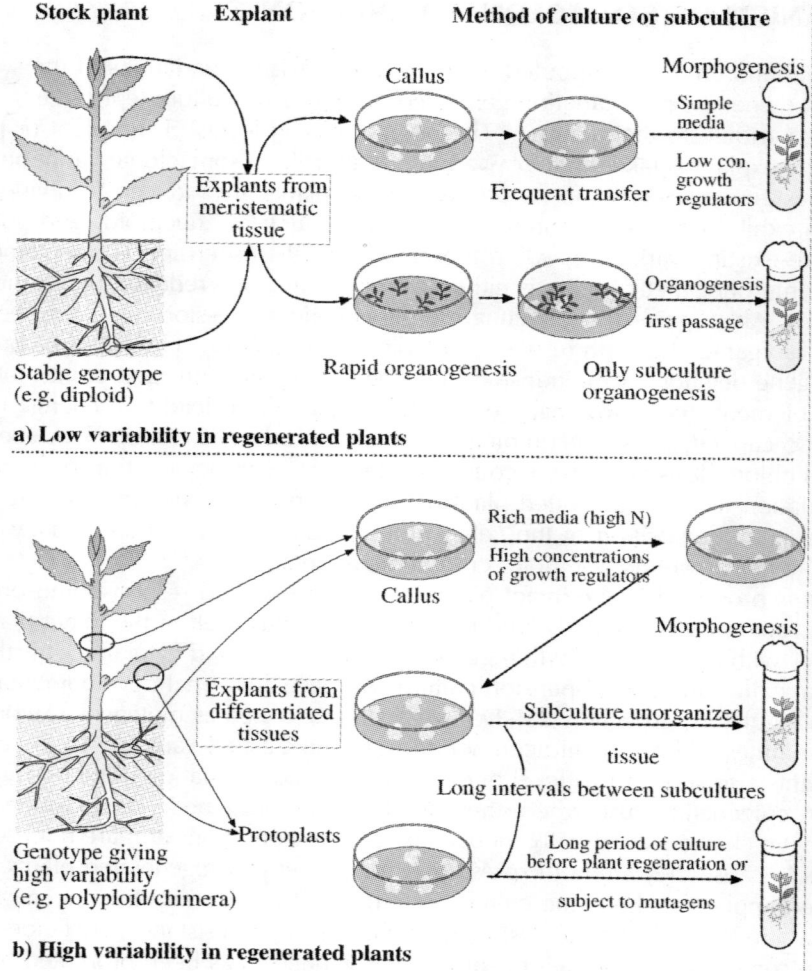

Figure 36.5 *Factors found to affect the genetic variability in callus cultures*

High proportions of growth regulators effect karyotypic alterations in cultured cells. 2,4–D, BAP, and others have been shown to induce chromosomal variability in cultured cells, leading to the formation of somaclones. Growth hormones are essential for the induction of organogenesis and shoot differentiation; however, high concentrations of these substances may not permit recovery of whole plants in tissue cultures and the proportion of hormones in the medium needs to be carefully monitored in establishing culture systems.

C. MOLECULAR BASIS OF VARIATION

Somaclonal variants may also arise as a result of more subtle changes due to single gene mutations in cultures, which have cells apparently showing no karyological changes. Recessive mutations are not detected in regenerated plants, but are expressed in progeny segregates of regenerated plants after selfing. Changes in the cytoplasmic genomes have also been observed in somaclones. Another aspect of single gene mutation responsible for somaclonal variation relates to transposable elements. Somaclonal variation may also be due to molecular changes caused by mitotic crossing over in regenerated plants. This could include both symmetric and asymmetric variation. Small changes in the structure of chromosomes could alter the expression and genetic transmission of specific genes, such as deletion or duplication of a copy of a gene, or gene conversion during repair processes. Further recombination, or chromosome breakage, at preferential regions or 'hot spots' of a particular chromosome affects the genome in a disproportionately high frequency, resulting in an altered phenotypic expression. Recent studies have demonstrated that changes in the organelle DNA, isoenzyme, and protein profiles correlate with the occurrence of somaclonal variation in many plants. Comprehensive information on various aspects of the molecular basis of somaclonal variation has been provided by Ball (1990).

D. ISOLATION OF SOMACLONAL VARIANTS

Mutants for several characters can be isolated from cell cultures either without selection pressure or with selection pressure. In the **program without selection pressure**, unorganized callus and cells are grown in cultures for various periods on a medium that contains no selective agent (toxin or inhibitory compounds); such calli are induced to differentiate into whole plants. The regenerated plants are ultimately transferred to the field and screened for somaclonal variations. Somaclonal variants of sugarcane, potato, tomato, geranium, wheat, maize, lucerne, and many other crops have been obtained through this approach. In the program **with selection pressure**, variant cell lines are screened from cultures by their ability to survive in the presence of a substance in the medium that may be toxic or inhibitory or may survive under conditions of environmental stress. Variants may be obtained by direct selection, indirect selection, or by individually testing single cell-derived colonies. The various approaches to the isolation of somaclonal variants can be grouped into two broad categories: (i) screening for spontaneous variants and (ii) selection of variants *in vitro*.

1. Screening for spontaneous variants

Unorganized callus and cells grown in cultures for various periods on a medium that contains no selective agent are induced to differentiate into whole plants. The existence of genetic variations in cell cultures makes it possible to screen for resistance to a selection pressure. Sometimes selection can be effective *in vivo* after a population of variable plants has been obtained. This technique involves the observation of a large number of cells or regenerated plants for the detection of variant individuals. This approach is the only feasible technique for the isolation of mutants for yield and yield traits, pathogen resistance, dwarf forms, etc. In general, R1 progeny (progeny of regenerated RO plants) are scored for the identification of variant plants, and their R2 progeny lines are evaluated for confirmation. Screening has been profitably and widely employed for the isolation of cell clones that produce higher quantities of certain biochemicals. Somaclonal variants of various crop species that have been obtained using this approach are: (1) Sugarcane: Sugarcane cv. Pindar for resistance to Fiji disease (virus) and to downy mildew. (2) Potato: Somaclonal variants resistant to early- and late-blight were isolated from potato mesophyll protopalsts. (3) Tomato: In tomato many somaclones were isolated, such as recessive mutations for male sterility, resistance to Fusarium oxysporium, tangerine virescent leaf, flower, and fruit color, etc. (4) Geranium: An improved scented geranium named 'Velvet Rose' was developed from somaclonal variation. (5) Cereals and Grasses: The callus-derived plants of cereals and grasses can be a rich source of somaclones. Variants developed were for male sterility, gliadin protein regulation, frost resistance in wheat, increased yield in rice, and earliness in maize. (5) Miscellaneous: Attempts have been made to induce somaclonal variation in lucerne, cucurbits, strawberry, peach, ornamental and forage legumes, Fuchsia, carnations, etc.

2. Selection of variants *in vitro* with selection pressure

A more efficient selection system can often result from the application of appropriate stress. In this approach, suitable selection pressure is applied which permits the preferential survival/growth of only variant cells. When selection pressure allows only the mutant cells to survive or divide, it is called positive selection. On the other hand, in the case of negative selection, the wild type cells divide normally and therefore are killed by a counter selection agent, e.g. arsenate; the mutant cells are unable to divide as a result of which they escape the counter selection agent. These cells are subsequently rescued by removal of the counter selection agent. The negative selection approach is utilized for the isolation of auxotrophic mutants.

The positive selection approach may be further subdivided into four categories: (*i*) direct selection, (*ii*) rescue method, (*iii*) step-wise selection, and (*iv*) double selection. In **direct selection**, the cells resistant to the selection pressure survive and divide to form colonies; the wild type cells are killed by the selection agent. This is the most common selection method; it is used for the isolation of cells resistant to toxins, herbicides, salt, antibiotics etc. In the **rescue method**, the wild type cells are killed by the selection-agent, while the variant cells remain alive but do not usually divide due to the unfavorable environment. The selection agent is then removed to recover the variant cells. This approach has been used to recover low temperature and aluminum resistant variant cells. In **step-wise selection**, the selection pressure may be gradually increased from a relatively low level to the cytotoxic level; the resistant clones isolated at each stage are subjected to the higher selection pressure. It may often favor gene amplification (generally unstable) or mutations in the organelle DNA. In some cases, it may be feasible to select for survival and/or growth on one hand and some other features reflecting resistance to the selection pressure on the other; this is called **double selection**. An example is selection for resistance to the antibiotic (streptomycin), which inhibits chlorophyll development. The selection is based on cell survival, i.e., colony formation as well as for the development of green color in these colonies in the presence of streptomycin.

Protocol-1: Selection of resistant somaclonal variants for glyphosate

Selection for resistance to a chemical is the most straight-forward selection method. In this exercise, selection for the glyphosate weedicide could be attempted.

Protocol

1. Take three 250 ml Erlenmeyer flasks, add aliquots (50 ml) of suspension culture and 50 ml of a liquid MS medium containing glyphosate in a sublethal or borderline lethal or lethal concentration to each flask.

2. Incubate the culture flasks for two weeks. Upon incubation, resistant cells can be observed as they undergo vigorous growth.
3. Following two weeks of incubation, transfer resistant cells onto the surface of a solid MS medium containing a borderline lethal concentration of glyphosate.
4. The resistant cells will form a callus and if subcultured onto a regeneration medium, they will probably develop plants which are glyphosate tolerant.

3. Disease resistance

Improved disease-resistance is a major goal in the breeding programs of many crop plants. Tissue culture techniques offer plant breeders a variety of new ways to identify, select, and transfer the genes involved in disease resistance. These include: (1) Use of cultured cells and protoplasts to study disease mechanisms; (2) production of haploid tissues and/or plants in which recessive genes for disease resistance can be more readily detected; (3) transfer of resistance via the wide crosses made possible by embryo culture, protoplast fusion, and other novel genetic manipulations; (4) regeneration from culture of large populations of plants, some of which may show enhanced disease resistance; (5) *in vitro* selection of material that can be regenerated into disease-resistant plants. *In vitro* selection is an appealing strategy. It takes advantage of many of the most distinctive features of plant cell cultures – growing large numbers of single cells or cell clumps in a small space, exposing plant populations from resistant material.

Types of selective agents: The agents can be pathogens, host-specific toxins, or non-specific toxins. A study of phytotoxins helps in understanding the mode and site of action at cellular and subcellular levels. Protoplasts can be exposed to physical or chemical mutagens to induce mutations in cells for resistance to phytotoxin. Selection for resistance to *Helminthosporium* T-toxin in maize cultures was developed and in this study, the embryogenic maize calli were exposed to the toxin in cultures. In the process, they obtained resistant and fertile plants. The calli were derived from plants inheriting toxin-suceptible and male sterile traits maternally. This indicates that the variation is of cytoplasmic origin.

4. Environmental stress tolerance

In many parts of the world, environmental stress due to high salt levels in soil is a major constraint in the development of agriculture (south and southeast Asia, an estimated 54 million hectares are saline soils). Since the first report of sodium chlorite tolerant tobacco plants, a number of cell lines and plants resistant to rigorous salinity have been developed. Screening existing material for salt tolerance has met with some success. However, the need is to increase the salt tolerance of rice beyond the maximum levels now possible. Tissue culture is recognized as a novel means to generate genetic variability and is possibly a new tool for improving salt tolerance in crop species.

Attempts to develop chilling tolerant cell lines were made in tobacco by exposing cell lines to lower temperatures.

5. Antibiotic resistance

Somaclonal variants resistant to the antibiotic streptomycin, linomycin, kanamycin, chloramphenicol, and cycloheximide have been developed from various plant species. Streptomycin resistance can be controlled by cytoplasmic (*Nicotiana tabacum*) or recessive nuclear genes (*N. sylvestris*). In *N sylvestris*, resistance to linomycin has been identified as a chloroplast-controlled trait and inherits resistance maternally.

IV NATURE OF GAMETOCLONAL VARIATION

Contrary to somatic cells, the gametes, which are products of meiosis consist of only half-genetic complements of parents. To distinguish somatic-derived somaclones from gametic-derived gametoclones three parameters are considered. First, both dominant and recessive mutant genes induced by gametoclonal variation will express directly in haploid plants regenerated from the microspores of diploid anthers. This facilitates the study of gametoclones (RO) directly to identify new variants. Second, recombinants recovered in gametoclones would be the result of meiotic crossing over. Third, the gametoclones can be used only after having been stabilized by doubling its chromosome number. The value of gametoclonal variation in crop improvement is evident from the development of the double-haploid line by anther culture of F1 hybrid plants. Several double-haploid plants have been recovered that expressed mixed characters of both parents. Variation in the chromosome number of gametes or gametophytic tissue plays an important role in gametoclonal variation.

V SUGGESTED READING

Auriol P, Strobel G, Betran JP, and Gray G (1978) Rhynchosporoside, A Host Selective Toxin Produced by *Rhynchosporium secalis*, The Causal Agent of Scald Disease of Barley, *Proc. Natl. Acad. Sci. USA*, **75**:4339–4343.

Bajaj YPS, Phul PS and Sharma SK (1980) Differential Tolerance of Tissue Cultures of Pearl millet to Ergot Extract, *Indian J. Exp. Biol.*, **18**:429–432.

Bhojwani SS and Razdan MK (1983) *Plant Tissue Culture: Theory and Practice*, Elsevier.

Cocking (1960) *Nature*, **187**:927–929.

Earle ED and Gracen VE (1981) The Role of Protoplasts and Cell Cultures in Plant Disease Research, pp285–297, In: RC Staples and GH Toenniessen (Eds) *Plant Disease Control*, John Wiley and Sons, New York.

Finch RP (1994) An Introduction to Molecular Technology, In: *Molecular Biology in Crop Protection*, (Eds) G Marshall and D Walters. Chapman and Hall.

Gengenbach BG, Green CE, and Donovan CM (1977) Inheritance of Selected Pathotoxin Resistance in Maize Plants Regenerated from Cell Cultures, *Proc. Natl. Acad. Sci. USA*, **74**:5113–5117.

Ignacimuthu S (1998) *Plant Biotechnology*, Oxford and IBH Publishing Co. Pvt. Ltd. New Delhi, Calcutta.

Morrish F, Vasil V, and Vasil K (1987) Developmental Morphogenesis and Genetic Manipulation in Tissue and Cell Cultures of the *Graminae*, In: *Molecular Genetics of Development* (Eds) Scandalis JG. Advances in Genetics, **24**:431–499.

Nabors MW, Gibbs SE, Bernstein CS and Mein ME (1980) NaCl–tolerant Tobacco Plants from Cultured Cells, *Z. Pflanzenphysiol*, **97**:13–17.

Power JB, Berry SF, Chapman JV, and Cocking EC (1980) Somatic Hybridization of Sexually Incompatible Petunias: *Petunia parodii, P. parviflora, Theor. Appl. Genet.,* **57**:1–4.

Razdan MK (2000) *An Introduction to Plant Tissue Culture,* Oxford & IBH Publishing Co. Pvt. Ltd., New Delhi, Kolkata.

Uchimiya H (1982) Somatic Hybridization between Male Sterile *Nicotiana tabacum* and *N. glutinosa* through Protoplast Fusion, *Theor. Appl. Genet.,* **61**:69–72.

Wright WE (1978) The Isolation of Heterokaryons and Hybrids by a Selective System using Irreversible Biochemical Inhibitors, *Exp. Cell Res.,* **112**:395–407.

Zimmermann U and Scheurich P (1981) High Frequency Fusion of Plant Protoplasts by Electric Fields, *Planta (Berl),* **151**:26–32.

37

Application of Plant Cell, Tissue and Organ Culture

I INTRODUCTION

The applications of plant cell and tissue culture in plant science are vast and varied. The discussion on applications of cell culture in this chapter is restricted only to non-breeding techniques. The applications in plant breeding have been discussed in previous chapters. Particular genetic traits can be introduced into modern cultivars to improve nutritional quality, herbicide resistance, disease resistance, and plant yield. The establishment of micropropagation for rapid propagation, root tip culture for virus free plants, and anther culture for haploid plant development are some of the applications extensively used for plant improvement. Plants are also a source of high added value, low volume compounds for the pharmaceutical, perfume, and fine chemical industries. About 25% of all drugs prescribed are derived from plants. Almost 1500 new plant compounds are reported in literature each year. The majority of compounds of commercial importance are secondary metabolites. Large amounts of plant tissue need to be processed in order to isolate a small amount of drug from the wild plant. The potential consequences of this action is that without clonal propagation and plant breeding programs, the demand for the plants or plant products cannot be met and also may lead to the loss of some rare plant species.

II LARGE-SCALE PLANT PROPAGATION

If a plant can only be propagated from seed, then increasing plant numbers can be a lengthy process. This is particularly true in case of perennials where flowering takes several years after seed germination. In such cases, the use of *in vitro* methods of propagation can be invaluable. In practice, the explant may be taken and induced to form a callus on solid media and then a suspension culture derived by inoculating the callus into a liquid medium. After the cells have increased in number, they are plated out and somatic embryoids selected. On cultivation, embryoids will develop into regular plants. Another method of micropropagation is clonal propagation by tissue culture. In this chapter, as a case study, clonal propagation of orchids is discussed in detail.

A. CLONAL PROPAGATION OF ORCHIDS

Initially, *in vitro* techniques were used for germination of orchid seeds shed from mature fruits of various species and hybrids. However, recently, the practice of harvesting immature (green) fruits and aseptically culturing the contents has been used to overcome some barriers of incompatibility and produce some unique hybrids. Asexual reproduction gives rise to plants, that are genetically identical to the parent plant. A single plant with desirable characteristics can be selected from a breeding program and propagated so that further trials and selections can be carried out faster by *in vitro* clonal propagation than *in vivo* seed production. The undesirable juvenile phase associated with seed-rised plants in some varieties does not appear in the vegetatively propagated plants from adult material. For orchids, *in vitro* clonal propagation is the only commercially viable method of micropropagation. Clonal multiplication of the cultivar is very important in horticulture and silviculture.

Since Morel (1960) demonstrated the applicability of tissue culture for rapid clonal propagation of orchids, this technology has been successfully applied for commercial production of selected orchid plants. Large quantities of quality plants, establishment of new hybrids and multiplication and preservation of the germplasm are some of the benefits that have been derived from the application of this technology. As orchids are extremely heterozygous, a diversity of response to tissue culture is inevitable depending on genotype, species, and genus. Furthermore, since the time from initiation of tissue culture to maturity of plant may range from one to several years, data on uniformity, stability, and fidelity of end products is scarce.

Materials

Culture medium: The formulations for media to successfully grow orchids *in vitro* are relatively simple. Two commonly used basic media are Kundson C (1946) and Vacin and Went (1949) in which growth regulators or organic additives are incorporated. Coconut water (15% v/v) is added to the medium for initiation and multiplication while homogenized green banana (50–100 g/l) is added to the medium for rooting. Supplemental sucrose is either reduced or deleted in the media for shoot promotion and rooting.

Explants: Explants are excised from vegetative shoots, young inflorescence, leaves, and roots. Terminal and axillary buds from healthy plants are most preferred. Apical bud explants include the apical meristem with 3 to 4 nodes with attached young leaves. Axillary bud explants consist of the cube with axillary bud attached to the node. Root explants consist of 3 to 5 mm root tips.

Disinfestation: Disinfestation is accomplished in either 2 or 3 steps using household bleach in descending concentrations (20, 15, 10%) and time (20, 15, 10 min.). Tween-20 (2–3 drops/100 ml), a wetting agent, is usually added to the initial disinfesting solution. Sterile deionized water is used for rinsing after each treatment. More drastic treatment would be rinsing shoots in 70% ethanol for one min. followed by rinsing with sterile deionized water prior to disinfestation with household bleach.

Culture conditions: Aseptic cultures are maintained under continuous or interrupted light from cool white or power groove fluorescent lamps at approximately 25°C to 28°C. Light may be cycled to provide a 16 hour day. The growth habit of orchids is either monopodial or sympodial. Monopodial orchids have a single upright indeterminate axis of growth; the inflorescences are axillary (e.g. Ascocentrum, Doritis, Vanda, Phalaenopsis). Sympodial orchids have two axes of growth – a horizontal axis which is indeterminate and a vertical axis which is determined and terminated by an inflorescence (e.g. Cattleya, Dendrobium, Cymbidium, Epidendrum, Miltonia, Oncidium). Recognition of the basic growth habit is important for selecting the appropriate procedure for initiation and multiplication.

Solutions and media

Use deionized water for preparation of stock solutions and media. Use sterile deionized water for preparation of disinfesting solution and rinsing during disinfestation. The concentration of Tween-20 wetting agent used is 2 to 3 drops per 100 ml.

Modified Vacin and Went medium

	Amount/litre
$Ca_2(PO_4)_2$	0.20 g
KNO_3	0.525 g
$(NH_4)_2SO_4$	0.50 g
$MgSO_4.7H_2O$	0.25 g
KH_2PO_4	0.25 g
$MnSO_4.4H_2O$	20.00 g
Sucrose	20.00 g
Agar	8.00 g
Water	845.00 ml
Coconut water	150 ml
$FeSO_4.7H_2O$	27.8 mg
$Na_2.EDTA$	37.3 mg
pH 4.8–5.0	

Other natural complex additive: Green banana
1. Harvest, wash, and store mature green bananas in a freezer.
2. Just prior to use, defrost the banana at room temperature or in a microwave oven and peel. Fresh green bananas are very difficult to peel.
3. Weigh amount of banana to be added (50–100 g/l).
4. Homogenize in a blender for 30 to 60 seconds at high speed and add to the medium.

Procedure

Initiation of sympodial orchids

1. Select a young vegetative shoot without expanded leaves. Sever the shoot from the rhizome with a razor blade.
2. Remove all papery scale-like leaves from the shoot and rinse in running water for a minimum of 30 min.
3. Remove in sequence all exposed scale-like leaves with forceps until the first visible axillary bud is exposed. Make sure that each leaf is torn off to the node. It is better to tear the leaf off than to cut it.

4. Disinfect in 15% household bleach with Tween-20 for 15 min.; shake occasionally.
5. Rinse in sterile deionized water.
6. Remove all but 2 or 3 of the leaves, which cover the apical meristem with a sterile pair of forceps.
7. Disinfect with 10% household bleach for 10 min.; shake occasionally.
8. Rinse in sterile deionized water.
9. For apical shoot explant cut transversely across the shoot tip to obtain a piece with an apical meristem and 2 to 3 nodes.
10. For axillary shoot explants, isolate the cube with an axillary bud attached to the node by slicing on each side of the bud, across young leaves, and below the node.
11. Culture both apical and axillary bud explants in sterile liquid media.
12. Place cultures in liquid medium on rotary shaker.
13. If liquid or solid medium discolors, change the medium daily or as necessary until the medium remains clear.
14. The first sign of growth appears within 4 to 6 weeks after culture.

Initiation of monopodial orchid

1. Sever 10 to 15 cm of shoot from a stock plant.
2. Trim expanded leaves flush to the stem axis.
3. Cut the stem into segments in 20% (v/v) commercial bleach with Tween-20 for 10 min.; shake occasionally.
5. In a sterile transfer hood, tear the leaf base from the stem to expose the axillary bud, using a sterile pair of forceps.
6. Disinfect the stem in 5% commercial bleach for 10 min.; shake occasionally.
7. Rinse in sterile deionized water.
8. For apical shoot explants, cut transversely across shoot tip to obtain a piece including the apical meristem and 2 to 3 nodes.
9. For axillary shoot explants, isolate the cube which includes a bud attached to a node.
10. Transfer the explant into a liquid medium without supplemental sucrose and place on a gyrotary shaker (approx., 100 rpm).
11. If the liquid medium discolors, change the medium daily or as necessary until the liquid remains clear.
12. The first sign of growth appears within 6 to 8 weeks after culture.

Initiation from young influorescence

1. Carefully collect young influorescence which are still enclosed in bracts.
2. Disinfect in 10% commercial bleach with Tween-20 for 15 min.; shake occasionally.
3. In a transfer hood, remove bracts by tearing with a sterile pair of forceps.
4. Disinfect in 5% commercial bleach for 10 min.; shake occasionally.
5. Rinse in sterile deionized water.
6. Culture the inflorescence in liquid medium.

Multiplication and rooting

1. For multiplication of liquid cultures, subculture clusters into separate flasks with the same medium; protocorm-like bodies or shoots break apart with time.

2. For multiplication of cultures on a solid medium, separate protocorm-like bodies or shoots and subculture into flasks with the same medium.
3. For rooting, transfer shoots to a solid agar medium with sucrose. Homogenized green banana (50–100 g/l) is used as an additive to promote rooting.

Transfer to greenhouse

1. When plantlets have roots about 1–2 cm in length, flush plants out of the container with water, rinse agar off, and plant in media appropriate for growing orchid seedlings.
2. Place pots with under low-light and high humidity for a few weeks; do not over-water.
3. When active root growth begins, move pots to higher light conditions.

B. MICROPROPAGATION BY PRODUCTION OF MINITUBERS

Regeneration of a whole plant through tissue culture is popularly known as micropropagation. Within a short time and space, a large number of plantlets can be produced from callus tissue. It is possible to make a large number of callus pieces from the original stock culture during subculturing. From each one of the callus pieces, it is possible to produce hundreds of plantlets. For example, the micropropagation method used commercially for the production of callus pieces that can develop into plantlets is production of potato minitubers (Figure 37.1). When axillary shoot cultures of potato are cultivated in the presence of appropriate levels of cytokinins and gibberellic acid they form large numbers of very small tubers (**minitubers**). These minitubers can be sown directly in the field and generate normal plants. Compared with the traditional way of planting, minituber appear to be very economical besides which the plant material will be disease-free.

C. PRODUCTION OF GENETICALLY VARIABLE PLANTS

In *in vitro* protoplast culture, there is a tendency for the callus tissue to exhibit a numerical variation of the chromosomes in the cells after a number of serial subcultures. The chromosomal instability and subsequent structural changes in cultured cells play an important role in polyploidization of cells and genetically viable plants can be raised from such cells. Thus protoplast culture is proving to be a rich and novel source of variability with a great potential in crop improvement without having to resort to mutation or hybridization. Variants selected through tissue culture have been variously termed **calliclones** (plants from callus) or **protoclones** (plants from protoplasts) or together as **somaclones**. Such variants show some useful characteristics such as resistance to diseases, herbicide resistance, stress tolerance, etc.

D. DISEASE-FREE PLANTS FROM TISSUE CULTURE

Plant tissue culture has made many valuable contributions to problems concerning plant diseases. One outstanding success is the virus eradication by apical meristem culture. It is well known that apical meristems are generally either free or carry a very low concentration of viruses. Apical meristem culture is the only way to obtain a clone of virus-free plant, which can be multiplied vegetatively under control conditions that protect it from the chance of re-infec-

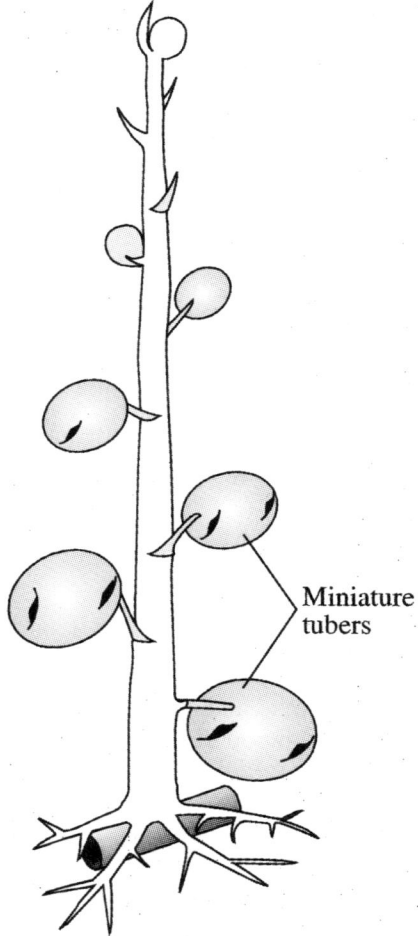

Figure 37.1 *Miniature tubers formed on an in vitro school of potato grown in a medium containing high cytokinin levels*

tion. Virus eradication by apical meristem culture has enormous horticultural and agricultural value. These virus-free stocks could provide ideal material for the national and international distribution of plants, either for future propagation or for use as breeding material.

III MASS PLANT CELL CULTURE FOR SECONDARY METABOLITES

Callus is too slow-growing and heterogenous to be useful for commercial production purposes. Suspension cultures are therefore used for mass culture. The ability to grow substantial volumes of plant cells under conditions similar to those used for microbial fermentation has enabled plant cell culture to be an alternative to field-based agricultural production. Field-grown product formation is fraught with many difficulties such as erratic supply due to the weather and the local political environment, effects of pests and diseases, and seasonal supply, which can vary in both quality and quantity.

In the 1970s, Japanese scientists grew tobacco cells in 6500 litres of modified stirred-tank bioreactors. Air-lift bioreactors have been used to grow up to 1500 litres of plant cell suspensions. An ideal bioreactor must have a moderate aeration rate, low shear, fine control of gaseous regime, and provide a good mixing environment. Plant cells grow more slowly with doubling times of at least two days resulting in long fermentation times and therefore increasing the infection. The plant cells can be immobilized so as to be protected from cell shearing. However, immobilization may be too severe for plant cells and plant secondary products are often stored in the vacuole and not released into the medium. The accumulation and excretion of secondary products may be controlled by the external pH. High-yielding undifferentiated cell lines also overcome both seasonal and tissue-specific production and can be selected to produce a significantly higher concentration of a particular secondary product than is found in the intact plant. In addition, specialized plant cells such as oil glands are required to synthesize some secondary products.

A. SECONDARY PRODUCTS

Since the early days of modern agriculture, plants have been an important source of precursors and products used in a variety of industries including those dealing with pharmaceuticals, food, cosmetics, and agriculture. More recently, in spite of many cases where natural plant products have been superseded by synthetic compounds, the plant kingdom is still a major source of specialty products with a market value in billions of dollars. There is tremendous potential for the synthesis of high-value natural products from plant cell cultures. The plant kingdom has the most varied biochemistry in terms of types of molecular structures that are synthesized, coupled with a highly specific enzymology. Many phytochemicals (plant metabolites) such as alkaloids, terpenoids, steroids, anthocyanins, anthraquinones, and polyphenols are used as drugs, food flavors, pigments, perfumes, and agrochemicals. It is worth remembering that if the chemical compound is structurally more complex, in particular the number of stereochemical centers in the molecule, the less likely it is that it is a synthetic one. The demand for plant-derived chemicals is largely from the developed countries but they are produced in third world countries, characterized by political instability, lack of funds, and interruption of supplies by many plant diseases. Plant cell cultures certainly provide an alternative source for production for such compounds. Over 1500 new compounds are identified in plants every year and it has been suggested that some of these could be developed into products produced in cell culture.

Table 37.1 *Major commercial phytochemicals*

Product	Use	Source
Atropine	Treatment of cardiac arrhythmias. Dilation of the pupil of the eye	*Atropa belladonna*
Crocin	Headache, general pain	*Gardenia jasminoides*

Continues...

... *Continued*

Codeine	Analgesic, Expectorant	*Papaver soniniferum*
Digoxin	Treatment of cardiovascular disorders, Cardiotonic	*Digitalis* sp.
Diosgenin and Sitosterol	Antifertility agent, Raw material for production of pharmacologically active steroids	*Zea mays*
Hyoscyamine	Muscle relaxants	*Hyoscyamus* spp.
Hyoscine	Muscle relaxants	*Datura* spp.
Jasmine	Perfumery	*Jasminum* sp.
Vanillin	Fragrance	*Vanilla*
Menthol	Flavoring	*Mentha piperita*
Morphine	Analgesics	*Papaaver* spp.
Pyrethrin	Insecticide	*Chrysanthemum* sp.
Quinine	Antimalarial; embittering agent for food and drink	*Cinchona ledgeriana*
Reserpine	Treatment of hypertension	*Rauwolfia serpentina*
Saffron	Food colourant and flavouring agent	*Crocus sativus*
Scopolamine (hyoscine)	Treatment of nausea, especially motion sickness	*Datura stramonium*
Podophyllotoxin	Anti-cancer	*P. hexandrum*
Thebaine	Analgesics	*Papaver* spp.
Ajmalicine	Anti-hypertension	*Catharanthus roseus*
VinblastineVincristine	Treatment of certain cancers, Leukemia	*Catharanthus roseus*
Taxol	Treatment of Ovarian and breast cancer	*Taxus brevifolia*

Data collected from different sources.

Plants are not as complacent and vulnerable as they may appear. They respond to elicitors by producing secondary metabolites. Metabolites are chemicals essential to metabolism, but secondary metabolites are not necessarily required for life. Secondary metabolites often provide the plant's defense and individual response to the environment. Some of these products are *allelopathic*, referring to the phenomenon of chemicals emitted by plants that influence or inhibit the growth of other plants. Perhaps best known is juglone, a toxin produced by black walnut trees that is leached from the trees into the ground, where it prevents the growth of some broad-leaved plants.

Many secondary metabolites are of research value or commercial importance, such as pharmaceuticals, medicinals (Figure 37.2), dyes, food additives, natural flavors, fragrances, gums, and pesticides. The products are conventionally extracted from whole plant or parts of plants. Traditionally the plants are field grown or are collected in the wild where they are often in short supply, limited by season and weather, of unreliable yield, and of questionable quality. The application of plant tissue culture to growing cells, callus, or plantlets for the purpose of extracting secondary products is the object of intensive research. Tissue culture production provides the opportunity to develop higher yields than is possible from field-grown plants. Higher yields in cell cultures are achieved by adjusting media, selection of high-yield cell lines, genetic engineering, or the application of elicitors, factors that help induce the sought-after chemicals.

The first step toward obtaining secondary products *in vitro* is to grow explants on a solid medium to produce the callus. The second phase is to use cells from the callus to proliferate in liquid cell cultures. It is necessary to grow large quantities of cells in tanks (reactors or ferment-

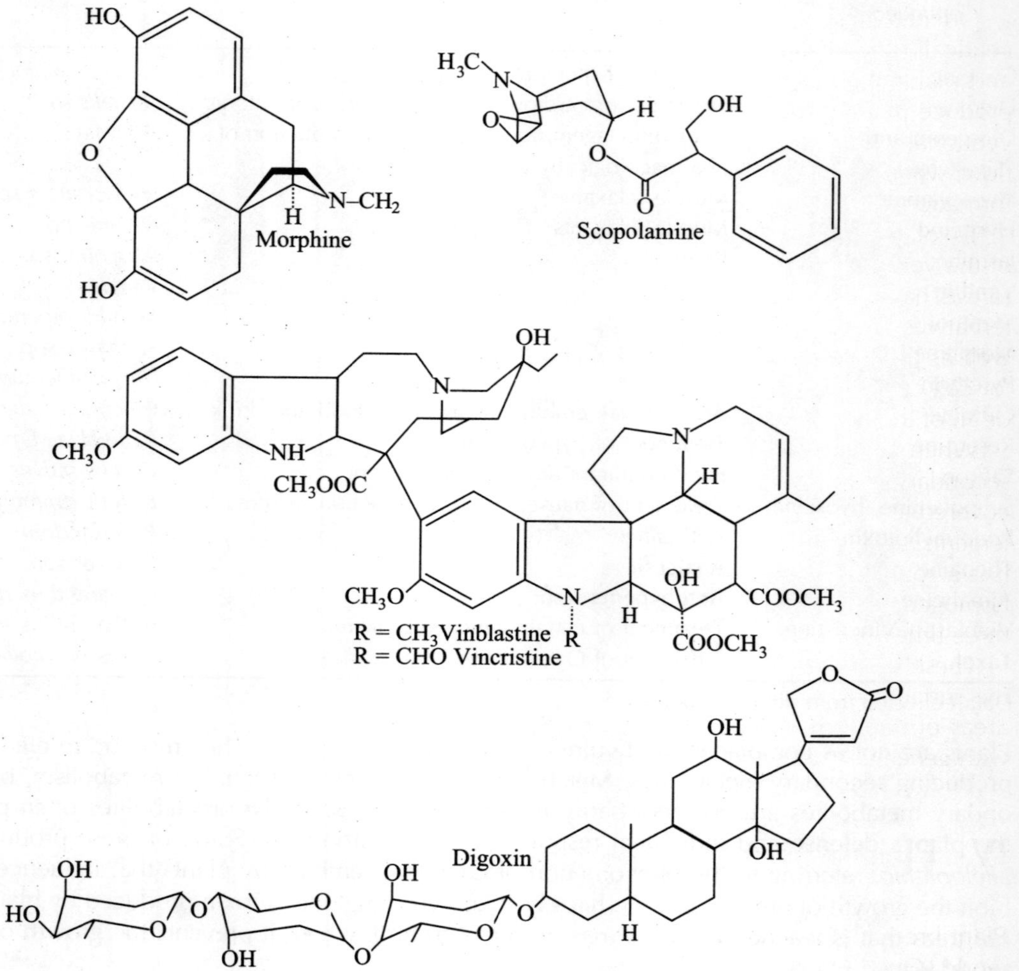

Figure 37.2 *Few important plant-derived drugs*

ers) in order to produce profitable volumes. The optimal medium, environment, treatment, and reactor must be carefully determined. One of the first secondary products to be successfully produced commercially by cell culture was shikonin (shikon), a secondary metabolite from the root of *Lithospermum erythrorhizon* that is used to remedy inflammation, to treat burns and hemorrhoids, as an antibiotic, and as a red dye. *Lithospermum* cannot be grown commercially, However, Mitsui Petrochemical Industries in Japan was determined not to import shikonin from China and Korea. By conventional methods it takes at least five years before the roots of *Lithospermum* can produce a mere 2% shikonin. With careful selection of cell lines, Mitsui was able to obtain 15% of dry weight within a matter of months; their production was estimated at 65 kilograms per year.

Oriental ginseng and American ginseng are widely used in health foods, cosmetics, pharmaceuticals, and other products, especially in the Orient. Seed germination in these species is poor and plants require five years to produce seed. But explants from roots, seeds, or flowers can produce plantlets *in vitro* in 3 months. Tissue cultured ginseng compares favorably with conventionally produced ginseng with respect to treating gastric ailments, hypoglycemia, and blood flow, when tested in rats. Furuya and Ushiyama states that tissue cultured ginseng inhibits ulcer formation, although ginseng from cultivated plants does not.

Taxol, the most-discussed drug for cancer treatment, was traditionally extracted from the bark of the yew (Taxus) trees. Thus thousands of regal giants from the Pacific Northwest old growth forests have been destroyed. Following great political and environmental pressure, several alternatives have been developed. Other parts of the trees have been found to produce taxol as well as the bark. Several laboratories are now using tissue culturing needles to produce callus for cell culture production of taxol. Meanwhile, other laboratories are striving to synthesize part or all of the complex compound.

When certain cell cultures are treated with elicitors, especially in combination, the yield of secondary metabolites can be increased significantly. Among the elicitors that have been discovered to serve this purpose are fungi, UV radiation, hormone deletion, heavy metal ions, detergents, agar, and heat. Periwinkle (*Catharanthus roseus*) is a plant of particular interest because it is a source of vinblastine and vincristine, alkaloids used in cancer treatment. When a fungal homogenate was added to cell cultures of *Catharanthus roseus*, the production of alkaloids was three times that of cultures lacking the fungal elicitor.

These brief examples of applying tissue culture to obtain secondary products barely touch, the surface of this exciting subject. This application of cell culture holds great promise in the areas of nutrition, pesticides, and pharmaceuticals, among others, especially when combined with genetic engineering (Refer to Chapter 22 for more details).

IV PLANT GERMPLASM BANKS

Plant breeders have made enormous contributions to increase food production throughout the world. One important part of their work is the introduction of genetic diversity by inter-crossing or mating selected germplasm with outstanding characters. One source of this genetic diversity is from related species occurring in the wild, frequently in the less well-developed countries of the world. Due to the growing urbanization and an increase in the amount of land being cultivated, the number of wild species and their natural habitats are rapidly disappearing in these countries. There is a general fear that potentially valuable germplasm is being irretrievably lost. Seed banks alone may not solve the long-term storage of these plants. First, it cannot be used with so-called recalcitrant seeds. Such seeds are large and succulent, lack the dormancy phase, and are incapable of surviving low temperatures and/or dehydration. Second, seed storage is not applicable for plants, that are propagated vegetatively. As a result of recent advances, cell tissue and organ culture are being considered as a means of germplasm storage. (Refer to Chapter 34 for more information on various methods and procedure for cryopreservation.)

V TISSUE CULTURE IN BIOTECHNOLOGY

Plants synthesize many metabolic intermediate or secondary metabolites that are very useful to humans. Biotechnologists are trying to increase the synthesis of natural compounds or new compounds by higher plant cell culture and transformation or by the conversion of added precursors in the culture medium (biotransformation). Biotechnologists are also trying to modify the genetics of the cultured cells in three ways. (1) Mutagenesis and selection of cell lines in cell culture; (2) introduction (transplantation) of foreign genes into protoplasts by means of genetic engineering; (3) somatic hybridization by the fusion of distantly related plant protoplasts (e.g. hybrid potato by cell fusion between *Solanum tuberosum* and *S. chacoense*). Plant scientists are also developing plants which produce improved proteins for human diet. The last two methods are very effective in stable transformation of plant cells in culture.

VI THE CONSTRAINTS AND RISK FACTORS ASSOCIATED WITH PLANT CELL AND TISSUE CULTURES

1. Plant cells have slow growth compared to microorganisms, with a doubling time in the range 20–60 hours (at optimal conditions bacteria can have a doubling time of just about 20 min.). As a result, the cultures are vulnerable to microbial contamination and a great deal of expenditure goes toward maintaining aseptic conditions.
2. There is genetic instability of plant cells in culture. The mutation frequency caused by the culture conditions, stress, and phytohormones poses problems for the maintenance of stability of cultures and cell lines.
3. There is difficulty in getting suitable cell lines. Highly productive cell lines occur only rarely, and are generally obtained through selection.
4. There is a difficulty in inducing the cells to produce the desired compound, and often, gene coding for products is expressed only in certain conditions.
5. There is a laborious excretion of secondary products by plant cells.
6. The technology development costs are high, the gestation period generally being three to four years.
7. There is a need for strong research and development support.

VII SUGGESTED READING

Ellis BR, Kuroki GW, and Stafford HA (Ed) *Genomic Engineering of Plant Secondary Metabolism*, Plenum Press New York and London.

Hasezawa H, Nagata T, and Syono K (1981) Transformation of *Vinca* Protoplasts Mediated by *Agrobacterium* spheroplasts, *Molec. Gen. Genet.*, **182**:206–210.

Ignacimuthu S (1998) *Plant Biotechnology*, Oxford and IBH Publishing Co. Pvt. Ltd., New Delhi, Kolkata.

Khanna, V K (1998) *Plant Tissue Culture Practice*, Kalyani Publishers, Ludhiana, pp1–20.

Kite Land Kleyn J (1996) *Plants from Test Tubes: An Introduction to Micropropagation*, Timber Press.

Sagawa Y (1991) Clonal Propagation of Orchids, In: *Plant Tissue Culture Manual* (Ed) K Lindsey, Kluwer Academic Publishers.

Wullems GJ, Molendijk L, and Schilperoort RA (1980) The Expression of Tumor Markers in Intraspecific Somatic Hybrids of Normal and Crown Gall Cells from *N. tabacum.*, Theor. Appl. Genet., **56**:203–208.

PART – IX
Environmental Biotechnology

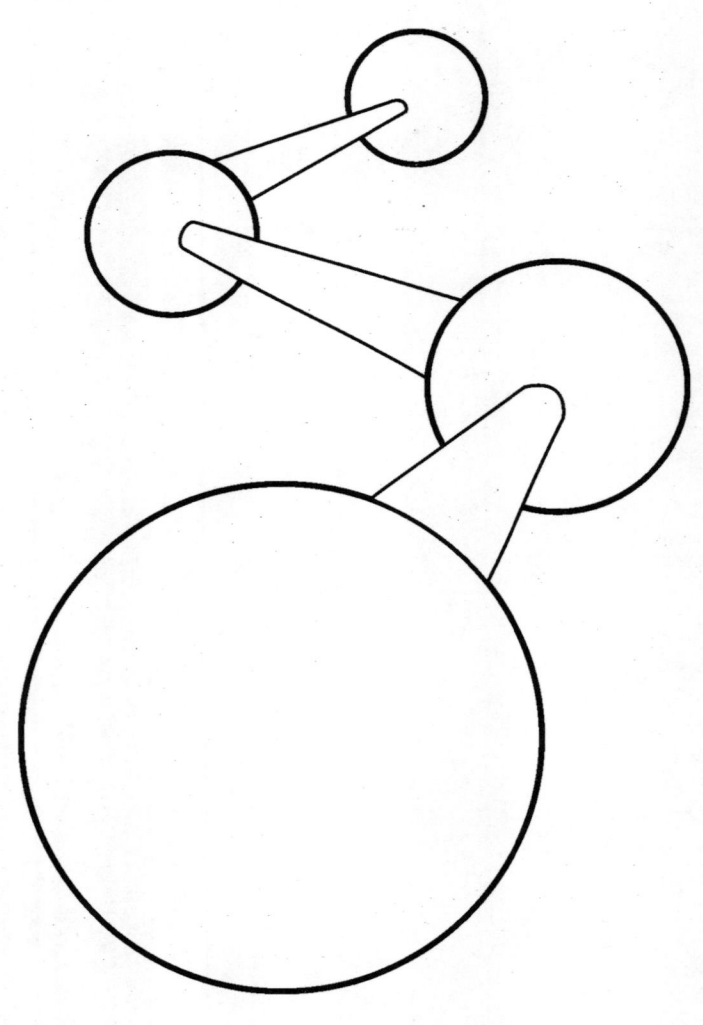

38

Biotechnology in Pollution Control

I INTRODUCTION

Polluted area (air, soil and water) is expanding day by day and the chemical and physical properties of the atmosphere are drastically changing as a consequence of anthropogenic activities. Annually, about 100 million people are added to the world population. Consequently, agriculture is expanding both vertically and horizontally to meet the demand for food and other agricultural products. More and more industries are coming up, and the number of automobiles is doubling. New and large quantities of health care products and medicines are being manufactured to meet people's demand. Consequently, an enormous quantity of wastes and pollutants are piling up at population centers, industries, and in farming. Notable examples are increases in tropospheric ozone (O_3), carbon dioxide (CO_2), oxides of nitrogen (NO_x) and sulfur (SO_x), ammonia (NH_3), particles, heavy metals, ultraviolet radiation, multiple ions in wet deposition, and organic vapors. From an ecological perspective, there is concern that many of these changes will influence terrestrial landscapes through their effects on plant growth and development or modification of ecosystem processes. People from all walks of life are voicing their concern about environmental hazards and the immediate need to develop technologies for minimizing waste generation on one hand and maximizing waste degradation and recycling on the other.

Today, there are two fundamental problems. First, how do we dispose of the large quantities of wastes that are continually being produced? Second, how do we remove the toxic compounds that have been accumulating at dump-sites, in the soil, and in water systems over the decades. With the discovery in the mid-1960s of a number of soil microorganisms that are capable of degrading xenobiotic chemicals such as herbicides, pesticides, refrigerants, solvents, and other organic compounds, disposal of toxic chemical wastes by microbial degradation gained credence — and funding. The biodegradation of complex organic molecules generally requires the concerted efforts of several different enzymes.

II ENVIRONMENT

The organisms and their environment together constitute an ecosystem. Balancing of both the components is essential for maintenance of a normal and sustainable ecosystem. Larger changes occurring in the ecosystem destroy the balance and lead to the destruction of ecosystems. Environmental disturbances (pollution) may drastically affect ecological balance. The environment consists of the sum of all the factors outside an organism. The environment in which organisms live will be determined by a range of biotic (other organisms) and physical or abiotic factors such as light, temperature, air, and moisture. All organisms and all features of the physical environment are necessary for the system to be maintained and to flourish. Changes move the system away from equilibrium in a minor or major way. Until the last 10,000 years or so, living systems evolved in response to changes in the abiotic environment, unaffected by human activity. Since the development of agriculture and technology, there has been an increasing human impact on the environment. In the last two centuries especially, widespread industrialization has led to potentially damaging environmental pollution.

III WASTES AND POLLUTION

Any product, by-product, or residue that cannot be used profitably is called a **waste**. A waste product is regarded as a pollutant when it damages the environment. Thus, **pollution** may be defined as the release into the environment of substances or energy in such quantities and for such duration that they cause harm to people or other organisms or their environment. Since pollutants are released into the atmosphere in a short time and in high amounts, there are no adaptive mechanisms that can deal with them. Pollution can affect all aspects of environment, human-made and natural, abiotic and biotic, and may be readily transferred between components of the life support system. Pollutants stimulate, initiate, or terminate the vital reactions for an organism and may also bring about changes in the organisms and thus modification of the entire ecosystem. Often wastes and pollutants are intricately linked. Wastes may be (i) biological, (ii) chemical, or (iii) physical in nature, and may originate from the following activities: (i) mines (ii) manufacturing, (iii) agriculture and dairy, (iv) energy production, (v) transport, and (vi) house building and house-keeping.

A. SOURCE OF WASTES AND POLLUTANTS

Wastes and pollutants are generated from various natural and human activities. Some of the sources due to human interference are summarized as below.

1. Manufacturing

Manufacturing industrial units generate a wide variety of wastes. The kind and the amount depend on the nature of raw materials and products, as well as the design and the operation of the involved processes. Manufacturing units processing biological materials (e.g. breweries, food processing, dairy etc.) generate wastes, that are biodegradable. Normally, these wastes can be utilized and recycled by microorganisms and plants. In contrast, industries based on non-biological raw materials and processes generate wastes, that are not easily biodegradable. Consequently, such wastes may pile up and persist for a long time in the environment until they have been modified/degraded by chemical/physical forces.

2. Energy production plants

Energy production by using fossil fuels generates large quantities of CO_2, and significant amounts of CO, various oxides of sulfur (SO_x) and nitrogen (NO_x), water vapor, and heat. Energy production by nuclear fuels, in particular, necessitates the maintenance and safe disposal of radioactive wastes, which have a profound effect on human health. The increase of CO_2 in the atmosphere due to constant burning of fossil fuels results in global warming. Different chemical species of SOx and NOx cause acid rain, which can affect the natural ecosystem substantially.

3. Agriculture

Agriculture and dairy activities produce crop residues and manure, but these are biodegradable. However, others such as pesticides, plastics, metal additives of feed, and wastes arising from burning of fossil fuels are recalcitrant to biodegradation. A substantial amount of ammonia is released from manure and fertilizers, and contributes to acid rain. Other chemical contaminants such as NO_3^- and PO_4^- are released in surface water causing massive blooms of algae; they also contaminate groundwater making it unfit for drinking.

4. Automobiles

Due to the revolution in the automobile industry, transportation activity has increased enormously in recent years. This is the major contributor to atmospheric pollution. Automobiles generate diverse air pollutants such as CO, SO_x, NO_x, chlorofluorocarbons (CFCs), volatile hydrocarbons, soot, and lead; they also contaminate surface waters and soil/underground water with oil and oil products. These pollutants can severely damage naturally occurring gases in the atmosphere, such as ozone (O_3).

5. Urban dwellings and domestic activities

The enormous increase in construction of buildings and domestic activities has generated a large amount of wastes. These activities generate both inorganic/non-biodegradable (e.g. stone, asbestos, synthetics, fly ash etc,) and biodegradable (sewage and many components of garbage) wastes. The chief wastes generated by domestic activities are human feces and urine, and garbage (food scraps, plastics, cardboard, tins, bottles, etc.). While sewage is biodegradable, its discharge into water bodies without proper treatment leads to the spread of many human diseases. It also causes reduction in oxygen tension and even induces anoxia, in the water, which has adverse effects on aquatic fauna (e.g. fish and other life forms).

B. HAZARDS FROM WASTES AND POLLUTANTS

The kind and magnitude of hazard produced by wastes and pollutants vary depending on their type and nature. These hazards are discussed under three broad groups (i) biological, (ii) chemical, and (iii) physical.

1. Biological agents

The main hazard from wastes related to biological agents are spread of diseases. This may occur due to a direct contamination of food or water by the waste (e.g. amoebiosis, cholera, diarrhea, typhoid, and hepatitis), or may be mediated by vectors, e.g. mosquitoes, flies, fleas, etc. Generally, it causes decrease in sanitation and hygiene and increases vectors and diseases.

2. Chemical hazards

Chemical agents are in wide forms, which include solid, liquid, and gas. They may be toxic or relatively non-toxic. These compounds are generally degraded into non-toxic form and assimilated by the microorganisms. However, sometimes, biodegradation can lead to conversion of a low-toxic form into a highly toxic one (e.g. production of highly toxic methyl mercury from relatively low toxic mercury compounds). Most **xenobiotic** (man-made compounds) agents are recalcitrant and some of them are **biomagnified** (progressive increase in concentration as they pass through the food chain) to dangerous/toxic levels. Generally these compounds are not found in nature (e.g. DDT, BHC, organophosphates, etc.,) If at all they are found, they are in lower concentrations than that liberated by man (e.g. phenols, aromatic hydrocarbons, metals etc.). Several features like aquatic and animal toxicity, mutagenicity, carcinogenicity, bioaccumulation, persistence, reactivity, eco-degradation, etc., are considered while evaluating the potential hazards from a chemical pollutant.

(a) **Planetary greenhouse effect**: Gases occurring naturally in the atmosphere may be affected seriously by chemical pollutants (**air pollution**). The burning of coal or fossil fuels in industrialized cities release airborne pollutants, including dust, smoke, soot, ashes, asbestos, oil, bits of lead and other heavy metals, and sulfur dioxides. In warm climates, photochemical smog develops as a brown, smelly haze over large cities. The main oxidants in smog are ozone and peroxyacyl nitrates (PANs). Traces of these compounds in air can make the eyes sting, irritate the lungs, and damage crops. Carbon dioxide is normally present in the lower atmosphere (the **troposphere**) in very small amounts, about 300 ppm or 0.03% by volume. Its importance, other than its being required for photosynthesis, lies in its contribution to the **planetary greenhouse effect**. Carbon dioxide is transparent to incoming short-wave radiation from the sun, but it strongly absorbs the long-wave radiation, which the earth re-radiates into space. It therefore 'traps' outgoing radiation, warming the lower atmosphere, which, in turn, radiates energy back to the surface of the earth. The contemporary concern lies with the clear evidence that carbon dioxide levels (CO, CH_4, CFCs, and other greenhouse gases) are rising at a rate unprecedented in recent earth history and their increased presence may lead to an increasingly warmer surface environment, namely an **enhanced greenhouse effect**. This may in turn, lead to increased evaporation and a greater atmospheric water vapor content, further increasing surface temperatures, since water vapor itself is a powerful long-wave absorber. Consequently, rise in surface temperature would cause changes in the distribution patterns and intensity of the major planetary weather systems, seriously affecting human activity and the ecological equilibrium.

(b) **Ozone depletion** is another serious problem being faced in recent years due to air pollution. The atmosphere provides a thermal blanket and radiation shield to the earth. In the upper atmosphere 15–50 km above the earth, oxygen and ozone absorb much of the incoming shortwave radiation, which is mostly harmful to living organisms. Radiation absorption by ozone high up in the atmosphere, warms these higher levels and creates a **deep temperature inversion layer** (where the highest temperature is at the greatest altitude). This effectively restricts the movement of air in the atmosphere by convection. Any disturbances of this inversion layer would profoundly alter global weather patterns and hence, earth surface climates. Ozone is produced high in the atmosphere by the action of sunlight on oxygen molecules. **Chlorofluorocarbons (CFCs)**, a group of chemicals which are commonly used as solvents, aerosol propellants, and refrigerator coolants, not only increase atmospheric temperature but also release chlorine and fluorine when broken down by sunlight. These react with ozone and break it down into oxygen faster than it can be reformed from oxygen into ozone, upsetting the oxygen–ozone equilibrium.

(c) **Acid rain**: Acid rain is due to complex factors. Acid gases, sulfur dioxide (SO_2) and oxides of nitrogen (NO_x), are produced by burning fossil fuels. Incomplete combustion of these fuels also releases hydrocarbons. Depending on climatic conditions, particles of oxides may stay airborne for a while, then fall to earth as *dry acid deposition*. Most sulfur and nitrogen dioxides dissolve in atmospheric water to form a weak solution of sulfuric and nitric acids. Winds may distribute them over a large area before they fall on earth in rain and snow. This is wet acid deposition (or **acid rain**). Many industrialized countries have all experienced rainfall with a pH well below 4.0 (normal rainwater has a pH ~5.0). Acid rainfall (pH<5) often causes major changes in ecosystems and damage to buildings. The acids chemically attack marble buildings, metals, mortar, rubber, plastic, even nylon stockings. They can also disrupt the physiology of organisms and the chemistry of ecosystems. Acid rainfall in central Sweden and southern Norway has affected salmon and trout fisheries and damaged forests. Acid rain leaches magnesium and calcium from soils and from damaged leaves. Eventually aluminum, manganese, and heavy metals such as iron and cadmium come into solution, causing damage to plant roots and the breakdown of mycorrhizas. These changes induce diseases due to the mineral deficiencies.

(d) **Water pollution**: The world has a tremendous supply of water, but most of it is too salty for human consumption or for agriculture. Only rain and a small portion of the groundwater is useful for terrestrial dwellings. Until recently, water pollution was a relatively local problem of the developed world. Human sewage, animal wastes, and toxic chemicals make water unfit to drink. Most river waters are polluted with nitrogen and phosphorus run-offs from fertilizers used in intensive agriculture and discharge of phosphate-rich sewage effluents. Such problems are now increasingly occurring on a worldwide basis. Sewage from coastal settlements are discharged, sometimes untreated, into coastal waters where it generates a direct health hazard for recreational bathers as well as marine organisms. Pollutants encourage contamination by pathogens. Agricultural run-off pollutes water with sediments, pesticides, and plant nutrients. Power plants and factories pollute water with chemicals, radioactive materials, and excess heat (thermal pollution). Another major problem is caused by excessive soil erosion on the land surface. This increases the silt load of rivers and coastal waters, which may enrich fisheries, but prove destructive to coral reefs.

(e) **Eutrophication**: Eutrophication means nutrient enrichment. Over a long period of time, typically several thousand years, lake ecosystems classically show a natural progression from an *oligotrophic* (few nutrients) to a *eutrophic* or even *dystrophic* (rich in nutrients) state. Pollutants collect in lakes, rivers, and bays before reaching the oceans. Many cities throughout the world dump untreated sewage into coastal waters. Cities along rivers and harbors maintain shipping channels by dredging the polluted muck and barging it out to sea. The main factors that cause eutrophication are heavy use of nitrogen fertilizers on agricultural land and the increased discharge of phosphates from sewage works. Eutrophication generates acute economic as well as ecological problems. Nitrates and particularly phosphates are the nutrients most commonly limiting primary productivity in aquatic ecosystems. Additional nitrate and phosphate, therefore, favor an increase in rapidly growing competitive planktonic species. But the consumer organisms grow slowly. Consequently, not all the increased primary production is eaten by the consumer organisms. This makes excess material enter the decomposition pathway, which is a more oxygen demanding process. Thus dissolved oxygen may be reduced to below the level necessary for the growth and reproduction of other species, making the situation increasingly worse. Changes associated with eutrophication can be monitored biologically and chemically. Changes in phytoplankton species present may help to indicate eutrophication (e.g. blue-green bacteria blooms). A useful chemical indicator of eutrophication is the biochemical oxygen demand (BOD), which measures the rate of oxygen depletion by organisms.

(f) **Oil pollution**: Oil pollution is a major hazard for marine and coastal environments. Each year, about 10 million tonnes of crude oil are spilt into the oceans in different incidents (damage to oil tankers, seepage from offshore installations, and flushing of tanker holds). Crude oil kills sea-weeds, mollusks, and crustaceans when washed onto rocky shores. Marine mammals may also be affected by oil spills. Oil spills are more severe in cold seasons as bacterial activity, which is responsible for oil degradation, is low then.

3. Hazards from physical pollutants

The most common physical pollutants are dust, humidity, radiation, fires, mechanical stress, sound pollution, electrical storms, etc. These hazards may be reduced using cleaner biotechnologies. However, the widespread use of biotechnological processes could generate its own wastes and pollutants, which may need to be managed.

4. Degradation of terrestrial ecosystems

Since the beginning of human civilization, humans have manipulated ecosystems in their search for food, shelter, fuel, and other resources and through the impact of discarded waste. The recent population boom increased the destruction of woodlands and forest, as more land is needed for growing crops and rearing livestock, and this has resulted in the creation of a completely new ecosystem. Mismanagement of this has further been aggravated with soil erosion, ***desertification*** and adverse effects associated with the widespread use of new synthetic organic compounds as pesticides. In temperate forests, there is no significant deforestation, but the tropical forests, on the other hand, are declining at a rate of 15% of global land surface in 1950 to 7% in the year 2000. It has been estimated that twelve million hectares of forest (the size of England) are disappearing annually. As the human population increases, larger areas of land must be cultivated to supply food. Fuel wood gathering will lead to further deforestation.

(a) **Genetic losses**: One individual is not capable of carrying all the alleles of its species. Each one carries a different sample of alleles, forming its genetic identity and causing the genetic source of phenotypic variation in response to the environment. Since the alleles are not randomly distributed over the range of the species, provenance variation is formed. It is, therefore, obvious that with the extinction of a species, its gene pool is lost. It is also obvious that if certain alleles are only present in one provenance, they will be lost if that provenance goes extinct as a result of the effects of air pollutants. Beyond these more obvious cases, a loss of genetic diversity within surviving populations is of more general interest. Adaptation of surviving populations can be accompanied by gene loss when a pollution regime causes viability and/or fertility selection. Genecology studies show that adaptation to different environments does not only include changes in allele frequencies but also differences in their presence. Loss of genetic multiplicity, however, is a threat to the adaptability of the species.

(b) **Deforestation** is the removal of all trees from large tracts of land for logging, agricultural, or grazing operations. Deforestation is serious for many reasons. (1) Traditionally harvested product (fuelwood, herbs, honey, fruit, animals etc.,) supply will decrease; (2) softwood timber used for making paper, for building, for furniture is threatened; (3) forests are often on uplands and on watersheds. The forest canopy softens the impact of intense tropical rainfall. If the forest canopy is removed, the soil surface bakes hard in the intense heat, rainfall cannot easily penetrate and this increases the surface run-off; (4) rapid run-off of rainwater results in soil erosion. This can remove the topsoil, making the soil unsuitable for cultivation; (5) deforestation increases global carbon dioxide, which has long-term effects on the global climate; (6) forests have several species and diverse wildlife communities. Their destruction will lead to the extinction of innumerable little-known forms of life with consequent loss of genetic diversity.

C. CONTROL OF WASTES AND POLLUTANTS

Modern human activities have not only increased the amount of wastes and pollutants they have also created a variety of new and complex pollutants. Therefore, there is an immediate need for the development of treatment methods to control the wastes and pollutants to save the ecosystem, including mankind. The main objective of biotechnology in this regard is to develop processes and products that minimize damage to the environment.

Waste and pollutants are different in nature, they can be gas, liquid, or solid and may occur in a concentrated and localized or dilute and dispersed form. In addition, wastes may be biodegradable, recalcitrant, or a mixture of both. Bioremediation technology also varies accordingly to whether the waste material involved is in its natural setting (*in situ bioremediation*) or is removed and transported into a fermenter or bioreactor (*ex situ bioremediation*). Various approaches to the *ex situ* waste treatment may be discussed under the following groups. (1) **Gases**: biofilters, (2) **Solids**: landfill, burning, or incineration and (3) **Liquids**: aerobic digestion and anaerobic digestion.

1. *In situ* bioremediation

In situ bioremediation is the clean-up approach, which directly involves the contact between microorganisms and the dissolved and sorbed contaminants for bio-transformation at waste sites. Biotransformation of waste in the *in situ* environment is a very complex process. The main advantages of *in situ* bioremediation methods are minimal site disruption, simultaneous treatment of contaminated soil and groundwater, minimal exposure of public and personnel, and low costs. However, there are many disadvantages also which include: it is time-consuming, there is seasonal variation of microbial activity resulting from direct exposure to environmental factors and the problematic application of treatment supplements (nutrients, surfactants and oxygen). The microorganisms are active only when the waste material helps them to generate energy and nutrients. If biodegradation by native flora is found low/slow, the degradation process can be enhanced by the addition of genetically engineered microorganisms (GEMs). *In situ* bioremediation is of two types, intrinsic and engineered *in situ* bioremediation.

(a) *Intrinsic in situ* **bioremediation**: Conversion of wastes into harmless forms through the innate capabilities of the naturally existing microbial population is called intrinsic *in situ* bioremediation. There is increasing interest in intrinsic bioremediation for control of pollutants at waste sites. Site monitoring programs should be carried on and recorded from time to time. The conditions that facilitate intrinsic bioremediation are groundwater flow, carbonate minerals to buffer pH change, supply of electron acceptors, nutrient availability, and the concentration for microbial growth, temperature, and the absence of toxic compounds (e.g. waste mixtures containing metals such as mercury, lead, arsenic, cyanide, etc.).

(b) *Engineered in situ* **bioremediation**: Intrinsic bioremediation works satisfactorily at some places, but it is a slow process due to poor microbial growth, limited ability of electron acceptors and nutrients, low temperature, and the high percent of contaminants. When site conditions are not favorable, bioremediation requires construction of engineered systems to supply materials that stimulate microorganisms. Engineered *in situ* bioremediation accelerates the desired biodegradation process by encouraging the growth of more microbes by optimizing both physical and chemical conditions. Optimizing the supply of oxygen and electron acceptors (e.g. NO_3^- and SO_4^{2-}) and nutrients (e.g. nitrogen and phosphorus) stimulates microbial growth. In some *in situ* bioremediation systems, deeply contaminated groundwater can be controlled by injecting amended-water through wells.

2. *Ex situ* bioremediation

Ex situ bioremediation involves the removal of waste materials and their collection at a place to facilitate microbial degradation. This process has many disadvantages. It suffers from high operational costs, e.g. for excavation, screening and fractionation, mixing, homogenizing, and final disposal.

(a) Gases

Biofilters: Various methods for treating toxic gaseous compounds have been developed. These devices consist of either a solid support or a two-phase (gas/liquid) system and appropriate biological agents to convert gaseous wastes to non-hazardous compounds. Generally, in a **solid support system**, peat or other solid material suitable for the growth of microorganisms is used in the biofilter. The gas to be treated is passed through the solid support layer, and the pollutants are broken down/converted by the biological agents present in the solid support. Due to the limitation in control of biomass in this method the bioconversion of gases is variable and unpredictable. The two-phase (gas/liquid) biofilters are an improvement, over the solid-support biofilters. In these filters, the liquid phase is separated from the gas phase by a membrane, and the biological agents are immobilized on the membrane on the side of the liquid phase. The gaseous pollutants present in the gas phase enter the liquid phase where they are consumed/broken/converted by the biological agents.

(b) Treatment of solid wastes

Solid waste includes organic wastes (e.g. leaves, animal manure and agricultural wastes), and problematic wastes (e.g. domestic and industrial wastes, sewage, sludge, and municipal solid wastes). Solid wastes can be treated first by sorting out the waste into biodegradable and non-biodegradable components; later the biodegradable components can be recycled while the non-biodegradable components can be (i) either incinerated or (ii) used for anaerobic digestion.

However, due to high costs of sorting, the entire solid waste is usually treated by conventional *composting* or dumped into pits (the process is called *landfill*).

(i) Composting: Composting is self-heating and substrate-dense and is processed by a naturally occurring microbial system. It is a single solid-phase biological treatment technology which is suitable for the treatment of a large amount of contaminated solid materials. Some of the limitations of this method are that many hazardous compounds are recalcitrant to biodegradation, due to their complex chemical structure, toxicity, and compound concentration. Microbial growth is also affected by moisture, pH, inorganic nutrients, and particle size. Composting can be done in an open system, i.e., land treatment, and in the closed system (Figure 38.1a–c). The *open land system* can be inexpensive, but the temperature fluctuates with seasons. Therefore, the biodegradation process declines. Oxygen limitation can be overcome by passing air (Figure

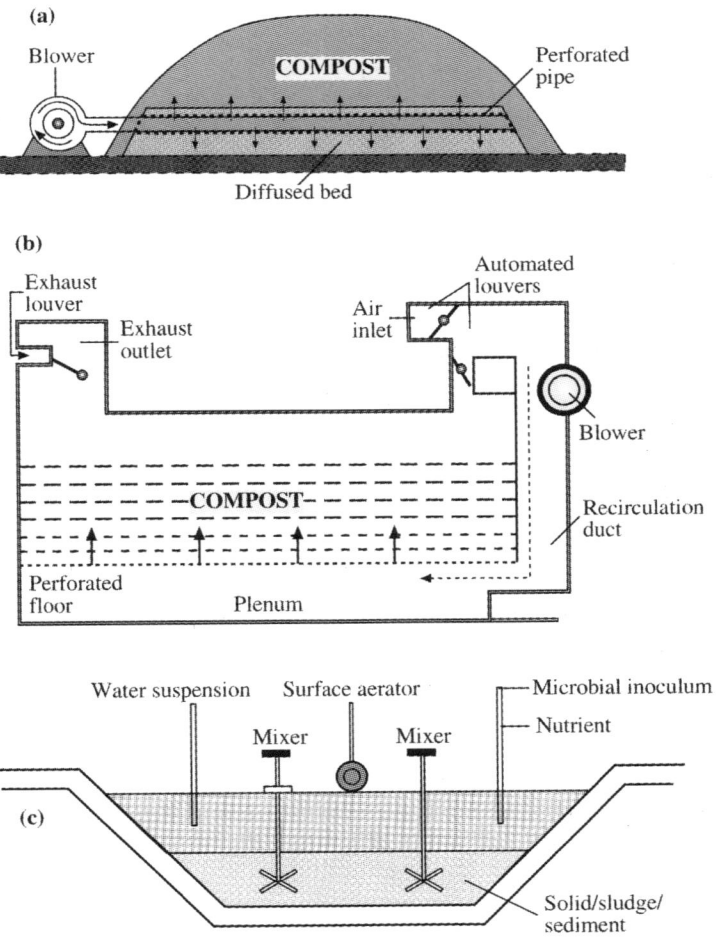

Figure 38.1 *Physical configuration of (a) open composting and (b) closed composting systems (c) slurry phase lagoon*

38.1a). This approach is referred to as **engineered soil piles** and **forced aeration treatment**. The **closed treatment system** is preferred to the open land treatment system because controlled air is supplied to maintain microbial activity. Rise in microbial growth results in a volatilization of hazardous compounds and the gradual rise in internal temperature. Therefore, blowers for air circulation and exhaust for removal of toxic-volatiles are set up in the closed treatment system.

Since composting is a solid-phase biological treatment, waste materials should be solids and hazardous compounds should be biologically transformed. The waste material should be suitably prepared so that biological treatment can be achieved. This is done by adjustment of several physical, chemical, and biological factors, including solubilization of hazardous waste. The hazardous compounds and soil organic matters provide the source of carbon and energy for microbial growth. Microbial enzymes secreted during the growth phase degrade most toxic compounds. Organic amendment also stabilizes microbial population in an adverse environment. The optimal amount of water and the addition of inorganic nutrients greatly influence microbial growth.

(ii) Landfill: Landfill is a natural or man-made pit or hollow in which waste is filled and covered with soil and often landscaped. The site of the landfill is usually in an unused area. The waste placed in the landfill is generally pre-treated. The pre-treatment may be sorting of the wastes, mechanical pulverization, or even incineration. This method is followed in two ways on the basis of the type of pit used, **cell emplacement** and **trench method** or **'cut and fill'** method. In the cell emplacement method, cells to a depth of 2.5 m, are excavated at the site of the landfill. The size depends upon the amount of garbage deposited each day. Every day the waste dumped in the cell is compacted and covered with about a 20 cm deep layer of soil. The cell may be designed to be single or multi-layered. In the trench method, long trenches are dug, filled with waste, and covered with soil. In both the trench and emplacement method, the soil for covering the waste is dug from the next trench/cell.

Landfill sites can be useful in two ways – as a source of biogas and for reclamation of derelict sites, to develop landscaped gardens, etc. Landfill sites generate considerable amount of methane, which leaks from the soil cover. This can be collected and used as biogas. The sites may be landscaped and planted with vegetation. But problems may arise due to the toxic substances present in the waste or which are produced due to degradation of the wastes, and many plant species may not survive. The disposal of wastes in landfills presents several hazards including fires in the waste materials, increase in the population of disease vectors like flies, offensive odors, methane leakage, and leaching of toxic and corrosive materials into surface and underground water.

(c) Treatment of liquid wastes

Liquid wastes are treated by either aerobic or anaerobic digestion in huge reclamation or sewage treatment plants. Particulate matter (e.g. grit and fibers) present in liquid water is first removed by sieving or settling down. The solid component that settles down is called **sludge**. The purpose of liquid waste treatment is to reduce the amount of organic material remaining in the solution phase and to inactivate pathogenic organisms that may be present in the waste. However, the non-degradable compounds pass through the process without being modified. Sludge can be used variously: (i) as fertilizer in agriculture (limited by presence of metals, xenobiotics, and pathogen), (ii) disposed of into the sea (this method is questioned by many authorities), (iii) in landfills along with other solid wastes, (iv) incinerated, and (v) disposed of on land following sanitation by pasteurization, thermophilic digestion, or irradiation.

(i) **Aerobic waste water treatment**: Most waste waters contribute organic materials and inorganic nutrients to river waters (called eutrophication). This can promote microbial and plankton growth which depletes oxygen in the water, resulting in the death of fish and other fauna. Waste water may also add inert particulate matter, which settles in the river; sometimes it may be hot and raise the river water temperatures. The amount of pollution can be measured by a number of techniques, particularly, the amount of organic matter content and the concentration of inorganic nutrients (e.g. nitrogen phosphorous). The important measurements are (i) biochemical oxygen demand (BOD), (ii) chemical oxygen demand (COD), (iii) suspended solids, (iv) ammonical nitrogen content, (v) phosphate content, and (vi) biological indicators.

(ii) **The process of waste water treatment**: Once the level of pollutants in waste water is determined, further treatment of polluted water proceeds through the following stages.

Preliminary treatment: In this preliminary stage, grit and heavy solids are removed by settling and floating solids are separated by screening. Finally, the particulate matter may be subjected to grinding. This process helps to prevent clogging of pipelines, damage to pumps, and interference with the subsequent processes.

Primary treatment: This treatment utilizes settling tanks of upward, horizontal, or radial flow type. The water flow is reduced to allow 15–45 min. (residence time), during which about 70% of the suspended solids either settle down at the bottom or float as scum on the top. The scum portion is removed by a scum baffle. This process reduces BOD by about 40%. To promote the settling of particles often, a chemical coagulant is added. The effluent obtained from primary treatment is called **settled sewage**.

Secondary treatment: Digestion of the organic matter dissolved/suspended in waste water by aerobic/anaerobic processes constitutes secondary treatment.

Tertiary treatment: This stage aims to remove suspended matter and to reduce BOD and to remove specific materials (e.g. phosphates) from the biologically treated water. The processes used are either physical or chemical. Suspended matter is removed by means of (i) lagooning (settlement of particles), (ii) grass-plot irrigation and recovery of under-drained water (filtration), (iii) micro-screening using large filtration devices, (iv) slow sand filters, (v) rapid downward flow filter, (v) up-flow filter, and (vii) upward flow clarified. Phosphates are usually removed by using alum or lime and the precipitate is removed after it settles down in the tanks.

Sludge treatment: Sludge is treated mainly anaerobically to make it safe before it is disposed of.

(iii) **Aerobic reactors**: The biological treatment of primary treated sewage may be based on a fixed film digester or a dispersal growth digester under aerobic conditions.

Fixed film digester: In these digesters, biological components or microorganisms, are present in the form of a film on filter particles or large discs. These are basically of two types, trickling filter digesters and rotating biological containers.

Trickling filter digesters: These are either rectangular or circular and consist of 3 m deep filter beds made up of 25–100 mm boulders or chunks of crushed rock, ceramic bricks, blast furnace slag, etc. The microorganisms are present as a layer called biofilm on the surface of the filter particles (Figure 38.2a). The biofilm grows thick around filter particles and peels off with time. A new biofilm soon forms due to fresh microbial growth. Thus cyclical death and detachment of biofilm prevents excess growth of microorganisms and the consequent clogging of the filters. The peeled off biofilm flows out with the treated sewage which is taken to a sedimentation tank to allow all types of solids from the effluent to settle down (the settled solids are called **humus sludge**). The settled sewage is redistributed mechanically over the filter surface from where

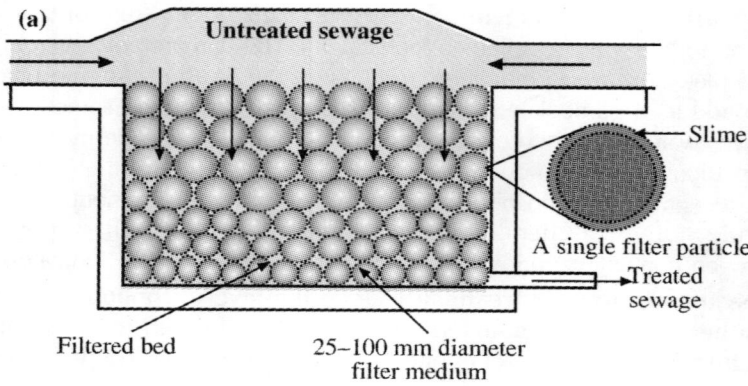

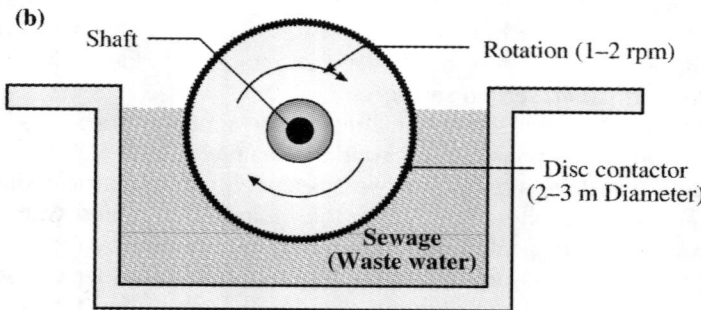

Figure 38.2 *A schematic representation of (a) a filter bed used in trickling filter method and (b) a rotating biological contactor, the biofilm is present on the surface of the disc*

it trickles down the filter. The organic materials dissolved in sewage water come into contact with the biofilm and are aerobically degraded. Various mechanical devices are used to ensure uniform sewage distribution over the filter surface. The permissible organic loading (BOD/m^3/day) depends on filter design. Trickling filters have been used for over a century and are usually made of inorganic filter media. Recently, synthetic supports like plastics have been used, which have several advantages over conventional supports.

Rotating biological contactors: These devices consist of a set of discs about 2–3 m in diameter, oriented vertically to the direction of waste water flow (Figure 38.2b). The biofilm is adhered to the disc surface, which keeps on rotating at the speed of 1–2 rpm. As shown in Figure 38.2b, the disc is partly submerged in the waste water and partly exposed in air. The rotation of the disc provides the aeration necessary for aerobic digestion. As the biofilm on the rotating disc comes into contact with sewage it digests the dissolved organic materials. The main advantages of these devices are low land requirement and very low maintenance. They are highly suited for small-scale applications.

Dispersed growth digesters: In this process, the microbial population is dispersed throughout the sewage being treated. The activated sludge process is a typical example. It uses a large aerated vessel for a large-scale oxidation of liquid wastes. The flocks of microorganisms are distributed throughout the waste water being treated. Waste water continuously flows into the vessel and treated water flows out at a predetermined rate to obtain optimal digestion of the dissolved organics. The microbial flocks flow out with the treated water and are carried into ponds with large surface areas. From here, the water flows out at very low flow rates. As a result, the suspended solids including the microbial flocks settle down (the sediment is called **activated sludge**) as it contains microbes for aerobic digestion. The aeration in the digester vessel is achieved by mechanical stirring, injecting a fine stream of air bubbles through porous ceramic rocks, generating a course of air bubble stream from a system of pipes, or introducing pure O_2 as a stream of bubbles. The quality of treated effluent from the digester will depend mainly on the organic loading rate, the residence time, and the sludge-loading rate. The sludge-loading rate is easy to control and can be varied to achieve the maximum possible reduction in BOD.

(iv) Microorganisms involved in aerobic treatment: Microbial digestion is a natural aerobic process. During this process, the organic material (biodegradable component) oxidized to CO_2 and H_2O, and nitrogenous compounds are produced. Since waste water contains more organic matter, it demands increased microbial activity. The microbial population and activities increase accordingly by providing a large area for biofilm formation and O_2 exchange and also returning a part of the activated sludge to the digester vessel and making effective aeration arrangements. A variety of microorganisms occur in the aerobic digestion systems, including bacteria, protozoa, fungi, viruses, cyanobacteria, and algae (the latter two need sunlight for photosynthetic activity).

Bacteria are the most common of the organisms and their number may be more than 10^{12} cells/l. Bacteria responsible for the removal of about 85–90% of the BOD remain after primary treatment. Bacteria can contribute variously in biodegradation. For example, the bacterium *Zoogloea ramigera* secretes a mucous like polysaccharide, which is involved in the attachment of various organisms (bacteria, fungi, algae, insect larvae, and nematodes) to the biofilm and also leads to floc formation. This complex community degrades carbohydrates, proteins, lipids, etc., into CO_2, NO_3^-, SO_4^{2-} and PO_4^{3-}. Many heterotrophic bacteria (e.g. *Sarcina, Pseudomonas, Escherichia, Staphylococcus, Streptococcus, Salmonella, Shigella, Aerobacter*, etc.,) are responsible for aerobic oxidation of organic molecules. Ammonia released from protein/amino acid degradation is converted to NO_3^- by nitrifying bacteria (*Nitrosomonas* and *Nitrobacter*). The excess of NO_3^- in drinking water may lead to a condition called 'blue babies' in the very young. The NO_3^- is ultimately removed from the water by the action of denitrifying bacteria e.g. *Alcaligenes, Achromobacter, Micrococcus, Pseudomonas* etc. These bacteria convert NO_3^- into N_2, which is liberated into the atmosphere. Denitrification is an anaerobic process. Therefore, it is achieved by alternating aerobic and anaerobic conditions.

Fungi usually occur as an external biofilm. They may help in the removal of nitrogen, phosphorous, and other nutrients. Protozoa are represented in waste water by flagellates, ciliates, and amoebal forms. The protozoa feed on nutrients and bacteria, and thereby convert smaller particles into larger ones, which may help in removal by sedimentation. Viruses also occur in waste water treatment processes, and may cause a decline in the populations of their host bacteria. Sewage harbors many pathogenic organisms. Most of these pathogens are contributed by feces and urine. Therefore, specific treatments are needed to eliminate pathogenic organisms from waste water.

(v) Anaerobic waste water treatment: Sewage treatment by anaerobic digestion has been practiced for over 100 years in Europe using specific tanks. A single large septic tank was earlier used for both sedimentation and anaerobic digestion. Later many improvements were made which include addition of baffles to septic tanks to aid sedimentation and for easy separation of sedimentation from digestion. Later, a variety of digesters were developed to operate at elevated temperatures of 35°C. The efficiency of digestion was also enhanced by increasing the population density of microorganisms by using solid support to retain biomass, recycling the active biomass, or limiting the biomass loss to a rate compatible with high population density in the digester.

Microbial digestion: The anaerobic digestion process involves a wide variety of organisms of which bacteria are the most predominant. These microorganisms digest the organic molecules like lipids, carbohydrates, proteins etc., into mainly methane and CO_2. In addition, if the waste water contains sulfate (SO_4^{2-}) and/or nitrate (NO_3^-) the following microbial activities also occur. (1) If SO_4^{2-} is used as electron acceptor by bacteria like *Desulphovibrio* during oxidation of organic compounds, they reduce SO_4^{2-} to S_2^-. (2) If denitrifying bacteria oxidize organic substrates and use NO_3^- as electron acceptor, they liberate N_2 in the process. At neutral pH, N_2 is the major product of this process. (3) Methanogenic bacteria contain several cofactors not found in other bacteria. Three such cofactors are involved in the reduction of CO_2 to CH_4 in a stepwise fashion beginning with methanopterin (MP), followed by methanofuran (MF) and co-enzyme M (CoM) in the end; the last reaction is catalyzed by Factor$_{430}$ (F_{430}), the prosthetic group of CoM. (4) The ATP generation in methanogens is assumed to involve a proton motive force. According to one model, H_2 oxidized by hydrogenase on the surface of the plasma membrane to generate H^+ is used for reduction of CO_2 inside the cell. This process also uses up the electrons generated during H_2 oxidation by hydrogenase.

Sludge treatment: The sewage or municipal sludge is treated to reduce odors and pathogens, and to stabilize the solids before final dispersal to landfills, as compost for agriculture. The sludge used for treatment may be the sediment from primary treatment or the activated sludge recovered after secondary aerobic treatment or scum from primary treatment tanks. The most common digester used for the purpose has two tanks, a digester, and a storage tank. The sludge is fed into the digester tank, which is maintained at 30–40°C (mesophilic treatment). Later, the stabilized sludge is taken to the second tank which is unheated and where the sludge is allowed to compact. The performance of sludge digesters cannot be controlled, and is markedly affected by the presence of toxins, climate, seasonal changes, presence of recalcitrant organic compounds, etc.

D. BIOREMEDIATION OF XENOBIOTIC COMPOUNDS

Xenobiotic compounds are man-made chemicals that are present in the environment in unnaturally high concentrations. They are not normally produced naturally, or if at all they are, they are in much lower concentrations than that in the case of man-made ones. Microorganisms play an important role in degrading all naturally occurring and a few xenobiotic compounds, but they are unable to degrade many others. The compounds that resist biodegradation and thereby persist in the environment are called **recalcitrant** (Table 38.1). The reasons for recalcitrance may be many. (i) They are not recognized as substrates by the existing degradative enzymes; (ii) they are highly stable; (iii) they are insoluble in water, or are adsorbed to external matrices like soil; (iv) they are highly toxic or give rise to toxic products due to microbial activity; (v) their large molecular size prevents entry into microbial cells; (vi) the compounds are unable to induce the synthesis of degrading enzymes; (vii) there is a lack of the permease needed for their transport into the microbial cells.

Table 38.1 *Long term persistence of insecticides and herbicides in soil*

Biocides	Time taken for 75 to 100% disappearance
A. Chlorinated insecticides	
DDT (1,1,1-trichloro-2, 2-bis-(p-chlorophenyl) ethane)	4 years
Aldrin	3 years
Chlordane	5 years
Heptachlor	2 years
Lindane (hexachloro-cyclohexane)	3 years
B. Organophosphate Insecticides	
Diazinon	12 years
Malathion	1 week
Parathion	1 week
C. Herbicides	
2,4-D (2,4-dichlorophenoxyacetic acid)	4 weeks
2,4,5-T	30 weeks
Atrazine	40 weeks
Simazine	48 weeks
Propazine	1.5 years

Source: Madigan et al. (1997)

1. Types of xenobiotic compounds

In recalcitrant xenobiotic compounds, the structural features that make these compounds resistant to microbial degradation are many. (i) The presence of halogens in the place of hydrogen. The carbon–halogen bond is highly stable and it requires a considerable amount of energy for bond cleavage; (ii) substitution of hydrogen by other groups like nitro-, sulphonate, methoxy-, amino- and carbonyl groups; (iii) cyclic structures, aromatic compounds, cycloalkanes, and heterocyclic compounds are more recalcitrant than linear chain or aliphatic compounds; (iv) branched linear chains resist biodegradation, etc. In general, the more complex the structure of a xenobiotic compound, the more resistant it is to biodegradation. Many other xenobiotics resist biodegradation due to their large molecular size and insolubility in water. The recalcitrant xenobiotic compounds are discussed under six groups.

(a) **Halocarbons**: These compounds contain different numbers of halogen (e.g. Cl, Br, F, I) atoms in place of H atoms. More than 200 halogenated natural products have been identified. Seventy five per cent of them contain chlorine and many originate in the salt water of the oceans. Most halocarbons are used as solvents (chloroform, $CHCl_3$), as propellants in spray cans of cosmetics, paints, etc., in condenser units of cooling systems (Freons, CCl_3F, CCl_2F_2, $CClF_3$, CF_4), and as insecticides (DDT, BHC, lindane, etc.) and herbicides (dalapon, 2,4–D, 2,4,5–T, etc.). The $C1$–C_2 haloalkanes like chloroform, freons, etc., are volatile and escape into the atmosphere, damaging the ozone layer. Pesticides are applied to crops from where they leach into water bodies. Many of them are subject to biomagnification.

(b) Polychlorinated biphenyls (PCBs): These compounds have two covalently linked benzene rings having halogens substituting for H. PCBs are used as plasticisers, insulator coolants in transformers, and as heat exchange fluids. They are both biologically and chemically inert to various degrees, which increases with the number of chlorine atoms present in the molecule. The resistant nature of the above compounds is due to their halogenation and as well as to their cyclic structure (PCBs).

(c) Synthetic polymers: These compounds are produced as plastics (e.g. polyethylene, polystyrene, polyvinyl chloride, etc.) and nylons which are used as garments, wrapping materials, etc. They are recalcitrant mainly due to their insolubility in water and molecular size.

(d) Alkylbenzyl sulphonates: These are surface-active detergents superior to soaps. The sulphonate ($-SO_3^-$) group present at one end resists microbial degradation, while the other end (non-polar alkyl end) becomes recalcitrant if it is branched. Resistance increases with the degree of branching.

(e) Oil mixtures: Oil is a natural product generated from fossil fuel. Oil wastes contain many components and are biodegradable, the different components being degraded at different rates. Biodegradation can handle small oil seepages, but when large spills occur the problem of pollution becomes acute. Oil is recalcitrant to biodegradation mainly because of its insolubility in water and because some components are toxic to microorganisms. Microorganisms which are capable of degrading petroleum include *Pseudomonads*, various cyanobacteria, and some yeast.

2. Hazards from xenobiotics

Xenobiotics pose a number of potential hazards to humans and the environment. These are briefly listed below.

(a) Toxicity: Many xenobiotics like halogenated and aromatic hydrocarbons are toxic to bacteria, lower eukaryotes, and humans. At low concentrations, they may cause various skin problems and reduce reproductive potential.

(b) Carcinogenecity: Certain halogenated hydrocarbons have been shown to be carcinogenic.

(c) Residual effect: Many xenobiotics are recalcitrant and persist in the environment, leading to a continuous accumulation of xenobiotics.

(d) Bioaccumulation or biomagnification: Many xenobiotics including DDT and PCBs are recalcitrant and lipophilic; as a consequence, they promote bioaccumulation often by a factor of 10^4–10^6. Biomagnification occurs mainly because these compounds are continuously taken up from the environment and accumulated in the lipid deposits of body. For example, there is a 100-fold accumulation of DDT by plankton from water. Also these organisms are consumed by other organisms in a sequential manner, constituting the food chain (e.g. plankton→small fish→large fish→sea-eagles). This build-up continues as we move up the food chain. In the case of DDT a 10^5–fold increase in DDT was observed in sea-eagles as compared to the concentration present in the aqueous environment. As a result of this, the sea-eagles laid fragile eggs. DDT and PCBs have been found in human tissues in high but sublethal concentrations in the regions where they have been used. They are produced and used in large quantities, which favors their accumulation in nature.

3. General features of xenobiotic biodegradation

Xenobiotics are a mixture of a wide variety of compounds. Their degradation occurs via a large number of metabolic pathways. Degradation of alkanes and aromatic hydrocarbons generally occurs in sequential steps. An oxygenase first introduces a hydroxyl group to make the compound reactive. The hydroxyl group is then oxidized to a carboxyl group. The ring structure is opened up and the linear molecule is degraded by β-oxidation to yield acetyl CoA, which is metabolized in the usual manner. Similarly, an alicyclic hydrocarbon e.g. cyclohexane, is oxidized as first an oxygenase adds an –OH group in the ring, then another oxygenase forms an ester in the form of a lactone, which is then hydrolyzed to open the ring structure to yield a linear molecule. In both these oxidations mono-oxygenases are involved which add oxygen to a single position in the molecule. In contrast, oxidation of the benzene ring may involve a di-oxygenase, which adds oxygen at two positions in the molecule in a single step. Both mono- and di-oxygenases are of a variety of types which catalyse specific groups of short chain or cyclic alkanes. Therefore, xenobiotics are degraded by a wide variety of microorganisms, which work in sequence. Often biodegradation involves cytochrome P_{450} or rubredoxin. In addition, the halogens and/or other substituent groups are either modified or removed usually as one of the initial reactions are achieved.

4. Bioremediation of hydrocarbons

Petroleum and its products are hydrocarbons. Biological degradation of hydrocarbons is one of the important biodegradation processes in pollution control. Some of the important hydrocarbons are briefly discussed below.

(a) Halomethanes are transferred into methanol by the enzyme methane monoxygenase, which uses them as substrate and it exists in a number of methylotrophs. Alternatively, a glutathione-dependent hydrolase catalyzes oxidative dechlorination of halomethanes into methanol; this reaction is anaerobic and uses oxygen derived from water. Methanol is oxidized to CO_2+H_2O via formaldehyde and formic acid.

(b) Cyanide (HCN): HCN is toxic to biological systems and not even microorganisms that degrade HCN can withstand it at higher concentrations. Therefore, disposal of cyanide is strictly controlled. Some microorganisms degrade HCN. For example, fungal hydratase converts HCN into $HCONH_2$ and bacteria *Pseudomonas fluorescens* converts HCN into CO_2 and NH_2.

(c) Aliphatic hydrocarbons: These compounds may be saturated or unsaturated. N-Alkanes of 10–24 carbons are the most readily biodegraded. Biodegradation of n-alkane is catalyzed by oxygenases to produce carboxylic acid, which is then degraded by β-oxidation. Oxidation may involve the methane group at one end of the n-alkane molecule, or it may occur in a β-methylene group. Sometimes, both terminal methyl groups are oxidized to yield a dicarboxylic acid. This reaction is used by many microorganisms for the biodegradation of branched chain n-alkanes.

(d) Alicyclic hydrocarbons are present naturally in waxes from plants, crude oil, microbial lipids etc., and are represented by xenobiotics used as pesticides and also in petroleum products.

(e) Aromatic hydrocarbons are oxidized by di-oxygenases to catechol, which is further metabolized by two separate pathways. In the case of the ortho-ring cleavage pathway, a 1,2–dioxygenase cleaves the ring between the two adjacent hydroxyl groups and the sequential catabolism of the product *cis, cis*-muconate yields succinate + acetyl-CoA. Alternatively, the enzyme 2,3-di-oxygenase cleaves the ring between the carbon atom having an OH group and an adjacent carbon lacking an OH group (*meta*-cleavage). The products at the end of reaction sequence are acetaldehyde and pyruvate. Both *ortho*- and *meta*-pathways are involved in degradation of aromatic hydrocarbons. Benzene is degraded by the *meta*-pathway.

(f) **Polycyclic hydrocarbons** contain two or more rings. Generally, one of the terminal rings is attacked by a di-oxygenase, leading to ring cleavage and degradation so that in the end, a single ring remains which is catabolised in a manner similar to that described above. Degradation of complex molecules containing aliphatic, aromatic, alicyclic, or heterocyclic components is difficult to generalize but the following features are observed. Amide-, ester- or ether bonds are first attacked and the further degradation of the products so generated takes place. If these bonds are absent or inaccessible, the aliphatic chains are degraded; if the aliphatic chains are branched, the aromatic component of complex molecules may be attacked. The site and mode of attack depends on the molecular structure, the microorganism involved, and the environmental conditions. In general, recalcitrance of various benzene derivatives increases with the substituent groups (at *meta*-position) as follows: $COOH=OH \leftarrow NH_2 \leftarrow O-CH_3 \leftarrow SO_3^- \leftarrow NO_2^-$. Further, the greater the number of substituent groups on the benzene ring, the higher the degree of recalcitrance. The position of substitution also affects recalcitrance as *meta* > *ortho* > *para* in recalcitrance.

(g) **Co-Metabolism and gratutious metabolism:** Some xenobiotic compounds (e.g. cyclohexane, halogenated compounds etc.) are degraded by microbes, but these compounds are not used as a source of energy and carbon. Degradation of such compounds, therefore, depends on the presence of other compounds which induce the required enzymes, and the metabolism of which provides both energy and reducing agents for metabolism of xenobiotics. These compounds are called **co-metabolites**, and such a degradation is called **co-metabolism**. In contrast, many xenobiotics are degraded by an existing pathway and are used by microbes as a source of energy and reducing equivalents. This is known as **gratuitous metabolism**. In this process, the enzyme needed are induced by another compound which is not needed as a co-metabolite. Often a xenobiotic compound may not be completely degraded by gratuitous metabolism, but the product may be less polluting or may be used as substrate by some other organisms.

5. Biodegradation of halogenated compounds

In general, two types of microbial degradation exist. In one, compounds are broken down to support the growth of an organism and to provide essential nutrients. In the second type of metabolism, which is commonly known as 'co-metabolism', the compounds neither support the organism's growth nor serve as nutrient sources. Biodegradation of halogenated compounds involves elimination of the halogen groups, and degradation of the non-halogenated product. Removal of halogen molecules may occur either directly, involving the removal of hydrogen halide (e.g. HCl), or it may involve the substitution of halogen by –H, –OH or a –thio group. The direct halogen removal produces a double bond and is less frequent. The mechanism involving halogen substitution, especially by OH, is far more common, particularly for fully reduced aliphatics or aromatics.

Aerobic degradation of halogenated aromatic compounds usually involves first the addition of the –OH group by a di-oxygenase to yield chlorinated catechols, then the cleavage of the ring by *ortho* or *meta* cleavage. It follows elimination of the halogen from the straight chain (alophatic) product and finally, the degradation of the aliphatic hydrocarbon so produced. In case of phenols, the step 1 reaction is catalyzed by a hydroxylase, which adds another –OH group to yield the catechols.

6. Bioremediation of heavy metals

Bacteria, algae, fungi, actinomycetes, and higher plants can metabolize high amounts of metals.

(a) **Algae**: The species of *Chlorella, Anabaena inaequalis, Westiellopsis prolifica*, Stigeoclonium tenue, *Synechococcus* sp., tolerate heavy metals. Several species of Chlorella and Anabaena have been used for removal of heavy metals. But the operational conditions limit the practical application of these organisms.

(b) **Fungi**: Fungi can accumulate large quantities of heavy metals in their cells. In fungi, several mechanisms operate for metal accumulation. Metabolism-independent accumulation: In this process, the positively charged ions in the solution are attracted to negatively charged ligands in cell membranes. Biosorption of metal ions is affected by the composition of biomass and other factors. The second mechanism is metabolism-dependent accumulation. In fungi and yeast, heavy metal ions are transported into the cells through the cell membrane. However, as a result of metabolic processes, ions are precipitated around the cells, and synthesized intracellularly as metal-binding proteins. The third mechanism is extracellular precipitation and complexation. Fungi produce several extracellular products, which can complex or precipitate heavy metals. For example, some fungi and yeast release high affinity Fe-binding compounds that chelate iron. They are called siderophores. The Fe^{3+} chelates are taken up by the fungal cells.

7. Bioremediation of coal wastes through VAM fungi

Recently coal mine areas are being encouraged to introduce selected VAM (Vesicular and Arbuscular Mycorrhiza) fungi for bioremediation. Extensive infection of most plant species colonizing coal wastes has been observed in India and other countries. It has been reported that VAM improves the growth and survival of desirable species of plants used for re-vegetation. It has also been reported that there has been increased growth of red maple, maize, alfalfa and several other plants inoculated with VAM growing in coal mine soils.

8. Evolution of metabolic processes to degrade xenobiotics

The continued exposure of microbes to xenobiotics can often induce the evolution of metabolic processes needed to wholly or partly degrade the xenobiotics. These novel capabilities may arise due to mutations or transfer of plasmid (vector) borne genes. Gene mutations occur spontaneously at a low frequency. These mutations can modify the active site of an enzyme so that it has an increased affinity for the xenobiotic, or it can eliminate regulatory controls and enhance its production. Such changes only enhance the rate of degradation of a xenobiotic. Very frequently, the new enzyme activities in bacteria are acquired by plasmid transfer (usually through conjugation), since many of the key enzymes conserved with xenobiotic metabolism are plasmid borne. Most bacteria containing such plasmids are gram-negative aerobes mainly from the genes *Pseudomonas*. Some plasmids encode the entire pathway for xenobiotic degradation. But many plasmids encode only some of the degradative enzymes. In such cases, the remaining enzymes involved in the degradation must be provided either by the chromosomal genes of the cell or by another microorganism. Plasmid transfers allow a microorganism to acquire the genes needed to complete the pathway for a xenobiotic metabolism, and/or to gain genes which improve the rate and/or the nature of degradation.

The mechanism of plasmid transfer can be exploited to create microbes with novel characteristics. For example, *Alcaligenes* sp. degrades 4-chlorophenol to 5-chloro-2-hydroxymuconic semialdehyde, which is toxic. When the concentration of 4-chlorophenol is higher than 5 m mol/l, the level of this intermediate becomes toxic preventing its further degradation. However, *Pseudomonas* strain B13 has a plasmid-borne gene, which encodes the enzyme 1,2-di-oxygenase. This enzyme cleaves 4-chlorophenol by the *ortho*-pathway, which prevents the production of toxic intermediates.

Attempts are being made to develop genetically engineered bacteria with recombinant pathways for xenobiotic compound degradation. Those strategies most likely to succeed are transferring genes encoding enzymes with a wider specificity for substrates, and modifying regulatory elements with a view to enhancing enzyme synthesis and promoting assimilation of the xenobiotic compound into the cell.

9. Use of mixed microbial populations

The use of mixed populations of microbes for degradation of xenobiotics is an effective strategy due to the following reasons. Two different microbes can together degrade a xenobiotic completely, while either of them alone is incapable of this feat. In such a case, the product of degradation by one microorganism serves as the substrate for the other. For example, *Acinetobacter* sp. has plasmid-borne genes for dihydroxylation of one of the rings of 4-chlorobiphenyl, its *meta*-ring cleavage, and subsequent degradation to produce 4-chlorobenzoate. However, it cannot degrade this product any further. Another bacteria *Pseudomonas putida* strain can cleave the ring of 4-chlorobenzoate by *ortho*-pathway, to ultimately generate acetyl-CoA and succinate, but this bacterium cannot utilize 4-chlorobiphenyl. Thus *Acinetobacter* and *P. putida* act in tandem to completely degrade the xenobiotic 4-chlorobiphenyl.

In a mixed population of microorganisms, one microorganism may produce the growth factor/nutrient required by the other. For example, *Nocardia* sp. degrades cyclohexane but is unable to produce biotin. A *Pseudomonas* sp. strain produces biotin but cannot degrade cyclohexane. *Nocardia* sp. can break down cyclohexane and the break-down products and *Nocardia* cell lysis products are used by the *Pseudomonas* strain which grows and releases biotin. The biotin, in turn, promotes *Nocardia* growth and cyclohexane break-down. Thus a mixture of these two strains would break down cyclohexane but neither of them can do it alone.

A co-culture of more microbes leads to plasmid transfer among the mixed culture, thereby creating a faster growing species capable of degrading the xenobiotic. An example is the transfer of plasmid from the *Pseudomonas* sp. strain B13 into *Alcaligenes* sp., which is faster growing than *Pseudomonas* sp. B13. Mixtures of xenobiotic compounds are found in natural environments. Use of mixed cultures increases the likelihood of the microbial components of the mixed culture being able to degrade all the xenobiotic compounds present. The biological treatment system, i.e., the microbial community used for degrading xenobiotics, is more stable and can withstand occasional shock loadings. In mixed cultures, the biodegradation rates are higher due to the microbial interactions described.

10. Practical solutions to xenobiotic degradation

Efficient biodegradation of xenobiotics depends on many factors such as their concentration (high concentrations are toxic), the pH of the medium, temperature, availability of water, nutrients, and the presence of organic compounds. Optimal concentration and non-toxicity is very important for efficient biodegradation. In a treatment system, continuous supply of the compound should be available for selective maintenance of biodegrading microbes. Inhibitory compounds should not be present in the environment. Biodegradation of xenobiotics is facilitated by the supply of sufficient nutrients or co-metabolites, maintenance of the xenobiotic compounds at non-toxic levels, and provision of a microbial population or inoculum.

(a) Nutrients and co-metabolites: All the nutrients required for growth and development of microbial cells must be available in the solution containing the xenobiotics. Often, the addition of nitrogen, phosphorus, and sulfur enhances xenobiotic degradation. Adequate oxygen supply must be ensured by aeration of the medium where xenobiotics are degraded aerobically. For some microbes, the addition of some compounds (co-metabolites) may be required to induce the necessary enzymes and to provide energy and the reducing equivalents.

(b) Xenobiotic concentration: If the xenobiotic concentration is high, it should be optimized by appropriate dilution, usually with water. At the same time, the xenobiotic should always be present in the effluent in order to maintain the necessary microbial population. In a medium containing a low concentration of the xenobiotic, a higher density of the degradative microbes will be needed for the degradation process.

(c) Microbial inoculum: In nature, microbial populations evolve naturally but it will take time to reach optimal density. Therefore, it is necessary to inoculate the system with a suitable population of microbes having the necessary degradative capabilities. This is called ***bioaugmentation***. A number of inoculants designed to degrade various xenobiotics are commercially available. The commercial inoculum is prepared by selecting the strains for specific xenobiotic degradation and then blending them with bulking agents, dispersing chemicals, wetting agents, and nutrients. These inoculants are usually available as dry powders.

IV WATER QUALITY

Water is the most important commodity on planet earth. Without water there will be no life. In all life forms, more than 80% of body weight is water. Therefore, clean and sufficient water is the key factor in functioning and sustainability of life. Modernization and increase in activities that increase the pollution of water by adding more organic and inorganic compounds, are causing a change in the physical and chemical properties of water. Water pollution can have serious effects on living systems and thus reduce the usefulness of water. Some of the important water polluting agents are household detergents, sewage and other wastes, industrial wastes, agricultural wastes, release from nuclear reactors, etc.

A. WATER QUALITY

Water quality is determined as a set of chemical, physical, and biological features of the water, which determines the suitability of the water for various uses. The analysis of water quality is very important for biosafety. Water analysis permits determination of the suitability of water for various uses. The important measures are initiation of steps for maintenance and improvement of water quality, determination of polluting effects of various effluents discharged into water courses, setting up of safe standards for discharge of effluents into water bodies, monitoring of discharges to ensure that the discharge standards are met, evaluation of the efficiency of effluent/waste treatment processes and the efficiency of drinking water treatment processes, etc.

B. WATER QUALITY PARAMETERS

A number of quality parameters are measured to determine water quality. Some of these are listed in Table 38.2. These parameters include **physical properties** like pH, color, turbidity, suspended solids, temperature, conductivity, odor, etc., **chemical properties**, such as, COD, BOD, total nitrogen, total phosphate, dissolved oxygen (DO), total pesticides, etc., and **biological properties** e.g., total coliform bacteria, fecal coliform counts, fecal streptococci counts, *Salmonella* counts, etc.

Table 38.2 *Quality parameters for water intended for human consumption*

Parameter	Maximum conn. /value	Parameter	Maximum conn. /value
Part A		Chromium (µg/l)	50
Color units	20	Mercury (µg/l)	1
Turbidity units	4	Lead (µg/l)	50
Odor (dilution #)	3 at 25°C	Nickel (µg/l)	50
Taste (dilution #)	3 at 25°C	Pesticides, total (µg/l)	0.5
pH value	9.5	Polycyclic aromatic hydrocarbons, total (µg/l)	0.2
Temperature	25°C	**Part C**	
Sulfate (mg/l)	250	Total coliforms	0/100 ml
Magnesium (mg/l)	50	Fecal coliforms	0/100 ml
Sodium (mg/l)	150	Fecal streptococci	0/100
Potassium (mg/l)	12	Sulphite reducing clostridia	≤1/100 ml
Dry residues (mg/l)	1500	**Part D**	
Nitrate (mg/l)	50	Chloride (mg/l)	400
Nitrite (mg/l)	0.1	Calcium (mg/l)	250
Ammonium (mg/l)	0.5	Substances extractable in chloroform (mg/l dry residue)	1
Dissolved or emulsified hydrocarbons (µg/l)	10	Boron (µg/l)	2000
Phenols (µg/l)	0.5	Barium (µg/l)	1000
Surfactants µg/l	200	**Part E**	
Part B		Total hardness (mg Ca/l)	60
Arsenic (µg/l)	50	Alkalinity (mg HCO_3/l)	30
Cyanide (µg/l)	50		

(UK water supply (water quality) regulations, 1989; Singh BD, 1998)

The determination of physical parameters is routine and simple. Inorganic parameters are estimated using specific colorimetric reactions or ion-specific electrodes. Now-a-days, there are specific kits available for several parameters simply to reduce time, labor, and cost of estimation and to have high specificity and reliability. However, the determination of organic compounds is relatively more difficult. It usually involves a step to increase their concentration. They are often determined by chromatographic characteristics (e.g. HPLC). This makes the estimations costly and time and labor intensive. Techniques for rapid, efficient, and low cost, are in progress by using biosensors, enzymes etc. Biosensors using microorganisms have been developed for detection of ammonia, BOD, nitrates, phosphate, etc. Similarly, the enzyme acetyl cholinestrase is used to detect organophosphate insecticides/nerve poisons since they inhibit the activity of this enzyme. The use of enzymes and biosensors for water analysis is expected to gain significance in future. The biological parameters are measured as numbers of colonies formed following culture on specific media at specific temperatures. The presence of coliform bacteria in water indicates the contamination by mammalian excreta and is suggestive of the presence of pathogenic bacteria like *Salmonella*, *Shigella*, *Vibrio cholera* etc. Microbiological tests can differentiate between fecal and non-fecal forms of coliforms. The ELISA has also been recently adapted for rapid detection of bacteria and viruses.

C. METAL TOXICITY OF WATER

Metal toxicity of water is a serious problem in many places and it can cause many diseases and many deaths. Some of the important metals found in polluted water are:

Mercury: This is generally released by chlorine and caustic soda manufacturing industries. It is one of the main contaminators of the food chain, particularly, by aquatic organisms such as fishes and also the biomagnification process. Mercury pollution can cause nerve defects and death in humans.

Lead: It is released by the outlet pipes carrying effluents and accumulates in animal tissues. Continuous accumulation of lead in humans can affect the gastrointestinal tract, neuromuscles, and the central nervous system.

Copper and zinc: Metal pollution is a big problem for aquatic animals, which can affect their growth, metabolism, and reproduction.

Cadmium and chromium: These metals can accumulate in marine organisms, and cause their death.

Fluoride pollution: Fluoride is known to cause dental fluorosis. Water contaminated with fluorine causes stiffening of bones, mottling of teeth, outward bending of legs from knees, etc.

The only effective method to control metal pollution of water is the treatment of factory effluents by detoxification treatments. This demands the enforcement of strict and appropriate laws.

V BIOTECHNOLOGY IN WASTE TREATMENT AND ENVIRONMENTAL MANAGEMENT

Biotechnology has made several contributions to waste treatment and environmental management. Environmental biotechnology employs the application of genetic engineering to improve efficiency and decrease costs for efficient management of pollution. It is hoped that in the future, the application of genetically modified microorganisms (GMMs) coupled with biotechnology techniques will make a major contribution to improving the quality of our environment. However, the risks associated with the use of GMMs are discussed in Chapter 39. The use of microorganisms in environmental clean-up with special reference to bioremediation types, processes and methods with several examples, utilization of sewage and agro-wastes, and the benefits associated with the release of GMMs for environmental clean-up have been discussed in the following sections.

A. GENE MANIPULATION OF BIODEGRADING MICROORGANISMS

The term *bioremediation* has been introduced to describe the process of using biological agents to remove toxic wastes from the environment. The basis for bioremediation is the enormous natural capacity of microorganisms to degrade organic compounds. This capacity can be improved by applying GMMs. The application of the new methods to constructing novel microbial strains that have improved capacities for degrading various synthetic compounds is currently an active area of research. Members of the genus *Pseudomonas* are the most predominant group of soil microorganisms that degrade xenobiotic compounds. The biodegradation of complex organic molecules generally requires the concerted efforts of several different enzymes. The genes that code for the enzymes of these biodegradative pathways are located mostly on the host chromosome, but in a few cases, these genes are also found on large plasmids, or on both.

Biodegradative pathways can be genetically engineered. The degradative capability of a particular bacterial strain is often restricted to a single class of chemical compounds. Oil spills, chemical dump-sites, and other wastes, however, contain mixtures of chemicals. Therefore, it is desirable to expand the degradative potential of a candidate strain. The simplest way to do this is to transfer by conjugation, into the recipient strain, plasmids that carry genes for different degradative pathways. If two resident plasmids contain homologous regions of DNA, recombination can occur, and a single, large "fusion" plasmid with combined functions can be created. Alternatively, it may also be possible to extend the degradative capability of a strain by altering the genes of the degradative pathway. The feasibility of this approach was examined for the toluene/xylene-degrading pathway of the plasmid pWWO. This plasmid encodes a "meta-cleavage" pathway that consists of 12 different genes and enables *Pseudomonads* carrying the plasmid to utilize various alkylbenzoates as carbon sources. The genes in the toluene/xylene pathway of pWWO are part of a single operon, called the *xyl* operon, under the control of the P_m promoter. Detailed biochemical and genetic analyses showed that bacteria carrying plasmid pWWO could degrade 4-ethylbenzoate to 4-ethylcatechol but not further. To solve this problem, mutants were developed which could grow and degrade 4-ethylbenzoate. This study demonstrates that by combining recombinant DNA technology, mutagenesis, and the appropriate selection protocols, novel properties can be added to a degradative pathway for a wide variety of waste products.

The feasibility of pesticide degradation through plasmid-mediated genetically engineered microorganisms was reported by Chakrabarty et al. (1981). The *opd* gene, isolated from *Flaavobacterium* sp. ATCC27551 and *Pseudomonas diminuta*, has been shown to degrade pesticides such as parathion, methylparathion, etc. Sims et al. (1990) transferred a recombinant

DNA plasmid containing the *opd* gene into a fungus, *Gliocladium virens*, which is a useful saprophyte showing a strong mycoparasitic activity against many fungal pathogens. These strains have potential for use in the bioremediation of contaminated soil. Nagata et al. (1993) cloned and sequenced two genes involved in early Y–HCH degradation in UT26. This *lin*A gene encodes the Y-HCH dehydrochlorinase which converts Y-CHC to 1,2,4TCB via Y-PCCH and 1,4-TCDN. The *lin*B gene encodes the 1,4-TCDN chlorohydrolase, which converts 1,4-TCDN to 2,4-DDOL via 2,4,5-DNOL. This gene belongs to the haloalkanedehalogenase family with a broad range specificity for substrate. The GMM *P. putida* comprises both *lin*A and *lin*B genes.

VI UTILIZATION OF SEWAGE AND AGRICULTURAL WASTES

A. PRODUCTION OF SINGLE CELL PROTEIN (SCP) ON SEWAGE

A relatively new field is the production of 'single cell protein' (SCP). The proteins from dried cells of microorganisms used as food or feed are collectively called microbial proteins. Microorganisms constitute a major part of our diet, as fermented yeast (*Saccharomyces* sp.) in bread, lactic acid bacteria (*Lactobacillus* and *Streptococcus*) in fermented milk and cheese, fungus from mushrooms, and blue-green algae (e.g. *Spirulina*) from water bodies. The term microbial protein was replaced by a new term 'single cell protein' (SCP) at the "First International Conference on Microbial Protein", held in 1967 at the Massachusetts Institute of Technology (MIT). The criterion for coining this term, was the single-celled habit of microorganisms used as food and feed. The term refers to protein obtained from the large-scale growth of microorganisms such as bacteria, yeasts and other fungi, and algae. The importance of mass production of microorganisms as a direct source of microbial protein was realized during World War I in Germany. Since then baker's yeast (*S. cerevisiae*) has been produced in an aerated molasses medium supplemented by ammonium salts. During World War II, the aerobic yeasts (e.g. *Candida utilis*) were produced for food and feed in Germany. In 1959, British Petroleum produced the SCP from hydrocarbons and established the first large-scale plant. It had a capacity of 1,00,000 tons of SCP per annum. In the USSR, SCP production reached 1.1 million tons per annum. SCP can provide a good source of protein and can replace, partially or totally, the protein requirements for various purposes. For the development of technologies to produce SCP on large scale and at cheaper price, the utilization of waste products will play an important role.

There are several advantages in using microorganisms (SCP) as food source. (i) They occupy less room than conventional crops and animals; (ii) they grow rapidly; (iii) they are easy to modify genetically; (iv) they have high protein content (range 43–85); (v) they can grow on a broad spectrum of waste products (e.g. agriculture and industrial wastes, petroleum products, methane, methanol, ethanol, sugar, molasses, dairy products, waste from pulp and paper mills), they have the secondary advantage that they can help to recycle materials and clean up waste; (vi) high reproducibility of quality and quantity since they can be continuously cultured and are independent of climate; (vii) there are fewer ethical issues associated with their exploitation and no animal rights issues. They have many other advantages, such as high efficiency in solar energy conversion per unit area, easy regulation of environmental factors, etc. Many groups of microorganisms are used as sources of proteins.

Substrates used for SCP Production: A variety of substrates from diverse sources are used for SCP production. The major components of the substrates are the raw materials which contain sugars (sugarcane, sugar beet etc.), starch (cereals, tapioca, potato, and their by-products), lignocelluloses from woody plants and herbs and other raw materials (whey and refuses from processed food). Organic wastes generated by industries, hydrocarbons, agricultural wastes, and any other organic materials can be used for SCP production. A range of substrate materials have been used to produce SCP (Table 38.3).

Nutritional value of SCP: The commercial value of SCP depends on their nutritional value. SCP may be used for human consumption or animal feed. SCP are a rich source of proteins. In addition, the high levels of methionine, lysine, vitamins, and essential minerals often make SCP more nutritious than some plant and animal foods. The composition of the growth medium governs the protein and lipid contents of the microorganisms. For example, yeast, moulds, and higher fungi have higher cellular lipid content and lower nitrogen and protein content, when grown in media having a high amount of available carbon as source and low nitrogen. Except in a few cases, the mean crude protein in dry matter of algae and yeast on conventional substrates ranges between 50 to 60%. For alkane yeasts, it will be between 55 and 65% and bacteria can have up to 89%. A high content of nucleic acid free protein is extremely important for the economic efficiency of the procedure in SCP production. However, there are some problems with the use of SCP products as human foods such as high content of nucleic acids can cause kidney stones and gout and there is the possibility of the presence of toxic metabolites and poor digestibility and stimulation of gastrointestinal and skin reactions.

Table 38.3 *Substrates and organisms used for SCP production*

Raw material	Type of microorganism
Carbon dioxide	*Spirulina maxima*, Cyanobacterium
Whey (lactose)	*Kluyveromyces fragilis*, Yeast
Petroleum alkanes	*Candida lipolytica*, Yeast
Cellulosic wastes	*Chaetomium cellulolyticum*, Fungus
Methane (methanol)	*Methylophilus methylotrophus*, Bacterium

Genetic improvement in microbial cells: SCP production by mass culture of microorganisms is not on a full-scale. One of the ways to enhance productivity and the quality of SCP products is the genetic improvement of microorganisms. Transfer and expression of beneficial genes in the microorganisms have started a new era. For example, genes responsible for dehalogenase can be introduced into organisms that grow in sewage polluted with halogenated chemicals. Rochix and van Dillerviger (1982) have successfully introduced genes of *S. cerevisiae* into *Chlamydomonas reinhardii* cells and got expression of fungal genes in algal cell. Imperial Chemical Industries (ICI) in England successfully developed a continuous methanol fermentation process for the commercial production of SCP from the bacterium *Methylophilus methylotrophus* (called Pruteen). Pruteen manufacture is a classic example of continuous batch culture. *M. methylotrophus* can use methane as a substrate for growth although in practice, the methane is converted into methanol, which is used as the principal substrate.

SCP from photosynthetic organisms: Both photosynthetic bacteria, such as blue-green bacteria and algae are used to make SCP. Algae, unicellular eukaryotes belonging to the genus *cyanobacteria*, grow autotrophically and synthesize their food by taking energy from sunlight or artificial light, carbon source from carbon dioxide, and nutrients from the carbohydrates present in the growth medium (Figure 38.3). *Chlorella* strains are being used for a variety of applications due to their very high protein contents, which can be used for improving the nutrition of animal feeds. In some countries, *Chlorella* are utilized for sewage oxidation and waste-water treatment. For cultivation of algae on sewage wastes, oxidation ponds have to be constructed, where sewage is allowed to accumulate and in which mixed cultures of algae will be grown. Mass cultivation of algae has been started in many countries. Harvesting of algal biomass becomes problematic because the cells settles down at the bottom and the algal cultures mix. The cells are recovered by concentration, de-watering, and drying. Methods of separation and concentration also follow centrifugation, flocculation, and centrifugation plus flocculation. Harvesting the *cyanobacteria*, e.g. Spirulina sp., is less troublesome as their spiral filaments float on the surface of the water. Cells are able to fix atmospheric nitrogen. Algal mats are filtered and the suspension of *Spirulina* is dried with hot air to get a fine powder. Algal yield can be

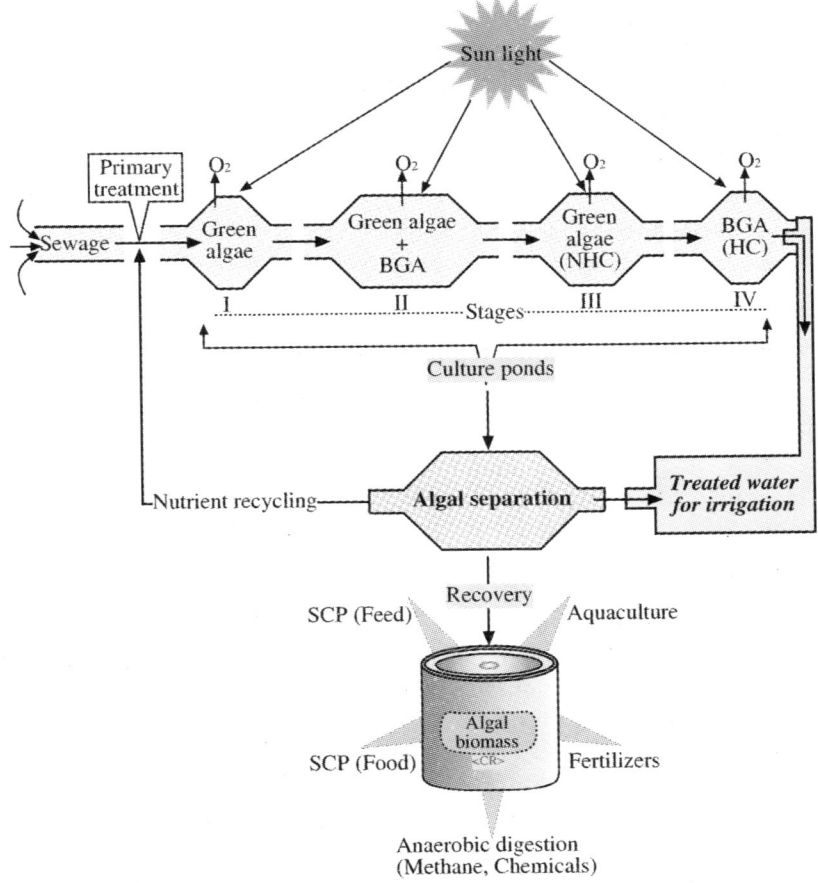

Figure 38.3 *Steps in cultivation of algae in sewage oxidation ponds and possible application of algal biomass*

from 114 to 170 tons/hectare/year. *Spirulina* powder can be used for various purposes, to make lozenges, capsules, biscuits, and confectionery with a high protein content.

Many types of farms for mass cultivation of *Spirulina* SCP are under operation. The **Seminatural lake system** is an ideal environment for the natural growth of *Spirulina*. Due to contamination problems in this system, SCP are used for fish and animal feed. The second type of cultivation is the **artificially built cultivation system**. Based on the water quality and nutritional status, this system can be grouped into **clean water system** and **waste water system**. Clean water system is generally more expensive due to the construction of artificial cultivation farms. These have shallow raceway ponds circulated by paddle wheels and high quality nutrients. For quick growth $NaNO_3$ and $NaHCO_3$ are added. The pH of the water must be maintained initially at 8.5 then alga self elevates to 10–10.5. The second method is waste water system, which is applicable in highly populated countries, where wastes are generated in higher quantities and pose environmental problems. In this system, human and animal wastes and sewage are used for the growth of *Spirulina*.

Requirements for growth of *Spirulina*: (i) **Algal tanks**: These are generally circular or rectangular cemented tanks. The circular ones are generally preferred and the size depends on convenience and yield needed. (ii) **Light**: Initially low light intensity is required in order to avoid photolysis. (iii) **Temperature**: Optimum temperature for growth should be between 35 and 40°C. (iv) **pH**: *Spirulina* grows at high pH ranging from 8.5 to 10.5. (v) **Agitation**: Agitation of culture by brush, paddle power, pipe pumps rotators, etc., is very necessary to get good quality and better yield. (vi) **Harvesting**: The filaments of *Spirulina* float on the surface of the water forming a thick mat. Therefore, they can be easily harvested by a fine mesh. (vii) **Drying**: Since *Spirulina* cells have thin cell walls, sun drying is the most suitable and economical. (viii) **Yield**: An average yield of 8–12 g *Spirulina* powder/m^2/day (which is equivalent to 20 tons/ha/annum) can be harvested. (ix) **Avoiding contamination**: Regular monitoring for chance contamination is necessary.

Uses of *Spirulina* SCP: Uses of *Spirulina* SCP are many: (i) as protein supplemented foods; (ii) as health food; (iii) in therapeutic and natural medicine; (iv) in cosmetics etc.

B. BACTERIA AND ACTINOMYCETOUS BIOMASS

Bacteria and actinomycetous biomass are widely used as a source of SCP because of their short life cycle and capacity to utilize a wide range of organic substrates. Since the establishment of British Petroleum in 1960, significant progress has been made in the production of microbial SCP using gaseous and liquid hydrocarbons and the chemicals derived from them (methanol, ethanol etc.). Shell Research Limited, UK, conducted research on the pilot plant scale process for the production of bacterial SCP from methane by using *Methylococcus capsulatus* or a mixed culture of *Pseudomonas* sp., *Hyphomicrobium* sp., *Acinetobacter* sp., and *Flavobacterium*. *Streptomyces* sp. is capable of growing on methanol. *Thermonospora fusca*, a thermophilic species, degrades 60–65% paper mill fines, resulting in a 30% protein product.

Harvesting: However, there are many problems related to the recovery of bacterial cells, since they are very small in size and cell density is in the order of 10–20 g/litre. The centrifugation cost is also high. Many pilot plants use flocculents and have set up a decanter type centrifuge. The cells are washed and the spent medium is again treated by the conventional treatment process as it contains inorganic salts and a small amount of cells. The cell biomass is then spray-dried.

C. MYCOPROTEIN

Fungi are another source of SCP. Yeast and molds, for example, are used for both human and animal consumption in processes similar to those for bacteria. A good example for use of molds is the manufacture of mycoprotein (**Quorn**).

1. Production of yeast biomass

Yeast cells as SCP was started during World War I in Germany. Since then many improvements have taken place in biotechnological applications of yeast, as far as culture development, process optimization, and scale up of products are concerned. Yeast biomass production in the world is 0.4 million metric tons per annum which includes 0.2 million metric tons of baker's yeast alone. Yeast synthesizes amino acids from inorganic acids and sulfur supplemented in the form of salts. They get carbon and energy sources from the organic wastes, e.g. molasses, starchy materials, milk whey, fruit pulp, wood pulp, and sulphite liquor. Yield corresponds to nutrients in the growth medium, temperature, culture, oxygen, etc. Yeast cells are recovered by decantation–centrifugation including washing and drying treatment methods. After harvesting by vacuum filter, a cake containing 20–40% dry matter is obtained which is then dried to get a product with 6–10% moisture.

Yeast extract: Yeast cells are rich in B vitamins, particularly niacin (B_6), riboflavin (B_2), thiamin (B_1), folic acid, and B_{12}. They can be dried and made into vitamin-rich tablets or converted into products like marmite. This process involves heating the yeast in large vats to 50°C and adding salt to encourage the process of autolysis. Autolysis is self-digestion and is carried out by enzymes in dying cells. The autolysed products are filtered and centrifuged to remove the cell walls and then concentrated into a thick paste. Vegetable extract is added to marmite. Alternatively, hydrolysis with hydrochloric acid can digest the cells. The HCl is later neutralized with sodium hydroxide.

2. Production of fungal biomass

As stated earlier, fungi are another source of SCP. Initially cultures of *Fusarium* and *Rhizopus* were grown in fermentation as protein food (**mycoproteins**). These organisms were chosen because they are non-toxic and grown on complex organic compounds. The doubling time of the fungus in culture was 5.5 hours, which is slower than bacteria, and it used glucose as a carbon and energy source. This comes from any cheap source of starch such as corn, wheat, rice, potato, or molasses. It produces 0.5 kg dry biomass per kilogram of sugar used. Fungi have the advantage that they can be grown at a pH which is acidic enough to inhibit bacterial growth, thus reducing the risk of contamination. *Fusarium* is grown at 30°C in continuous culture. Ammonium salts are the nitrogen source and mineral salts are also added to maintain growth. Agitation of the medium is very important because the hyphae tend to become tangled with stirrers and not uniformly distributed inside the fermenter. This is achieved by a special aeration mechanism called 'air-lift', because the culture is circulated continuously, once every two min. by the air.

Eukaryotic cells contain a higher proportion of nucleic acid than prokaryotic cells, so the mycoprotein product may contain 5–15% (dry mass) of nucleic acids. This is mostly RNA and must be reduced because consumption of more than 2 g a day can lead to gout and kidney stones in humans. It is removed by a simple procedure, i.e., heating the culture to 64°C for 20–30 min. in a separate steam-heated vessel. This inactivates fungal proteases, so proteins are not affected, but RNA is reduced to about 1%, well below the recommended limit of 2% set by the WHO.

As a result of rapid growth, a high amount of fungal biomass is produced. An advantage of using fungi is that the mycelium is easier to separate from the medium than are bacterial cells, and simple filtration is sufficient. Mycelial yield varies widely depending on organisms and substrates. Filtration and drying leaves a thin flexible sheet of Quorn. At this stage, it looks and tastes a little like raw pastry. Flavors, e.g. vegetable and egg white, can be added. It is then sliced, diced, or shredded for use. The fungus is already fibrous so it is easy to give it a meaty texture. For human consumption, factors other than just economics are important. Very strict safety guidelines must be adhered to, and the nutritional value of the fungal SCP must be acceptable. Trials on rodents and bovine have shown no long-term harmful effects. These products have several health advantages over meat. They are cholesterol-free and high in fiber, low in fat and calories with a good poly-unsaturated: saturated fatty acid ratio. They are also a good source of vitamin B_{12} and zinc.

However, strains of some species of molds, for example, *Aspergillus niger*, *A. fumigatus*, and *Fusarium graminearum* are very hazardous to humans, therefore use of those species should be avoided. Fungi can grow on many sources of organic wastes. Large quantities of cellulosic and lignocellulosic materials are present in agricultural, forestry, and industry wastes. There are many traditional fungi fermented foods that are in use throughout the world.

D. VERMICOMPOSTING

Vermicomposting is the phenomenon of compost formation by earthworms. Earthworms play an important role in the cycling of plant nutrients, turnover of organic matter, and maintenance of soil structure. The most important activity of earthworms in agro-ecosystems is the increase in nutrient cycling, particularly of nitrogen. They consume organic matter with a relatively wide C:N ratio and convert it into earthworm tissue with a lower C:N ratio. They can consume 10–20% of their own biomass per day. Therefore, earthworms can play a important role in the physico-chemical properties of soil.

VII BIOFERTILIZERS ARE AN ALTERNATIVE TO INORGANIC FERTILIZERS

The extensive use of inorganic fertilizers and pesticides to produce more food has had harmful effects on the ecosystem. Many of these pollutants have long-term residual effects and also lead to an increased cost of production. Therefore, the search for alternatives is going on throughout the world. There are many reports that certain plants can be used as fertilizers and pesticides. Since the plant products are of organic origin, they will easily be degraded in natural conditions. Use of biofertilizers is economical as well as environmental-friendly. Biofertilizers are of two main categories: Green manure and biofertilizers.

A. GREEN MANURES

Manures are organic materials mixed with soil. Manure supplies most of the nutrients necessary for the growth of crop plants. Thus manure applications increase crop productivity. Manures are of three types.

(i) Farmyard manure (**FYM**): This is a mixture of cattle dung and various forms of agricultural residues, unused parts of straw, and plant stalk fed to the cattle.

(ii) Composited manure: This consists of a mixture of partially or fully decomposed animal and plant residues.

(iii) Green manure: This is a herbaceous crop (such as clover), leaf branches (e.g. pongamea, glyricedia, etc.), ploughed under soil and mixed well with the soil while still green to enrich the soil fertility. The plants used as green manure are often quick growing, not used as fodder, and generally belong to legumes. Green manure not only enriches the soil but also gives very good texture to the soil so that aeration and water holding capacity is increased. Nutrients will be released slowly and microbial activity is enhanced.

B. BIOFERTILIZERS

Biofertilizers are generally defined as biologically active products consisting of bacteria, algae, and fungi (single or in combination) which supplement soil nutrients by biological activity. These mostly include nitrogen-fixing microorganisms, phosphate solubilizers, etc. Some of the biofertilizers are: (1) symbiotic nitrogen-fixing bacteria in legumes; (2) nitrogen-fixing fern Azolla Anabaena symbiosis; (3) nitrogen-fixing free living bacteria such as azospirillum; (4) cyanobacteria; (5) mycorrhiza phosphate solubilizers, etc.

VIII RESOURCES RECOVERY OR MICROBIAL LEACHING AND BIO-MINING

Useful metals such as copper, iron, uranium, gold, lead, nickel, and cobalt are found naturally as minerals (known as ores). The ores, which are sufficiently concentrated minerals are mined and the metals are extracted. Microorganisms have been forming and decomposing minerals in the earth's crust since geologically ancient times. Recently, the potential role of microbes in mining of minerals has been realized. Microbes can be used to recover valuable minerals such as uranium and silver from low-grade ores. Bacterial leaching is now used throughout the world as an additional technique for extracting metals from ores (Figure 38.4). Suitable combinations of bacteria are used, each making its own unique contribution. Many bacteria harbor plasmids that carry genes for transformation of metals, aromatics, and other hydrocarbons. Since they are located on plasmids, such genes can be easily isolated and then introduced into organisms like *Azotobacter*. *Azotobacter* can fix nitrogen and utilize the excessive carbon present in municipal sewage. The microbially treated sewage can be used as fertilizer.

Microbial leaching is the process by which metals are dissolved from bearing rocks using microorganisms. At present, a number of ores cannot be economically extracted using chemical methods due to their low metal content. In addition, large quantities of low grade ores are

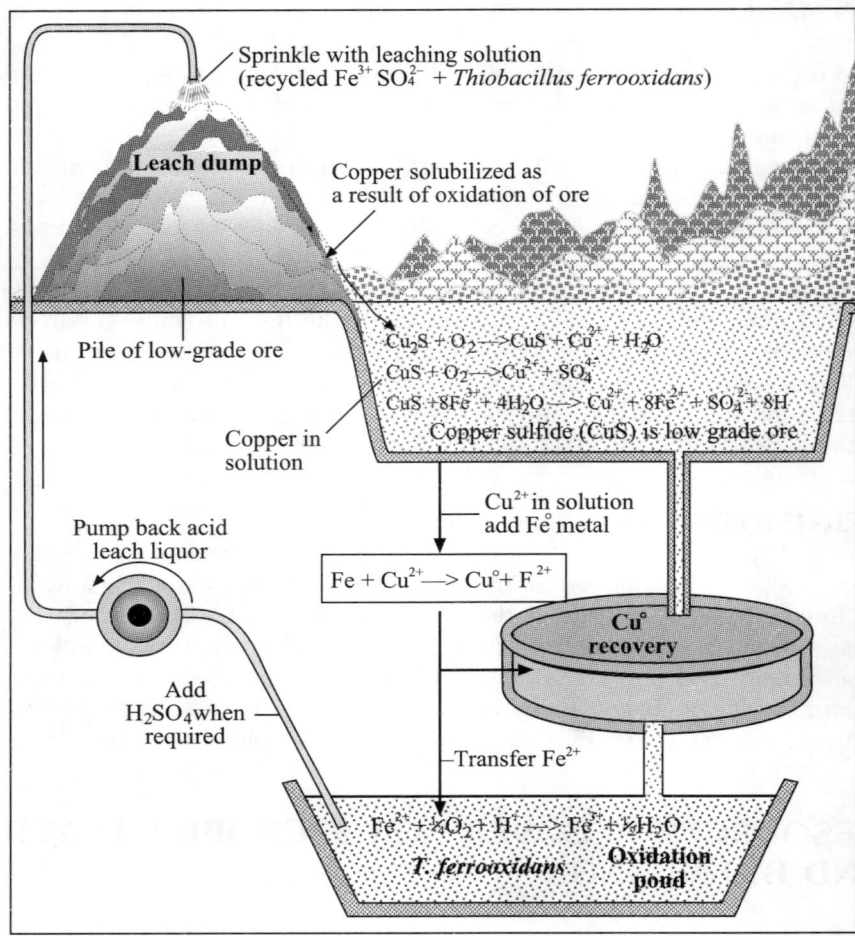

Figure 38.4 *Microbial leaching of copper (heap leaching)*

produced during separation of higher-grade ores and are generally discarded in waste heaps. Throughout the world, there are vast quantities of such low-grade copper ores that cannot be profitably purified by conventional chemical methods, but are profitably processed by microbial leaching. Thus bio-mining has emerged as an important branch of biotechnology.

Bio-leaching bacteria: Several different microorganisms have been isolated from the sites where bio-leaching takes place. Two most commonly used organisms in microbial leaching are *Thiobacillus thioxidans* and *T. ferrooxidans*. A number of others may also be used (e.g. *Bacillus licheniformis*, *B. luteus*, *B. megaterium*, *B. polymyxa*, *Leptospirillum ferrooxidaans*, *Pseudomonas fluorescens*, *Sulfolobus acidocaldarius*, *Thermothrix thioparus* etc. All these organisms are adapted to living in the harsh conditions (they thrive in acid solutions, high temperature, etc.) of their natural environment.

Leaching reactions: Leaching reactions generally involve the conversion of insoluble metal ores, which are often sulfides, to soluble compounds from which the desired metal can be more readily isolated. The chemistry of microbial leaching, for example for autotrophic aerobes, obtains its energy from the oxidation of Fe^{2+} to Fe^{3+} or from the oxidation of elemental sulfur and reduced sulfur compounds to sulfate. The oxidation of insoluble sulfur to sulfuric acid, which is also performed by *T. thiooxidans*, occurs in the periplasmic space. Copper leaching and uranium leaching are others successfully exploited. In bio-leaching, two reaction mechanisms take place (**direct** and **indirect bacterial leaching**). In ***direct bacterial leaching***, a physical contact exists between bacteria and ores and oxidation of minerals takes place through several enzymatically catalyzed steps. For example, pyrite is oxidized to ferric sulfate. In ***indirect bacterial leaching***, the microbes are not in direct contact with minerals but leaching agents are produced by the microorganisms that oxidize them.

Leaching techniques: Bio-leaching can be applied to the recovery of valuable metals from low-grade mineral deposits, vast tonnages of which are available. According to one estimate, more than 33 billion kilograms of copper are located in mine dumps in western USA. There are four commercial methods used in bio-leaching. (i) ***Slope leaching***: Large amounts (about 10,000 tonns) of ores are ground to fine pieces. It is dumped in large piles down a mountain-side. Water containing an inoculum of *Thiobacillus* is continuously sprinkled over the pile. Water is collected at the bottom. It is used to extract metals and generate bacteria in an oxidation pond. (ii) ***Heap leaching***: The ore is dumped in large heaps called dump and further steps of treatment are as described above. (iii) ***Vat leaching***: High-grade ores or concentrates are treated in aerated tanks. This is even more efficient than the heap methods and can be completed in a matter of days. Vat leaching has been successfully used to decompose gold-containing arsenopyrite ores before gold recovery. (iv) ***In situ leaching***: In this process, the ores remain in their original position inside the earth. Surface blasting of rocks is done just to increase the permeability of water. Water containing *Thiobacillus* is pumped through a drilled passage to the ores. Acidic water seeps through the rocks and collects at the bottom. Then the bottom water is pumped and the mineral is extracted.

Genetic improvement of leaching bacteria: Non-biological methods of mining and metal processing will continue to dominate in the immediate future. However, the economic competitiveness of bio-leaching for recovering large quantities of metals from low-grade ores has been amply demonstrated. Investigators are now turning their attention to the improvement of the bacterial strains used for leaching by the use of recombinant DNA technology. Genetic manipulation of the bacteria may permit the development of new strains with better growth rates, increased leaching efficiencies, and improved resistance to metals that have toxic effects on the organisms. *T. ferrooxidans* is a prime target for genetic improvement because of its role in bio-leaching. Metal ion tolerance is a particularly attractive marker for genetic studies of *T. ferrooxidans* because increased metal resistance has the potential for conferring an industrially significant characteristic on the bacterium. Genes coding for resistance to arsenic and mercury have been identified in other bacteria. If such genes could be introduced into *T. ferooxidans*, they might improve its leaching capabilities.

IX MANAGED ECOSYSTEMS

Ample evidence indicates that plants vary markedly in their response to air pollutants. Air pollution exerts a selective force on plants at both the sporophytic and gametic levels. The application of genetic knowledge regarding air pollution and the ecosystem focuses on two central concepts. (i) Since plants vary extremely from being highly susceptible to highly resistant to air pollutants, it is possible to select and breed for plants with either high or low sensitivity for use in managed ecosystems; (ii) the maintenance of genetic diversity in plant populations in both unmanaged and managed ecosystems should be given more serious attention.

Developing pollution-resistant plants: Many scientists have reported that it is possible to select and breed plants with improved air pollution resistance through traditional genetic approaches. Pollution-resistant plants could be useful in re-vegetating local areas affected by point-source pollutants (e.g. SO_x, heavy metals), particularly when the pollution control methods are not developed or are costly. It is also useful where the political or economic situation in the country prevents clean-up at the source and in urban areas where accumulated effects of multiple pollutants create very complex conditions.

Developing pollution-sensitive bio-indicator plants: Monitoring multiple pollutants with equipment designed to accurately characterize ambient air pollution loading is a costly and rigorous exercise. Monitoring requires large capital expense and intensive and skilled labor to calibrate, maintain, and operate air-monitoring equipment; so alternative methods are desirable. One alternative is to use biological-sensitive indicator plants or biomonitors. Bel W–3 tobacco is probably the best-known bioindicator plant, having been extensively used to monitor O_3 levels. Plants that respond predictably to air pollutants can be used independently or in conjunction with mechanical monitors to biomonitor the presence and amount of air pollution. Bioindicators have many advantages over instruments such as rapid response of plants to air pollution and lower cost. Besides, biomonitors can be used in remote sites.

Germplasm preservation and maintenance of diversity: The evidence of the impact of air-pollution on the gene pool of many plant species has been reported by many scientists. Numerous population studies have been conducted to demonstrate that gene frequencies can be changed by pollution-induced natural selection. This warrants immediate action to conserve genetic diversity by *in situ* or *ex situ* conservation methods.

(For conservation of **unmanaged ecosystems** Refer to the next chapter.)

X SUGGESTED READING

Bernarde MA (1992) *Global Warming—Global Warning*, John Wiley and Sons, Chichester.

CAMLAB (1991) *Design of Municipal Waste Water Treatment Plants*, CAMLAB, Cambridge.

Gilpin A (1994) *Environmental Impact Assessment*, Cambridge University Press, Cambridge.

Leach C K and Van Dam–Mieras MCE (1994) *Biotechnological Innovations in Environmental Management*, Butterworth–Heinemann, Oxford (BIOTOL series).

Miller MC and Powell W (1994) A Commercial View of Biotechnology in Crop Protection, *Molecular Biology in Crop Protection*, pp225–245. (Eds) Marshal G and Walters D, Chapman and Hall.

Singh BD (1998) *Biotechnology*, Kalyani Publishers.

Taylor DJ, Green NPO and Stout GW (1998) *Biological Science*, Cambridge University Press, Cambridge.

Taylor GE, Pitelka LF and Clegg MT (Eds) (1991) *Ecological Genetics and Air Pollution*, Springer-Verlag. New York Inc.

39

Biodiversity and Genetic Conservation

I INTRODUCTION

All forms of life on planet earth are part of one great interdependent system. This system interacts with, and depends on the non-living components of the earth: environment, atmosphere, oceans, water, rocks, and soils. Humans are part of this biosphere and totally depend on this community of life. The diversity, within or between these organisms forms valuable genetic resources on the earth. However, the explosion in human population have resulted in the expansion of agriculture and industrial activities, which has had an enormous and adverse impact on biodiversity and the ecosystem. Therefore, the extinction of wild species and the deterioration of the ecosystem have been areas of major concern for policy makers and biotechnologists. Development has to be both people-centered and conservation-based. Unless we protect the structure, functions, and diversity of the world's ecosystem, development will diminish and fail irreversibly. The earth's resources should be prudently used and development must not come at the expense of other groups or later generations, nor threaten the survival of other species. The conservation of biodiversity is fundamental to the success of the development process.

The steady erosion in the diversity of genes, and ecosystems that is taking place today will undermine progress toward a sustainable society. There have been major efforts made to conduct a survey and conserve the country's biodiversity, so as to save wild species from extinction. At the international level, under the auspices of the United Nations, funds are being established and many other efforts are being made for the conservation of germplasm. Biodiversity studies thus include a comprehensive and systematic examination of the array of organisms on earth and establishment of different strategies for the maintenance of biodiversity. In this regard, several biodiversity conservation meetings were held in 1992 for discussions on measures to be taken by the developing and developed countries to preserve biodiversity at the global level. Although most countries agree to the need for preserving biodiversity, there is still disagreement in relation to the financing and policy making.

Conservation is about maintaining the biosphere, i.e. taking action to avoid species decline and extinction and permanent detrimental change to the environment. To achieve the goal of conservation we need to understand how the biosphere functions or how organisms and the environment interact. The conservation of genetic diversity requires the application of knowledge of ecology and the related environmental sciences and mainly public and governmental support.

II BIODIVERSITY

'**Biological diversity**' (or biodiversity) has a long history of being used in a variety of contexts. Biodiversity in a broad sense, is the number, variety, and variability of all living organisms. A simple definition of ***biodiversity*** is the totality of the varieties of genes, organisms (species), and ecosystems (ecological) in a region or on earth. These three components at different levels are all considered in the Global Biodiversity Assessment currently being prepared by the United Nations Environment Program with funding through the Global Environment Facility (GEF) administered by the United Nations Development Program. The wealth of life on earth today is the product of hundreds of millions of years of evolutionary history. However, biodiversity can also be viewed on a much smaller or larger scale. Biodiversity can be divided into three hierarchical categories: genes, species, and ecosystems (Figure 39.1).

(i) Genetic diversity: Genetic diversity is the basis of biodiversity and only a small fraction (< 1%) of genetic material in higher organisms is outwardly expressed in the form and function of the organism. Genetic diversity refers mainly to the variation of genes within a species. This covers distinct populations of the same species or genetic variation within a population.

(ii) Organismal diversity: Organismal is preferred to 'species' so as to embrace taxonomic categories above the species rank. Organismal diversity refers to the variety of species within the region. Species diversity is commonly considered the measure of biodiversity (1.7 million species have been described although estimates for the total number of extant species range from 5 to 100 million, mainly consisting of insects and microorganisms). Organismal diversity can be measured in many ways. The number of species in a region is one often-used to measure. Taxonomic diversity also considers the relationship of species to each other. For example,

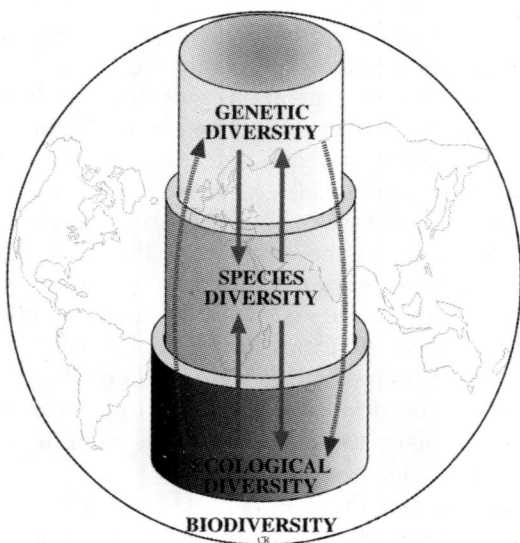

Figure 39.1 *A definition of biodiversity based on its components and their interactions (indicated by the arrows) and the trilogy of biodiversity presented as a hierarchical zoom (the three overlapping cylinders) (Source: Castri and Younes, 1996, CAB International)*

a region with three species of monkeys and two species of birds has greater taxonomic diversity than a region with five species of monkeys but no birds.

(iii) Ecosystem (Ecological) diversity: This is harder to measure than species or genetic diversity because the "boundaries" of communities and ecosystems are elusive. Nevertheless, as long as a consistent set of criteria is used to define communities and ecosystems, their number and distribution can be measured.

A. GEOGRAPHICAL PATTERN OF BIODIVERSITY

The geographical pattern of biodiversity distribution has been considered based on the species richness. The effect of latitude on species richness varies from one group of plants to the other and from one region to the other. Tree species richness has been shown to increase with decrease in latitude and vary from one region to the other. But it decreases more steeply from the equatorial regions into the southern temperate zones than it does to the north, indicating the variability in latitude effect in different geographical regions. Varied patterns of diversity are reported along altitudinal gradients; as altitude increases, the richness of woody species decreases. Many economically important families are entirely restricted to the tropics. However, for biodiversity, the main factor correlated was with evapotranspiration and ecosystem productivity.

1. Centers of biodiversity

Centers of biodiversity are the areas of exceptionally high levels of diversity or those having a unique assemblage of species such as endemism (product of isolation). Diversity is achieved through isolation, allowing genetic differentiation between populations and exchange of populations between semi-isolated areas, allowing new allopatrically formed species into sympatry. Centers of diversity can be recognized at the global, regional (e.g. endemics of Western Ghats, India) and local scale. The centers would contain more species or more of some other measures of biodiversity than would be expected from their environmental predictor variables.

2. Endemics of the Western Ghats (India)

The Western ghats, which run parallel to the west coast of India, form a chain of hills about 1600 km long. These reliefs are still largely covered by vegetation, though many stretches have been exploited and some parts continue to be under intense biotic pressure, leading to the degradation of the forests and fragmentation of the forest continuum. The vegetation of the ghats is relatively well known compared to other tropical areas. Despite the numerous surveys, new species continue to be discovered and new records for districts and states are still being added (Ramesh and Pascal, 1993). Many ecologically oriented studies on the Western Ghats have been done on the associations between species, the relationships between the vegetation types, the species and their natural environmental conditions. Many of these studies reveal the fact that there is a high rate of endemism – one of the most interesting aspects of the floristics of the Western Ghats. Nearly 63% of the tree species of the low and the medium elevation evergreen forests are endemic (implying that they are confined only to the Western Ghats). This high level of diversity and endemism in the Western Ghats has conferred on them the status of being one of the biodiversity "**hotspots**" of the world (Nayar, 1996).

Besides ecosystem diversity, many other expressions of biodiversity are important. These include the relative abundance of the species, the age structure of populations, the pattern of communities in a region, changes in community composition and structure over time, and also ecological processes such as predation, parasitism, and mutualism. Human cultural diversity could also be considered a part of biodiversity.

B. COMPONENTS OF BIODIVERSITY

Humans derive all their food, medicines, and industrial products from both wild and domesticated components of biodiversity. Economic benefits from wild species alone in the US made up an estimated worth of $87 billion annually in the late 1970s. Timber, ornamental plants, oils, proteins, gums, and many fibers also come from the wild. The current economic value of domesticated species is even greater. Trade in agricultural products amounted to $3 trillion in 1989. Once nearly all medicines came from plants and animals. Even today, they remain a vital and important source. More than 5,100 species are used in Chinese traditional medicine alone, and traditional medicine is now encouraged by many countries and also by the WHO. One-fourth of modern pharmaceuticals dispensed in the USA contain active ingredients extracted from plants. Compounds extracted from plants, microbes, and animals are involved in the development of all of the twenty best-selling drugs in the USA.

C. THE VALUE OF BIODIVERSITY

The variety of distinctive species, ecosystems, and habitats influences the productivity and services provided by ecosystems. If the diversity of species in an ecosystem changes, the ecosystem's ability to absorb pollution, maintain soil fertility and micro-climates, cleanse water, and provide other valuable services also changes drastically. The value of biodiversity is particularly apparent in agricultural crop plants. For generations, people have raised a wide range of crops and livestock to stabilize and enhance productivity. The genetic diversity found within individual crops is also of tremendous value. Genetic diversity provides an edge in the constant evolutionary battle between crops and livestock and the pests and diseases that prey on them. In the olden days, farmers grew several genetically distinct varieties of crops planted together as a hedge against crop failure. Breeders and farmers also draw on the genetic diversity of crops and livestock to increase yields and to respond to changing environmental conditions.

Over time, the greatest value of the variety of life may be found in the opportunities it provides to humanity to adapt to local and global changes. The unknown potential of genes, species, and ecosystems represents a never-ending biological frontier of inestimable but certainly high value. Genetic diversity will enable breeders to tailor crops to new climatic conditions. Earth's biota (a biochemical laboratory unmatched in size and innovation) hold still secret cures for emerging diseases. Life on earth still remains largely unexplored at the species and intraspecies levels. Therefore, complete knowledge of the world fauna and flora is not available even to the systematists. One estimate suggests that the total number of species that exist on earth may be close to 100 million (10^8). Out of this large number, only about 1.4 million species of organisms have so far been given scientific names. Studies also report that the diversity of fauna and flora varies in different ecosystems, habitats, geographical regions, and also among different taxonomic groups. There is more diversity in the tropics than elsewhere. For instance, about 300 tree species occur in hectare plots in Peru, while only 700 tree species have been identified in the whole of North America. Again in Peru, a single tree yields about 43 ant species belonging to 26 genera, a level of diversity found in the whole of the British Isles. **Hyperdiversity** can occur in certain habitats or geographical regions or among certain taxonomic groups depending on special adaptations, which allow penetration of multiple niches by them.

D. FACTOR OF SPATIAL SCALE IN DIVERSITY ASSESSMENTS

Biodiversity cannot be easily replaced because the species becomes adapted to a given habitat after a long course of time and through interaction with numerous biotic and abiotic factors. Whittaker (1972) characterized diversity on different scales: ***alpha***, ***beta***, and ***gamma*** components. (i) ***Point*** or ***alpha diversity*** is represented by the number of species in a specific area. This measure of diversity would be most valid when sampling is done in ecosystems on the finest spatial scale. (ii) ***Beta diversity*** is represented by the turnover of species across space. Alpha diversity increases with the total number of individuals encompassed and thus with the increase in the area sampled and the productivity per unit area. Generally, alpha diversity is less on remote islands and increases toward the equator. Not much is known about beta diversity to predict its patterns or its future when the natural areas will be surrounded by highly modified habitats. Beta diversity depends on how large species' ranges are. If the range is large, alpha diversity is independent of the area sampled, so that a national park to protect diversity can be placed anywhere. If alpha diversity is low with species' ranges being small, despite high total diversity and the fact that they are non-overlapping, many parks will be needed to protect diversity. (iii) ***Gamma diversity*** is the overall diversity within a large region, usually a country or region but rarely a continent.

E. ECOLOGICAL THEORIES OF SPECIES DIVERSITY

Species diversity is a product of histories of species, accumulation, and disappearance. In a given region, a species is gained by migration from outside or from production of new species within the region. Species loss could be due to extinction, unpredictable drastic changes in physical conditions, stochastic events, and biological factors, including competition, predation, pathogens, and dispersal agents. A number of hypotheses have been conceived to explain biodiversity patterns.

1. Equilibrium versus non-equilibrium theories: The equilibrium theory suggests that habitats become saturated with species and the saturation limit is determined by the outcomes of local interactions of species. The system tends to compensate for species losses due to extinction through gains as a result of speciation. Non-equilibrium theories suggest that communities never reach a state of equilibrium of diversity, and change in species richness at a given point or time or place is determined by the effect of the environment on species production, exchange, and extinction process. Equilibrium theory would assume that rare species would always be rare. The non-equilibrium theory would assume that rarity or abundance could be a transient unstable phase. The two theories are alternative explanations.

2. Spatial-temporal heterogeneity and niche segregation: Niche segregation could be related to evolution of diverse patterns of growth, reproductive strategies, photosynthetic pathways, and nutritional relationships. For example, the stable coexistence of trees and grasses in the Savanna ecosystem is attributed to exploitation of deep soils by trees and surface soil by the grasses under water-stress conditions. The co-existence of grass and legumes constitutes a benefit for each other in grassland ecosystems such that mixtures of the two use resources more effectively than one of them alone. Since trade-offs exist in species' abilities to compete for different resources, species differ sufficiently in their ability to tolerate different limiting resources, such that competitive exclusion does not occur.

3. Neighborhood effect theory: Pacala (1986) postulated that plant spatial distribution is typically intraspecifically clumped and sometimes interspecifically segregated. Intraspecific aggregation is likely to result in a higher intensity of intra-rather than interspecific competition, enabling species co-existence.

4. Neutral or lottery models: Neutral or lottery models rule out any significant role of niche segregation or competitive co-existence in supporting diversity. Fugitive species adapted to large-scale production of long dispersal mechanisms could survive with superior competitors by continually colonizing unoccupied space.

5. Community drift model: This model assumes that a large number of competitively identical species with a limited dispersal range change in abundance due to random drift. Rarity and narrow ranges guarantee that a species-accumulation curve would continue to increase at all scales until the boundary of the continent is reached. A ceiling value (asymptote) for the upper limit of local species richness is more likely when competition and other interactions determine community organization. Neutral and drift models have been supported by the co-existence of hundreds of tree species in a plot of one hectare (Condit et al, 1996).

6. High probability of predation near conspecific adults: According to this model, seeds are more likely to escape predation in the vicinity of adult consepecifics because the activities of seed herbivores tend to be concentrated here. This could lead to a possible bias in favor of rare species.

7. Gap dynamics: Gap dynamics has been found to be an important mechanism supporting high species richness in tropical rain forests. Gaps develop as a result of the natural death of old trees, which could be influenced by stochastic environmental processes. The gaps could be as small as 30 m^2 and the turnover rate could vary between 60 and 159 years. Species differ in their response to gaps.

8. Landscape ecology approach: The conventional approach of looking at ecosystems as self-contained ecological units is being increasingly elaborated different ecosystems to considering as interconnected spatial units in a landscape. Extinctions on a small scale could be inescapable but may be prevented on a large scale. Variability between different ecosystem types and their spatial configurations and interconnections together would determine the extinction, migration, and speciation in a landscape. Because of variation in topography, climate, disturbance regime, and resource-use history, the landscape could be highly complex.

9. Welfare ecology: The welfare economics of humans has to be backed by welfare ecology, since most of the economy and well-being springs from the use of ecological resources either directly or indirectly (e.g. atmosphere, hydrosphere, lithosphere, and biosphere) coupled with human technological adventures. As previously mentioned, ecology is an integral part of human welfare and there is a deep interconnection between human needs, wants, and aspirations. Therefore, there is an immediate need for humans to address the problem of how comfortable a lifestyle they can have at the cost of ecology. Welfare ecology is relevant and eventually the reality of all living organisms on this earth. It includes all the biota, since the health of the whole will determine the health of the part, and vice-versa. Therefore, welfare ecology has a wide meaning, and application, and has detrimental effects on the sustainability of all organisms on earth, including humans.

III DIFFERENT LEVELS OF BIODIVERSITY MEASUREMENT

Biological diversity exists at many different levels, from the genetic diversity within local populations of a species, or between geographically distinct populations of the same species, all the way up to communities or ecosystems. Depending on the context, any one of this nested hierarchy of levels can be of predominant importance. Three approaches are mainly considered by scientists to measure biodiversity.

1. Taxic measures

Some conservationists consider that the aim of measuring biodiversity is primarily to conserve the maximum number of species by measuring a number of higher taxa which is convenient and getting a quick estimate of the number of species in an area. Since, complete counts of organism are impractical at present, indirect solutions are needed that are both cheap and quick. Thus data from families of seed plants are used to make world-wide maps of the regional distribution of family richness and endemism. However, others argue that measurement of species is more appropriate since many species belong to a few phyla or families.

2. Molecular measure

An alternative and very sensitive method for the measurement of biodiversity is to use divergences in molecular characters, especially the percentage of either nucleic acid homology or base sequence differences. Unlike higher taxa, the DNA and RNA found in all living cells can provide a basis on which to make direct comparisons between diverse organisms. This makes sense, since the biodiversity of a community is expressed as the sum of the variety of genetic information coded within the genotypes of the inhabitants. A biodiversity estimation of species will be based on how many new base sequences contribute the genetic variability of a whole community.

3. Phylogenetic measures

Cladistics can be used to give an objective measure of taxonomic distance or 'independent evolutionary history (IEH) using methods pioneered by Vane-Wright et al. (1991). The approach provides information that is of special value for the conservation of target groups and in selecting areas appropriate for their conservation. But it is difficult to see how sufficient phylogenetic data could be generated in the near future for this method to be used to compare the diversity of whole countries. However, an ideal index of biodiversity is perhaps to be obtained by proceeding in the opposite direction, asking first the biggest question about diversity, how many kingdoms are represented on a site? Then, how many orders per phylum? And so on.

IV HUMAN USE OF BIODIVERSITY

From the human use perspective, four types of values are attached to biodiversity – direct, indirect, optional, and intrinsic values. Biodiversity is concentrated in areas inhabited by traditional societies who were presumed to be the major threats to the sustained use of biodiversity. This assumption led to conventional management of the allocation of land-agricultural area for direct uses of biodiversity to meet the immediate needs of humans and protected areas for indirect, optional, and intrinsic values of biodiversity. However, the recent concept of biosphere reserve is to reconcile the two extreme hypothetical views: (i) local people and their resource use practices threaten conservation objectives and so should be kept away from the conservation area, and (ii) local people have a major role to play in achieving the objectives of nature conservation as they have a rich empirical knowledge of sustainable resource uses.

Our immediate concern and need is to conserve as much genetic diversity as possible within each species as an insurance against future environmental change or new human uses of these resources. Therefore, our attention has to be focussed on genetically distinct sub-populations or sub-species, as for example, pest resistance and stress-tolerant plants and heavy metal tolerant strains of common grasses. Conservation of close living relatives can be avoided based on phylogenetic distinctiveness. Our focus can also be on conservation of biotic communities and ecosystems. We should try to maintain full representation of all earth's ecosystems. This should help to ensure the survival of species.

A. TROPICAL FORESTS AS 'SOURCES' OF GLOBAL BIODIVERSITY

The importance of tropical moist forests to the maintenance and use of biodiversity is enormous. As a resource and environmental stabilizer, a tropical moist forest is the most important ecosystem. The important resources it provides include timber, non-timber products, tourism income, plant genetic material, raw-materials for biotechnology, medicines, pharmaceuticals, and food and shelter for its people. Where environmental issues are concerned, it prevents erosion, maintains soil fertility, regulates water run-off, moderates climate, reduces stream and river sedimentation which kills fish, stops nutrient wash-out which pollutes waterways, and fixes carbon dioxide. The most important aspect of it is that it is a 'goldmine' of biodiversity. Tropical forest timber is a very valuable commodity; it brings in any enormous amount of foreign exchange, and is a major contributor to the GNP of most developing countries. Recently many environmental economists and conservation biologists have been promoting the non-timber product utilization of tropical forests which is not only sustainable and uses low technology but also involves local people in co-operative arrangements which provide a stable long-term future for both the people and the ecosystem. Particularly, tourism, centered on foreigners' fascination with tropical forests, has become feasible and commercially very attractive to local and foreign businessmen.

Forest microorganisms, so tractable to biotechnological development, play a major role in the evolution and diversification of nutrient supply such as through the breakdown of cellulose and lignins, and existence of mycorrhiza around the roots of 85% of all vascular plants being crucial to the absorption of growth-limiting nutrients. Food networks of all life are dependent on microorganisms, which also play a role in the maintenance of soil structures, biodegradation, biocontrol such as plant pathogenic microorganisms which can limit plants and entomogenous microorganisms which can limit insect populations. Microbial biodiversity in tropical forests will in the future be used by biotechnologists to improve nitrogen-fixing in legumes, enhance nitrogen-fixing using cyanobacteria, for biocontrol of insects, plants, diseases, and weeds; finally mycorrhizal fungi will be used for improved plant growth and stress tolerance. The medicines from tropical forest species are common, and require no biotechnological modification. Tropical forests support unique cultural systems, societies of gatherers, which are an integral part of the forests, in balance with the stability of the ecosystem. Tropical forest ecosystems surprisingly have very few recorded extinction rates (2–3% per decade), because habitat loss is faster than any other ecosystem. However, there are **lag time effects** not only between habitat loss and actual extinction, but also in the detection and reporting of such events. Another reason could be that only 4% of species within tropical forests have been identified. It is possible that extinction rates are much higher than currently recorded.

B. ISSUES CONCERNING BIODIVERSITY AND TROPICAL FOREST

The central issue involving biodiversity and tropical forest centers is the right of use of this common resource. Who has the right to exploit the forest resources, in which way, and how much? The local people, the government, the businessmen, the international community, or the scientists? But the overwhelming issue arising out of the uses and values of biodiversity in tropical forests, the politico-economic market, will never be resolved. The uses of tropical forests are enormous (Swingland, 1993).

1. Food crops (200–3000 spp).
2. Genetic resources for plant crop improvement (wild genes resistant to pathogens and insects, rapid growth, higher yields).
3. Timber, world export 1989 exceeds US$ six billion.
4. Rattans (90% extracted from the wild, industry employs 0.5 million people, >90% spp., vulnerable or endangered).
5. Medicinal plants and animals (90 plant spp. in worldwide use, US alone annually imports US$ 20 million animal spp. and annual world trade crosses US$ 350–500 million).
6. Ornamental use of plants and animals (species orchids, cacti, insects, skins, feathers, worldwide trade over US$ three billions).
7. Meat and eggs, hunting or gathering from the forests (no estimates available).
8. Working animals (16,000 working Indian elephants in captivity).
9. Sports, hunting (annual worldwide turnover in tropical forests >US$ 43 million).
10. Tourism, recreational benefits (annual worldwide turnover US$ 1.3 billion).
11. Captive animals as pets or for display (annual worldwide turnover US$ 1.6–2.3 billion).
12. Animal domestication (crib-biting wear found on horses' teeth 27–29,000 BP).
13. Selective breeding, livestock improvement and farming (quicker to import genes from the wild than selection within a breed).

V LOSSES OF BIODIVERSITY AND THEIR CAUSES

Biological diversity is being eroded as fast today as at any time (Table 39.1). The crucible of extinction is believed to be in the tropical forests, since these forests house 90% of the around

10 million species that live on earth. The biodiversity of marine and freshwater systems faces serious loss and degradation. The number of documented extinctions of species over the past century is small compared to those predicted for the coming decades due to the acceleration in the rates of habitat loss. Habitat loss not only precipitates species extinctions, it also represents a loss of biodiversity in its own right. The rapid losses of species and ecosystems obscure the loss of genetic diversity. Loss of genetic diversity could imperil agriculture. The current losses of biodiversity have both direct and indirect causes.

Table 39.1 *Estimated rates of species extinction based on forest area losses*

% Global loss per decade	Period	Estimation Method	Authority
4	1975–2000	Extrapolation	Myers, 1979
2–5	By 2000	Species-area projection	Reid and Miller, 1989
1–5	1990–2015	Species-area	Reid, 1992

Source: Swingland, 1993.

A. DIRECT MECHANISM OF BIODIVERSITY LOSS

 (i) Habitat loss and fragmentation: A sudden explosion in the human population and the consequent increase in resource consumption, have led to relatively undisturbed ecosystems shrinking dramatically in area over the past few decades. About 98% of the tropical dry forest along Central America's Pacific coast and 22% of Thailand's mangroves have disappeared in the past few decades. The river and stream habitat of freshwater ecosystems have been destroyed due to dams and reefs and near-shore communities of marine ecosystems wiped out due to coastal developments. In tropical forests, a major cause of forest losses is the expansion of marginal agriculture and timber harvest.

 (ii) Introduced species: Introduced species are responsible for the extinction of many species. Especially, in isolated ecosystems, the entry of a new organism (predator, competitor, or pathogen) can rapidly replace some native species, which cannot co-evolve with the newcomer.

 (iii) Over-exploitation of plant and animal species: Numerous natural resources (forest, fisheries, and wildlife) have been over-exploited, often to the point of extinction. Many species have become extinct because of over-harvest for human food; the search for precious commodities (e.g. ivory, pets, curiosities, etc.), and collector's items have also impinged on some populations and severely damaged others.

 (iv) Environmental pollution: Pollution of soil, water and the atmosphere strains ecosystems and may reduce or eliminate populations of sensitive species. Contamination may cause damage along the food chain. Pesticides, herbicides, and other poisons have caused hundreds of species to become extinct. Soil microorganisms have also suffered from pollution as industry sheds heavy metals and irrigated agriculture brings on salinization. Acid rain can make lakes and pools lifeless and also affects forests.

 (v) Global climate change: A massive effect of air pollution is global warming, which can play havoc with the greenhouse gases in the atmosphere. Consequently, global temperature can go up (1° to 3°C) during the next century, with an associated rise in sea level of 1 to 2 metres. Many of the world's islands may be completely submerged by the more extreme projections of sea level rise, killing an enormous number of fauna and flora. Protected areas themselves will be placed under stress as environmental conditions deteriorate.

(vi) Industrial agriculture and forestry: Since prehistoric times, farmers bred and maintained a tremendous diversity of crop and livestock varieties around the world. However, recently, on-farm diversity has been shrinking fast due to the growing of fewer varieties that respond better to water, fertilizers, and pests. Similar trends are transforming diverse forest ecosystems into high-yielding monocultural tree plantations.

B. INDIRECT CAUSES OF BIODIVERSITY LOSS

1. The un-sustainably high rate of human population growth and natural resource consumption: Modern agriculture and industry have had a positive effect on population growth. Another one billion people are likely to be added to the world population in each of the next three decades. The rates and magnitude of this growth and the eventual size at which the global population will stabilize will prove to be a critical consideration for biodiversity. Critical environmental resources are now under stress. The emissions of pollutants, including greenhouse gases, are already overtaxing the tolerance of the ecosystems. Ozone depletion, acid rain, and air pollution are all taking a toll on biodiversity today and may threaten it more severely in the future. Excessive consumption of minerals and other non-renewable resources aggravates these problems.

2. Steadily narrowing spectrum of traded products from agriculture, forestry, and fisheries: Modern agriculture produces now specialize in the relatively few crops that provide an edge in the world economy. As the number of crop species declines, many other species that have co-evolved over the centuries with traditional agricultural systems are dying out. In forest areas, the rapid and total conversion of forests (to monocultural cash crops) is widespread.

3. Economic systems and policies that fail to value the environment and its resources: Many conversions of natural systems such as forests or wetlands to farmlands and range-lands, are economically and biologically inefficient. There are several reasons for the mis-valuation of biological resources. (i) Many biological resources are consumed directly and never enter the markets. Accordingly, the economic values of logging and other potentially exhaustive uses are overestimated while the sustainable uses are underestimated. (ii) Biodiversity benefits are in large part "public goods" that no single owner can claim. For example, wetland protection benefits are so diffuse that no market incentives for wetland conservation ever develop. (iii) Property rights are more likely to be granted to those who clear and settle forests and other lands covered with natural vegetation than to forest dwellers living by the sustainable harvest for natural products. Correctly valued, biologically diverse natural systems are major economic assets. But because such systems are commonly undervalued, biodiversity conservation is seen as a cost rather than an investment. Correcting this attitude is essential for conserving global and national biodiversity.

4. Inequalities in the ownership, management, and flow of benefits from both the use and conservation of biological resources: In most countries, ownership and control of land and biotic resources are distributed in ways that work against biodiversity conservation and sustainable living. Quick profits from excessive logging or over-fishing, for example, lead to rapid depletion of species and the destruction of habitats. Another problem is that the concentration of resource control and responsibility for environmental policy decisions are primarily in the hands of urban men. The international trade, debt, and technology transfer policies and practices foster inequalities that resemble and reinforce those found within nations.

5. Deficiencies in knowledge and its application: We still do not have complete knowledge on natural ecosystems and their innumerable components. The lack of a link between traditional cultures and policy makers has resulted in the failure to develop policies that reflect the scientific, economic, social, and ethical values of biodiversity. The public are reluctant to accept policies that reduce excessive resource consumption, no matter how logical or necessary such policies may be, and this also creates difficulties in understanding biodiversity.

6. Legal and institutional systems that promote unsustainable exploitation: Ecological and economic realities, operation of international institutions along rigidly sectoral lines, and many environmental institutions being too small and having too meager resources create difficulties in adopting a cross-sectoral approach to biodiversity conservation and management. Over-centralization of government and corporate planning hinders local implementation, discourages local participation, and closes the process to citizen's groups and non-governmental organizations (NGOs). Many conservation agencies and organizations lack the financial and personnel resources needed even to support minimal programs and often their efforts are fragmented and overlapping; they do not integrate *in situ* and *ex situ* conservation tools and technologies. In addition, many countries do not have a proper system of environmental laws and other instruments to safeguard the environment and the sustainable use of its resources. In spite of many protected habitats for a limited number of species, land-use practices outside of the protected area can alter water supplies, introduce pollutants, and change micro-climates. Region-wide management approaches are needed to address the habitat needs of whole biotic communities and to integrate conservation with regional development. But the lack of the integrated expertise and authority needed to manage a mix of developed and wild ecosystems impedes sound regional management. Often regions can come under the jurisdiction of various local, state, or provincial governments, and some involve two or more nations, thus making administration complicated.

VI THE IMPORTANCE OF CONSERVATION

The Global Biodiversity Assessment warns, 'unless actions are taken to protect biodiversity, we may lose the valuable plant wealth forever'. Due to the financial limitations and lack of justification for financial investment in species and countryside, conservation has been challenged by many. Policy decisions are overwhelmingly matters of economics. Thus standard economics assigned a nominal value to ecological resources. Most justifications for nature conservation center on the human benefits that will accrue from maintaining a full range of biodiversity and avoiding environmental degradation. These can be broadly grouped into aesthetic, utilitarian, and ecological or scientific reasons for conservation. Recently one more category, ethical reasons for conservation has been added.

A. ETHICAL REASONS

The ecological question, which includes biodiversity, cuts across all fields of knowledge and action, from experimental sciences to philosophy, politics, economy, religion, and ethics. The ethical, philosophical, religious, anthropological, cultural, and legal implications of human responsibility toward earth have to be considered in a holistic and sustainable effort of protection and conservation of earth and its natural resources. In a broad sense, ethics at a global level means all living organisms have equal rights to live and perpetuate on earth. Each and every organism either directly or indirectly depends on each other for food, space, shelter, transportation, resources recycling, energy, etc. This implies that co-existence is the means of living and an inevitable activity also. The economic condition, cultural tradition, religious beliefs, political persuasion, and other similar concepts all shape our attitudes toward nature. The respect of the biological universe calls for a set of ethical norms including a high consideration for diversity. The opinion of some is that nature does not exist simply for humans to transform and modify in such a way as to satisfy their own needs. They further argue that all living organisms have a right to co-exist with human beings on earth, and humans have no right to cause the extinction or diminish the quality of life of any organism. Sometimes it is linked with the idea of the concept of **custodianship** which challenges us to pass on to future generations all the diversity of life and quality of environment that we have inherited.

Even though we cannot separate the socio-cultural systems of humanity from the biological universe of which they are a part, the whole game of power consists of imposing one's own system or values and weakening, if not overtaking, the values of others, thereby reducing cultural diversity. We are thus confronted with a delicate ethical problem because all research today is demonstrating the very close links between ecological systems and cultural systems in operational terms. Diversity has positive as well as negative aspects. When we look at the diversity in the quality of life within and between countries, we are displeased by the huge inequalities. These inequalities make it very difficult to preserve biological diversity. How can one ethically give priority to biodiversity as long as the problem of poverty has not been solved through the partnership of science and development?

B. AESTHETIC REASONS

Humans derive pleasure from natural environments and the presence of other living organisms. Even though this is absolutely true, it is hard to measure it objectively. Hobbies like gardening, pet-keeping, bird watching, enjoying the circus, animal games, etc., suggest that human beings do derive an aesthetic reward from these companions. Our appreciation of nature permeates art, design, literature, and music and influences our recreational pursuits. Many consider that contact with nature is essential for human well-being and even for spiritual reasons. For this reason, numerous local, national, and international organizations exist worldwide to promote wildlife and countryside conservation.

C. UTILITARIAN REASONS

Wild and wonderful nature contributes to our immediate needs in many fundamental ways (food, cloth, medicines, industrial products, etc.). Obvious examples are agriculture and forestry and fisheries. We direct crop benefits, we make use of pollinating insects and beneficial predators in pest control. Many plant species have important medicinal and nutritional uses. A number of industrial processes depend solely on plant and animal materials and, microorganisms are increasingly being used industrially, for example to concentrate valuable metals from low-grade metals ores. Recent developments in biotechnology facilitate the transfer of genes across the kingdoms, since it breaks the species or even kingdom barrier for gene transfer. Therefore, conservation of genetic diversity can act as a natural gene bank for current and future use.

D. ECOLOGICAL OR SCIENTIFIC REASONS

The direct benefits of species and their genetic resources are not only important to humans, the well-being of humans also depends on the maintenance of a fully functional biosphere. The maintenance of balanced geo-biochemical cycles is vital for the avoidance of pollution and regulation of the earth's climatic systems. Loss of vegetation cover can have profound effects on soil erosion, lead to siltation of rivers and coasts, and may even result in changes in rainfall and climate patterns. Biodiversity also plays an important role in the food chain and species adaptation to changes in environment.

VII STRATEGIES FOR BIODIVERSITY CONSERVATION

A. AIM OF BIODIVERSITY CONSERVATION

A successful strategy to conserve biodiversity must address the full range of causes of its current loss and exploit the opportunities that genes, species, and ecosystems provide for sustainable development. This is very important because biodiversity conservation supports sustainable development by protecting and using biological resources in ways that do not diminish the world's variety of genes, species, and ecosystems. Biodiversity conservation strategy must have a broad scope, but the approach can be saving biodiversity, studying it, and using it sustainably and equitably.

(i) Saving biodiversity means developing strategies to protect genes, species, habitats, and ecosystems. Therefore, it often involves efforts to prevent the degradation of key natural ecosystems and to manage and protect them effectively. The simplest way to maintain genetic diversity is to maintain their habitats. Programs must include measures to maintain the habitats, which have been modified due to human activities. It also includes restoration of lost species to their original habitats and preserving species in gene banks, zoos, botanic gardens, and other off-site facilities.

(ii) Studying biodiversity means documenting its composition, distribution, structure, and function; undertaking the roles and functions of genes, species, and ecosystems; understanding the complex links between modified and natural systems, and utilizing this knowledge for sustainable development (Figure 39.2). It also means creating awareness of biodiversity values, identifying people's appreciation of nature's variety, including biodiversity issues into school and college syllabi, and facilitating the access to information on biodiversity and conservation developments to the public.

(iii) Using biodiversity sustainability and equitably means managing biological resources to last longer and making sure that biodiversity is used to improve the human condition, and is shared equally.

The biodiversity conservation program must be broad-based and sustainable. The biodiversity conservation strategy must establish contacts and partnerships within communities, bringing

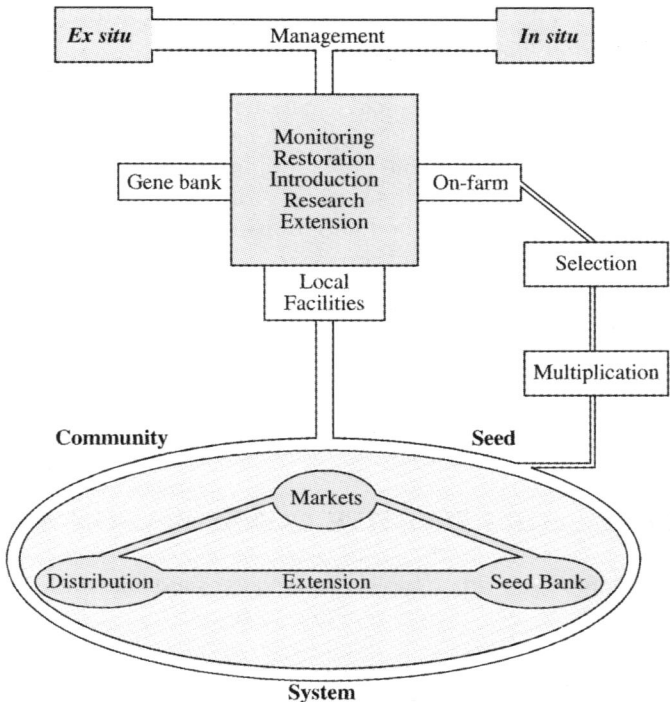

Figure 39.2 *Conservation–utilization link (Redrawn from Brush 2000, IPGRI, IDRC, Lewis Publishers)*

biologists and resource managers together with other members of the community (e.g. social scientists, political leaders, businessmen, religious leaders, farmers, journalists, artists, planners, etc.). There must be a good communication between central and local governments, industry and other citizen's groups. Conservation activities include activities at the individual level to global and in between. Many essential elements of biodiversity conservation require sustained commitment, wherein the results will be slow. However, those genes, species, and ecosystems that are disappearing faster, need immediate action.

B. PRINCIPLES FOR CONSERVING BIODIVERSITY

The following principles have guided individuals and institutions involved in the development of biodiversity conservation strategies.
 (i) Every living form of life is unique, and warrants respect from humanity.
 (ii) Biodiversity conservation is a long-term investment that yields substantial local, national, and global benefits.
 (iii) The financial burden and benefits of biodiversity conservation should be equitable among and within nations.
 (iv) To achieve sustainable development, conserving biodiversity requires fundamental changes in pattern and practices of economic development worldwide.

(v) In response to increased funding, policy and institutional reforms are needed to create the conditions for effective utilization of funding.
(vi) The priorities for biodiversity conservation must be legitimate and the focus should be on all species. It must be viewed at the local, national, and global levels and all countries and communities must show a vested interest in conserving their biodiversity.
(vii) Biodiversity conservation is sustainable only if public awareness and concern are substantially heightened, and policy makers have access to reliable information.
(viii) Conservation activities must be planned and implemented on a scale determined by ecological and social criteria and focused on where people live and work, as well as in protected wild land areas.
(ix) Cultural diversity and biodiversity are closely linked and often conserving biodiversity strengthens human cultural integrity and values.
(x) Biodiversity conservation is highly influenced by increased public participation, respect for basic human rights, improved access to education and information, and greater institutional accountability.

C. KEY STRATEGIES FOR EFFECTIVE CONSERVATION ACTION

The demands and expectations placed on biological resources are high and widely varied, calling for new approaches that go beyond merely reacting to resource crises and concerns. The limited conservation resources available must be focused on opportunities likely to yield the greatest benefits. The following five key strategic objectives offer significant possibilities for effective action.

1. Development of national and international policies that foster the sustainable use of biological resources and the maintenance of biodiversity

The legal and economic policies established by national governments create the incentives and obstacles that influence decisions about how to utilize and manage biological resources. Biotechnology is radically altering the market value of genetic resources. If right policies are established, countries rich in species and genetic resources stand to benefit substantially. Trade patterns and practices greatly influence what individuals and nations grow, harvest, buy, and sell. In many developing countries, public resources are spent to produce cash commodities that generate foreign exchange, and biodiversity conservation is neglected in the process.

2. Create conditions and incentives for effective conservation by local communities

Biodiversity conservation is effective only where people live and work. Thus, unless local communities have the incentive, the capacity, and the latitude to manage biodiversity sustainably, national and international actions will be ineffective. Local biodiversity conservation cannot succeed unless communities receive a fair share of the benefits, and assume a greater role in managing their biotic resources. The role of local communities in the management of wild lands as well as in stewardship of their natural resources as a whole is very important.

3. Tools for conserving biodiversity must be strengthened and applied more broadly

The world's protected areas (e.g. zoos, botanic gardens, and seed banks) are vital tools for conserving biodiversity. But these conservation tools may not do their work if they remain underfunded and understaffed. Conservation plans must be drawn up and implemented "bioregionally" to reflect both ecological and social realities. Protected areas will retain their central importance if planning is done bioregionally. Additionally, national networks of protected areas must be strengthened and expanded to cover all key *biomes* (a major portion of the living environment of a particular region, characterized by its distinctive vegetation and maintained by local climatic conditions) and ecosystems. The management objectives of protected areas must be harmonized with those for the surrounding ecosystems and human communities. The best means of strengthening protected areas is to better integrate them with local social and economic needs. This strategy emphasizes mechanisms for increasing benefits to local communities through ecotourism and sustainable use of non-timber forest products, the establishment of buffer zones between protected areas and surrounding communities, compensation to local communities for lost resources, and the use of integrated conservation strategies in establishing protected areas.

4. The human capacity for conserving and using biodiversity sustainability must be greatly strengthened, particularly in developing countries

Conservation actions are successful only if people understand the distribution and value of biodiversity, and its influence on their own lives and aspirations, and have the experience to manage areas to meet their needs without diminishing biodiversity. Unfortunately, this capacity is very little at present due to fewer taxonomists who have specialized in tropical species. No country has a complete list of species and for most ecosystems, little or no information is available on indicator and keystone species. Under investment, the notion that action on saving and studying biodiversity is wasteful expenditure is the main reason for the above gaps. Committed, skilled workers and experts in the biological and social sciences, economics, law, policy analysis, ethics, and community organizations are all required more in all the countries. The key to conserving genes, species and ecosystems is increasing awareness on the biodiversity and its role in human society. Research must be closely linked to national and local resource and development needs. The priorities for research and information systems should grow out of consultation with those who need and will use new data and analysis. The best option for helping to mobilize biodiversity to meet national needs is to establish institutions such as national biodiversity institutes to catalogue and explore a nation's biotic wealth.

5. Conservation action must be catalyzed through international cooperation and national planning

International cooperation plays a major role in conservation of biodiversity on earth. For effective conservation, international laws and institutions must be able to establish widely accepted norms of conduct, elicit firm commitments to action from governments, mobilize financial resources, develop accurate and timely information, and invite broad participation from scientific and non-governmental sectors. National or regional planning processes are also a key mechanism for catalyzing and focusing policy reforms to ensure sustainable resource use and support biodiversity conservation. The amendments needed to slow down biodiversity loss will involve policy adjustments. In case any disputes are anticipated and mechanisms for resolving them are established now, any hardships born out of changes can be minimized.

6. Preventing genetic homogeneity

Genetic homogeneity is associated with the introduction of high-yielding varieties and expansion of a single crop in a large area (mono-cropping). Consequently, it is leading to the replacement of large numbers of local varieties with a few high-yielding strains. However, the dangerous consequences of covering large contiguous areas with one or two genetic strains are now surfacing. Varietal diversification and crop rotations involving crops with non-overlapping pest sensitivity and nutritional depletion are important for sustainable agriculture. Mono-cropping can cause a lot of damage to agricultural lands and biodiversity in toto.

7. Reducing the effects of alien species invasion

Historically as well as in the recent past, the increasing liberalization of trade in food grains and other agricultural commodities has led to an alarming increase in an invasion of alien species into new regions. Most of these alien species get acclimatized to their new habitats very quickly and pose a threat to native biodiversity. A global strategy for the prevention of such invasions is, therefore, necessary. Some of the collaborative steps include: (a) improving sanitary and phytosanitary measures, (b) introducing effective methods of biological control, (c) finding economic uses (or incentives) for the removal of invasive species by the local communities, and (d) mobilizing new tools of genetic engineering to render such aggressive weeds sterile and thereby incapable of sexual propagation, etc. However, an integrated strategy may work more effectively than a single approach in most cases.

VIII IDENTIFYING THE RISKS TO BIODIVERSITY

A. RARE AND ENDANGERED SPECIES

Human population pressure and many scientific activities in the biosphere have caused the extinction of many species. For example, the woolly mammoth, dodo, great auk, and Bengal tiger are all examples of species made extinct by humans. It is estimated that, at present, one species is lost every day. If we want to prevent or stop this kind of continued extinction, we must identify the species at greatest risk, investigate why they have become vulnerable, and attempt to remedy the problem. The IUCN publishes detailed lists of species at risk of extinction in a series known as the **Red Data Books**.

B. CATEGORIES OF RISK

Four categories of risk are identified.
   1. Rare: Species with small populations either restricted geographically with localized habitats or with widely scattered individuals. These species are at risk of becoming more rare, but they are not in immediate danger of extinction.

2. Vulnerable: Species under threat of or actually declining in number, or species that have been seriously depleted in the past and have not yet recovered.

3. Endangered: Species with low population numbers that are in considerable danger of becoming extinct.

4. Extinct: Species which cannot be found in areas they recently inhabited nor in other likely habitats.

The Red Data Books for vertebrates list all known species in these four categories. This is impossible to do for plants since it is estimated that over 10% of all known plants, up to 60,000 species, are either rare or in danger of genetic erosion or extinction in the next 30 to 40 years.

C. Extinction

The process of **extinction** has always been part of the story of life on earth, as indicated by fossil records. However, since the emergence of humans, many new causes of extinction have arisen which are directly attributed to exploitation of biosphere resources. The extinction of these species may be natural or man-made. **Natural extinction** can be seen in species of small size breeding populations associated with a high coefficient of variation. When the size of the breeding population drops to lower than hundred, the likelihood of extinction is further enhanced due to inbreeding depression. The natural extinction rate estimated from fossil data is one species per 10^7 species per year. This rate refers to **true-extinction**, which means extinction of a species and its descendents. It differs from **'pseudo-extinction'**, which is caused by the evolution of one species into another. The longevity of a species defined as above varies from 1–2 million years to as much as 10 million years.

Human-caused extinction: Due to the destruction of natural habitats by human interference, biodiversity is being lost at a rapid rate. The loss of biodiversity is very high in tropical regions, where biodiversity is high and there is also high human interference. Loss of biodiversity due to human interference can be attributable to many factors. (i) Over-hunting; (ii) pollution of land, air and water bodies; (iii) the widespread use of pesticides and other organic chemicals; (iv) habitat fragmentation and loss; (v) agricultural intensification causing agricultural land expansion and monoculture cropping pattern; (vi) deforestation and mismanagement of soil and other natural resources; (vii) industrial revolution. Research results show that many species are getting rare and their long-term survival is also in doubt. Rare species are typically **genetically eroded**, which means that their gene pool is reduced and they also have reduced adaptability to the changed environment. Sudden change or adversity of environment can easily bring about the extinction of a rare species.

IX STEPS TO PRESERVE BIODIVERSITY

The dangers of depending on uniform genetic material in crop production have been amply demonstrated and the need to maintain the genetic diversity of crop and pasture species, and their wild relatives, firmly established. The past four decades have seen a steady increase in national and international activities to ensure that crop diversity is conserved for the benefit of future generations. More recently, the importance of conserving plant genetic resources of crop species was confirmed by the development of the program outlined in Agenda 21 and the coming entry into force of the Convention on Biological Diversity (**CBD**). CBD was signed by over 150 heads of states in Rio de Janeiro in June 1992. This is the only international legal instrument that addresses biological diversity in a comprehensive way. It offers a sound base for implementation of a sustainable agricultural system. The CBD has three stated objectives. (i) Conservation of biological diversity; (ii) sustainable use of its components; (iii) fair and equitable sharing of benefits, resulting from the utilization of genetic resources: This can take place through appropriate access to genetic resources, and transfer of relevant technologies, taking into account all rights over those resources and technologies, and funding.

The process of extinction can be prevented by taking appropriate restoration and rehabilitation measures in time. For example, the giant panda (*Ailuropoda melanoleuca*) found in eastern Tibet and southwest China, is a symbol of the Worldwide Fund for Nature (WWF). At one time, the species was very much endangered since its habitat, the bamboo forest, was increasingly being encroached upon by the human population. This is an example of an endangered species whose extinction has so far been prevented by a combination of habitat restoration measures and a captive breeding program. We can help to prevent extinction by several ways.

1. The first step in preventing extinction is protecting and restoring habitats. No undisturbed land should be used for human developments.
2. Catalogues of genetic resources and national biological inventories should be prepared, so that the threatened and endangered species may be protected against extinction.
3. Human population should be lowered and sustainable, high yielding agricultural systems must be developed, so that preservation and sustainable exploitation of biodiversity go hand in hand.
4. Large-scale conservation activities such as establishing game parks, national parks, nature reserves, and similar protected areas should be undertaken.
5. The impact of modern intensive agriculture should be assessed and the impact should be controlled and/or reduced by appropriate technology and policies.
6. The use of bio-poisons such as pesticides, insecticides, herbicides, xenobiotic compounds, etc., should be abandoned or reduced.
7. Measures should be taken to reduce environmental pollutants, especially emission of greenhouse gases and ozone-destroying chemicals.
8. The trade of endangered species should be restricted by strong national and international laws.
9. Sufficient refuge should be provided and the breeding programs for endangered species, facilitated in, for example, zoos and botanic gardens.
10. For restoration and revival for long-term needs, sperm banks and seed stores should be established to maintain the full range of genetic diversity of species.

A. GENETIC RESOURCES FOR HUMAN USE

The value and worldwide significance of the genetic variation present in traditional varieties of crops and their wild relatives was clearly established by Vavilov and his collaborators as a result of their extensive studies and collective programs (Table 39.2; Wilkes, 1977). Genetic resources are valuable assets for humans and have many uses. Some of the important areas for human uses are briefly described below.

Table 39.2 *Origin of the world's basic food plants*

WORLD CENTRES	WORLD'S BASIC FOOD PLANTS
OLD WORLD CENTRES	
1. ETHIOPIA	Banana (endemic), Barley, Caster bean, Coffee, Flax, Khat, Okra, Onion, Sesame. Sorghum, Wheat
2. MEDITERRANEAN	Asparagus, Beets, Cabbage, Carob, Chicory, Hops, Lettuce, Oats, Olive, Parsnip, Rhubarb, Wheat
3. ASIA MINOR	Alfalfa, Almond (wild), Apricot (secondary), Cabbage, Cherry, Date palm, Carrots, Fig, Flax, Grapes, Lentils, Oats, Onions (secondary), Opium poppy, Pea, Pear, Pistachio, Pomegranate, Rye, Wheat
4. CENTRAL ASIATIC (Afghanistan-Turkestan)	Almond, Apple (wild), Apricot, Broad bean, Cantaloupe, Carrots, Chick pea, Cotton (*G. vinifera*), Flax, Grapes, (*V. vinifera*), Hemp, Lentils, Mustard, Onion, Pea, Pear (wild), Sesame, Spinach, Turnips, Wheat
5. INDO-BURMA	Amaranths, Betel nut, Betel pepper, Chick pea, Cotton (*G. arboreum*), Cowpea, Cucumber, Egg plant, Hemp, Jute, Lemon, Mango, Millets, Orange, Pepper (black), Rice, Sugarcane (wild), Yam
6. SIAM, MALAYA, JAVA	Banana, Betel palm, Breadfruit, Coconut, Ginger, Grapefruit, Sugarcane (wild), Tung, Yam
7. CHINA	Adzuki bean, Apricot, Buckwheat, Chinese cabbage, Cowpea (secondary), Kaoliang (sorghum), Millets, Oats (secondary) Orange (secondary), Paper mulberry, Peach, Radish, Rhubarb, Soybean, Sugarcane (endemic), Tea
NEW WORLD CENTRES	
8. MEXICO-GUATEMALA	Amaranths, Bean (*P. vulgaris*), Bean (*P. lunatus*), Bean (*P. acutifolius*), Corn, Cacao, Cashew, Cotton (*G. hirsutum*), Guava, Papaya, Pepper (red), Sapodilla, Sisal, Squash, Sweet potato, Tobacco (*N. rustica*), Tomato
9. PERU-ECUADOR-BOLIVIA	Bean (*P. vulgaris*), Bean (*P. lunatus*), Cacao, Corn (secondary), Cotton, Edible roots (oca, ullucu, arracacha, anu), Guava, Papaya, Pepper (red), Potato (many species), Quinine, Quinoa, Squash (*C. maxima*), Tobacco (*N. tabacum*), Tomato
Minor Centres	
10. SOUTHERN CHILE	Potato, Strawberry (Chilean)
11. BRAZIL-PARAGUAY	Brazil nut, Cacao (Secondary), Cashew, Cassava, Mate, Para rubber, Peanut, Pineapple
12. UNITED STATES	Sunflower, Blueberry, Cranberry, Jerusalem artichoke

Source: Wilkes (1977)

1. Economical considerations

Biological resources represent a significant contribution to economic activity and in the USA alone between 1976 and 1989 wild species contributed about 4.5% to the gross domestic product (GDP). In less developed nations, the wild element in economic productivity probably looms larger. Another development in the contribution of biodiversity to economic development in recent years has been nature tourism.

2. Agriculture and pest management

Modern crops are the result of thousands of years of evolutionary processes. Like all biological evolution, crop evolution involves two fundamental processes: the creation of diversity and selection (both natural and artificial or conscious). These evolutionary processes must continue in order for agriculture, a living and evolving system, to remain viable. Therefore, an essential criterion of crop evolution is the availability of genetic diversity (Table 39.2). Farmers and scientists have both relied on the store of genetic diversity present in crop plants that has been accumulated by hundreds of generations who have observed, selected, multiplied, traded, and kept variants of crop plants. The result is a legacy of genetic resources that, today, feeds billions of humans throughout the world.

One surprising fact about human food plants is that only 30 of the 250,000 known higher plant species currently account for 95% of human nutrition. In developed countries where agricultural operations are mechanized, it is common for just a few varieties of these species to be used. For example, half of all the wheat grown in the Canadian wheat-lands is of one variety, "Neepawa". The greatest danger of this is that if major physical or biotic environmental change occurs, the variety in use will not continue to thrive. This discourages the practice of the monoculture crop system. A clear-cut example for this is the catastrophic spread of potato blight and the resulting crop failure that led to a widespread famine in Ireland in the 1940s. The solutions for this kind of a problem is to find a wild relative that is naturally resistant to the pest, or better adapted to the new climatic conditions. The desired genes from the new stock can then be transferred into the cultivar by a careful program of crossbreeding. Similarly, disease resistance in other organisms can be improved by incorporating genes from wild relatives.

3. Pharmaceuticals and other compounds

These options clearly emphasize the importance of wild relatives and the need for conservation of germplasm for future use. Conservation in seed stores, sperm banks, field gene banks and cryopreservation have obvious importance for domesticated plant and animal species (Refer to Chapter 34 for more details). Maintaining genetic diversity in non-domesticated species has become important as they are the source of many drugs used in medicine. **Aspirin**, for example, was originally derived from the leaves of a species of willow, *Salix alba*, whilst the rosy perwinkle, *Catharanthus roseus*, has yielded potent anti-cancer drugs (e.g. vincristin sulfate, vinblastin) for the treatment of Hodgkin's disease and some types of lymphatic cancer. In 1982, the world value of plant-based prescription drugs was estimated at $40 billion. These drugs came from just 41 species out of the 5000 plants tested. Given a world total of 250,000 known plant species, it is quite possible that more useful plant-based drugs have yet to be discovered. In the light of that potential, any future success depends on the conservation of maximum plant genetic diversity.

4. Environmental applications

The value of biodiversity to waste management and environmental clean-up problems, through a technique known as bioremediation, is increasing rapidly (Refer to Chapter 38 for more details on bioremediation).

5. Molecular level benefits

We have reason to be optimistic that major benefits will accrue to society and that sustainable development from wild resources is possible because of the wealth generated at the level of the molecule. Genetic engineering now makes it possible to introduce desirable genetic traits from one species into another, which are not closely related. For example, pest resistant genes from *Bacillus thuringiensis* have been transferred to a variety of crop species. Probably the most significant example of the power of a single molecule from the wild to generate new possibilities and economic development involves the polymerase chain reaction (PCR). The PCR method can be used effectively to construct phylogenetic maps and to measure the amount of biodiversity in a population.

B. CONSERVATION OF DEGRADED TERRESTRIAL ECOSYSTEM

The earth's ecosystem has significantly changed from prehistoric times as a natural process of geo-chemical and biological evolution. In addition to this, humans have manipulated the ecosystems in their search for food, shelter, fuel, and other resources. Discarded waste materials have also had an effect on the ecosystems. Deforestation on a global scale is now an issue of greater concern. The destruction of woodlands and forests is usually in response to the need for more land for agriculture and it results in the creation of completely new, managed ecosystems.

Global deforestation: Forests are the source of natural vegetation for many parts of the world covering one-third of the land surface. Recent reports say that the temperate forests are not significantly decreasing in area though in the past, large-scale deforestation did occur. Tropical forests, on the other hand, are declining at such a rate that in the last 50 years, they have decreased from 15% to 7% of global land surface. One estimate suggests that twelve million hectares of forest (the size of England) are disappearing annually. As human population size increases, larger areas of land may be cultivated to supply food. In addition, fuel, wood gathering will lead to further deforestation. In some forests, commercial logging for tropical hardwoods such as teak and mahogany is another major cause of deforestation. Loss of forests is serious for many reasons.
1. There is significant loss of traditionally harvested products such as timber, poles, twine, fuelwood, organic manure, honey, fruits, game animals, and herbs, that at one time supplied local people with their needs.
2. The demand for softwood timber, pulpwood and hardwood is rising globally. Long-term supplies are threatened.
3. Forests are unique watersheds and catch large amounts of rain.
4. Deforestation can increase rapid run-off of rainwater resulting in soil erosion.
5. It increases global carbon dioxide, which has a long-term effect on global climate.

6. Forests are species-rich and house diverse wildlife communities. Their destruction will lead to innumerable extinctions of small forms of life with the consequent loss of genetic variety and potential resources. Tropical forests have already given us anti-cancer and anti-malaria drugs.

C. *IN SITU* AND *EX SITU* CONSERVATION

The importance of crop genetic resources and the threats that they face have led to the creation of conservation programs to preserve crop resources for future generations. The conservation of biodiversity can be achieved in a number of ways. All these methods fall within the broader concept of '**gene banks**' and can be broadly classified into two categories. (i) ***in situ* conservation**: conservation of genetic resources in their natural habitat (on farm or natural habitats). (ii) ***ex situ* conservation**: conservation of genetic resources taken away from their natural habitat (i.e. gene banks, botanical gardens, and agricultural research stations).

1. *In situ* conservation

Plants are best conserved in their natural habitats (i.e. *in situ* conservation). *In situ* conservation applies only to wild fauna and flora and not to domesticated animals and plants, since conservation is achieved by protection of populations in nature. *In situ* conservation is the best way to conserve plants since it is possible to maintain a large number of individuals with minimum management, effort and cost. Species protected in their natural habitat can continue to evolve alongside their pollinators, symbionts, competitors and predators. This will ensure not only their conservation but also retain their 'fitness' and adaptiveness. In actuality, two types of *in situ* conservation can be distinguished. The first refers to the persistence of genetic resources in their natural habitats, including areas where farmers maintain genetic diversity on their farms. The second refers to specific projects and programs to support and promote the maintenance of crop diversity, sponsored by national governments, international programs, and private organizations. Efforts are being planned by the **Consultative Group on International Agriculture Research's** (CGIARs) latest international agricultural research center, CIFOR (Center for International Forestry Research) to conduct research for *in situ* conservation of forest trees. *In situ* conservation areas of different categories include national parks, sanctuaries, nature reserves, natural monuments, cultural landscapes, biosphere resources, etc.

The goal of *in situ* conservation is to encourage farmers to continue to select and manage local crop populations. These embody not only diverse alleles and genotypes but also evolutionary processes such as gene flow between different populations and local knowledge systems such as folk taxonomies and information about heterogeneous environments. There are several reasons for promoting *in situ* conservation of crop genetic resources: (1) conservation of indigenous knowledge; (2) conservation linked with use; (3) avoidance of regeneration; (4) the fact that key elements of crop genetic resources cannot be captured and stored off-site; (5) the fact that agro-ecosystems continue to generate new genetic resources; (6) because a backup to gene bank collection is necessary; (7) the fact that agro-ecosystems in centers of crop diversity/evolution provide natural laboratories for agricultural research; (8) because the convention of biological diversity mandates *in situ* conservation.

2. *Ex situ* conservation

However, where natural habitats are particularly fragmented and vulnerable to exploitation, *in situ* conservation may not be the solution. This is more applicable to many tropical species where rates of habitat loss are very high. *Ex situ* conservation, using sample populations, is done through establishment of 'gene banks', which include genetic resource centers, zoos, botanical gardens, culture collections, etc. *Ex situ* conservation in botanic gardens and special collection areas, called arboreta, is a possible solution to this problem. Worldwide there are about 1500 botanic gardens but these are concentrated in North America and the European countries, whereas plant diversity is greatest in the tropics and subtropics. Because many of these plants have different day-length and temperature regimens, reproduction and long-term survival of the species limit mere preservation as examples of living organisms. These difficulties can be overcome to some extent by combining the cultivation expertise of botanic gardens with the more innovative methods of germplasm conservation based on establishment of seed banks, field gene banks, and cryopreservation.

***Ex situ* conservation efforts at international level**: Major efforts in the *ex situ* conservation of crop genetic resources also became possible due to the support provided to the crop-based research center of CGIAR, which is a broadly based consortium supporting a worldwide network of 17 international agricultural research centers (IARCs). These IARCs (Figure 39.3) have built up the world's largest *ex situ* collection of crop gene pools, approaching as many as 600,000 individual accessions. These centers have been established to safeguard these resources for the use of present and future generations of research workers throughout the world. The CGIAR is also committed to strengthening national agricultural research systems in genetic resource programs, which have now been established in more than 100 countries.

With the support of CGIAR, an international research center exclusively devoted to plant genetic resources (PGRs) has also been established, named the **International Board for Plant**

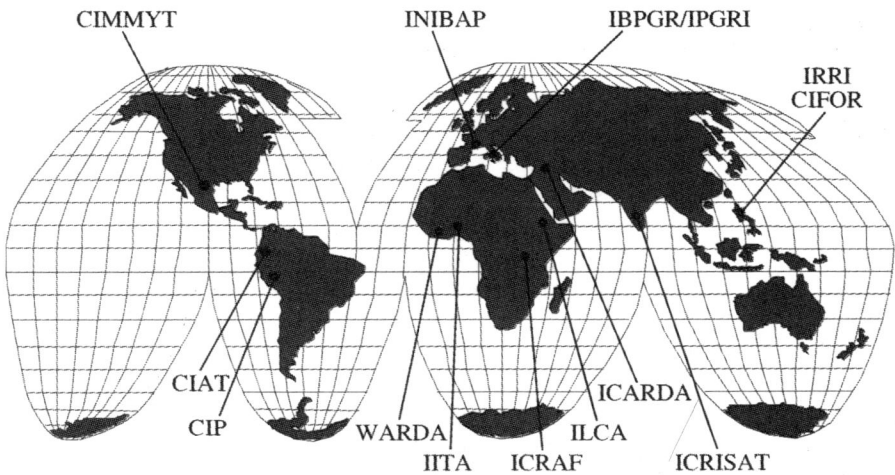

Figure 39.3 *Different International Agricultural Research Centers (IARCs) supported by Consultative Group on International Agricultural Research (CGIAR)*

Genetic Resources (IBPGR). This institute has four main objectives: (i) to assist countries, particularly developing nations, to assess and meet their needs for plant genetic resources; (ii) to strengthen and contribute to international collaboration in the conservation and use of plant genetic resources; (iii) to develop and promote improved strategies and technologies for plant genetic resources conservation; (iv) to provide an international information service on plant genetic resources. Recently, global interest in conservation has been encouraged by the "**Botanic Gardens Conservation Secretariat**" and the "**Center for Plant Conservation**".

Ex situ **conservation efforts in India**: India is rich in biodiversity of different species on the earth. In India, over 115,000 species of plants and animals have already been identified and described. The Indian subcontinet is an important center of diversity and origin of over 167 important cultivated plant species and domestic animals. Biodiversity conservation activities in India come under the Ministry of Environment and Forest, Agriculture, and Science and Technology. In India, *ex situ* conservation efforts were started in 1988, under a Indo-US project with an outlay of US$ 23.95 million. This project involves the construction of a **National Gene Bank** consisting of a seed repository (storage at $-20°C$), cryo-bank (storage at $-196°C$) and tissue culture repository ($10–25°C$). *In vitro* conservation techniques including tissue culture and cryo-preservation offer distinct advantages over other methods. Because of this reason, in 1986 with the financial support from the Department of Biotechnology, the National Facility for Plant Tissue Culture Repository (NFPTCR) was established at NBPGR (National Bureau of Plant Genetic Repository) located in New Delhi. This national facility aims at developing suitable *in vitro* conservation technologies for medium and long-term preservation of clonally propagated agri-horticultural crops including wild relatives. The program also aims at developing techniques for cryo-preservation of seeds, pollen, and *in vitro* cultures. A range of crops like millets (pearl millet, minor millet), oil seeds (*Brassica* sp., sunflower, sesame, etc.), vegetables (onion, carrot, chilli, amaranth, turnip, radish, tomato, etc.), pulses (peas, cow pea sp.) and narcotics (tobacco, poppy) have been maintained for several years in liquid nitrogen. The strategy of desiccating excised embryonic axes has led to the successful cryopreservation of even recalcitrant seed species (e.g. tea, orange) and is being tried for other crops.

D. SEED BANKS

The establishment of seed banks is the most convenient and space-saving method of conserving germplasm. Originally it was restricted to germplasm food crops, but now it includes stores of many endangered wild species. Characteristically, the seeds of many species can remain dormant for thousands of years if kept at low humidity (5–10%) and low temperature ($-20°C$). Seeds that can be stored in this way are termed **orthodox** and include major crop plants such as cereals, soybean, cotton, and many vegetable species. One of the limitations of this method is in ensuring seed viability. Seed viability needs to be frequently checked. Sub-samples are regularly tested for germination and if viability falls below acceptable levels – about 95% – new seeds for storage are generated by growing the stored seed and harvesting the seed. This can be a costly exercise and enforced inbreeding may lead to a loss of vigor in the stored seed. The **recalcitrant** seeds are damaged by drying and cannot be conserved for long periods in a seed bank. Some crops like potatoes mainly reproduce vegetatively. They therefore require an alternative long-term conservation strategy.

E. CRYOPRESERVATION

Cryopreservation involves storage of cells from embryos and shoot tips in liquid nitrogen at $-196°C$. This stops all metabolic processes, which means that the material can be preserved indefinitely assuming no mechanical failure of the storage system occurs. The main biological drawback of this approach is that since the germplasm is evolutionarily frozen, there is decreased viability and increased mutation frequencies (Refer to Chapter 34 for more details).

F. FIELD GENE BANKS

Field gene banks are permanent living plant collections. In this method, a small plot or strip of each variety of a particular species is grown. For example, the International Cocoa Genebank in Trinidad, which specializes in cocoa types from Latin America, grows 16 trees of each of 2500 types of *Theobroma cacao* that it holds. The main limitations are the space and susceptibility to pests and natural disasters such as fire and flood.

G. ZOOS

Zoos were originally established as collections of animals for curiosity's sake but later, their aim became captive breeding. The objectives of captive breeding programs is to preserve the genetic stocks of threatened species so that they can be reintroduced into the wild when conditions permit. Most zoos have a limited number of individuals of each type. To avoid inbreeding and weakening of stock, cooperation with other zoos is essential. A means of breeding through artificial insemination is used to avoid transportation around the world. The method of captive breeding has been very successful, and in some zoos, overpopulation problems have resulted, especially where a return to the wild has not been possible. In zoos, there is a decrease in mortality rate and increase in longevity, since captivity poses fewer threats than in the wild. For example, lions rarely live more than 7 years in the wild but, in zoos, life expectancy is about 20 years.

H. SUSTAINABLE USE OF PLANT AND ANIMAL RESOURCES

Ecological studies have revealed that the environment is deteriorating and resources on earth are diminishing. They also bring out the immediate need to reduce pollution and conserve resources. The Earth Summit held at Rio de Janeiro in 1992 was an important event in raising political and public awareness on environmental concerns and the need to promote **sustainable development**. The Earth Summit identified 27 principles for environmental and social development in the 21st century. Together these initiatives are known as **Agenda 21**. Local communities in each participating country were asked to consult the people they represented to formulate a **Local Agenda 21** by 1996.

For example, food crops and some other resources are easily renewable in short-time scales (which are called as **renewable resources**), provided we do not catastrophically damage our environment. Other resources, notably mineral resources and fossil fuels, cannot be replenished on human time-scales (these are called **non-renewable resources**). Deforestation and the extinction of many animals are indicators of how valuable resources are diminishing due to human ignorance or greed. In contrast, management based on ecological knowledge brings continued benefits, though short-term profits may be less. The attempt to regulate a state of chronic over-fishing now exists in the world's oceans, with an agreed international quota system. This is one example of ecological management at the international level. Some ecosystems are very difficult to manage in a way that provides sustainable resources. Long-living species and ecosystems such as elephants and forests also require sensitive management if future stocks are to be maintained. Productive and sustainable management of woodland ecosystems are long-established concerns of human populations. There are very few woodlands that are primary forest remnants and even these have been modified by human activities. Ancient woodlands (before 1600 AD) have the highest conservation value and often contain great species diversity. Many people believe that hedgerows have high value to wildlife conservation. Conservation of forests and the judicious management of both renewable and non-renewable resources play an important role in ecological balancing.

I. RESTORATION OF DEGRADED LANDS AND REFORESTATION

The increase in human population and modern agricultural activities have affected the ecosystems in a variety of ways. This ecological damage has led to the degradation of habitats, including cultivated land. A significant area of land in the world is arid, with problems of salinity, acidity, or aluminum toxicity. Restoration of degraded land needs immediate action to save the environment and the ecosystem. The conventional methods available for reconstruction are inadequate. However, with the advent of biotechnology, ample opportunities for the restoration of the ecosystem have become available through manipulation of biological systems. Some of the biotechnological methods established for the restoration of degraded land are discussed in the following sections.

Reforestation through micropropagation and use of mycorrhizae

Of the total land area on earth, 29–30% are forests. Forests provide various benefits to human beings (firewood, fodder, pulp, medicinal plants, animal products, etc.) and to the ecosystem (climatic stability and conservation of water and soil) on earth. Today, there is a widening gap between the supply and demand for forest-based materials and a large part of forests is degraded. Degraded lands can be effectively restored by planting trees and conserving the endangered species. Restoration of degraded lands not only provides biomass, it also helps to balance the ecosystem. For effective restoration, it is advisable to use a genetically modified or clonal propagation of superior genetic stocks rather than seeds of uncertain genetic quality.

In tropical and subtropical regions, where desert encroachment and soil degradation are major concerns, different species of *Casuarina* are recommended. These plants can harbor bacteria for nitrogen fixation. Consequently, these trees in addition to giving wood, also enrich the soil with nitrogen fertilizers. *Casuarina* grows faster and provides excellent firewood with high calorific value. Efforts are also under way for the development of plants resistant to abiotic stresses, through the use of techniques of tissue culture and genetic engineering. Salinity, acidity, and aluminum toxicity are the most important abiotic stresses. Use of mycorrhizae in reforestation has been recommended. In poor soils, mycorrhizae can improve seedling survival and growth by enhancing uptake of nutrients (particularly, phosphorous) and water, by extending the root life and surface area, and by providing protection against pathogens. These fungi can be used as inocula to be applied to roots of seedlings, to allow formation of mycorrhiza. Inoculants like vesicular arbuscular mycorrhizae (VAM) fungi may be supplemented with Rhizobium/Azotobacter and phosphorous solubilizing microbes (PSM) to allow better establishment and growth of the seedlings. Inoculation of hardwood and conferring seedlings with specific fungi in nurseries and glasshouses has shown dramatic improvement in growth and survival following transplantation.

X IMPACT OF BIOTECHNOLOGY ON BIOLOGICAL DIVERSITY

The rapid advancement of biotechnology and its applications in agriculture, health, industry, and environmental pollution have aroused tremendous and wide-ranging expectations. These technologies have the potential to provide more abundant and nutritious food, medicines, and environmentally friendly products, beside the means to clean up pollution. These technologies can also be very effectively used for assessment, monitoring, management, and sustainable utilization of biological diversity (Table 39.3).

Biotechnology provides a range of tools and methods for assessment, monitoring, and managing biological diversity, such as clarifying taxonomic and evolutionary relationships among groups of organisms and assessing the effects of ecosystem disturbance on components of biological diversity and biological processes. In addition to the above, biotechnology can also be effectively used for *in situ* and *ex situ* conservation.

Table 39.3 *Areas of application of biotechnology in biodiversity conservation*

Using technologies as source of: Proteins and peptides, lipids and fatty acids, carbohydrates, secondary metabolites for pharmaceuticals, food additives, and bio-pesticides.

Genetic engineering, breeding, and *in vitro* culture systems can be used to enhance agronomic performance.

Improving environmental conditions through: Identification of soil microorganisms and determination of best combinations for soil rehabilitation; use of plants to mitigate heavy pollution, engineering key genes in bacteria for pollutant degradation; improving plant microbe symbiotic systems for waste water treatments; production of biosurfactants (e.g. bio-plastics, and other biodegradable products).

Enhancing the efficiency of microorganisms in industrial processes such as: Microbial-enhanced secondary recovery of oil from reservoirs; bio-leaching (microbiological extraction of metals from low-grade ores), production of industrial enzymes; production of endogenous products (e.g. antibiotics, alcohol, organic acids).

(Source: Montaagu et al. 1995; Chauhan KPS, 1999)

A. IMPACT OF BIOTECHNOLOGY ON BIODIVERSITY

The applications of biotechnology in different areas using the components of biodiversity can have direct and indirect impact. This impact mainly arises from the release of genetically modified organisms and their interactions with the environment.

Some kinds of direct impact (undesirable) on biological community are: (i) displacement or destruction of indigenous/endangered or endemic species; (ii) exposure of species to new pathogenic or toxic agents; (iii) pollution of the gene pool; (iv) loss of species diversity; (v) disruption of energy and nutrient cycling. These kinds of impact are generally ecological and evolutionary and can be assessed scientifically and tested through simulated conditions.

The indirect impact of biotechnology are many and most of them are socio-economic in nature. They can be of major importance and have secondary or tertiary effects. Some of the indirect effects are: (i) pressure on natural habitats because of the increasing value of genetic resources; (ii) lack of immediately perceivable incentives for conservation; (iii) moral/ethical problems of ownership of genetic resources and benefit-sharing; (iv) increase in agricultural productivity; (v) replacement of traditional land races; (vi) decline or loss of opportunities for disadvantaged groups in areas of marginal production.

B. USE OF PLANT GENETIC RESOURCES FOR AGRICULTURE AND INDUSTRIAL PURPOSES

1. Agro-ecosystems

Agro-ecosystems are ecological systems transformed for the purpose of agriculture. For example, rice-fields are created out of a swamp. Each field is formed by building up a bund that defines the biophysical boundary (Figure 39.4). Within the boundary the great diversity of the original wildlife is reduced to a restricted assemblage of a few crops, and associated pests and weeds. The basic, renewable ecological processes remain: competition between the rice and the weeds, herbivory of the rice by the pests, and predation of the pests by their natural enemies. But these ecological processes are now overlain and controlled by the agricultural processes of cultivation, subsidy (with fertilizers), control (of water, pests and diseases), and harvesting operations. The result is an agricultural ecosystem (Lowrance et al., 1984 and Spedding, 1975). However, this is only a partial picture of the transformation. The agricultural process in turn is regulated by economic and social decisions. Rice-growing farmers interact with one another and the market, and exchange or consume their produce. The resulting system is as much a socio-economic system as it is an ecological system, and has a socio-economic boundary. This new complex, agro-socio-economic-ecological system, bounded in by several dimensions, is called an agroecosystem (Conway, 1993). More formally, an agro-ecosystem is "an ecological and socio-economic system, comprising domesticated plants and/or animals and the people who husband them, intended for the purpose of producing food, fiber, or other agricultural products".

2. Utilization of genetic resources in agriculture

The major centers of plant biological diversity are found in tropical regions, where two-thirds (250,000–300,000 species) of the world's resources lie. In the tropics, plant genetic resources are threatened by the destruction of the natural environment and are indirectly threatened by the spread of monocultures; high-yielding varieties resulting from conventional crop breeding or biotechnological processes replace the sturdy traditional cultivars that are less yielding but more resistant to various biotic and abiotic stresses. However, plant biotechnologies can play a key role in the massive production of improved crop varieties as well as in their genetic improvement. They can also help in propagating plant species, which contain useful and

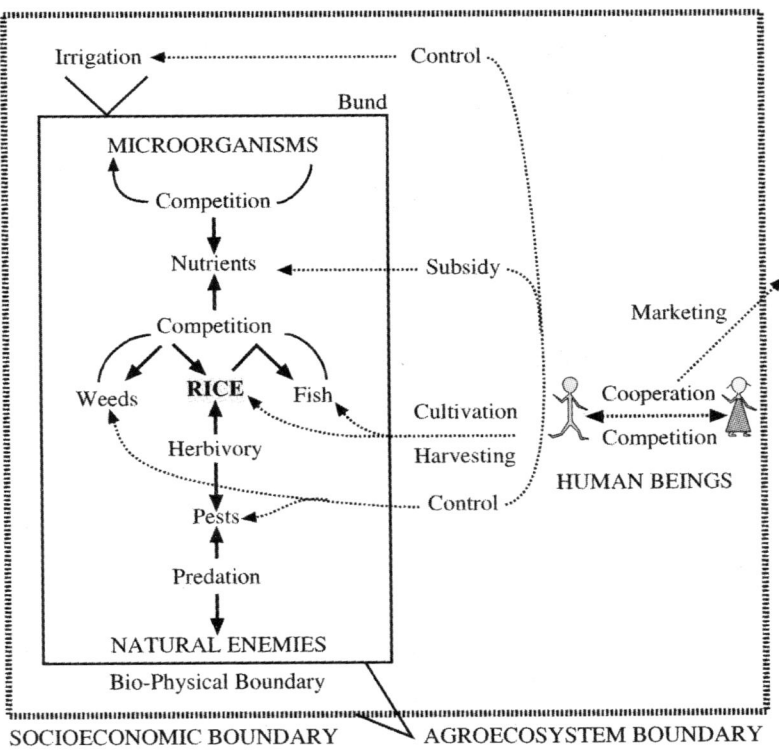

Figure 39.4 *The ricefield as an agro-ecosystem. Solid lines represent ecological processes while dotted lines indicate the overlying agricultural and socioeconomic processes (Redrawn from Conway, 1987)*

biologically active substances. Developing countries can benefit from these applications; an increasing number of them are screening their plant genetic resources for this purpose, sometimes with the help of corporations from the industrialized countries.

Plant genetic resources can be used in conventional breeding. For example, the Japanese primitive dwarf wheat variety Norin 10, which was introduced into America in 1946, played an important role in efforts to improve wheat varieties. The selected dwarf cultivars reacted well to fertilizers, giving higher yields. The primitive rice cultivars in north-western India have been particularly useful in breeding rice varieties resistant to several diseases and parasites. In the case of corn, six hardy varieties identified in Mexico and Guatemala contributed to introducing a much higher resistance to viruses and to parasitic insects in varieties that were being cultivated. In using wild groundnuts found in the Amazonian forest for crossbreeding purposes, resistance to a leaf disease of this legume species was improved, the resulting increase in harvest being evaluated at some $500 million by ICRISAT. Similar examples concern fodder species. Varieties of *Lolium multiflorum*, collected in Uruguay have been used in selection schemes aimed at increasing the resistance of this species to crown rust. A variety of *Bromus biebersteinii* collected in Turkey has been used for the transfer of good agronomic traits and increased vigor in the Ragar variety grown in the US. The wild species of the genus *Lycopersicon* which cross-breeds well with the cultivated tomato, *Lycopersicon esculentum*, has been used as gene pool for the selection of varieties resistant to parasitic fungi, viruses, nematodes, insects, for the improvement of fruit quality, and adaptation to unfavorable environmental conditions. One-third of the rubber used throughout the world was of natural origin and 99% of this was extracted from plantations of a few selected varieties of *Hevea*. Regarding resistance to pathogens and pests, the genetic sources of resistance to the golden nematode in potatoes came from Peru.

The conservation of sufficient genetic diversity within the same species was indispensable to the future utilization of its full genetic potential. Access to abundant plant genetic resources has therefore been an important factor for any work on improving and selecting crop varieties, either through conventional cross-breeding and hybridization techniques, or cultures of plant cells, tissue and organs, or gene transfer.

3. Production of plant-derived drugs and phytochemicals

The market potential for herbal drugs in the "western" world exceeded $49 billion in the year 2002. This growth was due to consumer preferences shifting away from chemicals to plant-derived drugs and the fact that pharmaceutical companies are continuously seeking new compounds for medicines to improve their competitiveness. The World Health Organization (WHO) estimates that about 80% of people in developing countries still rely on traditional plant-derived drugs, the main reason being their low price. Thus, plant-derived drugs offer an interesting potential as resources for local industry and a substitute for costly pharmaceutical imports. Most of the world's medicinal plants are located in the tropics. Of the 250,000 to 300,000 plant species, only about 5000 have been exhaustively studied for possible medical applications. Plant tissue culture could be used for the production of pharmaceuticals, food additives, perfumes, and biopesticides. It has been observed that molecular, cellular, and organ differentiation influences product biosynthesis. Selecting cell lines with high productivity can be simply done on the basis of color as in the case of *Plumbago*, *Nicotiana*, and *Lithospermum*, by plating the cell suspensions and/or protoplasts. A large number of compounds are the products of organogenesis. However, cultures could produce secondary metabolites even in the absence of organogenesis. (Refer to Chapter 37 for more details on the commercial exploitation of tissue culture products.)

C. PUBLIC PERCEPTION

Public concerns on the applications of biotechnology and their impact on ecosystems have been debated throughout the world. An intense debate has been going on between molecular biologists and ecologists on controversial issues regarding risk assessment related to the release of new biotechnology products into the environment. Several surveys conducted to analyze public perceptions have revealed that they were very similar in spite of geographical situations.

XI FURTHER READING

Abramovitz J N (1991) *Investing in Biological Diversity: US Research and Conservation Efforts in Developing Countries*, World Resource Institute, Washington, D.C.

Bernarde MA (1992) *Global Warming—Global Warning*, John Wiley and Sons, Chichester.

Caldecott J (1988) Climbing Towards Extinction, *New Scientist,* **118**:62–66.

CAMLAB (1991) *Design of Municipal Waste Water Treatment Plants*, CAMLAB, Cambridge.

Castri F. di and T Younes (Eds) (1996) *Biodiversity, Science and Development: Towards a New Partnership*, CAB International, UK.

Chauhan K P S (1999) Aspects of Biosafety in the Conservation of Biological Diversity, In: *Biotechnology, Biosafety, and Biodiversity: Scientific and ethical issues for sustainable development* (Eds) Shantharam S and Montgomery JF; Science Publishers, Inc. USA.

Conway GR (1993) *Sustainable Agriculture: The Trade-offs with Productivity, Stability and Equitability*.

Diamond JM (1989) The Present, Past and Future of Human-caused Extinctions, *Phil. Trans. R. Soc. Lond. B,* **325**:469–477.

Ford Foundation (1990) *Joint Management of Forest Lands: Experiences from South Asia*, Ford Foundation, New Delhi, India.

Gilpin A (1994) *Environmental Impact Assessment*, Cambridge University Press. Cambridge.

Grassle JF (1989) Species Diversity in Deep-sea Communities, *Trends in Ecology and Evolution,* **4(1)**:12–15.

Hawksworth DL (Ed) (1996) *Biodiversity Measurement and Estimation*, Chapman and Hall.

Hoyt E (1988) *Conserving the Wild Relatives of Crops*, International Board for Plant Genetic Resources, World Conservation Union (IUCN), and World Wide Fund for Nature, Rome, Italy and Gland, Switzerland.

IUCN and WWF (1989) *The Botanic Gardens Conservation Strategy*, IUCN, Gland, Switzerland.

IUCN, UNEP and WWF (1980) *Caring for the Earth: A Strategy for Sustainable Living*, IUCN, Gland, Switzerland.

IUCN, UNEP and WWF (1980) *World Conservation Strategy: Living Resource Conservation for Sustainable Development*, IUCN, Gland, Switzerland.

Khoshoo TN (1999) The Dharma of Ecology, *Current Science,* **77**:1147–1153.

Knutson L and Stoner AK (Eds) (1989) *Biotic Diversity and Germplasm Preservation, Global Imperatives*, Kluwer Academic Publishers, The Netherlands.

Leach CK and Van Dam-Mieras MCE (1994) *Biotechnological Innovations in Environmental Management*, Butterworth-Heinemann. Oxford (BIOTOL series).

Lowarance R, Stinner BR and House GJ (Eds) (1984) *Agricultural Eco-systems: Unifying Concepts*, John Wiley, New York.

Myers N (1979) *The Sinking Ark: A New Look at the Problem of Disappearing Species*, Pergamon Press, Oxford, UK.

Nayar MP (1996) *Hot Spots of Endemic Plants of India, Nepal and Bhutan*, Tropical Botanical Garden and Research Institute, Thiruvanthapuram, 252p.

NRC (National Research Council). (1980) *Research Priorities in Tropical Biology*, National Academy of Sciences, Washington, D.C., USA.

NRC (National Research Council) (1991) *Managing Global Genetic Resources: Forest Trees*, National Academy of Sciences, Washington, D.C., USA.

Ramesh BR and Pascal JP (1993) Five New Additions to the Flora of Karnataka, *J. Bombay Nat. Hist. Soc.* **90**:323–326.

Reid WV (1992) How Many Species Will There Be? In: *Tropical Deforestation and Species Extinction* (Eds) T.C Whitmore and JA Sayer) Chapman and Hall, London.

Reid WV and Miller KR (1989) *Keeping Options Alive: The Scientific Basis for Conserving Biodiversity*, World Resources Institute, Washington, USA.

Saxena KG, Rao KS and Ramakrishnan PS (1999) Ecological Context of Biodiversity, In: Shantharam S and Montgomery JF (Eds) *Biotechnology, Biosafety, and Biodiversity: Scientific and Ethical Issues for Sustainable Development*, Science Publishers, Inc. USA.

Singh R and Lal M (2000) Sustainable Forestry in India for Carbon Mitigation, *Current Science* **78**:563–567.

Spedding CRW (1975) *The Biology of Agricultural Systems*, Academic Press, London.

Swaminathan MS (2000) Government-industry-civil Society Partnerships in Integrated Gene Management, *Current Science,* **78**:555–562.

Swingland IR (1993) Tropical Forests and Biodiversity Conservation: A New Ecological Inperative, In: *Economics and Ecology: New Frontiers and Sustainable Development* (Ed), Barbier EBs. Chapman and Hall, London.

Wilkes G (1977) The World's Crop Plant Germplasm—An Endangered Resource, *Bull. Atomic Sci.,* **33**:8–16.

Wilson EO and Peter FM (Eds) (1988) *Biodiversity*, National Academy Press, Washington, D.C. USA.

WRI, IUCN and UNEP (1992) *Global Biodiversity Strategy*, WRI, IUCN, UNEP.

40

Bioenergy Fuel from Biomass

I INTRODUCTION

Energy is the pervasive element of a modern industrial economy and a substantial portion of our present day energy needs is met through fossil fuels derived from ultimately finite reserves. The fuel sources and fuel consumption of a country are linked with the economic prosperity and industrial superiority of the nation. The industrial and automobile revolution in the twentieth century caused a quantum jump in fossil fuel use (non-renewable energy sources that include coal, petroleum, and gas). These fossil fuels are depleting faster than they are being formed and are therefore, classified as **non-renewable energy sources**. The rapid depletion of energy sources coupled with the increased demand for energy are posing a threat to the global ecosystem in terms of human health and climatic changes. The deleterious effects of excessive consumption of carbonaceous fuels on the economy and ecology of a large part of the world are already apparent. As a consequence of this, the main focus now is on the utilization of fully renewable energy sources like solar, hydro, biomass, etc. Natural gas is another source of fossil fuel energy, and is emerging as an important substitute for oil. Natural gas is preferred over oil because it burns cleaner and offers great potential for conservation of energy. At the present rate of production and consumption, our oil reserves may hardly last for another 20–30 years and natural gas for about 150 years. In view of the very short-term supply of hydrocarbon reserves, it is important to have strategies for utilization of alternative energy sources.

II SOURCES OF ENERGY

Energy forms an important requirement for development and for all human activities. Energy sources have been variously classified. Energy can be placed into three broad categories: animal, electricity, and biofuel, based on energy forms. But depending on the sustainability of energy sources, energy source can be classified into two categories: (i) renewable and (ii) non-renewable.

A. FORMS OF ENERGY

1. Animal energy

In general, animal energy comes in two forms, (i) human muscle power (HMP) and (ii) draught animal power (DAP). HMP is used throughout the world. This relates to physical work done by human beings. HMP forms one of the major energy sources in some countries, especially in developing countries. In India, HMP constitutes more than 20% of the energy of total electricity produced in one year. HMP is used maximally in agriculture and transport.

2. Electricity

Energy from electricity, can be obtained in different forms such as nuclear energy (by nuclear fission and fusion) and other sources such as solar (solar thermal, solar photovoltic), geothermal (earth), tidal waves (due to moon), water (streams and falls) and wind.

3. Biofuels

Energy from biofuels include fossil fuel energy (coal, oil, and gas), oil from plants, bio-gas, and other sources such as solar-energy trapping by photosynthetic plants, etc.

B. FORMS OF ENERGY BASED ON SUSTAINABILITY

1. Non-renewable energy source

(a) **Fossil fuels**: Forests that existed millions of years ago gave us fossil fuels. Living plants buried during the carboniferous period (about 330–350 million years ago) have been the source of fossil fuels (coal, oil and gases). Their carbon-containing remains were buried and compressed in the sediments, then were transformed into coal, petroleum (oil), and natural gas. They are non-renewable resources.

(b) **Nuclear energy**: Nuclear energy can be generated through two processes: (a) nuclear fission, where a nucleus of an element is broken down into two nuclei or more, releasing a sufficient amount of energy, and (b) nuclear fusion, in which energy is released as a result of two small nuclei joining together. As nuclear fuel breaks down, it releases considerable heat. Typically, water circulating over the fuel absorbs the heat and produces steam that drives electricity-generating turbines. About 10% of electrical energy is generated from nuclear power, which apart from its inherent dangers and the problems of nuclear waste disposal and decommissioning of obsolete nuclear plants, is also dependent on a non-renewable resource, uranium.

An awareness about the rate of depletion of fossil fuels and the pollution caused by their use, coupled with the unease about the use of nuclear power, has prompted many people to think creatively about alternative sources of energy.

2. Renewable energy source

Hydropower is the most widely used form of renewable energy. It currently produces energy equivalent to 500,000 MW worldwide, supplying about 23% of the world's electricity. Most remaining potential hydro resources are concentrated in developing countries.

Solar energy can be harvested in the form of light, heat, electricity, and mainly as biomass through the use of waste material and by growing energy crops (**petrocrops**). It can also be harvested through the use of microorganisms, which can convert solar energy into chemical energy in the form of molecules like glycerol, from minimal nutrients (inorganic salts, water, and atmospheric CO_2). In California, USA, and in many cities in India, the use of solar energy for space and water heating has become a regular feature of modern house design.

Tidal power has been successfully harnessed in La Rance, France. However, though several designs have been proposed for using wave power to generate electricity, none has, as yet, proved commercially successful. **Wind-generated electricity** is already used in many countries.

These energy sources are renewable and offer less polluting alternatives, since petroleum consumption and also production of diverse organic substances by fermentation lead to environmental pollution by production of CH_4 and CO_2. Renewable energy sources are ideally suited for supplying small-scale dispersed power needs. They are particularly useful in developing countries where sunshine is plentiful, providing energy for local industries, schools, health clinics, water pumping stations, and may other small-scale projects. Biotechnology can also be used to produce alternative sources of energy based on naturally produced organic materials. These and other aspects are an important component of biotechnology programs and recognize the role of biotechnology for solving many environmental problems.

C. BIOMASS

Globally, biomass energy is the most important fuel energy source aside from fossil fuel. This source supplies 14% of the world's energy needs and is the main source of energy in the developing world, supplying 35% of its needs. Many new methods are being explored for using living organisms and biological processes as sources of alternative fuel. The main source of biomass can be classified into three groups. Waste materials including those derived from (i) agriculture, (ii) forestry and municipal wastes and growing energy crops involving short rotation forestry plantations and (iii) ***artificial photosynthesis***, a process of producing hydrogen gas as a fuel from water, which is one of the most sustainable possibilities. Another basic approach is to change the energy trapped in biomass into another form, which can be utilized as fuel. Among the raw materials being investigated are ***waste materials*** such as animal manure, sewage sludge, domestic wastes, food wastes, paper wastes, spoilt crops, sugarcane tops, and molasses. Two processes currently dominate – production of biogas (methane) by bacteria, and production of ethanol by yeast. Both processes are anaerobic.

D. BIOMASS AS A SOURCE OF RENEWABLE ENERGY

The unique feature of plants is the presence of the photosynthetic apparatus (green pigment chlorophyll in chloroplast organelle) in the cells. During the photosynthetic process, chlorophyll converts CO_2 into carbohydrates with the evolution of oxygen. Photosynthesis is a unique process, since solar energy is trapped into light harvesting molecules in the chloroplast; this then reduces CO_2 into carbohydrates, fats, and proteins. Radiant energy stored in plants is known as **primary production**, which later on creates plant biomass or bio-material. The rate of storage of photosynthetic products is known as *'primary productivity'*. The energy remaining as organic matter (after respiration) is called **net primary productivity** (**NPP**). NPP is expressed as kCal or $g/m^2/yr$. Biomass exists in many forms and is derived from biological activities present on the surface of soil or at different depths of the vast body of water, lakes, rivers, seas, and ocean. Photosynthetic efficiency depends mainly on the plant efficiency and light intensity, which also affects the accumulation of biomass. Net primary productivity and biomass accumulation are high when the rate of CO_2 assimilation is high and photo-respiration is low (between 25–30ºC). (Photo-respiration accounts for a 50% reduction in the efficiency of the process, as it has a rapid rate of photo-respiration and low rate of CO_2 fixation). In recent years, attempts have been made to convert C_3 plants into C_4 by the introduction of the characteristics of C_4 plants into C_3.

E. COMPOSITION OF BIOMASS

Biomass is living matter (grass, herbs, microorganisms, etc) or its residues (e.g. traditional fuel in the form of products like coal, gas, oil, wood, bagasse, peat, and dried animal dung), which is a renewable source. Plant cells, for example, are a major component of biomass and have many components: (i) cellulose, (ii) hemicellulose, (iii) lignin, (iv) water soluble sugars, amino acids and aliphatic acids, (v) ether and alcohol-soluble constituents (e.g. fats, oils, waxes, resin, and many pigments), and (vi) proteins. These components build up the major portion of plant biomass. The concentrations of these constituents vary with plant group and species.

Waste materials for energy include a variety of wastes originating from plants (wood, green plant matter, straw, paddy husk, rice bran, sawdust, etc.), animal and domestic or municipal wastes.

III CONVERSION OF BIOMASS INTO ENERGY

Since biomass is a complex material, it can be converted into energy by different ways.

A. NON-BIOLOGICAL PROCESS

Non-biologically, (**thermo-chemical process**) biomass can be converted into energy by different routes such as direct combustion, gasification, pyrolysis, and liquefaction.

1. Direct combustion

Biomass from plant or animal origin, are directly burnt for cooking and other purposes. Recently, "hog fuel" production technology has been developed in the US. Hog fuel is a mixture of wood and bark waste burnt directly. This fuel is produced in large sized boilers made up of steel. In the US a co-generation technology has been developed to generate electricity from hog fuel, and to use the exhaust heat in the form of process steam for manufacturing operations. In India, paddy husk (obtained during milling of paddy) can be converted into smokeless solid fuel **briquettes** suitable for use in domestic cooking, hotels, kilns, and boilers. This process is based on the principle of pyrolysis and involves heating the raw material at 250–300°C in the absence of air.

Pyrolysis: Pyrolysis is defined as the destructive distillation or decomposition of organic matter, for example, solid residues, wastes (sawdust, wood chips, and pieces) in an oxygen low atmosphere or in the absence of oxygen at high temperature (200–500°C or rarely 900°C). The type and amount of products from pyrolysis varies with the nature of wood, type of equipment, and system employed. For example, low temperature favors liquids and char, low heating rates favor gas and char, and short gas-residence time favors liquids. In contrast, a high heating rate, favors liquids and the long gas-residence time favors gas. Thus, the products of pyrolysis are gases, organic liquids, and chars, depending on the pyrolysis process and temperature of reaction. The condensable liquids separate into aqueous (pyroligneous acid), oil and tar fraction (if the substrate is wood). The composition of gas is carbon monoxide (28–33%), methane (3–18%), higher hydrocarbons (1–3%) and hydrogen (1–3%). During pyrolysis, the hydrogen content of gas increases as the temperature is increased. Pyrolysis has been employed to produce charcoal for the last few decades. Charcoal is a smokeless and low sulfur fuel used mostly for cooking purposes.

2. Gasification

Gasification is a process of thermal degradation of carbonaceous material under controlled conditions of air or pure oxygen, and high temperature of up to around 1,000°C. As a result of gasification, a high amount of gases is produced. Gasification of biomass is done in a gasifier designed in various ways. The success of the gasification process is based on its designing. Gas is used in a controlled manner for irrigation, pumping and electricity generation. When gasification of farm wastes (manure) takes place, the phenomenon is known as hydrogasification, because gasification of organic wastes occurs in the presence of hydrogen at 500–600°C.

3. Liquefaction

Liquefaction involves the production of oils for energy from wood or agriculture and carbon residues by making them react with carbon monoxide and water/steam at high pressure (4,000 lb/in^2) and temperature (350–400°C) in the presence of catalysts. By this technique, about 40–50% oil can be obtained from wood. This oil serves as a good sources of fuel.

B. BIOLOGICAL PROCESS (BIOCONVERSION)

Bioconversion involves the conversion of organic materials into energy, fertilizer, food, and chemicals through biological agents (i.e. bacteria, actinomycetes, fungi and algae). In a broad sense, bioconversion takes place in two steps, photosynthetic production of biomass, and its subsequent conversion into more useful energy forms (gaseous, liquid or solid fuel; heat and electricity).

1. Biogas

Biogas is a term used to represent a mixture of different gases produced as a result of anaerobic microbial action on domestic and agricultural wastes. The entire process involves three steps: (i) **Solubilization/hydrolysis**: The feedstock is solubilized by water and enzymes. The complex polymers hydrolyze and convert organic polymers into monomers with the help of hydrolytic bacteria. (ii) **Acidogenesis/acid formation**: This involves conversion of monomers into simple compounds such as acetate, CO_2, NH_3, and H_2, using a group of acid forming bacteria (e.g. acetogenic bacteria). (iii) **Methanogenesis/methane formation**: This involves conversion of simple compounds (acetate, H_2, and CO_2) into methane, carbon dioxide, and other products with the help of anaerobic bacteria (e.g. *methanogenic archaebacteria*).

Overall equation: $C_6H_{12}O_6 \rightarrow 3CH_4 + 3CO_2$
Glucose   methane   carbon dioxide

Energy value: $16 kJ g^{-1}$ $56 kJ^{-1}$

Biogas is about 50–70% methane, most of the rest is carbon dioxide (25–35%), with traces of nitrogen (2–7%), hydrogen (1–5%), oxygen (0–0.1%) and other gases (whereas natural gas is about 80% methane). In this process, a mixture of microorganisms is used in the fermentation, including a group of bacteria called **methanogens**, e.g. *Methanobacterium*, which can produce methane from carbon dioxide and hydrogen. These are **archaebacteria**, an ancient group of organisms closely related to the true bacteria. These microorganisms can use a wide range of waste materials or plant products as a substrate for fermentation. In USA, India, China and other parts of the world, the water hyacinth (*Eicchornia crassipes*), a vigorously growing plant, has been used.

Biogas production is a very economically feasible technology. The manure from one cow in one year can be converted to an amount of methane, which is equivalent to over 227 litres (dm^3) of petrol. For example, 0.5 kg of cow manure could generate enough gas to cook a family's meals for a day. In China, over 18 million family-scale digesters have been built. Similarly, in India, a large number of '**Gobar gas plants**' are being used in the villages, which run on farm manure and agricultural wastes produced in people's homes. These Gobar gas plants are not only a cheap source of energy, but are also used as a measure for pollution control. Biogas plants also provide highly enriched and contamination (e.g. insect pests, pathogens, weed seeds, etc.) free organic fertilizer. The gas is typically used for cooking, lighting, tractor or car fuel, and for running electricity generators. On a large scale, the gas can be produced as a byproduct of landfill, sewage, or waste from factories such as sugar factories and distilleries. It can be used to drive generators in sewage works and waste treatment plants (up to 20 dm^3 from one kilogram of rubbish can be obtained). It is important to know that even if the conventional fuels are cheaper than biogas production, it may still be worth it to develop processors for fermenting products like paper and cardboard, because there is a shortage of landfill sites and because of the nuisance they cause. Particularly for developing countries, this technology is more important because they lack their own fossil fuel reserves and have dwindling supplies of timber. Raw sewage and dried animal dung can also be used in a fermenter. The fuel value of fermented dung is six times greater than that of dried dung.

The important steps involved in biogas production using mainly cattle waste (dung) in Gobar gas plants generally followed in India are.

1. First, mix organic matter (e.g. cow dung, night soil, silkworm excretes, and other organic materials) and water in a 1:1 proportion (which makes an organic slurry) so that inorganic particles are maintained at 10% level (Cattle dung usually contains 20% inorganic particles). Combination of other inputs can raise biogas production from 38.6 litres/kg of input (when dung alone) to 59.6 litres/kg (when it combines with 3% night soil, silkworm excretes, 1% of calcium ammonium nitrate).
2. Feed the organic slurry into the biogas plant inlet as shown in Figure 40.1. Make sure sand particles or stone do not get in. These can get deposited at the bottom of the plant. Therefore, frequent cleaning needs to be done.

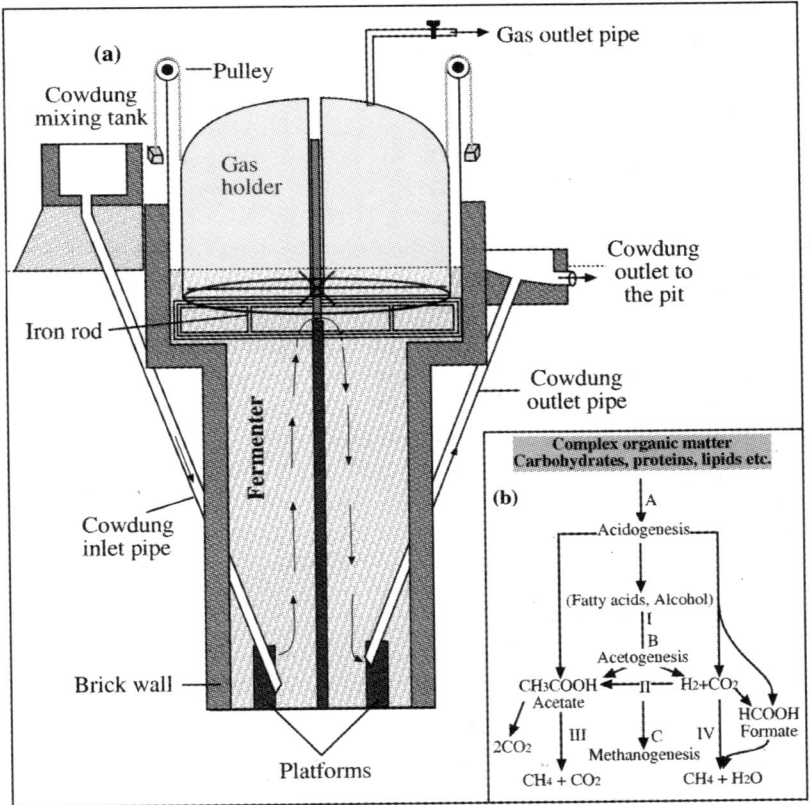

Figure 40.1 *Metabolic stages in the anaerobic degradation of organic matter (a) Diagram of a Biogas (gobar gas) production unit, (b) Anaerobic digestion of organic matter and production of methane. A:hydrolytic and fermentative bacterial; B:acetogenic bacteria (I-acetogenic dehydrogenation by proton reducing acetogenic bacteria; II-acetogenic hydrogenation by acetogenic bacteria); C:methanogenesis by acetoclastic methanogens (acetate respiratory bacteria) (III) and, hydrogen oxidizing methanogens (IV)*

3. The dung-feeding rate varies depending upon the plant (digester) size. In a medium-sized digester, the feeding rate is maintained at 3,500 kg/day.
4. The outlet will generally be open, but to prevent falling of heavier particles from falling, it is to be covered loosely with a wooden peg, which should be opened at the time of feeding to facilitate the slurry getting in.
5. The optimum temperature for biogas production ranges from 35–38°C. During winter seasons, this can be achieved by using hot water, solar panels, or other heating systems. The pH of the slurry should be maintained around neutral pH (7.00) for optimal biogas production.

2. Biodiesel

Diesel-like liquid generated from a biological material is called **biodiesel**. Biodiesel can be obtained from two sources, from lipids accumulated in plants and algae or from hydrocarbons produced by some plants and algae.

(a) Lipids as a source of biodiesel: Lipids are accumulated in the seeds of many oilseed plant species like sunflower, rapeseed, linseed, soybean, safflower, peanut, olive, etc., and by some algae. The lipids have high-energy value and can be burnt to heat boilers or used as diesel engine fuel. Due to high viscosity, these oils are not injectable into the engine combustion chamber. However, this problem is overcome by producing esters of the lipid fatty acids. The esterified lipid fatty acids constitute 'biodiesel' and can be used in unblended form in normal diesel engines with little or no modification.

Some algae may also accumulate up to 60% of their biomass with lipids when they are grown under mineral nutrient limitations. The reduced electron carriers used for the production of new cells under conditions of nutrient sufficiency are diverted to lipids under nutrient deficit conditions.

(b) Biodiesel from hydrocarbon producing crops: A number of **hydrocarbons** (an energy source) producing plants have been identified from different species of plants. These plants can accumulate the photosynthetic products (hydrocarbons) of high molecular weight (10,000). They are commonly known as **petroplants** or **petroleum plants**. Some of them include a large number of *Euphorbia* and *Asclepias* species. The petroplants have lactiferous canals in their stem and secrete milky latex. The latex and 35% of their dry weight can be extracted as organic extracts. The latex is rich in hydrocrackable hydrocarbons, which is called '**biocrude**'. Biocrude yields about 70.6% energy, out of which 22% is kerosene and 44.6% is gasoline. After drying, the latex gives rise to liquid oil. Chemical analyses of extracts, for example, *E. lathyris* show that 5% of its dry weight is a mixture of terpenoids and 20% is a simple sugar (hexose). The terpenoids can be converted into a gasoline like product and hexose may be fermented to ethanol. This results in conversion of 8–12% of its dry weight as oil, giving 20 barrels of oil/acre year.

Sugar producing crops (e.g., sugarcane and sugar beet), and starch crops (e.g., corn, potato, tapioca, etc.) are also a valuable source of energy. An efficient use of these renewable resources leads to conversion of several products that can be used as liquid fuel, chemicals, and other products. Sugarcane can give both liquid fuel (conversion of sugar into ethanol) and also solid fuel (bagasse).

3. Ethanol from biomass

Ethanol is produced by chemicals as well as biological routes. Through the chemical route, synthetic alcohol is produced by catalytic hydration of ethylene with water, using phosphoric acid at 70 atmoshperic pressure and 300°C. The biological route is an alternative way of producing alcohol. Ethanol for human consumption has been manufactured as a component of alcoholic fermentation since ancient times (Figure 40.2). Recently, it has also been recognized as a useful alternative to fossil fuel and an important chemical feed stock.

Overall equation:

$$\text{Carbohydrate} \rightarrow \text{Glucose} \rightarrow \text{Pyruvate} \rightarrow \text{Acetaldehyde} \rightarrow 2\ \text{Ethanol} + 2CO_2$$

Energy value: $16 kJg^{-1}$ $\qquad\qquad\qquad\qquad\qquad\qquad\qquad\qquad 30 kJ^{-1}$

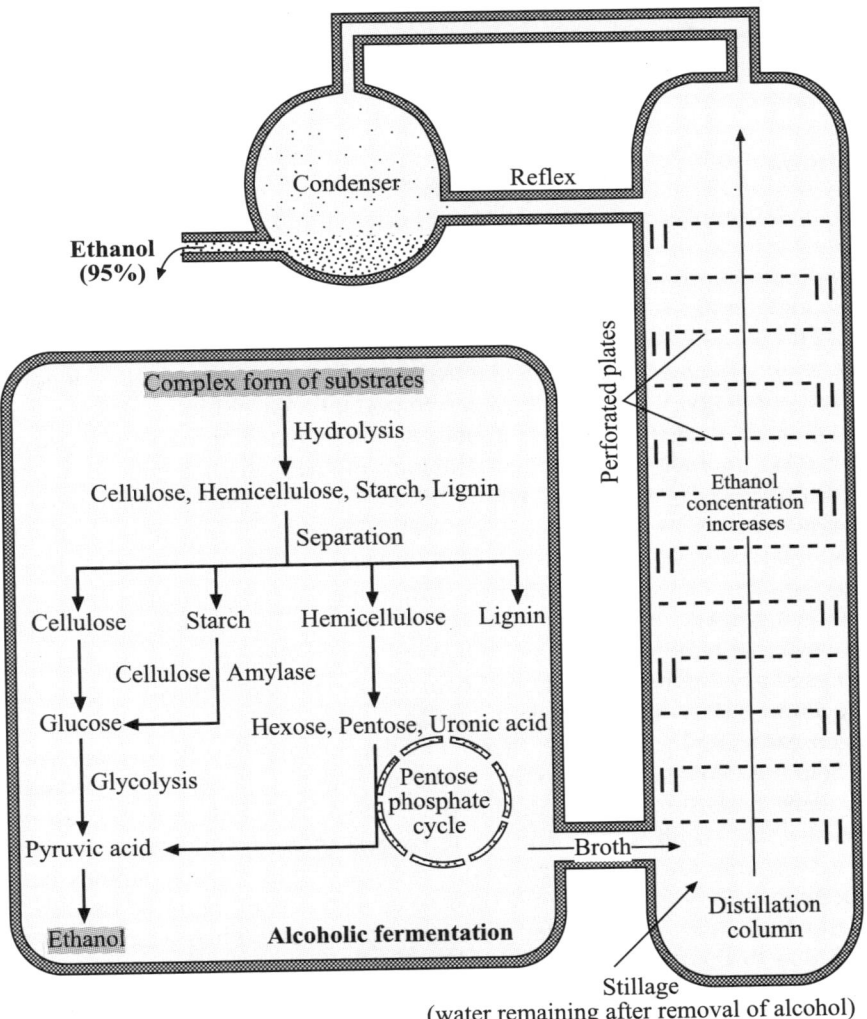

Figure 40.2 *Ethanol formation from the fermentable complex substrates and a schematic design of a cylindrical column distillation unit used for sequential distillation of ethanol during downstream processing*

Commercial production of ethanol involves three steps: (i) preparation of substrate, (ii) fermentation, and (iii) distillation. Ethanol has been produced successfully for use as a fuel in Brazil since 1975. Sugarcane is the starting material. Sugar is extracted from sugarcane juice as a commercial product, but this leaves a syrup called molasses, which contains glucose and fructose. The molasses can be used as material for fermentation by the yeast *Saccharomyces cerevisiae* or sometimes with *Kluyveromyces fragilis*. Several other organisms (fungi and bacteria) are also known to produce small quantities of alcohol. Fermentation is often most active under anaerobic conditions, when carbohydrates are converted into ethanol via pyruvate and acetaldehyde. Under aerobic conditions very little alcohol is produced, since most of the carbohydrates get oxidized to CO_2+H_2O. Ethanol is distilled to separate it from the other fermentation products. Some Brazilian cars are adapted to run on pure alcohol, although the ethanol is mixed with a little petrol to stop people drinking it! Over 11 billion dm^3 were produced in 1985. Since then, oil consumption has been cut by 20%. Some cars run on a mixture of ethanol and petrol called *gasohol*.

4. Hydrogen gas production

Hydrogen (H) is the simplest and lightest element present in the universe. Hydrogen gas (H_2) is one of the cheapest and most sustainable forms of fuel because the raw material used for the hydrogen gas production is water, which is abundant and the energy used for splitting the water to obtain hydrogen gas is solar, which is also available in abundance. It has a high energy to mass ratio. Furthermore, when used as a fuel, hydrogen gas causes no pollution and forms water, thus renewing the raw material. Since hydrogen fuel is environmental-friendly and a renewable source of energy, it is emerging as an important area of biotechnology. The production of H_2 depends on the principle used in photosynthetic machinery by the green plants. The chlorophyll helps in trapping solar energy in the form of ATP, which is utilized in photosynthesis. During this process, water is subjected to photolysis, which leads to splitting of water molecules into oxygen, electrons and hydrogen ions (H^+). The water molecules can also be split by electrolysis (splitting of water by electricity), thermolysis (splitting of water by heat), thermochemical lysis (splitting of water by both heat and chemical catalysis). The hydrogen ions do not form hydrogen gas, but are used to form energy rich compounds like glucose. However, if by some means, these hydrogen ions are converted into hydrogen gas (H_2), the latter can be collected and used as a biofuel. Enzyme ***hydrogenase*** and ***nitrogenase*** have been used to try to convert hydrogen ions (H^+) into hydrogen gas (H_2).

(a) **Hydrogen gas using hydrogenase**: Many microbes possess the enzyme hydrogenase. This enzyme can help the two electrons join two hydrogen ions to produce one molecule of hydrogen gas (H_2). Some of these microbes include different species of algae (e.g. *Chlamydomonas, Dunaliella, Oscillatoria, Porphyridium,* and *Scenedesmus*) and some photosynthetic bacteria (e.g. *Chromatium, Clostridium Rhodospirillium,* and *Thiocapsa*) which can be gainfully employed for hydrogen production. These microbes may be used as a source for the isolation of hydrogenase enzyme, which can be used with isolated chloroplasts, so that the hydrogen ion (H^+) and electrons liberated due to the use of solar energy by the chloroplasts are made to combine into hydrogen molecules, with the help of hydrogenase enzyme. The hydrogen gas thus formed will bubble out of the solution and can be easily collected.

(b) **Hydrogen gas using nitrogenase**: In cyanobacteria (blue green algae), the **'nitrogenase'** enzyme is the chief hydrogen-producing enzyme. It is considered better than hydrogenase because it is less sensitive to oxygen. And this enzyme is present in the heterocysts of cyanobacteria, where oxygen concentration is very low or completely absent, so that hydrogen gas production is very high. The nitrogenase reduces atmospheric nitrogen into NH_3 with the associated release of hydrogen gas. However, hydrogen evolution depends on light, which provides electron donors and also activates nitrogenase and ATP production in heterocysts.

IV SUGGESTED READING

Bernarde MA (1992) *Global Warming—Global Warning*, John Wiley and Sons, Chichester.

CAMLAB (1991) *Design of Municipal Wastewater Treatment Plants*, CAMLAB, Cambridge.

Gilpin A (1994) *Environmental Impact Assessment*, Cambridge University Press. Cambridge.

Hobson PN (1993) *Anaerobic Digestion: Modern Theory and Practice*, Elsevier, London.

Klasson KT, Ackerson MD, Clausen EC and Gaddy JL (1992) Bioconversion of Synthesis Gas into Liquid or Gaseous Fuels. *Enzyme Microb. Technol.*, **4**:602–608.

Leach CK and Van Dam-Mieras MCE (1994) Biotechnological Innovations, In: *Environmental Management*, Butterworth-Heinemann, Oxford (BIOTOL series).

Lynd LR, Cushman JH, Nichols RJ and Wyman CE (1991) Fuel Ethanol from Cellulosic Biomass, *Science* **251**:1318–1323.

Miller MC and Powell W (1994) A Commercial View of Biotechnology in Crop Protection, In: *Molecular Biology in Crop Protection*, pp225–245. (Eds) Marshal G and Walters D Chapman and Hall.

Singh BD (1998) *Biotechnology*, Kalyani Publishers.

41

Regulatory Aspects of Using Genetically-Modified Organisms

I INTRODUCTION

Now it has become evident that the application of molecular biotechnology offers tremendous potential benefits. It will facilitate the tackling of the pressing challenges of increasing crop productivity, sustaining biodiversity, and ameliorating environmental degradation by integrating the modern technology-based production processes with the traditional life support systems based on sound ecological principles. There are exciting prospects of elaborating products that were hitherto obtained from bacteria, fungi, fish, animals, plants, etc. However, this technology raises moral, ethical, and social questions besides issues relevant to the environment, human and animal health safety. In particular, molecular biotechnology is seldom implemented without controversy. Some people think that the new technology is a boon for mankind whereas others believe it may be the bane of society. Because molecular biotechnology can potentially affect many aspects of modern society – including agricultural production and medicine, it raises ethical, legal, economic, and social issues that need to be considered. Ecologists and other scientists have long expressed concerns about the potential impacts of releasing genetically engineered organisms (GEOs) into the environment. These concerns have prompted scientists to declare a self-imposed moratorium on certain types of recombinant DNA experiments until the adoption of official regulatory guidelines.

The successful utilization of recent advances in biotechnology for the benefit of mankind requires wise and balanced legislation. This is dependent on an informed public with a sensitivity to and knowledge and understanding of, the issues at stake. While most people do not doubt that genetic engineering will lead to significant benefits, there is considerable concern that it may lead to the production of organizations that turn out to have serious ill-effects. Public attitude poses a special obstacle in Europe where the focus is on the ethical, economic, political, and cultural implications of manipulating plant genetic material.

Public opposition to biotechnology is spreading throughout the world and there is increasing pressure on governments and communities to restrict commercial activities. In Europe, for example, Germany has already lost much of its biotechnology research due to strict regulations. In the Netherlands, underground groups of environmental activists have claimed numerous attacks on the biotechnology industry over the past three years. They have destroyed laboratories, glasshouses and novel bio-engineered crops, inflicting millions of pounds worth of damage. The attacks have become a major problem for the Dutch biotechnology industry. The stance of the European governments is in contrast to the US where the Bush administration has strongly supported the efforts of industry to develop high yielding disease resistant crops. Another advantage in the US is the adoption, by the US Food and Drug Administration, of rules that would make it easier to market bio-engineered foods in the US than would be permitted in the European Community.

The hazards and risks associated with the release of new organisms into the environment and also the hazards of traditional biotechnology have not been fully understood. In numerous cases, many of the important data on environmental interactions are not known. This article summarizes the ecological consequences and potential risks of the release of traditional and genetically modified organisms into the environment. Finally, the structure and some of the problems are outlined.

II POTENTIAL RISKS

Biotechnology is unique among the technologies developed so far and it will remain so in the future in the sense that its potential negative aspects are debated publicly even before the products reach the market place and the benefits can reach people of all sectors of society. However, the first biosafety concerns were raised from within the scientific community that pioneered genetic engineering during the 1970s. The unfortunate consequence was that it sowed in the public mind doubts and apprehension that the new science could spell doom to all life forms. A plethora of possibilities as projected in the media, was sufficient to construct the doomsday scenario. Some of the major issues of concern to scientists and other sectors of society are briefly discussed below.

A. RISK OF INVASIVENESS

The release of genetically engineered organisms (GEOs) can cause alarm as there is difficulty in predicting the occurrence and extent of long-term environmental effects when nonnative organisms are introduced into ecosystems. There are possibilities of genetic modifications through traditional breeding or genetic engineering of crops or other species to create changes that enhance an organism's ability to become an invasive species. Although genetic engineering transfers only short sequences of DNA relative to a host's entire genome, the resulting phenotype, which includes the transgenic trait and possibly accompanying changes in traits, can produce an organism novel to the existing network of ecological relationships. Potential ecological impacts through invasiveness depend on existing opportunities for unintended establishment, persistence, and gene flow of an introduced organism. Each of these, in turn, depends on various components of survival and reproduction of an organism or its hybrids. Even though only a few introduced organisms become invasive, an issue for the management of all introduced organisms is how to identify those modifications that may lead to or augment invasive characteristics.

For GEOs, one approach has compared the likelihood that transgenic organisms or their hybrids would persist outside of cultivation compared to non-transgenic controls. Some evidence indicates that under experimental conditions, transgenic crops can hybridize with closely related species or subspecies, a prerequisite for gene introgression. Natural hybridization occurs between 12 of the world's 13 most important food crops, as well as numerous other crop species, and some wild relatives. Large areas of cultivation may increase the opportunity for range overlap with compatible relatives. Consequently, introgression into wild relatives may increase as particular cultivars are more widely adapted. Ecological impacts of pollen transfer, a reproductive mechanism through which introgression might occur, depend on whether hybrids survive or not. Reproduction between transgenics and controls suggest, but do not indicate, the opportunity for introgression of transgenes into natural populations, depending on subsequent gene flow and selective pressures. The complex nature of biological invasions means that simple comparisons of fecundity and survival will not adequately predict invasiveness. Variations in the competitive environment and timing of introductions can confound predictions. A thorough understanding of factors, such as viral infections, insect predators, competition, or human-mediated controls that limit reproduction will highlight how transgenic traits affect the reproductive ability of GEOs and their wild relatives in different ways.

B. DIRECT NON-TARGET EFFECTS ON BENEFICIAL AND NATIVE ORGANISMS

Plants engineered to produce proteins with pesticidal properties such as *Bacillus thuringiensis* (Bt) toxin, may have both direct and indirect effects on populations of non-target species. For example, one group of Bt toxins primarily targets *Lepidoptera* (butterflies and moths) and another mostly affects beetles (*Coleopteran*). Effects on non-target species of these groups can vary depending on differences in sensitivity among species and the concentration of Bt toxin produced by transgenic lines. But how these potential risks compare with those of chemical pest control remains critical to understanding the net effect of Bt crops on non-target populations. Some genetically engineered crops affect soil ecosystems, but the long-term significance of any of these changes is unclear.

C. INDIRECT EFFECTS

GEOs may have indirect impacts on populations of species that depend on the pests controlled for survival or reproduction. Population models suggest that more effective control of weeds by using herbicide-tolerant crops could lead to lower food availability for seed specialists. For example, effective control of the Colorado potato beetle in transgenic fields showed a decrease in a predatory specialist on it but in plots of Bt and non-transgenic corn, the predatory population was same. The pesticidal proteins affect indirectly by bioaccumulation, if predators consume the prey that contain pesticidal proteins. The rate of persistence of pesticidal proteins may affect the probability of non-target effects.

D. NEW VIRAL DISEASES

Viruses with new biological characteristics could potentially arise in trangenic viral-related plants through **recombination** and **heteroencapsidation** (phenotypic mixing). New viral strains can evolve through recombination between closely related strains. Both events have resulted in the production of a recombinant virus from a transgenic plant under experimental conditions. However, we lack empirical evidence to understand the likelihood of this transference under natural circumstances. Strategies to reduce the biological risk of heteroencapsidation and accompanying changes in transmissibility are under investigation. **Synergism** is the result of more than the additive effect of infections by two viruses. In transgenic plants, the expression of certain genes from one of the synergistic "pair" can result in a synergistic reaction upon infection by the other member of the pair.

E. VARIABILITY AND UNEXPECTED RESULTS

Ecosystems are complex, and not every risk associated with the release of new organisms, including transgenics, can be identified. Unknown risks may surface as the frequency and scale of the introduction increases. Because some consequences, such as the probability of gene flow, are a function of the spatial scale of the introduction, limited field experiments do not always sufficiently mimic future reality prior to widespread planting. Ecological relationships include many cascading and higher order interactions that are intrinsically difficult to test and evaluate for significance on limited temporal and spatial scales. On a larger spatial scale, there is a greater possibility for contact with sensitive species or habitats or for landscape-level changes because on a larger scale, more ecosystems could be altered. In addition to this, environmental and cultivar variability complicates the task of assessing risk. Transgenic organisms, such as genetically engineered crops, released into the environment will potentially interact with a diversity of habitats in time and in space, and the potential risks from a single type of transgenic organism may vary accordingly. Therefore, risk assessments will need to be especially sensitive to temporal and spatial factors.

III   REGULATING RECOMBINANT DNA TECHNOLOGY

In the early 1970s, when it was realized that recombinant DNA technology could be used to manipulate its genetic constitution and engineer organisms with novel genes, concerns about safety, ethics, and unforeseen consequences were raised by scientists, the public, and government officials. The major apprehension was that, either inadvertently or possibly deliberately for the purposes of warfare, unique microorganisms that had not previously existed would be developed and would cause epidemics or environmental catastrophes. Thus the biosafety issue has to be addressed at all stages of development and release of GMOs or their derived products, starting with research laboratory to controlled glasshouses, to limited field trials, to permitted release in environment, and culminating in free-release without any regulatory oversight. Safety is achieved by the analysis of risks involved and by devising appropriate management practices to mitigate possible negative impacts. Since GMOs can move beyond political boundaries on a geographic continuum, safety regulations need to be based on internationally accepted principles.

In 1976, the US National Institute of Health (NIH), the primary US research grant agency in the medical and health sciences, issued a set of guidelines for the conduct of NIH-supported research using recombinant DNA (rDNA) technology. These rules and regulations rigorously defined laboratory containment levels for the conduct of rDNA experiments. They also required that biological containment be a component of any rDNA research (Chapter 2) and the preferred hosts for foreign DNA would be those microorganisms considered least likely to proliferate outside of the laboratory or to transfer their DNA to other microorganisms. For research with known pathogenic organisms, elaborate negative-pressure controlled, self-contained rooms were recommended. The initial NIH guidelines were very stringent, and many scientists thought that the regulations were excessive. In anticipation of a need to modify the original guidelines, the NIH Recombinant DNA Advisory Committee (NIH-RAC) was created. This committee was charged with overseeing the developments in rDNA research and, if necessary, refining the regulations. The NIH-RAC had to hold open meetings to discuss its decisions, and it had to publish and distribute the min. of its meetings. These guidelines were later adapted and applied in many countries all over the world.

As part of the original NIH guidelines, one specific class of experiments that was "not to be initiated at the present time" under any circumstances was the "deliberate release into the environment of any organism containing a rDNA molecule". However, it was inevitable that genetically engineered organisms (GEOs) would be developed that could function in natural settings. By 1980, the original NIH guidelines were relaxed considerably by the NIH-RAC as result of experience and specific experimental data from studies that the host organism e.g. *E. coli* K–12, most commonly used in rDNA experiments, was unable to proliferate to any significant extent outside the laboratory. Finally, it was conceded that it was extremely unlikely that a pathogenic organism would be created if the cloned gene had nothing to do with pathogenesis in the original organism. As a result of the easing of the containment requirements for more routine experiments, the use of rDNA technology became more prevalent and flourished.

A. APPROACHING THE BIOHAZARD PROBLEM

Biosafety can be achieved through the analysis and management of risks. Risk analysis generally involves identification of hazards, factors that would influence the occurrence of the hazard, and consequences of the occurrence of the predicted hazard. Management of risk on the other hand involves taking steps to reduce the probability of occurrence of the hazard, and to contain or neutralize the effect should the hazard actually occur. One of the major concerns with GMOs is that once released into the environment, it is difficult to fully monitor their spread and an impossible task to contain them if they are later found hazardous. As mentioned earlier, biosafety issues of GMOs always addressed the product rather than the process by which the product has been developed. The framework developed by the NRC of the US National Academy of Sciences has set out the following factors that are important for risk assessment: (1) characteristic of the organism contributing the gene; (2) the gene or sequence that is being mobilized; (3) the recipient of the genetic material; (4) the trait conferred by the gene in the new host; (5) the environment into which the GMOs is to be introduced; (6) interaction between the organism and the environment; (7) the application to which the product is to be put.

Information related to these factors is very critical for risk assessment. If the donor proves either pathogenic or there is lack of information about it, it should be treated very cautiously. Based on the 'familiarity' of the donor, the recipient, the environment, and the gene product, a scientific risk assessment of GMOs can be done. The intrinsic risk for each of the above factors can be categorized to determine the overall level of safety concern. After constructing the risk profile, risk reduction can be considered for those risks that exceed the determined protection level. Since the magnitude of the risk is the product of the probability of occurrence of the hazard and the severity of the hazard, the risk can be lowered by reducing one or both of the above factors. Risk assessment thus becomes critical to determine what to regulate and how much to regulate. It also identifies gaps in our measures and helps in deciding on the type of experiments to be conducted and kind of data to be gathered for future use.

B. ENVIRONMENTAL AND REGULATORY ASPECTS OF USING GENETICALLY-MODIFIED PLANTS

The techniques of genetic engineering have advanced to the point where products from recombinant organisms will soon be available to consumers. Speaking specifically of plants, almost all agronomically important crops have been engineered in an effort to improve upon one or more characteristics. Crops that produce better quality of food are more resistant to pests and environmental stress, and those that are tolerant to environmentally innocuous herbicides are very close to commercial reality. Such products have the potential to bring significant benefits to food processors, consumers, and agriculture. They will be more cost-effective to produce and process, of higher nutritional quality, and will give farmers more flexibility in environmentally sound cultural practices. Looking further, plants may be used to produce large quantities of therapeutically useful peptides and otherwise scarce chemotherapy agents like taxol at a lower cost.

The plant biotechnology industry is now at the point where field testing of genetically-modified plants (GMPs) is acknowledged as a critical and realistic step in assessing the safety, commercial potential, and environmental impact of each product. Literally hundreds of field tests have occurred throughout the world without incident and the results have been reported to the public (Duesing and Goy, 1996). Getting to this step in the commercialization process requires working with governmental bodies that have the authority to approve field tests of GMPs. An application for a field release permit requires that data and information be provided on the organisms, the transformation method, the gene and its origin, expression products, and their potential environmental impact. Such data are acquired from studies under controlled conditions such as glasshouses and growth chambers.

All studies with GMPs must be undertaken with the proper regulatory oversight by the government presiding over the location in which the studies are done. As GMPs will be used globally, it is critical to work with, and understand the regulations and approval processes applicable to field releases in place today as they relate to field release of GMPs in Canada, the US, Europe, and Japan. A discussion of environmental aspects of GMPs is also given. Environmental assessment is then presented as case studies: a herbicide tolerant canola or rapeseed, and insect resistant cotton.

Regulatory Aspects: Plant breeders have been genetically-modifying plants for centuries to improve the agronomic characteristics of crops and ornamentals. In Canada and Europe, there is an approval process for all new plant varieties, while the US has no such requirement. Concern over the uses and potential of genetic engineering for plant improvement has brought about more involvement of governmental bodies in a regulatory role. In Canada, the US, Europe, and Japan, certain public agencies that regulate agricultural and food industries under the jurisdiction of national laws will be responsible for approving the release of GMPs into the environment. With the help of the academia, industry, and a group of international experts, and the

Organization for Economic Cooperative Development (**OECD**), guidelines are being developed to regulate GMPs (Bureau of National Affairs, 1992).

C. RELEVANCE OF MULTIPLE ENVIRONMENT TESTING

The influence of the environment on transgene expression is recognized as important for shaping field phenotypic manifestation. Biosafety testing in multiple environments is expensive, but it is important to recognize that the environment has many components. The components relevant to biosafety tests on transgenics include – the people, agricultural practices, diseases, pests, abiotic stress, wild species, pollen dispersal agents, and pollution levels. It is well known that the pest/pathogen structure is highly variable between different locations and the expression of specific host genes is variable in relation to inciting stimuli. Therefore, testing transgenics in a new location is necessary. However, when the ambient environment at a new location is substantially similar to a previously tested location, there is no need for fresh testing.

The Canadian government is strongly supportive of the development of biotechnology in Canada as a means of strengthening economic competitiveness. Field testing of GMPs in Canada requires permits from Agriculture Canada, which is also responsible for environmental safety assessments of field trials under the Environmental Assessmental and Review Process Guidelines. Agriculture Canada is the lead agency when considering the regulatory aspects of using GMPs in the field. They have developed a flexible approach to regulation built on existing laws. Their approach is a product-based, case-by-case assessment with a science-based evaluation of the risk related to each field release. Based on that risk, a four-stage system of regulatory oversight is used. Contained glasshouse or lab research (**Stage 1**) is regulated as minimum risk and involves no regulation. Of incrementally greater perceived risk and regulatory involvement are confined field research trials (**Stage 2**), unconfined field research trials (**Stage 3**) and commercial release (**Stage 4**). Since 1988, there have been over 300 field trials at the stage 2 level, but only a few at stage 3. Agriculture Canada approves important movement of seed or plant parts and all field release of GMPs, issues safety requirements for field tests, and inspects release sites during the course of a test to ensure that the conditions of the permit are being followed. Before an approval is issued, an Environmental Assessment and Review Process (EARP) must be completed.

D. ENVIRONMENTAL AND REGULATORY ASPECTS OF USING GMMS

Advances in the techniques of molecular biology over the last two decades are such that genetically-modified microorganisms (GMMs), a category which includes viruses, bacteria, and lower eukaryotes, are being used commercially in contained systems and are also being developed for intentional release into the environment. Despite its initial prohibition, by 1982, it became clear that the NIH-RAC would have to cope with requests for open field testing of GMMs, i.e. for their deliberate release into the environment. There are diverse applications of GMMs in the environment and these include crop protection, biological control, bioremediation (degradation of chemical waste), extraction of metals from ores, enhancement of plant nutrient availability, etc. However, the deliberate release of GMMs into the environment has to be very much cautioned against due to the difficulty in assessing their impact on the environment and also the many inherent problems involved in monitoring the organisms in the environment and assessing their interaction with other biota. Indeed, if an unforeseen hazard is realized after the deliberate release of GMMs, it would be a major and perhaps impossible task to implement a recovery operation. This fundamental difference has had important implications for the formulation of guidelines by the regulatory authorities responsible for granting licenses for the release of GMMs.

In order that the risks of releasing GMMs into the environment can be assessed, we need to be able to address some fundamental questions about the ecology of microorganisms in the environment. We need to know what factors affect their growth, survival, activity, movement, and impact on indigenous microflora through the environment. To answer these questions, it is necessary to be able to detect and quantify reliably concentrations of total, viable, culturable, and non-culturable GMMs, to determine the presence and expression of recombinant DNA both in the introduced and indigenous populations. There are two main areas of concern when introducing a new species into the environment or into a part of the environment where it has not previously been present. The first concern is that the new species will display traits that **negatively affect the environment** of other species within it. The second concern is that **genetic traits in the released organism will be transferred to other organisms**, which will further disrupt the ecological balance. Most environmental releases follow the pattern of release, dispersion, survival, and then interaction with the environment. The interaction can be on a physical or genetic level.

1. Dispersion and Survival of Microorganisms in the Environment

The three major pathways of environmental dispersion are air, water, and vectors. Vectors are moving solid surfaces and may be either biological or physical. Viable airborne microorganisms can be transported by wind currents over a long distance and their survival may be prolonged through association with aerosols, soil, or other particles. Microbial transport in surface water or groundwater is difficult to predict because of its dependence on complex physical mechanisms, such as attachment. Biological dispersion is either in the intestinal tract, in the haemolymph or salivary glands, or on the body surface of a variety of vertebrate and invertebrate animals. Physical dispersion is on the surface or in internal spaces of man-made material, equipment, or vehicles.

When a species is introduced into a new ecosystem in which no ecological barrier exists, its population reaches a considerable level in a short time. The reasons for such population explosions can be found in the factors that affect organism survival in the environment. Some of the factors that affect survival of all microorganisms are for example, temperature, sunlight, moisture, nutrients, and biological interactions. Other factors include competition among microorganisms sharing the same ecosystem, and dispersion of the microorganism. The following mechanisms and requirements are necessary for the survival of native and released GMMs into a new environment: (1) adherence to host tissue; (2) colonization on a host body; (3) availability of essential nutrient elements; (4) resistance to antimicrobial activity; and (5) formation of a protective coat to resist adverse conditions in the new environment.

Once it establishes itself in the new environment, the organism is capable of transferring its genetic traits to other members of the ecosystem. There are four ways of gene transfer in a microbial ecosystem viz., transformation by free DNAs released, transduction by bacteriophage DNAs, conjugation by plasmids, and conjugal transfer by conjugative transposons. Although this exchange is not preferred, it cannot be guaranteed, as there are numerous examples of genetic exchange within the environment. Gene transfer can occur within or between the microorganisms, between microorganisms and animals and more significantly, between microorganisms and man. The most significant negative effect of gene exchange in the environment to date has been the transfer of antibiotic-resistant genes to organisms harmful to man. Bacteria resistant to antibiotics have been isolated in numerous places, for example, in hospital waste, raw sewage, water recovering sewage effluent, sediments, marine water, estuaries and rivers, fisheries, animal feed, abattoirs, plants and soil, drinking water, etc. There are an abundance of plasmids in aquatic microbial systems. In some lake waters, up to 46% of heterotrops contain plasmids. The majority of these plasmids are conjugative and transfer frequency varies from 5×10^8 to 2.5×10^{-2} conjugants per donor.

2. Monitoring of Introduced Microorganisms

In nature, the microbial community contains diverse groups of microorganisms. Monitoring introduced GMMs is not an easy task. There is no sole detection system which is perfectly suited to comprehensive risk assessment studies and so the combined use of traditional and molecular techniques is essential. Traditional microbial detection techniques which mainly require cultivation on laboratory media, are inadequate as they cannot detect non-cultivable cells and are very selective depending on the choice of culture media and growth conditions. In addition, the efficiency of extraction of GMMs from environmental samples varies between organisms and sample types and the harsh methods required often reduce cell viability. These deficiencies have stimulated the development of new methodologies, in particular molecular techniques, which offer the sensitivity and selectivity required to monitor GMMs and their recombinant DNA in the environment. The molecular techniques fall into three distinct groups: ***immunological methods***, ***nucleic acid probing***, and ***molecular markers***.

A variety of methodologies have been employed to introduce marker genes into the host GMM. The most straightforward is the introduction of marker genes on the plasmid, e.g. the introduction of *lac ZY* genes into fluorescent *Pseudomonads*. Multicopy plasmids can be used with higher sensitivity of detection. The use of antibiotic resistance markers has been very common as they allow direct selection of the marked population by plating on an antibiotic containing medium, which will select against background populations. However, there is a strong objection to the use of antibiotic resistance genes as concern has been voiced over the introduction of antibiotic resistance genes into the environment. All these molecular markers have been used successfully to monitor the presence of GMMs in environmental samples and the details of each system can be found in some excellent review articles (Pickup, 1991; Prosser 1994).

3. Ecological Impact on GMM Release

There are a number of ecological concerns raised when considering the proposed release of a GMM. These include: whether the microbial community structure and function will be distributed by the introduced inoculum and whether the recombinant DNA will be transferred into the indigenous community. Our knowledge of microbial ecology is limited although it is improving all the time. With the molecular tools described there in place, meaningful studies can be made on the impact an introduced inoculum will have on the indigenous microbial community. This has been tackled both by the use of contained microorganisms, greenhouse experiments, and limited field introductions.

4. Risk assessment

Risk may be defined as a measure of the likelihood and severity of harm and is generally assessed through three types of investigations (Al Bourquin and Seidler, 1986): (1) exposure assessment (determining the conditions of exposure); (2) hazard identification (attributing adverse effects to the hazard), including dose-response assessment (relating exposure to effects); (3) risk characterization (estimating overall release).

A useful descriptive model for the risk assessment of a biotechnology application was presented in a report by the US office of Technology Assessment (Fiksel and Corello, 1986).
1. *Formation*: The creation of GMMs which may be deliberate or accidental.
2. *Release*: The release of GMMs into the environment, which may be deliberate or accidental.
3. *Proliferation*: The subsequent multiplication, genetic reconstruction, growth, transport, modification, and death of these microorganisms in the environment including possible transfer of genetic material to other microorganisms
4. *Establishment*: The establishment of these microorganisms within an ecosystem niche, including possible colonization in human or other biota.
5. *Effect on humans and ecological effects*: Human or ecological effects occuring, due to the interaction of the organisms with some host or environmental factors.

The hazardous situations that may arise due to GMMs have been described by Krimsky and Fracknel (1985) and are as follows. (1) A novel organism might be released into the environment with unpredictable and possible irreversible effects on the environment. (2) The release of a new microbial agent may infect humans or animals. (3) A conventional pathogen might have its host range broadened. (4) A rapid-rise in application of large-scale biotechnological processes could result in bioeffluents that would place additional stress on the quality of land and water resources. (5) Organisms engineered to perform useful functions in the environment might produce adverse secondary effects of an unanticipated nature.

5. Direct risk to humans

To evaluate the risk potential to humans, we must first take into account how these microorganisms gain entry into humans. Humans have a natural protection against most bacterial infection i.e. the skin and stomach. Bacteria gain entry through abrasions and cuts in the skin and stomach. Naturally occurring bacteria have intrinsic traits that aid this ability to colonize, and protect themselves against competitive colonization by invading potential pathogens. Human infection depends on four major interacting factors: (1) the intrinsic traits, including pathogenicity of the organisms; (2) the health of the individual; (3) the number of invasive organisms; and (4) the first and second lines of antimicrobial defense. There are several lists that classify microorganisms based on "Safety in Biotechnology of the European Federation of Biotechnology".

6. Regulations governing GMM releases

Due to growing concerns over GMOs throughout the world, the UNIDO/WHO/FAO/UNEP has formulated an Informal Working Group on Biosafety. In 1991, this group prepared the "***Voluntary Code of Conduct for the Release of Organisms into the Environment***". The ICGEB also plays an important role in issues related to biosafety and the environmentally sustainable use of biotechnology. It organizes annual workshops on biosafety and on risk assessment for the release of GMOs. It also provides an on-line bibliographic database on biosafety and risk assessment for the environmental release of GMOs. Besides it collaborates with other organizations such as UNIDO's Biosafety Information Network and Advisory Service (BINAS), aimed at monitoring the global development in regulatory issues in biotechnology.

The potential hazards resulting from the release of genetically engineered organisms into the environment were initially raised at the Gordon Research Conference on Nucleic Acids in July 1973 (Hartl, 1985). One of the early local initiatives for regulation of recombinant microorganisms was a 3-week moratorium by Cambridge, Massachusetts, on recombinant DNA research at the MIT and Harvard University. Later in 1976, Cambridge passed an ordinance for the safe use of microorganisms. Today there is an extensive network of agencies and regulations governing the use of GMMs. In several cases, regulatory authorities worldwide have adopted a case-by-case approach to each application for permission for release although the regulations remain different for individual countries. The main safety considerations are different from those already in place to deal with the contained use of GMMs as, by their very nature, deliberately released microorganisms will be designated to persist in the environment. There have been a number of international conferences held to address the issues surrounding the deliberate release of GMOs, such as Release of Genetically Engineered Microorganisms (REGM 1) in 1988 (Sussman et al., 1988), and REGEM 2 in 1991 (Tull and Sussman 1992), as well as a number of research programs (BAP) (Economics, 1990) and the PROSAMO initiative (Killham, 1992). The tendency of regulatory authorities to adopt a cautious approach to deliberate release has been due part to the inability of scientists to predict the exact outcome of a release experiment and also general public concern that the release of GMMs may have long-term environmental or health risks.

7. Problems of regulation of biotechnology

The regulations governing biotechnology are complex and government approval is required for field testing genetically novel organisms. These regulations have led to high costs and long delays for companies wishing to test new products in the environment. The regulations attempt to strike a balance between protecting the environment and mankind while at the same time not stifling new advances and benefits to society. One of the consequences of the comprehensive regulation of biotechnology in some countries is that these regulations are often bypassed by conducting trials in other countries, where the regulations are not as strict.

8. Case studies of GMMs used in bioprocesses

Genetically engineered organisms are being exploited to develop processes for the production of many important biologicals. Some of the bioprocesses that might use genetically engineered organisms are:

(a) **Food bioprocesses**: A wide variety of food products are made by utilizing the action of microorganisms. Many of these processes have been practiced for thousands of years, for example, for the production of cheese, beer, wine, and vinegar. The main hazard involved in the use of microorganisms in food bioprocesses is the contamination of the food product with toxic chemicals. Though contamination of food products can be avoided, many industrial strains produce toxic compounds (aflatoxin-related compounds). The use of genetically engineered yeast has shown some problems in the food industry.

In the USA, the Food and Drug Administration (FDA) is responsible for regulating the introduction of foods, drugs, pharmaceuticals, and medical devices into the marketplace. Both the FDA and the food industry, which is represented by the International Food Biotechnology Council (IFBC), are convinced that new regulations are not required for foods and foodstuffs that are developed by recombinant DNA technology because any unlicensed food or food ingredient, regardless of how it is produced, must be assessed for safety by toxicity, allergenicity, and impurity testing. If genetic modifications by either selective breeding or recombinant DNA technology change the composition of an accepted food item or foodstuff, then once the safety of the product has been proved, the company, must inform the consumer about it through labeling as the new product differs from the traditional one.

(b) Pharmaceutical and enzyme bioprocesses: Biotechnology enables the manufacture of a wide range of drugs, therapeutics, and enzymes. Most enzyme products cannot be made 100% pure and may contain potentially harmful compounds from the fermentation broth. Like all proteins, enzymes are antigenic and may cause allergic reactions in some people. Historically, the most visible problems involving enzymes and human health have been in the production of detergent proteases. These problems, involving enzyme dusts, have been solved by encapsulation technology. Another problem occurs in the area of vaccine production. Extreme care is required in the inactivation of the virus to ensure that no residual live virus is present in the vaccine. All products especially those for human use should be free of bacterial DNA, and the complete removal of DNA from many products is a problem.

(c) Waste treatment: In waste-water treatment, organisms are used in an effluent stream that enhances their growth, thereby helping them break down organic material. Microorganisms and their activity are used to detoxify and degrade sewage and industrial wastes. Efforts are underway to improve the efficiency of treatment plants by adding specially cultured GMMs. *Pseudomonas putida* endowed with multiple compatible plasmids from other strains, giving unique multi substrate-utilizing capabilities, has been patented (Chakrabarty, 1974). The specific plasmids were for octane, xylene, m-xylene, camphor, and salicylate degradation. The degradation of many xenobiotic compounds is often the result of concerted metabolic activity by different members of a stable, highly interactive microbial community. In case of waste-water treatment, the microorganisms may have a direct manmade pathway into the environment. Some policy makers are concerned that novel organisms developed for sewage treatment may constitute a new form of pollution.

(d) Chemical bioprocesses: Various acids, solvents, and other chemicals have been manufactured by microbial fermentation. According to a recent report, more than 200 microorganisms currently produce substances of commercial value. Problems may arise if the microorganism used to produce a valuable product is a pathogen. Production of gibberellic acid in a solid medium by *Gibberella fujikuroi*, which is a pathogen to the rice plant, may pose a problem to rice crops if proper care has not been taken to remove the plant pathogen in downstream processing.

(e) Mining: Bacteria have been used to leach metals from low-grade ores of copper and uranium. The genus *Thiobacillus* is commonly used for this purpose. However, they pose some potential hazards in the mining industry: (1) Bacterial leaching operations that generate large quantities of sulfuric acid acidify water supplies; (2) *Thiobacillus* and related species may acquire the ability to infect humans by natural interaction with pathogenic organisms; (3) metals concentrated by bacteria from dilute mine water can accumulate in the food chain.

(f) Agriculture: The deliberate release of microorganisms to control plants and insects is of great significance, because widespread dispersion of the microorganisms is practiced. Currently, 13 microbial pesticide agents are approved and registered with the EPA. These 13 organisms are marked in 75 different products for use in agriculture, forestry, and insect control. *Bacillus thuringiensis* carries a gene that codes for a protein (endotoxin) called Bt toxin. The toxin breaks down the gut of some insects disrupting the gut membrane and eventually killing the insect. The release of the plant pathogen *Puccinia chondrillina* has been used successfully to control skeleton weed in Australia.

IV INTERNATIONAL BIOSAFETY PROTOCOL

Genetically modified organisms (GMOs) were first developed and released for trial in the early 1980s. Since then, many GMOs have been released for field trial in the United States, Japan, China, and countries belonging to the European Community (EC). More than 2,500 cases of experimental release of genetically engineered organisms were known in the US by May 1995. Fifteen GMPs and four GMMs have already received marketing permission. Many GMOs are waiting for permission. A legally binding international biosafety protocol is desirable for a number of reasons. The most important of these are: (i) it will ensure that countries do not neglect biosafety requirements in order to promote their biotechnology industries, (ii) it will prevent the testing and use of GMOs in developing countries with weak or no regulations, and (iii) it will help developing countries to formulate national biosafety regulations. The need for international biosafety guidelines was accepted at the 1992 Rio Earth Summit. After initial opposition from some developed countries the need for an international protocol was widely accepted. Due to some differences between developed countries and developing countries about the scope of the protocol, no comprehensive international protocol on the use of GMOs has yet emerged. In the absence of an international protocol, the use of GMOs will need to be regulated by national biosafety regulations.

V INTERNATIONAL ACTIVITIES IN BIOSAFETY CAPACITY BUILDING

The sophisticated techniques used in modern biotechnology offer innovative solutions to today's pressing need for sustainable development in agricultural, environmental, and energy applications. Recently, countries throughout the world, including many developing countries, have shown great interest in establishing research and development programs in genetic engineering. The rate of development and the level of success are dependent not only on acquiring the scientific and technical capabilities of resident scientists, but also on having a supportive infrastructure and an accepting environment in which to introduce and reap the benefits of biotechnologies. A key component in the formulation of a "biotechnology-accepting" environment is the establishment of a biosafety regulatory oversight infrastructure (Brenner, 1995; Virgin *et al* 1999). Apart from farmers, industry, and **non-governmental organizations** (NGO)s, the **Convention on Biological Diversity and Agenda 21** program extensions will also apply pressure on developing countries to adopt the technology. To facilitate the safe transfer of technologies and fully benefit from genetic resources and the rapid advances in modern biotechnology, developing countries need to formulate biosafety regulations. Concomitantly, the countries must acquire the capacity to implement these regulations via scientifically based, environmental impact assessments. To achieve these objectives, several things are needed, including training at all levels to address shortage of human resources, access to information, and international expertise.

The **United Nations Environmental Program** (UNEP) definition of capacity building is "the strengthening and/or development of human resources and institutional capacities. It involves the transfer of know-how, the development of appropriate facilities, and training in sciences related to safety in biotechnology and in the use of risk-assessment and risk-management" (UNEP, 1996). Even though much of the support to deal with many shortcomings has to come from the developing countries themselves, there is a clear need for continuing assistance from international organizations. To support capacity building in biosafety, one can distinguish the following two areas of international activities, both of which should be developed, but not necessarily simultaneously, (i) short-term training in risk assessment and risk management and (ii) sustainable strengthening of infrastructure and expertise in biotechnology. There are today a substantial number of international organizations that are directly or indirectly providing assistance to developing countries within these fields of biosafety capacity building (Table 41.1).

Table 41.1 *Some international organizations involved in biosafety activities*

Intergovernmental:
- Organization for Economic Co-operation and Development (OECD)
- Commission for the European Communities (EC)
- Consultative Group in International Agricultural Research (CGIAR)
- Technical Center for Agricultural and Rural Co-operation (CTA)

UN Organizations:
- United Nations Environmental Program (UNEP)
- United Nations Industrial Development Organization (UNIDO)
- World Health Organization (WHO)
- Food and Agriculture Organization (FAO)
- Parties to the Convention on Biological Diversity

Independent/non-profit:
- Biotechnology Advisory Commission (BAC)
- International Service for the Acquisition of Agri-biotech Applications (ISAAA)
- International Service for National Agricultural Research (ISNAR)
- International Academy of the Environment, Geneva (IAE)
- Agriculture Biotechnology for Sustainable Productivity (ABSP)

Donor organizations:
- Netherlands Ministry of Foreign Affairs (DGIS)
- Royal Danish Ministry for Foreign Affairs/ Department for International Development
- Deutsche Gesellschaft fur Technische Zusammenarbeit (GTZ)

... *Continued*

>   Rockefeller Foundation
>   Swedish International Development Cooperation Agency (SIDA)
>   The World Bank
>   United States Agency for International Development (USAID)
>
> Environmental NGOs:
>   Genetic Resource Action International (GRAIN)
>   Greenpeace International
>   Rural Advancement Fund International (RAFI)
>   Union of concerned Scientist (UCS)
>   Third World Network (TWN)
>   The World Conservation Union (IUCN)
>   World Wildlife Fund for Nature (WWF)

Source: Virgin et al. (1999) Shantharam and Montgomery (Eds.), Science Publishers, Inc.

VI BIOSAFETY REGULATIONS IN INDIA

In India, biosafety regulations were first prepared in 1990. These were revised in 1994. The regulations cover both the research and field trials of GMOs imported or locally developed. The regulations are implemented by three committees. These are as follows:

1. Institutional Biosafety Committee (IBSC): All institutions interested in undertaking research activities involving genetic manipulation of living forms are required to set up IBSCs. IBSC includes representatives from the concerned institution, the Department of Biotechnology (DBT) and a doctor. At present, about 69 research institutes have IBSCs. The researchers are required to inform the institute's IBSC of any experiment with biosafety implications. In addition to looking after the biosafety aspects of experiments, the IBSC is also responsible for the training of researchers on biosafety and for monitoring the health of researchers.

2. Review Committee on Genetic Manipulation (RCGM): The RCGM functions under the DBT and monitors all ongoing research with biosafety implications. The committee also issues clearance for import and export of genetic material required for research and training purposes.

3. Genetic Engineering Approval Committee (GEAC): GEAC functions under the Ministry of Environment and is responsible for biosafety in the cases of: (a) field trials and large-scale (commercial) release of GMOs; (b) production, sale, import, or use of genetically engineered products; (c) import, export, transport, handling, and use of GMOs.

In addition to these, there are also committees at the state and district level.

According to the perceived risks involved, genetic experiments are classified into three catagories. The experiments considered to have the highest risk include the use of toxin genes, cloning of genes for vaccine production, transfer of antibiotic resistant genes to pathogenic organisms not normally carrying such resistance, manipulation involving plant and animal viruses, and gene transfer to whole plants and animals. These experiments require clearance by the highest expert committee on a case-by-case basis.

A complete list of research with biosafety implications being carried out in India is not available publicly at present. It is therefore not possible to make definite comments on the potential risks associated with research on GMOs in India. In India most of the work on GMOs is in the early stages. Some of them include the following: (i) tagging and transfer of blight resistant genes from wild to cultivated species of rice; (ii) cloning of genes responsible for resistance to gall midges in rice; (iii) development of transgenic rice with Bt genes; (iv) development of transgenic rice with fungicide tolerance; (v) development of transgenic rice with herbicide resistance; (vi) development of transgenic chickpea with Bt genes; (vii) development of transgenic cotton with Bt genes.

While locally developed GMOs are unlikely to be ready for field trial, more imported transgenics may enter India in the near future.

VII BIOLOGICAL WEAPONS

Though the term **biological warfare**, appears very recent, historically, it existed even in early civilizations. The ancient Romans, and others before them, threw carrion into wells to poison their adversaries' drinking water. In the 14th century, the Tatars catapulted the bodies of bubonic plague victims over the city walls of Kaffa, a Black Sea port that served as a gateway to the Silk Road trade route. People inside the city soon came down with the disease, which suggests that the maneuver may have worked. But its effect became more severe when some of the city's inhabitants escaped in sailing ships, which happened to be infested with rats, carrying fleas infected with the causative agent of plague (the bacterium *Yersinia pestis*). The escaping ships entered various Italian ports that subsequently served as foci for the spread of the disease. After about three years, the bubonic plague (Black Death) raged northward, wiping out nearly a third of Western Europe. It was not until the 19th century that the microbial basis for the infectious disease was understood.

One of the first illnesses to be explained by the new germ theory was anthrax (*Bacillus anthracis*), an infectious disease common to sheep and cattle. Anthrax is only weakly communicable in humans and rarely causes disease, unless the bacterium comes into contact with the bloodstream through a wound (**cutaneous anthrax**), or is ingested through contaminated meat (**intestinal anthrax**) and it can enter through breathing causing **inhalation anthrax** in humans. Inhalation anthrax is a very deadly disease in humans and has 80% mortality. Because of this, it became the agent of choice for most biological warfare programs. It produces a large number of variants and great quantities of spores can be readily prepared from liquid cultures. Once desiccated and stabilized, the hardy spores have a long shelf life (more than 100 years) and are well suited to weaponization in a device that can deliver a widespread aerosol.

Table 41.2 *Biological agents that could be used in a biowarfare weapons*

Bacterial disease	Causative agent
Anthrax	*Bacillus anthraci*
Brucellosis	*Brucella suis, B. melitensis, B. abortus*
Glanders	*Burkholeria mallei, B. pseudomallei*
Plague	*Yersinia pestis*

Continues...

... *Continued*

Q fever	*Coxiella burnetis*
Typhus	*Rickettsia prowazeki*
Viral disease	**Causative agent**
Smallpox	*Variola major*
Viral encephalitis	*Venezuelan equine, eastern equine, tick bornEncephalitis virus*
African hemorrhagic fever	*Ebola, Marburg, Congo-Crimean virus*
South American hemorrhagic fever	*Junin, Machupo, Sabia, Flexal, yellow fever virus*
Fungal diseases (or crops)	**Causative agent**
Rice blast	*Magnaporthe grisea*
Rye stem rust	*Puccinia graminis*
Wheat stem rust	*Puccinia graminis*
Biological toxin	**Source**
Botulinum toxin	*Clostridium botulinum*
Enterotoxin B	*Staphylococcus aureus*
Epsilon toxin	*Clostridium perfringens*
Ricin	*Ricinus communis* (Castor bean)
Shiga toxin	*Shigella dysenteriae, S. flexneri*

Source: Block SM, 2001

The first World War (WWI) saw one of the first attempts to use anthrax during warfare, directed ineffectively against animal populations. Coincidentally, WWI also overlapped with a deadly outbreak of influenza, the Great Pandemic of 1918, which eventually killed more people than the Great War itself. Biological agents that could be used in a weapon include various bacteria, viruses, fungi, and toxins (Table 41.2). Other than these conventional biowarfare organisms, what worries people most nowadays is the impact of modern biotechnology. For better or worse, the world is in the midst of a stunning revolution in the life sciences. Scientists have already determined the complete genomic sequences of more than 30 microbes and even more viruses. The DNA code for the cholera pathogen (*Vibrio cholerae*) was recently published, and the genomes of more than 100 other microorganisms are now being sequenced. Although the new technology critical for answering fundamental and practical questions in biology and medicine, is what about "**black biology**"? Could biotechnology be used to produce a new generation of biowarfare agents with unprecedented power to destroy? Or is just hype? No one can say for sure, at least for the time being.

VIII SUGGESTED READING

Block SM (2001) The Growing Threat of Biological Weapons, *American Scientist*, **89**:28–37.

Brenner, C (1995) Technology Transfer: Public and Private Sector Roles, In: J Komen, J Cohen, and SK Lee (Eds), Turning priorities into feasible programs: proceedings of a regional seminar on planning, priorities and policies for agricultural biotechnology in Southeast Asia, 134pp. The Hague/Singapore: Intermediary Biotechnology Service/Nanyang Technological University.

Chopra VL (Ed) (2000) *Plant Breeding Theory and Practice* (Second Edition) Oxford & IBH Publishing Co. Pvt. Ltd.

Drell SD, Sofaer AD, and Wilson GD (1999) *The New Terror: Facing the Threat of Biological and Chemical Weapons*, Standard, Calif,: Hoover Institution Press.

Duesing JH and Goy PA (1996) Assessing the Environmental Impact of Gene Transfer to Wild Relatives, *Biotechnology*, **14**:39–40.

Ellstrand NC, Prentice HC, and Hancock JF (1999) Gene Flow and Introgression from Domestic Plants into Their Wild Relatives, *Ann. Rev. Ecol. System.*, **30**:539–563.

Hoyle T (1996) Taking the Hex off Transgenic Plant Export, *Nature Biotechnology*, **14**:1628.

James C (2000) Global Status of Transgenic Crops: Challenges and Opportunities. In: *Plant Genetic Engineering: Towards the Third Millennium* (Ed) AD Arencibia, New York: Elsevier. pp1–6.

Krimsky S and Wrubel R (1996) *Agricultural Biotechnology and the Environment: Science, Policy and Social Issues*, University of Illinois Press, Urbana, Chicago.

Marvier M (2001) Ecology of Transgenic Crops, *American Scientist,* **89**:160–167.

Miller MC and Powell W (1994) A Commercial View of Biotechnology in Crop Protection, In: *Molecular Biology in Crop Protection*, pp225–245 (Eds) Marshal G and Walters D, Chapman and Hall.

Moffat AS (2000) Can Genetically Modified Crops Go 'Greener'? *Science,* **290**:253–254.

Phifer PR and Wolfenbarger LL (2000) The Ecological Risks and Benefits of Genetically Engineered Plants, *Science,* **290**:2088–2092.

Poppy G (2000) GM crops: Environmental Risks and Non-target Effects, *Tends Plant Sci.*, **5**:4–6.

Shantharam S and Montgomery JF (Eds) (1999) *Biotechnology, Biosafety, and Biodiversity: Scientific and Ethical Issues for Sustainable Development*, Science Publishers, Inc. USA.

Somerville C (2000) The Genetically Modified Organism Conflict, *Plant Physiology*, **123**:1201–1202.

Stam M, Mol JNM and Kooter JM (1997) The Silence of Genes in Transgenic Plants, *Ann. Bot.*, **79**:3–12.

Tucker JB (2000) *Toxic Terror: Assessing Terrorist Use of Chemical and Biological Weapons*, Cambridge:MIT Press.

UNIDO (1995) *Environmental Sound Management of Biotechnology*, Vienna.

Virgin I, Frederick RJ, and Ramachandran S (1999) Biosafety Training Programs and Their Importance in Capacity Building and Technology Assessment, In: Shantharam S and Montgomery JF (Eds) *Biotechnology, Biosafety, and Biodiversity: Scientific and Ethical Issues for sustainable development*, Science Publishers, Inc. USA.

42

Intellectual Property Rights and Socio-Legal Aspects of Biotechnology

I INTRODUCTION

Biotechnology has been defined in many ways, all of which reflects its diverse range of applications. Historically, man has been harnessing biological processes for thousands of years. The ability of yeast to make alcohol in the form of beer, the traditional way of combining the genetic content of animals and plants by cross-breeding and using bacterial cultures to make cheese and yogurt are a few examples. However, it was not till the last century that mankind learned to change purposefully the genetic content of organisms in a direct way without depending on the process of reproduction. The quantum jump in man's ability to manipulate the genetic content of an organism, with the first successful experiment in gene cloning occurred in the 1970s.

As mentioned in the previous chapters, the potential benefits of genetic engineering are numerous. In medicine, the development of human insulin, human growth hormone, tissue plasminogen activator, and numerous vaccines stands out. In agriculture, there is interest in developing plants resistant to heat, cold, frost, insects and pests, herbicides, metals, salt, drought, etc. Some biotechnology products, such as bovine growth hormone, are of dubious benefit. There are many reports of overproduction of milk or meat by the use of genetically engineered animal growth hormones. Many microorganisms that produce a novel or high quantity and quality of secondary products have been genetically engineered and are in the phase of commercialization. Thus the advances in genetic engineering have created the potential for large benefits to society, but they have also meant that many genetically novel organisms are interacting with the environment.

Intellectual Property Rights: One of the most important issues, which has been raised due to the emergence of modern biotechnology, is the legal characterization and treatment of trade related biotechnological processes and products, popularly described as **Intellectual Property**. *Intellectual property rights* (**IPR**) *refer to exclusive authority provided by the government to the first innovator for manufacturing and marketing the innovation, prohibiting other parties from doing so unless licensed by the IPR holder, on payment of fees or royalty*. Intellectual property protection (**IPP**) and **IPR** have been the subject of discussion in recent years. The term, property, refers here to intangible property and includes '*patents*', '*trade secrets*', *copyrights*' and '*trademarks*'. The right to protect the property prohibits others from unauthorized use. Another form of IPR is the development of crop varieties, 'which are protected through '*plant breeders' rights*' (**PBRs**). More recently, **utility patents** for both plant and animal genetic material have been formulated in some countries, so that the patented material can neither be used for further breeding, nor will farmers be allowed to be the patent holder. Patents, the most relevant form of IPRs for pharmaceuticals, are primarily used to protect industrial innovations.

II A COMMERCIAL VIEW OF BIOTECHNOLOGY

Biotechnology is a rapidly expanding sector with potential academic as well as commercial benefits, particularly in the area of medicine and crop improvement, and is of great economic and social importance. Therefore, the principle objective of biotechnology is to produce commercial products for economic gain. The development of techniques such as genetic engineering, bioprocessing, monoclonal antibodies, vaccines, protein engineering, tissue culture, and protoplast fusion in the last decade promised to open up a new world of opportunities for increasing profits in biotechnology. In agriculture a number of areas were identified as having potential for the application of biotechnology. Improving plant breeding through genetic engineering was expected to provide new transgenic crops with insect and disease resistance. The crop protection area would see new biological control agents and herbicide resistant crops. These promised to bring good returns on investment and, as a result, money poured into the industry.

The extent to which scientific and technical advances become translated into new varieties will be influenced by many factors, with industry structure, technical progress, property protection, and the regulatory system being of particular importance. Research and development in the area of agriculture biotechnology is costly and time-consuming. The protection of this new technology, to ensure that the producer makes a profit out of this investment, is of great importance. No industry will initiate high-risk, long-term projects without the guarantee that the results of its research efforts will be legally protected from competitors. At the same time, society at large has a stake in encouraging industrial innovation. A strategy that meets both of these objectives is for the government to grant inventors exclusive rights to the novel products or processes that they develop. The developed countries have evolved legal systems which reward invention of products and processes by the granting of rights of limited monopoly for a defined period of time. These rights take a variety of forms and are generally called intellectual property rights (IPRs). Patents, trademarks, trade secrets, copyrights and plant variety rights are all forms of IPRs. For the seeds industry, the most important forms of IPRs are plant variety rights (PVRs) [also known as plant breeders' rights (PBRs)] and patents. These two systems were designated to be mutually exclusive, with each having a particular type of subject matter. Recent developments in biotechnology have, however, blurred the distinction between them.

Public perception: Public perception of the new technologies is also an important factor to be considered in the introduction of plant varieties produced by biotechnology. Fears about the safety of genetically engineered crops may severely disrupt the introduction of genetically engineered varieties, while concerns about the effects of intellectual property protection on the use of new varieties and genes and therein in further breeding programs, is creating disruption within the industry itself. Many research scientists still regard the patent law as a "jungle" and find many of its concepts strange. In this chapter, a brief outline of the current methods available for the protection of biotechnology developments will be presented and the main issues facing the introduction of biotechnology in crop protection will be outlined.

A. INTELLECTUAL PROPERTY RIGHTS AND PATENTING

Intellectual property rights (IPRs) encompass trade secrets, know-how, rights in design and copyright, confidential information, patents, and plant breeders' rights. All afford protection to varying degrees. The technical developments in agricultural biotechnology have focused on patents and plant breeders' rights for intellectual protection. The legislation, which protects this intellectual property, must stimulate industrial developments and encourage cooperation between plant breeders, the biotechnology industry, farmers and processors. The advent of the new technologies require harmonization between the PBRs and patents, and the system of laws that provide them. Legal rights or patents provide an inventor only a temporary monopoly on the use of an invention, in return for disclosing the knowledge to the others in a specification that is intended to be both comprehensive to, and experimentally reproducible by a person skilled in the art. Others in the society may use the knowledge to develop further inventions and innovations (Bull et al., 1993). The laws are formulated time to time at the national and international levels. Some useful websites for patent literature is given in Table 42.1.

Table 42.1 *Some useful websites for patent literature*

US Patent Office	http://www.uspto.gov
European Patent Office	http://www.epo.co.at
Internet patent news service	http://www.sunsite.unc.edu
Intellectual property & Cyberspace	http://www.fplc.edu
Patent bibliographic data-free access	http://patents.cnidr.org/
IBM patent server	http://patent.womplex.ibm.com
Biotechnology patents	http://www.inform.umd.edu:8080/EdRes/topic/AgrEnv/Biotech
General information about patents	http://www.patents.com
Patents on Internet	http://www.aber.ac.uk
An overview of intellectual property	http://www.aipla.com
Patent resources on Internet	http://members.tripod.com/~vkv/patent0.htm

B. FORMS OF PROTECTION

The IPR is protected by different ways: patents, copyrights, trade secrets and trademarks.

1. Patents

A patent is a property right granted to the inventor by the state authority through legislation, which excludes others from the use or benefit of the protected invention, without the consent of the patentee. Patents are granted to individuals and companies who can lay claim to a new product or manufacturing process, or to an improvement of an existing product or process which was not previously known. Patent laws vary considerably from country to country, although there are ongoing attempts to develop international standards. The grant of a patent gives the patentee a monopoly to make use of or sell the invention for a fixed period of time (usually 17 years in the USA) from the date of filling of the application. In India, the Indian Patent Act (1970) allows 'process patents' but not 'product patents', and the maximum duration of the patent is for five years from the date of grant, and seven years from the date of filing the patent application. In return for this right, the patentee pays a fee to cover the costs of processing the patent and publicly discloses the details of the invention. Eighty per cent of the information these patents contain is not published elsewhere.

The patent consists of three parts – the grant, specifications, and claims. The **grant** is filed at the patent office, and is not published. Generally, it will be a signed document, which is actually the agreement that grants patent right to the inventor. However, the specification and claims are published as a single document, which is made public at a minimum charge from the patent office. The **specification** is a narrative in which the subject matter of the invention and how the invention was carried out is described. On the other hand, the **claim** specifically defines the scope of the invention to be protected by the patent, which the others may not practice.

(a) Conditions for patenting

Patentable articles and conditions related with them are under serious debate. Not all discoveries are granted patents. Discovery cannot be patented because the discovered article is the product of nature. But the process or techniques used to discover nature's product may be granted patents. Therefore, patent laws differentiate between discovery and invention. The European Patent Office (EPO) has given suggestions that the process developed to isolate the products from nature is patentable. If the product is new and does not have previously existing recognition (e.g. microbial metabolites, alcohols, organic acids, vitamins, enzymes, etc.,) it is patentable.

The question of what is patentable under the European Patent Convention (EPC) has to be answered on the basis of articles 52 and 53 EPC which read as follows:

"Article 52"

Patentable inventions: European patents shall be granted for any inventions which are susceptible of industrial application, which are new, and which involve an innovative step.

"Article 53"

Exemptions to patentability
1. Inventions, publication or exploitation of which would be contrary to 'order public' or morality, provided that the exploitation shall not be deemed to be so contrary merely because it is prohibited by law or regulation in some or all of the contracting states;
2. Plant or animal varieties or essentially biological processes for the production of plants or animals; this provision does not apply to microbiological processes or products thereof. Thus in order for a patent application to become a valid patent, it must meet several criteria (Van Dullen, 1992).

Criteria for patenting
1. It must be 'novel'. This means that it must be original as a patent, and indeed new in any published format. In addition, the application is unlikely to be accepted if the applicant

describes it (except in confidence) or, if it is manufactured before the application is submitted, so it is vital that it is kept secret until this time. Otherwise the invention is open to anyone to manufacture it.

2. It must not be obvious. This means that it must not be a predictable improvement of something already in existence or described in published literature. Theoretically if an inventive person who knows all prior art thinks that an idea is an innovative step, then it is not obvious.
3. It must be useful. It must do or be something of practical benefit, rather than being a scientific observation, or a work of art.
4. It must be capable of being industrially reproduced. This criterion is applied very loosely: many chemical patents, for instance, refer to substances that cannot be reproduced in a factory environment.
5. It must not be illegal or immoral. Examples of illegal patents would be ones for mantraps or counterfeiting machinery. However, some countries allow 'illegal patents' if the applicant intends to export the product to countries where they are legal.
6. It must be detailed. The patent must be detailed enough so that someone skilled in the art can reconstruct the invention. This is fundamental and the failure to give sufficient details can lead to refusal of a patent.

Some categories are generally unpatentable. These include computer software and higher life forms (such as the genetically altered mouse), although the latter at least may soon be patentable. Plant patents have evolved alongside the plant variety rights. The US has allowed plant patents since its Plant Protection Act of 1930 and has issued more than 6500 patents. The Japanese have allowed plant patents since 1970. The European Patent Office (EPO) excludes plant varieties from patenting but the EPO board of appeal ruled in 1991 that plant patents may be awarded for useful and inventive genetic modifications, for example, a variety of corn having an additional, useful gene. The first patent issued in this category covered forage crops such as alfalfa (*Medicago sativa*), with increased protein content.

Given the complexity and rapid technical advances in plant biotechnology, there are strong reasons to suppose that the criteria for obtaining patent protection can be met. The types of inventions for which patents have been filed include tissue culture and micropropagation methods, methods for protoplast fusion, techniques for gene insertion, vectors, isolated genes, and gene promoter and terminator sequences. The claims usually extend to plants developed using such methods or containing such genes or parts of genes. Thus several patent applications have claims of the type: *Plants of species A having resistance to insect pest B through genetic transfer of gene C.* This claim is related to a plant, not a plant variety.

(b) Procedure

There is no simple, immediate system for the granting of a patent. The application must be prepared by an expert – normally a patent lawyer – and is organized following a defined pattern (Figure 42.1). Patents are granted after submission of an application, fulfilling certain statutory requirements. When granted, a patent is also published (Patent and Trademark Office, PTO, in USA). Prior to filing the patent application, the inventor must deposit a sample of officially approved material declaring that it is free from dispute of novelty and can be used by others when it becomes legally free (i.e. **utility, novelty,** and **statutory subject matter**). The procedure for filing a patent is an involved, lengthy process and according to many, drafting a good patent claim in one of the most difficult legal tasks. It usually needs collaboration between the inven-

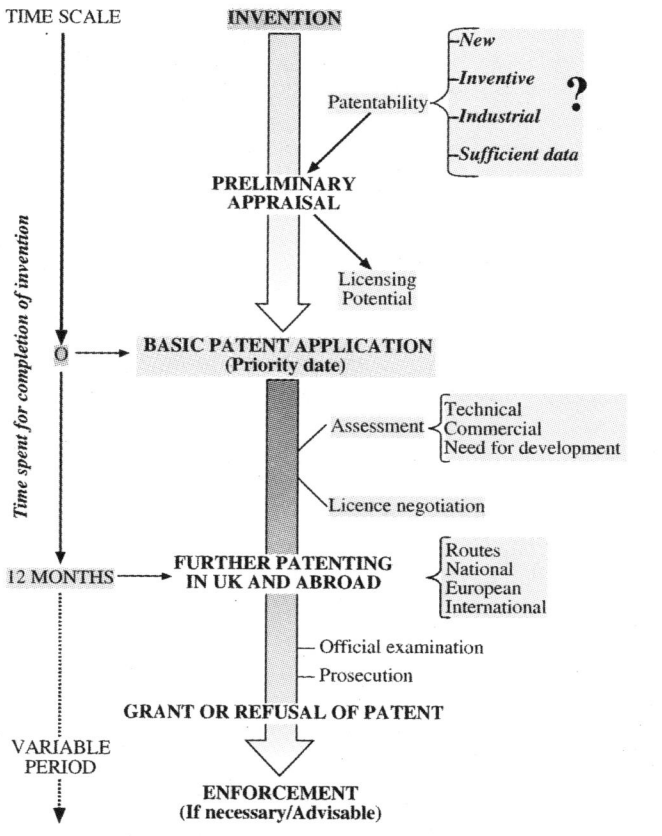

Figure 42.1 *Flow chart of patenting procedure (Redrawn from DBT, WIPO, 1999)*

tors and a patent attorney. An application for patent protection is usually first made in the country of residence or place of business of the applicant. This establishes a priority date, which will be recognized in most of the other countries of the world under the provisions of an international convention known as the Paris Convention. This postpones the major expenses of a foreign patenting program until toward the end of the first year after the initial filing date in the home country. An application for a European patent is on the same level as national applications in other countries. The one year interim period is very valuable both to industry and other organizations which have the problem of assessing the potential industrial applicability of the new research results. The other advantage is that the inventor can publish details of his invention without determent to his patent prospects once his priority date has been established. Thus the invention should be clearly defined and well supported by data by the date in the first application and the foreign filing should take place no later than one year from the first application.

(c) The patent application

To obtain a patent, an application must be filed with the relevant national authority (Patent office) and will be examined for compliance with the legal requirements. The basic procedure for applying patents is regulated by the **Paris Convention for the Protection of Industrial Property** of 1883, to which most countries belong. Separate patent applications are usually necessary in each country where protection is sought but a single application in the European Patent Office can cover a number of European countries up to the point at which rights are granted. This occurs under the **European Patent Convention** (**EPC**). In contrast, the **Patent Cooperation Treaty** (**PCT**) provides, on the basis of a single international application in one language, for an international filing and the search which will be effective in any one of the countries which are party to the treaty. This has worldwide membership. The PCT is administered by the World Intellectual Property Organization (WIPO) in Geneva and the first PCT applications were filed in June 1978.

The invention is defined in the 'claims', which form part of the specifications. Claims are a guide to the scope of the protection conferred by the patent. Patent claims are expressed in the broadest and most general language which the patent attorney can devise to avoid loopholes which competitors can exploit (Crespi, 1988). The patent office will carry out a search of previously published documents including the scientific and patent literature to determine the relevant 'prior art'. Subsequently, the patent application will be examined in light of the search results. This usually involves an argument about the specifications, especially the scope of the claims, and may take considerable time to settle. Even after acceptance by the office, a patent application or granted patent can, in most countries, be opposed by third parties who may raise objections and prior art similar to or additional to those already overcome by the applicant. This is usually termed Opposition and involves an argument between the applicant/patentee and Opposition who have equal status as contending parties. The US patent law does not provide for opposition in this sense but allows a third party to request official re-examination of the patent in light of the prior art which has not already been considered.

(d) Patenting in different countries

The patenting process is inconsistent, varying from country to country. Even though there are some differences between the patent laws in various countries one of the most important involves the first to file principle.

(a) The first to file principle in UK law: Priority in patent law is decided not only by the date of filing an application but also by the content of what is filed, i.e. the application must be a proper disclosure of what is claimed. For example, in the case of Dainippon in Japan, the question arose whether the earlier application by Dainippon was entitled to its date for priority purposes (Crespi, 1972). The application had disclosed the DNA sequence without providing experimental information on the preparation of the material. Asahi's application had an enabling disclosure and claimed that their patent should get precedence. The law concluded that the Dainippon application was not an enabling disclosure. This would seem to encourage the filing of applications for 'inventions' which may or may not be capable of commercial exploitation, in the hope that in the ensuing 12 months, something would turn up to render them patentable. Such applications would have the effect of preventing anyone else who may be better informed from obtaining a patent.

(b) The first to invent principle of US law: The US patent system also recognizes the distinction between bare and enabling disclosure. However, in relation to an inventor's own enabling publications, the US law is more generous than the European one. In deciding the question of priority, the actual date of invention is taken into account, proven from laboratory notebook records. An inventor's own prior publications will not ruin the chances of obtaining a US patent provided the application is filed within a 'grace period' of one year from the publication date.

(c) IPR in India: India does not currently recognize IPRs in agriculture. However, since India became a signatory to the GATT, it is required to provide for the protection of microorganisms and microbiological processes. Plant varieties are to be protected either through patents or through some other **sui generis** system (of his, her, its, or their own kind) to be based on the International Plant Breeders' Rights Convention for the Protection of New Plant Varieties (the UPOV convention). In response to the UPOV convention, the Indian government accordingly drafted a Plant Variety Act (PVA) in 1993. The provisions of the PVA are similar to those of the 1978 UPOV convention and its 1991 revisions and are:

(d) Financial gain from patenting

Once a patent has been granted, the next step is to begin to retrieve some of the investment in the biotechnology and cost of protection. This can be achieved in a number of ways but the most common is through licensing agreements. These can be exclusive or non-exclusive and in the 1990s this trading began in earnest. For example, Bio-Rad was given an exclusive license for manufacturing a 'gene gun' developed by DuPont together with associated consumables. Asgrow Seed purchased non-exclusive rights to Mogen International's fungal resistance research in horticultural crops. The two companies will jointly develop fungal resistant plant varieties, with Mogen receiving a royalty from Asgrow on the sales of improved varieties.

If a patent holder refuses to permit the use of the invention by another person, when this person offers to pay a reasonable compensation and thus to furnish security, that person can be given a **compulsory license** if permission is indispensable in the public interest.

2. Copyrights

Copyright protection is only a form of expression of ideas. It protects artistic expressions for about 50 years. One of the best examples of copyrights is the authored and cited books or audio- and video-cassettes. The authors, editors, publishers or both publisher and author/editor can have copyrights. The copyright laws provide that the materials of the book cannot be reprinted or reproduced without written permission from the copyright holders. Copyright rules may be modified from time to time as has been done to allow copy rightability to computer software. Biotechnological materials subject to copyright include databases of DNA sequence or any published forms, photographs, etc. The protection to copyright is limited, because although one may not be allowed to photocopy the present book on biotechnology due to copyright, the ideas given in the book can be used for any purpose.

3. Trade secrets (protection of undisclosed information, know-how)

The private proprietary information that benefits the owners is called trade secret. It covers all forms from process to product yield and protects undisclosed information like recipes (e.g. Coca Cola brand's syrup formula). Generally, renewable after every 7 years, a patent expires in 10–20 years, but under the law of trade secrets, a company will have no obligation to reveal the trade secrets. If the patent becomes public before the granted period, the intellectual is paid compensation and unauthorized users are punished by the court of law. However, if a trade secret becomes public knowledge by independent discovery or other means, is in no longer protectable. The trade secrets in the area of biotechnology may comprise hybridization conditions, cell lines, processing, designing, consumer's list, etc.

4. Trade mark

This is an identification symbol (e.g. ™), which is used in the course of trade to enable the public to distinguish a trader's (manufacturers or merchants) goods from similar goods of other traders. The public makes use of these trademarks in order to choose whose they will buy. If they are satisfied with the purchase, they can simply repeat their order by using the trademark, for example, KODAK for photography goods, IBM and Apple for computers, laboratory equipment (thermocycler, centrifuges etc), certain vectors used in molecular cloning, etc.

5. Other forms of protection

(i) Geographical indications, (ii) Industrial designs and (iii) Layout designs (Topographics) of integrated ciruits.

C. CHOICE OF IPP

Out of the four IPRs discussed above, selection of the most appropriate mode is a business judgement based on several factors which differ from case to case. These include the following: (1) pace of technological development: if it is rapid then a trade secret approach may be preferable to patenting; (2) associated costs: if the cost of secrecy exceeds the cost of obtaining a patent, trade secrets will be a better option; (3) security considerations: it may be impossible to prevent disclosure of a trade secret, so one prefers other forms; (4) need to show patents: patents may have to be shown to investors as a measure of success. The choice of protection also depends on the nature of the subject matter that needs to be protected. Patent information can be obtained on Internet websites (Table 42.2).

Table 42.2 *Patent information on Internet: some important websites*

Subject	Prominent patents related websites
Biotechnology patents: a service of the NAL, USDA	http://www.nal.usda.gov/bic/Biotech_Patents/
Patent and Know-how Information Division, National Informatics Center, India. Provides online search to database, EPIDOS	http://pk2id.delhi.nic.in
AIDS Patent Project	http://patents.uspto.gov
STO Patent Search System	http://sunsite.unc.edu/patents/intropat.html
EDS Shadow Ptatent Office	http://www.spo.eds.com
Community of Science US Patent Citation Database	http://cos.gdb.org/repos/pat/pat–intro.html
Gene Therapy Patent Library	http://www.-neca.com/~breffni/genetitle.txt
Derwent Scientific and Patent Searching Services	http://www.derwent.com
CNIDR/USPTO Patent Abstracts Searching System	http://patents.cnidr.org/access.html
Chemical Abstracts Service (CAS/STN)	http://www.cas/
IBM Patent Server	http://patent.womplex.ibm.com
The Dialog Corporation	http://www.dialog.com
MicroPatent	http://www.micropat.com

Source: DBT-BPFC, 1999.

III PATENTING BIOLOGICAL MATERIAL

Patents are granted on novel, non-obvious, and useful innovations. Knowledge already in the public domain (e.g. Ayurvedic formulations) is not considered novel and hence, if the innovation is a mere discovery (e.g. new species) and does not involve an inventive step, it is also not patentable. Further, just the knowledge of using a plant or a mixture of herbal extracts to cure disease is not patentable in India. However, screening, isolation of active ingredients, and demonstrable market potential may render the invention patentable.

A. PATENTING OF LIVE FORMS

Initially, there was strong resistance to the patenting of living organisms. However, now this has changed due to strong pressure from industrialized countries. The decision to patent life forms was initially influenced by the patent given to a bacterium belonging to *Pseudomonas* filed by Chakrabarty from USA in 1980. "A bacterium from the genus *Pseudomonas* containing at least two energy generating plasmids, each of said providing a separate hydrogen degradative pathway". The claim for patenting was that the bacterium was more effective in treating oil spills by manipulating natural *Pseudomonas*. The argument against awarding the patent was whether something already existing in nature could be patented. Patent law distinguishes discovery from invention and does not allow patent for a mere discovery. However, if a substance needs to be isolated from its surroundings and a process is developed to recover it from nature, then the process (**Process inventions**) may be patentable. Furthermore, if the substance is characterized and is 'new' having 'no previously recognized existence', then the substance per se (**Product inventions**) may also be patentable. Biological molecules such as antibiotics and other microbial metabolites, preparation of pure proteins, vitamin B12 by sophisticated techniques or by recombinant DNA technology or newly isolated microorganisms are examples of such patentable substances. Microorganisms such as *E. coli*, in which human growth hormone, human tissue plasminogen activator (t–PA), etc., have been recognized for patents in USA. A protein may be modified by altering the gene sequences responsible for its synthesis. This modification may or may not amount to infringement of a protein patent.

B. PATENTING OF GENES AND DNA SEQUENCES

Nucleotide sequences or genes that are artificially synthesized normally fulfill the requirements of patents. If the protein of an artificial gene and the organism into which the gene is inserted are also novel having desirable attributes, the patent may be extended to the protein and the organism having the artificial genes. The only concern about gene patenting is about naturally occurring useful genes, because they do not fulfill the requirement of novelty since they are just discovered and are not an invention. In the UK, the courts recently held a discussion on naturally occurring gene sequences and decided to not allow patents on them. In contrast, in the USA and some other developed countries, patents will be allowed in future for genes isolated and cloned from nature. An example of such a patent granted involved cloning of a DNA fragment, which originated due to mutation in a microorganism and imparted resistance to the herbicide 'glyphosate'. This gene was patented by Calgene Inc., USA, in terms of a DNA sequence containing the relevant structural gene. In 1992, the question of patenting of isolated genes or DNA and cDNA sequences once again come up for discussion in the USA. The main issue for discussion was whether these sequences were '**useful**' as required in the patent statute (35 USC 101). This requirement (Brenner vs Manson, US Supreme Court, 148 USPQ 689, 1966) was interpreted to mean that if the process or product was to be patentable, there should exist a '*specific benefit in currently available form*'. Subsequently, the word 'useful' was interpreted to mean '*practical utility*'. Despite the above requirement, isolated genes, vectors, and transformed cells expressing the hormone 'angiogenesis factor (AGF)', which increases vascularization, has been allowed a patent. AGF (in 1985), did not really have practical utility for therapy or for diagnostic tests. Patenting of gene sequences whether they have practical utility or not is necessary to encourage research leading to isolation of genes, which if patented, will be available to other researchers and if not patented, will be kept in secret.

C. PLANT BREEDERS' RIGHTS AND FARMERS' RIGHTS

In most industrialized countries, crop varieties are subject to intellectual property rights in the form of **PBR** (**Plant Breeders' Rights**). But in India, plant varieties are not generally protected through PBR. In the UK, 1964 marked the introduction of plant breeder's rights, which gave breeders a major incentive to invest in new varieties and allowed plant breeding companies to expand and prosper. It also benefited farmers and consumers through the introduction of new varieties with better yields and disease resistance. To qualify for PBRs, the new variety must be distinct, uniform, and stable (**DUS**). PBRs are protected under the **UPOV** convention (***International Union for the Protection of New Varieties and Seeds Act 1964***). Plant breeders – conventional ones and users of the modern biotechnologies, take the risk that their new variety stands only a slim chance of success. A new variety passes from the seven or so years of a breeding program to exhaustive statutory evaluation tests before finally being approved by the authorities for sale as a certified seed to the farmer. The new plant varieties are protected, allowing the breeder or their agent to sell the seeds. However, there are a number of important exemptions, e.g. other breeders could use the protected material in further breeding programs. Farmers could save the seed from the protected crop for re-sowing (**plantback**). In 1991, the UK signed a revised version of the UPOV convention, which maintains the monopoly for the original breeder and provides greater protection of protected varieties. In this patent law, the 1991 convention put certain acts beyond the breeders. A right in the original convention that farmers could save and re-use seed has been delegated to the individual governments to decide in the future. The revised treaty no longer prohibits dual protection, the use of both patents and breeder's rights to protect intellectual property on plants. The present IPR systems, however, do not promote, nor were they designed to promote, the protection of biodiversity of whole ecosystems or unmodified plants.

Dependent varieties: In the 1991 convention, the topic of ***dependent varieties*** was addressed. This arose due to fear that the new technologies would result in plagiarism of new varieties with genetic engineers able to insert genes to confer traits like, for example, insect and virus resistance. Articles 14 (1)–(4) cover varieties which are essentially derived from a protected variety or are not clearly distinguishable therefrom, or whose production involves the repeated use of the protected variety. In the 1991 test, an essentially derived variety is a variety that is predominantly derived, but clearly distinguishable from, the 'initial variety' and conforms to genotype or a combination of genotypes of that variety. The derived variety retains almost the totality of the initial variety's genotype yet is distinguishable from that variety by a limited number of characters (typically one). This could arise via genetic engineering where a gene is inserted into an existing protected variety. The repeated use of a protected variety in the production of a third variety (e.g. an F1 hybrid) requires the authorization of the holder of the breeders rights. Classical breeding methods are not regarded as giving rise to derived varieties.

Farm saved seed: The rate at which farmers saved seed from one harvest for planting to give the next commercial crop, was estimated in 1990 for small grain crops at 89% in Australia, 85% in Spain, 70% in Canada and the US, 45% in France, and 30% in the UK. These constitute very large losses in financial terms for the plant breeders and the seed trade. Under the 1991 convention, the person who saves seeds is infringing the Plant Breeder's Rights. In the US, the Plant Varieties Act 1970, amended in 1980, states that 'a person does not infringe on the breeder's rights by saving seeds produced by him from seeds obtained or descended from seeds obtained by authority of the owner of the variety for seedling purposes.'

Patents verses plant breeders' rights: There are many differences between the two main forms of intellectual property protection (patent and PBR) ranging from cost to legal complexity. Patents offer greater protection than PBRs since the rights extend (for the lifetime of the patent) to all subsequent generations of the plant. New varieties made by crossing a patented parent with another parent could be eligible for both patents and PBRs. Breeders see patent rights as potentially inhibiting the free exchange and use of genetic material which breeder's rights have sought to maintain. Companies claim that it is not their intention to restrict access to germplasm but that if a breeder subsequently markets a variety, which includes the patented trait, e.g. insect resistance, he should have to obtain the patent holder's permission. Enforcing patent rights through successive generations of breeding could be difficult and suggestions have been made that patent rights be limited to the plant varieties initially used. The breeder would pay a reasonable royalty and then be free to use the patented plant in subsequent breeding programs without repetitive royalties. The need to recoup investment in research and development from a single payment could make the charge prohibitive and provide a disincentive to innovation.

In India, new varieties are developed at state agriculture universities and at state department of agriculture. The seeds of new crop varieties are given to farmers and to the private companies without royalty. This service encouraged the farmers to adapt new varieties in the past, leading to the green revolution. If multinational companies are allowed to invest in India for seed production and development, they will have monopoly through PBR. At present conditions, this may not be in the interests of Indian agriculture at least for several years to come. It is possible to introduce some sort of PBR in India for crop varieties to encourage private companies to enter plant improvement programs.

D. FARMERS' RIGHTS

Farmers' rights are a concept developed and adopted by the FAO as a resolution and endorsed by all member countries. The understanding of farmers' rights is that farmers and rural communities have greatly contributed to the generation, conservation, exchange, and knowledge of genetic diversity for centuries. Therefore, the world community is obliged to help them to carry out the task and also help them in utilizing the biodiversity available with them. No IPR system or other mechanism exists for compensation or reward to the farmers for their contributions.

Efforts, therefore, have been made to give a concrete legal form to these farmers. In this regard, in a document published in 1989 (Rural Advancement Fund International 'Farmers rights', **RAFI**), it is suggested that a tax should be payable on commercialized biological material derived from developing countries, but used in developed countries. The idea of informal innovation eligible for farmers' rights is based on the following criteria. (i) Farmer's land races, plants (used as medicinal and other biological products), processes, and innovations are the result of human ingenuity and represent immediate inventions; (ii) most of these inventions arise from informal efforts, which are purposeful and creative, (ii) collectors gather present-day improved germplasm and the knowledge about its breeding and discovery. These aspects in relation to farmers' rights are not covered under TRIPS within the GATT negotiations.

An international debate on biotechnology and farmers' rights: "Opportunities and Threats for Small Scale Farmers in Developing Countries" was organized in 1991 at Amsterdam, the Netherlands. Some of the recommendations made in this debate are: (1) Plant breeders' rights should be maintained. (2) Patent legislation should not be extended to genetic material of plants and animals and there should be legal sanctions prescribed in this connection. (3) An "***International Gene Fund***" should be established, and used to compensate the local communities in the Third World including farmers and farm communities for their past and present contributions. Payments for this purpose must be met by the member countries and the amount will be based on the degree to which the users of the genetic material profit from it. The United Nations Conference on Environment and Development (**UNCED**) held in 1992 took further measures to establish an international fund, which gave substance to the principle of 'farmers' rights' and also to the genetic resource as a *'heritage of mankind'*, embodied in the "**FAO Undertaking**" formulated in 1990.

IV INTERNATIONAL CONVENTIONS AND COOPERATION

A. BIOPIRACY

Many industrialists in the developed countries argue that developing a new drug takes 10–12 years and nearly US $400 million. They advocate strong and longer IPR protection to their investment, as drug manufacturing is easy to copy and manipulate. This argument, however, overlooks the valuable diversity of crop plants, medicinal plants, and the people's knowledge that the developing countries have been freely providing to the developed world. The developed world, has made huge profits from the resultant new drugs, crops, and cosmetics, by patenting these products and selling them at exorbitantly high prices. Since IPRS protect only commercial inventions, day-to-day domestic use of bioresources are not prevented. Our folklore and ayurvedic *vaidyas* or pharmacies can continue to sell their powders and syrups unabated, but they cannot today claim a share in the profits generated from a derived drug and other valuable information. Such injustice in sharing of benefits led to a growing discontent and eventually gave birth to the International Convention on Biological Diversity (**CBD**) 1992.

CBD, signed by 170 countries, reaffirms the sovereign rights of the nations. Article 15 requires member countries to transfer genetic resources on the basis of prior consent and mutually agreed terms. These conditions, however, do not apply to genetic resources obtained prior to the 1992 CBD convention. Thus, biodiversity rich southern countries can set terms of benefit sharing only in future transactions, if any. Further, biodiversity rich countries can dictate terms only if the genetic resource is endemic (see Chapter 39).

Article 8J of the CBD requires member nations to respect and preserve the knowledge of local people and apply it only with their approval, involvement, and equitable sharing of benefits. Article 19c requires member nations to protect customary usage of biological resources. Article 16(5) mandates nations to ensure that IPRS are supportive of and do not run counter to the CBD objectives. Unfortunately, the approvals apply only to undisclosed information. Much of the traditional knowledge is already in the public domain – in daily use or in the form of a computerized database with quick search facilities, often housed in developed countries. Industries access this information without any consultation or benefit sharing with the original contributors.

B. INTERNATIONAL CONVENTIONS

Patent law has a long tradition of international co-operation to solve problems, which are not confined to one or a few countries. In relation to intellectual properties dealing with biological material, international conventions, do not allow patents for products or processes dealing with alleviation of human diseases: (i) related to all kinds of surgery, (ii) the use of drugs, antibiotics, or vaccines for any form of diagnosis, prevention or cure of diseases, (iii) artificial insemination, and (iv) *in vitro* fertilization and embryo transfer, etc. In case of plants, live plants (not transgenic plants), naturally occurring microorganisms, micropropagation, and all kinds of tissue and organ culture techniques, biological control of pests or hybrid varieties cannot be patented. But transgenic animals and plants can be. In case of microorganisms, those selected for production of antibiotics, amino acids, enzymes, alcohol, etc., cannot be patented except in certain situations, protection for microorganisms modified for commercial use may be allowed by contracts involving transfer of technology through restrictions on further transfer to third parties (i.e. trade secrets).

C. INTERNATIONAL COOPERATION

International cooperation is essential for solving problems related to patents, which are not confined to one or a few countries. In an international convention called "**Paris Convention of 1998**", the basic principles of equal treatment for domestic and foreign inventors were established. The Paris Convention now has more than 100 member states. The Convention allows inventors to claim international priority by filling of a patent application initially in one of the member-states and subsequently claiming in others. Where inventors working in different countries seek to patent the same invention, the Convention allows an international priority to be claimed based on the filing of a patent application initially in one member state and subsequently in others. The notion of priority date obtained in this way is very important in patent law because for almost all countries, the party with the earliest date wins the contest, subject to certain provisos.

D. THE WORLD INTELLECTUAL PROPERTY ORGANIZATION

The main instrument of international collaboration in these matters is the World Intellectual Property Organization (**WIPO**), based in Geneva. The WIPO is one of the specialized agencies of the United Nations. It administers (but does not enforce) all the conventions. WIPO asks the member states to ratify a convention and to introduce the agreed basic principles into their national laws. It provides the following intellectual property rights: (i) Literary, artistic, and scientific works, performance of artists, phonograms, broadcasts, innovations in all fields of human endeavor, scientific discoveries, trademarks, service marks and commercial names, industrial designs, protection against unfair competition and all other rights resulting from intellectual expression. (ii) Intellectual property is protected by and governed by appropriate national legislation. The national legislation specifically describes the inventions, which are the subject matter of protection and those which are excluded from protection (e.g. those that are contrary to law and injurious to public health). The European Patent Convention (**EPC**) of 1973, another international convention, began operation in 1978 and has 14 member states. EPC was the first to introduce patent statutes for biotechnology inventions. These include (i) need of 'culture collections' as patent depositories for the placement of microorganisms referred to in patent applications. This is because the complete description of living material is difficult to give and may not be fully reproducible. (ii) Exclusion of certain plant related inventions from the list of those which can be patented e.g. plant and animal varieties bred through classical methods.

The major study of international patent protection for biotechnology was published in 1985 by the ***Organization for Economic Cooperation and Development*** (**OECD**). In 1988, taking EPC as a model, the European Commission (EC) formulated a ***European Commission Directive***, to help the member states to modify their national laws to have uniform laws in European countries. This newly formulated law says, (i) an invention should not be refused to be given protection simply because it involves living matter, and (ii) the scope of the patent should extend to all progeny produced by multiplication of parental material, provided it retains the characteristics of this patented material. It also deleted the provisions of 'exhaustion of rights' and the 'experimental use' granted by the patent laws. Thus if this directive is accepted patented living forms will neither be allowed multiplication by the purchaser, nor will they be allowed to be used for experiments without paying royalty being paid for the patent. (iii) The microorganisms to be patented will have to be deposited in culture collections, officially recognized under the treaty as 'International Depository Authority' (IDAs). Any IDA in any member state can be selected by the patent applicant for the deposit of the relevant biological material, and this deposit will suffice for all member states in which the applicant files for patent protection.

E. GENERAL AGREEMENT OF TARIFFS AND TRADE (GATT) AND TRADE RELATED IPRS (TRIPS)

Scope, application procedure, and period of IPR protection varies across countries. Further, protection needs to be separately obtained in each country. The General Agreement on Tariffs and Trade (GATT) was framed in 1948 by developed countries to settle the disputes among the countries regarding share of world trade. It is decided based on the tariff rates and quantitative restrictions on imports and exports. Ever since its establishment, GATT has remained a provisional treaty – a sort of contractual agreement. The terms of agreements are modified by GATT from time to time and are placed for consent from countries, which are contracting parties of GATT, but not its members. Consequently, the developed countries could achieve the long time benefits from GATT. In 1988, the US Congress enacted a law 'the Omnibus Trade and Competitiveness Act (OTCA). As a result of this the USA became powerful to investigate the laws related to trade and to check them if they were not beneficial to its interest.

In the seventh round of negotiations, which concluded in 1994, GATT finalized a multilateral treaty signed by 117 member nations, including India. This treaty will be administered by the WTO. A requirement of TRIPs is that member nations provide for, beside other things, plant variety protection by patents or by an effective *sui generis* system or by any combination of the two.

V IMPLICATIONS OF PATENTING

The patent application with accompanying description and the deposited material, will be kept secret initially for a short time, after which the application must be published. However, the application can be withdrawn before publication, if the assessment suggests that the patent may not be granted. The publication gives advantage to an applicant's competitors, who can use this information for developing an improved process or product. Disclosure of the application also makes the deposited material available to the public. This law in Europe has serious drawbacks. Following are some of the serious consequences of the easy availability of the deposited materials to the third parties. (i) The loss of effective control on the material and its uses; (ii) irrevocable loss of the option to revert back to trade secrecy, if patent is not granted; (iii) possibility of circumvention of the patent protection by other competitors of the applicant through genetic modification of the deposited microorganism. However, some efforts are being made to overcome this problem associated with the European Patent Law.

A. BIOSAFETY AND ETHICAL ISSUES

1. Ethics

The patenting of whole organisms, plant, or animal raises a number of issues. From the legal/scientific standpoint, there are questions over how far the requirements for a patent can be applied to a whole organism. To date, most patents have been awarded for a plant or animal species, which has been genetically-modified by the insertion of an additional gene which confers the desired characteristic. The patenting of whole genes has been challenged on the grounds that the genes are naturally occurring entities and the procedures for their insertion into plants are well known and straightforward. Any such addition of a gene will affect only a tiny proportion of the organism; over 99.99% of the genetic make-up will remain unchanged.

In the UK, the 'Council of Bioethics' was set up to address questions raised by research on animals and plants. In 1992, the minister of agriculture commissioned an *ad hoc* study of the ethical issues that may arise from the potential consumption of genetically-modified organisms, their products, and related safety issues. The study was limited to the food use of organisms from genetic modification programs and not the ethics of genetic modification *per se*. The advisory group was to consider likely future trends in the use of genetic modification in food production and concerns which might arise from the consumption of food from such programs, and to recommend how the concerns might be addressed.

Patent Concern, a coalition of 31 consumer, environmental, animal rights, and Third World development organizations, is campaigning against the draft directive on patents put forward by the Commission of European Communities. The directive would allow the patenting of animal and plants that had been genetically engineered. The Patent Concern Coalition has called for a moratorium on patents on plants and animals until socioeconomic and ethical issues have been properly debated (King, 1991).

2. Genetic resources

One argument against biotechnology is that valuable genetic material, often in developing countries, will be destroyed as a result of this search for new genes with unique properties, which will be exploited in developed countries. The issue here is that the genetic material, if patented, is no longer freely available. However, until these genes are isolated they are only available in the original plant, and so little has changed. The resultant new plants will be freely available when the patent expires and a new resource will have been created.

Another argument is that agricultural production will focus on the patented lines leaving the earlier varieties behind and lost for posterity. This ignores the possibility of coexistence and competition on price, which will exist between the patented varieties and the existing free varieties. Large gene banks throughout the world have been set up to preserve the plants for generations. This need has been identified in advance of gene technology.

3. Financial aspects

In 1988, few biotechnology companies were making a profit, according to a report by the Arthur Young, High Technology Group (1988). Also, despite many successes in cloning genes and transforming and regenerating plants, laboratory techniques have been used to introduce a new trait. The pace and direction of innovation is the outcome of many factors, both external and internal, which influence a company. In the case of biotechnology, a number of these influences such as the cost of patents, regulations, the acceptability of the techniques and products, are making themselves felt as constants in the biotechnology industry.

Of considerable importance to the industry is whether or not they can get any financial return on plant biotechnology. The current economics of many companies is that they cannot, the reason being the length of time it takes to introduce a product in the market, the expenditure of meeting regulatory requirements, and the overall level of public acceptance of the products. A further consideration is the level of return from the farmers and/or consumer that is required to meet the cost of developing the technology.

To ensure a fair return on investment, there needs to be a secure structure of intellectual property protection. The industrial argument in favor of patents is an economic one. Many companies claim that they would not invest in research without the protection of a patent guaranteeing financial returns for innovation, as many of them have conducted expensive research before it was clear that they would gain patents on their work (Watts, 1991).

B. INVESTMENT

Uncertainties over patent rights and safety regulations are deterring investment by biotechnology companies. In 1991 venture capital organizations invested £ 63 million in biotechnology, down from £ 101 million in 1989. In Britain, only £ 15 million was invested, a decrease from the figure of £ 35 million in 1989. The US agricultural biotechnology companies increased research and development spending by 5% from fiscal 1990 to fiscal 1991 on the investment by seed and pesticide companies. There was a 2.5% reduction in research and development spending over the same period.

C. VALUE OF THE MARKET

In 1992, Mycogen Corporation received an experiment use permit (EUP) from the US Environmental Protection Agency (EPA) to begin large-scale field testing of its third genetically engineered product, M-Peril bioinsecticides, currently approved by the EPA for commercial sale. The construction of M-Peril involves the transfer and expression of a Bt toxin-genes in *Pseudomonas fluorescens* cells, which are then killed and the dead cells used. M-Peril controls the European corn borer (*Ostrinia nubilalis*), a major pest of maize in mid-western USA. The company believes the market for control of the European corn borer exceeds US $20 millions annually. Only a few of the bioinsecticides have been developed into products and today it is estimated that Bt accounts for 90–95% of the insect biocontrol market, the latter estimated at US$105 million (Rigby, 1991).

VI OPPOSITION TO PATENTS IN BIOTECHNOLOGY

The obligations related to patenting set by the CBD, IPR, and other regimes are being widely opposed through various seemingly conflicting strategies. Environmental and social activists reject patenting of life, citing three reasons – monopolies are socially unjust as they lead to exorbitant prices, monopolies often lead to monocultures, which are unsustainable in the long run, and monopolies over life are unethical, as humans have not created life. However, most governments in the world do not seem to buy these views today and have signed GATT.

A highly vocal challenge to this assumption has come particularly from the Animal Rights and Green movements and their supporters in the political arena and elsewhere. These oppositions were extended to any significant structural change in the agricultural industry which might stem from biotechnology, and especially from the acquisition in the hands of the larger Corporations of monopoly rights on the advances that are being made. These arguments apply to both plants and animals. Regarding animals, in addition to the above, a moral objection is also raised against interference with the assumed right to integrity of the species. These oppositions are mainly targeted against the patenting of these inventions, no less than against the research itself.

VII SUGGESTED READING

Crespi RS (1991) Biotechnology and Intellectual Property, Part 2: Microorgansim Deposit Questions and Agricultural Biotechnology, *TIBTECH*:**9**:151–157.

Crespi RS (1991) Biotechnology and Intellectual Property, Part 1: Patenting in Biotechnology, *TIBTECH*:**9**:117–122.

DBT, Biotechnology Patent Facilitating Cell, Ministry of Science and Technology, New Delhi, In: *Patenting in Biotechnology*, Patenting information on Internet, 1999.

DBT, WIPO, (1999) *National Roving Seminar on Patenting in Biotechnology*, 5 November 1999, CIMAP, CSIR, Lucknow.

Erbisch FH and Maredia KM (2000) *Intellectual Property Rights in Agricultural Biotechnology*. Universities Press.

Gupta PK (1997) *Elements of Biotechnology*, Rastogi Publications, Meerut.

Miller MC and Powell W (1994) A Commercial View of Biotechnology in Crop Protection, *Molecular Biology in Crop Protection*, pp225–245, (Eds) Marshal G and Walters D Chapman and Hall.

Utkarsh G (1999) What is Patenting?, Amruth, February 1999.

Appendices

Appendix-1: Bacterial Media

LIQUID MEDIA

1. YT and LB media and plates

The YT (yeast extract and tryptone) and LB media are two of the most useful media for growing bacteria. Almost all *E. coli* mutants grow on them. These two commonly used media differ slightly in the amount of yeast extract and are probably interchangeable in all cases. Most *E. coli* strains will attain densities of near 1 g/litre of medium when grown on YT. Usually between 20 and 30 ml are poured into 100 × 15 mm plastic Petri plates. In humid weather, these plates, like all others, require 'curing' for a day or two until faint wrinkles appear on the surface. The plates should not be more than two days old for plating phage.

LB medium (Luria–Bertani medium) (per litre)

This is the standard medium for liquid bacteriological culture. Some bacteria will grow more rapidly in a more concentrated medium, termed super LB broth.

To 950 ml of deionized H_2O, add:
Bacto-tryptone	10 g
Bacto-yeast extract	5 g
NaCl	10 g

Super LB broth (per litre)

To 950 ml of deionized H_2O, add:
Bacto-tryptone	20 g
Bacto-yeast extract	10 g
NaCl	5 g
Glucose	2 g

Shake until the solutes have dissolved. Adjust the pH to 7.0 with 5 N NaOH. Adjust the volume of the solution to 1 litre with deionized H_2O. Sterilize by autoclaving for 20 min. at 15 lb/sq. in. on liquid cycle.

For growing bacteria on agar plates, make up bottom agar LB containing 15 g/L of Bacto-agar, autoclave, and pour out while still hot into Petri dishes. These can be stored for some time at 4°C. Bacteria are added by mixture with heated top agar at 42°C (take care not to exceed this temperature), made up of 7 g/L Bacto-agar in LB plates. Add 15 g agar (Difco) per litre.

A plus B Minimal Medium for Plates:

This is a minimal medium to which the desired carbon-source, amino acids, vitamins, and antibiotics may be added.

Flask A (1-litre flask)		Flask B (2-litre flask)	
$Na_2HPO_4.7H_2O$	7 g	Agar (Difco)	15 g
KH_2PO_4	3 g	Distilled water	500 ml
NH_4Cl	1 g		
Na_2SO_4	0.8 g		
Distilled water	500 ml		

a. Autoclave.
b. Mix flasks A and B.
c. Add sugar to 2 g/litre, 10 ml of vitamin B1 at 1 mg/ml, and desired amino acids at 20 µg/ml.

M9 minimal medium (per litre)

To 750 ml of sterile deionized H_2O (≤ 50°C), add:

5X M9 salts	200 ml
Sterile deionized H_2O to 1 litre	
$MgSO_4$	2 ml
1 M $CaCl_2$	0.1 ml

20% solution of appropriate carbon source (e.g. 20% glucose)

If necessary, supplement the M9 medium with stock solutions of the appropriate amino acids.

5X M9 salts is made by dissolving the following salts in deionized H_2O to a final volume of 1 litre:

$Na_2HPO_4.7H_2O$	64 g
KH_2PO_4	15 g
NH_4Cl	5.0 g
NaCl	2.5 g

The salt solution is divided into 200-ml aliquots and sterilized by autoclaving for 15 min. at 15 lb/sq. in. on liquid cycle.

The $MgSO_4$ and $CaCl_2$ solutions should be prepared separately, sterilized by autoclaving, and added after diluting the 5X M9 salts to 1 litre with sterile H_2O. Glucose should be sterilized by filtration before it is added to the diluted M9 salts.

SOC medium

Add the following to 950ml deionized H_2O.

Bacto-tryptone 20g, bacto-yeast extract 5g, NaCl 0.5g. Dissolve by shaking. Then add 10 ml of a 250 mM KCl solution. Adjust the pH to 7.0 with 5 N NaOH. Sterilize by autoclaving. Cool to 60°C then add glucose to get final concentration of 20mM. Just before use, add 5ml of a sterile solution of 2M $MgCl_2$.

MEDIA CONTAINING AGAR OR AGAROSE

Prepare liquid media according to the recipes in the previous section. Just before autoclaving, add one of the following:

Bacto-agar	(for plates)	15 g/litre
Bacto-agar	(for top agar)	7 g/litre
Agarose	(for plates)	15 g/litre
Agarose	(for top agarose)	15 g/litre

Sterilize by autoclaving for 20 min. at 15 lb/sq. in. on liquid cycle. When the medium is removed from the autoclave, swirl it gently to distribute the melted agar or agarose evenly throughout the solution. Allow the medium to cool to 50°C before adding thermolabile substances (e.g. antibiotics). The plates can then be poured directly from the flask; allow about 30–35 ml of medium per 90-mm plate. Remove the air bubbles if any by flaming the surface of the medium before agar or agarose hardens. When the medium has hardened completely, invert the plates and store them at 4°C until needed. The plates should be removed from storage 1–2 hours before they are used.

Phage R17 and P1 Plates

Agar (Difco)	15 g
NaCl	5 g
Bacto-tryptone (Difco)	
8 g Yeast extract (Difco)	1 g
Distilled water	1 litre

Autoclave and when cool add the following sterile solutions:
$CaCl_2$	1 ml of 1 M
Glucose	2.5 ml of 20% (wt/vol)

Phage P1 Top Agar

Bacto-tryptone (Difco)	8 g
NaCl	5 g
Agar (Difco)	6.5 g
Distilled water	1 litre

Just before use, add $CaCl_2$ to make 0.025 M.

Phage T4 Plates

	Bottom Agar	Top Agar
Bacto-agar (Difco)	12g	6g
Bactotryptone (Difco)	13g	10g
NaCl	8g	8g
Na^+-citrate	2g	2g
Glucose	1.3g	3g
Distilled water	1 litre	1 litre

Phage Mu Plates

Bacto-agar (Difco)	10g
Yeast extract (Difco)	5g
NaCl	10g
Agar (Difco)	10g
Distilled water	1 litre

Make the plates 1 mM $CaCl_2$ and 2.5 mM $MgSO_4$ by addition of sterile salts after autoclaving. For top agar, reduce the agar concentration by half.

X-gal (5-bromo-4-chloro-3-indolyl-β-D-galactoside) Plates

An insoluble blue dye is produced when X-gal is cleaved by β-galactosidase. This is a particularly sensitive indicator for low levels of the enzyme.

Shortly before use, dissolve X-gal at 20 mg/ml in N, N-dimethyl formamide. Add this to any desired plate medium just before pouring to make a concentration of 40 μg/ml.

Stab-Agar for Storing Strains

Bacto-tryptone (Difco)	10g
NaCl	8g
Agar (Difco)	6g
Cysteine-HCl	20g
Distilled water	1 litre

Phage Lambda Suspension Medium and Storage Buffer

Tris-HCl, pH 7.5	10 ml, 1M
$MgCl_2$	10 ml, 1M
Distilled water	961 ml

For storage buffer, include gelatin at 0.05%.

Appendix-2: Some Selected Antibiotics[a] and their Mode of Action

Actinomycin D: Binds to DNA and blocks the movement of RNA polymerase (prevents RNA synthesis).

α-Amanitin: Blocks mRNA synthesis by binding preferentially to RNA polymerase II.

Ampicillin: Bactericidal; only kills growing *E. coli*; inhibits cell wall synthesis by inhibiting formation of the peptidoglycan cross-link. Resistance gene is β-lactamase (amp^r); gene product is secreted and hydrolyzes ampicillin. Application: amp^r gene is included on plasmid vectors as a positive selection marker.

Anisomycin: Blocks the peptidyl transferase reaction on ribosomes.

Bleomycin: Inhibits translation in prokaryotes by interfering with ribosome translocation. Resistance gene is the *bla*^{r} gene product. It binds to bleomycin and prevents it from binding to DNA. Application: *bla*^{r} gene is a positive selection marker on plasmids and also used as a marker in eukaryotic cells (zeo).

Chloramphenicol, in ethanol: Bacteriostatic; inhibits protein synthesis by interacting with the 50S ribosomal subunit and inhibiting the peptidyltransferase reaction. Resistance gene is chloramphenicol acetyl transferase (CAT or CM^r); gene product metabolizes chloramphenicol in the presence of acetyl CoA. Application: CAT/CM^r gene is used as a selectable marker, and as a transcriptional reporter gene of promoter activity in eukaryotic cells.

Cycloheximide: Blocks the translocation reaction on ribosomes. Acts only on eukaryotes.

D-Cycloserine, in 0.1 M sodium phosphate buffer, pH 8: Bacteriocidal; only kills growing *E. coli*; inhibits cell wall synthesis by preventing formation of D-alanine from L-alanine and formation of peptide bonds involving D-alanine.

Erythromycin: Blocks the translocation reaction on ribosomes.

Gentamycin: Bacteriocidal; inhibits protein synthesis by binding to the L6 protein of the 50S ribosomal subunit

Hygromycin B: Inhibits protein synthesis in prokaryotes and eukaryotes by interfering with ribosome translocation. Resistance gene is hygromycin-B-phosphotransferase (hph or hug^r); gene product inactivates hygromycin B by phosphorylation. Application: hug^r gene is used as a positive selection marker in eukaryotic cells that are sensitive to hygromycin B.

Kanamycin: Bacteriocidal; inhibits protein synthesis by interfering with ribosome function; inhibits translocation and elicits miscoding. Resistance gene is Neomycin or aminoglycoside phosphotransferase (neo^r); gene product inactivates kanamycin by phosphorylation. Application: neo^r gene is a positive selection marker on plasmids commonly used in eukaryotic molecular genetics.

Kasugamycin: Bacteriocidal; inhibits protein synthesis by altering the methylation of the 16S RNA and thus an altered 30S ribosomal subunit.

Nalidixic acid, pH to 11 with NaOH: Bacteriostatic; inhibits DNA synthesis by inhibiting DNA gyrase.

Penicillin G: Inhibits cell wall synthesis, inhibits final step in the synthesis of the peptidoglycan of the bacterial cell wall. It is very effective only in growing cells. Application: penicillin has a high therapeutic index particularly for treatment of most cases of gonococcal and streptococcal infections.

Puromycin: Causes the premature release of nascent polypeptide chains by its addition to growing chain end.

Rifampicin[b], in methanol: Bacteriostatic; inhibits RNA synthesis by binding to and inhibiting the β subunit of RNA polymerase; rifampicin sensitivity is dominant.

Spectinomycin: Bacteriostatic; inhibits translocation of peptidyl tRNA from the A site to the P site.

Streptomycin: Bacteriostatic; inhibits protein synthesis by binding to the S12 protein of the 30S ribosomal subunit and inhibiting proper translation; streptomycin sensitivity is dominant.

Tetracyclin[b, c] in 70% ethanol: Bacteriostatic; inhibits protein synthesis by preventing binding of aminoacyl tRNA to the ribosome A site. Resistance gene is tet^r; gene product is membrane bound and prevents tetracycline accumulation by an efflux mechanism. Application: tet^r gene is a positive selection marker on plasmids commonly used in eukaryotic molecular genetics.

---

[a]All antibiotics should be stored at 4°C, except tetracycline, which should be stored at –20°C. All antibiotics should be dissolved in sterile-distilled water unless otherwise indicated and be sterilized by filtration through a 0.22-micron filter.

[b]Light sensitive, store stock solution and plates in the dark.

[c]Magnesium ions are antagonists of tetracycline. Use media without magnesium salts for selection of bacteria resistant to tetracycline.

Appendix-3: Buffers and Standard Solutions

Most of the recipes given below are commonly used standard buffers. All volumes should be made up with the purest available water, either double-distilled, distilled-deionized, or reverse osmosis grade water.

Concentrations of acids and bases

Table 2.1. *pKa's of commonly used buffers*

Buffer	Mol. weight	pKa	Bufferring range
Tris[a]	121.1	8.08	7.1–8.9
HEPES[b]	238.3	7.47	7.2–8.2
MOPS[c]	209.3	7.15	6.6–7.8
PIPES[d]	304.3	6.76	6.2–7.3
MES[e]	195.2	6.09	5.4–6.8

[a]Tris(hydroxymethyl)aminomethane.
[b]N-2-hydroxyethylpiperazine-N'-2-ethanesulfonic acid.
[c]3-(N-morpholino)propanesulfonic acid.
[d]Piperazine-N,N'-bis(2-ethanesulfonic acid).
[e]2-(N-morpholino)ethanesulfonic acid.

Acetate buffers

To make 1 litre of a 0.1 M acetate buffer pH 4.0:

Dissolve 13.6 g sodium acetate (CH_3COONa–$3H_2O$) and 22 ml glacial acetic acid in one litre of distilled water.

To make 1 litre of 0.1 M acetate-citrate pH 5.0 or pH 6.0:

Dissolve 8.2 g sodium acetate in one litre of distilled water and adjust pH with 1 M (19.2%) citric acid.

Note that buffers of ammonium acetate, made up with acetic acid and ammonia solution, are useful because the salts can be completely removed by lyophilization.

Borate buffer

To make 1 litre of a 0.1 M borate buffer pH 8.0:

Dissolve 23.33 g of boric acid (H_3BO_3) and 4 g NaOH in one litre distilled water and adjust pH with saturated solution of H_3BO_3 (26.6 g/L).

For 0.2 M borate buffer, pH 8.5, start with double the quantity of H_3BO_3 and NaOH and add less H_3BO_3. Other formulations use sodium tetraborate (borax, $Na_2B_4O_7$) as a starting solution and adjust pH with boric acid.

Carbonate buffer

To make carbonate buffers of various molarities and pH, first make up two stock 1 M solutions of sodium carbonate (Na_2CO_3, 106 g/L) and sodium bicarbonate ($NaHCO_3$, 84 g/L).

To make 1 M carbonate buffer, pH 9.5, take 75 ml of 1 M sodium bicarbonate and raise pH to required value with 1 M sodium carbonate: expect to add about 25 ml. For more dilute carbonate buffers, use appropriate dilutions of stock solutions or add the required volume of water to a 1 M buffer and re-adjust pH if necessary.

Cacodylate buffer (2.0 M)

Dissolve 214 g sodium cacodylate ~$3H_2O$ in 500 ml distilled water and adjust pH to 7.3 with HCl

Glycine buffer

To make 1 litre of a 0.2 M glycine buffer pH 2.8:

Dissolve 5.14 g of glycine in one litre of distilled water and adjust pH with concentrated HCl.

Phage storage buffer (PSB)

This buffer is also known as SM (Storage medium), or simply as phage buffer.

To make 1 litre of PSB: add

5.84 g NaCl
2.03 g $MgCl_2$–$6H_2O$
50 ml 1 M Tris pH 7.4
10 ml 1 % Gelatin

Tris (tris[hydroxymethyl]aminomethane) buffer:

Tris buffers are convenient to make and use but possess several drawbacks. They chelate some ions. They also possess a rather large temperature coefficient, so it is necessary to be alert to the pH change upon cooling or heating a Tris buffer from room temperature. Most often, a given pH refers to a measurement made at room temperature even though the buffer may be at different temperatures. Another drawback of Tris-buffered solutions is that their pH varies considerably with concentration. Usually 1 M stocks are made by dissolving the Tris base in water, stirring in the desired acid, often HCl, until the desired pH is obtained, and then adjusting the volume to give a concentration of 1 M.

Phosphate-buffered saline (PBS):

PBS has several variant compositions, which are all suitable for most protein and antibody work; for functional assays, pay attention to whether Mg^{2+} and Ca^{2+} ions are present in the formula. PBS are frequently used to buffer enzymatic reactions that require a pH between 6 and 8. As with Tris buffers, the pH varies appreciably with concentration. Thus, PBS are conveniently made by diluting stocks of monobasic and dibasic phosphates to the desired concentration, then stirring one into the other until the desired pH is attained. The recipes here are for general purpose PBS pH 7.2 and 0.15 M.

Phosphate-buffered solutions support the growth of microorganisms to densities high enough to generate serious problems from contaminating nucleases, but so low that the contaminating growth cannot be seen by the eye. Hence, caution is necessary. Sodium phosphate stocks of 1 M do not become contaminated with bacterial growth. Potassium phosphate stocks of 0.2 M are near the limit of useful solubility and are best stored frozen. Rapid thawing before use is easily done in a microwave oven.

	Per L of 1X	Per 5 L of 10X
NaCl	8.00 g	400 g
Na_2HPO_4	1.15 g	57.5 g
KCl	0.20 g	10 g
KH_2PO_4	0.20 g	10 g

More complete PBS formulations include $CaCl_2$ and $MgCl_2$ at concentrations of 190 mg/L. These can be added to the above solutions, or, as an alternative, PBS tablets can be purchased to make small volumes of a complete buffer.

20X SSC Buffer for DNA

NaCl	175 g
Na_2-citrate	88 g
H_2O	1 litre

Concentrated HCl, approximately 1 ml, is added to give pH 7.0.

50X TE Buffer for DNA

The TE buffer is 10 mM Tris, 1 mM EDTA, pH 8.0. It is convenient to make the following 50X stock that will be 8.0 when diluted with distilled water.

Tris-base	60.55 g
EDTA	14.61 g
Distilled water	800 ml

Adjust to pH 8.2 with HCl and fill to 1 litre with distilled water.

TAE: Tris-Acetate-EDTA

To make 1 litre of a 50X concentrate stock:

Tris-base	54 g
Boric acid	27.5 g

500 mM EDTA	20 ml
Distilled water	800 ml

Adjust to pH 8.0 with HCl and fill to 1 litre with distilled water.

TBS: Tris-Buffered saline

To make 1 litre of a TBS pH 7.4

NaCl	8.75 g
Tris-base	2.42 g

Adjust pH with concentrated HCl.

2X Freezing medium for cells:

K_2HPO_4	12.6 g
Na^+-citrate	0.9 g
$MgSO_4$-$7H_2O$	0.18 g
$(NH_4)_2SO_4$	1.8 g
KH_2PO_4	3.6 g
Glycerol	88 g

Fill to 1 litre with distilled water and sterilize by autoclaving.

Appendix-4: Molarities of Common Reagents

		Molarity (M)	To make 1 M	Density (g/ml)
2-Mercaptoethanol	100%	14.26 M	70.1 ml/L	1.11
Conc. HCl	38%	12.04 M	83.0 ml/L	1.20
Glacial acetic acid	99%	17.57 M	56.9 ml/L	1.05
Conc. H_2SO_4	96%	35.21 M	28.4 ml/L	1.84
Conc. ammonium hydroxide	38%	15.03 M	66.5 ml/L	0.90±

Appendix-5: Physical characteristics of the nucleotides

Nucleotide	Mol.wt. (g/mol)	λ_{max} (nm)	λ_{min} (nm)	ε_{max} (mM^{-1}cm^{-1})	A_{280}/A_{260}	TLC mobility[a] A	B	C
ATP	507.2	259	227	15.4	0.15	0	6	34
ADP	427.2	259	227	15.4	0.16	0	26	54
AMP	347.2	259	227	15.4	0.16	11	52	65
Adenosine	267.2	260	227	14.9	0.14	–	–	–
dATP	491.2	259	226	15.4	0.15	0	–	35
dAMP	331.2	259	226	15.2	0.15	11	52	–
dA	251.2	260	225	15.2	0.15	–	–	–
CTP	483.2	271	249	9.0	0.97	0	11	41
CDP	403.2	271	249	9.1	0.98	0	33	64
CMP	232.2	271	249	9.1	0.98	15	64	75
Cytidine	243.2	271	250	9.1	0.93	–	–	–
dCTP	467.2	272	–	9.1	0.98	0	–	43
dCMP	307.2	271	249	9.3	0.99	18	65	–
dC	227.2	271	250	9.0	0.97	–	–	–
GTP	523.2	253	223	13.7	0.66	0	5	25
GDP	443.2	253	224	13.7	0.66	0	17	45
GMP	363.2	252	224	13.7	0.66	6	40	51
Guanosine	283.2	253	223	13.6	0.67	–	–	–
dGTP	507.2	252	222	13.7	0.66	0	–	26
dGMP	347.2	253	222	13.7	0.67	6	41	–
dG	267.2	254	223	13.0	0.68	–	–	–
UTP	484.2	262	230	10.0	0.38	0	14	49
UDP	404.2	262	230	10.0	0.39	0	41	80
UMP	324.2	262	230	10.0	0.39	20	75	80
Uridine	244.2	262	230	10.1	0.35	–	–	–
TTP	482.2	267	–	9.6	0.73	0	–	52
TMP	322.2	267	234	9.6	0.73	24	74	–
Thymidine	242.2	267	235	9.7	0.70	–	–	–

[a]TLC mobility is expressed as the % distance a given spot migrates relative to the solution front (R_f) in three different TLC systems using 0.5 mm polyethylenimine cellulose plates: "A" is 0.25 M, "B" is 1.0 M and "C" is 1.6 M LiCl.

Appendix-6: Properties of Amino Acids

1-letter code	3-letter code	Amino acid	Mol. wt. (g/mol)	Side chain	Group
A	Ala	Alanine	89.1	CH_3	Non-polar, aliphatic
C	Cys	Cysteine	121.2	CH_2-SH	Polar, uncharged
D	Asp	Aspartate	133.1	CH_2-COOH	Acidic
E	Glu	Glutamate	147.1	CH_2-CH_2-COOH	Acidic
F	Phe	Phenylalanine	165.2	CH_2-C_6H_6	Non-polar, aromatic
G	Gly	Glycine	75.1	None	Aliphatic
H	His	Histidine	155.2	CH_2-$C_3N_2H_3$ (ring)	Basic
I	Ile	Isoleucine	131.2	CH-(CH_3)CH_2-CH_3	Non-polar, aliphatic
K	Lys	Lysine	146.2	CH_2-CH_2-CH_2-CH_2-NH_2	Basic
L	Leu	Leucine	131.2	CH_2-CH_2-(CH_3)CH_3	Non-polar, aliphatic
M	Met	Methionine	149.2	CH_2-CH_2-S-CH_3	Non-polar
N	Asn	Asparagine	132.1	CH_2-CO.NH_2	Polar, uncharged
P	Pro	Proline	115.1	N-CH_2-CH_2-CH_2-C*	Non-polar (imino ring)
Q	Gln	Glutamine	146.2	CH_2-CH_2-CO.NH_2	Polar, uncharged
R	Arg	Arginine	174.2	CH_2-CH_2-CH_2-NH-C=(NH)NH_2	Basic
S	Ser	Serine	105.1	CH_2-OH	Polar, uncharged
T	Thr	Threonine	119.1	CH-(CH_3)OH	Polar, uncharged
V	Val	Valine	117.1	CH-(CH_3)CH_3	Non-polar, aliphatic
W	Trp	Tryptophan	204.2	CH_2-C_2HNH-C_6H_6	Non-polar, aromatic
Y	Tyr	Tyrosine	181.2	CH_2-C_6H_5-OH	Polar, uncharged, aromatic

*N and C of the peptide backbone

Appendix-7: Preparation of Commonly Used Stock Solutions

30% Acrylamide: Dissolve 29 g of acrylamide and 1 g of N,N"–methylenebisacrylamide in a total volume of 60 ml of H_2O. Heat the solution to 37°C to dissolve the chemicals. Adjust the volume to 100 ml with H_2O. Sterilize the solution by filtration through a filter (0.45-micro pore). Check that pH is 7.0 or less, and store the solution in dark bottles at room temperature.

40% Acrylamide: Dissolve 38 g of acrylamide and 2 g of N,N"–methylenebisacrylamide in a total volume of 600 ml of H_2O. Continue to prepare the solution as described above for 30% acrylamide, except adjusting the volume to 1 litre.

Actinomycin D: Dissolve 20 mg of actinomycin D in 4 ml of 100% ethanol. Read the OD_{440} of a 1:10 dilution of the stock solution in ethanol, using 100% ethanol as a blank. The absorbance at 440 nm of a solution containing 1 mg/ml of the drug is therefore 0.182. Stock solutions of actinomycin D are stored at –20°C in foil-wrapped tubes.

0.1 M Adenosine triphosphate (ATP): Dissolve 60 mg of ATP in 0.8 ml of H_2O. Adjust the pH to 7.0 with 0.1 N NaOH. Adjust the volume to 1 ml with distilled H_2O. Dispense the solution into small aliquots and store at –70°C.

10 M Ammonium acetate: Dissolve 770 g of ammonium acetate in 800 ml of H_2O. Adjust the volume to 1 litre with H_2O. Sterilize by filtration.

Appendix-8: Useful Numbers and Conversions

Table 1 *Molar solutions*

A molar (mole, molecular) solution is one in which 1 litre (1000 ml) of the solution contains the number of grams of the solute equal to its molecular weight (sum of atomic weights).

Example of molar solution, to make up a 1 M solution of sodium chloride.

1. Obtain the atomic weights of the elements (see Periodic Table).
   Sodium (Na) = 22.997
   Chloride (Cl) = 35.4592.

2. Obtain the sum of atomic weights from the formula for molecular weight.
   Molecular weight of NaCl = 22.997 + 35.459 = 58.456

3. Weigh 58.46 g of NaCl and make up to a total volume of 1000 ml with distilled water.

Table 2 *Metric prefixes*

milli (m) = 10^{-3} kilo (k) = 10^{3}
micro (μ) = 10^{-6} mega (M) = 10^{6}
nano (n) = 10^{-9} giga (G) = 10^{9}
pico (p) = 10^{-12} tera (T) = 10^{12}
femto (f) = 10^{-15} atto (a) = 10^{-18}

Table 3 *Isotope data*

Isotope	Stable or radioactive	Emission	Half-life
^{2}H	Stable		
^{3}H	Radioactive	β	12.1
^{14}C	Radioactive	β	5,730 years
^{15}N	Stable		
^{18}O	Stable		
^{24}Na	Radioactive	β (and γ)	15 hours
^{32}P	Radioactive	β	14.3 days
^{35}S	Radioactive	β	87.4 days
^{45}Ca	Radioactive	β	164 days
^{59}Fe	Radioactive	β (and γ)	45 days
^{125}I	Radioactive	γ	60 days
^{131}I	Radioactive	β (and γ)	8.1 days

1 Ci = 1,000 mCi
1 mCi = 1,000 µCi
1 µCi = 3.70 × 10^4 disintegrations/second
1 µCi = 2.2 × 10^6 disintegrations/min.
1 Becquerel = 1 disintegrations/min.
1 µCi = 3.70 × 10^4 Becquerels
1 Becquerel = 2.70 × 10^{-5} µCi

Table 4 *Spectrophotometric conversion*

1 A$_{260}$ unit of double-stranded DNA = 50 µg/ml
1 A$_{260}$ unit of single-stranded DNA = 33 µg/ml
1 A$_{260}$ unit of single-stranded RNA = 40 µg/ml

Table 5 *DNA molar conversions*

Average molecular weight of deoxynucleotide base pair = 660 Da
Molecular weight of an average base is approximately 330 daltons

$$\text{Number of moles of dsDNA} = \frac{\text{Weight of dsDNA in g}}{\text{Length in base pairs} \times 649}$$

1 kb of nucleic acid = 6.5 × 10^5 daltons dsDNA (sodium salt)
$\qquad\qquad\qquad\quad$ = 3.3 × 10^5 daltons single-stranded DNA (sodium salt)
$\qquad\qquad\qquad\quad$ = 3.4 × 10^5 daltons single-stranded RNA (sodium salt)
1 µg/ml of a 1 kb linear dsDNA = 3.08 nM 5'–ends.
10^6 daltons of double-stranded DNA is approximately 0.5 µm long
10^6 daltons of double-stranded DNA has approximately 1.5 kb
3 nmol of bases weigh approximately 1 µg
1 µg of 1,000 bp DNA = 1.52 pmol (3.03 pmol ends)
1 pmol of 1,000 bp DNA = 0.66 µg
One picogram of DNA = 0.965 × 10^9 bp = 6.1 × 10^{11} dalton = 29 cm

Table 6 *Protein molar conversions*

100 pmol of 100 kDa protein = 10 µg
100 pmol of 50 kDa protein = 5 µg
100 pmol of 10 kDa protein = 1 µg

Table 7 *DNA/Protein conversion*

1 kb of DNA = 333 amino acids = 37 kDa of protein
Avogadro's number = 6 × 10^{23}

Table 8 *Concentrations of standard commercial reagents*

HCl	= 11.6 M (36% HCl)
Nitric acid	= 16.4 M
H_2SO_4	= 17.8 M
H_3PO_4	= 14.7 M (ortho; 85%)
Glacial acetic acid	= 17.4 M
2-mercaptoethanol	= 15.6 M

Table 9 *Interconversion of mol, mmol, and µmol in different volumes to give different concentrations.*

M	mM	µM
1 mol dm^{-3}	1 mmol dm^{-3}	1 µmol dm^{-3}
1 mmol cm^{-3}	1 µmol cm^{-3}	1 nmol cm^{-3}
1 µmol mm^{-3}	1 nmol mm^{-3}	1 pmol mm^{-3}

Table 10 *The DNA content of various cells and viruses*

Species	Base pairs in Haploid genome	pg DNA per Cell (or virion)	(G+C) content (%)	Approximate number of genes
Flowering plants	$0.3–100 \times 10^9$	0.6–24	37–50	<50,000
Mammals	$2–4 \times 10^9$	4–9		<25,000
Human (*Homo sapiens*)	3.9×10^9		59%	
Mouse	3×10^9		58%	
Amphibia	$0.9–80 \times 10^9$	2–75		
Fish	$0.3–3 \times 10^9$	0.6–6	38–47	
Crustacea	$0.1–5 \times 10^9$	0.2–5.5	31–47	
Nematodes	1×10^8		36%	
Yeast (*S. cerevisiae*)	1.3×10^7		39%	
Fungi	$1–3 \times 10^7$	$1.1–3.3 \times 10^{-2}$	28–63	~4,000
Bacteria	$2–9 \times 10^6$	$2.2–10 \times 10^{-3}$	50%	~2,000
Mycoplasma	$0.6–2 \times 10^6$	$0.7–2.2 \times 10^{-3}$	24–75	~750
dsDNA viruses	$5–200 \times 10^3$	$0.6–22 \times 10^{-5}$	28–74	6–300
ssDNA viruses	$~5 \times 10^3$	$~6 \times 10^{-6}$		5–12
dsRNA viruses	$~20 \times 10^3$	$~2 \times 10^{-5}$		~20
ssRNA viruses	$1–20 \times 10^3$	$0.1–2.2 \times 10^{-5}$		1–2

pg=picogram

Table 11 *Sizes of various DNAs*

Source	Molecular weight	Base pairs	Length
Subcellular genetic systems:			
SV40 (mammalian tumor virus)	3.5×10^6	5,226	1.7 µm
Bacteriophage øX174 (double-stranded form)	3.2×10^6	5,386	1.8 µm

Continues...

... Continued

Bacteriophage lambda	3.3×10^6	5×10^4	13 µm
Bacteriophage T2 or T4	1.3×10^8	2×10^3	50 µm
Human mitochondria	9.5×10^6	16,596	5 µm
Prokaryotes:			
Hemophilus influenzae	8×10^8	1.2×10^6	300 µm
Escherichia coli	2.8×10^9	4.2×10^6	1.4 mm
Salmonella typhimurium	8×10^9	1.1×10^7	3.8 mm
Eukaryotes (content per haploid nucleus):			
Saccharomyces cervisiae (yeast)	1.2×10^{10}	1.8×10^7	6.0 mm
Neurospora crassa (pink bread mold)	1.9×10^{10}	2.7×10^7	9.2 mm
Drosophila melanogaster (fruit fly)	1.2×10^{11}	1.8×10^8	6.0 cm
Rana pipiens (frog)	1.4×10^{13}	2.3×10^{10}	7.7 m
Mus musculus (mouse)	1.5×10^{12}	2.2×10^9	75 cm
Homo sapiens (human)	1.9×10^{12}	3.8×10^9	94 cm
Zea mays (maize)	4.4×10^{12}	6.6×10^9	2.2 m
Lilium longiflorum (lily)	2×10^{14}	3×10^{11}	100 m

Table 12 *Typical E. coli*

10^{-12} g wet weight per cell
10^{-13} g protein per cell
2×10^{-13} g dry weight per cell
OD_{500} (Zeiss) of 1.0 is about 4×10^8 cells/ml
One molecule per *E. coli* cell = 1×10^{-9} M

Table 13 *RNA facts*

~1 g of eukaryotic cells yields → ~2 mg of total RNA yields → ~60 µg poly A+ mRNA (~5×10^9 cells)

Total RNA consists of approximately 2% poly A$^+$ RNA, 80% ribosomal RNA (rRNA), and ~18% tRNA and small nuclear RNAs (snRNA)

The ratio of the absorbance of purified RNA at 260 nm to its absorbance at 280 nm should be in the range of 1.7–2.1. Lower values may indicate protein contamination of the RNA sample.

The ratio of the absorbance of purified RNA at 260 nm to its absorbance at 230 nm should be about 2.0. Lower values may indicate carbohydrate contamination of RNA sample.

Appendix-9: Preparation of Organic Reagents

Distilling phenol

Phenol extraction is an excellent method for deproteinizing DNA because it is rapid and nearly foolproof. However, the oxidation products of phenol are reputed to produce cross-links between DNA strands. Thus, many laboratories repurify phenol by distillation. This is most conveniently done in sizable quantities and the aliquots of the phenol are then stored frozen.

Requirements

2000 ml round-bottom flask with ground-glass taper joint.
Heating mantle for flask
Distilled adapter with top joint for thermometer
Thermometer with ground-glass joint
Condenser, about 18 in. long
Variable voltage source, for example, Variac
Glass wool to pack around top of flask

Procedure

1. Set up the distilling apparatus in a chemical hood. Add 200 ml of water to 2 kg of phenol and melt in a 65°C water bath. Pour the phenol into the distilling flask. Add approximately 50 boiling chips.
2. Adjust the airflow to achieve a moderate flow through the condenser. Begin heating.
3. Pack the glass wool around the top of the flask.
4. Begin collecting the distilled phenol when the thermometer indicates about 160°C. It is most convenient to collect into a 3 litre bottle containing 200 ml of water. The water greatly lowers the melting temperature so that the phenol may later be poured into milk dilution bottles and then stored frozen at –20°C. Phenol is stable for years when stored at 20°C.

Equilibration of phenol

Most batches of commercial liquefied phenol are clear, colorless, and can be used in molecular cloning without re-distillation. Phenol, if it is pink or yellow in color, should not be used. Phenol is highly corrosive and can cause severe burns. Gloves should be worn while handling it. Before use, phenol must be equilibrated to a pH > 7.8 because DNA will partition into the organic phase at acid pH.

Procedure

1. Liquefied phenol should be stored at –20°C. As needed, remove the phenol from the freezer, allow it to warm to room temperature and then melt it at 68°C. Add hydroxyquinoline to a final concentration of 0.1%.

2. To the melted phenol, add an equal volume of buffer (0.5 M Tris.Cl, pH 8.0) at room temperature. Stir the mixture on a magnetic stirrer for 15 min. Remove the upper aqueous phase as much as possible.
3. Add an equal volume of 0.1 M Tris.Cl (pH 8.0) to the phenol. Stir the mixture with magnetic stirrer and then remove the upper aqueous phase. Repeat the extractions until the pH of the phenolic phase is >7.8 (as measured with pH paper).
4. After the phenol is equilibrated, add 0.1 volume of 0.1 M Tris.Cl (pH 8.0) containing 0.2% β-mercaptoethanol. The phenol solution may be stored in this form under 100 mM Tris.Cl (pH 8.0) in light-tight bottle at 4°C for periods of up to 1 month.

Phenol:Chloroform:Isoamyl Alcohol (25:24:1)

A mixture consisting of equal parts of equilibrated phenol and chloroform:isoamyl alcohol (24:1) is frequently used to remove proteins from preparations of nucleic acids. The chloroform denaturates proteins and facilitates the separation of the aqueous and organic phases, and the isoamyl alcohol reduces foaming during extraction. The phenol:chloroform:isoamyl alcohol mixture may be stored under 100 mM Tris.Cl (pH 8.0) in a light-tight bottle at 4°C for periods of up to 1 month.

Appendix-10: Restriction Enzymes and Reaction Conditions

Table 1 Restriction enzyme reaction conditions

Enzyme	MgCl$_2$ (mM)	Tris (mM)	pH (mM)	NaCl (mM)	KCl (mM)[a]	Dithiothreitol (µg/ml)	BSA
Alu I	6	6	7.6	60	–	1	50
Ava I	6	6	7.4	60	–	1	50
BamH I	6	6	7.5	60	–	1	50
Bgl II	6	6	7.4	60	–	1	50
BstE II	6	6	7.9	60	–	1	50
EcoR	6	100	7.5	60	–	–[b]	50
Hae III	6	6	7.5	6 (or 60)	–	1	50
Hha I	6	6	7.4	60	–	1	50
Hind III	6	6	7.4	60	–	–[b]	50
Hinf I	6	6	7.4	60	–	1	50
Hpa I	6	6	7.4	60	–	1	50
Kpn I	6	6	7.5	6	–	1	50
Msp I (Hpa II)	6	6	7.4	60	–	1	50
Pst I	6	6	7.4	60	–	1	50
Pvu II	6	6	7.5	6 (or 60)	–	1	50
Sac I	6	6	7.4	–	–	1	50
Sal I	6	6	7.5	150	–	1	50
Sau3A I	6	6	7.5	60	20[c]	1	50
Sma I	6	6	8.0	–	–	1	50
Xba I	6	6	7.4	150	–	1	50
Xho I	6	6	7.4	150	–	1	50

[a] 10 mM 2-mercaptoethanol can be substituted for 1 mM dithiothreitol.
[b] It does no harm to include dithiothreitol or 2-mercaptoethanol in these reactions.
[c] It may be possible to substitute NaCl for KCl in this reaction.

Table 2 *Some important restriction endonucleases*

The following restriction enzymes belong to the class II. The complete list of Kessler et al. (1985), comprise 626 entries.

Bacteria	Restriction endonuclease	Sequence recognized	Ends
Bacillus amyloliquefaciens H	*Bam* HI	G^GATCC	Cohesive
Bacillus globigii	*Bgl*II	A^GATCT	"
Escherichia coli RY13	*Eco*RI	G^AATTC	"
Escherichia coli R245	*Eco*RII	^CCTGG	"
Haemophilus influenzae Rd	*Hind*II	GTPy^PuAC	"
Haemophilus influenzae Rd	*Hind*III	A^AGCTT	"
Haemophilus parainfluenzae	*Hpa*II	C^CGG	"
Moraxella spp.	*Msp*I	C^GGG	"
Nocardia otitidis-caviarum	*Not*I	GC^GGCCGC	"
Providencia stuartii 164	*Pst*I	CTGCA^G	"
Proteus vulgaris	*Pvu*I	CGAT^CG	"
Bacillus globigii	*Bgl*I	GCC(N)$_4$^NGGC	"
Bacillus amyloliquefaciens	*Bam*HI	G^GATCC	"
Haemophilus haemolyticus	*Hha*I	GCG^C	"
Moraxella bovis	*Mbo*I	^GATC	"
Streptomyces albus	*Sal*I	G^TCGAC	"
Thermus aquaticus	*Taq*I	T^CGA	"
Xanthomonas badrii	*Xba*I	T^CTAGA	"
Xanthomonas holcicola	*Xho*I	C^TCGAC	"
Arthrobacter luteus	*Alu*I	AG^CT	Blunt
Serratia marcescens Sb	*Sma*I	CCC^GGG	"
Haemophilus aegyptius	*Hae*III	GG^CC	"
Haemophilus parainfluenzae	*Hpa*I	GTT^AAC	"
Brevibacterium albidum	*Bal*I	TGG^CCA	"
Haemophilus influenzae	*Hinc*II	GTPy^PuAC	"
Deinococcus radiophilus	*Dra*I	TTT^AAA	"
Aphanothece halophytica	*Aha*III	TTT^AAA	"
Microcoleus species	*Mst*I	TGC/GCA	"
Haemophilus aegyptius	*Hae*III	GG^CC	"
Nocardia aerocolonigenes	*Nae*I	GCC^GGC	"

^ = indicates enzyme cleaving site

Appendix-11: Commonly used Genetic Markers and How to Test them

Nutritional markers

Streak or replica plate colonies of the strain onto plates with and without the nutrient to be tested, but which contain all other necessary nutrients.

Antibiotic resistance markers

Streak or replica plate colonies of the strain onto plates with and without the antibiotic.

Other markers

lac Z^+

Streak strain on an LB plate with Xgal and IPTG. Colonies should turn blue. Colonies of control *lac Z⁻* strain should not turn blue.

lac Z"M15b

Transform strain with pUC plasmid and with control plasmid such as pBR322. Streak transform onto LB/ampicillin plate with Xgal and IPTG. Colonies bearing pUC plasmid should turn blue, while colonies bearing pBR322 should not.

F^+ or F'

Spot M13 phage onto a lawn of the cells. Small plaques should appear.

recA

Using a toothpick, make a horizontal stripe of cells across an LB plate. Also make a stripe of $recA^+$ control cells. Cover half of the plate with a piece of cardboard, and irradiate the plate with 300 ergs/cm^2 of 254 nm UV light from a hand-held UV source (20 sec). recA⁻ cells are very sensitive to UV light, and the recA⁻ cells in the unshielded part of the plate should be killed by this level of irradiation.

recBCD

Spot dilutions of lambda gam⁻ on a lawn of cells side by side with dilutions of lambda gam⁺. The gam-plaques should be almost as big as the gam⁺ plaques.

hisdS⁻

(1) Use the strain and a wild type strain to plate out serial dilution of a lambda-like phage stock grown on an hsdS⁻ or hsdR⁻ host. If the phage stock came from an hsdS⁻ host, then it should make plaques with 10^4 to 10^6 higher efficiency on the putative hsdS⁻ host than on a wild type host. If the plate stock came from an hsdR⁻ (hsdS⁺ hsdM⁺) host, it should make plaques with the same efficiency on both strains.

(2) Suspend one of the fresh plaques from the putative hsdS⁻ host in 1 ml lambda dilution buffer. Titer this suspension on the putative hsdS⁻ strain and on a wild type strain. The suspension should make plaques at 10^4 to 10^6 higher efficiency on the hsdS⁻ strain than on the wild type strain. One plaque contains ~10^7 phage.

hsdR⁻ (hsdS⁺ hsdM⁺)

(1) Perform step 1 described above, using a plate stock made on an hsdS⁻ host.

(2) Suspend one of the fresh plaques in 1 ml lambda dilution buffer. Titer this suspension on the putative hsdR⁻ strain and on a wild type strain. This suspension should make plaques with the same efficiency on the hsdR⁻ as on a wild type strain.

dam

Transform the strain and a wild type strain with a plasmid that contains recognition sites for the enzymes MboI or BclI. Prepare plasmid DNA from both strains and verify that plasmid DNA isolated from the dam-strain is sensitive to digestion by the enzyme.

dcm

Transform the strain and a wild type strain with a plasmid that contains recognition sites for ScrFI. Prepare plasmid DNA from both strains to verify that only plasmid DNA from the dcm strain is fully sensitive to digestion by the enzyme. Half of the ScrFI sites will be cut even when the DNA is dcm-methylated.

lon

Streak LB plate for single colonies. Also streak a control plate of a wild type strain. Incubate at 37°C. Colonies of the lon-strain should be larger, glittering, and mucoidal.

Appendix 12: Basal Media for Plant Cell and Tissue Culture

Table 1 Composition of some plants tissue culture media

Components	Concentration (mg per litre) in medium*							
	MS	B5	N6	NN	AA**	White	ER	SH
Macronutrients:								
KNO_3	1900	2500	28030	950	-	80	1900	2500
NH_4NO_3	1650	-	-	720	-	-	1200	300
$MgSO_4.7H_2O$	370	250	185	185	250	720	180	400
KH_2PO_4	170	-	400	68	-	-	340	-
$NaH_2PO_4.H_2O$	-	150	-	-	150	16.5	-	-
$CaCl_2.2H_2O$	440	150	166	166	150	-	440	200
$Ca(NO_3)_2.4H_2O$	-	-	-	-	-	300	-	-
$(NH_4)_2.SO_4$	-	134	463	-	-	-	-	-
KCl	-	-	-	-	2950	-	-	-
Na_2SO_4	-	-	-	-	-	65	-	-
Micronutrients:								
H_3BO_3	6.2	3	1.6	10	3.0	1.50	0.63	5.0
$MnSO_4.H_2O$	16.9	10	3.3	19.0	10.0	7.00	2,23	10
$ZnSO_4.7H_2O$	8.6	2	1.5	10.0	2	2.60	8.6	1.0
$Na_2MoO_4.2H_2O$	0.25	0.25	0.25	0.25	0.25	-	0.025	0.1
$CuSO_4.5H_2O$	0.025	0.025	0.025	0.025	0.025	-	0.0025	0.2
$CoCl_2.6H_2O$	0.025	0.025	-	0.025	0.025	-	0.0025	0.1
KI	0.83	0.75	0.8	-	-	0.75	0.83	1.0
$FeSO_4.7H_2O$	27.8	-	27.8	-	-	-	27.8	15
$Na_2.EDTA$	37.3	-	37.3	-	-	-	37.3	20
NaFe.EDTA salt	-	40	-	100	40	-	-	-
Vitamins and organics:								
Thiamine HCl	0.5	10	1	0.5	10	0.1	0.5	5.0
Pyridoxine HCl	0.5	1	0.5	0.5	1	0.1	0.5	0.5
Nicotinic acid	0.5	1	0.5	5.0	1	0.5	0.5	5.0
Myo-inositol	100	100	-	100	100	-	100	100
Glycine	-	-	40	5	-	3	2.0	-
Ca Panthothenate	-	-	-	-	-	1	-	-
Cysteine HCl	-	-	-	-	-	1	-	-
Sucrose	30×0^3	20×10^3	50×10^3	20×10^3	20×10^3	-	40×10^3	25×10^3
PH	5.8	5.5	5.8	5.5	5.5	-	5.8	5.8

*Source: MS=Murashige and Skoog (2); B5=Gamborg et al. (1); N6=Chu (7); NN=Nitsch and Nitsch (5); AA=Toriyama et al. (15); ER=Eriksson; SH=Schenk and Hildebrandt.
**The nitrogen source is mg/l, L-glutamine 730; L-aspartate 200; L-arginine 176; Glycine 7.5.

Table 2 Some commonly used sugars in tissue culture medium

Sugars	Mol. wt.	Sugars	Mol.wt.
Fructose	180.16	Mannitol	182.17
Galactose	180.16	Ribose	150.13
Glucose	180.16	Sorbitol	182.17
Lactose	360.30	Sucrose	342.30
Maltose	360.13	Xylose	150.13

Table 3 Preparation of growth hormone stock solutions

Compounds	Mol. wt.	Preparation
Cytokinins		
6-Benzyladenine (BA or BAP)	225.2	Dissolve in 25 ml of 0.5 N HCl; heat slightly, make to volume, adjust pH 5.0.
Isopentenyl adenine (2–iP)	203.2	
Kinetin (K, KIN)	215.2	
Zeatin (Z, ZEA)	219.2	
Auxins		
2,4-Dichlorophenoxyacetic acid (2,4–D)	221	Dissolve in 2 to 5 ml of ethanol, gradually add water; heat slightly; make to volume; adjust to pH 5
2,4,5-Trichlorophenoxyacetic acid (2,3,5–T)	255.5	
1-Napththaleneacetic acid (NAA)	186.2	
Indolbutyric acid (IBA)	203.2	
Indolacetic acid (IAA)	175.2	
Miscellaneous		
Gibberellic acid (GA)	330.0	Dissolve in 2 to 5 ml of 0.2 M KOH; adjust to pH 5.0
Picloram (PIC)	241.2	
Abscisic acid (ABA)	264	Distilled water, Store in dark
Adenine	135.1	
Dicamba (DCA)	221.0	Dissolve in 1 M NaOH
p-Chlorophenoxyacetic acid (CPA)	186.6	
Jasmonic acid (JA)	210.3	
Thiadiazuron (TDZ)	220.2	Dissolve in 5 ml of 95% ethanol, stir, heat gently, filter sterilize
Silver nitrate (AgNO$_3$)	169.9	

All the stock solutions should be stored in refrigerator; IBA, IAA, GA, and ABA should be filter sterilized.

Table 4 MS medium for sugarcane callus initiation

Add the following to MS medium from Table 1 (Appendix 12) + Growth regulators

Growth regulators	Concentrations (mg/litre)
IBA	0.01
GA3	0.10
Coconut milk (10%)	
Sucrose	20,000
Agar not added, liquid medium	

Table 5 Modified White's medium (1954) for sugarcane rooting

Macronutrients	Mg/litre	Vitamins and organics	Mg/litre
$Ca(NO_3)_2 \cdot 4H_2O$	288.0	Nicotinic acid	0.05
KNO_3	80	Thiamine HCl	0.01
KCl	65	Pyridoxine HCl	0.01
Na_2SO_4	200	Glycine	0.03
$NaH_2PO_4 \cdot H_2O$	20		
$MgSO_4 \cdot 7H_2O$	360	IBA	0.01
$ZnSO_4 \cdot 7H_2O$	2.64		
H_3BO_3	1.50	Sucrose	20×10^3
$MnSO_4 \cdot H_2O$	5.00		
$CuCl_2$	0.33	pH	5.8
$Na_2MoO_4 \cdot 2H_2O$	0.28		
KI	0.09		
Ferric citrate	1.60		

Table 6 Media for Papaya (Carica papaya)

Add the following to White (1954) macro- and microelements (as in Table 1, Appendix 12)

Constituents	Concentrations (mg/litre)	
A. Culture media:		**B. Rooting media:**
Nicotinic acid	5.0	5.0
Thiamine HCl	1.0	1.0
Pyridoxine HCl	1.0	1.0
Glycine	30.0	30.0
Sucrose	20×10^3	20×10^3
Agar	8,000	-
pH	5.8	5.8

Table 7 MS medium as in Table 1 of Appendix 12

Add the following to MS medium from Table 1 (Appendix 12) + Growth regulators

Constituents	Concentrations (mg/litre)
Kinetin	0.10
Benzyl aminopurine	0.20
Coconut milk (10%)	100
Sucrose	20,000
Agar	8,000

Table 8 Composition of the modified Nitsch's (1951) medium used to culture in vitro pollinated ovules

Constituents	Amount (mg 1^{-1})
$CaNO_3.4H_2O$	500
KNO_3	125
KH_2PO_4	125
$MgSO_4.7H_2O$	125
$CuSO_4.5H_2O$	0.025
Na_2MoO_4	0.025
$ZnSO_4.7H_2O$	0.5
H_3BO_3	0.5
$MnSO_4.H_2O$	3.0
$FeC_6O_5H_7.5H_2O$	10.0
Glycine	7.5
Ca-Pantothenate	0.25
Thymine HCl	0.25
Pyridoxine HCl	0.25
Niacin	1.25
Sucrose	50×10^3
Agar	7000

REFERENCES

Kessler, Ch., Neumaier PS and Wolf W (1985) Recognition Sequences of Restriction Endonucleases and Methylases–A Review, *Gene* **33**:1–102.

Roberts RJ (1984) Restriction and Modification Enzymes and Their Recognition Sequences, *Nucleic Acids Res*, **12**:167–204.

Index

% recovery of DNA 168
2,4–D 623
2D electrophoresis 254
2D-gel electrophoresis 718
^{32}P 5'-end-labeling 188
^{32}P-labelled mRNA 522
35S RNA promoter 403
3-D map 116
3-D molecular structures 103
3-D structures 111, 112
3D-immobilization 86
4-methyl-umbelliferyl-b-D-glucuronide (MUG) 544
5'–leader sequence 403
5-enolpyruvylshikimic acid-3-phosphate synthase 624
7-deaza-dGTP 237
α-1-Antitrypsin 601
α-carboxyl 680
β-galactosidase 557
β-galactosidase gene 566
β-globin DNA probe 569
β-glucoronidase activity 544
β-glucuronidase (GUS) gene 542, 544
β-lactamase gene 566, 567
β-rays 513
λ DNA 264
λ exonuclease digestion 306
λZAPII libraries 462

A

abrasives 654
absorbance 166, 666
    at 280 and 260 nm 668
    at 280 nm 667
    spectroscopy 166
absorption maximum 166
absorption spectroscopy 666
Ac and Dc (maize transposons) 437
Ac/Ds family of transposable elements 438
ACC oxidase 630
ACC-deaminase 631
acclimatization 903
    chamber 903
    room 877
acetate 47
acetic acid 47
aceto-carmine 997
aceto-orcein 997
acetonitrile 66
acetylation 325
acid rain 1043
acidic amino acids 681
acidity 40
acidogenesis 1112
acquired immunity 734
acrylamide 202, 205, 1168
    gel solutions 718
    solution 238
    /urea top solution 239
    /urea bottom solution 239
acrylic resin 323
actinomycetous biomass 1066
actinomycin D 1168
activated 3-aminopropyltriethoxy-saline (APES) 319
activated alumina column 66
activated charcoal 892
activated sludge 1051
activation vessel 64
adaptive enzymes 813
adaptive immune systems 734, 735
addition of (dT)n tails 397
addition of synthetic DNA linkers 461
adenosine triphosphate (ATP) 1168
adherent cell lines 848
adjuvants 755
ADP-glucose pyrophosphorylase 626
adsorbents 38
advantage of DNA chips 89
adventitious embryogenesis 971
aerobic reactors 1049

affinity 683, 743
affinity chromatography 704, 707
affinity cross-flow ultrafiltration 705
affinity electrophoresis 705
affinity escort 705
affinity precipitation 705
affymetrix gene-expression chips 80
AFLP mapping 283
AG medium 963
agar gel 840
agarase enzyme 212
agarose
    concentration 195
    gel electrophoresis 170
    gels 193
    plate 171
    aging 991
agriculture 1041
agro-ecosystems 1102, 1103
agrobacterium transfers T-DNA 482
Agrobacterium tumefaciens 426, 479
agrobacterium-mediated transformation 481
agroinfection 488
agroinfection with a geminivirus 477
air pollution 1042
air-lift bioreactors 1032
Alcaligenes sp. 1058
alcoholic beverages 555
alcohols 34
algae 1057
algal biomass 1065
algal tanks 1066
alginate 553
algorithms 94
alicyclic hydrocarbons 1055
alien introgression 286
alien species invasion 1090
aliphatic hydrocarbons 1055
Aljanabi and Martinez 148
alkaline lysis 137
alkaline phosphatase 331, 384, 399
alkyl amine spacer-chain 61
alkylbenzyl sulphonates 1054
allele-specific associated primers (ASAPs) 272
allelic exclusion 751
allergen 15
allotypes 740
allotypic characteristics 741
alpha diversity 1077
alu-repeats 274

amides 891
amino acid biosynthesis 625
amino acids 891, 1167
aminopterin 765
ammonium acetate 176, 1168
ammonium persulfate 202, 205, 239
ammonium sulfate precipitation 690
ampholytes 720, 721
ampicillin 137
amplification
    fragment length polymorphism (AFLP) 282
    of arbitrary sequences 289
    of related sequences 289
    of unrelated sequences 289
    reaction 305
AmpliTaq™ 295, 395
ampliTaq DNA polymerase 84
AMV-RT 384
amylase 558
amylopectin 626
amylose 626
anabolic pathways 572
anaerobic digestion 1052
analogs of dNTPs 237
analysis of whole genome 537
analyte detection 711
analytical centrifugation techniques 49
anchorage dependent 846
anchored PCR 301, 303
androgenesis 964, 969
animal
    biotechnology 11
    energy 1108
    pharming 600
    vectors 429
annealing buffer 188
annotation of the genome 119
anther culture 11, 963, 964, 966
anthocyanin biosynthesis 632
anti-idiotype 740
anti-idiotypic antibodies 771
anti-isotype Abs 741
anti-lipogenic action 594
antibiotic biosynthesis genes 555
antibiotic resistance 1023
antibiotics 540, 555, 762, 1159
antibody
    diversity 748
    producing lymphocyte 761
    reaction 227

screening assay 763
antigen (Ag) 742
    internalization and processing 738
    presenting cells (APCs) 737
    produced in transgenic plants 642, 643
    purification 754
    triggering of B cell 739
antimetabolites 1013
antimicrobial properties 812
antisense
    DNA/RNA 71
    RNA 617, 630
    technology 632
    therapy 610
antiseptic 34
antisera 753
antithrombin III 601
aphidicolin 492
aphid-transmission factor 617
apical-tip culture 926
application
    of 2D-PAGE 719
    of ISH 360
    of microchip technology 89
    of proteomics 118
    of radioisotopes 513
    of RFLPs 257
applied voltage 196
aqueous two-phase affinity partitioning 705
arbitrary PCR 303
arbitrary DNA sequence 275
arbitrary primed polymerase chain reaction AP-PCR 281
arbitrary sequence markers 275
archaebacteria 1112
Archie 99
argon 65
aromatic hydrocarbons 1055
array hybridization 75
array printing 85
arraying 77
artificial photosynthesis 1109
artificial recombinant molecules 363
artificial seeds 976
artificial skin 842
ascitic fluid 769
aseptic culture 917
aseptic technique 794
aspartate (acidic) protease inhibitors 651
aspergillus oryzae 558

asymmetric end labeling 354
asymmetric hybrids 1015
asymmetric PCR (single-stranded PCR) 303, 313
asymmetrical hybridization 1004
atomic structures 512
atrazine 623
attack by force 473
authoring tools 121
autoclaves 31, 882
autoclaving 26, 32
Autographa californica nuclear polyhedrosis virus (AcMNPV) 435
automated amplification 305
automated DNA sequencing 244
automated DNA synthesis 64, 68
automated fluorescence-based DNA sequencer 244
automobile industry 1041
automutagenesis 1019
autonomously replicating sequence or ARS 423
autoplastic connections 1004
autoradiographic techniques 184, 208, 242, 338, 513, 514, 779
auxins 1180
avidin-HRP 228
avidity 744
Azospirullum 633

B
B–DNA 128
Bacillus thuringiensis 12, 552, 620, 1120
BACs 422
bacterial
    alkaline phosphatase (BAP) 342
    chromosomal DNA 142
    chromosome 823
    culture 135
    genomic DNA 142
    growth 804, 805
    growth curves 810
    media 1155
    staining techniques 815
    transformation 504, 824
bacteriophage 823, 828
    based library 449
    lambda 414
    SP6 396
    T4 RNA ligase 393
    vectors 414
    T7 and T3 396

bacteriostatic agent 652
bacteroids 633
baculovirus 403, 435, 436, 579, 622
baculovirus expression systems 437
balancing tubes 54
ballistic device 494
banana plug connector 199
bands in the blank 302
bandshift 530
bandshift assay 530, 535
bar coding 93
barnase 634
barstar 635
base 40
   composition 196
   modifying reagent 233
   specific chemical cleavage method 232
basic amino acid 681
Basic Local Alignment Search Tool (BLAST) 104
batch culture 564, 950
beads-on-a-string motif 132
Beer–Lambert law 665
benedict's reagent 545
Bergmann's plating technique 955
beta diversity 1077
bialaphos 624
bicarbonate ion 41
binary vector system 430
binary vectors 428, 481
bio-indicator plants 1072
bio-leaching bacteria 1070
bio-mining 1070
bio-poisons 1092
bioaccumulation 1054
bioaugmentation 1059
biocatalysts 559
biochemical oxygen demand (BOD) 1049
biochemical role of micronutrients 889
biochip 74
biocontrol 14
bioconversion 1111
biocrude 1114
biodegradation 13, 1055
biodegradation of halogenated compounds 1056
biodegradative pathways 1062
biodiesel 1114
biodiversity 12, 1073, 1074
   conservation strategies 1087
   "hotspots" 1075
bioeffluents 1127

biofertilizers 1068, 1069
biofilm 1050
biofilters 1046
biofuels 10, 1108
biogas 1112
biogas (gobar gas) production unit 1113
bioinformatics 94
bioinformatics access tools 99
bioleaching 10
biolistics 493
biological
   agents 1042, 1133
   containment 21
   control 14
   dispersion 1125
   hazard 26
   warfare 1133
   weapons 1133
biologically active molecules 468
biomagnification 13, 1054
biomass 1109, 1110
   production 10
biomes 1089
biomining 10
bionet news groups 107
biopharming technology 642
biopiracy 1148
biopolymers 556
bioremediation 1055
   of heavy metals 1057
biosafety 17
   concerns 1119
   information network and advisory service 1128
   issues of GMOs 1122
   regulations 1131
   regulations in India 1132
BIOSCI 107
BIOSCI newsgroups 107
biotech food 15
biotechnological applications of plant tissue culture 871
biotechnology 1036, 1101
   accepting environment 1130
   companies 551
biotransformation 10
bitmapped images 121
Biuret method 669
Biuret reagents 669
black biology 1134

BLAST 104
BLASTN 105
BLASTP 105
BLASTX 105
bleeding from ear 756
blood collection 756
blood proteins 557
Bloom's syndrome 393
blotting matrices 225
blotting techniques 528
blue colonies 412
blue or white selection 544
blunt end 385
   ligation 388
   ligation using T4 DNA ligase 459
   linkers 378
boiling "miniprep" 138
bovine
   papilloma virus (BPV) 431
   serum albumin 225
   somatotropin (BST) 594
Bradford reagent 671
brain 98
Brazil nut 15
breed true 966
breeding strategies 595
bright bands in the well of the gel 302
bright-field microscopy 332
briquettes 1111
Brome mosaic virus 477
bromophenol blue 137, 170, 200
bromoxynil 623, 624
Bronstead 40
browning of medium 924
Bt corn 15
Bt subsp. *Israelensis* insecticidal protein 579
Bt toxins 578
buffer 46, 1161
   adjust 45
   gradient polyacrylamide gels 238
   reservoir 206
   solution 41
   used for PCR 294
bulldog clips 206
buoyant densities 52, 159

C

$CaCl_2$ precipitation 506
calcium alginate beads 560
calcium phosphate precipitation method 503

calcofluor 960
calibration 45
   curve method 668
calliclones 1017, 1030
callus culture 944
CaMV
   35S promoter 619
   as a vector 476
   genome 476
candidate gene 599
canola oil 627
cap site 403
capillary
   electrophoretic separations 729
   transfer 216
   transfer of DNA 219
capping 61, 63
CAPS 272
capture PCR 313
carbencillin 483
carbol fuchsin 997
carbon sources 890
carcinogenecity 1054
cardiac puncture 757
care of centrifuges 56
Carica papaya 938
carmine 998
carrot 974
casein 553
catabolic pathways 571
cation exchange column 699
cauliflower mosaic virus (CaMV) 476
caulimovirus 150
caulogenesis 909
CCD camera 87
cDNA 370
   array 77
   cloning 463, 464
   library 441, 452, 453
   microarrays 83
celite 682
   abrasive 150
cell
   based molecular cloning 365
   count by measuring turbidity 821
   culture laboratory 871
   cultures 836
   cycle 492
   debris remover (CDR) 682
   disruption 37, 653

disruption methods 38
emplacement 1048
fractionation 37
free system 180
fusion 764
immobilization 560
lines 843
mapping 117
mediated immunity 739
smears 322
specific gene knockouts 587
spreading 320
viability 993
cellular totipotency 905 906
cellulase R-10 959
cellulysin 1005
centers of biodiversity 1075
centimorgan (cM) 250
centiRays 248
centrifugation 37, 48
techniques 49
centrifuge tubes 57
centrifuges 56
centrosymmetric "palindromic" sequences 659
cesium chloride 50
cesium chloride density gradients 159
cetyltrimethylammonium bromide (CTAB) 146, 653
CGIARs 1096
chain termination 232
reaction 237
Chakrabarty et al. (1981) 1062
chalcone synthase (CHS) 371, 632
chaotropic anions 40
Chargaff 128
Chargwin 154
chelating agents 38
chemical
bioprocesses 1129
degradation technique 232
factories 549
fume hood 22
hazards 22, 1042
labeling 350
methods of cell lysis 654
modifications 233
oxygen demand (COD) 1049
synthesis of genes 374
synthesis of oligonucleotides 58
chemically synthesized genes 72

chicken embryo fibroblasts 849
chimeras 363
chimeric gene 483
chimeric plasmids 410
chip formats 77
chitinase genes 603
chloramphenicol acetyl transferase (CAT) 542
chlorella 1065
chlorinated insecticides 1053
chlorofluorocarbons (CFCs) 1043
chloroplast 133
extracts 656
lysate 657
choice of animals 754
choice of matrix 706
Chomczynski and Sacchi 154
choosing IEF type 723
chromatin 132
chromatographic techniques 692
chromium trioxide 319
chromosomal mosaicism 1019
chromosome
complements 1015
doubling 968
library 441, 451
segments 265
walking 285
cibacron blue 659
cibacron blue sepharose 661
CIFOR 1096
citric acid 553
Ck gene segment 746
cladistics 1079
class II restriction enzymes 377
class switching 747
classification of proteins 647
claviceps purpurea 558
cleaning solution 28
cleaved amplified polymorphic sequences (CAPs) 282
cleaved radioactivity of phosphopeptides 101
clonal
culture 858
multiplication 1027
propagation 11, 913
restriction 752
selection 750
selection theory 750
clone 407
cloned nucleic acids 336

cloning 407
  a gene by complementation of mutations 469
  vector 365, 379, 406, 407
clonogenic assay 859
  in methylcellulose 861
closed composting 1047
closed treatment system 1048
CM–Sephadex column 701
Cn3D 106
co-dominant markers 266, 276
co-inheritance 599
co-integrate disarmed Ti-plasmid 429
co-integrate pTi vectors 428
co-integrative vector systems 481
co-metabolism 1056
co-metabolites 1056, 1059
co-segregation 268
co-suppression 616, 631
co-transfection 483
coat protein 616
  mediated transgenics 616
coding region identification 110
codon choice 433
cohesive termini 386
colchicine 968, 999
cold boiling 33
cold conditioning 986
cold storage 988
colony hybridization 449, 450, 466, 521, 522
colorimeter 666
colorimetric 43
colorimetric assays 669
column chromatography 171, 693
comb 198
commensalism 792
commercial view of biotechnology 1137
common marker genes 540
common promoters 404
community drift model 1078
comparative mapping 285
compartmentalization 120
compatible cohesive termini 386
competence 504, 909
competent cell preparation 486
competitive ELISA 775
complement-fixing antibodies 767
complementary determining regions (CDRs) 741
complementary DNA (cDNA) 340
complementary RNA 370

complementation 516
  methods 1011
  of mutants 468
  test 566
complete Freund's adjuvant 755
complete linkage 251
complex medium 791
  organics 892
components of biodiversity 1076
composited manure 1069
composition of the electrophoresis buffer 196
composting 1047
compulsory license 1143
computational biology 102
computer aided pattern analysis, CAPA 728
computer databases 94
computer software for analysis of 2-D gels 729
computer-based translation of genomic data 103
concentration 684
conditions for patenting 1139
confocal microscopy 334
conformation of hybridity 1014
conformation of the nucleic acids 195
conformational epitope 743
conjugate base 41
conjugation 824
  in bacteria 826
conservation 13, 1073
  tools 1089
  utilization link 1087
conspecific adults 1078
constant regions 739, 741
constitutive enzymes 813
construction of the matrix 261
contact printing 85
containment level 19
contaminants 22
contamination 301
contig 423
contiguous epitopes 743
continuous
  affinity recycle extraction 705
  buffer systems 711
  culture 564, 951
  streak method 800
control room 877
convention on biological diversity (CBD) 1092, 1148
conventional genetic markers 266
conventional marker 268

conversion adapter 379
Coomassie
   blue 671
   blue staining 675
   Brilliant Blue 671
   Brilliant Blue R-250 675
copper sulfate 34
copyrights 1143
cornucopia 6
cos site 417
cosmid vectors 417, 418
Coulter counter 822
Council of Bioethics 1151
counter-staining 333
coupling 67
   or *cis* configuration 250
   reaction 63
cpDNA 133
cross-hybridization 83
cross-protection 475, 615
cross-reactive idiotypes (CRIs) 741
crown galls 479
crown gall tumour 482
"crush and soak" technique 211
cryopreservation 769, 980, 981, 985, 1099
cryoprotectants 324, 986
cryosections 322
cryptotopes 743
crystalline (cry) proteins 620
CsCl
   centrifugation 145
   gradients 146
   –ethidium bromide 135
CTAB 146
cultural characteristics 801
culture
   fresh weight 995
   media 761, 788, 789, 846
   of bone marrow cells 851
   of embryos 975
   of spleen cells 852
   of organised tissues 886
   of unorganized tissues 886
cumulina 602
custodianship 1085
'cut and fill' method 1048
cutaneous anthrax 1133
cuvettes 673
cyanide (HCN) 1055
cyanobacteria 1065, 1117

cyanocobalamin (vit B12) 555
cybrid formation 1016
cybrids 1015
cycling parameters 296
cycling procedure 298
cylindrical rod gels 204
cysteine (thiol) protease inhibitors 651
cytodifferentiation 907, 908
cytogenetic maps 251
cytokinins 1180
cytological techiques 996
cytomegalovirus (CMV) 471
cytoplasm transfer 1017

D
DAF-PCR 303
dam methylase 394
data mining 103
Daucus carota 945
David Lipman 104
DbdbEST 101
DBT 19
DDBJ 100
ddNTPs 238
DE-52 column 703
de-salting 684
de-salting methods 685
DEAE-
   cellulose 172, 703
   cellulose resin column 702
   Sephracel 210
death phase 551
deblocking 63
decline or death phase 807
Decon–90 25
dedifferentiation 905
deep temperature inversion layer 1042
defective interfering RNAs 617
defective interfering DNA 617
defined medium 791
deflasking 920
deforestation 1045
degenerate PCR 303
delivery valves 65
delivery vector 508
dem methylase 394
denaturant 327
denaturation 327, 682
denaturation conditions 352
denaturing polyacrylamide gels 205, 238

densitometer 514
densitometry trance 710
density 902
density gradients 51
deoxyribonuclease 1 391
deoxyribonucleotide triphosphates 296
deoxyribose 127
Department of Biotechnology (DBT) 1132
DEPC-treated SW40 tube 177
DEPC-treated water 153
dependent varieties 1146
deprotection 63, 67
desertification 1044
desiccation 38
design of expression cassettes 434
detecting protein-DNA binding 532
detection of peptides 663
detection of proteins in gels 674
detergents 34
determinate organs 925
determining generation time 808
detritylation 63, 67
Dewards et al. 148
diagnostic
    applications 8
    medium 791
diagnostics 11
dialysis membranes 685
dialysis techniques 685
diazophenylthioether cellulose 225
dideoxy method 232
dideoxy sequencing method 235
diethylpyrocarbonate 177
different forms of immunoglobulin 742
different types of media 789
differential
    centrifugation 50
    display technique 374
    growth 1013
    hybridization 466
    medium 791
    screening 372, 522, 523
differentiation 907, 928
digital imaging 335
digital video 121
digoxigenin-labeled DNA probes 356
dihydrofolate reductase (DHFR) 516
dilution rate, D 551
dimethoxytrityl 62
dimethyl sulfate 233

dimethyldioctadecyl ammonium bromide (DDA) 756
dinucleotide/trinucleotide sticky-end cloning (DISEC/TRISEC) 289
diploidization 969
direct amplification of minisatellite DNA markers 274
direct bacterial leaching 1071
direct cloning of a gene by cDNA library 452
direct combustion 1110
direct DNA uptake by imbibation 502
direct ELISA 775
direct gene transfer 488
direct selection 1022
direct staining of target cells 780
direction of the electric field 196
directional cloning 378
disarming 428
disc-electrophoresis 713
discontinous buffer systems 711
disease resistance 12, 1023
disinfecting the explants 899
disinfection 34
disomic 966
dispersed growth digesters 1051
distant recombinant form 364
distillation of ethanol 1115
distilling phenol 1173
distribution coefficient (K_d) 692
divalent cations 38
diversity (D) segments 744
DMS footprinting 530, 532
DNA
    amplification 70
    amplification fingerprinting (DAF) 281
    binding proteins 384, 399
    chip manufacturers 92
    chip technology 71
    concentration 167
    content of various cells 1171
    counter-staining 330
    deamination 174
    delivery pathways 474
    delivery via growing pollen tubes 502
    double helix 129
    extraction 147
    fingerprinting 276
    fingerprinting of crop germplasm 285
    fingerprinting of pathogen populations 286
    fragmentation 84

isolation   126
library   365, 368, 441, 442
ligases   384, 388, 393
linkers   378
methylases   384, 394
microarray   74
microchip   74
microextraction   278
microinjection   587
molar conversions   1170
polymerase I (Pol-I)   383, 394
polymerases   236, 383, 394
print   520
protein conversion   1170
sequence databases   245
sequencing   70, 232
synthesizers   64, 65
RNA synthesizer   83
typing   270
DNase   142
   footprinting   530
   I   382, 449
   I cleavage   449
dNTPs   238
Dolly   602
domestic activities   1041
dominant selectable markers   541
DOP-PCR   289
dormancy   916
dot blot hybridization   228
dot hybridizations   184
double immunodiffusion (DID)   774
double mutants   1012
double selection   1022
double-beam   175
double-haploid plants   1024
Dounce homogenizer   39, 40
downstream
   applications   289
   oligonucleotide primer   307
   processing   9
   processing steps   858
drag force   49
dried gels   208
Drosophila development   582
drought and salt tolerance   637
drug-response curves   862
dry acid deposition   1043
dry mounts   781
dry weight   823

dual promoter   403
dual-labeling system   1014
Dynabeads   164
Dynal   164
Dynal MPC magnet   164
dystrophic   1043

E
E-mail   107
E. coli   1172
   DNA   143
   strains DH5α and MC 1061   136
Earth Summit   1099
eco-tourism   1089
ecological impact on GMM release   1126
ecological theories   1077
EcoRI   386
ecosystem (ecological) diversity   1075
ecosystems   1076
edible interferons   642
edible vaccines   641
Edman sequencing   116
EDTA   134, 171, 652
effecter function domain   740
EGTA   651
electric field strength   491
electric oven   881
electrical pulse treatments   1013
electricity   1108
electro blotting   224
electrode   42
electroelution   209
electrolytes   42
electron microscopy   369
electronic communication system   98
electromotive force   43
electrophoresis   181, 191
   buffers   196
   of RNA   175
   unit   193
electrophoretic analysis   280
   methods   711
   mobility   192, 195
   transfer   215
   transfer apparatus   226
electroporation   487, 490, 507
   media   492
electrostatic device   494
electrostimulation   1014

elongation 923
elution
　buffer 162, 211
　by diffusion 730
　by electrophoresis 730
EM grids 324
Emax 338
EMBL 100, 104, 245
EMBL 3 448
EMBL 3 library 448
embryo
　cloning 597
　culture 11, 935, 938, 942
　culture of papaya 937
　development 936
　splitting 596
　transfer 11, 595
　transfer techniques 597
　transformation 505
embryogenesis 978
embryogenic clumps 973
embryogenic potential 973
embryonal-suspensor mass (ESM) 971
embryonic stem (ES) cells 589
encapsulation of somatic embryos 976
end-labeled cDNA 187
end-labeling
　of 3'-OH termini 343
　of nucleic acids 341
　reaction 343
end-modification 378
endangered species 1090, 1091
endopolyploidization 907
endosperm development 937
endotoxin gene 620
energy production plants 1041
energy sources 1107
enforced self-fertilization 936
engineered
　genetic male sterility 635
　in situ bioremediation 1046
　soil piles 1048
　vectors 417
engineering
　for extended shelf-life 629
　herbicide resistant plants 623
　modified reproductive systems 634
　of disease resistance in plants 615
　protein for purification 683
　proteins 10

enhanced greenhouse effect 1042
enhancer 404
enrichment (enriched) medium 789
entrapment technique 559
entrapped biocatalyst 560
enucleated egg cells 603
enucleated oocytes 598
enumeration of bacteriophages 833
environment 1039
environmental
　biotechnology 13, 1061
　effect 266
　effects 16
　growth cabinets 883
　stress 1023
enzymatic digestion 40
enzymatic labeling 339
enzyme
　bioprocesses 1129
　engineering 558
　for protoplast isolation 957
　immobilization 560, 562
　induction 813
　linked immunoabsorbant assay (ELISA) 775
　mediated reporter systems 331
enzymic methods of cell disruption 654
epidermal growth 7
epifluorescence microscopy 333, 334
epigenetic changes 1019
epigenetic programming 838
episomal vectors 421
epitope 740
EPSP synthase 624
equilibration of phenol 1173
equilibrium constant 41
equilibrium theory 1077
equipment for media preparation 892
erusic acid 627
Escherichia coli (*E. coli*) 787
essential amino acids 625, 626
establishment of human peripheral blood
　　lymphocyte 853
EtBr drop 170
ethanol 1114
ethanol precipitation 135
ethical reasons 1084
ethics 1151
ethidium bromide 87, 169, 176, 196, 201, 207
ethidium bromide–CsCl 140
ethylene 630

eukaryotic nuclear gene 402
eukaryotic vectors 420
European Commission Directive 1150
European Patent Convention (EPC) 1142, 1150
European Patent Office (EPO) 1140
eutrophication 1043, 1049
evolution 13
evolutionary genomics 102
ex situ bioremediation 1045, 1046
ex situ conservation 980, 1096, 1097
ex vivo gene therapy 610
excised plant tissues (explants) 887, 897
excitation filters 781
exemptions to patentability 1139
exon amplification 538
exon trapping 538
exonuclease Bal31 382
exonuclease HI 382
exonuclease III 391
explant 897
   cleaning 898
   size 922
   terminology 898
   transfer 900
exponential phase 807
expressed sequence tag markers (EST) 272
expression
   cassettes 433, 609
   libraries 470
   of satellite RNA 617
   vectors 433, 434, 470, 575
external jugular vein 757
extinct 1091
extinction 13, 1090, 1091
   coefficient 166, 667
extra-somatic evolution 98
extracting DNA fragments 211
extraction buffer 145

F
F factor 824
Fab fragments 741
factor IX 601, 602
factors affecting the immune response 752
false color image 88
FAO undertaking 1148
farm saved seed 1147
farmer's rights 14, 1147
farmyard manure (FYM) 1069
FASTA algorithm 109

FASTA search programs 100
Fc fragment 741
fed-batch culture 551, 564
feeder
   effect 896
   layer 846, 954
   layer cells 770
FEP–Teflon diaphragms 65
fermentation 9, 1109
   medium 562
fermenter design 564
fermenters 562
fertilizer management 633
feulgen 996, 997
ficoll 50
field gene banks 1099
filamentous fungi 426
filamentous fungi vectors 425
file transfer protocol (FTP) 104
fill-in reaction 460
filter binding 532
filter paper raft-nurse tissue culture 954
filter-sterilization equipment 882
filtration 30
   manifold 185
financial gain from patenting 1143
FindMod 100
finger-printing 9, 252, 538
first to file principle in UK law 1142
first to invent principle of US law 1142
first-strand cDNA synthesis 455
FITC 781
fixative 318
fixed angle rotors 55
fixed film digester 1049
fixed immobilized beds 857
flavonoid biosynthesis pathway 631
FLAVR SAVR 631
fleshly beans 656
flow cytometry 782, 783
flower bud culture 943
fluidized bed systems 857
fluorescein diacetate (FDA) 993
   staining 987
fluorescence 169
   signals 87
fluorescent staining methods 676
fluorescently-labeled DNA 87
fluoride pollution 1061
fluorinated alkyl silanes 80

greenhouse 904
greenhouse effect 1042
grid method 841
gridding robots 81
growing phage stocks 525
growing room shelves 882
growth analysis 992
growth curve 992
growth curve for a bacteria 806
growth hormone stock solutions 1180
growth regulators 890
guanidine isothiocyante (GITC) 154
gyrase 384
gyratory shakers 883

H
habitat loss 1082
habitats 1076
haemacytometer 866
Hae III 389
hairpin amplification method 453
halocarbons 1053
halomethanes 1055
handling restriction enzymes 254
Hank's minimum essential medium (HMEM) 849
haploid genomes 168
haploid plants 967
haploid production 936
HAPPY mapping 248
hardening 931
hardening off 903
HAT medium 765
HAT selection 765, 766
heap leaching 1071
heat sterilization 31
heavy chains 746
helix axis 129
helper component 617
hemacytometer 993
hemopoietic cells 858
Henderson–Hasselbatch equation 42
HEPA 32
hepatitis B virus 576
HEPES 46
herbicide
  glyphosate (roundup) 624
  resistance 12, 623
herbicides 540, 623, 1053
heritage of mankind 1148

heteroduplex 252
  analysis 369
  mapping 252
heteroencapsidation 1120
heterokaryons 1007, 1014
heterologous expression vectors 434
heterologous probes 371, 468
heteroplastic plasmodesmata 1004
high capacity cloning vectors 408
high-density chips 78
high-density reactors 858
high-molecular-weight compounds 556
high-molecular-weight DNA 143
high-performance liquid chromatography (HPLC) 709
high-value proteins 648
*Hind*III 387
*Hind*III reaction buffer 260
Hindu gods 363
histones 132
HML and HMR loci 569
hog fuel 1110
hollow fiber systems 857
home page 99
homogenization 37
  buffer 649, 655, 654, 656
  medium 37, 38
homogenous gels 204
homologous probes 465
homologues in human 599
homopolymer tailing 378
  tails 397
homopolymeric run 243
Honolulu nuclear transfer technique 604
Honolulu technique 602
horizontal gel electrophonesis apparatus 199
horizontal slab gel 197
hormone gene 600
host-encoded resistance 615
hot start PCR 299
human
  body 98
  caused extinction 1091
  diagnostics 90
  genome mapping 538
  growth hormone (HGH) 376, 575, 606
  hemoglobin 602
  hormones 557
  insulin (Humulin) 573

interface 121
interferon (IFN–l) 642
    proteins 557, 573
    use of biodiversity 1080
    vaccines 576
humus sludge 1049
hyaluronidase 558
hybrid
    arrest translation 527
    dysgenesis 582
    embryo 941
    protein 433
    release translation 527
    select (release) translation 527
    selection 468
    sorting 966
hybridization 182, 214
    arrest 370
    buffer 87
    of radio-labeled oligonucleotides 222
    of radio-labeled probes 220
    solution 221
    specificity 83
    stringency 326
hybridoma growth factors (HGF) 770
    technology 8, 759, 867
hydrazine 233
hydrocarbons 1114
hydrogen 43, 614, 1116
hydrogen peroxide 34
hydrogen-bonded 129
hydrogenase 1116
hydrophobic interaction chromatography (HIC) 683
hydrophobic polyvinylidene difluoride (PVDF) 225
hydrophobicity 683
    patterns 114
hydropower 1109
hydroxyapatite (HAP) 173
hydroxyapatite chromatography 172
hyper-diversity 1077
hyper-immune serum 757
hyperchromacity 166
hypermedia 121
hypersensitive reaction 618
hypertext file-linking 106
hypervariable regions 740, 741
hypotonic lysis 657
hypoxanthine, aminopterin, and thymidine (HAT) 762

hypoxanthine guanine phosphoribosyl transferase (HGPRT) 761

I

IARCs 1097
IBPGR 1098
IDA-agarose 708
identifications of protein spots 728
identity testing 252
idiotypes 740
IEF electrophoresis buffer 724
IEF gel mixture 724
illegal patents 1140
imaging systems 332
imidazolinones 623
iminodiacetate (IDA) 707
immobilization
    of cells 560, 856
    of enzymes 10
    techniques 707
immune modulators 558
    system 734
immunization 576, 755
    schedule 764
immuno-staining of proteins 227
immunoblotting 777, 779
immunochemical detection 517
immunochemistry 329
immunodeletion 224
immunoelectrophoresis (IEP) 717, 773
immunofixation 224
immunological
    detection 466
    purification of polysomes 455
immunoprecipitation 719, 778
    techniques 768
    test 518
immunoradiometric assay (IRMA) 776
immunostimulants 734
immunotherapeutic drugs 640
Immunotherapy 640
Imperial Chemical Industries (ICI) 1064
importance of conservation 1084
important plant tissue culture types 885
improving plant nutritional value 625
in planta transformation 483, 486
in situ bioremediation 1045
    conservation 980, 1096
    digestion 261
    hybridization 71, 189, 317, 368, 519

immunoassay 368
leaching 1071
PCR 358
polymerase chain reaction (ISPCR) 358
RNA hybridization 189
synthesis 78
in vitro cell culture 842
   culture of embryos 596
   fertilization 11
   localized mutagenesis 568
   packaging 416
   pollination 940, 941
   radiolabeling of DNA and RNA 522
   transcription 178, 339
   transcription reaction 179
   translation of mRNA 180
in vivo gene therapy 610
incentives for effective conservation 1088
incompatible cohesive termini 387
incomplete adjuvant 755
incomplete linkage 251
incubation conditions 808
incubation room 876
Incyte pharmaceuticals 80, 81
independent assortment 250
indeterminate organs 925
indirect bacterial leaching 1071
indirect causes of biodiversity loss 1083
indirect ELISA 775
indoleacelae 479
induced fusion 1004
inducing parthenocarpy 636
induction 813
industrial agriculture 1083
   biotechnology 9
   effluents 13
infection thread 633
infectious microorganisms 21
informatics 94
information-processing abilities 98
inhalation anthrax 1133
inheritance of transgenes 509
inhibitor mix 654
initial variety 1146
injection methods 756
injection schedule 756
ink-jet printing technology 79
ink-jetting 77
innate immunity 734, 735
inner and outer shell membranes 605

inner cell mass (ICM) 596
inoculants 552
inoculation cabinet 875
inorganic salts 888
insect resistance 12
insecticides 552, 578
insertion vectors 415
Institutional Biosafety Committee (IBSC) 1132
insulin-like growth factor (IGF-1) 594
integrated suspension culture 855
integrating vectors 421
integrative stable transformation 474
intellectual property rights (IPRs) 14, 1136, 1138
intensifying screen 513
inter simple sequence repeat markers (ISSR) 274
inter-atomic interactions 96
inter-cellular interactions 98
interactive multimedia 121
intercalating dyes 196
interleukin-1 (IL-1) 737
intermicrobial relationships 792
International Agricultural Research Centers (IARCs) 1097
International Biosafety Protocol 1130
international conventions 1149
international cooperation 1089, 1149
International Depository Authority (IDAs) 1150
International Food Biotechnology Council (IFBC) 1129
International Gene Fund 1148
international organizations involved in biosafety 1131
internet service provider 101
internode 948
intervening sequences 569
intestinal anthrax 1133
intra-atomic interactions 96
intrinsic *in situ* bioremediation 1046
inverse PCR 300, 301, 303, 304
invertase 558
investment by biotechnology companies 1152
ion product of water 40, 42
ion-exchange chromatography 683, 698, 700, 773
ionic balance and concentration 888
ionic properties of amino acids 679
ionic strength 192
ionization of acids 40
IPR 14
IPR in India 1143
IPTG 412, 461

irradiation 33
isoelectric
    focusing (IEF) 720, 721, 722
    point (PI) 680
isoenzyme analysis 1015
isoionic point 680, 681
isolating animal genes 598
isolating DNA from a single colony 526
isolation
    of enzymes 658, 660
    of nuclear DNA 144
    of nucleic acids 134
    of restriction enzymes 658
isopentenyl adenine 479
isopropanol 143
isopycnic density gradient ultracentrifugation 53
isopycnic sedimentation 51
isopycnic centrifugation 52
isoschizomers 389, 659
isotachophoresis 212, 213, 713
isotope 338
    data 1169
isotype determination 770
isotypes 740
isozyme analysis 247

J
Java QuickPDB 111
Joe Felsenstein 110
juvenile hormone esterase 579

K
k-casein 598
K3M basal 962
kalparuksha 6
Kalrez O-ring 65
kanamycin 483
Kappa (K) light chains 746, 752
Kauravas 836
ketolactose test 545
key strategies for effective conservation action 1088
kinase buffer 188
kinase reaction 342
kinases 398
Kjeldahl analysis 672
Klenow fragment 383
knockout (KO) mice 535, 589
knockout (KO) transgenic mice 589
knockouts 535

L
laboratory guidelines 17
laboratory-acquired infections 18
lac Z operon 461
lactic acid 553
lactoferrin 601
lactose (lac) operon 814
lactose medium 545
lag-phase 551, 807
laissez-faire 983
lambda cloning vectors 414
lambda light chain 746, 747, 752
lambdoid phage 829
laminar
    air flow system 878
    airflow cabinets 883
    flow cabinet 879
landfill 1047, 1048
landscape ecology 1078
large plasmid 479
large-scale preparation of plasmid DNA 140
laser 33
laser-induced DNA delivery 501
layered ink-jet printing 78
LB medium 135, 1155
leaching techniques 1071
lead 1061
leaf-disc transformation 485
leaf-disk technique 483
Leeuwenhoek 2
left-handed helix 131
libraries of unfractionated genomic DNA 446
life cycle of bacteriophage M13 416
ligase chain reaction (LCR) 312
ligases 384
ligation 380
    amplification reaction (LAR) 312
    with T4 DNA ligase 393
light responsive element (LRE) 405
light-directed deprotection 78
light-sensitive mutant 1012
limiting dilutions 822
linear map of λ phage 265
LINES 274
lingo-hemicellulose complexes 595
linkage
    analysis 268
    groups 249
    mapping 599
    −1 268

linked
  high-speed computers 101
  markers 599
  RFLP marker 269
linker molecule 87
Linsmaier and Skoog (LS) salt composition 888
lipid vesicles (bags) 497
lipoinfection 497
liposome
  formation 499
  structures 498
  mediated DNA delivery 497
liquefaction 1111
liquid
  batch cultures 805
  media 895
  scintillation counting 515
  shear 38, 39, 654
  wastes 1048
Lithospermum 1034
living skin equivalent, LSE 839
Lk gene segments 746
loading
  buffer 137
  rotors 54
  samples 201
local
  alignment 109
  biodiversity conservation 1088
  communities 1088
log or exponential phase 551
logarithmic (log) phase 807
long-term storage 985
loop fermenters 564
losses of biodiversity 1081
low-abundance mRNAs 184, 454
low-melting-temperature agarose 210
Lowry method 670
luc (luciferase) 542
lymphoid system 736, 737
lyophilization 689
lysine 85, 554, 626
lysis 39
lysogenic 414
lysogenic cycle 829
lysogenization 831
lysogeny 414
lysozyme 654
lytic cycle 829
lZAPII libraries 462

M

M13 vectors 417
M9 medium 793
M9 minimal medium 1156
macerase 1005
macerozyme R-10 959
macroelements 963
macroelements K3 963
macroinjection 502
macronutrients 888
MACs 422
magnetic beads 163
magnetic stirrer 177, 880
Mahabharat 836
maintaining genetic diversity 1095
maintenance of biodiversity 1088
major histocompatibility molecules (MHC) 737
making a knockout mouse 536, 537
male sterility 634
mammalian cell culture 844
mammary gland 601
mammary-specific genes 601
man-made supercomputer 98
managed ecosystems 1071
mannopine 479
map distance 249
map-based cloning of genes 285
Mapmaker software 268
mapped RAPD markers 277
mapping population 267
mapping transcripts 528
Marine Leukemia virus 432
marker gene 540
marker-assisted selection 272, 284
markers 183
mass production 6
mating type 569
matrix's beads 695
Maxam and Gilbert 232
Maxam and Gilbert method 233, 234
Maximow's single slide technique 841
maximum likelihood 110
maximum parsimony 110
MEA buffer 182
measurement of cell growth 992
measuring bacterial growth 820
mechanical microspotting 77
media
  dispensor 881
  for papaya 1181

preparation room 876
storage room 877
medical biotechnology 8
meiotic chromosome preparations 1000
melting temperature (T_m) 326
MEM 846
membrane
   affinity filtration 705
   blotting 214
   bound proteins 647
   filtration 31, 32
   proteases 38
   proteins 653
memory cells 751
Mendel 4
   laws of inheritance 247
mercuration 351
mercury 1061
mericloning 871
meristem culture 925, 926
meristem tip culture 927, 928
meristemming 871
merodiploids 566
mesophiles 809
mesophyll cells 1005
metal receptor systems 331
metal toxicity of water 1061
metalloprotease inhibitors 651
methanogenesis 1112
methanogens 1112
methionine 626
methods
   for the microinjection 501
   of enzyme immobilization 561
   of sterilization 30
methylation reaction 459
methylcellulose (Methocel) 862
methylmercuric hydroxide 175
metric prefixes 1169
micro-Ti 428
microarray hybridization 82
microbial
   biomass 552, 553
   digestion 1051, 1052
   enzymes 558
   fermentations 552
   growth 551
   inoculum 1059
   insecticides 552, 578
   leaching 1069, 1070
   metabolites 549
   polysaccharides 556
microchamber 955
microchips 76
micrococcal nuclease 383
microdialyzer devices 687
microdrop method 955
microelements MS 962
microfuge 137
microinjection 499
microinjection method 588
micronutrients 889
micropipetting 35
micropipettor 35
microprojectile 493, 494, 495
   acceleration devices 494
   bombardment 495
micropropagation 905, 913, 914, 920, 932, 933, 1026
microsatellites 273
microsporogenesis 965
microwave 33
milk-borne pharmaceuticals 600
mini-gel 170
mini-Ti 428
miniature tubers 1031
minimal medium 792, 1156
minimum inoculation density 896
mining 1129
miniprep 137
minisatellites 273
minitubers 1030
minor groove 129
mitochondria 132
mitotic chromosome
   analysis 999
   preparation 320
mixed cultures 1058
mixed-bed ion exchange resins 702
modern biotechnology 4
modern brewing vessel 563
modern science 2
modification
   of fatty acid composition 627
   of functional groups 571
   of the PCR 303
modified
   carbol fuchsin 997
   Lowery's method 670
   Nitsch's 1182

nucleotides 349
White's medium 1181
molar concentration 42
molar solutions 1169
molarities of common reagents 1165
molecular
   analysis 8, 1015
   biology 102
   blueprints 2
   interactions 96
   mapping 12
   markers 248
   measure 1079
   size 195
   taxonomy 277
   weight cutoff (MWCO) 685
   weights of protein standards 715
Molly and Polly 602
Moloney murine leukemia virus 504
monellin 639
monitoring introduced GMMs 1126
mono-cropping 1090
monoclonal antibodies (Mabs) 759
monolayer cultures 155
monomeric proteins 647
monopodial orchid 1029
monopolies 1153
Morel-Vitamins 962
morphine 1034
morphogenesis 909
morphological (traits) 247
morphological characteristics 1014
morphological markers 247
mouse embryonic stem (ES) cells 590
mouse polyclonal antisera 758
movement proteins 616
MOWSE 100
MR1 medium 963
MS inorganic salts 496
MS medium 1182
MS micronutrients 496
MS salts 962
MS vitamins 496
MS-identification 120
MSI's quanta 111
MSI's insight 111
mtDNA 132
multi-point attachment 706
multicellular organisms 95, 96, 98
multilingual video on demand 123

multimedia 121
multimedia-based internet delivery 122
multinucleate protoplasts 1016
multiple bands 302
multiple environment testing 1124
multiple gradient mini-gels 716
multiple ovulation with embryo transfer (MOET) 595
multiple sequence alignment 109
multiple-sequence detection 332
multiplex PCR 290
Mung bean nucleases 186, 391
Murashige–Skoog (MS) salt composition 888
mutagenesis 311
mutualism 792
mycobacterium 756
mycoprotein (Quorn) 1067
mycoproteins 1067
myeloma cells 761

N

N-acetylmuramyl-L-alanyl-D-isoglutamine 756
N. tabacum 492
napin 628
national biodiversity institutes 1089
National Gene Bank 1098
natural extinction 1091
natural hybridization 1120
natural plant protoplast fusion 1003
natural rubber 557
natural variation 266
nature tourism 1094
NBPGR 1098
NCBI 104
NCBI databases 106
near-isogenic lines (NILs) 284
Needleman–Wunsch algorithm 109
negative control 301
negative selection markers 541
negative staining 818
negative staining procedure 819
neighbor joining 110
neighborhood effect theory 1078
neomycin phosphotransferase II (NPTII) gene 542
neotope 743
nested PCR 299, 300
nested primers 84
net primary productivity (NPP) 1110
neutral or lottery models 1078
neutralization 673

new products in bacteria 550
niche segregation 1077
nick translation 345
    reaction 456
Nicotiana tobaccum 959
NIH 19
NIH Recombinant DNA Advisory Committee
    (NIH-RAC) 1122
nitrate reductase-deficient (NR⁻) 1012
nitrilase 624
nitrocellulose filter binding essay 534
nitrocellulose filters 182, 222
nitrogen fixation 577, 633
nitrogen-fixing symbiosis 552
nitrogenase 1116
    complex 633
    reductase 633
Nitsch and Nitsch (NN) media 888
N,N'-Methylene-bis-acrylamide 202
no PCR product 301
nodule 633
non-biological process 1110
non-breeding strategies 594
non-contact printing 85
non-denaturing
    agarose gel electrophoresis 181
    polyacrylamide gels 204
non-equilibrium theory 1077
non-fat dried milk 226
non-governmental organizations (NGOs) 1084,
    1130
non-integrative stable transformation 474
non-linear multi-media projects 121
non-radioactive
    labels 338
    tracers 515
non-renewable energy sources 1099, 1107
non-self 734
non-specific sites 466
non-target animals 620
non-translated region 403
nonoccluded viruses 435
nopaline 479
nopaline synthase (nos) promoter 403
Norin 10 1103
Northern Blotting 223
Northern hybridization 182
novel genes 571
novel products 7
novelty 6

NPTII assay 546
NRL-3D 101
nuclear
    energy 1108
    poly-hedrosis viruses (NPV) 435
    transfer 598, 603
nuclease BAL 31 390
nuclease S1 390
nucleases 382, 385, 390
nucleic acid 126
    hybridization 465, 519
    modifying enzymes 384
    probe 520
nucleoids 134
nucleoside 128
nucleosome 132
nucleotides 128, 1166
nucleotide-pair substitutions 750
nutrient agar plates 803
nutrient agar slants 803
nutrient broth cultures 803
nylon membranes 182, 218, 223

O

obtaining explant 897
occluded viruses 435
oil mixtures 1054
oil pollution 1044
oligo (dT)-cellulose 153, 161
oligomeric identical subunits 647
    subunits 647
oligonucleotide arrays 77
    based microarrays 88
    hybridization probes 87
    microarrays 78
    primers 293
    probes 71, 469
oligonucleotides 59, 70, 350
oligotrophic 1043
Omnibus Trade and Competitiveness Act (OTCA)
    1150
on-plate quantification 711
on-site DNA sequence algorithms 104
oncogenesis 431
one gene, one protein 120
one-dimensional electrophoresis 260
one-pin robot 86
ONPG 813
open composting 1047
open land system 1047

operon fusions 566
operons 367, 813
opines 480
opposition to patents 1153
optical density (OD) 666
optical methods 166
optimization of PCR reaction 299
optimizing the IEF 723
optimizing the separation 713
orcein 998
orchids 1027
organ culture 836, 840, 841
organellar DNA 132
organic
   acids 892
   compounds 9
   floatation 653
   solvents 40
organismal diversity 1074
Organization for Economic Cooperative Development 1124
organized
   cells 474
   growth 885
   tissues 925
organophosphate insecticides 1053
origin of replication 401, 570
osmoticum 37
ovary culture 939, 941
over-production 7
overlapping genes 569
overlapping oligonucleotides 71
overnight culture 805
oxidation 67
ozone depletion 1042, 1083

P

P element
   transposase 583
   enhancer trap vectors 583
   mediated gene transfer 584
   mediated transformation 582
packed cell volume (PCV) 995
PACs 422
PAGE 202
pair-wise distance method 110
pair-wise sequence alignment 109
pair-wise sequence comparison 108
palindromic symmetry 385
panda 1092
panhandle PCR 313
paraffin sections 322
paraffin wax 323
paraffin-embedding 322
paraformaldehyde 319, 781
parallel computing 97
parallel two-color monitoring 88
parasitism 792
paratope 740
Paris Convention of 1998 1149
partial metabolic pathways 571
partially homologous probes 465
particle bombardment 493
Pasteur pipette 148
pasteurization 33
patent
   application 1141
   Cooperation Treaty (PCT) 1142
   information on Internet 1144
patentable inventions 1139
patenting
   biological material 1145
   of genes 1145
   of live forms 1145
   procedure 1141
patents 1138
   verses plant breeders' rights 1147
pathogenesis-related (PR) proteins 615
pattern recognition 107
pBR322 409
PBS 32
PCR 365
   amplification 296
   based DNA mapping 270
   generated probes 337
   mutagenesis 289
   reaction with RAPD primers 279
   RFLP 303
   screening of gene libraries 528
   technique(s) 291
   versus genomic and cDNA libraries 464
PDB 101
PDB code 112
pectinase 1005
PEG 141
pelargonidin-3-glucoside 632
pEMBL plasmids 413
penicillin 555
Penicillium notatum 555
pentose sugar 127

PeptIdent 100
percoll 50
perfume 553
periwinkle 1035
permeabilization 325
petrocrops 1109
petroleum plants 1114
petroplants 1114
Petunia hybrida 632
pGEM–1 136
pH 41, 42
   electrodes 43
   gradient engineering 721
   meters 45, 881
   scale 44
phage 133
   based libraries 449
   contamination 795
   f1 420
   lambda suspension medium 1158
   Mu plates 1158
   P1 top agar 1157
   P2 interference (spi) 448
   R17 and P1 Plates 1157
   recovery procedure 525
   T4 plates 1158
phagemids 419
pharmaceutical-producing animals 11
pharming 600
phase variation 569
phaseolin 626
phenol
   chloroform 137
   chloroform lysis extraction 139
   chloroform: isoamyl alcohol (25:24:1) 1174
   rich leaves 656
phenolic compounds 34
   oxidation 896
phenotypic
   analysis of regenerated plant (RI) progeny 543
   effect 266
   mixing 1120
phenylmethylsulfonyl fluoride (PMSF) 651
phosphatases 398
phosphinothricin (PPT) 623, 624
phosphodiester bonds 128
phosphoramidite nucleosides 60
phosphoramidites 59
phosphorimager 515
phosphorimaging 514

phosphorous solubilizing microbes (PSM) 1101
phosphorylation 435
photo pigment genes 614
photo-deprotection 78
photographic
   camera systems 335
   emulsion 513
photographs of gels 202
photolabeling 351
photolithographic patterning 78
photolithography 77, 78
photomask 77
photometry 665
photomultiplier 334
photopolymerization 202
photorespiration 614
photosynthesis 614, 1109
photosynthetic bacteria 1064
   efficiency 614
PHYLIP 110
phylogenetic
   analysis 110
   measures 1079
physical containment procedures 20
physical genome mapping 538
physical mapping of genomes 284
physical maps 252
physical methods of cell lysis 654
physical pollutants 1044
physical protection 19
physical stress 392
phytoalexins 618
phytochemicals 1032, 1104
phytohormones 577
picking off of colonies 863
piezoelectric ink-jet printers 79
pigmentation in transgenic plants 12
pin and ring system 85
pinda 837
piperdine formate 233
PIPES 46
PIR 100
pith tissue 947
pK_a 47
pK_a's of commonly used buffers 1161
planetary greenhouse effect 1042
plant biopolymer 557
plant biotechnology 11, 613
Plant Breeders' Rights 1146
plant cell culture 11

plant conservation 13
plant growth
    enhancement 577
    hormones 891
plant pathogens 615
Plant Protection Act of 1930 1140
plant RNA viruses 476
plant tissue culture 870
    laboratory design 874
Plant Variety Act (PVA) 1143
plant vectors 426
plant viral nucleic acid 149
plant viral vectors 475
plant viruses 438, 439
plant-based drugs 1095
plant-derived drugs 1034, 1104
plant-genomic DNA 148
plant-nuclear DNA 145, 146
plantback 1146
plantibodies 618, 640
planting out 920
plants tissue culture media 1179
plaque assay 832
plaque formation assay 834
plaque hybridization 466
plaque-forming unit (PFU) 832
plaque-lift 521
plasma cells 751
plasma clot 840
plasmid 133, 368
    based library 449
    DNA 135
    isolation 137
    pBR322 411
    preparation 136
    pWWO 1062
    vectors 409
plasticware 28
plastocytes 960
plate
    counts 822
    cultures 805
    development 710
plating efficiency (PE) 862
plexiglas shields 25
plugging room 876
pluripotent ES cells 589
pollen
    culture 969
    transformation 501

pollution 1040
pollution control 1039
pollution-resistant plants 1072
poly(A)+ RNA 155, 161
poly-L-lysine 319, 324
polyacrylamide gel 675
polyacrylamide gel electrophoresis 202, 712
polyadenylation signal
    sequences 620
    site 403
polychlorinated biphenyls (PCBs) 1054
polycistronic gene 402
polyclonal antibody (antisera) 753
polycyclic hydrocarbons 1056
polyethylene glycol (PEG) 141, 173, 489, 763, 1007
polygalcturonase gene 631
polygenic traits 1018
polyhedrin 435
polyhedron promoter 403
polylinker 418
polymerase chain reaction (PCR) 243, 288, 349, 376, 471
polymerase enzymes 295
polymerization catalysts 202
polypeptide-encoding genes 402
polyphenols 650
polysomal RNA 156
polysome buffer 160
polysomes 160, 371
polyvinyl pyrrolidone (PVP) 650
pomato/topato 1018
population genetics 277
porcine growth hormone gene 600
pore gradient gels 713
pore limit 713
pore size 203
position-specific iterated BLAST (PSI-BLAST) 105
positive-negative selection procedure 592
post-harvest quality improvement 629
post-translational modifications 114, 120
potato 1031
potentiometric 43
Potter–Elvehjem homogenizer 39, 40
pouring the gel 205
power pack 193
practical utility 1146
pre-hybridization solutions 220
pre-immune serum 758
precipitating nucleic acids 135

precipitation
   in the medium 896
   of antiserum 690
   of immunoglobulins 691
precipitin 518
predicting heterosis 286
preformed adapters 379
pregnant mare serum gonadotropin (PMSG) 595
prehybridization fixation 325
prenatal diagnosis 8
preparation
   of antigen 756
   of dialysis membrane 685
   of extracts 649
   of lysates 832
   of primers 237
   of SAS 691
   of separating gel 714
   of single-stranded DNA 237
   of slides 780
   of stains 996
   of thin layer 709
preparative centrifugation 48, 49
preparing ion exchange resins 702
pressure-cycle fermenter 564
pretreatments 353
primary antibody 226
primary cell cultures 836, 849
primary immune response 751
primary production 1110
primary reaction vessel 64
primary response 751
primary structure 114
primed *in situ* labeling (PRINS) 351
primed synthesis of the second-strand of cDNA 459
primer extension 187, 531
   for location of the 5' end 533
   reactions 188
primers 235, 288
prior art 1142
private idiotopes (IDI) 741
probe 75, 78, 336, 352
probe labeling methods 337
process inventions 1145
product formation 857
   inventions 1145
   patents 1139
production level vectors 433, 434
production of monoclonal antibodies (Mabs) 760

profiling 252
ProFound 100
proinsulin 574
promoter finders 434
promoters 402, 404
pronculear injection methodology 601
PROSITE 114
prostaglandin F2a (PGF2a) 595
protease inhibitors 622, 650, 651, 732
proteases 558
protected nucleosides 66
protection assays 530
protein
   A 753
   analysis 112
   A-Sepharose column chromatography 771
   based markers 247
   C 601
   complex identification 120
   concentration (mg/ml) 668
   databank 111
   database 100
   detection and quantitation 728
   expression 471
   expression mapping 119
   expression proteomics 117
   function 120
   G 753
   inactivation 731
   localization 120
   modifications 120
   molar conversions 1170
   purification stages 648
   purification techniques 646
   recovery 729
   sequence 100
   standards 673
   structure databases 101
proteinase digestion buffer 310
proteolysis 732
proteome 115, 117
proteomics 115, 118
protoclonal variation 1018
protoclones 1017, 1030
Protogene 81
proton acceptor 46
proton donor 46
protoplast
   co-cultivation technique 484
   culture 955, 960

fusion 1003, 1007, 1008
isolation 956, 1005
preparation 492
transformation 489
protoplasts 902
pRT series of expression cassettes 413
Pruteen 1064
pSC101 410
pseudo-extinction 1091
pseudo-first-order reaction kinetics 82
pseudo-pregnant 587
Pseudomonas 1145
pseudopregnant females 589
pSP65 136
PstI 387
psychrophiles 809
public idiotopes (Idx) 741
pUC family of plasmids 412
pUC plasmids 412
pulsed-field gel electrophoresis (PFGE) 230, 231
pure culture 789, 796
purification of IgG 773
purine 127
purity of acrylamide 202
pyrimidine 127
pyrolysis 1111

Q
QB replicase 383
QBLAST 105
quadrant streak method 799
quality indicator 896
quantitating viral DNA molecules 310
quantitation
  of DNA concentration 309
  of mRNA 304
  of signal 335
quantitative
  trait loci mapping 284
  trait loci, QTL 247, 285
quantitative traits 269
quartz cuvettes 167
Quillaja saponaria 756
quinhydrone 44

R
R-loop hybridization 370
rabbit
  albumin (RSA) 708
  polyclonal antisera 757

reticulocyte lysate 180
radiation 30
  hazards 24
  hybrid mapping 248
  hybrid maps 248
radio-labeled
  dNTPs 236
  RNA probes 186
radioactive
  antibody test 517, 518
  contamination 24
  detection methods 677
  ink 208
  labels 333, 337
  tracers 23, 511
  waste 25
radioactively labeled probe 214
radioactivity 512
radioimmunoassay (RIA) 776
radioisotopes 23
RAFI 1148
raft methods 840
random-primed probe labeling 347
randomly amplified microsatellite polymorphisms
    (RAMPO) 282
RAPD analysis 275
RAPD-PCR 304
rape seed 627
rapid amplification of cDNA ends (RACE) 313
rapid salt extraction 148
rare mRNAs 454
rate-zonal density-gradient centrifugation 50, 52
re-culture 988
reading the sequences 243
real-time and simultaneous monitoring 88
RecA protein 384, 399
recalcitrant 1052
recipes for polyacrylamide 715
recipes for tricine peptide separation gels 715
recombinant
  baculoviruses 436
  DNA 6, 365, 369, 1121
  DNA technology 6
  microorganisms 566
  microorganisms in agriculture 577
  microorganisms in environmental protection 579
  microorganisms in medicine 572
  phage screening 524
  vaccines 576
  vector 365

recombination 823
recombination selection 380
Red Data Books 1090
redifferentiation 905
reducing agents 649
reflection-contrast microscopy (RCM) 333
reforestation 1100
regeneration of protoplasts 960
regulating recombinant DNA technology 1121
regulation of biotechnology 1128
regulatory aspects of using GMMs 1124
regulatory sequences 405
regulon 367
release of genetically engineered microorganisms 1128
removal of interfering substitutes 673
renaturation (hybridization) 328
renewable resources 1099
repeat complementary primers 274
repetitive DNA 273
replacement
   or supportive therapy 608
   synthesis of double-stranded cDNA 457
   synthesis of second-strand cDNA 456
   vectors 415
replica plating 800
   techniques 796
replicase 616
replication of plasmids 409
replicon 409
reporter gene 539, 541, 542
repulsion or *trans* configuration 250
rescue method 1022
research tissue culture laboratory 873
residual effect 1054
resin sections 322
resistance
   genes from animals 618
   markers 1012
   to bacterial pathogens 619
   to insects 619
   using virus-encoded genes 616
resistant insects 15
restriction
   digestion 253
   digestion analysis 252
   endonucleases 365, 382, 385, 1176
   enzyme buffers 255
   enzyme digestion 255
   enzyme half-sites 462

   enzyme mapping 450
   enzyme reaction conditions 1175
   fragment maps 262
   fragment-length polymorphisms (RFLP) 252
   fragments 214
   landmark genomic scanning (RLGS) 270
   reticulocyte extract 179
retro-orbital bleeding 757
retrotransposition-mediated fingerprinting 274
retroviral
   genome 503
   mediated gene transfer 503, 504
   vectors 432
retrovirus-like vectors 438
retroviruses 432
reverse
   genetics 371
   phase HPLC 709
   transcriptase 359, 370, 384, 396, 455
   transcriptase amplification reaction 307
reversed micellar extraction 705
Review Committee on Genetic Manipulation (RCGM) 1132
RFLP 257
   genetic maps 265
   mapping 268
   maps 267
   markers 266, 268
Rhizobium 633
Rhizobium species 552
Rhizoctonia solani 619
rhizogenesis 909
Ri plasmid 427
Ribi adjuvant system (RAS) 755
ribonuclease 385, 391
   A 391
   protection assay 186
   T1 391
ribose 127
ribozyme-mediated virus protection 618
right-handed helix 131
Rio Earth Summit 1130
risk assessment 1127
risk potential to humans 1127
RNA
   amplification 304
   analysis 174
   blots 223
   blotting 182
   dependent DNA polymerases 396

expression 470
extraction 153
extraction buffer 160
facts 1172
loading buffer 182
polymerases 383
probes 351
size fractionation 177
splicing 748
RNase 142, 186
  A 383
  contamination 154
  protection assay 187
  treatment 325
RNaseH 383
robot spotting 78
rodent bone marrow 851
root culture 933, 934
root tips 999
rotor speed 56
RPMI-1640 763
RT-PCR 306
RuBISCO 405, 614
ruminant nutrition 595
Rural Advancement Fund International 1148

S

S1 mapping 529
S1 nuclease mapping 529
S1 nucleases 383, 529
safety 17
safety cabinets 19
safety of biotech products 14
saII 389
salt homogenizing buffer 149
sample buffer 181
Sandwich ELISA 775
Sanger sequencing 232
Sarkosyl 142
saturated ammonium sulfate (SAS) 690
saturated genetic maps 284
saving biodiversity 1086
SCAR 281
Schleif and Wensink 143
scintillation 515
  counting 181
  fluid 515
scopolamine 1034
scorable reporter genes 539
scoring colonies 863

SCP production 1064
scrapefection 503
screening
  a genomic library 450
  colonies 525
  genomic libraries 449, 464
  libraries 442
  of cDNA libraries 467
  of phage plaques 523
  strategies 380
scripting 121
SDS 140
SDS solubilization solution 720
SEAQUEST 100
second-dimension gels 726
second-order hybridization kinetics 82
second-strand cDNA synthesis 456
  by self-priming 458
secondary antibody 226
secondary metabolism 554
secondary metabolites 10, 1031, 1033
secondary response 751
secondary structure 114
secretion of proteins 683
secretion vectors 434
sedimentation 178
  coefficient 49, 54
  rate 49
  velocity 49
Seed banks 1098
seed germination *in vitro* 544
seed storage proteins 625
Seedlessness 636
segregation analysis 280
selectable markers 410
selectable reporter genes 539
selection
  for presence of vector 517
  of inserted sequences 517
  of somatic hybrids 1011
  of transformed cells 542
selective agents 1023
selective medium 790
self 734
self–inactivating (SIN) 433
semi solid methylcellulose-medium 860
semi-continuous culture 564
semi-dry blotting 224
semi-natural lake system 1066
semipermeable membrane 688

semi-rigid gels 694
semi-synthetic medium 792
semisolid agar medium 859
semisolid culture 805, 859
senescence 630
senescence-tolerant Plants 636
sense RNA 630
separating gel 713, 715
Sephacryl 697
Sephadex 697
Sephadex chromatography 172
Sephadex G-25 687
Sepharose 697
sequenase 383
sequence
    alignment 109
    analysis 108
    characterized amplified regions (SCARs) 276
    composition 83
    matrices 263
    pattern 108
    reading 108
    tagged microsatellite site markers (STMS) 273
    tagged sites (STSs) 271, 276
sequencing 312
sequencing by hybridization 93
sequencing gel 240
sequential epitope 743
sera 762
serial dilution–agar plate procedure 797
serial dilution (pour-plate) 796
serine protease inhibitors 651
serum substitutes 762
serum-free mediums 762, 771
Serva Blue G 672
settled sewage 1049
sewage oxidation ponds 1065
sex factor 824
sex selection 596
sexual incompatibility 1017
SH 34
shake culture 805
shaker room 877
shelving unit 883
shikimic acid pathway 624
shikonin 1034
shoot
    bud differentiation 912
    culture 931
    multiplication 918

regeneration 968
short-term storage 983
shotgun cloning 443, 451
shuttle vectors 421, 425
sib selection 369, 465
sib selection of cDNA clones 468
sieving effect 713
signal
    generating system 329
    noise ratio 328
    peptide 403
    sequences 749, 750
signatures 243
silane 85
silencers 404
silico-biology 116
silicon carbide fiber-mediated DNA delivery 500
silver stain kit 676
silver staining 676
simian vacuolating virus 430
simple medium 792
simple staining 818
simplest method of dialysis 687
SINES 274
single cell clones 953
single cell protein (SCP) 552, 553, 1063
single primer re-amplification 84
single-base mismatches 83
single-concentration mini-gels 715
single-strand conformation polymorphism (SSCP) 272
single-stranded adapter 379
single-stranded DNA-binding protein (SSB) 399
single-stranded hairpin loop 456
site-directed mutagenesis (SDM) 377
size-exclusion chromatography 694, 696
sizes of various DNAs 1171
slab gels 204
slope leaching 1071
slot hybridization 185
sludge 1048
sludge treatment 1052
slurry absorbents 689
slurry phase lagoon 1047
small biological molecules 553
smaller local area networks 101
smart polymers 705
smearing of PCR products 302
smiling bands of DNA 205
Smith–Waterman algorithm 109

SOC medium 487
sodium azide 652
sodium hypochlorite 885
sodium p-hydroxymercuribenzoate 651
solar energy 1109
solenoid 132
solid shear 38, 39
solid support 212
solid support system 1046
solid waste 1046
solid–substrate fermentation (SSF) 564
solid-phase phosphoramide chemistry 58
solid-phase synthesis 58
solidified media 895
solubility 682
somaclonal variants 1021
somaclonal variation 12, 474, 838, 923, 1017, 1019
somaclones 1017, 1030
somatic
   back hybridization 1015
   embryogenesis 970, 971, 972, 974, 975
   embryos 971
   evolution 98
   gene therapy 608
   hybridization 1003
   hypermutation 750
   incompatibility 1015
   mutations 750
   recombination 747
   variants 1018
somatostatin 7, 376, 567
   hormone 566
Sorenson 41
source of wastes 1040
South-western procedure 539
Southern blotting 212
Southern blot technique 214
spacer 61
speciality chemicals 640
specialty products 1032
species diversity 1077
species extinction 1082
specific activity 730
spectrofluorometer 169
spectrophotometer 167, 666, 821
spectrophotometric conversion 1170
spectrophotometry 151, 175
sperm as vector of DNA 502
spirulina 1066

spirulina SCP 1066
spleen
   cells 761
   phosphodiesterase 383
   suspension cell cultures 851
splice junctions 187
splicing 747
spontaneous fusion 1004
spooling DNA 143
spread-plate method 796, 800
squashing 321
SSCP 277
SSP-PCR 304
stab cultures 805
Stab-Agar 1158
stability 682
stabilizing proteins 731
stacking gel 713, 715
stages of micropropagation 914
staining DNA 201
staining of bacterial spores 820
staining of protoplasts 1011
standard PCR 303
standard amplification reaction 297
standard protocol for ELISA 776
starch biosynthesis 626
starter cultures 552
stationary phase 551, 807
stealth strategy 473
stearoyl-ACP desaturate 628
step-wise cooling 987
step-wise selection 1022
sterile plants 1017
sterile plating 798
sterile technique (aseptic technique) 845, 879, 900
sterile technique for explant transfer 901
sterilization 30, 34
sterilization room 875, 877
sticky ends 385
stock solutions 893
Stokes' law 49
storage of artificial seeds 977
storage of strains 795
storing DNA 174
story-boarding 121
strain purity 795
strategies for cDNA cloning 454
strategy to conserve biodiversity 1086
streak-plate method 796

strength of acid 40
streptavidin 329
streptokinase 558
stringency 325, 326, 352
strong acids 41
structural genomics 102
structural proteomics 117
structure
    of genes 103
    of plasmid pBR322 411
    of T-DNA 427
studying biodiversity 1086
subatomic particles 96
subcloning 451
subcloning hybridomas 767
subculturing 897
subculturing of monolayers 847
subprotoplasts 1011
substractive hybridization 373, 375
subtilisin 559
subtracted cDNA probes 465
subtractive cDNA cloning 522
sucrose gradients 177
sugarcane 926, 1116
    callus initiation 1181
    explant 929
sugars in tissue culture medium 1180
sui generis system 1143
suicide vector 508
sulphonation 351
sulphonylurea 623
super LB broth 1155
super-secondary structures 115
supercoiled 131
superhelical DNA 130
superior antibiotic 572
supermice 7
superovulated 595
superovulation 594
superserver 98
support matrix 85
support medium 193
surface sterilizing agents 880
surrogate mothers 594
suspension cell transformation 485
suspension cultures 155, 949, 950, 952
sustainable development 1099
SV 40 vectors 430
SV40 virus 431
Swiss-Prot 100

symbiosis 792
symbiosis genes of rhizobia 634
symmetrical hybridization 1004
sympodial orchids 1028
synchronization 854
    methods 855
synergism 792, 1121
synteny relationship 285
synthesis
    of double-stranded cDNA by the self-prim 458
    of human growth hormone (HGH) 575
    of insulin 574
    of oligonucleotides 83
    of the first strand of cDNA 455
synthesizer design 64
synthesizer operation 66
synthetic
    DNA molecules 350, 376
    gene 70, 71
    linkers 461
    medium 792, 887
    nucleotide primers 370
    oligonucleotide probes 466
    oligonucleotides 336
    polymers 1054
    probe 373
    seeds 976, 978

T

T-DNA 427, 480
T-DNA border sequences 479
T3 RNA polymerases 186
T4 DNA ligase 71, 384
T4 DNA pol 383
T4 DNA polymerase 395
T4 polymerase buffer 460
T4 polynucleotide kinase 384
T7 DNA pol 383
T7 DNA polymerase 395
tabtoxin 619
Taq DNA polymerase 236, 306, 395
Taq polymerase 383
target 75
    element 75
    sequences 296, 307
    sites for herbicide action 625
taxic measures 1079
taxol 1035
TBE 238
TBLASTN 105

TBLASTX 105
TE buffer 145, 188
technology 2
tele-teaching 123
Telnet 99
TEMED 202
temperate phage 830
template 235, 292
termed exon trapping 538
terminal deoxynucleotidyl transferase 83, 378, 396
terrestrial ecosystem 1095
terrific broth medium 140
tetramethylammonium chloride 83
tetramethylrhodamine 83
tetrazole 67
thawing and recovering cells 864
therapeutic techniques 9
thermo-chemical process 1110
thermocycle profile 293
thermocycler setting 279
thermophiles 809
Thermus aquaticus 236, 295, 395
thin layer chromatograph 710
thin layer chromatography 709
thiophenoxide ions 67
thylakoid extracts 657
thylakoid homogenization buffer 657
thymidine kinase 593
Ti plasmid vector systems 427
Ti plasmids 427, 480
tidal power 1109
tissue
    culture laboratory 872
    fixation 318
    plasminogen activator (t-PA) 604
    sectioning 322
    specific gene expression 91
TMV 133, 151
tobacco mosaic virus (TMV) 439, 478
tobacco shoot culture 931
tobamovirus 151
topoisomerase I 384, 399
topoisomerases 399
total acidity 40
total cDNA probes 465
total RNA 154
totipotency 613, 870
totipotent 474
touchdown PCR 300

tower fermenters 564
toxicity 1054
tracer laboratory 24
tracking dye 137
trade mark 1144
trade secrets 1143
traditional fermenters 562
transcript mapping 182
transcription buffer 178
transcription initiation complex 405
transcriptional start-points 187
transdetermination 909
transducing phage lines 829
transduction 827
transfection 416, 473
transfer
    buffers 225
    chamber or hood 884
    instruments 790
    room 876
transformation 9, 474, 824, 846
    by ultrasonication 503
    methods 380
    of *E. coli* 507
    of monocot plants by Agrobacterium 486
    of tobacco leaves 496
    process 559
transformed cell lines 1013
transgenesis 7
transgenic
    organisms 4
    animals 11, 21, 535, 582
    chickens 605, 607
    fish 11
    food crops 12
    founder animals 591
    goats 604
    livestock 593, 600
    mice 586, 593
    pigs 602
    plants 7, 12, 21, 612, 621
    poultry 605
    salmon 607
    sheep 602
transilluminator 169
transit peptide 624
translation 178
transmission
    electron microscopy (TEM) 335
    protein 617

transposable elements 274
transposons 437
TREMBL 100
tri-parental mating 481
triacylglycerides in plants 628
trichloroacetic acid (TCA) 674
trickling filter method 1050
trickling filter digesters 1049
triphenyl tetrazolium chloride method 988
Tripos' SYBYL 111
TRIPs 1150
tris-acetate (TAE) 196
tris-borate (TBE) 197
tris-phosphate (TPE) 197
tris-tricine method 714
TRITC 781
tropical forest 1081
tropical moist forests 1080
troposphere 1042
true-extinction 1091
trypsinizing 847
Tth DNA pol 383
tumorous growth 1013
two-colored probes 90
two-dimensional (2D) electrophoresis 258, 261, 727
two-dimensional gel electrophoresis 717
two-enzyme–2D analysis 259
two-plasmid strategy 484
two-site ELISA assay 775
two-wavelength fluorescence microscope 87
Ty element 438
Ty retrotransposon 425
type I integration 421
type II integration 421
Type II restriction enzymes 385
Type III integration 421
type of extract 649
types of selective agents 1023
typical plant gene 401
typical prokaryotic cell 788
typing 252

U
ultracentrifugation 54
ultrasonic vibrations 33
ultraviolet light 33, 166
United Nations Conference on Environment and Development 1148
United Nations Environmental Program (UNEP) 1131
universal buffer 255
universal enzyme 959
universal primer 235
unorganized cells 474
unorganized growth 885
unorganized tissues 944
upstream oligonucleotide primer 307
upstream processing 9
URL (uniform resource locator) 100
US Environmental Protection Agency (EPA) 1153
using bicdiversity sustainability 1086
ustilago zeae 553
ultraviolet spectrophotometer 167

V
vaccination 615
vaccine approaches 784
vaccine development 784
vaccines 9, 558
vaccines from plants 642
Vacin and Went medium 1028
vacuum manifold 229
vacuum transfer 216
value of biodiversity 1076
variable joining sites 750
variable region 739
variable region genes (V) 744
variation of expression 509
vat leaching 1071
vector-mediated DNA transformation 475
vectorette PCR 313
vectors 6, 407, 408
Veda Vyasa 837
vermicomposting 1068
Veronica 99
versions of protein liquid chromatography 694
vertical slab gel electrophoresis apparatus 199
vertical tube rotors 57
vesicular arbuscular mycorrhizae (VAM) 1057, 1101
viability testing 987
viable count 806, 820
vinblastine 1034
vincristine 1034
vir genes 428
viral DNA 134
viral genomes 133, 149

Virion extraction buffer 151
viroids 478
virtual classroom 123
virulence 479
virulent phage 831
virus 438
   free plants 11
   purification 152
   resistant plants 615
visual counts 822
visual selection 1014
vitamins 555, 890
Vk gene segment 746
VNTRs 270, 271

W
washing pipettes 29
waste 1040
   materials 1109
   treatment 1061, 1129
   water treatment 1049
water pollution 1043
water purification equipment 881
water quality 1059
Watson and Crick 128
weak acid 46
weakly dissociated 42
web browser 122
web-based protein structure tools 111
websites for patent literature 1138
welfare ecology 1078
Western blot 224
Western blotting 777
Western Ghats (India) 1075
wet gels 208
wet mounts 780
wheat germ 179
whole–cell extracts 656
whole-mount preparations 324
Wide Area Information Servers 99
wind-generated electricity 1109

world centres 1093
World Intellectual Property Organization (WIPO) 1142, 1149
World Wide Web 99
world's basic food plants 1093

X
X-gal 412, 461, 1158
X-ray film 208
xanthan gum 557
Xanthomonas 557
*Xba*I 387
xenobiotic compounds 1052
xenobiotic concentration 1059
xenobiotics 13, 1042
*Xho*I 389
XO-monosomic condition 838
xylene-cyanol FF 200

Y
YAC cloning vectors 424
YACs 422, 423
yeast
   artificial chromosomes 423
   biomass 1067
   cloning vectors 423
   extract 1067
   plasmid vectors 423
YEPD 792
YNB medium 793
yolk sac 606

Z
zeatin 479
zein 626
Zn-finger protein genes 274
zone electrophoresis 192
zone sharpening techniques 713
zoos 1099
zwitterion 679, 681
zygotic embryos 971